Modern Charge-Density Analysis

Carlo Gatti • Piero Macchi
Editors

Modern Charge-Density Analysis

Editors
Carlo Gatti
CNR-ISTM, Istituto di Scienze e Tecnologie Molecolari del CNR
c/o Dipartimento di Chimica Fisica ed Elettrochimica
Università degli Studi di Milano
Via Golgi 19
20133 Milano
Italy
carlo.gatti@istm.cnr.it

Piero Macchi
Department of Chemistry and Biochemistry
University of Bern
Freiestrasse 3
CH3012 Bern
Switzerland
piero.macchi@dcb.unibe.ch

ISBN 978-90-481-3835-7 e-ISBN 978-90-481-3836-4
DOI 10.1007/978-90-481-3836-4
Springer Dordrecht Heidelberg London New York

Library of Congress Control Number: 2011941749

Printed on acid-free paper

Springer is part of Springer Science+Business Media (www.springer.com)

This book is dedicated to
Carla Roetti, Miguel Álvarez Blanco,
Andrés Goeta and Cesare Pisani
Their valuable contribution in
the field of charge density has been
an inspiration for many of us and
will remain for the future generations
of scientists

Preface

The contributions in this book manifest the tremendous developments that have taken place in the last 10–20 years in electron density research in the broadest sense. Although the possibility of measuring the distribution of electrons in crystalline solids with X-rays was realized soon after the discovery of X-ray diffraction early in the twentieth century it was not until half a century later that the first systematic attempts started to bear fruit. This was of course well after the advent of quantum mechanics but before the development of fast computers made the calculation of electronic structure a tool employed by a large research community in the physical sciences.

Already in 1922 Bragg realized that measurement of the net charges in NaCl was as yet impossible, but not beyond reach, when he wrote (with James and Bosanquest) that 'it seems that crystal analysis must be pushed to a far greater degree of refinement before it can settle the point' (Phil. Mag44, 433 (1922)). The 'greater degree of refinement' has now been achieved to a degree well beyond what could be envisioned in 1922. This applies to both experiment and theory and to the broad field of charge spin and momentum densities, all three of which are discussed in the current volume. The development of radiation sources, detectors and automation of data collection equipment has been nothing short of dramatic. New theoretical methods and in particular Bader's development of the density-based Theory of Atoms in Molecules, combined with the exponential increase in computing power have created a link between theory and experiment which opened a new phase in electron density research in which the emphasis is very much on interpretation of the results and their application in the understanding of chemical, physical and biological phenomena. The current volume is highly timely in covering the advances. Its chapters survey the state of the art and point the way to further exciting developments.

Williamsville, NY Philip Coppens

Contents

Contributors

János G. Ángyán CRM2, CNRS and Nancy-University, B.P. 239, F-54506 Vandœuvre-lès-Nancy, France, angyan@crm2.uhp-nancy.fr

Paul W. Ayers Department of Chemistry, McMaster University, Hamilton, ON, Canada L8S4M1, ayers@mcmaster.ca

Libero J. Bartolotti Department of Chemistry, East Carolina University, Greenville, NC 27858, USA, bartolottil@mail.edu

Pierre Becker Structures, Properties and Modelling of Solids (SPMS) Laboratory, Ecole Centrale Paris, Grande Voie des Vignes, 92295, Chatenay-Malabry, France, pierre.becker@ecp.fr

Miguel Alvarez Blanco Dpto. Química Física y Analítica, Universidad de Oviedo, 33006 Oviedo, Spain, miguel@carbono.quimica.uniovi.es

Elena V. Boldyreva REC-008 Novosibirsk State University and Institute of Solid State Chemistry and Mechanochemistry SB RAS, ul. Kutateladze, 18, Novosibirsk 128, Russia, eboldyreva@yahoo.com

Julia Contreras-García MALTA-Consolider Team and Departamento de Química Física, Universidad de Oviedo, E-33006 Oviedo, Spain, contrera@lct.jussieu.fr

Malcom J. Cooper Department of Physics, University of Warwick, Coventry, CV4 7AL, UK, m.j.cooper@warwick.ac.uk

Riccardo Destro Department of Physical Chemistry and Electrochemistry, Università degli Studi di Milano, Via Golgi 19, 20133 Milan, Italy, riccardo.destro@unimi.it

Sławomir Domagała Laboratoire de Cristallographie Résonance Magnétique et Modélisations (CRM2) CNRS, UMR 7036, Faculté des Sciences et Techniques, Nancy University. Institut Jean Barriol, BP 70239, 54506 Vandoeuvre-lès-Nancy, Cedex, France, SLAWDOM@chem.uw.edu.pl

Paulina M. Dominiak Department of Chemistry, University of Warsaw, ul. Pasteura 1, 02-093 Warszawa, Poland, pdomin@chem.uw.edu.pl

Roberto Dovesi Dipartimento di Chimica IFM, and Centre of Excellence NIS (Nanostructured Interfaces and Surfaces), Università di Torino, via Giuria 5, I-10125 Torino, Italy, roberto.dovesi@unito.it

Jonathan A. Duffy Department of Physics, University of Warwick, Coventry, CV4 7AL, UK, j.a.duffy@warwick.ac.uk

Georg Eickerling Institut für Physik, Universität Augsburg, Universitätsstrasse 1, 86159 Augsburg, Germany, georg.eickerling@physik.uni-augsburg.de

Thomas Elsaesser Max-Born-Institut für Nichtlineare Optik und Kurzzeitspektroskopie, D-12489 Berlin, Germany, elsasser@mbi-berlin.de

Alessandro Erba Dipartimento di Chimica IFM, and Centre of Excellence NIS (Nanostructured Interfaces and Surfaces), Università di Torino, via Giuria 5, I-10125 Torino, Italy, alessandro_erba@virgilio.it

Enrique Espinosa CRM2, CNRS and Nancy-University, B.P. 239, F-54506 Vandœuvre-lès-Nancy, France, enrique.espinosa@crm2.uhp-nancy.fr

Volker Eyert Institut für Physik, Universität Augsburg, Universitätsstrasse 1, 86159 Augsburg, Germany, veyert@materialsdesign.com

Louis J. Farrugia School of Chemistry, University of Glasgow, G12 8QQ Scotland, UK, louis.farrugia@glasgow.ac.uk

Tim Fievez Eenheid Algemene Chemie (ALGC), Vrije Universiteit Brussel (VUB), Pleinlaan 2, 1050 Brussel, Belgium, tfievez@rub.vub.ac.be

Ulrike Flierler Institut für Anorganische Chemie, Universität Göttingen, Tammannstrasse 4, 37077 Göttingen, Germany, uflierler@chemie.uni-goettingen.de

Bertrand Fournier Laboratoire de Cristallographie Résonance Magnétique et Modélisations (CRM2) CNRS, UMR 7036, Faculté des Sciences et Techniques, Nancy University, Institut Jean Barriol, BP 70239, 54506 Vandoeuvre-lès-Nancy, Cedex, France, bertrand.fournier@crm2.uhp-nancy.fr

Evelio Francisco Dpto. Química Física y Analítica, Universidad de Oviedo, 33006 Oviedo, Spain, evelio@carbono.quimica.uniovi.es

Carlo Gatti Istituto di Scienze e Tecnologie Molecolari del CNR (CNR-ISTM) e Dipartimento di Chimica Fisica ed Elettrochimica, Università di Milano, Via Golgi 19, I-20133 Milan, Italy

Center for Materials Crystallography, Aarhus University, Langelandsgade 140, DK-8000 Aarhus C, Denmark, carlo.gatti@istm.cnr.it

Paul Geerlings Eenheid Algemene Chemie (ALGC), Vrije Universiteit Brussel (VUB), Pleinlaan 2, 1050 Brussel, Belgium, pgeerlin@vub.ac.be

Paolo Giannozzi Democritos Simulation Center, CNR-IOM Istituto Officina dei Materiali, I-34151 Trieste, Italy

Dipartimento di Chimica, Fisica e Ambiente, Università di Udine, via delle Scienze 208, I-33100 Udine, Italy, paolo.giannozzi@uniud.it

Jean-Michel Gillet Structures, Propriétés et Modélisation des Solides, UMR8580, Ecole Centrale Paris, Grande Voie des Vignes, 92295 Chatenay-Malabry Cedex, France, jean-michel.gillet@ecp.fr

Béatrice Gillon Laboratoire Léon Brillouin (CEA-CNRS), Centre d'Etudes de Saclay, 91191 Gif-sur-Yvette, Cedex, France, Beatrice.gillon@cea.fr

Andrés E. Goeta Chemistry Department, Durham University, South Road, Durham, DH1 3LE UK, a.e.goeta@durham.ac.uk

Yuri Grin Max-Planck-Institut für Chemische Physik fester Stoffe, D-01187 Dresden, Germany, grin@cpfs.mpg.de

Benoît Guillot Laboratoire de Cristallographie Résonance Magnétique et Modélisations (CRM2) CNRS, UMR 7036, Faculté des Sciences et Techniques, Nancy University, Institut Jean Barriol, BP 70239, 54506 Vandoeuvre-lès-Nancy, Cedex, France, benoit.guillot@lcm3b.uhp-nancy.fr

Christoph Hauf Institut für Physik, Universität Augsburg, Universitätsstrasse 1, 86159 Augsburg, Germany, christoph.hauf@physik.uni-augsburg.de

Bo B. Iversen Department of Chemistry and iNANO, Aarhus University, DK-8000 Arhus C, Denmark, bo@chem.au.dk

Dylan Jayatilaka Chemistry, School of Biomedical, Biological and Chemical Sciences, University of Western Australia, 35 Stirling Highway, Nedlands, 6009 WA, Australia, dylan.jayatilaka@uwa.edu.au

Christian Jelsch Laboratoire de Cristallographie Résonance Magnétique et Modélisations (CRM2) CNRS, UMR 7036, Faculté des Sciences et Techniques, Nancy University, Institut Jean Barriol, BP 70239, 54506 Vandoeuvre-lès-Nancy, Cedex, France, christian.jelsch@crm2.uhp-nancy.fr

Paul A. Johnson Department of Chemistry, McMaster University, Hamilton, ON, Canada L8S4M1, johnsopa@mcmaster.ca

Miroslav Kohout Max Planck Institute for Chemical Physics of Solids, Nöthnitzer Str. 40, 01187 Dresden, Germany, kohout@cpfs.mpg.de

Tibor Koritsanszky Department of Chemistry, Computational Science Program, Middle Tennessee State University, Murfreesboro, TN 37132, USA, tkoritsa@gmail.com

Claude Lecomte Laboratoire de Cristallographie Résonance Magnétique et Modélisations (CRM2) CNRS, UMR 7036, Faculté des Sciences et Techniques,

Nancy University, Institut Jean Barriol, BP 70239, 54506 Vandoeuvre-lès-Nancy, Cedex, France, claude.lecomte@crm2.uhp-nancy.fr

Dorothee Liebschner Laboratoire de Cristallographie Résonance Magnétique et Modélisations (CRM2) CNRS, UMR 7036, Faculté des Sciences et Techniques, Nancy University, Institut Jean Barriol, BP 70239, 54506 Vandoeuvre-lès-Nancy, Cedex, France, dorothee.liebschner@crm2.uhp-nancy.fr

Leonardo Lo Presti Department of Physical Chemistry and Electrochemistry, Università degli Studi di Milano, Via Golgi 19, 20133 Milan, Italy, leonardo.lopresti@unimi.it

Piero Macchi Department of Chemistry and Biochemistry, University of Bern, Freiestrasse 3, CH3012 Bern, Switzerland, piero.macchi@dcb.unibe.it

Anders Ø. Madsen Department of Chemistry, University of Copenhagen, Universitetsparken 5, DK-2100 Copenhagen Ø, Denmark, madsen@chem.ku.dk

Miriam Marqués SUPA, School of Physics and Centre for Science at Extreme Conditions, The University of Edinburgh, Mayfield Road, Edinburgh EH9 3JZ, UK, mmarques@staffmail.ed.ac.uk

Jacob Overgaard Department of Chemistry and iNANO, Aarhus University, DK-8000 Arhus C, Denmark, jacobo@chem.au.dk

Angel Martín Pendás Dpto. Química Física y Analítica, Universidad de Oviedo, 33006 Oviedo, Spain, angel@fluor.quimica.uniovi.es

Virginie Pichon-Pesme Laboratoire de Cristallographie Résonance Magnétique et Modélisations (CRM2) CNRS, UMR 7036, Faculté des Sciences et Techniques, Nancy University, Institut Jean Barriol, BP 70239, 54506 Vandoeuvre-lès-Nancy, Cedex, France, Virginie.Pichon@crm2.uhp-nancy.fr

Cesare Pisani Dipartimento di Chimica IFM, and Centre of Excellence NIS (Nanostructured Interfaces and Surfaces), Università di Torino, via Giuria 5, I-10125 Torino, Italy, cesare.pisani@unito.it

Paul L.A. Popelier Manchester Interdisciplinary Biocentre (MIB), 131 Princess Street, Manchester M1 7DN, UK

School of Chemistry, University of Manchester, Oxford Road, Manchester M13 9PL, UK, pla@manchester.ac.uk

Rainer Pöttgen Institut für Anorganische und Analytische Chemie, Universität Münster, Corrensstrasse 30, 48149 Münster, Germany, pottgen@uni-muenster.de

Manuel Presnitz Institut für Physik, Universität Augsburg, Universitätsstrasse 1, 86159 Augsburg, Germany, manuel.presnitz@physik.uni-augsburg.de

José M. Recio MALTA-Consolider Team and Departamento de Química Física, Universidad de Oviedo, E-33006 Oviedo, Spain, mateo@fluor.quimica.uniovi.es

Ernst-Wilhelm Scheidt Institut für Physik, Universität Augsburg, Universitätsstrasse 1, 86159 Augsburg, Germany, ernst-wilhelm.scheidt@physik.uni-augsburg.de

Wolfgang Scherer Institut für Physik, Universität Augsburg, Universitätsstrasse 1, 86159 Augsburg, Germany, wolfgang.scherer@physik.uni-augsburg.de

Bernard Silvi Laboratoire de Chimie Théorique (UMR-CNRS 7616), Université Pierre et Marie Curie, 3 rue Galilée, 94200 Ivry sur Seine, France, silvi@lct.jussieu.fr

Raffaella Soave Istituto di Scienze e Tecnologie Molecolari (ISTM), Italian National Reseearch Council (CNR), Via Golgi 19, 20133 Milan, Italy, raffaella.soave@istm.cnr.it

Mark A. Spackman School of Biomedical, Biomolecular & Chemical Sciences, University of Western Australia, Crawley, WA 6009, Australia, mark.spackman@uwa.edu.au

Dietmar Stalke Institut für Anorganische Chemie, Universität Göttingen, Tammannstrasse 4, 37077 Göttingen, Germany, dstalke@chemie.uni-goettingen.de

Masaki Takata SPring8 Synchrotron Facility, Koto 1-1-1, Sayo-cho, Sayo, Hyougo 679-5148, Japan, takatama@spring8.or.jp

John S. Tse Department of Physics and Engineering Physics, University of Saskatchewan, Saskatoon, SK, Canada S7N 5E2, John.Tse@usask.ca

Michael Woerner Max-Born-Institut für Nichtlineare Optik und Kurzzeitspektroskopie, D-12489 Berlin, Germany, woerner@mbi-berlin.de

Abbreviations

1-RDM	1-electron Reduced Density Matrix
2-RDM	2-electron Reduced Density Matrix
3D	three dimensional
AcG	N-acetylglycine
ACP	Anisotropic Compton Profile
ADF	Amsterdam Density Functional ab-initio package
ADF2004	Amsterdam Density Functional ab-initio package (2004 version)
ADPs	Anisotropic Displacement Parameters
ADPs	Atomic Displacement Parameters
AE	All-Electron
AF	antiferromagnetic
AFM	Antiferromagnetic
AIM	Atom's in Molecules (analogous to QTAIM, Quantum Theory of Atoms in Molecules, in other chapters of the book)
AMM	Anions in metallic matrix
AO	Atomic Orbital
AS	Ammonium Sulfate
A–TCNB	anthracene-tectracyanobenzene
aug-cc-pVTZ	Dunning's correlation consistent triple-zeta basis sets with added diffuse functions
B3LYP	Becke's 3 parameter functional (in density functional theory calculations)
bcc	body centered cubic
bcp, BCP	bond critical point
bct	body centered tetragonal
BDC	benzene dicarboxylate
BF	Bloch Function
BLYP	Becke's 1988 exchange functional and Lee, Yang and Parr correlation functional (in density functional theory calculations)
BOMD	Born-Oppenheimer Molecular Dynamics
BP	bond path

bpe	(E)-1,2-bis(4-pyridyl)ethylene
bpNO	4,4′-dipyridyl-*N*,*N*′-dioxide
BS	Basis Set
BSSE	Basis Set Superposition Error
btr	4,4′-bis-1,2,4-triazole
BvK	Born-vonKarman (cyclic conditions)
BZ	Brillouin Zone
CA	chloranil
CASSCF	Complete Active Space Self-Consistent method
CBED	Convergent Beam Electron Diffraction
CCD	Charge Coupled Device
CCSD	Coupled Cluster calculation with both Single and Double Substitutions from the Hartree-Fock determinant
CCSD(T)	coupled cluster method with single and double excitations with singles/triples coupling term
CD	Charge (nuclear+electronic) density
CDASE	Comprehensive Decomposition Analysis of Stabilization Energy
CDM	charge density maps
CHD	1,3-cyclohexanedione
CI	Configuration Interaction
CIF	Crystallographic Information File
CISD	Configuration Interaction with Single and Double substitutions from the Hartree-Fock reference determinant
CMOS	Complementary Metal Oxides Semiconductors
CO	Crystalline Orbital
CP,CPF	Compton Profile (Function)
CPHF	Coupled Perturbed Hartree-Fock
CPL-1	Coordination Polymer 1
CRYSTAL	Crystal ab initio package (any version)
CRYSTAL03	Crystal03 ab initio package
CRYSTAL-XX	ab-initio package for the calculation of the electronic structure of periodic systems (XX=98, 03, 09, according to 1998, 2003, 2009 version)
CSD	Cambridge Structural Database
CSO	Crystalline Spin-Orbital
CWM	Constrained Wavefunction Method
DAFH	Domain Averaged Fermi Hole
DCD	Dewar-Chatt-Duncanson Model
DCH	Dirac-Coulomb Hamiltonian
DCP	Directional Compton Profile
DEA	DiEthylAmine
DEF	DiEthylFormamide
DF	Density Function (charge, spin or momentum density)
DFPT	Density Functional Perturbation Theory
DFT	Density Functional Theory

dhcp	double hexagonal closed packed
DIABN	4-(diisopropylamino)benzonitrile
DKH	Douglas, Kroll and Hess (DKH) (relativistic Hamiltonian approach)
DM	Density Matrix
DMA	DiMethylAcetamide
DMA	Distributed Multipole Analysis
DMF	DiMethylFormamide
dmpe	dimethyl-phosphinoethane
DM-TTF	2,6-dimethyltetrathiofulvalene
DNA	deoxyribonucleic acid
DOS	Density of States
dppen	cis-1,2-bis(diphenylphosphino)ethylene
DZP	Double-zeta plus Polarization (basis set)
DZPT	Double-Zeta plus Triple Polarization (basis set)
ECP	Effective Core Potential
ED	Electron Density
ESD	Electron Spin Density
EDD	Electron density distribution
EDF	Electron population Distribution Function
EDI	ELF delocalization index
EF	Electric Field
EFG	Electric Field Gradient
ELF	Electron Localization Function
ELI	Electron Localizability Indicator
ELMAM	Experimental Library of Multipolar Atom Model
EM	Electromagnetic
EMAC	Extended Metal Atom Chain (compounds)
EMD	Electron Momentum Density
EP	Exact Potential (Exact evaluation of the electrostatic interaction energy)
EPMM	Exact Potential and Multipole Moment (Volkov's method for evaluating the electrostatic interaction energy of a pair of molecules)
EPR	Electron Pair Repulsion
ESD	Electron Spin Density
esd	estimated standard deviations
ESP	ElectroStatic Potential
ESRF	European Synchrotron Radiation Facility
F_4DBB	1,4-dibromotetrafluorobenzene, $C_6F_4Br_2$
F_4DIB	1,4-diiodotetrafluorobenzene, $C_6F_4I_2$
fcc,FCC	Face Centered Cubic
FLAPW	Full Potential Linearized Augmented PlaneWave (code, method)
FM	Ferromagnetic
FMO	Frontier Molecular Orbital
FMs	Ferromagnets

FT	Fourier Transform
FFT	Fast Fourier transform
Full-CI	Full Configuration Interaction method
FWHM	Full width at half maximum
G03	Gaussian-03 (ab-initio program)
G98	Gaussian-98 (ab-initio program)
GGA	Generalized Gradient Approximation
Gly	α-glycine
GTF	Gaussian-Type-Function
GTO	Gaussian Type Orbital
hAR	human aldose reductase
HB	Hydrogen Bond
HC	Hansen-Coppens formalism (in multipole models refinements)
hcp	hexagonal close packing
HF	Hartree-Fock
UHF	Unrestricted Hartree-Fock
HIV	Human Immunodeficiency Virus
HM	Halet's and Mingos' model on metal carbides
HMs	Half-Metals
HOMO	Highest Occupied Molecular Orbital
HS	High spin
HS	Hirshfeld surface
IA	Impulse Approximation (in inelastic scattering)
IAM	Independent Atom Model
Invariom	*Invari*ant at*om* (database)
IQA	Interacting Quantum Atoms
IUCr	International Union of Crystallography
KS	Kohn-Sham
UKS	Unrestricted Kohn-Sham
KS-DFT	Kohn-Sham Density Functional Theory approach
Lac	L-(+)-lactic acid
LCAO	Linear Combination of Atomic Orbitals
LDA	Local Density Approximation
LICC	Ligand Induced Charge Concentration
LIESST	Light-Induced Excited-Spin-State Trapping
LMP2	Local Møller Plesset perturbation theory truncated at second order
LMTO	Linearized Muffin-Tin Orbital
LOCCC	Ligand Opposed Core Charge Concentration
LOL	Localized Orbital Locator
LS	Least-Squares (procedure)
LS	Local Source
LS	Low spin
LUMO	Lowest Unoccupied Molecular Orbital
MCP	Magnetic Compton Profile
MCPD	2-methyl-1,3-cyclopentanedione

MCS	Magnetic Compton Scattering
MCSCF	Multi Configuration Self Consistent Field
MD	Magnetization density
MD	Molecular Dynamics
MEM	Maximum Entropy Method
MESP	Molecular ElectroStatic Potential
MK	Merz-Kollman
MM	Multipole Model
MMFF94	Molecular Mechanics Force Field package (1994 version)
MNA	2-methul-4-nitroaniline
MO	Molecular Orbital
MOF	Metal Organic Framework
MOON	Molecular Orbitals with variable Occupation Numbers (refinement)
MOON	Molecular Orbitals with variable Occupation Numbers (method)
MP2	Møller-Plesset perturbation theory truncated at second order
MP4SDQ	Møller-Plesset correlation energy correction truncated at fourth-order in the space of single, double and quadruple substitutions from the Hartree-Fock determinant
MQM	Molecular Quantum Mechanics
MSD	Mean Square Displacement
MSO	Molecular Spin-Orbital
MSR	Mean-Square Residual
MTA	Multi-Temperature Analysis
MV	Mixed Valence
NADP+	Nicotinamide Adenine Dinucleotide Phosphate (oxidized form)
NADPH	Nicotinamide Adenine Dinucleotide Phosphate (reduced form)
NBO	natural bond orbital
NCI	NonCovalent Interaction
ND	Neutron Diffraction
NLO	Non Linear Optical
NMR	Nuclear Magnetic Resonance
NPA	Natural Population Analysis
NPP	*N*-4-nitrophenyl-L-prolinol
n-RDM	*n*-electron Reduced Density Matrix
NRT	natural resonance theory
NSO	Natural Spin Orbitals
OCM	Outer Core Maxima
p.d.f.	probability density function
PAW	Projected Augmented Waves
PBC	Periodic Boundary Conditions
PCAR	Point Contact Andreev Reflection
pdf	probability distribution function

PDZ2	Second PDZ domain. PDZ is a common structural domain of 80–90 amino-acids found in the signaling proteins of bacteria, yeast, plant, viruses and animals
PEECM	Periodic Electrostatic Embedded Cluster Model
PES	Potential Energy Surface
PGEC	Phonon-Glass – Electron Crystal
PIXEL	Gavezzotti's method for evaluating the electrostatic interaction energy of a pair of molecules
pNA	4-nitroaniline
PND	Polarised Neutron Diffraction
PNP	2-(*N*-(L)-prolinol)-5-nitropyridine
POM	3-methyl-4-nitropyridine-*N*-oxide
PP	Pseudo-Potential
PSD	Position Sensitive Detector
PU	Partition of the Unit methodology
PW	Plane Waves
QCISD	Quadratic Configuration Interaction calculation including single and double substitutions
QCT	Quantum Chemical Topology
QM	Quantum Mechanics or Quantum-Mechanical
QSAR	Quantitative Structure Activity Relationships
QSPR	Quantitative Structure Property Relationships
QTAIM	Quantum Theory of Atoms in Molecules
QZ4P	core triple zeta, valence quadruple zeta, plus four polarization functions basis set (in the ADF, Amsterdam Density Functional ab-initio package)
r.m.s.	root mean square
r.m.s.d.	root mean square deviation
RBFNN	radial basis function neural networks
rcp	ring critical point
RDF	RaDial Functions (in the multipole models pseudoatom formalism)
RFF	Reciprocal Form Factor
RIET	Redox Induced Electron Transfer
RLFT	Rigorous Local Field Theory
RT	Room Temperature
SAPT	Symmetry Adapted Perturbation Theory
SAPT2002	Ab initio program based on Symmetry Adapted Perturbation Theory (2002 version)
SCDS	SemiClassic Density Sums (in Gavezzotti's pixel method)
SCMP	Self Consistent Madelung Potential
SCO	Spin Cross Over
SDS	H-atom form factor developed by Stewart, Davidson and Simpson
SF	Source Function
sh	simple hexagonal

SIBFA	Sum of Interactions Between Fragments computed Ab Initio
SP	Spin Polarisation
SPDFG	Volkov, King and Coppens code for evaluating the electrostatic repulsion in terms of the monomer charge distributions
STF	Slater-Type-Function
TAAD	Transferred Aspherical Atom Model
*t*Bu	tert-butyl
TE	ThermoElectric
TLS	Translation/Libration/Screw model (rigid body model for thermal motion)
TM	transition metal
TPEP	Topological Partitioning of Electronic Properties
TS	Transition State
TTF	tetrathiofulvalene
T_{tr}	transition temperature
TZP	Triple-zeta plus Polarization (basis set)
UBDB	University at Buffalo DataBank
USPP	Ultra-Soft Pseudo-Potential
VB	Valence Bond
VE	Valence electrons
VOM	Valence Orbital Model
VSCC	Valence Shell Charge Concentration
VSCD	Valence Shell Charge Depletion
VSCP	Valence Shell Charge Polarization
VSEPR	Valence Shell Electron Pair Repulsion
WF	Wannier Function
XAO	X-ray Atomic Orbital method
XCDFT	X-ray Constrained DFT
XCHF	X-ray Constrained Hartree-Fock
XCW	X-Ray Constrained (Hartree-Fock) Wavefunction
XMCD	X-ray Magnetic Circular Dichroism
XMD	X-ray Magnetic Diffraction
XRD	X-Ray Diffraction
ZORA	quasi-relativistic two-component Zeroth-Order Regular Approximation

Chapter 1
A Guided Tour Through Modern Charge Density Analysis

Carlo Gatti and Piero Macchi

1.1 Introduction

Back in 1990, Philip Coppens – the author of the preface to this book and a reckoned authority in the field of X-ray charge-density (CD) studies, – wrote (with Dirk Feil) that "at present, charge density analysis is far from being a routine technique" [1]. However, less than 10 years later, in a paper significantly entitled *Charge-Density Analysis at the Turn of the Century*, he could be far more optimistic, and both about the easiness and the potential role of such an investigative tool [2]. According to Coppens, the CD analysis had eventually "developed into a technique now ready for application to a broad range of problems of interest in chemistry, biology and solid state physics" [2]. The present book is aimed at showing how such a prophecy has indeed been realized during the first decade of the new century, offering a balanced exposition of the most relevant recent applications of CD analysis, along with those theoretical, methodological and instrumental developments that have made them possible. Although this book is focused more on the experimental facets of the charge density analysis, a large emphasis is also given to those progresses of theoretical and computational quantum theory that allow useful and most necessary comparisons between theory and experiment [3], very often complementary. The book also covers, from many points of view and interest, aspects that go far beyond

C. Gatti (✉)
Istituto di Scienze e Tecnologie Molecolari del CNR (CNR-ISTM) e Dipartimento di Chimica Fisica ed Elettrochimica, Università di Milano, Via Golgi 19, I-20133 Milan, Italy

Center for Materials Crystallography, Aarhus University, Langelandsgade 140, DK-8000 Aarhus C, Denmark
e-mail: carlo.gatti@istm.cnr.it

P. Macchi (✉)
Department of Chemistry and Biochemistry, University of Bern, Freiestrasse 3, CH3012 Bern, Switzerland
e-mail: piero.macchi@dcb.unibe.it

C. Gatti and P. Macchi (eds.), *Modern Charge-Density Analysis*,
DOI 10.1007/978-90-481-3836-4_1,

the "simple" charge density analysis, and highlight the relevance of joint charge, spin and momentum densities studies, as well as of the role of one- and two-electron density matrices for discussing how atoms get bonded to one another in molecules and crystals.

Even though CD *has now come of age* – in keeping with the title of a recent highlight published by Philip Coppens on Angewandte Chemie [4] – a comprehensive updated treatment of this broad field was definitely lacking. We hope that the over 20 chapters of this book might adequately satisfy this need. This chapter is simply aimed at introducing the reader to the whole book through a concise summary of basic aspects of CD analysis and a general overview of the CD research developments which took place over the last 10 years. While offering a *coup d'oeil* to those issues which are then developed in details in the remaining chapters, this first one also briefly highlights those few topics that despite their relevance could not find place in the book.

This book is a comprehensive update of research on charge density analysis. On the other hand, a textbook like *X-Ray Charge Densities and Chemical Bonding* [5] by Philip Coppens and *Atoms in Molecules, A Quantum Theory* [6] by Richard Bader are more adequate for educational purposes and for a general introduction to this field. A recent book, edited by Cherif Matta and Russel Boyd [7] gives an update on advanced research, though limited to application of the Quantum Theory of Atoms in Molecules. A number of excellent reviews dealing with the diverse aspects of modern charge density analysis have also appeared during the last 10 years. They are not listed here, but properly referred to in the various chapters of this book.

1.2 From Electron and Charge Density to Physical Properties and Chemical Bond

1.2.1 Charge, Electron and Electron Spin Densities

Let's consider a molecular system made of N electrons and M nuclei. The *probability* of finding any of its electrons at $\mathbf{r}_1$ regardless of the exact position of the other electrons is equal to $\rho(\mathbf{r}_1)d\mathbf{r}_1$ where the corresponding *probability density* is the *position electron density* defined as [8]

$$\rho(\mathbf{r}) = N \int \Psi_{el}(\mathbf{r},\mathbf{r_2},...\mathbf{r}_N;\mathbf{R}) \cdot \Psi^*_{el}(\mathbf{r},\mathbf{r_2},...\mathbf{r}_N;\mathbf{R})d\mathbf{r_2},...d\mathbf{r}_N \tag{1.1}$$

In Eq. 1.1 Ψ_{el} is the stationary wavefunction for fixed nuclear space coordinates (Born-Oppenheimer approximation, with $\mathbf{R}$ denoting the ensemble of nuclear coordinates for the M nuclei in an arbitrary reference frame).

The *charge density* is instead the sum of electron and nuclear densities in position space [5, 9]. Note that in literature the terms "charge density" and "electron density" are often used interchangeably [9] even when talking about the electron distribution alone and so are they used in the present book. The charge density term is often adopted to emphasize those cases where the determination of the distribution of both positive (nuclear) and negative (electronic) charge has taken place simultaneously, as it happens in the X-ray diffraction experiments (*see infra*) or in theoretical approaches like Car-Parrinello Molecular Dynamics [10].

The electron density may be obtained either from experiment (Sect. 1.2.2) or from *ab-initio* calculations (Sect. 1.2.3). According to the first Hohenberg-Kohn theorem, the electron density uniquely determines the Hamiltonian operator and thus, at least formally if not operationally, all properties of a system (Sect. 1.2.4), including its reactivity (Chap. 21). Since electrons interact only in pairs, the Hamiltonian contains just one-body and two-body terms (i.e. there are no specifically many-body effects) and all 1-electron or 2-electron properties may be evaluated as expectation values of the corresponding operators, using the so-called 1-electron and 2-electron reduced density matrices (Sects. 1.2.3, 1.2.4 and Ref. [8]). A subset of 1-electron properties may be then obtained from the electron density alone (Sects. 1.2.3 and 1.2.4).

The electron density and other scalar or vector functions derived from the density itself or from the 1-electron or 2-electron density matrices, form the basic ingredients of the so-called topological (or quantum chemical topology) approaches to chemical bonding (Sects. 1.2.4, 1.3.6 and Chap. 9).

Figure 1.1 summarizes how information coming from theory and experiment is combined in modern charge density analysis, as detailed throughout this section.

The position space electron density (and derived functions) is the main density function discussed in this book, though other functions are also introduced in view of their complementary aspects for the analysis of physical properties and chemical bonding (Sects. 1.2.3, 1.3.6, Chaps. 2, 4, 7 and 9). The electron density in the momentum space, as obtained from a six-dimensional Fourier Transform (*FT*) of the first order density matrix in position space (Chap. 2) or through reconstruction from directional Compton profiles measured for multiple crystal directions (Chap. 4) is one of such functions. Emphasis is also given to the spin-resolved counterparts of the electron density, both in position (Chaps. 2 and 8) and momentum spaces (Chaps. 2 and 4), which may be obtained either theoretically (Chap. 2) or from a variety of experimental techniques. Polarized neutron diffraction (PND), Chap. 8, is the primary experimental source for position electron spin density, while magnetic Compton scattering (Chap. 4) allows for the reconstruction of electron spin density in momentum space. Non-resonant X-ray diffraction experiments enable the study of both spin and orbital magnetization for antiferromagnets (Chap. 4).

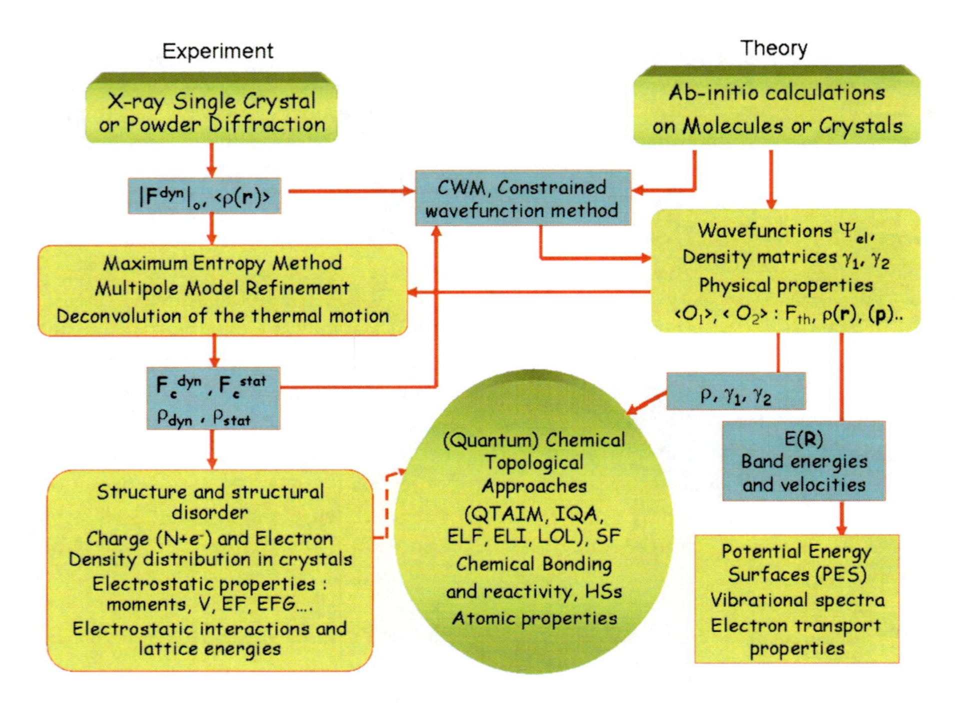

Fig. 1.1 Interplay of experiment and theory in modern charge density analysis. γ_1 and γ_2 are 1-electron and 2-electron density matrices; $\mathbf{F}_o^{dyn}$, $\mathbf{F}_c^{dyn}$ and $\mathbf{F}_c^{stat}$ are observed (o) or model calculated (c) dynamical (dyn) or static (stat) structure factors; $\mathbf{F}_{th}$ are structure factors from static *ab-initio* ED

1.2.2 Electron Density and Charge Density from X-Ray Diffraction

X-rays play a crucial role across the field of chemistry, as illustrated by a dedicated issue published some time ago on Chemical Reviews [11]. Upon interaction with the matter, X-rays can be either scattered or adsorbed, with both mechanisms being particularly relevant to chemistry. Due to their typical wavelengths (~1 Å), comparable to the distance between two bonded atoms, X-ray scattering or, for samples with long range order, X-ray diffraction, yields information about the size and shape of atomic systems, whereas the X-ray energies, which are of similar magnitude to the core electron binding energies, probe core electron energies and provide atom-specific spectral data through X-ray absorption.

The total scattering of X-rays contains both elastic and inelastic components. The intensities of elastic scattering are essentially determined by the electron density distribution since electrons scatter X-ray much more efficiently than nuclei [5]. The opposite is true for neutron scattering and related diffraction. Maxima in the electron density occur at the nuclear positions and elastic scattering from a periodic array of atoms like in crystals lead to significant intensities in discrete directions due to the destructive and constructive interference (diffraction) of scattered waves, as expressed in Laue equations or Bragg's law.

The coherent X-ray scattering amplitude (F) of the diffracted beam is the *FT* of the thermally averaged electron density $<\rho(\mathbf{r})>$ in the unit cell,

$$\mathrm{F}(\mathbf{H}) = \int_V <\rho(\mathbf{r})> e^{2\pi i \mathbf{H}\bullet\mathbf{r}} d\mathbf{r} = FT<\rho(\mathbf{r})> \tag{1.2}$$

where $\mathbf{H} = h\mathbf{a}^* + k\mathbf{b}^* + l\mathbf{c}^*$ is the scattering vector in the reciprocal space and V is the unit cell volume. The *FT* of the unit cell density is referred to as the *structure factor* F_{hkl}, which denotes the scattering amplitude obtained by sampling ρ over the reciprocal lattice point defined by the h, k, and l Miller indices. The structure factors are defined only at the discrete set of reciprocal lattice points therefore the electron density in the unit cell may then be obtained through an inverse *FT* summation of an infinite number of structure factors

$$\rho(\mathbf{r}) = 1/V \sum_{h,k,l} F_{hkl} e^{-2\pi i(hx+ky+lz)} \tag{1.3}$$

where x, y and z are fractional coordinates defining $\mathbf{r} = x\mathbf{a} + y\mathbf{b} + z\mathbf{c}$. Though Eq. 1.3 would in principle suffice to obtain the electron density in a crystal, this is not the case for a number of very well-known reasons [5]. In particular, (*a*) the number of available reflections is finite and limited by the wavelength λ of the employed radiation ($|\mathrm{H}|_{\max} = 2\sin\theta_{\max}/\lambda$) and the geometry of the diffractometer; (*b*) the intensities at high angle are much weaker, even though collecting data at very low temperature partly alleviates this problem (see Chaps. 3 and 19) and (*c*) only the magnitudes of F(**H**) are known experimentally while their phases, which are also needed to reconstruct $\rho(\mathbf{r})$ through Eq. 1.3, are not available from the diffraction experiment, under kinematic approximation.

1.2.2.1 Modeling and Refining the Charge Density

The problem of determining $\rho(\mathbf{r})$ is solved by introducing a model yielding an ample, theoretically unlimited, set of computed structure factors, known both in magnitude and phase. The model is *refined* (adjusted) and validated by evaluating and minimizing the deviations of the calculated from the experimental structure factor magnitudes, suitably weighted with the corresponding experimental uncertainties.

The electron density in the unit cell is taken [2, 5] as a sum over the convolution (symbol *) of "atomic" densities contributions ρ^{at} (whose expression may vary from model to model, Sect. 1.3.4 and Chap. 5) and delta functions centered at the nuclear average positions within the cell

$$\rho_{unit\ cell}(\mathbf{r}) = \sum_j \rho_j^{at}(\mathbf{r}) * \delta(\mathbf{r} - \mathbf{r}_j) \tag{1.4}$$

As said earlier, the actually observable electron density (Eq. 1.1) is the result of a thermal average over the vibrational states. In the usual models adopted for CD analysis, one does not use a lattice dynamical approach to treat thermal motion, but simply considers the atoms as individual harmonic oscillators, each described by a probability density function $p_j(\mathbf{r})$ giving the probability of atom j to be displaced from its reference position $\mathbf{r}_j$ (Chap. 3). The thermally averaged electron density is therefore expressed in terms of a superposition of "atomic" densities contributions each of which rigidly follows the motion of the nucleus it is attached to:

$$\rho_j^{at}(\mathbf{r}) = \rho_j^{at,\,static}(\mathbf{r}) * p_j(\mathbf{r}) = \int \rho_j^{at,\,static}(\mathbf{r} - \mathbf{r}') * p_j(\mathbf{r}')d\mathbf{r}'$$

$$\rho_{unit\ cell}(\mathbf{r}) = \sum_j \left\{ \left[\rho_j^{at,\,static}(\mathbf{r}) * p_j(\mathbf{r}) \right] * \delta(\mathbf{r} - \mathbf{r}_j) \right\} \tag{1.5}$$

This (harmonic) convolution approximation is generally termed as the *rigid pseudo atom approximation* (Chap. 3). By applying the Fourier convolution theorem the structure factor, F(**H**), is given by

$$F(\mathbf{H}) = FT\left[\rho_{unit\ cell}(\mathbf{r})\right] = \sum_{\mathrm{j}} \left\{ FT[\rho_j^{at,\,static}(\mathbf{r}) * p_j(\mathbf{r})] \cdot FT[\delta\,(\mathbf{r} - \mathbf{r}_j]\right\}$$

$$F(\mathbf{H}) = \sum_{\mathrm{j}} f_j(\mathbf{H}) \cdot \mathrm{T}_j(\mathbf{H}) \cdot e^{2\pi i \mathbf{H} \bullet \mathbf{r}_j}$$

$$f_j(\mathbf{H}) = \int \rho_j^{at,\,static}(\mathbf{r})\, e^{2\pi i \mathbf{H} \bullet \mathbf{r}} d\mathbf{r}; \quad T_j(\mathbf{H}) = \int p_j(\mathbf{r}')\, e^{i \mathbf{H} \bullet \mathbf{r}'} d\mathbf{r}' \tag{1.6}$$

where f_j and T_j are, respectively, the static "atomic" structure factor and the atomic temperature factor, often referred to as Debye-Waller factor. Models differ from each other in the form adopted for f_j and T_j and in the number of the associated free parameters to be refined (the so-called *X-ray data refinement*). The refinement is often carried out using the procedure of least square errors, but other techniques to search the minimal residuals are possible, for example the conjugate gradient method [12]. The parameters defining f_j and T_j include the electronic ones (functions and their expansion coefficients in the expression yielding $\rho_j^{at,\,static}$) and those related to the thermal motion (isotropic or anisotropic atomic displacement parameters). Positions of atoms are the center for both $\rho_j^{at,\,static}(\mathbf{r})$ and $p_j(\mathbf{r})$. In normal structure refinement procedures, and also in most CD studies, the probability density function for atomic displacements $p_j(\mathbf{r})$ is approximated by an isotropic or anisotropic Gaussian function (*harmonic motion approximation*, Chap. 3). The FT of a Gaussian function is also a Gaussian and T_j results defined from the elements (*atomic anisotropic displacement parameters*, ADPs) of a symmetric 3×3 matrix, with one or six free parameters to be determined for isotropic and anisotropic motion, respectively (Chap. 3). High-resolution data allow going beyond the harmonic motion approximation. Formalisms like the Gram Charlier expansion developed in statistics to describe non Gaussian distributions are often adopted to treat anharmonicity (Chap. 3). It seems necessary to remark that unfortunately very rarely the physical meaning of the refined parameters is carefully checked. In fact, it might well occur that experimental errors, totally unrelated with thermal motion, affect the Debye Waller factors, especially when the parameterization of thermal motion is extended without checking the hypothesis that the probability density function is consistent (for example that it is positive in space). In facts, going beyond harmonic approximation, regions of negative nuclear probability might be produced, besides lowering the residuals (*cf.* Sect. 1.3.3)

In the simplest approach, the static "atomic" electron densities, $\rho_j^{at,static}$, are taken as the Hartree-Fock or Dirac-Fock functions for neutral spherically symmetric atoms and only atomic coordinates and thermal parameters are refined (isolated-spherical atom model or Independent Atom Model, IAM) [5]. The model reproduces the structural results of a conventional least square refinement, but is clearly unable to describe any electron redistribution due to bonding. The next step is to distinguish the atomic core density, kept fixed with respect to the neutral atom, from the atomic valence density which is instead allowed to be non neutral and to expand or contract, reflecting the bonding environment. This model corresponds to the so-called "kappa" refinement [5], representing a substantial improvement over IAM, despite the refinement of only two more parameters per atom (the valence population and the k-factor scaling the radial coordinate of the valence function). The k-refinement allows a crude representation of the charge density and it is no longer used in modern CD analysis, except as a possible useful intermediate step in the course of a multi-step least squares refinement procedure. Most of the recent CD studies adopt a more flexible model, where each "atomic" unit is described by multipole expansion, composed by a set of monopole terms, yielding the population of the atom, and a number of higher poles allowing for deformation with respect to spherical symmetry (*multipole model*, MM, Sect. 1.3.4 and Chap. 5). This aspherical-atom formalism, which usually gives the best possible fit to the data within the pseudo-atomic model (Eq. 1.4), clearly involves a largely increased number of parameters to be refined, typically >20 per atom, and calls for X-ray data of very good quality (the so called "charge-density" quality data) and higher resolution (going beyond the data available from Cu Kα radiation). Although the MM represents the standard tool in nowadays experimental CD studies, the model is known to be yet not flexible enough and to bias in some way the resulting electron density determination. This becomes clearly evident when comparing a theoretical ED and the one obtained by applying the MM to the theoretical structure factors. Thus, lots of efforts are currently taking place to improve the pseudoatom model performance (Chap. 5). The various attempts are mostly targeted at determining optimal, still reasonably transferable, radial functions for the various poles of the model by exploiting *ab-initio* electron density calculations on the whole molecule under exam or on carefully chosen pieces of it (Chap. 5).

The experimental ED implicitly includes subtle effects like electron correlation or relativistic corrections, which are often difficult to treat theoretically. It also inherently incorporates the *matrix effects* due to the crystalline environment (sometime improperly called "crystal field"), which may be accounted for in theory through periodic *ab-initio* calculations (Chap. 2). Recently, imprints on the experimental ED of the mixing of certain excited states into the ground state have also been detected [13] and rough charge density maps for transient species begin to become available using femtosecond X-ray diffraction (Chap. 20). However, despite these inherent pros, the quality of experimental CDs, even under equilibrium conditions, is dampened by the accuracy and uncertainties of structure factors, by the deficiencies of current MMs and by the intrinsic limit of the expansion of the density in terms of pseudo-atom nucleus-centered contributions only (Eq. 1.4). Such a constraint rules automatically out all 2-centre terms, whose importance is instead evident from the expansion of the electron density or of the 1-electron density matrix in terms of one-electron atomic orbitals (Sect. 1.2.3, Chap. 2, and in particular Chaps. 5 and 7). Indeed, one may envisage to determine such matrix also from experiment by combining in a unique refinement diverse experimental data, coming from different kinds of experiments, in position and momentum space, and from elastic and inelastic scattering (Chaps. 5 and 7). Despite several attempts in the past 30 years, these studies are still in their infancy, but several groups in the world are now competing in this area and substantial progress towards the Holy Grail of trustable, experimental density matrices is soon expected. Another easier and promising, though less fundamental route towards an experimental "model wavefunction" (hence to an experimental "model 1-electron density matrix"), exploits the so-called *constrained wavefunction method* (CWM) [14] where the standard *ab-initio* variational procedure minimizes both the wavefunction energy and the deviation of experimental structure factor magnitudes from those calculated from the wavefunction being optimized (Chap. 6). Clearly, more than providing an experimental one-electron density matrix, the CWM approach yields a theoretical density matrix suitably modified to fit the results of (a number of) experiment(s).

Other methods have been proposed in the literature, for example the X-ray atomic orbital (XAO) approach [15], adopted to directly refine electronic states. The method is applicable mainly to analyses of the electron-density distribution in ionic solids, given that it is based on an atomic orbital assumption. The expansion coefficients of each atomic orbital are calculated within perturbation theory and the coefficients of each orbital are refined to fit to the observed structure factors keeping the orthonormal relationships among them. This model is somewhat similar to the valence orbital model (VOM), earlier introduced by Figgis et al. [16] to study transition metal complexes, within the Ligand field theory approach [17]. The VOM could be applied in such complexes, within the assumption that the metal and the ligands are linked by bonds with "low overlap" between the atomic orbitals, therefore the electron density around metals or in the ligand shell can be treated as a perturbation of the atomic density. This assumption is also adopted for the determination of orbital coefficients from multipolar model, as introduced by Coppens et al. [18].

All methods presented in this section are somehow based on atomic/molecular densities or on an orbital assumption (atomic or molecular) to model the electron density from diffraction data. Similarly to quantum chemistry, however, an atomic density or an orbital approach is not mandatory and alternative models can be adopted (for example, Monte Carlo wave functions). In experimental electron density determination, the alternatives are mostly related to maximum likelihood principle, therefore applying approaches within a statistical thermodynamic treatment of the data. The most popular method has been so far the Maximum Entropy Method, which is not discussed in a dedicated chapter of this book but will be briefly introduced in the next section. It could be possible that in the next future, methods like the so-called charge flipping [19], also based on pseudo-thermodynamic ideas, could be regarded with more attention by research groups active in accurate CD determination.

1.2.2.2 Maximum Entropy Method

The Maximum Entropy Method (MEM) provides an alternative to MM refinements for the determination of accurate CDs from X-ray diffraction data [5, 20]. It is a technique based on information theory, which was originally introduced in the field of radioastronomy to enhance the information from noisy data [21]. The MEM approach makes use of the concept of information entropy to find the most probable distribution of numerical quantities (in our case, the values of the ED) over the ensemble of equivalent pixels composing the system under exam (in our case, the unit cell). Within MEM, the most probable density is the result of a "compromise" between the one which best fits the diffraction data and that one which maximizes the information entropy S [22],

$$S = -\sum_{j=1}^{N_p} \rho_j \ln \left(\frac{\rho_j}{\rho_j^{prior}} \right) \qquad (1.7)$$

where N_p is the total number of pixels in the unit cell, $\rho_j = \rho(\mathbf{r}_j)$ is the value of the ED in the pixel j centered at $\mathbf{r}_j$ and $\rho_j^{prior} = \rho^{prior}(\mathbf{r}_j)$ is the corresponding value for a convenient, still arbitrary, "prior" reference density. Usually the prior is taken as the uniform distribution or as the ED corresponding to the IAM model. Without the constraint to fit the experimental structure factors, maximizing the entropy would invariably produce a uniform distribution and any deviation from such distribution has thus to be contained in the data.

This book does not deal with the MEM approach and its applications in great detail. MEM is introduced and discussed in the more general context of the so-called Bayesian approach to data interpretation in Chap. 5, while application of MEM to determine the spin density from PND experiments is presented in Chap. 8. Use of a combined and iterative MEM-Rietveld method to get structural information and imaging of the charge density of the guest molecules and of the host framework in metal organic frameworks (MOFs) from synchrotron powder diffraction data is illustrated in Chap. 13.

The basic aspects of the application of the MEM approach in crystallography are neatly and concisely summarized in the book by Coppens [5], while a discussion of most recent methodological improvements [23, 24] and applications of the MEM, in particular to molecular crystals of biological interest, may be found in Ref. [25]. MEM has indeed several important features. It does not suffer from the correlations among the parameters typical of the pseudoatom model, nor does it require applying restrictions among these parameters which go beyond those imposed by space-group symmetry. The problem is known to be particularly severe for systems with very large unit cells, like the proteins, where the number of free parameters of the MM becomes soon intractable, also in consideration of the generally low quality of the data. Several databases of transferable MM parameters have been proposed and adopted in such cases as a convenient way to drastically lower the number of parameters to be refined and their related correlations (see Chaps. 11 and 15). Another challenging advantage of MEM is that it treats on the same footing the ED both in the well-ordered regions and in those affected by static or dynamic disorder. All these features make MEM a very attractive method for structural and CD studies of protein crystals, which have a huge number of atoms in the cell and with only a fraction of them sitting in a well-ordered part of the protein. Static and dynamic disorder also typically occurs in materials of technological interest, with their peculiar properties being in most cases the result of such a disorder [26] (Sect. 1.5.1 and Chap. 13). Having a method able to detect the deviation from a reference ideal (symmetrical) structure and to provide a CD and bonding picture of the consequences of such a departure, is clearly of utmost technological importance. The structural information obtained from the MEM analysis may then be used as a convenient starting hypothesis for a conventional MM refinement using fractional occupancies for the disordered sites [26].

The MEM yields a map of the dynamic ED, ρ_{dyn}, while the MM provides both ρ_{dyn}, using the convolution approximation (Eq. 1.5), and the static ED, ρ_{stat}, from the sum of the refined static atomic densities $\rho_j^{at,\,static}$. The physical properties and the chemical bonding features (Sects. 1.2.4, 1.3.6 and Chaps. 9, 10, 12, and 13) are usually evaluated at the static level since, using ρ_{stat}, one affords them for a fixed and most probable nuclear arrangement, at the temperature of data collection. This is a most convenient choice, which allows for a direct comparison with ab-initio results within the Born-Oppenheimer approximation, but which also calls for an adequate deconvolution of ED and thermal motion parameters for its ultimate success (Chaps. 3 and 5). The chemical features (nuclear positions peaks, charge concentrations along the covalent bonds or in the lone pair regions) are generally flattened in the dynamic density and the study of bonding using ρ_{dyn} is occasionally performed only for comparison purposes between MEM and MM results (on ρ_{stat} and ρ_{dyn}) [25].

1.2.2.3 Improving the Thermal Motion Description

The electron and thermal motion parameters of the MM are often severely correlated. The problem becomes particularly serious in the case of H atoms (see Sect. 3.5.3), which should generally be modeled by combining information from X-ray diffraction with knowledge about their thermal motion from other independent sources [27]. The most common approach is that of using information from parallel neutron diffraction experiments [28, 29], but many alternatives routes have recently been suggested and their results compared to independently estimated ADPs. The new strategies include the use of spectroscopic data (IR and Raman) [30, 31] or of the (internal motion) vibrational spectra obtained either from *ab-initio* calculations on crystals [Chaps. 2 and 3] or on isolated crystal subunits in the gas-phase [32, 33]. The refinement of the overall rigid body motion and selected internal modes against the ADPs of structures investigated at several temperatures has also been proposed and explored [34]. A full account of the crucial progresses made during the last decade in treating the thermal motion and in so affording more accurate CDs is given in Chap. 3 along with the take-home message that the fact that X-ray diffraction experiment provides information about $<\rho>$ and not simply about ρ_{stat} should be more and more regarded as a true challenge for theory and experiment rather than a source of problems only. Efforts in this direction are currently targeted at analyzing and assessing the physical significance of the displacement parameters and at trying to start to move away from the independent atomic motion model towards a normal-mode based model of collective vibrations. With this latter improvement the number of parameters needed for describing thermal motion should eventually decrease.

1.2.3 *Electron Density and Beyond from ab-initio Calculations*

1.2.3.1 From the Wavefunction to the Electron Density

Equation 1.1 leads to different expressions for $\rho(\mathbf{r})$, depending on the functional form adopted for ψ_{el}. In the simplest case, the N-electron ground state wavefunction is approximated by an antisymmetrized product of N orthonormal one-electron functions ξ_i (spin orbitals) and is expressed in terms of a Slater determinant as: $|\psi_{el}\rangle = |\xi_1\xi_2.....\xi_N\rangle$ [35]. The electron density is the expectation value of the one-electron density operator $\hat{\rho}(\mathbf{r}) = \sum_i^N \delta(\mathbf{r}_i - \mathbf{r})$ and may then be obtained [35] from the general rules for the evaluation of the matrix elements of one-electron operators, $\hat{O}_1 = \sum_i^N O_1(\mathbf{r}_i)$, between N-electron Slater determinants $|K\rangle$:

$$\left\langle \hat{O}_1 \right\rangle = \langle K|\, \hat{O}_1\, |K\rangle = \sum_i^N \langle \xi_i|\, O_1\, |\xi_i\rangle$$

$$\rho(\mathbf{r}) = \langle K | \hat{\rho}(\mathbf{r}) | K \rangle = \sum_i^N \langle \xi_i | \delta(\mathbf{r}_i - \mathbf{r}) | \xi_i \rangle = \sum_i^N \xi_i^*(\mathbf{r}) \xi_i(\mathbf{r}) \tag{1.8}$$

By applying the variational principle under the only constraints of the double occupation for each spatial orbital, ϕ_i (with $\phi_i = \xi_i \sigma,\ \sigma = \alpha\ or\ \beta$), and of the orthogonality among them, $\langle \phi_i \mid \phi_j \rangle = \delta_{ij}$, the optimum wavefunction of type $|K\rangle$ is obtained. The variational flexibility lies in the form of the orbitals and the minimal energy wavefunction corresponds to the Hartree-Fock (HF) wavefunction [35]. With the constraint of double occupation of the ϕ_i (closed shell systems), Eq. 1.8 takes the form

$$\rho(\mathbf{r}) = 2 \cdot \sum_i^{N/2} \phi_i^*(\mathbf{r}) \phi_i(\mathbf{r}) \tag{1.9}$$

where the ϕ_i are the so called molecular orbitals (MOs). Within the MO-LCAO (Molecular Orbital as Linear Combination of Atomic Orbitals) approximation put forth by Roothaan,

$$\phi_i = \sum_j^m c_{ij} \chi_j \tag{1.10}$$

the variational flexibility in determining ϕ_i becomes limited to their expansion coefficients in terms of a fixed and suitable set of m basis functions $\{\chi\}$ (*basis set*), generally centered on the various nuclei of the system. For a closed-shell system, the electron density in the HF-MO-LCAO approach is then given by

$$\begin{aligned} \rho(\mathbf{r}) &= 2 \cdot \sum_i^{N/2} \sum_{j=1}^m c_{ji}^* \chi_j^*(\mathbf{r}) \sum_{k=1}^m c_{ki} \chi_k(\mathbf{r}) \\ &= \sum_{k,j=1}^m \left(2 \cdot \sum_{i=1}^{N/2} c_{ki} c_{ji}^* \right) \chi_k(\mathbf{r}) \chi_j^*(\mathbf{r}) = \sum_{k,j=1}^m \mathbf{P}_{kj}\, \chi_k(\mathbf{r}) \chi_j^*(\mathbf{r}) \end{aligned} \tag{1.11}$$

where the matrix **P** is the representation over $\{\chi\}$ of the first-order or one-electron *density matrix* γ_1. Note that **P** fully specifies $\rho(\mathbf{r})$ once $\{\chi\}$ and the point **r** are selected. A similar expression for the electron density holds within the Density Functional Theory (DFT) [36] Kohn-Sham approach [37], which in its standard implementation [38, 39] is, formally, also a single-determinant method. Extension of Eq. (1.11) and of first order density matrices to the regularly periodic systems, like are the ordered crystals, is detailed in Chap. 2.

1.2.3.2 Density Matrices and One- and Two-Electron Properties

Density matrices are a convenient mathematical device [8] for evaluating the expectation values of operators corresponding to physical observables (*see infra* and Chaps. 2, 5, 7 and 9). A density matrix of order p for a pure state of an N-electron system is given by [36]

$$\gamma_p(\mathbf{x}_1\mathbf{x}_2...\mathbf{x}_p;\mathbf{x}'_1\mathbf{x}'_2...\mathbf{x}'_p) = \binom{N}{p}\int \psi_{el}(\mathbf{x}_1\mathbf{x}_2...\mathbf{x}_p\mathbf{x}_{p+1}...\mathbf{x}_N)\psi^*_{el}(\mathbf{x}'_1\mathbf{x}'_2...\mathbf{x}'_p\mathbf{x}'_{p+1}...\mathbf{x}'_N)d\mathbf{x}_{p+1}...d\mathbf{x}_N \tag{1.12}$$

where $\{\mathbf{x}\}$ and $\{\mathbf{x}'\}$ are two sets of independent space ($\mathbf{r}$) and spin ($\mathbf{s}$) variables ($\mathbf{x}_i \equiv \mathbf{r}_i\mathbf{s}_i$; $\mathbf{x}'_i \equiv \mathbf{r}'_i\mathbf{s}'_i$). A numerical value is assigned to Eq. 1.12 by two sets of indices. γ_p may thus be seen as an element of a matrix and $\gamma_p(\{\mathbf{x}\}\ ;\{\mathbf{x}'\})$ as the corresponding *matrix* with infinite elements. The diagonal elements ($\mathbf{x}_i \equiv \mathbf{x}'_i$) of this matrix correspond to the probability of finding p electrons with given space and spin coordinates and regardless of those of the remaining (N-p) electrons. This is a p-particles density, motivating why $\gamma_p(\{\mathbf{x}\}\ ;\{\mathbf{x}'\})$ is called a *density matrix* of order p. $\binom{N}{p}$ is a binomial coefficient ensuring proper normalization. For instance, the second-order, γ_2,

$$\gamma_2(\mathbf{x}_1\mathbf{x}_2;\mathbf{x}'_1\mathbf{x}'_2) = \frac{N(N-1)}{2}\int \psi_{el}(\mathbf{x}_1\mathbf{x}_2\mathbf{x}_3..\mathbf{x}_N)\psi^*_{el}(\mathbf{x}'_1\mathbf{x}'_2\mathbf{x}'_3..\mathbf{x}'_N)d\mathbf{x}_3...d\mathbf{x}_N \tag{1.13}$$

and the first order, γ_1, density matrices

$$\gamma_1(\mathbf{x}_1;\mathbf{x}'_1) = N\int \psi_{el}(\mathbf{x}_1\mathbf{x}_2..\mathbf{x}_N)\psi^*_{el}(\mathbf{x}'_1\mathbf{x}'_2..\mathbf{x}'_N)d\mathbf{x}_2...d\mathbf{x}_N \tag{1.14}$$

turn out to be correctly normalized to the number of electron pairs (N(N-1)/2), and to the number of electrons N, respectively [36]. A density matrix of order p, with $p < N$, is often called a *reduced* density matrix of the corresponding order because of integration over N-p of the $\psi_{el}\psi^*_{el}$ product variables [36].

Many operators do not involve spin coordinates (e.g. the Hamiltonian operators for atoms or molecules) and density matrices may be further reduced to their spinless counterparts P, by integration (summation) over the spin coordinates (spin up, s = 1/2 and spin down, s = −1/2):

$$P_p(\mathbf{r}_1\mathbf{r}_2...\mathbf{r}_p;\mathbf{r}'_1\mathbf{r}'_2...\mathbf{r}'_p) = \int \gamma_p(\mathbf{x}_1\mathbf{x}_2...\mathbf{x}_p;\mathbf{x}'_1\mathbf{x}'_2...\mathbf{x}'_p)d\mathbf{s}_1....d\mathbf{s}_p$$

$$P_2(\mathbf{r}_1\mathbf{r}_2;\mathbf{r}'_1\mathbf{r}'_2) = \int \gamma_2(\mathbf{x}_1\mathbf{x}_2;\mathbf{x}'_1\mathbf{x}'_2)d\mathbf{s}_1 d\mathbf{s}_2 \tag{1.15}$$

$$P_1(\mathbf{r}_1;\mathbf{r}'_1) = \int \gamma_1(\mathbf{x}_1;\mathbf{x}'_1)d\mathbf{s}_1 \tag{1.16}$$

Since only one- and two-body interactions tale place among electrons, we may from now on restrict our attention to γ_1 and γ_2 or to P_I and P_2 for spin-independent effects [8, 36].

It is easily shown [8, 36] that the expectation value of any one- and two-electron operators may be concisely expressed as

$$\left\langle \hat{O}_1 \right\rangle = \left\langle \sum_i^N O_1(\mathbf{x}_1) \right\rangle = \int_{\mathbf{x}'_1=\mathbf{x}_1} O_1(\mathbf{x}_1)\gamma_1(\mathbf{x}_1;\mathbf{x}'_1)d\mathbf{x}_1 \tag{1.17}$$

$$\left\langle \hat{O}_2 \right\rangle = \left\langle \sum_{i,j}^N {}' O_2(\mathbf{x}_i\mathbf{x}_j) \right\rangle = \int_{\mathbf{x}'_1=\mathbf{x}_1;\mathbf{x}'_2=\mathbf{x}_2} O_2(\mathbf{x}_1\mathbf{x}_2)\gamma_1(\mathbf{x}_1\mathbf{x}_2;\mathbf{x}'_1\mathbf{x}'_2)d\mathbf{x}_1 d\mathbf{x}_2 \tag{1.18}$$

using the convention that (*a*) the operators act only on functions of the unprimed variables in the expressions giving γ_1 and γ_2 (Eqs. 1.13 and 1.14), and that (*b*) $\mathbf{x}'$ is put equal to $\mathbf{x}$ after operating with the operators but before completing the integration. Analogous expressions hold for spin-independent operators by replacing γ_1 and γ_2 with P_I and P_2 and the $\mathbf{x}$, $\mathbf{x}'$ with the $\mathbf{r}$, $\mathbf{r}'$ variables. For operators that are just a multiplier (e.g. some function of the coordinates), like the operator related to the potential energy of an electron in the field of the nuclei or that associated to the electrostatic interaction between pair of electrons, the primes in Eqs. 1.17 and 1.18 and in those equations where P_I and P_2 replace γ_1 and γ_2 may be dropped at once [8].

In summary, all one-electron and all two-electron properties may be obtained from the first and second order density matrices. Knowing an electron density from experiment (Sect. 1.2.2) enables one to calculate *that subset* of one-electron properties which are expectation values of purely multiplicative operators. For this kind of operators, Eq. 1.17 implies an average of the operator just over the electron density. Indeed, from Eqs. 1.16 and 1.14, the diagonal element of P_I ($\mathbf{r} = \mathbf{r}'$) equals the electron density $\rho_1(\mathbf{r})$ (denoted as $\rho(\mathbf{r})$ throughout the book, for the sake of simplicity).

1.2.3.3 The Pair Function and Bond Descriptors Derived Thereof

The diagonal elements of P_2 yield the *pair function* $\rho_2(\mathbf{r}_1,\mathbf{r}_2)$ [using the shorthand notation $\rho_2(\mathbf{r}_1,\mathbf{r}_2) \equiv \rho_2(\mathbf{r}_1,\mathbf{r}_2;\mathbf{r}_1,\mathbf{r}_2)$] which tells us how the motions of two electrons are correlated. It may be written as ([36] and references therein)

$$\rho_2(\mathbf{r}_1, \mathbf{r}_2) = \frac{1}{2}\left[\rho(\mathbf{r}_1)\rho(\mathbf{r}_2) - \rho_{2,xc}(\mathbf{r}_1, \mathbf{r}_2)\right] \tag{1.19}$$

where $\rho_{2,xc}(\mathbf{r}_1, \mathbf{r}_2)$ is the so called exchange-correlation density,[1] which incorporates all non classical effects and measures to which degree the density is excluded at $\mathbf{r}_2$ because of the presence of an electron at $\mathbf{r}_1$. $\rho_{2,xc}(\mathbf{r}_1, \mathbf{r}_2)$ integrates to N to correct the $N/2$ pair in excess of the total number of distinct pairs, $N(N\text{-}1)/2$, which would be obtained by integrating over $\mathbf{r}_1$ and $\mathbf{r}_2$ only the first term in parenthesis in the right-hand side of Eq. 1.19. The probability of finding one electron at $\mathbf{r}_1$ and another one at $\mathbf{r}_2$ deviates from the purely classical description of a product of independent densities (the factor ½ preventing double counting) because of Coulomb and Fermi correlation of electron motions. The first type of correlation accounts for the instantaneous electrostatic repulsion between electrons regardless of their spins, while the second arises from the fermion nature of electron particles and the related antisymmetry constraint on the wavefunction. In the HF method, where a MO orbital describes the motion of an electron in the mean field due to the (fixed) nuclei and to the average distribution of the remaining electrons, electron's motion is not instantaneously correlated, apart from that contribution (Fermi correlation) deriving from the antisymmetry requirement on same spin-electrons [35].

The exchange-correlation density serves as a convenient tool for defining several important descriptors which are intimately related to electron correlation and are increasingly being used in chemical bond analysis (Sects. 1.2.4 and 1.3.6 and Chaps. 9–10, 12, 13 and 18). Consider partitioning the molecular space in a number of non overlapping regions A, B . . . , spanning the whole space. The *localization index* $\lambda(A)$

$$\lambda(A) = \int_A \int_A \rho_{2,xc}(\mathbf{r}_1, \mathbf{r}_2)d\mathbf{r}_1 d\mathbf{r}_2 \tag{1.20}$$

and the *delocalization index* $\delta(A,B)$

$$\delta(A, B) = 2 \cdot \int_A \int_B \rho_{2,xc}(\mathbf{r}_1, \mathbf{r}_2)d\mathbf{r}_1 d\mathbf{r}_2 \tag{1.21}$$

[1]The ½ multiplicative factor in Eq. 1.19 complies with the adopted normalization of ρ_2 to the number of distinct electron pairs $N(N\text{-}1)/2$, as used in Parr's book [36]. In the literature on the exchange-correlation density, the Mc Weeney [8] ρ_2 normalization to $N(N\text{-}1)$ electron pairs is often adopted. Though somewhat less physically meaningful, such a normalization has undoubtedly some advantages in the derivations of quantities related to $\rho_{2,xc}$. However, we prefer to retain our initial choice and be consistent with the normalization adopted in Eq. 1.13 throughout all this chapter. Moreover, some authors prefer to precede $\rho_{2,xc}$ by a plus rather than a minus sign in Eq. 1.19 and, within such definition, $\rho_{2,xc}$ will integrate to –N. The different normalization on ρ_2 (and the difference in sign for $\rho_{2,xc}$) have clearly a consequence on some of the formula reported from now on in this section.

yield the number of electrons localized in basin A and delocalized (or shared) between two basins A and B, respectively [40, 41]. Note that by definition:

$$\sum_A \left[\lambda(A) + \frac{1}{2} \sum_{B \neq A} \delta(A, B) \right] = N \tag{1.22}$$

since $\rho_{2,xc}(\mathbf{r}_1, \mathbf{r}_2)$ integrate to N. Usually the non overlapping regions are taken as the Quantum Theory of Atoms in Molecules (QTAIM) atomic basins, but definitions given in Eqs. 1.20 and 1.21 and property (1.22) are completely general as they apply to any non overlapping and exhaustive set of regions. Localization and delocalization indices also have a very interesting statistical interpretation (Ref. [41] and Chap. 9], since it is easy to demonstrate (*see infra*) that the variance $\sigma^2(N_A)$ and covariances cov(N_A,N_B) of the basin electron populations N_Ω ($\Omega = A,B..$)

$$N_\Omega = \int_\Omega \rho(\mathbf{r}_1) d\mathbf{r}_1 \tag{1.23}$$

are intimately related to those indices

$$\sigma^2(N_A) = N_A - \lambda_A \tag{1.24}$$

$$\text{cov}(N_A, N_B) = -\delta(A, B)/2 \tag{1.25}$$

The variance is a measure of the quantum mechanical uncertainty of the basin population or of the degree of fluctuation of the electron pair. It becomes zero only when electrons are entirely localized into the basin, a limit reached only for totally isolated basins (atoms). The covariance between different groups of data is given by

$$\text{cov}(A, B) = <AB> - <A> <B> \tag{1.26}$$

where $<A>$, $<B>$ and $<AB>$ are averages of the data values and of their product; the diagonal elements of the covariance matrix are the variances

$$\sigma^2(A) = <A^2> - <A>^2 \tag{1.27}$$

Given that $\rho_2(\mathbf{r}_1, \mathbf{r}_2)\, d\mathbf{r}_1 d\mathbf{r}_2$ yields the probability density of finding simultaneously electron 1 at position $\mathbf{r}_1$ and electron 2 at position $\mathbf{r}_2$, Eq. 1.26 becomes

$$\text{cov}(N_A, N_B) = \int_A \int_B \rho_2(\mathbf{r}_1, \mathbf{r}_2) d\mathbf{r}_1 d\mathbf{r}_2 + \int_A \int_B \rho_2(\mathbf{r}_2, \mathbf{r}_1) d\mathbf{r}_2 d\mathbf{r}_1 - N_A N_B \tag{1.28}$$

and using the expression of ρ_2 in terms of the exchange-correlation density (Eq. 1.19) we get

$$\begin{aligned} \mathrm{cov}(N_A, N_B) &= -\frac{1}{2}\left[\int\limits_A\int\limits_B \rho_{2,xc}(\mathbf{r}_1,\mathbf{r}_2)d\mathbf{r}_1 d\mathbf{r}_2 + \int\limits_A\int\limits_B \rho_{2,xc}(\mathbf{r}_2,\mathbf{r}_1)d\mathbf{r}_2 d\mathbf{r}_1\right] \\ &= -\frac{\delta(A,B)}{2} \end{aligned} \tag{1.29}$$

Analogously, by taking into account the expression of $< N_A^2 >$ in terms of the electron and electron pair densities, developed by Bader and Stephens [42]

$$< N_A^2 >= 2\cdot\int\limits_A\int\limits_A \rho_2(\mathbf{r}_1,\mathbf{r}_2)d\mathbf{r}_1 d\mathbf{r}_2 + \int\limits_A \rho(\mathbf{r}_1)d\mathbf{r}_1 \tag{1.30}$$

Eq. 1.27 yields

$$\begin{aligned} \sigma^2(N_A) &= 2\int\limits_A\int\limits_A \rho_2(\mathbf{r}_1,\mathbf{r}_2)d\mathbf{r}_1 d\mathbf{r}_2 + N_A - (N_A)^2 \\ &= -\int\limits_A\int\limits_A \rho_{2,xc}(\mathbf{r}_1,\mathbf{r}_2)d\mathbf{r}_1 d\mathbf{r}_2 + N_A = -\lambda_A + N_A \end{aligned} \tag{1.31}$$

This statistical interpretation for λ_A and $\delta(A, B)$ is completely independent of the functional form adopted for the wavefunction and of whether it accounts only for the Fermi correlation (HF) or also for (a part of) the Coulomb electron correlation. However, the numerical values of the localization and delocalization indices will be clearly affected by the extent to which electron correlation is included [43, 44].

Using the exchange-correlation density, one may also introduce the scalar function $G^{\Omega}(\mathbf{r}_2)$ [45, 46],

$$G^{\Omega}(\mathbf{r}_2) = \int\limits_{\Omega} \rho_{2,xc}(\mathbf{r}_1,\mathbf{r}_2)d\mathbf{r}_1 \tag{1.32}$$

which has a quite interesting interpretation in terms of the exchange-correlation hole density [47], ρ^{hole}

$$\rho^{hole}(\mathbf{r}_2 \mid \mathbf{r}_1) = \rho(\mathbf{r}_2) - 2\frac{\rho_2(\mathbf{r}_1,\mathbf{r}_2)}{\rho(\mathbf{r}_1)} \tag{1.33}$$

This hole density measures the difference between the full electron density at point $\mathbf{r}_2$ and the conditional density that an electron lies at $\mathbf{r}_2$ when another *reference* electron is sitting at $\mathbf{r}_1$. Although ρ^{hole} neatly summarizes the effects of electron correlation, a more realistic picture and a more physically meaningful measure of the hole can be obtained [45] when the reference electron is known to be localized within a given, arbitrary, region of space Ω rather than being constrained to a single point location. By averaging the position of the reference electron over Ω,

$$\rho_{\Omega}^{hole}(\mathbf{r}_2) = \int_{\Omega} \rho^{hole}(\mathbf{r}_2 \mid \mathbf{r}_1)\, d\mathbf{r}_1 = \rho(\mathbf{r}_2) - 2 \cdot \frac{\int_{\Omega} \rho_2(\mathbf{r}_1, \mathbf{r}_2) d\mathbf{r}_1}{\int_{\Omega} \rho(\mathbf{r}_1) d\mathbf{r}_1} \tag{1.34}$$

one gets a *domain-averaged* exchange-correlation hole which when multiplied by the electron population of the basin Ω yields $G^{\Omega}(\mathbf{r}_2)$, a *charge-weighted* measure of the hole

$$G^{\Omega}(\mathbf{r}_2) = N_{\Omega} \rho_{\Omega}^{hole}(\mathbf{r}_2) \tag{1.35}$$

By combining Eqs. 1.34 and 1.19, one may easily demonstrate [46] the expression which was given earlier in Eq. 1.32:

$$\begin{aligned} G^{\Omega}(\mathbf{r}_2) &= N_{\Omega} \left[\rho(r_2) - \frac{\int_{\Omega} [\rho(\mathbf{r}_1)\rho(\mathbf{r}_2) - \rho_{2,xc}(\mathbf{r}_1, \mathbf{r}_2)]\, d\mathbf{r}_1}{N_{\Omega}} \right] \\ &= N_{\Omega} \left[\rho(r_2) - \frac{N_{\Omega}\rho(\mathbf{r}_2) - \int_{\Omega} \rho_{2,xc}(\mathbf{r}_1, \mathbf{r}_2) d\mathbf{r}_1}{N_{\Omega}} \right] = \int_{\Omega} \rho_{2,xc}(\mathbf{r}_1, \mathbf{r}_2) d\mathbf{r}_1 \end{aligned} \tag{1.36}$$

$G^{\Omega}(\mathbf{r}_2)$ was introduced some time ago by Ponec [45] and it is now increasingly been used to discuss chemical bonding in molecules (Sect. 1.3.6). When applied to HF wavefunctions which describe only the electron correlation of the same spin electrons, it is referred to as the charge-weighted Domain Averaged Fermi Hole or simply and more customarily as the *Domain Averaged Fermi Hole* (DAFH).

It is worth noting that the localization and delocalization indices are intimately related to $G^{\Omega_1}(\mathbf{r}_2)$ since they may be obtained by integrating this scalar function over Ω_1 and Ω_2, respectively (*cf.* Eqs. 1.20 and 1.21 with Eq. 1.32).

1.2.3.4 Electron Density, Density Matrices and Bonding Descriptors at Different Electron Correlation Levels

We have previously shown how all one-electron properties and all two-electron properties may be obtained from γ_1 and γ_2 (or P_I and P_2), through Eqs. 1.17 and 1.18. These equations and all the following ones are completely general and do not depend on the functional form adopted for ψ_{el}.

The Coulomb electron correlation, neglected at HF level, may be accounted for in several ways and one of the most general approach is that of taking the wavefunction as a linear combination of Slater determinants, defined in terms of the occupied and unoccupied subsets of MOs obtained from a preliminary single determinantal calculation [35]. The variational flexibility lies either in the expansion coefficients of the linear combination of Slater determinants (giving rise to the set of Configuration Interaction methods, CI) or also in the form of their composing orbitals (which leads to the various Multi Configuration Self Consistent Field, MCSCF, methods). Regardless of their specific form, most of the current theoretical *ab-initio* methods expand the wavefunction in terms of one-electron functions (MOs, natural orbitals) and the first and second order spinless density matrices take the general expressions [48]:

$$P_1(\mathbf{r}_1;\mathbf{r}'_1) = \sum_{i,j=1}^{m} P_{ij}\,\chi_i(\mathbf{r}_1)\chi_j^*(\mathbf{r}'_1) \quad (1.37)$$

$$P_2(\mathbf{r}_1\mathbf{r}_2;\mathbf{r}'_1\mathbf{r}'_2) = \sum_{i,j,k,l=1}^{m} D_{ijkl}\,\chi_i(\mathbf{r}_1)\chi_k(\mathbf{r}_2)\chi_j^*(\mathbf{r}'_1)\chi_l^*(\mathbf{r}'_2) \quad (1.38)$$

where P_{ij} and D_{ijkl} are elements of the corresponding matrix representations over the basis functions $\{\chi\}$. For wavefunctions expanded in terms of Slater determinants, the elements of **P** and **D** are computed from the coefficients of the wavefunction expansion and from the first and second-order transition density matrices associated to each pair of determinants [48, 49]. Evaluation of properties through Eqs. 1.17 and 1.18 requires in both cases the knowledge of **P** and **D** and of the matrix representations over the basis functions $\{\chi\}$ of the operators associated to the desired properties.

We recall that all one-electron properties, including the energy and ρ, are correct to the second-order at the HF level (Brillouin and Møller-Plesset theorems, [50]). Single-excitation configurations, i.e. Slater determinants differing by only one spin orbital with respect to the HF solution, give a null contribution to the second-order correction to the energy, which is only determined by the double-excitation configurations. However, single excitations provide the dominant contribution to the second-order correction to the electron density [51] and studies of correlated electron densities should always include both single and double excitations, since the coupling of singly excited configurations with the ground state through their interactions with the doubly excited ones is of paramount importance [52–54].

It is worth noting that within the single determinantal approach, γ_2 and P_2 turn out to be factorized in terms of γ_1 and P_I, respectively [8, 36]

$$\gamma_2(\mathbf{x}_1\mathbf{x}_2;\mathbf{x}'_1\mathbf{x}'_2) = \frac{1}{2}[\gamma_1(\mathbf{x}_1;\mathbf{x}'_1)\cdot\gamma_1(\mathbf{x}_2;\mathbf{x}'_2) - \gamma_1(\mathbf{x}_2;\mathbf{x}'_1)\cdot\gamma_1(\mathbf{x}_1;\mathbf{x}'_2)] \quad (1.39)$$

$$P_2(\mathbf{r}_1\mathbf{r}_2;\mathbf{r}'_1\mathbf{r}'_2) = \frac{1}{2}[P_1(\mathbf{r}_1;\mathbf{r}'_1)\cdot P_1(\mathbf{r}_2;\mathbf{r}'_2)] - \frac{1}{4}\left[P_1(\mathbf{r}_2;\mathbf{r}'_1)\cdot P_1(\mathbf{r}_1;\mathbf{r}'_2)\right] \tag{1.40}$$

with the pair function ρ_2 taking the simple form :

$$\rho_2(\mathbf{r}_1,\mathbf{r}_2) = \frac{1}{2}\left[\rho(\mathbf{r}_1)\rho(\mathbf{r}_2)\right] - \frac{1}{4}\left[P_1(\mathbf{r}_2;\mathbf{r}_1)P_1(\mathbf{r}_1;\mathbf{r}_2)\right] \tag{1.41}$$

Equations 1.39 and 1.40 tell us that within the one determinant constraint everything is determined by γ_1 (or P_I) and that only these first order density matrices are needed to evaluate any one- and two-electron property in such a circumstance. The accuracies of the properties will however reflect the rather incomplete treatment of electron correlation inherent to the single determinantal approach (HF or Kohn-Sham DFT, KS-DFT). Besides, for the case of KS-DFT method, the situation is far more problematic (*see infra*).

For a closed-shell system, in the one determinant approach, the P_1 density matrices take the following expressions in terms of the $M = N/2$ doubly occupied molecular or KS ϕ_i orbitals,

$$P_1(\mathbf{r}_1;\mathbf{r}'_1) = 2\cdot\sum_{i=1}^{M=N/2}\phi_i(\mathbf{r}_1)\phi_i^*(\mathbf{r}'_1) \tag{1.42}$$

Accordingly, based on Eqs. 1.40, 1.41 and 1.19, the spinless second order density matrix P_2, the pair density, ρ_2, and the exchange correlation density are given by

$$\rho_2(\mathbf{r}_1\mathbf{r}_2;\mathbf{r}'_1\mathbf{r}'_2) = \sum_{i,j=1}^{M=N/2}\Big\{2\cdot\left[\phi_i(\mathbf{r}_1)\phi_i^*(\mathbf{r}'_1)\phi_j(\mathbf{r}_2)\phi_j^*(\mathbf{r}'_2)\right] - \left[\phi_i(\mathbf{r}_2)\phi_i^*(\mathbf{r}'_1)\phi_j(\mathbf{r}_1)\phi_j^*(\mathbf{r}'_2)\right]\Big\} \tag{1.43}$$

$$\rho_2(\mathbf{r}_1,\mathbf{r}_2) = \sum_{i,j=1}^{M=N/2}\Big\{2\cdot\left[\phi_i(\mathbf{r}_1)\phi_i^*(\mathbf{r}_1)\phi_j(\mathbf{r}_2)\phi_j^*(\mathbf{r}_2)\right] - \left[\phi_i(\mathbf{r}_2)\phi_i^*(\mathbf{r}_1)\phi_j(\mathbf{r}_1)\phi_j^*(\mathbf{r}_2)\right]\Big\} \tag{1.44}$$

$$\rho_{2,xc}(\mathbf{r}_1,\mathbf{r}_2) = \sum_{i,j=1}^{M=N/2} 2\left[\phi_i(\mathbf{r}_2)\phi_i^*(\mathbf{r}_1)\phi_j(\mathbf{r}_1)\phi_j^*(\mathbf{r}_2)\right] \tag{1.45}$$

By combining Eq. 1.45 with Eqs. 1.32, 1.20 and 1.21, the previously defined electron localization/delocalization descriptors may be easily expressed in terms of MOs or KS orbitals, as follows:

$$G^{\Omega_1}(\mathbf{r}_2) = 2 \cdot \sum_{i,j=1}^{M=N/2} \int_{\Omega_1} \phi_i(\mathbf{r}_2)\phi_i^*(\mathbf{r}_1)\phi_j(\mathbf{r}_1)\phi_j^*(\mathbf{r}_2)d\mathbf{r}_1$$

$$= 2 \cdot \sum_{i,j=1}^{M=N/2} S_{ij}^{\Omega_1}\phi_i(\mathbf{r}_2)\phi_j^*(\mathbf{r}_2) \quad (1.46)$$

$$\lambda(\Omega_1) = 2 \cdot \sum_{i,j=1}^{M=N/2} \int_{\Omega_1}\int_{\Omega_1} \phi_i(\mathbf{r}_2)\phi_i^*(\mathbf{r}_1)\phi_j(\mathbf{r}_1)\phi_j^*(\mathbf{r}_2)d\mathbf{r}_1 d\mathbf{r}_2 = 2 \cdot \sum_{i,j=1}^{M=N/2} \left(S_{ij}^{\Omega_1}\right)^2 \quad (1.47)$$

$$\delta(\Omega_1, \Omega_2) = 4 \cdot \sum_{i,j=1}^{M=N/2} \int_{\Omega_1}\int_{\Omega_2} \phi_i(\mathbf{r}_2)\phi_i^*(\mathbf{r}_1)\phi_j(\mathbf{r}_1)\phi_j^*(\mathbf{r}_2)d\mathbf{r}_1 d\mathbf{r}_2 = 4 \cdot \sum_{i,j=1}^{M=N/2} S_{ij}^{\Omega_1}S_{ij}^{\Omega_2} \quad (1.48)$$

where

$$S_{ij}^{\Omega_k} = \int_{\Omega_k} \phi_i^*(\mathbf{r}_1)\varphi_j(\mathbf{r}_1)d\mathbf{r}_1 \quad (1.49)$$

is the domain overlap matrix between the ϕ.

The localization and delocalization indices can, in principle, be exactly computed at correlated levels of theory such as CI, MCSCF, Møller-Plesset perturbation methods, coupled-cluster, etc. General expressions for these indices in terms of the second order density matrix representations over the MOs or over the basis functions $\{\chi\}$ may be derived from Eq. 1.38 and are explicitly reported in Ref. [43]. However, γ_2 (and P_2) are often computationally unavailable and in such cases it has become common practice to derive approximate λ and δ values by substituting an "HF-like" second order exchange density matrix (expressed in terms of γ_1 only, Eq. 1.39), for γ_2 (and P_2) [41, 43, 55]. Proceeding in this way, one takes into account the Coulomb electron correlation effects on the first-order density matrix but not those on the pair density. A detailed analysis of the differences on the calculated λ and δ arising from such approximation is reported in Ref. [43].

The situation is much less definite when dealing with a KS DFT wavefunction, despite its formal mono-determinantal nature. Use of an HF-like pair density from the KS orbitals to derive approximate λ and δ values has become quite usual, but this practice has not a true physical basis [43, 55]. Indeed, in the KS approach a computationally heavy many-electron problem is "magically" transformed into a computationally more tractable problem of noninteracting electrons moving in a self-consistent field in a way that is exact in principle for the ground-state energy and electron density [36, 39]. This latter is used as the prime variable to calculate (among other properties) the correlation energies. In DFT, the wavefunction is eliminated as the unknown quantity in favor of the electron density. Hence, density matrices of any order are neither available, nor even definable in the DFT context. Apart from those depending on ρ only, any other one- and two-electron property values obtained from DFT using HF-like density matrices should in principle be regarded as flawed. In practice, the situation is far less depressing than just outlined. It was found that the λ and δ indices obtained using KS orbitals in the HF-like pair density expression are often chemically reasonable and qualitatively similar to those calculated from an HF function with the same or similar basis set [43, 56]. In those cases, where the HF method is unable to yield a qualitatively correct geometry, like for many organometallic compounds with metal-metal bonding interactions, the λ and δ indices obtained from DFT may be even more reliable than those obtained from the HF method, because of the more correct geometry adopted for the former [56].

1.2.4 Obtaining and Comparing Physical Properties from Experiment and Theory

1.2.4.1 Properties from Theory as Expectation Values of Operators or as Responses to a Perturbation

In Sect. 1.2.3 we presented and discussed in some detail how to obtain one- and two-electron properties from *ab-initio* wavefunctions. We need now only to add that any property may be defined either as an expectation value of the associated operator (Sect. 1.2.3) or as a *response* of the system with respect to a perturbation (an energy derivative) ([54] and references therein, [44]). For exact wavefunctions and some approximate methods, like the self-consistent approach, both routes lead to the same result because of the Hellmann-Feynman theorem [50]. However, for most many-body methods, with the exception of those where the chosen active space is fully exploited (CASSCF and Full-CI calculations) such theorem is not fulfilled. As a consequence, the ambiguity arises of whether calculate a given property from the corresponding energy derivative (which is often easier for post-Hartree Fock wavefunctions) or as an expectation value. Analogous problems occur for the density matrices which can be obtained either as an expectation value of the proper Dirac's delta based operator or as a wavefunction response [44]. In the latter case the density matrix is known as a response or *relaxed density* since it includes also the effects due to orbital relaxation within the Coupled-Perturbed Hartree-Fock (CPHF) calculations. More technical and fundamental details on this delicate issue may be found, for instance, in Refs. [54, 44]. Section 2.5.3 of Chap. 2 in this book illustrates to some extent how such a problem has been afforded for estimating the correlated density matrices and physical properties within a local-MP2 approach to the *ab-initio* simulation of periodically ordered crystalline solids.

1.2.4.2 Properties from Bragg Data or Multipole-Model Pseudoatom Expansions

When dealing with Bragg diffraction data or computed structure factors, either derived from MM refinement of X-ray data or from *ab-initio* wavefunctions, electrostatic properties may be obtained through corresponding Fourier summations of the structure factors (see Eq. 1.3 for the electron density and Ref. [57] for a more complete list). Each property has a different dependency on $|\mathbf{H}|^n = \left(\frac{2\sin\vartheta}{\lambda}\right)^n$, with n being 0 for the electron density and the electric field gradient (EFG), -2 for the electrostatic potential (ESP), -1 for the electric field (EF), and $+1$ or $+2$ for the gradient and the Laplacian of the electron density, respectively. As a consequence, the convergence of the Fourier sum and the relative importance of the different sampled regions of reciprocal space depend on the property being evaluated. High angle data, which are experimentally more difficult to obtain (Chaps. 3 and 19), take an increasing emphasis with increasing n and become particularly necessary for evaluating the electron density gradient and Laplacian. Note that such functions, along with the electron density, have a prominent role in the Quantum Theory of Atoms in Molecules and that local concentrations of electronic charge upon (covalent) bonding yield to more diffuse distributions in reciprocal space, hence to significant contributions to the high angle data also from the valence electrons.

Fourier summations apply to an infinite Ewald sphere, but at $|\mathbf{H}|_{\max} = 2/\lambda$ they are necessarily terminated. Larger $|\mathbf{H}|$ would be available using harder radiations, like Ag Kα in laboratories or otherwise higher energies available at the Synchrotron stations. Nevertheless, some problems of termination will remain and the decreasing quality of data at higher resolution is unavoidable. A clever way to tackle the series termination is computing Fourier summation, using $\Delta F(\mathbf{H}) = F^{obs}(\mathbf{H}) - F^{mod}(\mathbf{H})$ as coefficients, thus deriving the "deformation" property, [57] i.e. the deviation from the pro-molecule behavior which can be easily calculated. The pro-molecule is obtained from the superposition of non-interacting atoms, so $F^{mod}(\mathbf{H})$ are calculated with the IAM model, using atomic positions and root-mean-square amplitudes of vibrations inferred from either X-ray or neutron diffraction experiments. F^{obs} may be phased by the IAM model or more precisely from aspherical models. If the $F^{mod}(\mathbf{H})$ accurately account for the missing data at large **H**, then the termination effects are eliminated. Most of the bonding effects are presumably well resolved by data within a Ewald sphere of $\approx$4 Å^{-1} (sin$\theta/\lambda \approx 2$ Å^{-1}) [57].

When using computed data from aspherical-model refinements, the series termination is apparently solved. However, one should never forget that data at high resolution have not been fitted. In addition, the accuracy of obtained properties largely decreases with increasing value of n. This may be a particularly serious problem for $n = 1$ or 2. Said in other words [3], "The multipole model represents an extrapolation to infinite resolution from a finite set of experimental data. Sharp features, even those induced in the valence density by bonding effects, may not be represented in the multipole model maps as they will modify the X-ray scattering at higher Bragg angles only" [3].

Using aspherical-model refinements of structure factors from experiment or *ab-initio* wavefunctions, properties may be also obtained by direct (real) space mapping via multipole functions. In such a case, a portion of the crystal lattice is constructed and the property is valuated at points of interest by summation over the contributions from all pseudoatoms of this cluster. The sum converges nicely for properties with $n > -1$, whereas for the electrostatic potential and the EF this is not so. For this reason, Ewald summation in reciprocal space (based on the *FT* of the aspherical density) can be used to map these scalar or vectorial properties in crystals, in combination with direct space evaluation of the electrostatic properties of the pro-molecule [58, 59].

The direct space reconstruction of the electron density and its properties from aspherical density offers a very appealing alternative, which is the calculation of a molecule "extracted" from the crystal, i.e. including in the calculation only pseudoatoms forming that molecule or molecular aggregate. The evaluation of integrals related with electrostatic potential is not trivial and it has been widely discussed in the literature, with different procedures [58–62]. When the molecule is extracted from the crystal there is no concern about the convergence as there is no summation to carry out. The electron density and derived properties in the fragment reflect however the matrix effect due to the crystalline environment, since pseudoatom densities are refined within such an environment. Therefore, strictly speaking, they cannot be compared against properties calculated on the same molecule in the gas phase, although this is a common procedure and an acceptable approximation for molecular crystals in the absence of very strong intermolecular interactions.

As mentioned before, all direct space properties are usually calculated at the static level. When derived from experimental data, by "static" it is meant an energy-weighted average over the static densities of each of the points along the vibrational surface of the molecule or crystal [2], rather than over the hypothetical stationary density obtained from a single point *ab-initio* calculation on a potential energy surface (minimum), performed within Born-Oppenheimer approximation.

1.2.4.3 Making Use of and Comparing Experimental and Theoretical Properties

As anticipated in Sect. 1.2.3, knowing an electron density from experiment enables one to calculate only *that subset* of one-electron properties which are expectation values of a purely multiplicative operator. Experimental charge densities studies have now advanced sufficiently to permit meaningful comparison with theoretical results [3, 63], and to possibly even surpass them in accuracy in selected cases [13, 64].

Typically comparison between experiment and theory involves the electron density itself, the various electrostatic properties derived from the electron density (see earlier) and the atomic and molecular electrostatic moments which may be obtained through one of the many available partitioning schemes, roughly classifiable in two main categories – those defining the atomic fragments in terms of *discrete* boundaries and those where boundaries are *fuzzy* and the "atoms" may overlap to some extent (Chaps. 9, 11 and 14). Electrostatic moments can be used for estimating the electrostatic contribution to intermolecular interactions and to lattice energies (Chaps. 11, 15). They were customarily evaluated using the Buckingham algorithm [65], based on the approximation that the densities of the interacting molecular fragments do not overlap. However, electric moments are unable to well reproduce the interaction between two atoms very close apart (as those forming a strong intermolecular interaction like an hydrogen bond, HB). In fact, the mutual penetration of the two electron distributions perturbs the interaction between atom centered electrostatic moments. For this reason, the actual space distribution of the two interacting electron densities should be properly considered. Spackman first recognized this [66], but the model was lacking the most important part of the penetration as later recognized [67]. Volkov [68] and Gavezzotti [69] proposed exact volume integration of the electron densities, as we will see in Sect. 1.3.7. Calculations could be very expensive if the *exact potential* is computed for all atom-atom interactions and at relatively large distances the Buckingham approach is sufficiently correct. Therefore the combined exact-potential (short atom-atom distances) multipole model (large atom-atom distances) EP/MM method [70] has now become the standard tool in the area. Intermolecular electrostatic interactions energies have been also compared [71] with those obtained from *ab-initio* wavefunctions within the Morokuma-Ziegler energy decomposition scheme (Chap. 11), confirming the need to use exact potential at short distances.

Competition among different philosophies and approaches to obtain electrostatic moment databanks and/or their expansion in terms of the multipole model contributions has been particularly vivid and thought provoking in the last 10 years (see e.g. [72, 73]). Approaches differ as of whether data are derived directly from experiment and averaged over a set of chemically related systems (Chap. 15), or obtained only from theory ([74], Chap. 11], or from exploiting *ab-initio* densities of molecules or molecular groups to get an improved expansion of experimental densities in terms of atomic moments or of multipole populations [75, 76]. Regardless of the way it is obtained, building a databank of transferable pseudoatoms is of particular relevance for calculating the interaction energies of very large systems (e.g. macromolecules), for which neither charge density quality data (if not the crystals themselves) nor accurate enough theoretical densities are available.

Electrostatic moments may then be used to improve the current force fields (Chaps. 11, and 14) or even to design new ones (Chap. 14) to be used in molecular dynamic simulations. Obtaining a good description of the electrostatic interaction between molecules is indeed one of the big problems in the development of current force fields [77].

Comparison between results derived from experiment or theory is often biased by the known inadequacies of the multipole models adopted in the aspherical refinements (Sect. 1.2.2), especially those deficiencies due to the choice of the radial fit functions, as extensively discussed in Chap. 5. The problem is particularly serious for those local properties of the electron density which involve higher derivatives of the density (*cf.* Sect. 1.2.4.1), like the density gradient, Laplacian and curvatures, these latter being the eigenvalues of the density Hessian matrix whose sum yields the ED Laplacian. It has so become an often common practice to filter the theoretical results by refining the theoretical structure factors through the MM before making any comparison between theory and experiment [3, 78–80]. Furthermore, as anticipated earlier, differences among the primary density properties and those obtained after filtering them through the MM, allows estimating the bias introduced by the MM on the various properties.

As for the electron density curvatures, it was also shown [70] that theory and experiment may be brought to a better agreement when Slater type functions (STF), which largely improve the description of the core density relative to the commonly adopted Gaussian type functions (GTF), are used in the wavefunction calculations. The Amsterdam Density Functional code, ADF [81], not only makes use of STFs, but also applies relativistic corrections, which are very important for heavier elements, and which also find correspondence in the theoretical scattering factors and functions (corrected for relativistic corrections,[82]) used in modern MM approaches [83].

1.2.4.4 Interaction Densities

Differences between experimental and theoretical densities also arise when the matrix effects are not taken into account in the latter. Chapter 2 summarizes the current status of the art of *ab-initio* calculations on systems with periodicity (so also on crystals), which by nature include such effects. The *interaction density*, that is the electron density of the crystal minus the superposition of the electron density of isolated molecules placed as in the crystal, provides an obvious measure of the matrix effects. It has often been considered as a quite elusive quantity and a number of studies have extensively discussed [84–87] whether it is amenable or not to experimental determination. Although ambiguities and practical difficulties arise for evaluating the interaction density [87], it is now essentially recognized [3, 88] that matrix effects can in principle be detected by both experimental and theoretical methods, especially so if analyzed in real space and possibly with the help of the electron density topological analysis (*see infra*) [89–91]. In particular molecular regions, the interaction density may be as large as 0.1 e$Å^{-3}$, which is well within the presently achievable experimental accuracy [3]. A large enhancement (even > 70%) of molecular dipole in the crystal with respect to its value in the vacuum has been repeatedly observed for several hydrogen-bonded molecular crystals [63, 79, 88, 90, 92, 93]. It provides a clear evidence of the relevance of the cumulative charge density rearrangement arising from the matrix effects.

By using the constrained wavefunction method (CWM, Chap. 6) it has now become possible to calculate properties which go beyond those definable in terms of the electron density alone and where the fundamental information from X-ray diffraction data is exploited to include effects, like those due to the crystalline environment and the electron correlation, which would otherwise difficult to incorporate in many interesting cases. The estimation of molecular polarizabilities and crystal refractive indices from X-ray diffraction data represents a very promising and interesting application of the CWM (see Ref. [94] and Chap. 6).

1.2.4.5 Dynamical Properties

Most of the properties discussed in this book concern equilibrium electronic states. However, some of the most fundamental and technologically interesting properties of extended solids, such as transport phenomena (e.g. electronic conductivity, thermoelectricity, superconductivity, band magnetism, etc.) relate to the *dynamical* properties of electrons, which depend on the energy dispersion of electronic bands in momentum space [95]. Bands energies and velocities may be easily calculated through periodic calculations on crystals (Fig. 1.1), but the study of how the characteristic features of these reciprocal space properties impact on the CD (a real space property), has seldom been undertaken. Chapter 10 pioneers such studies using the quasi-one dimensional organometallic carbides as a test case.

Electronic transport properties, which are of relevance *e.g.* for thermoelectricity may be estimated from the electronic band structure, within the semi-classical Boltzmann's transport theory [95]. Studies relating the promising experimental and theoretical transport properties of novel thermoelectric materials to their composition, geometrical structure and ensuing charge density features have start to appear in the last decade. Examples of them are illustrated in Chap. 13 and Refs. [26, 96, 97].

1.2.5 Chemical Bonding

1.2.5.1 Deformation Densities

Deformation density maps, which are electron density differences in given suitable molecular or crystalline planes relative to a non bonded state (typically the IAM model), have been largely investigated in the past [5]. They have primarily been adopted in the chemical bond studies since offered a natural and efficient way to overcome the problem of the finite set of experimental data in the Fourier density sums (Sect. 1.2.4.2).

Despite these maps usually allow for an easy detection of the most relevant charge rearrangements occurring when atoms get bonded to one another – electron density accumulation along covalent bonds, charge concentrations associated to lone electron pairs, etc. – they have well-known serious conceptual drawbacks, which may eventually lead to the lack of the expected density accumulations along the bonds in some circumstances [98]. Classical examples are provided by the cases of the F-F bond in the F_2 molecule [98] and of the O-O bond in H_2O_2 [98]: they both show depletion rather than accumulation of charge in the bonding region. To resolve these inadequacies, promolecular densities build up from *oriented* or *prepared* for bonding atoms or the use of the so-called *chemical deformations densities,* where the optimum shape and orientation of the electron density of the free atoms is determined by minimizing the integrated square difference density with respect to the elements of the atomic first order density matrix, have been proposed [98]. These recipes, although often successful, are yet still arbitrary, since they introduce an a priori knowledge about the mechanism of bonding before the bond is analyzed.

The deformation maps cannot produce quantitative information on the chemical bonding (especially if obtained from Fourier summations). In addition they lose qualitative information when the chemical bond under investigation is less conventional than that in simple organic molecules. Nevertheless, this approach remains interesting to check the quality of a refinement, to spot unusual features in a system and, more importantly nowadays, to visualize spin densities differences, because this function does not involve any arbitrary reference but it is the difference between two actual density distributions of opposite spin (See more discussion in Chap. 8).

1.2.5.2 Total Densities and (Quantum) Topological Approaches

In the last 10–15 years, the primary focus of experimental CD studies has progressively shifted from the deformation to the *total density*. This is certainly due to the enormous increase of the X-ray data quality (see Sect. 1.3.1) [3, 63] and to the parallel progress in both the methodological and computational possibilities of properly refining such data (Sect. 1.3.4, Chaps. 3, 5 and 19). In the meantime, the capabilities of the *ab-initio* periodic methods have also largely improved (Chap. 2) and being the electron density the common available observable in the experimental and theoretical approaches, it has become quite natural and a common practice to adopt *density-based* topological tools to confront and mutually validate their outcomes [3, 4, 63, 91, 99–101]. In this way, a *real space* quantitative description of the chemical bonding in crystals is enabled rather than the conventional bonding analysis in the space of the specific and different mathematical functions used to expand the wavefunction or the MM.

The Quantum Theory of Atoms in Molecules, QTAIM, [6, 7] is certainly the most complete density-based topological tool for chemical bonding studies and despite some general criticisms and vibrant discussions on the chemical meaning of the bond path – probably the most exploited cornerstone of QTAIM (see Sect. 1.3.6) – this theory has, within the X-ray ED community, progressively been selected as the paradigm for discussing how atoms get bonded to one another in crystals [4, 5, 9, 63, 91, 100–102]. QTAIM provides a bridge between chemistry, ED and quantum mechanics and it shows that there is not any specific need to invoke arbitrary reference densities to discuss features relevant to chemistry since most of them are already included in ρ.

The ED alone, however, does not describe chemical bonding in its entirety, especially the mechanism of electron pairing. It is well-known, for instance, that the structure of the whole one-electron density matrix is remarkably influenced by covalent chemical bonding (Chaps. 7 and 9 and Refs. [103, 104]). Yet, the obvious difficulty of visualizing this matrix, has led to propose new convenient functions, defined in the real space and intimately related to electron pairing. Examples are the Electron Localization Function, ELF, [105], the Electron Localizability Indicator (ELI), [106], the Localized Orbital Locator, LOL, [107], and all those functions easily expressible in terms of the exchange-correlation density matrix (see Sect. 1.2.3), like the Domain Averaged Fermi Hole [45], and, within QTAIM, the Fermi hole [6] and the delocalization indices [40].

Among those functions able to provide direct insights on electron pairing, the ELF has enjoyed an enormous popularity, due to its quite unique feature of offering immediately understandable and chemically informative patterns of bonding. The ELF has indeed been implemented within most of the plane-wave based *ab initio* codes for the study of the electronic structure in solids and it represents now the reckoned standard for bonding analysis in solid state chemistry and physics [91]. Chapter 18 briefly introduces ELF and, along with Chap. 17, illustrates in detail, several ELF applications to the realm of chemical bond and chemical bond evolution in systems under pressure. The ELI, which in the HF approximation can be related to the ELF without using an arbitrary reference to the uniform electron gas [106] is presented at length in Chap. 9.

Depending on the wavefunction model, the electron-pair related functions are obtainable through the pair or the first-order density matrices and can therefore be derived from "experiment" only by using the CWM (Chap. 6, [108, 109]) or, potentially, by combining in a unique refinement different experimental data, coming from different kinds of experiments (Chap. 7). A number of functions, like the ELF and LOL, may be also derived through approximate formulas [110] from DFT theory using the experimental density, its gradient and Laplacian [111–113] that are known analytically. Similar expressions hold for all those descriptors related to the kinetic and potential energy densities that, along with other descriptors, are customarily used in QTAIM for discussing the nature of chemical bonds [91, 114]. However, it is worth saying that the numerical values of the approximate "version" of all these descriptors should be taken with a special caution. Besides the fact that the approximate values can hardly be regarded as true experimental outcomes, their general reliability is rather questionable when molecular regions or bonding interactions characterized by a quite distinct physical nature, hence very different values for the corresponding descriptors, are investigated. When the "exact" quantum mechanical values are compared with the corresponding estimates from the approximate DFT formula, a very unbalanced replica of the exact values of the various descriptors is generally observed, since the accuracies of the approximated values are intimately related to the nature of the molecular region being sampled [89, 91, 110].

Traditional investigations of bonding, based on the LCAO approaches to periodic systems and on the breakdown into atomic orbital contributions of the ensuing band structure or density of states, retain an undisputed importance for many interpretive and predictive aspects of bonding in crystals [115, 116]. Use of the Wannier functions [117, 118] enables one to obtain a local picture of bonding even when using plane-wave approaches. The method has also been extended to and applied into the LCAO realm [119]. Use of these tools is largely documented in the literature but it is not discussed at length in this book. Although not yet common [91, 120–122], use of local-functions based analyses in combination with the tools presented in this book should be probably encouraged and more pursued in the next future. This combined approach is fully explored in Chap. 10 and in Ref. [122] which show how tiny differences in the electronic band structure of organometallic carbides, strongly affecting their physical properties, may be faithfully recovered in terms of the properties of the ED and in particular of its Laplacian distribution, whose qualitative features are in turn unveiled by simple orbital based analyses.

1.3 Technological and Methodological Developments

The possibility to probe the electronic structure of atoms by means of X-ray was recognized quite soon after the first diffraction experiments. In 1915, P. Debye wrote: "*It seems to me that experimental study of the scattered radiation, in particular from light atoms, should get more attention, since along this way it should be possible to determine the arrangement of the electrons in the atoms*" [123]. However, at that time measurements were not accurate enough to precisely detail the electron density distribution. It took several decades before what anticipated by Debye could in fact be realized. This became possible thanks to:

(1) improvements of the data quality, provided by semi-automatic diffractometers coupled with scintillation detectors;
(2) availability of high power sources, like synchrotron radiation storage rings, and of neutron diffraction (for an independent assessment of thermal motion and atomic positions);
(3) developments of models of electron density from measured Bragg intensities.

These three aspects more or less remained crucial also during the youth and maturity of the field and are still fundamental nowadays, when the level of sophistication and therefore the expectations from a charge density analysis are quite substantially higher. Indeed, the availability of high performance detectors, sources and models remain vital. Any attempt by the authors to photograph the state of the art of the technology is unavoidably expected to become out of date for the reader who approaches this book. Nevertheless, we dare pointing out the most important breakdowns occurred in the history of this discipline and possibly anticipating some of the future developments. Past experiences could certainly stimulate new progresses and new techniques, that's why it is always important to look backward and learn about past motivations in order to successfully improve the methodologies for future, higher quality performances.

1.3.1 Detecting the Diffracted Intensities

As mentioned above, one important aspect is the accurate measuring of Bragg intensities. After the introduction of scintillation detectors, able to measure with high accuracy only one reflection at the time, the most important breakthrough was the introduction of bi-dimensional digital detectors. This led back to the possibility of rapidly collecting images of the whole reciprocal space, as in the early days of photographic film, however with the advantage of much higher accuracy in the radiation detection. Among the many kinds of bi-dimensional detectors certainly the most diffuse (at least on a laboratory scale) are the charge coupled device (CCD) area detectors. They were introduced in the 1990s and their impact on the accurate charge density determination occurred at the end of the decade [124]. Among the papers published on charge density, the majority is now based on CCD area detectors. Other bi-dimensional detectors, like imaging plates, initially gave more problems for accurate measurements and, although introduced quite earlier than the CCD's, they became routinely and successfully employed in charge density analysis more recently. Nevertheless, excellent studies have become possible also based on this technology [125].

There had been many discussions over the years concerning the suitability of digital area detectors for accurate charge density measurements [126], however it is nowadays quite established that these systems are indeed very efficient, provided that data collections and data reduction are carried out with accuracy and all due corrections to the raw data are applied [127]. One major advantage of the area detector technology is the high redundancy of intensity measurements, which is easily obtained. This gives an enormous improvement to the statistics (*precision*), but it allows also more adequate data corrections (*accuracy*), like absorption, time decay, source instability, diffusion and absorption of the sample holders, etc. One major defect of most area detectors is the very low energy discrimination. Low or high energy photons cannot be filtered out by the phosphorous screens coupled with the CCD chips and therefore they are included in the recorded frame (beside the phosphorous sensitivity might substantially vary with the incoming radiation energy). This is of concern in accurate measurements, because perfect monochromatic waves are not possible and some contamination of photons with different energy is always expected. For example, traditional graphite monochromators are typically suffering some $\Delta\lambda$ dispersion around the selected wave length [128] and contamination from harmonics [126]. For this reason, the new technology based on complementary metal oxides semiconductors (CMOS) could represent the next breakthrough, because it gives the advantages of a position sensitive bi-dimensional detector and reasonable energy discrimination (in principle able to select a single spectral emission). However, at this time, it is not possible to further speculate on this possibility because lacking of sufficient testing. On the other hand, it clearly seems that the CCD and imaging plates technologies have reached their limit, and new investments for improving them are not expected by companies producing detectors.

1.3.2 X-Ray Sources

Concerning sources, instead, for many years 3rd generation synchrotron sources and rotating anodes have represented the frontier for large scale facility and normal research laboratories, respectively.

On the synchrotron site, the investments to build X-ray free electron laser (X-FEL) sources could guarantee enormous improvements, however it is not clear how much usage will be devoted to Bragg scattering measurements when these facilities will be operative, or otherwise how much charge density information could be obtained from non crystalline samples (a challenging frontier field that has not been much exploited so far).

On the laboratory scale, instead, the newest development is the more routine use of X-ray micro-sources, that guarantee a higher brilliance on a more focused beam, using much less power. Besides some very positive reports [129], we must say that this technology is not free of problems. In fact, multilayer mirrors used to monochromatize and focus a micro-source beam can be contaminated by low energy radiation, reflected because of the small incident angle and the radiation must be filtered [130]. In this book, we will not dedicate further attention to technical aspects of the X-ray diffraction experiments, concerning sources and detectors. The reader is referred to literature quoted in this chapter or in the chapters discussing the experimentally derived electron density measurements. This book also does not describe other methods to obtain experimental electron density, other than X-Ray Bragg scattering or Compton profiles (see in particular Chap. 4). Γ-rays can also be used to map the experimental electron density [131], however no examples are given in this book. To obtain a spin density map, instead, it is necessary collecting polarizing neutron diffraction data. Experimental details are given in Chaps. 4 and 8.

1.3.3 Low Temperature

As it will become clear especially after reading Chap. 3, experimental procedures require a fundamental condition, that atoms are as steady as possible about their equilibrium positions, for two main reasons: (a) the drop of scattered intensity induced by the atomic motion; (b) the more difficult separation between electron density and thermal parameters of a model, when the latter are large. This means that low temperature measurements are extremely important to obtain quality charge density maps. However, "low" is quite undefined, because it very much depends on the kind of material under investigation. Anyway, it should be considered that even if a data collection at 0 K were available, it would be impossible to exclude thermal motion completely, due to the zero point vibrations. Actually, this is an important component of the thermal motion and all modeling must consider the way to extract static electron density from the scattering experiments. An exception is provided by models like the MEM that returns a dynamic electron density, abandoning the possibility to extract a purely static model (*cf.* Sect. 1.2.2.2)

The cryogenic techniques adopted in many laboratories are nowadays quite standard and do not require more discussion. In particular stable and reliable equipment are typically employed to measure X-ray diffraction at temperatures around boiling point of liquid Nitrogen. More sophisticated and expensive is instead the technology necessary to reach lower temperature, down to the boiling point of Helium, for example. A seminal review nicely summarizes the necessities and pitfalls in this field [132].

The role of low temperature in an experimental determination of the electron density is multiple. The most important is certainly the reduction of thermal agitation of atoms, which makes the *pseudoatom* approach a more reliable approximation. As for normal structure refinements, smaller thermal motion means less correlation among parameters, hence higher reliability of the final model. Lower thermal motion also means that the harmonic approximation is more valid, as anticipated in Sect. 1.2.2.1. Although it is possible to go beyond the harmonic approximation, it should be considered that a model including for example a Gram Charlier expansion [133] would be extremely costly because of the very large number of parameters. Mallinson et al. [134] could show that residuals due to anharmonic motion are somewhat similar to residuals of deformation density, especially when dealing with transition metals. This would of course create confusion between true electron density features and residuals due to atomic displacements exceeding the model, with obvious consequences for the interpretation of the results. It is important to warn that the physical significance of the refined Gram Charlier parameters should be verified. In fact, it might easily occur that these coefficients are refined to nonsense values, implying for example negative nuclear probability at the equilibrium position. Additional advantages of low temperature in electron density refinements are connected with the higher accuracy of the measured intensities, in particular at high resolution. It is important to stress that features of the bonding electron density are very likely not recorded at such resolution, which is typically dominated by core electron scattering. However, the larger intensity at high angle can be very important to increase the precision of a refinement, including more reflections to refine atomic positions and thermal factors (apart for H atoms). As a matter of facts, a data/parameter ratio above 10 is often recommended, however in many cases the *effective* ratio is much smaller because variables introduced in Eq. 1.51 (see Sect. 1.3.4) are mainly refined from low angle data [135]. Thus, if some parameters of a model (positions and thermal motion) could insist more on high angle observations, the correlation among variables would be significantly reduced.

Another reason why high angle reflections are better measured at low temperature is the decreased thermal diffuse scattering which allows a more accurate integration of those intensities.

After listing all benefits of low temperature on accurate electron density refinement, the reader might ask what temperature is really necessary. Of course, the lower the better, however some electron density studies at room temperature or even above have been reported. For example, Tanaka et al. have investigated electron density of Lanthanide Borides [136] at variable temperatures using XAO method to determine the electronic configuration of the metal. Those studies addressed significant changes of the order and occupation of some electronic states (associated with 5d or 4f orbitals of Ce or Sm) as a function of the temperature. Experiments of this kind are not so frequent and they would be quite impossible on molecular crystals. On the other hand, results of Tanaka et al. show that it would be interesting investigating temperature induced changes of the electron density, even spanning the regime of high temperature.

More discussion on temperature dependent experiments is given in Chap. 19.

1.3.4 Modelling Electron Density from Experiments with Multipolar Expansion

As anticipated earlier, the possibility to model the electron density has been one of the fundamental breakthroughs, necessary to put into practice the prediction by Debye [123]. The main progresses occurred during the 1970s when, among many other approaches, the method of multipolar expansion of the electron density was recognized as the most applicable and accurate, hereinafter called the multipole model (MM). Several formulations were proposed [137], but the intuitive notation introduced by Hansen and Coppens [138] became afterwards the most popular. Within this method, the electron density of a crystal is expanded in atomic contributions, or better in terms of rigid *pseudoatoms*, *i.e.* atoms that behave structurally according to their electron charge distribution and rigidly follow the nuclear motion. A *pseudoatom* density is expanded according to its electronic structure, for simplicity reduced to the *core* and the *valence* electron densities (but in principle each atomic shell could be independently refined). Thus,

$$\rho(\mathbf{r}) = \sum_{atoms} \rho_i(\mathbf{r} - \mathbf{r}_i) \tag{1.50}$$

$$\rho_i(\mathbf{r}) = P_{i,core}\rho_{i,core}(\mathbf{r}) + \kappa_i^3 P_{i,valence}\rho_{i,valence}(\kappa\mathbf{r}) + \sum_{l=0,l\max} \left[\kappa_{il}'^3 R_{il}(\kappa'\mathbf{r}) \sum_{m=0,l} P_{ilm\pm} y_{lm\pm}(\mathbf{r}/r) \right] \tag{1.51}$$

The parameters $P_{lm\pm}$'s, P_{core}, $P_{valence}$ and κ's can be refined within a least square procedure, together with positional and thermal parameters of a normal refinement to obtain a crystal structure. In the Hansen and Coppens model, the valence shell is allowed to contract or expand and to assume an aspherical form (last term in Eq. 1.51), as it is conceivable when the atomic density is deformed by the chemical bonding. This is possible by refining the κ and κ' radial scaling parameters and population coefficients $P_{lm\pm}$ of the multipolar expansion. Spherical harmonics functions $y_{lm\pm}$ are used to describe the deformation part.

The radial function of the electron density is a crucial choice. Usually the core and spherical valence are taken from Roothan Hartree Fock atomic wave functions [139], Dirac Fock type [82, 140] if relativistic effects are taken into account. These wave functions are products of orbital functions φ_l, expanded in several Slater type functions

$$\varphi_l = \sum_{j=1}^{m} c_j \left[2n_j!\right]^{-1/2} \left(2\zeta_j\right)^{(n_j+1)/2} r^{n_j-1} \exp\left(-\zeta_j r\right) \tag{1.52}$$

The third term in Eq. 1.51 is instead typically constructed from single Slater functions, taken from the best single zeta function reproducing atomic orbitals [141]. This is not the only choice and a large variety of recipes has been proposed in the literature. In particular, it is notable that a pure "orbital assumption" would not be able to properly reproduce the atomic densities in molecules, especially for second row elements or in general main group elements that in ground state do not have other valence orbitals than *n*s and *n*p type. Given the properties of spherical harmonics, p orbitals could produce at most quadrupolar densities, whereas it is common to obtain significant octupolar functions for C, N, O atoms. As a matter of fact and as noted earlier, the *pseudoatom* expansion has an inherent limitation that the two center density cannot be exactly explained (at variance from the one center density). Thus higher multipoles are necessary to obtain satisfactory, though not exact and not unique, convergence of the expansion.

For transition metal atoms, the density is highly one center type, due to the low overlap with the ligand shell. Thus, the multipolar expansion can reproduce the metal density, but hexadecapolar functions are necessary. In fact (*n-1*)d orbitals produce hexadecapolar, quadrupolar and monopolar densities, whereas dipoles and octupoles are typically negligible. For transition metal atoms it is sometimes better using Roothan Hartree Fock orbital density for the higher multipoles as for the core and spherical valence. In fact, the rather contracted nature of the d orbitals makes a multi Slater function description more adequate. However, in the presence of experimental errors and thermal motion, this subtle difference may not be appreciated.

Simultaneous refinements of more valence shells have been proposed, either for transition metal atoms [142] or for main group elements [143]. Besides the improved accuracy of the model, it should be taken into account that the quality of the data, especially for larger systems, may not be sufficient to adopt systematically these advanced and very flexible models.

Other models differ from (1.51) in the treatment of the spherical valence. For example, in the multipolar model proposed by Stewart [137], there is no $\rho_{i,valence}(\mathbf{r})$ term, but only a monopole constructed with single Slater function, as the last term of Eq. 1.51.

Several software packages [83, 144] are available for multipolar refinement of the electron density and some of them ([83, 144b, 144d, 145, 146]) compute properties from the refined multipolar coefficients.

1.3.5 Recent Progresses in Quantum Modeling for Charge and Electron Density Studies in Position and Momentum Space

An overwhelming variety of efficient programs for the *ab-initio* treatment of solids have been recently made available. They either represent largely improved versions of previously existing codes or they have come to the fore for the first time during the last decade (see e.g. Refs. [1–3] in Chap. 2 and Refs. [147, 148]). The liveliness of the field is testified – just to take an example – by the fact that 8 out of a total of 16 papers of a dedicated Zeitschrift für Kristallographie issue on Computational Crystallography [149] were focused on the diverse features of codes for the (*ab-initio*) simulation of solids. Or, as a further case in point, by noting that as many as three versions of the CRYSTAL-XX code [150], one of the most successful *ab-initio* program for treating systems with periodic boundary conditions (PBCs), have been distributed since 2003, each of these versions providing an impressive improvement over the preceding release. For instance, among other relevant steps forward, CRYSTAL03 introduced an automated geometry optimizer using analytic or numerical gradients and CRYSTAL06 the calculation of the harmonic vibrational spectrum of crystalline compounds at Γ point. The most recent version, CRYSTAL09 [151], offers several new features of remarkable interest for modern charge and electron density studies. CRYSTAL program was taken here as an example, but similarly impressive progresses characterize other well-known packages for the electronic structure calculations of solids, like, for instance, Wien-2 k [152].[2]

Chapter 2 in this book overviews the state-of-the art of the *ab-initio* simulation of crystalline systems with a special emphasis on the evaluation of their position and momentum space density matrices and of the various functions one may derive thereof (electron and electron spin densities in position and momentum space, Compton Profiles, Auto correlation Functions, etc.). Approaches based on plane-waves or local functions are discussed, with comments on their main advantages and drawbacks, corroborated by examining their relative performance when applied to test cases of relevant interest to ED analysis. The authors of chapter 2 are among the developers of the two main codes – CRYSTAL09 and Quantum EXPRESSO [153], using a local AO basis and Plane Waves (PW), respectively – being illustrated and

[2]For instance, ten versions of the Wien-2 k code – one among the most accurate schemes for band structure calculations – have been released during the last decade and a separate version, WIEN-ncm able to handle the case of non collinear magnetism has also appeared.

tested throughout the chapter. To the best of our knowledge, this is the first time that authors of a package using the *de facto* standard within solid state physics modeling (DFT + PW) and those of a code adopting a local basis (the typical chemist's choice) accept to compare in detail these two alternative techniques on ED issues, and not just on the usual geometrical, energetic or computational aspects. Being the authors of the codes has enabled them to go deeper than any other could have done in analyzing the effect of the various computational parameters on the quality of the resulting density matrices and related properties. This clearly provides an unusual added value to the comparisons reported in the chapter. For instance, the problem of treating the core electrons when using valence-only pseudopotential approaches – a compulsory choice for PW codes – has been extensively addressed, both in the theoretical part and in the reported examples (Sects. 2.2.2.3, 2.3.3.1, 2.4). Core-electrons are usually of negligible importance when the emphasis is on energy, but they are not so for ED related analyses, if nothing else because the pseudo-orbitals are not orthogonal to the core and the "pseudo-density" summed to the core density is not the true position or momentum space ED. Chapter 2 also briefly outlines a number of ongoing research developments. They include the necessary steps leading to the thermally averaged estimate of density matrices and functions in crystals (Sect. 2.5.1) and to the evaluation of their changes induced either by external fields (Sect. 2.5.2) or by the inclusion of electron correlation within a post-HF perturbative approach (Sect. 2.5.3).

The last decade has also witnessed the further development and first practical implementation in solids of electron density linear response methods – the so-called density functional perturbation theory (DFPT) [154, 155] which focuses on the computation of the derivative of the DFT electronic energy with respect to different perturbations. The perturbations treated in DFPT might be external applied fields, as well as changes of potentials induced by nuclear displacements, or any type of perturbation of the equations that define the reference system. DFPT is clearly of great interest for ED related studies and it is implemented, in its KS form, in several computer packages, like in the Phonon code of the Quantum-EXPRESSO [153] package or in the ABINIT software package [156]. In quantum chemistry, when applied in the specific context of the HF approximation, the resulting algorithm is known as the CPHF method [157], which has also become recently available in CRYSTAL [158]. Use of such or similar response techniques to obtain perturbed density matrices and functions under an external field or due to local electron correlation corrections is briefly mentioned in Chap. 2. However, for the sake of space, the chapter neither deals with DFPT (or CPHF) methods nor with their implementation for PBC systems. The reader is referred to the relevant literature [154, 155].

It is worth mentioning two other recent developments of interest for charge density analysis, namely the proposal of embedding schemes to model crystal field effects and the increased availability of *ab-initio* codes able to introduce relativistic corrections at diverse levels of sophistication. Both themes are not treated in this book and are concisely summarized here.

As discussed in Sect. 1.2.4.4, EDs of molecules in *vacuo* differ from those in the related molecular crystals because of matrix effects due to the presence of neighbouring molecules in the crystal. The obvious way to properly account for such effects is that of performing *ab-initio* ED calculations with periodic boundary conditions (PBC), which relative to in-vacuo computations, have the *pros* of naturally including the matrix effect and of fully retaining the ideal crystal symmetry. They have, however, non negligible *cons:* PBC all-electron calculations are usually restricted to relatively small unit cells, HF or DFT level of Hamiltonian and limited basis sets in the case of atom-centered basis functions (as for instance in the CRYSTAL code), whereas quantum chemical methods for molecules provide much more flexibility. Calculations on molecules can be (much more easily) done, e.g., at post-HF *ab initio* levels, with extended basis sets including diffuse functions, or in relativistic two- and four-component frameworks [159] when dealing with compounds of heavier elements. Furthermore, the system size accessible is substantially larger. One way to avoid these cons while keeping the main effects inherent to the mentioned pros is that of adopting the so-called embedded cluster approaches [160–162], where one local region (cluster) is treated quantum-mechanically, while the crystal surroundings are described in a convenient approximate way. The performance of embedded cluster methods for modeling the matrix effects on the CD has not been tested until recently when two papers on subsystems of supermolecular complexes [163] and one on cluster models for the polar organometallic methyl lithium crystal [164] have appeared. The latter study by Götz et al. is of particular interest for CD analysis since a new, simple and very efficient embedding model is adopted, the so-called periodic electrostatic embedded cluster model (PEECM) [162], and also because the performance of this new approach is compared to the results of a corresponding, reference PBC computation.

The study of molecular compounds and crystals containing (heavy) transition-metal elements is still a challenge for experiment but also from a quantum chemical point of view. *Inter alias*, accurate calculations need the relativistic effects to be taken explicitly into account and especially so for the 4*d* and the 5*d* metal-elements. It is a common practice to include these effects only in an indirect way, by using relativistic effective core potentials (ECPs) in valence-only electron calculations, which allow for an approximate estimate of changes induced by relativity on the valence ED. However, ECPs have been calibrated with energy and pseudo-orbital shapes in mind and may not be necessarily the "best" solution to model such ED changes. In addition, the total electron density cannot be reconstructed, unless using some assumption, as for example summarized by Tiana et al. [165]. Conversely, in the last decade different variants of relativistic Hamiltonians have been developed and all-electron relativistic calculations with various electronic structure methods have now become feasible [159]. As a natural consequence, studies on the role of relativity on the topology of the complete CD have begun to appear [122, 166–169]. The reference *fully relativistic* theory is based on Dirac's theory of the electron and on the related Dirac Coulomb Hamiltonian (DCH) [159], which includes the four-component one-electron Dirac operator.[3] Due to the structure of the DCH, the one particle functions in the Slater determinant are no longer simple molecular spin-orbitals but become four-components molecular spinors, so that, in the HF approximation, for instance, the total ED turns out to be expressed in terms of a sum over all occupied *four-component* molecular spinors [159, 166]. The DCH explicitly treats spin, but unfortunately calculations at the DCH level are computationally very demanding since this Hamiltonian introduces besides the positive energy also the troublesome negative energy states [159]. A simplified and typically used approach is due to Douglas, Kroll and Hess (DKH) [170] where the DCH Hamiltonian is block-diagonalized using a unitary transformation U to decouple the positive and negative energy states and to lead to two *two-component* Hamiltonians, one having only positive and the other only negative energy eigenstates. The first, of primary interest to chemistry, is retained, whereas the second is simply ignored in the DKH approach. DKH Hamiltonians are customarily labeled as DKHn, where n corresponds to the number of terms in the product of unitary matrices defining U, the larger the value of n, the more complete being the decoupling of positive and negative energy eigenstates. DKH Hamiltonians can then be split into spin-free and spin-dependent terms, with retention of the first one only leading to the *one-component* or *scalar* or spin-free DHK Hamiltonian [159].

Eickerling et al. [166] have compared relativistic Hamiltonians with respect to their effect on the ED, in terms of a topological analysis along the series of $M(C_2H_4)$ complexes, with M = Ni, Pd, Pt. The investigated Hamiltonians were the four-component DCH, the quasi-relativistic two-component zeroth-order regular approximation (ZORA) [171],[4] the scalar relativistic DKHn ($n = 2, 10$) operators and, for the sake of assessing the magnitude of relativistic effects on the ED, the standard non-relativistic many-electron Hamiltonian. Since relativistic effects increase with the atomic number Z, one anticipates an increase of their importance through the investigated series of homologous complexes. Scalar relativistic effects are included through the chosen DKH Hamiltonian whereas spin-orbit effects, which, however, are known not to play a decisive role for all non-p block elements of the periodic Table of the elements with $Z < 100$ [159], are included in the ZORA framework. Results from scalar DHK and ZORA were compared by Eickerling et al. [166] to their limiting reference cases, the non-relativistic and the four-component results, respectively. It was found that for the Pt complex the $\rho(\mathbf{r})$ value at the M-C bcp is, relative to the four-component result, underestimated by 6% when a nonrelativistic Hamiltonian is used. Such a change is of significant magnitude when comparing theoretical and experimentally-derived EDs. Corresponding changes for the ED Laplacian values are much larger and, for instance, up to 90% for the Pt complex at the M-C bcp. This provided a further evidence of how the Laplacian

[3]Dirac's operator is defined in terms of the linear momentum operator and of a 3-vector whose components are (4×4) matrices built, on the off diagonal blocks, from Pauli's spin matrices.

[4]ZORA represents another approach for reducing the four-component Dirac equation to an effective two-component form.

enables one to detect subtle changes in the ED (see also Chap. 10). The relativistic contraction of the electronic core shells of the metal atoms leads to an increase in the magnitude of the ED Laplacian as well as in the ED at the position of local charge concentrations. In general, comparison of the EDs obtained from the various investigated models showed that the relativistic effects in the investigated complexes are already accounted for by a scalar relativistic approximation. As a consequence, the take-home message of Eickerling's paper was that computationally demanding two-component approaches, including spin-orbit effects, or even four-component calculations, seem not to be required for a meaningful ED topology analysis of complexes containing large Z *d*-block elements. Effects due to electron correlation on the topology of the ED, as introduced by state-of-the art DFT calculations and as assessed by comparison with HF results, were proved to be of similar order of magnitude as the relativistic effects in the analyzed complex [166]. Recently, Eickerling and Reiher [167] have used the fully-relativistic four-component approach and a multiconfigurational ansatz for the wave function to examine the effect of different electronic configurations on the shell structure of atoms and to evaluate whether the changes so observed in the Laplacian of the radial density are sufficiently large to be revealed by experimental studies of the ED topology.

Other kinds of bonding analyses have also been examined to test the effects of the relativistic corrections on the ED and wavefunctions. As an example, a very recent study by Ponec et al. [169] on the picture of the Re-Re bonding in the $Re_2Cl_8^{2-}$ ion, a well-known paradigm for quadruple metal-metal bonding, has shown that the DAFH picture of the bond is only marginally affected when the non relativistic Hamiltonian and a "relativistic" ECP basis set, are replaced by a scalar relativistic DKH Hamiltonian and the required all-electron basis set.

To conclude this point, we mention that the Constrained Wavefunction method (CWM) has been very recently extended by Hudák, Jayatilaka et al. [168] to include relativistic effects, using the scalar DKH2 approach (*cf* Chap. 6).

Clearly related to CD analysis are also two issues that are not treated in this book and are, therefore, only briefly mentioned here.

Standard Generalized Gradient Approximations (GGA) are commonly employed in condensed matter to model the exchange correlation energy in terms of the ED gradients. However, despite their widespread use, they do not always perform satisfactorily for solids [172,173], motivating why in the last decade several specifically designed new GGA's have been developed for them [172, 174–177], at the cost of a worsened accuracy for atoms and molecules. The main problem resides in the fact that the exchange-correlation hole cuts off rather sharply in an atom or in a small molecule because of the exponential decay of the electron density to zero away from the electron, whereas the hole can be more diffuse in a solid. The newly designed functionals for solids are aimed at comprising this important difference and the reader is addressed to two very recent review papers on the subject [173, 178]. The second of such papers deals with a new class of functionals developed at the Minnesota University. These functionals have been designed to give broad accuracy in chemistry and seem to perform well not only for molecules in vacuo, but also for quite diverse classes of solids.

Another field which has enjoyed a real blooming in the last decade is that of the computational recovery or even prediction of very large and complex crystal structures, at normal conditions or under stress (e.g. high pressure, Sect. 1.5.3). Evolutionary approaches, which combine specifically adapted genetic algorithms with standard *ab-initio* methods to find the most stable structures under given external constraints, have shown to be particularly powerful in this context [179, 180]. Crystal prediction methods have become of special interest in CD analysis. Indeed, by sampling potential energy surface (PES) regions thus far unexplored, they may lead to the discovery of new structures, often featuring exotic or unanticipated bonding patterns (Chaps. 17 and 18, [180, 121]).

1.3.6 Recent Progresses in the Analysis and Interpretation of the Electron Density and Chemical Bonding

In Sect. 1.2.5.2 we introduced the most adopted tools to study the molecular structure and chemical bonding in terms of the ED and of objects intimately related to electron pairing. Besides illustrating many examples of application of such interpretive techniques to the realm of materials science, of metallorganic chemistry, of life science (Chaps. 10–15) and of the matter under stress (high pressure, Chaps. 17–18), this book also offers, in Chap. 9, a detailed presentation of two emerging and very promising analyses. The first, developed by Pendás, Blanco and coworkers at the University of Oviedo, is based on QTAIM and the Interacting Quantum Atom (IQA) approach [181] and provides a rigorous, yet chemically intuitive energy partitioning in terms of atomic promotion energies and of interatomic, classic and non classic, interaction energies. IQA has been applied to shed light on a number of problems in chemical bonding, by combining this energy partitioning into 1- and 2-body contributions with the analysis of the statistics of the electron population among different atomic basins (*electron population distribution function,* EDF) [182]. The result is an extremely instructive picture of chemical bonding, one where its nature is understood in terms of how the relative weight of the various energy contributions reflects the kind and extent of the observed population fluctuations among basins. The other approach, illustrated in Chap. 9, is due to Kohout and coworkers and exploits a novel idea for creating families of new, interesting functionals [106, 183]. Properties, like the number of electrons, the number of the same-spin pair electrons, etc. are sampled (averaged out) over fine-grained, compact non-overlapping space filling regions (cells), characterized by having all a fixed, infinitesimally small integral value for a given control function ω. The integration of a sampling property over the cells results in a new discrete distribution that can be further examined by the 'usual' procedures of the topological analysis. The resulting distributions will clearly depend on the property being sampled, the control function ω and the chosen integral fixed value for ω. Out of the potentially infinite possibilities that may be explored using this approach, Chap. 9

discusses at length the electron localizability indicator (ELI) [106, 183] – an almost continuous distribution obtained by using the pair density as the control function and the electron density as the sampling property (or the other way around). The pair density may be defined for the same-spin or the opposite-spin electrons, and both pair and electron densities may be analyzed in position or momentum densities, thus leading to several kinds of ELIs, each of them carrying a peculiar information [183]. When the control function is the electron density and the sampled property is the same-spin electron pair density (both defined in position space), one may relate the Taylor expansion of ELI in the Hartree-Fock approximation to the ELF of Becke and Edgecombe without using any arbitrary reference to the uniform electron gas [106]. Furthermore, and differently from ELF, the ELI approach can be consistently applied to correlated wave functions.

It is worth noting that the two promising analyses discussed in Chap. 9 depend on quantum mechanical objects that cannot be derived from the electron density alone but require manipulating first and second-order density matrices, which, presently, makes the link with experiments rather weak. The increasing availability of methods able to derive these matrices from experiment (Chaps. 6 and 7) could to some extent fill the gap in the future.

One of the most intriguing and endless issues in ED analyses concerns the notion of chemical bond and the related choice of suitable criterions to establish which atom pairs are actually bonded to one another – eventually leading to the so-called "molecular structure" [6]. Once the criterions are selected, the problem arises of how to characterize and classify by nature the resulting chemical bonds [91]. Both aspects – individuation and characterization – become particularly delicate in the case of non conventional bonds, which are typical, though not exclusive, of organometallic compounds, and whose CD study in vacuum and in the solid state has been largely and increasingly pursued in the past 10–15 years [56, 91, 99, 101].

Defining and characterizing a chemical bond through the existence of and the properties at the associated bcp, is a cornerstone of the QTAIM study of chemical bonds in molecules and crystals [63, 91, 184]. This frequently used association has, however, been a continuous source of controversies and misinterpretations [91]. Recurrent criticisms in the last decade have in particular concerned the physical meaning of bond paths and bcps [185–187]. The presence of a line linking two atoms and along which the ED is a maximum with respect to any neighboring line, is, according to QTAIM, a necessary condition for the two atoms to be *bonded* to one another when the system is in a stationary state, i.e. an energy minimum at a given nuclear configuration. Persistence of this line when the system is also in a stable electrostatic equilibrium ensures both a necessary and a sufficient condition for bonding. Adoption of a single criterion to define a *bonding* interaction is challenging, since diverse kinds of interactions – covalent, ionic, metallic, van der Waals – may be all simultaneously present in many crystals and molecules, and performing their analysis on the same footing is a clear prerequisite for a meaningful comparison [91]. However, use of the bond path as a *universal indicator* of bonding poses questions that need to be answered and not simply ignored. Bond paths have been found where chemists will not find bonds [185, 186, 188] and *viceversa* [189], triggering waves of intense debate [185–187, 190]. A significant electron sharing between pair of atoms, as measured by their delocalization index, may occur despite the lack of a bond path linking those atoms. This is typically the case of systems, whose structure diagram evolves through a conflict mechanism [6] and where two alternative pairs of atoms are thus competing for a bond path [191]. According to the bond path criterion and for nuclear configurations in the neighborhood of the conflict catastrophe point, one pair of atoms would be termed as *bonded* and the other one as not bonded, irrespective of the almost indistinguishable extent of electron sharing for those two pairs of atoms. Another delicate situation is that of systems characterized by very soft potential energy surfaces and, hence, quite often extremely flat electron densities in their bonding regions. This is typically the case of metal-metal and metal-ligand interactions in organometallics, leading to structural diagrams exhibiting an enormous sensitivity to computational or experimental details [91, 99, 192]. It is clear that in such circumstances, the occurrence or lack of an atomic interaction appears as a rather subtle and controversial issue, if judged solely using the bond path criterion. Problems may arise even when the molecular structure is not so topologically unstable. For instance, electron sharing between the metal atoms in unsupported binuclear metal carbonyls can be comparable or even smaller than in the corresponding carbonyl bridged compounds, notwithstanding the metal atoms are found to be linked by a bond path in the former and unlinked by such path in the latter compounds [56]. This piece of evidence is of course related to the intrinsic incapability of the bond path criterion to directly detect *multicenter* bonding [193]. Using such criterion, bonding through the bridging ligands and direct M-M bond are generally alternative and competitive options [193], whereas they are no longer by necessity so, if continuous descriptors of bonding, like the delocalization indices or other tools mentioned below, are adopted. In summary, continuous descriptors rather than a single, *yes/no*, discontinuous bonding indicator (bond path), appear more suited in all these controversial cases or should at least be used to complement information arising from the bond path criterion alone.[5]

For instance, Macchi and Sironi [194] analyzed how the delocalization indices δ and other bonding properties evolve along the terminal-to-bridging CO reaction path for the gas-phase compound $[FeCo(CO)_8]^-$ and found that the discontinuities in the topological structure are neither mirrored in the delocalization indices, nor in the bcp density values of Fe-Co, Fe-C and Co-C interactions, which both exhibit continuous and generally smooth trends. It was shown that δ values characterize bonding in these kind of systems in terms of a mutual interplay of direct metal-metal and metal-CO and indirect M...M and M...C interactions. An overview on

[5]The fact that apparently contradictory descriptions of bonding may come out when using the bond path criterion or the delocalization index values should not be a source of particular concern, despite both these bonding indicators are defined within QTAIM. Indeed, information from different spaces is being analyzed and complementary information is thus obtained: bond path are made manifest in the 3D position space, whereas delocalization indices are defined in the 6D pair density space, where competition between electron localization within atomic basins and two-center or multi-center electron delocalization is observed and evaluated.

the use of delocalization indices in organometallic chemistry is provided in Chap. 12, along with a number of illuminating examples. Relationships between multi-center delocalization indices (including the standard two-center ones) and the EDFs are discussed in detail in Chap. 9.

Using IQA theory, Pendás et al. [191] have recently unveiled a profound physical aspect of bond paths – namely that they may be interpreted as privileged exchange energy channels, and without denying any of the already established, but less intuitive, physical tenets of those paths [184, 195]. It represents an important conceptual progress, one which easily explains why, under certain circumstances, two atoms turn out to be not linked by a bond path despite showing a significant electron sharing. By examining a number of interesting examples, Pendás et al. found that when two atomic pairs are competing for a bond path, the linked pair is always that having a larger exchange energy density [191]. Indeed, for systems evolving through conflict mechanisms, the exchange energy curves of the two competing atomic pairs are found to cross, almost exactly, at the conflict catastrophe point, namely where the bond path switches from one to the other pair of atoms. Much of the existing controversies on the bond path and on its relationship to chemical bonding might be probably calmed down if the extended vision on such a path offered through IQA would be embraced.

Though not explored in any of the following chapters, the domain averaged Fermi Hole (DAFH) analysis [45, 196, 197] introduced earlier in Sect. 1.2.3.3 represents another, very useful, continuous descriptor of bonding. Its capability to describe well-known standard chemical interactions has been firmly established and the DAFH tool is now increasingly being used for discussing bonding interactions in molecules with a non trivial bonding pattern, like hypervalence [196, 197], multicenter bonding [198, 199], metal-metal bonding ([169, 193, 199, 200] and references therein], etc. The approach has demonstrated its usefulness since it combines an appealing and highly visual description of the bonding interactions in terms close to classical chemical thinking with its clear physically-grounded basis. Analysis is performed by diagonalizing the matrix representation of the charge-weighted domain-averaged Fermi hole, $G^{\Omega}(\mathbf{r})$, (Eq. 1.35), in the basis of atomic orbitals. Eigenvalues and eigenvectors are then subjected to the so-called isopycnic transformation [201] to convert the original DAFH eigenvectors into more localized functions. Since the Fermi holes associated with a region Ω are predominantly localized in Ω, the transformed eigenvalues and eigenvectors provide specific information about the structure of that region. For the sake of ease in computations, regions Ω have been often defined in the basis set space, leading to the so-called Mulliken-like approximate formulation of the DAHF analysis, but definition in the position space (normally QTAIM basins) has now been established as a feasible and far less ambiguous procedure [169, 193]. When the domain is a single atom, the hole yields information about the valence state of the atom in the molecule. Conversely, if the domain is more complex and it is formed by the union of several atomic regions, the hole reveals the electron pairs (chemical bonds, lone pairs, etc.) that remain intact in that molecular fragment as well as the broken or dangling valences which have been formed by the formal splitting of the bonds required to isolate the fragment from the rest of the molecule. Structural information is extracted both from the numerical values of the eigenvalues and from the visual inspection of the corresponding localized eigenvectors. For instance, when applied to the highly debated case of the triply-bridged $Fe_2(CO)_9$ coordination complex, the DAFH approach rather than the direct Fe-Fe bond interaction predicted by the 18-electron rule, suggests [193, 199] a 3c-2e character for the bonding of the bridging ligands. This view nicely complies with the non negligible value found for δ(Fe,Fe), notwithstanding the two metal atoms are not linked by a bond path. Indeed, the existence of non-vanishing delocalization indices between all pair of atoms in an ABC fragment was proved to be a necessary requirement for the presence of 3c-2e bonding in the fragment [202]. Note that even when the Fe-Fe distance is decreased to the point where a Fe-Fe bcp appears, the DAFH picture of bonding remains almost unaffected, irrespective of the dramatic change in the ED topology [193]. Such an observation consistently shows that the bonding of bridging ligands exhibits the typical features of delocalized 3c-2e bonding.

Recently, Francisco et al. [46] have shown that the one-electron functions derived from the DAFH analysis on atomic domains are interestingly related to the electron number distribution functions (EDFs) and to the two-center or multicenter delocalization indices. In particular, when the eigenvalues of the DAFH spin orbitals are either close to 1.0 or 0.0, the orbitals are almost fully localized in Ω or in its complementary space, and do not contribute to $\delta(\Omega,\Omega')$ or may ignored in the determination of the EDF. Instead, eigenvalues close to 0.5 correspond to maximally delocalized orbitals that participate significantly to the bonding. A very promising, combined strategy that unifies IQA, DAFH and EDF to restoring orbital thinking from real space descriptions has been very recently proposed and applied to bonding in classical and non-classical transition metal carbonyls [203].

Multi-center bonding may also be analyzed through the ELF. The approach distinguishes among different bonding schemes by assigning a synaptic order to each of the recovered ELF valence basins and by finding the number and type of core basins with which they have a boundary [204, 205]. As an example, disynaptic valence basins are related to the conventional two-centre bonds and the trisynaptic ones to 3c-2e bonds. Electron populations of such ELF valence basins then allow ranking the relative importance of the associate bonding schemes. ELF is briefly introduced in Chap. 18 and, as mentioned earlier, amply used in Chaps. 18 and 17 to brilliantly highlight the relationships between structural and chemical bond evolution in crystals under pressure. Although published 6 years ago by now, the reader is also addressed to an extensive review [91] on chemical bond in crystals which, inter alia offers a detailed presentation of ELF, focusing on its physical and chemical interpretation, on its applications to several interesting classes of crystalline materials and, not less importantly, on the complementarities and differences with other topological approaches.

Another tool, able to overcome the problems inherent to the possibly discontinuous description of bonding provided by the ED topology and the bond path criterion, is the Source Function (SF), introduced sometime ago. Bader and Gatti [206] showed that the electron density at any point **r** within a system may be regarded as determined by the operation of a local source LS(**r**,**r**') at all other points of the space:

$$\rho(\mathbf{r}) = \int \mathrm{L}S(\mathbf{r}, \mathbf{r}') \cdot d\mathbf{r}' \qquad (1.53)$$

The local source (LS) is expressed as,$\mathrm{L}S(\mathbf{r}, \mathbf{r}') = -(4\pi \cdot |\mathbf{r} - \mathbf{r}'|)^{-1} \cdot \nabla^2\rho(\mathbf{r}')$, where the Green's or *influence* function [207],$(4\pi \cdot |\mathbf{r} - \mathbf{r}'|)^{-1}$, represents the effectiveness of the cause, the Laplacian of the density at **r**', $\nabla^2\rho(\mathbf{r}')$, to produce the effect, the electron density at **r**, $\rho(\mathbf{r})$. By integrating LS over the atomic basins of a system, one may equate $\rho(\mathbf{r})$ to a sum of atomic contributions $S(\mathbf{r};\Omega)$,

$$\rho(\mathbf{r}) = S(\mathbf{r}, \Omega) + \sum\nolimits_{\Omega' \neq \Omega} S(\mathbf{r}, \Omega') \ , \qquad (1.54)$$

each of which is termed as the *source function* of atom Ω to $\rho(\mathbf{r})$. Atomic basins may be taken as those of any mutually exclusive space partitioning or also as those of any conceivable fuzzy boundary partitioning scheme. However, they have usually been assumed as the basins of QTAIM to ensure an unbiased and quantum-mechanically rigorous association of $S(\mathbf{r};\Omega)$ to the atoms or group of atoms of "chemistry". The electron density at **r** may be seen (Eq. 1.54) as determined by an internal SF self-contribution and by a sum of SF contributions from the remaining atoms or groups of atoms within a molecule. Decomposition afforded by Eq. 1.54 enables one to view the properties of the density from an interesting perspective, one foreseeing the SF as a tool able to provide chemical insight. For instance, although a bond path is associated to the only two atoms it connects, the shape of the path and the values of the ED at any point along the path depend on the whole set of physical interactions present in a system and accounted for by its Hamiltonian operator. The SF approach *solves* this apparent inconsistency since the ED at any point of the path and in particular at bcp, assumed as the most representative density point along the bond path, turns out to be determined not only by the two connected atoms, but also, to some variable extent, by the remaining atoms in the system. It has been generally shown, [208], that the larger is the bond order value and the more covalently bonded are two atoms, the higher is their percentage contribution to the ED value at their bcp. For less localized bonding interactions, the SF contributions become much more delocalized throughout the molecule. The SF analysis has been applied to several classes of bonding types, including hydrogen-bonds, multi-center bonds, metal-metal and metal-ligand bonds in organometallic systems, and it has also been exploited for assessing typical chemical features, like chemical transferability, the effect of substituents, etc. [209].

Quite recently, the capability of the SF to neatly reveal π-electron conjugation directly from the electron distribution and independently from any MO scheme or decomposition, has been explored [209, 210]. Chemical insight may also be obtained by analyzing the LS portraits along the bond paths (or the internuclear axes) [56, 211] or by integrating $S(\mathbf{r},\Omega)$ over the same or a different Ω' basin to get a physically unbiased atomic population matrix **M**, {M(Ω,Ω')}, in terms of an observable [209] (this specific use of the SF presents however numerical difficulties).

A very comprehensive review on SF theory and applications as well as on recent and ongoing SF developments has been recently published by Gatti [209]. Vivid discussions on the interpretation of the SF can be found in a recent paper by Macchi and Farrugia [212], as well as in the review by Gatti [209]. Chapter 13 and in particular Chap. 12 in this book highlight the usefulness of the SF as a complementary tool for studying bonding in transition metal compounds, using both experimental and theoretical ED distributions.

Though the SF analysis lacks the direct physical connection to electron pairing associated to the delocalization indices as well as to the IQA, DAFH and ELF analyses, it still retains a quite important advantage over all these bond descriptors. Indeed, at variance with them, it does not require the pair density to be evaluated (or the first density matrix in single determinant theoretical approaches) and it is so applicable to both experimental and theoretical electron densities, on the same ground. This appears as a quite interesting pro of the SF analysis, since, in the experimental studies of organometallics compounds, the most salient features of bonding have been quite often derived from theoretical calculations – a fact that could raise some doubts on the real convenience of performing the more time consuming and costly experimental determinations for such systems [56].

As mentioned at the beginning of the present Subsection, the choice of a suitable scheme to categorize the relevant chemical interactions in a system is as important as or even more important than their detection [91, 99]. For instance, the dicothomous classification [6, 114] of bonding interactions based on the sign of the Laplacian of the electron density at the bcp, $\nabla^2\rho_b$, is known to be inappropriate [91, 99] for bonding between atoms whose atomic $\nabla^2\rho$ distributions lack the outermost regions of charge depletion and concentration. This typically occurs for most of the transition metal atoms. The review by Gatti [91] discusses at length the various types of bond classifications based on topological approaches that have been proposed in the literature until 2005, highlighting their pros and cons and their applicability to the various bonding situations. Both those tools mainly based on the ED properties at the bcps and those placing these properties on a less dominant role are discussed and compared in that review, along with classifications based on ELF, DAFH, LOL, etc. Categorization of bonding using IQA, and ELI is discussed in Chap. 9 of this book, whereas Chaps. 17 and 18 show the capabilities of the ELF approach to follow the evolution of bonding network and nature with change in pressure and structural phase. Chapter 12 focuses on the use of various and complementary bonding descriptors for organometallic and for main group element hypo- and hyper-valent molecules. Critical analyses on this issue may be also found in Refs. [56, 99, 209]. Relationships between interaction energies and topological properties at the intermolecular bcps are discussed in Sect. 11.3 of Chap. 11, whilst the problem of how actually informative are the properties at the intermolecular bcps as for the charge rearrangements due to molecular interactions is discussed in Ref.

[89–91, 213]. Chapter 16 compares the information obtained from local ED topological descriptors and from the use of the Hirshfeld surface analysis [214] in increasing our understanding of the nature of intermolecular interactions. A new promising approach, which reveals non covalent interactions (NCI) as continuous surfaces rather than close contacts between atom pairs (or possibly bcps in QTAIM language) and which is also based on the ED and its derivatives, has been very recently proposed by Johnson et al. [215].

1.3.7 Intermolecular Interactions

In the past decades chemistry has vigorously transformed from a molecular science into a molecular *and* supramolecular science. Molecular materials attract enormous attention, because of their poly-functional applications, thanks to specific functional groups of the constituting molecules and to their organization in the crystal state. This field encompasses inorganic, organic and bio-organic materials, with applications in energy storage or transformation, computing, communication, constructions, life science and medicine.

A typical issue for a material scientist is to understand how the (molecular) building blocks interact with each other to give a specific architecture and the desired properties. Again, the ED plays a dominant role in this context, because one of the most important interactions that occur between molecules is in fact the electrostatic attraction or repulsion, which can be interpreted in a semi-classical fashion. In fact, despite molecules are quantum mechanical objects, electrostatic interactions between them can be computed assuming quite simple laws of classical physics, once a quantum mechanical electron density distribution is known (at least with a good approximation). This subject has grown rapidly in the past decade and scientists more and more frequently investigate intermolecular interactions using ED analysis (as we will see in Chaps. 11, 14 and 15). The two kinds of applications mainly tackled by scientists are related with Crystal Engineering (see also Sect. 1.5.4) and Biomimetics (Sect. 1.5.2).

While electrostatics plays a fundamental role, it is clear that some more steps forward should be made. In facts, a classical interpretation of the ED distribution is useful to predict and explain interactions between molecules intended only in terms of *hard* nucleophiles or electrophiles, *i.e.* for so-called *charge controlled* interactions (which are sometime precursors of actual reactions). On the other hand, *soft* (or *orbital controlled*) interactions are not so easily predicted from the static electron density, unless heuristically, *i.e.* correlating some electron density feature with regions of soft nucleophilicity or electrophilicity in a molecule, see Sect. 1.5.6 and Chap. 21 for more details.

For these reasons, efforts to compute polarizabilities from ED (or to determine them from typical X-ray diffraction experiments) could be extremely interesting and important in this field. This remains, however, a new and yet unexplored frontier, but very promising for the next few years. Some study in this direction are illustrated in Chap. 10, where the investigation of direct and reciprocal space is used to extract information on material properties, especially related to conductivity.

The interaction between two electron density distributions A and B was originally proposed by Buckingham [65] in terms of atom centered electric moments (or even molecular centered moments), $E_{MM}(AB)$. An energy is easily calculated based on classical electrostatic and extension of Coulomb energy between two charges,

$$E_{MM}(AB) = \sum_{a \in A} \sum_{b \in B} Tq_a Tq_b + T_\alpha \left(q_a \mu_{\alpha,b} - q_b \mu_{\alpha,a}\right) + T_{\alpha\beta}\left(1/3 q_a \Theta_{\alpha\beta,b} + 1/3 q_b \Theta_{\alpha\beta,a} - \mu_{\alpha,a}\mu_{\beta,b}\right) + \ldots. \quad (1.55)$$

with implicit summations (Einstein convention) over dummy Greek indices. μ_α and $\Theta_{\alpha\beta}$ are electric dipole and quadrupole moment components, whereas $T_{\alpha\beta\gamma\ldots}$ are the symmetrical interaction tensors ($\nabla_\alpha \nabla_\beta \nabla_\gamma \ldots (\mathbf{r}_{ab})^{-1}$, $\mathbf{r}_{ab}$ being the vector from the origin of a to b). The sum extends up to the higher multipole-multipole interaction and it generally converges up to $l_a + l_b > 5$ (l being the order of the multipolar expansion on center a and b, respectively) (see Chaps. 11 and 14).

This approach was employed by Spackman who proposed an easy model for the hydrogen bond based on multipolar expanded electron density distributions [66b,c]. The total interaction energy between two molecules forming an adduct is evaluated not only in terms of electrostatic energy E_{es}, but also in terms of semi-empirical terms for dispersive and short range repulsive contributions. More generally, the interaction energy between two molecules can be expressed as

$$E_{int} = E_{es} + E_{rep} + E_{disp} + E_{pol} + E_{CT} \quad (1.56)$$

For repulsion and dispersion, Spackman [66a] proposed atomic pair parametrization for main group atoms of the second period, based on Gordon-Kim model for electron gas [216]. In this way, E_{rep} and E_{disp} energies are simple functions of the distance between two atoms.

$$E_{rep}(ab) = b_a b_b \exp\left[(c_a + c_b) r_{ab}\right]$$
$$E_{disp}(ab) = \frac{C_{ab}^{(6)}}{r_{ab}^{6}} + \frac{C_{ab}^{(8)}}{r_{ab}^{8}} + \frac{C_{ab}^{(10)}}{r_{ab}^{10}} + \ldots. \quad (1.57)$$

If E_{es} is taken just as E_{MM}, then an important effect is not accounted. In fact, when the electronic shells of two interacting atoms can "overlap", stabilization is produced because of the unshielded nuclear-electron attraction (called penetration energy, E_{pen}). A corrective term must be calculated and incorporated in E_{es} ($= E_{MM} + E_{pen}$). However, to correctly reproduce the energy of hydrogen bonded adducts, Spackman [66c] was forced to propose an empirical and quite mysterious assumption: the repulsive term H—X (where X = HB acceptor) should be ignored. Later on, it became clear that this was in fact the result of a mutual cancellation of errors: in fact the extra stabilization produced by neglecting E_{rep}(H–X) was compensating the missing stabilization produced by the mutual penetration of spherical electron density shells, which was not included. E_{pen} due to the aspherical shells was instead evaluated in Spackman model, but this contribution is generally less important because it involves a smaller fraction of the total electron density. For this reason, caution should be taken to choose appropriate atomic or atom pair parameters to describe the short range repulsion. Many other data banks accounts for the repulsion and dispersive term of Eq. 1.57, for example taking it proportional to the overlap between the molecular electron density distributions. An early work by Cox, Hsu and Williams [217] is still much used to estimate those contributions.

More recently, the approaches suggested by Gavezzotti [69] or by Volkov and Coppens [71] for E_{es} have been more widely adopted. Both evaluate numerically the electrostatic interaction between two electron density distributions using the whole three dimensional function $\rho(\mathbf{r})$, which therefore does not need further correction for E_{pen}. The difference between the two methods is mainly the type of grid used for this evaluation, an evenly spaced one by Gavezzotti, based on a more sophisticated quadrature scheme by Volkov and Coppens. The latter approach is quite faster and it is even speeded up by using Buckingham calculations for atoms at distances above a given threshold, where the exact potential or the multipolar expansion gives a very similar energy (but the latter method at much lower expense).

Gavezzotti has also suggested a more sophisticated way to evaluate the dispersion energy [218], by using a sort of atomic distributed polarizabilities, weighted on the actual electron density distribution and applying the London approximation.

More complicated terms are the charge transfer energy (E_{CT}) and the induced polarization (E_{pol}). As pointed out by Abramov et al. [219], if the electron density is obtained from the modeling of a molecule in the crystal, then E_{pol} is automatically included in E_{es}, otherwise it should be computed from the interaction between the electric field produced by a molecule and the polarizability of the other (and vice versa). In many cases the induced polarization is simply neglected, an approximation which could be quite severe if strong intermolecular interactions are taking place. The estimate of the charge transfer energy is also quite delicate. In principle all assumptions made so far are valid provided there are only classical electrostatic interactions between the molecules. However, charge transfer implies a certain degree of non-classical interactions (covalent bonding). In this respect, the discussion proposed in Chap. 9, on interacting atoms, is very important and has consequences also for intermolecular interactions.

Chapter 11 discusses in details the theoretical background of intermolecular interaction and the reader will find there more advanced discussion.

1.4 About the Structure of This Book

This book is divided in two parts. In the first nine Chapters (including the current) theories are exposed on the physics beyond electron density distribution, the machinery of quantum mechanical and pseudo-atomic models and the dynamical processes occurring in the solid state. In the second part (Chaps. 10–21), applications will be presented where the reader may become aware of the most important studies carried out on several kinds of chemical systems and materials using methods of charge density analysis.

This book is not intended to replace the fundamental texts referred earlier, where theories are explicitly and more didactically reported, for example.

On the other hand, the purpose of this book is to update the reader on the current status of the research in the field of charge density analysis, including more details of applicative studies, that are typically not included in more didactical textbooks. Although we are well aware that progresses in the next few years will certainly overwhelm some of the topics discussed in this book, we think that a snapshot is particularly important because it is taken after a very productive period, stimulated by many investments in this field.

1.5 Applications

1.5.1 Materials Science

Use of the charge density (CD) analysis in materials science has become increasingly pursued in the last decade, since this tool has a special ability to unveil, at an atomistic level, features that are notably important for material design and performance, while being difficult or often even impossible to be revealed with other approaches. Despite its usefulness, a CD study on materials is, however, recurrently a challenging task, one which requires using experimental and theoretical techniques at a sophisticated and at a cutting-edge level. Chapter 13 reviews recent applications of CD analysis in materials science with emphasis on thermoelectric, magnetic and porous materials. Many of the studies outlined in the chapter are indeed highly demanding and non-standard, since they include heavy atoms, large unit cells, high symmetry inorganic solids (with extinction problems), structural disorder, microcrystals and powders, and air sensitive compounds. A variety of techniques, such as multi-temperature approaches, combined theory and experiment, and synchrotron, neutron and conventional X-ray tube radiation had so to be employed in these studies to get a trustable CD (and wavefunction). Their analysis has then involved the study of thermally smeared densities, deformation densities, orbital populations and the exploitation of interpretive tools like QTAIM, ELI, Hirshfeld Surfaces (HS) and electrostatic properties.

Chapter 10, instead, focuses on quasi-one dimensional organometallic carbides, which exhibit interesting magnetic and electronic properties and represent textbook examples of extended systems displaying pronounced orbital interactions and anisotropic physical properties in real and reciprocal space. Thus, one expects that small changes in the chemical bonding and/or the electronic structures of these carbides should give rise to pronounced effects in the properties in both spaces. The carbides being investigated in the chapter, Sc_3TC_4 (T = Fe, Co, Ni), are structurally highly related and, going from T = Fe to T = Ni, allow for a stepwise increase of the d-electron count. This leads to a systematic variation of the states contributing to the Fermi level and, hence, to a change of the nature of the conduction bands, without any significant structural modifications. As already mentioned in Sects. 1.2.4.5 and 1.2.5.2, the aim of Chap. 10 is basically to explore, by using experimental and theoretical CD analyses, whether and how real space properties, such as the CD distribution or its Laplacian, are mirrored in reciprocal space properties in solids, like the electronic conductivity and superconductivity. The study convincingly shows that, despite structural similarities, even tiny differences in the electronic band structure of the systems are faithfully recovered in the properties of the Laplacian of the ED. The established link between, on the one side, chemical bonding, T metal d-orbital energies and populations, Laplacian distribution in the T metal valence shell charge concentration (VSCC) [6] and, on the other side, the resulting k-space properties of the carbide, is clearly of utmost interest in view of the design of new materials or of the improvement of the existing ones.

Rather than listing classes of materials to which CD analysis has been applied and presenting the results thereof obtained during the last decade, we prefer to conclude this Subsection by briefly mentioning a few aspects that make CD analysis on materials a difficult but worthwhile task. Materials are often characterized by very flat potential energy surfaces (PES) with (subset of their) atoms interacting through soft and feebly directed potentials and, even when not that soft, characterized by a number of energetically similar but geometrically quite distinct configurations. Flat PES means structural sensitivity to the various variables, like the synthetic route and conditions, the specific crystalline environment, the external parameters (T,P, fields), and so on and so forth. As a result, many important materials are only partially ordered and/or partially crystalline though, often, they owe their peculiar functionalities just to such departures from ideality. In most cases, a CD study of a material thus implies a detailed study of both its nuclear and electron densities (not to speak of all those important material properties, like electron transfer in proteins, phonon scattering and superconductivity in solids, that depend on the breakdown of the Born-Oppenheimer approximation and where the electron and nuclear motions are not separable, as, instead, usually assumed in CD crystallography and in *ab-initio* modeling). As already mentioned in Sect. 1.2.2.2, subtle structural disorder features can be revealed using the MEM that treats on the same footing both the ED in the well-ordered regions and in those affected by static or dynamic disorder. For instance, a combined Rietveld/MEM analysis of synchrotron powder diffraction data has revealed the presence of interstitial Zn atoms and partially occupied main Zn sites in the very promising Zn_4Sb_3 thermoelectric material [220]. By combining MEM information and extensive *ab-initio* theoretical modeling, the till then undiscovered interstitials were afterward demonstrated to play a fundamental role for thermoelectric performance since they act as electron suppliers and Seebeck coefficient enhancers, as well as lattice thermal conductivity suppressors [26]. Sometimes, structural disorder features are even more subtle and the thermally smeared electron distributions only show a very diffuse atomic density, with no peaks to define possible disorder. As shown in Chap. 13, in such cases, MEM electron deformation densities, given as differences between the observed ED and that of the promolecule are usefully investigated, showing the combined effects of chemical bond deformation, charge transfer and structural disorder. MEM deformation density studies may be also performed using neutron diffraction data, yielding a *nuclear* deformation density (see Chap. 13) with respect to an assumed more ordered structural model. Even when structural disorder is limited or absent, problems may arise. Typical is the case of materials where the "interesting atoms", as for assessing the material's performance, are numerically limited and also slightly scattering with respect to the other "unimportant" atoms in the unit cell. For instance, in the porous coordination polymers, which represent a promising solution to the issue of hydrogen storage, the weak X-ray scattering amplitude of hydrogen makes it difficult to determine the hydrogen position by X-ray structure analysis. However, Chap. 13 shows that the MEM/Rietveld analysis combined with high brilliance synchrotron powder diffraction made it possible to reveal the position of hydrogen atoms as well as the chemical bonding of hydrogen in those materials.

Ab-initio modeling of structural and compositional disorder is difficult. However, theoretical modeling may be as well very informative, just because it may be performed on systems with a known and assumed departure from an ideal reference system. It may so systematically explore structure and properties of materials with variable phase compositions, variable structure, most of which may be not (yet) experimentally accessible. A CD analysis on the obtained hypothetical materials may provide an atomistic insight about the relationships between material's composition, structure and properties, enabling predictions on how to improve a given class of materials or even propose a new hypothetical one.

Flat PES in materials also implies that the relationships between geometrical structure, electronic structure and material properties may often be very subtle and tricky to discover. However, this is namely a reason of why non-trivial CD analyses become so relevant in materials science. Which chemical interactions are present and which one are playing the major role? Which is their nature and how one may hope to modify them for improving material's performance? Are these interactions localized or not, i.e. are they or are they not affecting long-range properties as is, for instance, magnetic ordering in complex magnetic materials? To these issues, Chap. 13 presents a number of illuminating examples where CD studies have largely contributed to enhance understanding of technologically important materials. Examples include, among others, the case of skutterudites, for which ELI and QTAIM analyses discern the most probable bonding framework among the many alternative descriptions proposed by structural chemists and the case of CD studies on magnetic coordination polymers which provide insights into the pathways for the magnetic orderings occurring at low temperature.

1.5.2 Biomolecules

One of the major outcomes of research in electron density analysis has been the ability to model macro-molecules, especially those of interest in Biology and Biochemistry. There are several attractive topics in this field, although the work carried out so far has mainly focused on two major aspects: (a) improving the structural refinement by means of more accurate atomic models; (b) evaluating the nature and the strength of intermolecular interactions taking place in crystals.

Pioneering work was carried out by Lecomte and co-workers, who exploited one of the most attractive feature of the pseuodoatom multipolar expansion, namely the exportability of atomic parameters [138]. The principle is based on the possibility to define local coordinate systems for a *pseudoatom* and refine multipolar coefficients that could be sufficiently similar to those of another *pseudoatom*, in the same chemical environment. This idea, originally proposed by Brock et al. [221], is particularly efficient when a set of atomic multipoles is determined with high accuracy and exported to a structure determined from data of much lower accuracy, because of experimental limitations or because of the high complexity of the system.

There are several way to find out the best set of multipoles for a given atom type. Lecomte and co-workers have systematically investigated peptides using experimental methods (X-ray diffraction) and constructed two libraries of multipolar atom model refinement, ELMAM [222] and ELMAM2 [223] (where new consistent local axes systems are proposed), see also Chap. 15. A different approach was suggested by Volkov et al. [70] (UBDB library), based on theoretically calculated multipoles. Here a set of exportable multipoles is refined from theoretical electron density distribution on molecules containing the relevant atom types. A similar approach was proposed by Dittrich et al. [75], the so-called Invariom method, which is in the authors' intentions an unbiased way to define the reference atoms.

A recent article [224] compares the performances of the various databases and methods, in particular respect to structural refinement and ability to reproduce electrostatic properties (especially dipole moments) and electrostatic interactions, with respect to benchmark molecules in crystal forms. Electrostatic properties are difficult to reproduce with precision, mainly because of the well known weakness of the multipolar expansion to model the diffuse part of a molecular electron distribution. It is in fact quite common that molecular dipole moments calculated from refined MM coefficients are very much dependent on choices of radial density and the recipe used to refine the model [225]. The theoretical constraints or the average over several experimentally refined models temperate those problems and in general the electrostatic properties can be calculated from libraries with a sufficient accuracy (using theoretical data as benchmarks).

The exportability is very much favored by the multipolar model in Hansen and Coppens notation [138], however other atom-like partitioning of the electron density, in combination with traditional force field concepts are nowadays well established. Matta [226] discussed the possibility to export electron density of whole fragments, as defined by QTAIM partitioning. Force fields based on quantum chemical topology have been intensively investigated by Popelier et al. [227] and are discussed in details in Chap. 14. In these studies the focus is mainly on the electrostatic energy and on the polarizabilities [228], modeled within the QTAIM approach, and their merge with traditional force field based on molecular mechanics approaches.

1.5.3 Electron Density in Compounds Under Perturbation

The progresses of the experimental techniques make nowadays available data collections on compounds under conditions often called "extreme", that means under some kind of stress or excitation (mechanical, magnetic, electric, photonic etc.). "Extreme" indicates that the materials are far from their normal operative conditions. The aim of these studies is analysing the response upon the perturbation.

One example of the broad field of "extreme conditions" is the study of crystals under pressure, where the materials can substantially modify their electronic configuration and give rise to unprecedented properties. For example, in the "megabar regime" ($P > 100$ GPa), the PV term of the free energy might well exceed the dissociation energy of strong covalent bonds or binding energy of valence electrons. Under these circumstances, atoms are subjected to an additional confinement potential that could lead to unexpected new forms. In fact, the "chemical bonding" and the electronic properties might be very different from that of the solids at ambient conditions. We can mention some of the known effects: (a) core electrons could be involved in the bonding (together or even instead of valence electrons); (b) the distribution of valence electrons could resemble that of electron gases even for solids that are non-metallic at ambient conditions; (c) higher energy orbitals may become energetically available leading some elements to unexpected electronic configurations. These phenomena lead on one hand to the famous prediction by Bridgman [229] that all solids are expected to become metallic above a given pressure. On the other hand, however, other phenomena might occur, like the localization of valence electrons in metals, leading instead to post-metallic phases of some known metals, like in Sodium [230] and Potassium [231], that could open up a completely new interpretation of chemical bonding, going beyond the classical understanding.

In general, one can anticipate the breakdown of differences between elements in the periodic table and therefore an unusual behaviour of some elements (like alkali metals behaving as transition metals). A full discussion is presented in Chap. 17, whereas in Chap. 18, topological rules encompassing the solid-solid transformations are enounced and changes to chemical bonding are interpreted within the most modern schemes, in particular by means of electron localization function (ELF). In fact, the breakdown/formation of chemical bonds induced by modifications of the thermodynamic variables is central in the understanding of phase transformations. Usually, these transformations are described "statically", *i.e.* taking pictures of the structure before and after the event. The careful analysis of chemical bonding provides much more information on the mechanism. In particular, one may recognize cases where bonds are formed or destroyed, others where chemical bonds remain but change their nature (as for example the non-metallic to metallic or metallic to post-metallic transitions anticipated above).

The successful combination of X-ray techniques and laser excitements makes available the investigation of excited electronic states – a field which is inherently connected with time resolved crystallography given the very short lifetime of excited species. Again, these are "extreme conditions" because electronic states of materials that are available only for very short periods (down to femto-seconds) become the matter of detailed studies, intended to elucidate crucial steps of chemical reactions or transformations. In Chap. 20, laser driven X-ray diffraction experiments are described and their results are interpreted which allows to draw transient charge density maps, i.e. dynamical evolutions of charge distributions on a very short time scale, able to describe phenomena like those governing chemical reaction processes or phase transformations in solids.

In this book, other studies of species under perturbations are not reported, although it is extremely important to mention the interesting results obtained by scientists stressing crystals under electric field [232].

Another typical field of "extreme conditions" would be that of very high temperatures. However, studies of ED in solids at temperature well above ambient are very rare. As already mentioned in Sect. 1.3.3, Tanaka et al. have investigated electron density of Lanthanide Borides [136] at variable temperatures (including some measurements above ambient conditions). Of course, it is in general difficult to separate dynamical and static effects in X-ray diffraction experiments and experiments of this kind are therefore not so frequent and would be quite impossible on molecular crystals. The careful analyses in Chap. 3 and the examples in Chap. 19 illustrate clearly the reasons why, historically, experiments on solids (even relatively hard inorganic solids) have been carried out at conditions where atomic dynamics were minimal. This is of course a drawback for some applications, because it could be extremely interesting to study materials at high temperature where they exhibit in fact important properties, rather than at very low temperature where experiments are very accurate, but materials are less significant. Although it is very difficult to predict many more experimental studies at high temperatures, it is possible that combination of molecular dynamics and electron density analysis [233] might shed more light on the changes of electron distributions stimulated by thermal motions.

All the studies on materials under extreme conditions are of particular interest because they really bring to new fundamental questions on chemical bonding. Indeed, we have to take into account that all definitions of chemical bonds, all speculations on the different nature of chemical bonds are based on assumptions valid in "normal" conditions, that most of the time define precisely a material in a special region of the phase diagram, and in the long time scale requested by thermodynamical equilibrium conditions. However, the progresses in chemistry and physics of "extreme conditions" require that all these intuitive notions are revised and tuned for all regions of phase diagrams and at all time scales. It is clear that this task is extremely demanding and it would be impossible to obtain a generalised theory, able to provide rules and easy interpretative schemes as it has been in general true for chemistry in the XX century. Nevertheless this challenge is extremely fascinating and much progress is expected in the next decades.

1.5.4 Crystal Engineering

While in a not so remote past, crystal engineers made generally use of concepts and tools that were only indirectly related to the CD analysis [234], the situation has largely changed during the last decade, as brilliantly reviewed by Spackman in the Chap. 16 of this book, entirely focused on the interplay between charge densities and crystal engineering. According to Desiraju's, crystal engineering concerns *the understanding of intermolecular interactions in the context of crystal packing and in the utilization of such understanding in the design of new solids with desired physical and chemical properties* [235]. These new solids may be as complex and technologically important as those typical of host-guest materials, of nanoporous frameworks and of molecular crystals (Chap. 13). An understanding of intermolecular interactions in such complex entities can no longer be based (only) on the tabulation and analysis of close contact distances and geometries, but deserves new techniques and tools leading to more easily manageable quantities and immediately graspable pictures, while significantly increasing the amount of the available key structural information. This is namely the role which is being played in modern crystal engineering by the CD-based techniques which are outlined in Chap. 16 and either used in a complementary way, or compared among them, throughout the chapter. Techniques based on the (polarized) charge density of the interacting units enable the calculation of intermolecular, and hence crystal packing, energies (Chaps. 11, 14 and 16), and so pave the way to the rationalization of the observed packing motifs or even to their prediction, in some instances. Or, not less powerful, is the Hirshfeld Surface (HS) approach [214, 236]. Initially set up as a physically appealing and convenient way (it requires just the promolecular density) to define a molecular entity in a crystal and aimed at evaluating properties such as the molecular electrostatic moments in the crystal [214], it has since then evolved in a number of incredibly powerful graphic tools [237], able to directly visualize the interactions experienced by a molecule in a crystal as well as their different nature and effectiveness. The various HS tools provide an easy way to compare the packing of different molecules or of one given molecule in different environments, in terms of molecular shapes, of *fingerprint maps* [236] highlighting the distribution and nature of the intermolecular interactions a molecule is involved in, and in terms of HSs colours, these latter being obtained by mapping a suitably chosen property on the HSs. When the molecular electrostatic potential (ESP) is mapped, the electrostatic complementarity between interacting molecules in the crystal is, if any, immediately revealed and correlated to the estimates of the electrostatic component of the packing energy that become much more easily explainable, this way.

Section 11.3 of Chap. 11 illustrates how the properties at intermolecular bond critical points (bcps), especially the electron density, ρ_b, its Laplacian, $\nabla^2\rho_b$, and the energy densities G_b, V_b and H_b, can be used as a tool to reveal and classify intermolecular interactions in crystals in terms of their nature and strength. The appropriateness and soundness of such an approach for crystal engineering issues is analyzed in Chap. 16 on a number of halogen-bonded crystal structures and compared with the outcomes from a packing energy analysis on the same structures. Use of local density properties of intermolecular interactions in such area opens the more general question of whether intermolecular recognition and cohesion has to be seen more as driven by the atom-atom interactions involving the peripheral atoms of the molecules, or rather in terms of less localized molecule-molecule or even supramolecular interactions where the molecular distributions as a whole are really playing the game. This important question has been addressed some time ago, in particular by Dunitz and Gavezzotti [238] and by Gatti [89, 91], and it is also summarized in part in Chap. 16. Use of topological indices other than the only properties at bcps, is another way to go beyond the simple atom-atom view of intermolecular interaction, since a larger portion of the molecular space comes into play. In particular, it was shown that rather than in the local properties at its bcp, the sign of a weak intermolecular interaction appears to be more evident in the density rearrangements it provokes inside the basins of the interacting atoms [89, 91, 213]. For instance, for weak hydrogen bond interactions, the *deformation* and even the *interaction* densities have their minima at the intermolecular bcps and in the nearby regions, while these same densities have much larger magnitudes (well) inside the atomic basins [89, 91]. This observation clearly supports the good practice to take simultaneously into account (changes in) both atomic and local properties when analyzing the effect and the nature of intermolecular interactions, within a topological approach such as QTAIM. This practice is, e.g., at the basis of the well-known Koch's and Popelier's criteria [239] to establish hydrogen bonds and it has already been fully explored in the earlier applications of QTAIM to molecular crystals [90, 91, 93, 100]. By operating this way, significant changes, upon packing, in molecular properties like the molecular dipole, may be interestingly related to the local features of the intermolecular interactions which cause them.

Analyses based on the non covalent interactions (NCI) surfaces [215], mentioned in Sect. 1.3.6, go one step further the use of local bcp properties to reveal intermolecular interactions and are, clearly, of potential interest for crystal engineers. Although different from HSs in nature, NCI surfaces share similar purposes and investigating how their analysis relates to and complements with that of the former surfaces, appears a subject worth of future studies.

1.5.5 Stereochemistry

Chemistry is deeply connected with the three dimensional arrangement of atoms to form molecules, because chemical reactions are highly affected. It is interesting that in the history of Chemistry, some stereochemical problems were already addressed well before quantum mechanics, even ignoring the existence of electrons and their role in the chemical bonding [240]. Stereochemistry was first introduced to explain optical properties of substances [241] and then to rationalize the products resulting from reactions. The revolution introduced after knowing the electron probability distribution has widened this view, especially introducing the concept of electron pairs [242], thus adding one more type of objects to arrange in the molecular space. It is interesting that before the quantum mechanics, the atomic *valence* was supposed to be an intrinsic property of atoms, thus implying the number of connections to be assigned. After Lewis [243], the intrinsic properties were the number of electrons and the attitude to reach a noble gas configuration.

Stereochemistry remains extremely important in the modern chemistry, especially when the nature of chemical bonds in a molecule is somewhat ambiguous and the localization or delocalization of electron pairs is such that a simple Lewis-type model does not hold. For examples, bonds to transition metals are often subject to difficult interpretation [244], emphasized by a tendency of chemists toward *reductionism*. In this sense, the use of many mathematical and geometrical analyses inherent in the QTAIM sometimes has encouraged this tendency.

In this book, stereochemical problems are mainly discussed in Chap. 12, dedicated to transition metal complexes and to main group compounds with hypervalent atoms. Information from the electron density distribution is used to retrieve chemical concepts, not limited to bonded or non-bonded electron pairs.

Applications of ED analysis, especially within the QTAIM framework, have been quite numerous in the past two decades and proved the importance of this kind of studies. Macchi et al. [124a] used the bond path to rationalize the chemical bonding in metal olefin complexes, in relation with the famous dichotomy between metallacycles and olefin complexes (i.e. separate M-C bonds or Dewar-Chatt-Duncanson [245] donation and back donation mechanism). Those ideas were later used by Scherer et al. [246] when discussing similar systems. Analysis of bond paths was also successful for the metal-carbide bond in complex carbides with special conducting properties [247], as it will be further discussed in Chap. 10.

The rationalization of the polarizations around a metal atom, visualized by means of the VSCC distribution, is also very useful not only at qualitative level but also quantitatively. For example, in *agostic* interactions the ED analysis in connection with other spectroscopic and theoretical investigations could provide sufficient insight to re-define this kind of bond [248]. The presence or absence of bond paths connecting H and the metal is not the most important feature because the bonding mechanism does not depend on a direct M—H interaction.

As anticipated earlier, a similar discussion on the actual meaning of the bond path was proposed by Farrugia et al. [189] analyzing the Iron Trimethylenemethane complex. Interestingly, while the metal is connected through a bond path only with the C_α, the electron delocalization with the three C_β's is much larger. The result is provocatively summarized in the title of the paper: "Chemical Bonds without Chemical Bonding?".

The bond to hypervalent main group elements is also of interest within stereochemical discussions. Stalke et al. [249] analyzed several molecules with hypervalent atoms and addressed the problem that geometrical coordination alone is not enough to establish the augmented valence.

All these studies address the problem that stereochemistry cannot be analyzed just by means of the *molecular graph* as it would be natural within the QTAIM, because the bond path does not carry itself information on the number of electron pairs involved in the bonding. For this reason, the current consensus is that stereochemical issues can be solved within QTAIM by a combination of *molecular* and *atomic* graphs together with indicators of the electron pair delocalization between two or more atoms.

Another field of interest in organometallic chemistry is that of fluxional molecules, where one isomer rapidly converts into another (in solution). This is very much indicative of a relatively flat PES, and consequent equilibrium between more species. We have already discussed the complex $[FeCo(CO)_8]^-$ [194], where a carbonyl ligand is representative of the fluxional mechanism typically occurring on transition metal bonding. A more complex mechanism was investigated by Ortin et al. [250], who analyzed the experimental ED in a Mn ketene complex. These kind of studies proved that factors involved in dynamical mechanisms could be rationalized by ED analysis. An even more appealing kind of research is a combined QTAIM/IQA-molecular dynamic study [233], where the evolution of ED is analyzed along a path. This intuition is a natural consequence of the work proposed in ref. [194] and it could possibly find large application in the future, thanks also to the improved computational resources available in many laboratories.

1.5.6 Chemical Reactions

Molecular orbital theory and, in particular, its perturbational treatment – most familiar to organic chemists as the Frontier Molecular Orbital (FMO) theory of Fukui [251] – has since long time provided a very powerful basis for explaining, and often predicting, many aspects of chemical reactivity [252]. Rationalization of Woodward- Hoffmann rules [253], which elucidate the pattern of reactivity in pericyclic reactions, is just one, though very prominent example, of the ability of the theory [251]. Molecular orbitals are, however, unobservable and their phase, though very useful and largely exploited in the FMO theory, is totally arbitrary. Approaches based on CD and related quantities permit, instead, to firmly rest our understanding of chemical reactivity on quantum observables, thus avoiding any qualitative dependence of the derived concepts on both the type and the quality level of the computation, while always enabling a study to be performed, regardless of the experimental or theoretical origin of the CD. Molecular ElectroStatic Potential (MESP) contours plots or, more recently, mapping of MESP on molecular surfaces (*e.g.* the isodensity value or the Hirshfeld surfaces), have traditionally been investigated in this area. As mentioned in Sect. 1.5.4, the MESP mapped on HSs visibly reveals the electrostatic complementarity between interacting molecules and so provides a rationale for the intermolecular "reactions" leading to the observed packing in crystals. Being "long range" in nature, ESP has been amply used for exploring molecular recognition and describing reactivity in chemistry, and especially so when both the transfer of electrons between the reactants and the reactants' ED polarization are negligible (charge controlled reactions, cf. Sect. 1.3.7). Analysis of the so called ESP topography (or topology) has also been proposed, by Gadre and coworkers [254]. It enables to enhance our understanding of *molecular recognition* by signaling the point where two molecules start to "feel" each other [255], or, in some cases, it may also provide insights on chemical reactions pathways, by identifying and classifying each reaction step, through the analysis of the succeeding topology catastrophes along the reaction [256].

Other reactivity indicators are those related to the electron-pair structure, like, on an empirical basis, the Laplacian of the electron density, and, rigorously, but ahead of the properties directly obtainable from the CD, the ELI and the ELF. Likewise the ESP topography, but more generally applicable and powerful, the $\nabla^2\rho$ and ELF topologies are able to shed light on reaction mechanisms by revealing, and quite often also predicting, those bonds which are breaking and forming along a reaction coordinate. Early, pioneering examples on the interpretation of cycloaddition reactions pathways were given by Gatti et al. [257], using $\nabla^2\rho$, and by Silvi et al. [258], using the ELF. In the case of the transition state (TS) for the concerted synchronous approach of ethylene and butadiene in the Diels Alder reaction, it was shown [257], for example, that the reacting C atoms, despite being linked by atomic interaction lines at any approaching distance of the reactants, develop a fourth valence $-\nabla^2\rho$ bonded maximum concentration only far beyond the TS, thereby accounting for the ability of perturbational methods to predict the selectivity of such reactions. In the present book, Chap. 18 (and in part also Chap. 17) discusses at length the use of the ELF topology to disentangle the solid state reactions which take place under pressure and phase transformations.

Based on early studies by Bader et al. [6, 259], showing that the local charge concentrations and depletions of $\nabla^2\rho$ predict sites of electrophilic and nucleophilic attack, respectively, the Laplacian of the electron density has been repeatedly used to discuss molecular packing and reactivity in crystals [63, 90, 91]. For instance, concerning H-bond formation, it was shown that the approach of the acidic hydrogen to the base will be such as to align the (3,+3) $-\nabla^2\rho$ minimum in the valence shell charge concentration (VSCC) of the H with the most suitable (3,−3) $-\nabla^2\rho$ base maximum [6, 260]. Typically, an O atom can be involved in two H-bonds, by exploiting the two (3,−3) charge concentrations associated to its two lone pairs. However, in the crystal phase of urea, the O atom forms four similarly strong H-bonds, since by lengthening the C = O bond, the O VSCC distribution becomes more spherical and less asymmetric and two additional charge concentrations become available on the O atom, up and below the molecular plane, for interaction with hydrogens [90, 100]. It has been also shown that displaying the *reactive surface* of a molecule, defined as the $\nabla^2\rho = 0$ isosurface [6, 261], enables one to obtain an effective description of the qualitative affinity of the various portions of a molecule toward Lewis acids and bases. An earlier example of a reactive surface in a molecular organic crystal was obtained by Roversi et al. [135] from an X-ray diffraction study of the charge density in citrinin at 19 K. Many other interesting examples may be found in the more recent literature concerning this use of $\nabla^2\rho$ to rationalize crystal packing and/or reactivity, and a number of them were reviewed in Ref. [91]. The role of $\nabla^2\rho$ in revealing the interactions between a transition metal atom and its neighbouring ligands has been extensively investigated both experimentally and theoretically by Scherer et al. [246, 248, 262] during the last decade and it is summarized in Chap. 12.

Given that the mentioned reactivity descriptors are already fairly well-known to most of the CD practitioners, we thought it worth dedicating a full chapter of this book (Chap. 21) to the description of chemical reactivity through the so-called *conceptual DFT theory* [263]. There are two good reasons behind our choice. The first and most important one is that such theory provides a *unified view* of charge density and chemical reactivity, while bearing noteworthy relationships with quantities derived from the FMO picture of reactivity. Secondly, despite the profound physical grounds of the theory, conceptual DFT seems generally yet not so familiar to those working in the CD area, which makes a didactic, yet rigorous chapter on the argument likely most welcome.

Conceptual DFT theory provides a mathematical framework for using *changes* of the electron density to understand chemical reactions and chemical reactivity. In fact it is based on the key idea that reactivity preferences of a molecule or materials can be decoded by studying their responses to perturbations. Chapter 21, written by most prominent scientists in the field, provides a very comprehensive outline of the theory, by showing how the various reactivity indicators which have been proposed through the years, can all be rigorously derived within one of the alternative representations available for the system under investigation. Analogously to thermodynamics, where the state functions are obtained by differentiating with respect to the variables that define the state of the system, molecular changes associated with chemical reactivity are evaluated by differentiating with respect to the variables that define the molecular structure. And similarly to thermodynamics, where system's definition depends on the experimental setup, it is the chosen theoretical or experimental operative framework that delineates the most suited system's representation. As it is explained in Chap. 21, the quantum chemists, which specify their system based on the external potential (molecular geometry) and the number of electrons, will likely prefer the *closed-system* picture of conceptual DFT, which uses the number of electrons, whereas the *open-system* picture, using the external potential, will be likely adopted by the condensed-matter theorists, which normally specify their system on the basis of the external potential and the chemical potential (Fermi level). The X-ray spectroscopists, who measure the ED and determine the external potential from the nuclear peaks in the ED, will find the *electron-preceding* picture as probably the most suited for them. Rather than taking the external potential as a variable and observing how changes in the external potential induce changes in the ED (an *electron-following* picture common to both the open- and closed-system representations), are the changes in the ED that are used in the electron-preceding picture to infer the change in the external potential in response to the ED changes. Measuring such changes is, however, experimentally challenging and requires performing CD studies under non equilibrium conditions. Chapter 21 presents a number of examples, where various reactivity indicators are compared as for their pros and cons as a function of the case being discussed. The chapter concludes by examining a number of very provocative questions that are hardly ever addressed in the literature, despite or because of their fundamental and challenging nature. Among such questions, whether conceptual DFT is or is not a predictive theory and whether it may be used, like the simple MO-based models, for "Back of the Envelope" calculations. Or why conceptual DFT usually does not consider the attacking reagent explicitly, though a two-reagent picture of conceptual DFT has also been proposed and explored [264]. And, also, when should one use conceptual DFT or when should one rather exploit a simpler and more immediate approach like FMO.

Before concluding this Subsection, it is worth noting that the usual ingredients of CD studies of reactivity find their natural place in conceptual DFT. The first-order response of the energy to changes in external potential is none other than the electron density, while the MESP is a primary regioselectivity indicator in the theory (Sects. 21.6.1 and 21.7.1). Analogously to the Laplacian of the electron density, the local temperature (or nighness indicator) [265] and the local entropy [266] both unveil regions of charge concentration in atoms and molecules, related to the electron-pair structure. The link with FMO theory is instead disclosed (Sect. 21.8.4) by observing that the most commonly used reactivity indicators in FMO appear as zeroth-order approximations to reactivity indicators in conceptual DFT.

1.6 Conclusions and Outlook

This book is intended to present the current state of the art of CD analysis and its applications in many fields of chemistry, as well as materials and biomolecular sciences. We have summarized in this chapter some of the most important pieces of science appeared in the recent past and older seminal works that serve as base for the current research in the field. The list may obviously be incomplete and it only reflects the personal perspective of the authors. However, the reader will find more discussions and relevant literature in the other chapters of this book, reflecting the personal perspectives of the respective authors. Of course, this book cannot be fully comprehensive and some important fields regrettably could not find room here. In this chapter, we have anyway tried to summarize the important contributions from those areas otherwise not represented in the book.

The great potential of Electron Density analysis should easily emerge from this book, which contains contributions from many scientists who developed theories and methods or otherwise pursued applications of electron density distribution. This potential has not been fully explored yet, which justifies the broad and intense research activity currently ongoing in many research groups worldwide. The reader could probably appreciate that, just thinking at the enormous progresses made in the analysis of chemical bonding since the early days of charge density. It can be anticipated that methods and ideas presented in this and the other chapters will be further developed in the next future, because many methodological progresses and applications are already envisaged.

Acknowledgements This book was planned in autumn 2008, after the successful organization of the 5th European Charge Density Meeting (Gravedona, Italy, 6–11 June 2008). We accepted the invitation from Springer, by proposing a challenging project, that was to provide a broad overview of the many applications of charge density analysis, involving many scientists who were requested, whenever possible, to write joint contributions. We believe that this goal has been achieved, although with strong efforts, given its complexity.We are indebted to a number of colleagues, especially to all the authors of contributions reported in this book, for their excellent cooperation and many useful suggestions. We also thank those colleagues who acted as anonymous referees and largely contributed to the improvement of this project.We wish to dedicate this book to some colleagues who have unfortunately passed away during the production of this book, in particular Prof. M. A. Blanco (University of Oviedo), Prof. A. Goeta (University of Durham), Prof. C. Pisani and Prof. C. Roetti, both at University of Turin. P.M. thanks the Swiss National Science Foundation for financial support of his research (project 200021_125313).

References

1. Coppens P, Feil D (1991) The past and future of experimental charge density analysis. In: Jeffrey GA, Piniella JF (eds) The application of charge density research to chemistry and drug design, vol 250, NATO ASI series B. Plenum Publishing Corp., New York, pp 7–22
2. Coppens P (1998) Charge-density analysis at the turn of the century. Acta Crystallogr A 54:779–788
3. Coppens P, Volkov A (2004) The interplay between experiment and theory in charge-density analysis. Acta Crystallogr A 60:357–364
4. Coppens P (2005) Charge densities come of age. Angew Chem Int Ed 44:6810–6811
5. Coppens P (1997) X-ray charge densities and chemical bonding. IUCr texts on crystallography, vol 4. International Union of Crystallography/Oxford University Press, Oxford
6. Bader RFW (1990) Atoms in molecules: a quantum theory, vol 22, International series of monographs on chemistry. Oxford Science, Oxford
7. Matta CF, Boyd RJ (eds) (2007) The quantum theory of atoms in molecules: from solid state to DNA and drug design. Wiley-VCH, Weinheim
8. McWeeny R (1989) Methods of molecular quantum mechanics, 2nd edn. Academic, London
9. Spackman MA (1997) Charge densities from X-ray diffraction data. Annu Rep Progr Chem C Phys Chem 94:177–207
10. Car R, Parrinello M (1985) Unified approach for molecular dynamics and density functional theory. Phys Rev Lett 55:2471–2474
11. Coppens P, Penner-Hahn J (eds) (2001) X-rays in chemistry. (Dedicated issue on) Chem Rev 101(6):1567–1868
12. Hestenes MR, Stiefel EJ (1952) Methods of conjugate gradients for solving linear systems. Natl Bur Stand USA 49:409–436
13. Roquette P, Maronna A, Peters A, Kaifer E, Himmel HJ, Hauf C, Herz V, Scheidt EW, Scherer W (2010) On the electronic structure of Ni-II complexes that feature chelating bisguanidine ligands. Chem Eur J 16:1336–1350

14. Jayatilaka D, Grimwood DJ (2001) Wavefunctions derived from experiment. I. Motivation and theory. Acta Crystallogr A 57:76–86
15. Tanaka K, Makita R, Funahashi S, Komori T, Win Z (2008) X-ray atomic orbital analysis. I. Quantum mechanical and crystallographic framework of the method. Acta Crystallogr B 64:437–449
16. Figgis BN, Reynolds PA, Williams GA (1980) Spin-density and bonding in the $[CoCl_4]^{2-}$ ion in Cs_3CoCl_5. Part 2. Valence electron-distribution in the $CoCl_4^{2-}$ ion. J Chem Soc Dalton Trans 12:2339–2347
17. Figgis BN (2000) Ligand field theory and its applications. Wiley-VCH, New York
18. Holladay A, Leung PC, Coppens P (1983) Generalized relations between d-orbital cccupancies of transition-metal atoms and electron-density multipole population parameters from X-ray diffraction data. Acta Crystallogr A 39:377–387
19. Oszlányi G, Sütő A (2008) The charge flipping algorithm. Acta Crystallogr A 64:123–134
20. Collins DM (1982) Electron density images from imperfect data by iterative entropy maximization. Nature 298:49–51
21. Gull SF, Daniell GJ (1978) Image reconstruction from incomplete and noisy data. Nature 272:686–690
22. Jaynes ET (1968) Prior probabilities. IEEE Trans Syst Sci Cybern SSC-4:227–240
23. Roversi P, Irwin JJ, Bricogne G (1998) Accurate charge-density studies as an extension of Bayesian crystal structure determination. Acta Crystallogr A 54:971–996
24. (a) Palatinus L, van Smaalen S (2005) The prior-derived F constraints in the maximum-entropy method. Acta Crystallogr A 61:363–372; (b) van Smaalen S, Netzel J (2009) The maximum entropy method in accurate charge-density studies. Phys Scripta 79:048304
25. (a) Hofmann A, Netzel J, van Smaalen S (2007) Accurate charge density of trialanine: a comparison of the multipole formalism and the maximum entropy method (MEM). Acta Crystallogr B 63:285–295; (b) Netzel J, van Smaalen S (2009) Topological properties of hydrogen bonds and covalent bonds from charge densities obtained by the maximum entropy method (MEM). Acta Crystallogr B 65:624–638
26. Cargnoni F, Nishibori E, Rabiller P, Bertini L, Snyder GJ, Christensen M, Gatti C, Iversen BB (2004) Interstitial Zn atoms do the trick in thermoelectric zinc antimonide, Zn_4Sb_3: a combined maximum entropy method X-ray electron density and ab initio electronic structure study. Chem Eur J 10:3861–3870
27. Munshi P, Madsen AØ, Spackman MA, Larsen S, Destro R (2008) Estimated H-atom anisotropic displacement parameters: a comparison between different methods and with neutron diffraction results. Acta Crystallogr A 64:465–475
28. Johnson CK (1970) Generalized treatments for thermal motion, Chapter 9, pp 132–160. In: Willis BTM (ed) Thermal neutron diffraction. Oxford University Press, London
29. Madsen AØ (2006) SHADE web server for estimation of hydrogen anisotropic displacement parameters. J Appl Crystallogr 39:757–758
30. Hirshfeld FL (1976) Can X-ray data distinguish bonding effects from vibrational smearing? Acta Crystallogr A 32:239–244
31. Roversi P, Destro M (2004) Approximate anisotropic displacement parameters for H atoms in molecular crystals. Chem Phys Lett 386:472–478
32. Madsen AØ, Mason S, Larsen S (2003) A neutron diffraction study of xylitol: derivation of mean square internal vibrations for hydrogen atoms from a rigid-body description. Acta Crystallogr B 59:653–663
33. Whitten AE, Spackman MA (2006) Anisotropic displacement parameters for H atoms using an ONIOM approach. Acta Crystallogr B 62:875–888
34. Bürgi HB, Capelli SC (2000) Dynamics of molecules in crystals from multi-temperature anisotropic displacement parameters. I. Theory. Acta Crystallogr A 56:403–412
35. Szabo A, Ostlund NS (1982) Modern quantum chemistry: introduction to advanced electronic structure theory. Macmillan, New York
36. Parr RG, Yang W (1989) Density-functional theory of atoms and molecules, vol 16, International series of monographs on chemistry. Oxford Science, Oxford

37. Kohn W, Sham LJ (1965) Self-consistent equations including exchange and correlation effects. Phys Rev A 140:1133–1138
38. Koch W, Holthausen MC (2001) A chemist's guide to density functional theory, 2nd edn. Wiley-VCH, Weinheim
39. Perdew JP, Ruzsinszky A, Constantin LA, Sun J, Csonka GI (2009) Some fundamental issues in ground-state density functional theory: a guide for the perplexed. J Chem Theory Comput 5:902–908
40. (a) Ángyán JG, Loos M, Mayer I (1994) Covalent bond orders and atomic valence indices in the topological theory of atoms in molecules. J Phys Chem 98:5244–5248; (b) Fradera X, Austen MA, Bader RFW (1999) The Lewis model & beyond. J Phys Chem A 103:304–314
41. Poater J, Duran M, Solà M, Silvi B (2005) Theoretical evaluation of electron delocalization in aromatic molecules by means of atoms in molecules (AIM) and electron localization function (ELF) topological approaches. Chem Rev 105:3911–3947
42. Bader RFW, Stephens ME (1974) Fluctuation and correlation of electrons in molecular systems. Chem Phys Lett 26:445–449
43. Poater J, Solà M, Duran M, Fradera X (2002) The calculation of electron localization and delocalization indices at the Hartree-Fock, density functional and post-Hartree-Fock levels of theory. Theor Chem Acc 107:362–371
44. Matito E, Solà M, Salvador P, Duran M (2007) Electron sharing indexes at the correlated level. Applications to aromaticity calculations. Faraday Discuss 135:325–345
45. (a) Ponec R (1997) Electron pairing and chemical bonds. Chemical structure, valences and structural similarities from the analysis of the Fermi holes. J Math Chem 21:323; (b) Ponec R (1998) Electron pairing and chemical bonds. Molecular structure from the analysis of pair densities and related quantities. J Math Chem 23:85–103
46. Francisco E, Pendás AM, Blanco MA (2009) A connection between domain-averaged Fermi hole orbitals and electron number distribution functions in real space. J Chem Phys 131:124125
47. Seitz F (1987) The modern theory of solids. Dover, New York
48. Yamaguchi Y, Osamura Y, Goddard JD, Schäefer III HF (1994) In A new dimension to quantum chemistry. Analytic derivative methods in Ab initio molecular electronic structure theory. Chapter 2. International series of monographs on chemistry vol 29, Oxford Science Publications, Oxford
49. Davidson ER (1976) Reduced density matrices in quantum chemistry. Academic, New York
50. Pilar FL (1968) Elementary quantum chemistry. Mc Graw Hill, New York
51. Bender CF, Davidson ER (1968) Theoretical study of the LiH molecule. J Chem Phys 49:4222–4229
52. Gatti C, MacDougall PJ, Bader RFW (1988) Effect of electron correlation on the topological properties of molecular charge distributions. J Chem Phys 88:3792–3804
53. (a) Wang L-C, Boyd RJ (1989) The effect of electron correlation on the electron density distributions of molecules: comparison of perturbation and configuration interaction methods. J Chem Phys 90:1083–1090; (b) Boyd RJ, Wang L-C (1989) The effect of electron correlation on the topological and atomic properties of the electron density distributions of molecules. J Comp Chem10:367–375
54. Wiberg KB, Hadad CM, LePage TJ, Breneman CM, Frisch MJ (1992) Analysis of the effect of electron correlation on charge density distributions. J Phys Chem 96:671–679
55. Mayer I, Salvador P (2004) Overlap populations, bond orders and valences for "fuzzy" atoms. Chem Phys Lett 383:368–375
56. Gatti C, Lasi D (2007) Source function description of metal-metal bonding in d-block organometallic compounds. Faraday Discuss 135:55–78
57. Stewart RF (1979) On the mapping of electrostatic properties from Bragg diffraction data. Chem Phys Lett 65:335–342
58. (a) Stewart RF (1982) Mapping electrostatic potentials from diffraction data. God Jugosl Cent Kristalogr 17:1–24; (b) Stewart RF, Craven BM (1993) Molecular electrostatic potentials

from crystal diffraction – the neurotransmitter gamma-aminobutyric-acid. Biophys J 65: 998–1005; (c) Stewart RF (1991) Electrostatic properties of molecules from diffraction data. In: Jeffrey GA, Piniella JF (eds) The application of charge density research to chemistry and drug design, NATO ASI Series B, vol 250. Plenum Publishing Corp., New York, pp 63–101
59. Spackman MA (2007) Comment on On the calculation of the electrostatic potential, electric field and electric field gradient from the aspherical pseudoatom model by Volkov, King, Coppens & Farrugia. Acta Crystallogr A 63:198–200
60. Su Z, Coppens P (1992) On the mapping of electrostatic properties from the multipole description of the charge density. Acta Crystallogr A 48:188–197
61. Volkov A, Coppens P (2007) Response to Spackman's comment on On the calculation of the electrostatic potential, electric field and electric field gradient from the aspherical pseudoatom model. Acta Crystallogr A 63:201–203
62. Volkov A, King HF, Coppens P, Farrugia LJ (2006) On the calculation of the electrostatic potential, electric field and electric field gradient from the aspherical pseudoatom model. Acta Crystallogr 62:400–408
63. Koritsanszky T, Coppens P (2001) Chemical applications of X-ray charge density analysis. Chem Rev 101:1583–1627
64. Schiøtt B, Overgaard J, Larsen FK, Iversen BB (2004) Testing theory beyond molecular structure: electron density distributions of complex molecules. Int J Quantum Chem 96:23–31
65. Buckingham AD (1967) Permanent and induced molecular moments and long-range intermolecular forces. Adv Chem Phys 12:107–142
66. (a) Spackman MA (1986) Atom-atom potential via electron gas theory. J Chem Phys 85:6579–6586; (b) Spackman MA (1986) A simple quantitative model of hydrogen bonding. J Chem Phys 85:6587–6601; (c) Spackman MA (1987) A simple quantitative model of hydrogen bonding. Application to more complex systems. J Phys Chem 91:3179–3186
67. Spackman MA (2005) The use of promolecular density to approximate the penetration contribution to intermolecular energies. Chem Phys Lett 418:158–162
68. Volkov A, Koritsanszky T, Coppens P (2004) Combination of the exact potential and multipole methods (EP/MM) for evaluation of intermolecular electrostatic interaction energies with pseudoatom representation of molecular electron densities. Chem Phys Lett 391:170–175
69. Gavezzotti A (2002) Calculation of intermolecular interaction energies by direct numerical integration over electron densities. I. Electrostatic and polarization energies in molecular crystals. J Phys Chem 106:4145–4154
70. Volkov A, Li X, Koritsanszky T, Coppens P (2004) *Ab Initio* quality electrostatic atomic and molecular properties including intermolecular energies from a transferable theoretical pseudoatom databank. J Phys Chem A 108:4283–4300
71. Volkov A, Coppens P (2004) Calculation of electrostatic interaction energies in molecular dimers from atomic multipole moments obtained by different methods of electron density partitioning. J Comput Chem 25:921–934
72. Pichon-Pesme V, Jelsch C, Guillot B, Lecomte C (2004) A comparison between experimental and theoretical aspherical-atom scattering factors for charge-density refinement of large molecules. Acta Crystallogr A 60:204–208
73. Volkov A, Koritsanszky T, Li X, Coppens P (2004) Response to the paper "A comparison between experimental and theoretical aspherical-atom scattering factors for charge-density refinement of large molecules", by Pichon-Pesme, Jelsch, Guillot & Lecomte (2004). Acta Crystallogr A 60:638–639
74. Dominiak PM, Volkov A, Li X, Messerschmidt M, Coppens P (2007) A theoretical databank of transferable aspherical atoms and its application to electrostatic interaction energy calculations of macromolecules. J Chem Theory Comput 3:232–247
75. Dittrich B, Koritsanszky T, Luger P (2004) A simple approach to non-spherical electron densities by using invarioms. Angew Chem Int Ed Engl 43:2718–2721
76. Dittrich B, Hübschle CB, Luger P, Spackman MA (2006) Introduction and validation of an invariom database for amino-acid, peptide and protein molecules. Acta Crystallogr D 62:1325–1335

77. Jensen F (2007) Introduction to computational chemistry, 2nd edn. Wiley, Chichester
78. Volkov A, Abramov Y, Coppens P, Gatti C (2000) On the origin of topological differences between experimemtal and theoretical crystal charge densities. Acta Crystallogr A 56: 332–339
79. Volkov A, Gatti C, Abramov Y, Coppens P (2000) Evaluation of net atomic charges and atomic and molecular electrostatic moments through topological analysis of the experimental charge density. Acta Crystallogr A 56:252–258
80. Farrugia LJ, Cameron E (2009) The QTAIM approach to chemical bonding between transition metals and carbocyclic rings: a combined experimental and theoretical study of (η^5-C_5H_5)Mn(CO)$_3$, (η^6-C_6H_6)Cr(CO)$_3$, and (*E*)-{(η^5-C_5H_4)CF-CF(η^5-C_5H_4)}(η^5-C_5H_5)$_2$Fe$_2$. J Am Chem Soc 131:1251–1268
81. (a) Te Velde G, Bickelhaupt FM, van Gisbergen SJA, Fonseca Guerra C, Baerends EJ, Snijders JG, Ziegler T (2001) Chemistry with ADF. J Comput Chem 22:931–967; (b) ADF2007.01, SCM, (2007) Theoretical chemistry, Vrije Universiteit, Amsterdam. http://www.scm.com
82. Macchi P, Coppens P (2001) Relativistic analytical wave functions and scattering factors for neutral atoms beyond Kr and for all chemically important ions up to I. Acta Crystallogr A 57:656–662
83. Volkov A, Macchi P, Farrugia LJ, Gatti C, Mallinson P, Richter T, Koritsanszky T (2006) XD2006 – a computer program package for multipole refinement, topological analysis of charge densities and evaluation of intermolecular energies from experimental and theoretical structure factors
84. Krijn MPCM, Graafsma H, Feil D (1988) The influence of intermolecular interactions on the electron-density distribution. A comparison of experimental and theoretical results for -oxalic acid dehydrate. Acta Crystallogr B 44:609–616
85. Spackman MA, Byron PG, Alfredsson M, Hermansson K (1999) Influence of intermolecular interactions on multipole-refined electron densities. Acta Crystallogr A 55:30–47
86. de Vries RY, Feil D, Tsirelson VG (2000) Extracting charge density distributions from diffraction data: a model study on urea. Acta Crystallogr B 56:118–123
87. Dittrich B, Spackman MA (2007) Can the interaction density be measured? The example of the non-standard amino acid sarcosine. Acta Crystallogr A 63:426–436
88. Spackman MA, Munshi P, Dittrich B (2007) Dipole moment enhancement in molecular crystals from X-ray diffraction data. Chemphyschem 8:2051–2063
89. Gatti C, May E, Destro R, Cargnoni F (2002) Fundamental properties and nature of CH··O interactions in crystals on the basis of experimental and theoretical charge densities. The case of 3,4-Bis(dimethylamino)-3-cyclobutene-1,2-dione (DMACB) crystal. J Phys Chem A 106:2707–2720
90. Gatti C, Saunders VR, Roetti C (1994) Crystal field effects on the topological properties of the electron density in molecular crystals: the case of urea. J Chem Phys 101:10686–10696
91. Gatti C (2005) Chemical bonding in crystals: new directions. Z Krist 220:399–457
92. Gatti C, Silvi B, Colonna F (1995) Dipole moment of the water molecule in the condensed phase: a periodic Hartree-Fock estimate. Chem Phys Lett 247:135–141
93. May E, Destro R, Gatti C (2001) The unexpected and large enhancement of the dipole moment in the 3,4-Bis(dimethylamino)-3-cyclobutene-1,2-dione (DMACB) molecule upon crystallization: a new role of the intermolecular CH--O interactions. J Am Chem Soc 123:12248–12254
94. Whitten AE, Jayatilaka D, Spackman MA (2006) Effective molecular polarizabilities and crystal refractive indices estimated from X-ray diffraction data. J Chem Phys 125:174505
95. Elliott S (1998) The physics and chemistry of solids. Wiley, Chichester
96. Bertini L, Cargnoni F, Gatti C (2007) Chemical insight from electron density and wavefunctions: software developments and applications to crystals, molecular complexes and materials science. Theor Chem Acc 117:847–884
97. Bertini L, Cargnoni F, Gatti C (2006) A chemical approach to the first-principles modeling of novel thermoelectric materials, chapter 7. In: Rowe DM (ed) Thermoelectrics handbook: macro to nano. CRC Press/Taylor & Francis, Boca Raton

98. (a) Schwarz WHE, Valtzanos P, Ruedenberg K (1985) Electron difference densities and chemical bonding. Theor Chim Acta 68:471–506; (b) Schwarz WHE, Mensching L, Valtzanos P, Von Niessen W (1986) A chemically useful definition of electron difference densities. Int J Quant Chem 29:909–914; (c) Ruedenberg K, Schwarz WHE (1990) Nonspherical atomic ground-state densities and chemical deformation densities from x-ray scattering. J Chem Phys 92:4956–4969; (d) Wiberg KB, Bader RFW, Lau CDH (1987) Theoretical analysis of hydrocarbon properties. 1. Bonds, structures, charge concentrations, and charge relaxations. J Am Chem Soc 109:985–1001
99. Macchi P, Sironi A (2003) Chemical bonding in transition metal carbonyl clusters: complementary analysis of theoretical and experimental electron densities. Coord Chem Rev 238–239:383–412
100. Gatti C (2007) Solid state applications of QTAIM and the source function - molecular crystals, surfaces, host-guest systems and molecular complexes, chapter 7. In: Matta CF, Boyd RJ (eds) The quantum theory of atoms in molecules: from solid state to DNA and drug design. Wiley-VCH, Weinheim
101. Macchi P (2009) Electron density distribution in organometallic materials. Chimia 63:1–6
102. Tsirelson VG, Ozerov RP (1996) Electron density and bonding in crystals. Institute of Physics Publishing, Bristol
103. Schmider H, Edgecombe KE, Smith VH Jr, Weyrich W (1992) One-particle density matrices along the molecular-bonds in linear molecules. J Chem Phys 96:8411–8419
104. Asthalter T, Weyrich W (1997) On the chemical interpretation of the one-electron density matrix of some ionic solids: LiH, LiF, and LiFHF. Ber Bunsenges Phys Chem 101:11–22
105. Becke AD, Edgecombe KE (1990) A simple measure of electron localization in atomic and molecular systems. J Chem Phys 92:5397–5403
106. Kohout M (2004) A measure of electron localizability. Int J Quantum Chem 97:651–658
107. Schmider HL, Becke AD (2000) Chemical content of the kinetic energy density. J Mol Struct (Theochem) 527:51–61
108. Jayatilaka D, Grimwood J (2004) Electron localization functions obtained from X-ray constrained Hartree-Fock wavefunctions for molecular crystals of ammonia, urea and alloxan. Acta Crystallogr A 60:111–119
109. Grabowsky S, Jayatilaka D, Mebs S, Luger P (2010) The electron localizability indicator from X-ray diffraction data—a first application to a series of epoxide derivatives. Chem Eur J 16:12818–12821
110. Abramov YA (1997) On the possibility of kinetic energy density evaluation from the experimental electron-density distribution. Acta Crystallogr A 53:264–272
111. Tsirelson VG (2002) The mapping of electronic energy distributions using experimental electron density. Acta Crystallogr B 58:632–639
112. Tsirelson V, Stash A (2002) Determination of the electron localization function from electron density. Chem Phys Lett 351:142–148
113. Tsirelson V, Stash A (2002) Analyzing experimental electron density with the localized-orbital locator. Acta Crystallogr B 58:780–785
114. Bader RFW, Essén H (1984) The characterization of atomic interactions. J Chem Phys 80:1943–1960
115. Hoffmann R (1988) Solids and surfaces. A chemist's view of bonding in extended structures. VCH Publishers, Inc., New York
116. Dronskowski R, Blöchl PE (1993) Crystal orbital Hamilton populations (COHP): energy-resolved visualization of chemical bonding in solids based on density-functional calculations. J Phys Chem 97:8617–8624
117. Wannier GH (1937) The structure of electronic excitation levels in insulating crystals. Phys Rev 52:191–197
118. Marzari N, Vanderbilt D (1997) Maximally localized generalized Wannier functions for composite energy bands. Phys Rev B 56:12847–12865
119. (a) Zicovich-Wilson CM, Dovesi R, Saunders VR (2001) A general method to obtain well localized Wannier functions for composite energy bands in linear combination of atomic

orbital periodic calculations. J Chem Phys 115:9708–9719; (b) Casassa S, Zicovich-Wilson CM, Pisani C (2006) Symmetry-adapted localized Wannier functions suitable for periodic calculations. Theor Chem Acc 116:726–733

120. Gatti C, Bertini L, Blake NP, Iversen BB (2003) Guest-framework interaction in type I inorganic clathrates with promising thermoelectric properties: on the ionic versus neutral nature of the alkaline-earth metal guest A in $A_8Ga_{16}Ge_{30}$ (A = Sr, Ba). Chem Eur J 9:4556–4568
121. Oganov AR, Chen J, Gatti C, Ma YZ, Ma YM, Glass CW, Liu Z, Yu T, Kurakevych OO, Solozhenko VL (2009) Ionic high-pressure form of elemental boron. Nature 457:863–868
122. Vogt C, Hoffmann R-D, Rodewald UC, Eickerling G, Presnitz M, Eyert V, Scherer W, Pöttgen R (2009) High- and low-temperature modifications of Sc_3RuC_4 and Sc_3OsC_4 -relativistic effects, structure, and chemical bonding. Inorg Chem 48:6436–6451
123. Debye P (1915) Dispersion of Röntgen rays. Ann Phys 46:809–823
124. (a) Macchi P, Proserpio DM, Sironi A (1998) Experimental electron density studies for investigating the metal pi-ligand bond: the case of bis(1,5-cyclooctadiene)nickel. J Am Chem Soc 120:1447–1455; (b) Koritsanszky T, Flaig R, Zobel D, Krane HG, Morgenroth W, Luger P (1998) Accurate experimental electronic properties of DL-proline monohydrate obtained within 1 day. Science 279:356–358
125. (a) Zhurov VV, Zhurova EA, Pinkerton AA (2008) Optimization and evaluation of data quality for charge density studies. J Appl Cryst 41:340–349; (b) Zhurova EA, Zhurov VV, Pinkerton AA (2007) Structure and bonding in beta-HMX-characterization of a trans-annular N—N interaction. J Am Chem Soc 129:13887–13893
126. (a) Martin A, Pinkerton AA (1998) Charge density studies using CCD detectors: oxalic acid at 100 K revisited. Acta Cryst B54:471–477; (b) Macchi P, Proserpio DM, Sironi A, Soave R, Destro R (1998) A test of the suitability of CCD area detectors for accurate electron-density studies. J Appl Crystallogr 31:583–588
127. (a) Blessing RH (1987) Data reduction and error analysis for accurate single crystal diffraction intensities. Crystallogr Rev 1:3–58; (b) Blessing RH (1995) An empirical correction for absorption anisotropy. Acta Crystallogr A 51:33–38
128. Lenstra ATH, Kataeva ON (2001) Structures of copper(II) and manganese(II) di(hydrogen malonate) dihydrate; effects of intensity profile truncation and background modelling on structure models. Acta Crystallogr B 57:497–506; (b) Lenstra ATH, Van Loock JFJ, Rousseau B, Maes ST (2001) Systematic intensity errors caused by spectral truncation: origin and remedy. Acta Crystallogr A 57:629–641; (c) Rousseau B, Maes ST, Lenstra ATH (2000) Systematic intensity errors and model imperfection as the consequence of spectral truncation. Acta Crystallogr A 56:300–307; (d) Destro R, Marsh RE (1987) Scan-truncation corrections in single-crystal diffractometry – an empirical-method. Acta Crystallogr A 43:711–718; (e) Destro R (1988) Experimental-determination of scan-truncation losses from low-temperature (16 K) single-crystal x-ray measurements. Aust J Phys 41:503–510; (c) Destro R, Marsh RE (1993) On predicting scan profiles – the nature of the aberration function. Acta Crystallogr A 49:183–190
129. Schulz T, Meindl K, Leusser D, Stern D, Graf J, Michaelsen C, Ruf M, Sheldrick GM, Stalke D (2009) A comparison of a microfocus X-ray source and a conventional sealed tube for crystal structure determination. J Appl Crystallogr 42:885–891
130. Macchi P, Bürgi H-B, Chimpri AS, Hauser J, Gál Z (2011) Low energy contamination of Mo micro-source X-ray radiation: analysis and solution of the problem. J Appl Cryst 44:763–771
131. Jauch W, Reehius M, Schultz AJ (2004) γ-ray and neutron diffraction studies of CoF_2: magnetostriction, electron density and magnetic moments. Acta Crystallogr A 60:51–57
132. Larsen FK (1995) Diffraction studies of crystals at low temperatures - crystallography below 77 K. Acta Crystallogr B 51:468–482
133. Johnson CK (1969) Addition of higher cumulants to the crystallographic structure-factor equation: a generalized treatment for thermal-motion effects. Acta Crystallogr A 25:187–194

134. Mallinson PR, Koritsanszky T, Elkaim E, Li N, Coppens P (1988) The Gram-Charlier and multipole expansions in accurate X-ray diffraction studies: can they be distinguished? Acta Crystallogr A 44:336–343
135. Roversi P, Barzaghi M, Merati F, Destro R (1996) Charge density in crystalline citrinin from X-ray diffraction at 19 K. Can J Chem 74:1145–1161
136. (a) Makita R, Tanaka K, Onuki Y (2008) 5d and 4f electron configuration of CeB_6 at 340 and 535 K. Acta Crystallogr B 64:534–549; (b) Tanaka K, Onuki Y (2002) Observation of 4f electron transfer from Ce to B_6 in the Kondo crystal CeB_6 and its mechanism by multi-temperature X-ray diffraction. Acta Crystallogr B 58:423–436; Funahashi S, Tanaka K, Iga F (2010) X-ray atomic orbital analysis of 4f and 5d electron configuration of SmB_6 at 100, 165, 230, 298 K. Acta Crystallogr B 66:292–306
137. (a) Stewart RF, Bentley J, Goodman B (1975) Generalized X-ray scattering factors in diatomic molecules. J Chem Phys 63:3786–3793; (b) Kurki-Suonio K (1977) Charge density deformation models. Isr J Chem 16:115–123
138. Hansen NK, Coppens P (1978) Electron population analysis of accurate diffraction data. 6. Testing aspherical atom refinements on small-molecule data sets. Acta Crystallogr A 34: 909–921
139. Clementi E, Roetti C (1974) Roothaan-Hartree-Fock atomic wavefunctions. Basis functions and Their coefficients for ground and certain excited states of neutral and ionized atoms, $Z \leq 54$. Atom Data Nucl Data Tab 14:177
140. Su Z, Coppens P (1998) Nonlinear least-squares fitting of numerical relativistic atomic wave functions by a linear combination of Slater-type functions for atoms with $Z = 1$–36. Acta Crystallogr A 54:646–652
141. Clementi E, Raimondi DL (1963) Atomic screening constants from SCF functions. J Chem Phys 38:2686–2689
142. Iversen BB, Larsen FK, Figgis BN, Reynolds PA (1997) X-ray–neutron diffraction study of the electron-density distribution in trans-tetraaminedinitronickel(II) at 9 K: transition-metal bonding and topological analysis. J Chem Soc Dalton Trans 2227–2240
143. Volkov A, Coppens P (2001) Critical examination of the radial functions in the Hansen-Coppens multipole model through topological analysis of primary and refined theoretical densities. Acta Crystallogr A 57:395–405
144. (a) MOLLY, See ref. 138; (b) Stewart RF, Spackman MA, Flensburg C (2000) VALRAY – User's manual, 2.1 edn. Carnegie Mellon University/University of Copenhagen, Pittsburgh/Denmark; (c) Petricek V, Dusek M, Palatinus L (2006) JANA2006, Structure determination software programs. Institute of Physics, Praha; (d) Jelsch C, Guillot B, Lagoutte L, Lecomte C (2005) Advances in proteins and small molecules. Charge density refinement methods using software MoPro. J Appl Crystallogr 38:38–54
145. Ghermani NE, Bouhmaida N, Lecomte C (1992) ELECTROS: computer program to calculate electrostatic properties from high resolution X-ray diffraction. Universite´ de Nancy I, France
146. Tsirelson V, Stasch A (2002) WinXPRO: a program for calculating crystal and molecular properties using multipole parameters of the electron density. J Appl Crystallogr 35:371–373
147. Evarestov RA (2007) Quantum chemistry of solids. The LCAO first principle treatment of crystals. Springer, Berlin
148. Bredow T, Dronskowski R, Ebert H, Jug K (2009) Theory and computer simulation of perfect and defective solids. Prog Solid State Chem 37:70–80
149. Oganov AR (2005) Dedicated issue on computational crystallography. Z Krist 220:399–585
150. Dovesi R, Orlando R, Civalleri B, Roetti C, Saunders VR, Zicovich-Wilson CM (2005) CRYSTAL: a computational tool for the *ab initio* study of the electronic properties of crystals. Z Krist 220:571–573
151. Dovesi R, Saunders VR, Roetti C, Orlando R, Zicovich-Wilson CM, Pascale F, Civalleri B, Doll K, Harrison NM, Bush IJ, D'Arco Ph, Llunell M (2009) CRYSTAL09 user's manual, Università di Torino, Torino. http://www.CRYSTAL.unito.it

152. (a) Blaha P, Schwarz K, Madsen G, Kvasnicka D, Luitz J (2010) WIEN2k (current version WIEN2k_10.1), Institute für Materials Chemistry, TU Vienna. http://www.wien2k.at/; (b) Laskowski R (2008) WIENncm: a non-collinear magnetism version of WIEN2k. http://www.wien2k.at/reg_user/ncm/
153. Giannozzi P, Baroni S, Bonini N, Calandra M, Car R, Cavazzoni C, Ceresoli D, Chiarotti GL, Cococcioni M, Dabo I, Dal Corso A, Fabris S, FratesiG, de Gironcoli S, Gebauer R, Gerstmann U, Gougoussis C, Kokalj A, Lazzeri M, Martin-Samos L, Marzari N, Mauri F, Mazzarello R, Paolini S, Pasquarello A, Paulatto L, Sbraccia S, Scandolo C, Sclauzero G, Seitsonen AP, Smogunov A, Umari P, Wentzcovitch RM (2009) QUANTUM ESPRESSO: a modular and open-source software project for quantum simulations of materials. J Phys Condens Matter 21, 395502 (19 pp). http://www.quantumespresso.org/
154. Baroni S, Giannozzi P, Testa A (1987) Green's-function approach to linear response in solids. Phys Rev Lett 58:1861–1864; Gonze X, Vigneron J-P (1989) Density-functional approach to non-linear-response coefficients of solids. Phys Rev B 49:13120–13128; (c) Baroni S, de Gironcoli S, Dal Corso A, Giannozzi P (2001) Phonons and related crystal properties from density-functional perturbation theory. Rev Mod Phys 73:515–562
155. Gonze X, Rignanese G-M, Caracas R (2005) First principle studies of the lattice dynamic of crystals, and related properties. Z Krist 220:458–472
156. Gonze X, Rignanese G-M, Verstraete M, Beuken J-M, Pouillon Y, Caracas R, Jollet F, Torrent M, Zerah G, Mikami M, Ghosez P, Veithen M, Raty J-Y, Olevano V, Bruneval F, Reining L, Godby R, Onida G, Hamann DR, Allan DC (2005) A brief introduction to the ABINIT software package. Z Krist 220:558–562
157. Gerratt J, Mills I (1968) Force constants and dipole-moment derivatives of molecules from perturbed Hartree-Fock calculations. I. J Chem Phys 49:1719–1729
158. a) Ferrero M, Rérat M, Kirtman B, Dovesi R (2008) Calculation of first and second static hyperpolarizabilities of one- to three-dimensional periodic compounds. Implementation in the CRYSTAL code. J Chem Phys 129:244100; (b) Ferrero M, Rérat M, Orlando R, Dovesi R (2008) The calculation of static polarizabilities of 1-3D periodic compounds. The implementation in the crystal code. J Comp Chem 29:1450–1459
159. Reiher M, Wolf A (2009) Relativistic quantum theory: the fundamental theory of molecular science. Wiley-VCH Verlag Gmb&Co. KGaA, Weinheim
160. Voloshina E, Paulus B (2006) On the application of the incremental scheme to ionic solids: test of different embeddings. Theor Chem Acc 114:259–264
161. Wesolowski TA, Warshel A (1993) Frozen density functional approach for ab initio calculations of solvated molecules. J Phys Chem 97:8050
162. Burow A, Sierka M, Döbler J, Sauer J (2009) Point defects in CaF_2 and CeO_2 investigated by the periodic electrostatic embedded cluster method. J Chem Phys 130:174710
163. (a) Kiewisch K, Eickerling G, Reiher M, Neugebauer J (2008) Topological analysis of electron densities from Kohn-Sham and subsystem density functional theory. J Chem Phys 128:044114; (b) Fux S, Kiewisch K, Jacob CR, Neugebauer J, Reiher M (2008) Analysis of electron density distributions from subsystem density functional theory applied to coordination bonds. Chem Phys Lett 461:353–359
164. Götz K, Meier F, Gatti C, Burow AM, Sierka M, Sauer J, Kaupp M (2010) Modeling environmental effects on charge density distributions in polar crganometallics: validation of embedded cluster models for the methyl lithium crystal. J Comp Chem 31:2568–2576
165. Tiana D, Fancisco E, Blanco MA, Martín Pendás A (2009) Using pseudopotential within the interacting quantum atoms approach. J Phys Chem A 113:7963–7971
166. Eickerling G, Mastalerz R, Herz V, Scherer W, Himmel H-J, Reiher M (2007) Relativistic effects on the topology of the electron density. J Chem Theory Comput 3:2182–2197
167. Eickerling G, Reiher M (2008) The shell structure of atoms. J Chem Theory Comput 4: 286–296
168. Hudák M, Jayatilaka D, Perašínová L, Biskupič S, Kožíšek J, Bučinský L (2010) X-ray constrained unrestricted Hartree–Fock and Douglas–Kroll–Hess wavefunctions. Acta Crystallogr A 66:78–92

169. Ponec R, Bučinský L, Gatti C (2010) Relativistic effects of metal-metal bonding: comparison of the performance of ECP and scalar DKH Description on the picture of metal-metal bonding in $Re_2Cl_8^{2-}$. J Chem Theory Comput 6:3113–3121
170. (a) Douglas M, Kroll NM (1974) Quantum electrodynamical corrections to fine-structure of helium. Ann Phys 82:89–155; (b) Hess BA (1986) Relativistic electronic-structure calculations employing a two-component no-pair formalism with external-field projection operators. Phys Rev A 33:3742–3748; (c) Wolf A, Reiher M, Hess BA (2002) The generalized Douglas-Kroll transformation. J Chem Phys 117:9215–9226
171. (a) Chang C, Pelissier M, Durand P (1986) Regular two-component Pauli-like effective hamiltonians in Dirac theory. Phys Scr 34:394–404; (b) van Lenthe E, Baerends E-J, Snijders JG (1993) Relativistic regular two-component Hamiltonians. J Chem Phys 99:4597–4610; (c) van Lenthe E, Baerends E-J, Snijders JG (1994) Relativistic total energy using regular approximations. J Chem Phys 101:9783–9792
172. Wu ZG, Cohen RE (2006) More accurate generalized gradient approximation for solids. Phys Rev B 73:235116
173. Perdew JP, Ruzsinszky A (2010) Density functional theory of electronic structure: a short course for mineralogists and geophysicists. Rev Miner Geochem 71:1–18
174. Csonka GI, Vydrov OA, Scuseria GE, Ruzsinszky A, Perdew JP (2007) Diminished gradient dependence of density functionals: constraint satisfaction and self-interaction correction. J Chem Phys 126:244107
175. Perdew JP, Ruzsinszky A, Tao J, Csonka GI, Constantin LA, Zhou X, Vydrov OA, Scuseria GE, Burke K (2008) Restoring the gradient expansion for exchange in solids and surfaces. Phys Rev Lett 100:136406
176. Zhao Y, Truhlar DG (2008) Construction of a generalized gradient approximation by restoring the density-gradient expansion and enforcing a tight Lieb-Oxford bond. J Chem Phys 128:184109
177. Csonka GI, Perdew JP, Ruzsinszky A, Philipsen PHT, Lebegue S, Paier J, Vydrov OA, Angyan JG (2009) Assessing the performance of recent density functional for solids. Phys Rev B 79:155107
178. Zhao Y, Truhlar DG (2010) The Minnesota density functionals and their applications to problems in mineralogy and geochemistry. Rev Miner Geochem 71:19–37
179. Oganov AR, Glass CW (2006) Crystal structure prediction using ab initio evolutionary techniques: principles and applications. J Chem Phys 124:244704
180. Oganov AR, Ma Y, Lyakhov AO, Valle M, Gatti C (2010) Evolutionary crystal structure prediction as a method for the discovery of minerals and materials. Rev Miner Geochem 71:271–298
181. Blanco MA, Martín Pendás A, Francisco E (2005) Interacting quantum atoms: a correlated energy decomposition scheme based on the quantum theory of atoms in molecules. J Chem Theory Comput 1:1096–1109
182. Martín Pendás A, Francisco E, Blanco MA (2007) An electron number distribution view of chemical bonds in real space. Phys Chem Chem Phys 9:1087–1092
183. Kohout M (2007) Bonding indicators from electron pair density functionals. Faraday Discuss 135:43–54
184. (a) Bader RFW (1998) A bond path: a universal indicator of bonded interactions. J Phys Chem A 102:7314–7323; (b) Bader RFW (2009) Bond paths are not chemical bonds. J Phys Chem A 113:10391–10396; (c) Bader RFW (2010) Bond definition of molecular structure: by choice or by appeal to observation? J Phys Chem A 114:7431–7444
185. (a) Poater J, Sola M, Bickelhaupt FM (2006) Hydrogen–hydrogen bonding in planar biphenyl, predicted by atoms-in-molecules theory, does not exist. Chem Eur J 12:2889–2895; (b) Poater J, Sola M, Bickelhaupt FM (2006) A model of the chemical bond must be rooted in quantum mechanics, provide insight, and possess predictive power. Chem Eur J 12:2902–2905
186. Haaland A, Shorokhov DJ, Tverdova NV (2004) Topological analysis of electron densities: is the presence of an atomic interaction line in an equilibrium geometry a sufficient condition for the existence of a chemical bond? Chem Eur J 10:4416–4421

187. Grimme S, Mück-Lichtenfeld C, Erker G, Kehr G, Wang H, Beckers H, Willner H (2009) When do interacting atoms form a chemical bond? Spectroscopic measurements and theoretical analyses of dideuteriophenantrene. Angew Chem Int Ed 48:2592–2595
188. Cioslowski J, Mixon ST (1992) Topological properties of electron density in search of steric interactions in molecules: electronic structure calculations on ortho-substituted biphenyls. J Am Chem Soc 114:4382–4387
189. Farrugia LJ, Evans C, Tegel M (2006) Chemical bonds without "chemical bonding"? A combined experimental and theoretical charge density study on an iron trimethylenemethane complex. J Phys Chem A 110:7952–7961
190. Bader RFW (2006) Pauli Repulsions exist only in the eye of the beholder. Chem Eur J 12:2896–2901
191. Pendás AM, Francisco E, Blanco MA, Gatti C (2007) Bond paths as privileged exchange channels. Chem Eur J 12:9362–9371
192. Reinhold J, Kluge O, Mealli C (2007) Integration of electron density and molecular orbital techniques to reveal questionable bonds: the test case of the direct Fe—Fe bond in $Fe_2(CO)_9$. Inorg Chem 46:7142–7147
193. Ponec R, Gatti C (2009) Do the structural changes defined by the electron density topology necessarily affect the picture of the bonding? Inorg Chem 48:11024–11031
194. Macchi P, Garlaschelli L, Sironi A (2002) Electron density of semi-bridging carbonyls. Metamorphosis of CO ligands observed *via* experimental and theoretical investigations on $[FeCo(CO)_8]^-$. J Am Chem Soc 124:14173–14184
195. Bader RFW, Fang D-C, Properties of atoms in molecules: caged atoms and the Ehrenfest force. J Chem Theory Comput 1:403–414
196. Ponec R, Roithová J (2001) Domain-averaged Fermi holes - a new means of visualization of chemical bonds. Bonding in hypervalent molecules. Theor Chem Acc 105:383–392
197. Ponec R, Duben AJ (1999) Electron pairing and chemical bonds: bonding in hypervalent molecules from analysis of Fermi holes. J Comput Chem 20:760–771
198. Ponec R, Cooper DL, Savin A (2008) Analytic models of domain-averaged Fermi holes: a new tool for the study of the nature of chemical bonds. Chem Eur J 14:3338–3345
199. Ponec R, Lendvay G, Chaves J (2008) Structure and bonding in binuclear metal carbonyls from the analysis of domain averaged Fermi holes. I. $Fe_2(CO)_9$ and $Co_2(CO)_8$. J Comput Chem 29:1387–1398
200. Ponec R, Yuzhakov G, Carbó-Dorca R (2003) Chemical structures from the analysis of domain-averaged Fermi holes: multiple metal-metal bonding in transition metal compounds. J Comput Chem 24:1829–1838
201. Cioslowski J (1990) Isopycnic orbital transformations and localization of natural orbitals. Int J Quantum Chem S24:15–28
202. Ponec R, Uhlik F (1997) Electron pairing and chemical bonds. On the accuracy of the electron pair model of chemical bond. J Mol Struct (THEOCHEM) 391:159–168
203. Tiana D, Francisco E, Blanco AM, Macchi P, Sironi A, Pendás AM (2011) Restoring orbital thinking from real space descriptions: bonding in classical and non classical transition metal carbonyls. Phys Chem Chem Phys 13(11):5068–5077
204. Savin A, Silvi B, Colonna F (1996) Topological analysis of the electron localization function applied to delocalized bonds. Can J Chem 74:1088–1096
205. Silvi B (2002) The synaptic order: a key concept to understand multicenter bonding. J Mol Struct 614:3–10
206. Bader RFW, Gatti C (1998) A Green's function for the density. Chem Phys Lett 287:233–238
207. Arfken G (1985) Mathematical methods for physicists. Academic, Orlando
208. Gatti C, Cargnoni F, Bertini L (2003) Chemical information from the source function. J Comput Chem 24:422–436
209. Gatti C (2011) The source function descriptor as a tool to extract chemical information from theoretical and experimental electron densities. Struct Bond 1–93. doi:10.1007/430_2010_31
210. Monza E, Gatti C, Lo Presti L, Ortoleva E (2011) Revealing electron delocalization through the source function. J Phys Chem A 115:12864–12878

211. Gatti C, Bertini L (2004) The local form of the source function as a fingerprint of strong and weak intra- and intermolecular interactions. Acta Crystallogr A 60:438–449
212. Farrugia LJ, Macchi P (2009) On the interpretation of the source function. J Phys Chem A 113:10058–10067
213. Spackman MA (1999) Hydrogen bond energetics from topological analysis of experimental electron densities: recognising the importance of the promolecule. Chem Phys Lett 301: 425–429
214. Spackman MA, Byrom PG (1997) A novel definition of a molecule in a crystal. Chem Phys Lett 267:215–220
215. Johnson ER, Keinan S, Mori-Sánchez P, Contreras-García J, Cohen JA, Yang W (2010) Revealing noncovalent interactions. J Am Chem Soc 132:6498–6506
216. Gordon G, Kim YS (1972) Theory for the forces between closed-shell atoms and molecules. J Chem Phys 56:3122–3133
217. Cox SR, Hsu L-Y, Williams DE (1981) Nonbonded potential function models for crystalline oxohydrocarbons. Acta Crystallogr A 37:293–301
218. Gavezzotti A (2003) Calculation of intermolecular interaction energies by direct numerical integration over electron densities. 2. An improved polarization model and the evaluation of dispersion and repulsion energies. J Phys Chem B 107:2344–2353
219. Abramov YA, Volkov A, Wu G, Coppens P (2000) Use of X-ray charge density in the calculation of intermolecular interactions and lattice energies: application to glycylglycine, DL-histidine and DL-proline and comparison with theory. J Phys Chem B 104:2183–2188
220. Snyder J, Mogens C, Nishibori E, Caillat T, Iversen BB (2004) Disordered zinc in Zn_4Sb_3 with phonon glass, electron crystal thermoelectric properties. Nature Mater 3:458–463
221. Brock CP, Dunitz JD, Hirshfeld FL (1991) Transferability of deformation densities among related molecules - atomic multipole parameters from perylene for improved estimation of molecular vibrations in naphthalene and anthracene. Acta Crystallogr B 47:789–797
222. (a) Pichon-Pesme V, Lecomte C, Lachekar H (1995) On building a data bank of transferable experimental electron density parameters: application to polypeptides. J Phys Chem 99: 6242–6250; (b) Zarychta B, Pichon-Pesme V, Guillot B, Lecomte C, Jelsch C (2007) On the application of an experimental multipolar pseudo-atom library for accurate refinement of small-molecule and protein crystal structures. Acta Crystallogr A 63:108–125
223. Domagała S, Jelsch C (2008) Optimal local axes and symmetry assignment for charge-density refinement. J Appl Crystallogr 41:1140–1149
224. Bak JM, Domagala S, Hubschle C, Jelsch C, Dittrich B, Dominiak PM (2011) Verification of structural and electrostatic properties obtained by the use of different pseudoatom databases. Acta Crystallogr A67:141–153
225. (a) Abramov Yu A, Volkov AV, Coppens P (1999) On the evaluation of molecular dipole moments from multipole refinement of X-ray diffraction data. Chem Phys Lett 311:81–86; (b) Whitten AE, Turner P, Klooster WT, Piltz RO, Spackman MA (2006) Reassessment of large dipole moment enhancements in crystals: a detailed experimental and theoretical charge density analysis of 2-Methyl-4-nitroaniline. J Phys Chem A 110:8763–8776
226. Matta CF (2001) Theoretical reconstruction of the electron density of large molecules from fragments determined as proper open quantum systems: the properties of the oripavine PEO, enkephalins, and morphine. J Phys Chem A 105:11088–11101
227. (a) Rafat M, Shaik M, Popelier PLA (2006) Transferability of quantum topological atoms in terms of electrostatic interaction energy. J Phys Chem A 110:13578–13583; (b) Rafat M, Popelier PLA (2007) Atom-atom partitioning of total (super)molecular energy: the hidden terms of classical force fields. J Comput Chem 28:292–301; (c) Rafat M, Popelier PLA (2007) Long range behavior of high-rank topological multipole moments. J Comput Chem 28: 832–838
228. Houlding S, Liem SY, Popelier PLA (2007) A polarizable high-rank quantum topological electrostatic potential developed using neural networks: molecular dynamics simulations on the hydrogen fluoride dimer. Int J Quantum Chem 107:2817–2827
229. Bridgman PW (1931) The physics of high pressure. Bell and Sons, London

230. Ma Y, Eremets M, Oganov AR, Xie Y, Trojan I, Medvedev S, Lyakhov AO, Valle M, Prakapenka V (2009) Transparent dense sodium. Nature 458: 182–185
231. Marqués M, Ackland GJ, Lundegaard LF, Stinton G, Nelmes RJ, McMahon MI, Contreras-García J (2009) Potassium under pressure: a pseudobinary ionic compound. Phys Rev Lett 103:115501
232. (a) Pietsch U, Mahlberg J, Unger K (1985) Investigation of dynamical bond charge-transfer in GaAs by changing x-ray reflection power under high electric-field. Phys Status Solidi B 131:67–73; (b) Pietsch U, Unger K (1987) An experimental proof of the valence electron-density variation in silicon under high electric-field. Phys Status Solidi B 143:K95–K97; (c) Tsirelson VG, Gorfman SV, Pietsch U (2003) X-ray scattering amplitude of an atom in a permanent external electric field. Acta Crystallogr A 59:221–227; (d) Hansen NK, Fertey P, Guillot R (2004) Studies of electric field induced structural and electron-density modifications by X-ray diffraction. Acta Crystallogr A 60:465–471
233. Tiana D (2010) Organometallic chemistry from the interacting quantum atoms approach, PhD thesis, University of Milano
234. Seddon KR (1999) Crystal engineering. A case study. In: Seddon KR, Zaworotko M (eds) Crystal engineering. The design and application of functional solids. Kluwer, Amsterdam, pp 1–28
235. Desiraju GR (ed) (1989) Crystal engineering: the design of organic solids. Elsevier, Amsterdam
236. Spackman MA, Jayatilaka D (2009) Hirshfeld surface analysis. CrystEngComm 11:19–32
237. Wolff SK, Grimwood DJ, McKinnon JJ, Jayatilaka D, Spackman MA (2008) CrystalExplorer 2.1. http://hirshfeldsurface.net/CrystalExplorer
238. Dunitz JD, Gavezzotti A (2005) Molecular recognition in organic crystals: directed intermolecular bonds or nonlocalized bonding? Angew Chem Int Ed 44:1766–1787
239. Koch U, Popelier PLA (1995) Characterization of C-H-O hydrogen-bonds on the basis of the charge density. J Phys Chem 99:9747–9754
240. Ramberg PJ (2003) Chemical structure, spatial arrangements, the early history of stereochemistry 1874–1914. Ashgate, Aldershot
241. (a) Pasteur L (1848) Mémoire sur la relation qui peut exister entre la forme cristalline et la composition chimique, et sur la cause de la polarisation rotatoire. Comptes Rendus Ac. Sc. 26:535–538; (b) van't Hoff JH (1875) Sur les formules de structure dans l'espace. Bull Soc Chim France 23:295–301
242. Gillespie RJ, Hargittai I (1991) The VSEPR model of molecular geometry. Allyn & Bacon, Boston
243. Lewis GN (1916) The atom and the molecule. J Am Chem Soc 38:762–785
244. Macchi P, Sironi A (2007) Interactions involving metals: from "chemical categories" to QTAIM, and backwards, chapter 13. In: Matta CF, Boyd RJ (eds) The quantum theory of atoms in molecules: from solid state to DNA and drug design. Wiley-VCH, Weinheim
245. (a) Chatt J, Duncanson LA (1953) Olefin co-ordination compounds. Part III. Infra-red spectra and structure: attempted preparation of acetylene complexes. J Chem Soc 2939–2947; (b) Dewar JS (1951) A review of the pi-complex theory. Bull Soc Chim Fr 18:C71–C79
246. (a) Scherer W, Eickerling G, Shorokhov D, Gullo E, McGrady GS, Sirsch P (2006) Valence shell charge concentrations and the Dewar–Chatt–Duncanson bonding model. New J Chem 30:309–312; (b) Hebben N, Himmel HJ, Eickerling G, Herrmann C, Reiher M, Herz V, Presnitz M, Scherer W (2007) The electronic structure of the tris(ethylene) complexes $[M(C_2H_4)_3]$ (M = Ni, Pd, and Pt): a combined experimental and theoretical study. Chem Eur J 13:10078–10087; (c) Reisinger A, Trapp N, Krossing I, Altmannshofer S, Herz V, Presnitz M, Scherer W (2008) Homoleptic silver(I) acetylene complexes. Angew Chem Int Ed Engl 46:8295–8298
247. Rohrmoser B, Eickerling G, Presnitz M, Scherer W, Eyert V, Hoffmann R-D, Rodewald UC, Vogt C, Pöttgen R (2007) Experimental electron density of the complex carbides $Sc_3[Fe(C_2)_2]$ and $Sc_3[Co(C_2)_2]$. J Am Chem Soc 129:9356–9365

248. Scherer W, McGrady GS (2004) Agostic interactions in d^0 metal alkyl complexes. Angew Chem Int Ed Engl 43:1782–1806
249. (a) Leusser D, Walfort B, Stalke D (2002) Charge-density study of methane di(triimido)sulfonic acid $H_2C\{S(Nt\text{-}Bu)_2(NHt\text{-}Bu)\}_2$ – the NR analogue of $H_2C\{S(O)_2(OH)\}_2$. Angew Chem 41:2079–2082; (b) Leusser D, Henn J, Kocher N, Engels B, Stalke D (2004) S = N versus S^+-N^-: an experimental and theoretical charge density study. J Am Chem Soc 126:1781–1793; (c) Kocher N, Henn J, Gostevskii B, Kost D, Kalikhman I, Engels B, Stalke D (2004) Si-E (E = N, O, F) bonding in a hexacoordinated silicon complex: new facts from experimental and theoretical charge density studies. J Am Chem Soc 126:5563–5568
250. Ortin Y, Lugan N, Pillet S, Souhassou M, Lecomte C, Costuas K, Saillard J-Y (2005) A favorable case where an experimental electron density analysis offers a lead for understanding a specific fluxional process observed in solution. Inorg Chem 44:9607–9609
251. (a) Fukui K, Yonezawa T, Shingu H (1952) A molecular orbital theory of reactivity in aromatic hydrocarbons. J Chem Phys 20:722–725; (b) Fukui K (1971) Recognition of stereochemical paths by orbital interaction. Acc Chem Res 4:57–64
252. Fleming I (1976) Frontier orbitals and organic chemical reactions. Wiley, Chichester
253. Woodward RB, Hoffmann R (1965) Stereochemistry of electrocyclic reactions. J Am Chem Soc 87:395–397; Hoffmann R, Woodward RB (1970) Orbital symmetry control of chemical reactions. Science 167:825–831
254. (a) Pathak RK, Gadre SR (1990) Maximal and minimal characteristics of molecular electrostatic potentials. J Chem Phys 93:1770–1773; Gadre SR (1999) Topography of atomic and molecular scalar fields. In: Jerzy Leszczynski (ed) Computational chemistry: reviews of current trends, vol 4. World Scientific, Singapore, pp 1–53
255. Roy DK, Balanarayan P, Gadre SR (2009) Signatures of molecular recognition from the topography of electrostatic potential. J Chem Sci 121:815–821
256. Balanarayan P, Kavathekar R, Gadre SR (2007) Electrostatic potential topography for exploring electronic reorganizations in 1,3 dipolar cycloadditions. J Phys Chem A 111: 2733–2738
257. Gatti C, Barzaghi M, Bonati L, Pitea D (1989) On the chemical nature of transition states in cycloaddition reactions: a charge density topological approach. Application to the thermal cycloaddition of two ethylenes and to the Diels Alder reaction of butadiene and ethylene. In: Carbó R (ed) Quantum chemistry: basic aspects, actual trends, studies in physical and theoretical chemistry, vol 62. Elsevier Publishers, Amsterdam, pp 401–427
258. (a) Berski S, Andrés J, Silvi B, Domingo LR (2003) The joint use of catastrophe theory and electron localization function to characterize molecular mechanisms. A density functional study of the Diels – Alder reaction between ethylene and 1,3-butadiene. J Phys Chem A 107:6014–6024; (b) Polo V, Andres J, Berski S, Domingo LR, Silvi B (2008) Understanding reaction mechanisms in organic chemistry from catastrophe theory applied to the electron localization function topology. J Phys Chem A112:7128–7136
259. Bader RFW, MacDougall PJ, Lau CDH (1984) Bonded and nonbonded charge concentrations and their relation to molecular geometry and reactivity. J Am Chem Soc 106:1594–1605
260. Carrol MT, Chang C, Bader RFW (1988) Prediction of the structures of hydrogen-bonded complexes using the Laplacian of the charge density. Mol Phys 63:387–405
261. Bader RFW, Popelier PLA, Chang C (1992) Similarity and complementarity in chemistry. Theochem J Mol Struct 87:145–171
262. (a) Scherer W, Sirsch P, Shorokhov D, Tafipolsky M, McGrady GS, Gullo E (2003) Valence charge concentrations, electron delocalization and β-agostic bonding in d(0) metal alkyl complexes. Chem Eur J 9:6057–6070; (b) Himmel D, Trapp N, Krossing I, Altmannshofer S, Herz V, Eickerling G, SchererW (2008) Reply Angew Chem Int Ed 47:7798–7801; (c) Scherer W, Wolstenholme DJ, Herz V, Eickerling G, Brück A, Benndorf P, Roesky PW (2010) On the nature of agostic interactions in transition-metal amido complexes. Angew Chem Int Ed 49:2242–2246

263. (a) Parr RG, Yang W (1989) Density-functional theory of atoms and molecules, Oxford University Press, New York; (b) Geerlings P, De Proft, F, Langenaeker W (2003) Conceptual density functional theory. Chem Rev 103:1793–1873
264. Berkowitz M (1987) Density functional-approach to frontier controlled reactions. J Am Chem Soc 109:4823–4825
265. (a) Ayers PW, Parr RG, Nagy A (2002) Local kinetic energy and local temperature in the density – functional theory of electronic structure. Int J Quantum Chem 90:309–326; (b) Ayers PW (2005) Electron localization functions and local measures of the covariance. J Chem Sci 117:441–454
266. Chattaraj PK, Chamorro E, Fuentealba P (1999) Chemical bonding and reactivity: a local thermodynamic viewpoint. Chem Phys Lett 314:114–121

Chapter 2
Electron Densities and Related Properties from the *ab-initio* Simulation of Crystalline Solids*

Cesare Pisani, Roberto Dovesi, Alessandro Erba, and Paolo Giannozzi

2.1 Introduction

The aim of this chapter is to provide a general introduction to the *ab-initio* simulation of the density matrix (DM) of crystalline systems, and derived functions: charge, spin and momentum densities generally indicated as "density functions (DF)", and to describe by way of examples the performance of state-of-the-art computational tools in this respect.

Table 2.1 sketches possible standards of quality which may be pursued. Stepping down from one level to that below corresponds to missing some aspects of the physical description of the problem, and/or to adopting some simplifying assumptions as specified in the framed boxes in between.

We shall take as the center of our treatment the level marked "1" in that table, that is, our reference *exact* DM and DFs will be those obtained from the ground-state solution of the Schrödinger equation in the absence of external fields:

$$\widehat{H}_{el}\,\Psi_0 = E_0\,\Psi_0 \tag{2.1}$$

In Sect. 2.2 the formalism is introduced. In Sect. 2.2.1 we define the exact DM and DFs for periodic structures, establish their relationship with the *ideal*

---

*To their great sorrow, Roberto Dovesi, Alessandro Erba and Paolo Giannozzi announce the death of Professor Cesare Pisani on July 17th, 2011

C. Pisani • R. Dovesi (✉) • A. Erba
Dipartimento di Chimica IFM, and Centre of Excellence NIS (Nanostructured Interfaces and Surfaces), Università di Torino, via Giuria 5, I-10125 Torino, Italy
e-mail: cesare.pisani@unito.it; roberto.dovesi@unito.it; alessandro_erba@virgilio.it

P. Giannozzi
Democritos Simulation Center, CNR-IOM Istituto Officina dei Materiali, I-34151 Trieste, Italy

Dipartimento di Chimica, Fisica e Ambiente, Universitá di Udine, via delle Scienze 208, I-33100 Udine, Italy
e-mail: paolo.giannozzi@uniud.it

C. Gatti and P. Macchi (eds.), *Modern Charge-Density Analysis*,
DOI 10.1007/978-90-481-3836-4_2,

Table 2.1 Different levels of approximation in the theoretical simulation (see text for comments)

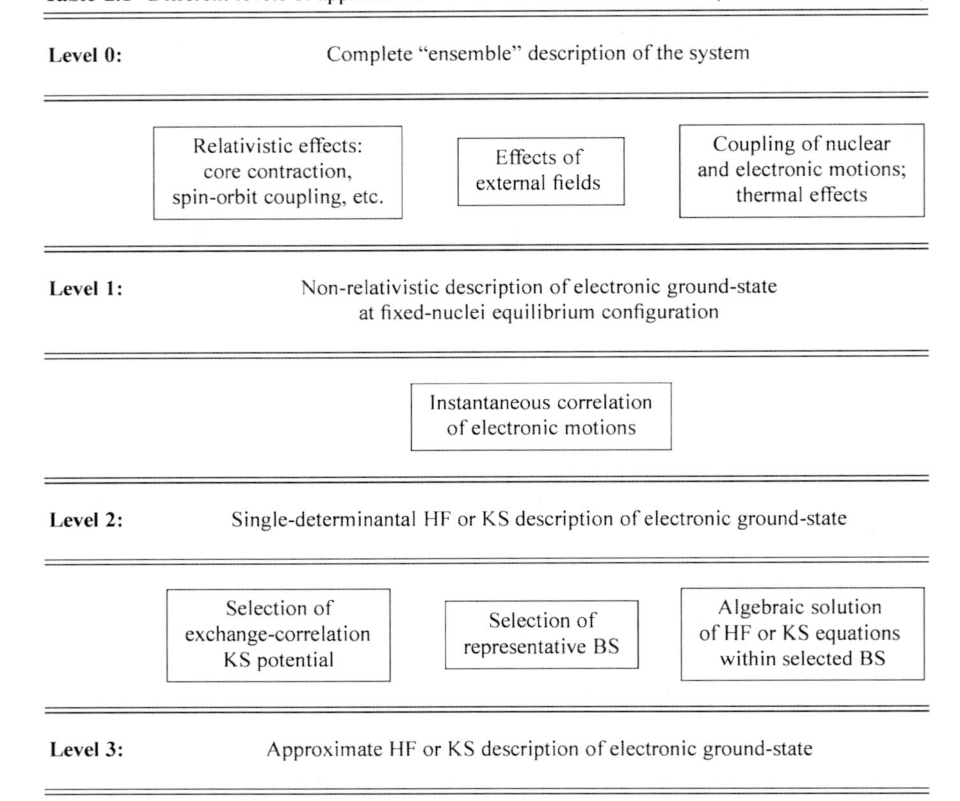

Level 0:	Complete "ensemble" description of the system
	Relativistic effects: core contraction, spin-orbit coupling, etc. — Effects of external fields — Coupling of nuclear and electronic motions; thermal effects
Level 1:	Non-relativistic description of electronic ground-state at fixed-nuclei equilibrium configuration
	Instantaneous correlation of electronic motions
Level 2:	Single-determinantal HF or KS description of electronic ground-state
	Selection of exchange-correlation KS potential — Selection of representative BS — Algebraic solution of HF or KS equations within selected BS
Level 3:	Approximate HF or KS description of electronic ground-state

experimental observables, and derive some general facts about them. The very fact that we are interested in "infinite" systems, makes the solution of Eq. 2.1 practically impossible at the present state-of-the-art of computational techniques, in spite of the simplifying constraint of translational symmetry. In order to obtain useful results, we are forced to adopt an "average-field" approximation of the electrostatic Hamiltonian that is, to step down to level 2 of Table 2.1, using either a Hartree-Fock (HF), or a Kohn-Sham (KS) approach. In Sect. 2.2.2 these one-electron schemes are considered. The general aspects of the corresponding single-determinant wavefunctions are reviewed, and some peculiar properties of the resulting DM and DFs are recalled. Specific problems related with the treatment of core electrons (the possible use of pseudopotentials, the need for relativistic corrections) are briefly discussed.

Of course, substantial effort and computational approximations are required even for obtaining a reasonable description of the single-determinantal solution of our problem. In practice, we are working at level 3 of Table 2.1, and our main aim will be to assess the importance of computational parameters in determining the distance with respect to level 2 quantities. Section 2.3 discusses explicitly the schemes of solution which are most commonly adopted for this case. It is shown that the very structure of the code and the technical peculiarities are to a large extent determined by the basis set (BS) adopted, either plane waves (PW) or local functions attached to the nuclei, conventionally indicated as atomic orbitals (AO). To be more specific, two codes are illustrated in some detail, Quantum ESPRESSO [1] and CRYSTAL [2], which adopt PWs and AOs respectively. It is only fair to state that a number of excellent periodic programs exist nowadays which are in widespread use in the scientific community (reference to some of them is provided in [3]). However, the fact that the Authors of this Chapter are among the developers of those two codes has permitted them to go deeper into the analysis of the effect of the various computational parameters on the quality of the resulting DM. As a matter of fact, this kind of analysis is relatively new, since the primary (often exclusive) object of interest for developers and users of the codes is the total energy of the system. On the other hand, this quantity is of fundamental importance also in the present context because it determines factors such as equilibrium geometries, pressure effects and normal vibrational modes, which directly affect calculated DFs.

Section 2.4 considers precisely by way of examples the influence of computational parameters (including the special form of the one-electron Hamiltonian and the BS) on charge, spin, momentum densities. Some simple systems are considered which cover a wide variety of electronic structures and types of bonding; for each of them a few selected aspects are analyzed in depth.

Once we have reached a solution close to our level-2 objective, it is worth trying to look back at the higher levels of the ladder of Table 2.1, in order to see what information has been missed, and possibly to partly recover it. Three aspects will be considered for this purpose in Sect. 2.5, with reference to ongoing research. In the first instance, the harmonic description of nuclear motions is shown to allow a statistical estimate of zero-point and thermal effects on DFs. Secondly, we discuss how to study the modifications induced by the presence of external fields. Thirdly, the *ab-initio* re-introduction of instantaneous electron correlation that is, the upgrade from level 2 to level 1, is entering into the realm of feasibility even for periodic systems, and it may be interesting to understand how much this aspect may influence DFs.

Some general conclusions are tentatively drawn in Sect. 2.6.

2.2 The Theoretical Frame

As anticipated in the Introduction, our reference level requires the solution of Eq. 2.1, where the non-relativistic electrostatic Hamiltonian for the N-electron system in the field of M nuclei of charge Z_A at $\mathbf{R}_A$ is given by the following expression:

$$\widehat{H}_{el} = \sum_{n=1}^{N} \frac{-\nabla_n^2}{2} + \sum_{n=1}^{N}\sum_{A=1}^{M} \frac{-Z_A}{r_{nA}} + \frac{1}{2}\sum_{n,m=1}^{N}{}' \frac{1}{r_{nm}} + \frac{1}{2}\sum_{A,B=1}^{M}{}' \frac{Z_A Z_B}{r_{AB}}. \quad (2.2)$$

Atomic units (au) are here used (see Appendix, Tables 2.5 and 2.6); the primed double sums exclude "diagonal" terms ($n = m$, $A = B$). $\hat{H}_{\rm el}$ depends parametrically on the sets $\{\mathbf{R}\}$, $\{Z\}$ of the positions $\mathbf{R_A}$ and charges Z_A of the M nuclei, so the same is true for E_0, Ψ_0.

Ψ_0 is an antisymmetric function of the space-spin coordinates $\mathbf{x}_n = \mathbf{r}_n, \sigma_n$ of the N electrons ($n = 1, \ldots, N$). The nuclear coordinates for which $E_0[\{\mathbf{R}\}, \{Z\}]$ has its minimum E_0^{eq} will be labelled $\{\mathbf{R}^{\mathrm{eq}}\}$. The corresponding eigenfunction, Ψ_0^{eq}, can then be written as follows:

$$\Psi_0^{\mathrm{eq}}(\ldots, \mathbf{x}_n, \ldots) \equiv \Psi_0(\ldots, \mathbf{x}_n, \ldots; [\{\mathbf{R}^{\mathrm{eq}}\}, \{Z\}]). \tag{2.3}$$

The special case of neutral periodic systems is the only one that will be treated here. Consider a lattice of vectors $\mathbf{T}_\mathbf{m}$ generated starting from D primitive linearly independent *basis vectors* $\mathbf{a}_i$ of ordinary space:

$$\mathbf{T}_\mathbf{m} = \sum_{i=1}^{D} m_i \mathbf{a}_i \tag{2.4}$$

where m_i are integers, and D is the number of periodic directions (three for bulk crystals, two for slabs, one for polymeric structures). The $\mathbf{a}_i$ vectors define (not univocally) the *unit cell* of the crystal. It is assumed that the coordinates and charges of all nuclei can be generated from those of a translationally irreducible finite set $\{\mathbf{R}_{A,\mathbf{0}}; Z_{A,\mathbf{0}}\}$ as follows:

$$\mathbf{R}_{A,\mathbf{m}} = \mathbf{R}_{A,\mathbf{0}} + \mathbf{T}_\mathbf{m} \quad ; \quad Z_{A,\mathbf{m}} = Z_{A,\mathbf{0}}. \tag{2.5}$$

For the solution of Eq. 2.1, the Born-von Kármán (BvK) cyclic conditions will be adopted. They correspond to assuming that Ψ_0 is cyclically periodic with respect to a super-lattice of vectors $\bar{\mathbf{W}}_\mathbf{m}$ defined as in Eq. 2.4, but starting from D *super-basis vectors* $\bar{\mathbf{A}}_i = w_i \mathbf{a}_i$:

$$\begin{aligned} \text{if}: \quad & \mathbf{x}_n = (\mathbf{r}_n, \sigma_n), \mathbf{x}'_n = (\mathbf{r}_n + \bar{\mathbf{W}}_\mathbf{m}, \sigma_n), \\ \text{then}: \quad & \Psi_0(\ldots, \mathbf{x}_n, \ldots) = \Psi_0(\ldots, \mathbf{x}'_n, \ldots) \quad \forall n, \mathbf{m} \qquad \{\text{BvK}\} \end{aligned} \tag{2.6}$$

(here and in the following {BvK} labels relationships that hold true owing to BvK conditions). The integers w_i define the *effective number* of electrons in the system: $N^{\mathrm{eff}} = W\, N_0$, with $W = \prod_i w_i$, and $N_0 = \sum_A Z_{A,\mathbf{0}}$ the number of electrons per cell. Provided the w_i's are large enough (see Sect. 2.3.1), their choice is scarcely influential on the quality of the results as concerns energy dependent quantities. On the other hand, account must be taken of the effect of the BvK conditions on DM and DFs in order to avoid misinterpretation of the results (see for instance Eq. 2.23).

2.2.1 Definition and Properties of Exact DM and DFs

Following McWeeny [4], we can define the one-particle generalized density function, our exact *position-spin* DM, as follows (by default, $\Psi_0 = \Psi_0^{\mathrm{eq}}$):

$$\gamma(\mathbf{x};\mathbf{x}') = N\int \Psi_0(\mathbf{x},\mathbf{x}_2,\ldots,\mathbf{x}_N)\times\left(\Psi_0(\mathbf{x}',\mathbf{x}_2,\ldots,\mathbf{x}_N)\right)^{*}\mathrm{d}\mathbf{x}_2\ldots\mathrm{d}\mathbf{x}_N. \tag{2.7}$$

Higher-order DMs can be usefully defined (see for instance Chaps. 5 and 9), but they don't matter in the present context. The content of information of $\gamma(\mathbf{x};\mathbf{x}')$ is very rich. This function of only six spatial and two spin coordinates provides the ground-state expectation value of any observable described by a one-electron operator $\widehat{F} = \sum_n \hat{f}[\mathbf{x}_n]$:

$$\left\langle \widehat{F}\right\rangle_0 = \int \{\hat{f}[\mathbf{x}]\,\gamma\,(\mathbf{x};\mathbf{x}')\}_{(\mathbf{x}'=\mathbf{x})}\mathrm{d}\mathbf{x}. \tag{2.8}$$

By making explicit the dependence of $\gamma(\mathbf{x};\mathbf{x}')$ on the two-valued spin coordinates with respect to the z direction $[\sigma,\ \sigma'=\alpha,\ \beta]$, we can define the *spin-projected components* of the DM:

$$\begin{gathered} P^{\sigma}(\mathbf{r};\mathbf{r}') = \gamma(\mathbf{r},\sigma;\mathbf{r}',\sigma) \\ P(\mathbf{r};\mathbf{r}') = P^{\alpha}(\mathbf{r};\mathbf{r}') + P^{\beta}(\mathbf{r};\mathbf{r}') \quad ; \ Q(\mathbf{r};\mathbf{r}') = P^{\alpha}(\mathbf{r};\mathbf{r}') - P^{\beta}(\mathbf{r};\mathbf{r}'). \end{gathered} \tag{2.9}$$

$P(\mathbf{r};\mathbf{r}')$ is called the *position* DM, obtained by integrating $\gamma(\mathbf{x};\mathbf{x}')$ over both spin components, while $Q(\mathbf{r};\mathbf{r}')$ is the *excess DM* of α- with respect to β-electrons.

The *momentum* DM is the six-dimensional Fourier transform (FT) of $P(\mathbf{r};\mathbf{r}')$:

$$\bar{P}(\mathbf{p};\mathbf{p}') = \int P(\mathbf{r};\mathbf{r}')\ \exp(-\imath\mathbf{r}\cdot\mathbf{p})\ \exp(\imath\mathbf{r}'\cdot\mathbf{p}')\,\mathrm{d}\mathbf{r}\,\mathrm{d}\mathbf{r}' \tag{2.10}$$

DFs are the "diagonal elements" of the DMs:

$$\begin{aligned} \rho(\mathbf{r}) &= P(\mathbf{r};\mathbf{r}) && \textbf{electron density (ED)} \\ \rho^{\sigma}(\mathbf{r}) &= P^{\sigma}(\mathbf{r};\mathbf{r}) && \textbf{electron } \sigma - \textbf{spin density} \\ \zeta(\mathbf{r}) = Q(\mathbf{r};\mathbf{r}) &= \rho^{\alpha}(\mathbf{r}) - \rho^{\beta}(\mathbf{r}) && \textbf{electron (net) spin density (ESD)} \text{ (along } z\text{)} \\ \pi(\mathbf{p}) &= \bar{P}(\mathbf{p};\mathbf{p}) && \textbf{electron momentum density (EMD)} \end{aligned} \tag{2.11}$$

They are *observable quantities* in a quantum-mechanical sense, since they are the ground-state expectation value of one-electron operators $\left[\widehat{F}^{\mathbf{y}} = \sum_n \delta(\mathbf{y}-\mathbf{y}_n)\right]$.

In a periodic system, the displacement of *all* $\mathbf{r}_n$ quantities by the same lattice vector $\mathbf{T}_m$ must just result in a change of the ground state wavefunction by a phase factor. From Eq. 2.7 we then have:

$$\gamma(\mathbf{r},\sigma;\mathbf{r}',\sigma') = \gamma(\mathbf{r}+\mathbf{T}_m,\sigma;\mathbf{r}'+\mathbf{T}_m,\sigma')\ ;\ P(\mathbf{r};\mathbf{r}') = P(\mathbf{r}+\mathbf{T}_m;\mathbf{r}'+\mathbf{T}_m). \tag{2.12}$$

All information about the position (−spin) DMs is thus obtained by confining the variable $\mathbf{r}$ to within the unit cell. In Eq. 2.10, the integral over $\mathbf{r}$ can then be limited to the unit cell. With this convention, which makes $\pi(\mathbf{p})$ independent of the "size" of the crystal: $\int \pi(\mathbf{p})\mathrm{d}\mathbf{p} = N_0$, the number of electrons per cell.

The BvK conditions (Eq. 2.6) entail a stronger consequence on DMs:

$$\gamma(\mathbf{r},\sigma;\mathbf{r}',\sigma') = \gamma(\mathbf{r},\sigma;\mathbf{r}' + \bar{\mathbf{W}}_m,\sigma')\ ;\ P(\mathbf{r};\mathbf{r}') = P(\mathbf{r};\mathbf{r}' + \bar{\mathbf{W}}_m) \quad \{\mathrm{BvK}\} \tag{2.13}$$

It is not our concern to discuss explicitly how information on these observables can be obtained from actual experiments: this question is dealt with elsewhere in this Book, see for instance Chaps. 3–5, 8 and 13. But while commenting below on some general properties of the exact DFs, with special attention to the consequences of translational periodicity, we will also establish their relationship with "ideal" measurable quantities such as structure factors, Compton profiles, etc.

Chapters 5–7 and in particular Chap. 9 provide a more complete analysis of the relationships among the various DMs and DFs, and between these and the chemical characteristics of the system.

2.2.1.1 Electron Density

In many respects, the ED $\rho(\mathbf{r})$ (also called electron charge density, ECD) and its spin components have the same properties; we consider here only the former quantity while we shall discuss some aspects of the ESD $\zeta(\mathbf{r})$ in Sect. 2.4.

$\rho(\mathbf{r})$ plays a specially important role in the characterization of the many-electron system. This is a simple function of the space coordinates (it is real, non-negative, finite everywhere and regular except for isolated cusps) which reflects faithfully the chemical composition and geometry of the system. In fact, the position $\mathbf{R}_A$ and the charge Z_A of all the nuclei are identified, respectively, through the location of the cusps and through the Kato's cusp condition [5]: $Z_A = -\overline{\rho'(\mathbf{R}_A)}/(2\rho(\mathbf{R}_A))$, where $\overline{\rho'(\mathbf{R}_A)}$ is the spherical average of the slope about the cusp. It therefore contains transparently all information about the Schrödinger Hamiltonian of the system. The recognition of this fact, although derived differently [6], is at the basis of density functional theory (DFT) which has enjoyed enormous success especially in its KS formulation [7] (see Sect. 2.2.2).

Bader's Atoms in Molecules theory [8] and its application to crystals [9, 10] allows a wealth of information about the chemical features of the system to be obtained through a topological analysis the ED and its derivatives (see also Chaps. 1, 9, 10 and 12–15 of this Book). Recent literature concerning the comparison between experimental and theoretical ED determinations is in fact often centered on the respective characterization of Bader's topological objects, in particular of the bond critical points [11].

Since a non-degenerate ground-state wavefunction [hence the corresponding $\rho(\mathbf{r})$] has the same symmetry as the Hamiltonian, the ED of a crystalline system is invariant to all operations of the related space group $\mathfrak{I}$:

$$\{V|T_{\mathbf{m}} + \mathbf{s}_V\}\rho(\mathbf{r}) \equiv \rho(V^{-1}\mathbf{r} - T_{\mathbf{m}} - \mathbf{s}_V) = \rho(\mathbf{r}). \tag{2.14}$$

The Seitz notation of the operators has been used [12], where V is a matrix representing a proper or improper rotation, and $\mathbf{s}_V$ the associated fractional translation (for the so-called symmorphic groups, $\forall V$: $\mathbf{s}_V = \mathbf{0}$). The rotations themselves form a group of order h, the *point group of the crystal*. Because of Eq. 2.14, all information about the ED is contained in a $1/h$-th wedge of the unit cell (the irreducible wedge). Integration of $\rho(\mathbf{r})$ over the irreducible wedge gives N_0/h.

A one-to-one correspondence exists between the ED and its three-dimensional FT, the *form factor*:

$$F(\kappa) = \int \rho(\mathbf{r})\ \exp(\iota\,\kappa \cdot \mathbf{r})\,\mathrm{d}\mathbf{r}. \tag{2.15}$$

Let us associate to the D unit vectors of direct lattice, $\mathbf{a}_i$, an equal number of unit vectors of reciprocal lattice, $\mathbf{B}_j$ defined by the relation: $\mathbf{a}_i \cdot \mathbf{B}_j = 2\pi\delta_{ij}$ $(i, j = 1, \ldots, D)$. The general point of reciprocal space can then be written as a sum of a "periodic part": $\kappa^{\parallel} = \sum_{i=1}^{D} \kappa_i\,\mathbf{B_i}$, and of a "non-periodic part", $\kappa^{\perp}$, perpendicular to the former. The symmetry properties of the ED (Eq. 2.14) entail the following consequences for the crystalline form factor (the case $D = 3$ is considered for definiteness, where $\kappa \equiv \kappa^{\parallel}$):

1. Due to translational invariance, $F(\kappa)$ is zero unless κ is a *reciprocal lattice vector* $\mathbf{G}_{hkl} \equiv h\mathbf{B}_1 + k\mathbf{B}_2 + l\mathbf{B}_3$, with h, k, l integers. The form factor is then defined by the discrete (but infinite) set of the *structure factors*:

   $$F_{hkl} = \int_{\text{unit cell}} \rho(\mathbf{r})\ \exp(\iota\,\mathbf{G}_{hkl} \cdot \mathbf{r})\,\mathrm{d}\mathbf{r}. \tag{2.16}$$

   By this convention, structure factors are normalized so that $F_{000} = N_0$, the number of electrons in the (conventional) crystallographic unit cell.
2. Due to rotational invariance, $F(\mathbf{G}) = \exp(-\iota\mathbf{G}\cdot\mathbf{s}_V)\,F(V\mathbf{G})$: that is, F_{hkl} is the same up to a phase factor for reciprocal lattice vectors which are obtained from each other by a point group operation (belong to the same *star*).
3. Depending on the space group, some structure factors are systematically zero ("*general*" *extinction conditions*); "*extra*" *extinction conditions* may apply to the triplet hkl, when $\rho(\mathbf{r})$ can be expressed as a sum of contributions from "spherical" atoms centered in some special Wyckoff positions: see e.g. reference [13].

In the dynamical limit and in the hypothesis of fixed nuclei, the X-ray structure factors provided by diffraction experiments are the *ideal* experimental counterpart of the theoretical ones. When trying to reconstruct EDs from actual diffraction data, apart from all the corrections that must be applied for obtaining a guess of the ideal structure factors [11], a *phase factor* problem exists since only the modulus of the latter is provided by the experiment. The comparison (Theory) → (Experiment) as concerns EDs is then more natural than the reciprocal one, because a number of corrections can easily be applied to the theoretical structure factors (including approximate account of nuclear thermal motion, see Sect. 2.5.1), so as to make the comparison with their experimental counterpart as justified as possible.

2.2.1.2 Momentum Density

The electron momentum density (EMD), $\pi(\mathbf{p})$, brings in complementary information with respect to the ED, as is discussed at length in Chap. 4 and, in particular, 5 and 7. In particular, the features of $\pi(\mathbf{p})$ near the origin are dominated by the contributions of the slow valence electrons, while those at large $|p|$ values reflect the properties of core electrons. The convention introduced at the beginning of this Section is adopted, according to which $\pi(\mathbf{p})$ in a crystal is normalized to N_0, the number of electrons per cell. Some basic facts about the EMD are here recalled.

$\pi(\mathbf{p})$ is a real, positive definite function which exhibits rotational invariance with respect to all operators of the point group of the crystal $[\pi(\mathbf{p}) = \pi(V\mathbf{p})]$.

It provides directly the ground-state expectation value of the total kinetic energy per cell of the electrons in the system:

$$\left\langle \widehat{T} \right\rangle_0 = \frac{1}{2} \int \pi(\mathbf{p}) p^2 \, \mathrm{d}\mathbf{p}. \tag{2.17}$$

Due to the virial theorem, which holds true for the electrostatic Hamiltonian of Eq. 2.2, the EMD also provides the total energy per cell at the equilibrium configuration:

$$\frac{1}{2} \int \pi(\mathbf{p})_{(\{\mathbf{R}\}=\{\mathbf{R}^{\mathrm{eq}}\})} p^2 \, \mathrm{d}\mathbf{p} = -E_0^{\mathrm{eq}}. \tag{2.18}$$

This important relationship, which derives from minimization of the expectation value of energy with respect to the scaling of *all* coordinates, is *not* valid in more general cases, for instance for volume-constrained optimized configurations.

Two interesting functions can be obtained from the EMD: the *Reciprocal Form Factor* (RFF), $B(\mathbf{r})$, and the *Compton Profile Function* (CPF), $J(\mathbf{q})$:

$$B(\mathbf{r}) = \int \pi(\mathbf{p}) \, \exp(-\imath \mathbf{p} \cdot \mathbf{r}) \, \mathrm{d}\mathbf{p} = \frac{1}{W} \int P(\mathbf{r}'; \mathbf{r} + \mathbf{r}') \, \mathrm{d}\mathbf{r}'$$

$$J(\mathbf{q}) = \int \pi(\mathbf{p}) \delta \left(\frac{\mathbf{p} \cdot \mathbf{q}}{|\mathbf{q}|} - |\mathbf{q}| \right) \, \mathrm{d}\mathbf{p}. \tag{2.19}$$

The two definitions of the RFF are easily seen to be equivalent; the second one justifies its alternative name of *Autocorrelation Function*. The normalizing factor W is the number of cells in the cyclic crystal; we then have, as expected: $B(\mathbf{0}) = N_0$.

The CPF results from the 2-D integration of $\pi(\mathbf{p})$ over a plane through $\mathbf{q}$ perpendicular to the $\mathbf{q}$ direction. Its main interest stems from the fact that it can be related to the *ideal* experimental Compton profiles (CP) [14]. Consider the *directional* CP:

$$J_{hkl}(q) = J(q\,\mathbf{e}_{hkl}) \qquad \left(\mathbf{e}_{hkl} = \frac{h\mathbf{a}_1 + k\mathbf{a}_2 + l\mathbf{a}_3}{|h\mathbf{a}_1 + k\mathbf{a}_2 + l\mathbf{a}_3|}\right). \tag{2.20}$$

In the sudden-impulse approximation, this function is proportional to the distribution of the loss, in the direction $\mathbf{e}_{hkl}$, of the momentum of scattered photons, as is measured in a Compton scattering experiment.

The *directional* RFF $[B_{hkl}(r) = B(r\mathbf{e}_{hkl})]$ and the corresponding directional CP are immediately seen to be related to each other by a *one-dimensional* FT:

$$B_{hkl}(r) = \int J_{hkl}(q)\ \exp(-\imath rq)\ \mathrm{d}q. \tag{2.21}$$

An interesting consequence of this relation can be cited. Due to limited resolution, experimental directional CPs $\left[J^{\exp}_{hkl}(q)\right]$ can be considered as the *convolution* of the ideal CP by the experimental resolution function $w(q)$ (usually a Gaussian):

$$J^{\exp}_{hkl}(q) \approx \int J_{hkl}(q')\, w(q' - q)\ \mathrm{d}q'. \tag{2.22}$$

It then follows that $B_{hkl}(r)$ is simply the FT of $J^{\exp}_{hkl}(q)$ *divided by* $\bar{w}(r)$, the FT of $w(q)$. The RFF is therefore easily accessible from the Compton scattering experiment. Furthermore, its fine structure is not affected very much by experimental errors; in particular its zeros can be located with relatively high precision.

A consequence of the BvK conditions on the calculated RFF can be noted: from Eq. 2.13 it immediately follows that the latter has (artificially) the periodicity of the superlattice:

$$B(\mathbf{r}) = B(\mathbf{r} + \bar{\mathbf{W}}_{\mathbf{m}}) \qquad \{\mathrm{BvK}\}. \tag{2.23}$$

The RFF conveys important information. Its oscillatory behaviour in direct space, the position of its nodal surfaces, the value of its maxima and minima are closely related to the chemical features of the system [15]: examples are provided in Chap. 7 of this Book.

2.2.2 HF and KS Schemes, and Related DM and DFs

The solution of Eq. 2.1 that is, the determination of the eigenfunction Ψ_0, is impossible for any system of real interest. Two simplified schemes, HF and KS, are usually adopted for describing periodic systems, which may provide valuable information on their DMs and DFs. Here we consider their exact formulation, and comment on some characteristic aspects of the resulting densities; in Sect. 2.3, we shall describe their actual implementation and the simplifying assumptions which must be adopted in order to obtain their approximate solution.

HF and KS have many features in common. In their *spin-unrestricted* formulation (UHF,UKS), both are intended to obtain a set of N one-electron functions, the *molecular spin-orbitals* (MSO) (or *crystalline spin-orbitals*, CSO, in the periodic case), $\psi_j^{\mathrm{X}}(\mathbf{x}) = \phi_j^{\mathrm{X},\sigma}(\mathbf{r})\omega(\sigma)$, with σ either α or β, which satisfy the equation:

$$\begin{aligned}\hat{h}^{\mathrm{X},\sigma}\phi_j^{\mathrm{X},\sigma}(\mathbf{r}) &= \left[-\frac{\nabla^2}{2} + \sum_A \frac{-Z_A}{|\mathbf{R}_A - \mathbf{r}|} + \int \frac{\rho^X(\mathbf{r}')}{|\mathbf{r}-\mathbf{r}'|}\mathrm{d}\mathbf{r}' + \widetilde{V}^{\mathrm{X},\sigma}\right]\phi_j^{\mathrm{X},\sigma}(\mathbf{r}) \\ &= \varepsilon_j^{\mathrm{X},\sigma}\phi_j^{\mathrm{X},\sigma}(\mathbf{r}).\end{aligned} \tag{2.24}$$

The effective Hamiltonian $\hat{h}^{\mathrm{X},\sigma}$ which acts on the individual MSO contains, apart from the kinetic, nuclear attraction and *Hartree* operators (the last one expressing the Coulomb repulsion with all the electrons in the system), a *corrective potential* operator, $\widetilde{V}^{\mathrm{X},\sigma}$, which differs in the two schemes (X = HF or KS), as is seen below.

A *single-determinant* N-electron function can be defined, after assigning the N electrons to the N MSOs corresponding to the lowest eigenvalues $\varepsilon_j^{\mathrm{X},\sigma}$ of Eq. 2.24, and antisymmetrizing their product:

$$\Psi_0^{\mathrm{X}} = N^{-1/2}\sum_P (-1)^{sp}\widehat{P}\left[\psi_1^{\mathrm{X}}(\mathbf{x}_1)\times\cdots\times\psi_N^{\mathrm{X}}(\mathbf{x}_N)\right] \equiv \|\cdots j \cdots\| . \tag{2.25}$$

Here $\widehat{P}$ is the general N-order permutation operator which acts on the electron co-ordinates and s_P the respective parity. The orthonormal MSOs which define Ψ_0^{X} are said to form the *occupied manifold*, all others belonging to the *virtual manifold*.

In the rest of this Section we shall assume, for simplicity, that we are describing a spinless system, where the occupied MSOs are in pairs having the same eigenvalue $\varepsilon_n^{\mathrm{X}}$, the same spatial part ϕ_n^{X} and α or β spin, to be labelled n^α and n^β, respectively, with $n = 1,\cdots,N/2$. In this case the spin index can be dropped from the effective Hamiltonian and from the corrective potential, since they are the same for α and β spin. Following the general definition (Eqs. 2.7 and 2.9), it is easily seen that the position DM and the ED associated with $\Psi_0^{\mathrm{X}} = \|\cdots n^\alpha n^\beta \cdots\|$ are simply:

$$P^X(\mathbf{r};\mathbf{r}') = 2\sum_{n=1}^{N/2}\phi_n^{\mathrm{X}}(\mathbf{r})(\phi_n^{\mathrm{X}}(\mathbf{r}'))^* \quad ; \quad \rho^X(\mathbf{r}) = P^X(\mathbf{r};\mathbf{r}) = 2\sum_{n=1}^{N/2}|\phi_n^{\mathrm{X}}(\mathbf{r})|^2. \tag{2.26}$$

In the HF scheme, the corrective potential $\widetilde{V}^{\mathrm{HF}}$ is defined by imposing that the HF energy E_0^{HF} that is, the Ψ_0^{HF}-expectation value of the electrostatic Hamiltonian (which cannot be less than the true ground state energy E_0) is a minimum with respect to any other single-determinant N-electron wavefunction. Therefore, the occupied manifold resulting from the solution of Eq. 2.24 defines the optimal (in a variational sense) single-determinant approximation of the true ground state Ψ_0. To achieve this goal, $\widetilde{V}^{\mathrm{HF}}$ must take the form of the *exact-exchange operator* $\widehat{V}_{\mathrm{exch}}$, whose action on the general function $\chi(\mathbf{r})$ is defined as follows:

$$\widehat{V}_{\mathrm{exch}}\chi(\mathbf{r}) = -\frac{1}{2}\int \frac{P^{\mathrm{HF}}(\mathbf{r};\mathbf{r}')\chi(\mathbf{r}')}{|\mathbf{r}-\mathbf{r}'|}d\mathbf{r}'. \tag{2.27}$$

We have, correspondingly:

$$E_0^{\mathrm{HF}} \equiv \left\langle \Psi_0^{\mathrm{HF}} | \widehat{H}_{\mathrm{el}} | \Psi_0^{\mathrm{HF}} \right\rangle = -\int \left[\frac{\nabla^2}{2} P^{\mathrm{HF}}(\mathbf{r};\mathbf{r}') \right]_{(\mathbf{r}'=\mathbf{r})} d\mathbf{r} - \sum_A Z_A \int \frac{\rho^{\mathrm{HF}}(\mathbf{r})}{|\mathbf{R}_A - \mathbf{r}|} d\mathbf{r} + \frac{1}{2}\int \frac{\rho^{\mathrm{HF}}(\mathbf{r})\rho^{\mathrm{HF}}(\mathbf{r}')}{|\mathbf{r}-\mathbf{r}'|} d\mathbf{r}\, d\mathbf{r}' - \frac{1}{4}\int \frac{|P^{\mathrm{HF}}(\mathbf{r};\mathbf{r}')|^2}{|\mathbf{r}-\mathbf{r}'|} d\mathbf{r}\, d\mathbf{r}' + \frac{1}{2}\sum_{A,B=1}^{M}{}' \frac{Z_A Z_B}{r_{AB}} \geq E_0. \tag{2.28}$$

The KS scheme, formulated in the frame of DFT [6, 7], introduces, for any given N-electron ED, $\rho(\mathbf{r})$, two *universal functionals*: $\varepsilon_{\mathrm{xc}}(\mathbf{r};[\rho])$ and $V_{\mathrm{xc}}(\mathbf{r};[\rho])$. The latter is obtained from the former via a functional derivative relationship: $V_{\mathrm{xc}} = \varepsilon_{\mathrm{xc}} + \rho(\delta\varepsilon_{\mathrm{xc}}/\delta\rho)$. When the *exchange-correlation potential* $V_{\mathrm{xc}}(\mathbf{r};[\rho]^{\mathrm{KS}})$ is used for $\widetilde{V}^{\mathrm{KS}}$ as a multiplicative operator in Eq. 2.24, the density from Eq. 2.26 coincides with the *exact* ground-state ED (Eq. 2.11):

$$\rho^{\mathrm{KS}}(\mathbf{r}) = \rho(\mathbf{r}). \tag{2.29}$$

The functional $\varepsilon_{\mathrm{xc}}(\mathbf{r};[\rho])$ allows the *exact* ground-state energy to be calculated, again with reference to the occupied KS manifold:

$$E_0^{\mathrm{KS}} = -\int \left[\frac{\nabla^2}{2} P^{\mathrm{KS}}(\mathbf{r};\mathbf{r}') \right]_{(\mathbf{r}'=\mathbf{r})} d\mathbf{r} - \sum_A Z_A \int \frac{\rho(\mathbf{r})}{|\mathbf{R}_A - \mathbf{r}|} d\mathbf{r} + \frac{1}{2}\int \frac{\rho(\mathbf{r})\rho(\mathbf{r}')}{|\mathbf{r}-\mathbf{r}'|} d\mathbf{r}\, d\mathbf{r}' + \int \rho(\mathbf{r})\,\varepsilon_{xc}(\mathbf{r};[\rho])\, d\mathbf{r} + \frac{1}{2}\sum_{A,B=1}^{M}{}' \frac{Z_A Z_B}{r_{AB}} = E_0. \tag{2.30}$$

Equation 2.24 must be solved self-consistently in both cases, because the Hartree and the corrective potential are defined in terms of the occupied manifold. Two important differences between the two schemes must be stressed, however.

1. In the HF case the corrective potential (the *non-local* operator of Eq. 2.27) is perfectly defined. On the contrary, no exact formula exists for the *local* exchange-correlation potential $V_{xc}(\mathbf{r}; [\rho])$ (or equivalently, for $\varepsilon_{xc}(\mathbf{r}; [\rho])$; we shall consider in Sect. 2.3.2 some powerful though approximate expressions that have been proposed for those functionals.
2. The single-determinant wavefunction Ψ_0^X defined in Eq. 2.25 has a different meaning in the two schemes. Ψ_0^{HF} may be considered as the zero-order approximation to the true ground-state wavefunction in a hierarchy of *post-HF* methods, which re-introduce the instantaneous electron correlation (see Sect. 2.5.3). Instead, Ψ_0^{KS} is in principle only a useful mathematical construction which describes a set of non-interacting electrons in an effective potential, whose ED is the same as that of the real system. It is however customary to use Ψ_0^{KS} *as though* it were representative of some properties of Ψ_0, for instance of the off diagonal terms of the position DM $[P^{KS}(\mathbf{r}; \mathbf{r}') \approx P(\mathbf{r}; \mathbf{r}')]$: this is what we shall do in the following. Some justifications for this assumption and some indications for correcting for its inadequacy have been provided, for instance, by Bauer [16].

2.2.2.1 HF and KS Solutions: Bloch Functions, Insulators, Metals

Some general facts about the solution of Eq. 2.24 in its spinless, periodic formulation are here recalled in order to fix notations and to prepare the discussion of the next sections.

Since the one-electron effective Hamiltonian $\hat{h}^X$ commutes with all operations of the space group $\Im$, in particular of the subgroup $\Im$ of the pure translations, its eigenfunctions, the COs, can be classified according to the irreducible representations of that group. As is shown in standard textbooks [12], they are then characterized by an index κ, a vector of reciprocal space, such that the corresponding COs are *Bloch functions* (BF), $\phi_n^X(\mathbf{r}; \kappa)$, which satisfy the property:

$$\phi_n^X(\mathbf{r} + \mathbf{T_m}; \kappa) = \phi_n^X(\mathbf{r}; \kappa)\ \exp(\iota\kappa \cdot \mathbf{T_m}). \tag{2.31}$$

Clearly, κ's differing by a reciprocal lattice vector $\mathbf{G}$ define the same irreducible representation. Among all equivalent κ's one can choose the one closest to the origin of the reciprocal space; this "minimal-length" set fills the so-called *(first) Brillouin zone* (BZ). The COs must also satisfy the BvK conditions $\left[\phi_n^X(\mathbf{r} + \overline{\mathbf{W}}_\mathbf{m}; \kappa) = \phi_n^X(\mathbf{r}; \kappa)\right]$, which means that $\exp(\iota\kappa \cdot \overline{\mathbf{W}}_\mathbf{m}) = 1$ or, otherwise stated, that the general κ must belong to a *Monkhorst grid* [17]:

$$\kappa_\mathbf{h} = \sum_{i=1}^{D} (h_i + s_i)\frac{\mathbf{B}_i}{w_i} \quad \left(\text{integer } h_i, s_i = 0 \text{ or } \frac{1}{2}\right) \qquad \{BvK\}. \tag{2.32}$$

This is our standard choice: the *exact* solutions (no BvK conditions imposed) can be obtained in the limit of infinite w_i's. It is customary and useful to choose the

w_i's such that the super-unit-vectors define the same Bravais lattice as the original one either "undisplaced" or "displaced" with respect to the origin according to the value of s_i. The $\kappa_{\mathbf{h}}$ vectors thus form a *contracted* reciprocal lattice with respect to the original one. The number of $\kappa_{\mathbf{h}}$ vectors in the BZ equals $W = \prod_i w_i$. As a consequence of rotational symmetry, if two sampling vectors are related to each other by a point group operator, the corresponding eigenvalues are the same, and the eigenfunctions coincide except for a rotation. This permits the determination of the solutions to be confined to $\kappa_{\mathbf{h}}$'s belonging to the irreducible wedge of the BZ.

As the effective number of electrons in the system is $N = W N_0$ (see Sect. 2.2), the number of occupied COs is $W N_0/2$ that is, on average, $N_0/2$ per sampling $\kappa_{\mathbf{h}}$ point. After ordering the eigenvalues by energy $\left[\varepsilon_n^X(\kappa_{\mathbf{h}}) \leq \varepsilon_{n+1}^X(\kappa_{\mathbf{h}})\right]$, a *Fermi energy* E_F can be defined such that there are exactly $W\ N_0/2$ eigenvalues $\varepsilon_n^X(\kappa_{\mathbf{h}}) \leq E_F$. A distinction can be made between insulators (including semiconductors) and metals.

With insulators, there are exactly $N_0/2$ eigenvalues below E_F at each $\kappa_{\mathbf{h}}$. The occupied manifold is then made of $N_0/2$ fully occupied *energy bands*. This is easily generalized to spin-polarized insulators, where the number of filled bands is different for the two spin subsystems. A unitary transformation is here feasible from the set of the occupied COs to an equivalent set of *Wannier functions* (WF) [18]:

$$\begin{aligned} \{\ldots,(n\kappa_{\mathbf{h}}),\ldots\} \stackrel{U}{\longleftrightarrow} \{\ldots,[\ell\mathbf{T_m}],\ldots\}\ [\ell\mathbf{T_m}] \equiv w_\ell(\mathbf{r}-\mathbf{T_m}); \\ \ell = 1, N_0/2 \quad , \quad 0 \leq m_i < w_i\,. \end{aligned} \tag{2.33}$$

The WFs $[\ell\mathbf{T_m}]$ are real, localized functions which can be assigned to the general cell $\mathbf{T_m}$: due to the BvK conditions, we need to consider just W inequivalent cells e.g., those contained in the Wigner-Seitz cell of the super-lattice. There are $N_0/2$ WFs in the reference cell, all the others are translationally equivalent. Since they are obtained through a unitary transformation U from the orthonormal set of the COs, they form themselves an orthonormal set; furthermore, U can be chosen such that they are well localized according to some localization criterion, while reflecting as far as possible the rotational symmetry properties of the system [19, 20]. The ground-state wavefunction Ψ_0^X (2.25) for insulating spinless crystals can then be recast in the form of an antisymmetrized product of spin-WFs:

$$\Psi_0^{\mathrm{X}} \stackrel{\text{(i)}}{=} \|\ldots[\ell\mathbf{T_m}]^\alpha[\ell\mathbf{T_m}]^\beta\ldots\|\,. \tag{2.34}$$

The symbol $\stackrel{\text{(i)}}{=}$ means that this formulation is possible only for spinless insulators.

With metals, instead, the number of eigenvalues below E_F is generally different at different $\kappa_{\mathbf{h}}$'s. Some bands are partially filled: in the limit of infinite w_i's, the surfaces in reciprocal space which separate regions with a different number of occupied COs constitute on the whole the *Fermi surface* of the metal.

2.2.2.2 Crystalline DM and DFs from Single-Determinant Wave-Functions

With the notations just introduced, the HF or KS DMs (Eqs. 2.9 and 2.10) can be written as follows, $\bar{f}(\mathbf{p})$ indicating the FT of $f(\mathbf{r})$:

$$P^{\mathrm{X},\sigma}(\mathbf{r};\mathbf{r}') = \sum_{\mathbf{h}}\sum_{n}{}'\phi_n^{\mathrm{X},\sigma}(\mathbf{r};\kappa_{\mathbf{h}})\left[\phi_n^{\mathrm{X},\sigma}(\mathbf{r}';\kappa_{\mathbf{h}})\right]^*;$$

$$P^{\mathrm{X}}(\mathbf{r};\mathbf{r}') = P^{\mathrm{X},\alpha}(\mathbf{r};\mathbf{r}') + P^{\mathrm{X},\beta}(\mathbf{r};\mathbf{r}')$$

$$\bar{P}^{\mathrm{X},\sigma}(\mathbf{p};\mathbf{p}') = \sum_{\mathbf{h}}\sum_{n}{}'\bar{\phi}_n^{\mathrm{X},\sigma}(\mathbf{p};\kappa_{\mathbf{h}})\left[\bar{\phi}_n^{\mathrm{X},\sigma}(\mathbf{p}';\kappa_{\mathbf{h}})\right]^*;$$

$$\bar{P}^{\mathrm{X}}(\mathbf{p};\mathbf{p}') = \bar{P}^{\mathrm{X},\alpha}(\mathbf{p};\mathbf{p}') + \bar{P}^{\mathrm{X},\beta}(\mathbf{p};\mathbf{p}') \quad (2.35)$$

where the primed sums are restricted to the occupied CSOs of σ spin $\left[\varepsilon_n^{\mathrm{X},\sigma}(\kappa_{\mathbf{h}})<E_F\right]$, and $\bar{\phi}_n^{\mathrm{X},\sigma}$ indicates the 3-D FT of the CO. The DM expressions simplify for spinless systems:

$$P^X(\mathbf{r};\mathbf{r}') = 2\sum_{\mathbf{h}}\sum_{n}{}'\phi_n^X(\mathbf{r};\kappa_{\mathbf{h}})\left[\phi_n^X(\mathbf{r}';\kappa_{\mathbf{h}})\right]^* \overset{(\mathrm{i})}{=} 2\sum_{\ell,\mathbf{m}} w_\ell^X(\mathbf{r}-\mathbf{T_m})w_\ell^X(\mathbf{r}'-\mathbf{T_m})$$

$$\bar{P}^X(\mathbf{p};\mathbf{p}') = \frac{2}{W}\sum_{\mathbf{h}}\sum_{n}{}'\bar{\phi}_n^X(\mathbf{p};\kappa_{\mathbf{h}})\left[\bar{\phi}_n^X(\mathbf{p}';\kappa_{\mathbf{h}})\right]^* \overset{(\mathrm{i})}{=} 2\sum_{\ell}\bar{w}_\ell^X(\mathbf{p})\bar{w}_\ell^X(\mathbf{p}'). \quad (2.36)$$

The various DFs and their properties can be obtained as discussed in Sect. 2.2.1. The following, however, should be noted.

While the KS ED, hence the corresponding structure factors, coincides in principle with the exact one (Eq. 2.29), the HF ED is expected to present systematic errors for instance, to overestimate the density in directed valence bonds (see Sect. 2.4).

HF and KS EMDs are both incorrect, but there are significant differences between the two cases. Consider Eqs. 2.17 and 2.18. The virial theorem is valid for HF, *not* for KS. We can therefore write, at the HF equilibrium configuration:

$$\frac{1}{2}\int \pi^{\mathrm{HF}}(\mathbf{p})p^2\,\mathrm{d}\mathbf{p} = -E_0^{\mathrm{HF}} < -E_0 \quad (\{\mathbf{R}\} = \{\mathbf{R}^{\mathrm{HF,eq}}\}) \quad (2.37)$$

which shows that the HF expectation value of the kinetic energy is systematically *underestimated*, but by an (approximately) known amount. The same cannot be said with the KS solution.

A notable property of the HF and KS RFF of insulators is obtained by using, in Eq. 2.19, the expression (2.36) of the DM in terms of WFs:

$$B^X(\mathbf{r}) \overset{\text{(i)}}{=} \frac{2}{W} \sum_{\ell,\mathbf{m}} \int w_\ell^X(\mathbf{r}' - \mathbf{T_m})\, w_\ell^X(\mathbf{r}' + \mathbf{r} - \mathbf{T_m})\, \mathrm{d}\mathbf{r}'$$

$$= 2 \sum_{\ell} \int w_\ell^X(\mathbf{r}')\, w_\ell^X(\mathbf{r}' + \mathbf{r})\, \mathrm{d}\mathbf{r}'. \tag{2.38}$$

If $\mathbf{r} = \mathbf{T_n} \neq \mathbf{0}$ the last integral vanishes, since it is the overlap between WFs belonging to different cells, meaning that $B^X(\mathbf{r})$ of insulators *must be zero* at all non-zero lattice points. If such condition is not met experimentally, this could happen because of inaccuracies in the CP measurement and/or because of inadequacy of the single-determinant description of the ground-state wavefunction (see Sect. 2.4.2).

2.2.2.3 The Problem of Core Electrons

A commonplace fact of chemistry is that only "valence electrons" are really involved in the formation of compounds (molecules, crystals), while "core electrons" are practically unaffected. This amounts to say that the wavefunction can be approximately written as an antisymmetrized product of $\Psi_v(\ldots, \mathbf{x}_n, \ldots)$, a wavefunction describing N_v valence electrons, times $\Psi_c(\ldots, \mathbf{x}_m, \ldots)$, which describes instead the remaining N_c core electrons; the latter can in turn be expressed as an antisymmetrized product of Ψ_c^A functions for the cores of the individual atoms A entering the compound: Ψ_c^A is obtained from the isolated atom solution except for a rigid displacement along with the nuclear coordinate. Using this core-valence separation Ansatz, which finds its fundamental justification in the prevailing importance of the nuclear attraction term in the proximity of nuclei, the problem is formally reduced to the determination of Ψ_v^X. This can be advantageous for different reasons.

1. Since HF and KS computational times scale rather rapidly with the number of electrons (typically, as N^3), getting rid of the core electrons may result in substantial savings, especially when heavy atoms are involved where the number of valence electrons is comparatively small.
2. The all-electron wavefunction has very sharp features in the proximity of the cores. Describing them with PWs would require extremely high energy cutoffs (see for details Sect. 2.3.3.1).
3. In the vicinity of nuclei, the speed of electrons is an appreciable fraction of the speed of light, and relativistic effects become important, the more so the higher the nuclear charge; the use of the non-relativistic Hamiltonian (2.2) can lead to serious errors in the description of the density in that region. For instance, the relativistic correction for the form factor of the Germanium atom at $\kappa = 0.75$ and $1.5\ \text{Å}^{-1}$ amounts to about 0.6% and 1% of the total, respectively [21]. Analytical expressions which permit the evaluation of relativistic atomic form factors up to $Z = 54$ have been provided by Coppens and coworkers [22]. In fact, in the frame of DFT, exchange-correlation potentials $V_{\mathrm{xc}}(\mathbf{r}; [\rho]^{\mathrm{KS}})$ have been proposed which include this kind of effects (see Sect. 2.3.2). The separation Ansatz offers an

easier solution for this problem, since it allows us to use different techniques for the two terms, by limiting the relativistic treatment to the simpler (central-field) core problem for each atom.

The easiest and most popular way to exploit the separation Ansatz in the frame of HF and KS schemes is to replace in Eq. 2.24 the nuclear Coulomb potential of atom A $(-Z_A/|\mathbf{R}_A - \mathbf{r}|)$ with a (generally non-local) operator $\widehat{V}_{\text{ps}}^{X,A}$, which is called the *pseudopotential* (PP) for atom A. From the solution of the modified equation, pseudo-orbitals $\psi_j^{\text{X,PP}}$ and pseudo-eigenvalues $\varepsilon_j^{\text{X,PP}}$ are obtained: the rest of the procedure is the same but only the N_v pseudo-orbitals lowest in energy are occupied.

It is not in the scopes of this Chapter to refer about the variety of PPs that have been proposed from their earliest formulations [23] to the present days. They can differ for the type of potential (local, semi-local, non-local), for the sub-division between core and valence electrons (large-core, small-core), for the criteria adopted for the optimization of the parameters involved. A number of families of PPs which cover a large part of the periodic system are included in most quantum chemical codes, and in particular in Quantum ESPRESSO and CRYSTAL.

In general, the generation of a PP for any given atom type starts from the corresponding isolated-atom solution, including relativistic corrections for its core electrons. A reference electronic configuration for the atom, typically the ground state, is chosen. The basic step in PP generation consists in replacing the "true" atomic valence orbitals with "pseudized" versions that are equal to the true ones in the outer (valence) region (i.e. for r larger than some suitable chosen *matching radius* r_c) and are smooth functions in the inner (core) region ($r \leq r_c$). In the simplest approach, the pseudized orbitals are nodeless and have the norm-conservation property, i.e. they contain the same amount of charge for $r \leq r_c$ as their atomic reference counter-parts. By inverting the radial KS equation at the same eigenvalues $\varepsilon_j^{\text{KS,PP}}$ of the corresponding true valence states, $\varepsilon_{N_c+j}^{\text{KS}}$, one obtains the so-called *Norm-Conserving* PPs [24]. By relaxing the norm-conservation property of the pseudized orbitals, one obtains the *Ultrasoft* PPs (USPP) [25], having better smoothness and lesser requirements in terms of number of PWs. The price to pay is the presence in the charge density of *augmentation* terms to compensate for the missing charge, and the loss of a simple orthonormality relation between orbitals.

The procedure outlined above is aimed at insuring that at a distance from any core region, the PP solution resembles the all-electron one, and permits the energy for any nuclear configuration to be obtained accurately and economically from that of the pseudo-system with reference to the energy of the constituent pseudo-atoms.

In the present context we are however interested in reconstructing DFs from the knowledge of ψ_v^{X} and of the atomic core solutions. The use of PPs has a serious drawback in this respect: since the pseudo-orbitals are not orthogonal to the core, the "pseudo-charge" one gets, summed to the core density, is not the true charge density. The same is true as concerns EMD. Special techniques for all-electron charge-density reconstruction must be applied, based for example on the Projector Augmented Waves (PAW) method [26]: see Sect. 2.3.3.1.

2.3 The Solution of the HF and KS Periodic Problem

2.3.1 General Solution Schemes

As shown in the previous Sections, both the HF and the KS problems can be recast under the form of single-particle Schrödinger equations under an effective self-consistent potential (Hartree and exchange potential for HF, Hartree and exchange-correlation potential for KS). The KS problem is simpler in this respect, since the effective potential depends only upon the charge density $\rho(\mathbf{r})$, while in HF it depends upon the DM $P(\mathbf{r},\mathbf{r}')$. The solution can be found using an iterative procedure to achieve self-consistency: starting from a suitable initial guess for the potential, single-particle orbitals are calculated, the effective potential is re-calculated with the new orbitals, and so on until self-consistency is achieved. Several well-established techniques for speeding up self-consistency are known. We remark however that the self-consistent solution of HF or KS equations is not the only possible way to find the HF or DFT ground state. In the *global minimization* approach, one directly minimizes the energy as a function of the orbitals. This approach is perfectly equivalent to the solution of HF or KS equations. In practice, it is used only in DFT with a PW BS, for aperiodic systems or systems described by a large unit cell, and typically in conjunction with *ab-initio* molecular dynamics [27].

In practical calculations, the orbitals must be expanded into some suitably chosen BS. In periodic systems, it is convenient to use a BS of BFs $f_\mu(\mathbf{r};\,\kappa)$ (see Sect. 2.2.2.1), so that determining the COs $\phi_n^X(\mathbf{r};\,\kappa)$ reduces to a secular problem that involves only basis functions of that given κ. The choice of the BS is crucial and determines the algorithms and numerical methods used in the actual solution. Most calculations use either PWs or atom-centered functions (AOs).

PWs are the traditional choice in solid state physics, reflecting the delocalized nature of valence and conduction electron states in crystals. PWs form an infinite *complete* BS uniquely determined by the crystal lattice:

$$f_{\mathrm{G}}(\mathbf{r};\kappa) = \frac{1}{\sqrt{\Omega}} \exp[\iota(\kappa + \mathbf{G}) \cdot \mathbf{r}], \tag{2.39}$$

where Ω is the volume of the BvK box, and $\mathbf{G}$ a reciprocal lattice vector (see Sect. 2.2.1.1). The Bloch condition (Eq. 2.31) is satisfied because $\mathbf{G} \cdot \mathbf{T_m} = 2n\pi$, with integer n. A finite set can be obtained by considering all PWs whose kinetic energy, $(\kappa + \mathbf{G})^2/2$, is below a given value E_c, the so-called *kinetic energy cutoff*. PWs present several advantages:

1. They are a numerically convenient, orthonormal set, allowing the usage of Fast FT (FFT) techniques.
2. They form an *unbiased* BS, since they do not depend upon which atoms are present and upon atomic positions: therefore, they do not suffer from incomplete

BS errors on forces (also known as Pulay forces [28]) or from BS superposition errors on energy (see Sect. 2.3.4.2) that affect calculations performed with AOs.
3. Convergence of the results can be evaluated by varying the single parameter E_c.

PWs have also some serious shortcomings, the most obvious being the inability to cope with the presence of core states in atoms. The standard solution is to introduce PPs, as was discussed in Sect. 2.2.2.3: in particular, USPPs [25] allow the practical usage of PWs in a large class of materials, including transition metals and first-row elements C, N, O, F. When the focus is on DFs, however, for the reasons explained there, a better solution is offered by the PAW technique, which allows the "true" all-electron charge density to be calculated, while retaining a PW BS of minimal size (see Sect. 2.3.3.1).

Even with the best USPP or PAW technology, the size of the PW BS vastly exceeds that of a well-designed basis of AOs: for typical systems, the average number of PWs per atom in the unit cell is in the order of a few hundreds. The advantages of PWs, coupled with algorithmic and numerical techniques specific for PWs, make however their usage interesting in spite of the large size of the BS, at least for DFT calculations. HF calculations in PWs are considerably slower than DFT calculations in the same system, though, and the size of the PW BS makes the expansion of the DM into PWs impractical for all but the simplest systems.

Localized BSs formed by atomic orbitals (AO), $\chi_\mu(\mathbf{r})$, are the traditional choice in quantum chemistry, reflecting the atomic composition of the matter. In periodic systems, one uses Bloch sums of AOs:

$$f_\mu(\mathbf{r};\kappa) = \frac{1}{\sqrt{W}} \sum_{\mathbf{T}} \exp[\iota\kappa \cdot \mathbf{T}]\, \chi_\mu(\mathbf{r}-\mathbf{T}). \tag{2.40}$$

AOs used in periodic systems include Linearized Muffin-Tin Orbitals, numerically defined Orbitals (including Slater-type orbitals) and Gaussian-type Orbitals (GTO). The CRYSTAL software described in this article shares with standard molecular quantum chemistry codes the use of GTOs: this technical similarity entails a number of useful consequences, as analyzed in more detail in Sect. 2.3.4.2. The other main advantage of AOs is the limited number of functions required for a good description of the COs. Their main disadvantage is that they do not form a complete BS: simply adding more AOs will eventually result in pseudo-overcompleteness (i.e. linear dependencies among basis functions). As a consequence there is no mathematically exact procedure to achieve convergence with respect to the BS. Since the actual degree of convergence will depend upon the atoms that are present in the cell and upon their positions, calculations with AOs suffer from BS superposition errors on energy and have Pulay terms in forces (see again Sect. 2.3.4.2 for a more complete discussion).

In the following we describe the two approaches, PWs and AOs, with reference to the two software implementations: Quantum ESPRESSO and CRYSTAL. In both cases, however, the expression of the corrective potential $\widetilde{V}^{X,\sigma}$ in Eq. 2.24 must be specified, which requires (apart from the HF case) selecting one in a variety of proposals. We consider preliminarily this question by briefly examining in the next section the choices available in the two codes.

2.3.2 The Exchange-Correlation Potential in KS Schemes

The *quality* of the KS Hamiltonian $\hat{h}^{KS}$, hence of the corresponding solution, depends primarily on the expression adopted for $V_{xc}(\mathbf{r}; [\rho])$, the exchange-correlation potential. Perdew [29] has suggested to classify the different proposals along a "Jacob's ladder", having at its summit the "true" potential, that is, such that Eqs. 2.29 and 2.30 are exactly satisfied. At the lowest rung of this ladder we find the local density approximation (LDA) [7], where the exchange-correlation at $\mathbf{r}$ is a *function* of the electron density at that point: $V^{LDA}(\mathbf{r}) = f^{LDA}(\rho(\mathbf{r}))$. The next rung is the generalized gradient approximation (GGA) which uses also the gradient of the electron density at $\mathbf{r}$ to improve upon LDA [30]: $V^{GGA}(\mathbf{r}) = f^{GGA}(\rho(\mathbf{r}), \nabla\rho(\mathbf{r}))$. The third rung, denoted meta-GGA, incorporates increasingly complex ingredients, such as the kinetic energy density, $\tau(\mathbf{r})$, or the Laplacian of the density $\nabla^2\rho(\mathbf{r})$ [31]. At higher levels of the ladder non-locality in both exchange and correlation components can be included (which represents a non-standard form of the KS Hamiltonian). Hybrid-exchange functionals that use a fraction of non-local HF exchange [32] can be considered as semi-empirical fourth-rung functionals. At all levels, relativistic effects may be taken into account [33]. All these potentials contain parameters which have been variously optimized to satisfy specific requirements; for instance, Zhao and Truhlar have recently proposed modified GGA-type functionals to recover the correct gradient expansion of slowly varying densities, a quite important condition for solids [34].

Molecular and crystalline codes in current use permit one in a variety of exchange-correlation functionals to be chosen from input. Table 2.2 lists some that are presently available in Quantum ESPRESSO and CRYSTAL (the list is continuously enlarged, and different combinations from those there reported can be created). Hundreds of papers have appeared where the performance of the different functionals has been tested with different families of compounds, molecules or crystals. It must be noted, however, that the basic quantity taken into account in such analyses is energy and energy derived quantities (formation energy, equilibrium geometry, vibrational frequencies . . .). Sometimes, other features of practical importance are considered, like the distribution of one-electron levels, etc., but almost never DFs. Suitably parameterized hybrid-exchange functionals are usually the best solution presently available in most respects: see for example reference [45]. In the following, the influence of the exchange-correlation functional on DFs is discussed by way of examples.

Table 2.2 Exchange-correlation functionals available in Quantum ESPRESSO (E) and CRYSTAL (C). Most of them are a combination of expressions for the exchange and correlation part, each described in the indicated reference

Type	Name	Exchange	[Ref.]	Correlation	[Ref.]	Availability	
LDA	SVWN	Slater	[35]	VWN	[36]	(E)	(C)
	SPWLSD	Slater	[35]	PWLSD	[30]	(E)	(C)
	SPZ	Slater	[35]	PZ	[37]	(E)	(C)
GGA	PBE	PBE	[38]	PBE	[38]	(E)	(C)
	PBEHCTH	PBE	[38]	HCTH	[39]	(E)	–
	PW91	PW91	[40]	PW91	[40]	(E)	(C)
	PBEsol	PBEsol	[41]	PBEsol	[41]	(E)	(C)
	SOGGA	SOGGA	[34]	PBE	[38]	–	(C)
	WC	WC	[42]	PBE	[38]	(E)	(C)
Hybrid	B3LYP	B/HF	[43]	LYP	[44]	(E)	(C)
	PBE0	PBE/HF	[38]	PBE	[38]	(E)	(C)
	B1WC	WC/HF	[42]	PW91	[40]	–	(C)

2.3.3 *The Quantum ESPRESSO Distribution*

The Quantum ESPRESSO distribution [1] is a rather large set of packages and utilities for electronic structure calculations using DFT and a PW BS. Quantum ESPRESSO is based on a panoply of codes and tools developed and used during many years by several research groups throughout the world. It is an open-source project, currently maintained by researchers at the DEMOCRITOS National Simulation Center of the Italian National Research Council with the strong support of several other institutions and of individual researchers interested in specific subjects or in implementing new developments, as specified in the web site [1].

The package that implements the most general approach to the calculation of the ED is `PWscf`. `PWscf` can perform single-point calculations, structural optimization (including crystal cell optimization) and molecular dynamics on the electronic ground state (including variable-cell dynamics), as well as the search for transition states and minimum energy pathways. Ab-initio Car-Parrinello molecular dynamics is instead performed by package `CP`. The basic ingredient of `PWscf` is the self-consistent solution of the KS Eq. 2.24, using mixing techniques (modified Broyden) to find the self-consistent charge, and iterative diagonalization (block Davidson) to determine the COs. Iterative diagonalization does not require to store the hamiltonian H as a matrix, since only $H\psi$ products are required. Such products, as well as the ED and the Hartree and exchange-correlation potential, are calculated by taking advantage of the "dual-space technique", i.e. the possibility to jump back and forth, using the FFT algorithm, from real to reciprocal space. This allows to perform the required operations in the space where it is more convenient.

Another important package is `PHonon`, allowing the calculation of dielectric properties and of the full phonon dispersions using Density-Functional Perturbation

Theory [46]. A series of tools and auxiliary codes allow the analysis of the data produced by `PWscf` and `CP`, including visualization and further processing of the ED and of the ESD.

2.3.3.1 The Treatment of Core Electrons in PW Codes

Let us consider the representation of the ED when simple norm-conserving PPs are used. The ED is simply given by Eq. 2.26, where the ϕ_n^X orbitals are replaced by the valence pseudo-orbitals $\phi_j^{X,PP}$. The ED thus contains Fourier components up to a maximum value of $|\mathbf{G}|$ such that $|\mathbf{G}|^2/2 \leq E_c^\rho$, where E_c^ρ is four times larger than the kinetic energy cutoff for PWs: $E_c^\rho = 4E_c$. A three-dimensional grid in reciprocal space ("FFT grid") is introduced:

$$\bar{\rho}_{\mathrm{FFT}}(h', k', l') \equiv \bar{\rho}(\mathbf{G}_{hkl}), \quad \mathbf{G}_{hkl} = h\mathbf{B}_1 + k\mathbf{B}_2 + l\mathbf{B}_3, \tag{2.41}$$

where $h' = 0, \ldots, N_1 - 1$ and $h = h'$ if $h' \leq N_1/2$ (N_1 even) or $h' \leq (N_1 - 1)/2$ (N_1 odd); $h = h' - N_1$ otherwise. The equivalent relations holds for k, k', N_2 and for l, l', N_3. The values of N_1, N_2, N_3 are determined by the condition that this grid must accommodate all $\mathbf{G}$ components of ED without any loss, i.e., components for which $h = h' - N_1$ and so on should not overlap components for which $h = h'$ and so on. A three-dimensional discrete FT with dimensions N_1,N_2,N_3 then directly yields the ED on the corresponding real-space FFT grid, spanning the unit cell of the crystal:

$$\rho_{\mathrm{FFT}}(m_1, m_2, m_3) \equiv \rho(\mathbf{r}_{m_1,m_2,m_3}), \quad \mathbf{r}_{m_1,m_2,m_3} = \frac{m_1}{N_1}\mathbf{a}_1 + \frac{m_2}{N_2}\mathbf{a}_2 + \frac{m_3}{N_3}\mathbf{a}_3, \tag{2.42}$$

where $m_1 = 0, \ldots, N_1 - 1$ and the equivalent for m_2 and m_3. As mentioned earlier, the ED so obtained is actually a pseudo-density.

Let us consider now the cases of USPPs and of PAWs, which can be treated in a unified framework [47]. The starting point is the introduction of a linear transformation, connecting the true orbitals $_n$ with pseudo-orbitals ϕ_n^{PP}:

$$\phi_n(\mathbf{r}) = \Im\phi_n^{PP}(\mathbf{r}) \equiv \phi_n^{PP}(\mathbf{r}) + \sum_i \left(\psi_i(\mathbf{r}) - \psi_i^{PP}(\mathbf{r})\right) \langle\beta_i|\phi_n^{PP}\rangle, \tag{2.43}$$

where the ψ_i are atomic reference orbitals (not necessarily bound states), the ψ_i^{PP} are the corresponding pseudized atomic orbitals, the β_i projectors are dual to the pseudized orbitals: $\left\langle\beta_i|\psi_j^{PP}\right\rangle = \delta_{ij}$. Both $\beta_i(\mathbf{r})$ and $\left(\psi_i(\mathbf{r}) - \psi_i^{PP}(\mathbf{r})\right)$ are nonzero by construction only in the core region. The sum over i runs over atoms and projectors for a given atom. Atomic functions are centered around the position of the corresponding atom.

Under suitable assumptions, one can show that the expectation value, $\sum_n \langle \phi_n | O | \phi_n \rangle$, of an operator O can be expressed as expectation value between pseudo-orbitals, $\sum_n \langle \phi_n^{\text{PP}} | \widetilde{O} | \phi_n^{\text{PP}} \rangle$, of an equivalent operator $\tilde{O}$ that can be written as:

$$\widetilde{O} \equiv \Im^{\dagger} O \Im = O + \sum_{ij} |\beta_i\rangle \left(\langle \psi_i | O | \psi_j \rangle - \left\langle \psi_i^{\text{PP}} \right| O \left| \psi_j^{\text{PP}} \right\rangle \right) \langle \beta_j |. \tag{2.44}$$

With this relation, one can express the total energy as a function of smooth pseudo-orbitals $\phi_n^{\text{PP}}(\mathbf{r})$, that can be easily expanded into PWs. The ED can be expressed via Eq. 2.44 by adding an "augmentation" term to the standard expression, Eq. 2.26:

$$\rho(\mathbf{r}) = 2 \sum_{n=1}^{N} \left(\left| \phi_n^{\text{PP}}(\mathbf{r}) \right|^2 + \sum_{ij} \langle \phi_n^{\text{PP}} | \beta_i \rangle \, Q_{ij}(\mathbf{r}) \, \langle \beta_j | \phi_n^{\text{PP}} \rangle \right). \tag{2.45}$$

The functions

$$Q_{ij}(\mathbf{r}) = \left(\psi_i(\mathbf{r}) \psi_j(\mathbf{r}) - \psi_i^{\text{PP}}(\mathbf{r}) \psi_j^{\text{PP}}(\mathbf{r}) \right) \tag{2.46}$$

are nonzero only in the core region(s) of the respective atom(s). These are however quickly oscillating functions, due to orthogonality to core states, and are as such unsuitable for Fourier expansion.

Within the USPP formalism, the Q functions are in turn pseudized and transformed into equivalent but smoother functions that can be safely expanded into Fourier components. The needed cutoff, however, often exceeds the cutoff $E_c^{\rho} = 4E_c$ that would be needed in the absence of the augmentation term. The solution that is typically adopted for USPP is the introduction of a second FFT grid, corresponding to a cutoff $E_c^{\rho} > 4E_c$. The ED is thus available, both in real and in reciprocal space on this grid. The analogy with PAW suggests that the true charge density can be reconstructed by replacing the pseudized Q functions with the original, unpseudized Q functions of Eq. 2.46.

In the PAW method, instead, the augmentation term is calculated and stored on radial grids centered around atomic positions. All energy and potential terms needed in the formalism are calculated – under suitable assumptions – using either the FFT grid or the radial grids; there are no "mixed" terms involving both grids. If the ED is desired e.g. for inspection or for visualization, however, both grids are needed.

2.3.3.2 The Problem of "Strongly Correlated" Systems (DFT+U)

The most commonly used DFT approximations, i.e. GGA (the "2nd rung of the ladder"), are notoriously unreliable for "strongly correlated" systems, i.e. those containing highly localized atomic-like orbitals. The big problem with DFT seems to be getting the correct occupancy of atomic-like orbitals: 3d, 4d, 5d for transition metals, 4f for rare earths, 5f for actinides, in a sea of delocalized band electrons. Current approximated exchange-correlation functionals tend to favor unphysical noninteger occupancies. Such behaviour can be traced to an important feature of the exact functional that is missing in approximate ones: a discontinuity, as a function of the number of electrons, when an integer number is crossed. This can in turn be traced to the incomplete cancellation of the self-energy, a problem that is absent by construction in HF, but is present to some extent in all approximate functionals.

DFT+U is a simple extension of conventional DFT that was devised to deal with highly correlated electrons. The basic idea of DFT+U (originally called LDA+U) is to add a Hubbard-like term for a suitably chosen subset of localized electron states [48]. The presence of (at least) an adjustable parameter U, of various possible choices for the manifold of localized states and for the Hubbard term itself, may induce to think that DFT+U is more akin to a semiempirical correction to DFT than to a real first-principle technique. Still, DFT+U is a very useful tool that has proven able to yield very good results in highly correlated materials at the price of a modest computational overhead.

Quantum ESPRESSO implements a simplified ("no-J") rotationally invariant form of the Hubbard term:

$$E_{Hub} = \frac{U}{2} \sum_{I,\sigma} \mathrm{Tr}[\mathbf{n}^{I\sigma}(1 - \mathbf{n}^{I\sigma})] \tag{2.47}$$

where $\mathbf{n}^{I\sigma}$ is the occupation matrix on the chosen manifold of localized states for atom I, for spin σ, and U is the Hubbard parameter [49]. The occupation matrix is defined as

$$n^{I\sigma}_{mm'} = \sum_{n} f^{\sigma}_{n} \langle \phi^{\sigma}_{n} | \, P^{I}_{mm'} \, | \phi^{\sigma}_{n} \rangle \tag{2.48}$$

where $P^{I}_{mm'}$ is the projector over the chosen manifold of localized states, f^{σ}_{n} the occupation (between 0 and 1) for electron orbital ϕ^{σ}_{n}. Typically the projector P simply projects over atomic states, and only on strongly localized ones (e.g. 3d in first-row transition metals and so on). The most delicate decision in a DFT+U calculation is probably the choice of the parameter U. For each atom in a given electronic configuration, experience and experiments indicate a typical range for U, usually a few eV. One can use U as an adjustable parameter; a more satisfactory procedure, described in Reference [49], allows a consistent value of U to be determined from first principles. The introduction of the U parameter may have profound effects on the electronic structure, and as a consequence, on the charge density of strongly correlated materials.

2.3.4 *The CRYSTAL Package*

2.3.4.1 General Features

CRYSTAL was conceived more than 30 years ago [50] as an extension to periodic systems of the powerful ab-initio molecular codes which were available at that time [51–53]. It has since been developed by researchers of the Theoretical Chemistry Group in Torino (Italy) and of the Computational Materials Science Group in Daresbury (UK), with important contributions from other scientists, as documented in the CRYSTAL site [2]. CRYSTAL solves the periodic HF and KS equations with a variety of exchange-correlation potentials (see Table 2.2). As in the molecular codes that served as its template, it adopts a BS of GTOs; this choice has some advantages and drawbacks as already anticipated in Sect. 2.3.1, and as analyzed in more detail in Sect. 2.3.4.2. An attractive feature related to the *local* character of the basis functions is that not only 3-dimensional crystals, but also structures periodic in 2 (slabs), 1 (polymers) and 0 (molecules) dimensions are treated by CRYSTAL with the same basic technology without any need of artificial replication of the subunits. In all cases, the symmetry of the system is fully exploited: for instance in carbon nanotubes (a 1-dimensional polymer), profit can be taken of the helicoidal symmetry with substantial time savings [54]. Among the many facilities embodied in CRYSTAL we list below some which are of interest in the present context.

1. Full geometry optimization is feasible with respect to both lattice parameters and atomic positions. It is also possible to perform *volume constrained* geometry optimization: that is, for a given crystalline structure and for a given cell volume V, the minimum energy configuration is determined. From the corresponding $E(V)$ curves, the effect of pressure on various properties of the system, including DFs, can be determined.
2. The vibrational frequencies at $\kappa = 0$ (Γ) and the corresponding infrared intensities are determined in the harmonic approximation; an anharmonic correction is performed for the stretching mode of X–H bonds. Each normal mode is classified by symmetry and can be visualized. The comparison with experimental vibrational data becomes easy, and often results in extremely good agreement [55]. Zero-point-motion and finite-temperature effects can be determined in the frame of the harmonic approximation (there may be the need of complementing the information at Γ with that at other points in the BZ, which may be obtained from supercell calculations). The knowledge of nuclear motions can be used for estimating Debye-Waller atomic factors or, more generally, for obtaining an ensemble description of electronic DFs at finite temperatures (see Sect. 2.5.1).
3. All-electron calculations are feasible in all cases, and are preferable to PP ones when one is interested in DFs for the reasons discussed in Sect. 2.2.2.3. Note however that no relativistic corrections for core electrons are yet implemented in CRYSTAL.
4. Various tools are available for representing the ED and calculating the static structure factors. Among the examples provided below of the use of these tools,

one concerns the "ED deformation map" of a molecular crystal, that is, the difference between the density in the actual system, and the superposition of the densities of the isolated molecules at the geometry they have in the crystal (see Sect. 2.4.2). In ED studies, this density is more customarily called interaction density; see later in this chapter and Refs. [9, 11].

5. An essential descriptor of EDs is Bader's topological analysis (see Sect. 2.2.1.1). CRYSTAL is connected to Gatti's TOPOND program [56] in the sense that it provides the latter with the information required for determining all topological properties of interest (see Refs. [9, 10]; for ED topological properties see Refs. [8–11] and use of them in Chaps. 1 and 9–16 of this Book).
6. The EMD, the directional CPs, the reciprocal form factor [$B(\mathbf{r})$], and their anisotropies can be calculated in different ways which provide results of different accuracy according to the nature of the system (insulator, conductor). Again, examples are provided below.
7. The effect of a static, uniform external electric field on the DM can be studied in CRYSTAL in two ways, either by superimposing a sawtooth potential (which preserves periodicity on a supercell scale), or by calculating the first and second derivatives of the DM with respect to the field components in the frame of coupled-perturbative HF or DFT: see Sect. 2.5.2 for more details.
8. In its most recent version CRYSTAL is connected to CRYSCOR, a post-HF code which permits the correction to the energy and to the DM of the crystalline system to be calculated at the lowest order of perturbation theory (see Sect. 2.5.3).

2.3.4.2 The Basis Set Problem

CRYSTAL shares with standard molecular codes the use of GTOs as basis functions. Each atom A carries p_A GTOs, each resulting from a "contraction" of M_{iA} Gaussian "primitives" of angular momentum components ℓ,m centered in $\mathbf{R}_A$:

$$\chi_{iA}(\mathbf{r}_A) = \sum_{j=1}^{M_{iA}} c_{iA,j} N^{\ell,m}(\alpha_{iA,j}) X^{\ell,m}(\mathbf{r}_A) \exp[-\alpha_{iA,j}\mathbf{r}_A^2].$$

Here $\mathbf{r}_A = \mathbf{r} - \mathbf{R}_A$, $X^{\ell,m}$ are real solid harmonics and $N^{\ell,m}$ normalization coefficients; $c_{iA,j}$ are known as "coefficients", $\alpha_{iA,j}$ as "exponents" of the GTO. As a rule, $\mathbf{R}_A$ are nuclear coordinates, but GTOs on "ghost atoms" at a general position can be added.

From its beginnings, CRYSTAL took over from molecular quantum chemistry the experience gained in the preparation of these sets and the extremely efficient algorithms already available for performing one- and two-electron GTO-integrals. During all these years other innovations were imported from computational quantum chemistry, related to the use of GTOs. Just to mention an example, density-fitting techniques, which permit incredible savings of time in the calculation of the integrals needed in post-HF schemes [57], can be transferred with few modifications to periodic GTO-based schemes (see Sect. 2.5.3). Another side advantage of the sharing of the same BS, is the possibility of one-to-one comparisons between the results of periodic and standard molecular calculations; examples thereof are provided below.

The evaluation of GTO-integrals in CRYSTAL entails problems related to the periodically infinite character of the system. Sophisticated techniques have been implemented which permit the truncation or the accurate approximation of lattice sums: Ewald techniques, multipolar treatment of non-overlapping distributions, bipolar expansion, etc. On the whole, many thousands lines of code have been developed for this purpose, which contrasts the amazing simplicity of the corresponding integral part in PW codes.

The real problem with GTO sets, however, is their inherent incompleteness. For each system, for each atomic species inside, a choice must be performed of the number (p_A), type (ℓ,m) and contraction scheme (M_{iA}, $c_{iA,j}$, $\alpha_{iA,j}$) of the χ_{iA} functions. Again, this contrasts with PW sets which are complete, in principle, and whose quality is determined by a single parameter (the energy cutoff). An enormous literature exists on how to set up GTO sets which perform efficiently for different types of system. The quality of the GTO set on atom A can in principle be improved at will by including more and more χ_{iA} functions, for instance following a precise strategy [58]. Techniques for extrapolating the computed energy to the *complete BS limit* have also been proposed [59]. For atoms in crystals, a number of AE or valence-only GTO sets are proposed in the CRYSTAL site, based on past experience [2]; those suggested for oxygen, for example, are different according to whether this species is present as an oxide ion (as in MgO) or is involved in semi-covalent bonds (Ice). A clever choice permits a very accurate representation of the ground state (HF, KS) wavefunction to be obtained with a surprisingly small number of functions, as compared to PWs.

Two additional problems can finally be mentioned.

1. The potential energy surface, $E_0(\{\mathbf{R}\})$, that is, the dependence of the calculated ground-state energy on the set of the nuclear coordinates, bears crucial importance because it determines not only reaction energies, but also equilibrium configuration and vibrational frequencies which directly influence DFs. In order to have reliable values for these quantities, all errors that affect $E_0(\{\mathbf{R}\})$ should be approximately constant over the range of nuclear coordinates considered. The part of this error related to the dependence of BS quality on $\{\mathbf{R}\}$ is known as BSSE (BS superposition error) [60]. It affects in particular schemes, such as CRYSTAL, using GTO functions centered in the nuclei: it is generally expected that the same GTO set describes better structures where atoms are close to each other than viceversa because in the former case the wavefunction in the interatomic regions can be represented more accurately using "redundant" functions from neighboring atoms. Many techniques have been proposed to estimate the BSSE and to correct for it [61]. The automated geometry optimization in

CRYSTAL, however, does not take BSSE into account, and may therefore result into too compact structures, for the reasons just explained.
2. The use of extended atomic sets comprising very diffuse (low exponent) primitives may lead to quasi-linear-dependence effects between functions centered in different atoms. With densely packed systems as are encountered in solid state problems, this type of "overcompleteness" easily results in catastrophic behaviour. As concerns the use of high angular momentum functions, the present version of CRYSTAL is limited to $\ell \leq 3$ (s, p, d, f GTOs).

2.4 Role of Computational Parameters on Density Functions

We provide below a few examples of simulated DFs for crystalline systems obtained with the two programs described in the previous Section. Different computational choices and their consequences on the results will be considered in the various cases. The objective is to provide an outlook of present capabilities and an indication of possible pitfalls due to unwise selection of the computational parameters. The cases of two typical covalent systems (Silicon, Diamond), of a molecular crystal (Urea) and of a metal (Aluminium) are considered. As concerns ionic systems (LiN_3, MgO, for instance), reference can be made to the existing literature [62, 63]. Finally, the effect of the Hamiltonian on spin localization in an ionic spin-polarized system ($KMnF_3$) is discussed.

2.4.1 ED of Silicon and Diamond: Basis Set and Hamiltonian Effects

Silicon and Diamond are two prototypical covalent systems: for this reason and for their intrinsic importance, they have been devoted enormous attention to. Among the many papers concerning the properties of their DM (see for instance the recent synchrotron-radiation experiment with powder samples [64]), we will refer in the following to the study by Lu et al. [21], who compared the EDs resulting from their all-electron LDA calculations for Diamond, Silicon and Germanium with those from experimental data, and in particular from accurately analyzed X-ray diffraction data for crystalline Silicon. Reference can also be made to the comprehensive studies performed with CRYSTAL about contemporarily [65, 66], and aimed at analyzing the quality of quasi-HF periodic solutions for this kind of systems. Here we consider the effect on the calculated ED of the quality of the representative BS and of the type of Hamiltonian adopted, by taking advantage of the availability of the two computational tools. Silicon is first considered in more detail; similar results are next reported for Diamond.

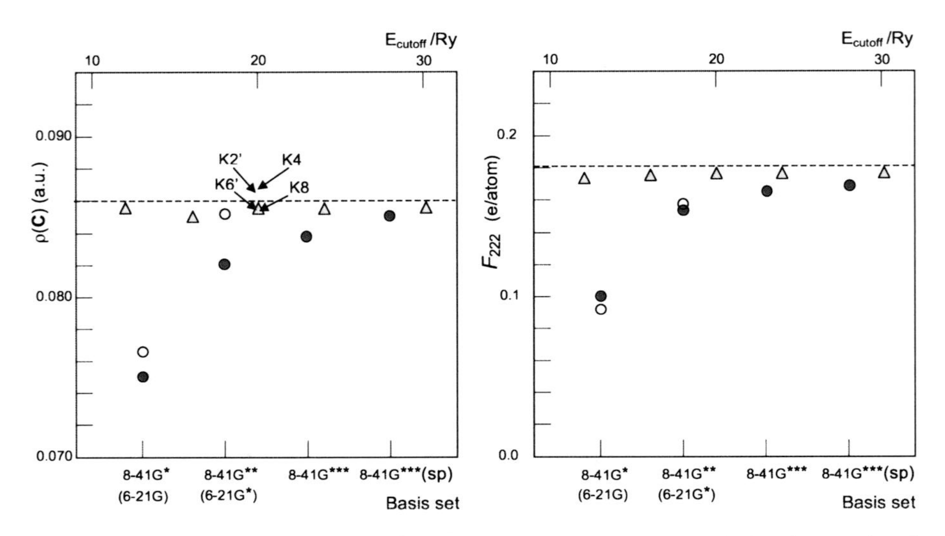

Fig. 2.1 Effect of basis set quality on $\rho(\delta)$, the calculated ED at the bond midpoint (*left panel*) and on the F_{222} structure factor (*right panel*) for crystalline Silicon, using the DFT-PBE Hamiltonian. The Quantum ESPRESSO results (*triangles*) are reported as a function of the cutoff energy E_c (*scale on top*); the CRYSTAL ones along an arbitrary scale (*at bottom*) corresponding to GTO sets of increasing quality: 8-41G^{n*} (*full circles*) or 6-21G^{n*} (*open circles*). The two horizontal *dashed lines* indicate the estimated experimental value [21] (See text for other details)

Figure 2.1 documents the BS dependence of two quantities related to the ED: its value at the Si-Si midpoint (δ) and the "forbidden" F_{222} structure factor, whose non-zero value is a measure of the asphericity of the ED about the individual atoms (see Sect. 2.2.1.1). All data here shown were obtained with the PBE choice for $V_{xc}(\mathbf{r}; [\rho])$ (see Sect. 2.3.2), but very similar trends were obtained with other choices; the lattice parameter was set at its experimental value, $a = 5.43$ Å. The Quantum ESPRESSO calculations were performed using the PAW technique (see Sect. 2.3.3.1; PAW pseudopotentials were generated using the parameters given in `paw_library`, contained in the Quantum ESPRESSO distribution). Different E_c values were tried as shown in the figure; correspondingly, the cutoff for the core contribution was varied from 48 to 120 Ry. As concerns the sampling in $\mathbf{k}$ space, a displaced Monkhorst grid with w=4, s=1/2 for all i's (see Eq. 2.32) was generally adopted, which may be labelled K4′; the adequacy of this choice is demonstrated in the left panel, which shows the effect of using different undisplaced (Kw) or displaced (Kw') k-grids with E_c=20 Ry. In the CRYSTAL all-electron calculations, a K8 sampling net was used, and a number of GTO sets were tried, which can be classified in two categories. The former category comprises double-zeta 8-41G sets as in reference [66], complemented with n (the number of asterisks) polarization functions of type d, d + d, d + d + f, respectively; the 8-41G***(sp) includes in addition a single-GTO sp set at the midpoint of each bond, with exponent 1.4 a.u. The latter category comprises the 6-21G and 6-21G* sets as in reference [66].

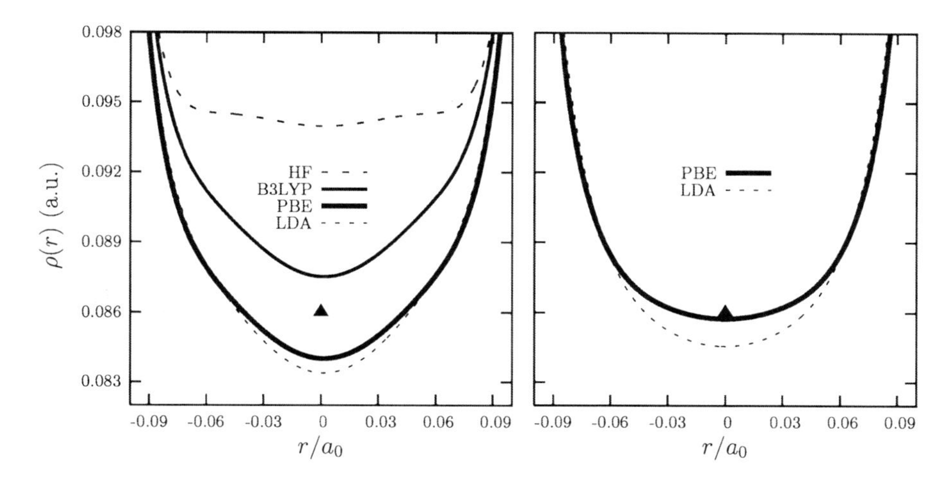

Fig. 2.2 Silicon ED along the Si-Si bond with different one-electron Hamiltonians, as indicated, and using either CRYSTAL (*left panel*) or Quantum ESPRESSO (*right panel*). The *triangles* are the experimental value [21]

The regular trend of the Quantum ESPRESSO results is clear and, as far as these quantities are concerned, the limit with respect to E_c seems reached. The CRYSTAL results are more scattered, as expected. Note in particular that the 6-21G* results are curiously very similar to the best PW results, which represents a warning against too hasty conclusions about the adequacy of the adopted BS. From the present data it appears in fact that convergence of the GTO sets towards the BS limit may be slower for DM related quantities than for energy: the PBE energy per cell for the three best GTO sets here used is -578.7798, -578.7821, -578.7826 E_h, respectively.

The converged results are also quite close to the data proposed by Lu et al. [21], by extrapolation to zero thermal motion of the best experimental results: $\rho(\delta) = 0.086$ a.u., $F_{222} = 0.181$ e/atom.

Figure 2.2 reports the ED along the Si–Si bond for various Hamiltonians (see Sect. 2.3.2), as a function of the distance r from the bond midpoint in units of the lattice parameter. For the CRYSTAL calculations (left panel), the 8-41G***(sp) GTO set was used. For the Quantum ESPRESSO calculations (right panel), a K4′ grid and a cutoff of 30 Ry were adopted; the core charge and the augmentation term in the valence charge were directly plotted in real space. Note that in the present version of this code, the PAW technique cannot be used with non-local exchange Hamiltonians. The LDA functionals SVWN and PZ have been used with the two codes, respectively. According to the CRYSTAL calculations, close to the bond midpoint the HF EC lies highest followed by the hybrid functional (B3LYP), the GGA (PBE), and the LDA ones; the Quantum ESPRESSO results with the pure DFT Hamiltonians are almost the same.

Results similar to those just presented were obtained with Diamond (see Fig. 2.3). Again, the experimental geometry was adopted ($a = 3.567$ Å); a 6-31G** GTO

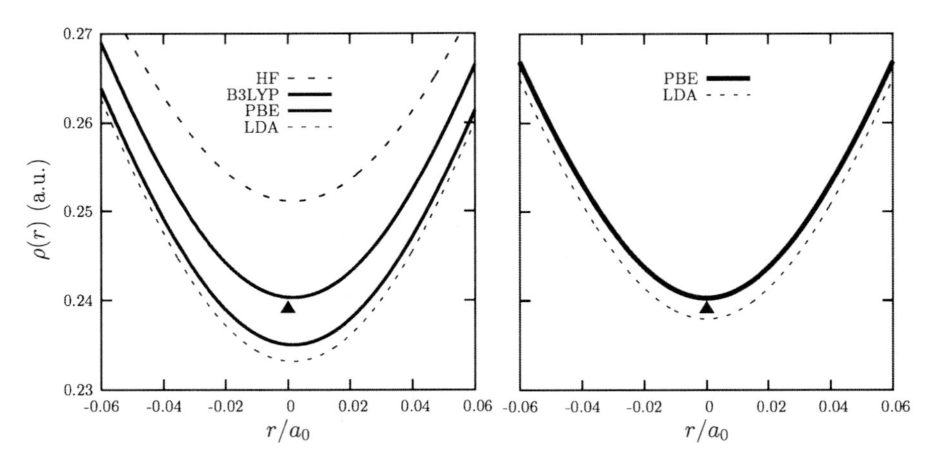

Fig. 2.3 Diamond ED. Symbols as in Fig. 2.2

set taken from the literature [65] complemented with an additional f-type polarization set was used for the CRYSTAL calculations (left panel); for the Quantum ESPRESSO ones (right panel), a K4′ grid and a cutoff of 45 Ry were adopted.

For the two systems, the ED presents a critical point [8] of type (3, −1) at the bond midpoint δ, and its value increases steadily moving away from it along the bond. The case is different if one considers only the contribution to the ED from valence electrons, $\rho_{\mathrm{val}}(\mathbf{r})$. An "experimental" determination of this non-observable quantity can be obtained by subtracting from the reconstructed ED the contribution of core electrons from accurate atomic calculations: both in silicon and in diamond ρ_{val} presents a relative maximum at two symmetry related points (γ) when moving from δ towards the cores [21]: in the case of silicon the difference between the two values is almost undetectable experimentally. It may be interesting to compare these experimental data with those from CRYSTAL all-electron calculations (but using only the valence bands in the primed sums of Eq. 2.35, or from Quantum ESPRESSO valence-only computations because the differences between the various schemes become so more evident (see Table 2.3). The computational conditions are as before; in particular, with CRYSTAL, two GTO sets of comparable quality: 8-41G*** and 6-31G**, have been used for silicon and diamond, respectively; in the PW calculations, norm-conserving pseudopotentials from the Quantum ESPRESSO library[1], a K4′ grid and cutoffs of 30 and 70 Ry for Silicon and Diamond, respectively, were adopted. The agreement with the experimental determinations is generally excellent, the hybrid B3LYP functional performing particularly well. The HF data appear to exaggerate the value of $\rho(\delta)$ for both systems.

[1]For Si: `Si.pbe-rrkj.UPF`, `Si.pz-vbc.UPF`, `Si.blyp-rrkj.UPF`; for C: `C.pbe-mt.UPF`, `C.pz-vbc.UPF`, `C.blyp-mt.UPF`.

Table 2.3 Valence electron density data (a.u.) for silicon and diamond at the experimental geometry: δ is the bond midpoint, γ the location of the maximum along the bond; $\Delta\rho_{val} = \rho_{val}(\gamma) - \rho_{val}(\delta)$ the depth of the minimum along the bond (See text for the computational conditions and other details)

		Silicon			Diamond		
Hamiltonian	BS	$\rho_{val}(\delta)$	$\rho_{val}(\gamma)$	$\Delta\rho_{val}$	$\rho_{val}(\delta)$	$\rho_{val}(\gamma)$	$\Delta\rho_{val}$
HF	GTO	0.093	0.094	0.001	0.251	0.285	0.034
	PW	0.095	0.095	0.001	0.261	0.300	0.039
B3LYP	GTO	0.087	0.088	0.001	0.240	0.285	0.045
	PW	0.089	0.091	0.002	0.247	0.307	0.060
PBE	GTO	0.084	0.085	0.001	0.235	0.281	0.044
	PW	0.085	0.086	0.001	0.240	0.300	0.061
LDA	GTO	0.083	0.084	0.001	0.233	0.280	0.047
	PW	0.084	0.086	0.002	0.242	0.295	0.053
Experiment [21]		0.086	0.086	0.000	0.240	0.287	0.048

2.4.2 *Environmental Effects on the DFs of a Molecular Crystal: Urea*

In a molecular crystal the constituent molecules are clearly identifiable even if in a geometry slightly different from the one they have in the gas phase. In the case of urea, the one treated in this Subsection, there are two symmetry-equivalent molecules per unit cell whose bond lengths and angles are modified to a small extent with respect to the free molecule (in particular, they take a planar configuration). On the whole, however, the weak interactions which set in between the molecules when the crystal is formed do not alter in any essential way their electronic structure. In order to make evident the role of intermolecular forces on DFs, it is then customary to consider so-called *interaction* DFs, $\Delta\rho(\mathbf{r})$ and $\Delta\pi(\mathbf{p})$. Reference is made for this purpose to a *procrystal* formed by $\bar{N}$ molecules in the unit cell (two in our case) and from all their translationally equivalent copies, in the same geometry and position they have in the crystal, but mutually independent. For each of those pseudo-molecules ($M = 1, \bar{N}$), the appropriate DF $[\rho^M(\mathbf{r}), \pi^M(\mathbf{p})]$, is computed using the same technique as for the crystal $[\rho^{cry}(\mathbf{r}), \pi^{cry}(\mathbf{p})]$. We can so define:

$$\Delta\rho(\mathbf{r}) = \rho^{cry}(\mathbf{r}) - \sum_{M,\mathbf{T}} \rho^M(\mathbf{r} - \mathbf{T}) \quad ; \quad \Delta\pi(\mathbf{p}) = \pi^{cry}(\mathbf{p}) - \sum_M \pi^M(\mathbf{p}). \quad (2.49)$$

In the second equation, we have exploited the independence of FTs from the origin, and used the normalization convention of Sect. 2.2.1.

The effect of the crystalline environment on the DFs of urea has been the object of intense experimental and theoretical work, because of the simple and at the same time intriguing structure of this system. In the crystal, the urea molecules are arranged top-to-tail to form two series of planar tapes, oppositely oriented and mutually orthogonal; hydrogen bonds are the main responsible for the links within the tape and between the tapes: the terminal oxygen of each molecule is thus forming four such bonds, an almost unique feature in molecular crystals.

Probably the most complete study to date of the ED of crystalline urea is the one by Birkedal et al. [67], who have reported synchrotron diffraction data of unprecedented precision, very accurately analyzed; the reconstructed experimental ED is there compared to that resulting from periodic HF calculations, by considering the respective characterization of the most important critical points owing to Bader's theory [8]. Gatti has recently reviewed the power of this theory using precisely urea as a test case [10] and extending the analysis performed in a pioneering ab-initio HF study [68]: in particular, he demonstrated the ability of topological analysis to describe quantitatively environmental effects on the ED, both as concerns the intermolecular and, indirectly, the intramolecular region (for instance, the appreciable change of the dipole moment of the molecule in the crystal with respect to its gas-phase value). The important discussion about the detectability of environmental effects from diffraction data by Spackman et al. [69] is also worth mentioning.

While in those studies only one type of periodic computation was considered (HF, with GTO BSs of rather good quality), we document here briefly the influence of Hamiltonian and BS on the description of $\rho(\mathbf{r})$ and $\Delta\rho(\mathbf{r})$, and extend the discussion to EMDs; for the latter case, which has been the object of a recent study by some of us [70], only results obtained with CRYSTAL are reported, because the analysis of EMDs is not yet feasible with the current version of Quantum ESPRESSO. In the following, all calculated data are referred to the experimental crystalline geometry [71] in order to make them comparable to each other and to the experimental data. Some of the GTO sets here used are taken from recent theoretical studies performed with CRYSTAL on urea and other molecular crystals (B Civalleri, M Ferrero, R Dovesi (private communication)) [72]; in order of increasing quality: 6-31 G(d,p) [here and in the following, the first set of polarization functions is assigned to first row atoms C,N,O, the second to H]; 6-311G(d,p); 6-311G(2df,2pd); TZPP; QZVPP. The last two GTO sets belong to a family devised by Ahlrichs and coworkers [73].

Figure 2.4 reports total and interaction ED maps obtained from HF CRYSTAL calculations with a very good (TZPP) GTO set. The picture is very similar to that provided by Spackman et al. [69]: with respect to the superposition of molecular densities the most notable feature is a build-up of charge in front of the terminal oxygen, due to the population of the hydrogen bonds with the neighboring molecule and to the corresponding de-population of the N–H intramolecular bonds.

Figure 2.5 describes the interaction ED along the line from the terminal O to the central C in the neighboring molecule, at a distance of 3.42 Å: this may not be the direction where environmental effects are the largest [10], but allows both inter- and intra-molecular changes of the ED to be recorded. The left panel shows the dependence on the Hamiltonian, as resulting from Quantum ESPRESSO calculations. In fact, in this case the procrystal was simulated by only two molecules, those containing the two selected atoms: therefore, $\Delta\rho(r)$ is not accurate beyond 2.4 Å from O, where the densities of other molecules become comparatively important. A cutoff E_c=50 Ry was used, practically corresponding to convergence:

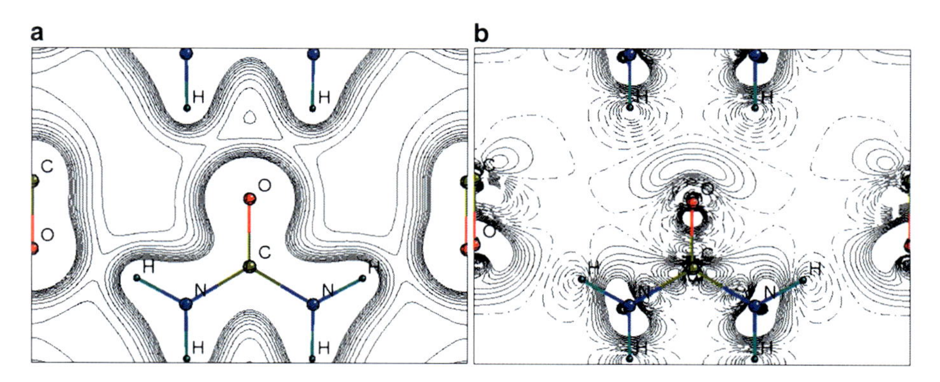

Fig. 2.4 Total (*left panel*) and interaction (*right panel*) charge density of crystalline urea in the (110) plane, resulting from HF calculations using a TZPP GTO set (see text). In the $\rho(\mathbf{r})$ map, the distance between consecutive lines is 0.01 a.u.; in the $\Delta\rho(\mathbf{r})$ map, it is 0.001 a.u.

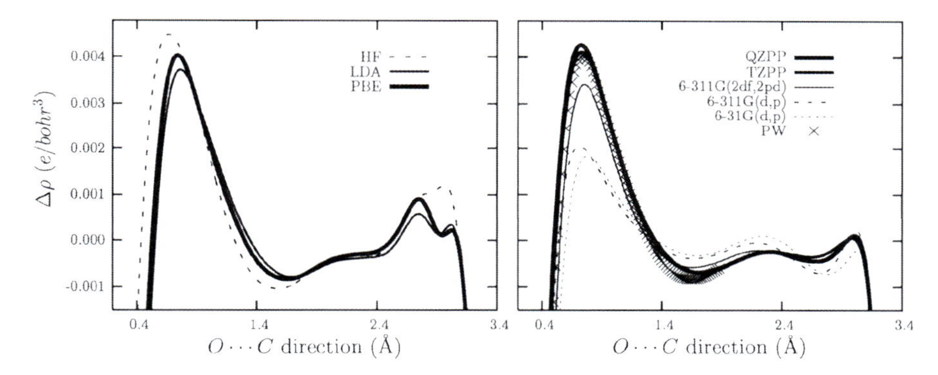

Fig. 2.5 Charge interaction density $[\Delta\rho(r)]$ along the $O \cdots C$ intermolecular direction (origin at O). The plots on the *left*, obtained with Quantum ESPRESSO, show the dependence on the Hamiltonian. Those on the *right*, the dependence on the BS for the PBE Hamiltonian: they were obtained with CRYSTAL except for the PW result, taken from the *left panel* (See text for details)

the HF and LDA solutions represent again two extremes as concerns the entity of the interaction density. The right panel, which refers to PBE calculations with different BSs, performed with CRYSTAL and using the complete procrystal, shows that convergence towards the PW solution in a vicinity of the O atom is achieved only with very sophisticated GTO sets. Interestingly, convergence of the calculated cohesive energy of the crystal to within 1 kJ/mol could be reached even when considering GTO sets of 6-311G(2df,2pd) quality (B Civalleri, M Ferrero, R Dovesi (private communication)) [72]: this shows again that DFs can be more sensitive to BS quality than energy related quantities.

The question has been raised by Spackman et al. [69] whether these environmental effects, unambiguously revealed by theoretical investigations, can be detected from experimental structure factors. In spite of the availability of new high quality

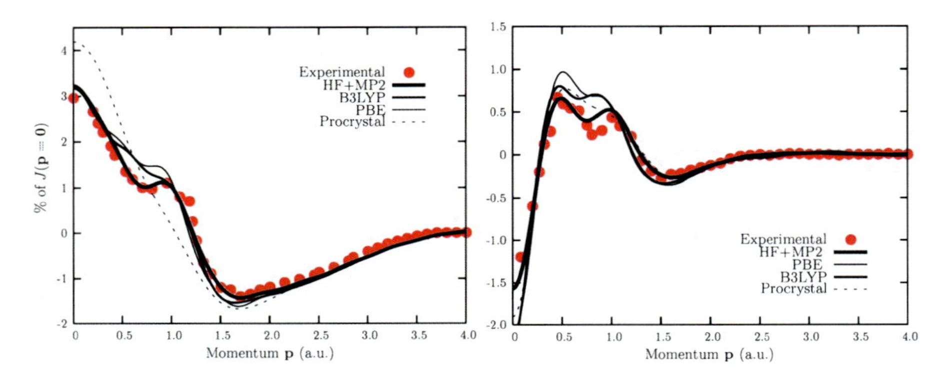

Fig. 2.6 CP anisotropies of urea: $[J_{001}(p) - J_{100}(p)]$ (*left panel*) and $[J_{110}(p) - J_{100}(p)]$ (*right panel*), obtained with different techniques (see inset), and given as percentage of the respective $J_{001}(0)$ value. All theoretical data were corrected for the experimental resolution

data obtained either via X-ray [74] or synchrotron diffraction [67] and of very sophisticated tools for their interpretation, the debate on this matter is still open.

Information from directional CPs is shown below to provide clearer evidence of the effect of the crystalline environment on the DM of urea; also the level adopted for its theoretical description appears to play here a more relevant role.

We have considered the CPs in the main three crystallographic directions: [001], along the tapes and perpendicular to the other two; [100], forming an angle of $\pi/4$ with the planes containing the tapes; [110], parallel to one set of tapes and perpendicular to the other. The experimental CP *anisotropies* $[J_{hkl}(p) - J_{h'k'l'}(p)]$, obtained from very accurate measurements using synchrotron radiation [75], can be compared directly with the calculated ones after correcting the latter for the limited experimental resolution (see Sect. 2.2.1.2). In examining the influence of the Hamiltonian on EMDs, we have also performed post-HF MP2 calculations, owing to the scheme (*b*) described in Sect. 2.5.3. We don't analyze here the important effects of BS quality, and report only the results obtained with the best BS feasible with all techniques, namely the 6-311G(d,p) BS previously introduced.

Two independent CP anisotropies, $[J_{001}(p) - J_{100}(p)]$ and $[J_{110}(p) - J_{100}(p)]$, are shown in Fig. 2.6. The first one is comparatively large, since it describes in a way the difference between the EMD in the direction of the tapes and perpendicular to them; the second one, between the two perpendicular directions, is much smaller, as expected, and still reveals the extreme sensitivity of CPs to directional effects. Comparison of the CP anisotropies for the crystal and the procrystal, the latter obtained following the scheme of Eq. 2.49, shows that environmental effects are clearly visible, both from the experimental and the theoretical viewpoint. Inspection of the results obtained with the different Hamiltonians, indicates that the HF+MP2 approach gives definitely better results than DFT, especially at intermediate moments. This is not unexpected (see Sect. 2.2.2): the MP2 method, taking advantage of the exact description of the electronic Fermi (exchange) correlation already provided

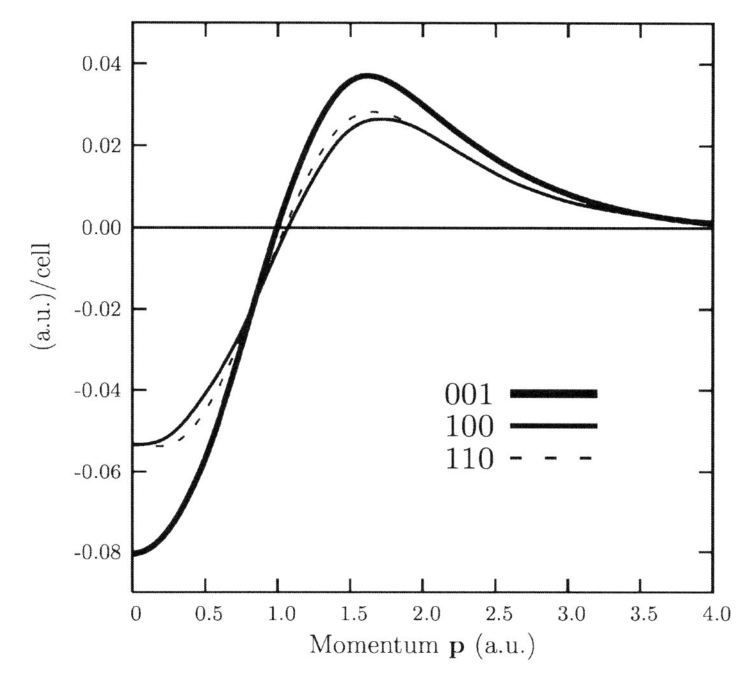

Fig. 2.7 Correlation contribution $J_{hkl}^{\mathrm{MP2}}(p)$ to the three CPs along the main crystallographic directions [001], [100], and [110]

by the reference HF method, can also recover a significant portion of the dynamic Coulomb correlation of the electronic motions. On the contrary, DFT describes both Fermi and Coulomb correlation as an average on the ground-state charge density and therefore cannot perform particularly well in predicting EMDs [76, 77].

Figure 2.7 shows $J_{hkl}^{\mathrm{MP2}}(p)$, representing the correlation correction to the HF CP, $J_{hkl}^{\mathrm{HF}}(p)$, in the three directions. This correction is small (well within 1% of $J_{001}(0)$, whose value is ≈24.70 a.u.) and would be hardly visible in the scale of Fig. 2.6. Two features can be noted, however: (i), the MP2 correction is negative at low momenta, positive at high momenta, corresponding to the higher average kinetic energy of correlated electrons and, (ii), this correction reduces the HF CP anisotropies.

More direct evidence of correlation effects is provided by the directional RFFs (see Sects. 2.2.1.2 and 2.2.2.3). Consider the data of Fig. 2.8. The RFFs along the [001] direction calculated with different Hamiltonians are there reported at an intermediate range as a function of the ratio r/R_L, R_L being the length of the first non-zero lattice vector in that direction. Equation 2.38 tells us that all RFFs obtained from single-determinant wavefunctions must be zero at $r/R_L=1$, independently of the BS used. This is in fact what is observed in the two examples reported (HF, PBE), and the same has been verified to happen in all cases and for all directions, provided that the calculations are numerically accurate. It is noteworthy that the

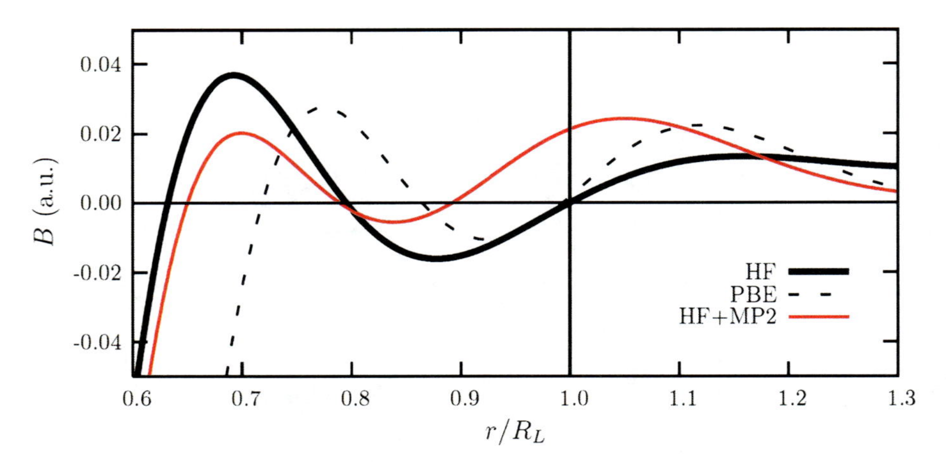

Fig. 2.8 Directional RFF $B_{001}(r/R_L)$ calculated with different techniques, as indicated. $R_L = 8.84$ a.u. is the length of the first non-zero lattice vector along the [001] direction

MP2 correction, though minute, results in a significant departure from such zero condition. This indication is susceptible of an easy experimental check, which is however not possible from the data provided in reference [75].

2.4.3 ED, EMD, CPF of Simple Metals: The Case of Aluminium

The accurate determination of the DM of metals involves a number of special problems, both fundamental and technical, which can be schematically stated as follows with reference to the prototypical metallic system, namely the *electron gas* [78].

1. The single-determinant description of this system is by necessity an antisymmetrized product of PWs: $\phi_n^X(\mathbf{r};\kappa) \propto \exp(\iota\kappa\cdot\mathbf{r})$ or $\phi_n^X(\mathbf{p};\kappa) \propto \delta(\mathbf{p}-\kappa)$. Since the associated eigenvalue $\varepsilon_n^X(\kappa)$ monotonously increases with $|\kappa|$ (in a way depending on X), the occupied ground-state manifold fills a sphere in reciprocal space centered in the origin and of radius $K_F = (3\rho/8\pi)^{1/3}$, where $\rho = N/V$ is the number of electrons per unit volume. As is clear from Eq. 2.36, the corresponding EMD has then the constant value $2/\rho$ for $|\mathbf{p}| < K_F$, zero otherwise. For real metallic systems, these features are obviously modified but the change in the number of occupied states at the two sides of the Fermi surface always results in a sharp discontinuity in the EMD at $\mathbf{p} = \kappa_F + \mathbf{G}$ where κ_F lies on the Fermi surface and $\mathbf{G}$ is a reciprocal lattice vector. This is an intrinsic deficiency of one-electron approximations. A many-body analysis of the interacting-electron-gas problem shows in fact that, due to electron correlation, the discontinuity of the

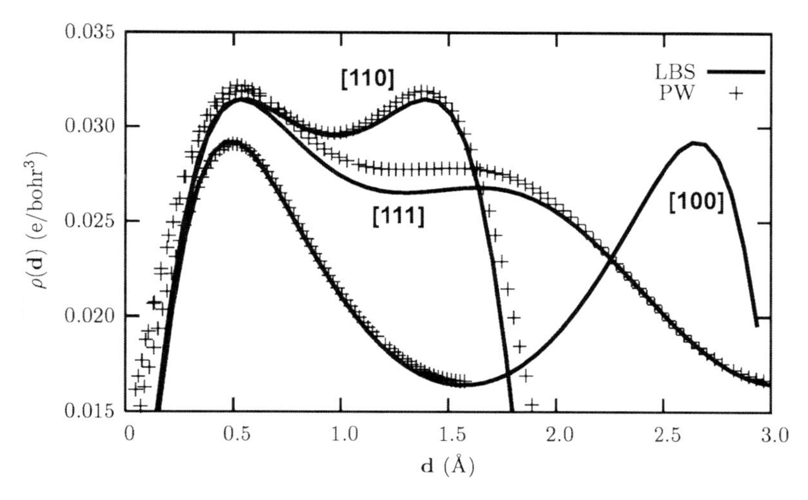

Fig. 2.9 Calculated valence ED of Al along the main directions using the PBE Hamiltonian. The Quantum-ESPRESSO results (*crosses*) were obtained using a PW cutoff of 25 Ry and a $20 \times 20 \times 20$ grid; for the CRYSTAL calculations (*continuous lines*) a 5-311G** BS (LBS) and a $16 \times 16 \times 16$ grid were adopted

EMD at K_F exists indeed, but is not as sharp: $\pi(\mathbf{p})$ has a non-zero tail also for $|\mathbf{p}| > K_F$, compensated for by a decrease of the EMD for $|\mathbf{p}|$ below K_F. Explicit expressions for correcting EMDs from one-electron approximations have been proposed by Lundqvist and Lydén [79] and by Lam and Platzman [80].

2. The HF description of the DM of the free electron gas coincides with that provided by any one-electron Hamiltonian. This is no longer true when real metals are considered, in particular as concerns the ED, for which the KS equation gives in principle the exact result. One may wonder how the two kinds of approximation perform when considering the EMD of metals. For this kind of systems HF is usually mistrusted because the κ dependence of the HF free-electron eigenvalues, $\varepsilon_n^{HF}(\kappa)$ has an unphysical logarithmic singularity at $|\kappa| = K_F$. While this fact has unpleasant consequences on band structures, it hardly influences DFs, as is shown below.
3. From a technical viewpoint, it is generally believed that GTF BSs (or any set of atom-centered local functions) cannot adequately describe conduction electrons, and that the use of PWs is nearly mandatory.

The case of Aluminium is here briefly considered in order to provide indications on the influence of these problems on calculated DFs of metals.

We first note that the third of the questions raised above is not fully justified. Figure 2.9 shows that a GTO BS of triple-zeta plus polarization quality quite accurately reproduces the valence ED obtained with a rich PW set and using norm

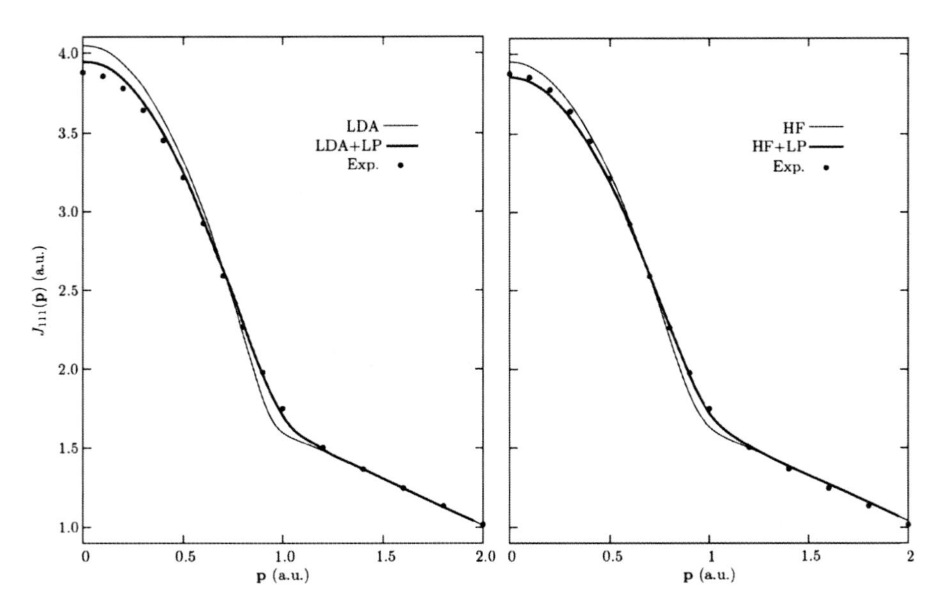

Fig. 2.10 CPs of Al in the [111] direction. LDA (*left panel*) and HF (*right panel*) all-electron results (*thin continuous lines*), were obtained with CRYSTAL using the same setup as in Fig. 2.9; the *thick continuous lines* show the Lam-Platzman-corrected data. Experimental data (*small full circles*) are from reference [81]. All calculated data were corrected for the experimental resolution

conserving PPs[2]. The results here reported are obtained with the PBE Hamiltonian. The effect of other choices for the exchange-correlation potential on the ED of Aluminium seems less relevant than in the case of the insulators considered in the previous sections. For instance, the value of the ED at the midpoint between two second-neighbor Al atoms is 0.0166 (0.0165); 0.0163 (0.0161); 0.0164 (0.0160) e Bohr^{-3} for PBE, LDA, HF, respectively, using the same computational conditions as in Fig. 2.9 (data in parentheses are those obtained with CRYSTAL).

While no accurate experimental determination of the Al ED seems to exist to compare with the present calculated results, the same is not true as concerns EMD data from directional CPs, for which very detailed experimental data and an accurate theoretical analysis exists [81, 82]. The case is more interesting here because, as stated at the beginning of this Subsection, the limitations of the one-electron description of metallic systems are more evident on EMD than on ED data.

As concerns the EMD, all the present one-electron calculations provide practically the same result, namely a quasi-free characterization of valence electrons, very much in line with that provided by Canney et al. a few years ago [83].

Figure 2.10 reports calculated and experimental CPs along the [111] direction; the following can be noted. The agreement with the experiment [81] of both

[2] `Al.pbe-rrkj.UPF` from the Quantum ESPRESSO web site.

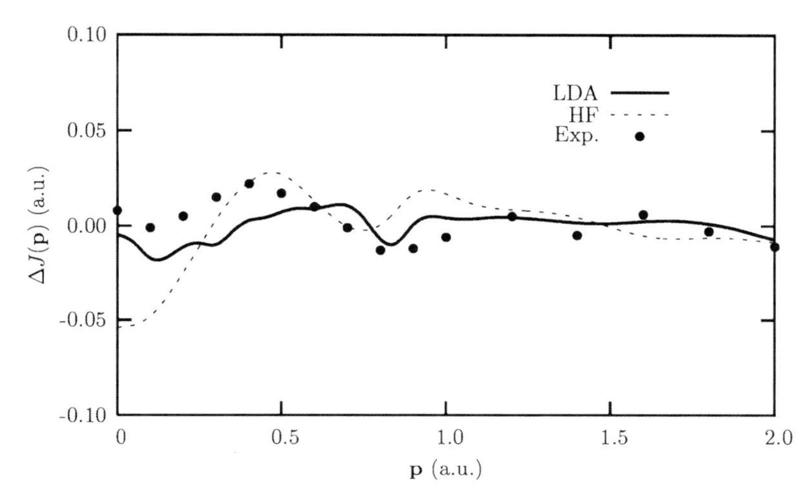

Fig. 2.11 CP anisotropy $[J_{110}(\mathbf{p}) - J_{111}(\mathbf{p})]$ for the Al crystal, at the HF (*dotted line*) and LDA (*continuous line*) level of theory. Experimental data (*small full circles*) are from reference [81]

LDA and HF results is significantly improved when the Lam-Platzman correlation correction [80] is included which brings in, as expected, an increase of momentum densities just above the Fermi momentum ($K_F \approx 0.92$ a.u.), and a decrease at low momenta. Apparently, HF performs slightly better than LDA.

CP anisotropies may provide additional information on the performance of various theoretical schemes, because many systematic errors which may be present in the individual experimental CPs are cancelled when performing the difference. Figure 2.11 reveals non negligible differences between the two calculated anisotropies: they are however quite small, of the same order as the experimental error (± 0.02 a.u.), so nothing definite can be inferred in the present case.

2.4.4 ESD: The Effect of the Hamiltonian on Spin Localization

An important sector of DM studies aims at the determination of accurate ESD data of spin-polarized crystals from diffraction experiments using polarized neutrons [84, 85] (see also Chaps. 4 and 8). The treatment of non-collinear magnetism in solids is very complex, and only recently some solutions have been proposed within KS-LDA approaches [86]. In many instances, however, when the spins are essentially aligned along one preferential direction, it may be expected that a single-determinant spin-unrestricted approach can provide a satisfactory description of ESD. These cases are ideally suited to ascertain the effect of the expression adopted for the corrective potential $\widetilde{V}^{X,\sigma}$ on the calculated ESD.

We consider here the case of a cubic perovskite ($KMnF_3$). The Mn^{+5} ions, formally in a d^5 configuration, are at the center of an octahedron of F^- ions, while

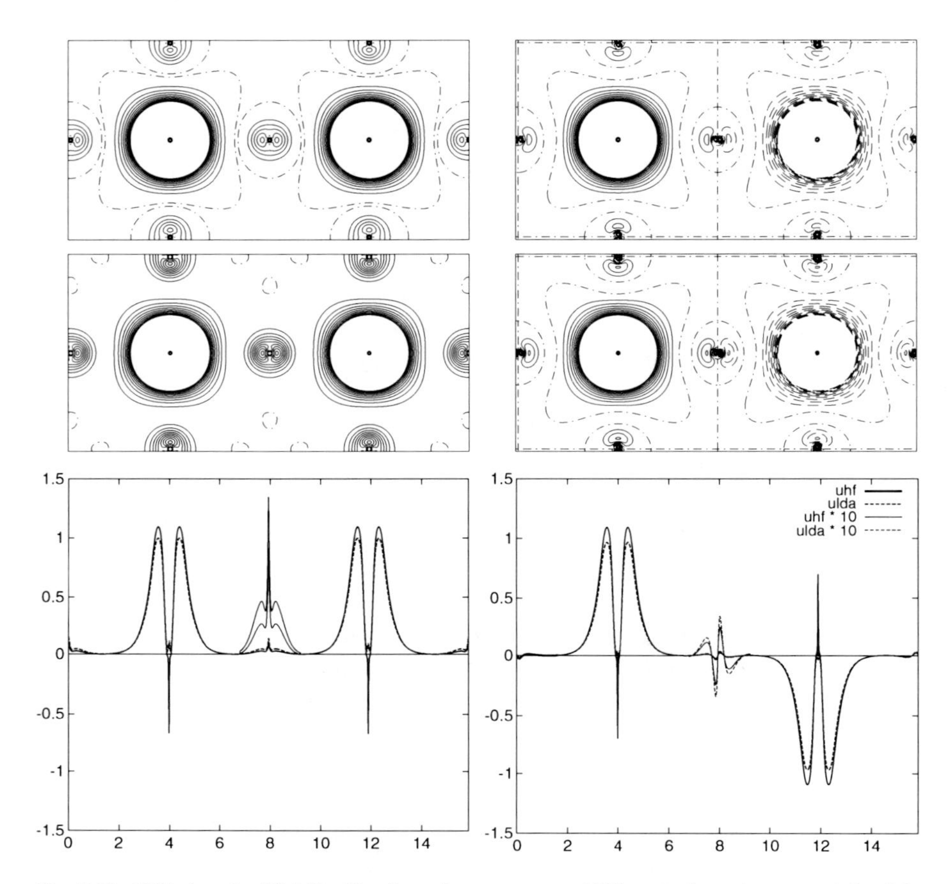

Fig. 2.12 ESD data for $KMnF_3$. The four plots on *top* are ESD maps in a plane containing Mn ions and four of their six nearest F ions, corresponding to the FM (*left*) and AFM (*right*) spin configurations, and are obtained using unrestricted HF (*above*) or LDA (*below*). *Continuous*, *dashed* and *dot-dash* isolines refer to positive, negative and zero ESD values, respectively; they are separated by 0.005 au; those whose value is larger than 0.05 a.u. (near the core of Mn) are not drawn. The two *bottom* plots report the FM- and AFM-ESD values along the line F-Mn-F-Mn-F (the distance from the F at the left is in Bohr). HF and LDA results are drawn as a *continuous* and a *dotted line*, respectively. The *thin lines* near the central F ion are a blow-up of the same data in that region by a factor 10

K^+ ions occupy the center of the vacancies between F^- octahedra. $KMnF_3$ can be found in a ferromagnetic (FM) configuration (all spins aligned) or an antiferromagnetic (AFM) one, where spins are opposite in sign on first neighbor Mn ions. The AFM configuration is more stable, due to a super-exchange effect mediated by the F ions midway between neighboring Mn ions. The magnetic coupling constant J (calculated from the difference in stability between the two configurations with reference to the Ising Hamiltonian) is experimentally known to be -3.65 K [87].

Table 2.4 ESD data for $KMnF_3$. For a variety of Hamiltonians, the calculated height of the FM- and AFM-ESD maxima in the valence region of Mn and F are reported, along with the calculated value of the spin coupling constant J (in Kelvin) whose experimental value is −3.65 K

	Maxima of ESD/(au)				
Hamiltonian	Mn-FM	Mn-AFM	F-FM	F-AFM	J/K
u-HF	1.095	1.094	0.027	0.025	−1.251
u-PBEO	1.039	1.033	0.040	0.034	−5.659
u-B3LYP	1.036	1.028	0.048	0.039	−7.879
u-PBE	1.005	0.980	0.047	0.037	−14.103
u-SVWN	0.998	0.967	0.047	0.035	−16.384

The unrestricted (u-) results reported below were obtained with the CRYSTAL program, by trying a variety of $\widetilde{V}^{X,\sigma}$ expressions (see Table 2.2), and using an all-electron BS of double-zeta quality (triple-zeta for fluorine). It is known that, while HF excessively favors spin localization (the exact-exchange operator $\widehat{V}_{exch}$ in Eq. 2.27 very effectively screens from each other electrons with the same spin), the opposite is true with expressions of the $\widetilde{V}^{KS}$ potential currently used; hybrid-exchange schemes are expected to work optimally in this respect.

Figure 2.12 shows ESD data obtained for the FM and AFM configurations using either the HF or an LDA (SVWN) Hamiltonian. Table 2.4 reports ESD and J data for a variety of one-electron Hamiltonians.

The general picture is similar in all cases. Apart from a fine structure near the nuclei, a very high ESD peak is observed at about 0.4 Bohr from the Mn nucleus, and a much smaller one at about 0.3 Bohr from the F nucleus; broadly speaking, the AFM picture can be obtained by reporting, in alternating cells, the opposite of the FM ESD. On closer inspection, it is found that the height of the Mn peak decreases and that of the F peak increases, when passing from HF, to the two hybrid schemes (PBE0 and B3LYP with a fraction of 25% and 20% exact exchange, respectively), to GGA (PBE), to LDA. This reflects the decreasing capability of localizing spin densities along this sequence. Higher values of the spin density on F entail larger super-exchange effects, which explains why the spin coupling constant is over-estimated with LDA, and under-estimated with HF. From a comparison with the experimental value of J, it can be argued that the ESD calculated with PBE0 is possibly the one closest to reality.

The alternative DFT+U scheme for favoring spin-localization in the frame of KS theory has been described in Sect. 2.3.3.2. The parallel application of the two techniques to the case of magnetic defects associated with O vacancies in metal-supported NiO and MgO monolayers has recently permitted their relative merits and problems to be analyzed [88].

2.5 Ongoing Developments

The subjects dealt with below are by no means exhaustive of ongoing theoretical research in the simulation of DMs and DFs in crystals. Rather, they serve the purpose of providing an indication of promising developments intended to overcome the limitations of existing software.

2.5.1 *The Harmonic Treatment of Thermal Effects*

It has been argued in Sect. 2.3.4.1 that a rather accurate description of the nuclear motions can often be obtained in the frame of the Born-Oppenheimer separation, by adopting the harmonic approximation. Vibrational modes in a crystal (phonons) can be described in terms of displacements of atoms in the unit cell and by a Bloch vector $\mathbf{q}$ in the Brillouin Zone. These modes can be conveniently calculated at any $\mathbf{q}$ using Density-Functional Perturbation Theory, as implemented in Quantum ESPRESSO [46].

In the following we instead suppose that the problem has been solved by using an F-fold supercell containing $F \times M$ nuclei. This amounts to defining $S = 3M \times F - 3$ normal modes each characterized by a normal coordinate Q_I and a characteristic angular frequency ω_I. The general eigenfunction for the nuclei can then be obtained by assigning an excitation level $n_I = 0,1,\ldots$ to each mode, and by multiplying by each other the S corresponding normalized eigenfunctions $H_{n_I}(Q_I)$ of the one-dimensional harmonic oscillator:

$$\Phi_{\mathbf{n}}^{\mathrm{Nuc}}(\{\mathbf{R}(\{\mathbf{Q}\})\}) = \prod_{I=1}^{S} H_{n_I}(Q_I); \tag{2.50}$$

the associated energy is simply given (in au) by $E_{\mathbf{n}} = \sum_I \omega_I(n_I + 1/2)$. Here $\{\mathbf{R}(\{\mathbf{Q}\})\}$ gives the position of the $F \times M$ nuclei in the supercell in terms of the set $\{\mathbf{Q}\}$ of the normal coordinates.

For a given temperature T, a Boltzmann weight $w_B(\mathbf{n}, T) = \exp(-E_{\mathbf{n}}/kT)/Z$ can be assigned to $\Phi_{\mathbf{n}}^{\mathrm{Nuc}}$. Owing to a statistical ensemble approach we can then define a joint-probability density for the set of the nuclei at the temperature T:

$$\rho^{\mathrm{Nuc}}(\{\mathbf{R}\}; T) = \sum_{\mathbf{n}} w_B(\mathbf{n}, T) \left| \Phi_{\mathbf{n}}^{\mathrm{Nuc}}(\{\mathbf{R}\}) \right|^2. \tag{2.51}$$

This function contains information on the *correlated* motion of all nuclei within the supercell. It can incidentally be noted that only the softest modes contribute significantly to the probability density at $\{\mathbf{R}\}$ coordinates not too close to $\{\mathbf{R}^{\mathrm{eq}}\}$: for them, the experimental information is usually insufficient to verify the adequacy of the theoretical harmonic description.

If we are interested in the probability distribution for a given nucleus (the Y-th, say), we have to integrate over the coordinates of all other nuclei:

$$\rho^{\mathrm{Y}}(\mathbf{R};T) = \int \rho^{\mathrm{Nuc}}(\{\mathbf{R}\};T)_{\mathbf{R}_Y=\mathbf{R}}\mathrm{d}\{\mathbf{R}\}_{A\neq Y}. \tag{2.52}$$

$\rho^{\mathrm{Y}}(\mathbf{R};\ T)$ has the full periodicity of the lattice. It can be used in principle for calculating iso-density surfaces enclosing a certain fraction (for instance 75%) of the occurrence probability of nucleus Y, to be compared to the anisotropic displacement parameters derived from X-ray and neutron diffraction experiments (especially important for hydrogens). In fact, simpler techniques are usually adopted for this purpose, which combine semi-empirical information from diffraction data with that resulting from normal mode analysis and take only approximately into account the correlation of nuclear motions [89] (*cf.* Chap. 3).

The joint nuclear probability distribution of Eq. 2.51 can be used in principle to obtain thermally averaged DFs, $\langle\langle\rho(\mathbf{y})\rangle\rangle_T$, where $\mathbf{y}$ can be $\mathbf{x}$, $\mathbf{r}$ or $\mathbf{p}$. Consider a sample of nuclear configurations in the supercell, $\{\mathbf{R}\}_j^T$, ($j = 1,\ldots,J$), which has been generated so as to reflect the distribution probability $\rho^{\mathrm{Nuc}}(\{\mathbf{R}\};\ T)$ (this may be accomplished, for instance, using a Metropolis algorithm [90, 91]). For each of them, the DF $\rho_j(\mathbf{y}) \equiv \rho\left(\mathbf{y};\{\mathbf{R}\}_j^T\right)$ can be evaluated as seen in Sect. 2.2. If it is assumed that the electron distribution follows instantaneously the nuclear motion, we can write, for large J:

$$\langle\langle\rho(\mathbf{y})\rangle\rangle_T; \approx \frac{1}{J}\sum_j \rho_j(\mathbf{y}). \tag{2.53}$$

While this equation seems at present exceedingly costly for practical applications, its use might be considered for benchmark studies with simple systems.

2.5.2 *The Effect of External Fields*

Among experimentalists, the effect of external fields on the properties of condensed matter, for instance, on their DFs is a subject of increasing interest [92]. We consider here the case where a constant electric field $\mathbf{E}$ acts on a finite (but macroscopic) crystalline sample; the field direction, $\mathbf{E}/|\mathbf{E}|$, perpendicular to the crystalline planes of Miller indices (h, k, ℓ), will be indicated conventionally as z. In the molecular case, the addition of an external potential $V^{\mathrm{ext}} = \mathbf{E}\cdot\mathbf{r} + c$ to the Hamiltonian of Eq. 2.2 introduces no essential complications for the solution of the corresponding problem (Eq. 2.1). The situation is different with (pseudo-)periodic systems, because the super-imposed potential destroys translational periodicity along z.

Two ways around this difficulty are implemented in CRYSTAL and in Quantum ESPRESSO. Both techniques are primarily intended to obtain the derivatives of energy with respect to the field components, which are essential for calculating the dielectric constant, the polarizability and hyperpolarizabilities of the crystal. These quantities are not only important on their own, but also because they are a necessary intermediate for evaluating the intensity of the vibrational excitations. A side benefit, however, which concerns us in the present context, is that information can be obtained on the change induced by the field on DM.

The first technique consists in superimposing to the crystal a "sawtooth" potential with periodicity D along z: $V^{\text{ext}}(z) = E\zeta$, where $\zeta = (-1)^n(z - nD/2)$ and $n = \text{int}[2z/D - 1/2]$. If $[\mathbf{a}'_1, \mathbf{a}'_2, \mathbf{a}'_3]$ is a "plane adapted" basis, with $\mathbf{a}'_1, \mathbf{a}'_2$ in the (h,k,ℓ) plane, and if D is a multiple of the interlayer distance d between those planes ($D = md$), then the crystal with unit supercell $[\mathbf{a}'_1, \mathbf{a}'_2, m\mathbf{a}'_3]$ preserves translational periodicity even in the presence of the added field, and its solution can be achieved according to the standard procedure [93]. In spite of the rather artificial device adopted (the derivative of the added potential is discontinuous at $z = nD/2 + D/4$), this technique performs surprisingly well [94] and, if m is large enough, in a region between the discontinuity planes the solution appears "quasi-periodic" with respect to translations $\mathbf{a}'_3$: this permits the changes in the DM to be analyzed by considering a cell in that region and its neighboring ones.

The second technique aims at calculating the zero-field derivatives of the energy ($\mathcal{E}_{abc\ldots}$) with respect to the field components. For this purpose, an extension to periodic systems of coupled-perturbed HF and DFT has been implemented in CRYSTAL [95, 96]. It provides first, second and third energy derivatives for systems periodic in 0, 1, 2, 3 dimensions. A by-product of these calculations are the derivatives of the DM elements in the GTO BS with respect to the field components: $\Im_{\mu\nu|abc\ldots}$. This allows in principle the estimate of the effect of a finite (but small) field in the general (z) direction on the DM, through a truncated Taylor expansion about its unperturbed value:

$$P_{\mu\nu}(E) = P_{\mu\nu}(0) + \Im_{\mu\nu|z}E + \frac{1}{2}\Im_{\mu\nu|zz}E^2 + \cdots. \qquad (2.54)$$

The results with the two techniques are in excellent agreement but the second one is easier to use, more accurate and cheaper.

The recent study of the inverse piezo-electric effect in α-quartz performed with the Wien2k code [3] provides a nice example of the use of the sawtooth technique [97]: both the atomic displacements and the change of the structure factors as a function of the field strength were calculated, in fair agreement with the experiment.

We finally remark that Quantum ESPRESSO also implements, in the framework of DFT, a third technique: finite macroscopic electric fields, treated via the modern theory of polarizability and the Berry phase concept [98, 99].

2.5.3 Post-HF Description of DMs: The Perturbative Approach

In the field of molecular studies, the use of post-HF techniques allows extremely accurate results to be collected, which is not possible using DFT-based schemes. While the main objective is usually the evaluation of energy eigenvalues, significant information can also be obtained as concerns wave-function related properties, in particular DMs and DFs. Among the many examples, we can cite the study on the ED of bullvalene and concerning the different types of C–C bonds in this molecule [100]: it is shown that both the use of a very good BS *and* the inclusion of correlation effects (at an MP2 level) are important to bring the theoretical results to close agreement with the experimental data. The advantage over DFT approaches should be even more evident when considering the EMD and related DFs, for the reasons expressed in Sect. 2.2.1 and 2.4.2. The cost of post-HF techniques in their standard formulation, which refers to the "canonical" HF COs delocalized over the whole system, ϕ_n^{HF}, scales however very rapidly with the number N of electrons, which prevents their use with "large" systems.

The *local-correlation* techniques proposed long ago by Pulay and others [101] provide a way out of this difficulty: they use as a reference a representation of the occupied HF manifold in terms of localized orthonormal functions like the WFs in Eq. 2.33, and exploit the short-range character of the inter-electronic correlation. Based on these ideas, N-scaling formulations of some of the most popular correlation techniques have been implemented, e.g., in the MOLPRO code [102]: Møller-Plesset perturbation techniques in second (MP2) and higher orders, coupled-cluster algorithms including triple corrections, etc.; very accurate correlated calculations for large molecules can thus be performed in reasonable times.

Quite recently, the CRYSCOR code has been developed [103] which implements an N_0-scaling local-correlation approach for crystals in an AO BS, though limited, for the time being, to an LMP2 level of approximation ("L" standing for "local") and to the case of spinless insulating systems. CRYSCOR uses as a reference the HF solution provided by CRYSTAL in the form of Eq. 2.34: $[\Psi_0^{\mathrm{HF}} = \| \ldots [\ell \mathbf{T_m}]^{\alpha} [\ell \mathbf{T_m}]^{\beta} \ \ldots \|]$; in addition, CRYSTAL provides all information concerning structure, symmetry and Fock matrix of the crystal. An essential feature of CRYSCOR which permits the LMP2 solution for relatively complicated periodic systems to be obtained in reasonable times, is the very efficient treatment of two-electron repulsion integrals, based on a combination of density-fitting and of multipolar techniques at short range and long range, respectively [104].

Energy related quantities are obviously the primary object of interest for such a code, since its use permits the importance of correlation effects to be estimated as concerns equilibrium geometries, elastic constants, vibrational spectra, cohesive energy, etc. However, as is clear from the preceding, DM related quantities in crystals represent a very rich ground for comparison between experiment and theory: it is therefore of interest to extract from a post-HF code also this kind of information. Two computational schemes have been implemented in CRYSCOR for this purpose, which provide an estimate $P^{(x)}(r, r')$ of the correlated DM: they can be related to the "expectation value" and the "response to an external perturbation" method, respectively, as are adopted in molecular calculations (see for example [105] and references therein).

The former technique ($x = a$) [106] defines a *locally-correlated wavefunction*, obtained by adding to the HF solution only those bi-excitations from two WFs to the unoccupied manifold where one of the two WFs is in the zero cell, but using for them the amplitudes from the periodic calculation. It can be viewed in a sense as the result of an "embedding" calculation, where electrons are allowed to correlate their motions in the zero cell and its neighborhood but are imposed to stay in their HF state far from it. There is more than that, however: for instance, the amplitudes include the effect of dispersive interactions up to infinite distance. A size-consistent periodic expression $P^{(a)}\ (r,\ r')$ can be obtained from there, which can be used for calculating the quantities of interest (ED, EMD, CPF, structure factors, etc.) by simply feeding this corrected DM instead of the HF one to the corresponding subroutines of CRYSTAL.

The second scheme ($x = b$) [107] is based on the calculation of the DM as the derivative of the LMP2 Lagrangian with respect to an external perturbation. In order to obtain a closed and simple expression, the response of the excitation amplitudes themselves to the external perturbation is not taken into account in this first implementation. An example of application of this approach has been provided in Sect. 2.4.2. Its generalization for including self-consistently orbital relaxation is the object of future work.

2.6 Final Considerations

Present-day ab-initio computer codes for periodic systems have been shown to provide valuable information on the DM and related functions. This may help experimentalists for the analysis and interpretation of their data concerning structure factors and CPs of crystals and their dependence on external conditions. Vice-versa, the availability of high quality experimental data concerning DMs is extremely important to assess the limitations of these powerful tools.

Two issues have been here devoted special attention to in this respect.

The Hamiltonian issue. Almost all present-day calculations of the electronic structure of crystalline systems are performed using one-electron approximations (DFT, HF or hybrid). Based on practical examples, it has been argued that very accurate DM predictions can be so obtained, probably hybrid Hamiltonians providing the best results on average. Definite discrepancies from the experiment, requiring more advanced theoretical tools, seem more likely to concern EMDs, both due to recent progress in the experimental determination of directional CPs, and to the fact that the instantaneous correlation of electronic motions affects more directly their momentum that their spatial distribution.

The basis set issue. It has been shown that the quality of the representative BS is extremely important not only as concerns energy, but also densities. PWs provide a reference in this respect, especially as concerns valence electrons. Calculations based on local functions (GTFs) can approach the PW results, but only using extensive sets allowing for large flexibility and including polarization functions.

We have finally tried to get a glimpse of the exciting new developments which are expected in a near future in this area of research: they will contribute to make the experiment-theory interaction even more fruitful in years to come.

Acknowledgements The Authors are grateful to Lorenzo Paulatto, Michele Catti, Piero Ugliengo, Marta Corno, Bartolomeo Civalleri, Mauro Ferrero for useful suggestions and for their contribution to performing some of the calculations whose results are here reported. They are also indebted to the Referees and to the Editors of this Book for their most helpful remarks and comments. Special and heartfelt thanks go to our young collaborator Alessio Meyer who unexpectedly died few months after the completion of the first draft of this manuscript, in December 2009.

References

1. Giannozzi P, Baroni S, Bonini N, Calandra M, Car R, Cavazzoni C, Ceresoli D, Chiarotti GL, Cococcioni M, Dabo I, Dal Corso A, Fabris S, Fratesi G, de Gironcoli S, Gebauer R, Gerstmann U, Gougoussis C, Kokalj A, Lazzeri M, Martin-Samos L, Marzari N, Mauri F, Mazzarello R, Paolini S, Pasquarello A, Paulatto L, Sbraccia S, Scandolo C, Sclauzero G, Seitsonen AP, Smogunov A, Umari P, Wentzcovitch RM (2009) QUANTUM ESPRESSO: a modular and open-source software project for quantum simulations of materials. J Phys Condens Matter 21:395502(19 pp). http://www.quantum-espresso.org/
2. Dovesi R, Orlando R, Civalleri B, Roetti C, Saunders VR, Zicovich-Wilson CM (2005) CRYSTAL: a computational tool for the *ab initio* study of the electronic properties of crystals. Z Kristallogr 220:571–573; Dovesi R, Saunders VR, Roetti C, Orlando R, Zicovich-Wilson CM, Pascale F, Civalleri B, Doll K, Harrison NM, Bush IJ, D'Arco Ph, Llunell M (2009) CRYSTAL09 user's manual. Università di Torino, Torino. http://www.CRYSTAL.unito.it
3. Gonze X, Beuken J-M, Caracas R, Detraux F, Fuchs M, Rignanese G-M, Sindic L. Verstraete M, Zerah G, Jollet F, Torrent M, Roy A, Mikami M, Ghosez Ph, Raty J-Y, Allan DC (2002) First-principle computation of material properties: the ABINIT software project. Comput Mater Sci 25:478. http://www.abinit.org; Kresse G, Furthmuller J (1996) Efficiency of ab-initio total energy calculations for metals and semiconductors using a plane-wave basis set. Comput Mater Sci 6:15. http://cmp.univie.ac.at/vasp; Artacho E, Anglada E, Diéguez O, Gale JD, García A, Junquera J, Martin RM, Ordejón P, Pruneda, JM, Sánchez-Portal D, Soler JM (2008) The SIESTA method; developments and applicability. J Phys Condens Matter 20:064208 (6 pp). http://www.icamb.es/siesta; Schwartz K (2003) DFT calculations of solids with LAPW and WIEN2k. J Solid State Chem 176:319–326. http://www.wien2k.at; Koepernik K, Eschrig H (1999) Full-potential nonorthogonal local-orbital minimum-basis band-structure scheme. Phys Rev B 59:1743–1757. http://www.fplo.de
4. McWeeny R, Sutcliffe BT (1969) Methods of molecular quantum mechanics. Academic, London
5. Kato T (1957) On the eigenfunctions of many particle systems in quantum mechanics. Commun Pure Appl Math 10:151–171
6. Hohenberg P, Kohn W (1964) Inhomogeneous electron gas. Phys Rev 136:B864–B871
7. Kohn W, Sham LJ (1965) Self-consistent equations including exchange and correlation effects. Phys Rev 140:A1133–A1138
8. Bader RFW (1990) Atoms in molecules – a quantum theory. Oxford University Press, Oxford
9. Gatti C (2005) Chemical bonding in crystals: new directions. Z Kristallogr 220:399–457

Appendix: Atomic Units, Glossary of Abbreviations

Table 2.5 Atomic units (au)

Quantity	Atomic unit	SI Equivalent	Notes
Mass	m_0	$9.1096\ 10^{-31}$ kg	The rest mass of the electron
Charge	e	$1.6022\ 10^{-18}$ C	The elementary charge
Angular momentum	$\hbar = h/(2\pi)$	$1.0546\ 10^{-34}$ J s	The reduced Planck constant: angular momentum operators have integer or semi-integer eigenvalues in au's
Length	$a_0 = 4\pi\varepsilon_0\hbar^2/(m_0e^2)$	$5.2918\ 10^{-11}$ m	The Bohr radius of H (also called "Bohr")
Permittivity	$4\pi\varepsilon_0$	$1.1126\ 10^{-10}$ $C^2m^{-1}\ J^{-1}$	The vacuum permittivity
Energy	$e^4 m_0/(4\pi\ \varepsilon_0\hbar)^2 = 1\ E_h$	$4.3598\ 10^{-18}$ J	The electrostatic repulsion energy between two electrons separated by 1 a_0 (also called "Hartree", abbreviated E_h: 1 E_h=2 Ry=2625.9 kJ/mol=27.21 eV)
Speed	$e^2/(4\pi\ \varepsilon_0\hbar) = c\alpha$	$2.1877\ 10^6$ m s^{-1}	The speed of the electron in the ground state of Bohr's H atom; c is the speed of light in vacuum, $\alpha = 137.036^{-1}$ the fine structure constant
Time	$\hbar/E_h$	$2.4189\ 10^{-17}$ s	The time taken to travel 1 a_0 at 1 au speed
E(S)D	$1/a_0^3$	$6.7482\ 10^{30}$ m^{-3}	N. of (unpaired) electrons per unit volume
EMD	$m_0\, c\alpha/a_0^3$	$1.3449\ 10^7$ kg m^{-2} s^{-1}	Electron momentum density

Table 2.6 Glossary of abbreviations

Acronym(s)	Meaning	Introduced in section:
AE	All-Electron	2.2.2.3
AFM	Antiferromagnetic	2.4.4
AO	Atomic Orbital	2.1
BF	Bloch Function	2.2.2.1
BS	Basis Set	2.1
BSSE	Basis Set Superposition Error	2.3.4.2
BvK	Born-von Kármán (cyclic conditions)	2.2
BZ	Brillouin Zone	2.2.2.1
CO,CSO	Crystalline (Spin-)Orbitals	2.2.2
CP,CPF	Compton Profile (Function)	2.2.1.2
DF	Density Function (charge, spin or momentum density)	2.1
DFT	Density Functional Theory	2.2.1.1
DM	Density Matrix	2.1
ED,ESD	Electron Density, Electron Spin Density	2.2.1
EMD	Electron Momentum Density	2.2.1
ESD	Electron (net) Spin Density along z	2.2.1
FM	Ferromagnetic	2.4.4
FT,FFT	Fourier Transform, Fast Fourier Transformation	2.2.1, 2.3.1
HF,UHF	Hartree-Fock, Unrestricted Hartree-Fock	2.1, 2.2.2
GTO	Gaussian Type Orbital	2.3.1, 2.3.4.2
KS,UKS	Kohn-Sham, Unrestricted Kohn-Sham	2.1, 2.2.2
LMP2	Local Møller Plesset perturbation theory at order 2	2.5.3
MO,MSO	Molecular (Spin-)Orbitals	2.2.2
PAW	Projected Augmented Waves	2.2.2.3, 2.3.3.1
PP	Pseudo-Potential	2.2.2.3
PW	Plane-Wave	2.1
RFF	Reciprocal Form Factor	2.2.1.2
USPP	Ultra-Soft Pseudo-Potential	2.2.2.3
WF	Wannier Function	2.2.2.1

10. Gatti C (2007) Solid state application of QTAIM and the source function – molecular crystals, host-guest systems and molecular complexes. In: Matta CF, Boyd RB (eds) The quantum theory of atoms in molecules: from solid state to DNA and drug design. Wiley-VCH Verlag, Weinheim, pp 165–206
11. Koritsanszky TS, Coppens Ph (2001) Chemical applications of X-ray charge density analysis: a review. Chem Rev 101:1583–1628
12. Tinkham M (1964) Group theory and quantum mechanics. McGraw-Hill, New York
13. Hahn T (ed) (1992) International tables for crystallography, 3rd Revised edn, Vol A (Section 2.13). Kluwer, Dordrecht
14. Williams B (ed) (1977) Compton scattering: the investigation of electron momentum distributions. McGraw-Hill, New York
15. Benesch R, Singh SR, Smith VH (1971) On the relationship of the X-ray form factor to the 1-matrix in momentum space. Chem Phys Lett 10:151; Schülke W (1977) The one-dimensional fourier transform of Compton profiles. Phys Status Sol B 82:229; Pattison P, Weyrich W, Williams B (1977) Observation of ionic deformation and bonding from Compton profiles. Solid State Commun 21:967–970; Weyrich W, Pattison P, Williams B (1979) Fourier analysis of the Compton profile: atoms and molecules. Chem Phys 41:271–284

16. Bauer GEW (1983) General operator ground-state expectation values in the Hohenberg-Kohn-Sham density-functional formalism. Phys Rev B 27:5912–5918
17. Monkhorst HJ, Pack JD (1976) Special points for Brillouin-zone integrations. Phys Rev B 13:5188–5192
18. Wannier GH (1937) The structure of electronic excitation levels in insulating crystals. Phys Rev 52:191–197
19. Marzari N, Vanderbilt D (1997) Maximally localized generalized Wannier functions for composite energy bands. Phys Rev B 56:12847–12865
20. Zicovich-Wilson CM, Dovesi R, Saunders VR (2001) A general method to obtain well localized Wannier functions for composite energy bands in linear combination of atomic orbital periodic calculations. J Chem Phys 115:9708–9719; Casassa S, Zicovich-Wilson CM, Pisani C (2006) Symmetry-adapted localized Wannier functions suitable for periodic calculations. Theor Chem Acc 116:726–733
21. Lu ZW, Zunger A, Deutsch M (1993) Electronic charge distribution in crystalline diamond, silicon, and germanium. Phys Rev B 47:9385–9410
22. Su Z, Coppens P (1998) Nonlinear least-squares fitting of numerical relativistic atomic wave functions by a linear combination of slater-type functions for atoms with $Z = 1$–36. Acta Crystallogr A 54:646–652; Macchi P, Coppens P (2001) Relativistic analytical wave functions and scattering factors for neutral atoms beyond Kr and for all chemically important ions up to I^-. Acta Crystallogr A 57:656–662
23. Phillips JC, Kleinmann L (1959) New method for calculating wave functions in crystals and molecules. Phys Rev 116:287–294
24. Hamann DR, Schlüter M, Chiang C (1979) Norm-conserving pseudopotentials. Phys Rev Lett 43:1494–1497
25. Vanderbilt D (1990) Soft self-consistent pseudopotentials in a generalized eigenvalue formalism. Phys Rev B 41:7892–7895. http://www.physics.rutgers.edu/dhv/uspp
26. Blöchl PE (1994) Projector augmented-wave method. Phys Rev B 50:17953–17979; Kresse G, Joubert D (1999) From ultrasoft pseudopotentials to the projector augmented-wave method. Phys Rev B 59:1758–1775
27. Car R, Parrinello M (1985) Unified approach for molecular dynamics and density-functional theory. Phys Rev Lett 55:2471–2474
28. Pulay P (1969) Ab initio calculation of force constants and equilibrium geometries. I. Theory. Mol Phys 17:197–204
29. Perdew JP, Schmidt K (2001) Jacob's ladder of density functional approximations for the exchange-correlation energy. In: Van Doren VE, Van Alsenoy C, Geerlings P (eds) Density functional theory and its applications to materials. AIP conference proceedings, vol 577. Melville, New York, pp 1–20
30. Perdew JP, Wang Y (1992) Accurate and simple analytic representation of the electron-gas correlation energy. Phys Rev B 45:13244–13249
31. Staroverov VN, Scuseria GE, Tao J, Perdew JP (2004) Tests of a ladder of density functionals for bulk solids and surfaces. Phys Rev B 69:075102 (11 pp)
32. Becke AD (1993) Density-functional thermochemistry. III. The role of exact exchange. J Chem Phys 98:5648–5652
33. MacDonald AH, Vosko SH (1979) A relativistic density functional formalism. J Phys C 12:2977–2990
34. Zhao Y, Truhlar DG (2008) Construction of a generalized gradient approximation by restoring the density-gradient expansion and enforcing a tight Lieb-Oxford bound. J Chem Phys 128:184109
35. Slater JC (1951) A simplification of the Hartree-Fock method. Phys Rev 81:385–390
36. Vosko SH, Wilk L, Nusair M (1980) Accurate spin-dependent electron liquid correlation energies for local spin density calculations: a critical analysis. Can J Phys 58:1200–1211
37. Perdew JP, Zunger A (1981) Self-interaction correction to density-functional approximations for many-electron systems. Phys Rev B 23:5048–5079

38. Perdew JP, Burke K, Ernzerhof M (1996) Generalized gradient approximation made simple. Phys Rev Lett 77:3865–3868
39. Hamprecht FA, Cohen AJ, Tozer DJ, Handy NC (1998) Development and assessment of new exchange-correlation functionals. J Chem Phys 109:6264–6271
40. Perdew JP, Chevary JA, Vosko SH, Jackson KA, Pederson MR, Singh D, Fiolhais C (1992) Atoms, molecules, solids and surfaces: applications of the generalized gradient approximation for exchange and correlation. Phys Rev B 46:6671–6687
41. Perdew JP, Ruzsinszky A, Csonka GI, Vydrov OA, Scuseria GE, Constantin LA, Zhou X, Burke K (2008) Restoring the density-gradient expansion for exchange in solids and surfaces. Phys Rev Lett 100:136406 (4 pp)
42. Wu Z, Cohen R (2006) More accurate generalized gradient approximation for solids. Phys Rev B 73:235116 (6 pp)
43. Becke AD (1988) Density-functional exchange-energy approximation with correct asymptotic behavior. Phys Rev A 38:3098–3100
44. Lee C, Yang W, Parr RG (1988) Development of the Colle-Salvetti correlation-energy formula into a functional of the electron density. Phys Rev B 37:785–789
45. Demichelis R, Civalleri B, Ferrabone M, Dovesi R (2010) On the performance of eleven DFT functionals in the description of the vibrational properties of aluminosilicates. Int J Quantum Chem 110:406–415
46. Baroni S, de Gironcoli S, Dal Corso A, Giannozzi P (2001) Phonons and related crystal properties from density-functional perturbation theory. Rev Mod Phys 73:515–562
47. Van de Walle CG, Blöchl PE (1993) First-principles calculations of hyperfine parameters. Phys Rev B 47:4244–4255; Hetényi B, de Angelis F, Giannozzi P, Car R (2001) Reconstruction of frozen-core all-electron orbitals from pseudo-orbitals. J Chem Phys 115:5791 (5 pp)
48. Anisimov VI, Zaanen J, Andersen OK (1991) Band theory and Mott insulators: Hubbard *U* instead of Stoner *I*. J Phys Rev B 44:943–954
49. Cococcioni M, de Gironcoli S (2005) Linear response approach to the calculation of the effective interaction parameters in the LDA+U method. Phys Rev B 71:035105 (16 pp)
50. Pisani C, Dovesi R (1980) Exact-exchange Hartree-Fock calculations for periodic systems. I. Illustration of the method. Int J Quantum Chem 17:501–516; Pisani C, Dovesi R, Roetti C (1988) Hartree-Fock ab-initio treatment of crystalline systems. Lecture notes in chemistry, vol 48. Springer, Heidelberg
51. Clementi E, Mehl J (1971) IBMOL 5 program user's guide, Publication RJ889, IBM Corporation
52. Hehre DJ, Lathan WA, Newton MD, Ditchfield R, Pople A (1972) GAUSSIAN70 program number 236. QCPE, Indiana University, Bloomington
53. Dupuis M, Spangler D, Wendoloski J (1980) NRCC software catalog, vol 1, program no. QG01 (GAMESS)
54. Noel Y, D'Arco P, Demichelis R, Zicovich-Wilson CM, Dovesi R (2010) On the use of symmetry in the ab initio quantum mechanical simulation of nanotubes and related materials. J Comput Chem 31:855–862
55. Zicovich-Wilson CM, Torres FJ, Pascale F, Valenzano L, Orlando R, Dovesi R (2008) Ab initio simulation of the IR spectra of pyrope, grossular, and andradite. J Comput Chem 29:2268–2278
56. Gatti C (1999) TOPOND-98 user's manual. CNR-CSRSRC, Milano. http://www.istm.cnr.it/gatti
57. Werner H-J, Knowles P, Manby F (2003) Fast linear scaling second-order Møller-Plesset perturbation theory (MP2) using local and density fitting approximations. J Chem Phys 118:8149–8160
58. Dunning TH (2000) A road map for the calculation of molecular binding energies. J Phys Chem A104:9062–9080
59. Jensen F (1999) The basis Set convergence of the Hartree-Fock energy for H_2. J Chem Phys 110:6601–6605

60. Boys SF, Bernardi F (1970) The calculations of small molecular interaction by the difference of separate total energies–some procedures with reduced error. Mol Phys 19:553–566
61. Kestner NR, Combariza JE (1999) Basis set superposition errors: theory and practice. Rev Comput Chem 13:99–132
62. Dovesi R, Pisani C, Ricca F, Roetti C, Saunders VR (1984) Hartree-Fock study of crystalline lithium nitride. Phys Rev B 30:972–979; Causà M, Dovesi R, Pisani C, Roetti C (1985) Ab initio study of the autocorrelation function for lithium nitride. Phys Rev B 32:1196–1202
63. Causà M, Dovesi R, Pisani C, Roetti C (1986) Electron charge density and electron momentum distribution in magnesium oxide. Acta Crystallogr B 42:247–253; Causà M, Dovesi R, Pisani C, Roetti C (1986) Directional Compton profiles and autocorrelation function of magnesium oxide. Phys Rev B 34:2939–2941
64. Nishibori E, Sunaoshi E, Yoshida A, Aoyagi S, Kato K, Takata M, Sakata M (2007) Accurate structure factors and experimental charge densities from synchrotron X-ray powder diffraction data at SPring-8. Acta Crystallogr A 63:43–52
65. Orlando R, Dovesi R, Roetti C, Saunders VR (1990) Ab initio Hartree-Fock calculations for periodic compounds: application to semiconductors. J Phys C: Condens Matter 2:7769–7789
66. Pisani C, Dovesi R, Orlando R (1992) Near-Hartree-Fock wave functions for solids: the case of crystalline silicon. Int J Quantum Chem 42:5–33
67. Birkedal H, Madsen D, Mathiesen RH, Knudsen K, Weber H-P, Pattison P, Schwarzenbach D (2004) The charge density of urea from synchrotron diffraction data. Acta Crystallogr A 60:371–381
68. Gatti C, Saunders VR, Roetti C (1994) Crystal field effects on the topological properties of the electron density in molecular crystals: the case of urea. J Chem Phys 101:10686–10696
69. Spackman MA, Byrom PG, Alfredsson M, Hermansson K (1999) Influence of intermolecular interactions on multipole-refined electron densities. Acta Crystallogr A 55:30–47
70. Erba A, Pisani C, Casassa S, Maschio L, Schütz M, Usvyat D (2010) MP2 versus density-functional theory study of the Compton profiles of crystalline urea. Phys Rev B 81:165108
71. Swaminathan S, Craven BM, Spackman MA, Stewart RF (1984) Theoretical and experimental studies of the charge density in urea. Acta Crystallogr B 40:398–404
72. Civalleri B, Doll K, Zicovich-Wilson CM (2007) Ab initio investigation of structure and cohesive energy of crystalline urea. J Phys Chem B 111:26–33
73. Schäfer A, Horn H, Ahlrichs R (1992) Fully optimized contracted Gaussian basis sets for atoms Li to Kr. J Chem Phys 97:2571–2577
74. Zavodnik VE, Stash AI, Tsirelson VG, de Vries RY, Feil D (1999) Electron density study of urea using TDS-corrected X-ray diffraction data: quantitative comparison of experimental and theoretical results. Acta Crystallogr B 55:45–54
75. Shukla A, Isaacs ED, Hamann DR, Platzman PM (2001) Hydrogen bonding in urea. Phys Rev B 64:052101 (4 pp)
76. Ragot S (2006) Exact Kohn-Sham versus Hartree-Fock in momentum space: examples of two-fermion systems. J Chem Phys 125:014106 (10 pp)
77. Thakkar AJ (2005) Electronic structure: the momentum perspective. In: Dykstra CE, Frenking G, Kim KS, Scuseria GE (eds) Theory and applications of computational chemistry: the first 40 years. Elsevier, Amsterdam, pp 483–505
78. Raimes S (1961) The wave mechanics of electrons in metals. North Holland, Amsterdam
79. Lundqvist SI, Lydén C (1971) Calculated momentum distributions and Compton profiles of interacting conduction electrons in lithium and sodium. Phys Rev B 4:3360–3370
80. Lam L, Platzman PM (1974) Momentum density and Compton profile of the inhomogeneous interacting electronic system. I. Formalism. Phys Rev B 9:5122–5127
81. Cardwell DA, Cooper MJ (1986) Directional Compton profile measurements of aluminium with 60 keV and 412 keV radiation. Philos Mag B 54:37–49
82. Cardwell DA, Cooper MJ (1989) The effect of exchange and correlation on the agreement between APW and LCAO Compton profiles and experiment. J Phys Condens Matter 1:9357–9368

83. Canney SA, Vos M, Kheifets AS, Clisby N, McCarthy IE, Weigold E (1997) Measured energy – momentum densities of the valence band of aluminium. J Phys Condens Matter 9:1931–1950
84. Lovesey SW (1984) Theory of neutron scattering from condensed matter, vols I, II. Clarendon Press, Oxford
85. Rousse G, Rodríguez-Carvajal J, Wurm C, Masquelier C (2001) Magnetic structural studies of the two polymorphs of $Li_3Fe_2(PO_4)_3$: analysis of the magnetic ground state from super-super exchange interactions. Chem Mater 13:4527–4532
86. Yamagami H (2000) Fully relativistic noncollinear magnetism in spin-density-functional theory: application to USb by means of the fully relativistic spin-polarized LAPW method. Phys Rev B 61:6246–6256; Hobbs D, Kresse G, Hafner J (2000) Fully unconstrained noncollinear magnetism within the projector augmented-wave method. Phys Rev B 62:11556–11570; Gebauer R, Baroni S (2000) Magnons in real materials from density-functional theory. Phys Rev B 61:R6459-R6462; Laskowski R, Madsen GKH, Blaha P, Schwarz K (2004) Magnetic structure and electric-field gradients of uranium dioxide: an ab initio study. Phys Rev B 69:140408 (4 pp); Dal Corso A, Mosca Conte A (2005) Spin-orbit coupling with ultrasoft pseudopotentials: application to Au and Pt. Phys Rev B 71:115106 (8 pp); Mosca Conte A (2007) SISSA/ISAS. PhD Thesis. http://www.sissa.it/cm/thesis/2007/moscaconte.pdf
87. de Jongh LJ, Block R (1975) On the exchange interactions in some 3d-metal ionic compounds: I. The 180° superexchange in the 3d-metal fluorides XMF_3 and X_2MF_4 (X=K, Rb, Tl; M=Mn, Co, Ni). Physica B 79:568–593
88. Ferrari AM, Pisani C, Cinquini F, Giordano L, Pacchioni G (2007) Cationic and anionic vacancies on the NiO(100) surface: DFT+U and hybrid functional density functional theory calculations. J Chem Phys 127:174711
89. Whitten AE, Spackman MA (2006) Anisotropic displacement parameters for H atoms using an ONIOM approach. Acta Crystallogr B 62:875–888
90. Metropolis N, Rosenbluth A, Rosenbluth M, Teller A, Teller E (1953) Equation of state calculations by fast computing machines. J Chem Phys 21:1087–1092
91. Koonin SE, Meredith DC (1990) Computational physics – Fortran version. Addison Wesley, Reading, Chapter 8
92. Pietsch U (2002) X-ray and visible light scattering from light-induced polymer gratings. Phys Rev B 66:155430 (9 pp)
93. Darrigan C, Rérat M, Mallia G, Dovesi R (2003) Implementation of the finite field perturbation method in the CRYSTAL program for calculating the dielectric constant of periodic systems. J Comput Chem 24:1305–1312
94. Rérat M, Ferrero M, Dovesi R (2006) Evolution of the (hyper)polarizability with the size and periodicity of the system. A model investigation from LiF molecule to the LiF 3D crystal. J Comput Methods Sci Eng 6:233–242
95. Ferrero M, Rérat M, Kirtman B, Dovesi R (2008) Calculation of first and second static hyperpolarizabilities of one- to three-dimensional periodic compounds. Implementation in the CRYSTAL code. J Chem Phys 129:244100
96. Ferrero M, Rérat M, Orlando R, Dovesi R (2008) The calculation of static polarizabilities of 1-3D periodic compounds. The implementation in the crystal code. J Comput Chem 29:1450–1459
97. Kochin V, Davaasambuu J, Pietsch U, Schwarz K, Blaha P (2004) The atomistic origin of the inverse piezoelectric effect in α-quartz. J Phys Chem Sol 65:1967–1972
98. Souza I, Íguez J, Vanderbilt D (2002) First-principles approach to insulators in finite electric fields. Phys Rev Lett 89:117602, (4 pp)
99. Umari P, Pasquarello A (2002) Ab initio molecular dynamics in a finite homogeneous electric field. Phys Rev Lett 89:157602 (4 pp)
100. Koritsanszky TS, Buschmann J, Luger P (1996) Topological analysis of experimental electron densities. 1. The different C–C bonds in bullvalene. J Phys Chem 100:10547–10553
101. Pulay P (1983) Localizability of dynamic electron correlation. Chem Phys Lett 100:151–154; Pulay P, Saebø S, Meyer W (1984) An efficient reformulation of the closed-shell self-

consistent electron pair theory. J Chem Phys 81:1901–1905; Saebø S, Pulay P (1985) Local configuration interaction: an efficient approach for larger molecules. Chem Phys Lett 113:13–18
102. Werner H-J, Knowles PJ, Lindh R, Manby FR, Schütz M, and others. MOLPRO version 2006.1, a package of ab initio programs. http://www.molpro.net
103. Pisani C, Maschio L, Casassa S, Halo M, Schütz M, Usvyat D (2008) Periodic local MP2 method for the study of electronic correlation in crystals: theory and preliminary applications. J Comput Chem 29:2113–2124; Erba A, Halo M (2009) CRYSCOR09 user's manual. Università di Torino, Torino. http://www.CRYSCOR.unito.it
104. Maschio L, Usvyat D (2008) Fitting of local densities in periodic systems. Phys Rev B 78:073102 (4 pp)
105. Wiberg KB, Hadad CM, LePage TJ, Breneman CM, Frisch MJ (1992) Analysis of the effect of electron correlation on charge density distributions. J Phys Chem 96:671–679
106. Pisani C, Casassa S, Maschio L (2006) On the prospective use of the one-electron density matrix as a test of the quality of post-Hartree-Fock schemes for crystals. Z Phys Chem 220:913–926
107. Usvyat D, Schütz M (2008) Orbital-unrelaxed Lagrangian density matrices for periodic systems at the local MP2 level. J Phys: Conf Ser 117:012027 (8 pp)

Chapter 3
Modeling and Analysing Thermal Motion in Experimental Charge Density Studies

Anders Ø. Madsen

3.1 Introduction

One of the greatest obstacles in the comparison of theoretical and experimental crystal electron densities is the modeling of thermal motion. Theoretical estimates of electron densities are based on *ab-initio* calculations that exclude nuclear motion, but atoms in crystals are always vibrating about their mean positions. Although this motion is lowered as the temperature is reduced, it is never completely absent due to the persistence of zero-point motion.

The atomic motion has important consequences for the scattering intensities of a diffraction experiment. The scattering of X-rays by atoms is a very fast process on the scale of about 10^{-18} s. This is much faster than the period of the atomic vibrations, where the period of lattice vibrations is of the order of 10^{-13} s. The X-ray scattering is recorded for several seconds, and is therefore an average of a large number of states in the crystal. Stewart and Feil [1] have shown that the scattering averaged over these instantaneous distributions is equivalent to the (hypothetical) scattering of the time-averaged electron distribution.

In order to compare theory and experiment it has become practice in most experimental charge density studies to investigate static charge densities, obtained by deconvoluting the nuclear motion. Because most investigators are interested in this static charge density, considerably less focus has been put on the modeling of atomic vibrations. This is a pity for two reasons: first of all, the atomic vibrations are interesting in themselves, and may tell a great deal about the potential energy hypersurface of the atoms and molecules in the crystal; secondly, the models of the static electron density and atomic vibrations are fitted against a common set

A.Ø. Madsen (✉)
Department of Chemistry, University of Copenhagen, Universitetsparken 5, DK-2100 Copenhagen Ø, Denmark
e-mail: madsen@chem.ku.dk

C. Gatti and P. Macchi (eds.), *Modern Charge-Density Analysis*,
DOI 10.1007/978-90-481-3836-4_3, © Springer Science+Business Media B.V. 2012

of diffraction data corresponding to the vibrationally averaged density; if either of these models is erroneous, correlations will introduce errors in the other too.

The present chapter is meant to present the scientist embarking on an experimental charge-density study with some background knowledge of atomic motion in crystals, an introduction to the way atomic motion is often modeled, as well as how this model can be validated, visualized and analyzed.

3.2 Lattice Dynamics

Some basic physics of atomic vibrations in crystals is needed in order to explain the models and methods often applied in experimental charge density studies. In this section we introduce the theory of lattice dynamics, used in subsequent sections. The theory of lattice dynamics is the basis for deriving the Debye-Waller factor, which describes the reduction of the Bragg diffraction due to atomic motion. The lattice dynamical model can be used to derive Debye-Waller factors from ab-initio calculations (Sects. 3.5.2 and 3.5.3) and to understand how to analyse the experimental Debye-Waller factors (Sect. 3.4).

The following section draws on the excellent book by Willis and Pryor [2]. Although the book is now more than 30 years old and out of print, it is still a most important reference for the crystallographer regarding atomic vibrations in crystals.

A crystal may be considered as one giant molecule. Each atom in the crystal has 3 degrees of freedom. With N unit cells and n atoms in each cell, the crystal has $3nN$ vibrational degrees of freedom. The vibrations of atoms are correlated and extend throughout the crystal in travelling waves, or *phonons*. In the theory of lattice dynamics developed by Born and von Kármán, and described in detail in the classical book by Born and Huang [3] the equations of motion of atoms are set up assuming periodic boundary conditions. In Chap. 2, similar Born and von Kármán conditions are applied to the electronic problem. The atoms at the surface of the crystals comprise a very small fraction of the total, and the periodic boundary conditions allow us to neglect these special surface conditions. Following the theory outlined by Willis and Pryor [2], the motion of each atom is described by

$$\mathbf{u}(kl, t, \mathbf{q}) = \mathbf{U}(k|\mathbf{q}) \exp[i(\mathbf{q} \cdot \mathbf{r}(kl) - \omega(\mathbf{q})t + \phi(k|\mathbf{q}))]. \tag{3.1}$$

k labels an atom in the unit cell, and l labels the unit cell. The displacement $\mathbf{u}(kl, t, \mathbf{q})$ of the (k, l) atom from its equilibrium position $\mathbf{r}(kl)$ depends on the *wave vector* $\mathbf{q}$ of the travelling wave. The displacement vector $\mathbf{U}(k|\mathbf{q})$ describes the maximum amplitude and the direction of motion of the (k, l) atom, as produced by the travelling wave of wave vector $\mathbf{q}$. $\mathbf{U}$ is independent of the unit cell l, because the motion of equivalent atoms (k) in different cells (l) have identical amplitude and direction and differ only in phase. This is *Bloch's theorem*, and introduces an enormous simplification as it allows us to restrict attention to the $3n$ equations of motion of the n atoms of just *one cell*, rather than the equations of motion of the $3nN$ atoms in the crystal. The frequency ω in Eq. 3.1 is a continuous function of $\mathbf{q}$, and the dependence of ω on $\mathbf{q}$ is called the *dispersion relation* for the propagation direction defined by $\mathbf{q}$. For a given $\mathbf{q}$, Eq. 3.1 describes $3n$ modes of vibration, so that we need N different $\mathbf{q}$-vectors to describe the atomic vibrations of the crystal. These $\mathbf{q}$ vectors can be chosen to be uniformly distributed within the so-called *first Brillouin zone*, or simply *the* Brillouin zone. This is the region, centered on the origin of reciprocal space and bound by planes drawn as perpendicular bisectors of the vectors joining the origin to the nearest reciprocal lattice points. Figure 3.1 shows the dispersion relation for the $3 \times 16 = 48$ modes of vibration for urea, which has $n = 16$ atoms in the unit cell. The dispersion relations are shown in the first Brillouin zone in the direction of the $\mathbf{q}$ vector $[0\ 0\ \xi]$, $0 < \xi < 0.5$. The dispersion curves are based on a force-constant model fitted against inelastic neutron scattering measurements [4] and *ab-initio* calculations [5].

The equations of motion within the harmonic and adiabatic approximation, for atom k in unit cell l are then

$$m(k)\ddot{\mathbf{u}}(kl,t) = -\sum_{k'l'} \Phi \begin{pmatrix} k & k' \\ l & l' \end{pmatrix} \mathbf{u}(k'l',t) \tag{3.2}$$

where Φ is a force-constant matrix related to the potential energy V of the crystal by

$$2V = \sum_{kl\alpha} \sum_{k'l'\alpha'} \Phi_{\alpha\alpha'} \begin{pmatrix} k & k' \\ l & l' \end{pmatrix} u_\alpha(kl) u_{\alpha'}(k'l'). \tag{3.3}$$

By substituting Eq. 3.1 in to Eq. 3.2 we obtain a set of equations which can be described using matrix algebra as:

$$\omega^2 \mathbf{U}_m = \mathbf{D}\mathbf{U}_m. \tag{3.4}$$

$\mathbf{U}_m$ is a mass-adjusted column matrix defined by

$$\mathbf{U}_m = \mathbf{m}^{1/2}\mathbf{U}_0 \tag{3.5}$$

where the mass matrix $\mathbf{m}$ is obtained by repeating the masses three times each along the diagonal. The elements of the column matrix $\mathbf{U}_0$ are taken from the elements of the atomic displacement vectors $\mathbf{U}(k|\mathbf{q})$:

$$\mathbf{U}_0(q) = \begin{pmatrix} U_1(1|\mathbf{q}) \\ U_2(1|\mathbf{q}) \\ U_3(1|\mathbf{q}) \\ U_1(2|\mathbf{q}) \\ \vdots \\ U_3(n|\mathbf{q}) \end{pmatrix}. \tag{3.6}$$

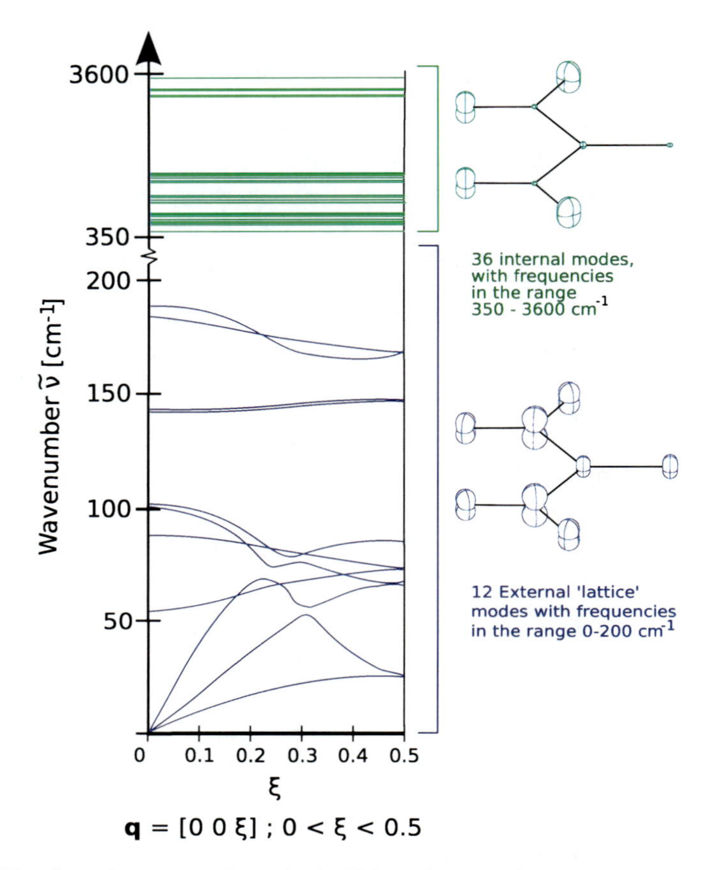

Fig. 3.1 The dispersion curves of urea in the [0 0 1] direction of the first Brillouin zone, given by the reduced wave vector coordinate ξ, $(0 < \xi < 0.5)$. Normally, the frequency on the ordinate is given in MHz. Here we have chosen to show the dispersion relation in terms of the wavenumber $\tilde{\nu}$ in order to relate them to Raman and IR spectroscopic measurements. The wavenumber can be converted into ordinary frequency by $\nu = c\tilde{\nu} = 2.9978 \times 10^{10}$ Hz cm $\times\, \tilde{\nu}$. The angular frequency is $\omega = 2\pi\nu$. Note the change of scale on the ordinate. The upper part (*coloured green*) of the diagram shows the (negligible) dispersion of the modes describing the internal vibrations of urea. The corresponding atomic mean square displacements of the atoms in urea at 123 K, obtained by summing over the internal modes using Eq. 3.9 are illustrated to the right of the dispersion curves. Similarly, the curves illustrating the dispersion of low-frequency external modes, as well as the corresponding atomic mean square displacements are shown in the lower part of the figure *coloured blue*

$\mathbf{D(q)}$ is the mass-adjusted $3n \times 3n$ dynamical matrix defined by

$$\mathbf{D} = \mathbf{m}^{-1/2}\mathbf{D}_0\mathbf{m}^{-1/2}. \tag{3.7}$$

The elements of $\mathbf{D}_0$ are obtained by calculating the forces between each atom in the unit cell and the interacting atoms in the same and neighbouring unit cells. An element of $\mathbf{D}$ is

$$D_{\alpha\alpha'}(kk'|\mathbf{q}) = (m(k)m(k'))^{-1/2} \sum_{l'} \Phi_{\alpha\alpha'} \begin{pmatrix} k & k' \\ 0 & l' \end{pmatrix} \exp[i\mathbf{q} \cdot \Delta\mathbf{r}(kk'l')], \quad (3.8)$$

where the $\Phi_{\alpha\alpha'}$'s are the corresponding components of force constants relative to the coordinates α and α', and $\Delta\mathbf{r}(kk'l')$ is the distance between the origins of asymmetric units. The solutions to the eigenvalue problem in Eq. 3.4 are eigenvectors $\mathbf{e}(j\mathbf{q})$ describing the atomic displacements of the mode $(j\mathbf{q})$. The eigenvalue corresponding to mode $(j\mathbf{q})$ is the (squared) normal mode frequency $\omega_j(\mathbf{q})^2$.

The forces necessary to set up $\mathbf{D}(\mathbf{q})$ can be obtained from force-field or *ab-initio* calculations, or they can be obtained by fitting an empirical model of forces against dispersion curves obtained from inelastic neutron scattering measurements, as well as frequencies obtained from Raman and IR spectroscopy.

Once the normal modes and frequencies have been obtained, they can be used to calculate the vibrational contributions to the thermodynamics of the crystal, i.e. the entropy and heat capacity. They can also be used to calculate the atomic mean square displacement tensors, and thereby the Anisotropic Displacement Parameters (ADPs), as further explained in Sect. 3.3.

The mean square displacement tensor of a vibrating atom k, may be written in terms of a summation of contributions from all the $3nN$ normal modes of vibration:

$$\mathbf{B}_{atom}(k) = \frac{1}{Nm_k} \sum_{j\mathbf{q}} \frac{E_j(\mathbf{q})}{\omega_j^2(\mathbf{q})} \mathbf{e}(k|j\mathbf{q})\mathbf{e}^*((k|j\mathbf{q}))^T \quad (3.9)$$

where $\mathbf{e}(k|j\mathbf{q})$ represents the kth component of a normalized complex eigenvector $\mathbf{e}(j\mathbf{q})$, and corresponds to atom k in normal mode j along the wavevector $\mathbf{q}$. ω_j is the frequency of mode j, m_k is the mass of atom k, and $E_j(\mathbf{q})$ is the energy of the mode, given by

$$E_j(\mathbf{q}) = \hbar\omega_j(\mathbf{q}) \left(\frac{1}{2} + \frac{1}{\exp(\hbar\omega_j(\mathbf{q})/k_B T) - 1} \right). \quad (3.10)$$

$\mathbf{B}_{atom}(k)$ in Eq. 3.9 is the *atomic* mean square displacement tensor. It is a symmetric 3×3 tensor equivalent to the tensor of ADPs, as further described in Sect. 3.3. The normal mode vectors $\mathbf{e}(j\mathbf{q})$ derived from the lattice dynamical model contain information about the correlation of atomic motion. This correlation is not directly available from a single temperature experiment, but some information can be derived using a rigid-body approximation and via multi-temperature experiments, as further explained in Sect. 3.4.2.1.

In Sect. 3.5.2 we will use Eqs. 3.10 and 3.9 above to obtain a theoretical estimate of the atomic Mean Square Displacement (MSD) tensors of crystalline urea.

3.3 Atomic Displacement Parameters

In the previous section we have sketched how the theory of lattice dynamics can be used to describe the atomic vibrations in a crystal in terms of normal modes of vibration. The usual models adopted for structure refinement and charge-density analysis in crystallography do not use a lattice dynamical approach, but consider the atoms as individual harmonic oscillators, moving in the mean field of the surrounding atoms. In this approach, the *mean thermal* electron density of an atom is considered to be the convolution of a static density $\rho_k(\mathbf{r})$ with the probability density function (p.d.f.) $p_k(\mathbf{r})$ describing the probability of having atom k displaced from its reference position $\mathbf{r}_{k0}$;

$$\langle \rho_k(\mathbf{r}) \rangle = \int \rho_k(\mathbf{r} - \mathbf{r}_k) p_k(\mathbf{r}_k - \mathbf{r}_{k0}) \, d\mathbf{r}_k. \tag{3.11}$$

It is important to realize that in this model, the static atomic electron density $\rho_k(\mathbf{r})$ is not deformable: As the atom is vibrating and thereby displaced from its equilibrium position, the electron density is rigidly following the nucleus. This approximation is called the *rigid pseudo atom approximation*. Whether this approximation is reasonable is difficult to investigate. The question is whether the perturbation of the atomic electron density caused by (vibrational) changes in the internuclear distances is of a magnitude that can be detected by experimental charge density studies. At least, the model is sufficient to provide displacement parameters in excellent agreement with parameters derived from inelastic neutron scattering measurements [6, 7].

The X-ray structure factor for the scattering vector $\mathbf{h}$ is given by the Fourier transform of the average electron density of the unit cell,

$$F(\mathbf{h}) = \int \langle \rho(\mathbf{r}) \rangle \, e^{2\pi i \mathbf{h} \cdot \mathbf{r}} d\mathbf{r}, \tag{3.12}$$

the average electron density in the unit cell is normally approximated as the sum of atomic contributions,

$$\langle \rho(\mathbf{r}) \rangle = \sum_{k=1}^{N} n_k \int \rho_k(\mathbf{r} - \mathbf{r}_k) p_k(\mathbf{r}_k - \mathbf{r}_{k0}) d\mathbf{r}_k \tag{3.13}$$

where n_k is the occupancy factor of the kth atom.

Inserting (3.13) into (3.12) and changing the order of the summation and integration, we obtain a summation of atomic contributions $F_k(\mathbf{h})$ to the structure factor,

$$F(\mathbf{h}) = \sum_{k=1}^{N} n_k F_k(\mathbf{h}). \tag{3.14}$$

Each atomic contribution consists of a form factor $f_k(\mathbf{h})$, which is the Fourier transform of the static density $\rho_k(\mathbf{r})$, multiplied by a term that is the Fourier transform of the atomic probability density function

$$F(\mathbf{h}) \approx \sum_{k=1}^{N} n_k f_k(\mathbf{h}) T_k(\mathbf{h}) \exp(2\pi i \mathbf{h} \cdot \mathbf{r}_{k0}). \tag{3.15}$$

Letting $\mathbf{v} = (\mathbf{r} - \mathbf{r}_k)$ and $\mathbf{u} = (\mathbf{r}_k - \mathbf{r}_{k0})$, we have the scattering factor, or atomic form factor, of atom k:

$$f_k(\mathbf{h}) = \int \rho_k(\mathbf{r}) \exp(2\pi i \mathbf{h} \cdot \mathbf{v}) d\mathbf{v} \tag{3.16}$$

and the Fourier transform of the atomic p.d.f.:

$$T_k(\mathbf{h}) = \int p_k(\mathbf{u}) \exp(2\pi i \mathbf{h} \cdot \mathbf{u}) d\mathbf{u}. \tag{3.17}$$

This last term contains the dependence of the structure factor on atomic displacements. Up to this point, we have not put any restrictions on the form of the p.d.f, and thus neither on $T_k(\mathbf{h})$. In normal structure refinement procedures, and also in most charge-density studies, the p.d.f. for atomic displacements is approximated by an isotropic or anisotropic Gaussian function. It can be shown, based on statistical physics, that in the harmonic approximation the probability density function is a Gaussian, as done for example by Born and Sarginson [8]. The Fourier transform of a Gaussian function is also a Gaussian; if the atomic subscript k is omitted, $T(\mathbf{h})$ from (3.17) can therefore be written as

$$T(\mathbf{h}) = \exp\left[-2\pi^2 \left\langle (\mathbf{h} \cdot \mathbf{u})^2 \right\rangle\right]. \tag{3.18}$$

The form that Eq. 3.18 takes depends on the basis vectors used to refer to the diffraction vectors ($\mathbf{h}$) and displacement vectors ($\mathbf{u}$). A confusing range of different bases have been used in the literature. A very thorough review of the different bases, their transformation properties and recommendations for the nomenclature has been given by a committee under the International Union of Crystallography (IUCr) [9]. Based on that nomenclature we present two important set of bases, customarily adopted today. In the most common representation used in reports on structure determination and charge density analysis the diffraction vector $\mathbf{h}$ is referred to the basis of the reciprocal lattice ($\mathbf{a}^*$, $\mathbf{b}^*$, $\mathbf{c}^*$) and the atomic displacement vector $\mathbf{u}$ to the basis of the direct lattice ($\mathbf{a}$, $\mathbf{b}$, $\mathbf{c}$), but normalized so that the components have dimension length;

$$\mathbf{h} = h\mathbf{a}^* + k\mathbf{b}^* + l\mathbf{c}^* \tag{3.19}$$

and

$$\mathbf{u} = \Delta\xi^1 a^* \mathbf{a} + \Delta\xi^2 b^* \mathbf{b} + \Delta\xi^3 c^* \mathbf{c}. \tag{3.20}$$

In this representation, we obtain for T

$$T = \exp\left(-2\pi^2 \sum_{j=1}^{3}\sum_{l=1}^{3} h_j a^j \left\langle \Delta\xi^j \Delta\xi^l \right\rangle a^l h_l\right) \tag{3.21}$$

$$\equiv \exp\left(-2\pi^2 \sum_{j=1}^{3}\sum_{l=1}^{3} h_j a^j U^{jl} a^l h_l\right) \tag{3.22}$$

with

$$U^{jl} = \left\langle \Delta\xi^j \Delta\xi^l \right\rangle. \tag{3.23}$$

The component U^{jl} is one form of the anisotropic displacement parameters. They have dimension (length)^2 and can be directly associated with the mean square displacements of the atom considered in the corresponding directions.

The symmetric 3×3 tensor $\mathbf{U}$ with anisotropic displacement parameters U^{jl} given by (3.23) is not generally expressed in a Cartesian coordinate system. When both the diffraction vector and the atomic displacement vector are referred to the same Cartesian basis, the hazards associated with calculations in oblique coordinate systems are avoided. Such a representation is used almost always in lattice-dynamical studies and thermal motion analysis. When the **h** and **u** vectors are expressed in a Cartesian representation;

$$\mathbf{h} = h_1^C \mathbf{e}_1 + h_2^C \mathbf{e}_2 + h_3^C \mathbf{e}_3 \tag{3.24}$$

and

$$\mathbf{u} = \xi_1^C \mathbf{e}_1 + \xi_2^C \mathbf{e}_2 + \xi_3^C \mathbf{e}_3 \tag{3.25}$$

we obtain for the Debye-Waller factor

$$T = \exp\left(-2\pi^2 \sum_{j=1}^{3}\sum_{l=1}^{3} h_j^C \left\langle \Delta\xi_j^C \Delta\xi_l^C \right\rangle h_l^C\right) \tag{3.26}$$

$$\equiv \exp\left(-2\pi^2 \sum_{j=1}^{3}\sum_{l=1}^{3} h_j^C U_{jl}^C h_l^C\right) \tag{3.27}$$

with

$$U_{jl}^{C} = \left\langle \Delta\xi_j^C \, \Delta\xi_l^C \right\rangle \tag{3.28}$$

The tensor of anisotropic displacement parameters expressed in Cartesian coordinates U_{jl}^{C} can be directly related to the tensor **B** of Eq. 3.9. Both are Cartesian representations of the mean square displacement tensor. If the coordinate systems are the same, they can be compared directly, otherwise a linear transformation of one of the tensors should be performed (see, e.g. Trueblood et al. [9]).

3.3.1 *Beyond the Gaussian Approximation*

The situation is more complicated if the atomic probability density function can not be considered Gaussian. There are several different formalisms for going beyond the harmonic approximation, and different varieties of nomenclature for similar formalisms. The most common method, recommended by the IUCr [9] is the Gram-Charlier (GC) expansion, which is a formalism developed in statistics to describe non-Gaussian distributions, based on a differential expansion of the Gaussian p.d.f. In terms of the Gaussian Debye-Waller factor $T_{\mathbf{h}}(\mathbf{h})$ it can be expressed as

$$T_{GC}(\mathbf{h}) = T_{\mathbf{h}}(\mathbf{h}) \left[1 + (2\pi i)^3 \gamma^{jkl} h_j h_k h_l / 3! + (2\pi i)^4 \delta^{jklm} h_j h_k h_l h_m / 4! + \ldots \right], \tag{3.29}$$

where the γ^{jkl}, δ^{jklm}, ... are the third-, fourth-, etc. order anharmonic tensorial coefficients. The Einstein summation over repeated indices is implied. There are in general 10 cubic, 15 quartic, ..., so-called quasi-moments that enter into the treatment and constitute the parameters of the refinement. The series is normally terminated to include only the third or fourth order cumulants. A very thorough summary of theoretical and experimental aspects of such *generalized* ADPs, i.e. displacement parameters beyond the harmonic approximation, is given by Kuhs [10].

Inclusion of a GC expansion requires high-resolution data. Based on the mean-square displacements $\langle u^2 \rangle$ for the atoms in the structure, the minimum resolution can be calculated from the equation

$$K_n = 2\left(\frac{n \ln 2}{\pi}\right)^{1/2} \frac{1}{\langle u^2 \rangle^{1/2}}, \tag{3.30}$$

where n is the order for the GC expansion [10]. In a charge-density study of tetrafluoroterephthalonitrile at 122 K, Sørensen et al. [11] found that only the N, F and cyano C atoms could be refined using third-order GC coefficients against the X-ray data, although the data extend to $\sin(\theta)/\lambda = 1.27\,\text{Å}^{-1}$.

Often refinements of anharmonic parameters are carried out in charge density studies, though very rarely the significance of these parameters is tested. A first indicator of the significance is to consider the standard uncertainties on these parameters, which are often found to be very high. Secondly, inspection of plots of the anharmonic p.d.f.s can reveal whether the deviations from the harmonic approximation are physically reasonable.

3.3.2 *Visualization*

The atomic mean square displacement tensors are normally visualized by means of ellipsoids of constant probability, defined by

$$\mathbf{u}^T \mathbf{B}^{-1} \mathbf{u} = c^2, \quad (3.31)$$

where c is a constant. For $c = 1.5382$ the atom has 50% probability of being inside the ellipsoid. An example is given in Fig. 3.2. These plots have been used in numerous papers, traditionally drawn by the ORTEP program [14], but today available from a range of programs. The ellipsoids give an impression of major directions of vibration of the atoms in a structure. An alternative is the surface, where vectors from the origin to the surface are proportional to the root mean square displacement in the vector direction. This surface is peanut-shaped (Fig. 3.2) and can be illustrated using the program PEANUT [13]. This type of illustration has the advantage that it can visualize difference-plots where negative mean square displacements can occur (Fig. 3.2).

3.4 Analysis of Atomic Motion

Once a model has been refined against the X-ray or neutron structure factors, it becomes important to test the physical validity of the model. It can also be desirable to analyse the model in order to obtain information about the molecular motion. After all, the atomic and molecular motion is one very important aspect of the physics of the crystal.

3.4.1 *Validation*

For the purpose of structural refinement an inspection of the equal-probability ellipsoids (Fig. 3.2) often suffices to elucidate problems. Abnormally oblate or prolate ellipsoids are usually caused by static disorder, where several conformations of molecules or parts of molecules occur in otherwise identical unit cells.

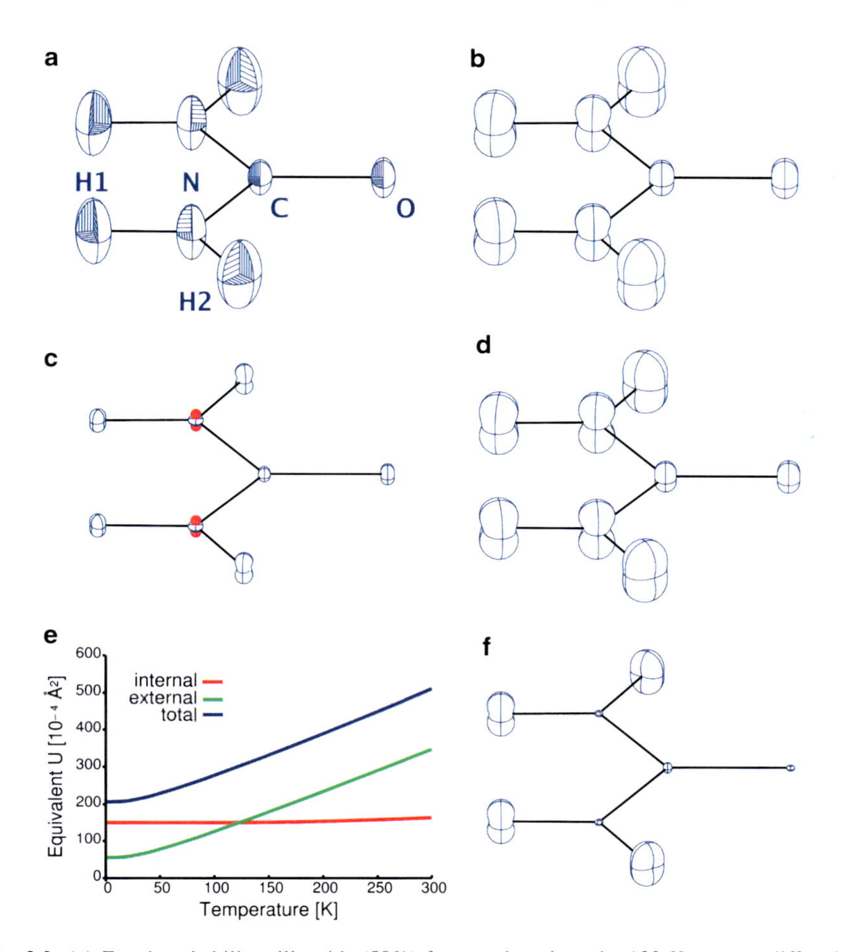

Fig. 3.2 (**a**) Equal probability ellipsoids (50%) for urea based on the 123 K neutron diffraction study by Swaminathan et al. [12]. (**b**) The root-mean-square surfaces of urea using the same model as in (**a**). (**c**) Rmsd difference surface plot showing the difference between the experimental (**b**) and *ab-initio* computed [5] (**d**) root mean square displacements at 123 K. The out-of-plane MSDs are slightly overestimated for N and underestimated for the C, O and H atoms. (**D**) Rmsd surface plot based on the *ab-initio* calculated mean square displacement tensors. (**e**) The temperature-evolution of the (isotropic) mean square displacement of hydrogen atom H1 of urea, based on *ab-initio* calculations. The mean square displacements corresponding to internal modes (*red line*) are almost temperature independent, whereas the low-frequency modes involving collective displacements of all atoms in the molecule depend on the temperature (*green line*). These contributions sum up to the total mean square displacement (*blue line*) which show a linear temperature dependence above ca. 30 K. (**f**) The internal mean-square displacements (based on *ab-initio* calculations) are virtually temperature-independent. The very small internal mean-square displacements of the non-hydrogen atoms is part of the reason for the success of the TLS rigid body analysis (Sect. 3.4.2). The plots were made using the Peanut program [13]

Static displacive disorder is quite common in crystals, however very few systems showing obvious signs of static disorder have been subjected to a charge-density analysis. Static disorder is temperature independent, as opposed to the thermal vibrations, and multi-temperature studies are an obvious way of distinguishing between the different types of contributions to the ADPs. Ellipsoids that are elongated in the same direction for all atoms are typically caused by a missing or erroneous absorption correction.

It is a necessary but not sufficient condition that the ADPs fulfill the *rigid bond test* proposed by Hirshfeld [15]; covalently bonded atoms of similar mass (e.g. second-row atoms in organic molecules) must have similar mean-square displacements in the direction of the bond. The mean square displacement of an atom k in the direction of the unit vector $\mathbf{v}$ is given by $\mathbf{vB}(k)\mathbf{v}^T$ where $\mathbf{B}(k)$ is the mean square displacement tensor (Cartesian representation as defined in Sect. 3.3) of atom k, and $\mathbf{v}^T$ is the transpose of $\mathbf{v}$. Differences of more than 10^{-4} Å^2 should be viewed with skepticism.

The rigid bond test can be extended to include non-bonded atoms in the structure [16]. If the mean square displacements between non-bonded atoms in a molecule are found to obey the rigid bond test, the entire molecule is probably vibrating as a rigid body. It may also be found that only a part of the molecule is moving as a rigid body. In either case, the ADPs of the structure can be subjected to a rigid body analysis or segmented rigid body analysis, as explained further below.

3.4.2 Rigid Body Analysis

Rigid body analysis is an attempt to analyze the atomic mean square displacements of a molecule as if the molecule was vibrating as a rigid unit, independent of the motion of the surrounding molecules in the crystal.

Following the pioneering work of Cruickshank [17, 18], researchers have analyzed the ADPs as if they originated from collective motion with a considerable amount of success [19–24]. The most well known model is the Translation/Libration/Screw (TLS) model developed by Schomaker and Trueblood [25]. The ADPs do not contain information about the correlation of motion between different atoms – however since the energy of the molecular modes depend on the temperature via Eq. 3.10, multitemperature experiments can recover part of this correlation, as shown by Bürgi and co-workers [26–28].

In a molecular crystal there is often a large gap between the strength of the inter- and intra-molecular forces. This implies that the modes of vibration may be separated into low-frequency external (rigid-body) and high-frequency internal (bond bending, stretching and torsional) modes. The situation for urea is depicted in Fig. 3.1, where the external modes have frequencies in the 0–200 cm^{-1} range, and the internal modes are in the 350–3,600 cm^{-1} range. However, unless the molecule is very rigid (e.g. benzene) there will be some mixing of the two types of motion. To understand the relative contribution of high- and low-frequency modes to the atomic mean square displacements it is instructive to consider the mean square displacement of a harmonic oscillator, which is given by

$$\langle u^2 \rangle = \frac{h}{4\pi^2 m\nu}\left(\frac{1}{2} + \frac{1}{\exp(h\nu/k_B T) - 1}\right), \tag{3.32}$$

where m is the reduced mass, ν is the frequency and T is the absolute temperature. Equation 3.32 is of course closely related to Eqs. 3.9 and 3.10 because each normal mode in the crystal is considered to be a harmonic oscillator. Since there is an inverse relationship between the frequency ν and the magnitude of vibration $\langle u^2 \rangle$ in Eq. 3.32, the mean thermal displacements are mainly a consequence of the low-frequency 'rigid body' vibrations, especially at elevated temperatures.

Figure 3.2e depicts the isotropic mean square displacements of a hydrogen atom in urea. The high-frequency 'internal' modes are important for hydrogen atoms because of their light mass, but are much smaller for heavier atoms. They are almost temperature-independent, whereas the magnitude of vibration for low-frequency 'external' modes increase approximately linearly with temperature. So the rigid body approximation becomes increasingly valid at higher temperatures, and the rigid body or TLS analysis, in its general formulation developed by Trueblood and Schomaker [25], ignores contributions from the internal modes.

The mean square displacement tensor of an atom can be obtained by summing the contributions from all normal modes of vibration, following Eq. 3.9. In the case of a rigid molecule we may treat the entire molecule as a single entity, and Eq. 3.9 can be replaced by

$$\mathbf{B}_{molecule}(k) = \sum_{j\mathbf{q}} \mathbf{e}_{molecule}(k\,|\,j\,\mathbf{q})\mathbf{e}^{*}_{molecule}((k\,|\,j\,\mathbf{q}))^{T} \tag{3.33}$$

where $\mathbf{e}_{molecule}(k|j\mathbf{q})$ describes the translational and librational displacements of the rigid molecule k in mode j along the wavevector $\mathbf{q}$:

$$\mathbf{e}_{molecule}(k\,|\,j\,\mathbf{q}) = \begin{bmatrix} U_1(k\,|\,j\,\mathbf{q}) \\ U_2(k\,|\,j\,\mathbf{q}) \\ U_3(k\,|\,j\,\mathbf{q}) \\ --- \\ \Theta_1(k\,|\,j\,\mathbf{q}) \\ \Theta_2(k\,|\,j\,\mathbf{q}) \\ \Theta_3(k\,|\,j\,\mathbf{q}) \end{bmatrix}. \tag{3.34}$$

Now k represents a molecule in the primitive unit cell, and the summation is over the $6nN$ external modes of vibration, where n is the number of molecules in the unit cell. U_i is the translational displacement along axis i, while Θ_i is the angular displacement about the same axis. $\mathbf{B}_{molecule}$ is a symmetric 6×6 matrix that may be partitioned into four 3×3 matrices describing the translational (**T**) and librational (**L**) displacements and their correlation (**S**):

$$\mathbf{B}_{molecule}(k) = \begin{bmatrix} \mathbf{T} & \vdots & \mathbf{S} \\ \cdots & & \cdots\cdots \\ (\mathbf{S}^*)^{\mathrm{T}} & \vdots & \mathbf{L} \end{bmatrix}. \tag{3.35}$$

Where

$$\mathbf{T}(k) = \sum_{j\mathbf{q}} \mathbf{U}(k|j\mathbf{q})\mathbf{U}^*(k|j\mathbf{q})^T \tag{3.36}$$

$$\mathbf{L}(k) \sum_{j\mathbf{q}} \Theta(k|j\mathbf{q})\Theta^*(k|j\mathbf{q})^T \tag{3.37}$$

$$\mathbf{S}(k) = \sum_{j\mathbf{q}} \mathbf{U}(k|j\mathbf{q})\Theta^*(k|j\mathbf{q})^T. \tag{3.38}$$

T is the rigid-body equivalent of the atomic mean square displacement tensor. The T, L and S components of $\mathbf{B}_{molecule}$ are obtained by a least-squares fit against the ADPs. Hydrogen atoms have very large internal motion, they do not obey the rigid-body condition and are omitted from the fit. For the use in a rigid-body analysis the ADPs are transformed to the Cartesian components of the mean square displacement tensor, giving a direct relation between the atomic and rigid-body displacements [2]:

$$\mathbf{B}_{atom}(k\alpha) = \mathbf{T}(k) + \mathbf{R}(k\alpha)\mathbf{L}(k)\mathbf{R}^T(k\alpha) + \mathbf{S}(k)\mathbf{R}^T(k\alpha) + \mathbf{R}(k\alpha)\mathbf{S}^T(k) \tag{3.39}$$

where **R** is an antisymmetric matrix of the cartesian components of $\mathbf{r}(k\alpha)$, the vector denoting the equilibrium position of atom α in the kth molecule:

$$\mathbf{R}(k\alpha) = \begin{bmatrix} 0 & r_3 & -r_2 \\ -r_3 & 0 & r_1 \\ r_2 & -r_1 & 0 \end{bmatrix}. \tag{3.40}$$

Only a part of the correlation tensor **S** can be determined, since the diagonal elements of **S** appear in the three combinations $(S_{22} - S_{11})$, $(S_{33} - S_{22})$ and $(S_{11} - S_{33})$. A constant may be added to each of these diagonal elements without changing the observational equations – the trace of **S** is indeterminate. It is normally set to zero. The quality of the TLS model as a description of the molecular motion in the crystal can be judged by computing the residual $R_w(U_{i,j}) = \sum(\Delta(wU_{i,j})^2)/\sum(wU_{i.j}^{obs})^2$ where $\Delta(wU_{i,j})$ is the weighted differences between the experimental ADPs and the ADPs computed from the TLS model (Eq. 3.39). For truly rigid molecules $R_w(U_{i,j})$ is typically found to be about 5% while for molecules that could be expected to have e.g. torsional vibrations values of 8–12% are common.

A range of computer programs have been developed to perform rigid body analysis, either using the TLS formalism (PLATON [29] and THMA11 [19]) or related models (EKRT [24] and NKA [30]).

3.4.2.1 Beyond the Rigid Body Approximation

The TLS analysis may be extended in several ways. The most common extension is the segmented rigid body analysis, which basically considers the molecule to consist of several connected rigid bodies, or as a rigid body with attached rigid groups. This approach was pioneered by Johnson [31] and by Hamilton and co-workers (e.g. [32]). A very thorough study was performed by Trueblood et al. [21], testing the ability of the segmented rigid body approach to estimate force constants, frequencies and barriers of rotation in crystals. A program commonly used for segmented rigid body analysis is the THMA program [19, 25].

Craven and co-workers [23, 24] have developed their own program and formalism and applied it to a range of neutron diffraction studies. In recent years, Bürgi and co-workers [26, 27] have developed a formalism where the overall rigid body motion and selected internal modes are refined against ADPs of structures investigated at several temperatures. When very low temperatures are included (for molecular crystals 20 K), this procedure allows a very good separation between internal and external vibrational modes.

3.4.2.2 Thermodynamics from Analysis of ADPs

The information about low-frequency molecular vibrations obtained from rigid body or segmented rigid body analysis of ADPs can be used to estimate the vibrational entropy and heat capacity of molecular crystals. Cruickshank [18] demonstrated in a study of naphthalene how the combination of rigid body analysis and information from infrared spectroscopy gave entropies in close agreement with calorimetric measurements. This approach has recently been adopted to study the thermodynamics of naphthalene, anthracene and hexamethylenetetramine [33] as well as ribitol and xylitol [34, 35].

3.5 Complementary Information on Thermal Motion in Crystals

The deconvolution of thermal motion from the static charge density model is of course not guaranteed by the refinement of a multipole-model against the structure factors derived from X-ray diffraction experiments. The parameters describing the static electron density and the atomic coordinates and anisotropic displacement parameters are correlated. Complementary information about the nuclear positions and motion is highly desirable. The most common way is to complement the X-ray diffraction experiment with a neutron-diffraction experiment. However, this is not always possible, due to the limited access to neutron-sources. Furthermore, it may not be possible to grow crystals of a sufficient size to perform a single-crystal neutron diffraction experiment.

Lattice dynamical calculations, based on the lattice-dynamical model described in Sect. 3.2, either using empirical force-fields or *ab-initio* calculations, can provide important information about the atomic motion. An example for urea is given in Sect. 3.5.2.

One of the most valuable outcomes of a neutron-diffraction experiment is the possibility to assess the positions and motion of the hydrogen atoms. However, in lack of neutron diffraction data a number of approaches to model the hydrogen atom positions and ADPs have been proposed, as described in Sect. 3.5.3.

3.5.1 Neutron Diffraction Studies

The interaction between neutrons and the atomic nuclei, which is the basis for the structure determination by neutron diffraction, leads directly to information on the positions and mean-square vibrations of the nuclei. This information is important for all charge-density studies, and especially for systems containing hydrogen atoms, because their lack of core electrons make them difficult to study using X-ray diffraction only, as further described below (Sect. 3.5.3).

3.5.1.1 Relating Data from X-Ray and Neutron Experiments

However, the combination of data from different techniques raises problems, because the experimental conditions are seldom identical. It is thus often found that there are systematic differences between the X-ray and neutron temperature factors. The spherical-atom approximation used in standard X-ray diffraction studies is contributing to this effect, because the anisotropic displacement parameters tend to fit not only the atomic mean square displacements, but also the nonspherical deformations of the valence electron density [36]. This contribution is practically removed in charge-density studies using the atom-centered multipole formalism. However, a range of experimental differences are also important, as discussed in detail by Blessing [37];

Temperature differences. The most obvious of these is differences in temperature. Mean square displacements due to thermal vibrations increase approximately linearly with temperature, as is evident from considering the low-frequency high-temperature limit ($h\nu \leq k_B T$) of Eq. 3.32;

$$\langle u^2 \rangle = h/4\pi^2 m\nu \left(\tfrac{1}{2} + k_B T/h\nu\right). \tag{3.41}$$

Temperature control is especially difficult for X-ray diffraction studies, because the exact temperature varies along the stream of cold nitrogen or helium often used to cool the crystal. This problem can be circumvented by calibrating the temperature using a reference crystal that has a phase transition in the desired temperature range. Potassium dihydrogen phosphate, with a phase transition at 122.4 K is often used for calibration of liquid nitrogen cooling devices.

Absorption. Absorption effects can be very different for the neutron and X-ray experiments. Although the absorption of neutrons by crystals is often small, the incoherent scattering of hydrogen causes an effect similar to absorption. Furthermore, the crystals used for neutron experiments are much larger than for X-rays (generally larger than 1 mm^3), and consequently the transmission factors become smaller. Because absorption attenuates the low-angle reflections more than the high-angle reflections, uncorrected absorption biases the displacement parameters toward values that are too small.

Extinction. Extinction effects are often pronounced in neutron diffraction experiments because of the larger crystals used. Extinction in X-ray diffraction is often attenuating strong low-order reflections, while for neutron diffraction, extinction may persist to higher scattering angles because neutron scattering lengths do not fall off with increasing scattering angles, as X-ray scattering factors do. Uncorrected extinction effects have a similar effect as uncorrected absorption, leading to mean-square displacement parameter values that are too small. The most common method for extinction correction is the formalism of Becker and Coppens [38], which is implemented in several common computer program packages for experimental charge density studies.

Thermal diffuse scattering. Thermal diffuse scattering (TDS) is caused by energy exchange between the scattered radiation and the low-frequency lattice vibrational modes. As the elastic Bragg scattering intensity falls off with increasing temperature due to atomic thermal vibrations, the inelastic TDS intensity builds up; the total X-ray scattering is independent of temperature. Like Bragg scattering, the TDS intensity peaks at reciprocal lattice points. Because TDS is inelastic, the peaks have a broader wavelength distribution and a broader intensity profile than the Bragg peaks, and it is therefore possible to make an empirical estimate based on an analysis of scan profiles. To calculate the TDS contributions it is necessary to know the elastic constants of the crystal. To the authors knowledge, there exists no up-to-date software that makes it easy to correct for TDS effects, and there seems to be very few investigators who attempt to correct their data for TDS. The best way to avoid TDS effects is to measure at low temperatures.

Multiple scattering. Multiple scattering effects occurs when several reciprocal lattice points intersect the Ewald sphere simultaneously, in which case the diffracted beams acts as primary beams and can undergo further diffraction corresponding to the other lattice points in diffraction conditions. This tends to cause strong reflections to become weaker, while weak reflections become stronger.

In effect this causes the ADPs to become smaller, similarly to uncorrected TDS, extinction and absorption effects. Computational corrections are relatively straightforward using the program UMWEG [39], although such corrections are not commonly attempted.

3.5.1.2 Scaling of ADPs

Blessing [37] found that in most cases, the differences between the X-ray and neutron displacement parameters could be described by a linear relation $\mathbf{U}_X = q\mathbf{U}_N + \Delta\mathbf{U}$. No matter what the cause of discrepancies between the two experiments is, it is practice to use the relation between non-hydrogen ADPs to scale the hydrogen ADPs from the neutron diffraction experiment in order to use them as fixed parameters in the model refined against the X-ray diffraction data. This is a practical work-around, but it is of course much better to track down the reason for the discrepancy and correct the data. The program UIJXN by Blessing [37] is available for the comparison and scaling of ADPs.

3.5.1.3 Deuteration of Crystals

Because of the differences in spin, the incoherent scattering is much smaller for deuterium than hydrogen. It is therefore often proposed to exchange hydrogen with deuterium to obtain a better signal-to-noise ratio. However, for the purpose of combining X-ray and neutron diffraction studies this is problematic. The mean square amplitudes of vibration of deuterium is different from hydrogen because of the difference in mass. Furthermore, deuteration is often fractional, which gives an extra ambiguity in the interpretation of the data, because the degree of deuteration can only be estimated by introducing extra parameters in the model fitting against the neutron diffraction data.

3.5.1.4 Comparison of ADPs: Agreement Indices

Comparison of ADPs for the same structure obtained from different sources has traditionally been made using least-squares statistics based on squares of differences, as in the *THMA11 program* [19] used for TLS rigid body analysis. Merritt [40], as well as Whitten and Spackman [41] have used an agreement index based on the overlap integral between the probability density function (p.d.f) p_1 and p_2 of the same atom in the molecule but obtained from different sources of data, which makes the statistic less dependent of the choice of cell axes. The index used by Whitten and Spackman is $S_{12} = 100(1 - R_{12})$ where

$$R_{12} = \int [p_1(\mathbf{x})p_2(\mathbf{x})]^{1/2} \mathrm{d}^3\mathbf{x} = \frac{2^{3/2}\left(\det \mathbf{U}_1^{-1}\mathbf{U}_2^{-1}\right)^{1/4}}{\left[\det\left(\mathbf{U}_1^{-1} + \mathbf{U}_2^{-1}\right)\right]^{1/2}}. \tag{3.42}$$

The S_{12} index describes a percentage difference between the two p.d.f.s. However, whether the difference is due to an isotropic difference in the size of the p.d.f.s, or whether it is due to differences in their anisotropy is not revealed by this index. To address this problem, Merritt [40] normalizes the index by scaling $\mathbf{U}_1$ and $\mathbf{U}_2$ so that their equivalent isotropic displacement parameters $U_{eq} = 1/3(U^{11} + U^{22} + U^{33})$ becomes the same. A related approach is to compare the U_{eq} values along with the S_{12} index [42]. The S_{12} index has been used to compare approaches to estimate ADPs for hydrogen atoms [42], as further discussed in Sect. 3.5.3.

3.5.2 Atomic Motion Derived from Force-Field or ab-initio Calculations

The ADPs provide three-dimensional information about the mean square displacements of atoms in crystals. However, they contain no information about the correlation of displacements of the individual atoms. Nevertheless, as described previously, it is well known that atoms and molecules are moving in collective modes that extend throughout the crystal (phonons), and that there are also intramolecular correlated modes. ADPs are commonly analyzed using the rigid-body or TLS approach, as described in Sect. 3.4.2. When going beyond the rigid-body assumption by using a segmented rigid-body approach or normal-mode coordinate analysis of multi-temperature data, *a-priori* knowledge or chemical intuition is crucial to propose reasonable models of motion that can then be refined against the experimentally determined ADPs. In contrast, *ab-initio* calculations requires no other knowledge than the coordinates of the atoms in the structure, and may therefore provide fruitful information about the most important modes of correlated motion in a given system. Moreover, if the experimental and *ab-initio* calculated ADPs are sufficiently close, we may argue that the dynamics derived from the calculations quantitatively reflects the dynamics in the real crystal.

Normal modes of vibration for *isolated* molecules can be obtained using common *ab-initio* computational packages. The procedure is essentially the same as for periodic systems, as outlined in Sect. 3.2, where the elements of the dynamical matrix (Eq. 3.8) are obtained from summing over intramolecular forces only, and without dispersion, i.e. without the $\mathbf{q}$ dependency.

In the past, empirical force-fields have been used to study the dynamics of crystals and derive mean square displacement parameters for the comparison with ADPs obtained from experiment, most notably by Gramaccioli and co-workers [43, 44]. Today, computer power is available to investigate crystal dynamics using *ab-initio* methods.

Whether the information about the interatomic force-constants are derived from force fields or from *ab-initio* calculations, the starting point for obtaining vibrational information is a geometry optimization of the structure with respect to the energy.

Once a stationary point on the potential energy surface has been found, the mass-weighted force constant matrix **D** (Eq. 3.7) can be obtained from the second partial derivatives of the potential energy with respect to the mass-weighted Cartesian coordinates.

Although the derivation of lattice dynamics of crystals from first principles is well established (*cf.* Sect. 2.5.1), the literature is sparse when it comes to examples of calculations of atomic temperature factors. Lee and Gonze [45] reports calculations on $SiO_2\alpha$-quartz and stishovite. Their results are in reasonable agreement with experimental results. The interatomic forces of extended solids are normally well described by DFT calculations and for these systems recent studies indicate that calculation of total ADPs from DFT calculations are within reach [45–47]. The real challenge is the description of intermolecular interactions in molecular crystals. These interactions can be dominated by dispersive forces, which are only partly described by the frequently used density functional theory calculations (see Chap. 2). In order to describe these dispersive forces it is necessary to apply post-Hartree-Fock techniques, such as second-order Møller-Plesset perturbation calculations (*cf.* Sect. 2.5.3). An alternative which is much cheaper but not *ab-initio* is to use an empirical correction.

3.5.2.1 An Example for the Urea Crystal

Ab-initio calculation of atomic mean square displacement tensors are presently confined to rather small molecules because of the massive computer power needed.

As an example we present estimates of ADPs for the urea crystal [5] based on calculations using the CRYSTAL 2009 program [48]. We used the B3LYP functional and tested a range of standard basis sets (e.g. 6-31G(d,p), TZP). Starting from the structure obtained from neutron diffraction experiments at 12 K [12], the coordinates and unit cell were optimized to a minimum on the potential energy surface.

The vibrational frequencies at the Γ point were obtained by diagonalizing the dynamical matrix (Eq. 3.7). The forces for the construction of the dynamical matrix elements (Eq. 3.8) were obtained by numerical calculation of the second partial derivatives of the potential energy with respect to the mass-weighted Cartesian coordinates by finite displacements of the atomic positions, using a three-point formula with a step amplitude of 0.001 Å. The **q** dependence of the low-frequency modes was obtained by performing Γ-point calculations on a $2 \times 2 \times 2$ supercell (*cf.* Sect. 2.5.1). The ADPs obtained after summation of contributions from all normal modes (by Eq. 3.9) at 123 K are given in Table 3.1 and illustrated in Fig. 3.2.

In order to obtain ADPs in agreement with experimental values it was necessary to use a rather large TZP basis set, and to avoid optimization of the cell parameters. Using the less elaborate basis set 6-31G(d,p) resulted in elongated equal-probability ellipsoids for nitrogen, indicating that the out-of-plane vibration of nitrogen were

Table 3.1 Urea: ADPs [10^{-4} Å^2] from Crystal06 calculations (first line) as compared to the results of neutron diffraction experiments (second line) [12]. The method was B3LYP/TZP supercell (222) calculations fixing the cell to the 12 K experimental cell parameters

	12 K				123 K			
	$U_{11} = U_{22}$	U_{33}	U_{12}	$U_{13} = U_{23}$	$U_{11} = U_{22}$	U_{33}	U_{12}	$U_{13} = U_{23}$
C	44	30	−4	0	115	49	−9	0
	63(4)	35(3)	2(3)	0	147(5)	65(3)	1(4)	0
O	65	30	−2	0	139	49	12	0
	87(5)	32(4)	2(4)	0	197(6)	63(4)	17(5)	0
N	101	44	−60	4	300	70	−200	2
	113(3)	59(2)	−52(2)	4(2)	286(4)	95(2)	−147(2)	2(3)
H1	227	170	−122	−23	351	205	−184	−27
	277(9)	185(7)	−143(6)	−29(8)	440(11)	216(7)	−222(8)	−31(9)
H2	263	89	−80	13	379	115	−136	15
	294(8)	116(6)	−94(7)	15(7)	430(10)	140(6)	−158(8)	19(8)

not well described by the calculations. Indeed, comparison with the normal-mode coordinate analyses based on Raman and inelastic neutron scattering experiments [4] indicated that modes of libration about the C-O axis had higher frequencies, and thus lower vibrational amplitudes, than indicated by the *ab-initio* calculations.

Although these results for urea indicate that total mean square displacement tensors from *ab-initio* calculations are within reach for molecular crystals, it remains to be investigated whether a similar level of accuracy can be obtained in general. Notice how the comparison of computed ADPs with experimental ADPs allow a validation of the *ab-initio* lattice dynamical model, a validation that would otherwise be done by e.g. comparison with inelastic neutron scattering measurements. The advantage is that the X-ray single crystal diffraction study is possible for a much larger group of systems, i.e. it can be performed for much larger unit cells and with ease and speed on standard home-laboratory equipment.

The lattice-dynamical approach described in the previous sections relies on the harmonic approximation, i.e. that the atoms are vibrating in a harmonic potential. For molecular crystals this model becomes increasingly inaccurate at higher temperatures, as evidenced by pronounced changes in the cell parameters with temperature. Molecular dynamics simulations (either based on DFT calculations or empirical force-fields) probe the potential energy hypersurface at many non-equilibrium positions and therefore has the potential to describe this anharmonicity. Such molecular dynamics calculations could in principle allow modeling of disorder and other crystal ‘defects’ and their impact on the experimental ADPs.

Whereas it is computationally demanding to obtain intermolecular forces with sufficient accuracy to obtain *total* MSD tensors, it is much less demanding to compute the contribution from the *internal* modes. This information can be combined with information about the external modes obtained from TLS analysis to get an estimate of the ADPs for hydrogen atoms, as further described below.

3.5.3 Estimating ADPs for Hydrogen Atoms

X-ray diffraction data cannot in general provide detailed information about the positions and thermal motion of hydrogen atoms. For the sake of normal structural refinement, the SDS form factor developed by Stewart, Davidson and Simpson [49] as implemented in most standard structural refinement programs, is adequate to obtain an approximate position and isotropic description of the thermal motion.

On the other hand, if the charge density of a hydrogen-containing molecule is to be studied, independent information on positions and thermal vibrations of the H atoms is invaluable: a number of recent studies [50–52] indicate that the use of isotropic displacement parameters for hydrogen atoms leads to considerable bias in the static charge density models; the topology of the electron density, as revealed by a QTAIM analysis, changes considerably even in areas remote from the H atoms.

In order to model the charge density of hydrogen it seems important to include quadrupole components in the multipole expansion, as noted by Chandler and Spackman [53] in studies of model densities – and as witnessed in several recent studies [28, 50] where the quadrupole components play an important role in order to obtain electric field gradients at the hydrogen nuclei in agreement with the results from nuclear quadrupole resonance spectroscopy. The quadrupole components on hydrogen also plays an important role for the topological properties of intermolecular hydrogen bonding [54]. To use isotropic displacement parameters is a severe approximation, not only because the isotropic displacement parameters correlate with the monopole parameters of the multipole model, but also because the quadrupole parameters will include dynamic effects because the quadrupole deformations will mimic dynamic effects, as the quadrupole terms have the same symmetry as the ADPs.

As described earlier, neutron-diffraction experiments are sometimes undertaken to supplement the X-ray charge density analysis with positional parameters and ADPs for the H atoms. However this is not always possible due to difficulties in obtaining sufficiently large crystals and the limited access to neutron sources, as well as the difficulties often observed in relating the ADPs from the two experiments (Sect. 3.5.1). Therefore, many studies will benefit from applying models for the H atoms combining information from the X-ray diffraction data with *a-priori* information from other systems studied by neutron diffraction experiments, or by using information from spectroscopy or *ab-initio* calculations.

Only 18% of present day charge-density studies of systems involving H atoms have used an anisotropic description of the H atom motion [42], either derived from neutron diffraction experiments (13%) or from one of the approximate approaches (5%) described in the following.

In the following we describe some of the approaches used in recent years to estimate the ADPs of hydrogen atoms[1]

---

[1] A new approach to obtain H ADPs by refinement against the X-ray scattering factors is to improve the aspherical scattering factors for hydrogen. Jayatilaka and Dittrich [55] used transferable atomic

3.5.3.1 Estimates Combining External and Internal Motion

These methods all rely on the combination of rigid body motion derived from a TLS analysis of the heavy-atom skeleton with some estimate of the internal motion of the hydrogen atom,

$$U^{ij} = U^{ij}_{internal} + U^{ij}_{external}. \tag{3.43}$$

As can be seen from Fig. 3.2e the internal motion of hydrogen gives an important contribution even at room temperature. Hirshfeld [15], from whom this method seems to spring, used internal motion derived from Raman or IR spectroscopic measurements. Others have used estimates of internal motion based on neutron diffraction studies of related compounds, or from *ab-initio* calculations.

3.5.3.2 Spectroscopic Evidence

In the original work by Hirshfeld, information from Raman and infra-red spectroscopy was used to assess the frequencies and corresponding mean square displacements of the bond-stretching and bond-bending modes of the X–H bonds. This approach has been used several times by Destro and co-workers [57–61], and has been implemented in the program ADPH, described and tested in detail by Roversi and Destro [50]

The ADPH approach is to use normal mode frequencies based on spectroscopic data. Each vibration is described by approximate vibrational coordinates. In the simplest case, three independent modes – one bond-stretching and two perpendicular to the bond – are used to construct the internal part $U^{ij}_{internal}$ to the internal mean square displacement tensor for the hydrogen atom, however there is no limitation on the number of normal modes that can be used. An advantage of this approach is that the estimates of internal ADPs can be based on spectroscopic measurements on the same compound that is studied by X-ray diffraction. However, for larger molecules, with several similar functional groups, it becomes impossible to assign the different spectroscopic frequencies to individual hydrogen atoms, and the approach has to rely on mean group frequencies.

---

densities defined in terms of 'Hirshfeld atoms'. For a more detailed presentation, see Chap. 6, Sect. 6.3. The atoms (and thereby the atomic form factors) are defined by using Hirshfelds stockholder partioning [56] of an electron density obtained from quantum mechanical calculations. The strategy has been tested against X-ray data for urea and benzene and benchmarked against neutron diffraction results. The C–H and N–H bond distances are remarkably well reproduced. The ADPs of carbon and hydrogen atoms in benzene are in excellent agreement with the neutron diffraction results, whereas the results for urea are more ambiguous, with some ADPs in good agreement, while others show deviations of more than 50%, and a large dependence on whether the applied electron density was obtained using HF or DFT methods. The method is promising but must be further validated before any conclusions can be drawn as to its general applicability.

3.5.3.3 Information from Neutron Diffraction Studies

It is possible to analyze the vibrational motion of hydrogen atoms in a similar vein as the statistical analysis of X–H bond lengths derived from neutron diffraction studies found in International Tables for Crystallography [62]. When the total atomic mean square displacement tensor U^{ij} has been determined from neutron diffraction experiments, and the rigid molecular motion U^{ij}_{rigid} has been determined from a rigid-body analysis of the non-hydrogen ADPs, it becomes possible to get an estimate of the internal motion of the hydrogen atoms by rearranging Eq. 3.43;

$$U^{ij}_{internal} = U^{ij} - U^{ij}_{rigid}. \tag{3.44}$$

It was noted by Johnson [31] that the mean square displacements derived from $U^{ij}_{internal}$ of hydrogen atoms was in good agreement with spectroscopic information, showing systematic trends corresponding to the functional group that hydrogen was part of. Similar observations were done by Craven and co-workers in the analysis of several systems [63–66]. The internal torsional motion of a range of librating groups, including methyl, carboxyl and amino groups was also thoroughly investigated by Trueblood and Dunitz [21] based on more than 125 neutron diffraction studies of molecular crystals from the literature.

Madsen and co-workers [67] investigated the internal mean square displacements of hydrogen atoms in xylitol and a range of related carbohydrate compounds found in the literature, and these estimates of internal modes were collected in a 'library' and later improved and enhanced with more statistical material [42]. The present library provides mean values of internal stretch modes as well as in-plane and out-of-plane bending modes for a range of chemical groups involving hydrogen bound to C, N and O, and forms the basis for assigning anisotropic displacement parameters to hydrogen atoms in the SHADE server, [68] which allows users to submit a CIF file [69] containing the atomic coordinates and the ADPs of the non-hydrogen atoms. The server performs a TLS analysis using the THMA11 program, and combines the rigid body motion with the internal motion obtained from analysis of neutron diffraction data. It is possible to perform a segmented rigid body analysis using the attached rigid group approach of the THMA11 program [19, 42]. The segmented rigid body approach seems to give marginally better results, as compared to neutron diffraction experiments, as judged from a few test cases [42] on non-rigid molecules. For adenosine we observed that despite the improved description of the motion of the heavy-atom skeleton, only small improvements were observed for the H atom ADPs. For some hydrogen atoms there was a substantial improvement, while for other we observed a worsening agreement. There is definitely room for further testing and improvement of the segmented rigid body analysis in this context.

The SHADE server is available at the web-address http://shade.ki.ku.dk.

3.5.3.4 Information from *ab-initio* Calculations

Estimates of interatomic force-constants obtained from *ab-initio* quantum mechanical calculations is today a straightforward way to build the dynamical matrix used in a normal-mode coordinate analysis. Several academic and commercial programs offer integrated normal-mode analysis. A program by Koritsanszky called XDVIB [70] is able to read the output from a normal-mode analysis from the program Gaussian [71] and to compute the ADPs corresponding to internal vibrations. This procedure was used successfully by Koritsanszky and co-workers to provide ADPs for hydrogen atoms in a range of studies of molecular crystals [72–76]. In these studies, the external contribution to the ADPs were based on a rigid-link refinement of the non-hydrogen ADPs, which essentially mimics a rigid-body type refinement. However, the *ab-initio* calculation of internal modes was performed on an isolated (gas-phase) molecules. This is not always sufficient to obtain reliable results. Results by Luo et al. [66] and Madsen et al. [67] shows that gas-phase calculations can lead to internal mean square displacements that are much larger than the total mean square displacements as derived from neutron diffraction experiments, because the intermolecular potential energy surface (PES) of an isolated molecule is very different from the PES of a molecule in a crystalline environment, especially for non-rigid systems with large amplitude torsional vibrations. The flat PES causes large amplitudes of some of the internal molecular vibrations, e.g. torsional modes. In these cases, it is necessary to take the intermolecular environment into account. For rigid molecules with weak intermolecular interactions it may be sufficient to use gas-phase calculations (e.g. the case of naphthalene [77]).

Whitten and Spackman [41] used ONIOM calculations – a procedure where the central molecule is treated using quantum mechanics and a cluster of surrounding molecules using classical molecular mechanics – to mimick the intermolecular environment with excellent results. This "TLS + ONIOM" approach differs slightly from the ADPH and SHADE approaches in that the internal mean square displacements are subtracted from the ADPs of the non-hydrogen atoms before the TLS analysis. Although this is a small correction, it seems to be an improvement as it diminishes the differences between the mean square displacements of bonded atoms in the direction of the bond (this so-called rigid bond test by Hirshfeld [15] is often used to test the reliability of ADPs derived from experiments).

Recently, the ADPH, SHADE and TLS + ONIOM approaches have been compared by Munshi et al. [42]. They differ primarily in the way the internal motion is estimated. The ADPs of hydrogen atoms in 1-methyl-uracil based on these approaches are compared in Fig. 3.3, and the mean similarity index, based on Eq. 3.42 is given. All models are in excellent agreement with the ADPs based on neutron diffraction experiments, and this was also the general conclusion in the comparison by Munshi et al. [42], where the SHADE server http://shade.ki.ku.dk was recommended as a routine procedure for deriving estimates of H-atom ADPs suitable for charge-density studies of molecular crystals.

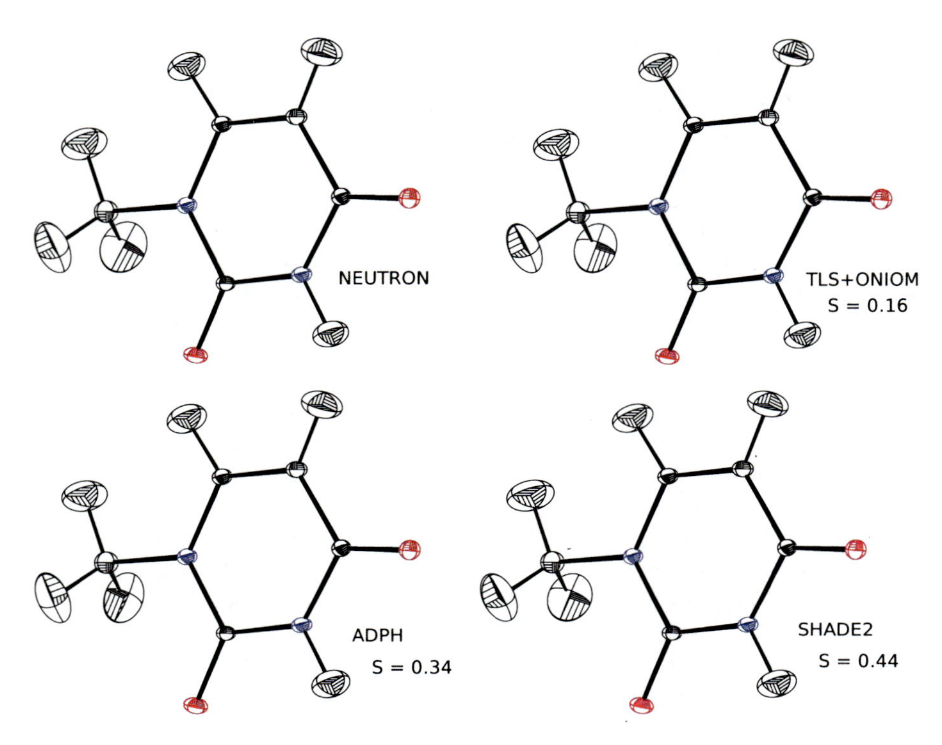

Fig. 3.3 A comparison of methods to estimate the hydrogen ADPs of 1-Methyl Uracil based on the work by Munshi et al. [42]. Equal-probability ellipsoids at the 70% probability level. The mean similarity index S is given (based on Eq. 3.42) for the hydrogen ADPs compared to the results derived from neutron diffraction data

3.6 Outlook

As outlined in this chapter the theory and procedures for modeling atomic thermal motion in standard crystallographic applications as well as in charge density studies is well established, and there is a range of possibilities for going beyond the standard model and for analyzing and assessing the physical significance of the displacement parameters.

Looking ahead, one experimental and one theoretical aspect will obviously affect the modeling and analysis of experimental charge densities.

The first is the advent of neutron spallation sources, which changes the size-requirements for the crystals used in neutron-diffraction experiments, and thereby enhances the applicability of neutron diffraction to a larger range of crystalline systems, providing information on thermal motion unbiased by inadequacies in the modeling of the static charge density. This will especially be an advantage for the modeling of hydrogen atoms, but may also be important if anharmonic atomic motion is significant. The smaller crystals will also diminish extinction effects.

The other important factor is the contribution from quantum-mechanical calculations. Crystallography has always benefitted from quantum mechanical calculations because they have provided the accurate wavefunctions needed to construct the spherical as well as aspherical atomic scattering factors. In recent years the advances in density functional theory, but especially the ever-increasing speed of computers has made it easier to estimate the dynamics of crystals based on ab-initio methods, as illustrated for urea in this chapter. The most obvious advantages are the estimation of hydrogen atom displacement parameters as well as the comparison with vibrational frequencies and normal modes derived from multi-temperature studies. The *ab-initio* approaches can also be used to estimate the atomic anharmonic motion via molecular dynamics simulations.

The incorporation of vibrational information from quantum mechanical calculations could pave the way for a better description of thermal motion in experimental charge density studies, moving away from the independent atomic motion towards a normal-mode based model of collective vibrations, which could diminish the number of parameters needed to describe the thermal motion, provide a framework for performing refinement of a charge density model against data from multiple temperatures, and allow for modeling the effects of thermal diffuse scattering.

In contrast to the maximum entropy method the multipole formalism provides a deconvolution of the thermal motion from a hypothetical static charge density, which can then be compared with quantum mechanical calculations. However, this popular approach with a focus on static densities has in many studies removed the focus from the importance of thermal motion.

The fact that X-ray diffraction experiments provide information about the thermally averaged charge density should be considered an advantage as well as a challenge, and I hope that further amalgamation of theory and experimental approaches will continue to explore the dynamical nature of crystals, for the mutual benefit of experimentalists and theoreticians.

References

1. Stewart RF, Feil D (1980) A theoretical study of elastic X-ray scattering. Acta Crystallogr A 36(4):503–509
2. Willis BTM, Pryor AW (1975) Thermal vibrations in crystallography. Cambridge University Press, London/New York
3. Born M, Huang K (1954) Dynamical theory of crystal lattices. Clarendon, Oxford
4. Lefebvre J, More M, Fouret R, Hennion B, Currat R (1975) Lattice-vibrations in deuterated urea. J Phys C 8(13):2011–2021
5. Madsen AØ, Civalleri B, Ferrabone M, Pascale F, Dovesi R (2011) Anisotropic displacement parameters for molecular crystals from periodic HF and DFT calculations. Acta Crystallogr B (submitted)
6. Willis BTM, Howard JAK (1975) Do the ellipsoids of thermal vibration mean anything? – analysis of neutron diffraction measurements on hexamethylenetetramine. Acta Crystallogr A 31:514–520

7. Flensburg C, Stewart RF (1999) Lattice dynamical Debye-Waller factor for silicon. Phys Rev B 60(1):284–290
8. Born M, Sarginson K (1941) The effect of thermal vibrations on the scattering of X-rays. Proc R Soc Lond A 179:69–93
9. Trueblood KN, Bürgi HB, Burzlaff H, Dunitz JD, Gramaccioli CM, Schulz HH, Shmueli U, Abrahams SC (1996) Atomic displacement parameter nomenclature. Report of a subcommittee on atomic displacement parameter nomenclature. Acta Crystallogr A 52(5):770–781
10. Kuhs WF (1992) Generalized atomic displacements in crystallographic structure analysis. Acta Crystallogr A 48:80–98
11. Sørensen HO, Stewart RF, McIntyre GJ, Larsen S (2003) Simultaneous variation of multipole parameters and Gram–Charlier coefficients in a charge-density study of tetrafluoroterephthalonitrile based on X-ray and neutron data. Acta Crystallogr A 59(6):540–550
12. Swaminathan S, Craven BN, McMullan RK (1984) The crystal structure and molecular thermal motion of urea at 12, 60 and 123 k from neutron diffraction. Acta Crystallogr B 40:300–306
13. Hummel W, Hauser J, Buergi HB (1990) PEANUT: computer graphics program to represent atomic displacement parameters. J Mol Graph 8:214–220
14. Johnson CK (1976) ORTEPII. Report ORNL-5138. Technical report. Oak Ridge National Laboratory, Tennessee
15. Hirshfeld FL (1976) Can X-ray data distinguish bonding effects from vibrational smearing? Acta Crystallogr A 32:239–244
16. Rosenfield RE Jr, Trueblood KN, Dunitz JD (1978) A test for rigid-body vibrations, based on a generalization of Hirshfeld's 'rigid-bond' postulate. Acta Crystallogr A 34:828–829
17. Cruickshank DWJ (1956) The analysis of the anisotropic thermal motion of molecules in crystals. Acta Crystallogr 9:754
18. Cruickshank DWJ (1956) The entropy of crystalline naphthalene. Acta Crystallogr 9(12):1010–1011
19. Schomaker V, Trueblood KN (1998) Correlation of internal torsional motion with overall molecular motion in crystals. Acta Crystallogr B 54:507–514
20. Dunitz JD, Maverick EF, Trueblood KN (1988) Atomic motions in molecular crystals from diffraction measurements. Angew Chem Int Ed Engl 27:880–895
21. Trueblood KN, Dunitz JD (1983) Internal molecular motions in crystals. The estimation of force constants, frequencies and barriers from diffraction data. A feasibility study. Acta Crystallogr B Acta Cryst. B 39:120–133
22. Dunitz JD, White DNJ (1973) Non-rigid-body thermal-motion analysis. Acta Crystallogr A 29:93
23. He XM, Craven BM (1993) Internal vibrations of a molecule consisting of rigid segments. I. Non-interacting internal vibrations. Acta Crystallogr A 49:10–22
24. He XM, Craven BM (1985) Internal molecular vibrations from crystal diffraction data by quasi-normal mode analysis. Acta Crystallogr A 41:244–251
25. Schomaker V, Trueblood KN (1968) On the rigid-body motion of molecules in crystals. Acta Crystallogr B 24:63–76
26. Bürgi HB, Capelli SC (2000) Dynamics of molecules in crystals from multi-temperature anisotropic displacement parameters. I. Theory. Acta Crystallogr A 56:403–412
27. Capelli SC, Förtsch M, Bürgi HB (2000) Dynamics of molecules in crystals from multi-temperature anisotropic displacement parameters. II. Application to benzene (C_6D_6) and urea [$OC(NH)_2$]. Acta Crystallogr A 56:413–424
28. Bürgi HB, Capelli SC, Goeta AE, Howard JAK, Spackman MA, Yufit DS (2002) Electron distribution and molecular motion in crystalline benzene: an accurate experimental study combining CCD X-ray data on C_6H_6 with multitemperature neutron-diffraction results on C_6H_6. Chem Eur J 8(15):3512–3521
29. Spek AL (1990) PLATON, an integrated tool for the analysis of the results of a single crystal structure determination. Acta Crystallogr A 46:C34

30. Capelli SC, Hauser J (2004) NKA user manual, 16.2.2004 edn. Universität Bern, Bern
31. Johnson CK (1970) Thermal neutron diffraction, Chap. 9. Oxford University Press, New York
32. Hamilton WC, Edmonds JW, Tippe A, Rush JJ (1969) Methyl group rotation and low temperature transition in hexamethylbenzene – a neutron diffraction study. Discuss Faraday Soc 48:192–204
33. Aree T, Buergi HB (2006) Specific heat of molecular crystals from atomic mean square displacements with the Einstein, Debye, and Nernst-Lindemann models. J Phys Chem B 110(51):26, 129–26, 134
34. Madsen AØ, Larsen S (2007) Insights into solid state thermodynamics from diffraction data. Angew Chem Int Ed 46:8609–8613
35. Madsen AØ, Mattson R, Larsen S (2011) Understanding thermodynamic properties at the molecular level: multiple temperature charge density study of ribitol and xylitol. J Phys Chem 115(26):7794–7804
36. Coppens P (1968) Evidence for systematic errors in X-ray temperature parameters resulting from bonding effects. Acta Crystallogr B 24(9):1272–1274
37. Blessing RH (1995) On the differences between X-ray and neutron thermal vibration parameters. Acta Crystallogr B 51:816–823
38. Becker PJ, Coppens P (1975) Extinction within the limit of validity of the Darwin transfer equations. I. General formalisms for primary and secondary extinction and their application to spherical crystals. Acta Crystallogr A 31:129–147
39. Rossmanith E (2003) UMWEG: a program for the calculation and graphical representation of multiple-diffraction patterns. J Appl Cryst 36(Part 6):1467–1474
40. Merritt E (1999) Comparing anisotropic displacement parameters in protein structures. Acta Crystallogr D 55:1997–2004
41. Whitten AE, Spackman MA (2006) Anisotropic displacement parameters for H atoms using an ONIOM approach. Acta Crystallogr B 62:875–888
42. Munshi P, Madsen AO, Spackman MA, Larsen S, Destro R (2008) Estimated H-atom anisotropic displacement parameters: a comparison between different methods and with neutron diffraction results. Acta Crystallogr A 64(Part 4):465–475
43. Gramaccioli CM, Filippini G, Simonetta M (1982) Lattice-dynamical evaluation of temperature factors for aromatic hydrocarbons, including internal molecular motion: a straightforward systematic procedure. Acta Crystallogr A 38:350–356
44. Gramaccioli CM, Filippini G (1983) Lattice-dynamical evaluation of temperature factors in non-rigid molecular crystals: a first application to aromatic hydrocarbons. Acta Crystallogr A 39:784–791
45. Lee C, Gonze X (1995) *Ab initio* calculation of the thermodynamic properties and atomic temperature factors of $SiO_2\alpha$-quarts and stishovite. Phys Rev B 51(13):8610–8613
46. Parlinski K, Li ZQ, Kawazoe Y (1997) First-principles determination of the soft mode in cubic ZrO_2. Phys Rev Lett 78(21):4063–4066
47. Schowalter M, Rosenauer A, Titantah JT, Lamoen D (2009) Temperature-dependent Debye-Waller factors for semiconductors with the wurtzite-type structure. Acta Crystallogr A 65: 227–231
48. Dovesi R, Saunders VR, Roetti C, Orlando R, Zicovich-Wilson CM, Pascale F, Civalleri B, Doll K, Harrison NM, Bush IJ, D'Arco P, Llunell M (2009) CRYSTAL09 user's manual. Universita di Torino, Torino
49. Stewart RF, Davidson ER, Simpson WT (1965) Coherent X-ray scattering for the hydrogen atom in the hydrogen molecule. J Chem Phys 42(9):3175–3187
50. Roversi P, Destro M (2004) Approximate anisotropic displacement parameters for H atoms in molecular crystals. Chem Phys Lett 386:472–478
51. Madsen AØ, Sørensen HO, Stewart RF, Flensburg C, Larsen S (2004) Modeling of nuclear parameters for hydrogen atoms in X-ray charge density studies. Acta Crystallogr A 60: 550–561
52. Hoser AA, Dominiak PM, Woźniak K (2009) Towards the best model for H atoms in experimental charge-density refinement. Acta Crystallogr A 65(4):300–311

53. Chandler GS, Spackman MA (1982) Pseudoatom expansions of the first-row diatomic hydride electron densities. Acta Crystallogr A 38:225–239
54. Mata I, Espinosa E, Molins E, Veintemillas S, Maniukiewicz W, Lecomte C, Cousson A, Paulus W (2006) Contributions to the application of the transferability principle and the multipolar modeling of H atoms: electron-density study of L-histidinium dihydrogen orthophosphate orthophosphoric acid. I. Acta Crystallogr A 62(5):365–378
55. Jayatilaka D, Dittrich B (2008) X-ray structure refinement using aspherical atomic density functions obtained from quantum-mechanical calculations. Acta Crystallogr A 64(Part 3): 383–393
56. Hirshfeld F (1977) Bonded-atom fragments for describing molecular charge-densities. Theor Chim Acta 44(2):129–138
57. Destro R, Roversi P, Barzaghi M, Marsh RE (2000) Experimental charge density of α-glycine at 23 K. J Phys Chem A 104:1047–1054
58. May E, Destro R, Gatti C (2001) The unexpected and large enhancement of the dipole moment in the 3,4-bis(dimethylamino)-3-cyclobutene-1,2-dione (DMACB) molecule upon crystallization: a new role of the intermolecular CH···O interactions. J Am Chem Soc 123(49):12248–12254
59. Forni A, Destro R (2003) Electron density investigation of a push-pull ethylene (C14H12NO2 ··· H2O) by x-ray diffraction at T = 21 K. Chem Eur J 9(22):5528–5537
60. Destro R, Soave R, Barzaghi M, Lo Presti L (2005) Progress in the understanding of drug-receptor interactions, part 1: experimental charge-density study of an angiotensin II receptor antagonist ($C_{30}H_{30}N_6O_3S$) at T = 17 K. Chem Eur J 11(16):4621–4634
61. Soave R, Barzaghi M, Destro R (2007) Progress in the understanding of drug-receptor interactions, part 2: experimental and theoretical electrostatic moments and interaction energies of an angiotensin II receptor antagonist $C_{30}H_{30}N_6O_3S$. Chem Eur J 13(24):6942–6956
62. Allen FH, Watson DG, Brammer L, Orpen AG, Taylor R (1999) Typical interatomic distances: organic compounds. In: Wilson AJC, Prince E (eds) International tables for crystallography, vol C. Kluwer Academic Publishers, Dordrecht/Boston/London, pp 782–803
63. Weber HP, Craven BM, Sawzip P, McMullan RK (1991) Crystal structure and thermal vibrations of cholesteryl acetate from neutron diffraction at 123 and 20 K. Acta Crystallogr B 47:116–127
64. Gao Q, Weber HP, Craven BM, McMullan RK (1994) Structure of suberic acid at 18.4, 75 and 123 K from neutron diffraction data. Acta Crystallogr B 50(6):695–703
65. Kampermann SP, Sabine TM, Craven BM, McMullan RK (1995) Hexamethylenetetramine: extinction and thermal vibrations from neutron diffraction at six temperatures. Acta Crystallogr A 51:489–497
66. Luo J, Ruble JR, Craven BM, McMullan RK (1996) Effects of H/D substitution on thermal vibrations in piperazinium hexanoate-H1 1, D1 1. Acta Crystallogr B 52(2):357–368
67. Madsen AØ, Mason S, Larsen S (2003) A neutron diffraction study of xylitol: derivation of mean square internal vibrations for hydrogen atoms from a rigid-body description. Acta Crystallogr B 59:653–663
68. Madsen AØ (2006) SHADE web server for estimation of hydrogen anisotropic displacement parameters. J Appl Cryst 39:757–758
69. Hall SR, Allen FH, Brown ID (1991) The crystallographic information file (CIF): a new standard archive file for crystallography. Acta Crystallogr A 47:655–685
70. Volkov A, Macchi P, Farrugia LJ, Gatti C, Mallinson P, Richter T, Koritsánszky T. (2006) XD2006 – A computer program package for multipole refinement, topological analysis of charge densities and evaluation of intermolecular energies from experimental and theoretical structure factors
71. Frisch MJ, Trucks GW, Schlegel HB, Scuseria GE, Robb MA, Cheeseman JR, Montgomery Jr JA, Vreven T, Kudin KN, Burant JC, Millam JM, Iyengar SS, Tomasi J, Barone V, Mennucci B, Cossi M, Scalmani G, Rega N, Petersson GA, Nakatsuji H, Hada M, Ehara M, Toyota K, Fukuda R, Hasegawa J, Ishida M, Nakajima T, Honda Y, Kitao O, Nakai H, Klene M, Li X,

Knox JE, Hratchian HP, Cross JB, Bakken V, Adamo C, Jaramillo J, Gomperts R, Stratmann RE, Yazyev O, Austin AJ, Cammi R, Pomelli C, Ochterski JW, Ayala PY, Morokuma K, Voth GA, Salvador P, Dannenberg JJ, Zakrzewski VG, Dapprich S, Daniels AD, Strain MC, Farkas O, Malick DK, Rabuck AD, Raghavachari K, Foresman JB, Ortiz JV, Cui Q, Baboul AG, Clifford S, Cioslowski J, Stefanov BB, Liu G, Liashenko A, Piskorz P, Komaromi I, Martin RL, Fox DJ, Keith T, Al-Laham MA, Peng CY, Nanayakkara A, Challacombe M, Gill PMW, Johnson B, Chen W, Wong MW, Gonzalez C, Pople JA (2004) Gaussian 03, revision C.02 (2003). Gaussian, Inc, Wallingford

72. Williams RV, Gadgil VR, Luger P, Koritsanszky T, Weber M (1999) The search for homoaromatic semibullvalenes. 6. X-ray structure and charge density studies of 1,5-dimethyl-2,4,6,8-semibullvalenetetracarboxylic dianhydride in the temperature range 123-15 K. J Org Chem 64:1180–1190
73. Buschmann J, Koritsanszky T, Lentz D, Luger P, Nickelt N, Willemsen S (2000) Structure and charge density studies on 1,1-difluoroallene and tetrafluoroallene. Z Kristallogr 215:487–494
74. Koritsanszky T, Buschmann J, Lentz D, Luger P, Perpetuo G, Röttger M (1999) Topological analysis of the experimental electron density of diisocyanomethane at 115 k. Chem Eur J 5:3413–3420
75. Koritsanszky T, Zobel D, Luger P (2000) Topological analysis of experimental electron densities. 3. potassium hydrogen(+)-tartrate at 15 K. J Phys Chem A 104:1549–1556
76. Flaig R, Koritsanszky T, Zobel D, Luger P (1998) Topological analysis of the experimental electron densities of amino acids. 1. D,L-Aspartic acid at 20 K. J Am Chem Soc 120: 2227–2238
77. Oddershede J, Larsen S (2004) Charge density study of naphthalene based on X-ray diffraction data at four different temperatures and theoretical calculations. J Phys Chem A 108:1057–1063

Chapter 4
Spin and the Complementary Worlds of Electron Position and Momentum Densities

Jonathan A. Duffy and Malcom J. Cooper

4.1 Introduction

The Compton effect is the very well-known phenomenon of inelastic scattering of a photon off a free, stationary electron. It is most familiar when written in terms of the wavelength shift, $\Delta\lambda$, which is simply fixed by the angle of scattering, ϕ, and the fundamental constants h, m and c, Planck's constant, the rest mass of the electron and the speed of light in vacuum, respectively.

$$\Delta\lambda = \frac{h}{mc}(1 - \cos\phi) \tag{4.1}$$

However, in order to understand how the motion of the electron affects the phenomenon it is more useful to write the formula in terms of photon energies:

$$E_2 = \frac{E_1}{[1 + (E_1/mc^2)(1 - \cos\phi)]} \tag{4.2}$$

Compton scattering from a "free" moving electron can them be simply understood as a Doppler effect, the motion of the electron along the x-ray scattering vector causing a frequency (and hence energy) shift proportional to the projection of the electron's momentum (velocity) onto the scattering direction which is defined as the scattering vector, $\boldsymbol{K} = \boldsymbol{k}_1 - \boldsymbol{k}_2$ [1]. If this direction is defined as the z-axis of a Cartesian coordinate system then the energy of the scattered photon E_2 is related to p_z the electron's momentum resolved along the scattering vector, $\boldsymbol{K}$

J.A. Duffy (✉) • M.J. Cooper
Department of Physics, University of Warwick, Coventry, CV4 7AL, UK
e-mail: j.a.duffy@warwick.ac.uk; m.j.cooper@warwick.ac.uk

C. Gatti and P. Macchi (eds.), *Modern Charge-Density Analysis*,
DOI 10.1007/978-90-481-3836-4_4,

$$\frac{p_z}{mc} = \frac{(E_2 - E_1) + E_1 E_2 (1 - \cos\varphi)\,/mc^2}{(E_1^2 + E_2^2 - 2E_1 E_2 \cos\varphi)^{\frac{1}{2}}}. \tag{4.3}$$

This results in a spectral line, which when plotted as a function of electron momentum, p_z, is symmetric about $p_z = 0$. It is the projection of the electron momentum distribution, $n(\boldsymbol{p})$ onto the scattering vector, is called the Compton profile and universally denoted $J(p_z)$. Thus the momentum density $n(\mathbf{p})$ is probed,

$$J(p_z) = \iint n(\mathbf{p}) dp_x dp_y. \tag{4.4}$$

The development of Compton scattering as a potent probe first of electron charge and then much later spin density owes much to the work of Richard J Weiss (1923–2008) whose book, *X-ray Determination of Electron Density Distributions* [2], introduced chemists and physicists to the possibilities of an alternative probe of charge density to x-ray diffraction. One which, due to its incoherent nature, avoided the big issues that then surrounded extinction in the interpretation of diffraction data. However, there were theoretical and experimental problems. In principle the full relativistic scattering cross section is itself dependent on the electron's momentum and the moving target electrons are bound, not free. In practice the incoherent scattering processes are inherently weak and so the low flux of scattered photons severely constrained statistical accuracy and resolution with which experiments could be performed. In fact the weakness of the scattering cross section, coupled with the limited x-ray flux of conventional x-ray tubes meant that subtleties associated with the momentum dependence of the cross section or binding effects destroying relationship between the measured spectra and the Compton profile, $J(p_z)$, were largely inconsequential. After much heart searching it has been well established that the Compton profile, as defined by Eq. 4.4 can be extracted from the cross section when the experiment is performed in the Impulse Approximation in which the energy transfer greatly exceeds the electron's binding energy and so it can be treated as if it were a free electron moving with the same velocity. Following Lovesey and Collins [3], the cross section can be most succinctly written as:

$$\frac{d^2\sigma}{d\Omega dE_2} = N\left(\frac{e^2}{mc^2}\right)\left(\frac{E_2}{E_1}\right)\left(\frac{m}{\hbar^2 K}\right) \times \left\{ \begin{array}{l} \left[1 + \cos^2\varphi + P_l \sin^2\varphi\right] \times \left[J\,(p_Z)\right] \\ +2\dfrac{E_1}{mc^2}\left[(\cos\varphi - 1)\, P_c \widehat{\sigma} \bullet (\mathrm{k}_1 \cos\varphi + \mathrm{k}_2)\right] \times \left[J_{mag}\,(p_Z)\right] \end{array} \right\} \tag{4.5}$$

The new symbols appearing in this expression are: P_l and P_C, which describe the linear and circular polarization of the incident photons respectively, $\hat{\sigma}$ which is a unit vector denoting the direction of the electron's spin moment and $J_{mag}(p_z)$. This last is usually called the magnetic Compton profile but more accurately would be termed the spin–dependent Compton profile. It is defined as the difference between the spin-up $n(\mathbf{p}\uparrow)$ and spin down $n(\mathbf{p}\downarrow)$momentum densities,

$$J_{mag}(p_z) = \iint (n(\mathbf{p}\uparrow) - n(\mathbf{p}\downarrow))\, d p_x d p_y. \tag{4.6}$$

Again it is the Impulse Approximation that is responsible for the absence of orbital magnetization: simplistically orbital motion is not sensed in an instantaneous interaction (i.e. one for which $\Delta E.\Delta t < \hbar$). This means that "magnetic" Compton scattering probes something different from neutron scattering where the total magnetization is always sampled by the neutron's magnetic moment. It was Platzman and Tzoar [4] who suggested that spin density could be probed specifically and would yield $J_{mag}(p_z)$. However, they left it to others to work out how to produce a polarised beam, of sufficient flux and with finite helicity!

4.2 Spin Densities in Momentum Space

The spin-dependent contribution to the scattering cross section is very small compared to the Compton cross section at conventional x-ray energies since, roughly speaking, it scales as (E_l/mc^2). Even at the higher energies available from gamma-ray sources ($E_l \sim mc^2$) it is unlikely that the spin dependent scattering from the small fraction of electrons with unpaired spins in a ferromagnet will be more than a few percent of the total Compton cross section from all the spin-paired electrons. For example in the archetypal ferromagnet, iron, just 10% of the electrons have unpaired spins and therefore at $E_l = mc^2$ the contribution to the cross section is only 1% and at 10 keV it would drop to below 0.05%.

In fact it is not obvious that charge and spin-dependent scattering can be separated because their scattering amplitudes are 90° out of phase. This is most easily understood classically by consideration of the effect of the electric field of an electromagnetic (EM) wave on an electron: its charge produces in-phase dipolar re-radiation but its effect upon the electron's magnetic (dipole) moment is to produce quadrupolar re-radiation. The factor of E/mc^2 also arises in this illustrative analysis, which was first presented by de Bergevin and Brunel [5] in their seminal papers on magnetic x-ray diffraction. The results carry across from diffraction to Compton scattering. Given this phase difference between the contributions to the scattering cross-section the only way to obtain a linear contribution to the cross section is to engineer a 90° phase shift. In diffraction there is the possibility of working at/near an absorption edge where the scattering factor has an appreciable imaginary component. At the high energies necessary for interpretable Compton scattering experiments this is not an option and the use of a complex polarization is the only practicable way to produce interference between the charge and spin-dependent scattering amplitudes. Then, as shown in Eq. 4.5, a first order term arises that is linear in spin and therefore switchable. It can be isolated by changing its sign. This in turn can be achieved either by reversing the hand of polarisation of the incident photon beam (i.e. flipping P_c), or by changing the direction of the unit spin vector, $\hat{\sigma}$, which is done by reversing the direction of the spin vector with an external magnetic field, always assuming that the material is not to magnetically hard for this. Now there is an interference term $\sim (E_1/mc^2)$ assuming a high degree of circular polarization can be achieved and a possibility that the magnetic Compton profile can be determined.

This was all known over 50 years ago. However, the interest from experimentalists was not at all in measurements of spin densities of moving, bound electrons, but from particle physicists intent on measuring photon helicity in beta decay in order to understand the neutrino. Thus Evan's review [6] refers to a Compton polarimeter in which the degree of photon helicity was determined from change in Compton scattered flux when a piece of iron was magnetized alternately parallel and antiparallel to the scattering vector. In 1970 Platzman and Tzoar [4] developed the cross section for a free moving electron, through expansions in powers of (E/mc^2) and showed that it would be possible to extract a quantity exactly akin to the Compton profile which is a projection of the spin-resolved momentum density $n(\mathbf{p}\uparrow) - n(\mathbf{p})\downarrow$. First Sakai and Ôno [7] then Sakai, Terashima and Sekizawa [8] turned the Compton polarimeter methodology on its head to use the now proven helicity of a cooled beta emitter to measure the magnetic Compton profile of ferromagnetic iron through the Compton scattering process: landmark results that defied not only the weakness of the "magnetic" scattering but also the weakness of the beta source. They demonstrated unequivocally that the spin density was being probed in an interpretable manner. It did not, however, offer an exploitable way forward: that had to wait until higher fluxes of circularly polarized photons could be extracted from synchrotron radiation beams.

Synchrotron radiation is habitually and correctly described as being linearly polarized in the orbital plane of the machine but it is often forgotten that it is elliptically polarized out of that plane, assuming the radiation originates from a single bend (or a dominant bend) of the electron beam (an undulator with alternate left and right hand bends of the electron trajectory will exhibit no degree of helicity out of the orbital plane, but a simple bending magnet or an asymmetric wiggler will). The degree of circular polarization that can be achieved out of the plane is limited by the decline in flux but the "inclined view method" was shown to be viable at the photon energies ($>$100 keV) needed for interpretable experiments [9]. Mills [10] demonstrated the use of a phase plate to convert linear to circular polarization and others [11] developed synchrotron insertion devices that produced circular polarization on axis. However, in the intervening two decades it is the "inclined view method" that has been generally adopted with asymmetric wigglers producing high energy flux at incident energies around 200 keV.

The pioneering of the experimental method stimulated further work on the relativistic cross section and an investigation of the assertion that orbital magnetization plays no role in the "Compton limit". This was shown experimentally [12] and explained for example in the chapter by Schülke in Cooper et al. [13]. Some treatments extend the quasi-relativistic Hamiltonian for bound electrons in a manner that does not preserve the direct relationship between the cross section and the Compton profile [14, 15]. However, the generally accepted treatment [16] follows the work of Ribberfors [17] in treating the electrons within the spirit of the impulse approximation as free but moving. In addition the initial electron momentum $\mathbf{p}$ is treated as small ($p <$ mc), which is a reasonable approximation for the unpaired spin electrons which are the slowly moving valence electrons in most ferromagnets. In the rare earths and transition metals they are the "inner shell" *4f* or *3d* electrons but nonetheless momenta still comfortably satisfy the above inequality. By this means the direct relationship between the cross section and the magnetic Compton profile is preserved. Tests on this cross section in terms of the symmetry of the profile (the Compton profile and the magnetic Compton profile are strictly symmetric because $n(\mathbf{p})$ is centro-symmetric) show that it provides a description of the scattering process that is perfectly adequate for the interpretation of spin-dependent scattering at the level of current experimental accuracy.

4.3 Experimental Approaches

Compton scattering studies of electron charge and spin densities have been thoroughly reviewed in the research monograph by Cooper et al. [13] to which the reader is referred for further reading. Here we will summarise the salient points and give a very limited number of examples of the "state-of the-art" in spin density studies. The production of the circular polarization itself is normally achieved by lining the experiment up at a slight angle (typically a few microradians) to the orbital plane of the synchrotron so that the incident radiation is elliptically polarized. The optimum conditions have been discussed by McCarthy et al. [18]. The spin dependent final term in the equation can then be isolated either by flipping the polarization of the photons, e.g. by moving from above to below the orbital plane, which is impracticable, using a phase plate which is equally problematic at such high energies (~200 keV), using a reversible polarization insertion device or reversing the magnetization of the sample. The last option is possible in soft magnetic materials but difficult in hard ones; indeed some studies of the latter have proceeded by physically rotating the permanent magnet. The detection and analysis of the small magnetic modulation of the charge scattering signal requires dispersive detectors with linearity over a large dynamic range. In fact Ge multi-head semiconductor detectors (typically 10–13 heads) are used to provide multiple parallel channels of data acquisition. The use of semiconductor detectors places a limit on the resolution of ~0.40 a.u., much inferior to what is possible in charge scattering Compton studies with dispersive spectrometers (<0.10 a.u., [19]) but the requirement for high statistical accuracy, driven by the need to separate the small magnetic modulation from the charge scattering dictates that the low resolution/high flux method has to be used. It is common practice to compare experiment and theory by only tail-stripping the data and then convoluting the theory with a Gaussian function, rather than attempting a full deconvolution of the Magnetic Compton Profile (MCP). Data need to be corrected for the energy dependence of the detector efficiency, sample absorption and of the relativistic cross-section. Although multiple scattering can affect the total profile significantly the MCP, isolated by field reversal, will typically contain only a negligible amount of multiple spin-dependent scattering and no correction need be applied.

It has always been clear that the direct interpretation of a Compton profile is almost impossible. The fact that the momentum density is projected down a line and that all the electrons contribute equally means that one of a number of strategies needs to be invoked in order to extract physically useful information: there is no way in which the line shape can be interpreted directly. This proviso is strengthened in practice by the fact that there are many corrections which need to be applied to the measured double differential cross section in order to extract the Compton profile. The energy dependence of the cross section has been alluded to above. In addition to that there are a number of energy dependent corrections for sample absorption, detector efficiency and detector resolution, which together with multiple scattering complicates the task enormously: readers are referred to the chapter by Zukowski in Cooper et al. [13] and elsewhere in this book. Suffice it to say that the correction algorithms are sufficient for measurements to pass the acid test of producing a profile $J(p_z)$ or $J_{mag}\ (p_z)$ that has symmetry about $p_z = 0$ within the statistical accuracy.

One way of minimizing systematic errors in the analysis of Compton profiles has been to look at directional difference profiles

$$\Delta J\ (p_Z) = J_{h',k',l'}\ (p_Z) - J_{h'',k'',l''}\ (p_Z)\,, \tag{4.7}$$

where h', k', l' and h'', k'', l'' represent different directions in a single crystal sample, i.e. the miller indices of the planes normal to the x-ray scattering vector. By this means energy–dependent systematic corrections are eliminated because they are common to both directions. Multiple scattering by its nature has very little dependence on the sample orientation and is therefore largely removed from the difference profile. The only problem remaining is to establish the correct scale for this difference and that does require knowledge (through Monte Carlo modeling) of the total amount of multiple scattering. Directional Compton profiles measured for multiple crystal directions can be used to reconstruct n($\mathbf{p}$) or the spin density, $n(\mathbf{p}\uparrow) - n(\mathbf{p})\downarrow$. Reconstruction algorithms are discussed in the contributions of Hansen and Dobrzynski in Cooper et al. [13]. The Fourier method is usually invoked in which the profile transforms (the so-called reciprocal form factor) is interpolated on a grid and then back transformed. Alternatively the maximum entropy can be used. The use of the Fourier method was established by Tanaka et al. [20] who reconstructed the spin density of iron (Fe stabilized with 3% Si) from 13 directional Compton profiles. More recently Nagao et al. [21] applied the same approach (again 13 directional profiles) to nickel, in a study which is discussed as an exemplar below.

Magnetic Compton profiles are naturally differences (see Eq. 4.6) obtained by flipping the magnetic field or possibly by reversing the helicity of the radiation.

Therefore they automatically benefit from relative freedom from systematic errors. Even the multiple scattering problem is less for the magnetic profile than the charge profile (by as much as an order of magnitude) as was pointed out by Sakai [22]. This is because there is a change of sign in the amplitude of scattering between small angle and high angle scattering. Thus Monte Carlo simulations provide a confident correction to the amount of multiple magnetic scattering. The estimation of multiple scattering is essential in data analysis because Compton profiles need to be normalized, viz:

$$\int J(p_z)dp_z = Z \qquad \int J_{mag}(p_z)dp_z = \mu_{spin}, \tag{4.8}$$

where Z and μ_{spin} are the number of electrons and the spin moment (in Bohr magnetons) respectively, either per formula unit or per unit cell. The choice is arbitrary, but the removal of multiple scattering must precede normalisation if the correct scales for J_{mag} (p_z) and ΔJ_{mag} (p_z) are to be established. Interestingly the fact that area under the magnetic Compton profile equates to the total spin moment allows the total orbital moment to be deduced by combining the Compton data with bulk magnetization measurements. In this limited sense both spin and orbital magnetic moments can be deduced experimentally.

Further analysis can be facilitated by noting that, in Eq. 4.4, n($\mathbf{p}$) can be written as the sum of the contributions from the majority and minority bands, $n(\mathbf{p}\uparrow) + n(\mathbf{p}\downarrow)$, and so majority and minority Compton profiles can be separated by combining J(p_z) with $J_{mag}(p_z)$ as was first shown by Collins et al. [23].

In the following section a number of exemplar studies are presented to illustrate what is currently attainable and in which directions future studies are likely to focus. The final comment that should be made here is simply to emphasize the need for some accompanying theory to facilitate interpretation, as illustrated by the examples below.

4.4 Recent Research

4.4.1 Combining Real and Momentum Space Studies

There has been little work combining real and momentum space studies of magnetic materials. Magnetic Compton scattering (MCS) is certainly complementary to polarised neutron diffraction (PND) (*cf.* Chap. 8), which is sensitive to the total moment, rather than just the spin, and indeed, is effectively substantially more sensitive than MCS [24]. Because of the fact that different atomic sites' contributions to different Bragg peaks vary with the structure factor, it is often possible to gain site specific information. The main disadvantage of PND is the requirement to make measurements at Bragg reflections, with the consequence that it can be insensitive to moments on itinerant electrons (for example the conduction electrons). Furthermore, in rare earth and actinide systems, the sensitivity to the total moment can be a disadvantage when the spin and orbital moments are antiparallel. A good example is $SmMn_2Ge_2$, where the *4f* spin and orbital moments nearly cancel. MCS was able to clarify the magnetism in this material [25], and was able to show that the Sm site ordering changes as a function of applied magnetic field, even though its total moment remains constant: the antiparallel spin and orbital moments increase at the same rate [26].

In recent work Qureshi et al. [27] combined polarised neutron diffraction and MCS in a single combined study of $Co_3V_2O_8$. Here results from both experiments were used together to achieve a stable refinement of the magnetic structure. As a result, the authors were able to determine the magnetic state of the different Co sites, which had been unclear from previous work, and reveal the involvement of the induced magnetism of the V and O sites. The authors note that there are discrepancies, and comment that these may be reduced if solid state effects were included in their calculations. Typically, PND experiments have been interpreted using orbital calculations: the sensitivity of MCS to itinerant moments has the consequence that the limitations of such an approach become apparent.

4.4.2 Spin Polarisation in Highly Spin-Polarised Materials

Spintronic materials are the subject of considerable research and rapid technological development [28]: indeed some commercial devices exist. Because of the potential for the development of novel devices using spin transport in semiconductors, so-called spin-injector materials, which would act as a source of highly spin-polarized (SP) electrons, currently attract considerable interest. Ferromagnets (FMs) are an obvious choice: they naturally have an imbalance of the electron spin population at the Fermi level and thus possess a degree of spin-polarization. Ideally, materials for applications would be fully spin-polarized 'half-metals' (HMs), where the density of states (DOS) at the Fermi level is finite for one spin, but zero for the other, such that carriers of only one spin exist at the Fermi level. Unfortunately, ferromagnets typically only possess partial spin-polarization, and much research effort is being applied to finding new candidates.

Spin polarisation is defined in terms of the spin-polarised density of states, $N_{\uparrow,\downarrow}$, at the Fermi level. Experimental methods are typically also sensitive to the spin polarised Fermi velocity v_F, and so the spin polarisation is normally defined as [29]:

$$P_n = \frac{N_\uparrow v^n_{F,\uparrow} - N_\downarrow v^n_{F,\downarrow}}{N_\uparrow v^n_{F,\uparrow} + N_\downarrow v^n_{F,\downarrow}}$$

here v_F is raised to the power n, where n = 0,1,2.

Experimental studies of P_n have proved to be problematic. Magneto-resistance studies are not sensitive to the sign of the effect. Point contact Andreev reflection (PCAR) is surface sensitive, and the dependence on the Fermi velocity is not always apparent: as shown in the case of $Co_{1-X}Fe_XS_2$ below, this appears to vary with the concentration x [30]. Another new method [31], which indirectly probes the polarisation via measurements of the magnetic relaxation time shows promise.

Magnetic Compton scattering has been used in conjunction with band structure calculations to determine the spin polarisation of Ni and $Co_{1-X}Fe_XS_2$ [32]. The method permits the determination of the bare spin polarisation and its Fermi velocity weighted values. CoS_2 has been predicted to have a high degree of spin polarisation, and, furthermore, the potential to tune the polarization via Fe doping. This makes it a good case for using this technique.

Band structure calculations, such as those using the LMTO method can be used to predict a number of experimental observables, including the electron momentum density, and hence the Compton profile, and parameters including the Fermi velocity. However, the bare *ab-initio* calculations, even using full potential techniques, are normally unable to reproduce the itinerant part of the MCP satisfactorily. This is particularly true in the case of transition metal systems. For example, see the work on nickel [33], where experimental data was compared with three *ab-initio* calculations. This is perhaps not surprising given that the density of states is usually sharply peaked near E_F for these flat bands. To address this, Major et al. [34] showed that applying a small energy shift to the relevant bands can effect a major improvement, demonstrating this on the same experimental nickel data [33]. Applying small energy shifts makes subtle differences to the calculated MCPs, and a fitting procedure is adopted to determine the shift values that give the best agreement with experiment.

In the new work, this has been used to determine the spin polarisation based on the fitted bands. This procedure has been tested using the nickel MCPs. Experimental transport measurements performed on Ni found $P_2 = 23\%$ [35]. However ab initio calculations by Mazin [29] predicted $P_2 = 0\%$. In the new work, the ab initio calculations were in agreement with this value. The fitting procedure was then applied, using experimental MCPs measured along four crystallographic directions. The refinement furnished a value of $P_2 = 20\%$, representing a dramatic improvement.

Turning to $Co_{1-x}Fe_xS_2$, a previous measurement of the spin polarisation has been made using PCAR. However, as indicated above, and discussed generally by Woods et al. [36], the relevant definition of P is not clear, and indeed is thought to change with Fe concentration, probably because the electron transport changes from ballistic to diffusive. Furthermore, the surface sensitivity may be problematic, as it is believed that S may diffuse out of the surface changing the stoichiometry.

The MCPs measured, together with the predicted profiles are shown in Fig. 4.1. The spin-polarization values, $P_{0,1,2}$ determined from the data are presented with the PCAR data in Fig. 4.2. The fitting procedure was applied to two samples, CoS_2 and $Co_{0.9}Fe_{0.1}S_2$. The results indicated that it is indeed difficult to be precise about the nature of the polarization as determined from the PCAR, as those experimental

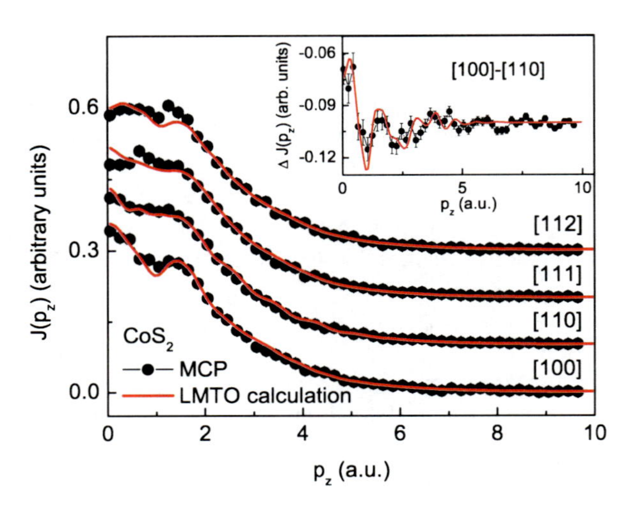

Fig. 4.1 The experimental MCPs for CoS_2, measured at 10K on beamline BL08w at SPring-8, together with the results of the LMTO calculations. The inset shows the anisotropy. Reproduced from Ref. [32] with kind permission of The American Physical Society

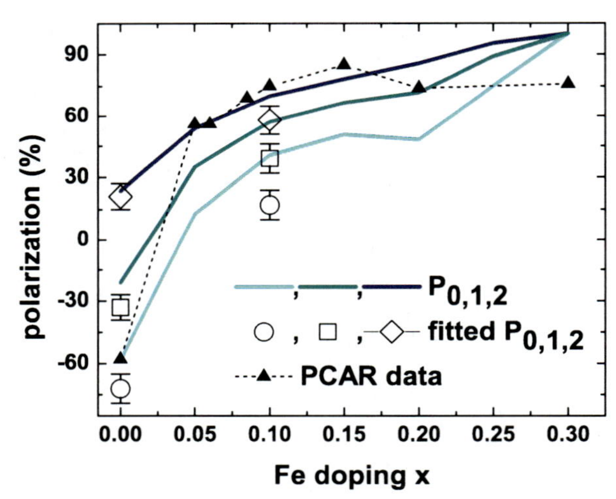

Fig. 4.2 Spin polarization in $Co_{1-x}Fe_xS_2$. The *filled triangles* represent the PCAR data reproduced from Wang et al. [30]. The *coloured lines* represent the unfitted ab initio calculations of $P_{0,1,2}$ by Utfeld et al. [32]. The *open symbols* show the fitted values using the experimental MCPs. Reproduced from Ref. [32] with kind permission of The American Physical Society

results appear to depend on the Fermi velocity in a non-trivial manner. Note that the PCAR polarization for CoS_2 was inferred to be negative from magnetoresistance measurements: the sign cannot be determined from the measurement.

4.4.3 Use of Reconstruction

Reconstruction of either a two-dimensional projection or the full three-dimensional momentum density has not been used extensively in MCS studies, although such methods are often used in related techniques such as PND and positron annihilation. Part of the reason is that many one-dimensional projections are required to produce a reliable reconstruction. However, such work is worth pursuing for the extra interpretation that may be achieved.

A study on nickel has been published by Nagao et al. [21], involving reconstruction of the three-dimensional spin density, and detailed comparison with a full potential calculation (FLAPW). As described in Sub-Sect. 4.4.2, nickel has been well-studied by MCS over recent years. This, together with its particular intricate and anisotropic spin density makes nickel an ideal test case. Back in 1998, Dixon et al. [33] compared four directional MCPs with three different ab-initio calculations, showing distinct discrepancies at low momentum. However, it was difficult to determine the cause of the discrepancy. It is worth noting that the four projections obtained were the maximum possible at that time and at the relatively good resolution at the ESRF. In 2004, Major et al. [34] showed how shifting certain bands could improve the agreement between experiment and theory, and that this also improved the predicted Fermi surface and calculated spin polarization.

In order to reconstruct the three-dimensional spin density, reproduced in Fig. 4.3, Nagao et al. measured MCPs along 13 directions in the crystal. The authors showed that the discrepancy between experiment and theory can be attributed to the 5th contributing band, which has d-like character, rather than the negatively polarised s and p-like bands. They also found negatively aligned polarisation near the first Brillouin zone (BZ) boundary, attributed to the 1st p-like band. Such reconstructions can clearly aid interpretation and provide a stringent test of theory.

4.5 X-ray Diffraction-Based Studies of Spin Density

Whereas Compton scattering studies of spin density are limited to ferromagnets by the incoherent nature of the scattering process, x-ray diffraction studies of spin and orbital magnetisation are predominantly reported for antiferromagnets because the very weak superlattice peaks associated with the magnetic lattice are not coincident with the charge peaks and therefore they may be observable against the background. Moreover they are observable with unpolarised as well as polarised beams. This field, which has been authoritatively reviewed by Lovesey and Collins [3], really began with the experiments of de Bergevin and Brunel [4, 37] in observing magnetic superlattice peaks from NiO with a standard x-ray tube source: a remarkable achievement. Later Gibbs et al. [38], using synchrotron radiation, revealed the spin-slip discommensurations in the spin structure of the rare earth holmium. They were also able to isolate the spin and orbital terms experimentally and thereby show that Hund's rules do indeed give an appropriate description of the Ho^{3+} ion.

The scattering cross section contains terms directly associated with both spin magnetization density, $\boldsymbol{S}$ and orbital magnetization density $\boldsymbol{L}$ through the term $\boldsymbol{K} \times \boldsymbol{p}$ [39]. Once again these terms scale as E/mc^2 compared to the charge scattering.

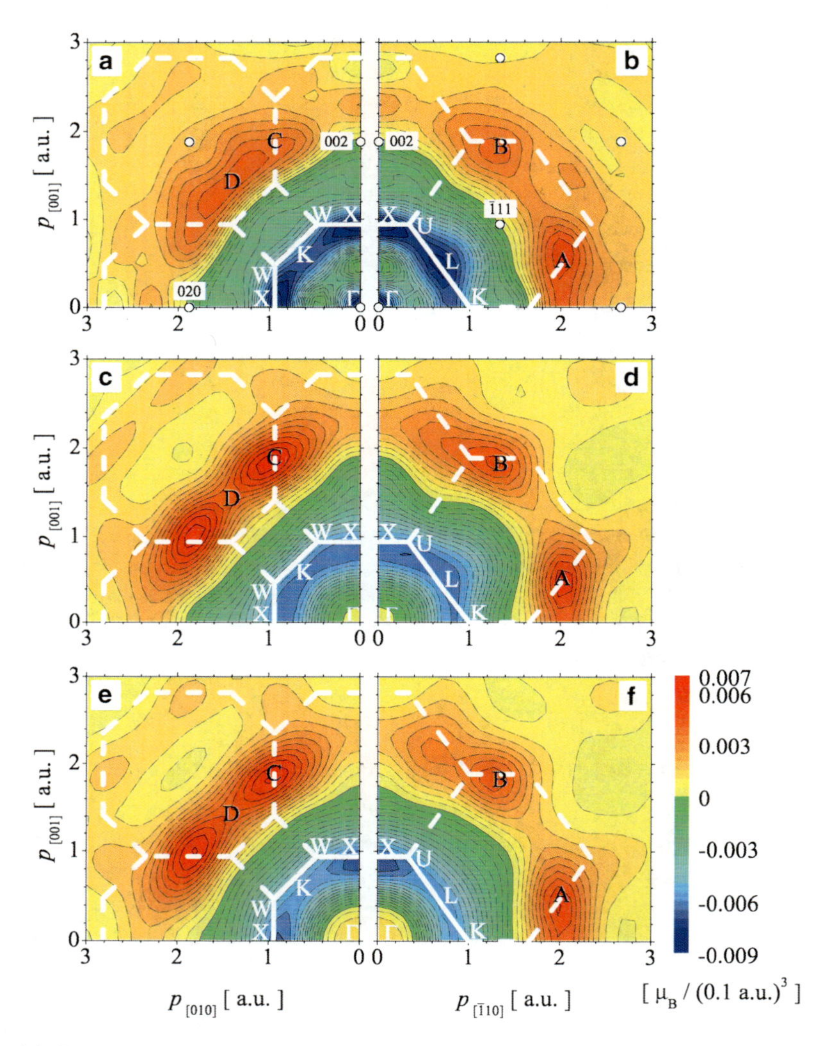

Fig. 4.3 Reconstructed momentum-space spin density in Ni. The experimental reconstructions in the [100] and [110] planes are shown in (**a**) and (**b**). The corresponding theoretical densities are shown in (**c**) and (**d**), and convoluted with the experimental resolution in (**e**) and (**f**). Reproduced from Ref. [21] with kind permission of IOP Publishing Ltd

$$\left(\frac{d^2\sigma}{d\Omega dE}\right) = \left(\frac{e^2}{mc^2}\right)^2 \left[\langle b| \sum_j e^{iK\cdot r_j} |a\rangle \widehat{\varepsilon} \bullet \widehat{\varepsilon}' - \frac{i\hbar\omega}{mc^2} \langle b| \sum_j e^{iK\cdot r_j} \left(\frac{iK \times p_j}{\hbar k^2} \bullet A + s_j \bullet B \right) |a\rangle \right]. \times \delta\left(E_a - E_b - \left(\hbar\omega'_{k_1} - \hbar\omega_{k_2}\right)\right) \tag{4.9}$$

For example for a centrosymmetric ferromagnet this reduces to:

$$\frac{d\sigma}{d\omega} = \frac{1}{2} r_e^2 F_C \left[F_C (1 + \cos^2\theta - P_L \sin^2\theta) \right] + 2\tau P_C (1 - \cos\theta) \left\{ \begin{array}{l} \mathbf{F}_S \cdot (\mathbf{k}_1 \cos\theta + \mathbf{k}_2) \\ + (1 + \cos\theta)\, \mathbf{F}_L \cdot (\mathbf{k}_1 + \mathbf{k}_2) \end{array} \right\} \tag{4.10}$$

where F_C , $\mathbf{F}_S$ and $\mathbf{F}_L$ are the Fourier transforms of the electron charge, spin and orbital densities, respectively, and r_e is the classical electron radius. The incident and scattered wave-vectors are denoted $\boldsymbol{k}_1$ and $\boldsymbol{k}_2$ and θ is the Bragg angle. The first term in Eq. 4.10 is the cross-section for charge scattering, which is proportional to P_l. The second term describes the magnetic scattering and changes sign if P_C or $\mathbf{F}_S$ and $\mathbf{F}_L$ are reversed. Also, the magnetic term is scaled by the factor τ $(=E/mc^2)$.

However, almost all studies of antiferromagnets are performed "at resonance" i.e. at an x-ray energy close to an absorption edge for which the selection rules permit transitions between the inner shell and the spin split valence band (so the L-edges for rare earths such as holmium and the M-edges of the actinides such as uranium or plutonium). The resonant enhancements are $\sim 10^2$ and $\sim 10^6$ respectively for these two classes of material. Clearly these massive enhancements make the observation of magnetic superlattice peaks much easier than ever imagined and much structural information can be determined simply from the observation of the position of the satellite reflections. However the cross section at resonance does not retain the simple relationship with $\mathbf{F}_S$ and $\mathbf{F}_L$ but is more to do with the densities of states in the sub bands and the nature (dipole, quadrupole, octupole etc.) of the transitions between filled and empty states. It is only recently that ab-initio calculations have been able to model this process [40].

Studies of ferromagnets suffer from the same fundamental problem as Compton studies of spin density, namely that the charge and spin terms are coincident. Once again the scattering amplitudes are in quadrature so that the use of a complex polarisation (elliptical or circular polarisation) produces a first order interference term whose sign can be reversed either by reversing the applied magnetic field or the helicity of the beam. Unlike Compton scattering, diffraction is generally always sensitive to some combination of spin and orbital magnetisation and the choice of geometry. In particular the total magnetisation ($\boldsymbol{L} + 2\boldsymbol{S}$) can be measured in an experiment with the field parallel to a beam diffracted through 90° whereas with the same scattering angle and the field parallel to the incident beam the orbital magnetisation, $\boldsymbol{L}$, is isolated (another geometry isolates the spin term). The fact that these geometries need a Bragg angle fixed at 45° fixes the wavelength that can be used to observe a specific *hkl* reflection. Alternatively if a white beam is used a wide range of harmonic reflections from the *hkl* face can be observed. It is this "white beam method" that was originally developed [41, 42]. The amplitude of the magnetic term depends on the ratio $P_C / 1 - P_L$ where P_C and P_L are the degrees of circular and linear polarisation. Thus there is an optimum "out of plane" angle of view to maximise this function for each synchrotron. It is also clear that electron beam source movements will lead deterioration in the ratio and lack of reproducibility. Whilst this cannot be circumvented in the white beam method there are strategies that can be adopted with monochromatic radiation on a synchrotron beamline. First of all if a wavelength can be chosen for which the monochromator has a Bragg angle of 45° then on emerging from the double crystal monochromator the beam will have a high degree of linear polarisation, P_L which can then be converted to an optimum mixture of P_C and P_L with a phase plate. Of course the fixed wavelength then severely limits the reflection that can be studied and indeed only one reflection can be studied. For example Ito and Hirano [43] were able to study the 220 reflection from Fe by this approach and were able to increase the magnetic effect to $\sim 2\%$ *cf* $\sim 1/2\%$ obtainable in the white beam method. Once again beam movements change the amount of linear, circular and unpolarised radiation transmitted by the monochromator. Bouchenoire et al. [44] essentially went one stage further by adding a second channel cut monochromator which ensured that the beam delivered to the phase plate retained a very high degree of linear polarisation despite beam movements, which only affected the flux transmitted. By this means the magnetic effect for the same 220 reflection from Fe was increased to 6%. It is however difficult to see how this method can be used to obtain a large data set on magnetic scattering amplitudes.

References

1. Klein O, Nishina Y (1929) Über die Streuung von Strahlung durch freie Elektronen nach der neuen relativistischen Quantendynamik von Dirac. Z Phys 52:853–868
2. Weiss RJ (1966) X-ray determination of electron density distributions. North Holland publishing company, Amsterdam
3. Lovesey SW, Collins SP (1996) X-Ray scattering and absorption by magnetic materials. Oxford University Press, Oxford
4. Platzman P, Tzoar N (1970) Magnetic scattering of X rays from electrons in molecules and solids. Phys Rev B 2:3556–3559
5. de Bergevin F, Brunel M (1981) Diffraction of X-rays by magnetic materials. I. General formulae and measurements on ferro- and ferrimagnetic compounds. Acta Crystallogr A 37:314–324; Brunel M, de Bergevin F (1981) Diffraction of X-rays by magnetic materials. II. Measurements on antiferromagnetic Fe_2O_3. Acta Crystallogr A 37:324–331
6. Evans RD (1958) The Compton effect. In: Flügge S (ed) Handbuch der Physik, vol 34. Springer, Berlin, pp 218–298
7. Sakai N, Ôno K (1976) Compton profile due to magnetic electrons in ferromagnetic iron measured with circularly polarized γ rays. Phys Rev Lett 37:351–353
8. Sakai N, Terashima O, Sekizawa H (1984) Magnetic Compton profile measurement using circularly polarized gamma-rays from oriented ^{191m}Ir nuclei. Nucl Instrum Methods Phys Res 221:419–426
9. Cooper MJ, Collins SP, Timms DN, Brahmia A, Kane PP, Holt RS, Laundy D (1988) Magnetic-anisotropy in the electron momentum density of iron. Nature 333:151–153
10. Mills DM (1987) Spin-aligned momentum distributions of transition-metal ferromagnets studied with circularly polarized synchrotron radiation. Phys Rev B 36:6178–6181
11. Yamamoto S, Kawata H, Kitamura H, Ando M, Sakai N, Shiotani N (1989) First production of intense circularly polarized hard x rays from a novel multipole wiggler in an accumulation ring. Phys Rev Lett 62:2672–2675

12. Cooper MJ, Zukowski E, Collins SP, Timms DN, Itoh F, Sakurai H (1992) Does Compton scattering only measure spin magnetisation? J Phys Condens Matter 4:L399–L404
13. Cooper MJ, Mijnarends PE, Shiotani N, Sakai N, Bansil A (2004) X-ray Compton scattering. Oxford University Press, Oxford
14. Grotch H, Kazes E, Bhatt G, Owen DA (1983) Spin-dependent Compton scattering from bound electrons: quasi-relativistic case. Phys Rev A 27:243–256
15. Bhatt G, Grotch H, Kazes E, Owen DA (1983) Relativistic spin-dependent Compton scattering from electrons. Phys Rev A 28:2195–2200
16. Bell F, Felsteiner J, Pitaevskii LP (1996) Cross section for Compton scattering by polarized bound electrons. Phys Rev A 53:R1213–R1215
17. Ribberfors R (1975) Relationship of the relativistic Compton cross section to the momentum distribution of bound electron states. Phys Rev B12:2067–2074
18. McCarthy JE, Cooper MJ, Honkimaki V, Tschentscher T, Suortti P, Gardelis S, Hamalainen K, Manninen SO, Timms DN (1997) The cross-section for magnetic Compton scattering up to 1 MeV. Nucl Instrum Methods A 401:463–475
19. Suortti P, Buslaps T, Fajardo P, Honkimaki V, Kretzschmer M, Lienert U, McCarthy JE, Renier M, Shukla A, Tschentscher T, Meinander T (1999) Scanning x-ray spectrometer for high-resolution Compton profile measurements at ESRF. J Synchrotron Radiat 6:69–80
20. Tanaka Y, Sakai N, Kubo Y, Kawata H (1993) Three-dimensional momentum density of magnetic electrons in ferromagnetic iron. Phys Rev Lett 70:1537–1540
21. Nagao T, Kubo Y, Koizumi A, Kobayashi H, Itou M, Sakai N (2008) Momentum–density distribution of magnetic electrons in ferromagnetic nickel. J Phys Condens Matter 20: 055201–055208
22. Sakai N (1987) Simulation of Compton double scatterings of circularly polarized gamma-rays by magnetic electrons. J Phys Soc Jpn 56:2477–2485
23. Collins SP, Cooper MJ, Timms DN, Brahmia A, Laundy D, Kane PP (1989) Minority and majority spin band directional Compton profiles of ferromagnetic iron. J Phys Condens Matter 1:9009–9020
24. Duffy JA, Hayden SM, Maeno Y, Mao Z, Kulda J, McIntyre GJ (2000) A polarized neutron scattering study of the Cooper-pair moment in Sr_2RuO_4. Phys Rev Lett 85:5412
25. McCarthy JE, Duffy JA, Detlefs C, Cooper MJ, Canfield P (2000) *4f* spin density in the reentrant ferromagnet $SmMn_2Ge_2$. Phys Rev B 62:r6073–r6076
26. Bebb AM, Taylor JW, Duffy JA, Banfield ZF, Cooper MJ, Lees MR, McCarthy JE, Timms DN (2005) Temperature and field dependence of the spin magnetization density in $SmMn_2Ge_2$. Phys Rev B 71:024407
27. Qureshi N, Zbiri M, Rodriguez-Carvajal J, Stunault A, Ressouche E, Hansen TC, Fernandez-Diaz MT, Johnson MR, Fuess H, Ehrenberg H, Sakurai Y, Itou M, Gillon B, Wolf Th, Rodriguez-Velamazan JA, Sanchez-Montero J (2009) Experimental magnetic form factors in $Co_3V_2O_8$: a combined study of ab initio calculations, magnetic Compton scattering, and polarized neutron diffraction. Phys Rev B 79:094417
28. Wolf SA, Awschalom DD, Buhrman RA, Daughton JM, von Molnar S, Roukes ML, Chtchelkanova AY, Treger DM (2001) A spin-based electronics vision for the future. Science 294:1488–1495
29. Mazin II (1999) How to define and calculate the degree of spin polarization in ferromagnets. Phys Rev Lett 83:1427–1430
30. Wang L, Umemoto K, Wentzcovitch RM, Chen TY, Chien CL, Checkelsky JG, Eckert JC, Dahlberg ED, Leighton C (2005) $Co_{1-x}Fe_xS_2$: a tunable source of highly spin-polarized electrons. Phys Rev Lett 94:056602
31. Müller GM, Walowski J, Djordjevic M, Miao GX, Gupta A, Ramos AV, Gehrke K, Moshnyaga V, Samwer K, Schmalhorst J, Thomas A, Hutten A, Reiss G, Moodera JS, Munzenberg M (2009) Spin polarization in half-metals probed by femtosecond spin excitation. Nat Mater 8:56–61
32. Utfeld C, Giblin SR, Taylor JW, Duffy JA, Shenton-Taylor C, Laverock J, Dugdale SB, Manno M, Leighton C, Itou M, Sakurai Y (2009) Bulk spin polarization of $Co_{(1-x)}Fe_xS_2$. Phys Rev Lett 103:226403

33. Dixon MAG, Duffy JA, Gardelis S, McCarthy JE, Cooper MJ, Dugdale SB, Jarlborg T, Timms DN (1998) Spin density in nickel: a magnetic Compton scattering study. J Phys Condens Matter 10:2759–2771
34. Major Z, Dugdale SB, Watts RJ, Laverock J, Kelly JJ, Hedley DCR, Alam MA (2004) Refining Fermi surface topologies from ab-initio calculations through momentum density spectroscopies. J Phys Chem Sol 65:2011–2016
35. Meservey R, Tedrow PM (1994) Spin-polarized electron tunnelling. Phys Rep 238:173–243
36. Woods GT, Soulen RJ Jr, Mazin I, Nadgorny B, Osofsky MS, Sanders J, Srikanth H, Egelhoff WF, Datla R (2004) Analysis of point-contact Andreev reflection spectra in spin polarization measurements. Phys Rev B 70:054416
37. de Bergevin F, Brunel M (1972) Observation of magnetic superlattice peaks by X-ray diffraction on an antiferromagnetic NiO crystal. Phys Lett A 39:141–142
38. Gibbs D, Moncton DE, D'Amico KL, Bohr J, Grier B (1985) Magnetic x-ray scattering studies of holmium using synchrotron radiation. Phys Rev Lett 55:234–237
39. Blume M (1985) Magnetic scattering of x-rays. J Appl Phys 57:3615–3618
40. Brown SD, Strange P, Bouchenoire L, Zarychta B, Thompson P, Mannix D, Stockton SJ, Horne M, Arola E, Ebert H, Szotek Z, Temmerman WM, Fort D (2007) Dipolar excitations at the L_3 x-ray absorption edges of the heavy rare earth metals. Phys Rev Lett 99:247401
41. Collins SP, Laundy D, Rollason AJ (1992) Magnetic form factors of ferromagnetic iron by x-ray diffraction. Philos Mag B65:37–46
42. Zukowski E, Cooper MJ, Armstrong R, Ito M, Collins SP, Laundy D, Andrejczuk A (1992) Magnetic form factor of nickel determined by white beam x-ray diffraction. J X-ray Sci Technol 3:300–310
43. Ito M, Hirano K (1997) Enhanced magnetic diffraction from ferromagnetic iron. J Phys Condens Matter 9:L613–L617
44. Bouchenoire L, Brown SD, Thompson P, Detlefs C, Cooper MJ (2006) Polarisation optimisation for ferromagnetic diffraction: a case study of iron. Nucl Instrum Methods A 566:733–738

Chapter 5
Past, Present and Future of Charge Density and Density Matrix Refinements

Jean-Michel Gillet and Tibor Koritsanszky

5.1 Introduction

Scattering experiments are extremely powerful techniques to study the structure and dynamics of solid-state materials. In the course of such experiments, a sample is irradiated and some characteristics of the emitted wave are detected. Quantitative details and the physical content of the information that can be extracted from the observations thus depend on the properties of the scattering system, the primary radiation and the detector. Assuming that the properties of the latter two are known or under control, certain properties of the sample can, in principle, be retrieved from the measured data based on first principles of quantum physics. In what follows, first we introduce the theory underling X-ray scattering processes and give some details on the nature of the physical information embedded in the observed signal depending on the scattering regime considered.

In the simplest case, when no photon-electron energy transfer occurs, the diffraction pattern is related to the Fourier image of the electron distribution averaged over nuclear vibrations. This relationship must be inverted to obtain the thermally smeared electron density. We refer to this transformation as the inverse problem of diffraction. The word 'problem' enters the terminology for several reasons: (i) Since usually only intensity data are measured, the lack of phase information prohibits the direct inverse Fourier transform; (ii) The data are limited in number (the diffraction pattern is discrete) and contaminated by experimental errors; (iii) The ultimate goal

J.-M. Gillet (✉)
Structures, Propriétés et Modélisation des Solides, UMR8580, Ecole Centrale Paris, Grande Voie des Vignes, 92295 Chatenay-Malabry Cedex, France
e-mail: jean-michel.gillet@ecp.fr

T. Koritsanszky
Department of Chemistry, Computational Science Program, Middle Tennessee State University, Murfreesboro, TN 37132, USA
e-mail: tkoritsa@gmail.com

C. Gatti and P. Macchi (eds.), *Modern Charge-Density Analysis*,
DOI 10.1007/978-90-481-3836-4_5, © Springer Science+Business Media B.V. 2012

is to derive the static rather than the smeared density because the former one is a more illuminating manifestation of electronic effects. The interpretation of the experiment thus necessarily involves a great deal of modelling to render the complex relation between the 'object and its image' into a relatively simple but closed analytic function of a limited number of parameters. The best estimates of these parameters are derived by achieving the possible 'best match' between the model-predicted and measured data. Since this fitting can be considered as a projection from the data to the parameter space, and since the dimensionality of both spaces is part of the model, the parameterized inverse problem is *ill-posed* and lacking of a unique solution. We find it important to devote a section of this chapter to elaborate on the mathematical aspects of data refinement and the simultaneous analysis of more data sets of different physical content. This is followed by a discussion of basic models that have been applied so far to recover density matrices from X-ray data.

The rest of this chapter is about X-ray charge density refinement, the most mature and credited technique with the widest range of applications whose number has grown exponentially over the past decades. This is mainly due to the significant improvements in experimental techniques allowing for the collection of high resolution, near-to-complete and highly redundant Bragg data almost routinely and within a short period of time. While there has been a promising progress in X-ray constrained wave-function methods (see Chap. 6), the vast majority of studies utilize the nucleus-centered pseudoatom model. The advantages and limitations of this formalism are discussed in detail and some practical aspects of the multipole refinement are outlined. Uncertainties of the pseudoatom density are highlighted by some of the controversial results of recent comparative experimental, as well as simulated studies. We conclude the chapter by an analysis applicable to access the absolute error of the pseudoatom formalism and propose a relatively simple route to upgrading it.

We do not attempt to cover the topic in its entirety. Owing to page limitations, important techniques, such as the maximum entropy method (Chaps 1, 7 and 13), are not discussed. For a comprehensive coverage of the research done up to the 1990s, we refer to popular monographs [1, 2]. Our goal is to pinpoint some general trends emerging from recent applications of X-ray methods, rather than to give a review of these applications [3].

5.2 Reconstruction of Density Matrices

It is described elsewhere in this book how densities in position or momentum representations are linked through the reduced density matrix (Chap. 2). Here, we give only a brief summary of definitions relevant to the subsequent discussion on scattering theory. The pure state N-electron density matrix stems from the N-electron wavefunction:

$$\Gamma^{(N)}\left(\mathbf{x}_1...\mathbf{x}_N;\mathbf{x}'_1...\mathbf{x}'_N\right) = \langle \mathbf{x}_1...\mathbf{x}_N \mid \Psi\rangle\left\langle\Psi \mid \mathbf{x}'_1...\mathbf{x}'_N\right\rangle, \tag{5.1}$$

where $\mathbf{x}_i$ stands for the generalized spin-position coordinates of the ith particle. In this chapter, we will consider spin independent properties and thus need only spin-traced density matrices obtained by summing over all spin coordinates (with prior setting $\sigma_i = \sigma'_i$ for all particles):

$$\gamma^{(N)}\left(\mathbf{r}_1...\mathbf{r}_N;\mathbf{r}'_1...\mathbf{r}'_N\right) = \sum_{\sigma_1,...\sigma_N}\Gamma^{(N)}\left(\mathbf{x}_1...\mathbf{x}_N;\mathbf{x}'_1...\mathbf{x}'_N\right). \tag{5.2}$$

On this basis, one defines the n-electron reduced density matrix (n-RDM):

$$\gamma^{(n)}\left(\mathbf{r}_1...\mathbf{r}_n;\mathbf{r}'_1...\mathbf{r}'_n\right) = \binom{N}{n}\int\gamma^{(N)}\left(\mathbf{r}_1...\mathbf{r}_N;\mathbf{r}'_1...\mathbf{r}'_N\right)d\mathbf{r}_{n+1}...d\mathbf{r}_N\,d\mathbf{r}'_{n+1}...d\mathbf{r}'_N. \tag{5.3}$$

The 1-RDM and 2-RDM are related, respectively, to the electron density$\rho(\mathbf{r}_1)$ and the pair distribution $P(\mathbf{r}_1,\mathbf{r}_2)$,

$$\rho(\mathbf{r}_1) = \gamma^{(1)}\left(\mathbf{r}_1;\mathbf{r}'_1 = \mathbf{r}_1\right) \tag{5.4}$$

and

$$P(\mathbf{r}_1,\mathbf{r}_2) = \gamma^{(2)}(\mathbf{r}_1,\mathbf{r}_2;\mathbf{r}'_1 = \mathbf{r}_1,\mathbf{r}'_2 = \mathbf{r}_2). \tag{5.5}$$

5.2.1 Experimental Information Relevant to Densities and Density Matrices

The information provided by an X-ray scattering experiment is contained in the dynamical structure factor $S(\boldsymbol{\kappa},\omega)$, the space-time Fourier transform of the correlation function $G(\mathbf{r},t)$ [4]

$$S(\boldsymbol{\kappa},\omega) = \frac{1}{2\pi}\iint G(\mathbf{r},t)\,e^{-i\boldsymbol{\kappa}.\mathbf{r}}e^{i\omega t}\,d\mathbf{r}dt, \tag{5.6}$$

where $\boldsymbol{\kappa}$ is the scattering vector, ω is the angular frequency associated with the energy transferred to the system ($\hbar\omega = \hbar(\omega_1 - \omega_2)$) and

$$G(\mathbf{r},t) = \int\left\langle\hat{\rho}\left(\mathbf{r}',t\right)\hat{\rho}\left(\mathbf{r}'-\mathbf{r},0\right)\right\rangle d\mathbf{r}'. \tag{5.7}$$

The symbol $\langle ... \rangle$ stands for the canonical ensemble average over all possible quantum states of the scattering system, each weighted by the Boltzmann probability factor, p_n

$$\langle ... \rangle = \sum_n p_n \langle \Psi_n | ... | \Psi_n \rangle \tag{5.8}$$

and $\hat{\rho}(\mathbf{r}, t)$ is the electron density operator

$$\hat{\rho}(\mathbf{r}, t) = \sum_i \delta(\mathbf{r} - \mathbf{r}_i(t)). \tag{5.9}$$

The information content of the observations depends on the scattering regimes identified on the basis of the energy transfer occurring in the photon-target process.

5.2.2 *Total Scattering*

If no energy analysis is performed, all photons are collected in a given scattering direction. Integration over all possible energy transfers yields the static structure factor

$$S(\boldsymbol{\kappa}) = \int S(\boldsymbol{\kappa}, \omega) d\omega = \int G(\mathbf{r}, t = 0) e^{-i\boldsymbol{\kappa}.\mathbf{r}} d\mathbf{r}. \tag{5.10}$$

This regime turns out to be useful for investigating the pair distribution function [5–7], since:

$$\begin{aligned} G(\mathbf{r}, t = 0) &= \int \sum_{i,j} \langle \delta(\mathbf{r}' - \mathbf{r}_i(0)) \, \delta(\mathbf{r}' - \mathbf{r} - \mathbf{r}_j(0)) \rangle \, d\mathbf{r}' \\ &= \int \langle \gamma^{(1)}(\mathbf{r}', \mathbf{r}' - \mathbf{r}) \rangle \, \delta(\mathbf{r}) \, d\mathbf{r}' + \int \langle P(\mathbf{r}', \mathbf{r}' - \mathbf{r}) \rangle \, d\mathbf{r}' \end{aligned} \tag{5.11}$$

5.2.3 *Elastic Scattering*

Elastic scattering occurs when there is no energy transfer between the incoming photon and the target system. The process is thus dominated by large time-scale effects

$$S(\boldsymbol{\kappa}, \omega) \approx \delta(\omega) \int G(\mathbf{r}, t = \infty) e^{-i\boldsymbol{\kappa}.\mathbf{r}} d^3\mathbf{r}. \tag{5.12}$$

In this particular situation, the motion of electrons is considered to be uncorrelated

$$G(\mathbf{r}, t=\infty) = \int \left\langle \hat{\rho}\left(\mathbf{r}', t=\infty\right) \hat{\rho}\left(\mathbf{r}'-\mathbf{r}, 0\right)\right\rangle d\mathbf{r}' \\ = \int \left\langle \hat{\rho}\left(\mathbf{r}', t=\infty\right)\right\rangle \left\langle \hat{\rho}\left(\mathbf{r}'-\mathbf{r}, 0\right)\right\rangle d\mathbf{r}' \tag{5.13}$$

and

$$S(\boldsymbol{\kappa}, \omega=0) \approx \left| \int \left\langle \gamma^{(1)}\left(\mathbf{r}, \mathbf{r}\right)\right\rangle e^{-i\boldsymbol{\kappa}.\mathbf{r}} d\mathbf{r} \right|^2 \\ = \left| \int \left\langle \hat{\rho}\left(\mathbf{r}\right)\right\rangle e^{-i\boldsymbol{\kappa}.\mathbf{r}} d\mathbf{r} \right|^2 = |F(\boldsymbol{\kappa})|^2 \tag{5.14}$$

When the Bragg condition is satisfied, $\boldsymbol{\kappa} = \mathbf{H}$ is a reciprocal lattice vector, X-ray scattering in the elastic regime yields a signal proportional to the squared modulus of the coherent elastic (Bragg) structure factor ($F(\mathbf{H})$). The latter is related to the ensemble average of the electron density distribution in position space, or the diagonal part of the 1-RDM (expressed in position space). An essential point to be emphasized is that there is no way to obtain the charge density corresponding to a pure state, since there is always some phonon contribution to the scattering. The deconvolution of lattice vibrations and the separation of static structural disorders from the electron density is one of the major issues in interpreting elastic scattering measurements.

5.2.4 Inelastic Scattering

From the preceding discussion, one is left to conclude that most of the inelastic scattering originates from short-time correlations

$$S(\boldsymbol{\kappa}, \omega) = \frac{1}{2\pi} \iint \left\langle \hat{\rho}\left(\mathbf{r}', t\right) \hat{\rho}\left(\mathbf{r}'-\mathbf{r}, 0\right)\right\rangle e^{-i\boldsymbol{\kappa}.\mathbf{r}} e^{i\omega t} d\mathbf{r}' d\mathbf{r} dt. \tag{5.15}$$

Using the density operator (5.9), one readily gets:

$$S(\boldsymbol{\kappa}, \omega) = \frac{1}{2\pi} \iint \sum_{k,j} \left\langle \delta\left(\mathbf{r}' - \mathbf{r}_j(t)\right) e^{-i\boldsymbol{\kappa}.\mathbf{r}'} \delta\left(\mathbf{r} - \mathbf{r}_k(0)\right) e^{-i\boldsymbol{\kappa}.\mathbf{r}} \right\rangle e^{i\omega t} d\mathbf{r}' d\mathbf{r} dt$$

This expression then becomes

$$S(\boldsymbol{\kappa}, \omega) = \frac{1}{2\pi} \int \sum_{k,j} \left\langle e^{-i\boldsymbol{\kappa}.\mathbf{r}_j(t)} e^{i\boldsymbol{\kappa}.\mathbf{r}_k(0)} \right\rangle e^{i\omega t} dt. \tag{5.16}$$

The dynamical behaviour of the electrons is not straightforward to express. However, if $\hat{T}$ and $\hat{V}$ are respectively the kinetic and potential energy operators, it is always possible to write the expression:

$$e^{-i\boldsymbol{\kappa}.\mathbf{r}_j(t)} = e^{i(\hat{T}+\hat{V})t/\hbar}e^{-i\boldsymbol{\kappa}.\mathbf{r}_j}e^{-i(\hat{T}+\hat{V})t/\hbar}$$

using the Zassenhaus expansion of non-commuting operators:

$$e^{i(\hat{T}+\hat{V})t/\hbar} = e^{i\hat{T}t/\hbar}e^{i\hat{V}t/\hbar}e^{\hat{C}_2(it/\hbar)^2}e^{\hat{C}_3(it/\hbar)^3}\dots,$$

where the first $\hat{C}_n$ coefficients are:

$$\hat{C}_2 = -\frac{1}{2}\left[\hat{T},\hat{V}\right] \quad \text{and} \quad \hat{C}_3 = \frac{1}{3}\left[\hat{T},\left[\hat{T},\hat{V}\right]\right].$$

If the target potential does not change significantly during the interaction time (i.e. the energy transfer $\hbar\omega$ is such that all $e^{\hat{C}_n(it/\hbar)^n}$ can be considered as static quantities compared with the phase factor $e^{i\omega t}$ in expression 5.16), then one can safely use the approximation:

$$\begin{aligned} e^{-i\boldsymbol{\kappa}.\mathbf{r}_j(t)} &\approx e^{i\hat{T}\frac{t}{\hbar}}e^{i\hat{V}\frac{t}{\hbar}}e^{-i\boldsymbol{\kappa}.\mathbf{r}_j}e^{-i\hat{V}\frac{t}{\hbar}}e^{-i\hat{T}\frac{t}{\hbar}} \\ &= e^{i\frac{\hat{p}_j^2}{2m}\frac{t}{\hbar}}e^{-i\boldsymbol{\kappa}.\mathbf{r}_j}e^{-i\frac{\hat{p}_j^2}{2m}\frac{t}{\hbar}} = e^{-i\boldsymbol{\kappa}.\mathbf{r}_j}e^{i\left(2\mathbf{p}_j\boldsymbol{\kappa}-\hbar\kappa^2\right)\frac{t}{2m}}. \end{aligned} \quad (5.17)$$

Within this so-called '*sudden*' or '*impulse*' approximation (IA) [8–10], one thus assumes a large energy transfer between the electrons and X-ray photons. The dynamic structure factor becomes

$$S_{IA}(\boldsymbol{\kappa},\omega) = \frac{1}{2\pi}\int\sum_{k,j}\left\langle e^{i\boldsymbol{\kappa}.\left(\mathbf{r}_j-\mathbf{r}_k\right)}e^{i\left(2\mathbf{p}_j\boldsymbol{\kappa}-\hbar\kappa^2\right)\frac{t}{2m}}\right\rangle e^{i\omega t}dt, \quad (5.18)$$

which can be separated into a single- ($S_{IA}^{(1)}$) and an electron-pair ($S_{IA}^{(2)}$) contribution. While the former does not involve the position operators of the electrons, the amplitude of the latter strongly depends on the amount of momentum transfer. If the momentum transfer is large, then the position dependent phase factor exhibits rapid oscillations thereby yielding a negligible contribution to the signal. The remaining contribution will hence imply superimposed single-electron components without interferences. This incoherent process, in the large energy- and momentum-transfer regime, is known as X-ray *Compton scattering*. Within the IA framework

$$S_{IA}^{(1)}(\boldsymbol{\kappa},\omega) = \frac{1}{2\pi}\int\sum_j \left\langle e^{i\left(2\mathbf{p}_j\boldsymbol{\kappa}-\hbar\kappa^2\right)\frac{t}{2m}}\right\rangle e^{i\omega t}dt$$
$$= \left\langle \sum_j \delta\left(\frac{2\mathbf{p}_j\boldsymbol{\kappa}-\hbar\kappa^2}{2m}+\omega\right)\right\rangle. \tag{5.19}$$

Introducing $q = \frac{\hbar\kappa}{2} - m\frac{\omega}{\kappa}$ for the magnitude of momentum and denoting the unit vector along the scattering vector by $\mathbf{u}$ ($\mathbf{q} = q\mathbf{u}$), one obtains:

$$S_{IA}^{(1)}(\boldsymbol{\kappa},\omega) = \frac{m}{\kappa}\left\langle\sum_j \delta\left(\mathbf{p}_j.\mathbf{u}-q\right)\right\rangle = \frac{m}{\kappa}J\left(\mathbf{q}\right). \tag{5.20}$$

The quantity $J\left(\mathbf{q}\right)$ is the so-called directional Compton profile (DCP) which is related to the behaviour of the electron cloud in the momentum space. Indeed, it is straightforward to see that $J\left(\mathbf{q}\right)$ is the projection of the momentum density ($\pi\left(\mathbf{p}\right)$) onto the scattering vector

$$J\left(q\mathbf{e}_z\right) = \iint \left\langle\pi\left(\mathbf{p}\right)\right\rangle dp_x dp_y, \tag{5.21}$$

if the z-axis is taken along the scattering vector. It will be useful to consider the Fourier transform of the DCP, the so called reciprocal form factor [11], $B(\mathbf{s})$

$$B(\mathbf{s}) = \int\left\langle\sum_j \delta\left(\mathbf{p}_j-\mathbf{p}\right)\right\rangle e^{-i\mathbf{p}.\mathbf{s}}d\mathbf{p} = \int\left\langle\pi\left(\mathbf{p}\right)\right\rangle e^{-i\mathbf{p}.\mathbf{s}}d\mathbf{p}, \tag{5.22}$$

or

$$B(s\mathbf{u}) = \int J\left(q\mathbf{u}\right)e^{-iqs}dq = \int\left\langle\sum_j \delta\left(\mathbf{p}_j.\mathbf{u}-q\right)\right\rangle e^{-iqs}dq. \tag{5.23}$$

This property is extensively used (*cf* Chaps. 2 and 4) to recover the 3D momentum density from a limited set of DCP [12–16]. Finally, we note that

$$B(\mathbf{s}) = \int\left\langle\gamma^{(1)}\left(\mathbf{r},\mathbf{r}'\right)\right\rangle e^{-i\mathbf{p}.(\mathbf{r}-\mathbf{r}')}e^{-i\mathbf{p}.\mathbf{s}}d\mathbf{r}'d\mathbf{r}d\mathbf{p} = \int\left\langle\gamma^{(1)}\left(\mathbf{r},\mathbf{r}+\mathbf{s}\right)\right\rangle d\mathbf{r}. \tag{5.24}$$

Therefore, while the coherent elastic X-ray scattering provides information on the diagonal part of the 1-RDM (i.e. the charge density in position space), the inelastic incoherent scattering (in the impulse approximation framework) yields a wealth of complementary data related mostly to the non diagonal part of this very same 1-RDM. When one is concerned with an accurate description of the charge density, Compton scattering can be regarded as an independent way to challenge the adequacy of the model wavefunction.

5.2.5 Reconstruction and Refinement Techniques

The reconstruction of 1-RDMs or charge densities in position or momentum space from a limited set of data (*cf* Chap. 7) requires a mathematical model, a function of a possible small set of parameters, built from physical considerations (prior information). The aim is to derive unbiased parameter estimates so that the model closely reproduces the data and also reliably predicts additional physical properties. Here we describe the so-called Bayesian approach [17] to data interpretation.

Let's consider N data points $\{D_{i=1,N}\}$ arranged in the N-vector $\mathbf{D}$, with a degree of confidence estimated from an experimental variance-covariance matrix ($\boldsymbol{\Sigma}$). Difficulties in accessing $\boldsymbol{\Sigma}$ experimentally usually enforce one to approximate it by its diagonal form containing the standard deviations ($\Sigma_{ii} = \sigma_i^2$). Starting from some prior information (a physical theory), a model (M) is formulated as an explicit function of a limited number of parameters $\{\alpha_n\}$ arranged in the n-vector $\boldsymbol{\alpha}$. The goal is to find the 'best' estimate of parameters ($\bar{\boldsymbol{\alpha}}$), for which the model-calculated data $\mathbf{M}(\bar{\boldsymbol{\alpha}})$ 'most closely' match those observed ($\mathbf{D}$), within a dispersion given by $\boldsymbol{\Sigma}$. An alternate way to formulate the problem is that, for a given $\mathbf{D}$ and $\boldsymbol{\Sigma}$, one seeks the most probable values for the model parameters. We therefore need an expression for the probability associated with $\boldsymbol{\alpha}$. According to Bayes theorem, the posterior probability distribution function (pdf) $p(\{\alpha\}|\mathbf{D}, \boldsymbol{\Sigma}, I)$, associated with a set of model parameters, experimental data, its covariance matrix and some external (independent) information (I) is:

$$p(\{\alpha\}|\mathbf{D}, \boldsymbol{\Sigma}, I) = p(\mathbf{D}|\{\alpha\}, \boldsymbol{\Sigma}, I) \times \frac{p(\{\alpha\}|\boldsymbol{\Sigma}, I)}{p(\mathbf{D}|\boldsymbol{\Sigma}, I)}, \tag{5.25}$$

where $p(\mathbf{D}|\{\alpha\}, \boldsymbol{\Sigma}, I)$ is the likelihood function, while $p(\{\alpha\}|\boldsymbol{\Sigma}, I)$ and $p(\mathbf{D}|\boldsymbol{\Sigma}, I)$ are the prior and evidence pdf, respectively. We usually have a fair preliminary knowledge on the range of physically acceptable parameter values and we can expect the prior pdf to be rather flat. As a matter of fact, one can usually consider it as a uniform distribution compared to the likelihood function which is expected to be the most sharply peaked. This is also true for the evidence pdf which hence, turns out to be a mere normalisation factor to the posterior pdf. To find the most probable set of parameters corresponding to a given experimental outcome, one needs to maximize expression (5.25).

Determination of the form of the likelihood function is thus essential and can be done by assuming that the proper pdf is the one that maximises the total information entropy [18] while requiring the model (with its "best" parameters) to yield the mean values for each data:

$$\langle \mathbf{D} \rangle = \int \mathbf{D} p \left(\mathbf{D} \,|\boldsymbol{\alpha}, \boldsymbol{\Sigma}, I \right) d^N D, \tag{5.26}$$

and the variance

$$\left\langle D_i D_j - \langle D_i \rangle \left\langle D_j \right\rangle \right\rangle = \Sigma_{ij} . \tag{5.27}$$

In other words, it is assumed that the model can adequately account for all physical effects associated with the measurement. Furthermore, the mean value of a data point obtained by a large number of repeated measurements should asymptotically approach to the value predicted by the model: $\langle \mathbf{D} \rangle = \mathbf{M}(\bar{\boldsymbol{\alpha}})$. If D_{ij} is a possible outcome of D_i with a probability of occurrence p_{ij}, the configuration entropy to be maximized is

$$S_i = -\sum_j p_{ij} \log \left(p_{ij} \right), \tag{5.28}$$

subject to the constraints

$$\langle D_i \rangle = M_i \quad \text{and} \quad \left\langle D_i^2 - \langle D_i \rangle^2 \right\rangle = \sigma_i^2 . \tag{5.29}$$

That is,

$$-\sum_j p_{ij} \log \left(p_{ij} \right) + \lambda \left[1 - \sum_j p_{ij} \right] + \mu \left[\sigma_i^2 - \sum_j p_{ij} \left(D_{ij} - M_i \right)^2 \right] \tag{5.30}$$

where λ and μ are Lagrange multipliers. It is straightforward to show that cancellation of derivatives of this quantity with respect to p_{ij}, λ, μ yields a Gaussian pdf

$$p_{ij} = \left(\pi \sigma_i^2 \right)^{-\frac{1}{2}} \exp \left(-\frac{\left(D_{ij} - M_i \left(\boldsymbol{\alpha} \right) \right)^2}{\sigma_i^2} \right), \tag{5.31}$$

or in the limit of continuous distribution of possible outcomes

$$p_i \left(D_i \right) = \left(\pi \sigma_i^2 \right)^{-\frac{1}{2}} \exp \left(-\frac{\left(D_i - M_i \left(\boldsymbol{\alpha} \right) \right)^2}{\sigma_i^2} \right). \tag{5.32}$$

For uncorrelated data, the global probability for $\mathbf{D}$ is thus

$$p \left(\mathbf{D} \,|\boldsymbol{\alpha}, \boldsymbol{\Sigma}, I \right) = \prod_i p_i \left(D_i \,|\boldsymbol{\alpha}, \sigma_i^2, I \right) = \prod_i \left(\pi \sigma_i^2 \right)^{-\frac{1}{2}} \exp \left(-\frac{\left(D_i - M_i \left(\boldsymbol{\alpha} \right) \right)^2}{\sigma_i^2} \right), \tag{5.33}$$

which is a pdf of $N - n$ degrees of freedom. Using this expression for the likelihood function, Bayes theorem implies that the most probable set of parameters for model M is the one that maximizes the quantity (Eq. 5.33) or, equivalently, that minimizes the so-called mean-square residual (MSR)

$$\chi^2 = \sum_i \frac{(D_i - M_i(\boldsymbol{\alpha}))^2}{\sigma_i^2}. \quad (5.34)$$

In the limit of a large number of data points, the expected value for χ^2 (corresponding to the most probable set of parameters), is $N - n$ with mean dispersion of $\sqrt{N}$. The sharper the minimum, the more precise the parameter estimates are. The variance-covariance matrix associated with these parameters is given by the inverse of $\nabla_\alpha \nabla_\alpha \chi^2$. Additional details on optimization techniques and algorithms can be found in reference [17].

5.2.6 Combining Different Experimental Data in a Unique Refinement

Since the outcomes of different scattering experiments performed on the same system are the manifestations of the same wavefunction, it is legitimate to expect that the validity of the model is not limited to a unique type of data. By simultaneous interpretation of two or more sets of observations one can improve the data-to-parameter ratio. An obvious example is the combined treatment of coherent elastic X-ray diffraction data and Compton profiles for non equivalent crystallographic directions. Both experiments provide information on the electron distribution in crystals, but the former one in the position space, while the latter one in the momentum space. These complementarities can be made use of also in the study of unpaired electrons of magnetic systems by combining polarized neutrons diffraction (Chap. 8) and magnetic Compton scattering (Chap. 4). Other examples include the joint interpretation of (i) convergent beam electron diffraction and positron annihilation, or (e,2e) and (γ, e γ) coincidence spectroscopy for obtaining 2D or 3D momentum densities and (ii) neutron, polarized neutron and X-ray data to derive the nuclear, magnetic and electron densities simultaneously.

Let's consider the case when one has to carry out a minimization with two sets of data ($\mathbf{D}_A$ and $\mathbf{D}_B$) pertaining to two different experiments (A and B). A plausible way to proceed is to minimize the combined MSR

$$\chi^2 = \chi_A^2 + \chi_B^2 = \sum_{E=A,B} \left[\sum_{i \in E}^{N_E} \frac{(D_{E,i} - M_{E,i}(\boldsymbol{\alpha}))^2}{\sigma_{E,i}^2} \right]. \quad (5.35)$$

We have to consider two important aspects of this approach; (i) If one data set (say A) is much larger than the other (B), then A can have an overwhelming contribution to χ^2, unless the balance is enforced by a proper weighting scheme. (ii) The change in the parameter vector is driven by the gradient vector, whose components are inversely proportional to the individual variances ($\sigma^2_{E,i}$):

$$\frac{\partial \chi^2}{\partial \alpha} = -2 \sum_{E=A,B} \sum_{i \in E}^{N_E} \left[\frac{D_{E,i} - M_{E,i}(\boldsymbol{\alpha})}{\sigma^2_{E,i}} \right] \frac{\partial M_{E,i}(\boldsymbol{\alpha})}{\partial \alpha}. \tag{5.36}$$

One usually has to compromise with a crude estimate for these quantities, for example, assuming a Poisson or Gaussian distribution law for the counting statistics. In combined refinements it is thus crucial to have reliable weights for all data sets, as well as for the data points in each set.

Here we suggest a rather simple procedure [17] to overcome the problems mentioned above. Let us assume that, for some unknown reasons, the individual $\sigma^2_{E,i}$ values (within one data set) cannot be trusted but only their ratios $\sigma^2_{E,i}/\sigma^2_{E,j}$. Therefore, it is to believe that the relative values $\sigma^2_{A,i}/\sigma^2_{B,j}$ (from different experiments) do not fairly reflect the relative degree of confidence one could have in the two data sets. This situation can be summarized by stating that there exists an overall and unknown scale factor to be applied to each set of variances:

$$\boldsymbol{\Sigma}_E \rightarrow \eta^2_E \boldsymbol{\Sigma}_E \quad \text{or} \quad \left\{\sigma^2_{E,i}\right\} \rightarrow \left\{\eta^2_E \times \sigma^2_{E,i}\right\}. \tag{5.37}$$

Therefore the posterior pdf for each set of observations is obtained by integration over all possible values of the scale factor drawn from its own pdf, $p(\eta_E)$

$$\begin{aligned} p(\mathbf{D}_E | \boldsymbol{\alpha}, \boldsymbol{\Sigma}_E, I) &= \int p(\mathbf{D}_E | \boldsymbol{\alpha}, \eta_E \boldsymbol{\Sigma}_E, I) p(\eta_E) \, d\eta_E \\ &= \int \prod_i \left(\pi \eta^2_E \sigma^2_{E,i}\right)^{-\frac{1}{2}} \exp\left(-\frac{(D_{E,i} - M_{E,i}(\boldsymbol{\alpha}))^2}{\eta^2_E \sigma^2_{E,i}} p(\eta_E) \, d\eta_E \right). \end{aligned} \tag{5.38}$$

Jeffreys has shown that the pdf which best translates our total ignorance is the distribution ($p(\eta_E) \propto \eta_E^{-1}$) [19].

Hence, the marginal probability becomes

$$\begin{aligned} p(\mathbf{D}_E | \boldsymbol{\alpha}, \boldsymbol{\Sigma}_E, I) &\propto \left[\prod_{i=1} \left(\pi \sigma^2_{E,i}\right)^{-\frac{1}{2}} \right] \int \left(\frac{1}{\eta_E} \right)^{N_E+1} e^{-\frac{\chi^2_E}{\eta^2_E}} \, d\eta_E \\ &= \left[\prod_{i=1} \left(\pi \sigma^2_{E,i}\right)^{-\frac{1}{2}} \right] \times \Gamma\left(\frac{N_E}{2} \right) \times \left(\chi^2_E\right)^{-\frac{N_E}{2}} \end{aligned} \tag{5.39}$$

and the overall probability for the parameters given the combined data is

$$p(\boldsymbol{\alpha}\,|\mathbf{D}_A, \boldsymbol{\Sigma}_A, \mathbf{D}_B, \boldsymbol{\Sigma}_B, I\,) \propto \left(\chi_A^2\right)^{-\frac{N_A}{2}} \left(\chi_B^2\right)^{-\frac{N_B}{2}}. \tag{5.40}$$

As a consequence, the most probable set of parameters can be searched by minimization of

$$L(\boldsymbol{\alpha}\,|\mathbf{D}_A, \boldsymbol{\Sigma}_A, \mathbf{D}_B, \boldsymbol{\Sigma}_B, I\,) \propto N_A \log\left(\chi_A^2\right) + N_B \log\left(\chi_B^2\right). \tag{5.41}$$

It should be noted that the gradient

$$\frac{\partial}{\partial\alpha} L\,(\boldsymbol{\alpha}\,|\mathbf{D}_A, \boldsymbol{\Sigma}_A, \mathbf{D}_B, \boldsymbol{\Sigma}_B, I\,) \propto \frac{N_A}{\chi_A^2}\frac{\partial\chi_A^2}{\partial\alpha} + \frac{N_B}{\chi_B^2}\frac{\partial\chi_B^2}{\partial\alpha} \tag{5.42}$$

is equivalent to the traditional gradient (minimization of $\chi_A^2 + \chi_B^2$) only in the vicinity of the full convergence where, $\chi_A^2 \approx N_A - n$ and $\chi_B^2 \approx N_B - n$ for $N_E \gg n$.

5.2.7 Refinement of One-Electron Reduced Density Matrices

The extraction of 1-RDM from experimental data is certainly a less straightforward procedure than that for the electron density. Its intrinsic complexity has long been an obstacle, just like the difficulties associated with combining different sources of information from different experimental origins. However, the property of 1-RDM, which allows for a unique junction between position and momentum spaces, provides the opportunity to build robust and physically sound models that can be challenged through numerous measurements of rather different natures. It has been shown (*cf* Chaps. 2 and 7) that bonding effects manifest themselves more obviously in the 1-RDM than in the charge density [20]. The knowledge of 1-RDM would also allow for an efficient calculation of the total energy within the mean field approximation limit. Given the current rate of increase in computing power and improvement in experimental techniques, the scientific area of 1-RDM reconstruction is expected to expand both in quantity and quality.

All efforts up to now were limited to simple systems [21–25] and based on simple parameterized models, such as the independent electron gas formalism. Within this framework, the 1-RDM is expressed in terms of natural spin orbitals (NSO), with integer occupancy numbers (n_i)

$$\gamma^{(1)}\left(\mathbf{r}', \mathbf{r}\right) = \sum_i n_i \Phi_i^*\left(\mathbf{r}'\right) \Phi_i\left(\mathbf{r}\right). \tag{5.43}$$

Expanding each NSO on an auxiliary atomic basis set

$$\Phi_i(\mathbf{r}) = \sum_j c_{ij}\chi_j(\mathbf{r}), \tag{5.44}$$

the 1-RDM takes the form

$$\begin{aligned}\gamma^{(1)}\left(\mathbf{r}',\mathbf{r}\right) &= \left\langle \mathbf{r}'\right| \left[\sum_{i,j,k} n_i c_{ij} c_{ik}^* \left|\chi_j\right\rangle \left\langle \chi_k\right|\right] \left|\mathbf{r}\right\rangle \\ &= \left\langle \mathbf{r}'\right| \left[\sum_{j,k} P_{jk} \left|\chi_j\right\rangle \left\langle \chi_k\right|\right] \left|\mathbf{r}\right\rangle\end{aligned} \tag{5.45}$$

In order to limit the computational time, almost all works aimed at the determination of the elements (P_{jk}) of the population matrix $\mathbf{P}$. It was only recently proposed to include screening parameters into the model to allow for expansion or contraction of the atomic orbitals [25, 26].

The N-representability requirement is the condition to be imposed on the 1-RDM in order to maintain its relationship to an N-electron wavefunction (pure-state case) or to a statistical mixture of N-representable pure-state 1-RDM's. Though it was shown to be "impossible to formulate constraints to ensure N-representability" in general [27], it was also recognized that the idempotency constraint would suffice within the Hartree-Fock (HF) scheme. Therefore the majority of 1-RDM reconstructions are carried out by restricting the population matrix to satisfy the equation

$$\mathbf{PSP} = 2\mathbf{P}, \tag{5.46}$$

where $\boldsymbol{S}$ is the overlap matrix of the basis functions and the factor 2 only applies to closed-shell systems. This condition, vital to the pertinence of the 1-RDM model, can be achieved by means of a 'purification scheme' suggested by Mc Weeny [28, 29], or by restricting the matrix to a specific analytic form. The former method makes use of an iterative approach. It assumes that the initial population matrix is close enough to be idempotent and applies the recurrence operation

$$\frac{3}{2}\mathbf{P}_n\mathbf{S}\mathbf{P}_n - \frac{1}{2}\mathbf{P}_n\mathbf{S}\mathbf{P}_n\mathbf{S}\mathbf{P}_n = \mathbf{P}_{n+1}. \tag{5.47}$$

The refinement thus starts with a theoretical initial guess for $\boldsymbol{P}$ and the purification is carried out with the constraint for the expectation values to match experimental outcomes [30, 31].

An alternate procedure was proposed by Schmider and collaborators [22]. It involves writing the NSO coefficient matrix, $\mathbf{C}$, as a product of a matrix $\mathbf{C}_0$ (derived from an ab-initio calculation, for example) and a unitary matrix $\mathbf{C}_R(\boldsymbol{\alpha})$, containing the parameters and taking the form of a product of planar rotation matrices

$$\mathbf{C}_R(\{\alpha\}) = \prod_{ij} \mathbf{C}_{ij}\left(\alpha_{ij}\right), \tag{5.48}$$

where the indices run over all pairs of basis functions. The $\mathbf{C}_{ij}\left(\alpha_{ij}\right)$ matrix corresponds to the transformations

$$\chi_i(\mathbf{r}) \rightarrow \cos\left(\alpha_{ij}\right)\chi_i(\mathbf{r}) + \sin\left(\alpha_{ij}\right)\chi_j(\mathbf{r}), \tag{5.49}$$

$$\chi_j(\mathbf{r}) \rightarrow -\sin\left(\alpha_{ij}\right)\chi_i(\mathbf{r}) + \cos\left(\alpha_{ij}\right)\chi_j(\mathbf{r}), \tag{5.50}$$

while leaving the remaining basis functions unchanged.

Though computationally far more demanding, this strategy is a real cornerstone since, in principle, it allows to account for electron correlation. This last point is proved to be essential, even for small systems, if more types of experiments are considered simultaneously. It turns out that the joint fit of 1-RDM's in both position and momentum spaces cannot be successfully carried out within the single determinant limit [22]. It is thus not sufficient to rely on a mere idempotency requirement to obtain a valuable 1-RDM, but information from both position and momentum spaces are essential [22, 25, 26]. Figures 5.1 display the first 1-RDM refinements (on an isolated atom) using both coherent ($F(\boldsymbol{H})$) and incoherent data ($B(\boldsymbol{s})$).

More recently, it was shown that 1-RDM elements related to bonding features could not be obtained from a unique source of information. Figure 5.2 illustrate this result for an isolated diatomic molecule. In particular, Fig. 5.2 shows that the full reconstruction of the off-diagonal part of the matrix was impossible even in the course of a combined treatment of high resolution Bragg diffraction and DCP data when an insufficient number of Compton profiles were included in the analysis [26].

5.3 Electron Density Reconstruction

The single-crystal X-ray diffraction pattern is usually interpreted as a coherent elastic scattering event. As discussed above, the property relevant to this process is the Bragg structure factor amplitude ($F(\boldsymbol{H})$) which is the Fourier transform of the mean electron density in the unit cell of volume V_c

$$F(\mathbf{H}) = \int_{V_c} \langle\rho(\mathbf{r})\rangle e^{i2\pi\mathbf{Hr}} d\mathbf{r}. \tag{5.51}$$

$\boldsymbol{H}$ is the scattering vector, with integral reciprocal-axis components (Miller indices h, k and l), satisfying the Laue conditions ($\boldsymbol{Ha}^* = h$, $\boldsymbol{Hb}^* = k$ and $\boldsymbol{Hc}^* = l$)

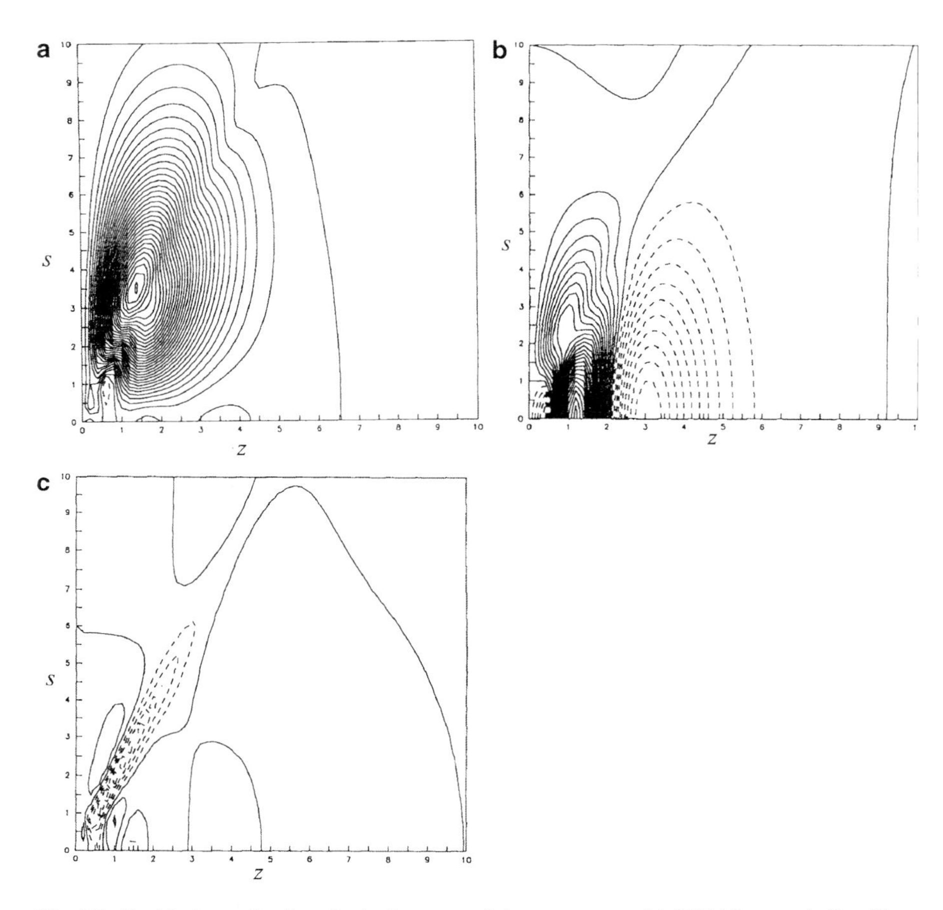

Fig. 5.1 Residual map for the spherical average of the reconstructed 1-RDM for atomic Beryllium with reference to the 1-RDM computed from correlated wavefunction by Weiss [32]. (**a**) Only from experimental structure factors (contour intervals: 0.001 a.u.); (**b**) Only from experimental $B(s)$ (contour intervals: 0.005 a.u.); (**c**) Using both structure factors and $B(s)$ data (contour intervals: 0.001 a.u.) (Reprinted from Ref. [22] with kind permission of The American Institute of Physics)

due to the translation symmetry of the crystal density. It is assumed that the system remains in the ground electronic state during the scattering process, that is, the averaging is taken over the vibration states. To bring Eq. 5.51 to an explicit parametric form, one needs an analytic representation for the static density and an analytically tractable treatment for thermal averaging. These requirements can ideally meet within the one-center density formalism and the harmonic convolution approximation to thermal smearing (*cf* Chaps. 1 and 3). Since molecular-orbital-based two-center models are described elsewhere in this book (*cf* Chaps. 2 and 7) our attention is focused on the one-center formalism, within which, the static

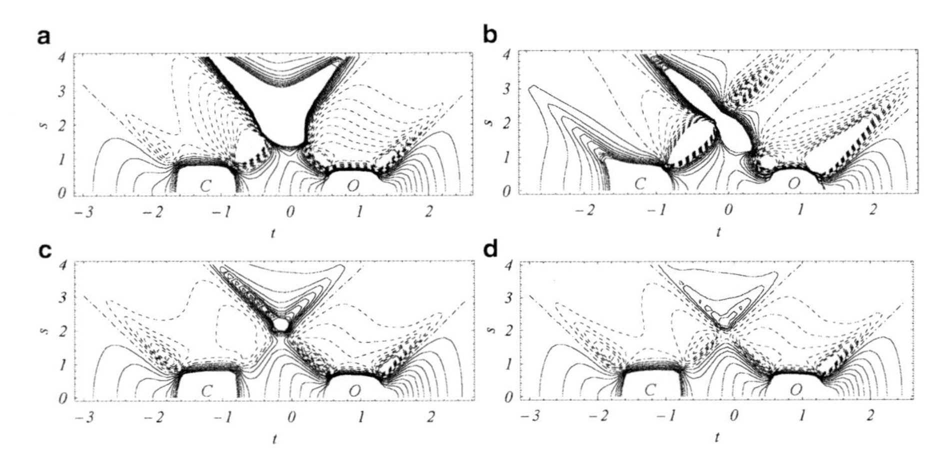

Fig. 5.2 Reconstructed 1-RDM for the CO molecule along the bond axis shown in terms of extracule-intracule coordinates ($t = (z + z')/2$ and $s = z - z'$); (**a**) Using theoretical structure factors generated from the "true" 1-RDM (see Fig. 5.2) with random noise added; (**b**) Reconstruction based on two DCPs only (70 grid points separated by 0.1 a. u.); (**c**) Reconstruction from structure factors and five DCPs; (**d**) The 'true' 1-RDM used as the target in the fit. Contour intervals: 0.1 a.u (Adapted from Ref. [25] with kind permission of Acta Crystallogr A)

crystalline density is partitioned into "atomic" densities (ρ_A), each assigned to a particular nucleus at $\boldsymbol{R}_A$:

$$\rho(\mathbf{r}) = \sum_A \rho_A(\mathbf{r}_A); \quad \mathbf{r}_A = \mathbf{r} - \mathbf{R}_A. \tag{5.52}$$

Within the harmonic-convolution approximation (a density unit rigidly follows the motion of its nucleus) the structure factor takes a closed analytic form:

$$F(\mathbf{H}) = \sum_A f_A(\mathbf{H}) t_A(\mathbf{H}) e^{i2\pi \mathbf{H}\mathbf{R}_A}, \tag{5.53}$$

where f_{A} is the complex atomic scattering factor, $\mathbf{R}_{\mathrm{A}}$ is the equilibrium nuclear position vector and t_{A} is the temperature factor

$$t_A(\mathbf{H}) = e^{-2\pi^2 \mathbf{H}' \mathbf{U}_A \mathbf{H}}, \tag{5.54}$$

which is the Fourier transform of the Gaussian pdf describing harmonic nuclear displacements with respect to the equilibrium configuration ($\mathbf{u}_{\mathrm{A}} = \mathbf{Q}_{\mathrm{A}} - \mathbf{R}_{\mathrm{A}}$):

$$P(\mathbf{u}_A) = (2\pi)^{-3/2} |\mathbf{U}_A|^{-1/2} e^{-\frac{1}{2}\mathbf{u}_A' \mathbf{U}_A^{-1} \mathbf{u}_A}, \quad \mathbf{U}_A = \langle \mathbf{u}_A \mathbf{u}_A' \rangle \tag{5.55}$$

The components of the mean-square displacement amplitude matrix ($\boldsymbol{U}$) are referred to as the anisotropic displacement parameters (ADP). A step-wise derivation of Eq. 5.53 is given in references [33, 34].

5.3.1 The Pseudoatom Formalism

The density ρ_A and its Fourier transform f_A in expressions (5.52 and 5.53) has been termed by Stewart as the A-centered 'rigid' pseudoatom and its generalized scattering factor, respectively [35]. While these equations are the key formulas for the interpretation of Bragg data, they contain a great deal of ambiguities as far as the definition of the atomic density/scattering unit is concerned. As spelled out by Stewart "a partitioning of ρ into constituent parts is an intellectual exercise that does not lend itself to unique measurement from elastic X-ray scattering experiments" [36]. What constitutes an 'atom' in a molecule or a crystal is obscure also from a theoretical point of view. Locality and compactness are often considered as guiding principles in designing fuzzy-type atomic decomposition schemes, though neither of these requirements is easy to impose and their validity is representation dependent. In other words, neither the pseudoatom nor the total static density (obtained through deconvolution of nuclear motion via pseudoatoms) can be considered physical observables.

The parameterization of the pseudoatom means choosing an appropriate basis set over which its density is expanded. The symmetry of the local Coulomb potential implies to adopt a nuclear centered multipole expansion, while the atomicity requirement enforces the expansion to be finite:

$$\rho(\mathbf{r}) = \sum_{l}^{L} \sum_{m=0}^{l} \rho_{lm\pm}(r) y_{lm\pm}(\vartheta, \phi), \tag{5.56}$$

where the index designating the nucleus is dropped. Since the angular part of the basis set is fixed to real spherical harmonics ($y_{lm\pm}$), only the radial functions (RDF: $\rho_{lm\pm}$) remain to be determined, usually by minimization of the MSR between the target and model densities or their Fourier transforms (direct- and reciprocal space optimizations are equivalent for infinite resolution). In what follows, we discuss how the RDFs can in principle be obtained if the target function is known (error-free data to infinite resolution). This derivation, detailed in the above-cited manuscript by Stewart [36] and referred here as the generalized multipole projection, has less of a practical than a theoretical significance. Nevertheless, we find it important to recast the main conclusion of the study because it clearly demonstrates the limitations of the model. Since the experimental data are always discrete and erroneous, 'model free' or unrestricted derivation of the RDFs is not a plausible approach in practice. One thus has to parameterize the RDFs using simple analytic functions or their combination [37, 38] and derive the best estimates of these parameters via the least-squares (LS) procedure. We refer to such a restricted treatment of data as a pseudoatom fit that leads to the model or multipole density consistent with the data in the LS sense.

Spherical harmonics, the eigenfunctions of the angular momentum operator, form a complete orthogonal basis. As a consequence, the product of any two spherical harmonics can be written as linear combination of real spherical harmonics. Since the angular part of the hydrogen-atom wave function is a spherical harmonics function, the density corresponding to any state has finite regular harmonics content. This is also true for a many-electron atom in a well-defined angular momentum state (L, M), since the square of the total angular momentum operator commutes with the spineless Hamiltonian [39]:

$$\rho(\mathbf{r}) = \sum_{l}^{L} \rho_{2l,0}(r) y_{2l,0}(\vartheta, \phi). \tag{5.57}$$

The quantum-chemical density of an atom obtained within the orbital approximation does not necessarily take the above form, yet its multipole expansion is finite. This can be easily seen for a single-determinant wave function composed of one-electron spin-functions, in which case the angular part of the density is also a combination of products of spherical harmonics. Within the linear combination of atomic orbitals (LCAO) approximation, the molecular density can be decomposed into one- and two-center orbital product. The multipole content of each one-center density, just like for an atom, is uniquely determined by the orbital basis. The projection of the two-center terms onto one-center functions however requires inclusion of higher-order spherical harmonics augmented with diffuse RDFs.

5.3.2 Generalized Multipole Projection

Given an analytic density (or its Fourier transform), the MSR (χ^2) is the functional of the pseudoatomic RDFs (ρ_{lm}^A)

$$\chi^2 = \int \left(\rho(\mathbf{r}) - \sum_A \sum_{l,m}^{L} \rho_{lm}^A(r_A) y_{lm}(\vartheta_A, \phi_A) \right)^2 d\mathbf{r} = \int_0^\infty \chi_0^2(r) r^2 dr \tag{5.58}$$

where χ_0^2 is a positive definite radial MSR (obtained by integrating χ^2 over the angular variables), whose functional derivative with respect to each RDF leads to an infinite set of inhomogeneous linear equations [36]:

$$\rho_{lm}(r_A) = \langle \rho \mid y_{lm}^A \rangle_\Omega = \rho_{lm}^A(r_A) + \sum_{B \neq A} \sum_{\lambda\mu}^{\Lambda} \rho_{\lambda\mu}^B(r_B) \left\langle y_{\lambda\mu}^B \;\middle|\; y_{lm}^A \right\rangle_\Omega. \tag{5.59}$$

The molecular RDF ($\rho_{lm}(r_A)$) of the A-centered multipole expansion (*l.h.s.*) at any radial grid point is given as a linear combination of pseudoatom RDFs (*r.h.s.*) taken at the same radial grid point. This projection, driven by the angular overlap integral, can yield 'exact' molecular moments up to Λ without necessarily reproducing the total density exactly. The procedure is computationally difficult, even for diatomic molecules, and leads to de-localized RDFs [40].

5.3.3 Restricted Multipole Formalism and Pseudoatom Fitting

In the most widely used parameterization (Hansen-Coppens or HC-formalism) [38], the pseudoatom density is written as

$$\rho(\mathbf{r}) = \sum_{l=0}^{L} R_l(\kappa_l r) \sum_{m=0}^{l} P_{lm\pm} d_{lm\pm}(\vartheta, \phi), \tag{5.60}$$

where d_{lm}'s are density normalized spherical harmonics

$$\int |d_{lm\pm}| \, d\Omega = 2 \quad \text{if } l > 0 \text{ and } \quad \int |d_{00}| \, d\Omega = 1, \tag{5.61}$$

expressed in a local coordinate system at each nucleus ($d\Omega$ is the angular volume element), allowing for an easy implementation of site-symmetry and chemical-equivalence constraints. The monopole term ($l = 0$) is composed of the core (ρ_c) and the spherically averaged and normalized valence density (ρ_v), both derived from the ground-state Hartree-Fock (HF) wave function of the isolated atom [41]:

$$R_0(\kappa_0 r) = \rho_c(r) + P_v \rho_v(\kappa_0 r). \tag{5.62}$$

A pseudoatom that looses/gains charge in a molecule is expected to get contracted/expanded. This isotropic deformation of the spherical valence density is accounted for by an expansion-contraction parameter ($\kappa_0 = \kappa$) applied to scaling the radial coordinate.

Since each deformation density unit ($l > 0$) of a pseudoatom integrates to zero, the model implicitly includes a definition of atomic charge ($q = N_c + P_v$ where N_c is the number of core electrons and P_v is the valence monopole population, P_{00}). Each RDF of the aspherical part is taken to be a Slater-type function

$$R_l(\kappa_l r) = \frac{(\eta_l \kappa_l)^{n_l+3}}{(n_l + 2)!} r^{n_l} \exp(-\eta_l \kappa_l r), \tag{5.63}$$

whose parameters are deduced from those of single-zeta HF Slater-orbitals of the corresponding ground-state atom [41]. Due to the Fourier-invariance of the spherical harmonics, the pseudoatom scattering factor takes the form

$$f(\mathbf{H}) = f_c(H) + P_v f_v(H/\kappa) + 4\pi \sum_l i^l J_l(H/\kappa_l) \sum_{m=0}^{l} P_{lm\pm} d_{lm}(\hat{\vartheta}, \hat{\phi}) \quad (5.64)$$

where J_l is the lth-order Fourier-Bessel transform of R_l. The static density parameters (P_v, $\kappa = \kappa_0$, $P_{lm\pm}$ and $\kappa' = \kappa_l$ for all l, usually) together with nuclear positions and ADP's are LS refined against the data. Obviously, the expansion terminated at the monopole level with $P_v = N_v$ (where N_v is the number of valence electrons of the isolated atom) and $\kappa = 1$, gives the spherical isolated-atom ($\rho_0 = \rho_c + N_v\rho_v$). The HC-formalism is thus built around the 'safe' concept of the so called promolecule used in conventional X-ray structure refinements. Density-related properties, such as electric moments, electrostatic potential, field, field-gradient and lattice energy can readily be obtained from the pseudoatomic density [1, 42–44].

5.3.4 *Multipole Refinement*

The pseudoatom-based data analysis is far from being a routine undertaking and the critical evaluation of the results is mandatory. Evidently, the physical content of the experimental pseudoatomic density is subject to data quality, approximations/inadequacies inherent to the scattering model and ambiguities associated with the refinement. It is crucial to measure the possible most extended and complete diffraction image with manifold redundancy. The sample must be cooled to the possible lowest temperature to minimize thermal motion and thermal diffuse scattering and thus increase the signal-to-noise ratio of high-order reflections. Accurate information on these regions of reciprocal space is needed to precisely locate core-density maxima and adequately deconvolute thermal smearing effects from density deformations due to chemical bonding. Ideally, the nuclear distribution (structure) and the electron density should emerge simultaneously and in a mutually consistent way in the course of data modeling. Since no direct relationship between the two distributions is incorporated into the scattering formalism, static and dynamic model parameters are often strongly correlated and their LS-estimate can be unavoidably and intractably biased [45]. Incorporation of temperature dependent data into the analysis can thus help monitor the adequacy of thermal motion models [46].

The HC representation does not adhere to basic requirements for a 'physically valid' density, such as positivity, nuclear cusp condition [47] and the proper asymptotic behavior [48]. The frozen core approximation can bias the valence density parameters, especially for elements beyond the second period [49]. The pseudoatom electronic moments ($l > 0$) obtained via a LS fit of a limited data do not necessarily add up to the 'true' molecular moments.

The inadequacy of the deformation RDFs is by far the weakest part of the formalism. The same Slater function is shared by all poles of a given l, and usually the same set of RDF's is assigned to identical nuclei, irrespective of their bonding environment. The location of the maxima of these functions is controlled by the parameters n_l and η_l, whose selection is not straightforward, even for quadrupolar atoms (*sp*-valence shell). The usual practice is to keep all exponents at the same value for a given pseudoatom and try to optimize a single radial screening ($\kappa' = \kappa_{l>0}$). This κ'-refinement is however notoriously unstable and often fails to converge. The expansion is usually terminated at the hexadecapolar level ($l = 4$), which obviously limits the description of diffuse density features [49]. Basis set problems are especially pronounced for elements of the third period and transition metals of diffuse valence shell. A further ambiguity for transition metals is the selection of starting electron configuration, that is, a neutral (d^n, s^2) or an ionic state (d^n).

Fitting a multivariate, nonlinear function to a limited number of data points is not a trivial exercise. However, it is routinely conducted nowadays [50]. A subset of variables can be equally dominating in the fit of a subset of data, while a single data point can be extremely influential to estimating a specific parameter. Since the basis functions of the structure factor expansion do not form an orthogonal set, parameter indeterminacies can occur due to ill-conditioned LS matrix. The latter problem, detectable via high correlations between LS variables, can be controlled to some extent by introducing pertinent constraints, or even eliminated by singular value decomposition. The physical origin of the bias however remains intractable because the 'authentic' solution, enforced or selected, is of purely mathematical in nature.

There is no general recipe for an ultimate refinement strategy, and the reliable tests to monitor the results are limited in number and performance. The complexity of the model is usually increased in a step-wise manner, proceeding from a more towards a less restricted parameter set, while observing changes in the goodness of fit and the correlations between parameter estimates. While a good fit and a flat residual map are necessary requirements for the density to be consistent with the data, they are, by no means, sufficient criteria for the results to be physically meaningful.

Adequate modeling of hydrogen atoms is of particular importance in X-ray only analyses. Because of their low scattering power (lack of core electrons and intense thermal motion), the contribution of H-atoms to diffraction intensities is marginal. In addition, essential density rearrangements and charge loss occur at these sites upon covalent bond formation. The H-pseudoatom is over-parameterized even at the dipolar level, in the sense that the joint refinement of multipole populations and conventional parameters (thermal displacement amplitudes and proton coordinates) is troublesome, if not impossible. There are 'tricks' of different levels of sophistication and rationale that apparently overcome the problem. However, the bias introduced in such ways, necessarily obstruct the results, especially if fine details are to be addressed, such as the topology of hydrogen bonds. Support from parallel neutron diffraction (ND) experiment appears to be essential. The X-(N + X) analysis makes a direct use of ND positional parameters and ADPs of protons [51]. The latter quantities usually have to be readjusted to account for different experimental conditions and different systematic errors affecting the ND and x-ray diffraction (XRD) measurements [52]. The most popular but least efficient practice is to fix the X–H bond distances to average ND-derived values and model the thermal motion of the protons at the isotropic level during the multipole refinement. There are several feasible approaches to estimate the ADPs of H-atoms in the absence of ND parameters (see Chap. 3). They are based on the assumption that external and internal modes are uncorrelated (the corresponding ADPs are additive) and combine the results of rigid-body analysis [53] with independent information on the internal modes. Such models include the use of average bond-parallel and perpendicular ADPs of H-atoms derived either from ND [54] or Infrared Spectroscopy [55, 56], the analysis of multi-temperature X-ray data [57], and normal coordinate analysis of the isolated molecule or clusters based on *ab initio* force fields [58]. A comparison of some of these methods is given in reference [59].

5.3.5 Uncertainties in Experimentally Determined Pseudoatom Densities

Overall reproducibility is an important requirement for evaluation of the physical significance of the results. Usual validation methods include comparative data analyses utilizing different constrained models, contrasting the outcomes with that obtained by theoretical methods or independent observations, systematic studies of chemically related systems (to reveal some expected chemical transferability), simultaneous analyses of temperature dependent data and the pseudoatom interpretation of error-free, theoretical data.

The number and complexity of experimental pseudoatom properties that are claimed to be routinely accessible has been increased significantly over the past years. The majority of recent papers address problems of chemical relevance through a complete topological analysis [60] of the model density and related properties whose spectrum has recently been extended to approximate energy densities [61]. Bond-critical-point (BCP) indices, in particular, are being reported with increasing confidence, in spite of a growing number of contradictive results that questions their experimental accessibility and reliance and that warns against their indiscriminate use as ultimate descriptors of chemical interactions in crystals [62].

The error distribution of the experimental model density, as calculated from the variance-covariance matrix, exhibits a topology similar to that of density; sharp peaks in the vicinity of the nuclei and minima in the internuclear regions. This general trend suggests that the density is more reliably accessible in the internuclear than in the nuclear regions and gives some credit to the widely-cited error-estimate of 0.05 e/Å^3 in the bond. Indeed, the experiment versus theory comparisons find ρ_{BCP} to be more reproducible than $\nabla^2\rho_{\mathrm{BCP}}$, especially for polar bonds, in which case the BCP is situated closer to the valence shell charge concentration of the less electronegative atom where the bond curvature changes rapidly. A large number of theoretical, experimental and combined studies have addressed this issue for carbonyl bonds [3]. Periodic quantum chemical methods [63] have revealed a strong effect of hydrogen bonding on intramolecular topology [64]. There are however quite inconsistent and even contradicting results concerning the experimental accessibility of crystal-field effects in general. The X-ray charge density analysis of famotidine polymorphs, for example, suggests that local topological properties are rather insensitive to intermolecular interactions [65]. The study finds good overall agreement in the BCP indices of chemically equivalent bonds for the two molecules in spite of their markedly different crystal environments, evidenced also by differences in the dipole moment, electrostatic potential and interaction energy of the two polymorphs. The results of a comparative study on two isomers of tetrafluorophthalonitrile, on the other hand, implies that even a weak crystal field can have experimentally detectable and distinct effects on the intramolecular topology [66]. While the B3LYP calculations predict the equivalent bonds in the isolated isomers to be identical, the experimental $\nabla^2\rho_{BCP}$ values for the polar bonds are significantly different for the two molecules in the two crystals. Another example that well illustrates the difficulties in detecting the weak signal of bulk effects is the study of 9-ethynyl-9-fluorenol that crystallizes with two molecules in the asymmetric unit. The unconstrained multipole refinement, treating the two symmetry-independent species autonomously to account for the different intermolecular environment, and the constrained refinement enforcing their static density to be identical, yields the same fit but statistically different BCP indices [67].

The comparison of experimental and theoretical integrated "Quantum Theory of Atoms in Molecules" (QTAIM) properties can also lead to rather controversial conclusions. The net charges obtained for naphthalene (weak intermolecular interactions) by the X-N analysis, for example, are consistent with those derived from periodic HF calculations [46]. For the crystalline L-tryptophan however, the experimental QTAIM charges and dipole moments are practically identical to those derived from the B3LYP/6-311++G (3d,3p) density of the isolated molecule, even for the H-atom involved in a strong ionic O–H$\cdots$O$^{(-)}$ hydrogen bond [68]. This is in contrast with the pronounced charge loss of H-atoms (0.1–0.3 e) in weak C–H$\cdots$O interactions found for crystalline coumarin and its derivatives [69].

5.3.6 Studies Using Simulated Data

In the course of simulated studies, structure factors, generated from a wave-function-based density corresponding to the periodic system or the isolated molecule, are targeted in the multipole fit to assess model inadequacies and derive 'theory-supported' pseudoatom parameters directly applicable to the interpretation of experimental data. Such analyses have the advantage that the target data can be designed so as to include distinct physical effect, separately or combined, in a controlled way. One of the earliest applications [70] validated the κ-formalism

[71] by showing that the monopole model was capable to account for the charge transfer in diatomics and that the pseudoatomic net charge correlated with the radial expansion/contraction of the valence shell in a plausible way. Simulations on non-centrosymmetric structures [72] have revealed that indeterminacies in odd-order multipole populations [73] can severely bias the model density.

It has been found that the experimental BCP indices are in a better agreement with the theoretical values if these are derived from the multipole-model density fitted to theoretical structure factors [74]. This is a clear indication that the limited multipole model biases the results. Volkov et al. [75] optimized the shape of deformation RDF's against simulated data and developed a databank of κ'-parameters for light atoms (H, C, N, O) in typical bonding situations. A throughout investigation of theoretical structure factor data (HF/6-311G**) of solid ammonia [76] clearly shows the inadequacy of the single Slater functions utilized in the HC-formalism.

Studies using periodic HF data of small model compounds [77] show that the interaction density ($\delta\rho = \rho(\text{crystal}) - \rho(\text{isolated molecule})$), in spite of its weak effect on the low-order structure factors, can be retrieved by multipole refinement.

Nevertheless, there is no doubt that pseudoatoms can selectively and efficiently absorb 'chemical signals' from accurate Bragg reflections. Indeed, several independent observations have found that the multipole refinement led to statistically equivalent set of deformation density parameters for chemically equivalent sites. The recognition of this 'chemical transferability' has stimulated efforts towards developing pseudoatom databanks (see Chaps. 1 and 11), either from experimental [78] or theoretical densities [79, 80] of small molecules, applicable to construct the densities of large systems. Several promising applications of both types of databanks have been reported [81–83].

5.3.7 Toward Improving the Pseudoatom Model

It is anticipated that wave-function-based models and 1-RDM refinement protocols are becoming increasingly applicable, as data access and quality continues to improve. Current applications however suffer from difficulties associated with simultaneous derivation of electronic and nuclear properties, that is, the nuclear geometry and ADPs enter into the model as fixed parameters that are usually derived via pseudoatom refinement of the same data. It is also evident that such methods will remain unaffordable for larger unit-cell systems. The pseudoatom formalism is thus likely to have a bright future, especially if efforts towards improving its physical soundness are successful.

A modified version of Stewart's multipole projection technique [36] offers a plausible route to such an upgrade [84]. An important objective is to derive RDFs that are flexible and structured enough to describe the total density in both nuclear and bonding regions but exhibit some degree of chemical transferability. More adequate basis sets can be developed by systematic analyses of quantum-chemical densities of analog systems in which 'chemical atom types' can be identified based on their similar bonding situation. The approach involves the derivation of the spherical harmonics content of atomic densities obtained via a fuzzy type fragmentation of a known molecular density (ρ). Since the target functions are atomic densities ($\rho(\mathbf{r}) = \sum_A \hat{\rho}_A(\mathbf{r}_A)$), the multicenter non-orthogonal projection (5.59) is reduced to a one-center orthogonal projection

$$\hat{\rho}_{lm}(r) = \langle \hat{\rho}(r) \mid d_{lm}(\vartheta, \phi) \rangle_{\Omega} \quad \text{and} \quad \int_0^{\infty} r^2 \hat{\rho}_{lm}(r) dr = \hat{P}_{lm}, \tag{5.65}$$

where the $\hat{P}_{lm}$'s are uncorrelated multipole populations. The angular integrals are numerically calculated on a fine radial grid and can be further projected onto an auxiliary radial basis. The analytic RDFs obtained in such a way are (l,m)-dependent, that is, an individual RDF is assigned to each $d_{lm\pm}$. Such a representation of atomic and molecular densities can be considered exact in the sense that it can, in principle, reproduce the target density within an arbitrary accuracy that depends only on the level of expansion (L). Optimal m-independent RDFs, directly applicable in the standard HC model, can also be derived by principal component analysis. The transformation between the two models is achieved by an LS-projection:

$$\sum_{m=-l}^{+l} \hat{\rho}_{lm} d_{lm} \Leftrightarrow \hat{\rho}_l \sum_{m=-l}^{+l} Q_{lm} d_{lm}, \tag{5.66}$$

for which the coefficients (C_{lm}) of the expansion,

$$\hat{\rho}_l = \sum_{m=-l}^{+l} C_{lm} \hat{\rho}_{lm}, \tag{5.67}$$

are the components of the eigenvector corresponding to the largest eigenvalue of the overlap matrix of the m-dependent RDFs. Fitting each $\hat{\rho}_l$ with a single Slater function ($\hat{R}_l$) yields the 'best-standard' pseudoatom. Applications of the above formalism to the *ab initio* density of α-glycine (BLYP/QZ4P) [85] partitioned with the stockholder scheme [86] leads to the following results.

The atomic core densities (derived from the core molecular orbitals) have non-spherical contributions of mainly dipolar ($l = 1$) and quadrupolar ($l = 2$) symmetry. The corresponding RDFs possess sharp features in the vicinity of the nucleus and are weakly, but not negligibly, populated. The RDFs of the higher-order terms in the expansion are far more structured than those utilized in the HC formalism and their fit needs a combination of several Slater functions.

The 'exact' model, using m-dependent RDFs ($\hat{\rho}_{lm}$) and terminated at the hexadecapolar level ($L = 4$), reconstructs the target density within an accuracy

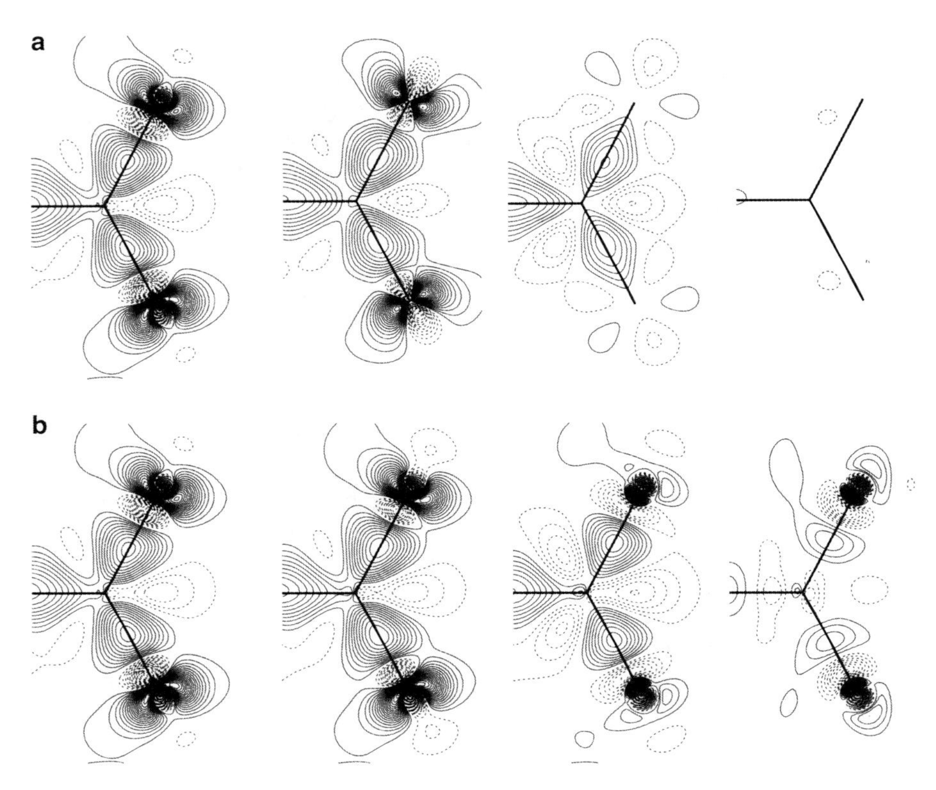

Fig. 5.3 Residual density maps in the plane of the carboxylate group in α-glycine: $\Delta\rho = \rho_\Psi - \hat{\rho}_L$ (from *left* to *right*, the level of expansion: L = 0, 1, 2, 3). (**a**) exact pseudoatom with numerical *m*-dependent RDFs ($\hat{\rho}_{lm}$). (**b**) the best standard model with *m*-independent RDFs, each fitted by a single Slater function ($\hat{R}_l$) (Contours: 0.05 e/Å^3; *solid* positive, *dotted* negative)

of 0.05 e/Å^3; the maximum/minimum residual density of 0.043/–0.047 e/Å^3 is located in the close vicinity of the oxygen nuclei and rapidly diminishes as the level of expansion increases (Fig. 5.3a). The inadequacy of the 'best-standard' model (with single Slater functions, $\hat{R}_l$, derived via principal component analysis) is well revealed by a strong residual peak of 0.15 e/Å^3 in the C=O bond and a maximum/minimum residual electron density of 0.163/–0.235 e/Å^3 near to the oxygen nuclei (Fig. 5.3b). These systematic features cannot be accounted for even with the inclusion of higher-order multipoles, since the convergence stops already at the octupolar level ($L = 3$). While the density is well described by the *exact* RDFs, the Laplacian at the BCP of the C=O bond is underestimated by 25%. A detailed analysis reveals that the expansion must be extended up to $L = 18$ to obtain the target value with four significant figures. An important observation is that the model with *m*-independent RDF's ($L = 4$) can occasionally perform better in predicting the bond-parallel curvature (λ_3) at the BCP than that with *m*-dependent RDFs.

This indicates that fortuitous cancellation of errors in the principal curvatures obtained by the restricted model can occur locally, yielding apparently correct values for the Laplacian. An even more important outcome of the analysis is that the best standard model can only modestly account for certain components of the molecular electrostatic moments. The relative error in the higher-order moments can exceed 100%.

The pseudoatom RDFs derived from theoretical molecular densities can significantly enhance the physical soundness of the model, because they are more adequate to describe density deformations due to bonding than those derived from atomic densities. As the above example clearly demonstrates, the standard model with m-independent functions fails to represent π-bonding. The introduction of the split-valence basis set into quantum chemistry was dictated by the recognition that those p-orbitals that are involved in π-bonding must have different radial behavior than those involved in σ-bonding. While the one-center molecular density corresponding to a minimal basis (all p-orbitals have the same RDF's) can be represented in terms of standard pseudoatoms, orbital splitting evidently requires m-dependent RDF's in the multipole expansion.

For molecular crystals, one can start the refinement with stockholder pseudoatoms consistent with a single-point wave function of the isolated molecule whose geometry is obtained via a standard HC-refinement. Subsequent refinements of the ADPs, using stockholder pseudoatom scattering factors, are expected to improve the physical significance of thermal parameters and lead to a static density which is less biased by thermal motion effects and systematic experimental errors. The analysis of the residual density, relative to a quantum chemical density of the isolated molecule, can help identify anharmonic sites and reveal crystal-field effects directly. This opens new revenues to extract the interaction density from high-quality data. The chemical transferability of the stockholder atoms provides an easy method for construction of relatively accurate density of larger systems of known structure.

Acknowledgements J.-M. Gillet would like to thank D. Sivia for fruitful discussions on combined data treatment. CNRS and Agence Nationale pour la Recherche (CEDA project) are also thanked for financial support. T. Koritsanszky acknowledges the support of the German Science Foundation (SPP 1178: "Experimental Electron Density as the Key to Understand Chemical Bonding").

References

1. Coppens P (1997) X-ray charge densities and chemical bonding. Oxford University Press, Oxford
2. Tsirelson VG, Ozerov RP (1996) Electron density and bonding in crystals. Institute of Physics Publishing, Bristol
3. Koritsanszky TS, Coppens P (2001) Chemical applications of X-ray charge density analysis. Chem Rev 101:1583–1628
4. Van Hove L (1954) Correlations in space and time and Born approximation scattering in systems of interacting particles. Phys Rev 95(1):249–262

5. Schülke W, Schmitz JR, Schulte-Schrepping H, Kaprolat A (1995) Dynamic and structure factor of electrons in Si: inelastic X-ray scattering results. Phys Rev B 52(16):11721–11732
6. Shukla A (1999) Ab initio Hartree-Fock computation of the electronic static structure factor for crystalline insulators: benchmark results on LiF. Phys Rev B 60(7):4539–4544
7. Watanabe N, Hayashi H, Udagawa Y, Ten-no S, Iwata S (1998) Static structure factor and electron correlation effects studied by inelastic X-ray scattering spectroscopy. J Chem Phys 108(11):4545–4553
8. Heisenberger P, Platzman PM (1970) Compton scattering of X-rays from bound electrons. Phys Rev A 2(2):415–423
9. Chew G (1950) The inelastic scattering of high energy neutrons by deuterons according to the impulse approximation. Phys Rev 80(2):186–202
10. Chew G, Wick GC (1952) The impulse approximation. Phys Rev 85(4):636–642
11. Pattison P, Weyrich W, Williams B (1977) Observation of ionic deformation and bonding from Compton profiles. Solid State Commun 21:967–970
12. Hansen NK (1980) Reports of Hahn-Meitner Institute HMI B342
13. Hansen NK, Pattison P, Schneider J (1987) Analysis of the 3-dimensional electron distribution in silicon using directional Compton profile measurements. Z Phys B 66:305–315
14. Gillet J-M, Fluteaux C, Becker PJ (1999) Analytical reconstruction of momentum density from directional Compton profiles. Phys Rev B 60(4):2345–2349
15. Kontrym-Sznajd G (1990) Three dimensional image reconstruction with application in positron annihilation. Phys Stat Solid A 117(1):227–240
16. Reiter G, Silver R (1985) Measurement of interionic potentials in solids using deep-inelastic neutron scattering. Phys Rev Lett 54(10):1047–1050
17. Sivia DS, Skilling J (2006) Data analysis. Oxford University Press, USA
18. Jaynes ET (2003) Probability theory: the logic of science. Cambridge University Press, Cambridge
19. Jeffreys H (1939) Theory of probability. Clarendon, Oxford
20. Schmider H, Edgecombe KE, Smith VH Jr (1992) One-particle matrices along the molecular bonds in linear molecules. J Chem Phys 96(11):8411–8419
21. Howard S, Hulke JP, Mallinson PR, Frampton CS (1994) Density matrix refinement for molecular crystals. Phys Rev B 49(11):7124–7136
22. Schmider H, Smith VH Jr, Weyrich W (1992) Reconstruction of the one particle density matrix from expectation values in position and momentum space. J Chem Phys 96(12):8986–8994
23. Schmider H, Smith VH Jr, Weyrich W (1993) On the inference of the one-particle density matrix from position and momentum-space form factors. Z Naturforsch 48A:211–220
24. Pecora LM (1986) Determination of the quantum density matrix from experiment: an application to positron annihilation. Phys Rev B 33(9):5987–5993
25. Gillet J-M (2007) Determination of a one-electron reduced density matrix using a coupled pseudoatom model and a set of complementary scattering data. Acta Crystallogr A 63: 234–238
26. Gillet J-M, Becker PJ, Cortona P (2001) Joint refinement of a local wave-function model from Compton and Bragg scattering data. Phys Rev B 63:235115
27. Kiang HS (1969) N-representability theorem for reduced density matrices. J Math Phys 10(10):1920–1921
28. McWeeny R (1959) Hartree-Fock theory with non-orthogonal basis functions. Phys Rev 114(6):1528–1529
29. McWeeny R (1960) Some recent advances in density matrix theory. Rev Mod Phys 32(2): 335–369
30. Clinton W, Galli A, Massa L (1969) Direct determination of pure-state density matrices. II. Construction of constrained idempotent one-body densities. Phys Rev 177(1):7–13
31. Clinton W, Massa L (1972) Determination of the electron density matrix from X-ray diffraction data. Phys Rev Lett 29(20):1363–1366
32. Weiss AW (1961) Configuration interaction in simple atomic systems. Phys Rev 122: 1826–1836

33. Stewart RF, Feil D (1980) A theoretical study of elastic X-ray scattering. Acta Crystallogr A 36:503–509
34. Stewart RF (1997) Vibrational averaging of X-ray-scattering intensities. Isr J Chem 16: 137–143
35. Stewart RF (1969) Generalized X-ray scattering factors. J Chem Phys 51:4569–4576
36. Stewart RF (1977) One-electron density functions and many-centered finite multipole expansions. Isr J Chem 16:124–131
37. Stewart RF (1976) Electron population analysis with rigid pseudoatoms. Acta Crystallogr A 32:565–574
38. Hansen NK, Coppens P (1978) Testing aspherical atom refinements on small molecule data sets. Acta Crystallogr A 34:909–921
39. Fertig HA, Kohn W (2000) Symmetry of atomic electron density in Hartree, Hartree-Fock and density-functional theories. Phys Rev A 62:052511–10
40. Stewart RF, Bentley J, Goodman B (1975) Generalized X-ray scattering factors in diatomic molecules. J Chem Phys 63:3786–3793
41. Clementi E, Roetti C (1974) Roothaan-Hartree-Fock atomic wavefunctions. Basis functions and their coefficients for ground and certain excited states of neutral and ionized atoms, $Z \leq 54$. Atom Data Nucl Data Tab 14:177
42. Spackman MA (1992) Molecular electric moments from X-ray diffraction data. Chem Rev 92:1769–1797
43. Volkov A, King HF, Coppens P, Farrugia LJ (2006) On the calculation of electrostatic potential, electric field and electric field gradient from the aspherical pseudoatom model. Acta Crystallogr A 62:400–408
44. Spackman MA (2007) Comments on On the calculation of electrostatic potential, electric field and electric field gradient from the aspherical pseudoatom model by Volkov, King, Coppens & Farrugia (2006). Acta Crystallogr A 63:198–200
45. Hirshfeld FL (1977) Charge deformation and vibrational smearing. Isr J Chem 16:168–174
46. Oddershede J, Larsen S (2004) Charge density study of naphthalene based on X-ray diffraction data at four different temperatures and theoretical calculations. J Phys Chem A 108:1057–1063
47. Kato T (1957) On the eigenfunctions of many-particle systems in quantum mechanics. Commun Pure Appl Math 10:151–177
48. Katriel J, Davidson ER (1980) Asymptotic behavior of atomic and molecular wave functions. Proc Natl Acad Sci USA 77:4403–4406
49. Pillet S, Souhassou M, Lecomte C, Schwarz K, Blaha P, Rerat M, Lichanot A, Roversi P (2001) Recovering experimental and theoretical electron densities in corundum using the multipolar model: IUCr multipole refinement project. Acta Crystallogr A 57:290–303
50. Volkov A, Macchi P, Farrugia LJ, Gatti C, Mallinson P, Richter T, Koritsanszky T (2006) Program manual, XD2006 – a computer program package for multipole refinement, topological analysis of charge densities and evaluation of intermolecular energies from experimental and theoretical structure factors. User's manual. http://xd.chem.buffalo.edu/docs/xdmanual.pdf
51. Coppens P, Boehme R, Price PF, Stevens ED (1981) Electron population analysis of accurate diffraction data. 10. Joint X-ray and neutron data refinement of structural and charge density parameters. Acta Crystallogr A 37:857–863
52. Blessing RH (1995) On the differences between X-ray and neutron thermal vibration parameters. Acta Crystallogr B 51:816–823
53. Schomaker V, Trueblood KN (1968) On the rigid-body motion of molecules in crystals. Acta Crystallogr B 24:63–76
54. Madsen AO, Sorensen HO, Flensburg C, Stewart RF, Larsen S (2004) Modeling of the nuclear parameters of H atoms in X-ray charge density studies. Acta Crystallogr A 60:550–561
55. Destro R, Roversi P, Barzaghi M, Marsh RE (2000) Experimental charge density of α-glycine at 23 K. J Phys Chem A 104:1047–1054
56. Roversi P, Destro R (2004) Approximate anisotropic displacement parameters for H atoms in molecular crystals. Chem Phys Lett 386:472–478

57. Bürgi HB, Capelli SC, Goeta AE, Howard JAK, Spackman MA, Yufit DS (2002) Electron distribution and molecular motion in crystalline benzene: an accurate experimental study combining CCD X-ray data on C_6H_6 with multi-temperature neutron-diffraction results on C_6D_6. Chem Eur J 8:3512–3521
58. Flaig R, Koritsanszky T, Zobel D, Luger P (1998) Topological analysis of experimental electron densities of amino acids: 1. D,L-Aspartic acid at 20 K. J Am Chem Soc 120:2227–2236
59. Munshi P, Madsen AO, Spackman MA, Larsen S, Destro R (2008) Estimated H-atom anisotropic displacement parameters: a comparison between different methods and with neutron diffraction results. Acta Crystallogr A 64:465–475
60. Bader RFW (1990) Atoms in molecules: a quantum theory. Oxford Science, Oxford
61. Tsirelson VT (2002) Mapping of electronic energy distributions using experimental electron density. Acta Crystallogr B 58:632–639
62. Gatti C (2005) Chemical bonding in crystals: new directions. Z Kristallogr 220:399–487
63. Saunders VR, Dovesi R, Roetti C, Causa M, Harrison NM, Orlando R, Zicovich-Wilson CM (1998) CRYSTAL98 user's manual. University of Turin, Turin
64. Gatti C, Saunders VR, Roetti C (1994) Crystal-field effect on the topological properties of the electron density in molecular-crystals – the case of urea. J Chem Phys 101:10686–10696
65. Ovegaard J, Hibbs DE (2004) The experimental electron density in polymorphs A and B of the anti-ulcer drug famotidine. Acta Crystallogr A 60:480–487
66. Hibbs DE, Overgaard J, Platts JA, Waller MP, Hursthouse MB (2004) Experimental and theoretical charge density studies of tetrafluoro-phthalonitrile and tetrafluoro-isophthalonitrile. J Phys Chem B 108:3663–3673
67. Overgaard J, Waller MP, Platts JA, Hibbs DE (2003) Influence of crystal effects on molecular densities in a study of 9-Ethynyl-9-fluorenol. J Phys Chem A 107:11201–11208
68. Scheins S, Dittrich B, Messerschmidt M, Paulmann C, Luger P (2004) Atomic volumes and charges in a system with a strong hydrogen bond: L-tryptophan formic acid. Acta Crystallogr B 60:184–190
69. Munshi P, Guru Row TN (2005) Exploring the lower limit in hydrogen bonds: analysis of weak C–H$\cdots$O and C–H$\cdots\pi$ interactions in substituted coumarins from charge density analysis. J Phys Chem A 109:659–672
70. Brown AS, Spackman MA (1991) A model study of κ-refinement procedure for fitting valence electron densities. Acta Crystallogr A 47:21–29
71. Coppens P, Guru Row TN, Leung P, Stevens ED, Becker PJ, Wang YW (1979) Net atomic charges and molecular dipole moments from spherical-atom X-ray refinements, and the relation between atomic charge and shape. Acta Crystallogr A 35:63–72
72. Spackman MA, Byrom PG (1997) Retrieval of structure-factor phases in non-centrosymmetric space group. Model studies using multipole refinement. Acta Crystallogr B 53:553–564
73. Haouzi AEl, Hansen NK, Hènass CLe, Protas J (1996) The phase problem in the analysis of X-ray diffraction data in terms of electron-density distributions. Acta Crystallogr A 52:291–301
74. Howard ST, Hursthouse MB, Lehmann CW (1995) Experimental and theoretical determination of electronic properties in L-dopa. Acta Crystallogr B 51:328–337
75. Volkov A, Abramov YA, Coppens P (2001) Density optimized radial exponents for X-ray charge density refinement from ab initio calculations. Acta Crystallogr A 57:272–282
76. Bytheway I, Chandler SG, Figgis BN (2002) Can a multipole analysis faithfully reproduce topological descriptors of a total charge density? Acta Crystallogr A 58:451–459
77. Spackman MA, Byrom PG, Alfredsson M, Hermansson K (1999) Influence of intermolecular interactions on multipole refined electron densities. Acta Crystallogr A 55:30–47
78. Pichon-Pesme V, Lecomte C, Lachekar H (1995) On building a data bank of transferable experimental electron density parameters: applications to polypeptides. J Phys Chem 99:6242–6250

79. Volkov A, Li X, Koritsanszky T, Coppens P (2004) Ab initio quality electro-static atomic and molecular properties from a transferable theoretical pseudoatom databank: comparison of electrostatic moments, topological properties, and interaction energies with theoretical and force-field results. J Phys Chem A 108:4283–4300
80. Dittrich B, Koritsanszky T, Luger P (2004) A simple approach to molecular densities with invarioms. Angew Chem Int Ed 43:2718–2721
81. Jelsch C, Pichon-Pesme V, Lecomte C, Aubry A (1998) Transferability of multipole charge-density parameters: application to very high resolution oligopeptide and protein structures. Acta Crystallogr D 54:1306–1318
82. Pichon-Pesme V, Zarychta B, Guillot B, Lecomte C, Jelsch C (2007) On the application of an experimental multipolar pseudo-atom library for accurate refinement of small-molecule and protein crystal structures. Acta Crystallogr A 63:108–125
83. Dominiak PM, Volkov A, Dominiak AP, Jarzembska KN, Coppens P (2009) Combining crystallographic information and an aspherical-atom data bank in the evaluation of the electrostatic interaction energy in an enzyme-substrate complex: influenza neuraminidase inhibition. Acta Crystallogr D 65:485–499
84. Koritsanszky T, Volkov A (2004) Density radial functions for bonded atoms. Chem Phys Lett 383:431–435
85. te Velde B, Bickelhaupt FM, van Gisbergen SJA, Fonseca Guerra C, Baerends EJ, Snijders JG, Ziegler TJ (2001) Chemistry with ADF. J Comput Chem 22:931–967
86. Hirshfeld FL (1977) Spatial partitioning of charge density. Theor Chim Acta 44:129–132

Chapter 6
Using Wavefunctions to Get More Information Out of Diffraction Experiments

Dylan Jayatilaka

6.1 Introduction

The goal of quantum chemistry is to determine the physical properties of molecules or larger aggregates of atoms such as crystalline solids by computation, using the fundamental laws of quantum mechanics. There are two successful approaches to obtaining such properties: the *ab initio* wavefunction approach (see, for example, [1]), and the density functional theory (DFT) approach (see e.g. [2]). The purpose of this article is to present the case for a third way: a method which synthesizes these two approaches and which makes use of experimental diffraction data.

In the *ab initio* paradigm advocated by Pople [3], physical properties are determined by making controlled and systematically improvable approximations to Schrödinger's wavefunction. For larger systems, *ab initio* methods become computationally unwieldy. Indeed, Walter Kohn, one of DFT's main champions has asserted that "the wavefunction is meaningless for systems of more than 1,000 electrons" [4]. In a landmark theorem, Kohn and Hohenberg proved that the electron density is sufficient to predict all chemical properties for the ground state [5]. Thus, DFT was born with the electron density as the prime variable for predicting all chemical properties.[1] In contrast to *ab initio* methods, DFT and its generalisations are characterised by a non-systematic approach: new methods are developed *ad hoc*, making use (where possible) of exact relations obeyed by the electron density and

[1]It is somewhat amusing to note that, only a year after this theorem was published, the notion of the wavefunction returns quickly via the Kohn-Sham method [6] because of the difficulty of computing a kinetic energy functional of the density. It is this latter 'wavefunction' version of DFT that is most commonly employed.

D. Jayatilaka (✉)
Chemistry, School of Biomedical, Biological and Chemical Sciences, University of Western Australia, 35 Stirling Highway, Nedlands, 6009 WA, Australia
e-mail: dylan.jayatilaka@uwa.edu.au

C. Gatti and P. Macchi (eds.), *Modern Charge-Density Analysis*,
DOI 10.1007/978-90-481-3836-4_6, © Springer Science+Business Media B.V. 2012

its generalisations. Typically these relations yield expressions which may be approximated, introducing adjustable parameters which are further fitted to the results of experiment (a semi-empirical approach), or which are fitted to the results from wavefunction calculations (a meta-semi-empirical approach). Curiously, experimentally measured properties of the electron density itself (through the X-ray or electron diffraction experiment, for example) are rarely used in density functional theory.

In this chapter, a conceptually different way to obtain chemical properties is proposed: the method of experimentally refined [7] and experimentally constrained wavefunctions [7–16]. In this semi-empirical method, reasonable model electronic wavefunctions are derived from the experimentally data pertaining to the electron density, specifically electronic information pertaining to periodic molecular crystals. Hence in this approach, wavefunctions and density functional theory techniques are merged together.

This chapter will review in the next section the concepts and practice of experimentally optimising and constraining model wavefunctions to experimental X-ray diffraction data. The optimisation process involves wavefunction parameters describing both electronic and nuclear (atomic positional) degrees of freedom. The optimisation of the nuclear degrees of freedom, although an important prerequisite for obtaining accurate electronic information, has been much less discussed. The reasons for this state of affairs has to do with the authors own initially too eager interest in electronic properties, coupled with the false assumption that the optimisation of geometric parameters from X-ray data is a solved and somewhat dead problem. Therefore, Sect. 6.2 discusses the relationship of electronic and nuclear parameter optimisation, before examining the techniques in detail. In Sect. 6.3 a review of the history of obtaining "experimental wavefunctions" is discussed, with emphasis on diffraction data. In Sect. 6.4 some of the more recent developments concerning "geometry optimisations" with wavefunction-derived electron densities are presented: namely the determination of hydrogen atom atomic anisotropic displacement parameters (ADPs). This goes some way toward obtaining dynamic information from the experimental X-ray diffraction data and more important ensuring the electronic parameters pertaining to these atoms are correct. In Sect. 6.5, several electronic properties obtained from the experimentally constrained wavefunctions are produced, such as deformation densities, dipole moments. More interesting response properties, such as refractive index and in-crystal hyperpolarisabilities are also produced. These are compared with independent experimental data and high-level calculations. A perspective of the future of the technique in given in the conclusions.

6.2 Can a Wavefunction Be Determined Experimentally?

The notion of fixing parameters in a model wavefunction from experiment raises the question about whether it makes sense to actually do this. This has been a somewhat controversial question, so it deserves some attention. The controversy is usually associated with those who approach quantum mechanics from an axiomatic standpoint; differences of opinion can usually be resolved by adopting acceptable terminology, specifically by being careful to use "experimental *model* wavefunctions" rather than "experimental wavefunctions" (we shall use both in this article as synonyms).[2] The determination of these experimental model wavefunctions has a long history, but it is more instructive to begin here with a didactic approach and defer to later the review of the literature.

6.2.1 Experimental Wavefunctions – Briefly

First, it is worth noting that the exact wavefunction is an experimentally derived entity. When the Schrödinger equation (or Dirac equation) is solved to obtain a wavefunction, the experimentally determined fundamental physical constants are introduced in order that experimental quantities can be produced from the wavefunction. In this sense then, the wavefunction is experimentally derived.

For many reading this article, however, the use of (experimentally determined) fundamental constants would not confer upon the wavefunction any experimental nature. Rather, those readers might expect that an experimental wavefunction is one whose values in some coordinate system are determined *directly* by experiment.[3] Rather than argue abstractly, it is easier to mention at least three experiments where wavefunctions or wavefunction-like entities have been determined.

The most elegant experiment is probably by Lohmann and Weigold [17], who have determined the ground state electronic wavefunction of the hydrogen atom using (e,2e) scattering experiments. These remarkable results, displayed in Fig. 6.1, should perhaps be more widely known. Although the hydrogen atom represents an optimum (one electron!) case, McCarthy and Weigold discuss other experimental techniques for mapping wavefunctions, or, more correctly, Dyson orbitals [18].

Recently, scanning tunneling microscopy has been used to obtain spatially distributed current-voltage transmission probabilities for single molecules adsorbed to surfaces [19, 20]. At certain energies, these current-voltage images correspond well to theoretically calculated molecular orbitals for the isolated molecule although theoretically the transition probabilities probably include other contributions which are not just molecular-orbital related.

[2] Some authors prefer the terminology "experimental density matrix reconstruction". Experimental wavefunctions are a restricted subclass of these methods, as explained later.

[3] Note that a wavefunction is only unique up to a unitary phase transformation of the basis in which it is determined. It is fallacious to use this fact to argue against the notion of an experimental wavefunction; the measurement of the position of a particle in space is also arbitrary up to an orthogonal rotation of the axes used to defined the coordinate system, yet few would argue that positions are unobtainable by experiment. In both cases it is simply necessary to specify the basis (axes) with which the measurement is made.

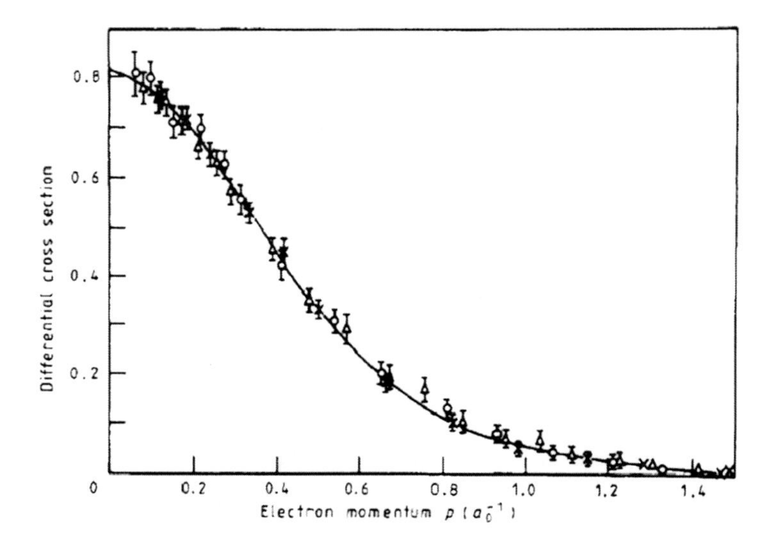

Fig. 6.1 Wavefunction for the H atom in momentum space, measured using (e,2e) spectroscopy [17]. *Data points* represent measurements at different total energy; *solid line* is the expected result $8\pi^{-1}(1+p^2)^{-4}$ (Reprinted from Ref. [17] with kind permission of Elsevier)

Very recently, the determination of a photon wavefunction has been reported using the concept of a "weak measurement" [21]. The technique involves few assumptions and is different to that employed here (the ideas used in this chapter are similar in spirit to "quantum state tomography" [21]).

Finally, a well-known example occurs in molecular ro-vibrational spectroscopy. In this field, much effort has been expended in determining parameters in effective Hamiltonians which describe an observed spectrum (see, for example, [22]). The eigenstates of these effective Hamiltonians are, essentially, experimentally determined rovibrational wavefunctions, typically expressed in a basis of a product of harmonic-oscillators rigid-rotor functions. This example highlights the fact that model Hamiltonians are typically used whenever there is a spectrum to be modelled, whereas *a model wavefunction would be used when focussing on a particular quantum state, typically the ground state*.

It should be clear from these examples that the reason one wants to derive a model wavefunction (or model Hamiltonian) is the same as for any model: to condense the observed data into a form of underlying physical significance. In the case of a wavefunction-based model, there is the tantalising chance to obtain other physical properties, from the model quite different from the data used to define the model.

Clearly, model wavefunctions (or components of a model wavefunction, such as molecular orbitals) are indeed related to and hence derivable from experimental measurements. The derivation of such model wavefunctions has been done for some time in various guises. Whether one regards this as an "experimental determination" of a wavefunction, or not, is largely a matter of taste [23]. Certainly, the theory

Fig. 6.2 Stages in a typical measurement. 1 Conception of a model *and* corresponding measurement process to determine unknown parameters. Here, a model wavefunction is used with parameter K. 2 Data is collected in accordance with the preconceived measurement scheme; here, we argue that X-ray diffraction data should be used to determine a model wavefunction. 3 Parameters (K) are determined according to the preconceived method

of quantum mechanics is set up so that observables are produced from the wavefunction by the recipe of expectation values; this recipe does not change, whether the (model) wavefunction is determined by *ab initio*, DFT, or by fitting to experimental data.

6.2.2 How to Determine Wavefunction Parameters?

Two questions should be addressed in this context: how, in practice, should it be done? And what experimental data could or should be used?

Since the wavefunction is in general a multidimensional quantity of high dimension practicality dictates that we should describe it not in coordinate space, but in terms of model parameters. In both *ab initio* and DFT methods, basis sets of functions are used to expand the wavefunction. For the wavefunctions considered here, the model wavefunction parameters are of the Hartree-Fock type and comprised of the linear combination of atomic orbitals (LCAO) molecular orbital (MO) coefficients in terms of a given basis set, and the positions of the atoms in the system (there are also other parameters required to describe the experimental data; this and related issues are discussed at more length below). So far, the scheme to determine model wavefunction parameters is no different to any other experiment; the procedure is illustrated in Fig. 6.2.

Although the representation of orbital-type wavefunctions in terms of basis sets is quite efficient, there are still many unknown parameters to determine in order to produce a reasonable model wavefunction. Therefore, a considerable amount of experimental data pertaining to the state under study is required.

6.2.2.1 Importance of Diffraction Experiments

In the case of electronic wavefunctions (assuming a Born-Oppenheimer separation between electronic and nuclear vibrational degrees of freedom) the Hohenberg-Kohn theorem already discussed makes it natural to turn to the *X-ray* or *electron diffraction experiments* which yield structure factor data with information pertaining to the electron density [24, 25]. It is fortunate that these experiments do indeed yield much data. However, the amount of structure factor data is usually still insufficient to determine all the parameters in even the simplest possible electronic wavefunction–that of the Hartree-Fock type–by means of a traditional least-squares procedure.[4] An alternative must be found.

6.2.2.2 Constrained Electronic Wavefunctions

In this article the constrained variational wavefunction approach is proposed as one alternative for deriving reasonable electronic wavefunctions from the diffraction data. It can be explained as follows as a hybrid of *ab initio* quantum mechanics and least-squares fitting.

In *ab initio* quantum mechanics, one of the ways to determine the parameters $\mathbf{p}$ in a wavefunction $\Psi(\mathbf{r}; \mathbf{p})$ is to use the variational theorem; namely to minimise the expression

$$E(\mathbf{p}) = \langle \Psi | \hat{H} | \Psi \rangle \tag{6.1}$$

with respect to the wavefunction parameters $\mathbf{p}$ (in DFT, there is an analogous energy expression which one minimises variationally). Note that the parameters $\mathbf{p} = (\mathbf{p}_e, \mathbf{p}_n)$ may be conceptually separated into two classes: those describing the electronic degrees of freedom $\mathbf{p}_e$, for example the molecular orbital coefficients, and those describing the nuclear degrees of freedom $\mathbf{p}_n$, for example the nuclear positions.

On the other hand, in traditional least squares theory one minimises the least-squares error

$$\chi^2(\mathbf{p}, \mathbf{q}; \mathbf{F}^{\text{exp}}) = \frac{1}{N_r - N_p} \sum_{k}^{N_r} \frac{\left(F_k(\mathbf{p}, \mathbf{q}) - F_k^{\text{exp}}\right)^2}{\sigma_k^2}. \tag{6.2}$$

[4] In fact, as discussed later, X-ray diffraction data is not sufficient on its own to obtain the exact one-particle density matrix for the system [26]. Later results concerning refractive indices show that the contribution of X-ray data is significant for particular response properties.

Here N_r is the number of experimental data, N_p is the number of parameters in the model (discussed below), the σ_k are the estimated errors in the measured data F_k^{exp}, and $F_k(\mathbf{p}, \mathbf{q})$ are the values of the experimental data calculated from the wavefunction $\Psi(\mathbf{r}; \mathbf{p})$ and from additional phenomenological parameters $\mathbf{q}$ required to model the experimental data properly e.g. for the X-ray experiment, atomic displacement parameters (ADPs) parameters, or extinction coefficients. These additional phenomenological parameters arise due to deficiencies or shortcomings in the electronic wavefunction model: specifically, that the crystal wavefunction, which is the one representing the actual experiment, is not a periodic replication of the molecular electronic wavefunction written above (this is discussed later on).

In the constrained wavefunction approach we combine the least-squares and variational ideas: a solution with desired least-squares error χ_0^2 is derived which at the same time minimises the wavefunction energy $E(\mathbf{p})$. This is equivalent to minimising either

$$L(\mathbf{p}, \mathbf{q}) = E(\mathbf{p}) + \lambda \chi^2(\mathbf{p}, \mathbf{q}; \mathbf{F}^{\text{exp}}), \quad \text{or} \tag{6.3}$$

$$\tilde{L}(\mathbf{p}, \mathbf{q}) = \mu \, E(\mathbf{p}) + \chi^2(\mathbf{p}, \mathbf{q}; \mathbf{F}^{\text{exp}}). \tag{6.4}$$

Minimising L or L is equivalent if $L = L/\lambda$ and $\mu = 1/\lambda$. The Lagrange multiplier λ (or μ) is chosen so that the desired value of χ^2 is obtained.

6.2.2.3 Refining Nuclear Parameters

Although this article is concerned with obtaining electronic wavefunctions, and we have argued that diffraction experiment are well suited to furnish relevant experiment information to determine such an electronic wavefunction, the development above makes it clear that the *nuclear degrees of freedom* $\mathbf{p}_n$ in any electronic wavefunction must also be determined.

However, an approach using constrained *nuclear* wavefunctions would not be appropriate for the nuclear parameters. That is because the diffraction experiments are performed on systems (crystals) where essentially all vibrational states (phonon modes) are occupied (according to a Boltzmann distribution); as already noted, a wavefunction approach requires a pure state. Another way to look at the mode problem in a crystal is that it amounts to a breakdown of the perfect-periodic approximation: one has to account for the relative displacements of *all* the atoms in the crystal away from their ideal position. This is true even at zero temperature, where all phonons are in the ground state and a pure-state (wavefunction) representation is appropriate; even in this case, there would be correlations between all the atoms in the crystal. Hence, dealing with the nuclear positions (if the nuclear parameters are chosen to be atomic positions) in practice requires some other *ad hoc* model with associated phenomenological nuclear parameters $\mathbf{q}_n$ which describe the non-periodic coupling between the nuclear position parameters. Couplings between nuclear positions in the same unit cell lead to Bragg scattering; those between atoms not in the same cell lead to thermal diffuse scattering. It is worth remarking that although correlations between an atom position and itself are taken into account (these are the so-called ADPs $\mathbf{q}_n \equiv \mathbf{U}$), between different atoms in the *same* unit cell are not, even though they should also contribute to Bragg scattering.[5]

There are other couplings that should be considered. For example, we know that the Born-Oppenheimer approximation is rather good. Therefore, the electronic parameters are dependent upon and thus totally correlated with the nuclear positions. Thus, there should some phenomenological description of $\mathbf{p}_e/\mathbf{p}_n$ and $\mathbf{p}_e/\mathbf{q}_n$ coupling. The phenomenological parameters need to describe both these effects in X-ray crystallography have not been much explored.[6]

These are important issues because in models where nuclear parameters are assumed independent of electronic parameters, their determination (in a least-squares sense) is often highly correlated with the electronic parameters, particularly for hydrogen atoms where changes in hydrogen mean atom position and its vibrational parameters are of the same order as the effects of chemical bonding on the diffraction data (cf. Chaps. 3 and 5).

6.2.2.4 Strategies for Finding Nuclear Position Parameters in the Electronic Wavefunction and the Associated Phenomenological Parameters

The simplest approach is to take the nuclear parameters from a refinement using the pseudoatom [29] multipole model (see, e.g. [24, 25]). In this model mean unit-cell positions and atomic anisotropic displacement parameters (ADPs), are least-squares fitted to the X-ray diffraction data. Then the position parameters are simply used in the constrained wavefunction without change. The ADP's of the atoms describe the cumulative effect of the phonons on each atom, and they are used to 'thermally smear' the partitioned atomic density by a convolution. Fortunately, there is usually enough data to determine the mean positions and ADPs by least-squares refinement in this way (though not usually the hydrogen atom positions and ADP's, unless the data is of exceptional quality).

An aesthetically much more pleasing method which is dealt with in detail later, has been described by Jayatilaka and Dittrich [7]. Here aspherical pseudoatoms are defined by an atomic partitioning of a molecular density. Specifically, a Hirshfeld atomic partitioning [30] is applied to an *ab initio* quantum mechanical calculation

---

[5] In Ref. [27] two-center Bragg scattering terms do not appear due to the application of the pseudoatom approximation. The terms we are discussing are not those called I_{elastic} in that paper

[6] See however the exciting work on polarizable multipoles by Schnieders et al. [28]

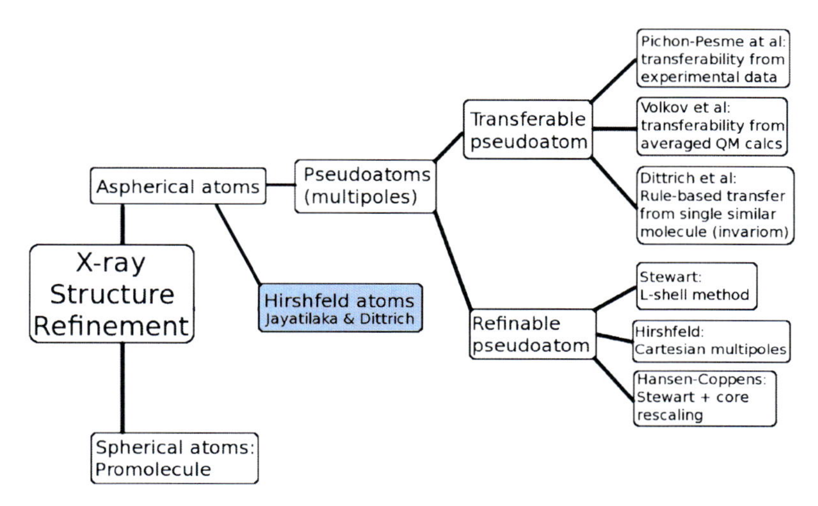

Fig. 6.3 Map of geometric refinement techniques in crystallography showing the positioning of the new wavefunction-based technique in relation to other methods

of the molecular electron density; the molecule itself possibly embedded in a local crystal field.[7,8] These aspherical "Hirshfeld atoms" are then used to obtain refined nuclear parameters.

The invariom method of Dittrich [32] can be considered an approximation of Hirshfeld atom refinement since it uses multipole-type pseudoatoms fitted to *ab initio* quantum mechanical data. The Hirshfeld atom refinement technique is also a generalization of the usual method which uses spherical atomic densities, also derived from quantum mechanical calculations.

Figure 6.3 puts the Hirshfeld atom refinement in perspective with others used in crystallography for atomic position determinations. Though aspherical atomic densities are used, Hirshfeld atom refinement is *not* a method for determining the charge density from experimental diffraction data: it is different from the pseudoatom-based methods because Hirshfeld atoms are defined from the underlying wavefunction. Neither is Hirshfeld atom refinement like those methods which use fixed transferable pseudoatoms: the Hirshfeld atoms change when the atoms change position, since wavefunction itself changes. The lineage of Hirshfeld atom refinement is best viewed as an extension of the spherical atom (pseudoatom) model, where the experimental structure factor data are modelled from *atomic* wavefunctions.

---

[7]Other partitioning schemes are conceivable. The Bader partitioning scheme would not be one such scheme since these atoms would have high frequency components in reciprocal space due to the sharp boundaries, making them highly unsuitable for structure refinement [31].

[8]The method could be applied to a molecular fragment from a *periodic* quantum mechanical calculation; this is a technical but important detail

When using constrained (electronic) wavefunction analysis, this "Hirshfeld atom refinement" procedure to obtain nuclear parameters is preferred because the same entity (the electronic wavefunction) is used in both nuclear and electronic refinement. The only difference is that, whereas the nuclear parameters are determined by least-squares, the electronic parameters are determined by contrained-least-squares. The use of the same underlying wavefunction model ensures that the features in the electronic density displayed by this reflect the underlying wavefunction and the quantum mechanical methods used to obtain it.

6.2.3 *Details of the Hirshfeld Atom Refinement Method*

Obtaining accurate nuclear positions and position-related information is a logical prerequisite to obtaining accurate electronic structure information from diffraction experiments. Therefore we begin with a description of the Hirshfeld atom refinement procedure.

6.2.3.1 Choice of Crystal Wavefunction

The method begins with a choice of wavefunction from which approximate in-crystal atomic electron densities (pseudoatoms) are obtained. The same choice is made for constrained Hartree-Fock wavefunctions, described later.

Since the diffraction data comes from a periodic crystal, we should use a wavefunction appropriate to a periodic system. In crystallography it is traditional to use the promolecule approximation, which is comprised of an (infinite and periodic) Hartree product of antisymmetric and non-interacting atomic wavefunctions

$$\Psi = \prod_A \Psi_A(\mathbf{x}; \mathbf{R}_A). \tag{6.5}$$

Each atomic wavefunction Ψ_A is centered at a nuclear coordinate $\mathbf{R}_A$ (the atom coordinates are symmetry- and translation-related to a unique set in the asymmetric unit) and is a function of several electronic space and spin coordinates $\mathbf{x}$ (nuclear spin coordinates are ignored for simplicity).

For the molecular systems considered in this article we use a minimal extension of this model: the so-called non-interacting fragment procrystal wavefunction, which is an (infinite and periodic) Hartree product of non-interacting molecular wavefunctions [33]

$$\Psi = \prod_M \Psi_M(\mathbf{x}; \mathbf{R}_M). \tag{6.6}$$

The molecules M (and the corresponding set of nuclear coordinates $\mathbf{R}_M$) may be symmetry- and translation-related to a unique set, but unlike the case of the promolecule wavefunction, the atoms in these molecules comprise *at least* an asymmetric unit but may be more, especially in the case where any individual molecule has a non-trivial symmetry. In this case a "whole" molecule is used rather than symmetry-unique portion.[9] Hence the above *ansatz* would be inappropriate for network-covalent or metallic crystals where there is no good concept of a separate "molecule", although we have nevertheless used it for both.

In this article the wavefunctions Ψ_M are chosen to be of Hartree-Fock or DFT form. Consequently the one-particle density matrix is idempotent and has orbital occupancies equal to one or zero; the real one-particle density matrix has no such restriction. However, a different choice of wavefunction Ψ_M can remove such restrictions.

For example, one could include exchange-correlation and coulomb interactions between the electrons in the molecule M and its neighbours, so that Ψ_M was not just a fragment wavefunction. By symmetry, such a wavefunction on its own would be sufficient to represent all the interactions in the crystal. An easier alternative would to neglect exchange-correlation interactions and approximate the periodic-coulomb interactions by embedding the molecular wavefunction within a one-electron self-consistent field of distributed point multipoles derived from the wavefunction itself (later, we report results using just such a method). In this case, the wavefunctions appearing in the Hartree product above are not strictly non-interacting.

It is also worth remarking that the fragment-based model *ansatz* can be extended to a full periodic Hartree-Fock wavefunction if one includes (i) a periodic coulomb potential, i.e. some form of Ewald summation, and (ii) short-range exchange interactions. Such periodic Hartree-Fock wavefunctions have not been used in this work. Others, though, have advocated such a real-space fragment-based approach to periodic electronic structure, sometimes also known as "divide and conquer" methods [34–36] (Ref. [34] is an excellent comparison of these local methods vs the usual Bloch-function based approaches). Such methods appear to be a very encouraging way to go beyond the Hartree-Fock approximation, and obtain correlated periodic wavefunctions; the developments in this field are still ongoing (cf. Chap. 1, Sect. 1.3.5 and Chap. 2, Sect. 2.5.3).

6.2.3.2 Expression for the Electron Density of Molecule *M*

All one-electron properties can be calculated from the wavefunction as an expectation value, which in turn can be expressed as a trace over the one particle density matrix $\mathbf{P}$. In a quantum mechanical calculation these density matrix elements will be a function of the wavefunction parameters $\mathbf{p}$ e.g. the molecular orbital

[9] It would not be a good approximation to model benzene by its symmetry-unique portion; the electron density at the bond-junctions would be very poorly modelled.

coefficients, and nuclear positions. If an atom-centered basis set is used to expand the wavefunction, as is often the case,[10] then the density for a single molecule M in the unit cell is

$$\rho_M(\mathbf{r}) = \sum_{\alpha,\beta} P_{\alpha\beta}\chi_\alpha(\mathbf{r} - \mathbf{R}_{A_\alpha})\chi_\beta(\mathbf{r} - \mathbf{R}_{A_\beta}). \tag{6.7}$$

Note that the expansion for the density involves terms where both basis functions are on the same center ($A_\alpha = A_\beta$) and two-center terms ($A_\alpha \neq A_\beta$). The above expansion can be made exact provided that at least one of the indices β extends over a significant portion of the infinite crystal; for the wavefunctions used here, however, they are restricted to the locality of the non-interacting molecule M.

6.2.3.3 Expression for the Unit-Cell Electron Density

We will see later that the structure factors depend on the (thermally averaged unit) cell density. Therefore we need to derive an expression for this quantity from our non-interaction fragment wavefunction.

If there are several molecules N_M in the unit cell all symmetry-equivalent to M (with wavefunction Ψ_M) their densities will be related by $\bar{\rho}_M(\mathbf{r}) = \bar{\rho}_m(\mathbf{S}_m\mathbf{r} + \mathbf{s}_m)$ where $(\mathbf{S}_m, \mathbf{s}_m)$ is a roto-translation (crystal symmetry) operation. Hence in this case[11] the unit cell electron density is given by

$$\rho_{\text{cell}}(\mathbf{r}) = \sum_m^{N_M} \rho_m(\mathbf{r}) = \sum_m^{N_M} \rho_M(\mathbf{S}_m^{-1}(\mathbf{r} - \mathbf{s}_m)). \tag{6.8}$$

This equation does not account for thermal averaging effects.

6.2.3.4 Hirshfeld Atom Partitioning

The basis of all real-space partitioning is the definition of an atomic weight function W_A with the property that

$$\sum_A w_A(\mathbf{r}) = 1. \tag{6.9}$$

[10]Although plane wave basis sets can be used, they are very inefficient in modelling the core density which is responsible for the bulk of the X-ray scattering. Plane-wave-type basis sets are often used with the pseudopotential approximation where the core electrons are "eliminated" from the calculation

[11]The case of several *different* molecules in the unit cell is a straightforward generalisation, but has not been used in our work yet.

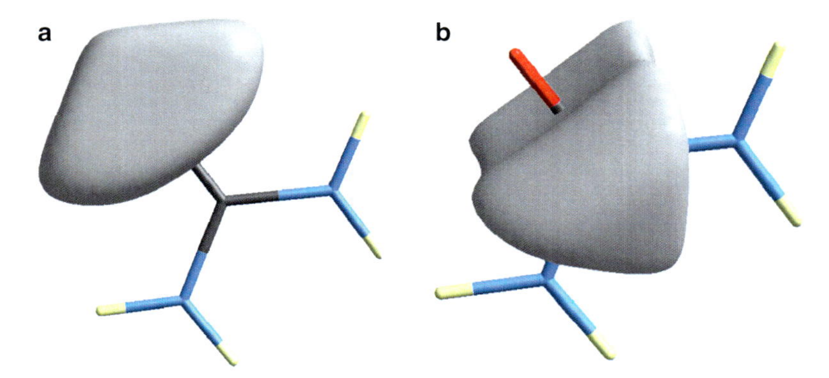

Fig. 6.4 Hirshfeld weight functions (plotted at the 50% contour level) associated with the Hirshfeld atoms for (**a**) hydrogen and (**b**) oxygen in the urea crystal for use in structure refinement

Hirshfeld defined his weight function by [30]

$$w_A(\mathbf{r}) = \rho_A^0(\mathbf{r} - \mathbf{R}_A) \Big/ \sum_{B \in M} \rho_B^0(\mathbf{r} - \mathbf{R}_B) \tag{6.10}$$

where ρ_A^0 are reference spherically averaged atomic electron densities. There are several other choices that could be made, and for our purposes this definition is not critical as long as the sum rule (6.9) is obeyed from which it follows that

$$\rho_M(\mathbf{r}) = \sum_A \rho_A(\mathbf{r}), \quad \text{where } \rho_A(\mathbf{r}) = w_A(\mathbf{r})\rho_M(\mathbf{r}). \tag{6.11}$$

The aspherical "Hirshfeld atoms" ρ_A are "cut out" of the molecular density ρ_M using Hirshfeld's partitioning technique.[12] A convenient way to visualize them is via a 50% contour level, as shown in Fig. 6.4 for the atoms in urea, a molecule considered later.

6.2.3.5 Structure Factors

Within the first Born approximation, and assuming ideal conditions, the X-ray diffraction experiment yields quantities $F_{\mathbf{k}}^{\text{exp}}$ (and associated errors $\sigma_{\mathbf{k}}^{\text{exp}}$) for a diffraction reflection labelled by scattering vector **k**; it is related to the Miller indices **h** by $\mathbf{k} = 2\pi\mathbf{B}\mathbf{h}$ where **B** is the reciprocal cell matrix itself related to the columns

[12]Note that in this and all published work by us, the Hirshfeld atom is "cut out" of the molecular density ρ_M and not the unit cell density.

of unit cell axis vectors, the so-called direct cell matrix $\mathbf{D}$ by $\mathbf{B} = (\mathbf{D}^{-1})^T$. The experimentally measured quantities are modelled by

$$F_{\mathbf{k}}^{\text{calc}} = \xi_{\mathbf{k}} |F_{\mathbf{k}}|, \tag{6.12}$$

where on the right hand side are the structure factor magnitude $|F_{\mathbf{k}}| = \left(F_{\mathbf{k}}^* F_{\mathbf{k}}\right)^{1/2}$, and $\xi_{\mathbf{k}}$, an experimental correction factor which accounts for an arbitrary scale factor and other sources of systematic error such as absorption, extinction, multiple reflection, thermal diffuse scattering and so on. The structure factors are related to the Fourier transform of the thermally smeared (i.e. averaged over the crystal phonons) unit cell electron density $\bar{\rho}_{\text{cell}}$ by

$$F_{\mathbf{k}} = \int \bar{\rho}_{\text{cell}}(\mathbf{r}) e^{i\mathbf{k}\cdot\mathbf{r}} d\mathbf{r}. \tag{6.13}$$

Substituting in the expression (6.8) and using the substitution $\mathbf{r}' = \mathbf{S}_m\mathbf{r} + \mathbf{s}_m$ for each term in the summation gives an expression in terms of the unique molecule density

$$F_k = \sum_m^{N_M} e^{i\mathbf{k}\cdot s_m} \int \bar{\rho}_M(\mathbf{r}') e^{i(S_m^t \mathbf{k})\cdot\mathbf{r}'} d\mathbf{r}'. \tag{6.14}$$

To get this result it was assumed that $\mathbf{k}\cdots(\mathbf{Sr}) = \mathbf{S}^t\mathbf{k} \cdot \mathbf{r}$ where superscript t is the transpose; and since $\mathbf{S}_m$ is a rotation, det $\mathbf{S}_m = \pm 1$ so that $d\mathbf{r} = d\mathbf{r}'$. The last integral is a Fourier transform of the density $\bar{\rho}_M$ evaluated at a symmetry-transformed reciprocal lattice point $\mathbf{S}_m^t\mathbf{k}$. Thus, although symmetry-unique reflections may be used as input data, the symmetry equivalents *are* generated and used to calculate the cell density if the molecule M has symmetry-equivalents.

6.2.3.6 Dealing with Thermal Smearing

To determine the structure factor we need to thermally average the density ρ_M to obtain $\bar{\rho}_M$.

To do this in the rigid Hirshfeld atom approximation, it is assumed that the pseudoatom $\rho_A(\mathbf{r})$ moves unchanged with the coordinate $\mathbf{R}_A$, and independent of the coordinates of the other atoms. This is actually not true, since w_A depends on all the coordinates in the molecule M. Furthermore, the density ρ_M also changes according to the nuclear positions. Finally, the atom motions are necessarily correlated with each other through the phonons. Nevertheless, with this assumption the thermal averaging of the density ρ_A is achieved by convolution with the nuclear-position probability distribution function for nucleus A. The Fourier transform of the averaged distribution $\bar{\rho}_A$, will, according to the convolution theorem, be the product of the Fourier transform of the functions being convoluted,

$$\bar{\rho}_A(\mathbf{k}) = T_A(\mathbf{k}) \times e^{i\mathbf{k}\cdot\mathbf{R}_A} \int \bar{\rho}_A(\mathbf{r} + \mathbf{R}_A) e^{i\mathbf{k}\cdot\mathbf{r}} d\mathbf{r}. \tag{6.15}$$

The translation factor $e^{i\mathbf{k}\cdot\mathbf{R}_A}$ has been explicitly introduced, by a simple substitution, to explicitly show the dependence on $\mathbf{R}_A$. The integral is the Hirshfeld atom form factor, and is assumed constant. T_A is the Fourier transform of the nuclear-positions probability distribution function, and it can be expressed in terms of the 3×3 ADP matrix $\mathbf{U}^{A_\alpha}$ parameters,

$$T_A(\mathbf{k}) = e^{-\frac{1}{2}\mathbf{k}^T\mathbf{U}^A\mathbf{k}}. \tag{6.16}$$

6.2.3.7 Final Expression for the Structure Factors and Refinement of Parameters

The expression for the structure factor is obtained by substituting $\bar{\rho}_M = \Sigma_A \bar{\rho}_A$ in Eq. (6.14), and then in (6.12), to give

$$F_{\mathbf{k}}^{\text{calc}} = \xi_{\mathbf{k}}(\mathbf{q}) \left| \sum_m^{N_M} e^{i\mathbf{k}\cdot\mathbf{s}_m} \sum_A e^{i\mathbf{k}\cdot\mathbf{R}_A} e^{-\frac{1}{2}\mathbf{k}^T\mathbf{U}^A\mathbf{k}} \cdot \int \bar{\rho}_A(\mathbf{r} + \mathbf{R}_A) e^{i(\mathbf{S}_m^t\mathbf{k})\cdot\mathbf{r}} d\mathbf{r} \right|. \tag{6.17}$$

The parameters to refine are $\mathbf{R}_A$, $\mathbf{U}^A$ and other phenomenological constants $\mathbf{q}$ such as scale constants and so on. Note that if the molecule has symmetry, not all the positions will be independent. Nevertheless, the calculation must still be done on a whole molecule, or a larger cluster of molecules; the structure factors for symmetry equivalent atoms occurring later in the list will be generated from those earlier.[13]

The expression above assumes that the Hirshfeld atoms are "rigid". To eliminate errors due to this assumption, after the parameters have been obtained by a least-squares refinement, one should consider recalculating the wavefunction Ψ_M at the new positions, and re-refining the parameters; this should be repeated until no change is observed.[14]

6.2.4 *Details of the Constrained-Wavefunction Method*

Much of the theory for constrained wavefunction optimisation has already been discussed when reviewing the Hirshfeld atom refinement procedure. For example, the choice of wavefunction is the same, leading to the same expressions for the final

[13]This can be quite useful to study, for example, the effect of intermolecular interactions on refined parameters. In this case a wavefunction for a cluster of molecules would used so that the Hirshfeld atoms include some of the effects of intermolecular binding

[14]The analytical "geometry optimization" of a molecule using this method would require the derivatives of the density matrices with respect to nuclear coordinates, and would be quite challenging to implement.

structure factor, Eq. 6.14. There are two major differences (i) there is an alternative way of dealing with thermal averaging, and (ii) the refinement involves derivatives of the density matrix **P** with respect to the wavefunction parameters.

6.2.4.1 Dealing with Thermal Averaging Effects: Alternatives

As already discussed, the treatment of thermal motion requires a knowledge of the phonon density of states, and the form of the phonons; or, what amounts to the same thing, the nuclear probability distribution function. Thermal averaging is then performed by a convolution of the density over the distribution of the nuclear coordinates.

Since the density expansion for molecule M, Eq. 6.7, involves both one- and two-center basis function products, thermal averaging requires parameters describing both the one- and two-nucleus probability distribution functions. The Fourier transform of the one-nucleus distribution for an atom A was already given in Eq. 6.16,

$$T_\alpha(\mathbf{k}) \equiv T_{A_\alpha}(\mathbf{k}). \tag{6.18}$$

Here, for later convenience, we abuse notation slightly and label the distribution by the basis function χ_α centered on atom A_α. On the other hand, the two-nucleus distributions are not available, so some approximation is required. Two approaches are available:

- Project ρ_M onto a one-center expansion e.g. as used for Hirshfeld refinement. The expressions for the structure factors have been derived. However, one could expand using fitting functions such as, for example, the multipole pseudoatoms [37]. In quantum chemistry the latter approach is sometimes referred to as "density fitting". A real-space atomic partitioning is recommended here because distributed atomic properties from such partitioning seem to show much more sensible behaviour.
- Recognize that a two-nucleus distribution is required, and approximate it from the one-nucleus parameters. Stewart [38], Coppens [39] and Tanaka [40] have proposed such approximations, e.g.

$$T_{\alpha\beta}(\mathbf{k}) = \left(e^{-\frac{1}{2}\mathbf{k}^T \mathbf{U}^{A_\alpha}\mathbf{k}} + e^{-\frac{1}{2}\mathbf{k}^T \mathbf{U}^{A_\beta}\mathbf{k}}\right)\Big/2 \qquad \text{(Coppens)} \tag{6.19}$$

$$T_{\alpha\beta}(\mathbf{k}) = \left(e^{-\frac{1}{2}g\mathbf{k}^T (\mathbf{U}^{A_\alpha}+\mathbf{U}^{B_\beta})\mathbf{k}}\right)\Big/2 \qquad \text{(Stewart)} \tag{6.20}$$

$$T_{\alpha\beta}(\mathbf{k}) = \left(e^{-\frac{1}{2}g\mathbf{k}^T (g_\alpha\mathbf{U}^{A_\alpha}+g_\beta\mathbf{U}^{A_\beta})\mathbf{k}}\right)\Big/2 \qquad \text{(Tanaka).} \tag{6.21}$$

All the formula satisfy the condition $T_{\alpha\beta} = T_\alpha$ when $A_\alpha = A_\beta$ i.e. when the basis functions are on the same center, the usual one-center distribution formula is used.[15] In Stewart's formula the factor g is taken to be 1/2 if the atom motions are correlated, or 1/4 if uncorrelated; in practice this means assigning the factor according to whether the atoms are considered bonded. The formula of Tanaka uses different distributions for different Gaussian basis functions: the factors g_α and g_β are the ratios of the exponents of the corresponding Gaussians to their sum [40]. The rationale for this is that, when Gaussian basis functions are used, the product (or overlap) distribution $\chi_\alpha \chi_\beta$ is another Gaussian centered at exactly these proportions along the line joining centers A and B.[16]

Note that, in both approaches described above only the nuclear-position dependence of the basis functions is considered; the position dependence of the density matrix has been ignored completely (this was discussed above as $\mathbf{q}_e/\mathbf{q}_n$ coupling). In the projecting-onto-pseudoatoms approach mentioned above, one also has to consider the position-dependence of the weight function w_A itself. Anharmonicity has also been ignored.

6.2.4.2 Structure Factors Formula Exhibiting Density Matrix Dependence

Using the results above, expression for the temperature dependent structure factor can be written exhibiting the dependence on the density matrix $\mathbf{P}$:

$$F_{\mathbf{k}} = \mathrm{Tr}(\mathbf{P}\bar{\mathbf{I}}_{\mathbf{k}}) = \sum_{\alpha\beta} P_{\alpha\beta} \bar{I}_{\mathbf{k}\alpha\beta}. \tag{6.22}$$

In the case where we approximate the two-nucleus distributions in terms of one-nucleus distribution parameters $\bar{I}_{\mathbf{k}}$ are the symmetry-averaged thermally smeared Fourier-transform basis-function pair integrals:

$$\bar{I}_{\mathbf{k}\alpha\beta} = \sum_m e^{i\mathbf{k}\cdot\mathbf{s}_m} T_{\alpha\beta}(\mathbf{k}) \int \chi_\alpha(\mathbf{r} - \mathbf{R}_{A_\alpha}) \chi_\beta(\mathbf{r} - \mathbf{R}_{A_\beta}) e^{i(\mathbf{S}_m^t \mathbf{k})\cdot\mathbf{r}} d\mathbf{r}. \tag{6.23}$$

In the case where we project the density ρ_M into a single-center expansion defined by weight functions w_A

$$\bar{I}_{\mathbf{k}\alpha\beta} = \sum_m e^{i\mathbf{k}\cdot\mathbf{s}_m} \sum_A T_A(\mathbf{k}) \int w_A(\mathbf{r}) \chi_\alpha(\mathbf{r} - \mathbf{R}_{A_\alpha}) \chi_\beta(\mathbf{r} - \mathbf{R}_{A_\beta}) e^{i(\mathbf{S}_m^t \mathbf{k})\cdot\mathbf{r}} d\mathbf{r}. \tag{6.24}$$

---

[15] There is actually no good justification for this; it only required that the marginal of the two center distributions should equal the one center distribution.

[16] When the Gaussians have exactly the same exponent the overlap distribution lies at the midpoint between A and B and the factors $g_\alpha = g_\beta = 1/2$, corresponding to Stewart's correlated case. In other words, the degree of correlation between the overlap distributions is governed by the relative distance of the overlap distribution to center A_α or B_α, respectively.

6.2.4.3 X-ray Constrained Hartree-Fock (and DFT) Wavefunctions

We are now in a position to present the equations for the X-ray constrained (non-interacting) Hartree-Fock wavefunction (XCHF). For this wavefunction, in the restricted case, the molecular wavefunction Ψ_M is made of a determinant of molecular spinorbitals whose spatial part ϕ_i is expanded in a basis

$$\phi_i = \sum_i c_{\alpha i} \chi_\alpha. \tag{6.25}$$

The wavefunction parameters $\mathbf{p} \equiv \mathbf{c}$ are the occupied molecular orbital coefficients. The molecular orbitals $\{\phi_i\}$ are required to be orthonormal to ensure determinant Ψ_M is normalized.

To obtain the orbitals in the absence of any experimental constraint we minimise (6.1) subject to the constraint that the orbitals remain orthonormal. Setting to zero the derivative of the Lagrange function,

$$L(\mathbf{c}) = E(\mathbf{c}) = \sum_{i,j} \varepsilon_{ij} \left(\int \phi_i \phi_j - \delta_{ij} \right), \tag{6.26}$$

yields the well-known Hartree-Fock equations:

$$\mathbf{Fc} = \mathbf{Sc}\varepsilon. \tag{6.27}$$

Here $\mathbf{F}$ is the Fock matrix, $\mathbf{S}$ is the overlap matrix of the basis functions, and ε is the symmetric matrix of Lagrange multipliers associated with the orthogonality constraint; all of these are defined in the usual way [1].

The application of the X-ray constraint requires that we minimise Eq. 6.1 with E on the right hand side replaced by L in Eq. 6.26. Thus the Lagrange functional to be minimised is identical to the Hartree-Fock one, except that there is the extra term involving $\lambda\chi^2(\mathbf{c}, \mathbf{q})$ (we show explicitly the dependence on the phenomenological parameters $\mathbf{q}$). In the restricted Hartree-Fock case the density matrix is given by $\mathbf{P} = 2\,\mathbf{cm}^{3\dagger}$ which allows us to evaluate (6.22) and hence χ^2 and its derivatives; since the structure factors are complex some care needs to be taken. After some algebra, the result is

$$\lambda \frac{\partial \chi^2}{\partial c^*_{\alpha i}} = \lambda \frac{2}{N_r - N_p} \sum_r^{N_r} \left[\frac{\left(F^{\text{calc}}_{\mathbf{k}_r} - F^{\text{exp}}_r\right) \xi_{\mathbf{k}_r}(\mathbf{q})}{\left(\sigma^{\text{exp}}_{\mathbf{k}_j}\right)^2 \left|F^{\text{calc}}_{\mathbf{k}_r}\right|} \right] \sum_\beta \text{Re}\left[F^*_{\mathbf{k}_r} \bar{I}_{\mathbf{k}_r\alpha\beta}\right] c_{\beta i}. \tag{6.28}$$

The right-hand side can be written as the matrix elements of an effective potential induced by the experimental data, $\lambda \sum_\beta v^{\text{exp}}_{\alpha\beta} c_{\beta i}$. This extra term is added to the usual Fock matrix to give the X-ray constrained Hartree-Fock equations

$$(\mathbf{F} + \lambda\mathbf{v})\mathbf{c} = \mathbf{Sc}\varepsilon \tag{6.29}$$

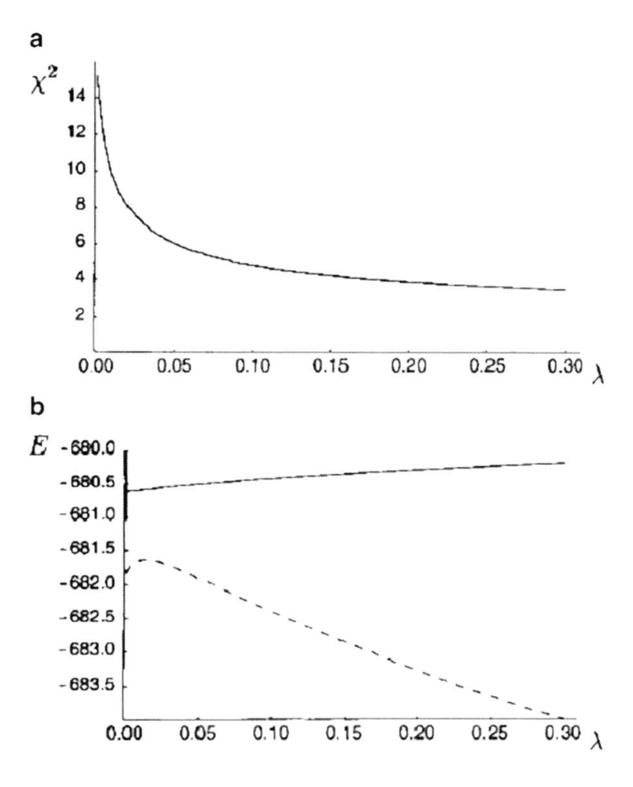

Fig. 6.5 Typical dependence of the (**a**) χ^2 versus the Langrange multiplier λ and (**b**) the energy E (*solid*) and the negative kinetic energy T (*dashed*) versus λ. Note that T does not have a predictable behaviour. The data is for α oxalic acid; see Ref. [11] for details (Reprinted from Ref. [11] with kind permission of Acta Crystallogr A)

The X-ray constrained DFT equations (XCDFT) are essentially the same as the above. The only difference is that **F** becomes the Kohn-Sham Fock matrix. An extension of the above technique to open-shell systems is also straightforward: the UHF and relativistic extensions have been reported in Ref. [12].

6.2.4.4 Solving the X-ray Data Constraint

In this work X-ray constrained Hartree-Fock equations are solved iteratively in the usual way for the modified molecular orbital coefficients **c**, for a chosen value of the λ parameter. As the λ parameter becomes larger the Lagrange function L becomes dominated by the χ^2 term, and the minimisation problem becomes more like a least-squares problem. Consequently, larger values of λ lead to lower values of the agreement statistic χ^2. This is illustrated in Fig. 6.5. Note that the SCF energy becomes larger, because with increasing λ one moves away from the energy-optimum SCF solution, and instead approaches the experimental-optimum (least-squares) solution. The ideal solution is obtained for the smallest value of λ where the χ^2 is approximately equal to one i.e. when the model data is within (roughly) one standard deviation of the experimentally measured values.

The values of the phenomenological parameters should be obtained by least-squares for each chosen value of λ; typically it is only the scale factor and extinction parameters which are treated this way, while the ADPs and nuclear position parameters are kept constant. The value of N_p, the number of parameters in the model, is chosen as one if only the λ parameter is varied to minimise χ^2.

6.2.5 *Discussion of the Constrained Wavefunction Method*

The Hirshfeld atom refinement technique is rather straightforward. By contrast, the constrained wavefunction method may be confusing. To get a clearer idea of the implications of the constrained wavefunction method, several issues are now discussed in a more informal way.

6.2.5.1 Terminology: Constraint or Restraint?

In crystallography, what we here call a constraint might be regarded a *restraint*: specifically, in Eq. 6.4, the term $\mu E(\mathbf{p})$ is regarded as a restraint or penalty term applied with strength μ to the parameters $\mathbf{p}$ during the minimisation of χ^2. If the restraint has zero strength (i.e. if $\mu = 0$) then this is a least squares minimisation. If, however, there are more parameters than data this will be an underdetermined problem. On the other hand if $\mu > 0$ is sufficiently large then the minimisation of L will be dominated by the restraint term E and the least-squares procedure will be well defined. The price to be paid for solving the ill-determined least squares problem is that the parameters are determined more by the restraint and less by the experimental data.[17] The extreme case of the restraint is when $\mu \to \infty$. This corresponds to Eq. 6.3, the minimisation of L, when $\lambda = 0$. In this case the parameters $\mathbf{p}[\lambda = 0]$ which minimise L are those from an (unconstrained) wavefunction minimisation of E. It follows that for $\lambda > 0$ a new set of wavefunction parameters $\mathbf{p}[\lambda]$ is obtained, and the value of $\chi^2(\mathbf{p}[\lambda])$ is reduced since it now appears in the expression for L with non zero weight λ. We now have a *constraint*: the value of λ is chosen to yield or constrain χ^2 to a desired value.

6.2.5.2 Interchangeable Role of E and χ^2; Interpretation of the λ Parameter

By contrast with $\chi^2(\mathbf{p}[\lambda])$ which is reduced as λ increases from 0, the energy $E(\mathbf{p}[\lambda])$ associated with the wavefunction is *increased* since we are no longer at the minimum of E which occurs for parameters $\mathbf{p}[\lambda = 0]$. All this was illustrated

[17] Though one must keep in mind that the restraint itself may be well motivated from experimental data, particularly if it comes from *ab initio* quantum chemistry

in Fig. 6.5. It is clear that as $\lambda \to \infty$ the χ^2 approaches limiting value which is dependent on the parameters used in the wavefunction calculation. Thus the range of values of χ^2 that can be obtained is limited by both the wavefunction method and the quality of the data.

The above analysis indicates that the roles of $E(\mathbf{p})$ and $\chi^2(\mathbf{p})$ are interchangeable: minimising either offers two competing ways of determining the model parameters $\mathbf{p}$. It follows from this that λ (or μ) is a switching parameter that smoothly moves between the least squares minimum for χ^2 and the quantum mechanical minimum for E. In the case when the minimisation of χ^2 is underdetermined, λ will be a switching parameter between the smallest value of χ^2 where the restrained problem is well defined.

6.2.5.3 Alternative Methods of Solving the Constrained-Least Squares Equations

One could solve for the minimum of the Lagrange function using second derivatives in a Newton-Raphson like approach reminiscent of least-squares. This would be a better approach because one could in principle solve directly for the required value of all the parameters at once. We have not done this yet because of the added computational effort involved.

6.2.5.4 Does the Magnitude of λ Have Any Significance?

The magnitude of λ has no physical significance because it is related to the errors in the measured data. To see this, note that in Eq. 6.2, scaling the σ_k by a constant s is the same as multiplying the λ value by $1/s^2$. Thus the choice of λ for a given problem to achieve a given χ^2 depends on the errors in the measurement, and these are system dependent and therefore not of universal significance. Because of this it is usually more informative to report the final agreement statistics rather than a particular value of λ.

6.2.5.5 Is a Constrained Wavefunction "Worse" than an Unconstrained One?

One must keep in mind two sources of error in the wavefunction. One source comes from the choice of the model wavefunction. This choice imposes an error at the outset, but it can be systematically improved by using a better model, just as in *ab intio* methodology. There is by now a well-known hierarchy for improved wavefunction models in quantum chemistry [1]. The second source of error comes from how the parameters in the model wavefunction are determined. Since the constrained wavefunction approach fits to experimental electron density data, it follows that one electron properties related to the density (e.g. multipole moments, electrostatic potentials) will be *improved*. On the other hand, since the total energy is worsened, and the total energy is a sum of one-electron and two-electron contributions, it seems likely that the magnitude of two-electron contribution to the energy will be *worsened*.[18] It is not clear from any argument such as this whether *derivatives* of the total energy with respect to perturbing parameters such as electric or magnetic fields will be improved or worsened. Such derivatives determine the response properties of a system.[19]

6.2.5.6 Constraint and Regularisation

The constrained approach here is often called "regularisation" in mathematics. Regularisation is an approach heavily used in situations where there is not enough data to obtain a unique solution to the problem, so called ill-posed problems [42]. Here the regularising term is the energy term E.

6.2.5.7 Constrained Wavefunctions vs. the Maximum Entropy Method (MEM)

The constrained approach described here is the same, in spirit, to the maximum entropy method (MEM) [43] (see also [42] for a good introduction, as well as Chaps. 1, 5 and 13, which include some applications). In the case of the maximum entropy approach, the regularisation function used is not the wavefunction energy E but a function called the entropy S which is maximised when the electron density becomes flat (featureless), or more often, when the density becomes equal to a "prior" density which represents the expected outcome. The expected outcome embodied by the function E is far more sophisticated than any prior used in the MEM approach; all the information pertaining to the assumed form of the wavefunction is embedded in the E functional.

6.2.5.8 Alternative Penalty Functions; More than One Set of Experimental Data

Alternative expressions for the Lagrangian penalty function may be considered. For example, we have also chosen to define the χ^2 with respect to the structure factor magnitudes; a modification to define it in terms of structure factor magnitudes *squared* would be straightforward.

---

[18] I am grateful to Dr. Carlo Gatti for discussing some of these matter with me

[19] A discussion of criteria for judging the quality of densities and density matrices is given by Kutzelnigg and Smith [41].

If one has data from multiple experiments (e.g. two different X-ray experiments, or one X-ray and one polarised neutron diffraction experiment) one might consider the penalty term $\lambda\chi^2$ to be a sum of the least squares errors from the combined data ([44] suggested this, using a perturbative treatment). One could also associate different Lagrange multipliers λ_1, $\lambda_2, \ldots$ with each experiments, leading to a penalty term $\sum_i \lambda_i \chi_i^2$; the λ_i would be chosen to achieve $\chi_i^2 = 1$ for each experiment (there are much better choices of penalty function; see Ref. [45] and Chap. 5, Sect. 5.2.6). One might extend this idea to the extreme and consider each datum $\mathbf{F}^{\mathrm{exp}}$ as having being obtained from a separate experiment.[20] We have not adopted any of this last approach here because of the extra complexity in fixing so many Lagrange multipliers, although it offers a promising area for further research.

6.2.5.9 Relationship of the Orbitals to those from DFT; Orbital Plots

If the experimental data were without error and complete, the orbitals obtained by minimising L are essentially the same as those that would be obtained from exact DFT theory.[21] However, it is important to note that the energy E obtained from the constrained orbitals is not the same as the energy produced from the exact DFT theory. The exact energy E could only be obtained with a knowledge of the unknown exact DFT energy functional. Levy and Goldstein [46] have described methods which could be used to obtain the exact DFT energy; we have not pursued their ideas.

There have been several publications in the literature where "orbitals" are claimed to have been measured experimentally (see [23] for a good review). Orbitals which are derived using some form of energy resolution, such as in the (e,2e) experiment, correspond to *Dyson orbitals* which can legitimately be produced. Also, the polarised neutron diffraction (PND) experiment produces data on the unpaired electron density which is more-or-less the singly occupied valence molecular orbital. On the other hand, the orbitals produced from an X-ray constrained wavefunction procedure involving data from a single state produce orbitals which are indeterminate up to a gauge transformation. Such orbitals should not be plotted unless appropriately canonicalized (canonicalization means to diagonalise a specified operator, usually the Fock operator in the basis of the occupied orbitals, in order to fix the non-degenerate orbitals to within a phase [13]).

---

[20] For a single X-ray experiment, the resulting λ_i values would constitute the DFT exchange correlation potential in a reciprocal space formulation, and the λ_i would be subject to continuity conditions

[21] The orbitals produced from the X-ray constrained method involve a non-local constraining potential whereas in DFT the constraining potential is required to be local in position space. In fact, the constraining potential in the X-ray method is local in reciprocal space

6.2.5.10 When to Terminate a Constrained Fitting procedure?

Ideally, the value of λ is chosen to obtain a χ^2 value equal to unity; this guarantees that, on average every predicted value if the experimental data is within one standard deviation of the experimental value. More correctly, one can select the appropriate value for χ^2 according the χ^2 distribution and a desired confidence interval.

Unfortunately, in practice, this value of $\chi^2 = 1$ is rarely a sensible choice because (a) the estimates for the errors in the measurements are rarely quantitative and (b) the number of parameters in the constrained model is difficult to estimate. Errors are difficult to estimate because they are often non-systematic, and further, all sources of error would have to be quantified. The number of parameters in the constrained model is difficult to estimate because in the case $\lambda \to \infty$ the constrained approach is essentially a (regularised) least-squares fit [47, 48] so that all the parameters become independent; on the other hand, when $\lambda \to 0$ the constrained approach becomes a "pure" wavefunction calculation and there are no adjustable parameters.

Different strategies are therefore needed to decide when to terminate the constraint procedure.

In the past we have advocated to use the largest possible value of λ to achieve the small possible χ^2. This is still one of the most common strategies, but it is somewhat unsatisfactory because the smallest value of χ^2 achievable depends strongly on the wavefunction parameters (particularly the basis set) used in quantum mechanical calculations.

Another strategy that could be used is to reserve a portion of the data for cross validation and terminate the procedure when a cross validation statistic such as R_{free} ceases to decrease. Our investigations have shown that in almost all cases R_{free} continues to decrease as λ gets larger. The cross validation technique seems unsatisfactory also because it requires reserving a portion of the data that could otherwise be used for fitting, and it is not clear if an increase in a cross validation statistic is the best way to assess the quality of a fit; very little is known about the distribution properties of R_{free}.

In recent work using X-ray data, we have terminated fitting when a figure of merit such as the R factor is the same as that from a multipole refinement using the same data. Although this procedure clearly prevents overfitting, it is also unsatisfactory because it means that the constrained wavefunction procedure is not self-contained but relies on another experimental technique. This deficiency could be rectified by using an agreement statistic such as R_σ which is based purely on the data itself, and which is not dependent on estimates of the errors, but it has not yet been tried.

In summary, although there are several well-defined protocols for terminating a constrained wavefunction procedure, it remains an area for research.

6.2.6 *Computer Implementation of the Constrained-Wavefunction and Hirshfeld Atom Refinement Techniques*

Over the past decade, the author and his colleagues have developed a computer package called Tonto [49] to perform the required calculations described in this article.[22] There were several motivations for doing this:

- Although *ab initio* quantum mechanical code was available in other packages, it was not always easy to graft the necessary crystallographic code onto the molecule-centric view of quantum chemistry.
- At the time of development, it was not certain if Fortran would remain the main computational language for numerical applications: therefore we decided to write our codes in a hybrid language called Foo with at least the capability of being translated into other languages.[23]
- The possibility of developing a new code base from the start, despite a very large initial outlay of effort, would allow good coding practices to be observed: clear and copious documentation, clear design, modular structure, with long and descriptive routine and variable names. We felt that this would make possible quick and easy future development by professionals and students.
- Very few of the computer codes were available under a free and unrestrictive license. Therefore, we were largely forced to start from scratch if we wanted to achieve the above goals.

Unless stated otherwise, all calculations were performed using the Tonto program package. The inputs to the program are usually a crystallographic information file to specify the starting geometry if a Hirshfeld atom refinement is performed, the basis set and associated options for the quantum mechanical calculations (which are essentially the same as for other *ab initio* packages) and several options describing exactly how the structure factors are to be calculated from the molecular fragment.

6.3 The History of Obtaining Experimental Wavefunctions

This chapter contribution has deliberately taken a pedagogic approach emphasizing the authors constrained-wavefunction approach for determining an experimental wavefunction; other attempts to do the same thing have only been briefly mentioned. In this section I correct this deficiency.

---

[22]Tonto is available under an open source license from the Sourceforge repository.

[23]So far, this has not been done, but it remains a possibility if the base-language needs to be changed. It is worth noting that many of the Foo-language facilities we have introduced (e.g. type-bound procedures) appear in the newer Fortran standards.

There are other measurements, including the (e,2e) and tunneling probe methods briefly mentioned at the start that provide relevant data for wavefunction mapping; these will not be discussed, except for the important Compton experiments.

6.3.1 Early Work

The earliest work on constrained wavefunctions is that by Mukherji and Karplus [50]. Using a perturbative solution technique these authors constrained a Slater-type Hartree-Fock wavefunction for HF to reproduce experimental values for the dipole moment and electric field gradient, and they also looked at the effects of doing this on other one electron properties such as the magnetic susceptibility and the forces on the nuclei. Rasiel and Whitman [51], and Chong and Byers Brown [44] performed similar perturbative calculations on LiH. This work was extended by Chong and Rasiel, who studied several non-perturbative methods to perform the constraint procedure on selected observables [52].

Chong and coworkers went on to constrain into an approximate wavefunction various known properties of the *exact* wavefunction, such as the cusp and coalescence conditions and off-diagonal hypervirial relations [53–56]. Weber and coworkers took a similar path, and developed an elegant technique for forcing linear homogenous constraints on approximate wavefunctions [57] which they applied to cusp conditions and long range wavefunction behaviour [58].

Clinton and coworkers have investigated issues related to constrained wavefunctions in great depth. The 1969 edition of the Physical Review presents a *tour de force* of five consecutive papers under the general heading "Direct determination of pure-state density matrices" [59–63]. The density matrix which they discuss in this series papers is a quantity derived from the square of the wavefunction, integrated over all but one or two electron coordinates (the former is the one-electron density matrix, while the latter is the two electron density matrix, which is, in principle, all that is required to obtain any physical observable [26]), although there is a strong focus on Hartree-Fock or idempotent density matrices fitted to (mostly theoretical) density data [64–74]. Perhaps the culmination of this work was the extraction of an idempotent density matrix for the Be atom from γ-ray diffraction data [73]. The work of Aleksandrov *et al.* in producing Hartree-Fock type fitted densities is also noteworthy [75]

A criticism of the work of Clinton and later workers is that the use of an idempotent density matrix (i.e. a single determinant wavefunction) fitted to density data does not necessarily yield unique results. This fact is admitted by, for example, Frishberg [69]; and in fact, the search over all wavefunctions which have a certain density is the basis of Levy's formulation of the unknown density functional [76]. There is, in fact, an infinity of Slater-determinant wavefunctions which can reproduce a given density [77, 78]. Henderson and Zimmerman suggested a solution to the uniqueness problem by combining the fitting procedure with an energy minimum criterion [79]. This is the earliest link to the authors own work.

In practice, with the use of a small basis it *is* possible to obtain unique results. This idea was discussed by Harriman who considered the problem of linearly-dependent basis-function products (recall that such product are used to represent the density in any basis-function quantum mechanical calculation [78]. However, Morrison demonstrated that, while the use of linearly independent basis function products might lead to excellent one electron properties, poor kinetic energies are obtained [80]). Schwarz and Muller also found that the (except for basis set comprised of *s* type basis functions) the linearly-independence criterion was not satisfied [81]. Indeed, Coppens and coworkers had much earlier found it difficult to refine density models based on product basis functions [39]. It is worth pointing out that these two center terms are easily distinguished using the Compton techniques discussed below.

The number of linearly independent basis function products to determine a given density was also considered by Levy and Goldstein [46], however, in their work are derived interesting relationships for the total ground state energy of a system from the kinetic energy and correlation energy functionals of DFT. These formula have not yet been applied to real systems.

6.3.2 Period 1990-

In this era, there were several developments which were significant, and a variety of new modelling techniques were developed.

Howard and coworkers have also applied Clintons method to real X-ray data, and to much larger systems [82]. This article generated considerable interest and excitement from the X-ray community. Likewise, Snyder and Stevens [83] obtained an idempotent density matrix (Slater determinant wavefunction) for the azide ion in potassium azide. Snyder and Stevens work demonstrated that within a finite basis set, unique results could be obtained, and the use of symmetry significantly reduced the number of free parameters. Their results indicated a closer agreement between *ab initio* electron density maps than multipole-derived maps.

Zhao, Parr and Morrison advocated obtaining Kohn-Sham wavefunctions (and effective potentials) directly from electron density data obtained from *theoretical* wavefunction or quantum Monte-Carlo sources [84–86]. That technique is widely used to generate new DFT functionals. The authors own work is, in fact, very closely related to their approach.

The authors own work has been discussed extensively already; it differs from the earlier contributions in that the density matrices obtained are unique and satisfy an energy minimum criterion. [7–16]. An interesting recent development is the use of open-shell (relativistic) constrained wavefunctions, which can potentially model magnetic diffraction data simultaneously with X-ray diffraction data [12].

Cassam-Chenai has stressed the importance of using ensemble representable density matrices, especially in connection with modelling polarised neutron diffraction data [87–89].

Hibbs, Waller and coworkers have proposed a somewhat similar density matrix fitting procedure, by using molecular orbitals with variable occupation numbers [90] (this is the MOON method). They have applied the method to and have pointed out that it can, like the authors own method, produce novel properties such as kinetic energy densities, lattice energies, etc. [91]. The problem with such methods is, as always, the data to parameter ratio and the selection of the orbitals used to do the fitting.

Tanaka and coworker have continued to develop their X-ray atomic orbital (XAO) method which is geared toward the modelling of crystal-field distorted atomic orbitals on heavy atoms [40, 92–94]. The method imposes an orthogonality constraint amongst the atomic orbitals and imposes relevant symmetry conditions particularly appropriate to the crystal field of the heavy atoms. Some interesting temperature and entropy-dependent electron transfer effects were observed in CeB_6 using the technique [92, 94].

6.3.3 Momentum Space Approaches

It would be remiss not to mention the importance of approaches that make use of *inelastic* (Compton) scattering data, since it is known that such data is highly sensitive to chemical bonding effects [95, 96] (see also Chaps. 2, 5 and 7). Indeed, it is possible to reconstruct the exact one-particle density matrix from such data, as has been demonstrated by the work of Weyrich and coworkers [97–100] and Gillet and coworkers [45, 101, 102]; in fact, these workers were the first to refine joint X-ray and Compton data experimental data to obtain "local wavefunctions" [102] (see also Chaps. 5 and 7). Importantly, Schmider et al. has shown that *it is not possible to obtain the correct one-particle density matrix if density data is used* [97]; in fact, the two center terms which are very hard to determine from X-ray diffraction methods clearly appear in the Compton scattering experiments [95, 96].

By contrast with the constrained wavefunction methods used in this work, where the density matrices have a restricted form dictated by the model wavefunction *ansatz* (see earlier), momentum-space density matrix reconstruction techniques are unhindered.[24] Unfortunately, the technique of Weyrich and coworkers has not, to the authors knowledge, been applied to real data, but results are imminent [100], and the methods deserve further investigation.

[24]Though, of course, whenever a model is built, underlying assumptions are made in representing the data e.g. the choice of fitting functions and so on. These assumptions can be progressively removed. Likewise, the choice of more flexible wavefunction *ansatz* can progressively remove the inflexibility in the one-particle density matrix. It is primarily for this reason that the author avoids using the term "experimental density matrix", and prefers to use "experimental model wavefunctions". The two approaches are complementary.

6.4 Atom Positional Information Derived from Wavefunctions

In this section we consider some representative results for benzene and urea.

6.4.1 Agreement Statistics

Table 6.1 shows results for agreement statistics between calculated and observed X-ray structure factor data after Hirshfeld atom refinement, using Hartree-Fock wavefunctions for the isolated urea and benzene molecules, as a function of basis set, compared with a more usual multipole refinement. The least squares error χ^2 is shown as well as the R factors commonly used in crystallography. The X-ray data are from accurate low temperature data. The refinements started from the neutron-refined geometries and the Hirshfeld atoms were calculated once and held constant during the refinement; this is called the rigid approximation. Further details of the refinements and data are provided elsewhere [7].

Since the multipole model adjusts the electron density parameters to fit the data, whereas Hirshfeld atom refinement does not, we expect that the agreement is better for the former model. In this context it is remarkable that for urea the results from Hirshfeld atom refinement yield better R factors. It is also apparent that the triple-zeta basis set cc-pVTZ is sufficient to obtain R factors converged to three decimal places, and results from the smaller double-zeta type basis cc-pVDZ are nearly as good.

The wavefunction for the isolated molecule is clearly not ideal for modelling the electron density of the molecule in a crystal. One can go some way towards ameliorating this deficiency by surrounding the isolated molecule with point charges at the crystalline lattice sites. This was done, and the charges were calculated using Hirshfeld's technique; a pair of charges at each atomic site represented the atomic

Table 6.1 Figures of merit for Hirshfeld atom refinement in comparison with multipole refinement, for various basis sets cc-pVnZ used to expand the molecular orbitals (Taken from [7])

	Multipole	Hirshfeld-atom refinement			With charges	BLYP	BLYP
	model	cc-pVDZ	cc-pVTZ	cc-pVQZ	cc-pVTZ+−	cc-pVTZ	cc-pVQZ+−
Benzene							
$R(F)$	1.89	2.17	2.16	2.16	2.16	2.15	2.15
$R_w(F)$	1.68	2.11	2.08	2.08	2.09	2.07	2.08
χ^2		1.64	1.59	1.59	1.60	1.57	1.59
Urea							
$R(F)$	2.02	1.71	1.66	1.66	1.54	1.53	1.40
$R_w(F)$	0.54	1.27	1.21	1.21	1.11	1.08	0.88
χ^2		9.40	8.51	8.53	7.14	6.82	4.51

The symbol '+−' refers to results for a molecule embedded in a self consistent field of charges dipoles. Charges on all molecules within a radius of 5 Å of the central molecule were used, and the charges were calculated self-consistently. The resulting agreement statistic are shown in the column cc-PVTZ+- of Table 6.1. For benzene, there is hardly any effect due to the surrounding crystal field, whereas for urea, there is an improvement. This is entirely understandable since there is a strong hydrogen bonding network present in urea crystal.

What about electron correlation? What effect does it have on the electron density? Are these effects experimentally detectable? In the last two columns data for Hirshfeld refinement using the BLYP density functional theory is presented, both for an isolated molecule and one embedded in a self-consistent field of point charges. For benzene, the effect of electron correlation is rather small, but for urea the results are systematically better when using BLYP for correlation. For urea, the crystal field seems to have a larger effect than electron correlation. At the very best level of theory, the agreement with experiment rivals that obtained from a multipole model.

The main conclusion to be drawn from this table is that *systematic effects of the crystal environment and electron correlation can be detected in the X-ray data and accounting for those effects leads to improvement in agreement with the experimental data.*

6.4.2 Bond Lengths

The differences in bond lengths obtained from Hirshfeld atom refinement and neutron diffraction are shown in Table 6.2. The agreement for heavy-atom bond lengths is about 0.002 Å and the results are just outside the error bars from each measurement. The uncertainty in the bond lengths from the Hirshfeld atom method for these heavy atoms is smaller than from the neutron diffraction experiment. The situation is reversed for the hydrogen bond lengths: here the uncertainty in the Hirshfeld atom refinement is larger. The agreement is within 0.002 Å and the agreement for benzene is within the sum, but the error for the CO bond in urea is well outside the error range. Nevertheless, the agreement between the parameters is very good .

The conclusion from these results is that *accurate atomic positions can be obtained, very nearly in agreement with neutron diffraction results, using good quality X-ray diffraction data and using the electrons densities from quantum mechanical wavefunctions as input data.*

6.4.3 ADP's

For heavy atoms, Fig. 6.7 shows that the ADPs are not greatly affected by the use of a surrounding cluster of charges, and for benzene there is agreement with the

Table 6.2 Differences between bond lengths obtained from Hirshfeld atom refinement and neutron diffraction ($\Delta_{\text{neutron}}/10^{-3}$ Å), for benzene and urea, using the BLYP/cc-pVTZ+- wavefunction. Uncertainties in the Hirshfeld atom refinement (*HA*) and neutron refinement (*N*) also reported (For details of atom numbering see Fig. 6.6 or Ref. [7])

Bond length	$\Delta_{\text{neutron}}/$ 10^{-3} Å	Uncertainty		Bond length	$\Delta_{\text{neutron}}/$ 10^{-3} Å	Uncertainty	
		HA	N			HA	N
Benzene							
C1–C2	1.4	0.4	1.1	C1–H1	−3.	5.	2.
C2–C3	0.7	0.4	1.1	C2–H2	3.	5.	2.
C1–C3a	2.5	0.3	1.1	C3–H3	−7.	5.	2.
Urea							
O1–C1	2.0	0.4	1.0	N1–H1	18.	3.	2.
N1–C1	0.0	0.2	1.0	N1–H2	−9.	5.	2.

multi-temperature analysis (MTA) X-ray measurements (note that the neutron and MTA results are in disagreement by up to 25×10^{-4} Å^{-2}). For urea the results agree within the error bars with the experimental values.

Now to the more interesting hydrogen atom ADP's. The general consensus amongst crystallographers is that hydrogen atom ADP's cannot be obtained from X-ray data (cf. Chaps. 3 and 1). Figure 6.8 presents data that puts this notion in doubt. Now there is a clear dependence of the ADP's on the use of correlation and embedding charges, and the use of both systematically improves agreement with experiment. The results are generally better for benzene than urea (disagreement within 50×10^{-4} vs 100×10^{-4}, respectively).

A more visual representation of the ADP information is shown in the peanut plots in Fig. 6.6. It is clear that both improvement in the correlation method and the inclusion of surrounding point charges improves the ADP's.

In summary, *reasonable hydrogen atom ADP's can be obtained from X-ray diffraction data, and the values are within by about 20% of those obtained from neutron diffraction experiments.*

6.5 Electronic Properties from X-ray Constrained Wavefunctions

The constrained wavefunction method is capable of producing all the properties obtained from an *ab initio* quantum mechanical calculation. Indeed, comparison with such calculations is facilitated by the fact that exactly the same basis set and level of theory can be used; the only difference is that in the constrained procedure one introduces the experimental data. In this section we focus on deformation densities, molecular dipole moments, and some new response properties such as in-crystal polarisabilities, refractive indices as a selection of what is possible.

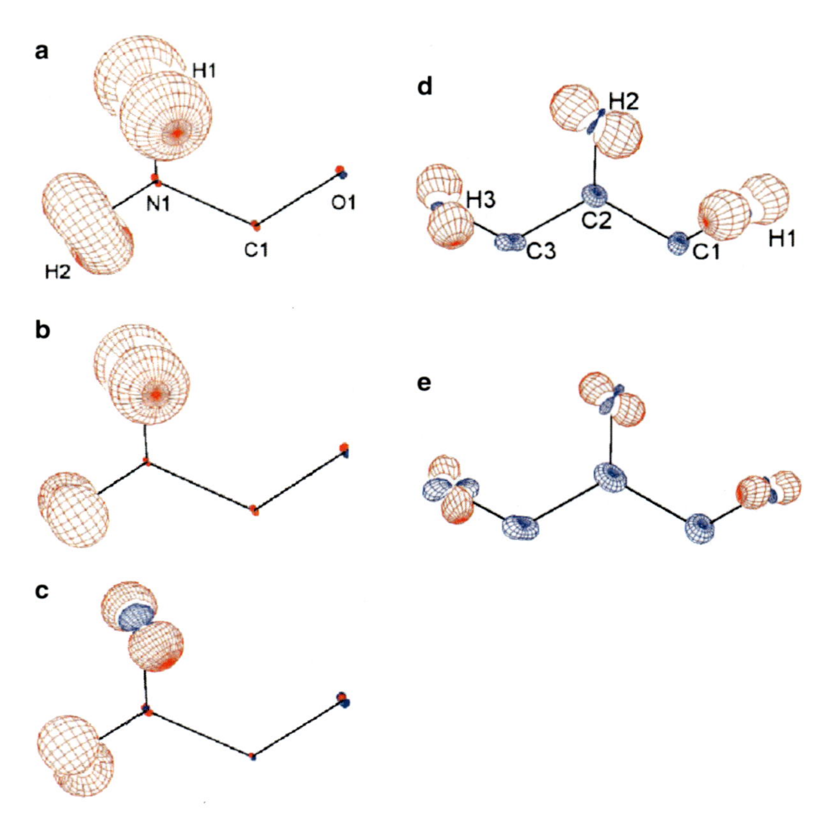

Fig. 6.6 For urea (*left*) and benzene (*right*), differences between ADPs from neutron experiments and those from Hirshfeld atom refinement displayed using the PEANUT program [103]. A scale of 6.15 was used for the RMSD surfaces. (**a**) neutron – HF/cc-pVTZ (**b**) neutron – HF/cc-pVTZ+- (**c**) neutron – BLYP/cc-pVTZ+- (**d**) neutron – HF/cc-pVTZ+- (**e**) neutron – BLYP/cc-pVTZ+- (Adapted from [7]. Reprinted from Ref. [11]. With kind permission of Acta Crystallogr A)

We have not discussed the possibility to produce experimentally derived electron localisation function (ELF) or related plots, or the possibility of producing spin densities from X-ray diffraction data.

6.5.1 Deformation Densities

Since X-ray diffraction data pertains to the electron density is used to obtained constrained wavefunctions, we start by comparing the electron densities themselves. Densities for α oxalic acid dihydrate are shown in Fig. 6.9. Most notable here is the close agreement between the BLYP and constrained wavefunction calculations.

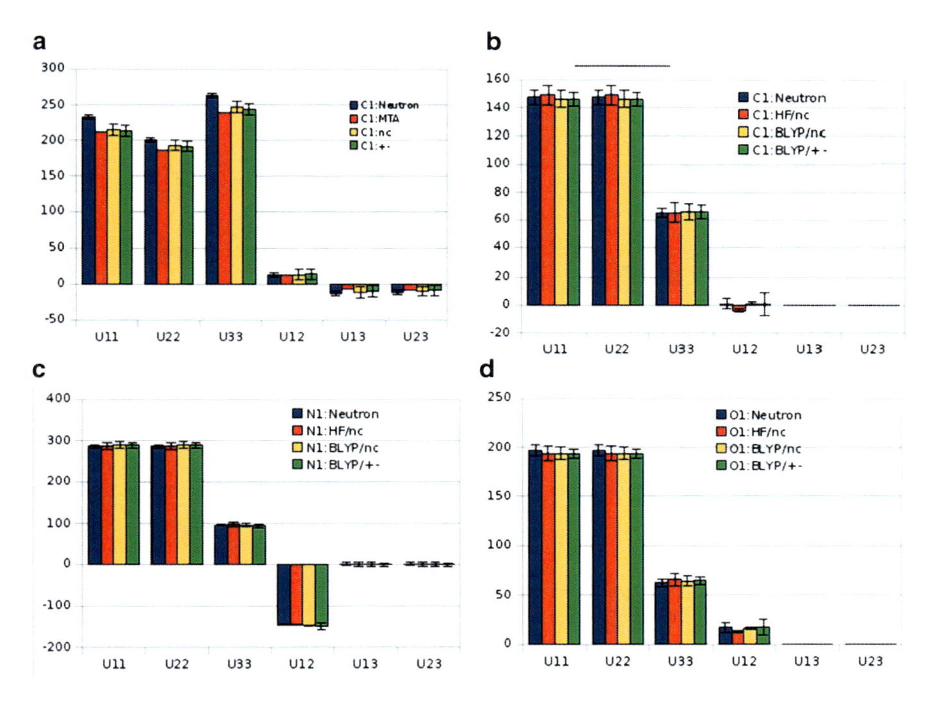

Fig. 6.7 ADP's for selected non-hydrogen atoms ($/10^4$ Å^2) from Hirshfeld atom refinement compared with independent neutron and multi-temperature analysis (*MTA*) X-ray experiments. For benzene (**a**) the C1 atom. In urea, the (**b**) C1 (**c**) N1 and (**d**) O1 atoms. The symbol 'nc' stands for no surrounding cluster of charges, while '+-' stands for self-consistent cluster-of-charges. BLYP or HF wavefunctions were used with a cc-pVTZ basis

Whitten and coworkers [105] have presented more examples for urea, benzene and the MNA molecule, shown in Fig. 6.10. The similarity between the features around the CO group in urea and oxalic acid, Fig. 6.9, are remarkable; as are similarities in the deformation density features around the C atoms in the rings of benzene and MNA, and the CH_2 groups in urea and MNA. It appears certain features are present across several molecules and several X-ray data sets. In Fig. 6.10b we see the results of crystalline Hartree-Fock calculations, taken from [104]. Qualitatively they are very similar, but it seems that the constrained wavefunction results are slightly more exaggerated; a detailed comparison of these effects is now warranted, since a possible explanation of these small discrepancies may be found in the use of geometries which have not been fully optimised to the X-ray data, in the sense of the previous section.

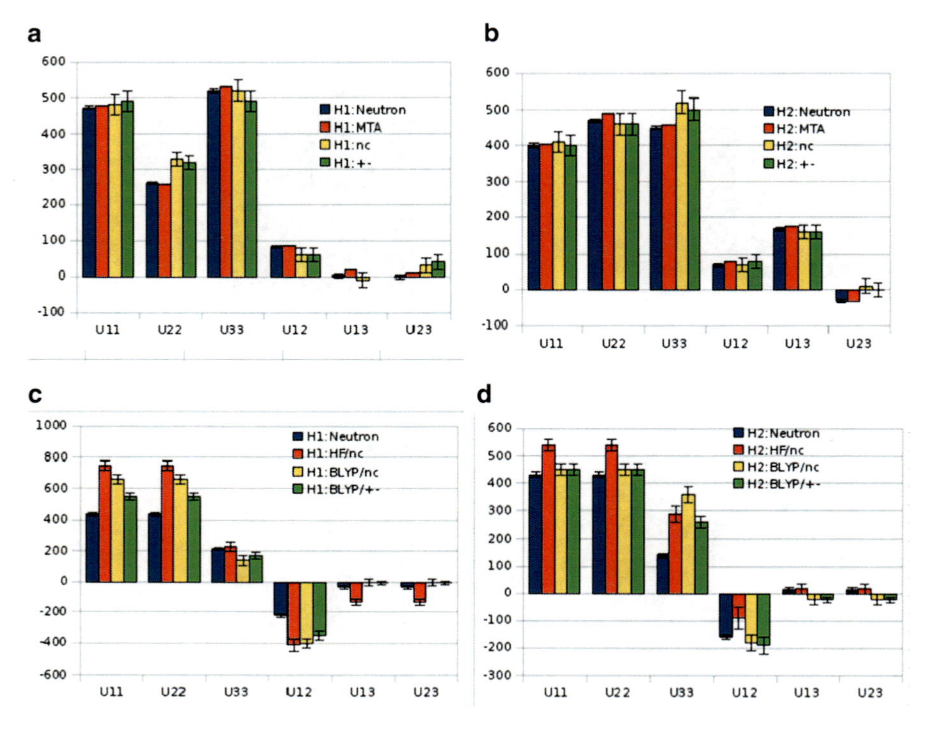

Fig. 6.8 ADP's for selected hydrogen atoms ($/10^4$ Å^2) from Hirshfeld atom refinement compared with independent neutron and multi-temperature analysis (*MTA*) X-ray experiments. For benzene (**a**) H1 and (**b**) H2 atoms. For urea (**c**) H1 and (**d**) H2 atoms. The symbol 'nc' stands for no surrounding cluster of charges, while '+-' stands for self-consistent cluster charges. BLYP or HF wavefunctions were used with a cc-pVTZ basis

6.5.2 In Crystal Dipole Moments

The calculation of in-crystal dipole moments has also been a somewhat controversial topic; the definition of the in-crystal dipole depends sensitively on how the molecules are partitioned from the crystalline density. Table 6.3 presents the in-crystal dipole moments for urea, and the POM, NPP, and PNP molecules (not shown; see [16]) derived from the constrained wavefunction approach and compared with experiment and various high level calculations. It is clear that the constrained wavefunction approach produces dipole moments in excellent agreement with *ab initio* MP2+field results (here, the molecule is embedded self-consistently in an static field calculated from an Ewald sum of the surrounding (self) dipole moments, as described earlier). It is worth mentioning that rather anomalous result for POM is due to the cancellation of large contributions around the ring. The conclusion from Ref. [16] is that "*despite the fact that the XCHF wavefunctions are only single*

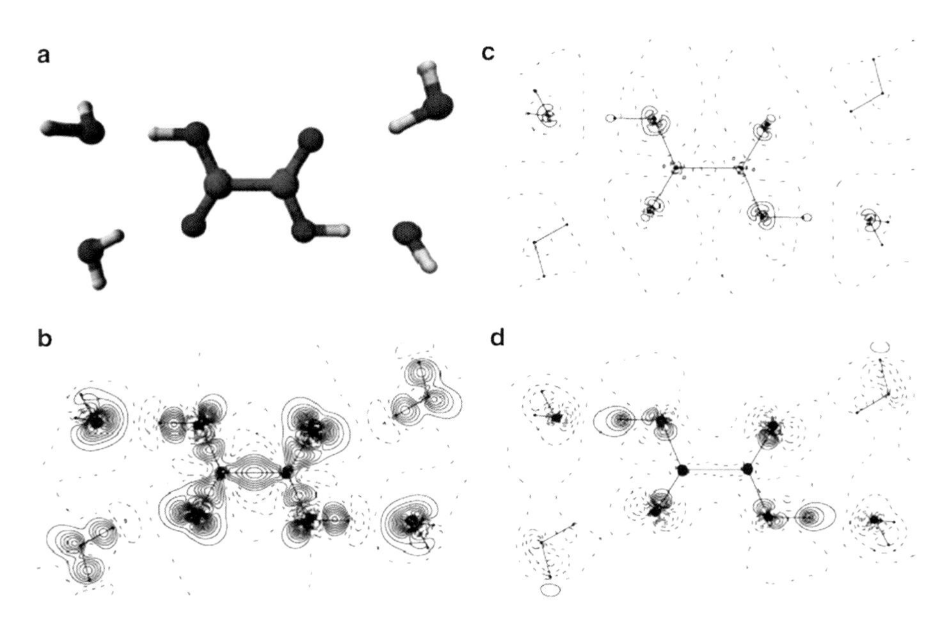

Fig. 6.9 The electron density from (**a**) a model of α oxalic acid dihydrate crystal compared with (**b**) the Hartree-Fock deformation density, $\rho_{HF} - \rho_{promolecule}$, (**c**) the correlation contribution to the electron density, $\rho_{BLYP} - \rho_{HF}$ [all from Ref. [11]] and (**d**) the X-ray constrained Hartree Fock wavefunction contribution to the electron density, $\rho_{XCHF} - \rho_{HF}$ [from Ref. [104]] All contours at 0.1 eÅ^{-3}. The BLYP and constrained X-ray Hartree-Fock contributions are remarkably similar, though the polarization on the hydrogen atoms is exaggerated in the latter case (Reprinted from Ref. [11]. With kind permission of Acta Crystallogr A)

Table 6.3 In-crystal dipole moments at various levels of theory of increasing accuracy (see Fig. 6.12 for explanation) compared with the X-ray constrained-Hartree-Fock results (*XCHF*) and with experimental results (*Expt*) for urea and the molecules POM, NPP and PNP (For details see [16])

MP2					
	HF	MP2	+field	XCHF	Expt
Urea	5.11	4.61	6.51	6.89	3.83(4)
POM	1.04	0.82	1.32	0.73	0.69(5)
NPP	7.52	6.82	8.70	8.99	7.3(5)
PNP	6.65	6.07	7.71	7.66	7.2(6)

determinant and minimize the Hartree-Fock energy, the additional constraint of fitting the X-ray structure factors results in a wavefunction that reflects the effects of both electron correlation and intermolecular interactions (i.e. hydrogen bonding and the crystalline electric field experienced by the molecule)"

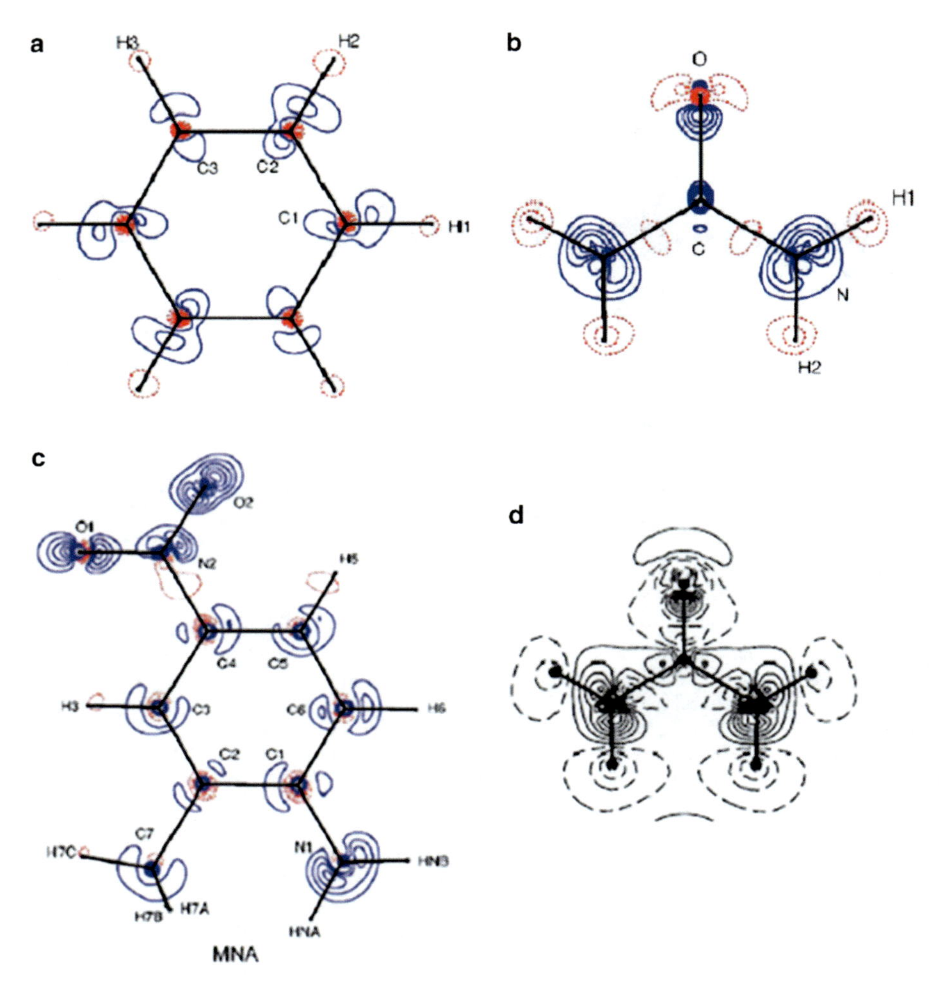

Fig. 6.10 The molecular Hartree-Fock deformation electron density for (**a**) benzene, (**b**) urea, and (**c**) MNA. Contour levels 0.1 eÅ^{-3} (Taken from [105]). In (**d**) the deformation densities from periodic crystal Hartree-Fock calculations for urea [104] (Reprinted from Ref. [104] with kind permission of Acta Crystallogr A and from Ref. [105] with kind permission of the American Institute of Physics ("AIP"))

6.5.3 Response Properties: In-Crystal Polarisabilities and Refractive Indices

The (bulk) electrical susceptibilities $\chi^{(1)}$ and $\chi^{(2)}$ of molecular crystals are of fundamental importance. For example, the first susceptibility determines the refractive indices and optic axes of the crystal, while the second order susceptibility determines the non-linear optical (NLO) properties, which are of increasing importance

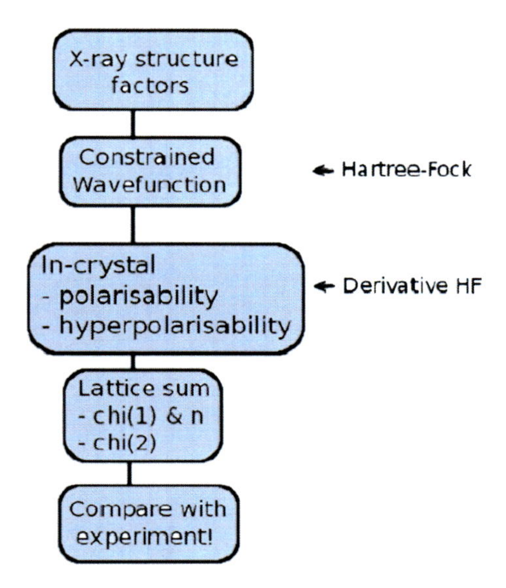

Fig. 6.11 Outline scheme for calculating refractive indices and NLO properties from X-ray data (See Ref. [16] for details)

in technological applications. It is remarkable that a constrained wavefunction approach can yield these quantities in much the same way that they would be calculated from an *ab initio* approach. The scheme is briefly outlined below; details for the refractive index calculations are provided in Ref. [16].

In terms of the effective first and second polarisabilities (respectively α_k and β_k) of the Z molecules which comprise the unit cell of the crystal,

$$\chi^{(1)} = (\varepsilon_0 V)^{-1} \sum_k^Z \alpha_k \cdot \mathbf{d}_k, \tag{6.30}$$

$$\chi^{(2)} = (2\varepsilon_0 V)^{-1} \sum_k^Z \beta_k \vdots [\mathbf{d}_k \mathbf{d}_k \mathbf{d}_k]. \tag{6.31}$$

Here V is the volume of the unit cell, ε_0 is the permittivity of free space, and $\mathbf{d}_k$ is a matrix which transforms an externally applied electric into a field actually experienced by the k-th molecule in the unit cell; according to RLFT theory (see [16] and references therein), it depends on a lattice sum of the first polarisability of all the molecules in the unit cell. The in-crystal polarisabilities are obtained as energy derivatives of the constrained wavefunction using coupled-perturbed Hartree-Fock (CPHF) theory in essentially the same way that they would be obtained *ab initio* for the isolated molecule. The steps in this scheme are summarised in Fig. 6.11.

Figure 6.12 shows the in-crystal polarisabilities for a series of selected NLO compounds relative to a series of purely *ab initio* models for the crystal; the model MP2+field includes the effect of electron correlation and a finite electric field representing the crystalline surroundings. The presentation of the data is supposed to

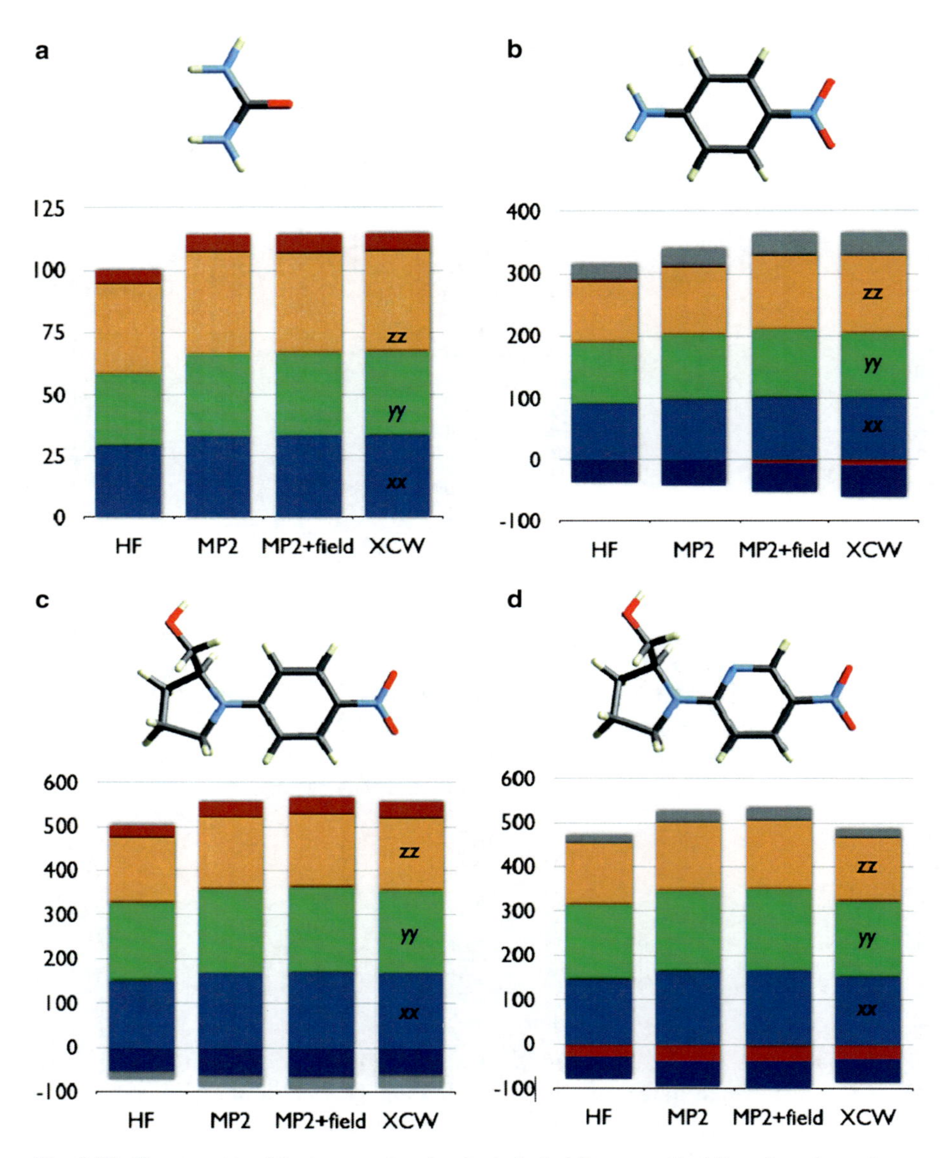

Fig. 6.12 Components of the in-crystal molecular polarisabily tensor (/au) for selected non-linear molecules (**a**) Urea (**b**) para-nitroaniline (**c**) NPP and (**d**) PNP for increasingly accurate levels of theory compared with the X-ray constrained Hartree-Fock (XCHF) wavefunction. Bar captions are; HF: Hartree-Fock, MP2: second order Møller-Plesset theory, MP2+field: an MP2 calculation in a static electric field calculated by Ewald summation of surrounding molecular dipole moments (For details of axis system see Ref. [16])

Table 6.4 Refractive indices for a series of NLO active molecules, at various levels of theory of increasing accuracy (see Fig. 6.12 for explanation) compared with the X-ray constrained-Hartree-Fock results (*XCHF*) and with experimental results (*Expt*) (For details see Ref. [16])

MP2						
	axis	HF	MP2	+field	XCHF	Expt
urea	$n_a = n_b$	1.39	1.44	1.45	1.45	1.477
	n_c	1.52	1.62	1.59	1.60	1.583
POM	n_a	1.55	1.58	1.58	1.55	1.625
	n_c	1.58	1.62	1.62	1.60	1.663
	n_b	1.71	1.82	1.81	1.78	1.829
NPP	n_X	1.42	1.45	1.44	1.45	1.457
	$n_Y = n_b$	1.63	1.69	1.70	1.67	1.774
	n_Z	1.71	1.80	1.82	1.79	1.829
PNP	n_X	1.43	1.45	1.45	1.43	1.456
	$n_Y = n_b$	1.62	1.68	1.69	1.61	1.732
	n_Z	1.70	1.80	1.82	1.74	1.880

facilitate a quick assessment of the agreement of a large amount of numerical data, which is presented in [16]. Table 6.4 shows the corresponding refractive indices derived from the in-crystal polarisabilities. *It is remarkable that the constrained Hartree-Fock results are in good agreement with the MP2 results.* Generally, the Hartree-Fock results are too small by a few percent, while by imposing an electric field representing the surrounding crystalline environment, become larger by a somewhat smaller margin.

The conclusion is that *first order electric-field response properties can be predicted for molecular crystals, by combining X-ray diffraction data and quantum mechanical techniques, and the results are very encouraging when compared to experimental results or higher level quantum mechanical calculations.*

6.6 Concluding Remarks, and the Future

There is now considerable evidence that meaningful "experimental" wavefunctions can be produced from diffraction data. The evidence for this comes from the comparison of geometric data, ADPs, dipole moments, polarisabilities and refractive indices with experimental data and with higher level calculations.

However, significant challenges remain. These can broadly be classified in *technical challenges* and *scientific opportunities*. On the technical level we can list the following issues:

- When should we terminate the constrained-fitting procedure? Fitting until the wavefunction ceases to converge is somewhat unsatisfactory even if the desired properties are not greatly affected by the final λ value, and even if such a procedure is supported by cross validation, using R_{free}. On the other hand, the idea to fit to a figure of merit (e.g. an R value from a multipole refinement) is also unsatisfactory because it means that the constrained wavefunction procedure is

not self-contained. It would seem that an approach based on cross validation or one related to the intrinsic quality of the data based on R_σ offers some merit.
- Constrained wavefunction methods will never be as fast as a least-squares refinement, but can they be made *faster*? Massively parallel computers will certainly alleviate this problem in the future – perhaps making possible constrained wavefunction calculations on small proteins. However, a better solution would be to expend effort optimising the code, and designing new algorithms. In this regard we are exploring the use of some kind of adjustable, polarisable model for an atom in a molecule based (of course) on an underlying wavefunction model.
- Is it possible to use a better model wavefunction, than just an single determinant one? Do the X-ray diffraction data warrant the added effort? The solution of the nuclear motion problem is a critical *scientific* prerequisite.

The most interesting scientific opportunities concern the interactions with other experiments, and the prediction of new properties:

- How far can the refinement of nuclear parameters be pushed? Can we incorporate, for example, temperature dependent ADP data to recover even more accurate positional information, and perhaps even normal mode information [106–108]?
- Will we be able to produce reasonable spin densities from X-ray diffraction results? Preliminary results not presented here are very encouraging [12].
- What about the combination of the constrained wavefunction technique with electron diffraction data? Since electron diffraction data usually pertains to hard inorganic crystalline materials, an essential component would be the use of crystalline-type wavefunctions. These crystalline wavefunctions might be approximated by using or extending the embedded molecule techniques described here.
- What about the prediction of other properties from the constrained wavefunction? For example, quadrupole coupling constants, electric fields at the nucleus, magnetisabilities, g-tensors, superconducting transition temperatures, and induced current densities, Hellmann-Feynman forces at the nuclei?
- What about charge density studies on heavy atom systems? The recent introduction of relativistic effects into the Tonto program package used to perform the calculations makes it possible to derive very high quality relativistic models for the heavy atoms in a *molecular* context, thus maximising the use of any relatively small valence electron signal seen in the X-ray diffraction data.
- Can we combine the results from different diffraction experiments? The answer is tentatively "yes" *if* the problem of obtaining results under exactly the same experimental conditions can be solved. We have already seen in this article that small geometric differences play an important role, and small differences in temperature can make these "joint" refinements very difficult.

The future is very promising for wavefunction-based techniques in diffraction science, with many opportunities for further work, a large body of free software being available.

Acknowledgements I am deeply indebted to several people who have made this work possible. Daniel Grimwood, Stephen Wolff, Anne Whitten, Andrew Whitton, Birger Dittrich, Martin Hudak, and Lukas Bucinsky have all helped to implement methods described, within the Tonto program system. Mark Spackman initiated the research into refractive indices. He has also provided several figures in this document. Mike Turner provided the first implementation of the coupled-perturbed Hartree-Fock equations for the response properties, while Andrew Whitten was involved in some earlier works. Parthapratim Munshi performed many of the calculations. Funding has been provided by the Australian Research Council over many years, and by the Centre National de la Recherche Scientifique. I am grateful to Sebastien Pillet and Claude Lecomte for arranging the latter funding and hosting an enjoyable visit. I am grateful to Carlo Gatti and Piero Macchi for the invitation to contribute, their discussions, and for their immense patience.

References

1. Helgaker T, Jorgensen P, Olsen J (2000) Molecular electronic structure theory. Wiley, Chichester
2. Parr RG, Yang W (1989) Density functional theory of atoms and molecules. Oxford University Press, New York
3. Pople JA (1999) Quantum chemical models (Nobel lecture). Angew Chem Int Ed 38:1894–1902
4. Kohn W (1999) Nobel lecture: electronic structure of matterwave functions and density functionals. Rev Mod Phys 71(5):1253–1266
5. Hohenberg P, Kohn W (1964) Inhomogeneous electron gas. Phys Rev B 136:B864–B871
6. Kohn W, Sham LJ (1965) Self-consistent equations including exchange and correlation effects. Phys Rev A 140:1133–1138
7. Jayatilaka D, Dittrich B (2008) X-ray structure refinement using aspherical atomic density functions obtained from quantum-mechanical calculations. Acta Crystallogr A 64:383–393
8. Bytheway I, Grimwood DJ, Figgis BN, Chandler GS, Jayatilaka D (2002) Wavefunctions derived from experiment. IV. Investigation of the crystal environment of ammonia. Acta Crystallogr A 58(3):244–251
9. Bytheway I, Grimwood DJ, Jayatilaka D (2002) Wavefunctions derived from experiment. III. Topological analysis of crystal fragments. Acta Crystallogr A 58:232–243
10. Grimwood DJ, Bytheway I, Jayatilaka D (2003) Wave functions derived from experiment. V. Investigation of electron densities, electrostatic potentials, and electron localization functions for noncentrosymmetric crystals. J Comput Chem 24:470–483
11. Grimwood DJ, Jayatilaka D (2001) Wavefunctions derived from experiment. II. A wavefunction for oxalic acid dihydrate. Acta Crystallogr A 57:87–100
12. Hudak M, Jayatilaka D, Perasinova L, Biskupic S, Kozisek J, Bucinsky L (2010) X-ray constrained unrestricted Hartree Fock and Douglas Kroll Hess wavefunctions. Acta Crystallogr A 66:78–92
13. Jayatilaka D (1998) Wave function for beryllium from X-ray diffraction data. Phys Rev Lett 80(4):798–801
14. Jayatilaka D, Grimwood DJ (2001) Wavefunctions derived from experiment. I. Motivation and theory. Acta Crystallogr A 57:76–86
15. Jayatilaka D, Grimwood DJ (2004) Electron localization functions obtained from X-ray constrained Hartree-Fock wavefunctions for molecular crystals of ammonia, urea and alloxan. Acta Crystallogr A 60(Pt 2):111–119
16. Jayatilaka D, Munshi P, Turner MJ, Howard JAK, Spackman MA (2009) Refractive indices for molecular crystals from the response of X-ray constrained Hartree Fock wavefunctions. Phys Chem Chem Phys 11:7209–7218

17. Lohmann B, Weigold E (1981) Direct measurement of the electron momentum probability distribution in atomic hydrogen. Phys Lett 86A(3):139–141
18. McCarthy IE, Weigold E (1988) Wavefunction mapping in collision experiments. Rep Prog Phys 51:299–392
19. Bellec A, Ample F, Riedel D, Dujardin G, Joachim C (2009) Imaging molecular orbitals by scanning tunneling microscopy on a passivated semiconductor. Nano Lett 9(1):144–147
20. Soe WH, Manzano C, De Sarkar A, Chandrasekhar N, Joachim C (2009) Direct observation of molecular orbitals of pentacene physisorbed on Au(111) by scanning tunneling microscope. Phys Rev Lett 102(17):100–103
21. Lundeen JS, Sutherland B, Patel A, Stewart C, Bamber C (2011) Direct measurement of the quantum wavefunction. Nature 474:188–191
22. Papousek D, Aliev MR (1982) Molecular vibrational-rotational spectra: theory and applications of high resolution infrared, microwave, and Raman spectroscopy of polyatomic molecules. Elsevier, Amsterdam
23. Schwarz WHE (2006) Measuring orbitals: provocation or reality? Angew Chem Int Ed 45(10):1508–1517
24. Coppens P (1997) X-Ray charge densities and chemical bonding. No. 4 in IUCr texts on crystallography. Oxford University Press, Oxford
25. Tsirelson VG, Ozerov RP (1996) Electron density and bonding in crystals: principles, theory, and X-ray diffraction experiments in solid state physics and chemistry. Institute of Physics Publishing, London
26. Davidson ER (1976) Reduced density matrices in quantum chemistry. Academic, New York
27. Stewart RF, Feil D (1980) A theoretical study of elastic X-ray scattering. Acta Crystallogr A 36:503–509
28. Schnieders MJ, Fenn TD, Pande VS, Brunger AT (2009) Polarizable atomic multipole X-ray refinement: application to peptide crystals. Acta Cryst D 65:952–965
29. Stewart RF (1976) Electron population analysis with rigid pseudoatoms. Acta Crystallogr A 32(4):565–574
30. Hirshfeld FL (1977) Bonded-atom fragments for describing molecular charge densities. Theor Chim Acta 44(2):129–138
31. Bruning H, Feil D (1992) Modeling the diffraction process of molecular crystals: computation of X-ray scattering intensities from ab initio electron densities. Acta Crystallogr A 48(6):865–872
32. Dittrich B, Koritsánszky T, Luger P (2004) A simple approach to nonspherical electron densities by using invarioms. Angew Chem Int Ed 43(20):2718–2721
33. McWeeny R (1992) Methods of molecular quantum mechanics, 2nd edn. Academic, San Diego
34. Hirata S (2009) Quantum chemistry of macromolecules and solids. Phys Chem Chem Phys 11(38):8397–8412
35. Shukla A, Dolg M, Fulde P, Stoll H (1998) Obtaining Wannier functions of a crystalline insulator within a Hartree-Fock approach: applications to LiF and LiCl. Phys Rev B 57:1471–1483
36. Shukla A, Dolg M, Stoll H, Fulde P (1996) An ab-initio embedded-cluster approach to electronic structure calculations on perfect solids: a Hartree-Fock study of lithium hydride. Chem Phys Lett 262:213–218
37. Koritsánszky T, Volkov A (2004) Atomic density radial functions from molecular densities. Chem Phys Lett 385:431–434
38. Stewart RF (1969) Generalized X-ray scattering factors. J Chem Phys 51(10):4569–4577
39. Coppens P, Willoughby TV, Csonka LN (1971) Electron population analysis of accurate diffraction data. I. Formalisms and restrictions. Acta Crystallogr A 27(3):248–256
40. Tanaka K (1988) X-ray analysis of wavefunctions by the least-squares method incorporating orthonormality. I. General formalism. Acta Crystallogr A 44:1002–1008
41. Kutzelnigg W, Smith VH (1969) On different criteria for the best independent-particle model approximation. J Chem Phys 41:896–897

42. Press WH, Teukolsky SA, Vetterling WT, Flannery BP (1992) Numerical recipes in Fortran – the art of scientific computing, 2nd edn. Cambridge University Press, Cambridge
43. Jaynes ET (1957) Information theory and statistical mechanics. Phys Rev 104:620–630
44. Chong DP, Byers-Brown W (1966) Perturbation theory of constraints: application to a lithium hydride calculation. J Chem Phys 45(1):392–395
45. Gillet JM, Becker PJ (2004) Position and momentum densities. Complementarity at work: refining a quantum model from different data sets. J Phys Chem Sol 65(12):2017–2023
46. Levy M, Goldstein JA (1987) Electron density-functional theory and x-ray structure factors. Phys Rev B 35(15):7887–7890
47. Tarantola A (2004) Inverse problem theory and methods for model parameter estimation. SIAM, Philadelphia
48. Tychonoff AN, Arsenin VY (1977) Solution of Ill-posed problems. Winston & Sons, Washington, DC
49. Jayatilaka D, Grimwood D (2003) Tonto: a Fortran based object-oriented system for quantum chemistry and crystallography. Comput Sci ICCS 2660:142–151
50. Mukherji A, Karplus M (1963) Constrained molecular wavefunctions: HF molecule. J Chem Phys 38:44–48
51. Rasiel Y, Whitman DR (1965) Constrained-variation method in molecular quantum mechanics. Application to lithium hydride. J Chem Phys 42:2124–2131
52. Chong DP, Rasiel Y (1966) Constrained-variation method in molecular quantum mechanics. Comparison of different approaches. J Chem Phys 44:1819–1823
53. Benston ML, Chong DP (1967) Off-diagonal constrained variations in open-shell SCF theory. Mol Phys 12:487–492
54. Chong DP (1967) Coalescence conditions as constraints in open-shell SCF theory. J Chem Phys 47(12):4907–4909
55. Chong DP, Benston ML (1968) Off-diagonal hypervirial theorems as constraints. J Chem Phys 49:1302–1306
56. Yue CP, Chong DP (1968) On the use of integral electron cusp conditions as constraints. Theor Chim Acta 12:431–435
57. Weber TA, Handy NC (1969) Linear homogeneous constrained variation procedure for molecular wavefunctions. J Chem Phys 50:2214–2215
58. Weber TA, Handy NC, Parr RG (1970) Self-consistent-field atomic wavefunctions from efficient nested basis sets. J Chem Phys 52:1501–1507
59. Clinton WL, Galli AJ, Henderson GA, Lamers GB, Massa LJ, Zarur J (1969) Direct determination of pure-state density matrices. V. Constrained eigenvalue problems. Phys Rev 177:27–33
60. Clinton WL, Galli AJ, Massa LJ (1969) Direct determination of pure-state density matrices. II. Construction of constrained idempotent one-body densities. Phys Rev 177:7–13
61. Clinton WL, Henderson GA, Prestia JV (1969) Direct determination of pure-state density matrices. III. Purely theoretical densities via an electrostatic-virial theorem. Phys Rev 177: 13–18
62. Clinton WL, Lamers GB (1969) Direct determination of pure-state density matrices. IV. Investigation of another constraint and another application of the P equations. Phys Rev 177:19–27
63. Clinton WL, Nakleh J, Wunderlich F (1969) Direct determination of pure-state density matrices. I. Some simple introductory calculations. Phys Rev 177:1–6
64. Clinton WL, Frishberg C, Massa LJ, Oldfield PA (1973) Methods for obtaining an electron-density matrix from X-ray diffraction data. Int J Quantum Chem Symp 7:505–514
65. Clinton WL, Massa LJ (1972) Determination of the electron density matrix from X-ray diffraction data. Phys Rev Lett 29:1363–1366
66. Clinton WL, Massa LJ (1972) The cusp condition: constraint on the electron density matrix. Int J Quantum Chem 6:519–523
67. Cohen L, Frishberg C (1976) On the errors in molecular dipole moments derived from accurate diffraction data. Phys Rev A 13:927–930

68. Cohen L, Frishberg C, Lee C, Massa LJ (1986) Correlation energy for a slater determinant fitted to the electron density. Int J Quantum Chem Symp 19:525–533
69. Frishberg C (1986) Slater determinant from atomic form factors. Int J Quantum Chem 30:1–5
70. Frishberg C, Massa LJ (1978) Notes on density matrix model for coherent X-ray diffraction. Int J Quantum Chem 13:801
71. Frishberg C, Massa LJ (1981) Idempotent density matrices for correlated systems from x-ray-diffraction structure factors. Phys Rev B24:7018–7024
72. Frishberg C, Massa LJ (1982) Numerical applications of a quantum model for the coherent diffraction experiment. Acta Crystallogr A 38:93–98
73. Massa LJ, Goldberg M, Frishberg C, Boehme RF, la Placa SJ (1985) Wave functions derived by quantum modelling of the electron density from coherent X-ray diffraction: beryllium metal. Phys Rev Lett 55(6):622–625
74. Pecora LM (1986) Determination of the quantum density matrix from experiment: an application to positron annihilation. Phys Rev B33:5987–5993
75. Aleksandrov YV, Tsirelson V, Reznik IM, Ozerov RP (1989) The crystal electron energy and Compton profile calculations from X-ray diffraction data. Phys Stat Sol B 155:201–207
76. Levy M (2001) Universal variational functionals of electron densities, first-order density matrices, and natural spin-orbitals and solution of the v-representability problem. Proc Natl Acad Sci USA 11(14):2549–2916
77. Gilbert TL (1975) Hohenberg-Kohn theorem for nonlocal external potentials. Phys Rev B 12(6):2111–2120
78. Harriman JE (1986) Densities, operators, and basis sets. Phys Rev A 34:29–39
79. Henderson GA, Zimmermann RK (1976) One-electron properties as variational parameters. J Chem Phys 65(2):619–622
80. Morrison RC (1988) Density and density matrix from optimized linearly independent product basis functions for Be. Int J Quantum Chem Symp 22:43–49
81. Schwarz WHE, Muller B (1990) Density matrices from densities. Chem Phys Lett 166(5):621–626
82. Howard ST, Huke JP, Mallinson PR, Frampton CS (1994) Density matrix refinement for molecular crystals. Phys Rev B 9:7124–7136
83. Snyder JA, Stevens ED (1999) A wavefunction and energy of the azide ion in potassium azide obtained by a quantum-mechanically constrained fit to X-ray diffraction data. Chem Phys Lett 313(1–2):293–298
84. Zhao Q, Morrison RC, Parr RG (1994) From electron densities to Kohn-Sham kinetic energies, orbital energies, exchange-correlation potentials, and exchange correlation energies. Phys Rev A 50(3):2138–2142
85. Zhao Q, Parr RG (1992) Quantities Tsn and Tcn in density functional theory. Phys Rev A 46(5):2337–2343
86. Zhao Q, Parr RG (1993) Constrained-search method to determine electronic wave functions from electronic densities. J Chem Phys 98(1):543–548
87. Cassam-Chenaï P, Wolff S, Chandler G, Figgis B (1996) Ensemble-representable densities for atoms and molecules. II. Application to $CoCl_4^{2-}$. Int J Quantum Chem 60:667–680
88. Cassam-Chenaï P (2002) Ensemble representable densities for atoms and molecules. III. Analysis of polarized neutron diffraction experiments when several Zeeman levels are populated. J Chem Phys 116(20):8677
89. Cassam-Chenaï P (1995) Ensemble representable densities for atoms and molecules. I. General theory. Int J Quantum Chem 54:201–210
90. Hibbs DE, Howard ST, Huke JP, Waller MP (2005) A new orbital-based model for the analysis of experimental molecular charge densities: an application to (Z)-N-methyl-C-phenylnitrone. Phys Chem Chem Phys 7:1772–1778
91. Waller MP, Howard ST, Platts JA, Piltz RO, Willock DJ, Hibbs DE (2006) Novel properties from experimental charge densities: an application to the zwitterionic neurotransmitter taurine. Chem Eur J 12(29):7603–7614

92. Tanaka K, Kato Y, Onuki Y (1997) 4f-electron density distribution in crystals of CeB6 at 165 K and its analysis based on the crystal field theory. Acta Crystallogr B 53:143–152
93. Tanaka K, Makita R, Funahashi S, Komori T, Win Z (2008) X-ray atomic orbital analysis. I. Quantum-mechanical and crystallographic framework of the method. Acta Crystallogr A 64(Pt 4):437–449
94. Tanaka K, Onuki Y (2002) Observation of 4f electron transfer from Ce to B6 in the Kondo crystal CeB6 and its mechanism by multitemperature X-ray diffraction. Acta Crystallogr B 58:423–436
95. Cooper M (1985) Compton scattering and electron momentum determination. Rep Prog Phys 48:415–481
96. Loupias G, Chomilier J (1986) Electron momentum density and Compton profiles: an accurate check of overlap models. Z Phys D Atoms Mol Clust 2:297–308
97. Schmider H, Smith VH, Weyrich W (1990) Determination of electron densities and one-matrices from experimental information. Trans Am Crystal Assoc 26:125–140
98. Schmider H, Smith VH, Weyrich W (1993) Z Naturforsch 48a:211–220
99. Schmider H, Smith VH, Weyrich W (1992) Reconstruction of the one-particle density matrix from expectation values in position and momentum space. J Chem Phys 96(5):8986–8994
100. Weyrich W (2006) An electronic position and momentum density study of chemical bonding in TiO_2 (Rutile). Lect Ser Comput Comput Sci 5:1–3
101. Gillet JM (2007) Determination of a one-electron reduced density matrix using a coupled pseudo-atom model and a set of complementary scattering data. Acta Crystallogr A 63 (Pt 3):234–238
102. Gillet JM, Becker P, Cortona P (2001) Joint refinement of a local wave-function model from Compton and Bragg scattering data. Phys Rev B 63(23):235, 115–7
103. Hummel W, Hauser J, Bürgi HB (1990) PEANUT: computer graphics program to represent atomic displacement parameters. J Mol Graph 8:214–218
104. Spackman MA, Byrom PG, Alfredsson M, Mermansson K (1999) Influence of intermolecular interactions on multipole-refined electron densities. Acta Crystallogr A 55:30–47
105. Whitten AE, Jayatilaka D, Spackman MA (2006) Effective molecular polarizabilities and crystal refractive indices estimated from x-ray diffraction data. J Chem Phys 125:174, 505
106. Bürgi HB, Capelli SC (2000) Dynamics of molecules in crystals from multitemperature anisotropic displacement parameters. I. Theory. Acta Crystallogr A 56:403–412
107. Bürgi HB, Capelli SC, Birkedal H (2000) Anharmonicity in anisotropic displacement parameters. Acta Crystallogr A 56:425–435
108. Capelli SC, Fortsch M, Bürgi HB (2000) Dynamics of molecules in crystals from multitemperature anisotropic displacement parameters. II. Application to benzene C_6D_6 and urea $OC(NH)_2$. Acta Crystallogr A 56:413–424

Chapter 7
Local Models for Joint Position and Momentum Density Studies

Jean-Michel Gillet

7.1 Introduction

The importance of electronic wave-functions and density matrices as well as their connection to charge density studies are explained in some detail in Chaps. 2, 5 and 6. Today, both can be computed with a flexible degree of accuracy for atoms, molecules and even infinite periodic systems. However, there are still limitations to find eigenstates of exact Hamiltonians. On the one hand, quantum computations necessitate simplifications, which can be relaxed later on, such as the Born Oppenheimer approximation, the mean field or the single determinant approach. On the other hand, in order to account for most of the remaining complexity, a large number of variational parameters are introduced in the model. When the calculation is carried out by means of a limited and fixed atomic basis set, the LCAO parameters constitute the so-called "population matrix".

As a reminder, we write here the expression for the N-electron wave function built from a single Slater determinant (the sum runs thru all possible permutations of particles):

$$\psi\left(\mathbf{r}_1, \mathbf{r}_2, ..., \mathbf{r}_N\right) = \frac{1}{\sqrt{N!}} \sum_p (-1)^p \hat{P}\left[\varphi_1\left(\mathbf{r}_1\right)\varphi_2\left(\mathbf{r}_2\right)\ldots\varphi_n\left(\mathbf{r}_n\right)\right]. \tag{7.1}$$

where, in the LCAO picture, each molecular orbital is written as a linear combination of atomic orbitals centred on nuclei located at $\mathbf{R}_A$ in the system:

$$\varphi_n\left(\mathbf{r}_1\right) = \sum_A \sum_{j\in A} c_{jn}^A \quad \varphi_{j\in A}(\mathbf{r}_1 - \mathbf{R}_A) \tag{7.2}$$

J.-M. Gillet (✉)
Structures, Propriétés et Modélisation des Solides, UMR8580, Ecole Centrale Paris, Grande Voie des Vignes, 92295 Chatenay-Malabry Cedex, France
e-mail: jean-michel.gillet@ecp.fr

C. Gatti and P. Macchi (eds.), *Modern Charge-Density Analysis*,
DOI 10.1007/978-90-481-3836-4_7,

The population matrix **P**, related to a given basis set of atomic functions, is obtained from its elements (see, also expressions 5.44 and 5.45 in Chap. 5):

$$P_{ij}^{AB} = \sum_n c_{in}^A c_{jn}^B.$$

The results from ab-initio computations are often very satisfactory when comparison is made with experimental data. They further allow for calculation of other properties that are difficult, if not impossible, to measure.

However, small discrepancies which cannot always be eliminated may occur. When they seem to correspond to physically significant features, it is usually assumed that improvement of the basis sets quality or a better account for the correlation effects, on a higher level of approximation, should solve the problem. Such improvement can be achieved but at the expense of a longer computing time with little gain in understanding the physical and chemical processes that are at work.

Another way to address the problem is to build models with few adjustable parameters and compute observables that are experimentally available. Such models are often oversimplification of reality. Nevertheless, in this case, the purpose is no longer to reach a high degree of accuracy but to gain insights into the basic mechanisms. This goal is often better reached with rather elementary pictures. However, there is a possibility that a particular "raw" model for a wave-function or a density matrix is suitable for reproducing (and explain) data from one particular experiment but fails to account for essential features of another kind of observable. The purpose of the model is usually to give a wider description of reality than that which a single experiment provides.

In a majority of cases, fine details of the wave-function were proved to have significant, but rather different, impacts on scattering spectra describing position or momentum spaces (see Chaps. 2 and 4). It is in this sense that both spaces complement one another.

Therefore, it is highly desirable to confront the model with a set of different experiments. In the case of a wave-function or a density matrix model, the least strategy is probably to make a simultaneous use of both position and momentum spaces observables. Attempts to combine experimental data from both spaces are rare and difficult because they require good data but also because they call for models that should easily be expressed in both spaces.

Schematic views of electron behaviours in matter can be built in a number of ways. A possible strategy is to focus on a local description of mechanisms whenever there is no strong electron delocalization.

This chapter explains how local models were used to jointly reproduce spectra from x-rays and electron scattering experiments.

First, a short range cluster approach to the reduced density matrix is proposed. Its adequacy to reproduce Compton scattering data is discussed using the example of the hydrogen bond in ice. Then, a simple local model, based on a "dressed ion orbital", is constructed for FCC ionic crystals. The first joint refinement of this model against electron and x-ray diffraction data and Compton spectra is given as an illustrative example of the possible quality of results that can be expected.

7.2 Local Models, A Progressive Cluster Approach to the 1-RDM

As discussed in Chaps. 2 and 5, and to follow "Coulson's challenge" [1], the determination of N-electron wave-function is usually unnecessary. The reason that is generally invoked is that all observables, from the electronic standpoint, are only one or two-particle properties. Therefore, it is often desirable, and sometimes inevitable, to elaborate models that involve at most a couple of electrons at a time. Such constructions require the use of two-electron reduced density matrices (2-RDM) and, within the Hartree-Fock framework, it is even possible to narrow down the description by means of a mere one-electron reduced density matrix (1-RDM). Indeed, in many cases, the 1-RDM is enough to provide a fair reproduction of experimental observations.

In fact, the 1-RDM, is the simplest mathematical object that is common to position and momentum spaces (*cf.* Chap. 2). Thus, it is commonly accepted that the 1-RDM is a convenient and economic way to model one electron properties. In particular, X-rays or electron scattering experiments are powerful techniques for investigating electron behaviour in atoms, molecules or solids. They offer a invaluable reference for checking the pertinence of a model (see Chap. 2) or determining its adjustable parameters.

The usual expansion of the 1-RDM on one-electron atomic orbitals is:

$$\gamma^{1}\left(\mathbf{r}'_{1};\mathbf{r}''_{2}\right)=\sum_{A,B}\sum_{\substack{i\in A\\ j\in B}}P_{ij}^{AB}\phi_{i\in A}(\mathbf{r}'_{1}-\mathbf{R}_{A})\phi_{j\in B}(\mathbf{r}''_{1}-\mathbf{R}_{B}) \tag{7.3}$$

where the first double sum runs through all atomic couples *(A,B)* in the system. The second sum, for a given couple of atoms, runs through all atomic orbitals $\varphi_i(\mathbf{r}_1-\mathbf{R}_A)$pertaining to each atom. For the sake of simplicity, orbitals are supposed to take only real values. $P_{ij}^{AB}$is the population matrix (see Sect. 7.1).

It is convenient to be able to partition the 1-RDM in terms of atomic contributions. A possible de-construction could be obtained in the following way:

$$\begin{aligned}\gamma^{(1)}\left(\mathbf{r}'_{1};\mathbf{r}''_{1}\right)&=\sum_{A}\gamma_{A}^{(1)}\left(\mathbf{r}'_{1};\mathbf{r}''_{1}\right)\\&=\sum_{A}\left[\gamma_{AA}^{(1)}\left(\mathbf{r}'_{1};\mathbf{r}''_{1}\right)+\frac{1}{2}\sum_{B\neq A}\left(\gamma_{AB}^{(1)}\left(\mathbf{r}'_{1};\mathbf{r}''_{1}\right)+\gamma_{BA}^{(1)}\left(\mathbf{r}'_{1};\mathbf{r}''_{1}\right)\right)\right]\end{aligned} \tag{7.4}$$

$\gamma_A^{(1)}$ can thus be interpreted here as the contribution of atom A to the total density matrix. It is constructed from a one-centre term $\gamma_{AA}^{(1)}$, i.e. involving only products of orbitals of the relevant atom, and two-centre terms, $\gamma_{AB}^{(1)}$, mixing orbitals from the current atom A and its neighbours:

$$\gamma^{(1)}_{AB}\left(\mathbf{r}'_1;\mathbf{r}''_1\right)=\sum_{\substack{i\in A\\ j\in B}}P^{AB}_{ij}\phi_{i\in A}(\mathbf{r}'_1-\mathbf{R}_A)\phi_{j\in B}(\mathbf{r}''_1-\mathbf{R}_B) \tag{7.5}$$

It is noteworthy that for a given couple of atomic centres, $\gamma^{(1)}_{AB}$ and $\gamma^{(1)}_{BA}$ should be introduced symmetrically to preserve the permutation invariance of electron coordinates $\mathbf{r}'_1 \Leftrightarrow \mathbf{r}''_1$.

In the case of an infinite perfect crystalline system, the construction must account for the translation invariance. This is best taken into account by means of the so-called intracular and extracular coordinates, respectively $\mathbf{s}_1 = \mathbf{r}'_1 - \mathbf{r}''_1$ and $\mathbf{r}_1 = \left(\mathbf{r}'_1 + \mathbf{r}''_1\right)/2$, such that using:

$$\gamma^{(1)}_{\mathbf{L},A}\left(\mathbf{r}_1-\mathbf{R}_A-\mathbf{L};\mathbf{s}_1\right)=\sum_{\mathbf{L}',B}P^{AB}_{ij}\left(\mathbf{L}'\right)\phi_{i\in A}(\mathbf{r}_1-\mathbf{L}-\left(\mathbf{s}_1+\mathbf{L}'\right)/2-\mathbf{R}_A)$$
$$\phi_{j\in B}(\mathbf{r}_1-\mathbf{L}+\left(\mathbf{s}_1+\mathbf{L}'\right)/2-\mathbf{R}_B) \tag{7.6}$$

the 1-RDM writes:

$$\gamma^{(1)}\left(\mathbf{r}_1;\mathbf{s}_1\right)=\sum_{\mathbf{L}}\sum_{A\in\mathbf{L}}\gamma^{(1)}_{\mathbf{L},A}\left(\mathbf{r}_1-\mathbf{R}_A-\mathbf{L};\mathbf{s}_1\right) \tag{7.7}$$

where $\mathbf{L}$ represents the lattice translations. This decomposition method emphasizes the predominant atomic (one centre) character of the 1-RDM though, so far, all two-centre terms are implicitly accounted for in each atomic contribution.

The pseudo-atom decomposition of the charge density as proposed by Hansen and Coppens [2] (see Chap. 5), can be considered as a limit case of such an expression in which all two-centre terms are neglected.

This pseudo-atomic model is still very successful because it is always limited to interpretations of electron density (charge or spin) in position space and allows for a convenient account of nuclear motions in terms of independent oscillators (*cf.* Chap. 3). Moreover, it is straightforward to check that two-centre contributions from distant atoms have a negligible impact on structure factors since they essentially decrease with the overlap between the associated orbitals:

$$\begin{aligned}F\left(\mathbf{Q}\right)&=\int\gamma^{(1)}\left(\mathbf{r}_1;0\right)e^{i\mathbf{Q}.\mathbf{r}_1}d\mathbf{r}_1\\&=\sum_A\left[\int\gamma^{(1)}_{AA}\left(\mathbf{r}_1;0\right)e^{i\mathbf{Q}.\mathbf{r}_1}d\mathbf{r}_1+\frac{1}{2}\sum_{B\neq A}\left(\int\gamma^{(1)}_{AB}\left(\mathbf{r}_1;0\right)e^{i\mathbf{Q}.\mathbf{r}_1}d\mathbf{r}_1\right.\right.\\&\qquad\left.\left.+\int\gamma^{(1)}_{BA}\left(\mathbf{r}_1;0\right)e^{i\mathbf{Q}.\mathbf{r}_1}d\mathbf{r}_1\right)\right]\end{aligned} \tag{7.8}$$

with a typical two-centre term:

$$\begin{aligned} F_{AB}(\mathbf{Q}) &= \int \gamma_{AB}^{(1)}(\mathbf{r}_1;0)\, e^{i\mathbf{Q}.\mathbf{r}_1} d\mathbf{r}_1 \\ &= e^{i\mathbf{Q}.\mathbf{R}_{AB}} \sum_{\substack{i\in A\\ j\in B}} P_{ij}^{AB} \int \phi_{i\in A}(\mathbf{r}-\mathbf{S}_{AB})\phi_{j\in B}(\mathbf{r}+\mathbf{S}_{AB}) e^{i\mathbf{Q}.\mathbf{r}} d\mathbf{r} \end{aligned} \tag{7.9}$$

with $\mathbf{R}_{AB} = (\mathbf{R}_A + \mathbf{R}_B)/2$ and $\mathbf{S}_{AB} = (\mathbf{R}_A - \mathbf{R}_B)/2$.

Hence, for an approximate computation of the density matrix, with connection to X-ray diffraction experiments, it is not required to use an extended number of neighbours for each atomic contribution.

On the contrary, it has been shown that two-centre terms [3] are of paramount importance when one seeks an accurate description of momentum space densities. Let us consider the charge density in momentum representation:

$$\begin{aligned} \pi(\mathbf{p}) &= \int \gamma^{(1)}(\mathbf{r}_1;\mathbf{s}_1)\, e^{i\mathbf{p}.\mathbf{s}_1} d\mathbf{r}_1 d\mathbf{s}_1 \\ &= \sum_A \left[\int \gamma_{AA}^{(1)}(\mathbf{r}_1;\mathbf{s}_1)\, e^{i\mathbf{p}.\mathbf{s}_1} d\mathbf{r}_1 d\mathbf{s}_1 + \frac{1}{2}\sum_{B\neq A}\left(\int \gamma_{AB}^{(1)}(\mathbf{r}_1;\mathbf{s}_1)\, e^{i\mathbf{p}.\mathbf{s}_1} d\mathbf{r}_1 d\mathbf{s}_1 \right.\right. \\ &\qquad \left.\left. + \int \gamma_{BA}^{(1)}(\mathbf{r}_1;\mathbf{s}_1)\, e^{i\mathbf{p}.\mathbf{s}_1} d\mathbf{r}_1 d\mathbf{s}_1 \right)\right] \end{aligned} \tag{7.10}$$

with a typical two-centre term:

$$\pi_{AB}(\mathbf{p}) = \int \gamma_{AB}^{(1)}(\mathbf{r}_1;\mathbf{s}_1)\, e^{i\mathbf{p}.\mathbf{s}_1} d\mathbf{r}_1 d\mathbf{s}_1 = \sum_{\substack{i\in A\\ j\in B}} P_{ij}^{AB} \tilde{\phi}_{i\in A}(\mathbf{p}) \tilde{\phi}_{j\in B}(\mathbf{p}) e^{2i\mathbf{p}.\mathbf{S}_{AB}} \tag{7.11}$$

Hence, in momentum space analysis, two-centre terms are mostly conditioned by the products of atomic orbitals in momentum representation. Of course, two atoms separated by a large distance will give a much too rapid oscillating contribution to be detected by a momentum resolution limited instrument $\sigma(p) > \pi / S_{AB}$.

7.3 From Computation to Refinement

The aforementioned decomposition of the 1-RDM is particularly well adapted to an approximate ab-initio computation of structure factors or deep inelastic X-ray scattering spectra (Compton regime). In particular, one should note that the

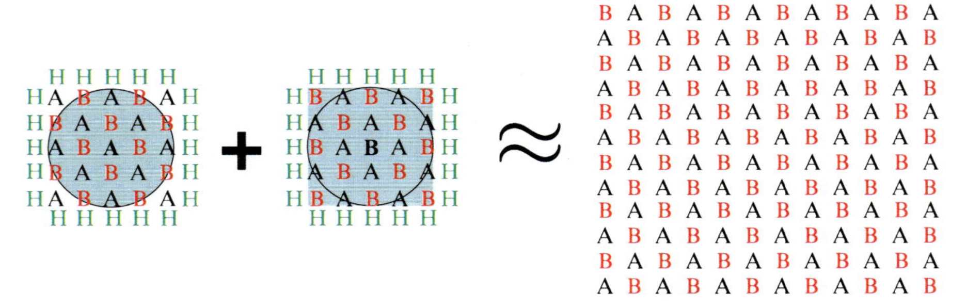

Fig. 7.1 Principle of reconstruction of the 1-RDM for an hypothetical diatomic A–B infinite crystal. Two molecular calculations are carried out. From the resulting population matrices, only the contributions involving the central atoms (*centre of blue circles*) are kept and recombined to approximate the actual content of a unit cell in the crystal

limitation of two-centre terms to a given shell of neighbours, for each atom, brings the computation of an infinite solid back to a set of molecular-like calculations. Moreover, because no long range order or periodicity is assumed, this technique is well adapted to amorphous and disordered systems.

As a schematic illustration (Fig. 7.1), approximate charge and momentum densities for a crystal with two atoms A and B per unit cell can be reduced to two ab-initio computations of clusters with, respectively, atoms A and B at the centres. Although it is not compulsory to conduct two different calculations, such a procedure has always been respected in order to limit artefacts from the asymmetry of the respective environment.

Moreover, to limit spurious finite size effects, dangling bonds appearing at the periphery of the cluster, are saturated by a set of additional hydrogen atoms.

In its principles, this approach is comparable to the "ONIOM method" now frequently used for the computation of large molecules [4] or the "divide and conquer" method proposed by Yang [5].

Obviously this cluster approach is not suitable for the computation of metallic compounds. However, it has been demonstrated [6] that with a moderate number of surrounding shells, not only ionic compounds (such as MgO, LiH ...) could be accurately calculated but also semi-conductors (Si).

In the next Subsection, we briefly report to what extent the above local model can be used to compute momentum space densities in relationship with x-ray scattering experiments.

7.3.1 Ice and the Hydrogen Bond Issue: An Example of a Moderately Disordered System

Since the early works by Pauling [7], the water molecule in its I_h solid form has received much attention. In particular, given the large number of H-bonds relative to the total number of atoms in the crystalline form, Ice I_h is usually considered as a good candidate for carrying out studies on hydrogen bonds.

To conduct an investigation that goes beyond the mere electrostatic interpretation of hydrogen bonds, Isaacs and co-workers [8] have performed an inelastic x-rays scattering experiment (Compton regime i.e. high momentum and energy transfer). Using a high purity Ice crystal in the I_h phase, they could then measure two projections of the electron momentum density (i.e. Compton profiles) respectively along the **c** crystallographic axis and along the perpendicular direction hereafter denoted $\overrightarrow{a/b}$.

Given the larger number of H bonds pointing in the **c** direction, anisotropic Compton profiles (ACP, difference between directional profiles; see also Chap. 2) could then be interpreted in terms of the electron delocalization on the hydrogen bonds. The Fourier transform of the ACP, i.e. the difference of directional auto-correlation functions (or "reciprocal form factors", see expressions 5.23 and 5.24 in Chap. 5 and 2.19 in Chap. 2), exhibits a peak corresponding to the O...H distance of the hydrogen bond (1.75 Å). Such a feature can unambiguously be attributed to a coherent contribution of the two atomic regions (O ... H). The authors further speculated that the feature was evidence for a partial covalence of the hydrogen bond.

In order to challenge this interpretation, it was essential to build a model for Ice that could not only accurately account for the local charge density behaviour, but also give a fair reproduction of the electron dynamics and the coherent contribution between atomic sites as described by Compton scattering data.

However, unlike other hydrogen bonded systems that show an almost perfect crystalline arrangement, ice in the I_h phase, has a disordered crystallographic structure. Following Bernal and Fowler "Ice-rule" [9], the "ideal" structure has the $P2_1$ symmetry such as displayed on Fig. 7.2. Therefore, the proton disorder makes it impossible to observe such a perfect arrangement and, according to Pauling [7], Owsten [10] and later Peterson and Levy [11], what can be observed is a statistical mixture of different arrangements leading to an average macroscopic crystal exhibiting a global $P6_3/mmc$ symmetry. Van Beek [12] showed that a superimposition of six locally ordered domains, each with a local $P2_1$ symmetry with **c** as a unique common axis would yield a satisfactory agreement with the measured structure factors amplitudes.

Therefore, very little information can be expected from elastic x-ray scattering concerning the nature of the hydrogen bond in ice I_h. For example, it is hardly conceivable to carry out charge density topology analysis in terms of properties (Laplacian, ellipticity) at critical points from high resolution x-ray diffraction at the level of accuracy reached by Espinosa and co-workers [13] on other crystalline systems.

Following Van Beek superposition, a set of cluster calculations was carried out. A fair convergence of the electronic properties was obtained for a cluster of 22 water molecules. As already observed in the literature both experimentally and theoretically [14–16], the molecular dipole values have been shown to increase with

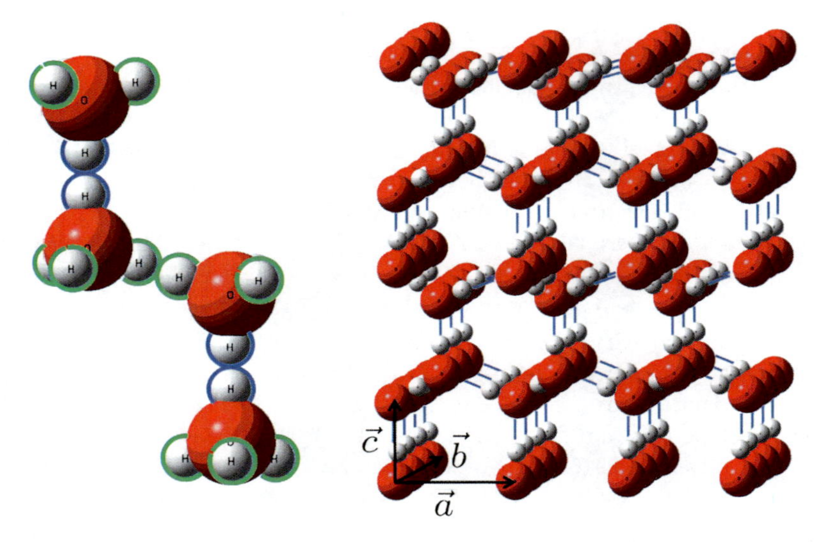

Fig. 7.2 Construction of the Ice crystal (Ih) according to the Bernal-Fowler ice-rules. Proton disorder is represented by the possible sites for each hydrogen atom H1 (*circled in blue*) and H2 (*circled in green*) with ½ occupancy probabilities

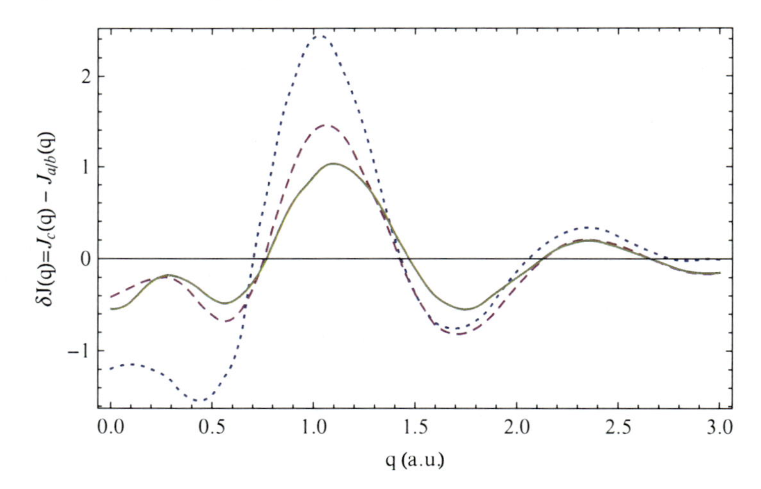

Fig. 7.3 Computed Compton anisotropy in Ice clusters. *Dotted*: dimer of water molecules. *Dashed*: 22 water molecules. *Solid*: 46 water molecules. The anisotropic Compton profiles (*ACPs*) are computed from a superimposition of six clusters configuration, each of P21 symmetry

the number of neighbours. In this study the criteria for checking the stability of the results were Compton anisotropies. They have been shown to be very stable for an ensemble of 46 molecules (Fig. 7.3).

On the basis of local model computations, it is thus possible to reproduce with an unprecedented quality the key features in Compton differences (Fig. 7.4). The

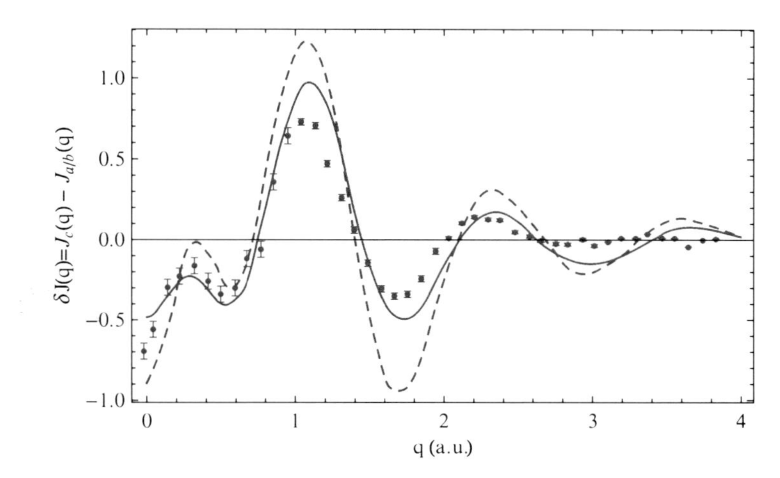

Fig. 7.4 Comparison between ACP for Ice Ih. *Solid*: 46 water molecules Hartree-Fock cluster based computation. *Dashed*: plane wave DFT based calculation (From [8]). *Dots* and *error bars*: data from experiment (From [8, 17]). For comparison, ab-initio results have been convoluted by the experimental resolution function (0.14 a.u.)

oscillations are believed to be characteristics of the bonding mechanism in ice but also in any other system. In order to check the pertinence of the "covalent" character of the hydrogen bond in ice, a similar calculation was carried out with the same structure and interatomic distances, but replacing oxygen atoms by neon and removing hydrogen atoms. In order to emphasize possible bond features arising from coherent contributions in the wave function, let us define the "power spectrum" as the squared modulus of the anisotropic autocorrelation function (Eq. 5.24):

$$Ps\,(\mathbf{s}) = |B\,(\mathbf{s}) - B_{\mathrm{iso}}\,(\mathbf{s})|^2.$$

Of course, this "Neon ice-like system" is not expected to exhibit any covalent bond. Nevertheless, a similar feature in the power spectrum and in the autocorrelation anisotropy could also be shown (Fig. 7.5). The peak at 1.75 Å (Fig. 7.6) in the power spectrum as a hint for a possible covalent effect in hydrogen bond could then be ruled out. As it was already mentioned by Ghanty and coworkers [18], the features can be also observed in antibonding systems and are now believed to be a side effect of the mere anti-symmetry of monomer wave-functions.

7.3.2 Refinement at the Lowest Order in Cluster Size

As specified earlier in the chapter, the cluster approach to the computation of the 1-RDM makes it possible to take into account two-centre contributions with a flexible

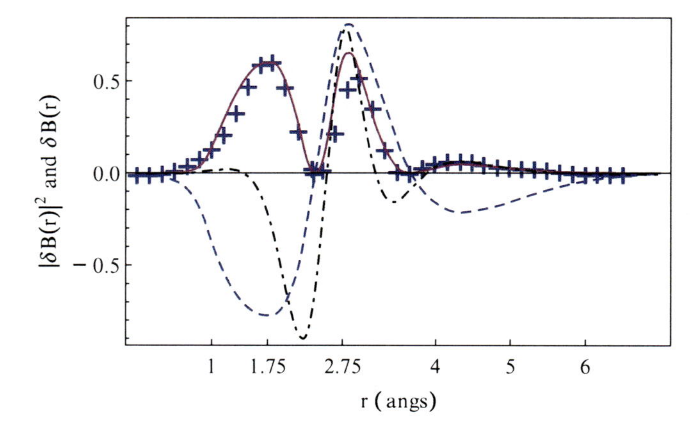

Fig. 7.5 Power spectrum (*solid*) and anisotropic autocorrelation function (*dashed*) as obtained from a 22 water molecules cluster based computation and Neon cluster (*dotted dashed*). For comparison, the experimental power spectrum is also plotted (*crosses*) (From [8, 17]). For a better display, multiplication factors of 3.5 and 20 were respectively applied to the theoretical anisotropic autocorrelation functions of ice and neon

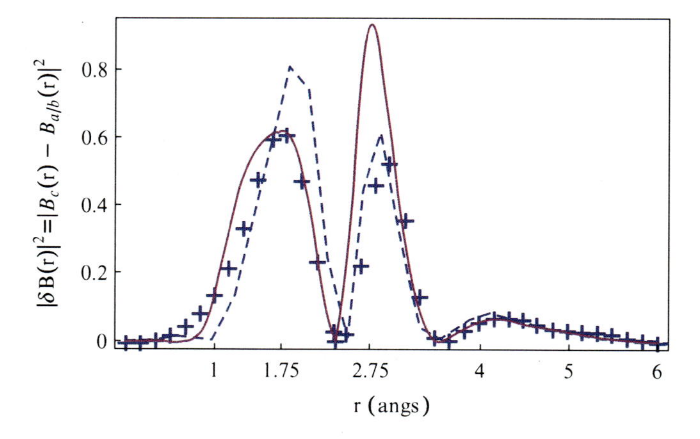

Fig. 7.6 Comparison between power spectra for Ice Ih. *Solid*: 46 water molecules cluster based Hartree-Fock computation. *Dashed*: plane wave DFT based calculation (From [8]). *Crosses*: data from experiment (From [8, 17])

level of accuracy. This property is also well adapted to the refinement of a 1-RDM model (with a limited number of parameters) against experimental data.

Ionic compounds are suitable for such a refinement approach thanks to the relatively short range of two centre contributions resulting from the limited delocalisation of valence electrons. Metals would definitely demand a different model.

Based on the sole input of experimental structure factors, from high resolution x-rays or convergent beam electron diffraction data, electron charge density in position space is usually reconstructed with a very good accuracy. It has been shown that, providing adapted atomic basis functions are used, multipoles on independent ions are enough to extract the best from available experimental data [19].

However, as it has been observed several times, the independent ions model no longer applies when electron dynamics come into play. Valence electrons behaviour retrieved from Compton scattering profiles analysis is a particular example of the inadequacy of one centre models. Therefore, more than a decade ago [20] the construction of Bloch functions from "dressed anions" was suggested. The purpose of this rather crude local model is to depart from the usual "bare anion" description of solids such as LiH, MgO, NaF [21] and include interaction between the central anion and its first neighbouring cations in order to build up a more flexible valence wave-function.

The physical argument underpinning this model is summarized by the following two step process. Firstly, a transfer of an electron from the cation to the anion is assumed. Secondly, it is expected that the newly populated valence orbital will tend to expand until overlaps with other surrounding anions become significant. Consequently, the Pauli and electrostatic repulsions will redistribute part of the charge towards available (i.e. not so jammed) directions. In an initial approximation, one could then consider that the valence electron will tend to be partially shared between the central ion and a surrounding cage constituted by the first cation neighbours. In a local description of the wave-function, it thus seems appropriate to describe the valence electrons not only by a mere anion orbital but with a mixture of orbitals pertaining to both the anion and the cation. Once more, this procedure allows for the inclusion of two-centre terms in the elaboration of an approximate 1-RDM.

Some details on the procedure that was adopted to jointly refine a local model from structure factors (x-rays and electron diffraction origin) [22, 23] and directional Compton profiles for crystalline FCC Magnesium Oxide [24] now follow.

Valence electrons are those commonly attributed to the 2s and 2p Oxygen orbitals. They are now described by Bloch functions respectively built from the dressed ion local functions with s and p-type symmetries:

$$\psi_0(\mathbf{r}) = N_0\left[\phi_0(\kappa_{O_{2s}}\mathbf{r}) + \lambda_s\chi_0\left(\kappa_{Mg_{3s}}\mathbf{r}\right)\right] \tag{7.12}$$

for the 2s-dressed orbital and

$$\psi_j(\mathbf{r}) = N_j\left[\phi_j\left(\kappa_{O_{2p}}\mathbf{r}\right) + \lambda_p\chi_j\left(\kappa_{Mg_{3s}}\mathbf{r}\right)\right] \quad \text{for } j = x, y, z \tag{7.13}$$

for the 2p-dressed orbitals. The N_0 and N_j are normalization constants. The κ and λ are parameters adjusted by fitting with experimental data. They respectively represent the necessity to rescale the extension of the original atomic functions and the amount of mixing between the anion function and its neighbours. This last parameter is thus expected to play a crucial role since it allows for fine tuning of two-centre term contributions to the model. The scaled ϕ_0 and ϕ_j oxygen orbitals are surrounded by the cation cage functions, respectively χ_0 and χ_j. The latter are built from the first Magnesium 3s valence orbitals φ_0 with appropriate symmetry:

$$\chi_0(\mathbf{r}) \propto \sum_{j=x,y,z} \varphi_0\left(\mathbf{r} - \frac{a}{2}\mathbf{e}_j\right) + \varphi_0\left(\mathbf{r} + \frac{a}{2}\mathbf{e}_j\right) \tag{7.14}$$

and

$$\chi_j(\mathbf{r}) \propto \varphi_0\left(\mathbf{r} - \frac{a}{2}\mathbf{e}_j\right) - \varphi_0\left(\mathbf{r} + \frac{a}{2}\mathbf{e}_j\right) \quad \text{with } j = x, y, z \tag{7.15}$$

All atomic functions were obtained by means of a density functional theory (DFT) program constructed by P. Cortona [25]. The virtue of such a program is that it is particularly well suited to crystalline systems with a cubic structure. The global system is partitioned into subsystems, typically each atom, placed into a spherically averaged potential created by the other subsystem. Kohn-Sham type orbitals are then obtained for each subsystems. Therefore, the net result is that these orbitals, by construction, already account for the effective surrounding potential and should exhibit a fair radial behaviour.

The valence model momentum density thus writes:

$$n_{\text{model}}(\mathbf{p}) = \sum_{\mu\nu} \left[\boldsymbol{\sigma}^{-1}(\mathbf{p})\right]_{\nu\mu} \tilde{\psi}_{\mu}^{*}(\mathbf{p})\, \tilde{\psi}_{\nu}(\mathbf{p}) \tag{7.16}$$

Where $\boldsymbol{\sigma}(\mathbf{p})$ is the overlap matrix in momentum space. It is constructed as:

$$\boldsymbol{\sigma}_{\nu\mu}(\mathbf{p}) = \sum_{\mathbf{R}} \mathbf{s}_{\nu\mu}(\mathbf{R})\, e^{i\mathbf{p}.\mathbf{R}} \tag{7.17}$$

where the sum runs over all the lattice vectors of direct space. The overlap in position space is:

$$\mathbf{s}_{\nu\mu}(\mathbf{R}) = \int \psi_{\mu}^{*}(\mathbf{r} - \mathbf{d}_{\mu})\, \psi_{\nu}(\mathbf{r} - \mathbf{d}_{\nu} - \mathbf{R})\, d\mathbf{r} \tag{7.18}$$

where explicit reference was made to the position of each function in the unit cell. Valence model directional Compton profiles (DCP) (*cf.* Chap. 2) can thus be obtained by numerical integration over the plane perpendicular to the scattering vector for a given momentum component in the crystallographic direction **u**:

$$J_{\text{model}}(\mathbf{u}, q) = \int n_{\text{model}}(\mathbf{p})\, \delta(\mathbf{u}.\mathbf{p} - q)\, d\mathbf{p} \tag{7.19}$$

Model structure factors are easier to compute, providing the assumption that overlaps between anions dressed orbitals have little impact on this type of observables.

Table 7.1 Numerical values for the parameters of the local model after refinement using different strategies

Strategy	κ_{O2s}	κ_{O2p}	λ_s	λ_p	$R_{wF}(\%)$	$R_{wJ}(\%)$
χ_J^2	0.86	1.08	−0.19	0.19	0.6	4.6
$L_{F,J}$	0.84	1.09	−0.17	0.17	0.2	4.8
$L_{F,REMD}$	0.90	1.07	−0.14	0.17	0.17	4.5

Owing to the fact that Mg and O bear similar isotropic Debye-Waller factors in addition to the expected limited weight of the cations in the local wave function, a unique B_O factor was used for the dressed anion, yielding:

$$F_{\text{model}}(\mathbf{Q}) = 4\left[f_{O_{\text{core}}}(\mathbf{Q}) + f_{\text{dress}}(\mathbf{Q})\right] e^{-B_O\left(\frac{Q}{4\pi}\right)^2} + 4(-1)^{h+k+l} f_{Mg_{\text{core}}}(\mathbf{Q})\, e^{-B_O\left(\frac{Q}{4\pi}\right)^2} \tag{7.20}$$

with the model valence form factor for the dressed anion:

$$f_{\text{dress}}(\mathbf{Q}) = 2 \sum_{j=0,x,y,z} \int \psi_j^*(\mathbf{r})\, \psi_j(\mathbf{r})\, e^{i\mathbf{Q}.\mathbf{r}} d\mathbf{r} \tag{7.21}$$

Given the limited number of available structure factors relevant to valence electrons and, on the contrary, the large amount of Compton data, the most probable values for $\kappa$and λ parameters are obtained following the procedure detailed in Chap. 5, by minimizing :

$$L = N_{\text{DCP}} \log\left[\sum_{\mathbf{u}} \sum_{j=1}^{N_{\mathbf{u}}} w_j \left|J_{\text{model}}(\mathbf{u}, q_j) - J_{\text{exp}}(\mathbf{u}, q_j)\right|^2\right] + N_{\text{SF}} \log\left[\sum_{\mathbf{Q}} w_{\mathbf{Q}} \left|F_{\text{model}}(\mathbf{Q}) - F_{\text{exp}}(\mathbf{Q})\right|^2\right] \tag{7.22}$$

where N_{DCP} andN_{SF} are respectively the total number of Compton and diffraction data points. $N_{\mathbf{u}}$ represents the number of uncorrelated experimental points in the Compton spectrum for the scattering vector along a crystallographic direction **u**.

Minimization of such a quantity has been shown to offer a fair balance between data of different experiments. In particular, it allows for an overall scale factor uncertainty regarding the weighting of data points.

Table 7.1 summarizes the results for different refinement strategies. Minimization of χ_J^2 refers to the use of three DCP and no structure factors whereas genuine joint refinements using DCPs and structure factors (x-rays and convergent beam electron

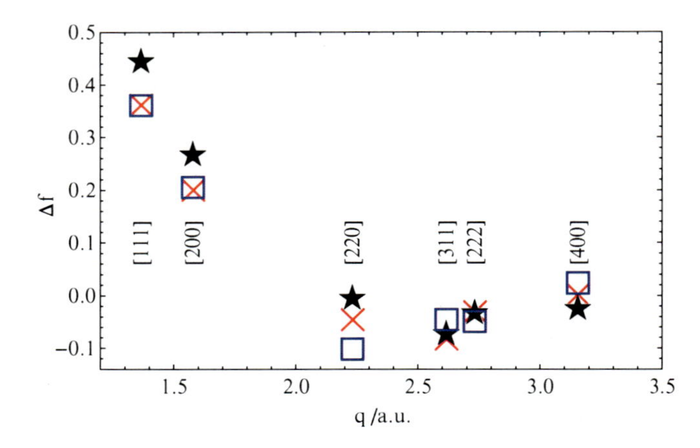

Fig. 7.7 Comparison between ab-initio (*stars*), refined (*crosses*) and experimental (*boxes*) structure factors in Magnesium Oxide

diffraction – CBED) were carried out either by numerical integration of the model in momentum space($L_{F,J}$) or by a prior reconstruction of experimental momentum density ($L_{F,REMD}$).

As expected, κ values, which drive the extension of individual atomic orbitals turn out to be mostly dependent on the diffraction data. Figure 7.7 illustrates the impact of the joint refinement on structure factors. In order to amplify small differences between structure factors, emphasis is put on valence electrons contribution by plotting:

$$\Delta f\,(\mathbf{Q}) = (-1)^{h+k+l}\left\{F\,(\mathbf{Q}) - 4\left[f_{\mathrm{Mg}^{2+}}\,(\mathbf{Q}) - (-1)^{h+k+l}f_{\mathrm{O}}\,(\mathbf{Q})\right]\right\}/4 \quad (7.23)$$

where the form factors for O and Mg^{2+} correspond to isolated systems.

A refinement using only inelastic scattering data did not significantly modify the κ values. However, the results for the mixing parameters λ proved to be strongly conditioned by the additional richness of Compton spectra. Indeed, although the mixing parameters appear to take rather weak values and can be interpreted (in a Mulliken point of view) as only 3% relative weight from the cation cage, they proved to play an essential role in the fair reproduction of Compton anisotropies for the (100) and (110) directions (Figs. 7.8 and 7.9). The simplicity of the model appears to be ineffective to account for the behaviour in the (111) direction (Fig. 7.10).

Owing to the weakness of the mixing effect, parameter κ_{Mg3s} could not be refined. Therefore, the value was kept equal to the unity.

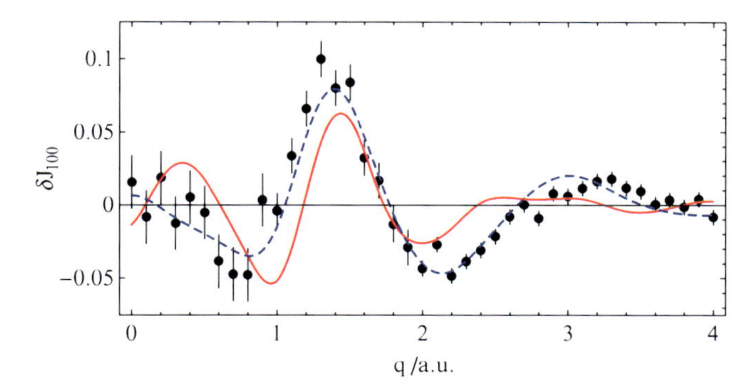

Fig. 7.8 Comparison between ab-initio (*dashed*), refined (*solid*) and experimental (*dots with error bars*) directional Compton profiles in Magnesium Oxide. In order to emphasize small differences between DCP, only absolute anisotropy (*ACP*) for direction (100) is plotted: $\delta J_u(q) = J\,(\mathbf{u}, q) - J_{\mathrm{iso}}(q)$

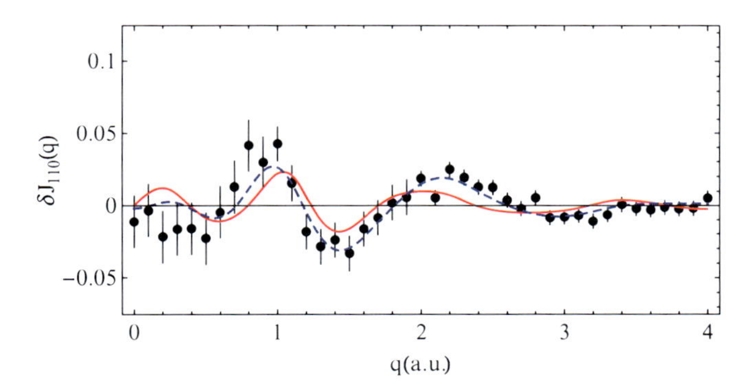

Fig. 7.9 Same as Fig. 7.8 for direction (110)

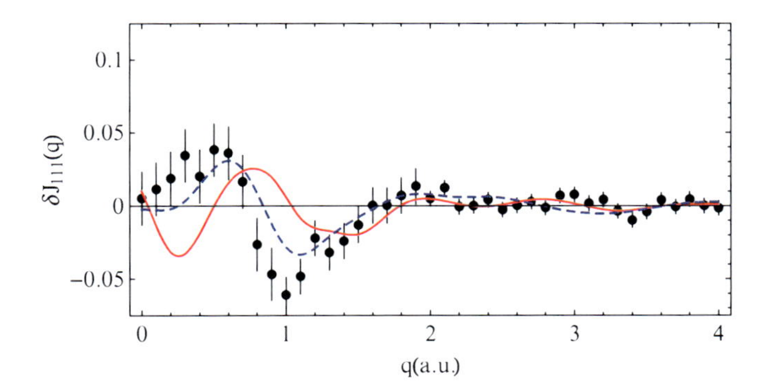

Fig. 7.10 Same as Fig. 7.8 for direction (111)

7.4 Conclusion

Local models for a wave-function have proved to be useful in at least two cases.

Firstly, they provide a convenient way to carry out computations on position and momentum densities for condensed systems when periodic codes are not fully adapted. Local models allow for a flexible account of the influence of the environment on a particular atom or molecule in the solid. This property is particularly valuable for disordered systems and very satisfactory results were obtained for Ice Ih, bringing some clarification on the character of hydrogen bond. Other studies [6] have demonstrated the ability of such method to take into account part of the correlation effects and, for example, improve significantly the agreement between total Compton profiles in ionic solids.

Secondly, a local model including adjustable parameters, with a very limited range of inter-atomic interaction, is also well suited to a refinement against a set of experimental data. In the case of ionic solids (MgO and LiH), the effect of two-centre contributions, that were often considered as negligible, actually confirmed to play a major role in the explanation of the oscillations observed in Compton anisotropies.

Acknowledgments The author warmly acknowledges P. J. Becker for his major contribution to this field and fruitful discussions over the years. The work of C. Fluteaux, S. Ragot and B. Courcot should over all be acknowledged. It was during their PhD time in our group that most of the reported results were obtained. P. Cortona is also thanked for stimulating exchanges and for providing high quality atomic functions.

References

1. Coleman AJ, Yukalov VL (2000) Reduced density matrices: Coulson's challenge. Springer, New York; Coulson C (1960) Present state of molecular structure calculations. Rev Mod Phys 32:170–177
2. Hansen NK, Coppens P (1978) Testing aspherical atom refinements on small molecule data sets. Acta Crystallogr A 34:909–921
3. Becker P, Gillet J-M, Cortona P, Ragot S (2001) Complementary aspects of charge and momentum densities for the study of the chemical bond. Theor Chem Acc 105:284–291
4. Dapprich S, Komáromi I, Byun KS, Morokuma K, Frisch MJ (1999) A new ONIOM implementation in Gaussian 98. Part 1. The calculation of energies, gradients and vibrational frequencies and electric field derivatives. J Mol Struct (Theochem) 462:1–21
5. Yang W, Zhou Z (1994) Electronic structure of solid-state systems via the divide and-conquer method. In: Ellis D (ed) Electronic functional theory of molecules, clusters, and solids. Kluwer Academic Publishers, Dordrecht, pp 177–188
6. Ragot S (2001) Matrices densité: modélisation des densités de charge et d'impulsion. Prédiction des propriétés des solides. Ph.D. Thesis, École Centrale Paris, France; Ragot S, Gillet JM, Becker PJ (2002) Local correlation mechanisms in ionic compounds: comparison with x-ray scattering experiments. J Chem Phys 117:6915–6922 and references therein
7. Pauling L (1935) The structure of entropy of Ice and of other crystals with some randomness of atomic arrangement. J Am Chem Soc 57:2680–2684

8. Isaacs ED, Shukla A, Platzman PM, Hamann DR, Barbiellini B, Tulk CA (1999) Covalency of the hydrogen bond in ice: a direct X-ray measurement. Phys Rev Lett 82:600–603
9. Bernal JD, Fowler RH (1933) A theory of water and ionic solution, with particular reference to hydrogen and hydroxyl ions. J Chem Phys 1:515–548
10. Owsten PG (1958) The structure of ice as determined by x-ray and neutron diffraction analysis. Adv Phys 7:171–188
11. Peterson SW, Levy HA (1957) A single-crystal neutron diffraction study of heavy ice. Acta Crystallogr 10:70–76
12. van Beek CG, Overeem J, Ruble JR, Craven BM (1996) Electrostatic properties of ammonium fluoride and deuterated ice-I_h. Can J Chem 74:943–950
13. Espinosa E, Molins E, Lecomte C (1998) Hydrogen bond strengths revealed by topological analysis of experimentally observed electron densities. Chem Phys Lett 285:170–173
14. Dyke TR, Muenter JS (1974) Microwave spectrum and structure of hydrogen bonded water dimer. J Chem Phys 60:2929–2930
15. Gatti C, Silvi B, Colonna F (1995) Dipole moment of the water molecule in the condensed phase: a periodic Hartree-Fock estimate. Chem Phys Lett 247:135–141
16. Batista ER, Xantheas SS, Jónsson H (1998) Molecular multipole moments of water molecules in ice Ih. J Chem Phys 109:4546–4551; Batista ER, Xantheas SS, Jónsson H (1999) Multipole moments of water molecules in clusters and ice Ih from first principles calculations. J Chem Phys 111:6011–6015; Batista ER, Xantheas SS, Jónsson H (2000) Electric field in ice and near water clusters. J Chem Phys 112:3285–3292
17. Ragot S, Gillet JM, Becker PJ (2002) Interpreting Compton anisotropy of ice Ih: a cluster partitioning method. Phys Rev B 65:235115
18. Ghanty TK, Staroverov VN, Koren PR, Davidson ER (2000) Is the hydrogen bond in water and ice covalent? J Am Chem Soc 122:1210–1214
19. Gillet JM, Cortona P (1999) Analysis of the MgO structure factors. Phys Rev B 60:8569–8574
20. Gillet JM, Becker PJ, Loupias G (1995) The refinement of anisotropic Compton profiles and of momentum densities. Acta Crystallogr A 51:405–413
21. Ramirez BI, McIntire WR, Matcha RL (1977) Theoretical Compton profile anisotropies in molecules and solids. II. Application of the MSC procedure to lithium hydride. J Chem Phys 66:373–376
22. Zuo JM, O'Keefe M, Rez P, Spence JCH (1997) Charge density of MgO: implications of precise new measurements for theory. Phys Rev Lett 78:4777–4780
23. Lawrence JL (1973) Debye-Waller factors for magnesium oxide. Acta Crystallogr A 29: 94–95
24. Fluteaux C (1999) Contribution to the study of electron density of ionic compounds by x-ray inelastic scattering. Ph.D. thesis, École Centrale Paris, Paris. Note: Valence profiles were evaluated through subtraction of a calculated Hartree-Fock core contribution from the total profiles; Gillet J-M, Fluteaux C, Becker PJ (1999) Analytical reconstruction of momentum density from directionl Compton profiles. Phys Rev B 60(4):2345–2349
25. Cortona P (1992) Direct determination of self-consistent total energies and charge densities of solids: a study of the cohesive properties of the alkali halides. Phys Rev B 46:2008–2014

Chapter 8
Magnetization Densities in Material Science

Béatrice Gillon and Pierre Becker

8.1 Introduction

Polarized neutron diffraction (PND) is the most widely used technique for magnetization density investigations as it gives access to the density at any point of the unit cell and can be applied for various magnetic crystalline materials including large molecular compounds [1]. The development of Magnetic Compton Scattering (MCS) (Chap. 4), Xray Magnetic Diffraction (XMD) and Xray Magnetic Circular Dichroism (XMCD) brings novel insights about the L/S separation in materials of interest in condensed matter physics, like on uranium intermetallic compounds [2–4], transition metal compounds [5, 6] and strongly correlated electrons systems [7, 8] also investigated by PND [9–11].

Molecular magnetic materials constitute a relatively recent field for PND studies. This rapidly evolving field receives growing attention because of the richness of the magnetic properties encountered in these new materials. In contrast with magnetic compounds like transition metal or rare earth intermetallic compounds and oxides, the spin distribution in molecule-based magnets is delocalized on a large number of atoms. MCS and XMD are not convenient for studying weak delocalized magnetization densities but XMCD and solid state NMR or Electron Paramagnetic Resonance are alternative methods to probe local moments in this type of materials [12, 13]. Magnetization density (MD) studies allow for exploring the electronic structure of open-shell molecular systems and the magnetic interaction mechanisms which may

B. Gillon (✉)
Laboratoire Léon Brillouin (CEA-CNRS), Centre d'Etudes de Saclay, 91191 Gif-sur-Yvette, Cedex, France
e-mail: Beatrice.gillon@cea.fr

P. Becker
Structures, Properties and Modelling of Solids (SPMS) Laboratory, Ecole Centrale Paris, Grande Voie des Vignes, 92295, Chatenay-Malabry, France
e-mail: pierre.becker@ecp.fr

C. Gatti and P. Macchi (eds.), *Modern Charge-Density Analysis*,
DOI 10.1007/978-90-481-3836-4_8, © Springer Science+Business Media B.V. 2012

lead to magnetic ordering. Extensive studies were devoted to paramagnetic organic radicals and organic magnets [14–16] as well as organometallic compounds forming extended magnetic structures or molecular clusters of nanometric size [17]. Charge density studies bring complementary outsights to MD studies of magnetic materials. Separate charge and spin densities studies were performed for various compounds [18–23].

The first section reviews the modern trends in the polarized neutron diffraction technique. In the second section, the methods for PND data analysis and magnetization densities reconstruction are described. We report in the third section some recent applications of PND to the field of molecule-based magnetism.

8.2 Polarized Neutron Diffraction

Magnetization density studies were traditionally performed in ferro- and ferrimagnetic states [24] but are now often realized in induced paramagnetic states or antiferromagnetic (AF) states. In this section we recall the principles of the flipping ratio method and overview the recent instrumental developments achieved in PND.

8.2.1 The Flipping Ratio Method

The experimental quantities measured by PND are the so-called flipping ratios $R(\boldsymbol{\kappa})$ between the respective diffracted intensities $I_+(\boldsymbol{\kappa})$ and $I_-(\boldsymbol{\kappa})$ by a magnetically ordered single crystal for a polarization vector $\mathbf{P}$ of the incident neutron beam parallel or antiparallel to the vertical applied magnetic field [25, 26]:

$$R(\boldsymbol{\kappa}) = \frac{I_+(\boldsymbol{\kappa})}{I_-(\boldsymbol{\kappa})} = \frac{F_N^2 + 2\mathbf{P}.(F_N'\mathbf{Q}_\perp' + F_N''\mathbf{Q}_\perp'') + Q_\perp^2}{F_N^2 - 2\mathbf{P}.(F_N'\mathbf{Q}_\perp' + F_N''\mathbf{Q}_\perp'') + Q_\perp^2} \tag{8.1}$$

$F_N(\boldsymbol{\kappa})$ is the nuclear structure factor and $\mathbf{Q}_\perp(\boldsymbol{\kappa})$ is the magnetic interaction vector: $\mathbf{Q}_\perp(\boldsymbol{\kappa}) = \hat{\boldsymbol{\kappa}} \times \mathbf{F_M}(\boldsymbol{\kappa}) \times \hat{\boldsymbol{\kappa}}$, where $\hat{\boldsymbol{\kappa}}$ is a unit vector parallel to the scattering vector $\boldsymbol{\kappa}$. These are complex quantities with real ($'$) and imaginary ($''$) parts.

The magnetic structure factor $\mathbf{F_M}(\boldsymbol{\kappa})$ is a vector, the direction of which is the magnetic moment direction $\boldsymbol{\mu}$ resulting from the sum of the atomic moments $\boldsymbol{\mu}_i$ due to spin and orbit in the unit cell. Its magnitude is related to the magnetization density $m(\mathbf{r})$ by Fourier transform:

$$\mathbf{F_M}(\boldsymbol{\kappa}) = \boldsymbol{\mu} \int_{\text{cell}} m(\mathbf{r})\, e^{i\boldsymbol{\kappa}\mathbf{r}} d\mathbf{r} \tag{8.2}$$

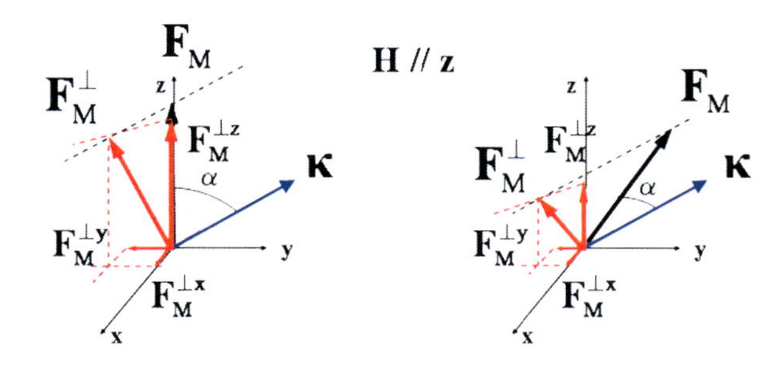

Fig. 8.1 *Left*: collinear case, *Right*: non-collinear case

$m(\mathbf{r})$ is the sum of the spin-only magnetization density $s(\mathbf{r})$ and the orbital magnetization density $L(\mathbf{r})$:

$$m(\mathbf{r}) = s(\mathbf{r}) + L(\mathbf{r}) \tag{8.3}$$

When atomic magnetic moments $\boldsymbol{\mu}_i$ are collinear with the applied magnetic field, the magnetic structure factor $\mathbf{F_M}(\boldsymbol{\kappa})$ is parallel to the vertical magnetic field (see Fig. 8.1) and the expression of the flipping ratio becomes:

$$R(\boldsymbol{\kappa}) = \frac{F_N^2 + 2Pq^2(F_N' F_M' + F_N'' F_M'') + q^2 F_M^2}{F_N^2 - 2Peq^2(F_N' F_M' + F_N'' F_M'') + q^2 F_M^2} \tag{8.4}$$

where e is the flipping efficiency and $q^2 = (\sin\alpha)^2$, α being the angle between the vertical magnetization direction and the scattering vector. In the case of centric space groups, that means real structure factors, the experimental $F_M(\boldsymbol{\kappa})$'s can directly be obtained from the flipping ratios.

In strongly anisotropic paramagnetic compounds, non collinear magnetic moments (Fig. 8.1 right) may exist on different atomic sites and the general expression (8.1) must be used. The local susceptibility tensor approach was developed to treat the PND measurements in such cases [27] (see Sect. 8.3.3).

8.2.2 *Recent Progress in PND*

The interest of PND for material science research was renewed by the important developments and improvements recently achieved in the field of neutron polarizers (i.e. ^{3}He filter), position sensitive detector for PND, pulsed sources, powder PND, irradiation in-situ. Up to now, the large majority of magnetization density studies using PND were performed on single crystals (see Sect. 8.4) but recent progress in the experimental devices will allow polarized neutron powder diffraction in a near future.

8.2.2.1 Instrumental Developments

In the two last decades most of PND studies were performed on one of the three polarized neutron diffractometers located at ILL (Institut Laue Langevin, Grenoble) or at LLB (Laboratoire Léon Brillouin, Saclay). These are two-axis diffractometers with lifting-arm detector using classical single crystal polarizers: D3 (ILL) installed on the hot source, D23 (ILL) on a thermal guide and 5C1 (LLB) on a hot channel. Two other two-axis thermal diffractometers are now available for PND: E4 at the BERII reactor in Berlin and recently SUPER6T2 [28] at LLB.

The most important progress realized in neutron polarization techniques in the past years is the development of ^{3}He spin filters [29, 30] at ILL, which led to implementation of polarized ^{3}He facilities in several neutron centres. The use of long-time relaxation polarized ^{3}He cells for obtaining polarized neutrons allows for relatively easily adding a polarized option to existing single crystal diffractometers, as for example POLI-HEiDi at the FRMII reactor in Münich [31]. For pulsed neutron beams classical single crystal polarizers are not of interest because of the large wavelength distribution versus time of the neutrons pulses, while ^{3}He spin filters are efficient [32]. New PND instruments are under construction on pulsed sources: TOPAZ at SNS (Spallation Neutron Source) in Oakridge and WISH at ISIS. Important improvement was also achieved concerning polarizing guides that can be used for both continuous and pulsed neutron beams. The implementation of a polarizing thermal guide on 6T2 at LLB provides benefit from the high thermal flux for PND. Moreover polarizing guides provide an interesting alternative to ^{3}He polarizer for pulsed neutron beams.

The spherical polarimetry technique developed at ILL [33, 34] to determine the absolute spin configuration in complex magnetic structures gives now access to AF form factors and magnetization density in AF materials [35].

Another new trend is the use of position sensitive detector (PSD) for flipping ratio measurements, like on E4 and SUPER6T2, which leads to an important gain in time in the PND starting procedure and data collection. The hot neutron diffractometer 5C1 at LLB will be equipped in a very next future with a curved bi-dimensional detector.

8.2.2.2 Recent Advances in Powder PND

Large single crystals for PND are not always possible to grow, particularly in the case of molecule-based magnets and compounds of biological interest. A method was proposed for determining the induced magnetization density on powder samples using unpolarized neutrons and applied to the determination of the Mn^{II}/Mn^{III} moment distribution in the paramagnetic single molecule magnet $[Mn_{12}]$ [36]. Promising polarized neutron powder diffraction measurements were recently realized at ILL thanks to the ^{3}He neutron spin filter which permits to achieve a good homogeneity of polarization (and magnetic field) over a large volume of powder. An experiment on the mixed-valence prussian blue compound $Fe_4^{III}[Fe^{II}(CN)],xD_2O$ on the powder diffractometer D1B permitted to evidence for the induced moments on the Fe^{II} and Fe^{III} sites on the magnetization density map [37]. Facilities for powder PND on new high-resolution powder diffractometer D20 at ILL will become soon available as well as on the future instruments TOPAZ and WISH on pulsed sources.

8.2.2.3 Light-Coupled PND

Irradiation of photomagnetic compounds by light of suitable wavelength induces a transition from the ground state towards a long-time living metastable magnetic state at low temperatures. An experimental setup allowing both in-situ light irradiation via an optical fibre and PND measurements was developed at LLB for light-coupled PND experiments devoted to the study of magnetization densities of photoexcited magnetic states in molecular compounds [38].

8.3 Modern Trends in Magnetization Density Reconstruction and Analysis

A major progress in PND data treatment was achieved by the application of the MEM to magnetization density mapping [39, 40]. Refinement methods based on density and atomic orbital models were generalized to non centric space groups [15]. The MEM and model refinement methods were until now applied to collinear magnetic moments only. Other important improvements were realized in PND data analysis of strongly anisotropic paramagnetic materials for which atomic magnetic moments are not collinear to the applied field [27, 41]. The method based on the local site magnetic susceptibility tensors approach provides a powerful tool for understanding the role played by the different sites carrying magnetism in the crystal cell.

8.3.1 Maximum of Entropy Method

The theoretical outlines of MEM charge density reconstruction from a limited set of N data points are described in Chap. 13. For charge density reconstruction [42], the density was modelled by a discrete distribution $\rho\left(x_j, y_j, z_j\right)$ using a grid of $M = m^3$ pixels dividing the crystallographic cell and the probability p_j is the density in pixel j normalized to 1:

$$p_j = \frac{\rho(x_j, y_j, z_j)}{\rho_{cell}} \; with \; \rho_{cell} = \sum_{j=1}^{M} \rho(x_j, y_j, z_j), \qquad (8.5)$$

At the difference to charge density which is always positive, magnetization density may be negative in some regions of the unit cell. Because the expression of the entropy given in (4.17) is valid for positive quantities p_j only, a two channels grid of 2M points is considered with $q_j = \rho_{\uparrow}(x_j, y_j, z_j)/\rho_{cell}^{\uparrow}$ for j = 1 to M and $q_j = \rho_{\downarrow}(x_j, y_j, z_j)/\rho_{cell}^{\downarrow}$ for j = M to 2M which leads to the expression of entropy:

$$S = -\sum_{j=1}^{2M} q_j \, Log\,(q_j) \tag{8.6}$$

The normalized MD, m_k, in pixel k, is the difference $m_k = q_k - q_{k+M}$(k = 1, M). The discrete distribution m is used to calculate the magnetic structure factors $F_M^{cal}(\boldsymbol{\kappa}_n)$ for N observations with scattering vectors $\boldsymbol{\kappa}_n$, n = 1 to N. This leads to an under-dimensioned system of N linear equations with M unknown quantities m_k, k = 1, M. The MEM method yields the solution m which maximizes the entropy (8.6) and minimizes the RMS by a least-square procedure, starting from a prior uniform density:

$$\chi^2(m) = \frac{1}{N}\sum_{i=1}^{N}\frac{1}{\sigma_i{}^2}\left|F_M(\boldsymbol{\kappa}_i) - F_M^{cal}(\boldsymbol{\kappa}_i)\right|^2 \tag{8.7}$$

The MEM method was also extended to non centric cases [40]. This method yields direct information on the location of the magnetic moments in the cell with their signs and on the nature of the orbitals involved in magnetism reflected by the shape of the distribution, as shown in Fig. 8.2 for the $Ca_{1.5}Sr_{0.5}RuO_4$ ruthenate compound [43]. The MEM magnetization density reconstruction reveals a strongly anisotropic density at the Ru site, consistent with the distribution of the d_{xy} (t_{2g} band) orbitals. The magnetization density on the in-plane oxygen arises from the strong coupling between the Ru 4*d* and O 2*p* orbitals.

8.3.2 Refinement Models

A major advantage of MEM is that it is a model free method. However when weak delocalized moments coexist beside strong moments, the reconstructed map is totally dominated by the large magnetization density peaks. Therefore model refinements remain necessary for a detailed analysis of the magnetization density in the case of delocalized densities like in molecule-based compounds.

The magnetization density in the unit cell is assumed to be equal to the sum of independent pseudo-atomic MD's, $m_i(\mathbf{r}_i)$, centered on atom i, in position $\mathbf{R}_i$, with $\mathbf{r}_i = \mathbf{r} - \mathbf{R}_i$. The magnetic structure factors are then written as a discrete sum on the n_a atoms in the cell [25]:

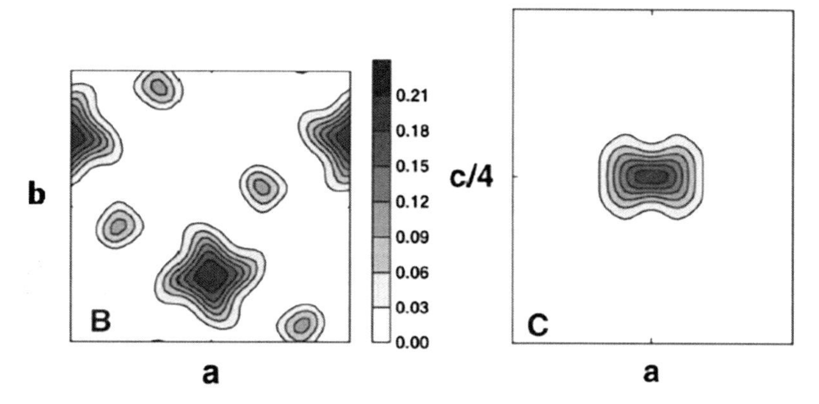

Fig. 8.2 Maximum entropy reconstruction of the magnetization distribution in $Ca_{1.5}Sr_{0.5}RuO_4$ at 1.6 K and 7 T. *Left*: section passing through the RuO_2 plane at z = 1/8, *right*: section perpendicular to the plane through the Ru position (Reprinted from Ref. [44] with kind permission of The American Physical Society)

$$\mathbf{F}_{\mathrm{M}}(\boldsymbol{\kappa}) = \sum_{\mathrm{i}=1}^{\mathrm{n_a}} \mu_i F^i_{mag}(\boldsymbol{\kappa})\, \mathrm{e}^{\mathrm{i}\boldsymbol{\kappa}\mathbf{R}_{\mathrm{i}}} \mathrm{e}^{-\mathrm{W}_{\mathrm{i}}} \tag{8.8}$$

where $F^i_{mag}(\boldsymbol{\kappa})$is the normalized magnetic form factor of pseudo-atom i, carrying a magnetic moment $\boldsymbol{\mu}_{\mathrm{i}}$ and $\mathrm{W_i}$ is the Debye-Waller factor of this pseudo-atom:

$$F^i_{mag}(\boldsymbol{\kappa}) = \frac{1}{\mu_i} \int m_i(\mathbf{r}) \mathrm{e}^{\mathrm{i}\boldsymbol{\kappa}r} \mathrm{d}\mathbf{r} \tag{8.9}$$

We consider the case of magnetic moments collinear to applied field in this section, for which expression (8.4) of the flipping ratio applies. Magnetization density refinement methods were first applied to magnetic structure factors data restricted to centric structures and later generalized to non centric structures by fitting the theoretical flipping ratios on the raw experimental data, which is now the standard procedure in all available refinement programs. In expression (8.4), the nuclear structure factor is known from the crystal structure and the magnetic structure factor is expressed by an analytical expression provided by the model. A least-squares refinement is performed by comparison between the calculated and the experimental flipping ratios. The magnetization density is then reconstructed using the refined model parameters. Two different models can be used depending on whether the model is applied to the unpaired electron AO or to MD. We first describe the models when MD is due to spin only, then when spin and orbital contributions coexist.

8.3.2.1 Pure Spin Contribution

The MD is the spin density when no orbital contribution exists. This is the case for organic radicals, as magnetism is due to 2p unpaired electrons and the orbital moment is zero. Some transition metal ions also present zero angular momentum like the Mn^{2+} ion with half filled 3d shell ($3d^5$, ${e_g}^2{t_{2g}}^3$, S = 5/2) or the Cr^{3+} ion in an octahedral field ($3d^3$, ${t_{2g}}^3$, S = 3/2), as well as the rare earth ion Gd^{3+} ($4f^7$, S = 7/2). The atomic magnetic moment is given by:

$$\boldsymbol{\mu}_i = g_s\, S_i \tag{8.10}$$

where g_s is the Landé factor and $\mathbf{S}_i$ the atomic spin operator.

Atomic Orbital Model

The general case consists in a molecular open-shell system of n_a atoms with n unpaired electrons. The UHF scheme has to be used (when remaining in the mean field approach) to account for the spin polarization effects which are responsible for negative spin density regions. Different $\phi_k^{\uparrow,\downarrow}(\mathbf{r})$ molecular orbitals (MO) describe the $n_\uparrow$ electrons with spin up and $n_\downarrow$ electrons with spin down respectively and the spin density is given by the difference [26]:

$$s(\mathbf{r}) = \rho^{\uparrow}(\mathbf{r}) - \rho^{\downarrow}(\mathbf{r}) = \sum_k^{n\uparrow} \left|\phi_k^{\uparrow}(\mathbf{r})\right|^2 - \sum_k^{n\downarrow} \left|\phi_k^{\downarrow}(\mathbf{r})\right|^2 \tag{8.11}$$

The MO's are linear combinations of n_a atomic orbitals $\psi_i(\mathbf{r_i})$ centred on atom i:

$$\phi_k^{\uparrow,\downarrow}(\mathbf{r}) = \sum_{i=1}^{n_a} c_i^{\uparrow,\downarrow}\, \psi_i(\mathbf{r}_i) \tag{8.12}$$

In the atom independent approximation (neglecting interatomic cross terms), the development of expression (8.11) of the spin density using expression (8.12) of the MO's leads to a sum of atomic spin densities $s_i(\mathbf{r}_i)$ localized on atom i:

$$s_i(\mathbf{r}_i) = p_i \left|\psi_i(\mathbf{r}_i)\right|^2 \tag{8.13}$$

p_i is the atomic spin population:

$$p_i = \sum_k^{n\uparrow} {c_i^{\uparrow}}^2 - \sum_k^{n\downarrow} {c_i^{\downarrow}}^2 \tag{8.14}$$

Atomic spin populations therefore may be positive or negative and satisfy:

$$\sum_{i}^{n_a} p_i = n_\uparrow - n_\downarrow = 2M_S \tag{8.15}$$

Assuming that magnetism on atom i only involves one atomic shell with atomic quantum numbers N and L (L = 0 to N − 1, L = 0 for s orbitals, L = 1 for p orbitals and L = 2 for d orbitals), the AO expression is:

$$\psi_i\left(\mathbf{r_i}\right) = \mathcal{N} r_i^{N-1} e^{-\xi_L^i r_i} \sum_{\mathrm{M}=-\mathrm{L}}^{\mathrm{L}} \alpha_{LM}^i \mathcal{N}_{ang} Y_{LM}\left(\theta_i, \varphi_i\right) \tag{8.16}$$

M is the magnetic quantum number (M = −L to L), $Y_{LM}\left(\theta_i, \varphi_i\right)$ is a usual spherical harmonics, ξ_L^i is the Slater radial exponent and α_{LM}^i are the atomic orbital coefficients with the normalization condition:

$$\sum_{\mathrm{M}=-\mathrm{L}}^{\mathrm{L}} \left|\alpha_{LM}^i\right|^2 = 1 \tag{8.17}$$

The following expression of the normalized magnetic form factor is obtained from (8.10) and (8.14) using description (8.17) of the AO $\psi_i\left(\mathbf{r_i}\right)$:

$$F_{mag}^i(\boldsymbol{\kappa}) = p_i \sum_{\ell=0}^{\infty} \left\langle j_\ell^i\left(\kappa\right)\right\rangle \sum_{\mathrm{m}=-\ell}^{\ell} C_{\ell \mathrm{m}}^i Y_{\ell m}^*\left(\theta_\kappa, \varphi_\kappa\right) \tag{8.18}$$

where $\left\langle j_\ell^i\left(\kappa\right)\right\rangle$ is the radial integral for atom i, involving the spherical Bessel function $j_\ell^i\left(\kappa r\right)$ of ℓth order:

$$\left\langle j_\ell^i\left(\kappa\right)\right\rangle = \int_0^\infty \mathrm{r}^2 \left(\mathcal{N} r_i^{N-1} e^{-\xi_L^i r_i}\right)^2 j_\ell^i(\kappa r) \mathrm{d}r \tag{8.19}$$

The coefficients $C_{\ell m}^i$ are linear combinations of products of orbital coefficients $\alpha_{\mathrm{LM}}^{\mathrm{i}} \alpha_{\mathrm{L'M'}}^{\mathrm{i}}$ and the ℓ and m values are restricted by the rules: $\ell \leq 2$ L and m = M−M′ [1].

The general analytical description of the magnetic structure factor in the AO model is from (8.8) and (8.18):

$$F_M(\boldsymbol{\kappa}) = \sum_{\mathrm{i}=1}^{\mathrm{n_a}} m_i \sum_{\ell=0}^{\infty} \left\langle j_\ell^i\left(\kappa\right)\right\rangle \sum_{\mathrm{m}=-\ell}^{\ell} C_{\ell \mathrm{m}}^i Y_{\ell m}^*\left(\theta_\kappa, \varphi_\kappa\right) \mathrm{e}^{\mathrm{i}\boldsymbol{\kappa}\mathbf{R}_\mathrm{i}} \mathrm{e}^{-\mathrm{W_i}} \tag{8.20}$$

where m_i are the atomic magnetic moments $m_i = \mu_i p_i$.

The AO $\psi_i\left(\mathbf{r_i}\right)$ may also be written in a basis of real spherical harmonics $y_{\ell m\pm}^i$ with atomic orbital coefficients a_{LM}^i obeying to a normalization condition. This leads to the following expression of $F_M^{cal}(\boldsymbol{\kappa})$ analogous to (8.20) but where $y^i_{\ell m}$ is a real spherical harmonics:

$$F_M(\boldsymbol{\kappa}) = \sum_{\mathrm{i}=1}^{\mathrm{n_a}} m_i \sum_{\ell=0}^{\infty} \mathrm{i}^{\ell} \left\langle j^i_{\ell}(\kappa)\right\rangle \sum_{\mathrm{m}=-\ell}^{\ell} A^i_{\ell m} y_{\ell m}(\theta_\kappa,\varphi_\kappa)\, \mathrm{e}^{\mathrm{i}\boldsymbol{\kappa}\mathbf{R}_\mathrm{i}} \mathrm{e}^{-\mathrm{W_i}} \qquad (8.21)$$

where $A^i_{\ell m}$ are linear combinations of products of orbital coefficients $a^i_{LM} a^i_{L'M'}$.

Two different refinement procedures are possible:

- The magnetic form factor is taken from literature and the atomic moments m_i are refined. The radial integrals $< j_\ell(\kappa) >$ are tabulated for 3d and 4f ions [44]. The spherical approximation ($\ell = 0$) is often used for the magnetic form factor:

$$F_M(\boldsymbol{\kappa}) = \sum_{\mathrm{i}=1}^{\mathrm{n_a}} m_i \left\langle j^i_0(\kappa)\right\rangle \qquad (8.22)$$

- In order to obtain information on the atomic orbitals, expression (8.21) is used: the AO coefficients a^i_{LM} involved in the $A^i_{\ell m}$ factors and the Slater exponent ξ^i_L in the radial integral are refined as well as the atomic moments m_i. The initial values of the Slater exponents are generally taken from atomic wave function calculations reported in literature [45].

The sum of the refined atomic moments m_i over the $\mathrm{n_a}$ atoms yields an experimental value of the induced magnetic moment M_{exp} per molecule:

$$\sum_{i=1}^{n_a} m_i = M_{\mathrm{exp}}$$

The normalized atomic spin populations p_i are then deduced from the refined atomic moments m_i using the relation:

$$p_i = m_i \frac{2M_S}{M_{\mathrm{exp}}} \qquad (8.23)$$

Multipole Refinement

The multipolar method (MM) in the Hansen-Coppens formalism [46] was first applied to spin density for centric structures [47] then for non centric structures [48]. The core term in the well-known multipolar expansion of the charge density [49] does not exist for spin density because only valence shells contribute to magnetism. The multipolar development of the atomic spin density is:

$$s_i(\mathbf{r}_i) = P^i_0 \mathcal{N} r_i^{n_0} e^{-k'\varsigma^i_0 r_i} + \sum_{\ell=0}^{\ell_{\max}=4} \mathcal{N} r_i^{n_\ell} e^{-k''_i \varsigma^i_\ell r_i} \sum_{\mathrm{m}=-\ell}^{\ell} P^i_{\ell m} y^i_{\ell m}(\theta_i,\varphi_i) \qquad (8.24)$$

Most of the time, the second term alone in (8.24) is used to describe the contribution resulting from p or d-type unpaired electron orbitals. The first monopole permits to refine an additional contribution due to a second type of orbital [50]. The expression of the magnetic structure factor obtained by Fourier transform of the second term of the multipolar development (8.24) is:

$$F_M(\kappa) = \sum_{i=1}^{n_a} \sum_{\ell=0}^{\ell_{max}} i^{\ell} \langle j_{\ell}^{i}(\kappa) \rangle \sum_{m=-\ell}^{\ell} P_{\ell m}^{i} y_{\ell m}(\theta_{\kappa}, \varphi_{\kappa}) e^{i\kappa \mathbf{R}_i} e^{-W_i} \tag{8.25}$$

where $\langle j_{\ell}^{i}(\kappa) \rangle$ is the Fourier-Bessel transform of ℓth order of the density radial Slater function:

$$\langle j_{\ell}^{i}(\kappa) \rangle = \int_{0}^{\infty} r^2 \left(\mathcal{N} r_i^{n_\ell} e^{-k''_i \varsigma_\ell^i r_i} \right) j_{\ell}^{i}(\kappa r)\, dr \tag{8.26}$$

Expressions (8.25) and (8.21) of $F_M(\kappa)$ are identical with $P_{\ell m}^{i} = m_i A_{\ell m}^{i}$, at the condition that a same radial function is taken for all multipoles $\ell = 1$ to $\ell_{max} = 4$, with exponents $n_\ell = 2(N-1)$ and $\zeta_\ell^i = 2\xi_L^i$. This leads to relations between the multipole populations $P_{\ell m}^{i}$ and the products of orbital coefficients $a_{LM}^{i} a_{L'M'}^{i}$ via $A_{\ell m}^{i}$ for different types of orbitals (2p, 3d). For 3d orbitals, these relations are the same as those described for charge density in 3d transition metal complexes [51]. Due to the limitation to $\ell \leq 4$, the multipole model is not convenient to treat 4f magnetism.

8.3.2.2 Spin and Orbital Contributions

PND yields information on the sum of the spin and orbital contributions. In situations where orbital magnetism is present, it is necessary to model both spin and orbital contribution in order to determine the ratio μ_L/μ_S. The treatment of the orbital contribution in PND data analysis depends on the relative strength of spin-orbit coupling and crystal field.

For transition metal ions and actinides, the magnetic moment of an atom or ion i can be written as:

$$\boldsymbol{\mu}_i = g_s S_i + g_L L_i \tag{8.27}$$

where g_s and g_L are respectively the spin and orbital components summing up to the Landé splitting factor g ($g = g_s + g_L$), $\boldsymbol{\mu}_i$ is thus the total atomic magnetic moment, equal to the sum of the spin and orbit magnetic moments ($\mu = \mu_s + \mu_L$). The magnetic form factor is the sum:

$$F_{mag}^{i}(\kappa) = \frac{\mu_S^i}{\mu_i} F_{mag}^{iS}(\kappa) + \frac{\mu_L^i}{\mu_i} F_{mag}^{iL}(\kappa) \tag{8.28}$$

Within the dipole approximation [52], using a spherical model for the spin part, the magnetic form factor is expressed by:

$$F^i_{mag}(\boldsymbol{\kappa}) = \left\langle j^i_0(\kappa)\right\rangle + c_2\left\langle j^i_2(\kappa)\right\rangle \tag{8.29}$$

with $c_2 = \mu_L/\mu$ and $\mu_L/\mu_S = c_2/(1 - c_2)$.

This model is used in most of the recent studies on spin and orbital moments in actinides [53–55]. Beyond the spherical approximation for the spin form factor, the AO model or multipole model were applied for transition metal compounds [10, 56].

In organometallic complexes, the orbital contribution due to the transition metal ion is often treated as a correction using an estimated value of μ_L in (8.28) and the dipolar approximation (8.29). Both refinement models described for the pure spin case in Sect. 8.3.2.1 can be applied to retrieve the pure spin density.

For rare earth, the spin-orbit coupling is much stronger than crystal field [57] and the $\mathbf{S}_i$ operator in (8.10) must be replaced by the $\mathbf{J}_i$ operator with eigenvalue $J = |L \pm S|$:

$$\boldsymbol{\mu}_i = g_j\, J_i \tag{8.30}$$

If the crystal field splitting is small compared to the energy of exchange and that due to applied magnetic field, the induced MD does not depend on the orientation of the magnetic field with respect to the crystal field principal axes. The total magnetic form factor is described in the spherical approximation by the following expression:

$$F^i_{mag}(\boldsymbol{\kappa}) = \langle j_0(\kappa)\rangle + c_2\langle j_2(\kappa)\rangle \;\; with\; c_2 = \frac{J(J+1) + L(L+1) - S(S+1)}{3J(J+1) - L(L+1) + S(S+1)} \tag{8.31}$$

8.3.3 *Local Site Susceptibility Tensor for Non-collinear Magnetization*

The local site susceptibility tensor approach [27] was developed to treat PND data for paramagnetic materials containing atoms on same crystallographic positions but having different local magnetic anisotropies. The expression (8.4) of the flipping ratio cannot be used and the general expression (8.1) has to be considered. In the weak field region, where non-linear effects due to applied magnetic field are not important, the magnetic moment induced on atom i by the applied field H can be written as:

$$\boldsymbol{\mu}_i = \boldsymbol{\chi}_i \mathbf{H} \tag{8.32}$$

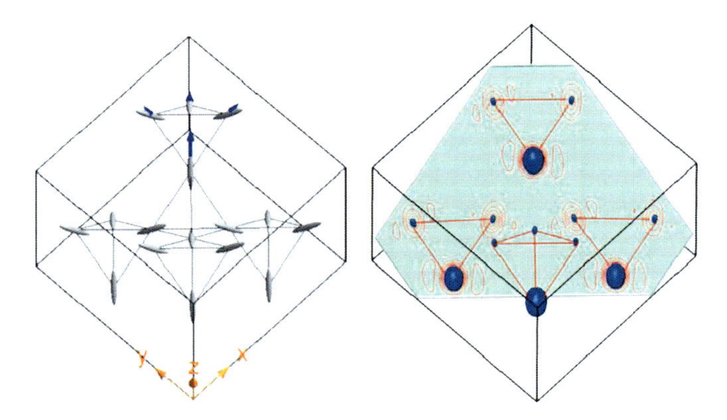

Fig. 8.3 (**a**) Magnetic structure and magnetic ellipsoids measured in a field H// [111] at 10 K (**b**) The corresponding 2D section (*red contours*) and 3D isosurface of the induced magnetization density (*blue*) (Reproduced from Ref. [58] with kind permission of Elsevier)

where $\boldsymbol{\mu}_i$ is the induced magnetic moment on atom i and χ_i the atomic site susceptibility tensor which accounts for the magnetic response of atom i to an external magnetic field. We consider the case where only one magnetic atom exists in the asymmetric unit. The magnetic structure factor can be written as:

$$\mathbf{F_M}(\boldsymbol{\kappa}) = F_{mag}(\boldsymbol{\kappa}) \sum_{i=1}^{n_a} \boldsymbol{\chi}_i \mathbf{H}.\mathrm{e}^{i\boldsymbol{\kappa}\mathbf{R}_i}\mathrm{e}^{-W_i} \tag{8.33}$$

where the summation runs on the crystallographically equivalent sites deduced by the symmetry operations of the space group from the initial atomic position. The symmetry operations apply to the tensor $\boldsymbol{\chi}$ in a very similar way as to the tensor $\boldsymbol{U}$ describing the thermal motion of atoms. Atomic susceptibility parameters χ_{ij} (ASPs), analogous with the atomic anisotropic displacement parameters U_{ij} (ADPs), can then be fully determined by least-square refinement providing that observed flipping ratios were collected for different directions of the applied field.

PND data collections on the cubic $Tb_2Ti_2O_7$ pyrochlore compound for two directions of field, [111] and [110], were analysed by this method. The magnetic ellipsoids in Fig. 8.3 visualize the local induced magnetization at low temperature on the Tb sites for a field along [111] [58]. The field induced magnetic structure resembles the so-called "one in-three out" found in spin ices. The magnetization density map in Fig. 8.3b is due to the induced component of the Tb moments along [111].

This method was up to now applied to inorganic ionic materials [59] but a very recent application to a Co(II) molecular complex gives very promising results for studying molecules with strong magnetic anisotropy [60]. In this tensor approach, a spherical magnetic form factor is assumed but the method could be extended to the atomic orbital model or to the multipole model.

8.3.4 PND Data Analysis Programs

Different MEM programs allowing for positive and negative densities were developed, as for example the free access MEND extended version (Tubingen University) of the MEED (Maximum-Entropy Electron Density) program [61].

The most recent programs for magnetic form factor and MD refinement work on experimental flipping ratios data. Multipole parameters or/and atomic orbital coefficients can be refined for atoms carrying unpaired electrons in p, sp or d-type orbitals. The nuclear structure factors are calculated from the crystal structure and all other quantities (polarization, flipping efficiency) are introduced as data. The program derived from the original MOLLY program written by N. Hansen [46] permits to refine orbital coefficients for 3d ions or 2p radical centres while monopole populations are refined for the rest of the molecule [62]. Another family of programs was developed by J. Brown using the CCSL library [63]. The MAGLSQ program [63] allows flipping ratio refinement using magnetic form factors, multipole or orbital models. The CHILSQ program [63] is devoted to paramagnetic non collinear magnetization using the local susceptibility tensors approach. The FULLPROF suite of programs for powder neutron diffraction data treatment was recently extended to single crystal structural analysis and PND data treatment for magnetization density determinations. [64]

8.4 Molecule-Based Magnetic Materials

Recent applications of PND to molecule-based compounds are presented in this section in order to illustrate the interest of magnetization density studies for a better understanding of the magnetism in these novel materials.

8.4.1 Structure of the Spin Ground State – Magnetic Interactions

Intra- and inter-molecular magnetic interactions in molecule-based materials can be due to magnetic (direct or indirect) exchange or to charge transfer [65]. PND brings information about the two main effects playing a key role in these magnetic interactions: spin delocalization and spin polarization. The first effect is responsible for a contribution on neighbouring atoms of same sign as the main magnetic centre while spin polarization yields an opposite sign contribution.

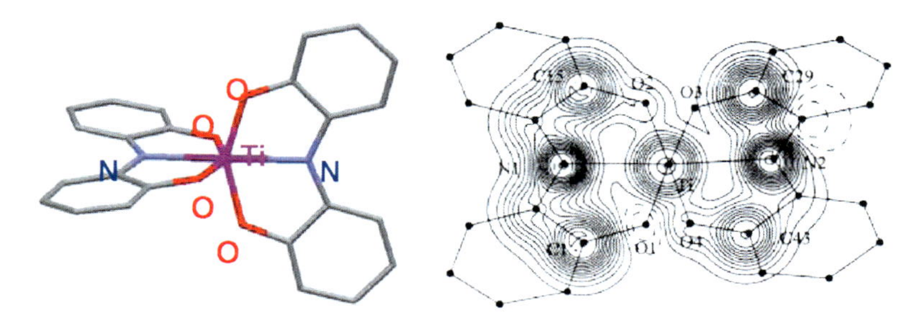

Fig. 8.4 *Left*: geometry of the TiL_2 molecule, right: induced spin density at 1.5 K, 7 T in projection along the direction bisecting the two radicals planes (step 0.008 $\mu_B/\text{Å}^2$) (Reprinted from Ref. [67] with kind permission of Wiley-VCH Verlag GmbH & Co. KGaA)

8.4.1.1 Understanding Exchange and Super-Exchange Mechanisms

The visualization of the magnetization density distribution in a paramagnetic molecule yields direct information about the magnetic interaction pathways and therefore enables to distinguish between direct or indirect exchange mechanisms. The role of superexchange in the intramolecular coupling between the two L^{2-} semiquinonato radical ligands[1] linked by a non magnetic Ti^{4+} ion in the paramagnetic TiL_2 complex was clearly demonstrated by PND [66]. The ferromagnetic interaction between the two radicals carrying one spin ½ each originates from the geometrical orthogonality of the magnetic orbitals describing the unpaired electrons. The induced spin density in the triplet ground state shows that a significant positive spin density is delocalized on the titanium atom while only a very weak spin density is carried by the oxygen atoms (Fig. 8.4).

This demonstrates that the ferromagnetic interaction occurs through indirect superexchange via the Ti^{4+} orbitals and not direct exchange due to the overlap of the radicals magnetic orbitals at the oxygen positions. Density Functional Theory (DFT) calculations are qualitatively in agreement with these results but underestimate the spin delocalization towards the titanium and carbon atoms.

8.4.1.2 Spin Polarization and Interplay with Quantum Mechanical Models

Two mechanisms were invoked to interpret the strong ferromagnetic coupling between the Cu^{2+} ions in all dinuclear Cu(II) complexes bridged by two symmetrical azido groups in end-on geometry $> N - N{-}N$. In the spin delocalization mechanism, the ferromagnetic coupling results from the accidental orthogonality of the copper magnetic orbitals delocalized on the bridging nitrogen atoms for a

[1] L = 3,5-di-ter-butyl-1,2-semiquinonato 1(2-hydroxy-3,5-di-ter-butylphenyl)imine.

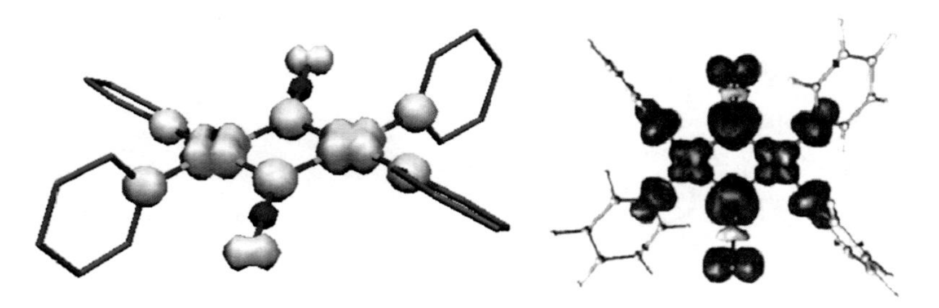

Fig. 8.5 *Left*: Induced spin density in $Cu_2(t\text{-bupy})_4(N_3)_2$ from PND using AO model reconstruction (isodensity 0.002 e.Å^{-3}), *Right*: Theoretical spin density (isodensity 0.001 e.Å^{-3}) (Reprinted from Ref. [68] with kind permission of The American Chemical Society)

bridging angle CuNCu = θ around 90°. AF coupling is predicted for angles larger than 100° but no example of singlet ground state was observed for such end-on complexes. The mechanism based on the spin polarization of the azido bridges (N_3^-) predicts a ferromagnetic coupling whatever the bridging angle is and implies the presence of negative spin density on the bridging nitrogen atoms. The spin density map in the triplet ground state of $Cu_2(t\text{-bupy})_4(N_3)_2(ClO_4)_2$[2] with $\theta = 100°$ [67] shows that a positive spin density region is located at the bridging nitrogen position (Fig. 8.5).

This positive density definitely rules out the spin polarization mechanism. However, the spin density map also reveals the existence of some negative spin density on the central nitrogen atom of each N_3 group, which can be assigned to the spin polarization of the doubly-occupied azido bridge π_g-type orbital, superimposed upon the main spin delocalization mechanism.

The theoretical DFT spin density in the triplet ground state in Fig. 8.5 (right) qualitatively agrees with observation regarding to the spin polarization of the N_3 group. However the spin delocalization from the Cu atoms towards the ligands is largely overestimated by the DFT calculations (58% instead of 22% as observed).

8.4.1.3 Role of H-Bonding in Magnetic Interactions: Charge Transfer Mechanism

The Nit (nitronyl-nitroxide) radical[3] carries one unpaired electron delocalized on a conjugated system of two equivalent NO groups. The NitSMePh[4] radical ferromagnetically orders at 0.2 K. The AO model refinement of PND data in the S = ½ state [21] gives evidence for a perturbation of the oxygen spin distribution of the two NO groups in comparison with the isolated Nit radical. Sp hybridization is

[2] t-bupy = p-tertbutylpyridine.

[3] Nit = $C_7H_{12}N_2O_2$.

[4] NitSMePh = p-(Methylthio)phenyl Nitronyl Nitroxide.

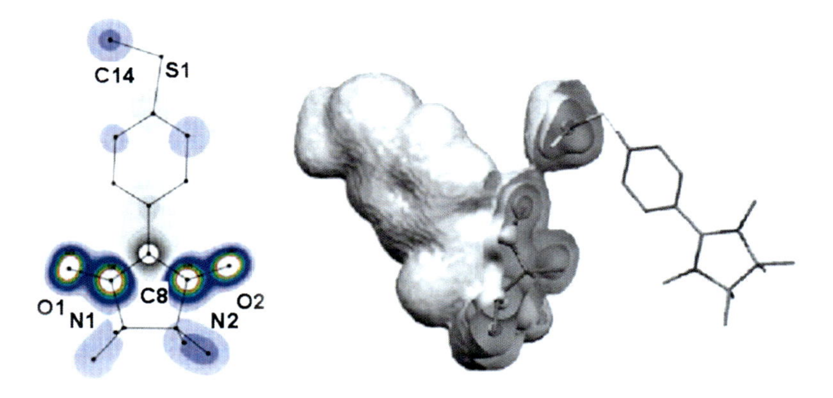

Fig. 8.6 *Left*: Projection of the induced spin density (T = 5.3 K, H = 8 T) on the Nit plane in the radical NitSMePh (isodensity levels, interval 0.006 $\mu_B.Å^{-2}$) (Reprinted with permission from the Ph.D. thesis of Dr. Y. Pontillon (1997)). *Right*: Total electron density along the intermolecular contact O2 ... H17–C14 for the NitSMePh radical (isodensity levels 0.03 $e.Å^{-3}$) (Reprinted from [21] with kind permission of The Royal Society of Chemistry)

observed together with a decrease of the oxygen spin populations. On another hand, a positive spin density is observed on the carbon atom of the SCH_3 group (Fig. 8.6 left). These two observations suggest the hypothesis that a charge transfer between the CH_3 and the oxygen atoms of the neighboring molecule is responsible for the intermolecular ferromagnetic interaction.

This hypothesis was confirmed by a charge density study as shown by the accumulation of electronic density along the hydrogen bond pathways O2 ... H17–C14 (Fig. 8.6 right) and O1 ... H18–C14 [21].

8.4.1.4 Quantum Size Effect in AF Coupled Systems

In a system of two AF coupled local spins $5/2$ and ½, the atomic spin populations differ from 5 and −1 because of the mixing of the spin configurations in the ground state wavefunction. This effect was observed on the experimental spin populations from PND for the paramagnetic molecular complex $Mn(cth)Cu(opxn)(CF_3SO_3)_2$[5] in which the Mn^{2+} and Cu^{2+} ions are bridged by a bidentate oxamido group $O_2C_2NX^{2-}$ (with X = N) [68]. The PND study of the oxamato bridged (X = O) compound $MnCu(pba)(H_2O)_3,2H_2O$[6] which forms ferrimagnetic chains $(MnCu)_n$ permitted to give evidence for the size effect on the spin distribution [23, 69]. The

[5](cth) = 5,7,7,12,14,14-hexamethyl-1,4,8,11-tetraazacyclotetradecane, (opxn) = N,N′-bis-(3-aminopropyl) oxamide.

[6](pba) = 1,3-propylenebis (oxamato).

Mn and Cu spin populations evolve from 4.32(2)/−0.47(1) in the (MnCu) molecule towards 4.93(3)/−0.75(2) in the $(MnCu)_n$ chain. The Cu population does not reach unity because of the strong ligand spin transfer. The DFT wavefunction is a single Slater determinant and therefore cannot correctly describe the triplet state (S = 2) but describes a broken symmetry state (BS). A correction must be applied to the BS spin populations in order to obtain the (S = 2) spin distribution. A good agreement is obtained for Mn but the Cu spin population is underestimated as already noticed for all other molecular Cu complexes. The observed trends due to the size effect are reproduced by DFT calculations on the molecule (DGauss code) and on the chain using periodic boundary conditions to treat the actual crystal structure (Dmol-Dsolid code) [23, 69].

8.4.1.5 Spin Ground State of Single Molecule Magnets

The field of single molecule magnets (SMM) recently emerged in molecular magnetism with the discovery of the mixed valence $Mn^{III,IV}$ cluster $[Mn_{12}]$ with ground state S = 10 which presents a very slow relaxation of the magnetization below the blocking temperature. The existence of a SMM behaviour requires a strong magnetic anisotropy and a large magnetic moment, therefore a high spin ground state. These conditions are realized for molecules where competing AF interactions exist. Classical magnetization measurements do not always permit to determine the spin arrangement in the spin ground state. PND was applied to elucidate the nature of the spin ground state for several SMM's $[Mn_{10}]$ [70], $[Mn_{12}]$ [71], $[Fe_8]$ [72, 73] and high spin clusters $[M_6W_9]$, with M = Mn^{2+}, Ni^{2+} [74, 75].

The $Fe_8{}^{II}$ molecular cluster presents a so-called butterfly structure with easy magnetization axis perpendicular to the molecular mean plane. The induced magnetization density in the S = 10 ground state of the $Fe_8{}^{II}$pcl[7] compound [73] confirms the relative spin arrangement proposed from classical magnetic measurements (Fig. 8.7).

8.4.2 *Excited States Under External Forces*

Excited magnetic states can be reached in molecule-based magnetic materials having a diamagnetic S = 0 ground state, by application of an external force as magnetic field [76] or light irradiation which produces long lifetime metastable states [77].

[7] $Fe_8{}^{II}$pcl = $[(tacn)_6Fe_8O_2(OH)_{12}]Br_{4.3}(ClO_4)_{3.7}$, $6H_2O$ with (tacn) = 1,4,7-triazacyclononane.

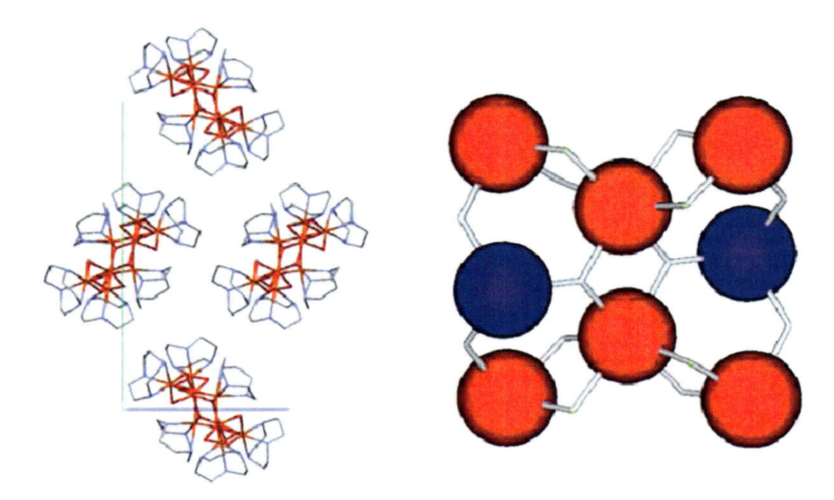

Fig. 8.7 Fe_8pcl compound: *Left*: Projection of the cell content along the **a** axis. *Right*: induced magnetization density reconstructed from a spherical model refinement (1.8 K, 7 T) (positive isosurface in *red*, negative in *blue*)

8.4.2.1 Field Induced Quantum Excited State

The F_2PNNNO[8] organic biradical presents a triplet ground state (S = 1) due to two ferromagnetically coupled spins ½ originating from a Nit radical and a tBuNO[9] radical group [76]. At low temperature, AF coupling between two neighbouring biradicals leads to the formation of AF spin tetramers with singlet ground state $S_t = 0$. This quantum size spin state cannot be described by a classical picture of two spin up on one radical and two spin down on the other. As a matter of fact, the spin density in the singlet ground state is zero in the entire space. In the presence of an external magnetic field applied along the z axis, non-zero spin states become populated due to Zeeman effect as shown in Fig. 8.8. At 10 K the lowest excited state only ($S_t = 1$, $S_{z,t} = +1$) is populated under a field of 7 Tesla. The wave function of this $|-1, +1>$ state is a linear combination of Slater determinants built on the four MO's (two Nit and two tBuNO) corresponding to the four possible spin configurations with three spin up and one spin down.

The induced magnetization in the tetramer displayed in Fig. 8.8b is not equally distributed on the four magnetic sites but the tBuNO population is reduced with respect to that of Nit due to AF coupling between dimers. The experimentally observed ratio between the Nit and tBuNO spin populations is in agreement with

[8] F_2PNNNO = 2-[2′,6′-difluoro-4′-(*N*-*tert*-butyl-*N*-oxyamino)phenyl]-4,4,5,5-tetramethyl-4,5-dihydro-1*H*-imidazol-1-oxyl.

[9] tBuNO = tertio-Butyl-nitroxide.

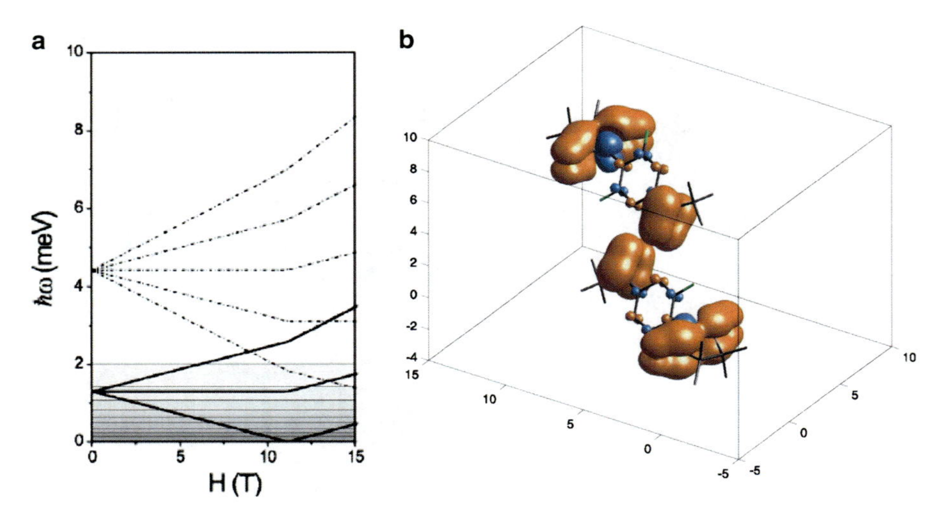

Fig. 8.8 The F_2PNNNO spin-tetramer (Reproduced from [77] with kind permission of The American Physical Society). (**a**) Calculated field dependence of the energy levels (*heavy solid lines* are the lowest energy $S = 0$ and $S = 1$ states, *dashed lines* are $S = 2$ states). (**b**) Experimental spin density at $H = 7$ T and $T = 10$ K, from AO model refinement (isosurfaces at 1.10^{-3} $\mu_B/\text{Å}^3$ (*orange*) and -1.10^{-3} $\mu_B/\text{Å}^3$ (*blue*) levels). The axes show Cartesian coordinates in Å

the ratio of 1.39 predicted from the $|-1, +1>$ wave function coefficients obtained by numerical diagonalization of the four-spin Heisenberg Hamiltonian.

8.4.2.2 Light-Induced Spin Crossover

Photoexcitation of spin crossover Fe^{II} complexes at low temperature, with light of suitable wavelength, induces a transition of the system in the LS ground state (S = 0) towards a HS meta-stable state (S = 2) having an extremely long lifetime at low temperatures. This effect is therefore called Light Induced Excited Spin State Trapping (LIESST) [77]. A single crystal of $[Fe(ptz)_6](BF_4)_2$ (ptz = 1-propyltetrazole) at 2 K installed on the 5C1 diffractometer (LLB) was exposed to light (3 mW/cm^2) produced by a laser at the optical wavelength 473 nm [38]. The time dependence of the photo-excitation process of the crystal was followed directly by PND in a constant magnetic field of 5T. For this purpose, the flipping ratio R of the (0 1 2) reflection was measured as a function of time. The time evolution of (1 – R) under illumination at 2 K, reported in Fig. 8.9 (left), shows that saturation is obtained after 2 h. The induced magnetization density in the photoexcited HS state is strongly localized on the Fe atoms (Fig. 8.9, right).

The moment of 4.05(7) μ_B on the iron site is very close to the theoretical value of the Fe^{2+} moment at saturation (S = 2). This provides a direct evidence of the complete phototransformation of the crystal.

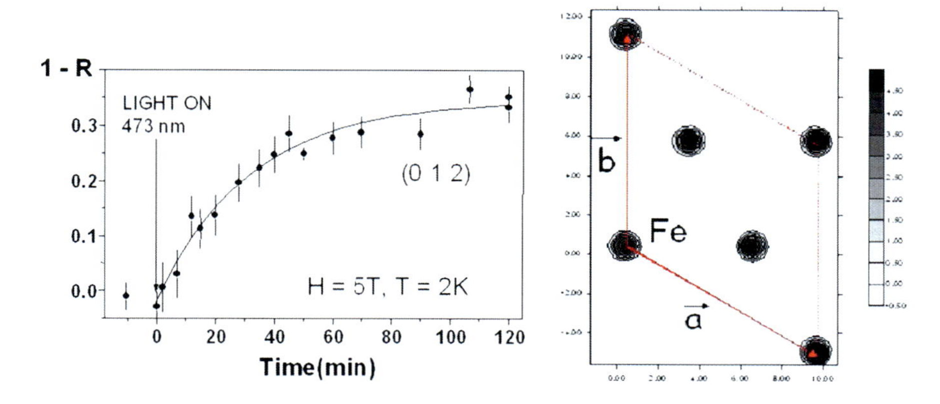

Fig. 8.9 *Left*: kinetics of the photexcitation process in $[Fe(ptz)_6](BF_4)_2$. *Right*: induced magnetization density in the photoexcited state at T = 2 K under 5 Tesla, in projection along the **c** axis

8.5 Conclusions

Progress achieved in Polarized Neutron Diffraction experimental facilities on both continuous and pulsed sources enlarge the domain of magnetization density studies. Powder PND could permit to investigate magnetism in samples of biological interest like proteins. The use of Position Sensitive Detectors considerably reduces the required single crystal size (less than 1 mm^3). Extreme conditions of very low T, very high magnetic field and high pressure are now accessible on several PND diffractometers and enable for full investigations of magnetic phase diagrams. High magnetic field and low temperature are for example necessary for the investigation of excited spin states in molecular clusters with AF zero ground spin state and the study of intermediate spin states in Single Molecule Magnet. The determination of magnetization densities under an applied electric field can be of special interest for multiferroics. Light-coupled PND has a great potential in studies of a variety of systems, like spin crossover complexes [78], molecule-based magnets [79] as well as doped manganites [80] where photoinduced magnetic effects have been evidenced.

Maximum of Entropy reconstruction of magnetization densities was proved to be a very powerful tool for studying the nature and origin of magnetism in various materials. The new approach based on local susceptibility tensor opens a wide domain of investigations on strongly magnetically anisotropic compounds. Atomic orbital and multipole model refinement methods are particularly worth for the analysis of the magnetic interaction mechanisms in molecule-based materials. Future improvements are expected from joint refinement methods based on data from different techniques like Xray diffraction and PND [81], PND and Magnetic Compton Scattering [82] or Xray diffraction, PND and MCS [83] (see Chaps. 5 and 7).

Acknowledgments The authors thank A. Gukasov for many helpful discussions.

References

1. Schweizer J (2006) Polarized neutrons and polarization analysis. In: Chatterji T (ed) Neutron scattering from magnetic crystals. Elsevier, Amsterdam, pp 153–213
2. Lawson PK, Cooper MJ, Dixon MAG, Timms DN, Zukowski E, Itoh F, Sakurai H (1997) Magnetic-Compton-scattering study of spin moments in UFe_2. Phys Rev B 56:3239–3243
3. Tsutsui S, Sakurai Y, Itou M, Matsuda TD, Haga Y, Onuki Y (2005) Magnetic Compton scattering study of UCoAl. Phys B 359–361:1117–1119
4. Kučera M, Kuneš J, Kolomiets A, Diviš M, Andreev AV, Sechovsky V, Kappler JP, Rogalev A (2002) X-ray magnetic circular dichroism studies of $5f$ magnetism in UCoAl and UPtAl. Phys Rev B 66:144405
5. Muro T, Shishidou T, Oda F, Fukawa T, Yamada H, Kimura A, Imada S, Suga S, Park SY, Miyahara T, Sato K (1996) Magnetic circular dichroism of the S $2p$, Co $2p$, and Co $3p$ core absorption and orbital angular momentum of the Co $3d$ state in low-spin CoS_2. Phys Rev B 53:7055
6. Neubeck W, Vettier C, de Bergevin F, Yakhou F, Mannix D, Ranno L, Chatterji T (2001) Orbital moment determination of simple transition metal oxides using magnetic X-ray diffraction. J Phys Chem Solid 62:2173–2180
7. Ito M, Tsuji N, Adachi H, Nakao H, Murakami Y, Taguchi Y (2005) Tokura, Y: direct observation of ordered orbital of $YTiO_3$ by the X-ray magnetic diffraction technique. Nucl Instrum Method Phys Res B 238:237–241
8. Tsuji N, Ito M, Sakurai H, Suzuki K, Tanaka K, Kitani K, Adachi H, Kawata H, Koizumi A, Nakao H, Murakami Y, Taguchi Y, Tokura Y (2008) Magnetic Compton profile study of orbital ordering state of 3d electrons in $YTiO_3$. J Phys Soc Jpn 77:023705–1–023705–4
9. Javorsky P, Sechovsky V, Schweizer J, Bourdarot F, Lelièvre-Berna E, Andreev AV, Shiokawa Y (2001) Magnetization densities in UCoAl studied by polarized neutron diffraction. Phys Rev B 63:064423
10. Kernavanois N, Ressouche E, Brown PJ, Henry JY, Lelièvre-Berna E (2003) Magnetization distribution in paramagnetic CoO: a polarized neutron diffraction study. J Phys Condens Matter 15:3433–3443
11. Akimitsu J, Ichikawa H, Eguchi N, Miyano T, Nishi M, Kakurai K (2001) Direct observation of orbital ordering in $YTiO_3$ by means of the polarized neutron diffraction technique. J Phys Soc Jpn 70:3475–3478
12. Champion G, Arrio MA, Sainctavit P, Zacchigna M, Zangrando M, Finazzi M, Parmigiani F, Villain F, Mathoniere C, Cartier dit Moulin C (2003) Size effect on local magnetic moments in ferrimagnetic molecular complexes: an XMCD investigation. Monatshefte fur Chem 134:277–284
13. Moroni R, Cartier dit Moulin Ch, Champion G, Arrio M-A, Sainctavit Ph, Verdaguer M, Gatteschi D (2003) X-ray magnetic circular dichroism investigation of magnetic contributions from Mn^{III} and Mn^{IV} ions in Mn_{12}-ac. Phys Rev B 68:064407
14. Schweizer J, Gillon B (1999) In: Lahti PM (ed) Magnetic properties of organic radicals. Marcel Dekker, Inc, New York, p. 449
15. Schweizer J, Ressouche E (2001) In: Miller J, Drillon M (eds) MagnetoScience – from molecules to materials. Wiley, Berlin, pp 325–357
16. Zheludev A, Chiarelli R, Delley B, Gillon B, Rassat A, Ressouche E, Schweizer J (1995) Spin density in an organic biradical ferromagnetic crystal. J Mag Mag Mater 140–144:1439–1440
17. Gillon B (2001) Spin distributions in molecular systems with interacting transition metal ions. In: Miller J, Drillon M (eds) MagnetoScience – from molecules to materials. Wiley, Berlin, pp 357–378
18. Reynolds PA, Figgis BN (1990) On the charge density distribution in cobalt(III) ammine hexacyanochromate(III) complexes. Aust J Chem 43:1929–1934
19. Figgis BN, Kucharski ES, Vrtis M (1993) Spin and charge transfer through hydrogen bonding in $[Co(NH_3)_5(OH_2)][Cr(CN)_6]$. J Am Chem Soc 115:176–181

20. Pontillon Y, Caneschi A, Gatteschi D, Grand A, Ressouche E, Sessoli R, Schweizer J (1999) Experimental spin density in a purely organic free radical: visualization of the ferromagnetic exchange pathway in p-(methylthio)phenyl nitronyl nitroxide, Nit(SMe)Ph. Chem Eur J 5:3616–3624
21. Pillet S, Souhassou M, Pontillon Y, Caneschi A, Gatteschi D, Lecomte C (2001) Investigation of magnetic interaction pathways by experimental electron density measurements: application to an organic free radical, ***p***(methylthio)phenyl nitronyl nitroxide. New J Chem 25:131–143
22. Baron V, Gillon B, Cousson A, Mathonière C, Kahn O, Grand A, Öhrström L, Delley B, Bonnet M, Boucherle JX (1997) Spin density maps for the ferrimagnetic chain compound $MnCu(pba)(H_2O)_3,2H_2O$, (pba = 1,2 propylènebis(oxamato)): polarized neutron diffraction and theoretical studies. J Am Chem Soc 119:3500–3506
23. Pillet S, Souhassou M, Mathonière C, Lecomte C (2004) Electron density distribution of an oxamato bridged Mn(II)-Cu(II) bimetallic chain and correlation to magnetic properties. J Am Chem Soc 126:1219–1228
24. Carlisle CJ, Willis BTM (eds) (2009) Magnetic neutron diffraction. In: Experimental neutron scattering. Oxford University Press, New York, pp 172–193
25. Brown PJ (1992) Magnetic scattering of neutrons. In: Wilson AJC (ed) International tables for crystallography, vol C. Kluwer, Dordrecht, pp 512–514
26. Coppens P, Becker PJ (1992) Analysis of charge and spin densities. In: Wilson AJC (ed) International tables of crystallography, vol C. Kluwer, Dordrecht, pp 627–645
27. Gukasov A, Brown PJ (2002) Determination of atomic site susceptibility tensors from polarized neutron diffraction data. J Phys Condens Matter 14:8831–8839
28. Gukasov A, Goujon A, Meuriot JL, Person C, Exil G, Koskas G (2007) Super-6T2, a new position-sensitive detector polarized neutron diffractometer. Phys B 397:131–134
29. Tasset F (1995) Towards helium-3 neutron polarizers. Phys B 213&214:935–938
30. Dreyer J, Regnault L-P, Bourgeat-Lami E, Lelièvre-Berna E, Pujol S, Thomas M, Thomas F, Tasset F (2000) Cryopol: a superconducting magnetostatic cavity for a ^{3}He neutron spin filter. Nucl Instrum Method Phys Res A 449:638
31. Hutanu V, Meven M, Lelièvre-Berna E, Heger G (2009) POLI-HEiDi: the new polarized neutron diffractometer at the hot source (SR9) at the FRMII – project status. Phys B 404:2633–2636
32. Anderson IS, Cook J, Felcher G, Gentile T, Greene G, Klose F, Koetzle T, Lelièvre-Berna E, Parizzi A, Pynn R, Zhao J (2005) Polarized neutrons for pulsed neutron sources. J Neutron Res 13:193–223
33. Tasset F, Brown PJ, Lelièvre-Berna E, Roberts T, Pujol S, Alibon J, Bourgeat-Lami E (1999) Spherical neutron polarimetry with Cryopad-II. Phys B 267–268:69
34. Brown PJ (2006) Spherical neutron polarimetry. In: Chatterji T (ed) Neutron scattering from magnetic crystals. Elsevier, Amsterdam, pp 215–244
35. Brown PJ, Forsyth JB, Lelièvre-Berna E, Tasset F (2002) Determination of the magnetization distribution in Cr_2O_3 using spherical neutron polarimetry. J Phys Condens Matter 14:1957–1966
36. Reynolds PA, Gilbert EP, Figgis BN (1996) Powder neutron diffraction in an applied magnetic field: a novel tool for transition metal chemistry. Inorg Chem 35:545–546
37. Wills AS, Lelièvre-Berna E, Tasset F, Schweizer J, Ballou R (2005) Magnetization distribution measurements from powders using a ^{3}He spin filter: a test experiment. Phys B 356:254–258
38. Goujon A, Gillon B, Gukasov A, Jeftic J, Nau Q, Codjovi E, Varret F (2003) Photoinduced molecular switching studied by polarized neutron diffraction. Phys Rev B 67:220401(R)
39. Papoular RJ, Gillon B (1990) Maximum entropy reconstruction of spin density maps in crystals from polarized neutron diffraction data. Europhys Lett 13:429–434
40. Schleger P, Puig-Molina A, Ressouche E, Schweizer J (1997) A general maximum-entropy method for model-free reconstructions of magnetization densities from polarized neutron diffraction data. Acta Crystallogr A 53:426–435
41. Brown PJ (2004) Experimental attempts to measure non-collinear local magnetization. J Phys Chem Solid 65:1977–1983

42. Sakata M, Sato M (1990) Accurate structure analysis by the maximum-entropy method. Acta Crystallogr A 46:263–270
43. Gukasov A, Braden M, Papoular RJ, Nakatsuji S, Maeno Y (2002) Anomalous spin-density distribution on oxygen and Ru in $Ca_{1.5}Sr_{0.5}RuO_4$: polarized neutron diffraction study. Phys Rev Lett 89:087202
44. Brown PJ (1992) Magnetic form factors. In: Wilson AJC (ed) International tables for crystallography, vol C. Kluwer, Dordrecht, pp 391–399
45. Clementi E, Roetti C (1974) Roothaan-Hartree-Fock atomic wavefunctions. Basis functions and their coefficients for ground and certain excited states of neutral and ionized atoms, $Z \leq 54$. Atom Data Nucl Data Tab 14:177
46. Hansen NK, Coppens P (1978) Testing aspherical atom refinements on small-molecule data sets. Acta Crystallogr A 34:909–921
47. Brown PJ, Capiomont A, Gillon B, Schweizer J (1979) Spin densities in free radicals. J Mag Mag Mater 14:289–294
48. Boucherle JX, Gillon B, Maruani J, Schweizer J (1982) Determination by polarized neutron diffraction of the spin density distribution in a non-centrosymmetrical crystal of DPPH:C_6H_6. J de Physique Colloque C7 43:227–234
49. Coppens P (1997) X-ray charge densities and chemical bonding. IUCr texts on crystallography 4. Oxford University Press, Oxford/New York
50. Claiser N, Souhassou M, Lecomte C, Gillon B, Carbonera C, Caneschi A, Dei A, Gatteschi D, Bencini A, Pontillon Y, Lelièvre-Berna E (2005) Combined charge and spin density experimental study of the yttrium(III) semiquinonato complex $Y(HBPz_3)_2$(DTBSQ) and DFT calculations. J Phys Chem B 109:2723
51. Holladay A, Leung P, Coppens P (1983) Generalized relations between d-orbital occupancies of transition metal atoms and electron-density multipole population parameters from X-ray diffraction data. Acta Crystallogr A 39:377
52. Squires GL (1978) Introduction to the theory of thermal neutron scattering. University Press, Cambridge, p. 139
53. Kernavanois N, Grenier B, Huxley A, Ressouche E, Sanchez JP, Flouquet J (2001) Neutron scattering study of the ferromagnetic superconductor UGe_2. Phys Rev B 64:174509
54. Hiess A, Bourdarot F, Coad S, Brown PJ, Burlet P, Lander GH, Brooks MSS, Kaczorowski D, Czopnik A, Troc R (2001) Spin and orbital moments in itinerant magnets. Europhys Lett 55:267–272
55. Prokes K, Gukasov A (2009) Magnetization densities in URhSi studied by polarized neutron diffraction. Phys Rev B 79:024406
56. Brown PJ, Neumann KU, Simon A, Ueno F, Ziebeck KRA (2005) Magnetization distribution in CoS_2; is it a half metallic ferromagnet? J Phys Condens Matter 17:1583–1592
57. Schweizer J (1980) Interpretation of magnetization densities. In: Becker P (ed) Electron and magnetization densities in molecules and crystals, vol 48, NATO advanced study institutes series, series B: physics. Plenum Press, New York, pp 501–519
58. Gukasov A, Cao H, Mirebeau I, Bonville P (2009) Spin density and non-collinear magnetization in frustrated pyrochlore $Tb_2Ti_2O_7$ from polarized neutron scattering. Phys B 404:2509–2512
59. Gukasov AG, Rogl P, Brown PJ, Mihalik M, Menovsky A (2002) Site susceptibility tensors and magnetic structure of $U_3Al_2Si_3$: a polarized neutron diffraction study. J Phys Condens Matter 14:8841–8851
60. Borta A, Gillon B, Gukasov A, Cousson A, Luneau D, Jeanneau E, Ciumacov I, Sakiyama H, Tone K, Mikuriya M (2011) Local magnetic moments in a dinuclear Co^{2+} complex as seen by polarized neutron diffraction: Beyond the effective spin-1/2 model. Phys Rev B 83:184429
61. Kumazawa S, Kubota Y, Takata M, Sakata M, Ishibashi Y (1993) *MEED*: a program package for electron-density-distribution calculation by the maximum-entropy method. J Appl Crystal 26:453–457

62. Ressouche E, Boucherle JX, Gillon B, Rey P, Schweizer J (1993) Spin density maps in nitroxide-Copper(II) complexes. A polarized neutron diffraction determination. J Am Chem Soc 115:3610–3617
63. Matthewman JC, Thompson P, Brown PJ (1982) The CCSL library. J Appl Crystallogr 15: 167–173; Brown PJ (2006) Cambridge crystallographic subroutine library. Documentation on https://www.ill.eu/sites/ccsl/html/ccsldoc.html
64. Frontera C, Rodriguez-Carvajal J (2003) FullProf as a new tool for flipping ratio analysis. Phys B 335:219–222
65. Kahn O (1993) Molecular magnetism. VVH Publishers, Inc., New York
66. Pontillon Y, Bencini A, Caneschi A, Dei A, Gatteschi D, Gillon B, Sangregorio C, Stride J, Totti F (2000) Spin density map of the triplet ground state of a Titanium(IV) complex with the Schiff base diquinone radical ligands: a polarized neutron diffraction and DFT investigation. Angew Chem Int Eng Ed 39:1786–1788
67. Aebersold M, Gillon B, Plantevin O, Pardi L, Kahn O, Bergerat P, von Seggern I, Tuczek F, Öhrström L, Grand A, Lelièvre-Berna E (1998) Spin density maps in the triplet ground state of $[Cu_2(t\text{-bupy})_4(N_3)_2](ClO_4)_2$ (t-bupy = p-tert-butylpyridine): a polarized neutron diffraction study. J Am Chem Soc 120:5238–5245
68. Baron V, Gillon B, Plantevin O, Cousson A, Mathonière C, Kahn O, Grand A, Öhrström L, Delley B (1996) Spin density maps in an oxamido-bridged Mn(II)Cu(II) binuclear compound; polarized neutron diffraction and theoretical studies. J Am Chem Soc 118:11822–11830
69. Kahn O, Mathonière C, Srinavasan B, Gillon B, Baron V, Grand A, Öhrström L, Ramashesha S (1997) Spin distributions in antiferromagnetically coupled $Mn^{2+}Cu^{2+}$ systems: from the pair to the infinite chain. New J Chem 21:1037–1045
70. Caneschi A, Gatteschi D, Sessoli R, Schweizer J (1998) Magnetization density in a Mn high-spin (S = 12) magnetic cluster. Phys B 241–243:600
71. Robinson RA, Brown PJ, Argyriou DN, Hendrickson DN, Aubin DMJ (2000) Internal magnetic structure of Mn_{12} acetate by polarized neutron diffraction. J Phys Condens Matter 12:2805–2810
72. Pontillon Y, Caneschi A, Gatteschi D, Sessoli R, Ressouche E, Schweizer J, Lelièvre-Berna E (1999) Magnetization density in an iron(III) magnetic cluster. A polarized neutron investigation. J Am Chem Soc 121:5342
73. Gillon B, Sangregorio C, Caneschi A, Gatteschi D, Sessoli R, Ressouche E, Pontillon Y (2007) Experimental spin density in the high spin ground state of the Fe_8pcl cluster. Inorg Chim Acta 360:3802–3806
74. Ruiz E, Rajaraman G, Alvarez S, Gillon B, Stride J, Clérac R, Larionova J, Decurtins S (2005) Symmetry and topology determine the Mo^{V}–CN–Mn^{II} exchange interactions in high-spin molecules. Angew Chem Int Ed 44:2711–2715
75. Gillon B, Larionova J, Ruiz E, Nau Q, Goujon A, Bonadio F, Decurtins S (2008) Experimental and theoretical study of the spin ground state of the high-spin molecular cluster $[Ni^{II}\{Ni^{II}(CH_3OH)_3\}_8(\mu\text{-CN})_{30}\{W^{V}(CN)_3\}_6]$,$15CH_3OH$ by polarized neutron diffraction and density functional theory calculations. Inorg Chem Acta 361:3609–3615
76. Zheludev A, Garlea VO, Nishina S, Hosokoshi Y, Cousson A, Gukasov A, Inoue K (2007) Spin-density distribution in the partially magnetized organic quantum magnet F_2PNNNO. Phys Rev B 75:104427
77. Decurtins S, Gütlich P, Hasselbach MK, Spiering H, Hauser A (1985) Light-induced excited-spin-state trapping in iron(II) spin-crossover systems. Optical spectroscopic and magnetic susceptibility study. Inorg Chem 24:2174
78. Ogawa Y, Koshihara S, Koshino K, Ogawa T, Urano C, Takagi H (2000) Dynamical aspects of the photoinduced phase transition in spin-crossover complexes. Phys Rev Lett 84:3181
79. Sato O, Iyoda T, Fujishima A, Hashimoto K (1996) Photoinduced magnetization of a cobalt-iron cyanide. Science 272:704
80. Matsuda K, Machida A, Moritomo Y, Nakamura A (1998) Photoinduced demagnetization and its dynamical behavior in a $(Nd_{0.5}Sm_{0.5})_{0.6}Sr_{0.4}MnO_3$ thin film. Phys Rev B 58:R4023

81. Becker PJ, Coppens P (1985) About the simultaneous interpretation of charge and spin density data. Acta Crystallogr A 41:177–182
82. Qureshi N, Zbiri M, Rodriguez-Carvajal J, Stunault A, Ressouche E, Hansen TC, Fernandez-Diaz T, Rodriguez-Valazaman J, Sanchez-Montero J, Johnson MR, Fuess H, Ehrenberg H, Sakurai Y, Itou M, Gillon B (2009) Experimental magnetic form factors in $Co_3V_2O_8$: A combined study of ab initio calculations, magnetic Compton scattering and polarized neutron diffraction. Phys Rev B 79:094417
83. Gillet JM (2007) Determination of a one-electron reduced density matrix using a coupled pseudo-atom model and a set of complementary scattering data. Acta Crystallogr A 63: 234–238

Chapter 9
Beyond Standard Charge Density Topological Analyses*

Angel Martín Pendás, Miroslav Kohout, Miguel Alvarez Blanco, and Evelio Francisco

9.1 Introduction

Many of the classical concepts exploited everyday in chemistry, such as atomic charges, chemical bonds, covalency, ionicity, resonance, bond energy, aromaticity, etc., are either difficult to recover from multi-electron wave functions, or not unequivocally defined. Moreover, many of the quantities that are widely used to quantify these concepts weaken, or even fade away, as the accuracy of the wave function increases. A breakthrough in this *status quo* was introduced when the scientific community recognized that sophisticated analyses of the electron density, ρ, might provide chemically appealing, yet physically founded answers to some of these fuzzy concepts. Hohenberg and Kohn's first theorem [1], on the one hand, and the development of the Quantum Theory of Atoms in Molecules (QTAIM) by Bader and coworkers [2], on the other, paved the way towards rigorous concepts in the theory of chemical bonding from electron densities. And the observable character of ρ, amenable to experimental determination, opened this field to a new audience that soon became one of its most faithful advocates, as this volume demonstrates.

---

*To their great sorrow, Angel Martín Pendás, Miroslav Kohout and Evelio Francisco announce the death of Professor Miguel Alvarez Blanco on April 4th, 2010.

A.M. Pendás (✉) • M.A. Blanco • E. Francisco
Dpto. Química Física y Analítica, Universidad de Oviedo, 33006 Oviedo, Spain
e-mail: angel@fluor.quimica.uniovi.es; miguel@carbono.quimica.uniovi.es; evelio@carbono.quimica.uniovi.es

M. Kohout (✉)
Max Planck Institute for Chemical Physics of Solids, Nöthnitzer Str. 40, 01187 Dresden, Germany
e-mail: kohout@cpfs.mpg.de

C. Gatti and P. Macchi (eds.), *Modern Charge-Density Analysis*,
DOI 10.1007/978-90-481-3836-4_9, © Springer Science+Business Media B.V. 2012

The topological analysis of scalar functions in real space, together with the associated 3D partitioning into disjoint regions that these analyses provide, are now consolidated techniques that have offered deep insights into the nature of our most beloved chemical notions. However, as the success of the ELF function [3] has taught us, not all these scalars may be obtained directly from the electron density. In the absence of the exact universal energy or electron repulsion density functionals, many of the actual concepts must unavoidably depend on quantities far more complex that the electron density.

It is the aim of this chapter to show how some progress is being done along these generalized analyses. Most of the concepts that we will introduce depend on quantum mechanical objects that cannot be derived from the electron density alone. Hence, our results will be based on the manipulations of density matrices. This does not mean that the methods here reviewed will remain uncoupled from experiments, for there are at least two ways to fill this gap. The first one is finding approximations to our more complex scalars that depend only on ρ, as made by Tsirelson [4] for the kinetic energy density, and the second, to experimentally adapt the Ansatz for wave functions, as pioneered by Jayatilaka [5].

Our procedures are all based on the partitioning of the physical 3 dimensional space into domains, cells or regions. Depending on the criteria used to define the latter, the partitioning acquires different physical meanings. Technical considerations aside, all of these methods are fairly independent of the accuracy and specific form of the wave function (as far as it captures the phenomena under scrutiny), and try to settle down and consolidate intuitive concepts of traditional chemistry by simply applying the basic rules of quantum-chemistry.

Generally, the partitioning either yields the relevant objects for the analysis, or the partitioning serves as the basis for the derivation of new distributions which subsequently can be partitioned. In this chapter, we are going to consider two different classes of space partitioning, which span two easy to understand limits. One in which the spatial regions are associated to chemical objects, typically atoms, and are thus very coarse-grained, and other in which they are extremely small and of changing size, such that a given quantum mechanical property is conserved in all of them.

Atomic domains can be chosen satisfying different conditions, and may be fuzzy or exhaustive. At each 3D point, atomic weight functions $w_A(\mathbf{r})$ are defined according to a well defined recipe such that $\sum_A w_A(\mathbf{r}) = 1$. In fuzzy partitions, all the $w_A(\mathbf{r})'$s are, in general, non-zero at every $\mathbf{r}$, while in exhaustive partitions 3D domains (Ω_A) are mutually exclusive and all but one of the $w_A(\mathbf{r})'$s is zero at any $\mathbf{r}$, $w_B(\mathbf{r} \in \Omega_A) = \delta_{AB}$. Although most of the concepts and equations considered in this chapter can be formulated for any space partitioning of this type, we will focus on the results corresponding to that provided by the Quantum Theory of Atoms in Molecules (QTAIM), developed by Bader et al. and induced by the topology of the electron density. Taking QTAIM atomic domains as the starting point, we will discuss recently developed theoretical tools of chemical bonding analysis in real space. Among them, the coarse graining of density matrices, the interacting quantum atoms (IQA) energy partition method, the statistics of electron population, and the concept of resonance structure in real space.

The other type of space partitioning, so called ω-restricted space partitioning, presented in this chapter is based on the space decomposition into a huge number of very small cells obeying certain rules. The partitioning is determined by a control function in such a way that the integral of the control function over any cell yields the same (infinitesimally) small fixed value. Then, the integration of a sampling property over the cells results in a new discrete distribution that can be further examined by the 'usual' procedures of the topological analysis. From the potentially infinite possibilities that can be explored by using this approach, the electron localizability indicator (ELI) will be described in detail. ELI is obtained as a quasi-continuous distribution where the pair density $\gamma^{(2)}(\mathbf{r}_1, \mathbf{r}_2)$ and the electron density $\rho(\mathbf{r})$ are utilised as the control function and the sampling property, respectively (and vice versa). This approach can be consistently applied to correlated wave functions as well as transformed into the momentum space representation without any change of the procedure.

9.2 IQA Partitioning

The success of conventional charge density analyses relies, in one way or another, on the existence of a chemically relevant partition of the electron density, ρ, into 3D regions (fuzzy or exhaustive) that carry chemical information. These may be atomic-like, as in the QTAIM, or electron-pair-like, as in the partitions provided by ELI or ELF functions. It is in this way that relations among the several regions in which a system is divided, e.g. bonds between atomic domains, or their properties, e.g. their electron populations, are obtained and compared to previous wisdom. As shown by Li and Parr [6], a wide formalism exists to deal with the energetics of general real space partitions, fuzzy or not. It is based on (i) any chosen spectral decomposition of the unity, and (ii) the equivalence principle that considers on the same footing any electron, be it considered either as a part of the whole system or of one of its 3D constituents. This partition of the unity (PU) methodology may be easily applied to general reduced density matrices (RDMs), and it provides a unified framework to deal with electron population statistics and energetic decompositions, as we are going to show in the following Subsections.

9.2.1 *Partitioning Density Matrices in Real Space*

Let us consider a partition of the unity in the physical space, that is, a set of independent weight functions $w_A(\mathbf{r})$, associated to the chemical objects A that we want to define, that add up to one at every point $\mathbf{r}$:

$$\sum_A w_A(\mathbf{r}) = 1(\mathbf{r}). \tag{9.1}$$

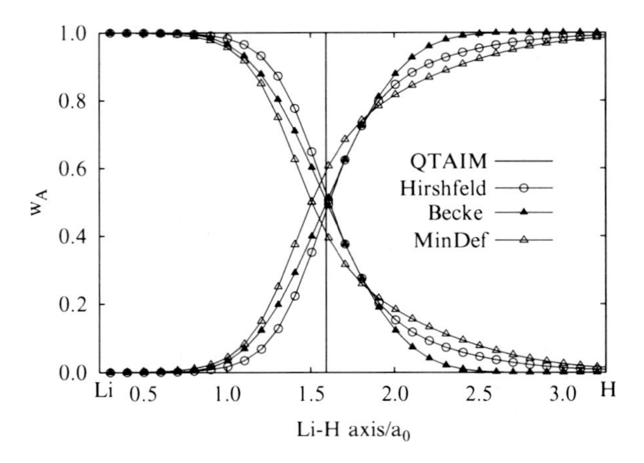

Fig. 9.1 Atomic weight functions, $w_A(r)$ ($A = \text{Li, H}$), for the QTAIM, Hirshfeld, Becke, and MinDef PUs obtained from a CAS[4,6]/6-311 G** calculation in LiH at the theoretical equilibrium distance, and plotted along the internuclear axis. The atomic radii for the Becke partition have been chosen equal to the QTAIM topological radii. The bond critical point coincides with the QTAIM weight function step

This framework is extremely general, since it unifies the description of fuzzy decompositions, defined by continuous weight functions, with that of exhaustive ones, characterized by a set of mutually exclusive 3D domains Ω_A in which all but one of the weight functions vanish, so that $w_A(\mathbf{r} \in \Omega_A) = 1$, $w_A(\mathbf{r} \notin \Omega_A) = 0$. The latter include all partitions induced by the topology of a scalar field, e.g. QTAIM and ELF partitions, and the former any of the fuzzy atomic decompositions proposed so far, e.g. *stockholder's* or Hirshfeld partitioning [7], Li and Parr's minimal self–energy atoms [6], Becke's decomposition [8], Fernández-Rico et al. minimum deformation (MinDef) atoms [9], and many others [10–13]. Figure 9.1 contains a plot of the w functions obtained for QTAIM, Hirshfeld, Becke (using the QTAIM topological radii to define the atomic sizes), and MinDef PUs in LiH. In every case, the weight for a given atom is almost equal to one far from the other nucleus, decreasing in a more or less smooth way to zero as we approach it. It is also interesting to notice that the $w_A = w_B = 0.5$ condition is satisfied in a narrow R window around the QTAIM bond critical point (exactly there by construction in the Becke case).

It is not too difficult to use PUs to partition RDMs. Let us take the first order, non-diagonal density $\gamma^{(1)}(\mathbf{r}'; \mathbf{r})$ as an example. Following Li and Parr, it is a reasonable assumption that

$$\left.\frac{\hat{T}\gamma^{(1)}(\mathbf{r}'; \mathbf{r})}{\gamma^{(1)}(\mathbf{r}'; \mathbf{r})}\right|_{\mathbf{r}' \to \mathbf{r}} = \left.\frac{\hat{T}\gamma_A^{(1)}(\mathbf{r}'; \mathbf{r})}{\gamma_A^{(1)}(\mathbf{r}'; \mathbf{r})}\right|_{\mathbf{r}' \to \mathbf{r}} \qquad \forall A, \tag{9.2}$$

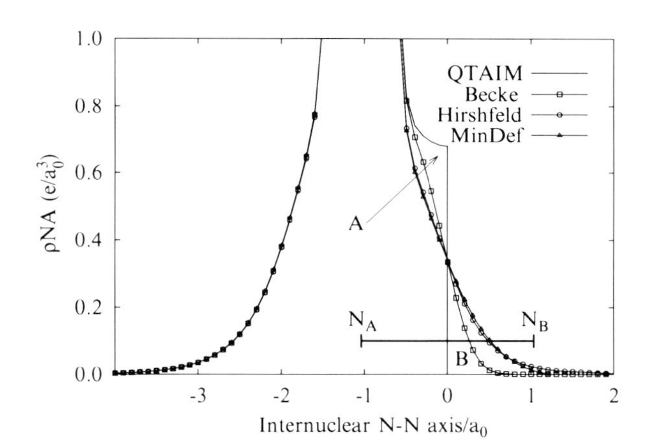

Fig. 9.2 Atomic densities of the left N atom in N_2, ρN_A, obtained at the theoretical equilibrium distance in a CAS[10,8]//6–311 G(d,p)++ calculation. The PUs are equivalent to those of Fig. 9.1. A and B refer to the two QTAIM basins. The QTAIM bond critical point is located at the origin of the x axis

so that the kinetic energy density per electron does not depend on whether we suppose the electron as part of the entire system or of subsystem A. This leads to considering

$$\gamma_A^{(1)}(\mathbf{r}';\mathbf{r}) = w_A(\mathbf{r}')\gamma^{(1)}(\mathbf{r}';\mathbf{r}) \tag{9.3}$$

or, equivalently, to

$$\gamma^{(1)}(\mathbf{r}';\mathbf{r}) = \sum_A \gamma_A^{(1)}(\mathbf{r}';\mathbf{r}) = 1(\mathbf{r}')\gamma^{(1)}(\mathbf{r}';\mathbf{r}). \tag{9.4}$$

The partition of ρ induced by the above argument is familiar, $\rho(\mathbf{r}) = \sum_A \rho_A(\mathbf{r})$. In Fig. 9.2 we plot the atomic densities of the N atom in N_2 induced by several PUs. It is clear from the Figure that the Hirshfeld and MinDef densities interpenetrate the other atomic domain considerably more than the Becke one or the non–interpenetrating QTAIM. This behavior is rather general, and its energetic and statistical consequences thus easily foreseen.

Similarly, $\gamma_{AB}^2(\mathbf{r'}_1,\mathbf{r'}_2;\mathbf{r}_1,\mathbf{r}_2)$ is written as

$$\gamma_{AB}^{(2)}\left(\mathbf{r'}_1,\mathbf{r'}_2;\mathbf{r}_1,\mathbf{r}_2\right) = w_A\left(\mathbf{r'}_1\right)w_B\left(\mathbf{r'}_2\right)\gamma^{(2)}\left(\mathbf{r'}_1,\mathbf{r'}_2;\mathbf{r}_1,\mathbf{r}_2\right), \tag{9.5}$$

that leads to the full 2-nd order RDM by summing over A and B:

$$\begin{aligned}\gamma^{(2)}\left(\mathbf{r'}_1,\mathbf{r'}_2;\mathbf{r}_1,\mathbf{r}_2\right) &= \sum_A\sum_B \gamma_{AB}^{(2)}\left(\mathbf{r'}_1,\mathbf{r'}_2;\mathbf{r}_1,\mathbf{r}_2\right) \\ &= 1\left(\mathbf{r'}_1\right)1(\mathbf{r'}_2)\gamma^{(2)}\left(\mathbf{r'}_1,\mathbf{r'}_2;\mathbf{r}_1,\mathbf{r}_2\right).\end{aligned} \tag{9.6}$$

The diagonal partitioning, $\gamma^{(2)}(\mathbf{r}_1,\mathbf{r}_2) = \sum_A \sum_B \gamma^{(2)}_{AB}(\mathbf{r}_1,\mathbf{r}_2)$, allows to write the total electron repulsion as a sum of one-domain contributions ($A = B$) and two-basin terms ($A \neq B$) (Sect. 9.2.2).

These ideas may be immediately generalized to p-th order RDMs, and

$$\gamma^{(p)}\left(\mathbf{r}'_1,\ldots,\mathbf{r}'_p;\mathbf{r}_1,\ldots\mathbf{r}_p\right) = 1\left(\mathbf{r}'_1\right)\ldots 1\left(\mathbf{r}'_p\right)\gamma^{(p)}\left(\mathbf{r}'_1,\ldots,\mathbf{r}'_p;\mathbf{r}_1,\ldots\mathbf{r}_p\right), \tag{9.7}$$

together with

$$1\left(\mathbf{r}'_1\right)\ldots 1\left(\mathbf{r}'_p\right) = \sum_A \cdots \sum_P w_A\left(\mathbf{r}'_1\right)\ldots w_P\left(\mathbf{r}'_p\right), \tag{9.8}$$

lead to

$$\gamma^{(p)}\left(\mathbf{r}'_1,\ldots,\mathbf{r}'_p;\mathbf{r}_1,\ldots\mathbf{r}_p\right) = \sum_A \cdots \sum_P \gamma^{(p)}_{A\ldots P}\left(\mathbf{r}'_1,\ldots,\mathbf{r}'_p;\mathbf{r}_1,\ldots\mathbf{r}_p\right), \tag{9.9}$$

where

$$\gamma^{(p)}_{A\ldots P}\left(\mathbf{r}'_1,\ldots,\mathbf{r}'_p;\mathbf{r}_1,\ldots\mathbf{r}_p\right) = w_A\left(\mathbf{r}'_1\right)\ldots w_P\left(\mathbf{r}'_p\right)\gamma^{(p)}\left(\mathbf{r}'_1,\ldots,\mathbf{r}'_p;\mathbf{r}_1,\ldots\mathbf{r}_p\right). \tag{9.10}$$

It should be clear from the preceding three equations that symbols $A,\ldots,P$, and electron coordinates $\mathbf{r}_1,\ldots,\mathbf{r}_p$ go in pairs: $(A,\mathbf{r}_1),\ldots,(P,\mathbf{r}_p)$. Questions related to the asymmetry of the above expressions are interesting [14], but out of the scope of this presentation.

Similarly, one may generalize these ideas and domain–average—i.e. w-weight and integrate—all of the electrons of the system. This leads to coarse-grained density matrices (CGDMs) [15]. Let us consider a N electron system with wave function Ψ, and a PU with m components (i.e. the summation in Eq. 9.1 contains m terms). We define a coarse-graining, domain-average, or condensation of an electron j over a domain Ω_k by $N\int d\mathbf{r}_j w_k(\mathbf{r}_j)\Psi^*\Psi$. The result is a function of $\mathbf{r}_1,\ldots,\mathbf{r}_{j-1},\mathbf{r}_{j+1},\ldots$, and $\mathbf{r}_N$, and the factor N accounts for the equivalence of the N electrons of the system. Analogously, a simultaneous condensation of electron j over Ω_k and of electron l over Ω'_k is defined by $N(N-1)\int d\mathbf{r}_j d\mathbf{r}_l w_k(\mathbf{r}_j) w'_k(\mathbf{r}_l)\Psi^*\Psi$ $(k \neq k')$ or $N(N-1)/2\int d\mathbf{r}_j d\mathbf{r}_l w_k(\mathbf{r}_j) w'_k(\mathbf{r}_l)\Psi^*\Psi (k = k')$, and the result is a function that depends on all except the coordinates of electrons j and l. By extension, we may domain–average (condense, coarse–grain) c_1 electrons over Ω_1, c_2 electrons over $\Omega_2,\ldots$, and c_m electrons over Ω_m. The total number of condensed electrons may be smaller than N, $c_1+\cdots+c_m=\mathcal{N}\leq N$, and the remaining $i=N-\mathcal{N}\geq 0$ electrons are called *free* or non–condensed electrons. Without loss of generality we can take electrons 1 to i as the free electrons, electrons $i+1$ to $i+c_1$ as the c_1 electrons averaged ver Ω_1, etc. A condensation C is understood as a partition of the $\mathcal{N}$ condensed electrons, and is specified by the ordered set $C=(c_1,\ldots,c_m)$. When the full set of N electrons is so divided, we speak of a *resonance structure* S in real space. Taking into account indistinguishability, we may define an i–th order CGDM associated to the C condensation as

$$\gamma^{(i)}(\mathbf{r}_1,\ldots,\mathbf{r}_i)[C] = I_C \int d\mathbf{r}_{i+1}\ldots d\mathbf{r}_N w_C \Psi^*\Psi, \tag{9.11}$$

where $I_C = N!/(i!c_1!\cdots c_m!) = \binom{N}{i}\mathcal{N}!/(c_1!\cdots c_m!)$ corrects for all permutations among equivalent groups of electrons, and $w_C = w_C(\mathbf{r}_{i+1},\ldots, \mathbf{r}_N)$ stands for the product of $\mathcal{N}$ weight functions such that electrons $(i+1)$ to $(i+c_1)$ are weighted through w_1, electrons $(i+c_1+1)$ to $(i+c_1+c_2)$ through w_2, etc., until each of the condensed electrons has been properly weighted. Transition CGDMs are also easily defined.

Particular cases of Eq. 9.11 are well known objects of Quantum Chemistry. For instance, when $\mathcal{N}=0$ and $i=N$, $\gamma^{(N)}(\mathbf{r}_1,\ldots, \mathbf{r}_N)[] = \Psi^*\Psi$, where "[]" means that none electron is condensed. When $\mathcal{N}=N$, $i=0$, and $m=1$, $\gamma^{(0)}()[C]$ represents the normalization of the wave function:

$$\gamma^{(0)}()[C;m=1] = \int d\mathbf{r}_1\ldots d\mathbf{r}_N \Psi^*\Psi = 1, \tag{9.12}$$

where "()" means that none electron is free. Finally, when $\mathcal{N}=N-i$ and $m=1$, $\gamma^{(i)}(\mathbf{r}_1,\ldots, \mathbf{r}_i)[C;\ m=1]$ becomes $\gamma^{(i)}(\mathbf{r}_1,\ldots, \mathbf{r}_i)$ the diagonal i-th order RDM. Using the property

$$\begin{aligned}1(\mathbf{r}_{i+1})\ldots 1(\mathbf{r}_{i+\mathcal{N}}) &= \sum_{A_1}\ldots\sum_{A_\mathcal{N}} w_{A_1}(\mathbf{r}_{i+1})\ldots w_{A_\mathcal{N}}(\mathbf{r}_{i+\mathcal{N}})\\ &= \sum_{\{C\}} \frac{\mathcal{N}!}{c_1!\ldots c_m!} w_C = \binom{N}{i}\sum_{\{C\}} I_C w_C,\end{aligned} \tag{9.13}$$

any traditional diagonal i-th order RDM may also be obtained by adding appropriate CGDMs over all possible condensations $\{C\}$ of the $\mathcal{N}$ electrons, so

$$\gamma^{(i)}(\mathbf{r}_1,\ldots,\mathbf{r}_i) = \binom{N}{i}\int d\mathbf{r}_{i+1}\ldots d\mathbf{r}_N \Psi^*\Psi = \sum_{\{C\}}\gamma^{(i)}(\mathbf{r}_1,\ldots,\mathbf{r}_i)[C]. \tag{9.14}$$

For compatibility with previous works, the $\gamma^{(i)}(\mathbf{r}_1,\ldots, \mathbf{r}_i)$ used in the following sections corresponds to that given by Eq. 9.14 multiplied by $i!$.

CGDMs allow a domain-based partition of any relevant quantum mechanical expectation value. It is particularly interesting that populations may be manipulated in a *per* electron basis using zeroth order densities. For instance, given a resonance structure $S=(n_1,\ldots, n_m)$, the probability of finding exactly n_1 electrons in domain

Ω_1, n_2 electrons in Ω_2, etc., is just $p(S) = p(n_1, \ldots, n_m) = \gamma^{(0)}()[S]$. The set of all of these probabilities forms the core of the electron distribution functions that will be discussed in Sect. 9.4.1. The application of Eq. 9.14 with $i = 0$ gives

$$\gamma^{(0)}() = \int d\mathbf{r}_1 \ldots d\mathbf{r}_N \Psi^* \Psi = \sum_{\{S\}} \gamma^{(0)}()[S] = \sum_{\{S\}} p(S) = 1, \tag{9.15}$$

as it should be since the sum of the probabilities for all resonance structures must be 1.

If we are only interested in a set of k domains $(\alpha, \ldots, \kappa)$, $k < m$, we may consider a system composed of $k + 1$ domains, the last one (ω) being described by the weight function $w_\omega = 1 - \sum_{i=1}^{k} w_i$. This is nothing but the rest of R^3 for exhaustive partitions. In this way we may correctly define objects like $p(n_\alpha, \ldots, n_\kappa) = p(n_\alpha, \ldots, n_\kappa, (N - \mathcal{N})_\omega)$, where $\mathcal{N} = n_\alpha + \cdots + n_\kappa \leq N$ in this condensation. This provides the probability of finding exactly $n_\alpha, \ldots, n_\kappa$ electrons in domains $\alpha, \ldots, \kappa$.

We can easily construct conditional RDMs [16] from the coarse–grained objects.

$$\gamma_A^{(1)}(\mathbf{r}_1|S) = w_A(\mathbf{r}_1) \frac{\gamma^{(1)}(\mathbf{r}_1)[C]}{p(S)}, \tag{9.16}$$

for instance, provides the conditional probability of finding an electron in domain A within a resonance structure S, C being obtained by subtracting one electron from domain A in S. This is easily shown to integrate to $n_A = c_A + 1$, the number of electrons in domain A within the resonance structure S, $\int d\mathbf{r}_1 \gamma_A^{(1)}(\mathbf{r}_1|S) = n_A$.

9.2.2 *Induced Energy Decomposition*

Once a PU has been chosen, any molecular observable may be decomposed into domain contributions. This may be done at the domain level, providing a coarse partition of one-particle observables into additive domain contributions, and of two-particle observables in a pairwise additive manner [17, 18]. But also at the resonance structure level, giving rise to a much thinner decomposition.

In the first case, let us write the electronic energy (under the Coulomb Hamiltonian approximation and the Born-Oppenheimer paradigm) in terms of the first and second order RDMs:

$$\begin{aligned} E_e = & \int d\mathbf{r}_1 \left(\hat{t} - \sum_\alpha \frac{Z_\alpha}{r_{1\alpha}} \right) 1\left(\mathbf{r}'_1\right) \gamma^{(1)}\left(\mathbf{r}'_1; \mathbf{r}_1\right) \\ & + \frac{1}{2} \int d\mathbf{r}_1\, d\mathbf{r}_2 \frac{1}{r_{12}} 1(\mathbf{r}_1) 1(\mathbf{r}_2) \gamma^{(2)}(\mathbf{r}_1, \mathbf{r}_2), \end{aligned} \tag{9.17}$$

where we have already introduced adequate PUs affecting the integrated electron coordinates. Upon expanding the PUs, the kinetic and electron-nucleus potential energies become a sum of one-domain contributions, and the electron repulsion a sum of two-basin terms. Using Eq. 9.1,

$$E_e = \sum_A \left(T^A + \sum_\alpha V_{\text{en}}^{A\alpha} \right) + \sum_A V_{\text{ee}}^{AA} + \sum_{A>B} V_{\text{ee}}^{AB}. \tag{9.18}$$

From this point on, we are going to use a simplified notation in which basins are specified as superindices. In this way, $T^A = T(\Omega_A)$ corresponds to the atomic kinetic energy, and $V_{\text{en}}^{A\alpha}$ stands, for instance, for the electron-nucleus interaction between the electrons of domain A and nucleus α. Notice that any PU will provide well–defined V_{en} and V_{ee} terms, but that only QTAIM partitions have a unique definition for the domain kinetic energies [2, 19]. In practice, one chooses a particular form for the one–electron kinetic energy operator, usually the positive definite one $\hat{t} = \hat{g} = (1/4)\nabla \cdot \nabla'$. When the PU is a full atomic partition, i.e. each domain A is associated to a given nucleus α, the total molecular energy may be rewritten as a sum of intra– and interatomic terms,

$$E = \sum_A \left(T^A + V_{\text{en}}^{AA} + V_{\text{ee}}^{AA} \right) + \sum_{A>B} \left(V_{\text{en}}^{AB} + V_{\text{en}}^{BA} + V_{\text{ee}}^{AB} + V_{\text{nn}}^{AB} \right), \tag{9.19}$$

where the standard nuclear repulsion energy, $V_{\text{nn}}^{AB} = Z_A Z_B / R_{AB}$, has been added. It is this partition which lies at the heart of the IQA success, since it separates the large intra–atomic energy contributions from the usually much smaller interaction terms. It was the computational complexity of the 6D numerical integrations needed to obtain the V_{ee} terms that precluded such treatments in the past until efficient algorithms were developed [20, 21]. Let us thus define the atomic self-energy of atom (domain) A as

$$E_{\text{self}}^A = T^A + V_{\text{en}}^{AA} + V_{\text{ee}}^{AA}. \tag{9.20}$$

E_{self}^A tends to the *in vacuo* atomic energy as the interactions vanish. We also define the interaction energy between atoms A and B,

$$E_{\text{int}}^{AB} = V_{\text{en}}^{AB} + V_{\text{en}}^{BA} + V_{\text{ee}}^{AB} + V_{\text{nn}}^{AB}, \tag{9.21}$$

so that

$$E = \sum_A E_{\text{self}}^A + \sum_{A>B} E_{\text{int}}^{AB}. \tag{9.22}$$

In this way, atoms (domains) behave as ordinary physical systems in real space: self–energies contain one–particle energy contributions as well as intra–system interparticle interactions, while interaction energies contain all two–particle inter–system interactions (i.e. the two particles "belong" to two different systems). Atoms are thus objects that can be philosophically considered as physical entities in real space.

Thus, if a domain is composed of several atoms, defining a group, the above expressions are immediately generalized, and the total energy becomes a sum of group self-energies and inter–group interaction contributions [14]. We have found that both self– and interaction energies are transferable for domains with chemical sense [22].

Up to this point, all terms entering the proposed partitioning have a clear physical, but not necessarily chemical meaning. After all, the theory of the chemical bond is currently understood with terms like covalency and ionicity, not electron–nucleus attractions or electron–electron repulsions. IQA allows us to extract the former from the latter in an elegant manner. This is done by noticing that save V_{ee}, all the terms in E_{int}^{AB} are classical in nature. The 2RDM may always be written as a sum of classical (or Coulombic), and non-classical (or exchange–correlation) terms,

$$\gamma^{(2)}(\mathbf{r}_1, \mathbf{r}_2) = \rho_2(\mathbf{r}_1, \mathbf{r}_2) = \rho(\mathbf{r}_1)\rho(\mathbf{r}_2) + \gamma_{xc}^{(2)}(\mathbf{r}_1, \mathbf{r}_2). \tag{9.23}$$

Introducing the above further decomposition into Eq. 9.17, we may write analogously the electron repulsion between domains *A* and *B*,

$$V_{ee}^{AB} = V_{C}^{AB} + V_{xc}^{AB}. \tag{9.24}$$

Gathering together all the classical interaction terms, V_{en}, V_{nn}, and V_{C},

$$V_{cl}^{AB} = V_{C}^{AB} + V_{en}^{AB} + V_{en}^{BA} + V_{nn}^{AB}, \tag{9.25}$$

so that the (almost) electroneutrality of the atoms introduces a tremendous cancellation of terms, usually larger than an order of magnitude. The interaction energy between two domains turns out to be composed of classical and non-classical components,

$$E_{int}^{AB} = V_{cl}^{AB} + V_{xc}^{AB}. \tag{9.26}$$

As we are going to show, they can be related to the concepts of ionicity and covalency in the *AB* chemical interaction. It is interesting to notice that all the terms of the above expression are symmetric with respect to the exchange of *A* and *B* although V_{en}^{AB} is not.

Usually, the large atomic self–energies E_{self}^{A} are not relevant themselves, and only their change with respect to a given reference state with energy E_0^A is interesting. We define the *deformation energy* of atom *A* with respect to this reference as

$$E_{def}^{A} = E_{self}^{A} - E_{0}^{A}. \tag{9.27}$$

This is the second (and largest) source of energy cancellation, and both deformation and interaction energies (together with its classical and exchange–correlation components) have thus entered the chemical energy scale. If the atomic reference states are used to define binding energies, the latter result from the balance between the atomic deformations that occur upon interaction, which are usually positive, and the interaction themselves, usually negative:

$$E_{\text{bind}} = \sum_{A} E_{\text{def}}^{A} + \sum_{A>B} E_{\text{int}}^{AB}. \tag{9.28}$$

This is again very close to conventional chemical thinking, with atoms that must promote to higher energetic states to guarantee optimal interactions, and total deformation $\left(E_{\text{def}} = \sum_A E_{\text{def}}^A\right)$ and binding energy scales several orders of magnitude smaller than both total (E) and atomic $\left(E_0^A\right)$ reference energies.

The energetic partition may also be carried out at the resonance structure level. To do that, we just need introduce a PU for each electronic coordinate, so that

$$E_e = \sum_{\{S\}} \int d\mathbf{r}_1 \ldots \int d\mathbf{r}_N w_S \Psi^* \hat{H}_e \Psi, \tag{9.29}$$

where w_S stands for a product of N weight functions associated to a resonance structure S which, in its turn, is defined by the number of electrons ascribed to each of the m domains, $S = (n_A, \ldots, n_M)$. Thanks to permutational symmetry, the m^N terms in the above sum could be grouped into resonance structure classes each containing $n_S = N!/(n_A!n_B! \ldots n_M!)$ identical terms. In this way, we can define an energetic contribution for each resonance structure, so $E_e = \sum_{\{S\}} \langle H_e \rangle_S$. It is not difficult to show [15] that $\langle H_e \rangle_S$ takes the following simplified form if we use CGDMs,

$$\begin{aligned}\langle H_e \rangle_S = & \sum_A \int d\mathbf{r}\, w_A(\mathbf{r}') \hat{h} \gamma^{(1)}(\mathbf{r}'; \mathbf{r})[C_A] \\ & + \sum_{A,B} \int d\mathbf{r}_1 \int d\mathbf{r}_2\, w_A(\mathbf{r}_1) w_B(\mathbf{r}_2) r_{12}^{-1} \gamma^{(2)}(\mathbf{r}_1, \mathbf{r}_2)[C_{AB}],\end{aligned} \tag{9.30}$$

where $\hat{h}$ is the one–electron hamiltonian, and C_A and C_{AB} the condensations stemming from S when one or two electrons are associated to domains A, or A and B.

If we take into account Eqs. 9.18, 9.19, 9.20, 9.21, 9.22, 9.23, 9.24, 9.25, 9.26, 9.27 and 9.28, it stands clear that the energy of any resonance structure may also be written in an IQA–like manner, so

$$\begin{aligned}\langle H_e \rangle_S = & \sum_A \left\{T^A(S) + V_{\text{ee}}^{AA}(S) + V_{\text{en}}^{AA}(S)\right\} \\ & + \sum_{A>B} \left\{V_{\text{ee}}^{AB}(S) + V_{\text{en}}^{AB}(S) + V_{\text{en}}^{BA}(S)\right\},\end{aligned} \tag{9.31}$$

and if we use the probability of finding structure S, $p(S)$, to construct the structure share in the interatomic repulsion energy, $V_{\rm nn}^{AB}(S) = p(S)Z_A Z_B/R_{AB}$, then

$$V_{\rm int}^{AB}(S) = V_{\rm ee}^{AB}(S) + V_{\rm en}^{AB}(S) + V_{\rm en}^{BA}(S) + V_{\rm nn}^{AB}(S). \tag{9.32}$$

Furthermore, we may build normalized energetic contributions for each resonance structure, that is, their contribution to the energy if $p(S)=1$. If we consider quantity $A(S)$, its normalized value is simply $\tilde{A}(S)=A(S)/p(S)$. Doing so, the total energy of a system may be written as a structure average of normalized structure energies,

$$E = \sum_{\{S\}} p(S)\tilde{E}(S), \tag{9.33}$$

and each of the latter may be IQA partitioned:

$$\tilde{E}(S) = \sum_{A} \tilde{E}_{\rm self}^{A}(S) + \sum_{A>B} \tilde{E}_{\rm int}^{AB}(S). \tag{9.34}$$

It is also clear that each structure interaction energy may be further partitioned into classical and exchange–correlation terms. Proceeding this way, a PU leads us as close to describe the energy of a stationary state like an ensemble average of structure energies as the rules of quantum mechanics allow for.

9.2.3 *Electron Population Statistics*

A rather interesting link among CGDMs, the fluctuations (thus the statistics) of coarse–grained electron populations, and many of the standard concepts of the theory of the chemical bond, like bond orders, exists and is being actively developed in the last years [15, 23–27]. A simple starting point is the generalized population analysis of one-determinant expansions [28–30], based on the idempotent nature of Hartree–Fock 1RDMs, $\mathrm{Tr}(PS)^i = N$. Here, P is the usual SCF charge density–bond order matrix in a spin–orbital basis, S the overlap matrix over the primitive function set, and i any positive integer. This expression allows an ith–order partition of the N electrons of a system into a sum of terms which are products of i factors:

$$N^{(i)} = \underbrace{\sum_{\alpha,\beta\ldots,\iota}}_{i} (PS)_{\alpha\beta}(PS)_{\beta\gamma}\ldots(PS)_{\iota\alpha}. \tag{9.35}$$

A Mulliken–like gathering of the $\alpha \ldots \iota$ indices of the primitive set into atomic centers leads immediately to a decomposition of the electron population into one–, two–, ..., and i–center terms,

$$N^{(i)} = \sum_A N_A^{(i)} + \sum_{A>B} N_{AB}^{(i)} + \cdots + \sum_{A>B>\ldots>I} N_{AB\ldots I}^{(i)}. \tag{9.36}$$

The first–order $N_A^{(i)}$ quantities are usually called i-th order localization indices, and the $N_{AB\ldots K}^{(i)}$ are known as i-th order, K–center delocalization indices. It is easy to show that $N_A^{(1)}$ coincides with the Mulliken population of center A, and that $N_{AB}^{(2)}$ is identical to the Wiberg bond order [31]. Multicenter delocalization indices have been shown to provide very fruitful measures of multicenter bonding [32–35]. These Fock–space implementation may be generalized by noticing that due to the circular property of matrix traces, Eq. 9.35 may be rewritten as

$$N^{(i)} = \int d\mathbf{r}_1 \int d\mathbf{r}_2 \ldots \int d\mathbf{r}_i \; \gamma^{(1)}(\mathbf{r}_1;\mathbf{r}_2)\gamma^{(1)}(\mathbf{r}_2;\mathbf{r}_3)\ldots\gamma^{(1)}(\mathbf{r}_i;\mathbf{r}_1) \tag{9.37}$$

for single–determinant wave functions. Actually, there are several non–equivalent permutations of the i electrons in Eq. 9.37 if $i > 3$, so the correct symmetric form for $N^{(i)}$ must include an average over all of them,

$$N^{(i)} = \frac{1}{i!}\sum_P \hat{P} \int d\mathbf{r}_1 \ldots d\mathbf{r}_i \; \gamma^{(1)}(\mathbf{r}_1;\mathbf{r}_2)\gamma^{(1)}(\mathbf{r}_2;\mathbf{r}_3)\ldots\gamma^{(1)}(\mathbf{r}_i;\mathbf{r}_1), \tag{9.38}$$

where $\hat{P}$ is the usual permutation operator that acts on the naturally ordered electronic coordinates. This form is more general, but its correct physical meaning is still only valid for single–determinants.

The extension to fully correlated wave functions is obtained through the cumulant expansion of general RDMs [36]. For instance, for a single–determinant wave function, $\gamma^{(1)}(\mathbf{r}_1;\, \mathbf{r}_2)\gamma^{(1)}(\mathbf{r}_2;\, \mathbf{r}_1) = \rho(\mathbf{r}_1)\rho(\mathbf{r}_2) - \gamma^{(2)}(\mathbf{r}_1;\, \mathbf{r}_2) = -\gamma_{\rm xc}^{(2)}(\mathbf{r}_1, \mathbf{r}_2)$. This is called the irreducible component of the 2RDM. Similarly,

$$\begin{aligned}\gamma_{\rm irr}^{(3)}(1,2,3) = \gamma^{(1)}(1)\gamma^{(1)}(2)\gamma^{(1)}(3) - \frac{1}{2}\gamma^{(1)}(1)\gamma^{(2)}(2,3) - \frac{1}{2}\gamma^{(1)}(2)\gamma^{(2)}(1,3)\\ - \frac{1}{2}\gamma^{(1)}(3)\gamma^{(2)}(1,2) + \frac{1}{2}\gamma^{(3)}(1,2,3),\end{aligned} \tag{9.39}$$

where a simplified notation for electron coordinates has been used (notice that only diagonal densities appear on the right hand side). After some algebraic manipulation, a completely general decomposition of N into i-th order localization and delocalization indices as in Eq. 9.36 is obtained if a product of i PUs is used to recover N by integration of $\gamma_{\rm irr}^{(i)}$,

$$N^{(i)} = \int d\mathbf{r}_1 \ldots d\mathbf{r}_i \; 1(\mathbf{r}_1) \ldots 1(\mathbf{r}_i) \gamma_{\text{irr}}^{(i)}(\mathbf{r}_1, \ldots, \mathbf{r}_i). \quad (9.40)$$

The explicit expressions of higher–order irreducible RDMs become more and more cumbersome, but cumulant expansions offer a systematic path to obtain them.

Our last ingredient comes from the expectation values of domain restricted electron number operators, $\hat{n}_A$ being, for instance, the operator corresponding to the number of electrons in domain A [36]:

$$\hat{n}_A = \int d\mathbf{r} \; w_A(\mathbf{r}) \sum_i^N \delta(\mathbf{r} - \mathbf{r}_i). \quad (9.41)$$

Taking Eq. 9.40 for $i = 2$, it is straightforward to show that the $N_{AB}^{(2)}$ term in Eq. 9.36 is exactly $-2\langle(\hat{n}_A - \bar{n}_A)(\hat{n}_B - \bar{n}_B)\rangle$, thus measuring the two–particle covariance of the A and B electron number distributions. Since $N_{AB}^{(2)}$, the standard two-center delocalization index, commonly written as $\delta(\Omega_A, \Omega_B)$ or simply δ^{AB} in our superindex notation, may be also understood as the number of shared pairs of electrons between domains A and B, all these relations connect the Mulliken–like Wiberg bond order, the usual QTAIM δ, and the fluctuations in the electron population in an extremely interesting way. Following these reasonings, it is also possible to show that

$$N_{AB\ldots I}^{(i)} \equiv i(-1)^{i-1} \left\langle (\hat{n}_A - \bar{n}_A)(\hat{n}_B - \bar{n}_B) \ldots (\hat{n}_I - \bar{n}_I) \right\rangle, \quad (9.42)$$

so that multicenter bonding is linked to the existence of multicenter circuits of electron population fluctuations. It is to be noticed that, in order to actually obtain multicenter indices with $i > 2$, either complete wave functions or high order density matrices are required. This is a task more complex than computing the energy, and electron correlation may play a dominant role. Only a little is known about the effect of correlation upon δ, and almost nothing about higher-order indices.

Knowledge of the set of probabilities of all the resonance structures induced by a given PU, $p(S)$ $\forall S$, provides what we have called the *electron population distribution function* (EDF) [15, 24–27]. It is clear that access to EDFs allows us to obtain, through the population fluctuation link, any localization or delocalization index. For instance,

$$\begin{aligned} \langle \hat{n}_A \rangle = \bar{n}_A &= N_A^{(1)}, \\ \bar{n}_A - \left\langle (\hat{n}_A - \bar{n}_A)^2 \right\rangle = \lambda^A &= N_A^{(2)}, \\ -2 \left\langle (\hat{n}_A - \bar{n}_A)(\hat{n}_B - \bar{n}_B) \right\rangle = \delta^{AB} &= N_{AB}^{(2)}, \end{aligned} \quad (9.43)$$

Table 9.1 6–311 G* RHF, CAS[2,2] and full–CI (FCI) $\delta^{\mathrm{H-H'}}$ in the H_2 molecule at $R_{\mathrm{H-H'}} = 1.40\ a_0$ with different PUs

PU	RHF	CAS[2,2]	FCI
QTAIM	1.000	0.834	0.849
Becke	1.000	0.850	0.864
Hirshfeld	1.000	0.920	0.927
MinDef	1.000	0.941	0.947

where $\lambda(\Omega_A) = \lambda^A$ is the standard localization index. In this way,

$$N_A^{(1)} = \sum_{n_A=0}^{N} n_A p(n_A),$$

$$N_A^{(2)} = N_A^{(1)} - \sum_{n_A=0}^{N} (n_A - \bar{n}_A)^2 p(n_A),$$

$$N_{AB}^{(2)} = -2 \sum_{n_A,n_B=0}^{N} (n_A - \bar{n}_A)(n_B - \bar{n}_B) p(n_A, n_B). \tag{9.44}$$

As relatively efficient methods have been devised to obtain EDFs for correlated wave functions [24, 27], higher order multicenter delocalization indices may now be explored.

Vivid images of bonding from this population fluctuation point of view are obtained if we take into account that at the one-determinant level, α spin and β spin electrons are statistically independent [27]. This means that the full EDF $p(S)$ is a direct product of up and down components, $p(S) = p^\alpha(S^\alpha) \otimes p^\beta(S^\beta)$. If we apply these ideas to H_2, we necessarily have to concur that, being any sensible atomic PU symmetric in both H atoms, the probability of finding the α electron in one of the two equivalent domains has to be 1/2. This means that the two-electron one-determinant EDF obtained by the above direct product is fixed, $p(2, 0) = p(0, 2) = 1/4$, $p(1, 1) = 1/2$. This is a two–body symmetric binomial distribution (BD), that provides $N_A^{(1)} = 1$, $N_A^{(2)} = 1/2$, $N_{AB}^{(2)} = 1$, independently of the PU. The bond order equal to one comes from two independent electrons (1–determinant model) being equally shared between both domains, contributing each of them with 0.5 to δ. We have shown [24] that despite their intrinsic complexity, EDFs are generally dominated by the direct product of a small number of BDs, supporting the usual assumption of independent bonds.

The main effect of α–β correlation on bonding is a decrease in the delocalization ability of opposite spin electrons. In our simple H_2 example, the (1, 1) resonance structure is enhanced, so that δ decreases, as shown in Table 9.1. Notice that the effects of correlation are almost saturated at the simplest CAS[2,2] level and that, as a rule, the more interpenetrating the PU, the larger the value of δ, as expected. In many electron systems, single determinant values for δ are usually not so close to nominal bond orders, on one hand, and correlation has a non–negligible effect, especially on multiple bonds, on the other. Hence, HF values of δ are approximations (usually from above) to correlated δ values, even for highly interpenetrating PUs.

9.3 IQA Energetics

Let us start this Section by considering the IQA energetic terms in a paradigmatic case, the strongly covalent N_2 molecule. We will then switch to examine the evolution of the several IQA energy terms, i.e. deformation, classical, and exchange–correlation energies, in different bonding regimes. At the end, we expect to have offered a fairly complete survey of the IQA image of chemical bonding.

9.3.1 *Global Approach: IQA Analysis of N_2*

Let us then consider the N_2 molecule, and examine its IQA energetic decomposition. Figure 9.3 contains the QTAIM and MinDef images, as extreme cases of non-interpenetrating and highly diffuse atomic densities. The gross behavior put forward in Sect. 9.2 is clearly seen. Binding is a trade–off between atomic deformation and interatomic interaction. Both tend to zero at large interatomic distances, increasing in absolute value when the atoms approach. However, the scales in both PUs are rather different. The QTAIM E_{def} of a N atom remains a relatively modest quantity except at very small distances. At R_e it is 159 kJ/mol and it is quite flat. On the contrary, the MinDef value is always much larger, rising to 1,393 kJ/mol at equilibrium. Since the energetic balance between atomic deformation and interatomic

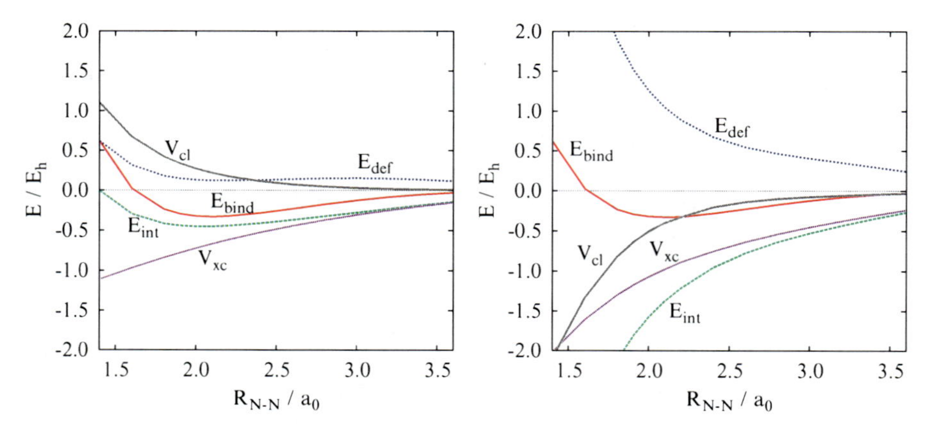

Fig. 9.3 IQA decomposition for N_2 as a function of the N–N distance for the QTAIM (*left*) and MinDef (*right*) PUs. Results from CAS[10,8]//6–311 G(d,p)++ calculations

interaction energy yields the same binding energy, this clearly means that $E_{\rm int}$ must be much more negative in the MinDef partition. In fact, it is equal to $-3{,}640$ kJ/mol, to be compared both with the $-1{,}184$ kJ/mol value obtained for the QTAIM PU, and with our calculated $E_{\rm bind} = -866$ kJ/mol. The reason for this behavior is rather clear. Interpenetrating densities exacerbate the interaction between the atoms, and this comes mainly from the classical interaction term, $V_{\rm cl}$ which, as we can see, has a completely different behavior in both cases. In the QTAIM partitioning it is continuously positive, and not too large. In the Mindef case it is attractive and extremely large. We have shown that a negative QTAIM $V_{\rm cl}$ value for homonuclear diatomic molecules is a necessary condition coming from basic electrostatics [14]. In our opinion, and as far as our goal is to study chemical bonding in real space, the assignment of electrons located very close to a given nucleus A to another one, let us say B, should be avoided. Only QTAIM or slightly interpenetrating PUs provide $E_{\rm def}$ values in the chemical binding scale, and thus give chemically appealing binding pictures in which atoms keep their energetic fingerprint upon interaction. In these IQA pictures, conventional covalent bonds (like that in N_2) are characterized by positive deformation energies which are small with respect to the large negative interactions, dominated by the quantum mechanical exchange–correlation term. If charge–transfer between two atoms is sizeable, their interaction energy will include a $V_{\rm cl}$ contribution, which may outweigh the covalent–like $V_{\rm xc}$ term, as it is found in LiF and many other molecules that are traditionally considered ionic, their $E_{\rm def}$'s also behaving differently, as will be shown shortly.

9.3.2 Deformation Energies

The self–energy of an atomic domain A, $E^A_{\rm self}$, may be written in terms of its *in vacuo* one–electron effective hamiltonian, $\hat{h}_A = \hat{t} - Z_A/r$, as

$$E^A_{\rm self} = \int d\mathbf{r}\, \hat{h}_A\, \gamma^{(1)}_A(\mathbf{r}';\mathbf{r}) + \frac{1}{2}\int d\mathbf{r}_1 d\mathbf{r}_2 \frac{\gamma^{(2)}_{AA}(\mathbf{r}_1,\mathbf{r}_2)}{r_{12}}, \tag{9.45}$$

where $\gamma^{(1)}_A$ and $\gamma^{(2)}_{AA}$ are the PU weighted densities introduced in Eqs. 9.7 and 9.10. If we set N–representability issues and other subtle questions aside, the above expression is the energy of an atomic open system (within the grand canonical ensemble scheme of the Statistical Mechanics of mixed quantum states [37]) with $N_A = \int d\mathbf{r}\rho_A(\mathbf{r})$ electrons and $N_{AA} = \int d\mathbf{r}_1 d\mathbf{r}_2 \gamma^{(2)}_{AA}$ distinguishable pairs. N_{AA} is only equal to $N_A(N_A - 1)$ for a closed system. The exact energy of a grand canonical ensemble of A atoms with average $N_A \in [n, n+1]$ electrons, n being an integer, is the linear interpolation between the energies of isolated A atoms with n and $n+1$ electrons, $E^A(N_A) = E^A(n) + (N_A - n)(E^A(n+1) - E^A(n))$ [22, 37]. It is then very likely that $E^A_{\rm self} \gtrsim E^A(N_A)$, although a rigorous demonstration is lacking due to the abovementioned subtleties. No violation of this relationship in actual chemical calculations has been found to date.

Table 9.2 Full valence CAS//TZV(2d,p) E^A_{def} values for first and second row homodiatomics A_2

E_{def}	H_2	He_2	Li_2	Be_2	B_2	C_2	N_2	O_2	F_2	Ne_2
QTAIM	33.5	1.3	70.3	110.4	199.1	202.5	149.3	213.8	190.8	5.4
MinDef	337.6	2.1	233.4	250.6	814.1	1390.1	1368.4	774.3	820.4	16.7

Energies in kJ/mol

When self–energies are measured with respect to the standard neutral atomic reference, so $E^A_0 = E^A(n = Z_A)$, it is clear that the deformation energy will have two components, one due to charge transfer (CT), and one due to electron reorganization (CR):

$$\begin{aligned} E^A_{\text{def}} &= E^A_{\text{self}} - E^A_0 = E^A_{\text{def}}(\text{CT}) + E^A_{\text{def}}(\text{CR}), \\ E^A_{\text{def}}(\text{CT}) &= E^A(N_A) - E^A_0, \\ E^A_{\text{def}}(\text{CR}) &= E^A_{\text{self}} - E^A(N_A). \end{aligned} \tag{9.46}$$

Deformation energies in homodiatomics lack the charge transfer term, and thus measure directly the effects of electron reorganization. For instance, using full–valence CAS//TZV(2d,p) wave functions in the first and second row homodiatomic molecules (i.e. active spaces including all the 2s, 2p orbitals plus the σ and p MOs of the M shell) provides the deformation energies of Table 9.2. As expected, the atomic deformation energies are much higher for the MinDef partition than for the QTAIM one. Moreover, in the MinDef PU total E_{def}'s are roughly twice the molecular binding energies, so the total interatomic interactions that must compensate this positive contributions must roughly be three times as large as the standard bond energies. This shows that the use of interpenetrating densities leads to interesting but not very chemical binding images.

The QTAIM partition, on the contrary, gives rise to an appealing picture of binding in these systems [38]. We should notice that we have used symmetrical topological atoms in all of our discussions. This means that half the central non-nuclear maximum found in Li_2, is associated to each of the Li basins. This symmetrical Li atoms also fullfils the basic QTAIM zero flux condition. The role of known sources of electron reorganization upon binding stands out clear. In this sense, we may see the multiconfigurational character (i.e. the need of promotion to higher electron states in chemical wisdom) of Be_2 and C_2, as well as the double–hump shape of E_{def} with Z, with a local minimum in N_2 which we have related to the particular stability of half-filled shells. It is also interesting to notice that E_{def}/Z is relatively constant. All these facts show that topological atoms keep track of their free-state electron structure when they participate in bonding events and that, grossly speaking, the CR component of E_{def} is actually measuring the energetic cost of the electron reorganization needed for accessing the bonded state. This is roughly proportional to the bonding strength, though other factors may be determinant.

In order to grasp the role of the charge transfer component of deformation energies it is necessary to turn to heteroatomic examples. Simple isoelectronic series of

Table 9.3 Topological charges $q_A = q(\Omega_A)$, in e, of the cationic species in the three isoelectronic series of heterodiatomics at the CAS[8,8] or CAS[10,10]//TZV(d) levels, as commented in the text

LiF	BeO	BN	BF	CO	NaF	MgO	AlN	SiC
0.9289	1.6305	1.2005	0.8858	1.1625	0.9317	1.0837	0.9345	0.9743

diatomic molecules serve well this purpose. Here we analyze the 12e series formed by the LiF, BeO, BN, and C_2 set; the 14e BF, CO, and N_2 one; and the 20e NaF, MgO, AlN, and SiC molecules, computed at the CAS[n,n]//TZV(d) level, where n is 8 for the iso12 and iso20 sets and 10 for the iso14 one. Since the calculation of electron affinities is notoriously difficult, the CT contributions to E_{def}'s will be discussed only for cations. Moreover, since the discussion of interpenetrating PUs follows the same patterns already commented, we restrict here to QTAIM data. Topological charges, contained in Table 9.3, allow us to distinguish rather clearly between traditionally ionic systems, like LiF, NaF, and BeO in a lesser extent, which show charge transfers very close to nominal oxidation states, and other highly polar molecules, like AlN or SiC, which remain quite far from closed shells. Notice also how difficult it is to transfer more than one electron beyond the second period.

Figure 9.4 contains a graphical summary of the results. The total deformation energies of cations may now acquire much larger values than those found in related homodiatomics. In the LiF, NaF, BeO, and MgO cases, E_{def} is completely dominated by the CT contribution. In other words, the Li, Na, Be, and Mg species within-the-molecule behave like grand canonical partially charged atoms, with small energetic losses due to promotion or delocalization. BN, BF, CO, AlN, and SiC, on the contrary, show the increasing role of the CR component on building the total E_{def}, although the CT one is still the largest contribution in every case.

The atomic, promotion–like nature of the CR term can also be grasped from the data, corroborating the ideas emerging from homodiatomics. $E^{\text{B}}_{\text{def}}(CR)$ is 577 kJ/mol in both the BN and BF molecules to within 0.5 kJ/mol. Moreover, the CR deformations of B and C in BN, BF, and CO slightly vary in the 577–644 kJ/mol range. Those of Al and Si in AlN and SiC are also similar and consistently smaller, 389 and 494 kJ/mol, respectively. This is the expected behavior of a promotion–like quantity upon changing the nuclear charge on rows and columns in the periodic table.

9.3.3 *Classical Interaction Terms*

Let us examine in this Subsection two atomic domains A and B, each of them associated to one nucleus, with charges Z_A and Z_B, respectively. The classical interaction energy between them may be compactly written as

$$V_{\text{cl}}^{AB} = \int d\mathbf{r}_1\, d\mathbf{r}_2 \frac{\rho_A^t(\mathbf{r}_1)\rho_B^t(\mathbf{r}_2)}{r_{12}}, \tag{9.47}$$

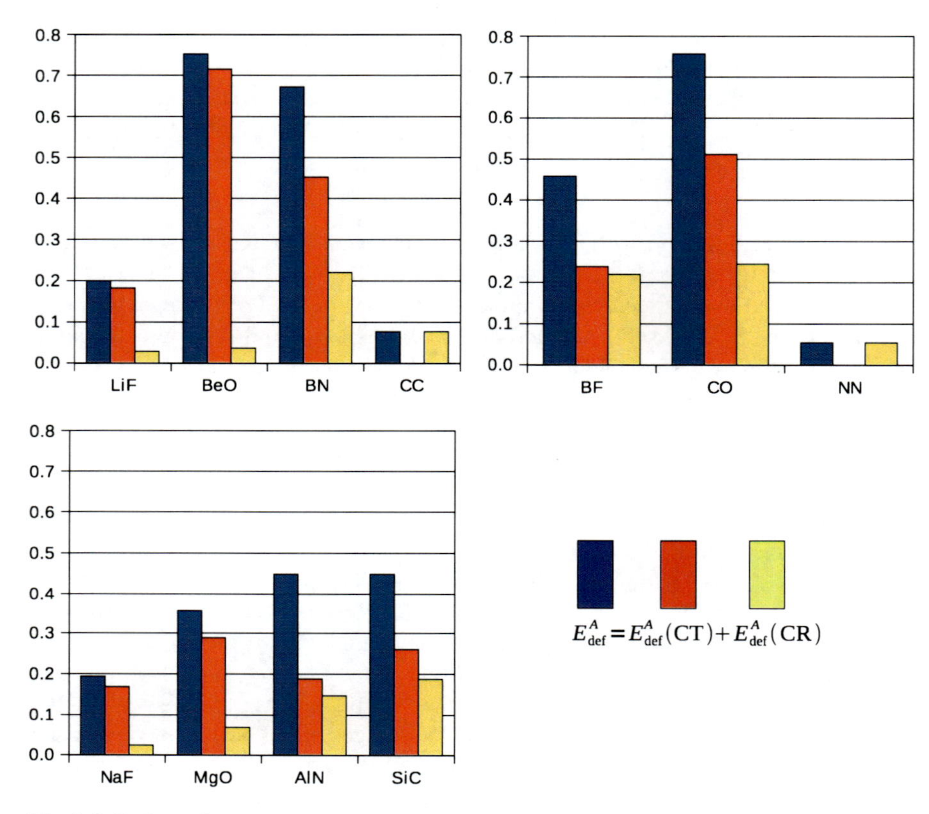

Fig. 9.4 Deformation energy components for the cationic species in three isoelectronic series of di-atomics. Results from CAS[8,8] or CAS[10,10]//TZV(d) calculations (see the text) for the QTAIM PU. All energies in E_h

where ρ^t_A is the total (electronic and nuclear) charge density of domain A: $\rho^t_A(\mathbf{r}) = -\rho_A(\mathbf{r}) + Z_A\delta(\mathbf{r} - \mathbf{R}_A)$, and $q_A = \int d\mathbf{r}\rho^t_A(\mathbf{r})$ its net charge. Deviations of $V^{AB}_{\rm cl}$ from $E^{AB}_Q = q_A q_B / R_{AB}$, its monopolar or first multipolar component [20, 21], measures the distortion of the atomic electronic charge from sphericity around each of the two nuclei, place the net charge of each atom on a single point and considering the centroid of its negative charge coincident with that of its positive charge (nucleus).

As commented before, it is easy to show from elementary electrostatics that two neutral non-interpenetrating symmetrical charge densities (as in a homodiatomic within a QTAIM PU) display a necessarily positive $V_{\rm cl}$. This ceases to be so if the densities interpenetrate each other or if they are not neutral.

Table 9.4 contains the QTAIM and MinDef $V^{AB}_{\rm cl}$ values for first and second row homodiatomics. As expected from the above comments, all QTAIM values are positive, and show the same triangular, rather symmetrical shape peaking in N_2 that was first seen in their deformation energies. When interpenetrating PUs are used,

Table 9.4 Full valence CAS//TZV(2d,p) classical interaction (in kJ/mol) between the two equivalent atomic domains in the first and second row homodiatomics A_2

V_{cl}^{AB}	H_2	He_2	Li_2	Be_2	B_2	C_2	N_2	O_2	F_2	Ne_2
QTAIM	110.4	0.0	3.3	9.2	124.7	365.6	557.6	360.6	138.5	0.0
MinDef	−273.2	−0.0	−157.3	−112.1	−615.0	−1160.1	−992.3	−194.9	−135.1	−0.4
Q_z^A										
QTAIM	−0.117	−0.006	0.482	−0.369	−0.365	−0.258	−0.639	−0.414	−0.374	−0.032
MinDef	0.245	0.003	1.310	−0.136	0.189	0.386	−0.207	−0.252	−0.110	−0.001

The z moment of the atomic electron distribution of the leftmost domain, $Q_z^A = \mu_z(\Omega_A) = \int d\mathbf{r}\, z\gamma_A^{(1)}(\mathbf{r})$ is also shown (in ea_0)

the classical interaction between the atoms becomes stabilizing, this time peaking in C_2. The different behavior may be understood if we examine the first non–trivial multipole of the electron distribution around a given nucleus, the leftmost one in the present case. As seen from the Table, the electron density of all the QTAIM atoms, except Li in Li_2, is back–polarized towards the rear part of the internuclear axis, a fact related with the position of the interatomic surface. The Li exception is well understood due to the formation of a non–nuclear maximum which we are not considering here [39]. The smaller the R_{AB} distance and the larger the total interaction, the greater the classical destabilization. MinDef atomic densities behave quite differently. H, Li, B, and C are forward–polarized, this explaining their large negative classical terms, while Be, N, O, and F are back–polarized.

When sufficiently large charge transfers occur, it turns out that the sign and magnitude of V_{cl} becomes dictated by the point charge contribution (the monopolar term) $E_Q^{AB} = q_A q_B / R_{AB}$. This may be clearly seen in Fig. 9.5, where QTAIM heterodiatomics are again analyzed. Notice how LiF, BeO, NaF, and MgO are nicely represented by just a classical point charge model in which nominal net charges have been substituted by topological ones. This is the origin of the success of the classical ionic model of bonding in ionic solids, where domain net charges are very close to nominal ones.

If charge redistribution is noticeable (see also Fig. 9.4), then non point charge contributions to V_{cl} may become as important as E_Q^{AB}. This is particularly clear in the CO and SiC cases, a fact that teaches us again how density distortions from sphericity lead to important energetic terms, even at the classical level.

The above comments are rather general for 1–2 interactions, where the multipolar expansion of V_{cl} usually fails to converge. V_{cl} may be formally written as the sum of multipolar and short-range components: $V_{cl} = V_{cl,mp} + V_{cl,sr}$. When $V_{cl,mp}$ is divergent, so it is $V_{cl,sr}$. If the multipolar term converges, as in 1–3 or more distant atomic pairs, its value may separate non-negligibly from the exact V_{cl} due to large short-range terms, and even in very favorable cases, asphericities may preclude the monopolar term to be dominant even for relatively remote centers.

For instance, let us take a HF//TZV++(d,p) calculation in the staggered conformation of methanol as an example. The QTAIM charges of the C, O, H_O, H_s, and H atoms, where H_O is the hydroxilic H, and H_s its corresponding staggered H in the methyl group, are 0.751, -1.246, 0.614, -0.015, and -0.052 e, respectively. The values of V_{cl} and E_Q^{AB} for the O–H_s pair are -12.6 and $+13.0$ kJ/mol, in that order. Notice how the asphericity of the electron distributions of both atoms is able to change the sign of the electrostatic interaction over that of the point charge model, even though both atoms are 3.79 a_0 apart and the multipolar expansion of the classical interaction has converged to 0.5 kJ/mol or better. Similarly, for the 1–4 H_O–H_s pair, $V_{cl} = 3.3$, while $E_Q = -4.6$ kJ/mol with $R_{AB} = 5.29$ a_0. The slightly different H_O–H pair shows V_{cl} and E_Q equal to -12.1 and -25.1 kJ/mol, respectively, with $R_{AB} = 4.42$ a_0, equal in sign and much larger in magnitude.

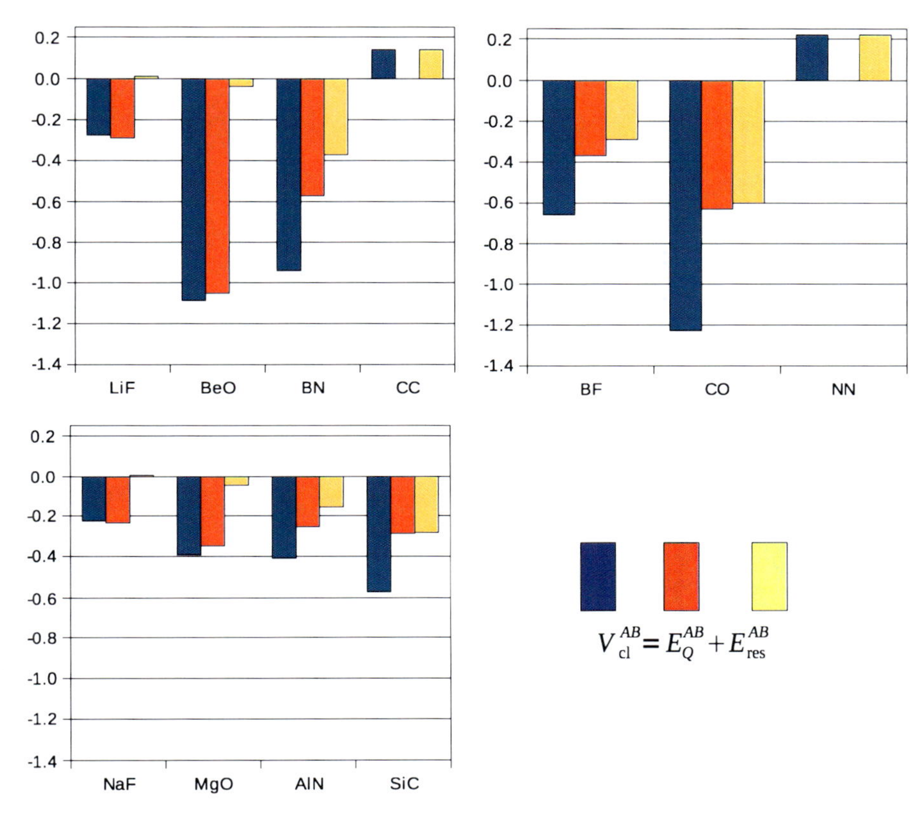

Fig. 9.5 Classical energy contributions to binding in three isoelectronic series of diatomics. $E_Q^{AB} = q_A q_B / R_{AB}$ is the zeroth order multipolar term to $V_{\rm cl}$, and $E_{\rm res}^{AB}$ the sum of all the residual multipole contributions. Results from CAS[8,8] or CAS[10,10]//TZV(d) calculations (see the text) for the QTAIM PU. All energies in E_h

9.3.4 *Quantum Mechanical Contributions*

The non–classical or exchange–correlation interaction energy between domains A and B,

$$V_{\rm xc}^{AB} = \int d\mathbf{r}_1\, d\mathbf{r}_2\, w_A w_B \frac{\gamma_{\rm xc}^{(2)}(\mathbf{r}_1, \mathbf{r}_2)}{r_{12}}, \tag{9.48}$$

contains the real space effect of antisymmetry, Fermi, and Coulomb correlation on the A, B pair. Recalling that $E_{\rm int}^{AB} = V_{\rm cl}^{AB} + V_{\rm xc}^{AB}$, and noticing how the classical term is mainly controlled by charge transfer, $V_{\rm xc}$ must underlie binding when CT is absent or small, thus providing a real space analogue of *covalency*. Table 9.5 justifies this idea showing the QTAIM and MinDef exchange–correlation energies

Table 9.5 Full valence CAS//TZV(2d,p) non–classical interaction (in kJ/mol) between the two equivalent atomic domains in the first and second row homodiatomics A_2

V_{xc}^{AB}	H_2	He_2	Li_2	Be_2	B_2	C_2	N_2	O_2	F_2	Ne_2
QTAIM	−620.8	−2.5	−259.4	−248.5	−827.9	−1387.6	−1805.6	−1296.9	−694.9	−13.8
MinDef	−853.8	−3.8	−425.0	−400.4	−1313.2	−2255.7	−2674.9	−1853.7	−1064.7	−20.9
δ^{AB}										
QTAIM	0.851	0.005	0.835	0.589	1.368	1.805	1.952	1.541	0.925	0.034
MinDef	0.952	0.008	1.014	0.786	1.765	2.393	2.439	1.875	1.171	0.045

Two-center delocalization indices are also shown

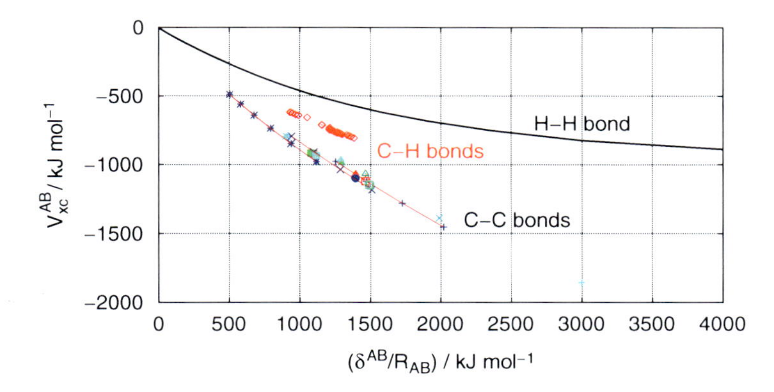

Fig. 9.6 Behavior of QTAIM V_{xc}^{AB} against δ^{AB}/R_{AB} for several CAS descriptions of the saturated and insaturated hydrocarbons described in the text at several geometries. A full–CI calculation in H_2 is also shown for comparison purposes

for our test homodiatomics. Notice how in either of the PUs their values are clearly proportional to the delocalization or bond order indices. It is also a general fact that delocalization indices coming from interpenetrating PUs resemble the nominal bond orders of simple MO theory more closely than those of the QTAIM partition.

The relation between V_{xc} and standard covalency deserves further analysis. To that end, it is reasonable to examine prototypical C–C and C–H interactions. We have obtained full valence CAS descriptions for ethane, ethene, and ethyne at different C–C internuclear distances; p–orbital CAS functions for the allyl cation, radical, and anion; and π–orbital CAS states for benzene and the cyclopropyl cation, radical, and anion. Figures 9.6 and 9.7 summarize some interesting results for QTAIM PUs. A mixed symbol, color code has been used to distinguish the systems: stars, crosses, and plus signs for ethane, ethene, and ethyne, respectively; squares, triangles, and filled circles for allyl, cyclopropyl, and benzene; red, yellow, and cyan for cationic, radical, and anionic species.

First, an interesting universality is found in the evolution of V_{xc} with δ^{AB}/R_{AB} for a fixed AB pair [40]. Its origin is rationalized by noticing that δ^{AB} (Eq. 9.43) differs from the form of Eq. 9.48 in the absence of the interelectron distance, $r_{12} \simeq R_{AB}$ in this regime. This means that the larger the number of shared electron pairs between the basins, the larger the stabilization due to non–classical effects. These type of plots contain information not only about the strength of a given interaction (i.e. bond), but also about its stiffness (i.e. the sensitivity of the bond strength to bond distance). Although the general validity of this assertion is still to be confirmed, other universal relationships displayed by topological indicators [41, 42] point towards its true generality. It is clear from these results that, at least in simple cases, bond order is a continuous parameter for a given pair of bonded atoms, and that V_{xc} depends smoothly on it.

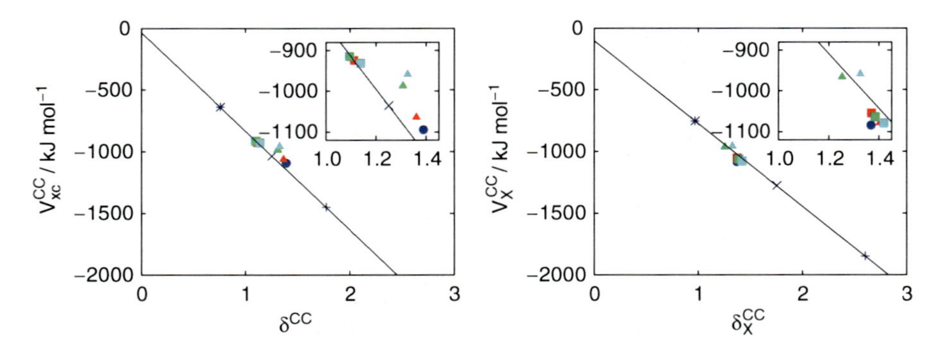

Fig. 9.7 QTAIM exchange–correlation (V_{xc}^{AB}, *left*) and exchange–only (V_{X}^{AB}, *right*) interaction energies for C–C bonds at equilibrium configurations versus the corresponding components of the C–C delocalization index, δ^{AB} and δ_{X}^{AB}. The systems are those discussed in the text. The insets expand interesting areas from the main plots

If we now constrain molecular geometries to equilibrium ones, as done in Fig. 9.7, we find a clearly linear relation between exchange–correlation energies and delocalization indices (left panel), with a correlation coefficient $r^2 = 0.9998$. This fact points again towards associating V_{xc} to covalent contributions, which in naïve descriptions are additive in bond order. Even more interesting, deviations from the trend line, as shown in the inset, may be used to find specially stable or unstable species. However, since the amount of correlation included in a given C–C bond depends on the quality of the CAS space (good for the saturated species, worse for allyl and cyclopropane, worst for benzene), this effect is best seen if we isolate the exchange–only component of V_{xc} and δ, which we will write V_{X}, and δ_X, respectively. When this is done (Fig. 9.7, right panel), the benzene and C_3H_3 moieties, which were lying above the trend in the left panel, now follow the trend much better. A stabilization energy of about 25.1 kJ/mol may be inferred for each C–C bond in benzene, a value well in line with standard measures of its resonance stabilization.

The above arguments do not only follow for mainly homopolar links. Figure 9.8 shows how the covalent-like contribution to bonding in the isoelectronic series of diatomics examined in this Section does also follow the δ/R proportionality. These results are remarkable from many points of view, but let us only notice here how V_{xc} becomes more negative on decreasing the electronegativity difference (i.e. on increasing homopolarity) in the AB pair, and how these data are even more universal than those found in hydrocarbons. Since in the present case the treatment of correlation is more uniform across the different systems, the possibility of a deeper V_{xc} versus δ/R universal relation should be clearly explored. We should also notice that V_{xc} seems also to control the appearance of bond critical points. We have shown that the latter seem to indicate the existence of a privileged exchange–correlation channel in molecules [43].

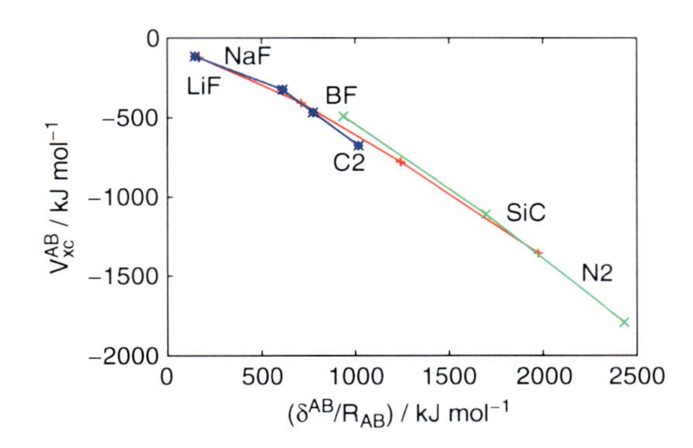

Fig. 9.8 QTAIM exchange–correlation interaction energies for the isoelectronic series of heterodiatomics of Sects. 9.3.2 and 9.3.3. In *red*, *green*, and *blue*, the 12, 14, and 20–electron series comprising LiF, BeO, BN, and C_2; BF, CO, and N_2; and NaF, MgO, AlN, and SiC, respectively

9.4 Understanding Chemistry with IQA

We will devote this Section to outline how the IQA approach together with a given PU (which will be generally, but not necessarily, a QTAIM partition) may be used to provide new insights in some problems of the theory of chemical bonding. We will first consider an outline of what image of bonding emerges from IQA thinking. To do so we will use EDFs to help us classify the electrons in a system into well localized and delocalized subsets. Only the latter participate in covalent–like bonding and contribute to V_{xc}. The former may, nevertheless, play a significant role determining the V_{cl} component of a given interatomic interaction. After this chemically appealing image has been presented, we will examine a couple of real world applications.

9.4.1 Real Space Picture of Bonding

Let us start by considering a simple dinitrogen molecule. With its 14 electrons and two atomic domains, its EDF only contains 8 independent components: $p(14, 0)$, $p(13, 1), \ldots, p(7, 7)$. This is so because, for any homonuclear diatomic molecule, $p(n_A, n_B)$, the probability of finding n_A electrons in the left A basin, and $n_B = 14 - n_A$ electrons in the right B domain, is symmetric in A and B. For N_2, we may thus restrict n_A to lie in the 0–7 range.

In a one determinant approximation, the full EDF $p(S)$ is the direct product of equivalent α and β distributions, $p(S) = p^{\alpha}(S^{\alpha}) \otimes p^{\beta}(S^{\beta})$, and the independent components of p^{α} (or p^{β}) are only 4: $p^{\alpha}(7,0), \ldots, p^{\alpha}(4,3)$. Figure 9.9 displays the p^{α} component. It is immediately seen that if we neglect the (7,0), (6,1) and their symmetric structures, only two independent $p^{\alpha\prime}$s arise, so that the initial 7 electron distribution is effectively generated by only three α electrons (this is shown in green). This means that the remaining four do not participate in delocalization events, i.e. they are completely localized. By symmetry, two of them lie in the

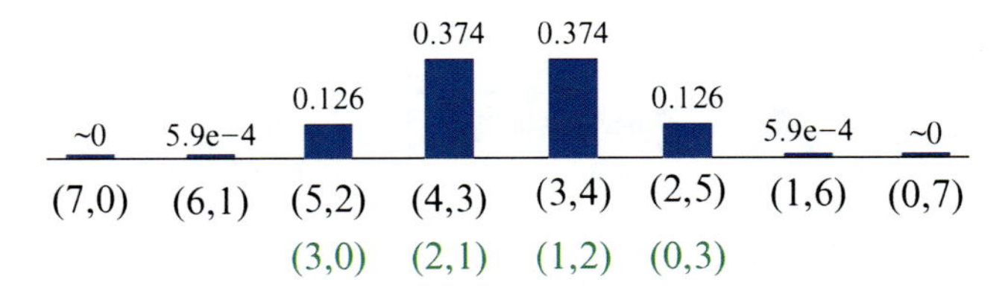

Fig. 9.9 $p^{\alpha}(n_A, n_B)$ EDF for the ground state of the N_2 molecule at the single determinant HF//6-311 G(d,p) level in a QTAIM partition. Only the p^{α} component is shown

left, two in the right domain. Moreover, the effective three electron distribution is extremely similar to the triple direct product of three one–electron distributions, $p_3^{\alpha} = p_1^{\alpha} \otimes p_1^{\alpha} \otimes p_1^{\alpha}$, with $p_1^{\alpha}(1,0) = 1/2$, and $p_1^{\alpha}(0,1) = 1/2$. These would lead to $p^{\alpha}(3,0) = p^{\alpha}(0,3) = 1/8 = 0.125$, and $p^{\alpha}(2,1) = p^{\alpha}(1,2) = 3/8 = 0.375$ [27]. Each of these independent electrons are completely delocalized, in the sense that the probability of finding them in each of the domains is equal. The α set of electrons is thus divided into two groups of two localized, and one group of three independent, delocalized electrons. Adding the equivalent β set, we arrive at an extremely familiar picture. Two cores made of two α, β pairs, and three pairs of independent α, β pairs, the standard triple bond. Using Eq. 9.44 we may use these data to obtain a delocalization index equal to 3.040.

Let us see how correlation alters this image. For instance, in LiH [24] a HF//6-311G(d,p) QTAIM calculation leads to $p^{\alpha}(0, 2) = p^{\beta}(0, 2) = 0.0034$, $p^{\alpha}(1, 1) = p^{\beta}(1, 1) = 0.9458$, and $p^{\alpha}(2, 0) = p^{\beta}(2, 0) = 0.0508$, where the first index refers to the Li domain. This provides a picture in which when one α (β) electron is found within the Li domain, the other $\alpha(\beta)$ electron is found in the H domain with a probability close to one (0.9458). Thus, the two same spin electrons avoid each other as much as possible, and Pauli's principle is clearly mapped to real space spin distributions. When obtaining the full four electron EDF by direct product, the dominant resonance structure corresponds to two very localized α, β pairs in each atomic domain with a probability $p(2, 2) = p^{\alpha}(1, 1) \times p^{\beta}(1, 1) + p^{\alpha}(0, 2) \times p^{\beta}(2, 0) + p^{\alpha}(2, 0) \times p^{\beta}(0, 2) = 0.8948$. Fermi correlation spatially segregates same spin electrons, which cannot be considered statistically independent anymore. When Coulomb correlation is added, a smaller similar effect is switched on for α, β pairs. In LiH, the two same–basin pairs try to move away from each other, and the ionic structure slightly drops to $p(2, 2) = 0.8907$ at the CAS[6,4] level. Similarly, Coulomb correlation in N_2 makes the three independent α, β pairs that form the standard triple bond to become statistically dependent. As a consequence, they try to localize in each basin, and the weight of localized structures increases, as shown in Fig. 9.10. For instance, $p(7, 7)$ and $p(8, 6) = p(6, 8)$ pass from 0.3110 and 0.2338, respectively, in the HF calculation to 0.3952 and 0.2438 at the correlated level. Since the delocalization index is a measure of the width of the EDF, this decreases sharply from the above 3.04 to the CAS 1.99 value [24]. This effect should be larger the larger the number of electrons involved, so the effective bond order of multiple bonds saturates quickly.

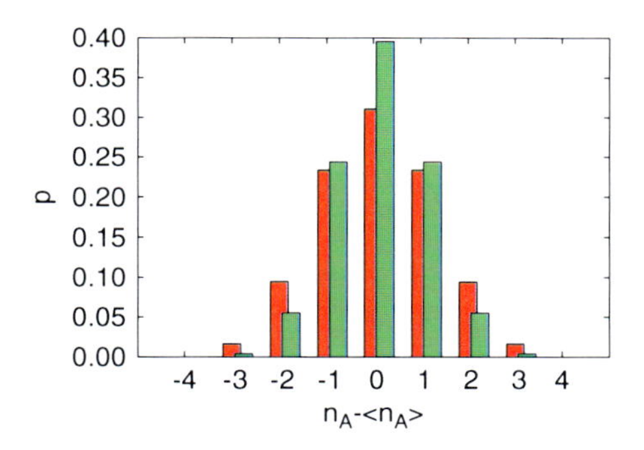

Fig. 9.10 $p(n_A)$ QTAIM EDF for the ground state of the N_2 molecule for TZV(2d,p) HF (*red*) and full valence CAS (*green*) descriptions, at their respective theoretical equilibrium geometries

It may further be shown that diagonalization of $\gamma_{\rm xc}^{(2)}$ in the occupied orbital basis leads to an approximate partition of $V_{\rm xc}^{AB}$ into a sum of contributions for effective one–electron states called domain natural orbitals [44]. Though we will not pursue this here any further, we will state that only delocalized electrons between domains A and B contribute non-negligibly to the final $V_{\rm xc}^{AB}$ value. This may be grasped from Eq. 9.48. Only situations in which $\gamma_{\rm xc}^{(2)}(\mathbf{r}_1, \mathbf{r}_2)$ is large for $\mathbf{r}_1$ in A and $\mathbf{r}_2$ in B (a delocalized state) will give a significant contribution to the exchange–correlation energy. Two-center covalent–like bonding in our IQA image is driven by electron delocalization between two domains. The larger the number of electrons delocalized (as well as its extent) the larger the exchange–correlation stabilization. Since Coulomb correlation tends to inhibit delocalization, correlated covalent–like bonds are generally weaker. When the dominant structures, as in LiH, contain very localized electrons well separated in space, and with a large CT, the covalent–like contribution is very small, and $V_{\rm cl}$ approaches the value of the naïve ionic model.

Full EDFs in polyatomic molecules allow us to access multicenter bonding features, even though their energetic significance is to be understood in terms of induced changes in the effective two–particle properties. Let us just show a couple of examples, the full valence CAS//6–311G(d,p) descriptions of BH_3 and NH_3 (Fig. 9.11). Notice the similarities and the differences in their behavior. In both cases, there are a number of parent structures, like (7, 1, 1, 1) in ammonia, followed by children ones in which the number of electrons in the central atom is unchanged but one electron in a H atom has jumped to another H. This is another way of talking about three-center delocalizations among the H_3 group.

Borane is clearly dominated by ionic contributions. The most probable one is the closed shell $(B^{+3})(H^-)_3$ structure. The probability of others quickly decreases as we approach the neutral one. Ammonia is much more covalent, on one hand, but the high electronegativity of the N atom provides a non negligible probability of finding the ten electrons on the molecule residing in the N domain. Actually, the (10, 0, 0, 0) structure is not much less probable than any of the three other dominating terms.

We can easily compute three– and four–center delocalization indices from our EDFs at the correlated level. With the normalization imposed by Eq. 9.42, some

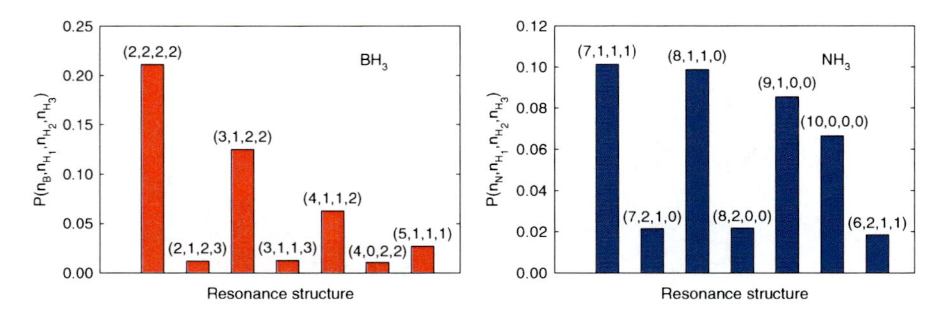

Fig. 9.11 QTAIM EDFs for full valence CAS//6–311 G(d,p) calculations in BH_3 and NH_3. Only structure classes with weights larger than 1% are shown. A structure class is made from all the structures equivalent by symmetry, e.g. $(3, 1, 2, 2) \equiv (3, 2, 1, 2) \equiv (3, 2, 2, 1)$

Table 9.6 Selected QTAIM CAS//6–311 G(d,p) two–, three–, and four–center delocalization indices for BH_3 and NH_3

System	$N^{(2)}_{AH_1}$	$N^{(2)}_{H_1H_2}$	$N^{(3)}_{AH_1H_2}$	$N^{(3)}_{H_1H_2H_3}$	$N^{(4)}_{AH_1H_2H_3}$
BH_3	0.523	0.123	0.048	0.011	−0.192
NH_3	0.756	0.054	0.068	0.004	−0.118

A is the central atom

of them may be found in Table 9.6. Notice that 3–center bonding among the H atoms is larger in BH_3, as expected, but that the opposite is true if the triad contains the central heteroatom, also in line with standard reasoning. Four–cycles contribute negatively to electron populations in both systems, more intensely in borane.

9.4.2 Solving Paradoxes in Hydrogen Bonding

The physical nature of hydrogen bonding is still today a source of debate. On the one hand, several electrostatic models [45–50] which originated in Pauling's view [51, 52] have been found to successfully predict the structure of many of these systems. On the other, a number of experimental signatures of hydrogen bond (HB) formation, like redshifts in the infrared *X*–H stretching bands [53–56], or large *J* couplings and chemical shifts [53], are interpreted by others as inequivocal signs of covalent contributions to bonding. From a fundamental perspective, these doubts stem from the inability to decompose total molecular, or supermolecular, energies into well–defined physical components. The application of IQA seems a natural way to deepen into these questions.

A systematic study in simple homo– and hetero–dimers of HF, H_2O, and NH_3 at the 6–311G + (d,p) IQA/QTAIM level both for single determinant and full valence CAS wave functions [57] reveals several interesting facts. First, in agreement with the seminal works by Umeyama and Morokuma [58], both the proton donor

(PD, AH) and proton acceptor (PA, B) fragments suffer important electron density rearrangements upon HB formation. By using an antisymmetrized state Ψ^0_{AH-B} constructed from the isolated fragments' wave functions at the single–determinant level, and using a point charge model to represent the electrostatic effect of each fragment onto the other, we have shown that the antisymmetrized and polarized density of the complexes, ρ^0_{pol}, is almost identical to the final density, and that the effect of Pauli exclusion is small and confined to the HB region. Since ρ determines the energy, and the latter is basically determined by polarization effects, electrostatic models should provide reasonable energetics and geometries.

Table 9.7 contains a summary of our results. Total deformation energies scale with HB strength, and as a rule, the acceptor fragment shows larger deformations than the donor one. It is also interesting to notice that if we fix the PD, then the PA deformation increases on going from N_2 to HF to NH_3, and does also correlate with the inter–fragment CT, and the electrostatic and binding energies of the molecular complexes. As we can see from the Table, deformation energies, inter–monomer exchange–correlation energies, total electrostatic interactions, binding energies, and charge transfer are stronger the larger the electronegativity of the central atom of the PD and the smaller that of the PA monomer.

We have shown [57, 59] that the sum of the total deformation energy between two interacting fragments and their mutual exchange–correlation energy, forms a short range quantity which corresponds, in many ways, to the exchange–repulsion terms of other energy partition techniques, like symmetry adapted perturbation theory (SAPT) [60] or the energy decomposition analysis (EDA) [61, 62]. We thus define XRC (exchange, repulsion, correlation) energy as: XRC $= V^{AB}_{\text{xc}} + E^A_{\text{def}} + E^B_{\text{def}}$. As we can see, XRCs are positive and rather small quantities in these weakly interacting systems, which may be understood to exist in a quasi–pertubative regime. This means that the main *covalent* contribution to HB bonding in IQA is basically canceled out by the deformation of the fragments upon interaction.

Moreover, it is found that $V^{\text{PD,PA}}_{\text{xc}}$ is basically saturated with the Ψ^0 antisymmetrized function, and that electron relaxation towards the final Ψ does only introduce a small extra stabilization. A similar result is found for E_{def}, but the correct V_{cl} term needs from polarization effects to fully develop. The fermionic character of electrons introduced by antisymmetrizing the fragments' wave functions turns out to be the root of the deformation and exchange–correlation energies of the fragments. The XRC sum of both effects is slightly repulsive, and were it not for the classical stabilization due to dipolar interaction and charge transfer, the complexes would not form.

The XRC cancellation allows us to view the origin of the electrostatic/covalent controversy from a real space perspective. For those advocates of the covalent picture, it is clear that $V^{\text{HB}}_{\text{xc}}$ itself, the exchange–correlation energy between the directly bonded HB atoms, provides a good correlation with the overall binding energies of the complexes. A similar good correlation is also obtained between the binding energy of the complexes and the delocalization index between the H and B atoms (δ^{HB} in Table 9.7), understood as a covalent bond order. Moreover, since the dependence of the V_{xc} terms as we explore different geometrical arrangement of

Table 9.7 IQA/QTAIM analysis for simple HB complexes (Adapted from Ref. [57] with kind permission of The American Institute of Physics)

System	E_{def}^{PD}	E_{def}^{PA}	E_{def}	$V_{cl}^{PD,PA}$	$V_{xc}^{PD,PA}$	E_{int}	XRC	E_{bind}	Q^{PD}	V_{cl}^{HB}	V_{xc}^{HB}	δ^{HB}
HF–HF	18.8	22.2	41.0	−20.5	−38.1	−58.6	2.9	−16.3	−5	−344.7	−21.3	0.036
	16.7	23.4	40.2	−23.4	−35.1	−58.6	5.0	−18.4	−5	−398.7	−17.2	0.029
HF–H_2O	26.8	34.7	61.1	−36.4	−60.7	−97.1	0.4	−34.7	−21	−635.5	−30.5	0.049
	28.4	38.9	67.4	−41.8	−63.2	−105.0	4.2	−36.4	−20	−702.4	−32.6	0.052
HF–NH_3	36.4	51.9	88.3	−47.3	−89.1	−136.4	−0.8	−48.1	−50	−621.2	−43.9	0.068
	38.5	56.9	95.4	−52.3	−92.5	−144.7	2.9	−44.8	−46	−678.5	−48.1	0.077
H_2O–H_2O	37.7	25.9	64.0	−25.5	−58.6	−84.1	5.4	−21.3	−15	−441.8	−37.7	0.066
	22.6	27.6	50.2	−26.8	−43.1	−69.9	7.1	−21.3	−9	−522.5	−25.5	0.045
H_2O–NH_3	26.4	40.6	66.9	−29.7	−59.4	−89.1	7.5	−23.8	−27	−453.5	−32.6	0.057
	25.9	36.8	62.3	−31.0	−54.4	−85.3	7.9	−24.3	−21	−487.4	−32.2	0.058
NH_3–H_2O	16.7	18.0	34.7	−13.4	−27.6	−41.0	7.1	−8.4	−3	−250.2	−17.6	0.036
	16.3	16.7	33.0	−13.8	−26.8	−40.2	6.3	−9.2	−3	−272.8	−17.6	0.035
NH_3–NH_3	18.8	28.4	47.3	−16.3	−38.5	−54.8	8.8	−12.1	−13	−237.2	−23.8	0.047
	18.8	22.6	41.4	−15.9	−31.8	−47.7	9.2	−10.9	−9	−251.0	−21.3	0.044
HF–N_2	7.1	18.4	25.5	−8.8	−23.8	−32.6	1.7	−7.5	−7	−74.0	−12.1	0.022
	6.7	16.7	23.4	−11.7	−21.8	−33.5	1.7	−10.5	−5	−70.7	−11.3	0.022
FHF^-	246.4	64.0	310.4	−284.9	−269.4	−554.3	41.0	−243.9	−124	−750.9	−182.8	0.219
	283.6	125.9	409.1	−329.7	−255.6	−585.3	153.5	−177.4	−105	−826.2	−159.0	0.192

Energetic quantities in kJ/mol, and charges in *me*. The last three columns contain the local V_{cl}, V_{xc}, and δ values for the hydrogen bond atomic pair. The first line for each system corresponds to the full–valence 6-311 + G(d,p) CAS optimized results, and the second to the Hartree–Fock level. E_{bind} values come from the analytic calculations, their deviations with respect to $E_{def} + E_{int} = V_{cl}^{PD,PA}$ + XRC give an indication of the (small) numerical integration errors the monomers is found to be similar to that displayed by the classical interactions, V_{xc} a may also be used to understand the equilibrium geometries of the complexes. Thus, covalency effects alone cleanly explain the observed energetics. Similarly, both $V_{cl}^{PD,PA}$ and a simple indicator of its strength, the charge transferred from the PA to the PD, show a very good correlation with the binding energies of the complexes, so binding may also be understood in terms of electrostatics alone. We should warn now, however, that this is a global effect, and that local electrostatic interactions (V_{cl}^{HB}, for instance) are much more intense and cannot be used to estimate the stabilization energy of a HB dimer.

It is the possibility of partitioning a final observed quantity, i.e. a binding energy, into multiple terms that originates different, many times irreconcilable positions. The fine graining that the IQA partitioning achieves, together with the clean physical meaning of all of its energetic quantities, allows us to observe chemical bonding from multiple perspectives. As the examples of this Section teach us, changing the focus from one to other of these energy terms may provide a completely different picture of the same underlying physics.

9.5 Restricted Space Partitioning

In the previous sections of this chapter approaches were discussed, that rely on the partitioning of the examined system into separate spatial fragments. The extent of the fragments, like for instance the atomic basins in case of QTAIM [2], were comparable with the inter-atomic distances (coarse-grained partitioning).

In the following sections another kind of space partitioning will be presented. This space partitioning results in huge number of extremely small regions in which the chosen property is evaluated, yielding a very dense distribution of values [63]. The individual spatial regions resulting from this approach are not aimed to represent some chemically relevant entities. Instead, the distribution based on such space partitioning can be subsequently used to decompose the system into possibly meaningful parts.

9.5.1 ω-Restricted Partitioning

To gain knowledge about a complex system it can be suitable to examine the desired properties for samples of same (chosen) quality, for instance, having the same volume, mass or speed. Then, the properties of all samples can be compared. This approach can also be used in case of continuous distributions, where the sample extent as well as the evaluated property are determined by integrals of particular function. More specifically, the sample volume V_i is given by region μ_i, centred around chosen position $\mathbf{a}_i$, which gains such extent that the integral of so called *control* function over this region attains the fixed value ω (i.e., the *control* function

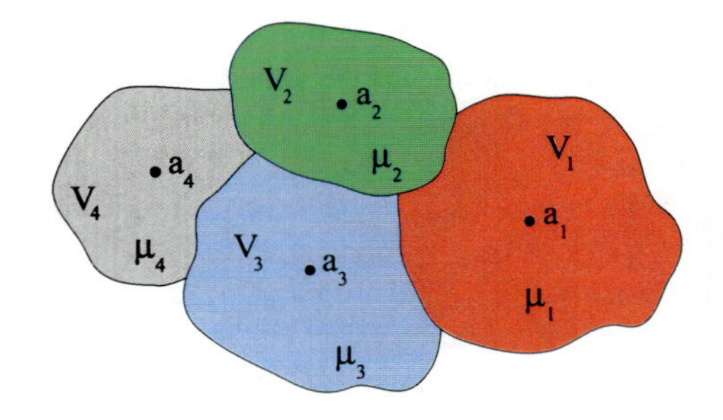

Fig. 9.12 Schematic representation of the space partitioning subject to the ω-restriction. The compact regions μ_i, centred at $\mathbf{a}_i$, have different volumes V_i but yield always the same integral ω

controls the extent of the regions μ_i and thus the partitioning of the space). This prescription would allow for large freedom in the shape of the sample region. Hence, we demand that the sample region describes the system as locally as possible. The positions in a sample volume must not spread too far away from the centre $\mathbf{a}_i$, i.e., the sample region should be as *compact* as possible. Clearly, with those conditions, a separate single sample would be a sphere around the chosen position with the integral of the control function over the sphere yielding the value ω [64, 65]. The approach described in this section follows another route [63].

Instead of separately probing different positions let us divide the whole system at once into so many samples that they form a compact non-overlapping and space filling set of regions fulfilling the restriction ω. The sum of all sample volumes yields the total volume of the system. For such space partitioning it is no more possible that all sample regions become spheres. Each sample region will adapt such form as to achieve the highest compactness subject to maximal compactness of the neighbouring sample regions, cf. Fig. 9.12. Though the freedom of the sample region shapes is somewhat reduced (due to the compactness) there are still possibly infinite many ways to perform such space partitioning for any chosen restriction ω. This is one significant difference to the partitioning scheme used in previous sections which was uniquely defined by the gradient of a scalar field.

The above described compact space partitioning crucially depends on the control function. Some general remarks regarding the number of sample regions and the ω values can be given for space partitioning based on control function $f_c(\mathbf{r})$ of single coordinate $\mathbf{r}$. The partitioning is contingent upon the value of the integral $F_c = \int f_c(\mathbf{r})\, d\mathbf{r}$ over the whole system:

- $F_c = \pm\infty$: for any real restriction ω (with appropriate sign) there will be infinite number of compact regions μ_i.
- $F_c = 0$: only the restriction $\omega = 0$ is allowed, because $F_c = \kappa\omega$ (with κ the total number of regions). The number of compact regions μ_i is undetermined (it can be the whole system or any decomposition into regions enclosing domains of positive and negative function values integrating to zero).
- $F_c \neq 0$: for any real restriction $\omega \neq 0$ (with appropriate sign) there will be finite number $\kappa = F_c/\omega$ of compact regions μ_i.

Relatively simple case for the first possibility is given when the control function is a non-zero constant $f_c \equiv \xi$. Then, all the compact regions μ_i will have identical volume $V_i = \omega/\xi$ and similar sample shape. If the control function is not a constant then the density of the compact regions (regions per volume) will also vary in the volume of the system.

Concerning the second possibility, $F_c = 0$, the control function integrates to zero over the whole system for instance (apart from the trivial case $f_c \equiv 0$) in case of the Laplacian of a scalar field. Thus, the space partitioning into basins describes (cf. previous sections), in certain sense, a space partitioning into regions controlled by the field Laplacian restricted to integrate to zero over the basin volumes. As mentioned above, these conditions are not sufficient to uniquely partition the system as there are many other decompositions conceivable where the control function integrates to zero.

The most interesting case is the one where the integral of the control function over the whole system yields a non-zero real value. Especially, if the control function does not change the sign in the system volume, because then the cancellation of values within the sample regions to fulfil the ω-restriction is not necessary. The number of sample regions is uniquely given and the local behaviour of the samples scales with the restriction ω. This scaling is manifested in the proportionality between the volume of the compact regions μ_i and the value of the restriction ω when reduced to very small value. For smooth control functions and fixed very small ω the volume V_i of the sample region μ_i will remain almost unchanged if the position of the centre $\mathbf{a}_i$ is slightly shifted. In other words, though the space partitioning is not determined uniquely with respect to the shape and volume of a particular sample region, for infinitesimally small restriction ω (here infinitesimally small means small enough) the density of sample regions will remain roughly constant for all the possible partitioning fulfilling the chosen fixed ω-restriction.

Definition 9.1. The ω-restricted space partitioning (ωRSP) is the decomposition of the volume of the analysed system into compact non-overlapping space filling regions such that the integral of chosen control function over each region yields the fixed value ω.

For sufficiently small ω the ωRSP regions become very small. Such regions will be termed *micro-cells*. The integrals of chosen functions over the *micro-cells* can safely be replaced by polynomials based on the first non-vanishing term of the corresponding Taylor expansions.

Let us apply the ωRSP idea using the electron density $\rho_1(\mathbf{r})$ as the control function. The integration of the molecular electron density over the whole space yields the total number of electrons N of the examined system. If the system is decomposed into non-overlapping space filling regions μ_i each enclosing the fixed charge:

$$q = \int_{\mu_i} \rho_1(\mathbf{r})\, d\mathbf{r} = \omega \tag{9.49}$$

then there will be $\kappa = N/q$ such regions. If the restriction ω is chosen to be small enough then the charge in the *micro-cell* volume V_i is approximately given by $q \approx \rho_1(\mathbf{a}_i)V_i$. Thus, the volume of the *micro-cell* centred at the position $\mathbf{a}_i$, cf. Fig. 9.12, and controlled by the restriction to enclose the charge q is inverse proportional to the electron density at the *micro-cell* centre $V_i \approx q/\rho_1(\mathbf{a}_i)$. This inverse proportionality of the *micro-cell* volume is valid for any positive control function $f_c(\mathbf{r})$ of single coordinate $\mathbf{r}$ normalised to $F_c = \int f_c(\mathbf{r})\,d\mathbf{r}$:

$$V_i \approx \omega / f_c(\mathbf{a}_i) \qquad F_c = \kappa\omega \tag{9.50}$$

with the restriction ω fixed at chosen value for all *micro-cells*. Of course, the exact *micro-cell* volume V_i is defined even for $f_c(\mathbf{a}_i) = 0$ (through the integral for ω, cf. Eq. 9.49) and for the approximate expression higher terms of the Taylor expansion need to be considered.

For the control function $f_c(\mathbf{r}_1, \ldots, \mathbf{r}_n)$ of n coordinates the situation is more subtle. In this case the ωRSP is performed in such a way that for each *micro-cell* the integral:

$$\omega = \int_{\mu_i} d\mathbf{r}_1 \ldots \int_{\mu_i} f_c(\mathbf{r}_1, \ldots, \mathbf{r}_n),\ d\mathbf{r}_n \tag{9.51}$$

where all coordinates are confined to the *micro-cell* region μ_i, yields the fixed value ω. Unlike the case of a single coordinate, the number κ of *micro-cells* cannot be simply inferred from the fact that the control function is normalised to the value F_c. For example, let us partition the space using the electron pair density $\rho_2(\mathbf{r}_1, \mathbf{r}_2)$ as the control function under the restriction that each *micro-cell* encloses a fixed number D of electron pairs. For any region Ω (not necessarily from a restricted partitioning) the integral:

$$D = \int\int_{\Omega} \rho_2(\mathbf{r}_1, \mathbf{r}_2)\, d\mathbf{r}_1 d\mathbf{r}_2 \tag{9.52}$$

is a positive number. For a given restriction ω the number κ of *micro-cells* μ_i fulfils $\kappa\omega \leq N(N-1)/2$, which just follows from the fact that the number of electron pairs in all the *micro-cells* cannot exceed the total number of pairs in the system. For sufficiently small restriction ω the volume V_i of the *micro-cell* region μ_i scales with ω. The actual expression for the inverse proportionality can be deduced from the Taylor expansion of the control function around the *micro-cell* centre $\mathbf{a}_i$.

In case of an atom or molecule $(\kappa - \iota)$ *micro-cells* of the ωRSP form for a given ω an envelope around the system, whereby ι *micro-cells* extend outside this envelope till infinity (because the system is defined in the whole space). This outermost *micro-cells*, each containing the quantity ω, cannot be made compact in the above sense. Of course, choosing sufficiently small ω will move the envelope to any desired distance, similarly to the molecular envelope defined by appropriate value for the electron density isosurface [2].

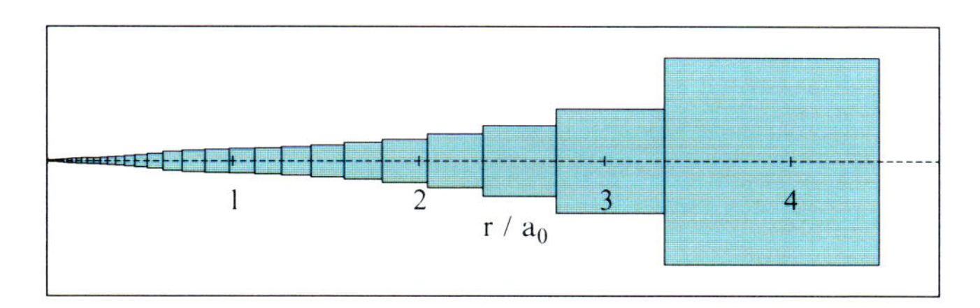

Fig. 9.13 ωRSP controlled by the electron density of the Ar atom (basis set of Clementi and Roetti) with $\omega = 10^{-3}$ electrons. Only *micro-cells* (approximated by cubes for simplicity) centred along the radius are shown

Figure 9.13 shows sequence of *micro-cells* of ωRSP, based on the electron density of the Ar atom (basis set of Clementi and Roetti [66]), in the radial direction. The *micro-cells*, each containing $\omega = 10^{-3}$ electrons, were approximated by cubes for simplicity (otherwise they will adopt such a form as to fill a sphere without gaps and overlaps, cf. Fig. 9.12). The outermost edge in the diagram is at 4.48 bohr from the nucleus. A sphere of this radius contains 17.9775 electrons. Thus, outside such sphere there are 22 *micro-cells* (the outermost one extending to infinity). Note, that the diagram is very approximate (just to show the principles of the distribution).

9.5.2 Restricted Populations

The ωRSP decomposition into *micro-cells* supply (virtually) the samples over which another function f_s, so called *sampling* property, can be evaluated. The idea behind the sampling over the *micro-cell* regions is as follows. The expectation value of an operator $\hat{A}$ acting on the wave function Ψ is given by the integral:

$$\left\langle \hat{A} \right\rangle = \int \Psi^* \hat{A} \Psi \, \mathrm{d}V \tag{9.53}$$

over all space. If $\hat{A}$ is a single-particle operator then the above equation can be reduced to (spin traced-out for simplicity):

$$\left\langle \hat{A} \right\rangle = \int \hat{A} \gamma^{(1)}(\mathbf{r}'; \mathbf{r}) \, \mathbf{dr} = \sum_{i=1}^{K} \int_{\mu_i} \hat{A} \gamma^{(1)}(\mathbf{r}'; \mathbf{r}) \, \mathbf{dr} \tag{9.54}$$

In this case the sum of the sampling values (for the sampling property $f_s(\mathbf{r}) = \hat{A} \gamma^{(1)}(\mathbf{r}'; \mathbf{r})|_{r' \to r}$) over all *micro-cells* μ_i yields the expectation value. Of course, for a two-particle operator (where the sampling property is similarly given by $f_s(\mathbf{r}_1, \mathbf{r}_2) = \hat{A} \gamma^{(2)} (\mathbf{r}'_1, \mathbf{r}'_2; \mathbf{r}_1, \mathbf{r}_2)|_{r' \to r}$) the total sum of the sampling values will recover only part of the expectation value.

The integration of the sampling property over the *micro-cells* of the ωRSP (based on chosen control function) will result in a discrete distribution of values $\{\zeta_i\}$ over the examined system. The distribution of the samples as well as the sampling values themselves will extremely depend on both the function f_c controlling the space partitioning and the chosen sampling property f_s. This can be easily seen by examining two special examples:

- $f_c \equiv f_s$: the function controlling the space partitioning is identical (up to a multiplication with a constant) with the sampling property. The sampling of f_s yields only one value, namely the chosen restriction ω, i.e., the result is a distribution of constant values. The density of sampling values can vary over the system, as the *micro-cells* of the partitioning will be smaller in regions of higher f_c values.
- $f_c \equiv \text{const}$: all *micro-cells* will have the same volume proportional to the restriction ω, cf. Sect. 9.5.1. For infinitesimally small ω the discrete distribution of samples $\{\zeta_i\}$ will mimic the (single coordinate) sampling property f_s and the samples will be equally distributed over the system.

An example of the sampling property is the diagonal part of the 1-matrix $\gamma^{(1)}(\mathbf{r}'; \mathbf{r})$, i.e., the electron density $\rho_1(\mathbf{r})$, which is normalised to the number of electrons N of the examined system. The integration of $\rho_1(\mathbf{r})$ over the *micro-cell* regions μ_i of an ωRSP yields a discrete distribution of populations (charges) with $\sum q_i = N$.

Another example is the diagonal part of the 2-matrix $\gamma^{(2)}(\mathbf{r}'_1, \mathbf{r}'_2; \mathbf{r}_1, \mathbf{r}_2)$, i.e., the electron pair density $\rho_2(\mathbf{r}_1, \mathbf{r}_2)$ normalised to the number of electron pairs $N(N-1)/2$, that can also be used as the sampling property. The integration of $\rho_2(\mathbf{r}_1, \mathbf{r}_2)$ over the *micro-cells* of an ωRSP yields a discrete distribution of electron pair populations $\{D_i\}$ in the respective regions. Clearly, now the individual pair populations will not sum up to the total number of pairs (for which also the pair populations D_{ij} over all region pairs μ_i and μ_j would be needed). The actual distribution (the topology) of the electron populations $\{q_i\}$, respectively pair populations $\{D_i\}$ in the *micro-cells* will of course strongly depend on the control function.

Definition 9.2. The discrete distribution of values generated by the integration of the n-th order electron density $\rho_n(\mathbf{r}_1, \ldots, \mathbf{r}_n)$ over the regions of an ωRSP is termed ω-restricted populations.

9.5.3 Quasi-Continuous Distributions

Sampling (integrating) the property f_s over the *micro-cells* of the ωRSP yields a discrete distribution of values $\{\zeta_i\}$. For properly behaving control function, cf. Sect. 9.5.1, the density of samples (samples per volume) will increase with decreasing ω. It is obvious that, at the same time, the absolute values of individual samples ζ_i will decrease. This is somewhat inconvenient situation, especially when the sum of all samples tends towards zero after such action (like for the restricted pair populations). Thus, it is necessary to extract the relevant information which is not affected by the change of ω. This can be achieved by suitable rescaling of the sampled values, such that the scaling depends only on the restriction ω and that the distribution converge to a unique function in the limiting case when $\omega \to 0$.

The scaling can be inferred from the behaviour of the integrals ζ_i. The control function $f_c(\mathbf{r}_1, \ldots, \mathbf{r}_m)$ of m coordinates can be approximated with the Taylor expansion around a point inside the *micro-cell* region μ_i, say (for convenience) the centre $\mathbf{a}_i$. Then, within μ_i, the integral for the restriction ω can be written as:

$$\omega = \int_{\mu_i} d\mathbf{r}_1 \ldots \int_{\mu_i} f_c(\mathbf{r}_1, \ldots, \mathbf{r}_m)\, d\mathbf{r}_m = t_c(\mathbf{a}_i) V_i^{\vartheta_c} + \varepsilon_c(\mathbf{a}_i) \tag{9.55}$$

where t_c and ϑ_c are function and parameter, respectively, determined by the first non-vanishing term of the Taylor expansion of the control function and V_i is the volume of the *micro-cell* μ_i (note that $t_c(\mathbf{a}_i)$ is the value at position $\mathbf{a}_i$ whereas ϑ_c is valid for the whole distribution). With the correction ε_c the exact value ω of the integral is obtained. ε_c accumulates all the higher terms from the Taylor expansion and becomes negligibly small with decreasing volume of the *micro-cell*. The volume V_i can be expressed as:

$$V_i = \left[\frac{\omega - \varepsilon_c(\mathbf{a}_i)}{t_c(\mathbf{a}_i)} \right]^{1/\vartheta_c}. \tag{9.56}$$

The same procedure can be applied to the sampling property $f_s(\mathbf{r}_1, \ldots, \mathbf{r}_n)$ of n coordinates (yielding the sampling values ζ_i for *micro-cells* μ_i):

$$\zeta_i = \int_{\mu_i} d\mathbf{r}_1 \ldots \int_{\mu_i} f_s(\mathbf{r}_1, \ldots, \mathbf{r}_n)\, d\mathbf{r}_n = t_s(\mathbf{a}_i) V_i^{\vartheta_s} + \varepsilon_s(\mathbf{a}_i). \tag{9.57}$$

The above expressions are exact due to the corrections ε_s. The volume V_i can be substituted from Eq. 9.56 giving:

$$\zeta_i = t_s(\mathbf{a}_i) \left[\frac{\omega - \varepsilon_c(\mathbf{a}_i)}{t_c(\mathbf{a}_i)} \right]^{\vartheta_s/\vartheta_c} + \varepsilon_s(\mathbf{a}_i) = t_s(\mathbf{a}_i) \left[\frac{\omega}{t_c(\mathbf{a}_i)} \right]^{\vartheta_s/\vartheta_c} + \varepsilon(\mathbf{a}_i) \tag{9.58}$$

where all corrections have been subsumed into $\varepsilon(\mathbf{a}_i)$. Let us rescale the sampling values $\{\zeta_i\}$ by the division with the parameter $\omega^{\vartheta_s/\vartheta_c}$. Then, with decreasing ω, the rescaled discrete distribution (with increasing number of members) will approach the values $\{t_s(\mathbf{a}_i)/t_c(\mathbf{a}_i)^{\vartheta_s/\vartheta_c}\}$ with the limit after rescaling given by the function:

$$\lim_{\omega \to 0} \{\zeta_i / \omega^{\vartheta_s/\vartheta_c}\} = t_s(\mathbf{r}) \left[\frac{1}{t_c(\mathbf{r})} \right]^{\vartheta_s/\vartheta_c} = t_s(\mathbf{r})\, \tilde{V}^{\vartheta_s}(\mathbf{r}) \tag{9.59}$$

where $\widetilde{V}(\mathbf{r})$ is the limit of the rescaled *micro-cell* volume, cf. Eq. 9.56, termed a volume function. Note that for given non-zero restriction ω the set $\{\zeta_i/\omega^{\vartheta_s/\vartheta_c}\}$ is a discrete distribution, whereas the limit after rescaling is a continuous function.

An interesting issue emerges concerning the rescaling procedure. There is a clear relationship between the value ζ_i for the *micro-cell* μ_i centred at $\mathbf{a}_i$ and the restriction ω, cf. Eq. 9.58. However, after the rescaling this relationship is no more explicitly given. The only possibility to conclude the value of ω from the rescaled distribution would be utilising the size of the sampling set. Then, at least in principle, such ω could be searched (or exactly determined) that yields an ωRSP of the same set size, i.e., the same number of *micro-cells*. An incomplete set of rescaled sampling values would hinder this intention. When reducing ω to an infinitesimally small value the distribution ζ_i remains discrete. With the proper choice of ω the distribution can be made as dense as desired, i.e., the centres $\mathbf{a}_i$ and $\mathbf{a}_j$ of two individual *micro-cells* can be moved to whatever non-zero distance, but not all the centres at once. Thus, one is allowed to compute any finite number of rescaled sampling values at any positions using Eq. 9.59 for the limit after rescaling knowing that, beside a small deviation, the results definitely will be members of a discrete distribution based on certain ω. Having a set of rescaled values one could compute additional ones which would just possibly change the (virtual) restriction. Although the distribution of the sampling values is discrete, the values can be determined at any chosen position (like for a continuous function). Let us term the set with such behaviour a *quasi-continuous* distribution [67].

9.6 Electron Localizability Indicator

The information about the chemical bonding (depending on the definition) within an N electron system is carried by the N-matrix $\Gamma^{(N)}$. As the energy is determined by 1 and 2-particle operators it can be assumed that the 2-matrix $\Gamma^{(2)}$ contains an important part of this chemically relevant information. In Sect. 9.5 the approach was described where on the basis of two functions new distributions are created that can be analysed. Utilising both, the electron density $\rho_1(\mathbf{r})$ and the electron pair density $\rho_2(\mathbf{r}_1, \mathbf{r}_2)$, offers the freedom to use one of the densities as the control function and the other as the sampling property. The resulting ω-restricted population can be seen as the charge needed to form a fixed fraction of an electron pair (D-restricted population) or, reverse, the number of electron pairs formed from a fixed electron population (q-restricted population).

The charge and the pair population in a *micro-cell* μ_i are given by Eqs. 9.49 and 9.52, respectively. If the q-restriction is chosen, then each *micro-cell* contains the same charge $q = \omega$. Sampling the pair density highlights actually the difference to the fixed value q^2 [63], because the pair density integral can be formally written as:

$$\frac{1}{2}\int\int_{\mu_i} \rho_1(\mathbf{r}_1)\rho_1(\mathbf{r}_2)[1 + f_{XC}(\mathbf{r}_1, \mathbf{r}_2)]\, d\mathbf{r}_1\, d\mathbf{r}_2 = \frac{1}{2}[q^2 + F_i] \tag{9.60}$$

with the correlation factor $f_{XC}(\mathbf{r}_1, \mathbf{r}_2)$. In certain sense, F_i can be seen as a measure of the correlation of electronic motion in the region μ_i [68, 69].

Because the electron density is 1-particle property, the integrals over the *micro-cells* can be evaluated, using the Taylor expansion, on the same footing irrespective whether the total $\rho_1(\mathbf{r})$ or the spin-resolved $\rho_1^\sigma(\mathbf{r})$ densities are used (with the spin σ set to α or β). This is not the case for the electron pair density (2-particle property). Due to the Pauli principle the value at the coalescence is zero for the same-spin pair density, $\rho_2^{\sigma\sigma}(\mathbf{r}, \mathbf{r}) = 0$, whereas the opposite-spin pair density can have nonzero on-top values, $\rho_2^{\alpha\beta}(\mathbf{r}, \mathbf{r}) \neq 0$ [70, 71]. This behaviour not only controls the first non-vanishing term of the Taylor expansion (which would be of interest only for the approximate evaluation of the distribution samples). It also shows that for sufficiently small *micro-cell* volumes the opposite-spin pair populations will dominate the total populations. The same-spin pair information can be analysed only if the opposite-spin pair density is not considered.

The interplay between the charge and pair populations within a *micro-cell* is connected with the 'refusal' of electrons to share the same region of space. D-restricted and q-restricted populations, respectively, can be used as a measure of electron localizability.

Definition 9.3. The electron localizability indicator (ELI) is the rescaled discrete distribution of electron populations (ELI-D, symbol Υ_D), respectively electron pair populations (ELI-q, symbol Υ_q) in *micro-cells* based on ωRSP with infinitesimally small restriction ω.

The abbreviation ELI-D emphasises that the ωRSP follows the restriction of fixed number $\omega = D$ of electron pairs (ELI-q for fixed charge $\omega = q$) in each *micro-cell*. Because the term ELI is used only for ωRSP functionals based explicitly on ρ_1 and ρ_2 it is clear that Υ_D samples the charge and thus the corresponding symbol needs not to be added. The utilised spin component should be highlighted by a superscript. Then, Υ_D^α stands for ELI-D sampling the α-spin charge in $\alpha\alpha$-pair restricted *micro-cells*. There are no approximations in the definition of ELI, as can be seen, e.g., for the Υ_D^α value corresponding to the *micro-cell* μ:

$$\Upsilon_D^\alpha(\mu) = \frac{1}{\omega^{3/8}} \int_\mu \rho_1^\alpha(\mathbf{r})\, d\mathbf{r}: \quad \omega = \int \int_{\mu_i} \rho_2^{\alpha\alpha}(\mathbf{r}_1, \mathbf{r}_2)\, d\mathbf{r}_1 d\mathbf{r}_2 = \text{const.} \tag{9.61}$$

The value 3/8 for the exponent in the rescaling factor will be derived in the next section. Of course, in the actual determination of the quasi-continuous ELI distribution the above integrals will be evaluated approximately, conveniently using the limit after rescaling, see Sect. 9.5.3. If, for any reasons, the discrete nature of ELI should not be taken into account the limit after rescaling will replace the quasi-continuous distribution (for instance, analysing the topology of ELI [72]).

9.6.1 Same-Spin Electron Pairs

The same-spin electron pair density $\rho_2^{\sigma\sigma}(\mathbf{r}_1, \mathbf{r}_2)$ can be either sampled over the *micro-cells* of the ωRSP or the same-spin pair population can be used as the restriction. In both cases the corresponding integrals over the *micro-cell* regions μ_i need to be computed. As mentioned in Sect. 9.5.3 the Taylor expansions around the centres $\mathbf{a}_i$ of the *micro-cells* can be utilised for that purpose. Obviously, due to the behaviour of $\rho_2^{\sigma\sigma}(\mathbf{r}_1, \mathbf{r}_2)$ at the electron-coalescence [73] the first non-vanishing term of the Taylor expansion for the coordinate $\mathbf{r}_2$ yields for the approximate same-spin pair population:

$$D_i^{\sigma\sigma} \approx \frac{1}{2} \int d\mathbf{r}_1 \int_{\mu} (\mathbf{s}_2 \cdot \nabla_{\mathbf{r}_2})^2 \rho_2^{\sigma\sigma}(\mathbf{r}_1, \mathbf{r}_2)\Big|_{\mathbf{r}\to\mathbf{a}} d\mathbf{r}_2 \tag{9.62}$$

with $\mathbf{s}_2 = \mathbf{r}_2 - \mathbf{a}$. After the expansion for $\mathbf{r}_2$ both coordinates $\mathbf{r}_1$ and $\mathbf{r}_2$ are set to $\mathbf{a}$ (i.e., zero-th term of the expansion for $\mathbf{r}_1$). Taking the $\rho_2^{\sigma\sigma}(\mathbf{r}_1, \mathbf{r}_2)$ component of the pair density expanded into Slater determinants (with orbitals ϕ):

$$\rho_2^{\sigma\sigma}(\mathbf{r}_1, \mathbf{r}_2) = \frac{1}{2} \sum_{i<j}^{\sigma} \sum_{k<l}^{\sigma} P_{ij,kl} |\phi_i(\mathbf{r}_1)\phi_j(\mathbf{r}_2)||\phi_k(\mathbf{r}_1)\phi_l(\mathbf{r}_2)| \tag{9.63}$$

gives after the application of the Taylor expansion and successive integration of the remaining variables in Eq. 9.62 the approximate value for the pair populations [74]:

$$D_i^{\sigma\sigma} \approx \frac{1}{12} V_i^{8/3} g^{\sigma}(\mathbf{a}_i) \tag{9.64}$$

where V_i is the volume of the *micro-cell* μ_i and $g^{\sigma}(\mathbf{a}_i)$ is the Fermi-hole curvature at the *micro-cell* centre:

$$g^{\sigma}(\mathbf{a}_i) = \sum_{i<j}^{\sigma} \sum_{k<l}^{\sigma} P_{ij,kl} [\phi_i(\mathbf{a}_i)\nabla\phi_j(\mathbf{a}_i) - \phi_j(\mathbf{a}_i)\nabla\phi_i(\mathbf{a}_i)] \times [\phi_k(\mathbf{a}_i)\nabla\phi_l(\mathbf{a}_i) - \phi_l(\mathbf{a}_i)\nabla\phi_k(\mathbf{a}_i)] \tag{9.65}$$

If the ωRSP is controlled by infinitesimally small $\omega = D^{\sigma\sigma}$, i.e., fixed same-spin pair population, then Eq. 9.64 determines approximately the volume V_i of the *micro-cell* μ_i:

$$V_i \approx \left[\frac{12\omega}{g^{\sigma}(\mathbf{a}_i)}\right]^{3/8} : \qquad \tilde{V}_D(\mathbf{r}) = \lim_{\omega\to 0} \left\{\frac{1}{\omega^{3/8}} V_i\right\} = \left[\frac{12}{g^{\sigma}(\mathbf{r})}\right]^{3/8}. \tag{9.66}$$

The limit after rescaling of all *micro-cell* volumes containing the fixed fraction ω of same-spin pair is the pair-volume function $\widetilde{V}_D(\mathbf{r})$ [67, 75]. Sampling the σ-spin

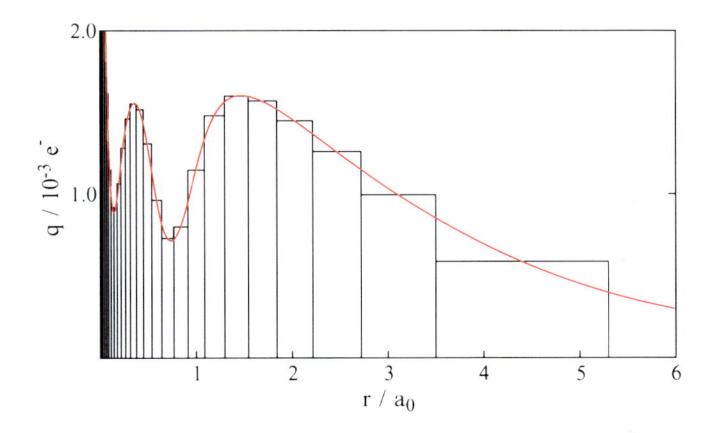

Fig. 9.14 *Red line*: Υ_D^α for the Ar atom using the basis of Clementi and Roetti; *black bars*: discrete distribution of charges from the integration of ρ_1^α over the *micro-cells* (approximated by cubes of the size given by the bar width) of the ωRSP restricted to enclose $10^{-8}\alpha\alpha$-pairs

electrons in the *micro-cells* μ_i of the above ωRSP centred around positions $\mathbf{a}_i$ yields the ELI-D values:

$$\Upsilon_D^\sigma(\mu_i) = \frac{1}{\omega^{3/8}} \int_\mu \rho_1^\sigma(\mathbf{r})\, d\mathbf{r} = \Upsilon_D^\sigma(\mathbf{a}_i) \approx \frac{1}{\omega^{3/8}} \rho_1^\sigma(\mathbf{a}_i)\; V_i \tag{9.67}$$

(the relation $\Upsilon_D^\sigma(\mu_i) = \Upsilon_D^\sigma(\mathbf{a}_i)$ is just a formal replacement) with the limit after rescaling:

$$\tilde{\Upsilon}_D^\sigma(\mathbf{r}) = \lim_{\omega \to 0} \{\Upsilon_D^\sigma(\mu_i)\} = \rho_1^\sigma(\mathbf{r})\; \tilde{V}_D(\mathbf{r}). \tag{9.68}$$

Because the rescaling factor $\omega^{3/8}$ is derived from the Taylor expansion it is clear, cf. Eqs. 9.66 and 9.67, that the exact ELI-D value will depend on the choice of ω, even if the centre of the region μ_i (occupying ω dependent volume) remains stationary. However, this rescaling ensures the limiting behaviour of the ELI-D distribution.

The red line in Fig. 9.14 shows Υ_D^α for the Ar atom [66]. Such diagram cannot clarify whether the continuous function $\tilde{\Upsilon}_D^\alpha(\mathbf{r})$ or the quasi-continuous distribution $\Upsilon_D^\sigma(\mathbf{r})$ (with isolated positions $\mathbf{r}$) for infinitesimally small ω, yielding regions smaller then a pixel, is examined. On the other hand the discrete form is immediately visible from the black steps in Fig. 9.14 where the height shows the α-spin charge in regions restricted to enclose $10^{-8}\alpha\alpha$-pairs (with the rescaling factor 1,000 for ELI-D the charges given in 10^{-3} electrons are identical to the ELI-D values). After the outermost edge at 5.3 bohr from the nucleus (just approximate value) only regions extending to infinity will follow. For quasi-continuous ELI-D distribution this edge can be located at any conceivable distance (not infinity) from the nucleus.

For the Ar atom 3 Υ_D^α maxima can be identified in Fig. 9.14 corresponding to 3 atomic shells separated from each other by Υ_D^α minima. The number of α-spin electrons (1.1, 3.9, 4.0) within the shells is close to the ones expected from the Periodic Table [76]. Using the event probabilities [77] it can be shown that high ELI-D values, i.e., high charges needed to form the fixed pair population, are proportional to the probability that the *micro-cell* is occupied by single electron. Thus, in *micro-cells* located around the Υ_D^α maximum it is difficult to form $\alpha\alpha$-pairs. In closed-shell system this is valid for the $\beta\beta$-pairs as well.

One could ask the question, what about the number of opposite-spin pairs in the *micro-cells* of the ωRSP with the same-spin restriction? At HF level this would be similar to one of the spin-pair compositions [78] defined, e.g., as the ratio of the number of same-spin pairs to the number of opposite-spin pairs in an arbitrary volume around the reference point. There, in principle, the q-restriction was used but ωRSP not explicitly utilised for the spin-pair composition. In case of Υ_D^α the pairing avoidance of the same-spin electrons, due to the Pauli principle, determines the α-spin charge q_α in a *micro-cell* (proportional to the ELI-D value). For closed-shell HF wave function the number of opposite-spin pairs equals q_α^2, cf. Eq. 9.60 with proper spin and zero correlation factor, which could lead to the persuasion that the pairing avoidance of the same-spin electrons favour the pairing of the opposite-spin electrons. However, the motion of the opposite-spin electrons is uncorrelated at HF level (in sense of Eq. 9.60). The increased number of opposite-spin pairs is just the consequence of the increased charge. It is like people having nothing in common, just being crowded around a bar to buy a drink. Using the q-restricted partitioning may illustrate it even better, because there the number of same-spin pairs in a *micro-cell* again depends on the correlation of motion (Fermi-hole), whereas the number of opposite-spin pairs remains constant (the squared restriction value) throughout the space [79].

9.6.2 Singlet and Triplet Pairs

In case of spin-polarised calculation two same-spin ELI-D data sets, one for each $\sigma\sigma$-spin component, need to be analysed. It could be desirable to have single result for such systems. Moreover, the decomposition into same-spin and opposite-spin contributions is not rotation invariant in the spin space [80, 81]. A possibility to combine the spin components is to form the symmetric and antisymmetric combinations of the reduced 2-matrix $\gamma^{(2)}(\mathbf{r}'_1, \mathbf{r}'_2; \mathbf{r}_1, \mathbf{r}_2)$ [82, 83]:

$$\gamma^{(s)}\left(\mathbf{r}'_1, \mathbf{r}'_2; \mathbf{r}_1, \mathbf{r}_2\right) = \frac{1}{2}\left[\gamma^{(2)}\left(\mathbf{r}'_1, \mathbf{r}'_2; \mathbf{r}_1, \mathbf{r}_2\right) + \gamma^{(2)}\left(\mathbf{r}'_1, \mathbf{r}'_2; \mathbf{r}_2, \mathbf{r}_1\right)\right]$$

$$\gamma^{(t)}\left(\mathbf{r}'_1, \mathbf{r}'_2; \mathbf{r}_1, \mathbf{r}_2\right) = \frac{1}{2}\left[\gamma^{(2)}\left(\mathbf{r}'_1, \mathbf{r}'_2; \mathbf{r}_1, \mathbf{r}_2\right) - \gamma^{(2)}\left(\mathbf{r}'_1, \mathbf{r}'_2; \mathbf{r}_2, \mathbf{r}_1\right)\right]. \quad (9.69)$$

The diagonal elements of the resulting matrices are the singlet and triplet pair densities $\rho_2^{(s)}(\mathbf{r}_1, \mathbf{r}_2)$ and $\rho_2^{(t)}(\mathbf{r}_1, \mathbf{r}_2)$, respectively. The integrals of those densities over a region yields the number $D^{(s)}$ of singlet-like, respectively $D^{(t)}$ of triplet-like electron pairs within. Whereas the same-spin pairs contribute only to the triplet pairs the opposite-spin pairs are participating on both singlet and triplet pairs. The reduction of the above pair densities yields the densities $\rho_1^{(s)}(\mathbf{r})$ and $\rho_1^{(t)}(\mathbf{r})$ of electrons coupled to a singlet and triplet, respectively.

The integration of $\rho_1^{(t)}(\mathbf{r})$ over the *micro-cells* μ_i of ωRSP based on the restriction $\omega = D^{(t)}$, i.e., fixed number of triplet-like pairs in each *micro-cell*, produces the discrete distribution of charges $\{q_i^{(t)}\}$ which after rescaling yields ELI-D for the triplet-coupled electrons $\Upsilon_D^{(t)}$ with the limit [67]:

$$\tilde{\Upsilon}_D^{(t)}(\mathbf{r}) = \lim_{\omega \to 0} \left\{ \Upsilon_D^{(t)}(\mu_i) \right\} = \rho_1^{(t)}(\mathbf{r})\, \tilde{V}_{D^{(t)}}(\mathbf{r}) \tag{9.70}$$

where the pair-volume function $\tilde{V}_{D^{(t)}}$ is computed from the curvature of the triplet-like pair density (with the contributions of the Fermi-hole curvatures) in similar fashion as in case of the same-spin pairs. For single-determinant closed-shell wave functions $\Upsilon_D^{(t)} \propto \Upsilon_D^{\alpha}$.

In Fig. 9.15 diagrams for the ROHF/6-311++G(3df,2p) calculation [84] of the CH_3 radical (CH bond length of 1.07 Å) are presented. The two left-side diagrams show the same-spin ELI-D, with the upper diagram for the majority spin displaying above and below the molecular plane the localisation domains that can be attributed to the presence of the 'radical' electron. Of course, those domains are not present in the minority channel (bottom left diagram). The upper right diagram with the $\Upsilon_D^{(t)}$ localisation domains unifies, in certain sense, the two left diagrams into a single one. The domains due to the 'radical' electron are still extant in this representation. The bottom right diagram shows the corresponding $\Upsilon_D^{(t)}$ basins [85] with the carbon core (black coloured, enclosing 2.1 electrons) and the 2 pink basins encompassing together roughly 1/3 electron.

In case of the singlet coupling the ωRSP based on the restriction $\omega = q^{(s)}$ of fixed number of singlet-like coupled electrons is used, i.e., the q-restricted partitioning (with this choice high ELI values correspond to high singlet-pair populations). The sampling of $\rho_2^{(s)}(\mathbf{r}_1, \mathbf{r}_2)$ over the *micro-cells* of the partitioning results in discrete distribution of singlet-like pair populations $\{D_i^{(s)}\}$, which after rescaling yields the ELI-q for the singlet coupled electrons $\Upsilon_q^{(s)}$. The exponent of the rescaling factor can be found from the Taylor expansion around the centre $\mathbf{a}_i$ of the *micro-cell* μ_i in the following way. The singlet-pair density comprises the opposite-spin pair densities, i.e., the expansion will be dominated by the on-top density $\rho_2^{(s)}(\mathbf{r}, \mathbf{r})$ and the singlet-like pair population can be approximated by:

$$D_i^{(s)} \approx \rho_2^{(s)}(\mathbf{a}_i, \mathbf{a}_i)\, V_i^2 \approx \rho_2^{(s)}(\mathbf{a}_i, \mathbf{a}_i) \left[\frac{\omega}{\rho_1^{(s)}(\mathbf{a}_i)} \right]^2 \tag{9.71}$$

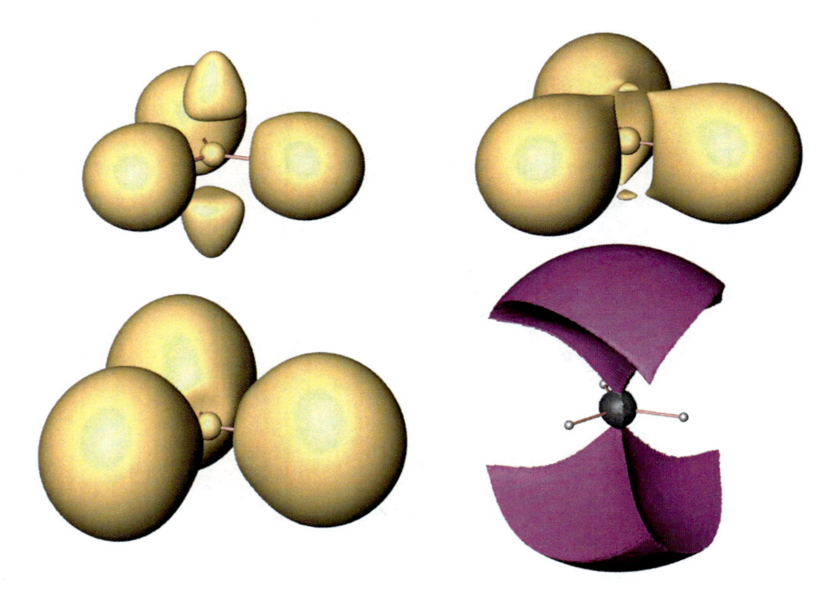

Fig. 9.15 ELI-D for the CH_3 radical from ROHF/6-311++G(3df,2p) calculation. *Upper left* and *bottom left* diagrams: 1.55-localisation domains of Υ_D^{α} and Υ_D^{β} for the majority and minority spin, respectively. *Upper right* diagram: 1.193-localisation domains of $\Upsilon_D^{(t)}$ for the triplet coupled electrons. *Bottom right*: core basin (*black coloured*) and the 'radical' basins (*pink*) for $\Upsilon_D^{(t)}$; the hydrogen positions are marked by *grey* spheres

where the *micro-cell* volume was approximated by $V_i \approx q^{(s)}/\rho_1^{(s)}$. With the scaling factor ω^2 the limit after rescaling for $\Upsilon_q^{(s)}$ is defined as [67]:

$$\tilde{\Upsilon}_q^{(s)}(\mathbf{r}) = \lim_{\omega \to 0} \left\{ \Upsilon_q^{(s)}(\mu_i) \right\} = \frac{1}{2}\left[\frac{N+2}{N-1}\right]^2 \frac{\rho_2^{(s)}(\mathbf{r},\mathbf{r})}{\left[\rho_1^{(s)}(\mathbf{r})\right]^2} \tag{9.72}$$

with the total number of electrons N. For closed-shell HF wave functions, where the motion of the opposite-spin electrons is, in sense of Eq. 9.60, uncorrelated we find $\Upsilon_q^{(s)}(\mathbf{r}) = const$ throughout the space. The N-dependent factor sets this constant to 1 (value previously chosen for the electron localizability indicator for antiparallel-spin pairs, the functional ELIA [79], using the charge product $q_\alpha q_\beta$ as the restriction for ωRSP). For spin-polarised HF wave functions $\Upsilon_q^{(s)}$ is non-uniform and shows certain modulation due to the spin density.

In case of DFT calculations the (correlated) electron density is given as the sum of squared Kohn-Sham (KS) orbitals. Although the density matrices constructed from the squared KS wave function do not truly represent the correlated density matrices they can, at least formally, be created. Once the (DFT) matrices are formed they cannot be immediately distinguished from the HF matrices (this could be done analysing different properties and representations). It follows that formally ELI can be derived from a DFT wave function and analysed [75, 86]. However, it cannot be *per se* expected to inspect here the influence of the correlated motion of electrons in all details. This is especially illustrated by $\Upsilon_q^{(s)}$ which is constant in all space for DFT wave function, because the opposite-spin pair population is given by the same expression like for a HF result.

9.6.3 ELI for Correlated Wave Functions

For explicitly correlated wave functions the whole formalism described in Sects. 9.5 and 9.6 remains unchanged. The correlated density matrices are simply fed into the expressions. The topology of both the same-spin and triplet-coupled ELI-D distributions is principally governed by the Fermi correlation, i.e., the structure is basically embossed already at the HF level. It seems that for atoms the account for the Coulomb correlation has only minor effect on ELI-D [87], whereby the changes due to the correlation are more pronounced for the pair-volume function than for the electron density (cf. Eq. 9.68). For molecules the overall topological ELI-D descriptors are not markedly influenced by the inclusion of Coulomb correlation, except in cases where the correlation plays an important role, like for the F_2 molecule, with correlated ELI-D yielding bifurcated bond attractor [79].

From Sect. 9.6.2 it is clear that reasonable $\Upsilon_q^{(s)}$ distributions can only arise if correlated density matrices are utilised [79]. Then, due to the correlation of electronic motion of the opposite spin electrons the number of opposite-spin pairs can significantly differ from the simple product of densities of the two spin components. The $\Upsilon_q^{(s)}$ distribution monitors how many singlet-like pairs are formed from the fixed amount ω of the density of singlet-coupled electrons. The degree to which the correlation reduces the singlet-pair population will depend on the position of the *micro-cell* in the examined system as well as on the effort to account for the correlation.

Often even relatively simple CAS calculations exhibit for the computed system patterns in $\Upsilon_q^{(s)}$ topologically similar to the ones from the ELI versions based on the Fermi correlation [79]. However, to reveal not only the core-valence separation but the full shell structure of atoms, respectively detailed bonding information of molecules, requires relatively high effort. For instance, in case of the Ar atom [87] and the first row dimers [88] MRCI calculations were necessary.

Figure 9.16 shows the 0.91-localisation domain for the CAS(8,14)/VDZ calculation of the CH_4 molecule (the grey spheres mark the position of the hydrogen atoms). The core region together with 4 domains, that can be attributed to the C–H bonds, are clearly identified. The core orbitals were not included into the CAS. Thus, the electron pairing remains HF-like with $\Upsilon_q^{(s)}$ values approaching 1 at the C nucleus (otherwise a non-nuclear $\Upsilon_q^{(s)}$ maximum for the first atomic shell is found [79, 87]). The core domain becomes visible, because at certain distance form the C nucleus the

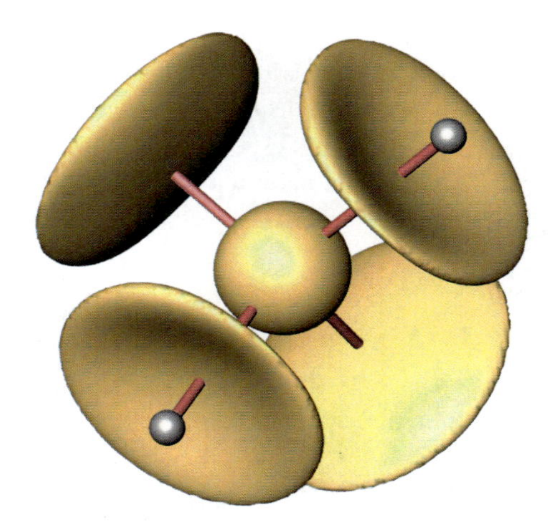

Fig. 9.16 ELI-q for the CAS(8,14)/VDZ calculation of CH_4. Shown are the 0.91-localisation domains of $\Upsilon_q^{(s)}$. The hydrogen positions are marked by grey spheres

reduction of the opposite-spin pairing starts to be effective. Everywhere outside the C–H bond region the correlation reduces the number of singlet-like pairs that could be formed from a fixed amount of charge. The performed CAS calculation involved one million determinants. For smaller calculations the bond pattern is not resolved. Instead, $\Upsilon_q^{(s)}$ maxima between the H atoms remain after the pairing reduction.

9.6.4 ELI in Momentum Space

The derivation of ELI is based on the ωRSP of the real space. This derivation can be consistently applied in the momentum space using exactly the same formalism with density matrices in the momentum space representation (cf. Chapter 6). The momentum space is partitioned in such a way that electrons having momenta **p** bounded by the *micro-cell* regions μ_i form a fixed number ω of electron or pair populations, respectively. In the *micro-cells* the electron density, respectively electron pair density is sampled and rescaled. Because in momentum space the behaviour with respect to the Fermi and Coulomb correlations is the same as for the position representation, i.e., vanishing same-spin pair density at the coalescence of the momenta and a non-zero cusp for the opposite-spins, the rescaling factors will be the same as given for the real space representation. The limit after rescaling for the same-spin ELI-D in the momentum space is defined by [89]:

$$\tilde{\Upsilon}_D^{\sigma}(\mathbf{p}) = \lim_{\omega \to 0} \{\Upsilon_D^{\sigma}(\mu_i)\} = \pi_1^{\sigma}(\mathbf{p})\ \tilde{V}_D(\mathbf{p}) \tag{9.73}$$

with the σ-spin momentum density $\pi_1^{\sigma}(\mathbf{p})$ and the momentum pair-volume $\widetilde{V}_D$ (and similarly for the triplet-coupled pairs, cf. Eq. 9.70 for the real space representation).

For the singlet-coupled pairs the same scaling factor and behaviour, with respect to the Coulomb correlation, is valid as in the real space.

ELI-D reveals the atomic shell structure in the momentum space. For the atom series Li to Ar the number of $\Upsilon_D^{\sigma}(\mathbf{p})$ maxima corresponds to the ones expected from the Periodic Table [89]. Additionally, the electron populations in spherical momentum space regions delimited by the $\Upsilon_D^{\sigma}(\mathbf{p})$ minima, yields reasonable shell occupations. For molecules the situation is different from that in the real space analysis, where ELI-D topology covers the whole molecular region, i.e., increases with the size of the system. In momentum space the individual atoms in the molecule are no more directly present. ELI-D is significantly structured only in the valence region which extends just few atomic units around zero momentum. All similar 'chemical objects', be it similar bonds or atomic cores, fall into similar momentum regions. Thus, for instance, in cores of identical atoms the electrons occupy near range of momenta. Because $\Upsilon_D^{\sigma}(\mathbf{p})$ describes the avoidance of electrons to gain nearby momenta, the ELI-D value can decrease to such extent that separate ELI-D maxima for the objects will not appear [90].

In Fig. 9.17 the same-spin ELI-D diagrams for the CH_4 molecule (computed with the HF/6-311++G(3df,2p) basis at the bond distance of 1.085 Å) are presented. In the real space (upper left diagram) the ELI-D localisation domains nicely represent the bonding situation in the molecule. The same-spin pairing is avoided in the region of carbon core and around the hydrogen atoms. The α-spin electrons are 'alone' in that regions, i.e., they are localised there. After the Dirac-Fock transformation of the HF Gaussian-type orbitals [91] the corresponding $\Upsilon_D^{\alpha}(\mathbf{p})$ diagram (upper right) shows 6 maxima (inversion symmetry) in the examined momentum region around the origin. How to find some correspondence between the two representations?

The ELI-D is the distribution of (rescaled) charges in the *micro-cells* sampled by the integration of electron density. The electron density can formally be decomposed into contributions from any parts used to construct the electron density. Then ELI-D can also be written as the sum of such contributions, e.g., based on the orbital densities [75, 90]:

$$\Upsilon_D(\mu_i) = \frac{1}{\omega^{3/8}} \int_{\mu_i} \pi(\mathbf{p})\, \mathrm{d}\mathbf{p} = \frac{1}{\omega^{3/8}} \int_{\mu_i} \sum_k \pi_k(\mathbf{p})\, \mathrm{d}\mathbf{p} = \sum_k \Upsilon_D(\mu_i|\phi_k). \quad (9.74)$$

The momentum density $\pi = \sum_k \pi_k$ is given as the sum of orbital contributions originating from the orbitals ϕ_k. The separate ELI-D contributions $\Upsilon_D(\mu_i|\phi_k)$, which exactly sum up to the total ELI-D, are termed partial ELI-D (pELI-D). Instead of the ϕ_k orbital there can be any other descriptor symbolically identifying the contributions. To locate the position of the *micro-cell* μ_i the centre $\mathbf{b}_i$ of the region μ_i can be used, as in the previous sections, i.e., $\Upsilon_D(\mathbf{b}_i|\phi_k)$ or simply $\Upsilon_D(\mathbf{p}|\phi_k)$, with isolated points $\mathbf{p}$.

The mid diagrams of Fig. 9.17 shows the pELI-D contributions computed from the localised orbitals. In the real space representation each of the bond orbital contribution (coloured red, green, blue and grey) almost fully reproduces the

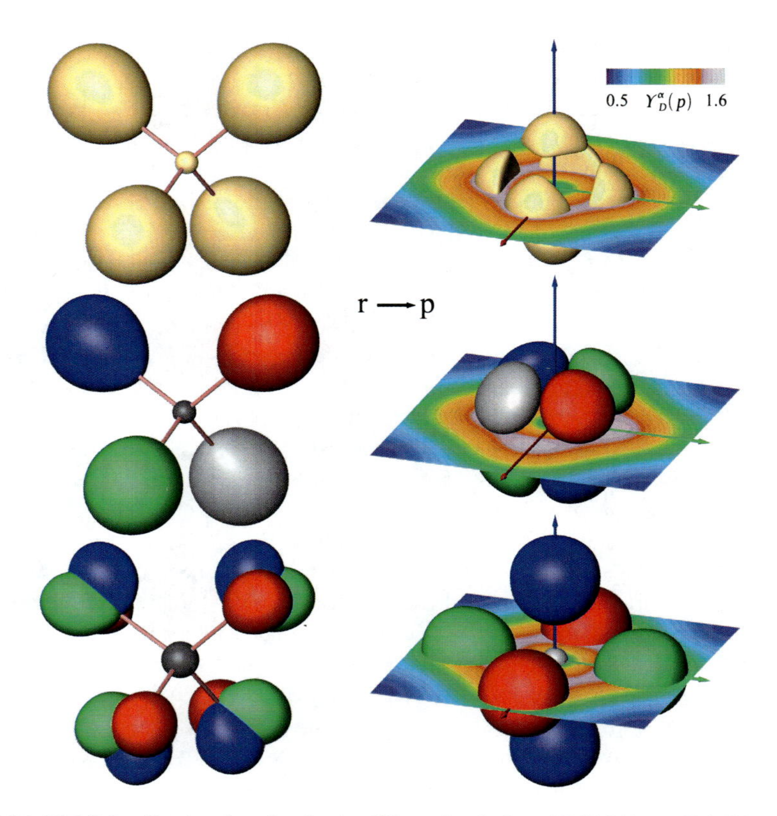

Fig. 9.17 ELI-D localisation domains for the CH_4 molecule from HF/6-311++G(3df,2p) calculation (*left*: real space representation; right: momentum space representation). *Upper*: $\Upsilon^\alpha_D(\mathbf{r}) = 2.5$ and $\Upsilon^\alpha_D(\mathbf{p}) = 1.4$ for the total density. *Mid*: $\Upsilon^\alpha_D(\mathbf{r}|\phi_l) = 2.5$ and $\Upsilon^\alpha_D(\mathbf{p}|\phi_l) = 0.5$ for the contributions of the localised orbitals ϕ_l. *Bottom*: $\Upsilon^\alpha_D(\mathbf{r}|\phi_k) = 1.0$ and $\Upsilon^\alpha_D(\mathbf{p}|\phi_k) = 0.8$ for the contributions of the canonical orbitals ϕ_k. The axes point in the p_x, p_y, and p_z directions. The scale applies to the planes

localisation domains of the total $\Upsilon^\alpha_D(\mathbf{r})$ around particular hydrogen atom, (the black coloured domain stems from the core orbital). However, the contributions of the localised orbitals transformed into the momentum space do not separately recover any of the localisation domains in total momentum space picture (as was the case for other analysed molecules [90]).

Taking instead of the localised orbitals the canonical ones has the opposite effect. Now the real space total $\Upsilon^\alpha_D(\mathbf{r})$ domains are not reproduced by the contributions of the canonical orbitals, cf. the bottom left diagram of Fig. 9.17 (the contributions from the 3 degenerate bond orbitals are coloured red, green, and blue; the contributions of the 4th, non-degenerate, bond orbital are not visible, because the domains are too small and close to the hydrogen atoms) as each of the pELI-D contributions is spread over all hydrogen atoms. In the momentum space each of the pELI-D contributions of the degenerate canonical orbitals nicely reproduces a pair of the localisation domains along the momentum space axes (bottom right diagram). The contribution from the non-degenerate canonical orbital (grey coloured domain) is located around zero momentum. It is not revealed in the total picture, the $\Upsilon_D(\mathbf{p}|\phi_2)$ contribution is concealed by the large contributions of the 3 degenerate bonding orbitals. Recapitulating, the ELI-D topology of CH_4 in the real space is reproduced by the pELI-D contributions of the localised orbitals (spatial viewpoint), whereas in the momentum space the ELI-D topology is controlled by the pELI-D contributions of the canonical orbitals (energetical viewpoint).

9.7 Beyond the Standard Analyses

We have tried to summarize in this chapter some of the directions along which the topological analyses of scalar fields related to the charge density are evolving in the last years. It is our opinion that both at the coarse (QTAIM/IQA), or fine-grained (ωRSP) levels reviewed here, vast territories remained unexplored.

From the coarse-grained viewpoints described in Sects. 9.2–9.4, many interesting new ideas stem from (atomic)-partitioning of higher order density matrices. Although the algebra and the basic insights that we expect to obtain from such procedures are already developed, efficient computational codes to exploit them are still to be published, particularly for extended systems. In their absence, energy decompositions at the resonance structure level will only be available for really toy systems. Similar comments apply to the emerging field of electron population statistics. Dealing with multicenter bonding issues from a purely real space point of view, a faraway prospect not long ago, is now an open possibility that fills a much criticized gap in the standard QTAIM.

An interesting outcome from studying electron statistics is a picture of chemical bonding in terms of population fluctuations among basins. This idea has been previously proposed [36], but the lack of examples and actual calculations has clearly precluded its spread. A much deeper analysis of the mutual relationships that exist among population fluctuation, correlation, delocalization, and bonding is urgently needed.

An important lesson to be learnt from our results is that once we allow further order (i.e. second order and beyond) reduced density matrices to enter our descriptors, the usual atomic QTAIM view starts including atomic pairs, trios, etc. It is in this way that some of the core concepts of chemistry find their natural way in real space topological analyses. Unfortunately, this richness cannot be easily obtained from experiments, which still mostly inform us about the diagonal part of the first order density. Much more effort should thus be devoted to finding adequate approximations to higher order quantities from density functional expansions [4] or to improve the refinement of suitable wavefunctions [5].

Turning to the fine-grained point of view, the method of ωRSP is applicable to a wide number of possible choices of *control* functions and *sampling* properties to create whole families of new functionals. Many of the resulting distributions will have the interesting feature that the sum of all values is proportional to the expectation value of the *sampling* property (respectively some specific part of it). The investigation of the ωRSP approach itself and the ELI distributions in particular is more or less at the initial stage. Thorough study of suitable examples is necessary to unveil possible connections to the bonding situation of the system.

Already now it is clear that the definition of ELI on the basis of ωRSP enables to apply the method transparently to any level of theory, in both real and momentum space, as long as the electron density and pair density are accessible. ELI-D is proportional to a discrete distribution of charges, thus, in a very loose sense, it could have relations to QTAIM, which deserves to be explored. Additionally, the viewpoint of the limit after rescaling makes it possible to analyse the ELI topology using derivatives of ELI (although this renounces the discrete character of the distribution). Because ELI-D in the limit after rescaling can be seen as the product of electron density and the pair-volume function, the Laplacian of ELI-D is connected with the electron density Laplacian [72]. Furthermore, there is the appealing challenge to reconstruct ELI-D from the experimental electron density using an approximate pair-volume function.

The availability of correlated wave functions enables to examine ELI for the opposite-spin electron pairs. The first analyses of such distributions showed [9, 10, 42] that the emergence of chemical descriptors (atomic shells, bonds or lone-pairs) is not so 'self-evident' as in case of same-spin pairs. What should an opposite-spin ELI diagram look like for 'perfectly' correlated wave function? Due to the non-zero on-top density for an opposite-spin pair the scaling of ELI (with respect to the restriction, cf. Sect. 9.5.3) for opposite-spin pairs is very different from the one for same-spin pairs. This also means that monitoring spinless pairs recovers only the features of the opposite-spin pairs (determined by the term of the Taylor expansion that governs the number of the respective pairs). Interestingly, if the on-top density would be zero over the whole system all ELI formulae would follow the same scaling.

The diagrams of ELI in the momentum space usually shows well pronounced structure only in relative small region of few atomic units around the origin. This is somewhat uncomfortable compared to the real space diagrams which reflect the size of the system. Possibly, at least for larger systems, the momentum space ELI is able to deliver just a kind of fingerprint. On the other hand, this could enhance the perception of similarities between systems differing in size and composition. ELI is connected to the local correlation of electronic motion, which in momentum space is related to the tendency to avoid similar momenta (irrespective of the real space positions), whereas in real space the avoidance concerns the position (irrespective of the momenta). This interplay needs to be examined in the future.

Acknowledgements Financial support from the Spanish MEC, Project No. CTQ2006-02976, the MALTA Consolider program (CSD2007-00045), and the ERDF of the European Union, is acknowledged.

References

1. Hohenberg P, Kohn W (1964) Inhomogeneous electron gas. Phys Rev 136:B864–B871
2. Bader RFW (1990) Atoms in molecules. Oxford University Press, Oxford
3. Becke AD, Edgecombe KE (1990) A simple measure of electron localization in atomic and molecular systems. J Chem Phys 92:5397–5403
4. Tsirelson V, Stash A (2002) Determination of the electron localization function from electron density. Chem Phys Lett 351:142–148
5. Jayatilaka D (1998) Wave function for beryllium from x-ray diffraction data. Phys Rev Lett 80:798–801
6. Li L, Parr RG (1986) The atom in a molecule: a density matrix approach. J Chem Phys 84:1704–1711
7. Hirshfeld FL (1977) Bonded-atom fragments for describing molecular charge densities. Theor Chim Acta 44:129–138
8. Becke AD (1988) A multicenter numerical integration scheme for polyatomic molecules. J Chem Phys 88:2547–2553
9. Fernández Rico J, López R, Ema I, Ludeña E (2004) Analytical representation and fast evaluation of density, electronic potential and field and forces on the nuclei. J Comput Chem 25:1355–1363
10. Alcoba DR, Lain L, Torre A, Bochicchio R (2005) A study of the partitioning of the first-order reduced density matrix according to the theory of atoms in molecules. J Chem Phys 123:144113
11. Mayer I (2003) An exact chemical decomposition scheme for the molecular energy. Chem Phys Lett 382:265–269
12. Salvador P, Mayer I (2004) Energy partitioning for fuzzy atoms. J Chem Phys 120:5046–5052
13. Sierraalta A, Frenking G (1997) Diatomic interaction energies in the topological theory of atoms in molecules. Theor Chim Acta 95:1–12
14. Martín Pendás A, Blanco MA, Francisco E (2007) Chemical fragments in real space: definitions, properties, and energetic decompositions. J Comput Chem 28:161–184
15. Martín Pendás A, Francisco E, Blanco MA (2007) Pauling resonant structures in real space through electron number probability distributions. J Phys Chem A 111:1084–1090
16. Martín Pendás A, Francisco E, Blanco MA (2007) Spatial localization, correlation, and statistical dependence of electrons in atomic domains: the $x^1\sigma_g^+$ and $b^3\sigma_u^+$ states of h_2. Chem Phys Lett 437:287
17. Blanco MA, Martín Pendás A, Francisco E (2005) Interacting quantum atoms: a correlated energy decomposition scheme based on the quantum theory of atoms in molecules. J Chem Theory Comput 1:1096–1109
18. Francisco E, Martín Pendás A, Blanco MA (2006) A molecular energy decomposition scheme for atoms in molecules. J Chem Theory Comput 2:90–102
19. Cohen L (1979) Local kinetic energy in quantum mechanics. J Chem Phys 70:788–789
20. Martín Pendás A, Blanco MA, Francisco E (2004) Two-electron integrations in the quantum theory of atoms in molecules. J Chem Phys 120:4581–4592
21. Martín Pendás A, Francisco E, Blanco MA (2005) Two-electron integrations in the quantum theory of atoms in molecules with correlated wavefunctions. J Comput Chem 26:344–351
22. Martín Pendás A, Francisco E, Blanco MA (2007) Charge transfer, chemical potentials, and the nature of functional groups: answers from a quantum chemical topology. Faraday Discuss 135:423–438
23. Chamorro E, Fuentealba P, Savin A (2003) Electron probability distribution in AIM and ELF basins. J Comput Chem 24:496–504
24. Francisco E, Martín Pendás A, Blanco MA (2007) Electron number probability distributions for correlated wave functions. J Chem Phys 126:094102
25. Francisco E, Martín Pendás A, Blanco MA (2008) EDF: computing electron number probability distribution functions in real space from molecular wave functions. Comput Phys Commun 178:621–634

26. Martín Pendás A, Francisco E, Blanco MA (2007) An electron number distribution view of chemical bonds in real space. Phys Chem Chem Phys 9:1087–1092
27. Martín Pendás A, Francisco E, Blanco MA (2007) Spin-resolved electron number distribution functions: how spins couple in real space. J Chem Phys 127:144103
28. Ponec R, Mayer I (1997) Investigation of some properties of multicenter bond indices. J Phys Chem A 101:1738–1741
29. Sannigrahi AB, Kar T (1990) Three-center bond index. Chem Phys Lett 173:569–572
30. Sannigrahi AB, Kar T (2000) Ab initio theoretical study of three-centre bonding on the basis of bond index. J Mol Struct Theochem 496:1–17
31. Wiberg KB (1968) Application of the pople-santry-segal cndo method to the cyclopropylcarbinyl and cyclobutyl cation and to bicyclobutane. Tetrahedron 24:1083–1096
32. Bochicchio R, Ponec R, Lain L, Torre A (2000) Pair population analysis within AIM theory. J Phys Chem A 104:9130–9135
33. Bochicchio R, Ponec R, Torre A, Lain L (2001) Multicenter bonding within the AIM theory. Theor Chem Acc 105:292–298
34. Giambiagi M, de Giambiagi MS, Mundim KC (1990) Definition of a multicenter bond index. Struct Chem 1:423–427
35. Mundim KC, Giambiagi M, de Giambiagi MS (1994) Multicenter bond index: Grassmann algebra and n-order density functional. J Phys Chem 98:6118–6119
36. Ziesche P (2000) Many-electron densities and reduced density matrices, chapter 3. Kluwer Academic, New York, p 33
37. Parr RG, Yang W (1989) Density-functional theory of atoms and molecules. Oxford University Press, New York
38. Martín Pendás A, Francisco E, Blanco MA (2006) Binding energies of first row diatomics in the light of the interacting quantum atoms approach. J Phys Chem A 110:12864–12869
39. Martín Pendás A, Blanco MA, Costales A, Mori-Sánchez P, Luaña V (1999) Non-nuclear maxima of the electron density. Phys Rev Lett 83:1930–1933
40. Rafat M, Popelier PLA (2007) The quantum theory of atoms in molecules, chapter 5. Wiley-VCH/GmbH & Co. KGaA, pp 121–140
41. Costales A, Blanco MA, Martín Pendás A, Mori-Sánchez P, Luaña V (2004) Universal features of the topological bond properties of the electron density. J Phys Chem A 108:2794–2801
42. Dominiak PM, Makal A, Mallinson PR, Trzcinska K, Eilmes J, Grech E, Chruszcz M, Minor W, Wozniak K (2006) Continua of interactions between pairs of atoms in molecular crystals. Chem Eur J 12:1941–1949
43. Martín Pendás A, Francisco E, Blanco MA, Gatti C (2007) Bond paths as privileged exchange channels. Chem Eur J 13:9362–9371
44. Ponec R, Cooper DL (2007) Anatomy of bond formation. Bond length dependence of the extent of electron sharing in chemical bonds from the analysis of domain-averaged Fermi holes. Faraday Discuss 135:31–42
45. Buckingham AD, Fowler PW (1983) Do electrostatic interactions predict structures of Van der Waals molecules? J Chem Phys 79:6426–6428
46. Castiglioni C, Gussoni M, Zerbi G (1984) Stabilization energies of weak hydrogen bonded molecular complexes. Comparison of simple models. J Chem Phys 80:3916–3918
47. Dykstra CE (1993) Electrostatic interaction potentials in molecular force fields. Chem Rev 93:2339–2353
48. Kollman P (1977) A general analysis of noncovalent intermolecular interactions. J Am Chem Soc 99:4875–4894
49. Rendell APL, Bacskay GB, Hush NS (1985) The validity of electrostatic predictions of the shapes of Van der Waals dimers. Chem Phys Lett 117:400–408
50. Spackman MA (1986) A simple quantitative model of hydrogen bonding. J Chem Phys 85:6587–6601
51. Pauling L (1928) The shared-electron chemical bond. Proc Natl Acad Sci USA 14:359–362
52. Pauling L (1960) The nature of the chemical bond, 3rd edn. Cornell University Press, Ithaca

53. Arnold WD, Oldfield E (2000) The chemical nature of hydrogen bonding in proteins via NMR: J-couplings, chemical shifts, and aim theory. J Am Chem Soc 122:12835–12841
54. Jeffrey GA, Saenger W (1991) Hydrogen bonding in biological structures. Springer, Heidelberg
55. Thompson WH, Hynes JT (2000) Frequency shifts in the hydrogen-bonded oh stretch in halide-water clusters. The importance of charge transfer. J Am Chem Soc 122:6278–6286
56. Zhang Y, Zhao GY, You XZ (1997) Systematic theoretical study of structures and bondings of the charge-transfer complexes of ammonia with HX, XY, and X_2 (X and Y are halogens). J Phys Chem A 101:2879–2885
57. Martín Pendás A, Blanco MA, Francisco E (2006) The nature of the hydrogen bond: a synthesis from the interacting quantum atoms picture. J Chem Phys 125:184112
58. Umeyama H, Morokuma K (1977) The origin of hydrogen bonding. An energy decomposition study. J Am Chem Soc 99:1316–1332
59. Martín Pendás A, Blanco MA, Francisco E (2009) Steric repulsions, rotation barriers, and stereoelectronic effects: a real space perspective. J Comput Chem 30:98
60. Jeziorski B, Moszynski R, Szalewicz K (1994) Perturbation theory approach to intermolecular potential energy surfaces of Van der Waals complexes. Chem Rev 94:1887–1930
61. Ziegler T, Rauk A (1979) Carbon monoxide, carbon monosulfide, molecular nitrogen, phosphorus trifluoride, and methyl isocyanide as .sigma. donors and .pi. acceptors. A theoretical study by the Hartree-Fock-Slater transition-state method. Inorg Chem 18:1755–1759
62. Ziegler T, Rauk A (1979) A theoretical study of the ethylene-metal bond in complexes between copper(1+), silver(1+), gold(1+), platinum(0) or platinum(2+) and ethylene, based on the Hartree-Fock-Slater transition-state method. Inorg Chem 18:1558–1565
63. Kohout M (2004) A measure of electron localizability. Int J Quantum Chem 97:651–658
64. Dobson JF (1991) Interpretation of the Fermi hole curvature. J Chem Phys 94:4328–4332
65. Savin A, Nesper R, Wengert S, Fässler TF (1997) Die Elektronenlokalierungsfunktion ELF, ELF: the electron localization function. Angew Chem Int Ed Engl 36:1808–1832
66. Clementi E, Roetti C (1974) Roothaan-Hartree-Fock atomic wavefunctions: basis functions and their coefficients for ground and certain excited states of neutral and ionized atoms, $Z \leq 54$. At Data Nucl Data Tables 14:177–478
67. Kohout M, Wagner FR, Grin Y (2008) Electron localizability indicator for correlated wavefunctions. III: singlet and triplet pairs. Theor Chem Acc 119:413–420
68. Bader RFW, Stephens ME (1974) Fluctuation and correlation of electrons in molecular systems. Chem Phys Lett 25:445–449
69. Bader RFW, Stephens ME (1975) Spatial localization of the electronic pair and number distributions in molecules. J A Chem Soc 97:7391–7399
70. Kato T (1957) On the eigenfunctions of many-particle systems in quantum mechanics. Commun Pure Appl Math 10:151–177
71. Smith VH Jr (1971) Cusp conditions for natural functions. Chem Phys Lett 9:365–371
72. Wagner FR, Kohout M, Grin Y (2008) Direct space decomposition of ELI-D: interplay of charge density and pair-volume function for different bonding situations. J Phys Chem A 112:9814–9828
73. Löwdin PO (1955) Quantum theory of many-particle systems. I: physical interpretations by means of density matrices, natural spin-orbitals, and convergence problems in the method of configurational interaction. Phys Rev 97:1474–1489
74. Kohout M, Pernal K, Wagner FR, Grin Y (2004) Electron localizability indicator for correlated wavefunctions. I: parallel-spin pairs. Theor Chem Acc 112:453–459
75. Wagner FR, Bezugly V, Kohout M, Grin Y (2007) Charge decomposition analysis of the electron localizability indicator: a bridge between the orbital and direct space representation of the chemical bond. Chem Eur J 13:5724–5741
76. Kohout M, Savin A (1996) Atomic shell structure and electron numbers. Int J Quantum Chem 60:875–882
77. Daudel R (1968) The fundamentals of theoretical chemistry. Pergamon Press, Oxford

78. Silvi B (2003) The spin-pair compositions as local indicators of the nature of the bonding. J Phys Chem A 107:3081–3085
79. Kohout M, Pernal K, Wagner FR, Grin Y (2005) Electron localizability indicator for correlated wavefunctions. II: antiparallel-spin pairs. Theor Chem Acc 113:287–293
80. Kutzelnigg W (1963) Über die Symmetrie-Eigenschaften der reduzierten Dichtematrizen. Z Naturforschg 18a:1058–1064
81. Kutzelnigg W, Mukherjee D (2002) Irreducible Brillouin conditions and contracted Schrödinger equations for n-electron systems. II: spin-free formulation. J Chem Phys 116:4787–4801
82. Cooper DL, Ponec R, Thorsteinssohn T, Raos G (1996) Pair populations and effective valencies from ab initio SCF and spin-coupled wavefunctions. Int J Quantum Chem 57:501–518
83. Kutzelnigg W (2002) In: Rychlewski J (ed) Explicitly correlated wave functions in chemistry and physics: theory and applications. Kluwer, Dordrecht, pp 14–17
84. Frisch MJ, Trucks GW, Schlegel HB, Scuseria GE, Robb MA, Cheeseman JR, Montgomery Jr JA, Vreven T, Kudin KN, Burant JC, Millam JM, Iyengar SS, Tomasi J, Barone V, Mennucci B, Cossi M, Scalmani G, Rega N, Petersson GA, Nakatsuji H, Hada M, Ehara M, Toyota K, Fukuda R, Hasegawa J, Ishida M, Nakajima T, Honda Y, Kitao O, Nakai H, Klene M, Li X, Knox JE, Hratchian HP, Cross JB, Bakken V, Adamo C, Jaramillo J,Gomperts R, Stratmann RE, Yazyev O, Austin AJ, Cammi R, Pomelli C, Ochterski JW, Ayala PY, Morokuma K, Voth GA, Salvador P, Dannenberg JJ, Zakrzewski VG, Dapprich S, Daniels AD, Strain MC, Farkas O, Malick DK, Rabuck AD, Raghavachari K, Foresman JB, Ortiz JV, Cui Q, Baboul AG, Clifford S, Cioslowski J, Stefanov BB, Liu G, Liashenko A, Piskorz P, Komaromi I, Martin RL, Fox DJ, Keith T, Al-Laham MA, Peng CY, Nanayakkara A, Challacombe M, Gill PMW, Johnson B, Chen W, Wong MW, Gonzalez C, Pople JA Gaussian 03, Revision C.02. Gaussian, Inc., Wallingford, CT, 2004
85. Kohout M (2009) DGrid 4.5. Radebeul
86. Baranov AI, Kohout M (2008) Electron localizability for hexagonal element structures. J Comput Chem 29:2161–2171
87. Bezugly V, Wielgus P, Wagner FR, Kohout M, Grin Y (2008) Electron localizability indicators ELI and ELIA: the case of highly correlated wavefunctions for the argon atom. J Comput Chem 29:1198–1207
88. Bezugly V, Wielgus P, Kohout M, Wagner FR (2010) Electron localizability indicators ELI-D and ELIA for highly correlated wavefunctions of homonuclear dimers. II. N_2, O_2, F_2, and Ne_2. J Comput Chem 31:2273–2285
89. Kohout M, Wagner FR, Grin Y (2006) Atomic shells from the electron localizability in momentum space. Int J Quantum Chem 106:1499–1507
90. Kohout M (2007) Bonding indicators from electron pair density functionals. Faraday Discuss 135:43–54
91. Kaijser P, Smith VH Jr (1977) Evaluation of momentum distributions and compton profiles for atomic and molecular systems. Adv Quantum Chem 10:37–76

Chapter 10
On the Interplay Between Real and Reciprocal Space Properties

Wolfgang Scherer, Georg Eickerling, Christoph Hauf, Manuel Presnitz, Ernst-Wilhelm Scheidt, Volker Eyert, and Rainer Pöttgen

10.1 Introduction

The charge density distribution – a real space property – is the inverse Fourier transform of the structure factors representing reciprocal space properties. The related linkage between the Patterson pair correlation function and the Bragg reflections – as intensity-weighted reciprocal space properties – provides another well-established example for such a direct and reciprocal space relationship. Even though these prototypes of real and reciprocal space correlations are well established and fundamental, we note that charge density and charge density matrix refinements from scattering experiments remain a true challenge (Chap. 5). This chapter, however, goes beyond these classical space correlations and addresses the problem how we can retrieve information from charge density distributions to understand and control the physical behavior of periodic solids. Many physical properties in extended solids such as transport phenomena (*e.g.* electronic conductivity, superconductivity, band magnetism) also depend directly on reciprocal space properties, namely, the energy dispersion of electronic bands in momentum or k-space. However, direct relationships with real space properties are less elaborated in these cases. It is therefore the major issue of this chapter to outline how characteristic features in the band structure of solids are reflected in charge density properties. To simplify this real and reciprocal space correlation we will focus on quasi-one dimensional

W. Scherer (✉) • G. Eickerling • C. Hauf • M. Presnitz • E.-W. Scheidt • V. Eyert
Institut für Physik, Universität Augsburg, Universitätsstrasse 1, 86159 Augsburg, Germany
e-mail: wolfgang.scherer@physik.uni-augsburg.de; georg.eickerling@physik.uni-augsburg.de; christoph.hauf@physik.uni-augsburg.de; manuel.presnitz@physik.uni-augsburg.de; ernst-wilhelm.scheidt@physik.uni-augsburg.de; veyert@materialsdesign.com

R. Pöttgen
Institut für Anorganische und Analytische Chemie, Universität Münster, Corrensstrasse 30, 48149 Münster, Germany
e-mail: pottgen@uni-muenster.de

C. Gatti and P. Macchi (eds.), *Modern Charge-Density Analysis*,
DOI 10.1007/978-90-481-3836-4_10, © Springer Science+Business Media B.V. 2012

case studies only. We therefore selected as benchmarks quasi one-dimensional organometallic carbides, which provide textbook examples of extended systems displaying pronounced orbital interactions and anisotropic physical properties in real and reciprocal space. Hence, small changes in the chemical bonding and/or their electronic structures should give rise to pronounced effects in both real and reciprocal space properties.

During the past three decades a great number of ternary rare-earth transition metal carbides $RE_xT_yC_z$ (RE = rare earth metal; T = transition metal) have been synthesized [1–14]. The $RE_xT_yC_z$ carbides display monoatomic C^{4-} units, C_2 pairs or in some cases linear C_3 moieties and are characterized by a large structural variety as exemplified by almost 60 different structure types [15]. Furthermore, these complex carbides display a great and puzzling variety of different chemical bonding scenarios as witnessed by a large number of theoretical studies in this field [15–19]. Since the pioneering theoretical studies by Burdett, Whangbo and Hoffmann on YCoC [16–18] – a textbook example displaying pronounced orbital interaction in an extended low-dimensional material – one might consider these species as anionic organometallic $[T_yC_z]^{\delta-}$ polymers embedded in an ionic matrix provided by the rare earth metal atoms [20–22]. The carbometallate YCoC also represents a prototype of a low-dimensional $RE_xT_yC_z$ carbide by forming infinite linear –Co-C-Co-C– chains (Fig. 10.1a) [23]. Besides its quasi one-dimensional character YCoC displays an unusual high electronic heat capacity with a Sommerfeld coefficient γ of 14.0 mJ/K^2 mol. Employing the free electron model, a density of states at the Fermi-level ($N(E_F)$ of 5.9 states/eV) was derived from the gamma value, suggesting the presence of narrow conduction bands [24]. This observation is supported by recent theoretical calculations predicting a weak ferromagnetic instability of YCoC via the Stoner mechanism, which postulates that itinerant ferromagnetic correlations are associated with an enhanced $N(E_F)$ [25].

Also other members of this family of low-dimensional carbides exhibit interesting magnetic and electronic properties. For example, if we interconnect the parallel –Co-C-Co-C– chains in YCoC by additional perpendicular Co–C–Co linkages, a hypothetical two-dimensional network composed of $[TC_4]^{\delta-}$ polyanions with formula "$YCoC_2$" is established (Scheme 10.1). However, as shown by Li and Hoffmann [19] this network is unstable with respect to the pairing of the carbon atoms leading to severe distortions of the square-planar network and finally to the $CeNiC_2$ type structure [1] (Scheme 10.1 and Fig. 10.1b). In this class of compounds, especially the ternary nickel carbides $RENiC_2$ exhibit a variety of magnetic ordering scenarios, which depend on the nature of the rare earth metal. For example $SmNiC_2$ displays an interesting interplay between charge density wave (CDW) formation and ferromagnetic order [26, 27]. However, the most prominent representative of this compound family is $LaNiC_2$, which exhibits superconductivity at $T_c = 2.7$ K despite the fact that the ferromagnetic element nickel usually has an adverse effect with respect to superconducting properties in solids [28]. However, the magnetism in the $RENiC_2$ phases appears to be solely due to the $4f$ electrons of the rare earth metal with almost no magnetic contribution from nickel [29]. Increasing the chemical pressure (reduced unit cell volume by replacing the La^{3+} by Th^{4+}) in

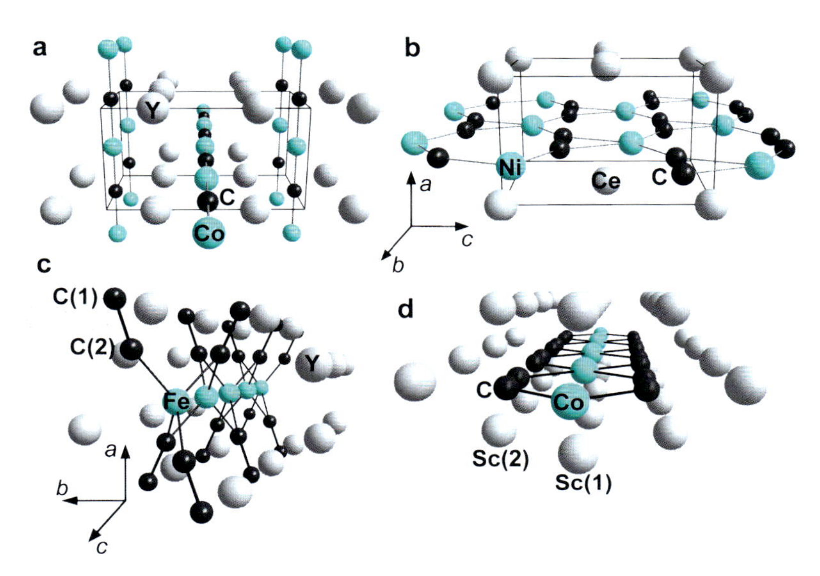

Fig. 10.1 Structural models of the ternary rare-earth transition metal carbides (**a**) YCoC, (**b**) $CeNiC_2$ and the embedded TC_4 ribbons in (**c**) Y_2FeC_4 and (**d**) Sc_3CoC_4. The coordinate system is identical for (**a**), (**b**) and (**d**)

the solid solution $La_{1-x}Th_xNiC_2$ [28, 30] leads to an even higher T_c of 7.9 K at $x = 0.5$. Despite its non-centrosymmetric structure, $LaNiC_2$ (space group *Amm*2; $CeNiC_2$ structure) has been recently classified as a conventional electron-phonon superconductor by density functional theory (DFT) calculations [29] in contrast to earlier suggestions based on muon spin-relaxation measurements [31]. Short C–C bond distances are another characteristic feature of the non-centrosymmetric $LaNiC_2$ structure and indicative for the presence of stiff ν(C–C) stretching modes. These modes and substantial contributions of antibonding π^*(C–C) states to the density of states at the Fermi level have been suggested to control the T_c in transition metal carbides displaying dicarbido moieties [29, 32].

Superconductivity has been also observed in Y_2FeC_4 (Fig. 10.1c) featuring quasi one-dimensional $[FeC_4]_\infty$ moieties with iron in a distorted tetrahedral coordination environment [33]. We note, that the superconductors $LaNiC_2$ and Y_2FeC_4 share the presence of dicarbido C_2 moieties as common structural features. This possibly further supports the idea that the presence of antibonding π^*(C–C) states near the Fermi level (which are absent in the carbometallate YCoC) might provide a prerequisite for the onset of superconductivity in the $RE_xT_yC_z$ carbides with $z \geq 2$ [32]. This is in line with DFT studies on the YC_2 carbides, which also show that antibonding π^*(C-C) states near the Fermi level might have substantial contributions to their electronic structures and physical properties [32]. The absence of paramagnetic *RE* cations in $LaNiC_2$ and Y_2FeC_4 might be another criterion while the presence of iron and nickel atoms has no adverse effect to the superconducting properties like in the iron arsenides [34].

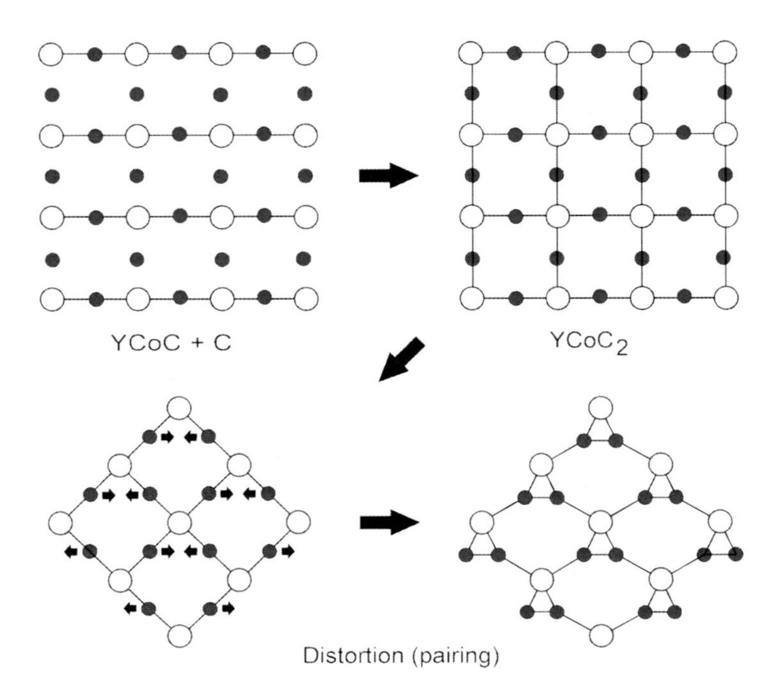

Scheme 10.1 (Adapted from Ref. [19] with kind permission of The American Chemical Society)

To gain a better understanding of the various control parameters on the physical properties in low-dimensional $RE_xT_yC_z$ carbides we focus in our studies on the electronic structure of the Sc_3TC_4 carbides (T = Fe, Co, Ni, Ru, Os) [35–42], which represent like Y_2FeC_4 quasi one-dimensional materials (Fig. 10.1d). However, in Y_2TC_4 two T-centers are bridged via one common carbon atom while in Sc_3TC_4 linear $[T(C_2)_2]_\infty$ ribbons with bridging μ-η^2-C_2 moieties are observed. Due to their isotypic character the Sc_3TC_4 carbides are structurally highly related and allow in the series T = Fe, Co, Ni a stepwise increase of the d-electron count and thus a systematic variation of the states contributing to the Fermi level. Accordingly, the selected systems allow a systematic design of their electronic structures and physical properties. For example, we will illustrate that substitution of the Group 8 element iron in Sc_3FeC_4 by its Group 9 congener cobalt enforces superconductivity ($T_c^{onset} = 4.5$ K) while the parent Sc_3FeC_4 carbide remains metallic above 1.9 K [40]. We will further outline that the changes in the electronic structures of the Sc_3TC_4 carbides can be clearly identified by analyses of the topology of their respective charge density distributions. Especially, we will employ the negative Laplacian of the charge density distributions, $-\nabla^2\rho(\mathbf{r}) = L(\mathbf{r})$, to trace and to quantify subtle electron localization phenomena and their influence on the electronic transport properties of these Sc_3TC_4 carbides [40–42].

10.2 Chemical Bonding in Quasi One-Dimensional Transition Metal Carbides

10.2.1 Structural Characteristics

Figure 10.1d displays a cutout of the structural motif of Sc_3CoC_4. The one-dimensional $[CoC_4]_\infty^{\delta-}$ polyanions extend along the *b* axis and are embedded in a scandium matrix. Within the $[CoC_4]_\infty^{\delta-}$ polyanion, each cobalt atom is coordinated by four C_2 pairs at Co–C distances of 2.0886(4) Å, significantly longer than the Co–C distance in the –Co-C-Co-C– chains of YCoC (1.8250(2) Å) or in the bulk transition metal carbide Co_2C (1.911(6)–1.918(5)) [18, 43]. Also cobalt carbonyls display significantly shorter Co–C bond lengths (*e.g.* 1.742(2) Å in $Li[Co(CO)_4]$) [18, 44] due to the pronounced σ-donor and π acceptor interaction. The dicarbido system $[Co_9(C_2)(CO)_{19}]^{2-}$ displaying C_2 moieties encapsulated in a tricapped trigonal prism of 9 cobalt atoms might act as a reference system for Co–C single bonds (Co–C 1.95(2)–1.97(2) and 2.04(2)–2.06(2) Å for the outer and central Co atoms, respectively; Fig. 10.2) [45]. Accordingly, Sc_3CoC_4 displays slightly elongated Co–C single bonds while the C_2 pairs are characterized by a C–C distance of 1.4539(8) Å, hinting for an intermediate bonding situation between a C–C single bond (1.54 Å) and a double bond (1.34 Å) [46–48].[1] The structural features of the 3d series Sc_3FeC_4 **1**, Sc_3CoC_4 **2**, Sc_3NiC_4 **3** are compared in Fig. 10.3 and Tables 10.1 and 10.2 which demonstrates that all three carbides are truly isotypic displaying only minute structural differences. The transition metal atom's site symmetry is close to D_{4h} (4/*mmm*) when limited to the first coordination sphere $[TC_4]$, but reduced to D_{2h} (*mmm*) for the whole chain $[T(C_2)_2]_\infty$.

Hence, the different physical properties of these three benchmarks will be virtually decoupled from structural parameters and will mainly depend on the local electronic structure/valence electron count of the transition metal atom. The Sc_3TC_4 (T = Fe, Co, Ni) compounds represent therefore ideal benchmarks to establish a relationship between local charge density features in real space and physical properties in reciprocal space.

---

[1] Indeed, the C–C bond distances in Sc_3TC_4 of 1.4498(11), 1.4539(8) and 1.4561(13) Å for T = Fe, Co and Ni, respectively, fall in a narrow range marked by (*i*) transition metal ethyl complexes with shortened C–C single bond distances as a consequence of negative (*M*–C) hyperconjugation (*e.g.* 1.5126(12) Å in the agostic $(C_2H_5)TiCl_3$(dmpe) complex; dmpe = $Me_2PCH_2CH_2PMe_2$) (Ref. [46]) and (*ii*) transition metal olefin complexes with weakened C = C double bonds due to pronounced $M \rightarrow \pi^*$(C–C) back donation (*e.g.* 1.4189(6) Å in $\eta^2(C_2H_4)$Ni(dbpe); dbpe = $Bu_2{}^tPCH_2CH_2PBu_2{}^t$) (Ref. [47]). However, the C–C bond distances clearly differ from those of (*iii*) transition metal acetylene complexes displaying C–C triple bonds (*e.g.* 1.209(1) Å in $[Ag(C_2H_2)Al(OCCH_3(CF_3)_2)_4]$; Ref. [48]). Note, that the C–C bond distances in all these molecular benchmark systems are directly comparable since they are based on multipolar refinements employing high-resolution X-ray data. Therefore, these distance parameters can be considered as being virtually deconvoluted from thermal smearing effects.

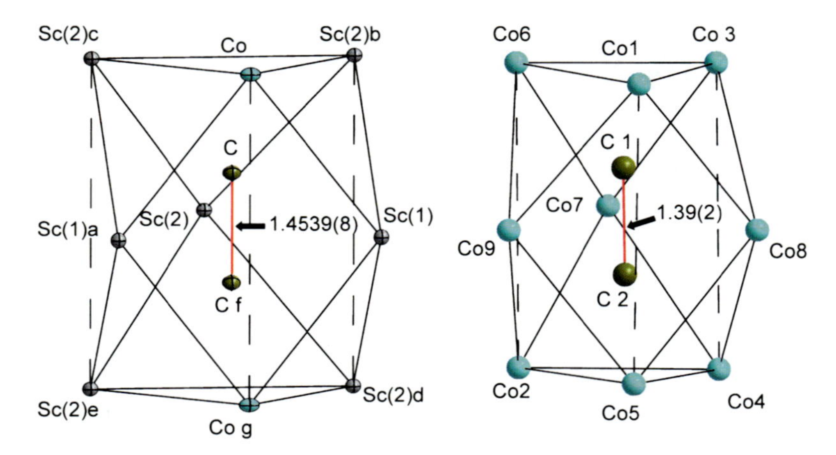

Fig. 10.2 Structural comparison of the dicarbido C_2 moieties encapsulated in a tricapped trigonal prism of 9 metal atoms in Sc_3CoC_4 (*left-hand side*) and the dicarbido cluster $[Co_9(C_2)(CO)_{19}]^{2-}$ (*right-hand side*). Symmetry related atomic coordinates: (*a*) $-1+x$, y, z; (*b*) $0.5+x$, $0.5-y$, $0.5-z$; (*c*) $-0.5+x$, $0.5-y$, $0.5-z$; (*d*) $0.5+x$, $-0.5-y$, $0.5-z$; (*e*) $-0.5+x$, $-0.5-y$, $0.5-z$; (*f*) $-x$, $-y$, z; and (*g*) x, $-1+y$, z. The C–C bond lengths of the dicarbido moieties are specified in [Å]

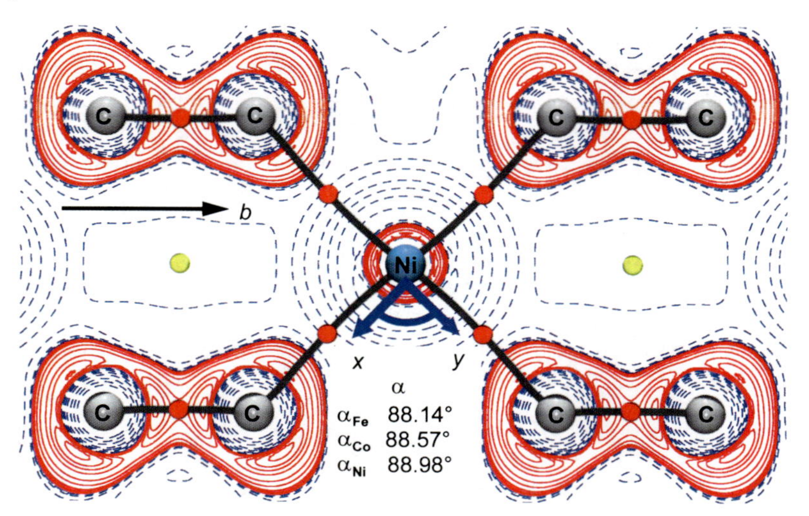

Fig. 10.3 $L(\mathbf{r})$ contour maps of the experimental electron density of **3** in the crystallographic b,c plane (adapted from Ref. [40]). Contour levels are drawn at 0.001, $\pm 2.0 \times 10^n$, $\pm 4.0 \times 10^n$, $\pm 8.0 \times 10^n$ eÅ^{-5}, where $n = 0, \pm 3, \pm 2, \pm 1$; extra levels at 11.5, 15.0, 1,200 and 1,500 eÅ^{-5}; positive and negative values are marked by *solid* and *dashed lines*, respectively. Bond paths (*solid lines*), bond critical points (bcps) and ring critical points (rcps) of **3** are marked by *red and yellow closed circles*, respectively. Note the definition of the local coordinate system and the transition metal atom's site symmetry which is close to D_{4h} ($4/mmm$) in the TC_4 moieties but reduced to D_{2h} (mmm) for the polyanions in the crystal

Table 10.1 Analysis of bond and ring CPs in **1–3** (ρ_b is given in eÅ^{-3}, L_b in eÅ^{-5}, $H(\mathbf{r}_c)$ in hartree Å^{-3} and G_b/ρ_b in hartree e^{-1}; all distances in Å)

Unit	Method[a,b]	Distance	ρ_b	L_b	Ellipticity ε	H_b	G_b/ρ_b[c]
1 (Sc_3FeC_4)							
Fe–C	Experiment	2.1074(6)	0.590	−5.3	0.01	−0.209	0.987
	Theory		0.553	−4.5	0.11	−0.194	0.923
C–C	Experiment	1.4498(11)	1.750	9.8	0.27	−2.272	0.902
	Theory		1.765	14.8	0.00	−2.421	0.776
Sc(1)–C[d]	Experiment	2.3789(4)	0.282	−3.9	2.75	−0.005	1.004
	Theory		0.301	−4.2	2.60	−0.010	1.018
Sc(2)–C[d]	Experiment	2.3631(6)	0.353	−3.6	1.41	−0.056	0.885
	Theory		0.332	−3.5	0.59	−0.046	0.880
[Sc(2)C_2]	Experiment	–	0.335	−4.0	–	−0.035	0.952
rcp	Theory	–	0.308	−4.6	–	−0.005	1.060
2 (Sc_3CoC_4)							
Co–C	Experiment	2.0886(4)	0.581	−5.4	0.17	−0.196	1.002
	Theory		0.572	−3.8	0.08	−0.226	0.869
C–C	Experiment	1.4539(8)	1.813	12.0	0.09	−2.449	0.881
	Theory		1.769	15.0	0.00	−2.433	0.773
Sc(1)–C[d]	Experiment	2.3761(3)	0.299	−3.4	10.71	−0.027	0.895
	Theory		0.308	−3.7	2.36	−0.025	0.940
Sc(2)–C[d]	Experiment	2.3576(4)	0.337	−4.0	2.35	−0.037	0.949
	Theory		0.333	−3.5	0.59	−0.046	0.883
[Sc(2)C_2]	Experiment	–	0.330	−4.0	–	−0.035	0.957
rcp	Theory	–	0.310	−4.5	–	−0.006	1.061
3 (Sc_3NiC_4)							
Ni–C	Experiment	2.094(1)	0.515	−5.1	0.09	−0.147	0.979
	Theory		0.563	−3.5	0.15	−0.227	0.838
C–C	Experiment	1.4561(13)	1.689	9.2	0.09	−2.141	0.886
	Theory		1.746	14.6	0.01	−2.376	0.776
Sc(1)–C[d]	Experiment	2.3813(8)	0.299	−3.7	5.41	−0.021	0.937
	Theory		0.306	−3.6	1.82	−0.028	0.914
Sc(2)–C[d]	Experiment	2.3692(7)	0.353	−3.0	11.26	−0.072	0.798
	Theory		0.337	−3.6	0.65	−0.047	0.888
[Sc(2)C_2]	Experiment	–	0.352	−3.0	–	−0.071	0.799
rcp	Theory	–	0.315	−4.6	–	−0.010	1.054

[a]Experimental fractional coordinates and lattice parameters were adopted and not further optimized during the band structure calculations
[b]Theoretical values are based on the WIEN2K (Ref. [51]) results
[c]Calculated according to Ref. [52][2]
[d]Bridging Sc-η^2(C_2) moieties

---

[2] Calculation were done using the WIEN2K, employing the gradient-corrected PBE functional and up to 512 k-points in the irreducible wedge of the body-centered orthorhombic Brillouin zone.

Table 10.2 Integrated atomic charges ($q(\Omega)$) of the atomic basins of the crystallographically independent atoms in **1**–**3**

	1 (Sc_3FeC_4)		**2** (Sc_3CoC_4)		**3** (Sc_3NiC_4)	
atomic charge	Experiment	Theory[a]	Experiment	Theory[a]	Experiment	Theory[a]
$q(\Omega)_T$	0.50	0.15	0.27	0.07	0.14	0.05
$q(\Omega)_C$	−0.99	−1.31	−1.06	−1.29	−1.02	−1.21
$q(\Omega)_{Sc(1)}$	1.47	1.60	1.33	1.61	1.12	1.58
$q(\Omega)_{Sc(2)}$	1.01	1.61	1.35	1.62	1.37	1.60

[a]Theoretical values are based on the WIEN2K (Ref. [51]) results

10.2.2 Topological Analysis of the Electronic Structures

Multipolar refinements on high resolution X-ray data of **1**–**3** yield monopole populations of 7.98(10), 9.07(6) and 10.01(1) considering the $4s^2 3d^6$, $4s^2 3d^7$ and $4s^2 3d^8$ electrons as valence for the Fe, Co and Ni atoms, respectively [40, 41]. Also the corresponding quantum theory of atoms in molecules (QTAIM) charges suggest that the electronic configuration of the transition metals is close to d^8 (Fe), d^9 (Co), and d^{10} (Ni) since only small positive QTAIM charges ($q(\Omega)$) were observed for the metal atoms. QTAIM analyses of the charges of the C_2 moieties suggests in all cases the presence of $(C_2)^{2-}$ dicarbido units (Table 10.2). However, hydrolysis in diluted hydrochloric acid releases only saturated hydrocarbons besides hydrogen [37]. This already suggests that the bonding between these formal $(C_2)^{2-}$ moieties and their tricapped trigonal-prismatic T_2Sc_7-matrix is complex and cannot be accounted for by ionic contributions only.

Indeed, band structure analyses [40–42] (see Sect. 10.2.4) suggest the presence of pronounced σ-donor and π-acceptor interaction between the transition metal and the dicarbido moiety. The π character of the T-C bonds is further signaled by non-zero ellipticities at the T–C bond critical points (bcps) (Table 10.1 and Fig. 10.3) based on the DFT calculations. Unfortunately, the respective experimental bond ellipticity values, ε, do not provide such a concise picture. We note, however, that the absolute values of bond ellipticities critically depend on the ratio of the curvatures λ_1 and λ_2, representing eigenvalues of the Hessian of $\rho(\mathbf{r})$. Hence, subtle modifications in the charge density maps (e.g. due to even minute insufficiencies in the experimental data) might enforce large variations in the ε-values. Considering a local coordinate system with the x and y axis parallel to the T–C bonds and the z-axis perpendicular to the molecular plane spanned by the TC_4 ribbons (Fig. 10.3) we can assign the σ-component to the in-plane C ($p_{x,y}$) → Fe ($d_{x^2-y^2}$) donation and the π-component to the out-of-plane Fe (d_{xz},d_{yz}) → C (p_z) back donation into the antibonding π* states of the C_2 moieties.

Furthermore, the QTAIM analyses of **1**–**3** reveal individual bond paths for the three bridging ScC_2 moieties (involving the Sc(1), Sc(2) and Sc(1)a atoms) and the four terminal Sc–C groups (involving the Sc(2)b-e atoms; Figs. 10.2 and 10.4). The terminal Sc–C groups and the bridging $Sc(1)C_2$ and $Sc(1a)C_2$ moieties are characterized by a T-shaped bond path topology while only the bridging $Sc(2)C_2$

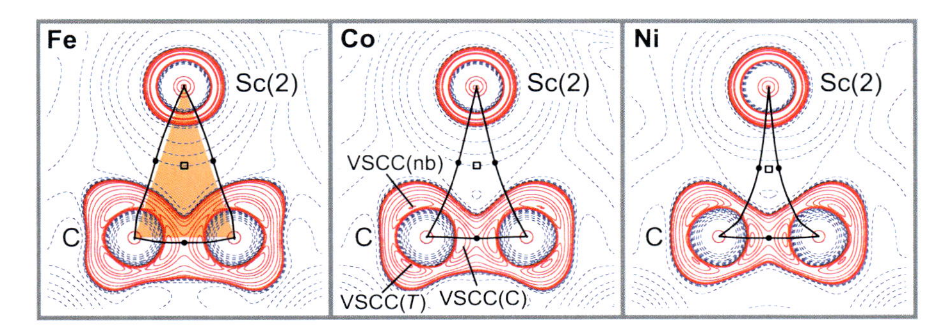

Fig. 10.4 Experimental $L(\mathbf{r})$ maps showing the separated Sc(2)–C bond paths in the $Sc(2)C_2$ planes of **1**–**3**, respectively. For a specification of the contour levels, see Fig. 10.3. Note, the presence of the three charge concentrations in the valence shell of the carbon atoms in the plane of the individual $[TC_4]$ units

moiety being located in the molecular plane of the $[TC_4]$ ribbons displays two separated Sc(2)–C bond paths. Such cyclic bond path topology suggests covalent metallacyclopropane-type bonding to occur in all bridging $Sc(2)C_2$ moieties of **1**–**3** [47, 53–57].[3,4] These cyclic bond paths are not well pronounced since the charge density accumulations at the Sc–C bcps and the corresponding ring critical points (rcps) are rather minute in **1**–**3** (Table 10.1). However, we note a distinct change of the bond path topologies in the $Sc(2)C_2$ moieties from exocyclic to endocyclic (Fig. 10.4) in the sequence Sc_3FeC_4, Sc_3CoC_4, Sc_3NiC_4. This might be due to a stepwise population of the antibonding π^* orbitals of the C_2 units with increasing d-electron count at the metal. Apparently, this trend is not only reflected in a lengthening of the corresponding C–C distances (Table 10.1) but also by a reduced T-C_2 bonding due to a stepwise enforced π back donation from the metal into the π^* orbitals of the dicarbido moieties in the series $Sc_3FeC_4 < Sc_3CoC_4 < Sc_3NiC_4$. Accordingly, this scenario indicates a crossover from a cyclic to a T-type bond path scenario[5] and is therefore indicative for an unstable bonding situation close

[3]Such bond path topology reflects the classical bonding scenario shown by early transition metal olefin complexes featuring a metallacyclopropane bonding mode.

[4]We therefore assume that the $\pi^*(C=C) \leftarrow Sc$ back donation component can be neglected due to the pronounced Lewis-acidity of the Sc atoms in **1**–**3** – in clear contrast to the Dewar-Chatt-Duncanson (DCD) bonding scenario displayed by late transition metal olefin complexes (Ref. [56, 57]).

[5]In case of the theoretical calculations a less consistent picture is derived. The WIEN2k calculations employing plane waves suggest in agreement with the experimental findings cyclic Sc(2)–C and T-shaped Sc(1)–C bcps bond path topologies for **1**–**3**. On contrast, CRYSTAL06 LCAO-DFT calculations employing limited basis sets predict cyclic bond paths also in case of the bridging $Sc(1)C_2$ moieties in Sc_3TC_4 (T = Fe, Co, Ni; Ref. [40–42]). Hence, subtle changes in the level of approximation employed might render the T-shaped bond paths in the bridging $Sc(1)C_2$ moieties into metalla-cyclic ones and *vice versa*. This renders the discrimination of both types of bridging ScC_2 bond path topologies somewhat difficult.

to a catastrophe point [58]. It should be noted at this point that such a *topological instability* can not be directly related to the *chemical stability* of these systems but rather refers to a change of the bonding scenario in these chemically stable compounds. Despite the fact that a T-shaped bond path does not necessarily rule out the presence of covalency, the crossover from a cyclic to a T-type bond path clearly signals a decrease of the interaction strength [48–50, 59]. We therefore suggest that the strength of the covalent bonding in these cyclic Sc(2)C_2 moieties decreases down the row $Sc_3FeC_4 < Sc_3CoC_4 < Sc_3NiC_4$. Indeed, the Sc-C bonds in the cyclic ScC_2 moieties appear to be somewhat stronger than the ones with T-shaped bond paths. Accordingly, a higher charge density accumulation is found at the Sc(2)–C bond critical points ($\rho_b = 0.35/0.34/0.35$ eÅ^{-3} in **1**, **2** and **3**, respectively) relative to the corresponding T-type bcp of the Sc(1)C_2 fragments ($\rho_b = 0.28/0.30/0.30$ eÅ^{-3} in **1**, **2** and **3**, respectively) [40–41] (Table 10.1). In addition, only in the iron carbide the Sc-C bond paths are directed towards the VSCC(nb) at the carbon atom which is another topological indicator for a more pronounced Sc-C covalency in **1** vs **2** and **3**. Accordingly, only in the Co and Ni carbides, the VSCC(nb) represents a truly non-bonding valence charge concentration.

Analysis of the total energy density [60, 61], ($H(\mathbf{r}) = G(\mathbf{r}) + V(\mathbf{r})$), at the cyclic Sc(2)–C and *T*–C bcps supports the partial covalent character of all these bonds but clearly reveals that the covalency is significantly stronger in the *T*–C case. The experimental $H(\mathbf{r})$ values [52][6] are therefore clearly negative (−0.209, −0.196 and −0.147 hartree·Å^{-3}) in case of the *T*–C bcps while close to zero at the Sc(2)–C bcps (Table 10.1). However, the $G(\mathbf{r})/\rho(\mathbf{r})$ ratios at the Sc(2)–C bcps are smaller than unity (0.89/0.95/0.80 Hartree·e^{-1}) for **1**, **2** and **3**, respectively, and support the noticeable covalent character of these bonds [53].

We therefore suggested that the bonding in the Sc_3TC_4 species is primarily controlled by covalent (*i*) $\sigma(T \leftarrow \text{C})$ donation, (*ii*) $T \rightarrow \pi^*$(C–C) back donation and (*iii*) partially covalent Sc–($\eta^2(C_2)$) bonding (Scheme 10.2) [41]. On the basis of the experimental QTAIM charges of **1**–**3** (Table 10.2) we classified the C_2 moieties as $(C_2)^{2-}$ dicarbido ligands while the transition metals appear virtually unoxidized (e.g. $q(\Omega)_{\text{exp}}(\text{Fe}) = +0.50$; $q(\Omega)_{\text{theor}}(\text{Fe}) = +0.15$; $q(\Omega)_{\text{exp}}(\text{Co}) = +0.27$; $q(\Omega)_{\text{exp}}$ $(\text{Ni}) = +0.14$; Table 10.2).[7] In first approximation the TC_4 ribbons are therefore composed of d^8, d^9 and d^{10} configurated *T* atoms (*T* = Fe, Co, Ni, respectively) which are coordinated by four bridging μ-η^2-$(C_2)^{2-}$ ligands in square-planar fashion. The $[TC_4]^{\delta-}$ units (where δ is approximately 4) can formally be considered as 16, 17 and 18 valence electron (VE) species, respectively, if we only consider

[6] In our experimental and theoretical studies we have used the approach developed by Abramov to derive $G(\mathbf{r})$ from the respective total charge density distributions.

[7] A referee of Ref. [42] pointed out, that alternatively the C_2 unit might be considered as $(C_2)^{4-}$ ethylene-like rather than $(C_2)^{2-}$ ethyne-like. This formulation assumes a formal charge distribution $(Sc^{3+})_3(T(C_2)_2)^{8-}$·[1e] and leads to the same VE count for the transition metals in **1**–**3** as in our bonding description. However, the multiple bond character of the *T*–C bond and the covalent Sc–C bonding are ignored in this ionic bonding picture. We therefore prefer to assume $(C_2)^{2-}$ units in agreement with the experimentally observed QTAIM charges (Table 10.2).

Scheme 10.2

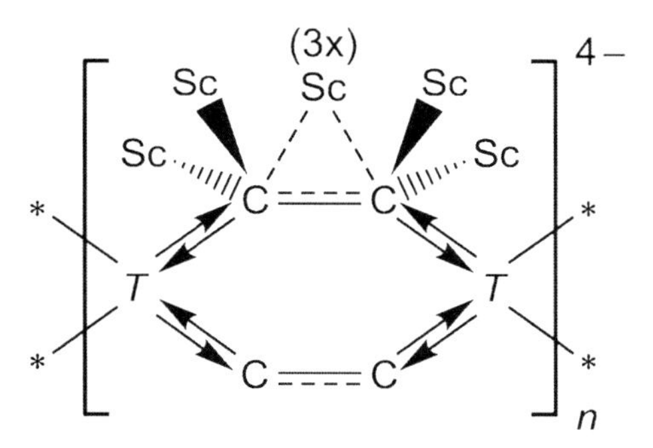

σ − type T-C bonding interactions. We note, that especially the stabilization of the $[FeC_4]^{4-}$ units in a scandium matrix is remarkable, since so far all attempts failed to isolate a corresponding molecular organometallic $[FeR_4]^{4-}$ (R = alkyl, phenyl) model complex [62, 63].[8] The additional valence electrons with respect to $[FeC_4]^{4-}$ will populate the conduction bands of **2** and **3** which display a $d_{xz/yz}$ character (Sect. 10.2.4).

We note that our bonding description for the rare earth transition metal carbides Sc_3TC_4 is related to Halet's and Mingos' (HM) model [64–66] if we also consider π-type bonding interactions between the dicarbido moieties and the transition metal atoms/scandium matrix. The HM model was originally developed for dicarbides fully encapsulated in metal clusters, where the C_2 units are formally considered as dianionic species acting as eight-electron σ/π donors. In line with our findings in **1**–**3**, the C–C bond lengths of the C_2 units depend in the HM model basically on the electronic effects coming from the metallic environment. Similar C–C distances can also be found for C_2 units encapsulated in metal clusters: *e.g.* $Co_6(C_2)(CO)_{18}$ (prismatic Co_6 coordination; C–C = 1.426(9) Å) [67] and $[Co_9(C_2)(CO_{19})]^{2-}$ (tricapped Co_9 prism; C–C = 1.39(2) Å; Fig. 10.2) [45]. Thus, simple bonding descriptions in the framework of the Lewis concept (Scheme 10.2) or on the basis of C–C bond orders will fail to quantify the extent of the various bonding interactions between the dicarbido units and the metal cluster atoms. We therefore analyzed the topology of the negative Laplacian in the valence shell of the carbon atoms of **1**–**3**. Figure 10.4 reveals the presence of three valence shell charge concentrations (VSCC) at the carbon atoms in the TC_4 plane in case of all systems. This demonstrates that the formal dianionic character of the $(C_2)^{2-}$ units as postulated by the QTAIM charges (Table 10.2) does not impose acetylenic bonding character – in line with the non-linear M–C–C coordination and the reduced C–C double bond character. Two of the VSCC are pointing approximately to the transition metal atom and the neighbouring carbon atom. These VSCC are denoted VSCC(T) and VSCC(C) in the following

[8] We note, that the existence of square planar organometallic $[FeR_4]^{4-}$ (R = alkyl, phenyl) complexes has been claimed for decades in the chemical literature. However, the most prominent example of such an Fe(d^8) species $[Li(Et_2O)]_4[FePh_4]$ turned out to be most likely a case of a mistaken identity ($[Li(Et_2O)]_4[FeH_2Ph_4]$).

and can be classified as bonded VSCCs in line with the presence of covalent *T*–C/C–C bonding. The third charge concentration at each carbon atom is, however, located in between terminal Sc–C bonds and represents a non-bonding VSCC(nb) which is only in case of the Fe carbide in close proximity to the Sc(2)-*T* bond path. A splitting of this charge concentration would lead to the typical *L*(**r**) pattern of four charge concentrations as expected for a sp^3 hybridized carbon atom. Hence, we suggest to interpret the presence of a non-bonding VSCC(nb) at each carbon atom as an indicator of a crossover from a sp^2 to a sp^3 hybridization scenario since these features are clearly elongated out of the [TC_4] planes. This is in line with a topological unstable situation leading to a crossover from three to four individual valence shell charge concentrations at the carbon atoms. This is also in agreement with the observation of a reduced C–C double bond character and the hydrolysis behaviour (no formation of unsaturated hydrocarbons) of the Sc_3TC_4 carbides [37].

10.2.3 Fine Structure of the Laplacian

In the next step we focus our analysis on the fine structure of *L*(**r**) at the transition metal centers in **1**–**3**. The polarization pattern of late transition metal complexes TL_n (L = ligand) reflects in many cases the occupation of the nonbonding *d*-orbitals [68–71]. For example, the presence and cubic arrangement of 8 VSCCs (Fig. 10.5) and 6 charge depletions (VSCDs) in d^6 configurated ML_6 complexes can be explained by the non-population of the d_{z^2} and $d_{x^2-y^2}$ orbitals of e_g symmetry [68–71]. Hence, the atomic graph of a d^6ML_6 complex usually displays a characteristic sets of critical points [*V* (vertexes) = 8, *E* (edges) = 12, *F* (faces) = 6]. Accordingly, we might be able to transfer this concept to discriminate the electronic situation in the square-planar $[TC_4]^{4-}$ moieties of **1**–**3** displaying d^8, d^9 and d^{10} configurated transition metal atoms, respectively. Indeed, Fig. 10.6 reveals that each transition metal center *T* in **1**–**3** displays an individual characteristic *L*(**r**) pattern which allows us to overcome the coloring problem (*vide supra*) and to distinguish the three carbides on the basis of their local electronic structures. As a common feature of all *L*(**r**) maps in **1**–**3** we find four equatorial charge depletion zones ($VSCD_{eq}$) located approximately on the *T*–C vectors. Four (*T* = Fe, Co) equatorial charge concentrations ($VSCC_{eq}$) in between the *T*–C bonds complement the equatorial polarization pattern. The same *L*(**r**) pattern is also displayed by the textbook example of a square planar d^8 complex $[Pd(NH_3)_4]^{2+}$ (Fig. 10.5) [72]. Hence, these *L*(**r**) features seem to be characteristic for (square-)planar TL_4 (L = ligand) moieties and might be due to the depopulation of the antibonding $d_{x^2-y^2}$ states [42]. Also the polarization of the nickel carbide can be derived from this global pattern by splitting two opposing equatorial $VSCC_{eq}$ which results in the formation of four charge concentrations above and below the [NiC_4] plane.

However, our simple crystal field approach as outlined in Fig. 10.5 does not explain the presence of two additional axial charge depletion zones ($VSCD_{ax}$) in *z*-direction at the d^8, d^9 and d^{10} configurated transition metal atoms in $[Pd(NH_3)_4]^{2+}$

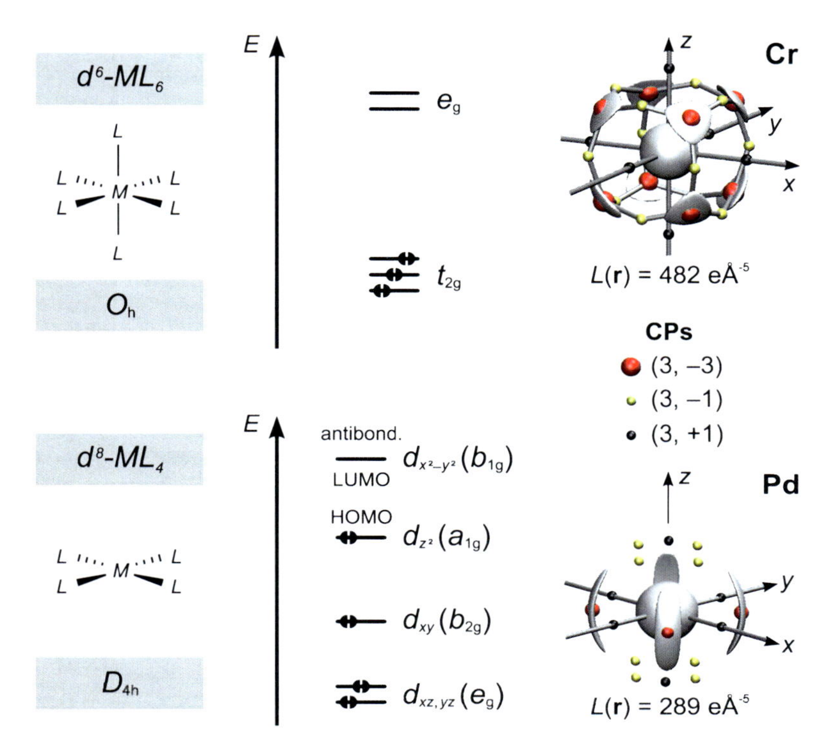

Fig. 10.5 Correlation between the occupation of *d*-orbitals and the resulting fine structure of $L(\mathbf{r})$ in d^6-ML_6 and d^8-ML_4 complexes. The d^6-ML_6 complex $Cr(CO)_6$ displays a cubic structure of local charge concentrations with a [8,12,6] set while the d^8-ML_4 complex $[Pd(NH_3)_4]^{2+}$ shows a characteristic [4,8,6] set (Adapted from Ref. [42])

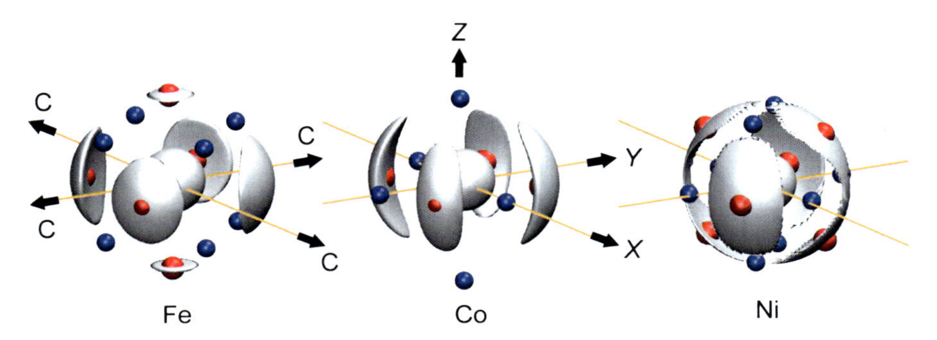

Fig. 10.6 Experimental $L(\mathbf{r})$ envelope map of **1**, **2** and **3**; $L(\mathbf{r}) = 777$, 1,400 and 1,277 $e\text{Å}^{-5}$, respectively (Adapted from Ref. [40]). Note the additional charge concentrations above and below the $[FeC_4]$ plane in the case of **1** which are lacking in **2** and **3**. The individual charge concentrations and depletions ((3, −3) and (3, +1) critical points) are marked by *red* and *blue spheres*, respectively

Scheme 10.3

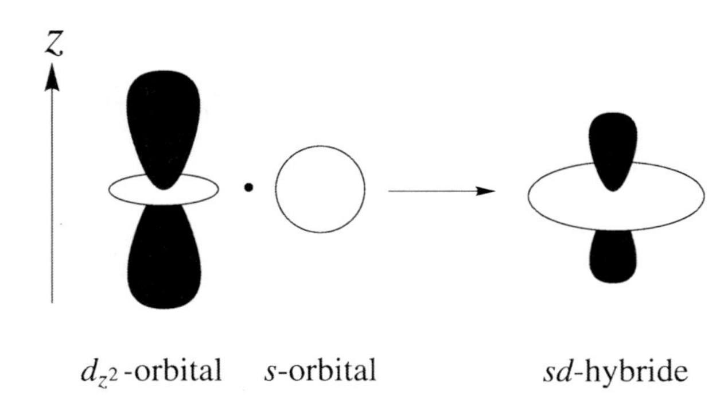

and the $[CoC_4]^{4-}$ and $[NiC_4]^{4-}$ moieties, respectively (Figs. 10.5 and 10.6). We suggest that these well-defined axial depletion zones are a signature of pronounced covalent *T*–C bonding in **1**–**3** [40–42].[9] In the case of **1**–**3** mixing of the $3d_z{}^2/s$ states can assist the strengthening of the *T*–C bonding by increasing the density in the TC_4 plane and by providing a better overlap between the $T(d_{x^2-y^2}, d_{z^2})$ and $C(p_{x,y})$ orbitals (Scheme 10.3). Hence, the presence of axial $VSCD_{ax}$ zones appears to be a natural consequence of such $3d_z{}^2/s$ mixing and its magnitude might be related to the extent of covalent *T*-C bonding [73][10].

If we ignore the additional splitting of two of the equatorial charge concentrations in **3**, all the d^8, d^9 and d^{10} configurated transition metal atoms in $[Pd(NH_3)_4]^{2+}$ and the $[CoC_4]^{4-}$ and $[NiC_4]^{4-}$ moieties, respectively, are characterized by the same set of critical points [4,8,6] (Figs. 10.5 and 10.6).[11,12] It is therefore surprising that only

---

[9]We note, that the *T*–C bond lenghts of **1**–**3** display a V-shaped pattern with the shortest distance shown by the cobalt carbide: *T*–C = 2.1074(6) Å (**1**), 2.0886(4) Å (**2**), 2.094(1) Å (**3**). Hence, the cobalt carbide seems to display the strongest *T*–C bond of these benchmarks – even under consideration of the general contraction of the atomic and ionic radii of these transition metal atoms along the periodic Table.

[10]The mixing of s/d_{z^2} states in $[Pd(NH_3)_4]^{2+}$ complexes enhances covalent bonding in the molecular plane at the expense of chemical interactions in axial direction. Pronounced s/dz^2 mixing, thus stabilizes square-planar coordination and the low-spin state in $[Pd(NH_3)_4]^{2+}$. This might explain that for Ni(II) amine complexes, displaying significantly weaker *T*-N bonds, relative to the corresponding Pd amine complexes, the alternative Jahn-Teller stable octahedral high-spin $^3A_{2g}$ state is observed instead.

[11]If we take the subtle splitting of two equatorial (3, −3) CPs of **3** into account the number of edges (12) and faces (8) would increase. This finally leads to a more complex atomic graph [6,12,8] which is, however, still closely related to the simplified [4,8,6] set.

[12]In general the *L*(**r**) pattern of **2** is significantly more pronounced than that of **3**. This clearly discriminates the electronic situation in both TL_4 units of **2** and **3**. Especially, the individual *L*(**r**) magnitudes of the VSCCs and VSCDs differ significantly in **2** while being rather similar in case of **3**. This simply shows in agreement with the derived monopole populations and QTAIM charges that the nickel atom in **3** is close to a formal d^{10} configuration. Accordingly, the rather complete filling of all five *d*-orbitals in **3** is in line with the rather unpronounced fine structure of *L*(**r**) at the nickel atom in **3**. Individual *L*(**r**) values in [eÅ^{-5}] of VSCCs and VSCDs: $VSCC_{eq}$ = 1002–1092, 1841–

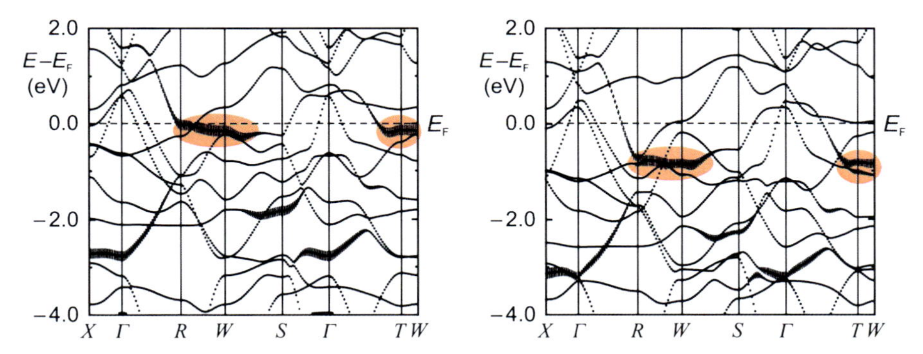

Fig. 10.7 Electronic bands of the ternary carbides **1** (*left*) and **2** (*right*) along selected symmetry lines within the first Brillouin zone of the body-centered orthorhombic unit cell (Adapted from Ref. [41] with kind permission of The American Chemical Society). The width of the bars given for each band indicates the contribution due to the $3d_{z^2}$ states of the Fe or Co atoms, respectively. For a definition of the Brillouin zone and other relevant orbital contributions, see Supporting Material of Ref. [41]

the iron carbide **1** – characterized by d^8 $[FeC_4]^{4-}$ units – displays two axial charge concentrations ($VSCC_{ax}$) instead of the expected depletion zones. The atomic graph of **1** is therefore defined by a [6,12,8] set of critical points. We will outline in Sect. 10.2.4 that these additional $VSCCs_{ax}$ in **1** signal a pronounced localization of itinerant (conducting) electrons in **1**. We will also show that the characteristic $L(\mathbf{r})$ features of **1** are reflected by the presence of localized states of predominant d_{z^2}-type character in the conduction band.

10.2.4 Band Structure Analysis

The close structural relationship of the isotypic carbides **1**–**3** is clearly reflected by their similar electronic structures. Accordingly, the individual electronic bands of **1** and **2** show highly related dispersions along selected symmetry lines within the first Brillouin zone of the body-centered orthorhombic unit cell (Fig. 10.7) [40–42].[13] Furthermore, the low-dimensional character of **1**–**3** due to the presence of the linear $[TC_4]$ ribbons is signaled in all three cases by reduced dispersions along the X–Γ and T–W lines (Fig. 10.7) [41]. The major difference between the band structures of **1**–**3** is therefore mainly due to an increase of the d-electron

2030, 1328–1391; $VSCD_{eq} = 588, 184, 1207$ for **1**, **2** and **3**, respectively; $VSCD_{ax} = 917, 1270$ in **2** and **3**, respectively; $VSCC_{ax} = 847$ in **1**.

[13]Note, that we used a rotated reference frame with the local z-axis perpendicular to the T–C-planes and the x-axis pointing along one of the T–C-bonds to assign the crystal orbitals. The y-axis closely approximates another T–C bond of the $[TC_4]$ units (see Fig. 10.3).

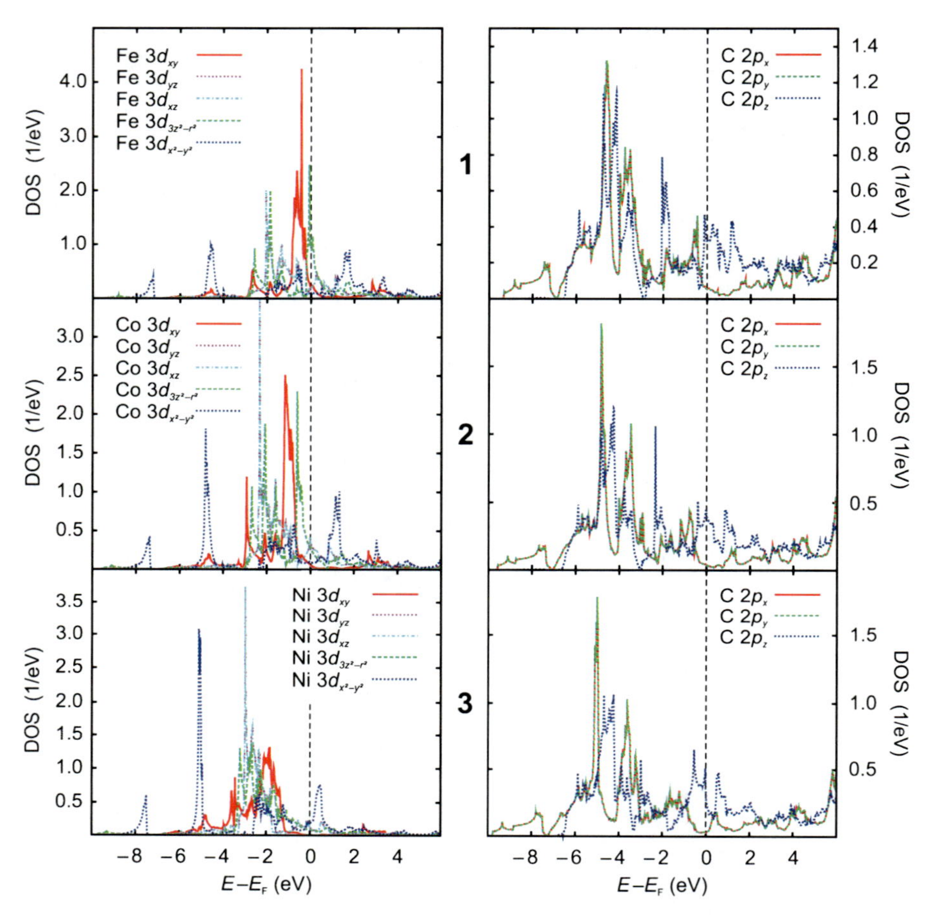

Fig. 10.8 Site and state projected partial densities of states (DOS) for the ternary carbides **1** (*top*), **2** (*middle*) and **3** (*bottom*) in the energy range between −10 and 6 eV (Adapted from Ref. [40, 41] with kind permission The American Chemical Society). The position of Fermi level (E_F) is specified by a *dashed line*

count in the formally 16 VE $[FeC_4]^{4-}$, the 17 VE $[CoC_4]^{4-}$, and 18 VE $[NiC_4]^{4-}$ moieties causing a subsequent lifting of the Fermi level. *Accordingly, the isotypic carbides **1–3** represent ideal model systems to study the electronic consequences of a stepwise population of higher energetic states – virtually decoupled from geometrical distortions.*

The major differences between **1** and **3** are therefore revealed by inspecting the site and state projected partial density of states (DOS) (Fig. 10.8). Only the iron carbide **1** displays a sharp and large DOS peak at the Fermi energy ($N(E_F) = 0.42$ states per eV and atom) which is due to Fe d_{z^2} states of a_{1g} symmetry and a minor contribution from C p_z at the Fermi-level. For symmetry reasons, these states represent basically nonbonding interactions in the $[FeC_4]_\infty$ ribbons (of approximate local D_{4h} symmetry) [41]. These states are also revealed in a "fat band" representation which correlates the contribution of d_{z^2}-type states with the width of the individual bands along a certain symmetry direction (Fig. 10.7). Here, the $d_z{}^2$ state contribution are dominant along the R–W and T–W lines, where the conduction band crossing the Fermi level is characterized by a rather weak dispersion – in line with the non-bonding character of these states. *We suggested earlier that these localized d_{z^2} states appear to be the origin of the axial VSCCs at the iron atom observed in our real space charge density picture* [40, 41]. We therefore aimed at another experimental evidence for the presence of these localized d_{z^2} states. Indeed, the presence of narrow conduction bands and the resulting high density of states lead to high electronic heat capacities in experimental specific heat studies as we will outline in the following.

In simple metals the specific heat C is given by the sum of the electronic, C_e, and phonon, C_{ph}, contribution

$$C = C_e + C_{ph} = \gamma T + \beta T^3 \tag{10.1}$$

for temperatures below $\Theta_D/50$ (and depending of the material even up to $\Theta_D/10$), where Θ_D is the Debye temperature. The Sommerfeld coefficient γ (in units of mJ/K^2mol) is proportional to $N(E_F)$, the bare (or band-structure) electronic DOS at the Fermi energy in units of states/(eV·atom) [74].

$$N(E_F)(1 + \lambda) = 0.4244\gamma/n \tag{10.2}$$

$$\beta = (1944 \times 10^3)n/\Theta^3{}_D \tag{10.3}$$

In this context, λ represents an interaction (electron-phonon and electron-electron) parameter and n the number of atoms of a formula unit (8 in case of Sc_3TC_4) [74]. Hence, the Sommerfeld coefficient $\gamma = 17$ mJ/K^2mol and the β-parameter (0.031 mJ/mol K^4) can be determined by the intercept and the slope of the fitted line in a C/T *vs.* T^2 plot of **1**, respectively (insert Fig. 10.9). Note, that the remarkably high DOS per unit formula (Table 10.3) estimated from the specific heat data between 7 and 14 K might cause a weak band-ferromagnetic instability via the Stoner mechanism as suggested for YCoC [25] displaying a γ of 14.0 mJ/K^2mol [24]. Such instability is supported by the DFT calculations which indicate that a ferromagnetic state is stabilized in **1** by 2.9 kJ/mol [40]. This might also explain the divergence of the C/T *vs.* T^2 plot in **1** at $T < 7$ K. A comparison of the experimental and theoretical Sommerfeld coefficient values of **1**–**3** (Table 10.3) underpins the unusual electronic situation of **1** in this series of carbides in line with its different $L(\mathbf{r})$ fine structure at the transition metal centre. The remarkably large gamma values obtained by our experimental studies of **1** hint for the presence of strong electronic correlation effects. These might not be fully recovered by DFT methods which again might explain the significant differences between the theoretical and experimentally obtained DOS at the Fermi energy of **1**. However, also in the calculations the presence of electronic correlations is supported by the presence of narrow bands

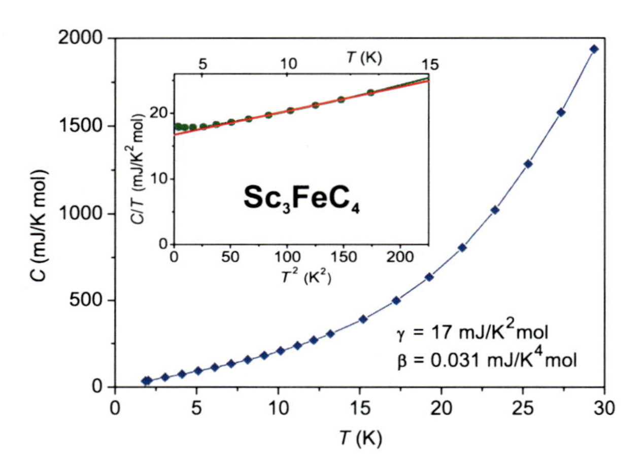

Fig. 10.9 Temperature dependence of the specific heat of Sc_3FeC_4. The *insert* shows the C/T vs. T^2 plot at low temperatures. Note the deviation of C/T from the linear fit following Eq. 10.1 in the temperature range between 7 and 14 K may be due to the presence of a weak ferromagnetic instability, or small amounts of Fe impurities in the sample

Table 10.3 Comparison of the Sommerfeld coefficient γ in mJ/K^{2}mol, the β-values in mJ/K^4 mol and the electronic DOS at the Fermi energy $N(E_F)$ reported in states/(eV atom) in **1**–**3**. Calculated values are based on WIEN2K DFT calculations (see Ref. [40])

Sc_3TC_4		$T = \text{Fe}$	$T = \text{Co}$	$T = \text{Ni}$
β	Experiment	0.031	0.035	0.066
γ	Experiment	17	5.7	7.7
	Theory	7.8	8.3	5.3
$N(E_F)(1+\lambda)$	Experiment	0.9	0.3	0.41
$N(E_F)$	Theory	0.42	0.44	0.28

at the Fermi-level. Indeed, the lack of a narrow conduction band (and axial VSCCs at the transition metal) might be correlated in **2** and **3** with their smaller Sommerfeld coefficients ($\gamma = 5.7$ [8.3] and 7.7 [5.3] mJ/K^2 mol; theoretical values are specified in square brackets).

In the next section, we will show that also the cobalt carbide **2** can be discriminated from its nickel congener **3** despite their similar γ-parameters by a comparison of the respective electronic conductivities. Here, only the cobalt compound displays superconducting behaviour below 4.5 K and a structural phase transition around 70 K.

10.3 Physical Properties

Figure 10.10a depicts the electronic specific heat capacity divided by temperature, $\Delta C/T$, of **2** for the temperatures between 0.1 and 300 K. The phonon contribution was subtracted from the specific heat data by fitting the C/T data, pictured in the insert of Fig. 10.10a, with a simple model assuming five phonon terms: one single Debye term and four Einstein modes accounting for the acoustic and optical

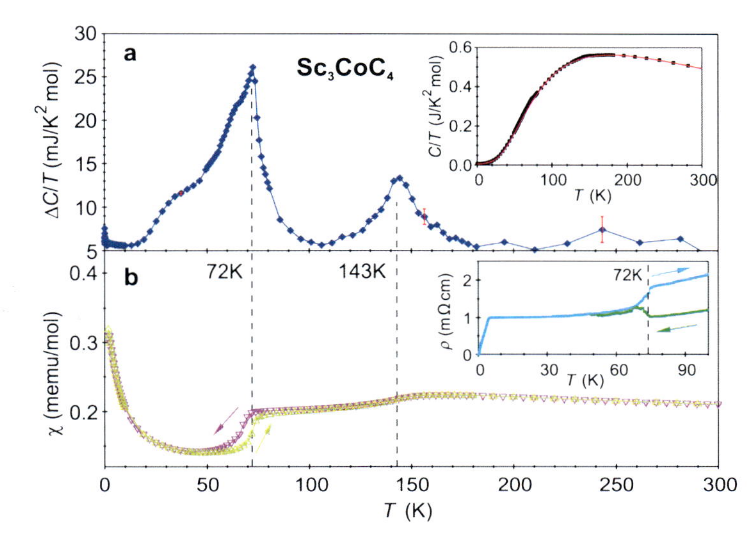

Fig. 10.10 (**a**) The electronic specific heat divided by temperature, $\Delta C/T$, of **2** in comparison with (**b**) the magnetic molar susceptibility, χ, at $B = 1$ T. The *upper insert* shows a C/T *vs.* T plot together with the calculated phonon contribution (*solid line*; see text). The *lower insert* depicts the electrical resistivity, ρ. The *arrows* specify the cooling and warming sequences of the resistivity and susceptibility measurements. The *dashed lines* mark the position of the two anomalies at 72 and 143 K on the temperature scale. The error bars are exemplarily drawn at three temperatures (~240, ~160, ~40 K) illustrating the decrease of the experimental errors upon cooling

modes, respectively [40]. In this calculation, we fixed the number of internal degrees of freedom to 24, according to the eight atoms in the asymmetric unit. Using a weight distribution of 1:1:2:2:2, we obtained a Debye temperature of $\Theta_D = 376$ K, two Einstein modes due to the presence of two different scandium sites of $\Theta_{E1} = 275$ K and $\Theta_{E2} = 447$ K, and two additional Einstein modes for the carbon atoms ($\Theta_{E3} = 589$ K and $\Theta_{E4} = 997$ K, respectively). Starting values for the two Einstein modes at the carbon atoms were taken from theoretical calculations [75, 76].[14] The observed two $\Delta C/T$ anomalies at 72 and 143 K are characterized by entropy changes of about 600 mJ/K mol and 200 mJ/K mol, respectively.

These features are also observed in the temperature-dependent magnetic *dc*-susceptibility, $\chi(T)$, and the electrical resistivity, $\rho(T)$ (Fig. 10.10b). The $\chi(T)$ data at the upper anomaly displays a subtle maximum, whereas the $\rho(T)$ values display a small shoulder at this temperature [40]. This might hint for an onset of a charge density wave formation [77]. The anomaly in the lower temperature

[14]These roughly estimated phonon modes are in line with calculated vibrational frequencies for the high temperature structural model (>72 K) of **2** (space group *Immm*) using the BAND program of the ADF package. These calculations predict for Sc a vibrational energy of 430 K and for the carbon atom two energies of 767 K and 1,760 K; BAND2008.01, SCM, Theoretical Chemistry, Vrije Universiteit, Amsterdam, The Netherlands, http://www.scm.com.

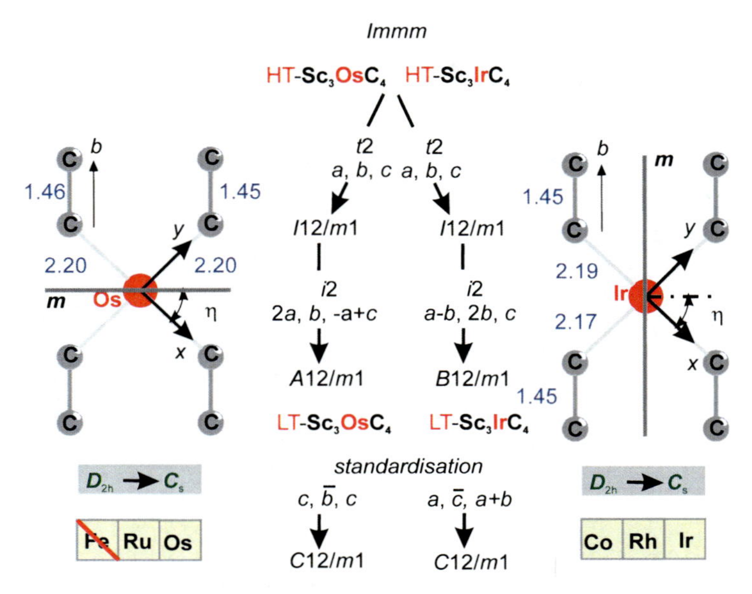

Fig. 10.11 Group-subgroup scheme in the Bärnighausen formalism (Ref. [78] and [79]) for the symmetry reduction displayed by the Group 8 and 9 Sc_3TC_4 carbides. Starting from space group *Immm* in both cases a *translationsgleiche* followed by an *isomorphic* transition leads to the loss of two crystallographic mirror planes. For the Group 8 carbides (*left*) only the mirror plane perpendicular to the TC_4 ribbons is retained while for the Group 9 carbides (*right*) the only remaining mirror plane in the low temperature space group *C*12/*m*1 is oriented parallel to the ribbon direction. Note, only for **1** no low-temperature (LT) modification is yet known. The indices for the *translationengleiche* (*t*), the *isomorphic* (*i*) symmetry reductions and the unit cell transformations are given

regime, however, is characterized by a distinct decrease of the susceptibility which might signal formation of antiferromagnetic correlations upon cooling below 72 K. We further note that $\chi(T)$ and the $\rho(T)$ values critically depend on the thermal history. This hysteretical behaviour between the cooling and warming cycles hints for a structural phase transition. Indeed, a single crystal X-ray diffraction study of Sc_3CoC_4 at 9 K reveals the formation of a low temperature modification of **2** below 72 K [40] via a *translationsgleiche* phase transition of index *t*2. Therefore, this transition involves pseudo-merohedral twinning of the crystals followed by an isomorphic transition of index *i*2 (Fig. 10.11). An isotypic low-temperature modification (space group *C*12/*m*1) has also been observed for the Group 9 carbides Sc_3RhC_4 and Sc_3IrC_4 [38].

This behaviour clearly allows a discrimination of **2** from **1** and **3**, which both do not display any structural phase transition down to 1.9 K. This is remarkable, since substitution of iron by the heavier Group 8 elements Ru and Os leads to the phases Sc_3RuC_4 (**4**) and Sc_3OsC_4 (**5**) which are isotypic to **1** at room temperature but show like **2** *translationsgleiche* phase transitions below 223 and 255 K,

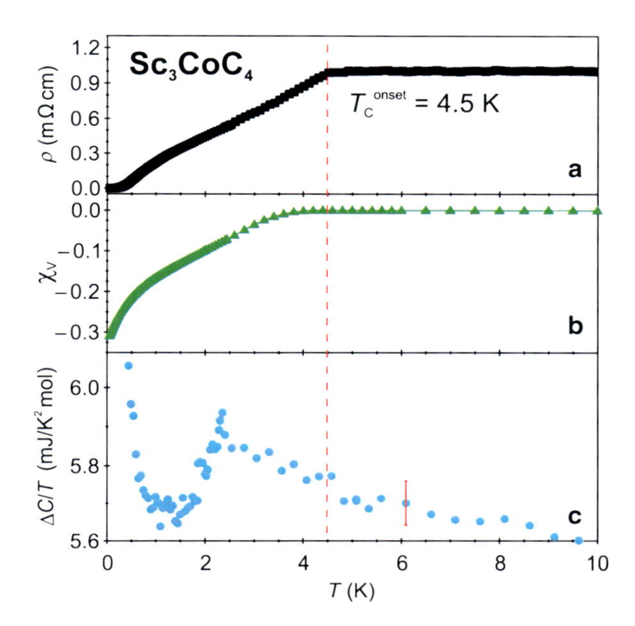

Fig. 10.12 Temperature dependency of (**a**) the electrical resistivity, ρ, (**b**) the volume susceptibility, χ_V, at $B = 0.1$ mT and (**c**) the electronic contribution of the specific heat divided by temperature, $\Delta C/T$, of **2** below 10 K (Adapted from Ref. [40] with kind permission of the John Wiley & Sons)

respectively [42].[15] We note that both possible *translationsgleiche* phase transition routes as illustrated in Fig. 10.11 lead to a symmetry reduction which results in two crystallographic independent TC_4 moieties. Hence, the antiferromagnetic correlations seen in the $\chi(T)$ behaviour of **2** below 72 K might be the true driving force behind the structural phase transition. The phase transitions in Sc_3RuC_4 **4** and Sc_3OsC_4 **5** are characterized by minute geometrical distortions of the TC_4 moieties and also yield significant antiferromagnetic correlations in the $\chi(T)$ cooling cycle [42]. Hence, the structural distortions in **2**, **4** and **5** appear to be rather the consequence of the antiferromagnetic correlations at low temperatures than the driving force of these transitions.

Below the structural phase transition of **2** we further observe a sudden drop in the resistivity at $T_c^{onset} = 4.5$ K. For a better understanding of the origin of this sudden resistivity reduction, we performed low temperature resistivity measurements down to 50 mK, plotted in Fig. 10.12a. At around 250 mK we observe a resistivity reduction to zero indicating the presence of superconductivity in the quasi one-dimensional carbide **2**. Low temperature susceptibility (Fig. 10.12b) and specific heat (Fig. 10.12c) investigations strongly corroborate this finding. Accordingly, we observe a lowering of the susceptibility response just below 4.4 K in parallel with

[15]The structural phase transitions observed for Sc_3RuC_4 and Sc_3OsC_4 at 223 and 255 K, respectively, are reflected by sharp peaks in specific heat measurements in the same temperature range. Furthermore, the resistivity curves, $\rho(T)$, reveal an increased metallic behaviour for both carbides which is paralleled by a sharp decrease of the magnetic susceptibility, $\chi(T)$, below the structural phase transition temperature.

the onset of the resistivity decrease [80].[16] The value of the volume susceptibility ($\chi_V \sim -0.3$ at 50 mK) and the large transition width of $\rho(T)$ may indicate a (cluster-) bulk superconductor which is approximately magnetically shielded by a factor of one third [81]. We further interpret the specific heat anomaly below 4 K (Fig. 10.12b), as another clear signature of a bulk superconductor.

In summary, the resistivity, susceptibility and specific heat measurements clearly characterize bulk superconductivity in the quasi one-dimensional structure of Sc_3CoC_4. In this respect, **2** represents a potential candidate to study superconductivity in quasi one-dimensional compounds.

10.4 Conclusions

The electronic transport properties of the conducting and superconducting carbides Sc_3TC_4 have been explored by physical measurements and interlinked with charge density properties derived by QTAIM analyses. Due to their isotypic character these organometallic carbides are structurally highly related and allow in the series $T = $ Fe, Co, Ni a stepwise increase of the d-electron count and thus a systematic variation of the states contributing to the Fermi level without any significant structural changes. Accordingly, these systems were found to allow a systematic alteration of their physical properties by controlling the nature of their conduction bands in k-space which again is reflected by subtle but significant changes in their underlying charge density distributions.

In the first step of our analyses we studied the electronic structures of the isotypic carbides Sc_3TC_4 ($T = $ Fe (**1**), Co (**2**), Ni (**3**)) by combined theoretical and experimental charge density studies. On the basis of topological analyses we showed that the bonding in the Sc_3TC_4 species is primarily controlled by covalent (*i*) $\sigma(T \leftarrow C)$ donation, (*ii*) $T \rightarrow \pi^*(C–C)$ back donation and (*iii*) partially covalent $Sc–(\eta^2(C_2))$ bonding. Furthermore, analyses of the atomic charges suggest that all $[TC_4]^{\delta-}_{\infty}$ polyanions carry approximately the same negative charge ($\delta \sim 4$). The transition metals in these $[TC_4]^{\delta-}$ units carry only small positive atomic charges (e.g. $q(\Omega)_{exp}(Fe) = +0.50$; $q(\Omega)_{exp}(Co) = +0.27$; $q(\Omega)_{exp}(Ni) = +0.14$) in line with small partial oxidation states. Hence, **1** can thus be formally considered as 16 valence electron (VE) species with a $Fe(d^8)$ center coordinated by four $(C_2)^{2-}$ ligands in a square-planar manner. Accordingly, **2** and **3** display 17/18VE $[TC_4]^{4-}$ polyanions, respectively.

Despite their highly similar crystal structures topological analyses of the fine structure of the negative Laplacian, $L(\mathbf{r})$, another real space property, allowed a clear discrimination of all three carbides on the basis of their different atomic graphs at the

[16] Hence, we can rule out the presence of surface superconductivity alone, because the $\rho(T)$ (a transport property) as well as the $\chi(T)$, behaviour, as a measure of the change of a thermodynamic property, exhibit both a kink at the same temperature.

transition metal atom. Only **1** displays two axial valence shell charge concentrations (VSCCs) pointing out of the molecular TC_4 plane at the iron metal centre. On the basis of band structure analyses we suggested that *the additional axial VSCCs at the iron atom (observed in our real space charge density picture) correlates with the presence of localized* d_{z^2} *states near the Fermi energy* which are lacking in case of the cobalt and nickel carbide. Indeed, the nature of the conduction bands in *k*- or reciprocal space in the cobalt and nickel species differ significantly from that of the iron species and display basically antibonding $T(d_{xz,}d_{yz})/\pi^*$(C-C) character. Accordingly, only the iron carbide **1** displays a sharp and large peak at the Fermi-level ($N(E_F) = 0.42$ states/eV atom) due to the presence of Fe($d_{3z^2-r^2}$) states and minor contributions from C(p_z) states. We therefore aimed at another experimental evidence for the presence of these localized $d_z{}^2$ states. Here, the presence of narrow conduction bands in combination with high density of states could be traced by an unusually high electronic heat capacity of **1** (Sommerfeld coefficient $\gamma = 17$ mJ/K^2 mol). On contrast, *the lack of a narrow conduction band (and axial VSCCs at the transition metal) could be correlated in* **2** *and* **3** *with their smaller Sommerfeld coefficients* ($\gamma = 5.7$ *and* 7.7 *mJ/K*^{2}*mol*). Finally, we demonstrated that also the cobalt carbide **2** can be discriminated from its nickel congener **3** despite their similar γ-parameters by transport studies of their individual electronic conductivities. Here, only the cobalt compound displays superconducting behaviour below 4.5 K. As a consequence of their highly related electronic structures and structural parameters the isotypic Sc_3TC_4 carbides provided excellent quasi one-dimensional experimental benchmarks to study the correlation between their physical reciprocal space properties and charge density distributions in real space.

References

1. Bodak OI, Marusin EP (1979) Crystal-structure of $CeNiC_2$, $LaNiC_2$, $PrNiC_2$ compounds. Dokl Akad Nauk Ukr SSR Ser A 12:1048–1050
2. Bodak OI, Marusin EP, Bruskov VA (1980) Crystal-structure of the compound $RCOC_2$ (R = Ce, La, Pr). Kristallografiya 25:617–619
3. Tsokol' AO, Bodak OI, Marusin EP, Zavodnik VE (1988) X-Ray-diffraction studies of ternary $RRhC_2$ (R = La, Ce, Pr, Nd, Sm) Compounds. Kristallografiya 33:345–348
4. Pecharskaya AO, Marusin EP, Bodak OI, Mazus MD (1990) Crystal-Structure of Sc_2CrC_3. Kristallografiya 35:47–49
5. Pöttgen R, Witte AM, Jeitschko W, Ebel T (1995) Preparation, properties, and crystal structures of alpha-$ScCrC_2$ and beta-$ScCrC_2$. J Solid State Chem 119:324–330
6. Reehuis M, Danebrock ME, Rodriguez-Carvajal J, Jeitschko W, Stüsser N, Hoffmann R-D (1996) Antiferromagnetic order of the lanthanoid moments in the carbides Ln_2ReC_2 with Ln = Tb, Dy, Ho and Er. J Magn Magn Mater 154:355–364
7. Böcker UA, Jeitschko W, Block G (1996) Preparation and crystal structure of the ternary carbides $Ln_{12}Mn_5C_{15}$ with Ln = Y, Pr, Nd, Sm, Gd, Tm, Lu. J Alloys Compd 236:58–62
8. Pöttgen R, Wachtmann KH, Jeitschko W, Lang A, Ebel T (1997) $Er_5Re_2C_7$, $Tm_5Re_2C_7$, and $Lu_5Re_2C_7$ with $Sc_5Re_2C_7$ type, and Yb_2ReC_2 with Pr_2ReC_2 type structures. Z Naturforsch 52b:231–236

9. Hoffmann R-D, Jeitschko W (1998) Carbides $Ln_{10}Ru_{10}C_{19}$ (Ln = Y, Gd-Lu): crystal structure of their subcells and the superstructures of $Er_{10}Ru_{10}C_{19}$. Acta Crystallogr B54:834–850
10. Hüfken Th, Wachtmann KH, Jeitschko W (1998) Preparation and crystal structure of the ternary carbides $R_{12}Os_5C_{15}$ with R = Y, Pr, Nd, Sm, Gd-Tm. J Alloys Compd 281:233–236
11. Reehuis M, Gerdes M, Jeitschko W, Ouladdiaf B, Stüsser N (1999) Crystal and magnetic structures of the ternary carbides $Ho_2Mo_2C_3$ and $Er_2Mo_2C_3$. J Magn Magn Mater 195:657–666
12. Pohlkamp MW, Hoffmann R-D, Kotzyba G, Jeitschko W (2001) Preparation, properties, and crystal structure of the rare earth ruthenium carbides $R_3Ru_2C_5$ (R = Y, Gd-Er). J Solid State Chem 160:77–87
13. Pohlkamp MW, Kotzyba G, Böcker UA, Gerdes MH, Wachtmann KH, Jeitschko W (2001) Preparation, crystal structure, and magnetic properties of rare earth ruthenium and osmium carbides with $La_{3.67}FeC_6$ type structure. Z Anorg Allg Chem 627:341–348
14. Adachi G-Y, Imanaka N, Fuzhong Zh (1991) Rare earth carbides. In: Gschneidner KA Jr, Eyring L (eds) Handbook on the physics and chemistry of rare earths. North-Holland, Amsterdam, pp 61–190
15. Dashjav E, Kreiner G, Schelle W, Wagner FR, Kniep R, Jeitschko W (2007) Ternary rare earth and actinoid transition metal carbides viewed as carbometalates. J Solid State Chem 180: 636–653
16. Burdett JK (1984) From bonds to bands and molecules to solids. Prog Solid State Chem 15:173–255
17. Whangbo M-H (1986) In: Rouxel J (ed) Crystal chemistry and properties of materials with quasi one dimensional structures. Reidel, Dordrecht, p 27
18. Hoffmann R, Li J, Wheeler RA (1987) YCoC - a simple organometallic polymer in the solid state with strong Co-C-π bonding. J Am Chem Soc 109:6600–6602
19. Li J, Hoffmann R (1989) How C-C bonds are formed and how they influence structural choices in some binary and ternary metal carbides carbides. Chem Mater 1:83–101
20. King RB (1993) 3-Dimensional aromaticity in deltahedral boranes and carboranes. Russian Chem Bull 42:1283–1291
21. King RB (1997) Organometallic structural units in solid state ternary transition metal carbides. J Organomet Chem 536:5–15
22. King RB (2000) Structure and bonding in ternary and quaternary transition metal carbides and carbide silicides containing linear M = C = M unit. J Indian Chem Soc 77:603–607
23. Gerss MH, Jeitschko W (1986) YCoC and isotypic carbides with a new, very simple structure type. Z Naturforsch 41b:946–950
24. Suzuki K, Murayama T, Eguchi M (2001) Magnetic heat capacity and electrical resistivity of RCoC (R = Dy, Ho, Er, Y). J Alloys Compd 317–318:306–310
25. Singh DJ (2002) Itinerant electrons and magnetism in the Co-C chain compound YCoC. Phys Rev B 66:132414
26. Shimomura S, Hayashi C, Asaka G, Wakabayashi N, Mizumaki M, Onodera H (2009) Charge-Density-Wave destruction and ferromagnetic order in $SmNiC_2$. Phys Rev Lett 102:076404
27. Laverock J, Hynes TD, Utfeld C, Dugdale SB (2009) Electronic structure of $RNiC_2$ (R = Sm, Gd, and Nd) intermetallic compounds. Phys Rev B 80:125111
28. Lee WH, Zheng HK (1997) Superconductivity in the series $(La_{1-x}Th_x)NiC_2$ (0 < =x < =0.8). Solid State Commun 101:323–326
29. Subedi A, Singh DJ (2009) Electron-phonon superconductivity in noncentrosymmetric $LaNiC_2$: First-principles calculations. Phys Rev B 80:092506
30. Lee WH, Zheng HK, Chen YY, Yao YD, Ho JC (1997) Calorimetric studies of superconducting $(La_{1-x}Th_x)NiC_2$ (x = 0.1-0.9). Solid State Commun 102:433–436
31. Hilier AD, Quintanilla J, Cywinski R (2009) Evidence for time-reversal symmetry breaking in the noncentrosymmetric superconductor $LaNiC_2$. Phys Rev Lett 102:117007
32. Gulden Th, Henn W, Jepsen O, Kremer RK, Schnelle W, Simon A, Felser C (1997) Electronic properties of the ytriumdicarbide superconductors YC_2, $Y_{1-x}Th_xC_2$, $Y_{1-x}Ca_xC_2$ ($0 < x < \sim 0.3$). Phys Rev B 56:90219029

33. Gerss MH, Jeitschko W, Boonk L, Nientiedt J, Grobe J (1987) Preparation and crystal structure of superconducting Y_2FeC_4 and isotypic lanthanoid iron carbides. J Solid State Chem 70: 19–28
34. Johrendt D, Pöttgen R (2008) Pnictide oxides: a new class of high-T_C superconductors. Angew Chem Int Ed 47:4782–4784
35. Tsokol' AO, Bodak OI, Marusin EP (1986) Crystal-structure of Sc_3CoC_4. Kristallografiya 31:788–790
36. Jeitschko W, Gerss M, Hoffmann R-D, Lee S (1989) Carbon pairs as structural elements of ternary carbides of the f elements with the late transition metals. J Less-Common Met 156: 397–412
37. Hoffmann R-D, Pöttgen R, Jeitschko W (1992) Scandium transition metal carbides Sc_3TC_4 with $T = $ Fe, Co, Ni, Ru, Rh, Os, Ir. J Solid State Chem 99:134–139
38. Vogt C, Hoffmann R-D, Pöttgen R (2005) The superstructure of Sc_3RhC_4 and Sc_3IrC_4. Solid State Sci 7:1003–1009
39. Zhang L, Fehse C, Eckert H, Vogt C, Hoffmann R-D, Pöttgen R (2007) Solid state NMR spectroscopy as a probe of structure and bonding in the carbides Sc_3TC_4 ($T = $ Co, Ni, Ru, Rh, Os, Ir). Solid State Sci 9:699–705
40. Scherer W, Hauf Ch, Presnitz M, Scheidt E-W, Eyert V, Eickerling G, Hoffmann R-D, Rodewald UCh, Hammerschmidt A, Vogt Ch, Pöttgen R (2010). Superconductivity in quasi one-dimensional carbides. Angew Chem Int Ed. 49:1578–1582
41. Rohrmoser B, Eickerling G, Presnitz M, Scherer W, Eyert V, Hoffmann R-D, Rodewald UCh, Vogt C, Pöttgen R (2007) Experimental electron density of the complex carbides $Sc_3[Fe(C_2)_2]$ and $Sc_3[Co(C_2)_2]$. J Am Chem Soc 129:9356–9365
42. Vogt Ch, Hoffmann R-D, Rodewald UCh, Eickerling G, Presnitz M, Eyert V, Scherer W, Pöttgen R (2009) High- and low-temperature modifications of Sc_3RuC_4 and Sc_3OsC_4—relativistic effects, structure, and chemical bonding. Inorg Chem 48:6436–6451
43. Clarke J, Jack KH (1951) The preparation and the crystal structures of Cobalt Nitride, Co_2N, of cobalt carbonitrides, $Co_2(C, N)$, and of Cobalt Carbide, Co_2C. Chem Ind 46:1004–1005
44. Klufers P (1984) The crystal-structure of $LiCo(CO)_4$ and $NaCo(CO)_4$. Z Kristallogr 167: 275–286
45. Martinengo S, Noziglia L, Fumagalli A, Albano VG, Braga D, Grepioni F (1998) Synthesis and structural characterisation of the dianion $[Co_9(C_2)(CO)_{19}]^{2-}$ as its tetramethylammonium salt. J Chem Soc Dalton Trans 2493–2496
46. Scherer W, Sirsch P, Shorokhov D, Tafipolsky M, McGrady GS, Gullo E (2003) Valence charge concentrations, electron delocalization and β-Agostic bonding in d^0 metal alkyl complexes. Chem Eur J 9:6057–6070
47. Scherer W, Eickerling G, Shorokhov D, Gullo E, McGrady GS, Sirsch P (2006) Valence shell charge concentrations and the Dewar–Chatt–Duncanson bonding model. New J Chem 30:309–312
48. Reisinger A, Trapp N, Krossing I, Altmannshofer S, Herz V, Presnitz M, Scherer W (2007) Homoleptic Silver(I) acetylene complexes. Angew Chem Int Ed 46:8295–8298
49. Himmel H.-J, Trapp N, Krossing I, Altmannshofer S, Herz V, Eickerling G, Scherer W (2008) Reply Angew Chem Int Ed Engl 47:7798–7801
50. Krapp A, Frenking G (2008) Comments on "Homoleptical Silver(I) acetylene complexes". Angew Chem Int Ed 47:7796–7797
51. Schwarz K, Blaha P, Madsen G, Kvasnicka D, Luitz J (2003) *WIEN2k*, an augmented plane wave + local orbitals program for calculating crystal properties, Technische Universität Wien (2003)
52. Abramov YA (1997) On the possibility of kinetic energy density evaluation from the experimental electron-density distribution. Acta Crystallogr A 53:264–272
53. Macchi P, Proserpio DM, Sironi A (1998) Experimental electron density in a transition metal dimer: metal – metal and metal – ligand bonds. J Am Chem Soc 120:13429–13435

54. Hertwig RH, Koch W, Schröder D, Schwarz H, Hrušák J, Schwertfeger P (1996) A comparative computational study of cationic coinage metal – ethylene complexes $(C_2H_4)M^+$ (M = Cu, Ag, and Au). J Phys Chem 100:12253–12260
55. Frenking G, Fröhlich N (2000) The nature of the bonding in transition-metal compounds. Chem Rev 100:717–774
56. Dewar MJS (1951) A review of the π-complex theory. Bull Soc Chim Fr 18:C71–C79
57. Chatt J, Duncanson LA (1953) Olefin co-ordination compounds. Part III. Infra-red spectra and structure: attempted preparation of acetylene complexes. J Chem Soc 2939–2947
58. Bader RFW, Nguyen-Dang TT, Tal Y (1981) A topological theory of molecular structure. Rep Prog Phys 44:893–948
59. Eickerling G, Masterlerz R, Herz V, Scherer W, Himmel H-J, Reiher M (2007) Relativistic effects on the topology of the electron density. J Chem Theory Comput 3:2182–2197
60. Cremer D, Kraka E (1984) Chemical bonds without bonding electron density - does the difference electron-density analysis suffice for a description of the chemical bond? Angew Chem Int Ed Engl 23:627–628
61. Cremer D, Kraka E (1984) A description of the chemical-bond in terms of local properties of electron-density and energy. Croat Chem Acta 57:1259–1281
62. Jefferis JM, Girolami GS (1998) Crystal structure of "$[Li(Et_2O)]_4[FePh_4]$": corrigendum and reformulation. A remarkable example of a false solution in a wrong space group Organometallics 17:3630–3632
63. Jefferis JM, Girolami GS (1999) Crystal structure of "$[Li(Et_2O)]_4[FePh_4]$". Organometallics 18:3768–3768
64. Halet J-F, Mingos DMP (1988) Molecular orbital analysis of dicarbido transition metal cluster compounds. Organometallics 7:51–58
65. Frapper G, Halet J-F (1995) Theoretical aspects of the bonding in organometallic clusters containing exposed dicarbon (C_2) entities. 1. Tetrametallic systems. Organometallics 14:5044–5053
66. Frapper G, Halet J-F, Bruce MI (1997) Theoretical aspects of the bonding in organometallic clusters containing exposed dicarbon (C_2) entities. 2. High-nuclearity systems. Organometallics 16:2590–2600
67. Geiser U, Kini AM (1993) Structure of the trigonal phase of bis(methinyltricobaltenneacarbonyl), $[CCo_3(CO)_9]_2$. Acta Crystallogr C 49:1322–1324
68. Bo C, Poblet JM, Bernard M (1990) Laplacian of charge density for binuclear complexes: the metal-metal bond in the $Rh_2{}^{4+}$ unit. Chem Phys Lett 169:89–92
69. Macchi P, Sironi A (2003) Chemical bonding in transition metal carbonyl clusters: complementary analysis of theoretical and experimental electron densities. Coord Chem Rev 238–239:383–412
70. Bianchi R, Gatti C, Adovasio V, Nardelli M (1996) Theoretical and experimental (113 K) electron-density study of lithium bis(tetramethylammonium) hexanitrocobaltate(III). Acta Crystallogr B52:471–478
71. Farrugia LJ, Evans C (2005) Experimental X-ray charge density studies on the binary Carbonyls $Cr(CO)_6$, $Fe(CO)_5$, and $Ni(CO)_4$. J Phys Chem A 109:8834–8848
72. Zadesenets AV, Filatov EYu, Yusenko KV, Shubin YuV, Korenev SV, Baidina IA (2008) Double complex salts of Pt and Pd ammines with Zn and Ni oxalates – promising precursors of nanosized alloy. Inorg Chim Acta 361:199–207
73. Essmann R, Kreiner G, Niemann A, Rechenbach D, Schmieding A, Sichla T, Zachwieja U, Jacobs H (1996) Isotype Strukturen einiger Hexaamminmetall(II)-halogenide von 3d-Metallen: $[V(NH_3)_6]I_2$, $[Cr(NH_3)_6]I_2$, $[Mn(NH_3)_6]Cl_2$, $[Fe(NH_3)_6]Cl_2$, $[Fe(NH_3)_6]Br_2$, $[Co(NH_3)_6]Br_2$ und $[Ni(NH_3)_6]Cl_2$. Z Anorg Allg Chem 622:1161–1166
74. Stewart GR (1984) Heavy-fermion systems. Rev Mod Phys 56:755–787
75. te Velde G, Baerends EJ (1991) Precise density-functional method for periodic structures. Phys Rev B 44:7888–7903
76. Wiesenekker G, Baerends EJ (1991) Quadratic integration over the three-dimensional Brillouin zone. J Phys Condens Matter 3:6721–6742

77. Srivastava SK, Avasthi BN (1992) Preparation, structure and properties of transition metal trichalcogenides. J Mater Sci 27:3693–3705
78. Bärnighausen H (1980) Group-subgroup relations between space groups: a useful tool in crystal chemistry. Commun Math Chem 9:139–175
79. Müller U (2004) Kristallographische Gruppe-Untergruppe-Beziehungen und ihre Anwendung in der Kristallchemie. Z Anorg Allg Chem 630:1519–1537
80. Saint-James D, Gennes PG (1963) Onset of superconductivity in decreasing fields. Phys Lett 7:306–308
81. Kuzmin YI (2000) Fractal geometry of normal phase clusters and magnetic flux trapping in high-T_c superconductors. Phys Lett A 267:66–70

Chapter 11
Intermolecular Interaction Energies from Experimental Charge Density Studies

Paulina M. Dominiak, Enrique Espinosa, and János G. Ángyán

11.1 Introduction

In the theory of intermolecular forces the most straightforward contribution is due to the classical electrostatic forces between the total charge distributions of the interacting units, molecules or ions. While often this is the most significant component of the binding energy in intermolecular complexes, strongly anisotropic electrostatic forces play an even more important role in determining the optimal structure of such systems. Progress in the last 30 years in high-resolution crystallography has made experimentally accessible the charge density distribution of crystalline solids at a subatomic resolution, and has raised the hope that experimental charge density data may provide access not only to the qualitative but also to the quantitative aspects of intermolecular forces responsible for crystal packing, polymorphism, phase transitions and many other phenomena. Although this hope has not been proven to be totally in vain, the initial expectations that precise energetic estimations could be done based on charge density information alone has not been completely fulfilled, at least for the time being. From a fundamental viewpoint, the precise prediction of intermolecular interaction energies would imply an operative knowledge of the exact total energy density functional. Furthermore, even if we knew such a functional, one should be ascertain that the experimentally attained precision in charge density determination is high enough to provide reasonable answers for the tiny energy differences arising in weak intermolecular interactions.

Taking into account the above general considerations, practical methods to assess intermolecular forces from (experimental) total charge densities usually take

P.M. Dominiak (✉)
Department of Chemistry, University of Warsaw, ul. Pasteura 1, 02-093 Warszawa, Poland
e-mail: pdomin@chem.uw.edu.pl

E. Espinosa (✉) • J.G. Ángyán (✉)
CRM2, CNRS and Nancy-University, B.P. 239, F-54506 Vandœuvre-lès-Nancy, France
e-mail: enrique.espinosa@crm2.uhp-nancy.fr; angyan@crm2.uhp-nancy.fr

C. Gatti and P. Macchi (eds.), *Modern Charge-Density Analysis*,
DOI 10.1007/978-90-481-3836-4_11, © Springer Science+Business Media B.V. 2012

advantage of complementary pieces of information coming either from electronic structure calculations, or from (semi-)empirical potential energy functions, in order to have a full energy expression which includes non-electrostatic contributions. A particularly delicate point in this latter case is to find appropriate van der Waals parameters that are compatible with the supposedly accurate electrostatic energy contribution deduced from the charge density. Another theoretically intricate point is the correct estimation of the induction/polarization energy, in the absence of the experimental knowledge of the non-polarized charge distributions. Most probably, this problem cannot be solved in a reliable way without accurate quantum chemical calculations to provide appropriate reference systems.

In the first part of this chapter we overview intermolecular interaction energy calculations based on separate evaluation of electrostatic, induction, dispersion and overlap-repulsion contributions, the last three ones often being represented semiempirically. The second part is devoted to the study of specific interactions, like H-bonds, which is pursued using a topological analysis of the total charge density.

11.2 Intermolecular Interaction Energies Calculated from Experimental Charge Densities

From the quantum chemical point of view, the interaction energy between molecules A, B, C, etc. is defined as the difference between the total energy of the complex $E_{ABC...}$ and the total energies of the monomers E_A, E_B, $E_C, \ldots$:

$$E_{\mathrm{int}}(\mathbf{R}) = E_{\mathrm{ABC...}}(\mathbf{R}) - [E_A + E_B + E_C + \ldots]. \tag{11.1}$$

It is usually supposed that the internal structural parameters of the monomers are kept fixed [1]. For a given structure of the monomers the interaction energy depends uniquely on the relative position of the centers of mass of the monomers and on their mutual orientation. This procedure, based on the notion of the potential energy surface, presupposes the validity of the Born-Oppenheimer approximation, i.e. the coupling of electronic and nuclear motions is neglected. The total energies of Eq. 11.1 are obtained from the electronic Schrödinger equation at fixed nuclear positions and the monomers are considered in a hypothetical, non-vibrating state.

Intermolecular interaction energies can be studied using either a super-molecule or a perturbation approach. In the supermolecule method $E_{\mathrm{int}}(\boldsymbol{R})$ is obtained from Eq. 11.1, by subtraction of total energies. Usually the exact interaction energy is four to seven orders of magnitude smaller then the terms subtracted in Eq. 11.1. Since for all many-electron systems, except very small ones, the errors in ab initio total energies are much larger than the interaction energy itself, the supermolecular methods rely heavily on a cancellation of these large errors [1, 2]. Moreover, these approaches do not give any physical insight into the nature of interaction.

An alternative to the supermolecular methods is the perturbation theory approach, in which the interaction energy is computed directly rather than by subtraction. The perturbation methods separate the overall energy into various physically meaningful contributions such as electrostatic, induction, dispersion and exchange-repulsion energies.

Electrostatic interaction energy, E_{es}, as defined within the perturbation approach, represents the energy of electrostatic (Coulombic) interaction between unperturbed monomer charge densities $\rho_{tot}^{A}(r)$ and $\rho_{tot}^{B}(r)$ [1]:

$$E_{es}^{AB}(\mathbf{R}) = \int_{A}\int_{B} \frac{\rho_{tot}^{A}(\boldsymbol{r}_1)\,\rho_{tot}^{B}(\boldsymbol{r}_2)}{|\boldsymbol{r}_1 - \boldsymbol{r}_2|} d\boldsymbol{r}_1\, d\boldsymbol{r}_2 \tag{11.2}$$

The charge density of the monomer A (and B, likewise) includes nuclear and electronic contributions:

$$\rho_{tot}^{A}(\boldsymbol{r}_1) = \sum_{i\in A} Z_i \delta(\boldsymbol{r}_1 - \boldsymbol{R}_i) - \rho_{elec}^{A}(\boldsymbol{r}_1) \tag{11.3}$$

where the first term containing Dirac's delta function represents positive charge Z_i at the position $\boldsymbol{R}_i$ of nucleus i and $\rho_{elec}^{A}(\boldsymbol{r})$ is the electron density.

The induction (sometimes called polarization) energy, always attractive, arises from the polarization of each molecule in the static electric field produced by the unperturbed charge density of the other. The dispersion energy, also always attractive, is caused by the instantaneous correlation in the fluctuations of the unperturbed charge density of each molecule. The exchange-repulsion energy arises from the Pauli principle, as the electrons of the same spin repel each other in the overlap region. There is also an attractive exchange energy, as the electron motion can extend over both molecules [3].

Since experimental techniques can nowadays provide quantitative information on charge densities of molecules constituting a crystal [4], quantitative estimation of intermolecular interactions from experimental data becomes possible. The commonly used models of the experimental charge density are based on a finite spherical harmonic expansion of the electronic part of the charge distribution about each atomic center, as introduced by Stewart [5]:

$$\rho_i(r) = \sum_{\ell=0}^{\ell_{max}} \sum_{m=0}^{\ell} \sum_{p} P_{\ell mp} N_{\ell mp} \rho_{\ell}(r) \mathcal{Y}_{\ell mp}(\theta, \phi) \tag{11.4}$$

where $\rho_{\ell}(r)$ is a radial function; P_{lmp} are electron population parameters; $N_{\ell mp}$ are normalization factors; $\mathcal{Y}_{\ell mp}(\theta,)$ are real spherical harmonic functions, often referred to as multipoles and the index $p = +$ or $-$ when m is different from 0. Such an atomic expansion is called a pseudoatom [5] and the crystal electron distribution can be written as a sum of pseudoatomic densities:

$$\rho_{elec}^{A}(\boldsymbol{r}) = \sum_{i} \rho_i(\boldsymbol{r} - \boldsymbol{R}_i) \tag{11.5}$$

where $\boldsymbol{R}_i$ represents the position of the nucleus i in the asymmetric part of the unit cell.

In the most commonly used variant of the above formalism, i.e. in the Hansen-Coppens [6] multipolar model, the pseudoatom electron density is defined by

$$\rho_i(\boldsymbol{r}) = P_c\rho_{core}(r) + P_v\kappa^3\rho_{valence}(\kappa r) + \sum_{\ell=0}^{\ell_{max}} \kappa'^3 R_\ell(\kappa'\zeta r) \sum_{m=0}^{m=\ell}\sum_{p} P_{\ell mp} d_{\ell mp}(\theta, \phi) \quad (11.6)$$

where $\rho_{core}(r)$ and $\rho_{valence}(r)$ are spherically averaged free-atom core and valence densities normalized to one electron, respectively; $R_\ell(\kappa'\zeta r)$ is a Slater-type radial function and $d_{\ell mp}(\theta, \phi)$ are density-normalized real spherical harmonic functions. The populations P_v and $P_{\ell mp}$, and the dimensionless expansion-contraction parameters. κ and κ' are refined against experimental data, while the population P_c of the core shell remains fixed.

As alternatives to multipole refinement, the X-ray constrained Hartree-Fock (XCHF) modeling [7, 8] and Molecular Orbitals with variable Occupation Numbers (MOON) refinement [9, 10] have been developed, both of which may offer in the future a possibility of directly applying quantum mechanical methods to calculate interaction energies (and other properties). It is also worth mentioning the maximum entropy method (MEM) of obtaining the electron density in the unit cell from phased X-ray diffraction data [11].

The experimentally attained crystal density should reflect all effects resulting from many-body interactions such as polarization, charge transfer, electron correlation and relativistic effects. The question arises whether the experimental density approach allows for observation of tiny density perturbation resulting from intermolecular interactions.

The process of experimentally determining the charge density is not a trivial task and often diffraction data are of not good enough quality to get reliable charge density results. The confidence in experimental charge density can be lowered not only by experimental errors [12] or multipolar pseudoatom model limitations (see, for example, Chap. 5 and Refs. [13–21]) but also by lack of phases [15] and large uncertainties of hydrogen atom position and thermal motion parameters [22–30]. A theoretical study based on crystal densities of ice VIII, acetylene, formamide and urea [17] have shown that the interaction density can be retrieved by multipolar model. However, addition of random errors to theoretical structure factors of urea obstructed the retrieval of the interaction density according to de Vries [15]. Nevertheless, Ditrich and Spackman have concluded recently [31] that an approximate interaction density is accessible by experiment from a highly restricted multipole refinement.

To compute intermolecular interaction energies within the experimental density approach, the crystal density is partitioned into the constituent molecules. Such a partition must rely on the assumption that molecules retain their identity upon bulk crystal formation [32, 33]. In a multipolar pseudoatom representation, the molecular density is usually obtained by simply summing densities of pseudoatoms belonging to a given molecule. This procedure is valid only if pseudoatom densities are sufficiently localized to avoid significant overlap with neighboring molecules. Abramov has shown [13] that the experimental basis set overlap error is greatly reduced by restricting the κ' values to those obtained from multipole refinement of static crystal-theoretical structure factors [18].

The experimental molecular densities are perturbed, i.e. deformed due to mutual polarization and compressed due to Pauli-repulsion. The enhancement of molecular dipole and higher electric moments in crystals has been pinpointed by several researchers [13, 20, 23, 34, 35]. For that reason, direct application of Coulomb's law, Eq. 11.2 to experimental molecular densities leads to an interaction energy which is not purely the E_{es} energy as defined in perturbation theory. It has been proposed that such energy equals the sum of E_{es} and twice the induction energy [36], which is a valid assumption as far as the polarization remains in the linear response regime. On the other hand, Ma and Politzer in their formulation of the interaction energy based upon the Hellmann–Feynman theorem [32] state that energy computed by application of Eq. 11.2 to the charge distributions of the components, as they are in the complex, is equal to the total interaction energy. Moreover, the induction, dispersion and exchange-repulsion energies defined within the perturbation approach are simply compensating for E_{es} being computed from unperturbed densities [37]. Clearly, the question whether the Columbic energy computed from experimental molecular densities should be treated in first approximation as a simple sum of electrostatic and induction energies, remains open.

11.2.1 Electrostatic Contribution to Interaction Energy from Experimental Charge Densities in Pseudoatom Representation

The exact expression (11.2) is really not suitable for practical electrostatic energy calculations because of the complicated 6D integration involved [38]. Several methods have been implemented in the experimental charge density field. Some of them have been designed *de novo* and extensively tested against theoretical results to avoid the influence of experimental errors.

11.2.1.1 Multipole Expansion Approximation

A practical method, avoiding heavy numerical or analytical integrations, consists in approximating Eq. 11.2 by a multipole series, i.e. through the expansion of $|\boldsymbol{r}_1 - \boldsymbol{r}_2|^{-1}$ in a Taylor series. If the series expansion is performed in Cartesian coordinates, the result can be expressed in terms of Buckingham's traceless multipole moments, $E_{es\,MM}$, and it reads as [39]:

$$
\begin{aligned}
E_{es\,MM}(\boldsymbol{R}) = {} & T\, q^A q^B + T_\alpha \left(\mu_\alpha^A q^B - q^A \mu_\alpha^B\right) \\
& + T_{\alpha\beta}\left(\frac{1}{3}\Theta_{\alpha\beta}^A q^B + \frac{1}{3} q^A \Theta_{\alpha\beta}^B - \mu_\alpha^A \mu_\beta^B\right) \\
& + T_{\alpha\beta\gamma}\left(\frac{1}{15} q^A \Omega_{\alpha\beta\gamma}^B - \frac{1}{15}\Omega_{\alpha\beta\gamma}^A q^B - \frac{1}{3}\mu_\alpha^A \Theta_{\beta\gamma}^B + \frac{1}{3}\Theta_{\beta\gamma}^A \mu_\alpha^B\right) \\
& + \frac{1}{9}\Theta_{\alpha\beta}^A T_{\alpha\beta\gamma\delta} \Theta_{\gamma\delta}^B + \ldots
\end{aligned}
\tag{11.7}
$$

where q, μ, $\Theta, \ldots$ are the permanent molecular electrostatic moments (monopole, dipole, quadrupole, ...), $T_{\alpha\beta\gamma\ldots} = \nabla_\alpha \nabla_\beta \nabla_\gamma \ldots (R^{AB})^{-1}$ are the Cartesian tensor, the indices α, β, $\gamma, \ldots$ denote radial vector components, and Einstein summation over repeated indices is implied. In Eq. 11.7 detailed knowledge of the two charge distributions, $\rho_{tot}^A(\boldsymbol{r})$ and $\rho_{tot}^B(\boldsymbol{r})$, is not needed and only permanent multipole moments of the molecular charge distributions are required. To obtain better representation of charge density (and better convergence), the molecular moments are usually replaced by a set of multipoles centered at the various atomic sites. A detailed argument in favor of such distributed multipole models to improve the convergence of electrostatic potentials and energies have been given by Stone [3, 40], who extended the concept of distributed multipoles to response (polarizability-dependent) properties as well [41].

Unfortunately, atomic moments cannot be determined uniquely as neither the wave function nor the electron density can be unambiguously partitioned into atomic fragments. Two types of electron density partitioning methods, referred to as fuzzy and discrete boundary partitionings, can be distinguished [23, 38, 42]. Examples of the former are the Hirshfeld's stockholder [43] and the pseudoatom partitioning [5, 6, 44], while the example of the latter is atoms-in-molecules (AIM, also called Quantum Theory of Atoms in Molecules, QTAIM) based topological partitioning [45]. The topological partitioning has been applied to the response properties [46], induction [47] and dispersion energies [48] as well. Chemically realistic atomic moments can also be determined from a least-squares fit to the total electrostatic potential [49], although the transferability of fitted atomic charges is usually not satisfactory [50].

Multipolar expansion of the electrostatic interaction energy is valid only when the electron densities of interacting atoms (or molecules) do not overlap. To be more precise, the expansion converges to the exact energy only if $|\boldsymbol{r}_1 - \boldsymbol{r}_2| < |\boldsymbol{R}_{ij}|$ where $\boldsymbol{r}_1$ and $\boldsymbol{r}_2$ are the coordinates that describe the charge density associated with atom i and j, respectively. $\boldsymbol{R}_{ij}$ is the vector linking the centers of the multipole expansion, which are usually placed at nuclear positions. To visualize this situation, a sphere can be constructed around the atom, centered on the expansion origin, with a radius that encompasses the whole electron density of the atom. If the spheres for the two atoms i and j intersect (overlap) then the expansion will diverge.

The molecular densities and the atomic densities resulting from fuzzy boundary partitioning extend to infinity, so by definition the multipole expansion of their electrostatic interaction will be divergent. However, for charge distributions approximated by finite-dimensional basis sets comprising a limited number of Gaussian (or Slater) type functions, convergence may be achieved [3, 51]. Electrostatic energies thus obtained systematically underestimate those derived by exact integration of Eq. 11.2 because of charge density penetration (overlap) i.e. incomplete shielding of the nuclear charges by the electron density. The difference, $E_{\mathrm{es}} - E_{\mathrm{esMM}}$, is called the penetration energy and can be significant. Note that the penetration energy defined in this manner depends also on the level of expansion in cases when, because of practical reasons, the expansion is truncated before the true convergence is achieved.

Atoms resulting from discrete boundary partitioning have a finite divergence sphere, therefore for them the convergence condition can be rigorously monitored and for most of the configuration space it can be exactly obeyed. Even for intersecting divergence spheres, if the intersection is small and involves regions of the atoms with low electron density, the expansion may still display pseudoconvergence [52]. Truncation of the expansion in the region of pseudoconvergence may lead to values acceptably close to the exact electrostatic energy.

The dependence of penetration energy on the partition scheme used to obtain multipolar moments has been studied, for example, by Volkov and Coppens [38]. For a total of 11 dimers of α-glycine (Gly, see Fig. 11.1), N-acetylglycine (AcG) and L-(+)-lactic acid (Lac) structures, they obtained the electrostatic interaction energies computed according to a Buckingham-type approximation with stockholder, atoms-in-molecules and pseudoatom (in Hansen-Coppens formalism) moments (E_{esMM}) and compared them with those evaluated directly from Eq. 11.2 and treated as exact electrostatic energies. The authors employed atomic moments based on theoretical single molecule electron densities calculated at various levels of theory and basis set expansion. They obtained atomic fragments from separate isolated monomers and then superimposed them according to the geometry of the dimer in the crystal structure. As a consequence of this procedure penetration energy may arise also in the case of AIM moments. The atomic moments were expanded up to hexadecapole level. In case of Slater-type base sets, they obtained the exact E_{es} from the Morokuma-Ziegler energy decomposition scheme as implemented in the ADF program (E_{esMZ}); and for Gauss-type base sets, they calculated the exact E_{es} using the new SPDFG program ($E_{esSPDFG}$) [53].

Their results have shown that for the same molecular density (ADF/BLYP/TZP or ADF/BLYP/DZP), the E_{esMM} calculated with AIM moments are usually closer to the exact E_{esMZ} values than those calculated with stockholder moments (Table 11.1). The average percentage difference between the exact E_{esMZ} and values calculated with AIM moments were 22% and 21% for ADF/BLYP/TZP and ADF/BLYP/DZP densities, respectively, while those calculated using stockholder moments were 30% and 28% for the same densities. The largest penetration energies (in absolute values) were -27 kJ·mol^{-1} in Gly4 dimer and -35 kJ·mol^{-1} in Gly3 and Gly4 for AIM

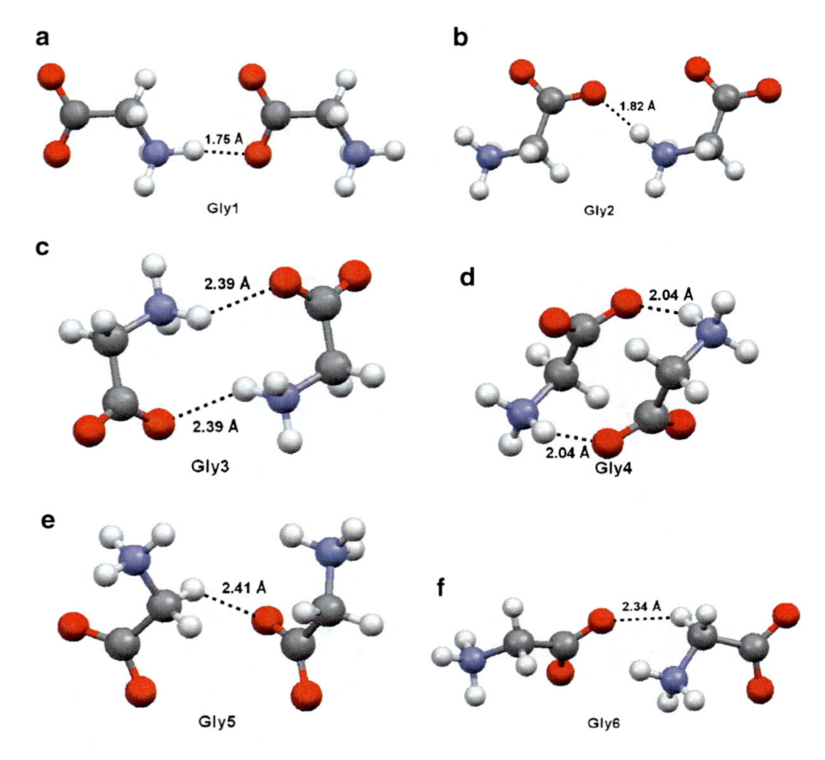

Fig. 11.1 Dimers in the crystal structure of α-glycine after Ref. [38]

and stockholder moments, respectively. For the G98/B3LYP/6-31G** calculation, the only one for which the pseudoatom moments have been also obtained, the discrepancies between the exact $E_{esSPDFG}$ and estimated E_{esMM} generally increased in the sequence AIM < stockholder < pseudoatoms.

From the first three columns of Table 11.1 it can be noted that there are important discrepancies in electrostatic interaction energies calculated directly from the wave function depending on the level of theory and the basis set. The root mean square (r.m.s.) deviation between the ADF/BLYP/TZP and ADF/BLYP/DZP energies is 5 kJ·mol^{-1}. More extensive analyses have been performed by Volkov et al. [53] using the above mentioned six dimers of α-glycine. For example, the r.m.s. deviation between G03/B3LYP/6-31G** and more advanced G03/CCSD/aug-cc-pVTZ energies is 14 kJ·mol^{-1}, and between the ADF/BLYP/QZ4P and G03/CCSD/aug-cc-pVTZ it is 9 kJ·mol^{-1}. The authors concluded that the main factor affecting E_{es} is the inclusion of diffuse functions in the basis set expansions. Advanced correlated methods, such as QCISD, CCSD, and MP4SDQ, and also MP2, show very consistent results for a given basis set, usually within a range of 1–2 kJ·mol^{-1}. Hartree-Fock and CISD methods usually overestimate the E_{es}, while DFT functionals tend to underestimate the magnitude of the electrostatic interaction.

Table 11.1 Electrostatic interaction energies (kJ·mol^{-1}) for isolated dimers of α-glycine (Gly), N-acetylglycine (AcG) and L-(+)-lactic acid (Lac) as computed from different methods. Geometries as in the crystal structures, see Fig. 11.1 [38, 53–55]

	Wave function			Stockholder			AIM		
	ADF	ADF	G03	ADF	ADF	G98	ADF	ADF	G98
	BLYP	BLYP	B3LYP	BLYP	BLYP	B3LYP	BLYP	BLYP	B3LYP
	TZP	DZP	6-31G**	TZP	DZP	6-31G**	TZP	DZP	6-31G**
Dimer	E_{esMZ}[a]	E_{esMZ}	$E_{esSPDFG}$[b]	E_{esMM}[c]	E_{esMM}	E_{esMM}	E_{esMM}	E_{esMM}	E_{esMM}
Gly1	−115	−108	−113	−86	−82	−92	−92	−88	−94
Gly2	−37	−35	−27	−15	−15	−14	−16	−15	−13
Gly3	−109	−102	−97	−74	−68	−82	−95	−88	−96
Gly4	−166	−165	−159	−134	−130	−143	−142	−138	−149
Gly5	43	35	49	47	44	52	48	45	52
Gly6	−26	−26	−23	−20	−20	−22	−24	−23	−24
AcG1	−48	−47	–	−27	−27	−27	−31	−31	−29
AcG2	−99	−90	–	−68	−63	−63	−92	−88	−69
Lac1	−70	−63	–	−46	−42	−45	−53	−48	−46
Lac2	−44	−42	–	−27	−26	−27	−25	−24	−26
Lac3	−13	−14	––	−9	−9	−9	−9	−9	−9

	Pseudoatom			Pseudoatom UBDB					
	G98			G98					
	B3LYP			B3LYP					
	631G**			631G**					
Dimer	E_{esMM}	E_{esEP}[d]	E_{esEPMM}[e]	E_{esMM}	E_{esEP}	E_{esEPMM}	$E_{pro\text{-}pro}$	$E_{pro\text{-}def}$	$E_{es\ SP}$[f]
Gly1	−83	−112	−112	−84	−115	−115	−33	3	−114
Gly2	−10	−31	−31	−5	−27	−27	−24	1	−28
Gly3	−78	−85	−85	−81	−88	−88	−11	0	−92
Gly4	−134	−167	−167	−129	−162	−162	−37	2	−164
Gly5	53	49	49	52	47	47	−6	0	46
Gly6	−20	−25	−25	−18	−23	−23	−5	0	−23
AcG1	−27	−53	−52	−29	−55	−54	−22	1	−50
AcG2	−30	−91	−91	−29	−93	−93	−66	2	−93
Lac1	−36	−83	−82	−26	−78	−78	−45	2	−69
Lac2	−21	−39	−39	−21	−41	−41	−24	1	−44
Lac3	−16	−18	−17	−15	−18	−18	−3	0	−18

[a] E_{esMZ} means exact E_{es} from Morokuma-Ziegler decomposition
[b] $E_{esSPDFG}$ means exact E_{es} calculated by the SPDFG program [56]
[c] $E_{esMM} - E_{es}$ within multipole moments expansion approximation
[d] E_{esEP} means exact E_{es} calculated from molecular densities in pseudoatom representation
[e] E_{esEPMM} means E_{es} from the EPMM method
[f] $E_{esSP} = E_{esMM}$ (pseudoatom UBDB) $+ E_{pro\text{-}pro} + E_{pro\text{-}def}$ with $E_{pro\text{-}pro}$ obtained from Clementi-Roetti atomic electron densities, and a ground-state H atom with standard molecular exponent $\zeta_H = 1.24$ a.u

Several strategies have been investigated to account for penetration energy in electrostatic interaction energy calculations. Some approaches use damping functions [57, 58], whereas others employ a simplified charge density treatment [54, 55, 59, 60]. Among the latter, the exact potential and multipole moment (EPMM) method [54], and the promolecular density approximation [55] have been recently explored in experimental charge density studies. Alternatively, there is always the possibility to completely resign from the multipole expansion approximation and perform time-consuming evaluation of the electrostatic interaction energy directly from the molecular charge density and the electrostatic potential derived from it [61, 62].

11.2.1.2 The Exact Potential and Multipole Model Method

The Exact Potential and Multipole Model (EPMM) method [54, 56, 63, 64] combines an exact evaluation of the electrostatic energy of interaction (EP) for the short-range pseudoatom-pseudoatom interactions with the Buckingham-type multipole approximation (MM) for the more numerous long-range interactions. The E_{esEP} between the two pseudoatom charge distributions $\rho_{tot,i}(\boldsymbol{r})$ and $\rho_{tot,j}(\boldsymbol{r})$ is evaluated exactly *via* numerical quadrature over the simpler three-dimensional integrals obtained by rewriting Eq. 11.2 in terms of the electrostatic potentials [54],

$$\begin{aligned} E_{esEP,ij} &= \int_i \rho_{tot,i}(r_1) V_{tot,j}(r_1) dr_1 = \int_j \rho_{tot,j}(r_2) V_{tot,i}(r_2) dr_2 \\ &= \frac{Z_i Z_j}{R_{ij}} + \int_i \rho_i(r_1) V_{nuc,j}(r_1) dr_1 + \int_j \rho_j(r_2) V_{nuc,i}(r_2) dr_2 \\ &\quad + \int_j \rho_j(r_2) V_{elec,i}(r_2) dr_2, \end{aligned} \tag{11.8}$$

where $V_{tot,i}$ and $V_{tot,j}$ are the total (nuclear plus electronic) electrostatic potentials of pseudoatoms i and j, respectively, Z_i and Z_j are the nuclear charges; R_{ij} is the distance between the nuclei; $\rho_i(\boldsymbol{r}_1)$ and $\rho_j(\boldsymbol{r}_2)$ are the pseudoatom electron densities; $V_{nuc,i}$ and $V_{nuc,i}$ are the nuclear potentials and $V_{elec,i}$ is the electronic potential of pseudoatom i.

In general, the electrostatic potential can be evaluated from the pseudoatom model in various ways which may employ Fourier-, direct-space or both types of summation [56, 63–67]. Volkov et al. opted for calculating the electrostatic potential from pseudoatoms by direct-space summation [56, 63, 64].

The EPMM method requires a user-supplied critical interatomic distance, R_{crit}, at which the calculation changes between the EP and MM. Volkov et al. [54] analyzed the differences in E_{es} between the EP and MM methods for all intermolecular interactions in the previously mentioned test of the 11 dimeric α-glycine structures (Gly, Fig. 11.1), N-acetylglycine (AcG) and L-(+)-lactic acid (Lac) structures. In case of O... H interatomic interactions the discrepancy increased exponentially up to 50 kJ.mol^{-1} as the distance approaches 1.5 Å (Fig. 11.2). Similar, but less

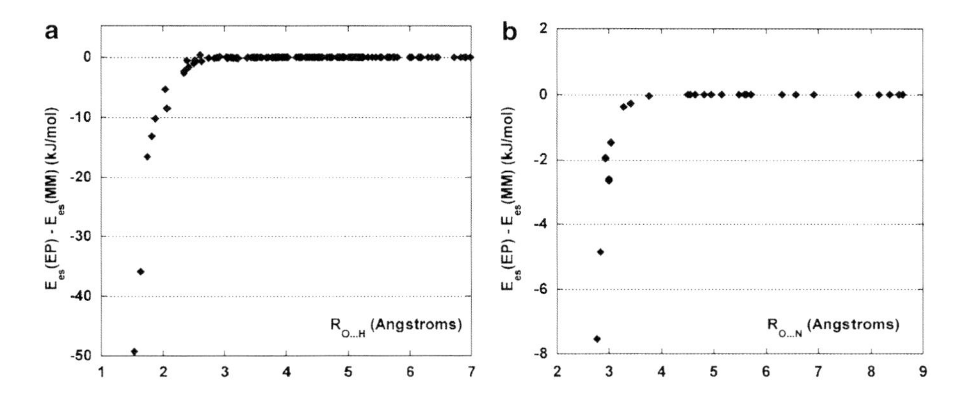

Fig. 11.2 Differences between electrostatic interaction energies for intermolecular: (**a**) O . . . H and (**b**) O . . . N interactions evaluated by the exact integration (EP) and multipole expansion (MM) (kJ.mol^{-1}) methods (Reprinted from Ref. [54] with kind permission of Elsevier)

dramatic differences were observed for other types of interactions, such as O . . . N (Fig. 11.2), O . . . C, etc. The authors concluded that the convergence of E_{esEPMM} to E_{esEP} is reached at around 4 Å.

The EPMM method results in an essentially exact evaluation of the electrostatic component of the interaction energy (see Table 11.1) at a considerable saving in computation time compared with full evaluation of all atom-atom pairs via Eq. 11.8. The speed of calculations does not increase much for more complex systems because the number of interatomic distances that are lower than R_{crit} increases slowly with the number of atoms in the complex.

The discrepancy between the E_{es} values obtained from the EPMM method and directly from the wave function (see columns $E_{esSPDFG}$ and E_{esEPMM} in Table 11.1) is not due to the electrostatic calculation which is exact, but attributed by the authors to the bias introduced by the pseudoatom model in the electron density upon projection via the X-ray structure factors into the Fourier space. The electrostatic interaction energy of Gly3 seems to be affected by this bias more than that of the other dimers. A direct-space pseudoatom partitioning, which is being investigated by Koritsanszky and co-workers (see Chap. 5 and Ref. [68]), may be needed to eliminate such a deviation.

11.2.1.3 Promolecule Approximation

Spackman proposed a different approach to approximate the electrostatic interaction energy [42, 55]. He partitioned each pseudoatom into a spherical atomic term $\rho_{atomic}(\boldsymbol{r})$ (a spherically averaged free-atom density) and a deformation term $\Delta\rho_{elec}(\boldsymbol{r})$:

$$\rho_{pseudoatom}(r) = \rho_{atomic}(r) + \Delta\rho_{elec}(r) \qquad (11.9)$$

and rewritten the integrand of Eq. 11.2:

$$\rho_{tot}^{A}(r_1)\rho_{tot}^{B}(r_2) = \sum_{i\in A}\sum_{j\in B}\Big[\rho_{atomic,i}^{A}(r_1)\rho_{atomic,j}^{B}(r_2) + \rho_{atomic,i}^{A}(r_1)\Delta\rho_{elec,j}^{B}(r_2) + \Delta\rho_{elec,i}^{A}(r_1)\rho_{atomic,j}^{B}(r_2) + \Delta\rho_{elec,i}^{A}(r_1)\Delta\rho_{elec,j}^{B}(r_2)\Big] \quad (11.10)$$

The sum of spherical atomic terms of pseudoatoms belonging to one molecule constitute the promolecule, therefore the electrostatic energy in Spackman formulation E_{esSP} can be expressed as a sum of promolecule–promolecule, promolecule–deformation and deformation–deformation contributions:

$$E_{esSP} = E_{pro-pro} + E_{pro-def} + E_{def-def} \quad (11.11)$$

By application of multipole expansion approximation to Eq. 11.11 only the $E_{def\text{–}def}$ term is approximated, the remaining two terms are zero as the spherical and neutral atoms have zero multipole moments. For short distances, however, multipole approximation may lead to incorrect values of $E_{def\text{–}def}$ because of deformation density overlap.

The first term in Eq. 11.11, $E_{pro\text{–}pro}$, involves a sum over electrostatic interactions between pairs of spherical atomic charge densities (including nuclei). Such interaction energy can be evaluated from the exact formula expressed in the reciprocal space [69]:

$$E_{es,spherical,ij} = \frac{2}{\pi}\int_0^{\infty}[Z_i - f_i(s)][Z_j - f_j(s)]\, j_0(sR_{ij})\, ds \quad (11.12)$$

where Z_i is the nuclear charge of atom i, $f_i(\boldsymbol{s})$ is the atomic scattering factor for atom i as a function of the scattering vector $\boldsymbol{s}$ ($\boldsymbol{s} = 4\pi\sin\theta/\lambda$), $j_0(\boldsymbol{s}R_{ij})$ is the spherical Bessel function of zeroth order and R_{ij} is the distance between the nuclei. For calculation simplification, the integrand is divided into four components, similarly to Eq. 11.8, and the nuclear term replaced by Z_iZ_j/R_{ij}. The remaining terms oscillate but decay quickly with increasing s, and are readily evaluated by one-dimensional numerical integration strategies. Further simplification of calculation might be achieved by tabulation of $E_{es,spherical,ij}$ as a function of R_{ij} for atom-atom pairs of interest and the particular charge density model under consideration (e.g. Clementi-Roetti Hartree-Fock wave functions).

The remaining term in Eq. 11.11, $E_{pro\text{–}def}$, involves pairwise interactions between a spherical atomic electron density in one molecule, $\rho_{atomic,i}$ and an atomic deformation term in another, $\Delta\rho_{elec,i}$. The deformation term can be approximated by the multipole moments derived from it. Again, for short distances, the multipole approximation may lead to incorrect values of $E_{pro\text{–}def}$.

Table 11.2 Electrostatic contribution (kJ·mol^{-1}) to the total interaction energies for the dimers in crystals of L-alanine and DL-alanine as obtained from different methods [70]

	Exp.		Theor.
dimer	E_{esSP} [a]	E_{esEPMM} [b]	E_{esMZ} [c]
L-alanine			
A	−135(9)	−141	−124
B	−155(8)	−171	−142
C	−23(4)	−29	−31
DL-alanine			
A	−153(20)	−171	−121
B	−5(9)	−34	−49
C	−15(5)	−28	−32

[a] E_{esSP} means E_{es} from Spackman's formulation
[b] E_{esEPMM} means E_{es} from the EPMM method
[c] E_{esEPMZ} means exact E_{es} from Morokuma-Ziegler decomposition using ADF2004/BLYP/TZP calculations

Spackman utilized a set of 11 dimers of α-glycine (Gly, Fig. 11.1), N-acetylglycine (AcG) and L-(+)-lactic acid (Lac) structures (the same as in Volkov et al. [54]) in order to estimate the closeness of his approximation to the exact electrostatic interaction energy and to assess the importance of particular terms. He approximated the $E_{def-def}$ by Volkov's E_{esMM}. The results show (Table 11.1) that Spackman's estimates of electrostatic energy $E_{esSP} = E_{esMM} + E_{pro-pro} + E_{pro-def}$ are very close to the reference exact values obtained from the EP method, the r.m.s. deviation being only 3 kJ·mol^{-1}. $E_{pro-pro}$ is always attractive for these dimers, often substantial, and in four cases is greater in magnitude than the contribution from the moment-moment expression (E_{esMM}). The contribution from the promolecule-deformation terms, $E_{pro-def}$, is small and positive, significantly smaller compared to $E_{pro-pro}$ and E_{esMM}, and has only a trivial effect on the agreement between the present E_{esSP} and the exact reference results. This shows that the penetration energy is largely accounted for by integration over promolecular charge densities. Thus Spackman suggested that an excellent and practical approximation to the exact electrostatic energy for intermolecular interactions is simply $E_{es} = E_{esMM} + E_{pro-pro}$.

The EPMM and Spackman's estimations of electrostatic interaction energy were also compared in case of experimental charge densities [70] determined for zwitterionic L- and DL-alanine crystals. The agreement between the experimental values of E_{esSP} and E_{esEPMM} is satisfactory (Table 11.2), although the r.m.s.d. is rather large, 17 kJ·mol^{-1}. Both methods equally deviate from the electrostatic energies computed with ADF2004/BLYP/TZP on isolated dimers. This may be due to crystal field effects included in electrostatic interaction energies derived from experimental data, wheras E_{es} from Morokuma-Ziegler decomposition refer to electrostatic interaction between unperturbed monomer charge distributions.

11.2.1.4 Pseudoatom Databanks

As previously mentioned, it is not straightforward to obtain experimentally reliable charge densities. On the other hand, it has been noted that pseudoatom parameters for atoms in chemically identical environments are transferable [71]. Therefore an idea arose to create a pseudoatom database to aid X-ray data refinement.

Three different databases are well established at the moment: the Experimental Library of Multipolar Atom Model (ELMAM) [72, 73], the Invariom database [74–76] and the University at Buffalo Databank (UBDB) [77–79]. The ELMAM database of experimental pseudoatom densities is constructed from a set of peptides refined within the Hansen-Coppens multipole formalism and covers all 20 amino acids present in proteins. Its main application is in the refinement of high resolution protein structures. A second generation ELMAM database is about to be released [80].

In the Invariom database the atomic electron density of an invariom (*invari*ant at*om*) is generated from a geometry-optimized small model compound containing the invariom, by quantum chemical calculations performed with the G98 program using the B3LYP/D95++(3df, 3pd) level of theory. Only the nearest or next-nearest neighbour influence is assumed when choosing a suitable model compound. Pseudoatom parameters of an invariom are obtained from Fourier-space fitting of the molecular density. This database was designed as a tool to improve structural parameters of small molecule or macromolecular crystals when high-resolution X-ray data are unavailable. It can be done via the transferred aspherical atom model (TAAM) refinement, in which pseudoatom parameters for each atom are transferred from the database and only coordinates and atomic anisotropic displacement parameters (ADPs) are refined.

In the UBDB, each atom type results from averaging over a family of chemically unique pseudoatoms derived from the theoretical density of a number of small molecules. The theoretical densities are obtained from G98, G03/B3LYP/6-31G** single-point calculations on the basis of experimental geometries taken from the CSD [81]. A statically controlled spawning procedure is used to ensure that close transferability is obeyed. Apart from its application to the refinement of macromolecular X-ray data [82], the UBDB is the only pseudoatom databank purposely designed for the evaluation of the electrostatic properties of large molecular complexes from the reconstructed molecular electron density [78].

An excellent reproduction of the electron density in a number of amino acids has been achieved from the UBDB when compared with densities calculated with conventional G98/B3LYP/6-31G** theoretical method [78]. The UBDB combined with the EPMM method (UBDB + EPMM) gives also very good predictions of the electrostatic interaction energies of amino acid dimers. For example, the r.m.s. deviation between the UBDB + EPMM and the G03/B3LYP/6-31G** results equals to 4 kJ·mol^{-1} for six α-glycine dimers, see Table 11.1 and [53]. In case of antibiotic vancomycine complexes [83], which is a much larger molecule, the accuracy is slightly worse. For seven complexes of vancomycine fragment with small ligands, including dipeptides, a depsipeptide and a tripeptide, the r.m.s. deviation between the UBDB + EPMM and G03/B3LYP/6-31G** results is 31 kJ·mol^{-1}, which is still only 5% of the averaged value of electrostatic energies of -500 kJ·mol^{-1}. Surprisingly, the UBDB + EPMM values for vancomycin complexes are closer to these from the higher level of computation (G03/B3LYP/DZP; r.m.s.d. = 20 kJ·mol^{-1}) than from the level used to generate the UBDB databank.

In a more recent studies, the UBDB + EPMM method has been used to evaluate the electrostatic interaction energy in case of the syntenin PDZ2 domain complexed with four-residue peptides and of the PDZ2 dimer [79], as well as for the influenza neuraminidase interacting with a series of inhibitors [84]. The quantitative analysis of the PDZ2 electrostatic interactions allowed the authors to identify the most important pairwise interactions in the complexes and to explain, at least to some extent, the degeneration of ligand binding specificity. For the wide range of complexes of influenza neuraminidases, the electrostatic component of inhibition have been discussed in terms of the contributions from different fragments of inhibitor molecule. Also, the importance of each amino acid residue, selected structural water molecules and proximal calcium ion have been analyzed. Comparative analysis of series of wild and mutated neuraminidase complexes allowed to understand better the molecular basis of the influenza virus resistance to neuraminidase inhibitors caused by the Arg292Lys mutation.

The advantages and disadvantages of experimental versus theoretical databases have been discussed [85, 86]. The major advantages of theoretical databases are the absence of experimental error in the construction process and the availability of unlimited types of pseudoatoms. On the other hand, theoretical databases does which is not take into account the influence of the crystal field on the charge density distribution. Very recently, Bak et al. [87] investigated the electrostatic interaction energies of paracetamol dimers obtained from the Invariom and UBDB databanks, and compared them with these obtained from periodic calculations (CRYSTAL03/B3LYP/6-31G**, Hansen-Coppens pseudoatoms fitted in the Fourier space) and from different strategies of multipolar refinement of experimental data. Significant differences with r.m.s.d.'s of 5–40 kJ·mol^{-1} have been found.

11.2.2 Non-electrostatic Contributions to Interaction Energy

Although very important and many times dominant, electrostatic contribution to interaction energy is not enough to fully describe the energetics of intermolecular interactions. As already discussed at the beginning of the Sect. 11.2, Coulumbic energy evaluated from the *experimental* molecular density includes already some induction contributions, beside the classic electrostatic term. For the exchange-repulsion and dispersion contributions, there is no method currently at hand to evaluate them directly from experimental charge densities in pseudoatom representation.

11.2.2.1 Gavezzotti's Semiclassical Density Sums Method

One way would be to utilize some semiempirical method developed for the calculation of intermolecular energies. An example of such is the Gavezzotti's semiclassical density sums (SCDS) method [88–91] used with success in crystal structure prediction research, although not implemented yet in experimental charge density studies. The method adapt theoretically determined valence electron densities of isolated molecules in geometries as found in the crystal lattice. It allows for intermolecular energies calculations partitioned over electrostatic, polarization, dispersion, and repulsion contributions. Electrostatic energies are calculated by direct numerical integration over charge densities calculated point by point (pixels), like in the EP method by Volkov [54]. However, the numerical integration method in SCDS is cruder. For the calculation of polarization energies, the total, many-body intermolecular electric field at points over molecular space is evaluated, and a semiempirical model for distributed polarizabilities is introduced. Dispersion is evaluated from atomic polarizabilities distributed over the electron density, using an average ionization potential taken as the energy of the highest occupied molecular orbital, in a London-type inverse sixth-power formulation. Repulsion is evaluated from the overlap between electron densities. The parameters involved in the evaluation of the polarization, dispersion and repulsion terms have been separately calibrated using physically consistent data (thermodynamic experimental data and results from high quality quantum mechanical calculations) [92]. It is an open question whether a combination of experimental molecular densities and the methodology used in the SCDS approach (especially the optimized parameters) would be able to lead to reasonable estimates of total interaction energies.

11.2.2.2 Isotropic Atom-Atom Potentials

It order to supplement electrostatic interaction energies obtained from experimental charge densities by repulsion and dispersion terms, several workers [93–95] suggested to use isotropic atom-atom potentials as parameterized by Williams and Cox [96, 97] or Spackman [42, 69, 93, 98]. Both types of parameterization result from the model in which the total intermolecular interaction energy of the AB dimer is approximated as a sum of repulsion, dispersion and electrostatic energies:

$$E_{int} = E_{rep} + E_{disp} + E_{es} \tag{11.13}$$

and the first two terms being expressed in the (exp-6) form (Buckingham potential):

$$E_{rep} + E_{disp} = \sum_{i \in A, j \in B} b_i b_j \exp[-(c_i + c_j) r_{ij}] - \sum_{i \in A, j \in B} a_i a_j r_{ij}^{-6} \tag{11.14}$$

where r_{ij} is a distance between non-bonded atoms i and j, and a, b and c are adjustable potential parameters defined separately for each element. The third term, resulting from classical electrostatic interactions between the monomer charge distributions, is approximated either from a set of site electric charges (Williams-Cox) or from the multipole moments and the derivatives of the electrostatic potential (Spackman) obtained specifically for a given molecule. The important details of both parameterizations are given below.

11.2.2.3 Cox-Williams Model

In the Cox-Williams model, the electrostatic energy is evaluated from Coulomb interactions between site electric charges q:

$$E_{es} = -\sum_{i \in A} \sum_{j \in B} q_i q_j r_{ij}^{-1} \tag{11.15}$$

Net atomic charges are obtained by fitting the calculated *ab initio* molecular electrostatic potential surrounding the molecules. Lone-pair site charges are necessary for aromatic heterocyclic N atoms.

The repulsion exponent c of Eq. 11.14 was estimated from theoretical calculations, whereas the remaining parameters a and b were empirically derived by fitting to 33 crystal structures which do not exhibit hydrogen bonding interactions. The force minimization method with full-matrix weights has been used for optimizing the potential parameters. The experimental heats of sublimation were used to scale the potential functions. The potentials for interactions between different elements were obtained from the geometric-mean combining law.

The semi-empirical Williams-Cox potentials were intended to describe nonbonding interactions between C, N, O, and HC (hydrogen bonded to carbon) atoms in a crystalline environment. However, they have to be complemented by the atom-atom potentials for polar hydrogen atoms, especially those involved in hydrogen bonding (like the Mitchell and Price potential optimized by Coombes et al. [99, 100], for example).

This model has been widely used in crystal structure prediction studies [101]. It has been also applied in some experimental charge density studies [94, 102]. However, it should be pointed out that the Williams-Cox parameterization absorbs some induction contributions into the isotropic parameters, as well as electrostatic penetration effects, and that the derived potentials are only available for a limited number of atoms.

11.2.2.4 The Spackman Model

Having the aim to be used together with experimentally obtained molecular electron densities, Spackman proposed a consistent set of nonbonding atom-atom potentials for a large number of atoms (up to Kr) [42, 69, 93, 98]. Contrary to the Williams-Cox parameterizations, the repulsion and dispersion parameters of Spackman potentials were essentially derived *ab initio*, independently from each other.

The repulsion parameters b and c of Eq. 11.14 have been obtained from the Gordon-Kim [103] electron-gas theory in which the repulsive interaction energy between molecular charge densities is expressed as a sum of the electrostatic (Coulomb), kinetic and exchange contributions:

$$E_{rep} = E_{Coul} + E_{kin} + E_{exc} \tag{11.16}$$

Following the assumptions of the Gordon-Kim model (no perturbation of density caused by interaction and uniform electron gas approximation) Spackman utilized spherically averaged atomic electron densities (promolecule approximation) to calculate the repulsion energy as a function of the interatomic distance. Hartree-Fock atomic wave functions of Clementi and Roetti [104] have been used. For hydrogen, a radially modified electron density function ($\zeta = 1.24$) has also been considered. The geometric mean combining rules have been applied.

The dispersion parameters a have been approximated by Spackman from the scaled theoretical (Hartree Fock level) atomic C_6 dispersion coefficients related to the square of the atomic dipole polarizability α.

To balance the model of the total interaction energy, Eq. 11.13, the expression for the electrostatic term E_{es} was derived from Eq. 11.11 by omitting the promolecule-promolecule component $E_{pro-pro}$, which has been already included in the E_{rep} term Eq. 11.16:

$$E_{es} = E_{pro-def} + E_{def-def} \tag{11.17}$$

For the details how $E_{pro-def}$ and $E_{def-def}$ are computed see Sect. 11.2.1. It is worth noticing here that electrostatic energy computed by the EPMM method (Sect. 11.2.1.2) should not be combined with the Spackman's E_{rep} and E_{disp} terms unless it is appropriately reduced to avoid double counting of $E_{pro-pro}$.

Again, special treatment is required for the hydrogen atoms involved in a hydrogen bond. Spackman recommends to use E_{rep} equal to zero for the interaction between the proton acceptor and the proton in the hydrogen bond. He showed [98] that such approximation is adequate for weak to medium hydrogen bonds, but worsens as the binding energy increases.

The Spackman model has already been applied in experimental charge density studies of α-glycine [105], glycylglycine, DL-histidine, DL-proline [106, 107], p-amino-p'nitrobiphenyl [107], p-nitroaniline, l-asparagine monohydrate, [95] the pentapeptide Boc-Gln-D-Iva-Hyp-Ala-Phol [108], two polymorphs of 3-acetylcoumarin [109], fungal secondary metabolite austdiol [110], angiotensin II receptor antagonist LR-B/081 [111], and zwitterionic L- and DL-alanine [70].

The group of Destro and co-workers [70, 105, 110, 111] applied multipole moments derived from pseudoatoms expressed in Stewart formalism and described a hydrogen bond by omitting the atom–atom potential terms between the proton

Table 11.3 Interaction energies (kJ·mol^{-1}) for dimers of α-glycine molecules in the crystal computed from the experimental charge density [105] by application of the Spackman model and from the SCDS method [89]. For the definition of dimers see Fig. 11.1

	Exp.						SCDS[a]			
Dimer	$E_{\text{def-def}}$	$E_{\text{pro-def}}$	E_{rep}	E_{dis}	E_{HB}[b]	E_{int}[c]	E_{es}[d]	E_{pol}	E_{disp}	E_{int}[c]
Gly1	−139(12)	3	120	−26	−76	−118	−123	−27	−11	−76
Gly2	2(8)	3	89	−21	−59	14	−30	−13	−10	1
Gly3	−149(6)	1	52	−21	0	−117	−103	−14	−13	−101
Gly4	−231(9)	3	125	−44	−53	−200	−170	−25	−24	−158
Gly5	94(5)	0	25	−15	0	104	51	−4	−10	51
Gly6	−38(8)	0	19	−10	0	−29	−24	−4	−6	−21

[a]Valence-only electron densities were calculated at the G98/MP2/6-31G** level
[b]Hydrogen bond energy. There are two identical H bonds in Gly4, each contributing to E_{HB} by −26.5 kJ·mol^{-1}; only one H bond occurs in Gly2 and Gly4
[c]Total interaction energy
[d]Electrostatic energy which from definition should include penetration energy

and its acceptor. Their results for α-glycine crystal, see Tab. 11.3, can be compared with these from theoretical calculations (Table 11.1) previously discussed. The experimental electrostatic interaction energies $E_{def-def}$ correlate remarkably well with the corresponding theoretical E_{esMM} energies. For example, in case of pseudoatom multipole moments computed at the G98/B3LYP/6-31G** level, the linear fit gives the equation: $E_{def-def} = 4(6) + 1.8(1)\ E_{esMM}$ with $R = 0.996$. However, for all eight sets of theoretical values, the theoretical results have to be up scaled by the factor of 1.8 to match experimental results. That might be due to polarization by the crystal field which is not taken into account in theoretical calculations and present in experimental electron densities. The total interaction energies E_{int} can be compared with the results from the Gavezzotti's SCDS method [89] (Table 11.3). Again, the values correlate very well ($R = 0.99$), but the theoretical method downscales the energies (E_{int}(Exp.) $= 15(9) + 1.4(1)$ *vs.* E_{int}(SCDS)). Similar situation is for the dispersive term alone ($E_{disp}(Exp.) = -0(4) + 1.8(3) E_{disp}(\text{SCDS})$, $R = 0.95$). The reasons of that have to be further investigated.

In other studies of Destro and co-workers [70, 110, 111], the result of comparison between total interaction energies estimated from experiment through Spackman model and from theoretical supramolecular calculations at ADF2004/BLYP/TZP level is less encouraging. For dimers of austdiol and LR-B/081, molecules which do not have any charged groups in contrast to α-glycine, the r.m.s.d. is 21 kJ.mol^{-1}, which is of the same magnitude as the estimated energy itself. Partially, the crystal field effect may be responsible for the above discrepancies. Also, deficiencies of both theoretical (like overestimated BSSE correction) and experimental (especially the repulsion term) approaches limits the possibility of determination of more realistic interaction energies.

11.2.2.5 Volkov Potentials

Recently, to complement electrostatic interaction energies calculated from the UBDB + EPMM method, Volkov and coworkers [83] have proposed to derive the remaining terms of the interaction energy from the symmetry-adapted perturbation theory (SAPT) [1, 112] and references cited therein. They approximated the interaction energy as a sum of electrostatic, exchange-repulsion, dispersion and induction contributions:

$$E_{int}^{AB} = E_{esEPMM} + \sum_{i \in A, j \in B} a_{ij} \exp(-b_{ij} r_{ij}) + \sum_{i \in A, j \in B} c_{ij} r_{ij}^{-6} + \sum_{i \in A, j \in B} d_{ij} \exp(-f_{ij} r_{ij}) \quad (11.18)$$

The pairwise coefficients a_{ij}, b_{ij}, c_{ij}, d_{ij} and f_{ij} for interactions between the atoms C, H, O and N have been obtained by simultaneous fits of the respective SAPT components computed for 138 small organic molecular complexes. The SAPT2002/6-31G**/DC + BS method [1, 113, 114] has been used for the energy calculations. Complexes included such molecules like L-serine, L-glutamine, α-glycine, DL-norleucine, L-(+)-lactic acid, benzene, and others. Several different fitting schemes of the dispersion term with the higher power of r_{ij} were also explored [115]. The r.m.s. discrepancies between the theoretical interaction energies and those from the fitted pairwise functions are 5.4, 1.0 and 5.6 kJ·mol^{-1} for the exchange-repulsion, dispersion and induction terms, respectively. The first application of the UBDB+EPMM+SAPT method to 7 complexes of vancomycine complexes [83] showed that the agreement between the UBDB+EPMM+SAPT and supermolecular theoretical results is of the order of 27 kJ·mol^{-1} (for both, G03/B3LYP/6-31G** and G03/B3LYP/DZP levels of theory), which is significantly better than for the MMFF94 molecular mechanics results compared with those from ab initio calculations (r.m.s.d. = 57 kJ·mol^{-1}).

11.2.3 Lattice Energies

Lattice energy (E_{latt}) may serve as a more global measure of intermolecular interactions in crystal lattice. It is generally defined as the energy of formation of a crystal from the isolated (gas phase) molecules. Typically, E_{latt} is evaluated as a difference between the intermolecular interaction (binding) energy in the crystal (E_{bind}) and the relaxation energy (E_{rel}):

$$E_{latt} = E_{bind} - E_{rel} \quad (11.19)$$

The crystal binding energy E_{bind} is defined as

$$E_{bind} = E_{cryst}/Z - E_{mol}(\text{cryst}) \tag{11.20}$$

where E_{cryst} is the crystal energy per unit cell, Z is the number of molecules per unit cell and E_{mol}(cryst) is the energy of an isolated molecule at the crystal geometry. Assuming additivity, E_{bin} can be evaluated by summation of the pairwise intermolecular interactions over an increasing number of neighboring cells until convergence is reached:

$$E_{bin} = \frac{1}{2}\sum_{B} E_{int}^{AB} \tag{11.21}$$

with E_{int}^{AB} being the interaction energy between an unique molecule A and the surrounding molecules B. Because the interactions are between two bodies, the factor of 1/2 arises to avoid double counting.

The relaxation energy E_{rel} is defined as a difference between the energies of an isolated molecule in its gas-phase and crystal geometries:

$$E_{rel} = E_{mol}(\text{gas}) - E_{mol}(\text{cryst}) \tag{11.22}$$

Besides conformational changes, the relaxation energy will also include proton transfer energy in cases, when a molecule crystallize as zwitterion and change to neutral form in the gas phase upon sublimation through intramolecular proton transfer [70].

The lattice energies E_{latt} are related to sublimation enthalpies ΔH_{sub} at some temperature T. In order to transform the former to sublimation enthalpies, the differences in the kinetic (translational, rotational, vibrational) contributions to the crystal and the gas-phase internal energies have to be taken into account. These differences are commonly approximated by the equipartition value of $-3RT$ [116]. This approximation is based on the assumption that internal vibrations between gas phase and crystal are invariant (see Chap. 3 and Refs. [116–119]) and that the temperature T is high enough to excite all librational and translational crystal modes [116, 117]. From the above and from the work of expansion of the gas ($pV = RT$ for an ideal gas), the relation between E_{latt} and ΔH_{sub} is given as

$$\Delta H_{sub}(T) = -E_{latt} - 2RT \tag{11.23}$$

Particularly interesting might be the sublimation enthalpy extrapolated to absolute zero $\left(\Delta H_{sub}^{0}\right)$ which equals to

$$\Delta H_{sub}^{0}(T) = -E_{latt} - \Delta E_{ZPE} \tag{11.24}$$

where ΔE_{ZPE} is the vibrational zero-point energy difference. For many crystals ΔE_{ZPE} is very small (<0.8 kJ·mol^{-1}) and often neglected [120]. However if one aims to estimate enthalpy difference between two polymorphic forms of organic

Table 11.4 Lattice energies (kJ·mol^{-1}) from X-ray charge densities (Exp.) and theoretical periodic calculations with CRYSTAL using Hartree-Fock (HF) or density functional (DFT) methods and 6-31G** basis set [95, 107, 109]. For the coumarine the DFT calculations used the B3LYP hybrid GGA functional, while for other systems Becke GGA exchange and Perdew-Wang GGA correlation functionals were used

		Theor.	
Crystal	Exp.	HF	DFT
DL-histidine	−131.2	−124.2	−154.3
DL-proline.H_2O	−172.5		−173.2
L-asparagine.H_2O	−98.7(16.7)	−122.8	
Glycylglycine	−274.9	−279.5	−254.6
p-Nitroalanine	−96.5(11.5)	+0.1	+12.2
p-Amino-p′-nitrobiphenyl	−88.0	+47.9	+32.1
3-Acetylcoumarin (form A)	−86.1	+31.8	+3.0
3-Acetylcoumarin (form B)	−78.7	+15.3	−7.1

crystals, the zero-point vibrational contributions cannot be ignored in cases where the pattern and strength of the hydrogen bonds differ [121, 122].

There have been several attempts to obtain reasonable estimates of the lattice energies on the basis of the experimental charge densities [36, 95, 102, 106, 107, 109, 123–126]. Most often, the crystal binding energies have been estimated on the basis of the Spackman's model of intermolecular interaction energy (see Sect. 11.2.2.4). In work of Abramov and co-workers [95, 106, 107] and of Munshi and Guru Row [109], the electrostatic term of Spackman's model have been evaluated solely from multipole moments derived from experimental pseudoatom densities. The Ewald lattice summation for the terms up to dipole-dipole interactions have been used in order to achieve faster convergence of crystal binding energy.

A comparison of charge-density-based lattice energies and results from theoretical periodic calculations, see Table 11.4, shows quite reasonable agreement for crystals in which electrostatic interactions dominate (DL-histidine, DL-proline.H_2O, L-asparagine.H_2O and Glycyl-Glycine). The remaining structures (p-nitroalanine, p-amino-p′-nitrobiphenyl and both forms of 3-acetylcoumarin) are not or hardly stable according to the periodic crystal calculations. Obviously, the theory at HF and DFT levels does not properly account for the stacking interaction between the aromatic rings. The sublimation enthalphy of p-nitroalanine of 98–109 kJ·mol^{-1} [107] indicates that experimental approach gives more reasonable results.

11.2.3.1 Electrostatic Lattice Sums

The calculation of the lattice energy consists in the evaluation of pair-interactions of the formula unit of the crystal with all symmetry-related, periodic images. Since the resulting lattice sums are either conditionally (typically the Coulomb interaction) or slowly convergent (e.g. dispersion interaction), specific lattice summation techniques, like the Ewald-method [127] are recommended to guarantee reliable results. In a similar manner, if one is interested in the interaction energy of a substrate, represented by its charge density, $\rho_{tot,A}(\boldsymbol{r})$ embedded in a periodic crystalline environment,

$$E_{int} = \int d\boldsymbol{r} \rho_{tot,A}(\boldsymbol{r}) V_{cryst}(\boldsymbol{r}) \tag{11.25}$$

where $V_{cryst}(\boldsymbol{r})$ is the electrostatic potential of the crystal, the calculation of the potential may require analogous treatment.

The conditional convergence of the lattice sum means that a finite sum corresponding to a finite sample depends on its size and shape [128, 129]. The principle of the Ewald-summation method, by far the most popular approach to accelerate the convergence of lattice sums, is to combine two, conditionally convergent lattice sums in such a way that when added together, they lead to an absolutely convergent result (*cf.* Chap. 2). In the practice this can be done by splitting the Coulomb interaction kernel to two components using the error function and the complementary error function. The short-range, complementary error function contribution can be safely summed in the direct space, while the contribution associated with the long-range error function converges rapidly in the reciprocal space [130, 131]:

$$G(\boldsymbol{r},\boldsymbol{r}') = \sum_{\boldsymbol{L}} \frac{\mathrm{erfc}(\mu|\boldsymbol{r}-\boldsymbol{r}'+\boldsymbol{L}|)}{|\boldsymbol{r}-\boldsymbol{r}'+\boldsymbol{L}|} + \frac{4\pi}{\Omega}\sum_{k\neq 0}\frac{\mathrm{e}^{-k^2/4\mu^2}}{k^2}\mathrm{e}^{i\boldsymbol{k}\cdot(\boldsymbol{r}-\boldsymbol{r}')}. \tag{11.26}$$

The parameter μ determines the relative weight of the direct and of the reciprocal space sums and its value should be chosen to optimize the overall convergence [132]. Although different variants of the Ewald summation method is widely used in solid state theory, as to our knowledge it has not yet been applied in the context of the Hansen-Coppens type multipolar charge distribution models.

A multipole-expanded variant of the above-defined Ewald kernel, Eq. 11.26 has been used in a quantum mechanical/molecular mechanical mixed approach, where the electrostatic effect of the periodic charge distribution of the crystal has been taken into account self-consistently in ab initio [133] or semi-empirical [134] electronic structure calculations on an appropriately selected subunit of a molecular crystal. This self-consistent Madelung potential (SCMP) method, when supplemented with a classical dispersion-repulsion potential, has been used quite successfully in predicting lattice energies [134, 135] and has proven to be appropriate to optimize the lattice structure [136] as well.

Most of the work involving lattice summations over experimental charge density models to obtain electrostatic potentials or interaction energies follow the approach originally proposed by Stewart [137–139]. This method consists in splitting the charge density to the superposition of spherical and neutral atomic densities, $\rho_{IAM}(\boldsymbol{r})$

and a multipolar deformation part, $\Delta\rho(\boldsymbol{r})$. The total electrostatic potential of the crystal, entering to Eq. 11.25 is

$$V_{cryst}(\boldsymbol{r}) = V_{IAM}(\boldsymbol{r}) + \Delta V(\boldsymbol{r}) \tag{11.27}$$

where $V_{IAM}(\boldsymbol{r})$ is the procrystal potential and $\Delta V(\boldsymbol{r})$ is the potential of the deformation density. While the potential of the procrystal density, $\rho_{IAM}(\boldsymbol{r})$, can be easily summed up in direct space,

$$V_{IAM}(\boldsymbol{r}) = \int dr' \frac{\rho_{IAM}(\boldsymbol{r}')}{|\boldsymbol{r} - \boldsymbol{r}'|} \tag{11.28}$$

the multipolar deformation contribution, $\Delta V(\boldsymbol{r})$ was found to converge rapidly in reciprocal space,

$$\Delta V(\boldsymbol{r}) = -\frac{1}{\pi\Omega} \sum_{\boldsymbol{k}(\neq 0)} \frac{F_{cryst}(\boldsymbol{k}) - F_{IAM}(\boldsymbol{k})}{k^2} \mathrm{e}^{-2\pi i \boldsymbol{k}\cdot\boldsymbol{r}}. \tag{11.29}$$

Such an approach has been successfully applied to estimate the interaction energies of water molecules, hydroxyl and ammonium ions embedded in an $AlPO_4$ molecular sieve [61]. Using experimental charge densities and a numerical integration technique inspired by Gavezzotti's pixel method [88], the calculated interaction energies could be related to experimental, calorimetric, desorption enthalpies as well.

11.3 Interaction Energy from Topological Analysis of Experimental Charge Densities

The modeling of crystalline electron density distributions $\rho(\boldsymbol{r})$ is essentially undertaken to extract information on chemical bonding and molecular interactions [4, 140]. In the study of interaction energies between molecular systems in crystals, the works developed by Gavezzotti [88–91, 141–144], Spackman [17, 55, 98, 145] and the group of Coppens [38, 54, 78, 79, 83, 95, 106, 107] are well-known and merit particular attention. Another approach to the same subject is coming from the topological analysis of experimental electron densities. Hereafter, we will focus on its contribution on the description of interaction energies and other related properties in intermolecular regions.

To analyze interatomic and intermolecular interactions, the Atoms in Molecules (AIM) methodology [45], which is based on the topological analysis of the electron density, is nowadays applied to almost any experimentally determined $\rho(\boldsymbol{r})$ model. However, in contrast to the AIM analysis of theoretical $\rho(\boldsymbol{r})$, which permits to retrieve all its topological and energetic properties from the calculated wavefunction, the experimental approach has only access to the topological ones from the multipolar $\rho(\boldsymbol{r})$ function. This problem was partially solved by Abramov [146], who proposed an estimation of the local electron kinetic energy density $G(\boldsymbol{r})$ for closed-shell interactions in terms of $\rho(\boldsymbol{r})$ and of its gradient $\nabla\rho(\boldsymbol{r})$ and Laplacian $\nabla^2\rho(\boldsymbol{r})$ functions based on the semiclassical Thomas-Fermi equation [147] with gradient quantum corrections [148, 149] (in a.u.),

$$G(\boldsymbol{r}) = \left(\frac{3}{10}\right)(3\pi^2)^{2/3}(\rho(\boldsymbol{r}))^{5/3} + \left(\frac{1}{72}\right)\left(\frac{(\nabla\rho(\boldsymbol{r}))^2}{\rho(\boldsymbol{r})}\right) + \left(\frac{1}{6}\right)\nabla^2\rho(\boldsymbol{r}). \tag{11.30}$$

The values of $G(\boldsymbol{r})$ estimated from Eq. 11.30 exhibit good agreement with Hartree-Fock calculations of the same quantity in the medium-range region (*i.e.* for distances 0.5–2.1 Å from the atomic nuclei), where the differences between them are within 4%. While expression (11.30) nicely works for closed-shell interactions, it only gives rough values of $G(\boldsymbol{r})$ for those of shared-shell nature. Thanks to this evaluation of $G(\boldsymbol{r})$, the AIM analysis of crystalline $\rho(\boldsymbol{r})$ has opened new possibilities to experimentalists, mainly for the analysis of intermolecular interactions. At (3, −1) bond critical points (BCP), where $\nabla\rho(\boldsymbol{r}_{\mathrm{BCP}}) = 0$, Eq. 11.30 leads to (hereafter, $\boldsymbol{r}_{\mathrm{BCP}}$ is omitted for clarity),

$$G = \left(\frac{3}{10}\right)(3\pi^2)^{2/3}\rho^{5/3} + \left(\frac{1}{6}\right)\nabla^2\rho. \tag{11.31}$$

Then, using Eq. 11.31 and the local form of the virial theorem (in a.u.) [45]

$$2G + V = \left(\frac{1}{4}\right)\nabla^2\rho \tag{11.32}$$

the local electron potential (V) and total (H) energy densities can be also estimated at intermolecular BCP's from (11.31),

$$V = \left(\frac{1}{4}\right)\nabla^2\rho - 2G \tag{11.33}$$

$$H = G + V. \tag{11.34}$$

This procedure was used for the first time in the analysis of a large set of 83 experimentally determined X-H···O (X = C, N, O) hydrogen bonding interactions (1.56 < d(H···O) <2.63 Å) [150, 151]. For this data set, the observed BCP's are situated at distances ranging 0.5–1.2 Å and 1.0–1.6 Å from the hydrogen and the oxygen nuclei, respectively. The validity of the estimated G, V and H values for closed-shell interactions ($\nabla^2\rho_{\mathrm{BCP}} > 0$, as characterized by the AIM theory) was tested against a set of 32 theoretically calculated X-H···F-Y interactions (1.30 < d(H···F) < 2.80 Å) at the MP2 level of theory with the 6-311++G** basis set [152].

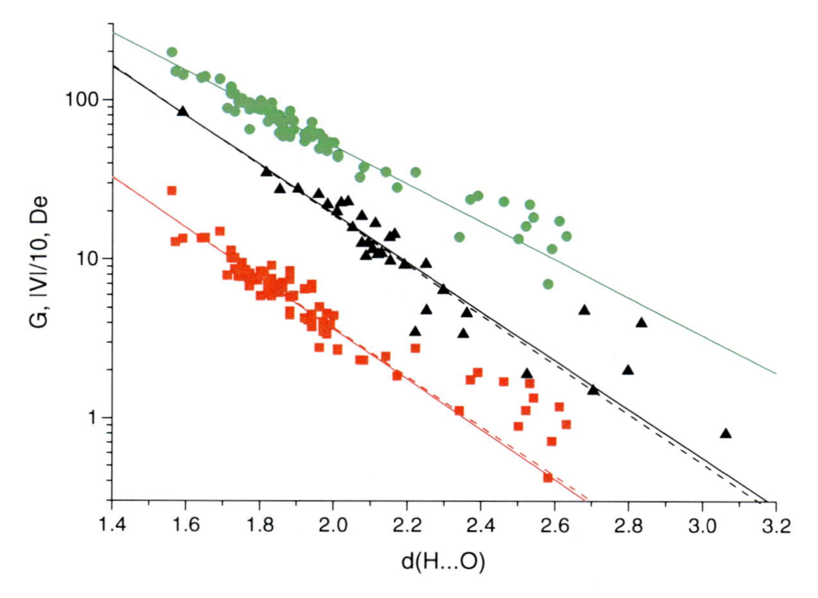

Fig. 11.3 Log-linear plots of experimentally estimated G (*green*) and |V|/10 (*red*) values (kJ·mol$^{-1}a_0^{-3}$), along with calculated D_e (*black*) dissociation energies (kJ·mol^{-1}) for several hydrogen bonded complexes, against internuclear distances d(H$\cdots$O). *Solid lines* represent the fitting curves for G, V and D_e. *Dashed lines* represent the V and D_e functions obtained by fitting new pre-exponential parameters while their exponential parameters are fixed to -3.6 Å^{-1} (Reproduced from Ref. [150] with kind permission of Elsevier)

Using the experimental data set, exponential dependences appear when plotting the estimated G and V values against the internuclear H$\cdots$O distance, as shown in the log-linear plot of Fig. 11.3. The data can be fitted to simple unweighted exponential functions,

$$G = a_G \cdot \exp[-b_G \cdot d(\mathrm{H}\cdots\mathrm{O})] \tag{11.35}$$

$$V = a_V \cdot \exp[-b_V \cdot d(\mathrm{H}\cdots\mathrm{O})] \tag{11.36}$$

where $a_G = 12(2) \cdot 10^3$ and $a_V = -54(18) \cdot 10^3$ are in kJ·mol$^{-1}a_0^{-3}$, and $b_G = 2.73(9)$ and $b_V = 3.65(18)$ are in Å^{-1}, showing nice distributions around the fitting curves for distances shorter than d(H$\cdots$O) $\approx$ 2.3 Å. For distances longer than this approximate threshold, where the overlap between the hydrogen and oxygen electron clouds is small and the interaction can be almost considered as pure electrostatic in nature, systematic deviations indicate that the exponential dependences are lost.

Local energy densities are dimensionally equivalent to a pressure. While V can be interpreted as the pressure exerted on the charge distribution to group electrons between the nuclei to strengthen the interaction, G is interpreted as the pressure exerted by these electrons on the surrounding charge. Thus, the electron kinetic energy density translates the repulsion between the grouped charges, such as simple electrostatic interactions and Pauli repulsion, as a reaction to V. As a consequence, the increase of both energy densities parallels the approach of the nuclei. For the same set of hydrogen bonding interactions, it has been shown that G and V at BCP are respectively reflected by the tightness of $\rho(\boldsymbol{r})$ along the bond path direction and in the perpendicular plane, *i.e.* by the three main curvatures (λ_i, $i = 1, 2, 3$) of the electron distribution [153]. The Laplacian components associated with the local concentration ($\lambda_1 + \lambda_2$) and depletion (λ_3) of the electron distribution have been thus found to follow proportional relationships with V and G, respectively. Similar dependences of the local energy densities with the internuclear distance, and with the topological curvatures of $\varrho(\boldsymbol{r})$, have been also shown with theoretical data involving X-H$\cdots$F-Y hydrogen bonding interactions in gas phase [154].

11.3.1 Dissociation Energies from the Charge Density at the Bond Critical Point

Along with the estimated G and V values at the BCPs of the experimentally characterized hydrogen bonds, calculated dissociation energies D_e for several H$\cdots$O hydrogen bonded complexes in gas phase are also represented in Fig. 11.3. As for G and V, D_e data can be also fitted to a simple unweighted exponential function ($D_e = a_{D_e} \cdot \exp[-b_{D_e} \cdot d(\mathrm{H}\cdots\mathrm{O})]$, where $a_{D_e} = 23(5) \cdot 10^3$ kJ·mol^{-1} and $b_{D_e} = 3.54(10)$ Å^{-1}). Although it is not straightforward to bridge the conceptual gap between local and integrated properties, a phenomenological relationship between V and D_e has been investigated as the former is related to the grouping of charge between the nuclei, strengthening the interaction. Using the fitting functions, the following expression relates V (at BCP) to D_e,

$$D_e = V \cdot \left(\frac{a_{D_e}}{a_V}\right) \cdot \exp[(b_V - b_{D_e}) \cdot d(\mathrm{H}\cdots\mathrm{O})] \tag{11.37}$$

where indexes V and D_e stand for the fitting parameters of the corresponding exponential functions. As the exponential fitting parameter of D_e is very close (and statistically equivalent) to that of V, equivalent exponential functions can be fitted without any statistically significant change for V and D_e by constraining their exponential term to an intermediate value ($b_V = b_{D_e} = -3.6$ Å^{-1}) and by fitting a new pre-exponential parameter ($a_V = -50.0(1.1) \cdot 10^3$ kJ/mol/a_0^3 and $a_{D_e} = 25.3(6) \cdot 10^3$ kJ·mol^{-1}). This procedure leads to fitting functions that are equivalent to the former (Fig. 11.3) and points to a proportionality relationship between the hydrogen bond energy ($E_{\mathrm{HB}} = -D_e$) and V,

$$E_{\mathrm{HB}} = -D_e \approx \frac{1}{2} \cdot V \tag{11.38}$$

the constant being in volume atomic units. A similar phenomenological correspondence was also observed with the X-H$\cdots$F-Y theoretical data, where the fitted proportionality factor between E_{HB} and V was 0.42(2)[154]. Equation 11.38 has been used in the investigation of weak bonding interactions in DNA [155], intramolecular hydrogen bonds (HBs) and π-stacking [156], hydrogen bonds between zwitterions [157], the role of HBs in charge transfer [158] and the role of oxonium in crystal phase [159], in the analysis of $N_3 \cdots N_3$[160], N...Si [161], anion interactions [162–164], and several other interactions involving Ru [165], Cr [166], Au [167], Pd [168] and Ln-complexes [169], and in the investigation of stereoelectronic effects in N-C-S and N-N-C systems [170]. Although last equation was derived for estimating dissociation energies of H$\cdots$O hydrogen bonded systems, its application to other types of closed-shell interactions has been shown in reasonable agreement with theoretical calculations and seems to provide suitable estimations of D_e for these cases (*cf.* Chap. 16).

The dependence of total energy density with the internuclear distance can be derived by using the fitting functions obtained for G and V,

$$H = G + V = (1.2 \cdot \mathrm{e}^{-2.73 \cdot d(\mathrm{H}\cdots\mathrm{O})} - 5.0 \cdot \mathrm{e}^{-3.6 \cdot d(\mathrm{H}\cdots\mathrm{O})}) \times 10^4 \qquad (11.39)$$

The plot of $-H$ shows a dependence with d(H$\cdots$O) that is very similar to which is expected for an interaction potential function. Thus, as Eq. 11.38 seems to point V as a contour condition at the interatomic surface for the hydrogen bond energy, a H$\cdots$O hydrogen bond interaction potential has been proposed in terms of the total energy density [171]

$$U = -\nu \cdot H \qquad (11.40)$$

where ν is a positive constant in volume atomic units. The three characteristic internuclear distances r defining the interaction potential $U(\boldsymbol{r})$ between two chemical entities (namely, the distance r_z vanishing U, the position r_0 of the potential well depth, and the geometry r_d corresponding to the inflexion of the curve) hold respectively the equations,

$$U = 0 \qquad (11.41)$$

$$\left(\frac{dU}{dr}\right) = 0 \qquad (11.42)$$

$$\left(\frac{d^2U}{dr^2}\right) = 0. \qquad (11.43)$$

From Eqs. 11.39 and 11.40, these distances are $r_z = 1.640$ Å, $r_0 = 1.958$ Å and $r_d = 2.276$ Å. The curvature of U at r_0 corresponds to the force constant k_0 of the interaction and is defined as,

$$\left(\frac{d^2U}{dr^2}\right)_{r_0} = k_0. \tag{11.44}$$

The value of the positive constant ν can be derived from Eqs. 11.39, 11.40 and 11.44,

$$\nu = -\frac{k_0 V_0}{b_G^2 H_0 G_0} \tag{11.45}$$

where b_G is the exponential parameter of G (c.f. Eq. 11.35), and G_0, V_0 and H_0 are the values of the local electron energy densities at r_0.

In order to check the model, ice was selected as a testing material. Among the known phases of ice, Ice-VIII is one of the simplest structures (it consists of two Ice-Ic structures that interpenetrate each other) that has been largely studied upon different conditions, either experimentally [172, 173] or theoretically [17, 174, 175]. Vibrational spectroscopy permits to determine k_0, and therefore to get ν through Eq. 11.45. Thus, using the force constant $k_{\mathrm{H}\cdots\mathrm{O}} = 22.7$ N/m derived from Raman data measured on Ice-VIII at T = 100 K [172] and the values of the energy densities calculated at the estimated equilibrium distance ($\mathrm{r}_{eq}^{100\mathrm{K}} \approx 1.996$ Å) [171, 172], Eq. 11.45 leads to $\nu = 0.982 a_0^3$ and therefore to a potential well depth of $U_0(r_0 = 1.958\,\text{Å}) = -13.6$ kJ·mol^{-1}. Then, using (11.39) and (11.40) with the calculated ν value, the interaction potential model of $U(\boldsymbol{r})$ is finally determined: $U(\boldsymbol{r}) = 49{,}100 \exp(-3.6 \cdot r) - 11{,}800 \cdot \exp(-2.73 \cdot r)$ (Fig. 11.4).

The potential well depth represents the energy necessary to break the interaction, formally bringing the entities from their equilibrium geometry to an infinite distance from each other. At the end of this process, the entities are not relaxed and remain polarized. Thus, the dissociation energy D_e at r_0 (*i.e.* D_{e0}, which is calculated as the difference between the total energy of the system at the equilibrium geometry and the sum of the energies corresponding to the isolated and relaxed entities) is bigger in magnitude than U_0 by a value $\left|U_0^{\mathrm{pol}}\right|$. The quantity $\left|U_0^{\mathrm{pol}}\right|$ is thus the polarization energy of the interaction when the entities are placed at their equilibrium distance r_0. This energy becomes $\left|U_0^{\mathrm{pol}} + U_0^{\mathrm{CT}}\right|$ if charge transfer takes place between molecules. As the latter is zero in Ice,

$$U_0 = -D_{e0} + \left|U_0^{\mathrm{pol}}\right|. \tag{11.46}$$

Thus, at $r_0 = 1.958$ Å, $U_0 = -13.6$ kJ·mol^{-1} and $D_{e0} = 21.7$ kJ·mol^{-1} (calculated from Eq. 11.38), and therefore the polarization energy is estimated at a value $\left|U_0^{\mathrm{pol}}\right| = 8.1$ kJ·mol^{-1}. It is obvious that $|U^{\mathrm{pol}}|$ depends on r, however close to r_0 (mainly in the condensed phase) its magnitude is expected to be approximately constant, *i.e.* $|U^{\mathrm{pol}}| \approx \left|U_0^{\mathrm{pol}}\right|$. According to this assumption, in Fig. 11.4 is also represented the function obtained by shifting $-D_e$ (Eq. 11.38) of a value $|U^{\mathrm{pol}}|$ (*i.e.*,

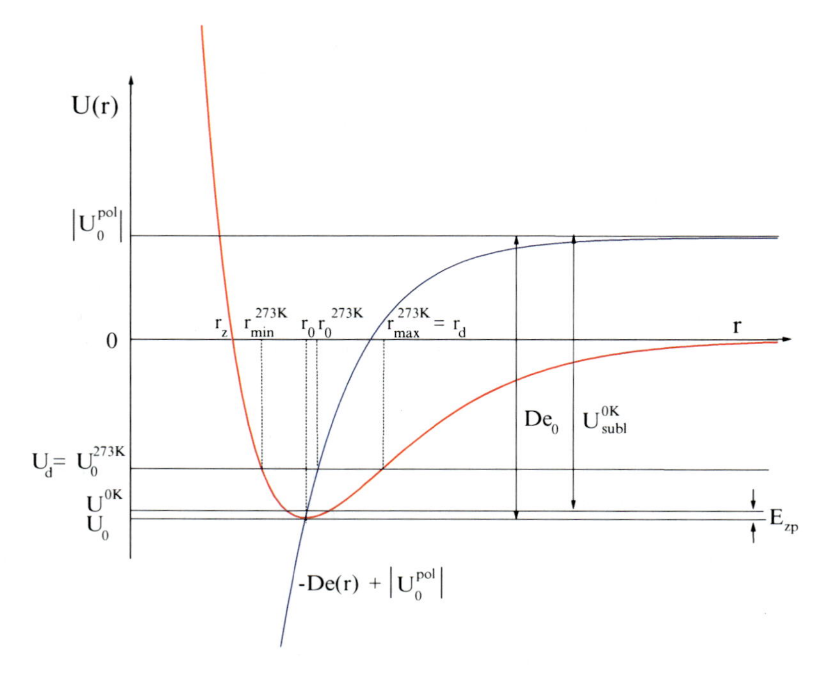

Fig. 11.4 Interaction potential curve $U(r_{H\cdots O})$ derived for H···O hydrogen bonds on Ice. At T = 273 K, the hydrogen bond distance varies from a minimum r_{min}^{273K} to a maximum $r_{max}^{273K} = r_d$ value, the mean thermal equilibrium distance being r_0^{273K}. Definition and values of the energetic properties and distances are given in the text (Reprinted from Ref. [171] with kind permission of The American Institute of Physics)

$-D_e(\boldsymbol{r}) + |U^{\mathrm{pol}}|$). The difference between the theoretically calculated quantity U_0 and the experimental energy of the system at $T = 0$ K $\left(U_0^{0K}\right)$ is called the zero-point energy E_{zp} (Fig. 11.4). The quantity $U_0^{0K} + |U^{\mathrm{pol}}|$ is related to the sublimation energy of the system at $T = 0\,K\left(U_{subl}^{0K}\right)$. In a molecular crystal, U_{subl}^{0K} differs from the lattice energy in the zero-point energies due to the intermolecular vibrations and to the difference between the intramolecular zero-point energies found in the crystal and in the gas phase. In the harmonic approximation, each of these vibrational modes contributes to E_{zp} with a magnitude equal to $1/2\hbar\omega$, ω being the oscillation frequency of the nuclei around the corresponding equilibrium distance r_0^{0K}.

11.3.1.1 Checking the Model

In order to check the interaction potential and the dissociation energy models derived from Eqs. 11.38 to 11.40, they were used to calculate several energetic properties that were compared to theoretical calculations of polarization [171, 175], lattice [174] and molecular binding [17] energies of Ice-VIII at experimental

Table 11.5 Energetic properties derived from the interaction potential model and from theoretical calculations on phases VIII and Ih of Ice, as well as from experimental sublimation energy of ice at 0 K

Energetic property	r_0	Model	Ice-type
Polarization energy		$U_0 + D_{e0}$	Ice-VIII
	$r_0^{model} = 1.958$ Å	8.1	5.9/9.0[a]
Lattice energy		D_{e0}	Ice-VIII
	$r_0^{10K,2.4GPa,exp} = 1.910$ Å	25.8	23.5[b]
	$r_0^{minimised,theo} = 1.928$ Å	24.2	22.1/24.3
Molecular binding energy (ignoring relaxation)		$D_{e0} - \left\|U_0^{pol}\right\|$	Ice-VIII
	$r_0^{10K,2.4GPa,exp} = 1.910$ Å	17.7	19.5[c]
Molecular binding energy		D_{e0}	Ice-Ih
	$r_0^{antiferroelectric,\ theo} = 1.813$ Å	36.6	33.0[d]
	$r_0^{ferroelectric,theo} = 1.878$ Å	29.0	29.9
Sublimation energy (0 K)		D_{e0}	Ice
	$r_0^{model} = 1.958$ Å	21.7	28.0[e]

Comparisons are done: (a) at the minimum of the model (r_0), (b) at the experimentally determined geometry of Ice-VIII at T = 10 K and P = 2.4 GPa $\left(r_0^{10K,2.4GPa,\ exp}\right)$ [173], (c) at minimised geometries of Ice-VIII $\left(r_0^{minimized,theo}\right)$ and (d) at theoretically calculated geometries for the antiferroelectric and ferroelectric phases of Ice-Ih $\left(r_0^{antiferrolectric,\ theo} \text{and } r_0^{ferroelectric,theo}\right)$. Two theoretical values at the same geometry indicate that two kind of calculations were carried out

[a]Ab initio periodic Hartree-Fock calculations (6-31G** basis set), Ref. [171, 175]

[b]Ab initio periodic Hartree-Fock calculations (6-31G** basis set), Ref. [174]. First line: without BSSE correction, second line: without and with BSSE correction, respectively

[c]Ab initio periodic Hartree-Fock calculations (DZPT wavefunction), Ref. [17], ignoring molecular relaxation in the gas phase, BSSE effects, electron correlation zero-point energies and temperature dependence

[d]Ab initio calculations based on two- and three-body analytical potentials, Refs. [176, 177]. First reference: based on the Matsuoka, Clementi and Yoshimine potential (Ref. [178]). Second reference: ab initio Hartree-Fock-Roothaan SCF method

[e]Experimental sublimation energy of Ice at 0 K. Ref. [179]

geometries, to binding energies of antiferroelectric and ferroelectric phases of Ice-Ih [176, 177], as well as to the experimentally determined sublimation energy of ice at 0 K [179]. All these comparisons are gathered in Table 11.5. While the agreements with theoretical calculations are found very good, the comparison with the sublimation energy at 0 K shows a larger difference. In this case, several factors should be however taken into account. Indeed, the energy difference observed between both quantities ($\Delta = 6.3$ kJ·mol^{-1}) is in part related to (i) the uncertainty in the zero-point rotational energy per molecule [179], (ii) the effect of anharmonicities, which were neglected [179], (iii) the r_0 value used in the calculation of D_e, which could differ to the unknown r_0^{0K} value corresponding to the experience of the sublimation energy determination, as well as to (iv) the missing contributions of O···O interactions to D_e, the dissociation energy of water molecules. The proposed model for D_e (Eq. 11.38) has been also used to investigate the lattice energies of sulflower $C_{16}S_8$

(S···S, S···C and C···C interactions) [180] and of [2.2] paracyclophane (H···H and C···H interactions) [181], the latter showing a reliable value compared to ΔH_{subl}.

The proposed models for U and D_e can be also used to predict the thermodynamic expansion of the mean equilibrium geometry with the temperature. Thus, provided $r_0 = r_0^{0K}$ and $U_0^{pol,\,0K} \approx U_0^{pol,\,T}$ in the condensed phase, the mean equilibrium distance at a given temperature $\left(r_0^T\right)$ can be derived from,

$$\Delta U(T) = U_0^T - U_0^{0K} = \Delta(-D_e + |U^{pol}|) \approx -\Delta D_e = \frac{a_0^3}{2}\left(V_0^T - V_0^{0K}\right) \quad (11.47)$$

and from (11.36), leading to

$$r_0^T = -\frac{1}{b_V} \cdot \ln\left(\frac{2a_0^{-3}\Delta U + V_0^{0K}}{a_V}\right). \quad (11.48)$$

Thanks to the large energy separation between the strongest intramolecular (>3,300 cm^{-1}) and intermolecular (translational 214 cm^{-1}, and rotational 494 cm^{-1}) modes (Raman spectrum of Ice-VIII [172]), it can be assumed that the former are inactive at $T = 273$ K ($k_B T/hc = 190$ cm^{-1}). Therefore, at $T = 273$ K, ΔU can be approximated to the thermal energy $kT = 2.27$ kJ·mol^{-1}, leading to $r_0^{273K} = 1.989$ Å. This value compares very well to that derived from the linear expansion coefficient of Ice ($\alpha = 5 \cdot 10^{-5}$ K^{-1}) [182],

$$r_0^{273K} = r_0^{0K} + \Delta r = r_0^{0K} + r_0^{0K} \cdot \Delta T \cdot \alpha \quad (11.49)$$

which leads to $r_0^{273K} = 1.985$ Å, provided $r_0^{0K} = r_0 = 1.958$ Å. In addition, as hydrogen bonds in Ice lose their elastic vibrations at $T = 273$ K (ice starts to break its crystalline structure), the total energy at this temperature U^{273K} should verify,

$$U^{273K} = U_0^{273K} = U_d \quad (11.50)$$

where U_0^{273K} (calculated at the equilibrium geometry r_0^{273K}) is only kinetic and U_d (calculated at the inflexion position r_d, where the interaction force can not restore anymore the elastic character of the vibrations in the condensed phase) is only potential energy. At a given T, the dissociation energy corresponds to the value calculated at the mean thermal equilibrium geometry r_0^T: $D_e = D_e\left(r_0^T\right)$. Thus, using $r_0^{273K} = 1.989$ Å from (11.48) and assuming $U_0^{pol,T} \approx U_0^{pol}$ in the condensed phase,

$$U_0^{273K} = -D_e\left(r_0^{273K}\right) + \left|U_0^{pol,273K}\right| \approx -D_e\left(r_0^{273K}\right) + \left|U_0^{pol}\right| \quad (11.51)$$

which leads to $U_0^{273K} = -11.3$ kJ·mol^{-1}. This value is in very good agreement with that obtained from the interaction potential function at $r_d = 2.276$ Å: $U_d = -\nu \cdot$

$H_d = -10.0$ kJ·mol^{-1}, indicating the internal coherence of the interaction potential and dissociation energy models.

To conclude this subsection, it is noteworthy that the equations derived from the model $U = -\nu \cdot H$ (Eq. 11.40) and Eqs. 11.35 and 11.36, are not sampling-depending. Indeed, the exponential dependencies of V and G on $d_{\mathrm{H}\cdots\mathrm{O}}$ in the experimental range of distances $1.5 < d_{\mathrm{H}\cdots\mathrm{O}} < 2.3$ Å have been corroborated in a recent theoretical study of $\rho(\mathbf{r})$ at the BCPs of 163 H···X (X = H, C, N, O, F, S, Cl, π) hydrogen bonded complexes, indicating universal dependences between the topological and the energetic properties of $\rho(\mathbf{r})$ at BCPs, as well as between each of them and the $d_{\mathrm{H}\cdots\mathrm{O}}$ distance [183]. In addition, from this theoretical study, the exponential and pre-exponential fitting parameters of G vs. $d_{\mathrm{H}\cdots\mathrm{O}}$ and V vs. $d_{\mathrm{H}\cdots\mathrm{O}}$ data were found very close to those of Eqs. 11.35 and 11.36 (almost statistically equivalent within standard deviations). Accordingly, while the values of the quantities here described could be in some measure sampling-depending, the equations describing the dependencies between these quantities are not. It should be however kept in mind that, outside the experimental range of distances ($d > 2.3$ Å), Eqs. 11.35 and 11.36 should not be used, as shown in Fig. 11.3. This is usually the case for weaker C-H···O interactions, as discussed in references [184] and [185].

11.3.1.2 Comparison to Other Interaction Potentials

Morse- and Buckingham-type interaction potentials are widely used in spectroscopy and atom-atom potential methods. The Morse potential is represented by the function,

$$U_{\mathrm{Morse}}(r) = U_0\left(1 - \exp\left[-\omega \cdot \left(\frac{\mu}{2U_0}\right)^{1/2} \cdot (r - r_0)\right]\right)^2 - U_0 \qquad (11.52)$$

where U_0 is the potential well depth at the minimum r_0, $\omega = (k_0/\mu)^{1/2}$ is the vibrational frequency of the mode, k_0 is the corresponding force constant at r_0 and μ the reduced mass of the system. The Buckingham potential takes the form,

$$U_{\mathrm{Buckingham}}(r) = -A \cdot r^{-6} + B \cdot \exp(-\alpha \cdot r) \qquad (11.53)$$

the first and the second terms representing the attractive dipole-dipole and the repulsive hard-spheres contributions in the interaction of two atoms separated by a distance r. Kitaigorodsky pointed out [186] that the Buckingham potential can be expressed in terms of three physically meaningful parameters, namely the equilibrium distance r_0, the potential well depth U_0 and the curvature k_0 of the potential at r_0. Indeed, knowing r_0, U_0 and k_0, the parameter α that satisfies the equation,

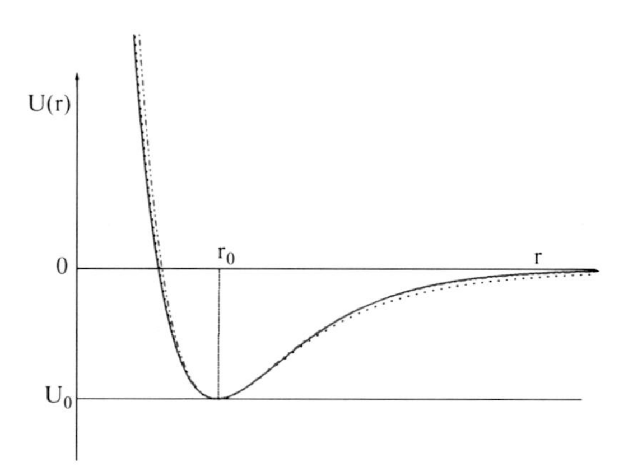

Fig. 11.5 Representation of $U = -\nu \cdot H$ (*solid blue line*), U_{Morse} (*dash-dot red line*) and $U_{\mathrm{Buckingham}}$ (*dot black line*) potentials while they are constrained to take the same characteristic values r_0, U_0 and k_0 at the minimum. Here, $r_0 = 1.958$ Å, $U_0 = -13.6$ kJ·mol^{-1} and $k_0 = 22.7$ N/m (Reprinted from Ref. [171] with kind permission of The American Institute of Physics)

$$\frac{k_0 \cdot r_0^2}{U_0} = \frac{6(\alpha \cdot r_0)^2 - 42\alpha \cdot r_0}{6 - \alpha \cdot r_0} \tag{11.54}$$

and leads to a negative constant for the attractive r^{-6} term and to a positive one for the repulsive exponential term, permits to express the Buckingham potential as,

$$U_{\mathrm{Buckingham}}(r) = \frac{U_0}{6 - \alpha \cdot r_0}\left(\frac{-\alpha \cdot r_0^7}{r^6} + 6 \cdot \exp[-\alpha \cdot (r - r_0)]\right). \tag{11.55}$$

The three physically meaningful parameters also define the Morse-type potential (11.52). Therefore, the model based on the experimental charge density (11.40) can be compared to both U_{Morse} and $U_{\mathrm{Buckingham}}$ potentials by constraining the two latter to have the same r_0, U_0 and k_0 values than those defining the former. Thus, representing (11.40, 11.52 and 11.55) altogether (Fig. 11.5), an almost perfect matching between the three potentials is observed throughout the full range of distances. Despite the applied constraints, this is a remarkable result when considering that they are defined by three different functions. Indeed, for all of them, the potential shape, and in particular the asymmetry around r_0, is similarly determined by only the characteristics of the corresponding function at the minimum (r_0, U_0, k_0). This result permits to apply closed-shell interaction potentials derived from the experimental charge density analysis to the domains of spectroscopy and of atom-atom potential methods.

11.3.1.3 Dependence of the Interaction Energy on the External Field

These results indicate that, derived from Eqs. 11.31 to 11.36, the dependence of the total energy density at BCP with the hydrogen bonding distance can be considered as a reference for H$\cdots$O interaction potentials. It has been shown to nicely work with Ice when the force constant $k_{\mathrm{H\cdots O}}$ determined from Raman data on Ice-VIII has been used to derive ν, and therefore $U_{\mathrm{H}\cdots\mathrm{O}}$ through (11.40) for this compound. However, it is obvious that the exact interaction potential for each particular X-H···O-Y case, involving different X- and Y-groups, should be different. Accordingly, the potential well depth U_0, the characteristic equilibrium distance r_0 and the corresponding force constant $k_{\mathrm{H}\cdots\mathrm{O}}$ should vary with the X- and Y-groups, therefore modifying the interaction potential shape as previously pointed out. With the aim to understand the influence of the environment on the electron properties associated to hydrogen bonding interactions, a first study on the effect of an external electric field $\boldsymbol{E}$ on the hydrogen bonded dimer HF···HF has been undertaken [187]. In this work, the applied field ε ($\boldsymbol{E} = \varepsilon \cdot \boldsymbol{u}$), $\boldsymbol{u}$ being the unit vector (along the F···H direction) ranges from −0.05 to +0.05 a.u. by steps of 0.01 a.u. (0.01 a.u. = 5.1422082 GV/m), taking as positive the field applied from H to F along F···H. Calculations were done by relaxing the F-H distances for each pair of parametrs ($d_{\mathrm{F}\cdots\mathrm{H}}$, ε) applied to the dimer and for each field ε applied to the monomers, while keeping the $C_{\infty v}$ symmetry in all cases. For $\varepsilon > 0$, both the integrated energy of the dimer (E_{dimer}) and the interaction energy U defined as

$$U(d_{\mathrm{F}\cdots\mathrm{H}}, \varepsilon) = E_{\mathrm{dimer}}(d_{\mathrm{F}\cdots\mathrm{H}}, \varepsilon) - 2 \cdot E_{\mathrm{monomer}}(\varepsilon) \tag{11.56}$$

shift to more negative values. Following the description given for the interaction potential (Eqs. 11.39 and 11.40), the dependence of U on both $d_{\mathrm{F}\cdots\mathrm{H}}$ and ε has been also fitted to a sum of two exponentials,

$$U(d_{\mathrm{F}\cdots\mathrm{H}}, \varepsilon) = -a_1 \cdot \exp(c_1 \cdot \varepsilon - b_1 \cdot d_{\mathrm{F}\cdots\mathrm{H}}) + a_2 \cdot \exp(-c_2 \cdot \varepsilon - b_2 \cdot d_{\mathrm{F}\cdots\mathrm{H}}) \tag{11.57}$$

where the fitting parameters are $a_i = 193(9)$ and $1.02(4) \cdot 10^4$ kJ.mol^{-1}, $b_i = 1.10(2)$ and 3.74(4) Å^{-1}, and $c_i = 13.9(2)$ and 5.3(2) inverse electric field a.u.'s for i = 1 and 2, respectively (correlation factor R = 0.9994). The fitted interaction energy surface $U(d_{\mathrm{F}\cdots\mathrm{H}}, \varepsilon)$ is represented in Fig. 11.6.

The dependence of the interaction energy on an external field presents a minimum for all values of ε. The minimum is displaced to larger distances and becomes shallower for negative fields. It is particularly interesting to determine the effect of ε on the equilibrium distance. This can be done by solving the equation,

$$\left.\frac{\partial E}{\partial d_{\mathrm{F}\cdots\mathrm{H}}}\right|_{\varepsilon} = \left.\frac{\partial U}{\partial d_{\mathrm{F}\cdots\mathrm{H}}}\right|_{\varepsilon} = 0. \tag{11.58}$$

Using (11.57), the following relationship is obtained,

$$d_{\mathrm{eq}} = d_0 - \frac{c_1 + c_2}{b_2 - b_1} \cdot \varepsilon \tag{11.59}$$

where

$$d_0 = \frac{1}{b_2 - b_1} \ln\left(\frac{a_2 \cdot b_2}{a_1 \cdot b_1}\right) \tag{11.60}$$

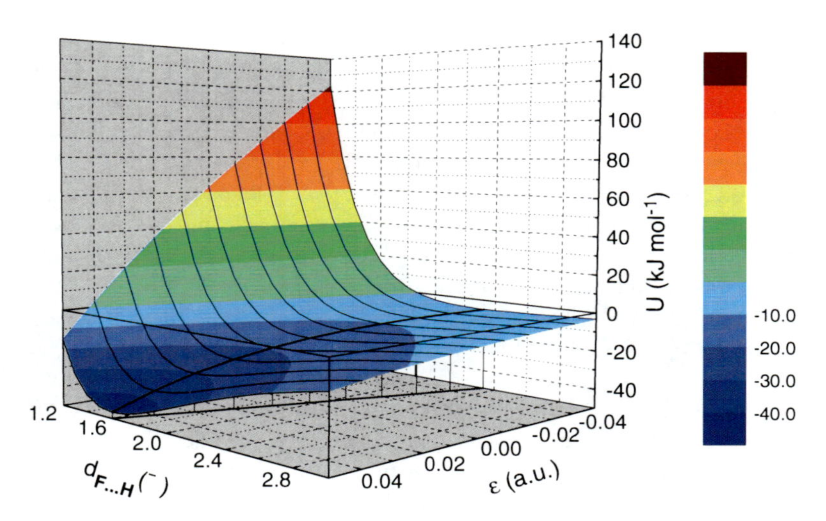

Fig. 11.6 Interaction energy surface $U(d_{\mathrm{F}\cdots\mathrm{H}}, \varepsilon)$ for the linear dimer HF···HF. The F···H distance is in Ångstrøm, the applied field is in atomic units, the interaction energy in kJ/mol. The thick line crossing the surface corresponds to the dependence of U at the equilibrium distance with the applied field ε, i.e. $U(d_{\mathrm{eq}}(\varepsilon), \varepsilon)$. The projection of the thick line on the $(d_{\mathrm{F}\cdots\mathrm{H}}, \varepsilon)$ surface corresponds to the intrinsic dependence between d_{eq} and ε analytically given by Eq. 11.59 (Reprinted from Ref. [187] with kind permission of The American Institute of Physics)

represents the equilibrium distance at $\varepsilon = 0$ ($d_0 = 1.966$ Å from the fitting parametrs, while $d_0 = 1.955$ Å for the linear dimer without applied field). The linear dependence of Eq. 11.59, represented in the projection of Fig. 11.6, can be easily inverted, leading to

$$\varepsilon = \frac{b_2 - b_1}{c_1 + c_2} \cdot (d_0 - d_{\mathrm{eq}}). \tag{11.61}$$

As far as the polarization induced by the external field provokes the variation of both d_{eq} and U, the dissociation energy of the system $D_e = U(d_{\mathrm{F}\cdots\mathrm{H}} = d_{\mathrm{eq}})$ can be determined through their dependences on ε given by (11.57) and (11.59). Thus, setting $d_{\mathrm{F}\cdots\mathrm{H}} = d_{\mathrm{eq}}$ and including (11.61) in (11.57),

$$D_e(d_{\mathrm{eq}}) = \zeta \cdot \exp(-\tau \cdot d_{\mathrm{eq}}) \tag{11.62}$$

where

$$\tau = \frac{b_1 \cdot c_2 + b_2 \cdot c_1}{c_1 + c_2} \tag{11.63}$$

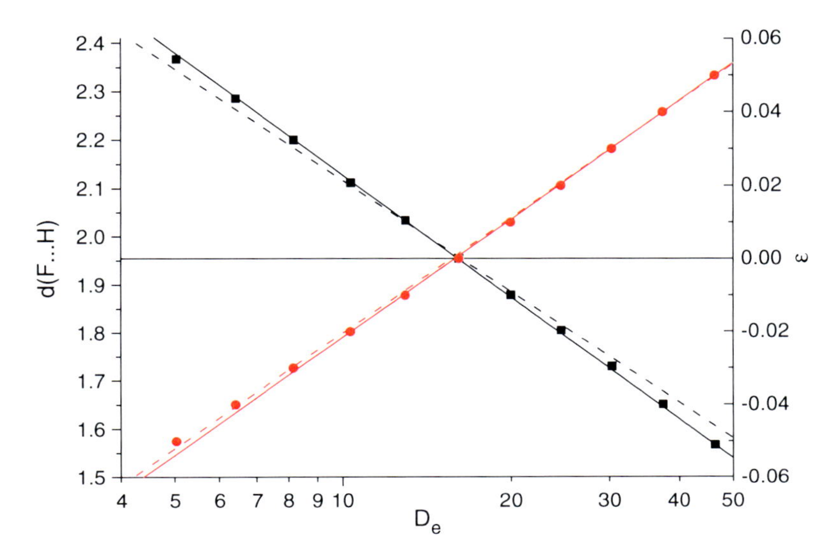

Fig. 11.7 Log-linear plots of the exponential dependence of D_e (kJ·mol^{-1}) with $d_{F\cdots H}$ measured in Å (*black lines*) and with ε in a.u. (*red lines*) at the HF···HF equilibrium configurations ($d_{F\cdots H} = d_{eq}(\varepsilon)$) calculated for $-0.05 < \varepsilon < +0.05$ a.u.. *Solid lines* correspond to the exponential fittings on the depicted data: $D_e = 3.4(1) \cdot 10^3 \cdot exp(-2.74(2) \cdot d_{F\cdots H})$ and $D_e = 15.92(9) \cdot \exp(21.5(1) \cdot \varepsilon)$, while *dashed lines* are obtained from the fitting parameters of Eq. 11.57 and Eqs. 11.59, 11.62–11.64: $D_e = 5841 \cdot \exp(-3.01 \cdot d_{F\cdots H})$ and $D_e = 15.66 \cdot \exp(21.9 \cdot \varepsilon)$ (Reprinted from Ref. [187] with kind permission of The American Institute of Physics)

$$\zeta = \left(a_1^{c_2} \cdot a_2^{c_1}\right)^{\frac{1}{(c_1+c_2)}} \cdot \left(\left(\frac{b_2}{b_1}\right)^{\frac{c_1}{(c_1+c_2)}} - \left(\frac{b_1}{b_2}\right)^{\frac{c_2}{(c_1+c_2)}}\right). \tag{11.64}$$

Using the values of the fitting parameters of (11.57) leads to $\tau = 3.01$ Å^{-1} and $\zeta = 5{,}841$ kJ·mol^{-1}. The exponential dependence of D_e on d_{eq} (similar to that previously found for the experimental H···O data) appears therefore as a direct consequence of the exponential dependence of U with both ε and $d_{F\cdots H}$. Due to the linear correspondence between d_{eq} and ε (Eq. 11.59), D_e can be also expressed in terms of an exponential dependence with the latter, where the parameters are $\tau_\varepsilon = -21.9$ electric field a.u.$^{-1}$ and $\zeta_\varepsilon = 15.66$ kJ·mol^{-1}. Figure 11.7 shows the log-linear plot of $D_e(d_{eq})$ and $D_e(\varepsilon)$ data along with the derived and the fitting functions.

The topological and energetic properties of $\rho(\boldsymbol{r})$ at the hydrogen bond BCP of the linear dimer HF···HF can be also described in terms of each pair of applied parameters ($d_{F\cdots H}$, ε)[187]. Therefore, following the same procedure than that described for D_e, *i.e.* setting $d_{F\cdots H} = d_{eq}$ and including (11.61) in the dependency of each BCP property with both $d_{F\cdots H}$ and ε parameters, it permits to retrieve the magnitude of the BCP property at the equilibrium configuration for each ε.

Accordingly, the value of ε fixes the interaction potential, the dissociation energy, the equilibrium distance and the BCP properties at this geometry, all of them appearing intrinsically related to each other.

As ε represents an external perturbation on the electron density distribution, it could be very adequate to interpret the effect of the atomic and molecular environment on the considered interaction in terms of an effective ε. However, it should be kept in mind that, in contrast to the applied ε, molecular environments induce non-homogeneous electric fields. In addition, it should be also noted that while interactions in gas phase can reach their equilibrium configurations, steric constraints could avoid full relaxations in crystal phases. With this respect, the scattering of values around the empirical dependencies that is observed for a family of intermolecular interactions, and in particular for a given family of X-H$\cdots$A-Y hydrogen bonds, can be due (besides the experimental noise) to the different electric field that each particular crystalline surrounding exerts on its corresponding member of the family and to a non-relaxed conformation resulting from steric constraints. Therefore, to characterize interaction potentials and dissociation energies from experimental charge densities, a good understanding of their relationships with the BCP properties is needed not only at equilibrium geometries but also out of equilibrium configurations. To end, the knowledge of electric fields miming atomic and molecular environments and their application to reference systems in gas phase seems an interesting issue to investigate the dependence of energetic properties of target systems embedded in different crystalline environments (*cf.* Chap. 6).

11.3.2 Concluding Remarks

To close this section, we would like to address a final remark. The information extracted from the analysis of the crystalline charge density $\varrho(\boldsymbol{r})$ provides deep insight on the physical and chemical properties of the system, including intermolecular cohesion. It is well-known that the $\varrho(\boldsymbol{r})$ function can be derived either from theoretical calculations of periodic wavefunctions or from high resolution X-ray diffraction experiments. Within these approaches, different decompositions of the total energy and different partitions of the $\varrho(\boldsymbol{r})$ function can be used to retrieve and to quantify crystal properties, as exemplified in the present chapter. Is it possible to decide whether one approach is better than another? It is difficult to find a satisfactory positive answer to this question. Provided the modeling of $\varrho(\boldsymbol{r})$ has been well undertaken in any case, the calculated crystal properties should converge to similar values independently of the approach, method or partition followed to derive it. Sometimes, a criticism has been raised concerning the ascription of the intermolecular cohesion to interactions between pairs of atoms linked by topological bond paths, rather than between charge density distributions [188], cf. also the discussion reported in Chap. 16. However, as far as the topological analysis of the intermolecular charge density in crystals is applied to the total electron distribution of the complete system, we think that both approaches should be fundamentally considered as valid. Indeed, while they are based on different representations that are made by pieces that can not be straightforwardly compared, both lead to similar quantifications for the inherent properties of the crystal.

References

1. Jeziorski B, Moszynski R, Szalewicz K (1994) Perturbation theory approach to intermolecular potential energy surfaces of van der Waals complexes. Chem Rev 94:1887
2. Moszynski R (2007) In: Sokalski WA (ed) Molecular materials with specific interactions. Springer, Dordrecht, pp 1–152
3. Stone AJ (1997) The theory of intermolecular forces. Oxford University Press, Oxford
4. Koritsanszky TS, Coppens P (2001) Chemical applications of X-ray charge-density analysis. Chem Rev 101:1583
5. Stewart RF (1976) Electron population analysis with rigid pseudoatoms. Acta Crystallogr A 32:565
6. Hansen N, Coppens P (1978) Testing aspherical atom refinements on small-molecule data sets. Acta Crystallogr A 34:909
7. Jayatilaka D (1998) Wave function for beryllium from X-Ray diffraction data. Phys Rev Lett 80:798
8. Jayatilaka D, Grimwood DJ (2001) Wavefunctions derived from experiment. I. Motivation and theory. Acta Crystallogr A 57:76
9. Hibbs DE, Howard ST, Huke JP, Waller MP (2005) A new orbital-based model for the analysis of experimental molecular charge densities: an application to (Z)-N-methyl-C-phenylnitrone. Phys Chem Chem Phys 7:1772
10. Waller MP, Howard ST, Platts JA, Piltz RO, Willock DJ, Hibbs DE (2006) Novel properties from experimental charge densities: an application to the zwitterionic neurotransmitter taurine. Chem Eur J 12:7603
11. van Smaalen S, Netzel J (2009) The maximum entropy method in accurate charge-density studies. Phys Scr 79:048304
12. Destro R, Loconte L, Presti LL, Roversi P, Soave R (2004) On the role of data quality in experimental charge-density studies. Acta Crystallogr A 60:365
13. Abramov Y, Volkov A, Coppens P (1999) On the evaluation of molecular dipole moments from multipole refinement of X-ray diffraction data. Chem Phys Lett 311:81
14. Bianchi R, Gatti C, Adovasio V, Nardelli M (1996) Theoretical and experimental (113 K) electron-density study of lithium bis(tetramethylammonium) hexanitrocobaltate(III). Acta Crystallogr B 52:471
15. de Vries RY, Feil D, Tsirelson VG (2000) Extracting charge density distributions from diffraction data: a model study on urea. Acta Crystallogr B 56:118
16. Spackman M, Byrom P (1996) Molecular electric moments and electric field gradients from X-ray diffraction data: model studies. Acta Crystallogr B 52:1023
17. Spackman MA, Byrom PG, Alfredsson M, Hermansson K (1999) Influence of intermolecular interactions on multipole-refined electron densities. Acta Crystallogr A 55:30
18. Volkov A, Abramov YA, Coppens P (2001) Density-optimized radial exponents for X-ray charge-density refinement from ab initio crystal calculations. Acta Crystallogr A 57:272
19. Volkov A, Coppens P (2001) Critical examination of the radial functions in the Hansen-Coppens multipole model through topological analysis of primary and refined theoretical densities. Acta Crystallogr A 57:395
20. Volkov A, Gatti C, Abramov YA, Coppens P (2000) Evaluation of net atomic charges and atomic and molecular electrostatic moments through topological analysis of the experimental charge density. Acta Crystallogr A 56:252

21. Swaminathan S, Craven BM, Spackman MA, Stewart RF (1984) Theoretical and experimental studies of the charge density in urea. Acta Crystallogr B 40:398
22. Stewart RF, Bentley J, Goodman B (1975) Generalized X-ray scattering factors in diatomic molecules. J Chem Phys 63:3786
23. Spackman MA (1992) Molecular electric moments from X-ray diffraction data. Chem Rev 92:1769
24. Flaig R, Koritsanszky T, Zobel D, Luger P (1998) Topological analysis of the experimental electron densities of amino acids. 1. D,L-aspartic acid at 20 K. J Am Chem Soc 120:2227
25. Sorensen H, Stewart R, McIntyre G, Larsen S (2003) Simultaneous variation of multipole parameters and Gram-Charlier coefficients in a charge-density study of tetrafluoroterephthalonitrile based on X-ray and neutron data. Acta Crystallogr A 59:540
26. Roversi P, Destro R (2004) Approximate anisotropic displacement parameters for H atoms in molecular crystals. Chem Phys Lett 386:472
27. Madsen AØ, Sørensen HO, Flensburg C, Stewart RF, Larsen S (2004) Modeling of the nuclear parameters for H atoms in X-ray charge-density studies. Acta Crystallogr A 60:550
28. Whitten AE, Spackman MA (2006) Anisotropic displacement parameters for H atoms using an ONIOM approach. Acta Crystallogr B 62:875
29. Jayatilaka D, Dittrich B (2008) X-ray structure refinement using aspherical atomic density functions obtained from quantum-mechanical calculations. Acta Crystallogr A 64:383
30. Hoser AA, Dominiak PM, Wozniak K (2009) Towards the best model for H atoms in experimental charge-density refinement. Acta Crystallogr A 65:300
31. Dittrich B, Spackman MA (2007) Can the interaction density be measured? The example of the non-standard amino acid sarcosine. Acta Crystallogr A 63:426
32. Ma Y, Politzer P (2004) Determination of noncovalent interaction energies from electronic densities. J Chem Phys 121:8955
33. Welch GWA, Karamertzanis PG, Misquitta AJ, Stone AJ, Price SL (2008) Is the induction energy important for modeling organic crystals? J Chem Theory Comput 4:522
34. Spackman MA, Munshi P, Dittrich B (2007) Dipole moment enhancement in molecular crystals from X-ray diffraction data. ChemPhysChem 8:2051
35. Whitten AE, Turner P, Klooster WT, Piltz RO, Spackman MA (2006) Reassessment of large dipole moment enhancements in crystals: a detailed experimental and theoretical charge density analysis of 2-methyl-4-nitroaniline. J Phys Chem A 110:8763
36. Suponitsky KY, Tsirelson VG, Feil D (1999) Electron-density-based calculations of intermolecular energy: case of urea. Acta Crystallogr A 55:821
37. Ma Y, Politzer P (2004) Electronic density approaches to the energetics of noncovalent interactions. Int J Mol Sci 05:130
38. Volkov A, Coppens P (2004) Calculation of electrostatic interaction energies in molecular dimers from atomic multipole moments obtained by different methods of electron density partitioning. J Comput Chem 25:921
39. Buckingham AD (1967) Advances in chemical physics. In: Hirschfelder J (ed) Intermolecular forces. Wiley, Madison, pp 107–142
40. Stone AJ, Price SL (1988) Some new ideas in the theory of intermolecular forces: anisotropic atom-atom potentials. J Phys Chem 92:3325
41. Stone AJ (1985) Distributed polarizabilities. Mol Phys 56:1065
42. Spackman MA (1986) A simple qualitative model of hydrogen bonding. J Chem Phys 85:6587
43. Hirshfeld FL (1977) Bonded-atom fragments for describing molecular charge densities. Theor Chim Acta 44:129
44. Hirshfeld FL (1971) Difference densities by least-squares refinement: fumaramic acid. Acta Crystallogr B 27:769
45. Bader RWF (1990) Atoms in molecules – a quantum theory. University of Oxford Press, Oxford
46. Ángyán JG, Jansen G, Loos M, Hättig C, Hess BA (1994) Distributed polarizabilities using the topological theory of atoms in molecules. Chem Phys Lett 219:267

47. Jansen G, Hättig C, Hess BA, Ángyán JG (1996) Intermolecular interaction energies by topologically partitioned electric properties. 1. Electrostatic and induction energies in one-centre and multicentre multipole expansions. Mol Phys 88:69
48. Hättig C, Jansen G, Hess BA, Ángyán JG (1997) Intermolecular interaction energies by topologically partitioned electric properties. II. Dispersion energies in one-centre and multicentre expansions. Mol Phys 91:145
49. Bouhmaida N, Ghermani N, Lecomte C, Thalal A (1997) Modelling electrostatic potential from experimentally determined charge densities. 2. Total potential. Acta Crystallogr A 53:556
50. Ángyán JG, Chipot C (1994) A comprehensive approach to molecular charge density models: from distributed multipoles to fitted atomic charges. Int J Quantum Chem 52:17
51. Jansen G (2000) Convergence of multipole expanded intermolecular interaction energies for Gaussian-type-function and Slater-type-function basis sets. Theor Chem Acc 104(6):499
52. Popelier PLA, Rafat M (2003) The electrostatic potential generated by topological atoms: a continuous multipole method leading to larger convergence regions. Chem Phys Lett 376:148
53. Volkov A, King HF, Coppens P (2006) Dependence of the intermolecular electrostatic interaction energy on the level of theory and the basis set. J Chem Theory Comput 2:81
54. Volkov A, Koritsanszky T, Coppens P (2004) Combination of the exact potential and multipole methods (EP/MM) for evaluation of intermolecular electrostatic interaction energies with pseudoatom representation of molecular electron densities. Chem Phys Lett 391:170–175
55. Spackman M (2006) The use of the promolecular charge density to approximate the penetration contribution to intermolecular electrostatic energies. Chem Phys Lett 418:158
56. Volkov A, King HF, Coppens P, Farrugia LJ (2006) On the calculation of the electrostatic potential, electric field and electric field gradient from the aspherical pseudoatom model. Acta Crystallogr A 62:400
57. Piquemal JP, Chevreau H, Gresh N (2007) Toward a separate reproduction of the contributions to the Hartree-Fock and DFT intermolecular interaction energies by polarizable molecular mechanics with the SIBFA potential. J Chem Theory Comput 3:824
58. Slipchenko LV, Gordon MS (2007) Electrostatic energy in the effective fragment potential method: theory and application to benzene dimer. J Comput Chem 28:276
59. Kairys V, Jensen J (1999) Evaluation of the charge penetration energy between non-orthogonal molecular orbitals using the Spherical Gaussian Overlap approximation. Chem Phys Lett 315:140
60. Qian W, Krimm S (2006) Charge density treatment of the molecule-charge interaction and its relation to the electrical component of hydrogen bonding: accuracy and distance dependence. J Mol Struct Theochem 766:93
61. Aubert E, Porcher F, Souhassou M, Lecomte C (2004) Electrostatic potential and interaction energies of molecular entities occluded in the $AlPO_4$-15 molecular sieve: determination from X-ray charge density analysis. J Phys Chem Solids 65:1943
62. Bouhmaida N, Bonhomme F, Guillot B, Jelsch C, Ghermani NE (2009) Charge density and electrostatic potential analyses in paracetamol. Acta Crystallogr B 65:363
63. Spackman MA (2007) Comment on On the calculation of the electrostatic potential, electric field and electric field gradient from the aspherical pseudoatom model by Volkov, King, Coppens & Farrugia (2006). Acta Crystallogr A 63:198
64. Volkov A, Coppens P (2007) Response to Spackman's comment on On the calculation of the electrostatic potential, electric field and electric field gradient from the aspherical pseudoatom model. Acta Crystallogr A 63:201
65. Su Z, Coppens P (1992) On the mapping of electrostatic properties from the multipole description of the charge-density. Acta Crystallogr A 48:188
66. Stewart R, Craven B (1993) Molecular electrostatic potentials from crystal diffraction: the neurotransmitter γ-aminobutyric acid. Biophys J 65:998
67. Ghermani N, Bouhmaida N, Lecomte C (1993) Modeling electrostatic potential from experimentally determined charge-densities. 1. Spherical-atom approximation. Acta Crystallogr A 49:781

68. Koritsanszky T, Volkov A (2004) Atomic density radial functions from molecular densities. Chem Phys Lett 385:431
69. Spackman M, Maslen E (1986) Chemical-properties from the promolecule. J Phys Chem 90:2020
70. Destro R, Soave R, Barzaghi M (2008) Physicochemical properties of zwitterionic L- and DL-alanine crystals from their experimental and theoretical charge densities. J Phys Chem B 112:5163
71. Brock C, Dunitz J, Hirshfeld F (1991) Transferability of deformation densities among related molecules - atomic multipole parameters from perylene for improved estimation of molecular vibrations in naphthalene and anthracene. Acta Crystallogr B 47:789
72. Pichon-Pesme V, Lecomte C, Lachekar H (1995) On building a data bank of transferable experimental electron density parameters: application to polypeptides. J Phys Chem 99:6242
73. Zarychta B, Pichon-Pesme V, Guillot B, Lecomte C, Jelsch C (2007) On the application of an experimental multipolar pseudo-atom library for accurate refinement of small-molecule and protein crystal structures. Acta Crystallogr A 63:108
74. Dittrich B, Koritsanszky T, Luger P (2004) A simple approach to nonspherical electron densities by using invarioms. Angew Chem Int Ed 43(20):2718
75. Dittrich B, Hübschle CB, Luger P, Spackman MA (2006) Introduction and validation of an invariom database for amino-acid, peptide and protein molecules. Acta Crystallogr D 62:1325
76. Hübschle CB, Luger P, Dittrich B (2007) Automation of invariom and of experimental charge density modelling of organic molecules with the preprocessor program InvariomTool. J Appl Crystallogr 40:623
77. Koritsanszky T, Volkov A, Coppens P (2002) Aspherical-atom scattering factors from molecular wave functions. 1. Transferability and conformation dependence of atomic electron densities of peptides within the multipole formalism. Acta Crystallogr A 58:464
78. Volkov A, Li X, Koritsanszky T, Coppens P (2004) Ab initio quality electrostatic atomic and molecular properties including intermolecular energies from a transferable theoretical pseudoatom databank. J Phys Chem A 108:4283
79. Dominiak PM, Volkov A, Li X, Messerschmidt M, Coppens P (2007) A theoretical databank of transferable aspherical atoms and its application to electrostatic interaction energy calculations of macromolecules. J Chem Theory Comput 3:232
80. Domagala S, Jelsch C (2008) Optimal local axes and symmetry assignment for charge-density refinement. J Appl Crystallogr 41:1140
81. Allen F (2002) The Cambridge Structural Database: a quarter of a million crystal structures and rising. Acta Crystallogr B 58:407
82. Volkov A, Messerschmidt M, Coppens P (2007) Improving the scattering-factor formalism in protein refinement: application of the University at Buffalo Aspherical-Atom Databank to polypeptide structures. Acta Crystallogr D 63:160
83. Li X, Volkov A, Szalewicz K, Coppens P (2006) Interaction energies between glycopeptide antibiotics and substrates in complexes determined by X-ray crystallography: application of a theoretical databank of aspherical atoms and a symmetry-adapted perturbation theory-based set of interatomic potentials. Acta Crystallogr D 62:639
84. Dominiak PM, Volkov A, Dominiak AP, Jarzembska KN, Coppens P (2009) Combining crystallographic information and an aspherical-atom data bank in the evaluation of the electrostatic interaction energy in an enzyme-substrate complex: influenza neuraminidase inhibition. Acta Crystallogr D 65:485
85. Pichon-Pesme V, Jelsch C, Guillot B, Lecomte C (2004) A comparison between experimental and theoretical aspherical-atom scattering factors for charge-density refinement of large molecules. Acta Crystallogr A 60:204
86. Volkov A, Koritsanszky T, Li X, Coppens P (2004) Response to the paper A comparison between experimental and theoretical aspherical-atom scattering factors for charge-density refinement of large molecules, by Pichon-Pesme, Jelsch, Guillot & Lecomte (2004). Acta Crystallogr A 60:638

87. Bak J, Dominiak PM, Wilson CC, Wozniak K (2009) Experimental charge density study of paracetamol – multipole refinement in the presence of a disordered methyl group. Acta Crystallogr A 65:490
88. Gavezzotti A (2002) Calculation of intermolecular interaction energies by direct numerical integration over electron densities. I. Electrostatic and polarization energies in molecular crystals. J Phys Chem B 106:4145
89. Gavezzotti A (2003) Towards a realistic model for the quantitative evaluation of intermolecular potentials and for the rationalization of organic crystal structures. Part II. Crystal energy landscapes. CrystEngComm 5:439
90. Gavezzotti A (2003) Calculation of intermolecular interaction energies by direct numerical integration over electron densities. 2. An improved polarization model and the evaluation of dispersion and repulsion energies. J Phys Chem B 107:2344
91. Gavezzotti A (2005) Quantitative ranking of crystal packing modes by systematic calculations on potential energies and vibrational amplitudes of molecular dimers. J Chem Theory Comp 1:834
92. Dunitz JD, Gavezzotti A (2009) How molecules stick together in organic crystals: weak intermolecular interactions. Chem Soc Rev 38:2622
93. Spackman MA (1986) Atom-atom potentials via electron gas theory. J Chem Phys 85:6579
94. Coppens P, Abramov Y, Carducci M, Korjov B, Novozhilova I, Alhambra C, Pressprich M (1999) Experimental charge densities and intermolecular interactions: electrostatic and topological analysis of DL-histidine. J Am Chem Soc 121:2585
95. Abramov Y, Volkov A, Wu G, Coppens P (2000) The experimental charge-density approach in the evaluation of intermolecular interactions. application of a new module of the XD programming package to several solids including a pentapeptide. Acta Crystallogr A 56:585
96. Cox S, Hsu L, Williams D (1981) Nonbonded potential function models for crystalline oxohydrocarbons. Acta Crystallogr A 37:293
97. Williams D, Cox S (1984) Nonbonded potentials for azahydrocarbons: the importance of the coulombic interaction. Acta Crystallogr B 40:404
98. Spackman M (1987) A simple quantitative model of hydrogen-bonding - application to more complex-systems. J Phys Chem 91:3179
99. Mitchell J, Price S (1990) The nature of the N-H ... O = C hydrogen bond: an intermolecular perturbation theory study of the formamide/formaldehyde complex. J Comput Chem 11:1217
100. Coombes DS, Price SL, Willock DJ, Leslie M (1996) Role of electrostatic interactions in determining the crystal structures of polar organic molecules. A distributed multipole study. J Phys Chem 100:7352
101. Price SL (2008) Computational prediction of organic crystal structures and polymorphism. Int Rev Phys Chem 27:541
102. Grabowsky S, Pfeuffer T, Morgenroth W, Paulmann C, Schirmeister T, Luger P (2008) A comparative study on the experimentally derived electron densities of three protease inhibitor model compounds. Org Biomol Chem 6:2295
103. Kim YS, Gordon RG (1974) Study of the electron gas approximation. J Chem Phys 60:1842
104. Clementi E, Roetti C (1974) Roothaan-Hartree-Fock atomic wavefunctions: basis functions and their coefficients for ground and certain excited states of neutral and ionized atoms, $Z \leq 54$. Data Nucl Data Tables 14:177
105. Destro R, Roversi P, Barzaghi M, Marsh RE (2000) Experimental charge density of α-glycine at 23 K. J Phys Chem A 104:1047
106. Abramov YA, Volkov A, Wu G, Coppens P (2000) Use of X-ray charge densities in the calculation of intermolecular interactions and lattice energies: application to glycylglycine, DL-histidine, and DL-proline and comparison with theory. J Phys Chem B 104:2183
107. Abramov Y, Volkov A, Coppens P (2000) Anisotropic atom-atom potentials from X-ray charge densities: application to intermolecular interactions and lattice energies in some biological and nonlinear optical materials. J Mol Struct Theochem 529:27
108. Li X, Wu G, Abramov Y, Volkov A, Coppens P (2002) Application of charge density methods to a protein model compound: calculation of Coulombic intermolecular interaction energies from the experimental charge density. Proc Natl Acad Sci 99:12132

109. Munshi P, Guru Row TN (2006) Topological analysis of charge density distribution in concomitant polymorphs of 3-acetylcoumarin, a case of packing polymorphism. Cryst Growth Design 6:708
110. Lo Presti L, Soave R, Destro R (2006) On the interplay between CH···O and OH···O interactions in determining crystal packing and molecular conformation: an experimental and theoretical charge density study of the fungal secondary metabolite austdiol ($C_{12}H_{12}O_5$). J Phys Chem B 110:6405
111. Soave R, Barzaghi M, Destro R (2007) Progress in the understanding of drug-receptor interactions, Part 2: experimental and theoretical electrostatic moments and interaction energies of an angiotensin II receptor antagonist ($C_{30}H_{30}N_6O_3S$). Chem Eur J 13:6942
112. Szalewicz K, Jeziorski B (1979) Symmetry-adapted double-perturbation analysis of intramolecular correlation effects in weak intermolecular interactions: the He-He interaction. Mol Phys 38:191
113. Williams HL, Mas EM, Szalewicz K, Jeziorski B (1995) On the effectiveness of monomer-, dimer-, and bond-centered basis functions in calculations of intermolecular interaction energies. J Chem Phys 103:7374
114. Bukowski R, Cencek W, Jankowski P, Jeziorski B, Jeziorska M, Kucharski SA, Misquitta AJ, Moszynski R, Patkowski K, Rybak S, Szalewicz K, Williams HL, Wormer PES, (2003) SAPT2002: an ab initio program for Many-Body Symmetry-Adapted Perturbation Theory calculations of intermolecular interaction energies. Technical report, University of Delaware/University of Warsaw
115. Szalewicz K, Patkowski K, Jeziorski B (2005) Intermolecular forces and clusters II. Struct Bond 116(Springer):43–117
116. Mirsky K (1976) Interatomic potential functions for hydrocarbons from crystal data: transferability of the empirical parameters. Acta Crystallogr A 32:199
117. Kitaigorodskii AI (1965) The principle of close packing and the condition of thermodynamic stability of organic crystals. Acta Crystallogr 18:585
118. Demartin F, Filippini G, Gavezzotti A, Rizzato S (2004) X-ray diffraction and packing analysis on vintage crystals: Wilhelm Koerner's nitrobenzene derivatives from the School of Agricultural Sciences in Milano. Acta Crystallogr B 60:609
119. Li T, Feng S (2006) Empirically augmented density functional theory for predicting lattice energies of aspirin, acetaminophen polymorphs, and ibuprofen homochiral and racemic crystals. Pharm Res 23:2326
120. Gilli G (2000) Molecules and molecular crystals. In: Giacovazzo C (ed) Fundamentals of crystaphy. Oxford University Press, New York, pp 465–534
121. Karamertzanis PG, Day GM, Welch GWA, Kendrick J, Leusen FJJ, Neumann MA, Price SL (2008) Modeling the interplay of inter- and intramolecular hydrogen bonding in conformational polymorphs. J Chem Phys 128:244708
122. Rivera SA, Allis DG, Hudson BS (2008) Importance of vibrational zero-point energy contribution to the relative polymorph energies of hydrogen-bonded species. Cryst Growth Des 8:3905
123. Tsirel'son VG, Kuleshova LN, Ozerov RP (1982) The electrostatic term in lattice-energy calculations for lithium formate monodeuterate: determination from the experimental electron density. Acta Crystallogr A 38:707
124. Su Z, Coppens P (1995) On the calculation of the lattice energy of ionic-crystals using the detailed electron-density distribution. 1. Treatment of spherical atomic distributions and application to NaF. Acta Crystallogr A 51:27
125. Overgaard J, Hibbs DE (2004) The experimental electron density in polymorphs a and b of the anti-ulcer drug famotidine. Acta Crystallogr A 60:480
126. Bianchi R, Gervasio G, Marabello D (2000) Experimental electron density analysis of $Mn_2(CO)_{10}$: metal-metal and metal-ligand bond characterization. Inorg Chem 39:2360
127. Cummins PG, Dunmur DA, Munn RW, Newham RJ (1976) Applications of the Ewald method. I. Calculation of multipole lattice sums. Acta Crystallogr A 32:847

128. De Leeuw SW, Perram JW, Smith ER (1980) Simulation of electrostatic systems in periodic boundary conditions. I. Lattice sums and dielectric constants. Proc R Soc A 373:27
129. De Leeuw SW, Perram JW, Smith ER (1980) Simulation of electrostatic systems in periodic boundary conditions. II. Equivalence of boundary conditions. Proc R Soc A 373:57
130. Makov G, Payne MC (1995) Periodic boundary conditions in ab initio calculations. Phys Rev B 51:4014
131. Marshall SL (2000) A periodic Green function for calculation of coloumbic lattice potentials. J Phys Condens Matter 12:4575
132. Catti M (1978) Electrostatic lattice energy in ionic crystals: optimization of the convergence of Ewald series. Acta Crystallogr A 34:974
133. Ángyán JG, Silvi B (1987) Electrostatic interactions in three-dimensional solids. Self-consistent Madelung potential (SCMP) approach. J Chem Phys 86:6957
134. Ferenczy GG, Csonka G, Náray-Szabó G, Ángyán JG (1998) Quantum mechanical/molecular mechanical self-consistent Madelung potential method for the treatment of molecular crystals. J Comput Chem 19(1):38
135. Ferenczy GG, Párkányi L, Ángyán JG, Kálmán A, Hegedüs B (2000) Crystal and electronic structure of two polymorphic modifications of famotidine. An experimental and theoretical study. J Mol Struct (Theochem) 503:73
136. Ferenczy GG, AÁngyán JG (2001) Intra- and intermolecular interactions of polar molecules. A study by the mixed quantum mechanical/molecular mechanical SCMP-NDDO method. J Comput Chem 22:1679
137. Stewart R (1979) On the mapping of electrostatic properties from Bragg diffraction data. Chem Phys Lett 65:335
138. Spackman MA, Stewart RF (1981) In: Politzer P, Truhlar DG (eds) Chemical applications of atomic and molecular electrostatic potentials. Plenum Press, New York, pp 407–425
139. Stewart R (1982) Mapping electrostatic potentials from diffraction data. God Jugosl Cent Kristalogr 17:1
140. Coppens P (1997) X-ray charge densities and chemical bonding. Oxford University Press, New York
141. Gavezzotti A (1994) Are crystal structures predictable? Acc Chem Res 27:309
142. Gavezzotti A (2002) Modeling hydrogen bonded crystals. J Mol Struct 615:5
143. Gavezzotti AT (2002) Ten years of experience in polymorph prediction: what next? CrystEngComm 4:343
144. Gavezzotti A (2005) Calculation of lattice energies of organic crystals: the pixel integration method in comparison with more traditional methods. Z Kristallogr 220:499
145. Spackman MA, Weber HP, Craven BM (1988) Energies of molecular interactions from Bragg diffraction data. J Am Chem Soc 110:775
146. Abramov YA (1997) On the possibility of kinetic energy density evaluation from the experimental electron-density distribution. Acta Crystallogr A 53:264
147. March NH (1957) The Thomas-Fermi approximation in quantum mechanics. Adv Phys 6:1
148. von Weizsäcker C (1935) Zur Theorie der Kernmassen. Z Phys 96:431
149. Kirzhnits DA (1957) Quantum corrections to the Thomas-Fermi equation. Soviet Phys JETP-USSR 5:64
150. Espinosa E, Molins E, Lecomte C (1998) Hydrogen bond strengths revealed by topological analyses of experimentally observed electron densities. Chem Phys Lett 285:170–173
151. Espinosa E, Souhassou M, Lachekar H, Lecomte C (1999) Topological analysis of the electron density in hydrogen bonds. Acta Crystallogr B 55:563
152. Espinosa E, Alkorta I, Rozas I, Elguero J, Molins E (2001) About the evaluation of the local kinetic, potential and total energy densities in closed-shell interactions. Chem Phys Lett 336:457
153. Espinosa E, Lecomte C, Molins E (1999) Experimental electron density overlapping in hydrogen bonds: topology vs. energetics. Chem Phys Lett 300:745

154. Espinosa E, Alkorta I, Elguero J, Molins E (2002) From weak to strong interactions: a comprehensive analysis of the topological and energetic properties of the electron density distribution involving X-H ... F-Y systems. J Chem Phys 117:5529
155. Matta CF, Castillo N, Boyd RJ (2006) Extended weak bonding interactions in DNA: π-stacking (base-base), base-backbone, and backbone-backbone interactions. J Phys Chem B 110:563
156. Lyssenko KA, Borissova AO, Burlov AS, Vasilchenko IS, Garnovskii AD, Minkin VI, Antipin MY (2007) Interplay of the intramolecular N-H ... N bond and π-stacking interaction in 2-(2-tosylaminophenyl)benzimidazoles. Mendeleev Commun 14:164
157. Nelyubina YV, Antipin MY, Lyssenko KA (2009) Hydrogen bonds between zwitterions: intermediate between classical and charge-assisted ones. a case study. J Phys Chem A 113:3615
158. Lyssenko KA, Barzilovich PY, Aldoshin SM, Antipin MY, Dobrovolsky YA (2008) The role of H-bonds in charge transfer in the crystal of 1,5-naphthalenedisulfonic acid tetrahydrate. Mendeleev Commun 18:312
159. Nelyubina YV, Troyanov SI, Antipin MY, Lyssenko KA (2009) Why oxonium cation in the crystal phase is a bad acceptor of hydrogen bonds: a charge density analysis of potassium oxonium bis(hydrogensulfate). J Phys Chem A 113:5151
160. Lyssenko KA, Nelubina YV, Safronov DV, Haustova OI, Kostyanovsky RG, Lenev DA, Antipin MY (2005) Intermolecular $N_3 \cdots N_3$ interactions in the crystal of pentaerythrityl tetraazide. Mendeleev Commun 15:32
161. Korlyukov AA, Lyssenko KA, Antipin MY, Grebneva EA, Albanov AI, Trofimova OM, Zel'bst EA, Voronkov MG (2009) Si-Fluoro substituted quasisilatranes $(N \rightarrow Si)FYSi(OCH_2CH_2)_2NR$. J Organomet Chem 694:607
162. Nelyubina YV, Lyssenko KA, Kostyanovsky RG, Bakulin DA, Antipin MY (2008) $ClO \cdots ClO_3^-$ interactions in crystalline sodium chlorate. Mendeleev Commun 18:29
163. Nelyubina YV, Lyssenko KA, Golovanov DG, Antipin MY (2007) $NO_3^- \cdots NO_3^-$ and $NO_3^- \cdots \pi$ interactions in the crystal of urea nitrate. CrystEngComm 9:991
164. Nelyubina YV, Antipin MY, Lyssenko KA (2007) Are Halide$\cdots$Halide contacts a feature of rock-salts only? J Phys Chem A 111:1091
165. Borissova AO, Antipin MY, Perekalin DS, Lyssenko KA (2008) Crucial role of Ru$\cdots$H interactions in the crystal packing of ruthenocene and its derivatives. CrystEngComm 10:827
166. Lyssenko KA, Korlyukov AA, Golovanov DG, Ketkov SY, Antipin MY (2006) Estimation of the barrier to rotation of benzene in the (η6–C6H6)2Cr crystal via topological analysis of the electron density distribution function. J Phys Chem A 110:6545
167. Borissova AO, Korlyukov AA, Antipin MY, Lyssenko KA (2008) Estimation of dissociation energy in donor-acceptor complex $AuCl\text{-}PPh_3$ via topological analysis of the experimental electron density distribution function. J Phys Chem A 112:11519
168. Peganova TA, Valyaeva A, Kalsin AM, Petrovskii PV, Borissova AO, Lyssenko KA, Ustynyuk NA (2009) Synthesis of aminoiminophosphoranate complexes of palladium and platinum and X-ray diffractional investigation of the weak C-H ... Pd interactions affecting the geometry of the PdNPN metallacycles. Organometallics 10:3021
169. Puntus LN, Lyssenko KA, Antipin MY, Bünzli JCG (2008) Role of inner- and outer-sphere bonding in the sensitization of Eu-III-luminescence deciphered by combined analysis of experimental electron density distribution function and photophysical data. Inorg Chem 47:11095
170. Bushmarinov IS, Antipin MY, Akhmetova VR, Nadyrgulova GR, Lyssenko KA (2008) Stereoelectronic effects in N-C-S and N-N-C systems: Experimental and ab initio AIM study. J Phys Chem A 112:5017
171. Espinosa E, Molins E (2000) Retrieving interaction potentials from the topology of the electron density distribution: the case of hydrogen bonds. J Chem Phys 113:5686
172. Wong PTT, Whalley E (1976) Raman-spectrum of ice-8. J Chem Phys 64:2359
173. Kuhs WF, Finney JL, Vettier C, Bliss DV (1984) Structure and hydrogen ordering in ice-VI, ice-VII, and ice-VIII by neutron powder diffraction. J Chem Phys 81:3612

174. Ojamäe L, Hermansson K, Dovesi R, Roetti C, Saunders VR (1994) Mechanical and molecular-properties of ice-VIII from crystal-orbital ab-initio calculations. J Chem Phys 100:2128
175. Gatti C, Silvi B, Colonna F (1995) Dipole moment of the water molecule in the condensed phase: a periodic Hartree-Fock estimate. Chem Phys Lett 247:135
176. Morse MD, Rice SA (1982) Tests of effective pair potentials for water. Predicted ice structures. J Chem Phys 76:650
177. Yoon BJ, Morokuma K, Davidson ER (1985) Structure of ice-Ih ab initio 2-body and 3-body water-water potentials and geometry optimization. J Chem Phys 83:1223
178. Matsuoka O, Clementi E, Yoshimine M (1976) CI study of the water dimer potential surface. J Chem Phys 64:1351
179. Whalley E (1957) The difference in the intermolecular forces of H_2O and D_2O. Trans Faraday Soc 53:1578
180. Bukalov SS, Leites LA, Lyssenko KA, Aysin RR, Korlyukov AA, Zubavichus JV, Chernichenko KY, Balenkova ES, Nenajdenko VG, Antipin MY (2008) Two modifications formed by sulflower $C_{16}S_8$ molecules, their study by XRD and optical spectroscopy (Raman, IR, UV-Vis) methods. J Phys Chem A 112:10949
181. Lyssenko KA, Korlyukov AA, Antipin MY (2005) The role of intermolecular H···H and C···H interactions in the ordering of [2.2]paracyclophane at 100 K: estimation of the sublimation energy from the experimental electron density function. Mendeleev Commun 15:90
182. Ashby MF (1992) Materials selection in mechanical design. Pergamon, Oxford
183. Mata I, Alkorta I, Molins E, Espinosa E (2010) Universal features of the electron density distribution in hydrogen-bonding regions: a comprehensive study involving H···X (X = H, C, N, O, F, S, Cl, π) interactions. Chem Eur J 16:2442
184. Gatti C, May E, Destro R, Cargnoni F (2002) Fundamental properties and nature of CH···O interactions in crystals on the basis of experimental and theoretical charge densities. The case of 3,4-bis(dimethylamino)-3-cyclobutene-1,2-dione (DMACB) crystal. J Phys Chem A 106:2707
185. Gatti C (2005) Chemical bonding in crystals: new directions. Z Kristallogr 220:399
186. Kitaigorodsky AI (1973) Molecular crystals and molecules. Academic, New York
187. Mata I, Molins E, Alkorta I, Espinosa E (2009) Effect of an external electric field on the dissociation energy and the electron density properties: the case of the hydrogen bonded dimer HF··· HF. J Chem Phys 130:044104
188. Dunitz J, Gavezzotti A (2005) Molecular recognition in organic crystals: directed intermolecular bonds or nonlocalized bonding? Angew Chem Int Ed 44(12):1766

Chapter 12
Chemical Information from Charge Density Studies

Ulrike Flierler, Dietmar Stalke, and Louis J. Farrugia

12.1 Introduction

Very often, the main purpose of undertaking a charge density analysis (either experimental or theoretical) is to shed some light on the chemical bonding or chemical reactivity of the system in question. If, as is usually the case these days, the Quantum Theory of Atoms in Molecules (QTAIM) [1] method is used to analyse the density, the direct results of such an analysis are local properties such as ρ_b, $\nabla^2\rho_b$ or derived energy terms like the total energy density H_b at the bond critical point (bcp). The relationship between these numerical data and commonly accepted chemical concepts is unfortunately not always straightforward, and often the researcher relies on heuristic connections, rather than rigorously defined ones. In this chapter we demonstrate how charge density analyses can shed light on aspects of chemical *bonding* and the chemical *reactivity* resulting from the determined bonding situation.

As originally suggested by Bader and Essén in 1984, the Laplacian of the charge density at the bcp $\nabla^2\rho_b$ provides a very useful indicator for atomic interactions [2]. A negative value of the function implies a local charge concentration, a factor easily linked with the chemical concept of covalency and a shared-shell interaction. Conversely, a positive Laplacian at the bcp indicates a local charge depletion, which may be associated with a closed-shell interaction, i.e. ionic bonding or with the weaker interatomic interactions such as hydrogen bonds or van der Waals bonds that are covered in Chap. 11 of this book. This very simple bipolar distinction is

U. Flierler • D. Stalke
Institut für Anorganische Chemie, Universität Göttingen, Tammannstrasse 4, 37077 Göttingen, Germany
e-mail: uflierler@chemie.uni-goettingen.de; dstalke@chemie.uni-goettingen.de

L.J. Farrugia (✉)
School of Chemistry, University of Glasgow, G12 8QQ, Scotland, UK
e-mail: louis.farrugia@glasgow.ac.uk

C. Gatti and P. Macchi (eds.), *Modern Charge-Density Analysis*,
DOI 10.1007/978-90-481-3836-4_12, © Springer Science+Business Media B.V. 2012

most useful for the majority of compounds between elements of the second and third periods. However, it has been shown, that already for very polar bonds this simple differentiation does not hold. Properties determined solely at the bcp are susceptible to misinterpretations as the bcps lie on a rising edge of the property functions. Additionally, other local properties are worthy of consideration. For instance, Cremer and Kraka [3] proposed using the local total electronic energy density $H(\mathbf{r}) = G(\mathbf{r}) + V(\mathbf{r})$ as an indicator of the extent of covalency. A negative value for $H(\mathbf{r})$ implies a dominating potential energy density $V(\mathbf{r})$ over the kinetic energy density $G(\mathbf{r})$, indicative of a stabilising covalent interaction. Several other more elaborate sets of indicators for weak interactions and hydrogen bonds are discussed in Chap. 11.

It was soon recognised that the simple bipolar model was not adequate to characterise bonds to heavier elements, such as the transition metals (TM) [4]. The reasons for this are discussed in considerable detail in recent reviews [5–7]. Space does not permit a full exposition of the arguments here, but in a nut-shell the lack of a distinct fourth shell of charge concentration for the first row transition metals and the quite contracted radial extension of the $3d$ orbitals, means that the bcp for almost any bonds to these metals will invariably fall in a zone of charge depletion, i.e. resulting in a positive Laplacian. However, it is quite unrealistic to induce from this that all bonds to transition metals are of the closed shell type! In octahedral organometallic or coordination compounds of the mid- to late-transition metals, the TM-ligand σ-bonding interaction is normally characterised by a (usually significant) charge concentration at the ligator atom, associated with charge depletion at the transition metal – the so-called "lock and key" mechanism [5–7]. A classic example of this is $Cr(CO)_6$ [7], where eight charge concentrations are observed, which maximally avoid the ligand charge concentrations, together with six charge depletions facing the ligands, consistent with the classic ligand field splitting of the d-orbitals. This pattern of polarisation at the TM often persists when the coordination geometry becomes less symmetric, especially if strong π-acceptor ligands are present [8]. However, for early transition metal complexes, particularly for formally d^0 alkyl compounds, the TM polarisation may result in the appearance of bonding charge concentrations along the TM–C bond, e.g. in $EtTiCl_3(dmpe)$, dmpe=dimethyl-phosphinoethane [9] or in ligand induced charge concentrations, e.g. in Me_3NbCl_2 and Me_2NbCl_3 [10].

For many transition metal-ligator bonds, the total electronic energy density H_b at the bcp is negative, providing an indicator of the covalency of the bond. Moreover, the diffuse character of many TM–X bonds means that ρ_{bcp} is generally much smaller than for "classical" covalent bonds formed by the elements of the second and third periods. As originally suggested by Cremer and Kraka [11] and as demonstrated by Macchi and Sironi [5], the integrated charge density over the shared interatomic surface $N\,(\mathrm{A,B}) = \oint_{A\cap B} \rho(\mathbf{r})$ may prove a better qualitative indicator of the bond strength in these circumstances. Two QTAIM indicators that are receiving increasing attention for chemical characterisation are the delocalisation index $\delta(\Omega_A, \Omega_B)$ which provides a measure of the Fermi correlation shared between two atomic basins A, B [12, 13] and the source function $S(\mathbf{r}, \Omega_A)$ which gives a measure of the contribution from atomic basin A to the density $\rho(\mathbf{r})$ [14]. Although no physical connection exists between these two indicators, it has been proposed [15] that a heuristic relationship may exist in some cases. This would be very useful, as the former indicator requires knowledge of the one and two-electron density matrices, while the latter requires only the density and its derivatives and is hence available from experimentally as well as theoretically obtained densities.

An even more fundamental indicator of a chemical interaction between atoms is the presence or absence of a path of maximal density linking two atomic nuclei through a (3, −1) cp, i.e. the atomic interaction line. For structures at equilibrium geometry, this line is known as the bond path (BP), as it is taken as an indicator of *chemical bonding* between two atomic centres [16]. The BP is often, though quite erroneously, taken to be synonymous with the well known concept of the *chemical bond*. This heuristic linkage has been the subject of controversy for many years, and continues to be so, but any linkage between a *chemical bond* and *chemical bonding* (as shown by the BP [16]) is to be discouraged. Recent work by Pendás et al. [17] provides an alternative, energetic interpretation of the BP as a preferred channel of quantum-mechanical exchange. For a recent example of this controversy in the literature, the readers attention is drawn to the case of the H ... H bond paths [18] observed between close H atoms such as the *ortho*-hydrogen atoms in biphenyl or the "bay" H atoms in phenanthrene. These have been described as either attractive [18] and locally energetically stabilising (though overall energetically destabilising [17]) or as purely repulsive [19] interactions. An interesting recent paper by Grimme et al. [20], drawing on experimental infra-red evidence, suggests the latter view may be correct, though this conclusion has since been vigorously disputed [21].

The reader will therefore be aware that there is currently no completely rigorous mapping of charge density indicators to established chemical concepts of bonding. Any interpretation of these relationships retains a degree of personal perspective. In this Chapter we describe how charge density studies have been used to investigate specific bonding issues in two areas of chemistry, namely main group chemistry and transition metal chemistry. In the former area, we focus on the traditional valence-bond approach to chemical structures, while in the latter one, the charge density indicators are related more to molecular orbital concepts of chemical bonding. Both approaches are equally valid.

12.2 Bonding in Main Group Element Compounds

12.2.1 Valence Expansion

When it comes to discussing the electronic structure of molecules the octet rule, as proposed by Lewis in 1916 [22] is one of the most important concepts. According to this rule, all atoms in a molecule possess eight electrons – or four (bonding or non-bonding) electron pairs – in their valence shell. Exceptions to this rule, such as BF_3, PCl_5 and SF_6 are frequently called hypo- or hypervalent molecules. In the valence shell of one (central) atom in hypovalent molecules there are less than eight electrons, while in hypervalent molecules there are more than eight electrons.

The bonding situation in hypervalent molecules [23, 24] is especially a topic of constant debate. In the framework of the Valence Bond (VB) theory, the bonding in hypervalent molecules is described in terms of $d^m sp^3$ hybrid orbitals pursuant to the sp^n hybridisation in molecules obeying the octet rule. However, *ab initio* calculations indicate that the *d*-orbitals merely serve as polarisation functions and do not contribute to the bonding [25]. Nevertheless, this description is still found in various textbooks. The observation that the majority of hypervalent molecules have substituents that are more electronegative than the central atom resulted in the attempt to formulate such molecules using resonance structures which avoid hypervalence, by involving ionic bonds. However, as these structures are intentionally written to be in accordance with the octet rule, they do not prove that the octet rule is obeyed. An alternative description of the bonding in "hypervalent" molecules has been suggested by Rundle [26]. He pointed out that in SO_x systems, the formation of a π-electron system is feasible, which leads to an *m*-centre-*n*-electron p_π-p_π-bonding, thus reducing the number of valence electrons around the central atom.

As electron density studies are able to shed light on the true nature of chemical bonds, they are definitely an appropriate tool to decide which of these interpretations of the hypervalent molecule is correct. In the hexacoordinate silicon complex difluoro-bis-[*N*-(dimethylamino) phenylacetimidato-*N*,*O*] silicon (**1**), the silicon atom is coordinated by two nitrogen, oxygen and fluorine atoms, respectively [27]. Thus, this compound contains three different sets of highly polar silicon–element bonds (Si–E, E=N, O, F) in the same molecule, resulting in a hexacoordinate, and thus formally hypervalent, silicon atom. Based on the nature of the silicon–element bonds, the hypervalent description of the whole molecule may be verified or falsified.

As early as 1939, on the basis of the difference in the electronegativity χ, Pauling [28] came to the conclusion that a Si–O bond has about 50% covalent character rather than being purely ionic. Nevertheless, while the nature of the Si–O bond still remains controversial [29], most studies mainly focused on silicates. Additionally, besides a theoretical study on the nature of the Si–N bond [30] and an experimental charge density study in K_2SiF_6 without a quantitative topological analysis [31] the nature of the Si–N and Si–F bonds had hardly been investigated.

An examination of the two-dimensional Laplacian distribution (Fig. 12.1) in the O_2SiF_2- and the SiN_2-plane of **1** reveals the nature of the bonding in the Si–E bonds. While the spatial distribution of the Laplacian is almost perfectly spherical around the silicon atom, it still shows a rather ionic-like behaviour around the oxygen and fluorine atoms. This indicates a distinct electronic depletion around the silicon atom and predominantly ionic Si–E bonding with differing – but always small – covalent bonding contributions. The deformations of the Laplacian around the substituents can be directly related to their electronegativity (χ) and thus to the degree of their respective bond polarisation – the lower χ of the bonding partner, the more the

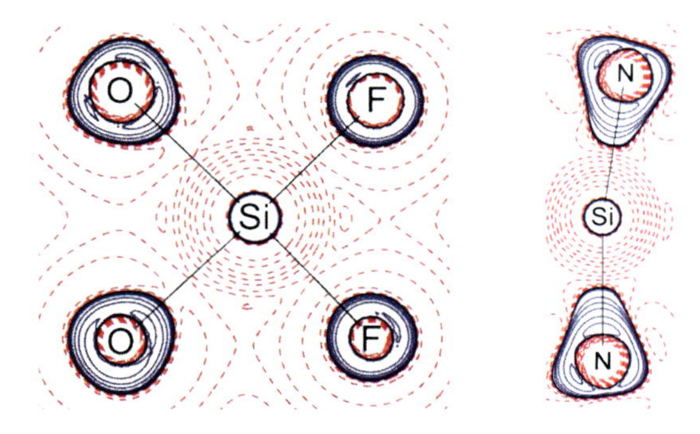

Fig. 12.1 Contour plot of the Laplacian distribution in the O_2SiF_2 (*left*) and the SiN_2 plane (*right*) of **1**. Positive values of $\nabla^2\rho(\mathbf{r})$ are depicted by *red*, negative by *blue lines* (Reprinted from Ref. [27] with kind permission of The American Chemical Society)

deformation is pronounced. These deformations are only marginal with the oxygen atoms, while with fluorine atoms they are virtually undetectable. The significant polarisation of the density at the nitrogen atoms towards the depleted silicon atom results from the lone pairs in the apical positions at the three-coordinate amine functions.

The dominating ionic character of the bonds is underlined by the integrated charges. A distinct positive charge of +2.78 e is found at the silicon atom and negative charges are found for the fluorine (–0.80 e), the oxygen (–1.21 e) and the nitrogen (–0.78 e) atoms. Thus, the experiment reveals predominantly ionic bonding and just a small covalent augmentation for all Si–E bonds. Therefore, the hypervalent description of the silicon atom is no longer required – due to the charge transfer from the silicon atom to its more electronegative bonding partners, the valence shell of the silicon atom has eight or less electrons and a Lewis formula that does not violate the octet rule, can and must be formulated. Notably however, in some transition metal complexes, a similar difference in charges along a bond implies a more distinct covalency than assumed here.

Sulfur-diimide $S(NR)_2$ and sulfur-triimide $S(NR)_3$ – imide analogues of SO_2 and SO_3 – are planar systems, which have been assumed to be examples of valence expansion at the sulfur atom. The structural characterisation of $S(NtBu)_2$ (**2**) [32] and $S(NtBu)_3$ (**3**) [33] revealed very short distances for the sulfur–nitrogen bonds of approximately 1.5 Å, which led to the formulation of S=N double bonds in those compounds [34]. This description avoids formal charges (Pauling's verdict) but implies valence expansion and *d*-orbital participation at the central sulfur atom. However, this formulation is in contrast to theoretical investigations from the mid 1980s, which verified that *d*-orbitals cannot participate in the sulfur–nitrogen bonding due to large energy differences between the sulfur *p*- and *d*-orbitals [25, 35–40]. Furthermore, these MO-calculations on second-row atoms in "hypervalent" molecules showed that the *d*-orbitals are mainly needed as polarisation functions rather than as bonding orbitals [41, 42]. Theoretical studies of SO_2 and SO_3 show that the S–O bonds have highly ionic character and bond orders close to one.

Additionally, several experimental observations in recent years [43–47] do not suit the idea of a classical S=N double bond, e.g. the reactivity of many polyimido

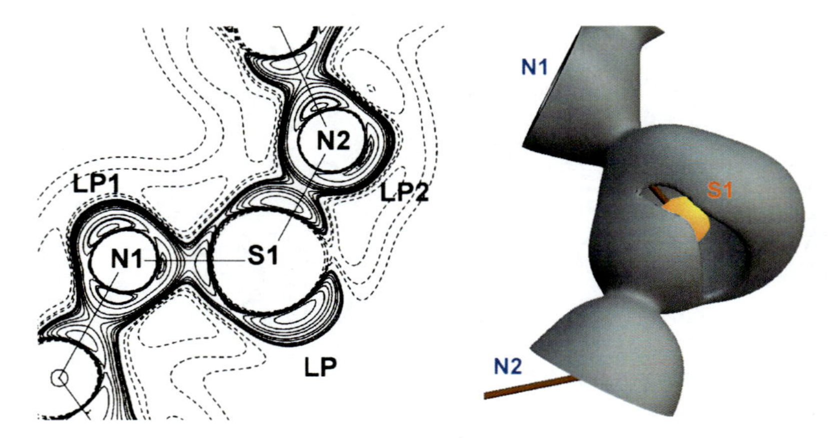

Fig. 12.2 Contour representation of $\nabla^2\rho(\mathbf{r})$ (*left*) and reactive surface ($\nabla^2\rho(\mathbf{r}) = 0$ eÅ^{-5}) around the sulfur atom (*right*) in the sulfur-diimide $S(NtBu)_2$ (**2**) (Reprinted from Ref. [54] with kind permission of The American Chemical Society)

sulfur species in polar media. They easily perform transimidation reactions [48] and generate diimides [49]. Since such reactions require facile S–N bond cleavage in polar media, the reactivities are a sign of quite polar bonding rather than of p_π–d_π double bonding. Furthermore, the reassignement of the SN stretching vibrations in the Raman spectroscopic experiment to much lower wavenumbers (640 and 920 cm^{-1} [49] instead of initially assumed 1200 cm^{-1} [50–53]) indicates a weaker bond and probably another bonding type rather than S=N. Indeed, the S–N bonds in both compounds were found to be polar in the topological analysis of the experimentally derived electron density distributions (EDD) [54].

Valence shell charge concentrations (VSCCs) can be utilised to determine the density-related bonding geometry of an atom. They depict the hybridisation better than do the traditional interatomic vectors. Experimental as well as theoretical Laplacian distributions in both planar $S(NtBu)_2$ (**2**) and $S(NtBu)_3$ (**3**) molecules reveal one single in-plane lone-pair VSCC at both nitrogen atoms. At the sulfur atom primarily sp^2-hybridisation is indicated (Figs. 12.2 and 12.3).

The sp^2-hybridisation is indicative of a π-system above and below the SN_x plane. Such a π-system is indeed also reflected in the corresponding π-orbitals and the leading resonance structures given by an NBO/NRT (natural bond orbital/natural resonance theory [55]) approach. This bonding type corresponds to 4-centre-6-electron bonding. As a consequence of the π-system, the redistribution of charge should be quite efficient. Indeed, the NBO/NRT analyses reveal increased covalent contributions to the SN bond orders, accompanied by decreased charges at the nitrogen atoms in **2** and **3**, compared to molecules in which the sulfur atom is sp^3-hybridised. However, from the shape of the orbitals and from the NBO/NRT resonance structures, it is obvious that the π-orbitals are polarised. Thus, the ionic contributions to the total bond orders are significant in the short SN bonds of **2** and **3**. Again, valence expansion at the sulfur atom can definitely be excluded.

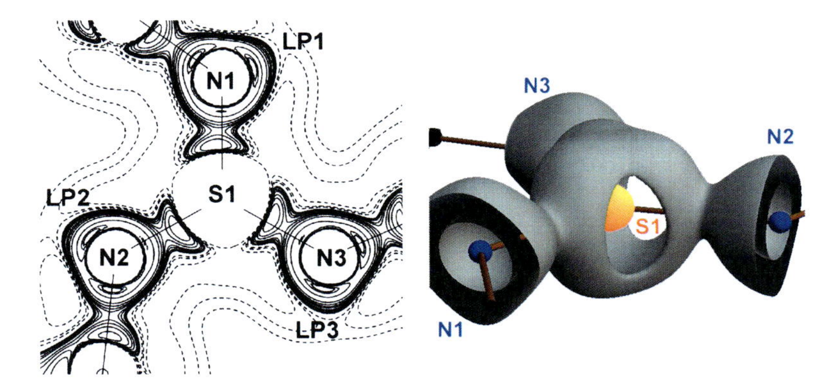

Fig. 12.3 Contour representation of $\nabla^2\rho(\mathbf{r})$ (*left*) and reactive surface ($\nabla^2\rho(\mathbf{r}) = 0$ eÅ^{-5}) around the sulfur atom (*right*) in the sulfur-triimide S(N*t*Bu)$_3$ (**3**) (Reprinted from Ref. [54] with kind permission of The American Chemical Society)

In addition to the bonding type, the investigations were also able to elucidate the experimentally observed reactivity by inspection of the reactive surfaces ($\nabla^2\rho(\mathbf{r}) = 0$ eÅ^{-5}; Figs. 12.2 and 12.3). S(N*t*Bu)$_3$ (**3**), for example, reacts smoothly with MeLi and PhCCLi but not with *n*BuLi or *t*BuLi. The topological analysis shows that this discrimination of large reactants can be related to small areas of strong charge depletion in the SN_3 plane at the bisections of the N–S–N angles. The carbanionic nucleophile has to approach the sulfur atom along the NSN bisection in the SN_3 plane or in an angle of less than about 45°, which is only feasible for small or planar carbanions. Bulky anions cannot reach the holes, due to the steric hindrance of the N*t*Bu groups. The steric argument would not be valid if a direct orthogonal attack above or underneath the SN_3 plane was favoured, as there is sufficient space in the planar molecule to reach the sulfur atom directly.

Electron density studies can thus not only be applied to understand bonding situations but from this knowledge also allow the deduction of the chemical reactivity. What is more, they can even be applied to predict reactivities and thus provide new synthetic reaction pathways.

12.2.2 *Reactivity of Double Bonds*

The stability and electric conductivity of polyphosphazenes and iminophosphoranes is, according to numerous textbooks, to be ascribed to the P=N double bond. The P=N bonds in iminophosphoranes are mostly described as a resonance hybrid between a double-bonded ylene $R_3P{=}NR$ and a dipolar ylidic form $R_3P^{\delta+}{-}N^{\delta-}R$ [56–58]. Therefore, iminophosphoranes are said to represent a thermodynamic sink, which is why attempts to reduce P^{V} to P^{III} were not undertaken synthetically.

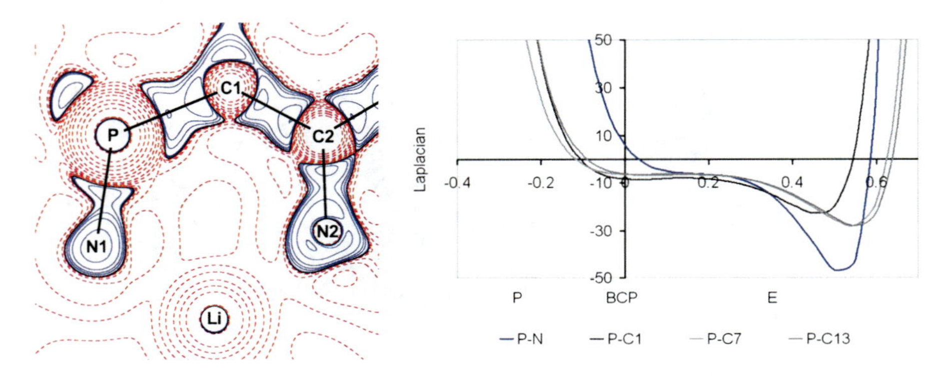

Fig. 12.4 Contour representation of $\nabla^2\rho(\mathbf{r})$ in the LiN_2C_2P-plane (*left*) of $[(Et_2O)Li\{Ph_2P(CHPy)(NSiMe_3)\}]$ (**4**) and the $\nabla^2\rho(\mathbf{r})$ distribution along the bond paths (*right*)

Only *via* an experimental electron density determination of $[(Et_2O)Li\{Ph_2P(CHPy)(NSiMe_3)\}]$ (**4**) has the picture of the bonding situation been utterly changed [59]. The assumed short P=N double bond should actually be described as a polar $P^{\delta+}$–$N^{\delta-}$ single bond, which is shortened only by ionic bond reinforcement. Additionally, it has been proposed, that negative hyperconjugation may be responsible for any π-character in the P–O, P–C, or P–N bonds [25, 36, 38, 60, 61]. These results have been substantiated by recent calculations dealing with the Wittig-type reactivity of phosphorous ylides [62–65] and iminophosphoranes [66–68] and by the experimental determination of the charge density in a phosphane ylide [69].

The experimental electron density determination of $[(Et_2O)Li\{Ph_2P(CHPy)(NSiMe_3)\}]$ (**4**) yields four VSCCs at the nitrogen atom, which indicates an sp^3-hybridisation (Fig. 12.4, left). For the formation of a P=N double bond an sp^2-hybridisation at both the nitrogen and the phosphorus atom would have been required.

The ionic bonding contributions are further supported by an inspection of the Laplacian along the P–C as well as the P–N bond paths (Fig. 12.4, right) in **4**. Along the P–C bond path, an almost equal charge distribution in the phosphorus basin is observed for the three P–C bonds considered, while the main difference concerning the Laplacian in the basin of the carbon atoms is related to the shorter distance between C1 and $bcp_{P\text{–}C1}$. This rather short P–C bond results from distinct electrostatic interactions between the negatively polarised deprotonated carbon and the electropositive phosphorus atom, as reflected in the charges of −0.52 e for C1 and +2.20 e for P from integration over the atomic basins.

The Laplacian distribution $\nabla^2\rho(\mathbf{r})$ along the P–N1 bond is completely different to that along the P–C_{ipso} bonds. The Laplacian at the bcp is positive and charge density is exclusively concentrated in the nitrogen basin. This indicates a severe contribution of electrostatic interaction to the bonding energy, further substantiated

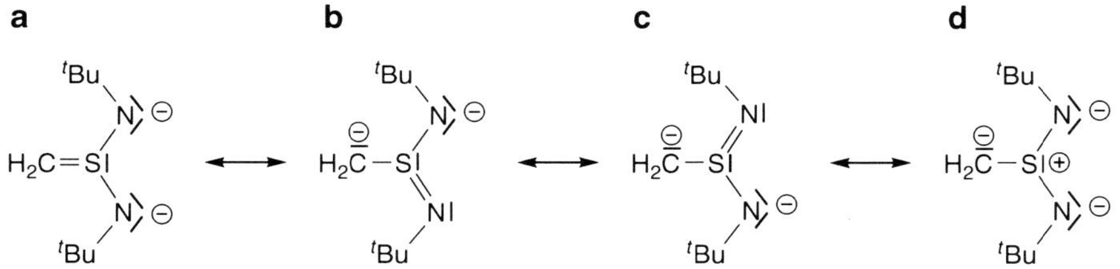

Fig. 12.5 Possible resonance formulas of the diimido sulfur ylide dianion, only the last not exceeding the octet at the central sulfur atom

by the integration of the atomic basins. The two basins related to the nitrogen atoms give distinctly negative values. The imino nitrogen atom N1 is bonded to three electropositive neighbours, and therefore the charge of −1.91 e is higher than that of −1.11 e for the ring nitrogen atom N2. The charge of the deprotonated carbon atom C1 is about 0.2 e higher than those of the two phosphorus bonded ipso carbon atoms (C7, −0.30 e; C13, −0.34 e) of the phenyl substituents. All these findings support an ylidic $P^{\delta+}$–$C^{\delta-}$ simultaneous with a $P^{\delta+}$–$N^{\delta-}$ bond.

Thus, the experimental charge-density distribution in $[(Et_2O)Li\{Ph_2P(CHPy)(NSiMe_3)\}]$ (**4**) clearly proves that the formal P=N imino double bond and the potential ylenic P=C double bond must be written as polar $P^{\delta+}$–$N^{\delta-}$ and $P^{\delta+}$–$C^{\delta-}$ single bonds, augmented by electrostatic contributions. With this knowledge, P^{V}–N molecules can be reduced with polar metallorganic bases in polar solvents. In this way, access to chiral phosphaneamines was opened up, which can hardly be synthesised by any other methods. Thus, an incorrect concept of the P=N double bond had blocked the synthetic access to these compounds for ages.

Sulfur ylides ($R_2S^{\delta+}$–${}^{\delta-}CR_2$) are widely used in organic synthesis for stereoselective epoxidations, cyclopropane formations and ring expansion reactions. [70–78] Nevertheless, their electronic properties are still under debate, because their ylenic text book formulation ($R_2S = CR_2$) contradicts their reactivity [79, 80].

The S–C as well as the S–N bond cleavages in molecules like $[(thf)Li_2\{H_2CS(N\textit{t}Bu)_2\}]_2$ (**5**) clearly contradict the classical Lewis notation of S=C or S=N double bonds (hypervalent ylenic form, Fig. 12.5a). Thus, an ylidic resonance form seems much more feasible (Fig. 12.5b–d) [81]. This fuels the dispute as to which extent sulfur ylides are dominated by ylidic or ylenic bonding. By analysing the topological properties of the experimental electron density distribution, it was possible to specify the S–N and S–C bonds in **5** as classical single bonds strengthened by electrostatic interactions ($S^{\delta+}$–$N^{\delta-}$ and $S^{\delta+}$–$C^{\delta-}$) [82]. This, together with the observation that four VSCCs are present at the sulfur atom, clearly rules out hypervalence, which is in good agreement with the above mentioned findings for other formally hypervalent SN-compounds (Fig. 12.6). Hence, the present compound should be formulated as ylidic rather than ylenic (Fig. 12.5d).

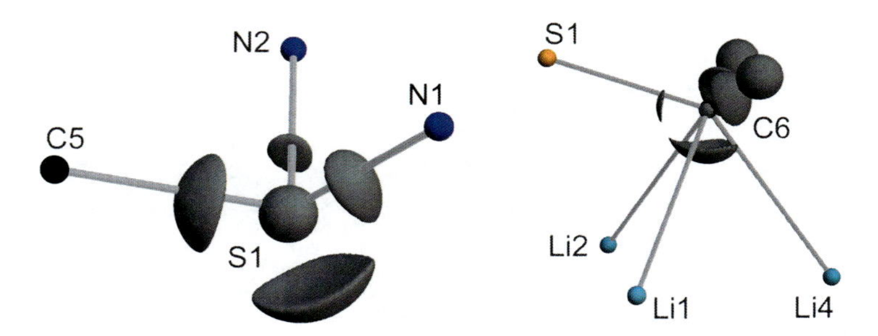

Fig. 12.6 Spatial distribution of the VSCCs around the sulfur atom (*left*) in $[(thf)Li_2\{H_2CS(NtBu)_2\}]_2$ (**5**) and at the carbanion, supporting the $H_2C^{\delta-}\cdots Li_3^{\delta+}$ 4c2e bond (Reprinted from Ref. [86] with kind permission of The American Chemical Society)

12.2.3 Multiple Centre Bonding

Compound **5** is especially interesting, as in addition to the formal hypervalency discussed above, the controversial interaction of a carbanion with a Li_3-triangle can be investigated. The coordination of a carbanion to a Li_3-triangle [83], a structural motif well-known throughout organolithium chemistry, is also present in **5.** Up to now the bonding mode and the forces that keep the highly charged Li^+ cations together are not fully understood. It is still an open question as to what extent the Li–C contacts can be regarded as mainly ionic or with appreciable covalent contributions [84, 85]. The interaction of the carbanion with the Li_3 triangle was determined to be a 4c-2e bond. The carbon atom forms a bond to each of the three lithium atoms, although one interaction is preferred. Thus, the VSCC at the carbanion that is representing the lone-pair is inclined away from the centre of the Li_3 triangle towards the lithium atom at the tip of the isosceles triangle (Fig. 12.6). This is also reflected by the characteristics of the bcps. No bcps and thus no direct bonding interactions between the lithium atoms were determined.

The knowledge gained should enable chemists to tailor target organolithium compounds. Furthermore, a better understanding of the reaction behaviour, especially of stereochemically active lithium-organic, can be envisaged [86].

This was tested in the complex of picoline ($(NC_5H_4)Me$, PicH) and the picolyl anion ($[(NC_5H_4)CH_2]^-$, Pic^-) coordinated to a lithium cation, forming the dimer $[LiPicPicH]_2$ (**6**) [87]. The coordination of the anion is of particular interest, as there is a diversity of feasible electronic configurations of Pic^-, which is reflected by the variety of different resonance formulas depicted in Fig. 12.7. In light of the solid-state structures, **A** and **D** appear to contribute most to the appropriate description of bonding. **A** emphasises the carbanionic form, while **D** interprets the anion as an enamide.

The decision as to which formulation best represents the molecule can be based on different levels of analysis. From bond-length considerations of the anion, the enamide resonance formula **D** seems most suitable to describe the electronic

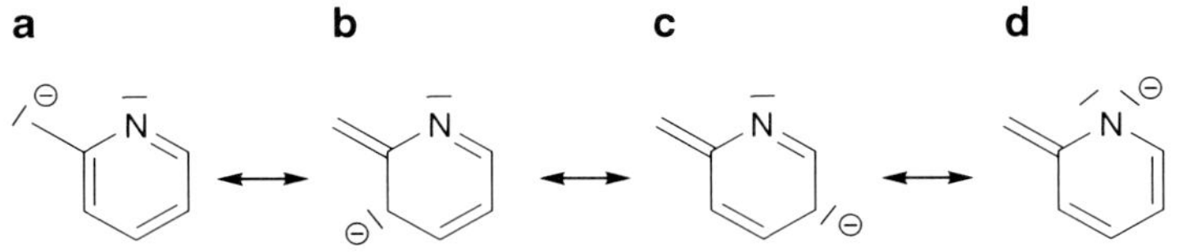

a b c d

Fig. 12.7 Resonance structures of the anion in $[LiPicPicH]_2$ (**6**)

situation in **6**. The topological analysis yields bond paths between the lithium and the nitrogen atoms, but none to the aza-allylic carbon atoms, thus supporting the idea of dominant Li–N bonds in the dimer and an auxiliary interaction with the formal anionic carbon atom C6.

An inspection of the atomic charges should help in choosing the correct formulation. The lithium cation exhibits a charge of +0.93 e, which is counterbalanced by the anion (sum of atomic contributions: −0.80 e) and the donor molecule (−0.13 e group charge). Thus, the main portion of the negative charge is located at the anion. Within the donor base, negative charge is exclusively concentrated at the ring nitrogen atom (−0.94 e). This situation results in a polarisation of the neighbouring carbon atoms (C7, +0.37; C11, +0.12 e) so that the negative charge is relativised. Interestingly, despite the deprotonation, the negative charges at N1 (−1.04 e) and the methylene group (−0.19 e) show just a marginal increase compared to the donor (N2, −0.94 e; methyl group, +0.02 e). Thus, the negative charge is distributed over the whole ring system with the largest increase at the methylene group ($\Delta = -0.21$ e). Hence, we suggest from the investigation of the group charges that the carbanionic resonance form **A** in Fig. 12.7 is suitable, even though the absolute values of the negative charges are relatively low.

A very helpful descriptor for distinguishing between different resonance forms has been provided from theoretical calculations. The electron delocalisation indices measure the number of electron pairs shared by two atomic basins. Macchi could show, that in picoline C1 and C6 share almost exactly one electron pair, while in Pic^- the delocalisation is about 1.5 larger than between two adjacent carbon atoms in the pyridine ring (1.3) [88]. However, this bond is not completely double. In fact, δ(C1, C6) is significantly smaller than for non-conjugated C–C double bonds [13]. This gives an idea of the interference caused by the resonance structure. The larger δ(C1, C6) causes a smaller δ(C1, N1) in Pic^-, through the conjugation with the pyridyl ring. It is also interesting, that upon complexation the delocalisation of electron pairs in the two ligands decrease, meaning that polarisation and partial charge transfer to Li cations affect the skeleton bonding.

Remarkably, electrophilic attack on 2-picolyllithium generally occurs at the methylene group. Can this be rationalised from the electron density distribution? The three-dimensional distribution of the electrostatic potential (ESP) displayed in Fig. 12.8 provides an answer. Opposite to the Li′–N bond, we find a vast region of negative ESP above the picolyl anion plane. The spatial distribution suggests that potential electrophiles are guided by the negative potential towards the nucleophilic

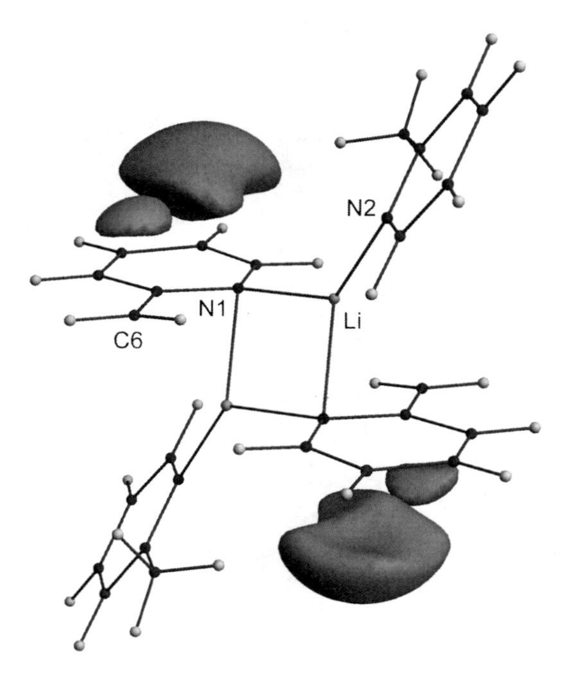

Fig. 12.8 Isosurface representation the ESP (-0.1 eÅ^{-1}) in **6** (Reprinted from Ref. [87] with kind permission of John Wiley and Sons)

deprotonated carbon atom C6. Most probably, it is the π-system of the anion as a whole that leads to the observed ESP, not a local charge concentration at the formal carbanion alone.

12.2.4 *Reactivity of Boranes*

As is known, electrophilic boranes are viable co-catalysts, which activate transition-metal alkyl or hydride complexes by abstracting alkyl anions and thereby generating transition-metal complex cations [89–91]. These play an important role in the catalysis of mainly polymerisation reactions. $[(C_6F_5)_2BNC_4H_4]$ (**7**) has proven to be a valuable alternative to the commonly employed Lewis acid $[(C_6F_5)_3B]$, while its saturated analogue, the $[(C_6F_5)_2BNC_4H_8]$ (**8**) does not show this advantage (Fig. 12.9) [92].

These differing reactivities could easily be deduced from the different electron density distributions [93]. The two-dimensional distribution of $\nabla^2\rho(\mathbf{r})$ in the principal mean plane of the heterocycle substituent including the boron atom reveals the expected features of shared and – at least for B–N – polarised bonds in both molecules. Charge concentrations in the bond from both atoms form a saddle shaped distribution with tailing perpendicular to the bonding vector (Fig. 12.10).

The polarisation is also reflected in the charges at the atoms involved. Insight into the electron shift within the whole molecules is provided by the group

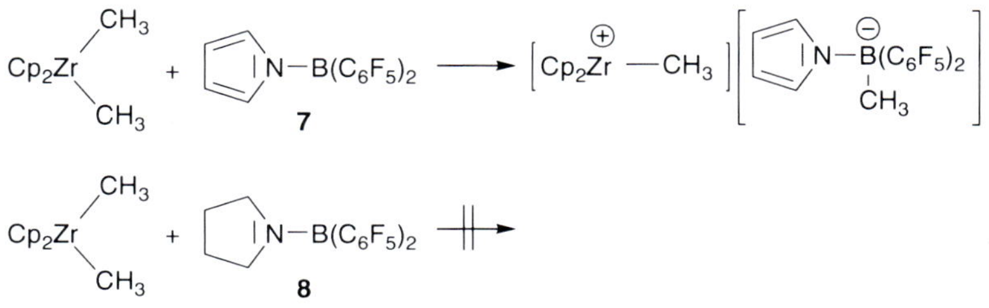

$Cp_2Zr(CH_3)_2$ + pyrrolyl–N–$B(C_6F_5)_2$ (**7**) ⟶ $[Cp_2Zr^{+}–CH_3]$ [pyrrolyl–N–$B^{-}(C_6F_5)_2(CH_3)$]

$Cp_2Zr(CH_3)_2$ + pyrrolidinyl–N–$B(C_6F_5)_2$ (**8**) ⟶|| (no reaction)

Fig. 12.9 Reaction scheme showing that **7** abstracts methanide from organometallics, while **8** does not

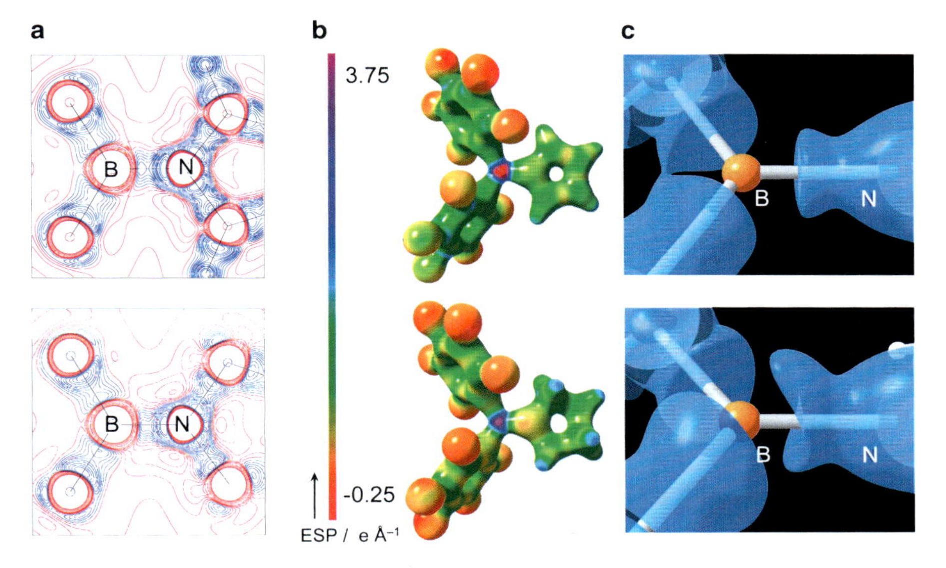

Fig. 12.10 Contour representation of $\nabla^2\rho(\mathbf{r})$ (**a**), isosurface representation of the electrostatic potential ranging from -0.25 to $+3.75$ eÅ^{-1} mapped on $\rho(\mathbf{r}) = 0.65$ eÅ^{-3} (**b**) and reactive surface ($\nabla^2\rho(\mathbf{r}) = 0$ eÅ^{-5}) (**c**) in the B–N–C_{ortho} planes of **7** (*top row*) and **8** (*bottom row*) (Reprinted from Ref. [93] with kind permission of John Wiley and Sons)

charges (summed over all atomic contributions of the defined group). The electron-withdrawing effect of the C_6F_5-groups leads to distinctly negative charges at the nitrogen atoms, which are counterbalanced by the boron atom *and* the heterocycle in **8**, but by the boron atom *alone* in **7**. The inability of the pyrrolyl ring to compensate for the electron depletion at the boron atom leaves the boron atom in **7** more positively charged than in **8**.

The consequence of the positive charge at the boron atom, also mirrored in the ESP, is echoed in the shape of the reactive surface (Fig. 12.10). In **7** the boron atom is only shielded towards the bonding partners, while above and below the B–N–C_{ortho} plane the room is wide open for a potential nucleophilic attack. The situation is different in **8**, where, in addition to the charge concentrations towards the bonding partners, claws are formed from the nitrogen atom, providing additional shielding.

In addition to the reactive surface, the spatial distribution of $\nabla^2\rho(\mathbf{r})$ and its local minima, the VSCCs also provide information about the electronic situation in the valence shells of the bonded atoms. Around N and B three VSCCs are found in each molecule. While the values at the boron atoms do not differ significantly (−10.9 to −12.3 eÅ^{-5}), the nitrogen VSCCs mirror the particular atomic contributions to the respective bonds. While in **7** the boron-directed VSCC is −67.7 and the two in-ring VSCCs are −73.1 eÅ^{-5}, the corresponding values in **8** are −70.2 (towards B), −58.3 (C5), and −65.2 eÅ^{-5} (C2). This indicates that the charge concentrations around the nitrogen atom in **7** are higher in the aromatic ring and, compared to that, slightly reduced towards the boron atom. In **8** we find the reverse situation – the most distinct concentration is towards the boron atom and the less pronounced VSCCs are directed into the heterocycle.

A conceivable interpretation of the bonding features found is the assumption of an electron density donation from the nitrogen atom of each heterocycle into the vacant *p*-orbital of the boron atom. This seems consistent in the case of **8**. Due to the (at first sight rather surprising) planarity of the system, the two electrons at the nitrogen atom reside in a non-hybridised *p*-orbital and can easily be donated from the nitrogen atom into the B–N bond by interaction with the vacant *p*-orbitals of the boron atom. In **7** this coupling has to be accomplished by π-density from the aromatic ring system. This picture is actually supported by the absolute values of the bond ellipticity ε. Even though a π-contribution is observed in both compounds, it is much more pronounced in **8** than in **7**.

12.3 Bonding in Transition Metal Compounds

As alluded to above, QTAIM investigations into compounds containing elements as heavy as the first row transition metals must take into account the differences between these elements and those of the second and third periods. A major difference is that formal outer valence shell may not be detectable in the Laplacian function $\nabla^2\rho(\mathbf{r})$, which has important implications in the interpretation of chemical bonding. These issues have been carefully discussed by Macchi and Sironi in a recent review [5] and monograph [94], and the interested reader is directed to these articles, which discuss several important aspects of QTAIM and chemical bonding involving transition metals, such as metal-carbonyl bonding. In this section, we will focus on more recent developments and elaborations in the interpretation of chemical bonding.

One issue that is increasingly evident is the realisation that, unfortunately for many of the more interesting types of bonds involving transition metals, the charge densities in the region of the bcps are very flat, and topologies containing ring structures in particular are susceptible to catastrophe situations. The value of the eigenvalue λ_2 of the Hessian for a pair of bcps and rcps vanishes as the two cps become physically close in space, resulting in a coalescence of both points into a (2,0) degenerate cp. This is the bifurcation catastrophe discussed by Bader [1], which leads to the opening of the ring structure. In this situation, the bond ellipticity ε ($\varepsilon = |\lambda_1/\lambda_2| - 1$) is very high, due to the small value of λ_2 and it does not provide any meaningful information on π-bonding. In the region of a catastrophe point, any small perturbation in the density model (e.g. a change of basis set or Hamiltonian for a theoretical study, or a change in a multipole modelling or experimental structure factors for an experimental study) may cause a change in topological structure. At present, there is no easy way of predicting on which side of the catastrophe point a particular structure will reside. A careful appraisal of the chemical bonding in terms of the charge density is required in such situations. It should also be pointed out that a topological catastrophe situation is not necessarily associated with chemical instability. In certain cases, it indeed might be true that the molecule has weak chemical bonding leading to instability. On the other hand in delocalised TM-π-carbocyclic systems, as discussed below, this is certainly not the case, though the catastrophe situation does lead to an ambiguity in the description of chemical bonding in terms of localised TM-C interactions.

12.3.1 Simple Alkene and Alkyne Compounds

An important area in which QTAIM techniques have been applied is in the understanding of transition metal–alkene and alkyne bonding. In an initial experimental study by Macchi et al. [95] on $Ni(1,5\text{-}C_8H_{12})_2$ (**9**) the Ni-olefin bonding was discussed in terms of the well known Dewar-Chatt-Duncanson (DCD) model. Bond paths from the Ni atom to all four pairs of olefinic carbon atoms were observed, and the inwardly curving, but well separated paths were taken to indicate agreement with the DCD model of σ-donation and π-back-bonding. The distinction between the DCD and metallocycle formalisms depends on subtle differences in the bond path curvatures [95]. Interestingly, when less elaborate experimental multipole models were used for **9**, the so-called "T-shaped" topology for the Ni-olefin interaction was obtained. This topology involves a single bond path from the metal atom to the bcp of the C–C bond and represents the conflict structure [1], in which the metal atom may be considered as equally bonded to both carbon atoms. This unstable conflict structure is maintained over a wider range of nuclear configurations than the bifurcation catastrophe, since a shift of the metal atom position along the bond path retains the conflict topology.

These typical bond paths are shown in Fig. 12.11 for theoretical charge distributions in the acetylene complexes of Pd^0, Ag^+ and Rb^+ [96]. The conflict "T-shaped" topology has in the past been specifically associated with closed shell "ionic" metal-alkene and -alkyne interactions [96, 97], but this direct association no longer seems tenable and has been recognised as a consequence of the catastrophe situation [96]. The bonding in the Pd^0 case may be assumed to be covalent and that for Rb^+ to be electrostatic, but in examining the case for covalency in the Ag^+ complex, Krossing,

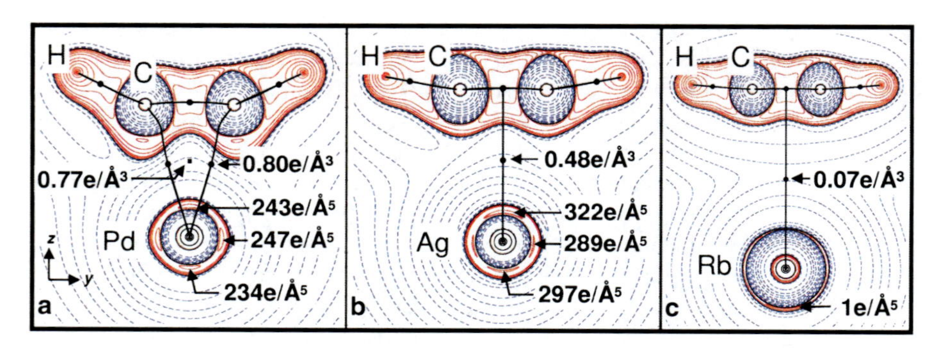

Fig. 12.11 Contour maps of the negative Laplacian $-\nabla^2\rho(\mathrm{r})$ for (**a**) $Pd(C_2H_2)$ (**b**) $[Ag(C_2H_2)]^+$ and (**c**) $[Rb(C_2H_2)]^+$, showing bond paths and critical points in $\rho(\mathrm{r})$ (Reproduced from Ref. [96] with kind permission of Wiley-VCH)

Scherer and coworkers argue that the valence shell charge polarisations (VSCP) on the metal also need to be considered as important indicators of the chemical bonding [96]. A combined experimental and theoretical study on the alkene complex $Ni(C_2H_4)(dbpe)$ (**10**) by Scherer et al. [98] has demonstrated the utility of this approach.

In the combined "AIM/MO" analysis shown in Fig. 12.12, the two most significant orbitals for σ-donation and π-back-bonding are HOMO-5 and HOMO-4, respectively. The effect of HOMO-4 in inducing the observed polarisation of the metal density is shown by subtracting its contribution to the total density. This operation clearly demonstrates the relationship between the charge transfer from the d_{xy} orbital of the nickel atom to the olefinic π^* orbitals, and the resultant polarisation of the metal atom. Using this approach, and/or other, non-discontinuous, topological indicators, such as the delocalisation index or the source function, as tools for the charge density analysis largely avoids the problems that can arise if solely focussing on the bond paths. It should be stressed that the accurate experimental determination of the atomic VSCP requires data of the highest quality, as this property is highly sensitive to experimental error [99].

The ubiquitous nature of the catastrophe problem in metal alkene and alkyne π-complexes is nicely illustrated by two recent studies. In $Co_2(CO)_6\{\mu\text{-}\eta^2\text{-}CH{\equiv}C(C_6H_{10}OH)\}$, experimental errors and inadequacies of the multipole model are assumed to be the reason for the differing topologies of the experimental and theoretical densities [100]. In the Rh^I norbornadiene complex $Rh(C_7H_8)\text{-}(PPh_3)Cl$ (**11**), only three of the four expected $Rh\text{–}C_{alkene}$ bond paths were detected [101]. Figure 12.13 shows there is considerably less polarisation of the alkene carbon atoms towards the metal than is observed in **10** [98]. The two Rh-alkene interactions are not equivalent, due to the *trans* influence of the phosphine ligand, with the Rh-C1/C2 distances $\sim$0.1Å longer than the corresponding C4/C5 distances.

This is evident in the topology, since only a single Rh–C bond path is observed for the former interaction. Interestingly, the path is highly curved, and is close to

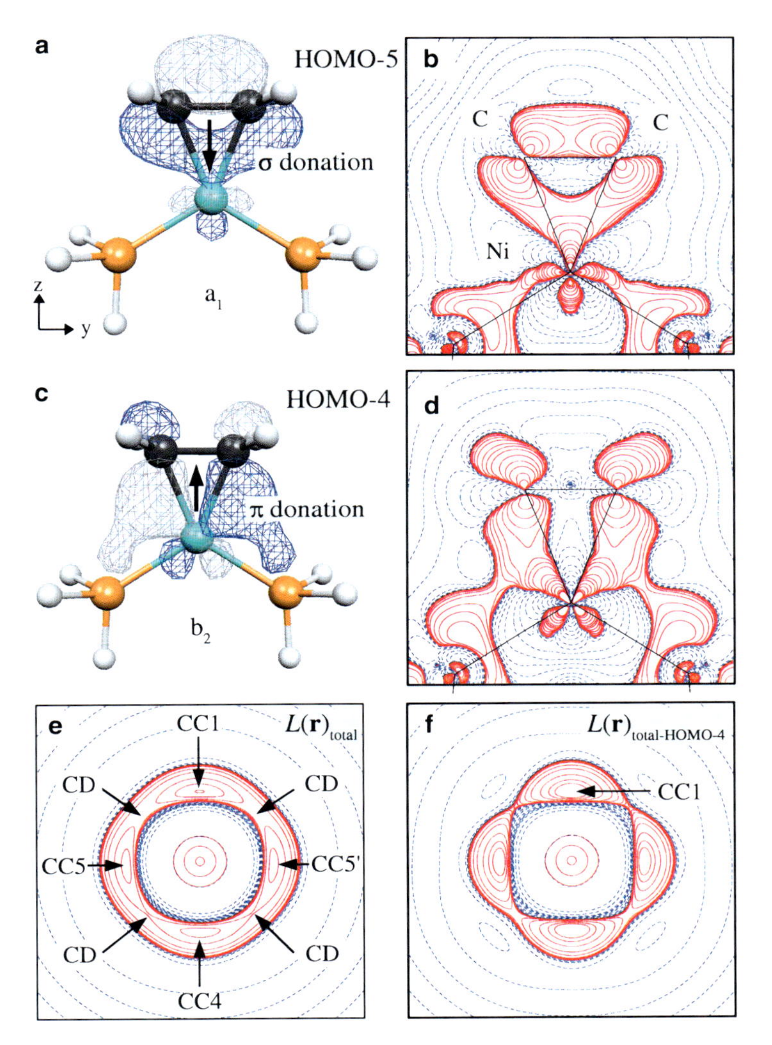

Fig. 12.12 (**a**, **c**) Kohn-Sham orbitals of HOMO-5 and HOMO-4 for **10**. (**b**, **d**) contour maps of $-\nabla^2\rho(\mathbf{r})$ for (**a**, **c**) respectively. (**e**, **f**) contour maps of $-\nabla^2\rho(\mathbf{r})$ at the Ni atom, for the total density and the density after subtracting contribution of HOMO-4, respectively (Reproduced from Ref. [98] with kind permission of The Royal Society of Chemistry)

becoming a "T-shaped" interaction. The distinction between the two Rh–alkene interactions is further evident in the ellipticity (ε) profiles shown in Fig. 12.14. These profiles provide a sensitive measure of the change in C–C π-bonding caused by interaction with the metal atom. The profile for C1–C2 is typical of unperturbed alkene or benzene C–C bonds [102], with a φ_{ref} angle close to zero along the central region, indicative of density accumulation perpendicular to the π-plane. The abrupt

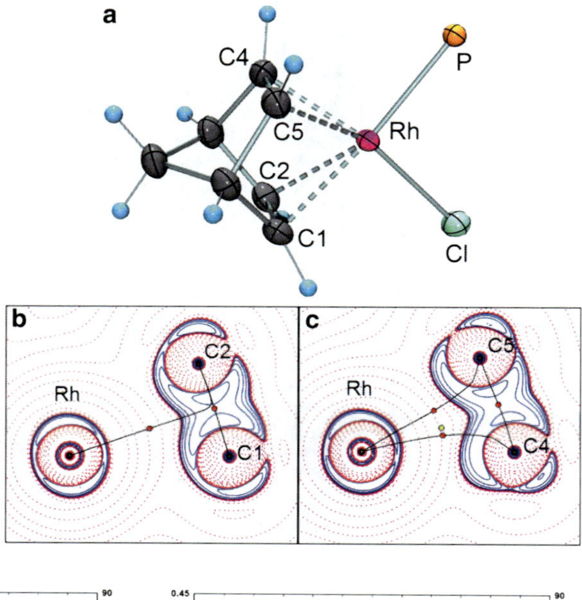

Fig. 12.13 (**a**) Molecular structure of **11**, with phenyl groups on the PPh_3 ligand omitted for clarity. (**b**, **c**) contour maps of the negative Laplacian $-\nabla^2\rho(\mathbf{r})$ with superimposed bond paths and cps in the planes of the Rh-alkene interactions

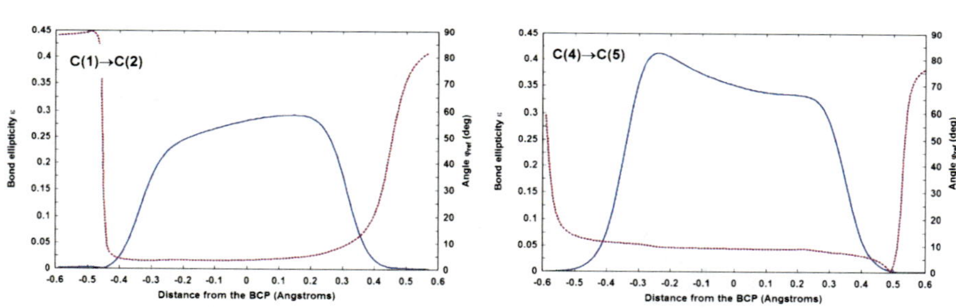

Fig. 12.14 Profile plots of the ellipticity (ε) (*solid blue line*) along the alkene C–C bonds in **11**. The φ_{ref} angle (*broken purple line*) is the angle between the normal to the π-plane (mean plane of olefinic carbon atoms and their attached carbon atoms) and the principal axis of ellipticity (λ_2 of the Hessian). A value of zero indicates that the preferential plane of accumulation is perpendicular to the π-plane

change in the φ_{ref} angle to nearly 90° occurs when ε becomes very small, near the charge concentrations in the VSCC [102]. The profile for C4–C5, with a central dip and a much larger value of ε, more closely resembles the profile for a typical strong metal–alkene interaction [9]. This indication of a stronger Rh–C4/C5 interaction agrees with the original conclusions of the authors [101].

12.3.2 Bonding in Conjugated π-Complexes

When the metal π-bonding is extended to more delocalised carbocyclic systems, such as η^5-C_5H_5 or η^6-C_6H_6, the number of observed TM–C bond paths very rarely

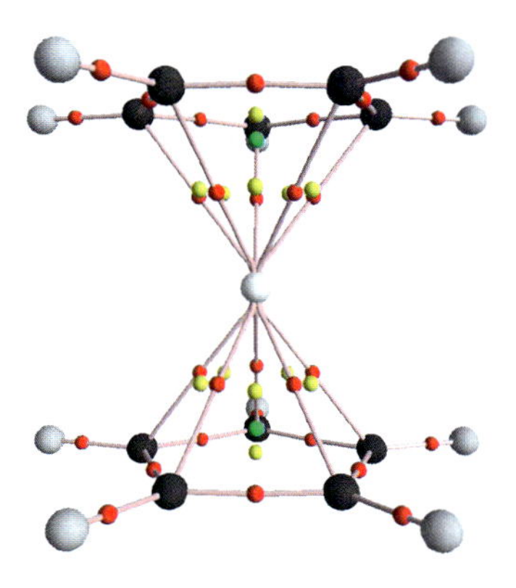

Fig. 12.15 Molecular graph of D_{5h} ferrocene (**12**). Colour coding for cps in $\rho(\mathbf{r})$, red = (3,−1), yellow = (3,+1), green = (3,+3)

matches the formal hapticity [8]. Figure 12.15 shows the molecular graph for $Fe(\eta^5$-$C_5H_5)_2$ (**12**) in the ground state eclipsed conformation – the slightly higher energy staggered D_{5d} conformation gives an essentially identical graph [7]. Each carbon atom of the ring is linked to the iron atom by a bond path, giving rise to a ring of five bcps, which are interspersed with another ring of rcps in virtually the same plane, nearly parallel to the C_5 ring plane. The bcps and rcps are in close physical proximity, with virtually identical electron densities, so presaging a catastrophe situation. It has been suggested [7] that "the interaction of a metal atom with the Cp ring is best viewed as involving an interaction with the delocalised density of the entire ring perimeter". Such a view is appealing, as it complies with chemical intuition and MO analyses regarding the delocalised nature of the metal-π-bond.

Consistent with the impending catastrophe situation implied in Fig. 12.15, in almost all reported examples where the exact fivefold symmetry of the TM-(η^5-C_5H_5) moiety is reduced, the number of bond paths no longer matches the formal hapticity. Amongst many examples, this has been shown for (E)-{(η^5-C_5H_4)-CF=CF(η^5-C_5H_4)}(η^5-C_5H_5)$_2$$Fe_2$ (**13**) [8]. The Fe–C distances are in a narrow range and essentially identical for both rings, with ranges of $Fe–C_{free} = 2.053 \pm 0.006$ Å and $Fe–C_{subst} = 2.050 \pm 0.004$ Å, so in terms of commonly accepted structural criteria, both rings would therefore be treated as idealised η^5. In fact only four bonds paths are found from the iron atom to either ring in the experimental study. The "missing" bond path in this case involves the longest Fe–C interaction, though this is not necessarily always found to be so. The delocalisation indices $\delta(\Omega_{Fe}, \Omega_C)$ fall in a narrow range of 0.43–0.47 and indicate essentially equivalent bonding of the iron atom to all five ring carbon atoms.

The same impression is given by the source function, which indicates an equivalent delocalised bonding, regardless of whether the reference point is taken at an Fe–C bcp or the Fe–C mid-point (for the "missing" bond paths). Figure 12.16 shows the related source function plots for the $Mn–C_{ring}$ interactions in (η^5-C_5H_5)$Mn(CO)_3$ (**14**) [8]. All basins act as sources for the density at reference points either at the $Mn–C_{ring}$ bcp or rcp, with relatively little difference between these two positions. The highly delocalised SF is quite typical of many $TM–C_{ring}$ interactions, and may be taken to indicate a strongly delocalised Mn–Cp bonding.

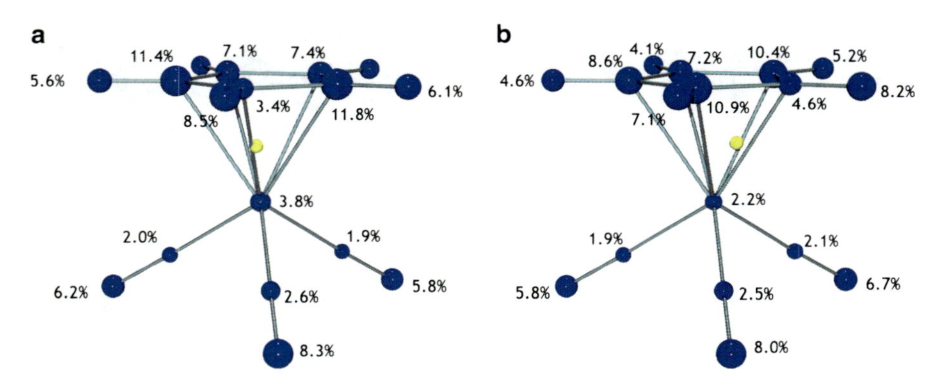

Fig. 12.16 Integrated source function plots for **14**, where the reference point is taken as (**a**) an Mn–C_{ring} bcp or (**b**) an Mn–C_{ring}-C_{ring} rcp. The volumes of the spheres on the atomic centres are proportional to the percentage contribution from the atomic basins and the reference points are shown as *yellow spheres*

A DFT investigation of the series of 18 electron half-sandwich complexes (η^n-C_nH_n)M(CO)$_3$ ($n = 3,8$; M=Co-Ti) showed that for rings with $n > 5$, the molecular graph was quite sensitive to the basis set used, and that fewer than n bond paths are obtained. A comparison of the molecular graphs, i.e. the atomic interaction lines and critical points in the scalar field of $\rho(\mathbf{r})$, with the virial graphs, i.e the atomic interaction lines and critical points in the scalar field of the potential energy density $V(\mathbf{r})$, for these series of compounds is illuminating. The virial and electron density fields are normally homeomorphic [103], and it has been argued [16] that "this implies that every bond path is mirrored by a virial path, a line linking the same neighbouring nuclei, along which the potential energy is maximally negative, i.e maximally stabilising with respect to any neighbouring line". This homeomorphism underlines the connection between a bond path and energetic stabilisation. As can be seen in Fig. 12.17 however, the two graphs for the complexes (η^n-C_nH_n)M(CO)$_3$ ($n = 5,8$; M=Mn-Ti) are not always structurally homeomorphic. This non-homeomorphic behaviour can arise when a structure is close to a catastrophe point [103], which of course is exactly the situation for these complexes. It can be seen, in fact, that the virial graphs show interaction lines between the metal and all the n carbon atoms of the rings, again emphasising the QTAIM view that the metal is equivalently bonded to all ring carbons. The "true" hapticity is thus recovered in the virial graph, which directly represents the energetic stabilisation on chemical bonding.

The usefulness of these QTAIM indicators in defining the hapticity in more complex systems is illustrated by the case of Fe(C_8H_8)$_2$ (**15**), which has one η^6-C_8H_8 ring and one η^4-C_8H_8 in the ground state structure as determined from X-ray diffraction [104] and dynamic ^{13}C NMR spectroscopy [105]. The plot of $|\nabla\rho(\mathbf{r})|$ through the plane of the Fe–C bcps for both rings, given in Fig. 12.18, shows open ended "horse-shoe" zones of very low values of $|\nabla\rho(\mathbf{r})|$, which indicate the regions

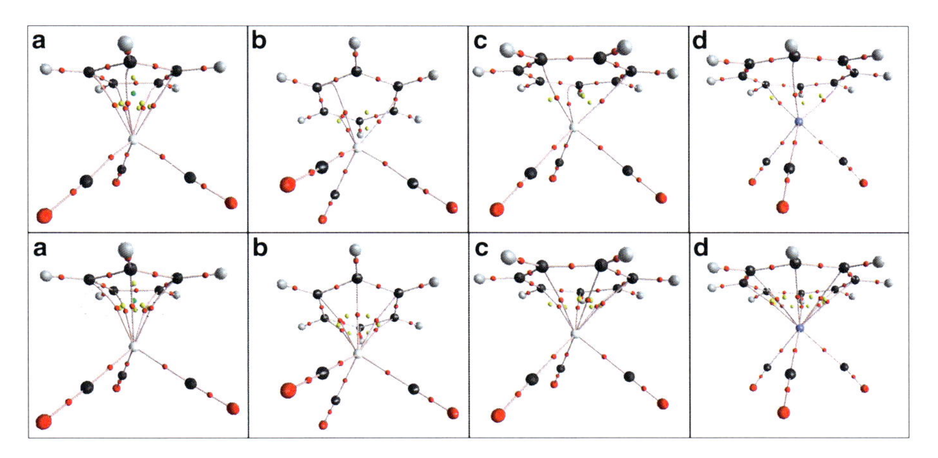

Fig. 12.17 Molecular graphs (*top row*) and virial graphs (*bottom row*) for $(\eta^n\text{-}C_nH_n)M(CO)_3$; (**a**) ($n = 5$; M=Mn), (**b**) ($n = 6$; M=Cr), (**c**) ($n = 7$; M=V), (**d**) ($n = 8$; M=Ti)

where the bcps between the ring carbon atoms and the iron atoms are located. The similarity in shape of these zones to the graphical symbols commonly used by chemists to represent the delocalised interactions between transition metal atoms and open or closed conjugated π-hydrocarbyl ligands is unmistakable, and shows they are highly appropriate to represent delocalised topological interactions. The delocalisation of electron density in the π-bonded ring is responsible to a large extent for the movement of the molecular graphs towards a catastrophe situation. Compared with the situation of the hypothetical pro-molecule, the electron density "smearing" of the ring due to covalent bonding results in an even flatter zone of density which interacts with the metal atom density [8].

Finally we note that the issues discussed above should also apply to metal π-hydrocarbyl complexes in general. One such example is the trimethylene(methane) complex $\{C(CH_2)_3\}Fe(CO)_3$ (**16**), which contains a stellated rather than ring conjugated π-system. A combined experimental and theoretical study showed that the molecular graph (Fig. 12.19) contains only one bond path between the Fe centre and the C_α of trimethylene(methane) ligand [106], despite considerable evidence for a strong π-interaction between the Fe atom and the ligand. This puzzling observation was further compounded by noticing that the delocalisation index associated with the non-existent bond path $\delta(\Omega_{Fe}, \Omega_{C\beta}) = 0.609$ was significantly greater than for $\delta(\Omega_{Fe}, \Omega_{C\alpha}) = 0.351$. The source function, shown in Fig. 12.19, implies a strongly delocalised interaction between the iron atom and the π-hydrocarbyl ligand, and the percentage basin contributions from C_β are larger than for C_α, regardless of where the reference point is taken. If the geometry is distorted by reducing the Fe-C_α-C_β angle from its equilibrium value of 76.6–73°, a catastrophe situation arises, and Fe–C_β bond paths appear [106].

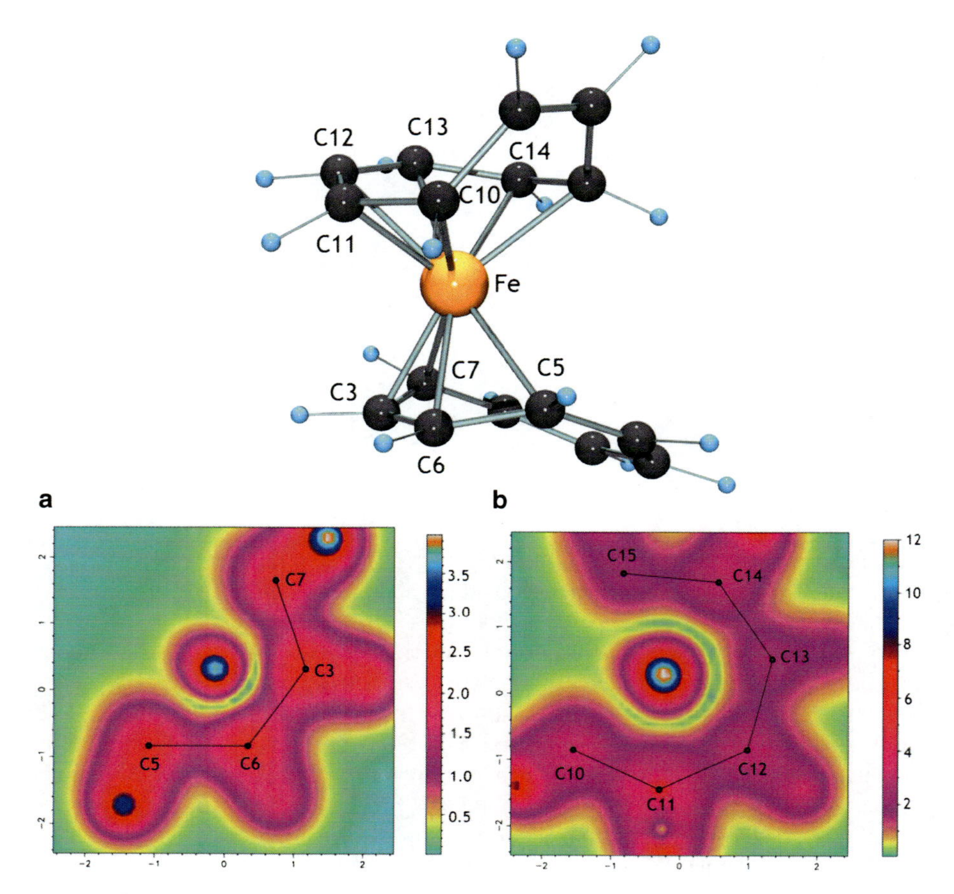

Fig. 12.18 Molecular structure of complex **15** and plots of $|\nabla\rho(\mathbf{r})|$ through the plane of the Fe-C bcps for (**a**) the η^4-C_8H_8 ring and (**b**) for the η^6-C_8H_8 ring. The positions of the ligated C atoms are projected onto the plane

12.3.3 Metal–Metal Bonding

The direct metal–metal bond, though known in organotransition complexes for many decades, is still a subject of interest and controversy. Silvi and Gatti [107] have discussed the metallic bond from an ELF and QTAIM perspective and various metallicity QTAIM indices have been proposed [108], though their relevance to molecular compounds containing isolated metal–metal bonds is disputable. The topological characteristics of metal–metal bonds in organometallic compounds are reasonably well established, with low values of $\rho(\mathbf{r})$ (~0.2–0.5 eÅ^{-3}), small positive $\nabla^2\rho(\mathbf{r})$ and very slightly negative H_b and delocalisation indices significantly less than one [5, 94, 109], indicating a sharing of less than one electron pair. Even for cases with unsupported metal–metal bonds such as $Mn_2(CO)_{10}$, there is

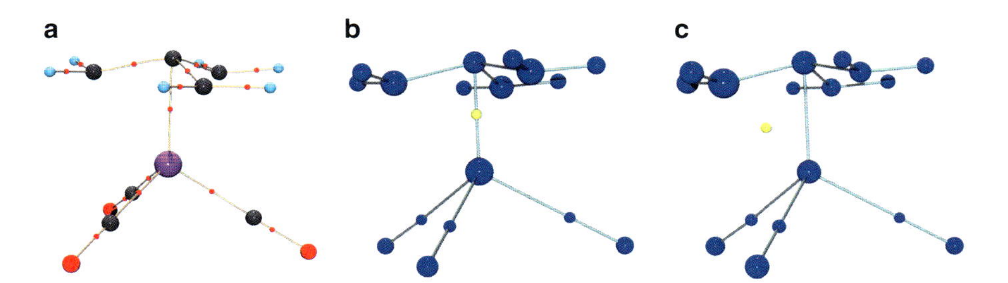

Fig. 12.19 (**a**) Molecular graph, (**b, c**) integrated source function plots for **16**, where the reference point is taken at (**b**) the Fe–C_α bcp or (**c**) the Fe–C_β midpoint. The volumes of the spheres on the atomic centres are proportional to the percentage contribution from the atomic basins and the reference points are shown as *yellow spheres*

evidence, from the domain averaged Fermi hole density, of electron sharing through interactions with the vicinal carbonyl ligands [110]. Metal–metal bcps are generally only observed for unsupported bonds – the densities in the central bonding regions are very flat, and a catastrophe situation is a constant possibility. As a consequence, despite often quite short metal–metal distances, and a requirement for a metal–metal bond through the 18-electron rule, a clear QTAIM confirmation of a direct bond through the appearance of a bond path and bcp is often elusive.

It is interesting to observe that arguments about metal–metal bonding in such well-known molecules as $Fe_2(CO)_9$ still occur. Reinhold et al. [111] have recently reported that, depending on the basis set, a bcp for the Fe–Fe interaction may be detected, despite several other studies that show a cage cp at the Fe–Fe midpoint. They argue, through a hybrid QTAIM/MO approach involving orbital partitioning of $\rho(\mathbf{r})$, $\nabla^2\rho(\mathbf{r})$ and H_b, that a residual σ-bonding combination of the t_{2g} results in a weak direct bond. Ponec et al. [112] on the other hand, through analysis of the domain averaged Fermi holes, arrive at the "classical" conclusion that there is no direct Fe–Fe bond, but that any metal–metal interactions are mediated through the bridging CO ligands.

An unusual example of a metal–metal bonded compound which has been recently examined by QTAIM techniques is $(\eta^5\text{-}C_5Me_5)_2Zn_2$ (**17**) [113], which was the first example of an organometallic compound containing a Zn–Zn bond (Fig. 12.20). This formally Zn^I complex has no bridging ligands and as a consequence a bond path and bcp are to be expected and are indeed observed. The theoretical topological properties, $\rho_b = 0.43$ eÅ^{-3}, $\nabla^2\rho_b = 1.6$ eÅ^{-5}, $H_b/\rho_b = -0.36$ He^{-1}, $\delta(\Omega_{Zn},\Omega_{Zn}) = 0.92$ and $\oint_{Zn\cap Zn}\rho(\mathbf{r}) = 1.25$ eÅ^{-1}, are suggestive of a reasonably strong bond with significant covalency. A number of theoretical studies on **17** agree that the formal Zn_2^{2+} unit is held together primarily (>90%) through a 4*s*-4*s* σ-bond, with much smaller contributions from 4*p*-4*p* and $3d_{z^2}$-$3d_{z^2}\sigma$-bonding. The bonding between the Zn_2^{2+} unit and the pentamethylcyclopentadiene ligands seems to be primarily ionic, and this compound presents yet another example of the catastrophe situation in M-(η^5-Cp) chemistry, as it is reported that only for some of

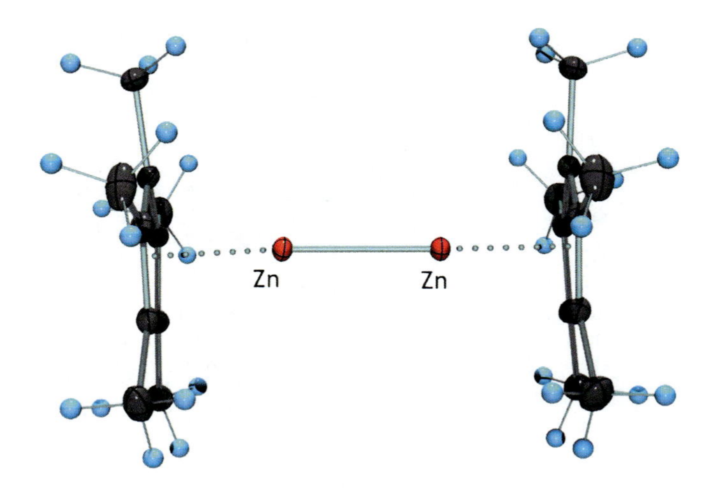

Fig. 12.20 Molecular structure of **17**

the multipole models tested are all ten potential bond paths between the Zn atoms and the Cp ligands actually observed [113]. Interest in the bonding in this unusual molecule remains high, though to date very few of its reactions have been reported where the Zn_2 unit remains intact [114].

More recently, in a related experimental topological study, the Mg–Mg bond in $Mg_2(\{(2,6\text{-}i\text{Pr}_2C_6H_3)NCMe\}_2CH)_2$ (**18**) has been examined by Overgaard et al. [115]. The Mg–Mg distance of 2.8456(2) Å in **18** is substantially longer than the Zn–Zn interaction in **17**, and the topological indicators $\rho_b = 0.10(2)$ eÅ^{-3}, $\nabla^2\rho_b = 10.26(1)$ eÅ^{-5}, $H_b = -0.01$ H are suggestive of a quite reduced covalency. The source function at the Mg–Mg bcp shows that the two Mg basins have only a modest influence in determining the density (17.4 and 13.7%), which again implies only a small covalency. Clearly more work is required to fully characterise these interesting and highly unusual metal–metal bonds.

12.3.4 Agostic Interactions

The characterisation of agostic interactions of coordinated alkyl C-H groups with transition metals is another area where QTAIM studies have had an impact. Studies were instigated by a theoretical paper in 1998 by Popelier and Logothesis on β- and γ-agostic interactions in Ti^{IV} alkyls [116]. They suggested that the presence of a bond path between the metal and interacting H atom was an indicator of the agostic interaction, which they found to be quite distinct from a H-bonding interaction. As is typical for TM compounds, a positive Laplacian is observed at the bcp. An alternative view of agostic bonding was expressed in an in-depth study by Scherer and coworkers [9], which included an experimental charge density analysis

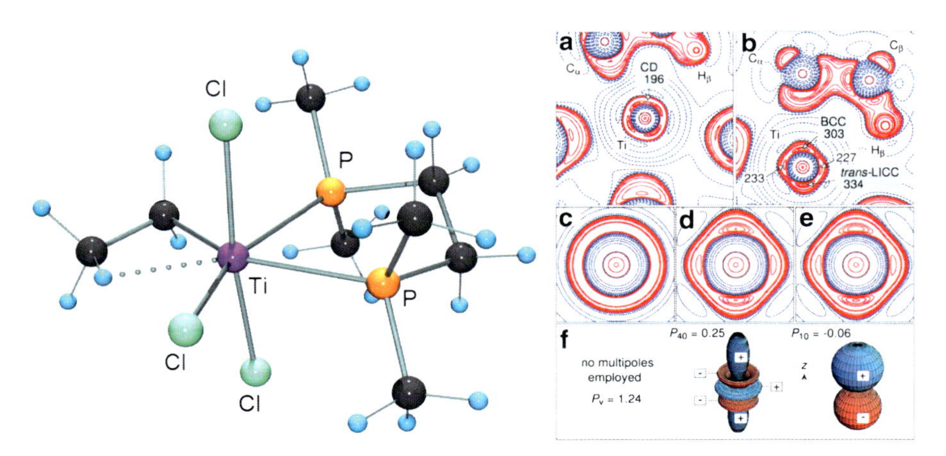

Fig. 12.21 Structure of **19** and contour plots of the negative Laplacian $-\nabla^2\rho(\mathbf{r})$ of (**a**) the theoretical and (**b**) the experimental densities in the Ti-C_α-C_β planes. The increasing polarisation of the Ti atom with differing levels of multipole modelling are shown in (**c–e**) and magnitudes of the LICC's (eÅ^{-5}) are indicated in (**a, b**) (Partly reproduced from Ref. [9] with kind permission of Wiley-VCH)

of $TiCl_3(C_2H_5)$(dmpe) (**19**). They propose that the agostic interaction in this and related species is derived primarily from a delocalisation of the M–C_α bonding electrons, with any M....H–C_β interactions playing a secondary role. The negative hyperconjugation which leads to this delocalisation is traced in the observable charge density through the bond ellipticity profiles and also the VSCCs in the metal and interacting carbon atoms. In contrast to the Popelier and Logothesis suggestion [116], no bond path is detected between the Ti and H_β atoms for the experimental density. As discussed by Scherer et al. [9], any M....H-C_β bond path occurs in a region of very flat density, and this provides yet one more example where interesting bonding interactions involving transition metals are plagued by the catastrophe problem. McGrady and coworkers [117] have confirmed that the bond path criterion is not a robust one as far as characterising the early transition metal agostic interaction. The complex $TpMe_2NbCl(MeC{\equiv}CPh)(i\text{Pr})$ (Tp=tri-pyrazolylborato) has two rotamers with an α- and a β-agostic *i*Pr group. The observation of a bond path for the Nb....H-C_β interaction is highly dependent on the DFT functional used, and one is never observed for the Nb....H-C_α interaction.

The ligand induced charge concentrations (LICC) for Ti in **19** are shown in Fig. 12.21. They are strongly influenced by the agostic alkyl ligand and can be adequately modelled by only two populated multipoles (P_{40} and P_{10}), Fig. 12.21. Moreover, the ellipticity along the C_α-C_β bond path shows a characteristic profile indicative of incipient π-bonding associated with the negative hyperconjugative delocalisation. Scherer et al. [9] conclude that in general, ellipticity profiles along C_α–X_β bonds (X=C, Si) can be used to reveal electronic features due to the delocalisation from M–C bonding density. The profiles for β-agostic interactions in

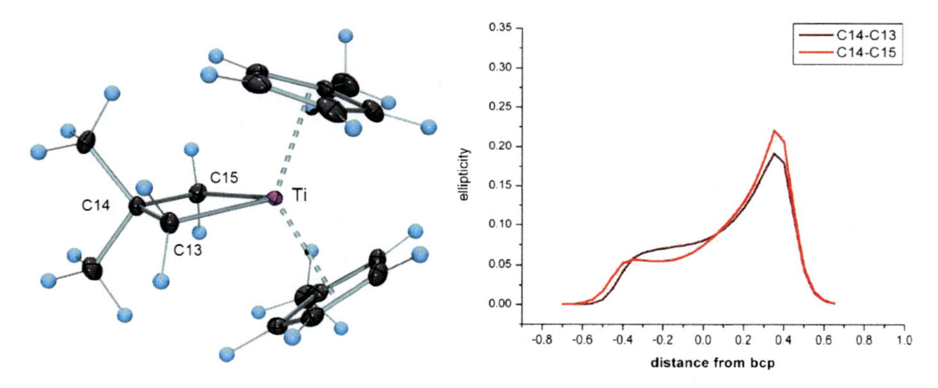

Fig. 12.22 Structure of **20** and ellipticity profiles along the Ti–C_α bonds (Partly reproduced from Ref. [120] with kind permission of The American Chemical Society)

late transition metal indicate a much greater degree of M–C electron delocalisation, consistent with the classical view that they are much closer to the alkene-hydrido β-elimination products [9]. Ellipticity profiles have also proved very useful topological indicators in other systems – the π-delocalisations in *N*-heterocyclic carbene [118] and in cyclopropenylidene [119] transition metal complexes are clearly revealed through their ellipticity profiles.

A further interesting example is the complex $(\eta^5\text{-}C_5H_5)_2Ti(CH_2CMe_2CH_2)$ (**20**), which contains a (C–C)–M agostic interaction [120]. This unusual type of agostic interaction may be implicated in olefin metathesis and alkene polymerisation reactions. The averaged experimental topological parameters for the Ti–C_α interactions, $\rho_b = 0.53$ eÅ^{-3}, $\nabla^2\rho_b = 5.9$ eÅ^{-5}, $H_b/\rho = -0.28$ au are indicative of reasonably strong covalent bonding, but a clear characterisation of the agostic interaction comes from the corresponding C_α-C_β ellipticity profiles, shown in Fig. 12.22. These show significant preferential accumulations in the metalla-cyclobutane ring plane close to C_α and strongly confirm the Scherer model [9]. Incidentally, this study also emphasises the issues raised above regarding M–C bond paths and formal hapticities, as only three and one bond paths are observed between the Ti atom and the two formally η^5-Cp rings.

Finally, we mention the σ-silane $HSiX_3$ complexes with the 16 electron fragment $CpMn(CO)_2$, in which the bonding has been controversial. The theoretical topology in the complex with $X_3 = Cl_3$ has been discussed by Bader et al. [121], while the complexes with $X_3 = HPh_2$, FPh_2 have been examined in combined experimental/theoretical work by Scherer and coworkers [122, 123]. The central question is whether the bonding is treated as an agostic type σ-interaction of the Mn centre with the Si–H bond, or whether an oxidative addition description, with Si–H bond cleavage is more appropriate. In all cases, residual Si–H bonding is observed in the electron density topology, which is found to involve an Mn–Si–H ring system, except for the SiH_2Ph_2 addition product. This is illustrated in the Laplacian distributions shown in Fig. 12.23.

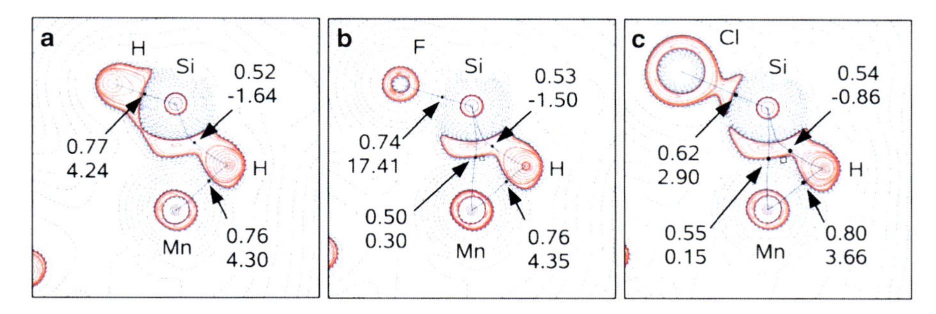

Fig. 12.23 Negative Laplacian plots in the Mn-H-Si planes for Cp'Mn(CO)$_2${η^2HSiX$_3$} for (**a**) X$_3$=HPh$_2$, (**b**) X$_3$=FPh$_2$ and (**c**) X$_3$=Cl$_3$. Value of the densities (e Å^{-3}, *above*) and Laplacian (e Å^{-5}, *below*) for indicated critical points are given (Reproduced from Ref. [122] with kind permission of The American Chemical Society)

Through a combined QTAIM/MO approach and an analysis of the source function for the Mn–Si–H ring system, the conclusion of Scherer et al. [122] is that the bonding is very similar in all three cases and is best described as an agostic type η^2-silane adduct. At the risk of tiring the reader, it is also worth mentioning that this ring system is close to a catastrophe situation, as indicated by Fig. 12.23. The implications of the catastrophe situation are discussed at some length by Bader et al. [121].

12.4 Conclusions

We have shown how QTAIM indicators obtained from experimental or theoretical charge density studies have been used to shed light on specific chemical bonding issues in two key areas of chemistry. In certain cases, they may provide a clear resolution of these issues, e.g. in hypervalent main group multiple bonding, the reactivity of boranes and in agostic interactions. In other areas such as metal-metal bonding, QTAIM studies rarely find evidence in the shape of a bond path for direct bonding (especially if there are bridging ligands present), and there remains a distinct conflict with established chemical concepts (in this case the 18 electron rule). Perhaps the extended interpretation of the bond path as proposed by Pendás et al. [17] will lead to a greater concensus. In any case, it is clear that the "orthodox" reliance on local topological properties at the bcp has proved too limited to be useful for a wide range of chemical interactions. Non-local and continuous QTAIM indicators such as the source function [14] may prove helpful, but they need to be used with thoughtful care, and interpretations may not always be straight forward, as shown by a recent study [124].

References

1. Bader RFW (1990) Atoms in molecules: a quantum theory. Oxford University Press, Oxford
2. Bader RFW, Essén H (1984) The characterisation of atomic interactions. J Chem Phys 80:1943–1960
3. Cremer D, Kraka E (1984) Chemical bonds without bonding electron density - does the difference electron-density analysis suffice for a description of the chemical bond? Angew Chem Int Ed Engl 23:627–628
4. Smith GT, Mallinson PR, Frampton CS, Farrugia LJ, Peacock RD, Howard JAK (1997) Experimental determination of the electron density topology in a non-centrosymmetric transition metal complex: [Ni(H3L)][NO3][PF6] [H3L=N, N′, N′′-tris(2-hydroxy-3-methylbutyl)-1,4,7-triazanonane]. J Am Chem Soc 119:5028–5034
5. Macchi P, Sironi A (2003) Chemical bonding in transition metal carbonyl clusters: complementary analysis of theoretical and experimental electron densities. Coord Chem Rev 238–239:383–412
6. Gatti C (2005) Chemical bonding in crystals: new directions. Z Kristallogr 220:399–457
7. Cortés-Guzmán F, Bader RFW (2005) Complementarity of QTAIM and MO theory in the study of bonding in donor-acceptor complexes. Coord Chem Rev 249:633–662
8. Farrugia LJ, Evans C, Lentz D, Roemer M (2009) The QTAIM approach to chemical bonding between transition metals and carbocyclic rings: a combined experimental and theoretical study of (η^5-C_5H_5)Mn$(CO)_3$, (η^6-C_6H_6)Cr$(CO)_3$ and (*E*)-{(η^5-C_5H_4)CF=CF(η^5-C_5H_4)}(η^5-C_5H_5)$_2$Fe$_2$. J Am Chem Soc 131:1251–1268
9. Scherer W, Sirsch P, Shorokhov D, Tafipolsky M, McGrady GS, Gullo E (2003) Valence charge concentrations, electron delocalisation and β-agostic bonding in d^{10} metal alkyl complexes. Chem Eur J 9:6057–6070
10. McGrady GS, Haaland A, Verne HP, Volden HV, Downs AJ, Shorokhov D, Eickerling G, Scherer W (2005) Valence shell charge concentrations at pentacoordinate d^{10} transition-metal centers: non-VSEPR structures of Me_2NbCl_3 and Me_3NbCl_2. Chem Eur J 11:4921–4934
11. Cremer D, Kraka E (1984) A description of the chemical bond in terms of local properties of electron density and energy. Croat Chem Acta 57:1259–1281
12. Bader RFW, Stephens ME (1975) Spatial localisation of the electron pair and number distribution in molecules. J Am Chem Soc 97:7391–7399
13. Fradera X, Austen MA, Bader RFW (1999) The Lewis model and beyond. J Phys Chem A 103:304–314
14. Bader RFW, Gatti C (1998) A Green's function for the density. Chem Phys Lett 287:233–238
15. Gatti C, Lasi D (2007) Source function description of metal-metal bonding in d-block organometallic compounds. Faraday Discuss 135:55–78
16. Bader RFW (1998) A bond path: a universal indicator of bonded interactions. J Phys Chem A 102:7314–7323
17. Pendás AM, Francisco E, Blanco MA, Gatti C (2007) Bond paths as privileged exchange channels. Chem Eur J 13:9362–9371
18. Matta CF, Hernández-Trujillo J, Tang T-H, Bader RFW (2003) Hydrogen-hydrogen bonding: a stabilising interaction in molecules and crystals. Chem Eur J 9:1940–1951
19. Poater J, Solá M, Bickelhaupt FM (2006) Hydrogen-hydrogen bonding in planar biphenyl, predicted by atoms-in-molecules theory, does not exist. Chem Eur J 12:2889–2895
20. Grimme S, Mück-Lichtenfeld C, Erker G, Kehr G, Wang H, Beckers H, Willner H (2009) When do interacting atoms form a chemical bond? Spectroscopic measurements and theoretical analysis of dideuteriophenanthrene. Angew Chem Int Ed 48:2592–2595
21. Bader RFW (2009) Bond paths are not chemical bonds. J Phys Chem A 113:10058–10067
22. Lewis GN (1916) The atom and the molecule. J Am Chem Soc 38:762–785
23. Gillespie RJ, Silvi B (2002) The octet rule and hypervalence: two misunderstood concepts. Coord Chem Rev 233–234:53–62

24. Noury S, Silvi B (2002) Chemical bonding in hypervalent molecules: is the octet rule relevant? Inorg Chem 41:2164–2172
25. Kutzelnigg W (1984) Chemical bonding in higher main group elements. Angew Chem Int Ed Engl 25:272–295
26. Rundle RE (1947) Electron deficient compounds. J Am Chem Soc 69:1327–1331
27. Kocher N, Henn J, Gostevskii B, Kost D, Kalikhman I, Engels B, Stalke D (2004) Si–E (E=N, O, F) Bonding in a hexacoordinated silicon complex – new facts from experimental and theoretical charge density studies. J Am Chem Soc 126:5563–5568
28. Pauling L (1939) The nature of the chemical bond. Cornell University Press, Ithaca
29. Gibbs GV, Downs JW, Boisen MB Jr (1994) The elusive SiO bond. Rev Mineral 29:331–368
30. Wang J, Eriksson LA, Boyd RJ, Shi Z, Johnson BG (1994) Diazasilene (SiNN): a comparative study of electron density distributions derived from Hartree-Fock, second-order Moller-Plesset perturbation theory, and density functional methods. J Phys Chem 98:1844–1850
31. Herster JR, Maslen EN (1995) Electron density - structure relationships in some perovskite-type compounds. Acta Crystallogr B 51:913–920
32. Herberhold M, Köhler C, Wrackmeyer B (1992) Sulfur diimides bearing bis(amino)phosphinyl substituents. Phosphorus, Sulfur, Silicon Relat Elem 68:219–222
33. Pohl S, Krebs B, Seyer U, Henkel G (1979) Trigonal planar koordinierter Schwefel(VI): Die kristall- und molekülstrukturen von $(R_3SiN)_3S$ und $(R_3CN)_3S$ ($R=CH_3$) bei –130°C. Chem Ber 112:1751–1755
34. Mayer I (1987) Bond orders and valences: role of *d*-orbitals for hypervalent sulphur. THEOCHEM 149:81–89
35. Bors DA, Streitwieser A (1986) Theoretical study of carbanions and lithium salts derived from dimethyl sulfone. J Am Chem Soc 108:1397–1404
36. Reed AE, Schleyer PVR (1990) Chemical bonding in hypervalent molecules. The dominance of ionic bonding and negative hyperconjugation over *d*-orbital participation. J Am Chem Soc 112:1434–1445
37. Salzner U, Schleyer PVR (1993) Generalized anomeric effects and hyperconjugation in $CH_2(OH)_2$, $CH_2(SH)_2$, $CH_2(SeH)_2$, and $CH_2(TeH)_2$. J Am Chem Soc 115:10231–10236
38. Gilheany DG (1994) No *d* orbitals but Walsh diagrams and maybe banana bonds: Chemical bonding in phosphines, phosphine oxides, and phosphonium ylides. Chem Rev 94:1339–1374
39. Dobado JA, Marinez-García H, Molina JM, Sundberg MR (1998) Chemical bonding in hypervalent molecules revised. Application of the atoms in molecules theory to Y_3X and Y_3XZ (Y=H or CH_3; X=N, P or As; Z=O or S) Compounds. J Am Chem Soc 120:8461–8471
40. Stefan T, Janoschek R (2000) How relevant are S=O and P=O double bonds for the description of the acid molecules H_2SO_3, H_2SO_4, and H_3PO_4, respectively? J Mol Struct 130:282–288
41. Cruickshank DWJ, Eisenstein M (1985) The role of *d* functions in ab-initio calculations: part 1. The deformation densities of H_3NSO_3 and SO_3^-. J Mol Struct 130:143–156
42. Cruickshank DWJ (1985) A reassessment of d_π–p_π bonding in the tetrahedral oxyanions of second-row atoms. J Mol Struct 130:177–191
43. Fleischer R, Freitag S, Pauer F, Stalke D (1997) $[S(N^tBu)_3]^{2-}$ – A cap-shaped dianion, isoelectronic with the sulfite ion and oxidizable to a stable radical anion. Angew Chem Int Ed Engl 35:204–206
44. Fleischer R, Rothenberger A, Stalke D (1997) $S(N^tBu)_4{}^{2-}$: a dianion isoelectronic to $SO_4{}^{2-}$ and the related $MeS(N^tBu)^{3-}$. Angew Chem Int Ed Engl 36:1105–1107
45. Ilge D, Wright DS, Stalke D (1998) Mixed alkali metal cages containing the cap-shaped $[S(NtBu)_3]^{2-}$ triazasulfite dianion. Chem Eur J 4:2275–2279
46. Walfort B, Lameyer L, Weiss W, Herbst-Irmer R, Bertermann R, Rocha J, Stalke D (2001) $[\{(MeLi)_4(dem)_{1.5}\}_1]$ and $[(thf)_3Li_3Me\{(N^tBu)_3S\}]$: how to reduce aggregation of parent methyllithium. Chem Eur J 7:1417–1423

47. Walfort B, Pandey SK, Stalke D (2001) The inverse podant $[Li_3(NBu^t)_3S\}]^+$ stabilises a single ethylene oxide $OCHNCH_2$ anion as a high- and low-temperature polymorph of $[(thf)_3Li_3(OCHNCH_2)\{(NBu^t)_3S\}]$. J Chem Soc Chem Commun 1640–1641
48. Fleischer R, Stalke D (1998) Syntheses and structures of main group metal complexes of the $S(NtBu)_3{}^{2-}$ dianion, an inorganic Y-conjugated tripod. Organometallics 17:832–838
49. Fleischer R, Walfort B, Gburek A, Scholz P, Kiefer W, Stalke D (1998) Raman spectroscopic investigation and coordination behavior of the polyimido S^{VI} anions $[RS(NR)_3]^-$ and $[S(NR)_4]^{2-}$. Chem Eur J 4:2266–2279
50. Glemser O, Pohl S, Tesky FM, Mews R (1977) Tris(tert-butylimino)sulfur(VI) and bis(tert-butylimino)(silylimino)sulfur(VI); The "Y-triene" structure Angew. Chem Int Ed Engl 16:789–790
51. Meij R, Oskam A, Stufkens DJ (1979) A conformational study of N-sulfinylanilines and the assignment of vibrations of N-sulfinylanilines and N, N´-di-arylsulfurdiimines using 15N enrichment and Raman resonance results. J Mol Struct 51:37–49
52. Herbrechtsmeier A, Schnepfel FM, Glemser O (1978) Vibrational studies of (n-trimethylsilylimido)sulfur compounds. J Mol Struct 50:43–63
53. Markowskii LN, Tovstenko VI, Pashinnik VE, Mel'nichuk EA, Makarenko AG, Shermolovich, Yu G (1991) Reactions of (α,α,ω -trihydropolyfluoroalkoxy)trifluorosulfuranes with primary amines and amides. Zh Org Khim 27:769–773
54. Leusser D, Henn J, Kocher N, Engels B, Stalke D (2004) S=N versus $S^+–N^-$ – an experimental and theoretical charge density study. J Am Chem Soc 126:1781–1793
55. Glendening ED, Badenhoop JK, Reed AE, Carpenter JE, Weinhold F (1999) NBO 4.M, Theoretical Chemical Institute, University of Wisconsin, Madison
56. Johnson AW (1993) Ylides and imines of phosphorous. Wiley, New York
57. Steiner A, Zacchini S, Richards PI (2002) From neutral iminophosphoranes to multianionic phosphazenates. The coordination chemistry of imino–aza-P(V) ligands. Coord Chem Rev 227:193–216
58. Bickley JF, Copsey MC, Jeffery FC, Leedham AP, Russell CA, Stalke D, Steiner A, Stey T, Zacchini S (2004) From the tetra(amino) phosphonium cation, $[P(NHPh)_4]^+$, to the tetra(imino) phosphate trianion, $[P(NPh)_4]^{3-}$, two-faced ligands that bind anions and cations. J Chem Soc, Dalton Trans 989–995
59. Kocher N, Leusser D, Murso A, Stalke D (2004) Metal coordination to the formal P=N Bond of an iminophosphorane and charge-density evidence against hypervalent phosphorus(V). Chem Eur J 10:3622–3631
60. Magnusson EJ (1993) The role of *d* functions in correlated wave functions: main group molecules. J Am Chem Soc 115:1051–1061
61. Chesnut DB (2003) Atoms-in-molecules and electron localization function study of the phosphoryl Bond. J Phys Chem A 107:4307–4313
62. Naito T, Nagase S, Yamataka H (1994) Theoretical study of the structure and reactivity of ylides of N, P, As, Sb, and B. J Am Chem Soc 116:10080–10088
63. Restrepo-Cossio AA, Gonzelez CA, Mari F (1998) Comparative ab initio treatment (Hartree – Fock, density functional theory, MP2, and quadratic configuration interactions) of the cycloaddition of phosphorus ylides with formaldehyde in the gas phase. J Phys Chem A 102:6993–7000
64. Yamataka H, Nagase S (1998) Theoretical calculations on the Wittig reaction revisited. J Am Chem Soc 120:7530–7536
65. Dobado JA, Martinez-Garcìa H, Molina JM, Sundberg MR (2000) Chemical bonding in hypervalent molecules revised. 3. Application of the atoms in molecules theory to Y_3X-CH_2 (X=N, P, or As; Y=H or F) and H_2X-CH_2 (X=O, S, or Se) Ylides. J Am Chem Soc 122:1144–1149
66. Lu CW, Liu CB, Sun CC (1999) Theoretical study of the $H_3PNH + H_2CO$ reaction mechanism via five reaction channels. J Phys Chem A 103:1078–1083

67. Lu CW, Sun CC, Zang QJ, Liu CB (1999) Theoretical study of the aza-Wittig reaction $X_3P{=}NH{+}O{=}CHCOOH \rightarrow X_3P{=}O{+}HN{=}CHCOOH$ for X=Cl, H and CH_3. Chem Phys Lett 311:491–498
68. Koketsu J, Ninomiya Y, Suzuki Y, Koga N (1997) Theoretical study on the structures of iminopnictoranes and their reactions with formaldehyde. Inorg Chem 36:694
69. Yufit DS, Howard JAK, Davidson MG (2000) Bonding in phosphorus ylides: topological analysis of experimental charge density distribution in triphenylphosphonium benzylide. J Chem Soc Perkin Trans 2:249–253
70. Aggarwal VK, Richardson J (2003) The complexity of catalysis: origins of enantio- and diastereocontrol in sulfur ylide mediated epoxidation reactions. Chem Commun 21:2644–2651
71. Aggarwal VK, Winn CL (2004) Catalytic, asymmetric sulfur ylide-mediated epoxidation of carbonyl compounds: scope, selectivity, and applications in synthesis. Acc Chem Res 37:611–620
72. Brandt S, Helquist P (1979) Cyclopropanation of olefins with a stable, iron-containing methylene transfer reagent. J Am Chem Soc 101:6473–6475
73. Kremer KAM, Helquist P, Kerber RC (1981) Ethylidenation of olefins using a convenient iron-containing cyclopropanation reagent. J Am Chem Soc 103:1862–1864
74. O'Connor EJ, Helquist P (1982) Stable precursors of transition-metal carbene complexes. Simplified preparation and crystal ctructure of (η^5-Cyclopentadienyl)[(dimethylsulfonium)-methyl]-dicarbonyliron(II) fluorosulfonate. J Am Chem Soc 104:1869–1874
75. Weber L (1983) Metal complexes of sulfur ylides: coordination chemistry, preparative organic chemistry, and biochemistry Angew. Chem Int Ed Engl 22:516–528
76. Aggarwal VK (1998) Catalytic asymmetric epoxidation and aziridination mediated by sulfur ylides. Evolution of a project. Synlett 4:329–336
77. Tewari RS, Awasthi AK, Awasthi A (1983) Phase-transfer catalyzed synthesis of 1,2-disubstituted aziridines from sulfuranes and Schiff bases or aldehyde arylhydrazones. Synthesis 4:330–331
78. Franzen V, Driesen H-E (1963) Umsetzung von sulfonium-yliden mit polaren doppelbindungen. Chem Ber 96:1881–1890
79. Walfort B, Leedham AP, Russell CR, Stalke D (2001) Triimidosulfonic acid and organometallic triimidosulfonates: S^+-N^- versus S=N bonding. Inorg Chem 40:5668–5674
80. Walfort B, Stalke D (2001) Methylenetriimidosulfate $H_2CS(N^tBu)_3{}^{2-}$ – the first dianionic sulfur(VI) ylide. Angew Chem Int Ed 40:3846–3849
81. Walfort B, Bertermann R, Stalke D (2001) A new class of dianionic sulfur-ylides: alkylenediazasulfites. Chem Eur J 7:1424–1430
82. Deuerlein S, Leusser D, Flierler U, Ott H, Stalke D (2008) $[(thf)Li_2\{H_2CS(NtBu)_2\}]_2$: synthesis, polymorphism, and experimental charge density to elucidate the bonding properties of a lithium sulfur ylide. Organometallics 27:2306–2315
83. Stey T, Stalke D (2004) Lead structures in lithium organic chemistry. In: Rappoport Z, Marek I (eds) The chemistry of organolithium compounds. Wiley, New York, pp 47–120
84. Bickelhaupt FM, Solà M, Guerra CF (2006) Covalency in highly polar bonds. Structure and bonding of methylalkalimetal oligomers $(CH_3M)_n$ (M=Li-Rb; n = 1, 4). J Chem Theory Comput 2:965–980
85. Matito E, Poater J, Bickelhaupt FM, Solà M (2006) Bonding in methylalkalimetals $(CH_3M)n$ (M) Li, Na, K; n) 1, 4). Agreement and divergences between AIM and ELF Analyses. J Phys Chem B 110:7189–7198
86. Ott H, Däschlein C, Leusser D, Schildbach D, Seibel T, Stalke D, Strohmann C (2008) Structure/reactivity studies on an α-Lithiated benzylsilane: chemical interpretation of experimental charge density. J Am Chem Soc 130:11901–11911
87. Ott H, Pieper U, Leusser D, Flierler U, Henn J, Stalke D (2008) Carbanion or amide? First charge density study of parent 2-picolyllithium. Angew Chem Int Ed 48:2978–2982
88. Macchi P (2009) Resonance structures and electron density analysis. Angew Chem Int Ed 48:5793–5795

89. Bochmann M (1992) “Non-coordinating” anions: underestimated ligands. Angew Chem Int Ed Engl 31:1181–1182
90. Piers WE, Chivers T (1997) Pentafluorophenylboranes: from obscurity to applications. Chem Soc Rev 26:345–354
91. Piers WE (1998) Zwitterionic metallocenes. Chem Eur J 4:13–18
92. Kehr G, Fröhlich R, Wibbeling B, Erker G (2000) (N-Pyrrolyl)B$(C_6F_5)_2$ – a new organometallic Lewis acid for the generation of group 4 metallocene cation complexes. Chem Eur J 6:258–266
93. Flierler U, Leusser D, Ott H, Kehr G, Erker G, Grimme S, Stalke D (2009) Catalytic abilities of $[(C_6F_5)_2BR]$ ($R{=}NC_4H_4$ and NC_4H_8) deduced from experimental and theoretical charge-density investigations. Chem Eur J 15:4595–4601
94. Macchi P, Sironi A (2007) Interactions involving metals: from “chemical categories” to QTAIM, and backwards. In: Matta CF, Boyd RJ (eds) The quantum theory of atoms in molecules. From solid state to DNA and drug design. Wiley-VCH, Weinheim, pp 345–374
95. Macchi P, Proserpio DM, Sironi A (1998) Experimental electron density studies for investigating the metal π-ligand bond: the case of bis(1,5-cyclooctadiene)nickel. J Am Chem Soc 120:1447–1455
96. Himmel D, Trapp N, Krossing I, Altmannshofer S, Herz V, Eickerling G, Scherer W (2008) Reply Angew Chem Int Ed 47:7798–7801
97. Reisinger A, Trapp N, Krossing I, Altmannshofer S, Herz V, Presnitz M, Scherer W (2007) Homoleptic silver(I) acetylene complexes. Angew Chem Int Ed 46:8295–8298
98. Scherer W, Eickerling G, Shorokhov D, Gullo E, McGrady GS, Sirsch P (2006) Valence shell charge concentrations and the Dewar-Chatt-Duncanson bonding model. New J Chem 30:309–312
99. Farrugia LJ, Evans C (2005) Experimental X-ray charge density studies on the binary carbonyls $Cr(CO)_6$, $Fe(CO)_5$ and $Ni(CO)_4$. J Phys Chem A 109:8834–8848
100. Overgaard J, Clausen HF, Platts JA, Iversen BB (2008) Experimental and theoretical charge density study of chemical bonding in a Co dimer complex. J Am Chem Soc 130:3834–3843
101. Sparkes HA, Brayshaw SK, Weller AS, Howard JAK (2008) $[Rh(C_7H_8)(PPh_3)Cl]$: an experimental charge-density study. Acta Crystallogr B 64:550–557
102. Cheeseman JR, Carroll MT, Bader RFW (1988) The mechanics of hydrogen bond formation in conjugated systems. Chem Phys Lett 143:450–548
103. Keith TA, Bader RFW, Aray Y (1996) Structural homeomorphism between the electron density and the virial field. Int J Quantum Chem 57:183–198
104. Allegra G, Colombo A, Mognaschi ER (1972) Bis(cyclooctatetraene) iron. Evidence of dynamic tautomerism in the crystalline state. Gazz Chim Ital 102:1060–1067
105. Mann BE (1978) A carbon-13 nuclear magnetic resonance investigation of the ring exchange in (1-4-η^4-cyclooctatetraene)(1-6-η^6-cyclooctatetraene) iron. J Chem Soc Dalton Trans 1761–1766
106. Farrugia LJ, Evans C, Tegel M (2006) Chemical bonds without “chemical bonding”? A combined experimental and theoretical charge density study on an iron trimethylenemethane complex. J Phys Chem A 110:7952–7961
107. Silvi B, Gatti C (2000) Direct space representation of the metallic bond. J Phys Chem A 104:947–953
108. Jenkins S, Ayers PW, Kirk SR, Mori-Sánchez P, Pendás AM (2009) Bond metallicity of materials from real space charge density distributions. Chem Phys Lett 471:174–177
109. Gervasio G, Bianchi R, Marabello D (2004) About the topological classification of the metal-metal bond. Chem Phys Lett 387:481–484
110. Ponec R, Yuzhakov G, Sundberg MR (2005) Chemical structures from the analysis of the domain-averaged Fermi holes. Nature of the Mn-Mn bond in bis(pentacarbonylmanganese). J Comput Chem 26:447–454
111. Reinhold J, Kluge O, Mealli C (2007) Integration of electron density and molecular orbital techniques to reveal questionable bonds: the test case of the direct Fe-Fe bonds in $Fe_2(CO)_9$. Inorg Chem 46:7142–7147

112. Ponec R, Lendvay G, Chaves J (2008) Structure and bonding in binuclear metal carbonyls from the analysis of domain averaged Fermi holes. I $Fe_2(CO)_9$ and $Co_2(CO)_8$. J Comput Chem 29:138–1398
113. van der Maelen J, Gutiérrez-Puebla E, Monge A, García-Granda S, Resa I, Carmona E, Fernández-Díaz MT, McIntyre GJ, Pattison P, Weber H-W (2007) Experimental and theoretical characterisation of the Zn-Zn bond in $[Zn_2(\eta^5\text{-}C_5Me_5)_2]$. Acta Crystallogr B 63:862–868
114. Schultz S, Schumann D, Westphal U, Bolte M (2009) Dizincocene as a building block for novel Zn-Zn bonded compounds? Organometallics 28:1590–1592
115. Overgaard J, Jones C, Stasch A, Iversen BB (2009) Experimental electron density study of the Mg-Mg bonding character in a magnesium(I) dimer. J Am Chem Soc 131:4208–4209
116. Popelier PLA, Logothesis G (1998) Characterisation of an agostic bond on the basis of the electron density. J Organomet Chem 555:101–111
117. Pantazis DA, McGrady JE, Besora M, Maseras F, Etienne M (2008) On the origin of the α- and β-agostic distortions in early-transition-metal alkyl complexes. Organometallics 27:1128–1134
118. Tafipolsky M, Scherer W, Öfele K, Artus G, Pedersen B, Herrmann WA, McGrady GS (2002) Electron delocalisation in acyclic and *N*-heterocyclic carbenes and their complexes. A combined experimental and theoretical charge-density study. J Am Chem Soc 124:5865–5880
119. Scherer W, Öfele K, Tafipolsky M (2008) On the electron delocalisation in cyclopropenylidenes – an experimental charge density approach. Inorg Chim Acta 361:513–520
120. Scheins S, Messerschmidt M, Gembicky M, Pitak M, Volkov A, Coppens P, Harvey BJ, Turpin GC, Arif AM, Ernst RD (2009) Charge density analysis of the (C-C) $\rightarrow$ Ti agostic interactions in a titanacyclobutane complex. J Am Chem Soc 131:6154–6160
121. Bader RFW, Matta CF, Cortés-Guzmán F (2004) Where to draw the line in defining a molecular structure. Organometallics 23:6253–6263
122. McGrady GS, Sirch P, Chatterton NP, Ostermann A, Gatti C, Altmannshofer S, Herz V, Eickerling G, Scherer W (2009) Nature of the bonding in metal-silane σ-complexes. Inorg Chem 48:1588–1598
123. Scherer W, Eickerling G, Tafipolsky M, McGrady G S, Sirch P, Chatterton NP (2006) Elucidation of the bonding in $Mn(\eta^2\text{-}SiH)$ complexes by charge density analysis and T_1 NMR measurements: asymmetric oxidative addition and anomeric effects at silicon. J Chem Soc Chem Commun 2986–2988
124. Farrugia LJ, Macchi P (2009) On the interpretation of the source function. J Phys Chem A 113:10058–10067

Chapter 13
Charge Density in Materials and Energy Science

Jacob Overgaard, Yuri Grin, Masaki Takata, and Bo B. Iversen

13.1 Introduction

The electron density of a molecular system is probably the most information-rich observable available to natural science. It can be obtained either from quantum mechanical calculations or estimated experimentally from accurate X-ray diffraction data [1]. Even though great advances have occurred in computational chemistry, it is still difficult to theoretically estimate accurate charge densities (CDs) for systems containing heavy atoms or for low symmetry systems containing many atoms. It is a strong challenge for theory to mimic the periodic and stochastic boundary conditions, which are vital for materials and always present in the experiment. The X-ray CD technique recently has been revitalized due to major advances in the experimental methods with the availability of fast area detectors, intense short wavelength synchrotron radiation and stable helium cooling devices [2]. In the most used experimental method X-ray structure factors are fitted with aspheric atomic models containing both density and thermal motion parameters, and this provides an analytical representation of the static electron density [1] (see Chaps. 3 and 5). Alternatively, one may use the model-free maximum entropy method (MEM) to obtain a thermally smeared electron density [3]. From the electron density a quantitative understanding of chemical bonding and molecular structure can be obtained using e.g. the Quantum Theory of Atoms In Molecules (QTAIM) [4]

J. Overgaard • B.B. Iversen (✉)
Department of Chemistry and iNANO, Aarhus University, DK-8000 Arhus C, Denmark
e-mail: jacobo@chem.au.dk; bo@chem.au.dk

Y. Grin
Max-Planck-Institut für Chemische Physik fester Stoffe, D-01187 Dresden, Germany
e-mail: grin@cpfs.mpg.de

M. Takata
SPring8 Synchrotron Facility, Koto 1-1-1, Sayo-cho, Sayo, Hyougo 679-5148, Japan
e-mail: takatama@spring8.or.jp

C. Gatti and P. Macchi (eds.), *Modern Charge-Density Analysis*,
DOI 10.1007/978-90-481-3836-4_13, © Springer Science+Business Media B.V. 2012

and derived properties such as the electrostatic potential or electrostatic moments provide fundamental insight into intermolecular interaction energies between atoms and molecules [5] (see Chaps. 11 and 14). For weaker interactions such as hydrogen bonds, more advanced descriptors like analysis of the source function [6] or the electron localizability function may be required [7] (see Chaps. 1 and 9). The present book documents the amazing versatility of modern charge density research. In this chapter the focus will be on applications in materials science. It is, however, important to first point out the term charge density actually covers two things: the electron density and the nuclear density. One of the most important approximations of quantum mechanics, the Born-Oppenheimer approximation, provides a separation of electron and nuclear dynamics by assuming that the faster moving electrons instantaneously adjust to a given nuclear skeleton. It is also this approximation, which allows a separation of electronic and thermal effects in the commonly used models of charge density crystallography [1]. However, many important phenomena are actually dependent on the breakdown of the Born-Oppenheimer approximation and examples range from the essential electron transfer reactions in proteins to phonon scattering in solids. In general it is the specific chemical bonding environment which sets up the potential felt by the nuclei and determines the lattice dynamics. The two are intimately related and our survey of charge density in materials and energy science will include discussion on nuclear densities.

13.2 Thermoelectric Materials

Thermoelectric (TE) materials are functional materials, which are attracting much attention due to their dual ability of electrical-thermal energy conversion [8]. They are used either for cooling (electrical energy is consumed) or for energy production (conversion of heat to electrical energy). The discovery of a series of promising new TE materials in the mid-nineties [9] has led to an explosion of interest [10]. The new materials are based on a design strategy coined the "phonon-glass – electron crystal" (PGEC) concept. Briefly, a good thermoelectric material should conduct heat like an amorphous material and electricity like a crystal. The basic hypothesis for the thermoelectric properties of the PGEC materials is that a semi-conducting host structure results in a high Seebeck coefficient (S) and electric conductivity (σ), while extreme thermal motion of loosely bound guest atoms located in cavities gives a reduction of the thermal conductivity (κ). This results in a large thermoelectric "figure of merit", $ZT = TS^2\sigma/\kappa$. Normally a good conductor of electricity (large σ) is also a good conductor of heat (large κ), and this makes the optimization of ZT extremely difficult. What makes the new materials stand out is that they minimize the lattice part of κ without destroying σ. The vibrational energy of the guest atoms and the coupling of these vibrations to the lattice phonons of the host structure can be manipulated by atomic scale engineering of the structure and chemical function of the cavity.

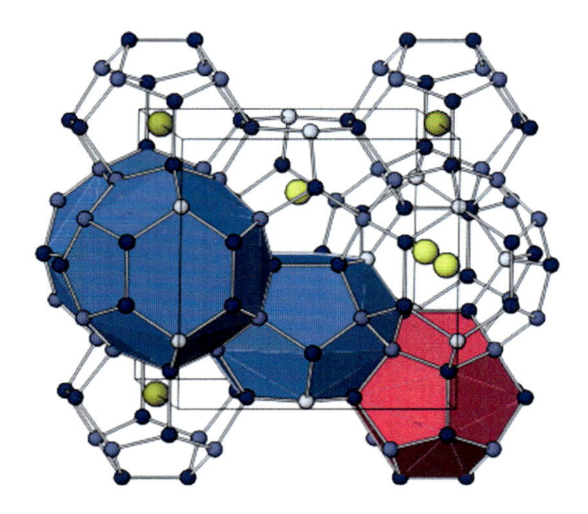

Fig. 13.1 The crystal structure of the clathrate type I with the large tetrakaidecahedral (24-atom) cage in *blue* and the dodecahedral (20-atom) cage in *red*

13.2.1 Inorganic Clathrates

One of the most important new classes of PGEC materials is the inorganic clathrates, which consists of a host framework of group III, IV or V atoms (or transition metals), while loosely bonded guest ions are trapped in the cages, Fig. 13.1 [11].

Most thermoelectric studies have concerned the type-I clathrate structure, $(M^{2+})_8 III_{16}IV_{30}$, Fig. 13.1. The primary guideline for understanding semiconducting clathrates has been the Zintl concept. In this model four electrons must be available for each tetradrahedrally bonded host atom. The guest atom is assumed to be ionic and to donate its valence electrons to the host structure, although in the case of the inverse clathrates the guest atoms accept electrons. Thus, in $Ba_8Ga_{16}Ge_{30}$ each of the eight Ba guest atoms donates two electrons to the host structure giving a total of 16 transferred electrons. The valences of the 16 Ga and 30 Ge atoms in the host structure are three and four, respectively. This gives a total of 184 valence electrons available for covalent bonding between the 46 host structure atoms, and the compound is expected to be a semiconductor. However, in fact most of the clathrate samples reported in the literature show metallic conduction. There are at least three potential explanations for this: (i) incomplete guest to host charge transfer, (ii) external doping by impurities and (iii) Non-stoichiometric composition. If the valences of the constituent atoms are not precise (case i), clearly it will have a strong influence on the physical properties such as the resistivity, where e.g. a slight lack of electron donation from the guest atoms will make the compound a p-type conductor. Clathrate studies were therefore from the beginning concerned with charge density issues [12]. With regard to case ii and iii detailed studies e.g. of zone melted purified samples have shown that the carrier concentration is determined by the overall stoichiometry and that external impurities have little influence on the transport properties [13]. There are many structural aspects, which are important for understanding the thermoelectric properties of clathrates. For the guest atoms the exact position and the atomic motion is important, i.e. are the atoms located in the centers of the cages or are they displaced towards the cage wall and statistically disordered (static or dynamic) over several sites. The guest atom position has a bearing on the nature of the host-guest interactions and thus the ability of the rattling guest atoms to scatter heat carrying phonons. Another question is whether these atoms are rattling, i.e. whether they exhibit low frequency motion that can hybridize with the acoustic phonons. Since anharmonicity is an important phonon scattering mechanism at high temperature it is also important to quantify the anharmonic contribution to the atomic motion.

The first series of clathrate diffraction studies addressed the location and atomic motion in the type I clathrates [14]. Chakoumakos et al. reported multi-temperature single crystal as well as powder neutron diffraction data on $Sr_8Ga_{16}Ge_{30}$ [14a]. The structural refinements with guest atoms centered in the cavities clearly showed that for the guest atom in the large cage, Sr(2), the mean square displacement, $\langle u^2 \rangle$, was an order of magnitude larger than for the framework atoms and the Sr atom in the small cage, Sr(1). Difference Fourier maps in the (100) plane through the Sr(2) site with the Sr(2) atom removed from the reference model, which in the case of neutron data gives the nuclear density, showed non-ellipsoidal features with peaks in the axes directions. This led Chakoumakos et al. to suggest a four site disorder model, but it was soon realized that this is too simple an approximation [15]. The complex nature of the disorder of the Sr(2) atom in $Sr_8Ga_{16}Ge_{30}$ and the Ba(2) atom in $Ba_8Ga_{16}Ge_{30}$ was directly revealed by MEM calculations based on single crystal X-ray diffraction data [15]. The advantage of the MEM electron density is that it is model free and the problem with Fourier termination ripples is minimized. However, a direct plot of the thermally smeared electron density distribution only showed a very diffuse atomic density with no peaks to define possible disorder. For this reason MEM deformation densities were introduced, which are differences between the observed electron density and the electron density of a superposition of free spherical atoms (promolecule). MEM difference densities show the combined effects of chemical bond deformation, charge transfer and structural disorder. The MEM deformation densities of the M(2) guest atom in the type I clathrates $Sr_8Ga_{16}Ge_{30}$ and $Ba_8Ga_{16}Ge_{30}$ have a negative central region surrounded by a positive region, Fig. 13.2. This was direct evidence for the structural disorder on the M(2) site.

Following the initial structural studies many other structural studies have been published on thermoelectric clathrates [16]. Sales et al. reported 15 K single crystal neutron data on $M_8Ga_{16}Ge_{30}$ with M = Ba, Sr and Eu [17]. Neutron diffraction data are preferred to X-ray diffraction data in analysis of disorder since they do not contain superimposed effects from chemical bonding. Based on difference Fourier nuclear densities for the three M(2) ions it was concluded that the Sr and Eu ions are positioned off center (0.3 Å for Sr and 0.4 Å for Eu), while for Ba "the nuclear density is broad, but centered at the center of the cage". Thus, the Ba atoms were believed to be without disorder and behaving fundamentally different from Sr and Eu. At the same time Sales et al. observed that in their samples there was a striking difference between the lattice thermal conductivity of $Ba_8Ga_{16}Ge_{30}$ and the lattice

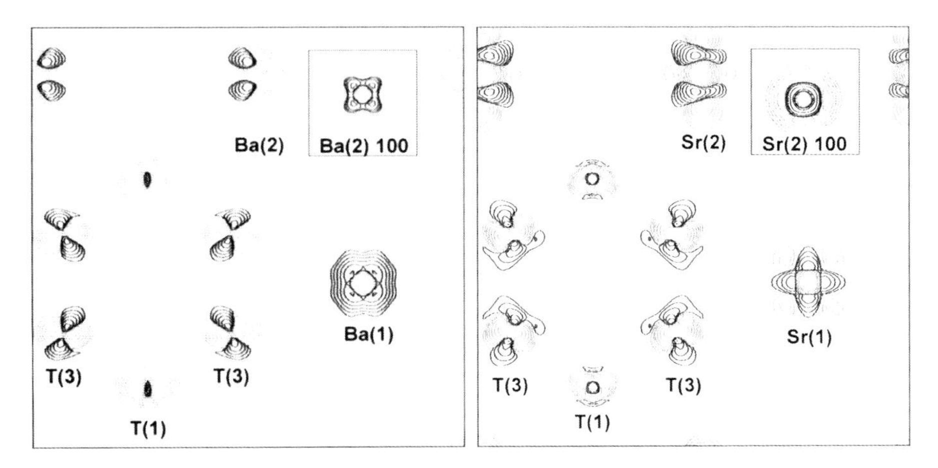

Fig. 13.2 MEM deformation electron density of $Sr_8Ga_{16}Ge_{30}$ and $Ba_8Ga_{16}Ge_{30}$. The *main panel* shows the (001) plane, while the *insert* shows the (100) plane for the M2 guest atom [15]. *T* denotes framework atoms, in this case the disordered Ge or Ga

thermal conductivities of $Sr_8Ga_{16}Ge_{30}$ and $Eu_8Ga_{16}Ge_{30}$. They suggested that for these compounds both rattling and tunneling states due to disorder are necessary to produce the glasslike thermal conductivity observed for the Sr and Eu samples. The nuclear densities of $Sr_8Ga_{16}Ge_{30}$ and $Eu_8Ga_{16}Ge_{30}$ were thought to represent a snapshot of the tunneling states.

The idea of disorder and tunneling states in the Sr and Eu systems but not in the Ba systems and the corroborative nature of the difference Fourier nuclear density and lattice thermal conductivity became established in the clathrate literature [18], even though evidence appeared against this model. Thus, very high resolution neutron diffraction studies revealed clear discrepancies with the disorder/tunneling picture. The most comprehensive study so far is a multi-temperature single crystal neutron diffraction study of $Ba_8Ga_{16}Si_{30}$ [19]. In this study data were collected at 15, 100, 150, 200, 250, 300, 450, 600, 900 K. In Fig. 13.3 the MEM nuclear *deformation density* in the (100) plane of Ba(2) is shown at 15, 300 and 900 K together with the total MEM nuclear density at 15 K. Just as the Fourier difference density of $Ba_8Ga_{16}Ge_{30}$ from Sales et al. [17] the MEM nuclear density of Ba(2) in $Ba_8Ga_{16}Si_{30}$ is broad and centered at the center of the cage. The Ba(2) nuclear density becomes much more diffuse with temperature, but there is no clear evidence of disorder. However, just as the subtle chemical bonding details of an electron density distribution can be revealed through *electron* deformation densities [1], subtle structural disorder features can be revealed by *nuclear* deformation densities. In the MEM nuclear deformation densities of Fig. 13.3 the reference model is based on a Ba(2) atom centered in the cage. This map provides direct evidence that the Ba(2) atom in $Ba_8Ga_{16}Si_{30}$ is disordered, and examination of the deformation density in three dimensions show that the disorder is similar to that observed in the MEM electron density deformation maps of $Sr_8Ga_{16}Ge_{30}$ and $Ba_8Ga_{16}Ge_{30}$, Fig. 13.2.

Fig. 13.3 MEM difference densities, ρ(MEM) – ρ(Non-Uniform Prior), in the (100) plane through Ba(2) in $Ba_8Ga_{16}Si_{30}$ at 15, 300, and 900 K. The difference densities are plotted on a logarithmic scale, 0.0125×2^N (n = 0, . . . ,6), and solid contours are positive, dotted contours are negative. The plot at the lower right is the total MEM nuclear density at 15 K

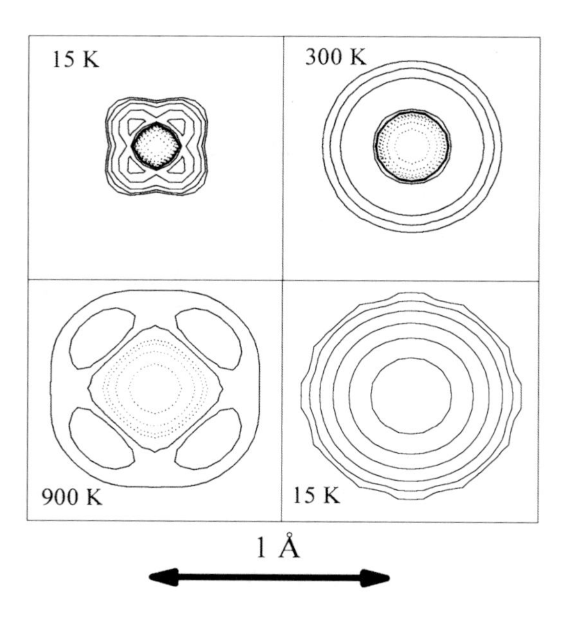

During the last 5 years single crystal neutron diffraction studies have been carried out on $Ba_8Ga_{16}Ge_{30}$, $Ba_8Al_{16}Ge_{30}$, and $Ba_8Zn_8Ge_{38}$ at the SCD instrument at the Intense Pulsed Neutron Source, Argonne National Laboratory in order to address the detailed nature of the clathrate host-guest chemistry [20]. These studies were strengthened relative to the initial $Ba_8Ga_{16}Si_{30}$ study by the availability of a second detector at the SCD instrument, which allowed data collection to extremely high resolution (>2 Å^{-1}). There are many aspects to discuss in connection with the Ba(2) dynamics in different clathrate framework, but here we will just point out one important point. In Fig. 13.4 the Ba(2) nuclear density is shown for two different $Ba_8Al_{16}Ge_{30}$ samples having slight differences in the framework stoichiometry and Al siting. However, even these slight differences in the framework lead to fundamental differences in the shape of the Ba(2) nuclear density. In one structure Ba(2) has a disorder similar to the $Ba_8Ga_{16}Ge_{30}$ and $Ba_8Ga_{16}Si_{30}$ structures discussed above, but in the other the nuclear density has a prolate shape. Overall, the Ba(2) nuclear density, and therefore the Ba(2) lattice dynamics, is found to be intimately related to the specific host structure, i.e. the specific distribution of the group III atoms. The nuclear density studies have shown that control of the host structure is the crucial element in controlling the thermoelectric properties of clathrates.

A lot of the discussions on clathrates revolve around the fact that they are host-guest systems, and that it is the nature of the host-guest interactions, which determine their thermo-electric properties. Transport properties are commonly understood in reciprocal space through band structure calculations, but chemistry takes place in real space, or at least in Hilbert space when considering orbital interactions [21]. So how can the host-guest interactions in clathrates be described

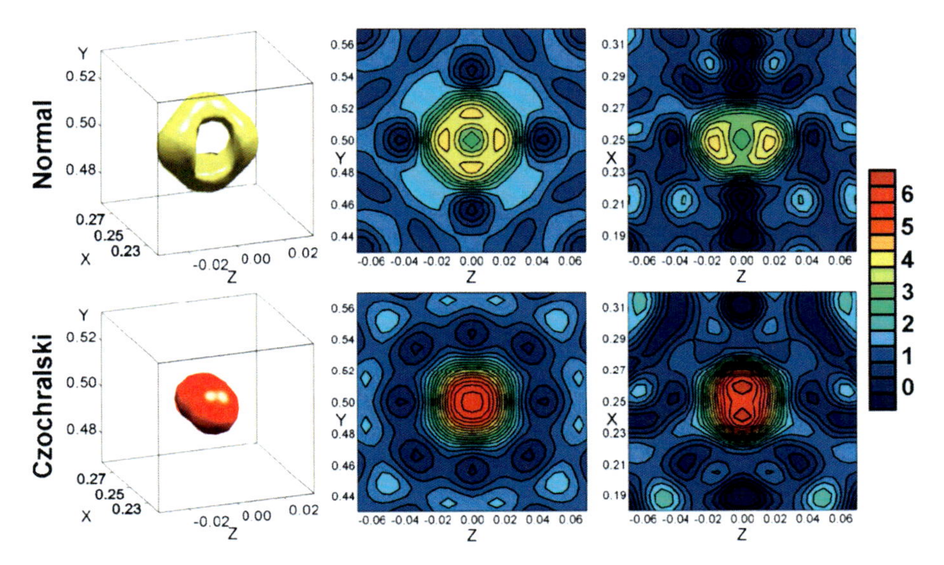

Fig. 13.4 Fourier nuclear difference density maps of the Ba(2) guest atom in the (100) plane obtained from refining neutron diffraction data with a model omitting Ba(2). (*top*) Normal stoichiometric synthesis $Ba_8Al_{16}Ge_{30}$ at 25 K, and (*bottom*) Czochralski synthesis $Ba_8Al_{16}Ge_{30}$ at 25 K

in terms of chemical bonding models? It turns out that this question is quite tricky. In one of the first theoretical papers on clathrate bonding Blake et al. showed that the guest atoms act as electron donors in accordance with the Zintl model [12]. However, when examining the electron densities in real space it was observed that the Ba(2) atom in $Ba_8Ga_{16}Ge_{30}$ had a very similar charge density to a free Ba atom. This indeed seemed contradictory, but Blake et al. concluded that "a charge density analysis reveals that Ba and Sr in clathrates are neutral". The theoretical charge densities were in agreement with the MEM X-ray charge densities [15]. Thus, it was experimentally observed that the difference between the guest atom charge density in the clathrate and the corresponding free atom charge density is very small. In simple terms it looked like the guest atoms are neutral and that the Zintl model may not be valid for clathrates.

It turns out that the mystery of the host-guest interactions in clathrates is similar to a famous discussion that took place almost 100 years earlier about the origin of the bond in ionic crystals such as NaCl [22]. It was early on observed that diffraction data of KCl showed the absence of even reflections, and this seemed to imply that K and Cl have the same number of electrons or, in other words, that the atoms of KCl are present as monovalent ions. In the crystal of NaF, consisting of much lighter atoms, the same reflections were not absent, owing to the different thermal motion of the two ions. It was difficult to make strong conclusions because the accuracy of X-ray diffraction measurements at that time was not deemed sufficient for determining such fine details. The real problem, however, was not only the treatment of the effect of the thermal motion (and extinction). The problem was that the chlorine acceptor orbital is located in the same spatial region as the Na donor orbital, and thus the crystal electron density looks very similar in the two cases. The NaCl example underlines that charge transfer cannot be defined from geometrical considerations only, and other criteria had to be developed. The main development was to introduce a quantum mechanical definition of an atom in a molecule [4]. Indeed, for the clathrates the spatial region of the Ba valence orbitals is almost the same as the spatial region of the host structure acceptor orbitals [23]. However, when the atomic volumes are defined from quantum mechanics the outcome is that the guest atoms in clathrates are highly ionic with charges close to +2 [23]. Thus, the dominant contribution to the host-guest bond is ionic. Nevertheless, the QTAIM analysis also shows that there are covalent effects, and the guest atoms form preferred directional bonds to specific host structure atoms. It is the relative strength of these directional bonds that e.g. makes Sr move far off-center in $Sr_8Ga_{16}Ga_{30}$ compared with Ba in $Ba_8Ga_{16}Ga_{30}$ [23].

13.2.2 Charge Density and Understanding of Atomic Interactions in Derivatives of Skutterudites

A second large group of cage compounds showing pronounced thermoelectric activity are derivatives of the mineral skutterudite ($CoAs_3$) which are often simply called *skutterudites* [24]. The general chemical formula for these compounds is $M_xT_4E_{12}$, where M can be either alkaline, alkaline earth, or the rare-earth metals uranium or thorium, T is a transition metal mainly belonging to groups 8–10, and E is a main group element mainly from groups 13–16. In this large family of materials the most promising thermoelectric materials are compounds with antimony in the place of arsenic. In particular they have been tested for use in space applications [ref [25] and references therein]. Thus, the understanding of the nature of the thermoelectric potential in this family of materials is driven not only by the basic research but also by commercial interests. In the following discussion we will focus mainly on the antimony-containing skutterudites.

The crystal structure of skutterudite and its derivatives can be represented in three different ways (Fig. 13.5). In the first view, the transition metal atoms are understood as forming a primitive cubic sub-lattice (edge length ca 4.5 Å). The square-shaped Sb_4 groups and, eventually, the filler atoms M occupy the $[T_8]$ cubes. In this representation, only the covalent interactions within the Sb_4 group are emphasized, while other interactions are not explicitly defined. The second view focuses on the vertices-condensed $[CoSb_{6/2}]$ octahedrons forming a 3D framework with M atoms in the larger icosahedral cavities. This representation suggests different interactions within the framework and between the framework and the filler atom. The third representation is similar to the second and has been developed rather recently, showing 3D T_4Sb_{12} framework formed by Sb-Sb and T-Sb bonds with the M atoms in the $[Sb_{12}]$ cages.

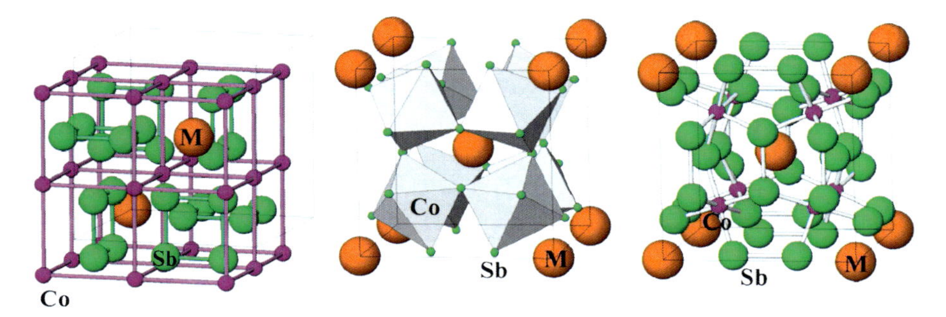

Fig. 13.5 Representations of the crystal structure of skutterudites $M_xCo_4Sb_{12}$: (*left*) Co atoms forming primitive cubic sublattice with the square Sb_4 groups and M atoms located in the centres of the $[Co_8]$ cubes. (*middle*) vertices-condensed $[CoSb_{6/2}]$ octahedrons forming a 3D framework with M atoms in the larger icosahedral cavities. (*right*) covalently bonded 3D framework Co_4Sb_{12} with the M atoms in the $[Sb_{12}]$ cages

In contrast to the intermetallic clathrates (see above), where the thermoelectric properties of the 'empty' frameworks are still not investigated (the first representative of this group – $\square_{24}Ge_{136}$ – was reported recently [26] with bulk preparation remaining a challenge), the non-filled skutterudite $CoSb_3$ reveals remarkable thermoelectric activity [27]. Even the 'empty' Skutterudite structure $CoSb_3$ has a relatively low thermal conductivity, which in this case cannot be caused by the phonon scattering on the filler atoms. Thus the question about the chemical bonding arises. Concerning the prototype Skutterudite $CoAs_3$ Pauling stated already in 1978 that '... in skutterudite ... the electronegativity of arsenic, 2.0, is so close to the electronegativity of cobalt, 1.9, that the bonds have very little ionic character ...' [28]. The first combined experimental (photoemission and photoabsorption spectra) and theoretical analysis of the bonding mechanism in $CoSb_3$ reveals, by exploiting the charge density maps, that the Sb-Sb and Co-Sb bonds are qualitatively similar [29]. Thus, the crystal structure of $CoSb_3$ may be described neither by means of Sb_4 rings inside the cubic cobalt sub-lattice nor by means of independent $CoSb_6$ units, but instead as a 3D covalently bonded network structure.

In order to obtain a direct confirmation of the character of the Co-Sb and Sb-Sb bonds, the charge density was experimentally examined [30] by the MEM/Rietveld method (see Sect. 13.4). The total charge density distributions are shown in Fig. 13.6, right. The two different Sb-Sb bonds within the Sb_4 ring have two different types of charge density overlaps yielding midpoint densities of 0.50 e/Å^3 for the shorter bond I and 0.35 e/Å^3 for the longer bond II. This lead to the suggestion, that the Sb_4 ring may be considered rather as two Sb_2 dumbbells. Moreover, the Co-Sb bond reveals an even larger density of 0.52 e/Å^3 at the midpoint, showing a direct similarity to the covalent Sb-Sb bonds. The results of the MEM charge density were in general confirmed by the calculations (LMTO method, [31] Fig. 13.6, left). Theory and experiment yield very similar isosurface of the total electron density at the value of 0.4 e/Å^3 and quite similar charge density distributions in the planes

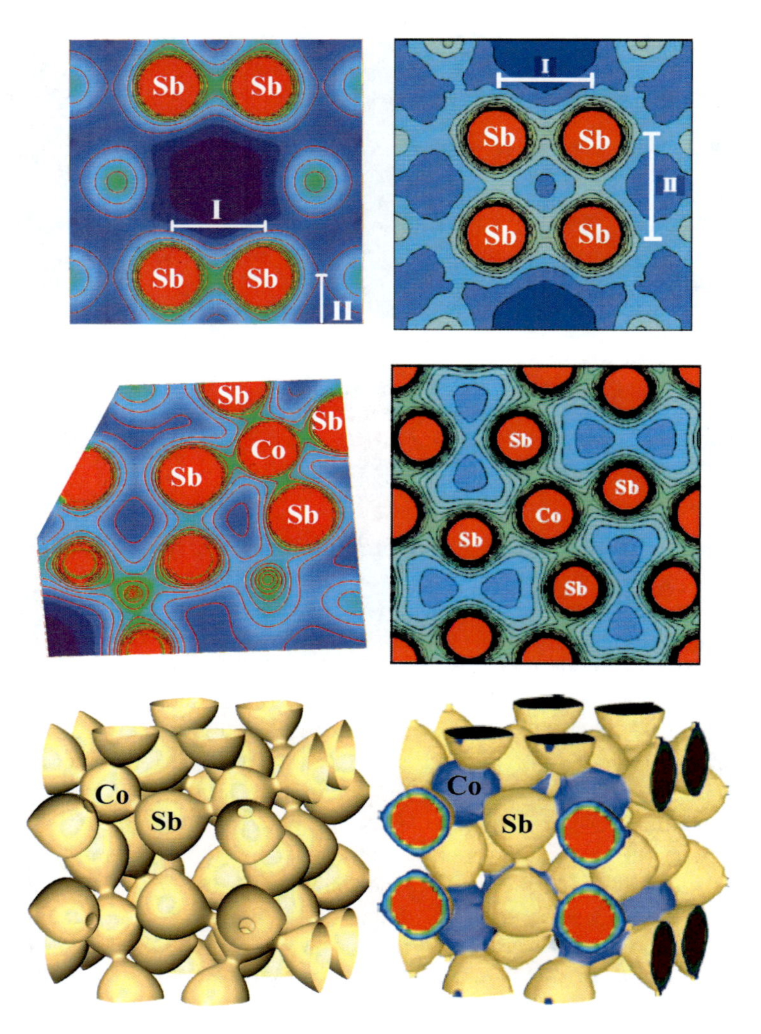

Fig. 13.6 Electron density in $CoSb_3$: (*left*) calculated by LMTO method; (*right*) experimental MEM charge density. (*top*) Density distribution in the plane of the square groups Sb_4. (*middle*) Density distribution in the equatorial plane of the $CoSb_6$ octahedron (calculated) and cumulated MEM density for $0.18 < z < 0.32$ projected along [001] in the same region. (*bottom*) calculated and experimental isosurface of total electron density at the density value of 0.4 e/A^3

of the Sb_4 rings and the equatorial planes of the $CoSb_6$ octahedrons. The essential differences are only in the density values in the bond midpoints. The calculations reproduce the experimental ratios but absolute values are remarkably smaller: Sb-Sb bond I 0.39 e/Å^3, Sb-Sb bond II 0.33 e/Å^3, Co-Sb bond 0.43 e/Å^3. Overall the obtained pattern is quite similar to the third crystal structure representation in Fig. 13.5 with the question remaining about the nature of the bonding in the Sb_4 square.

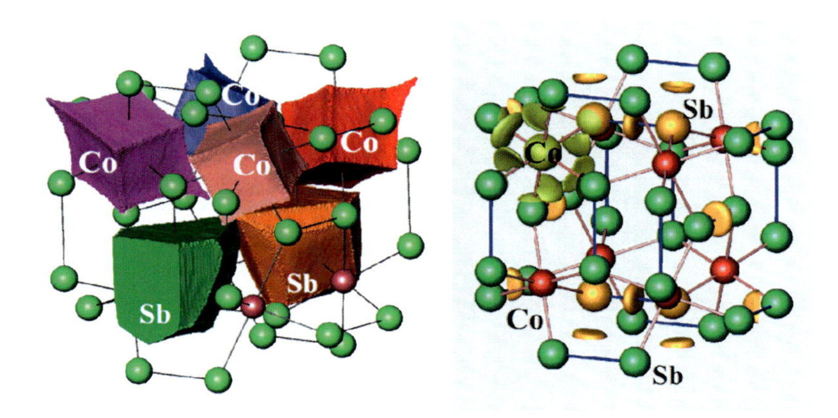

Fig. 13.7 (*left*) Cobalt and antimony QTAIM atoms in $CoSb_3$. (*right*) Isosurfaces of electron localizability indicator in $CoSb_3$ revealing the covalent interactions between Sb atoms and between Co and Sb atoms as well

This question was clarified by application of the techniques of the electron localizability. Calculation of the electron localizability indicator (ELI) [7] for $CoSb_3$ [31] and for $NaFe_4Sb_{12}$ [32] reveals clearly two topological features of the ELI in these compounds. First, the penultimate shell of the transition metal atoms (iron and cobalt) is highly anisotropic indicating the participation of the electrons of these shells in the valence region [33]. Further, clear maxima of ELI were found on both of the different Sb-Sb contacts and on the Co-Sb contacts (Fig. 13.7 right) confirming the covalent character of all three types of the atomic interactions in the non-filled and in the filled skutterudites. In the ELI representation, the sodium in $NaFe_4Sb_{12}$ shows a spherical distribution around the M position, but only with two shells, which clearly reveals the ionic interaction of sodium with the polyanionic framework [32]. For both compounds, the QTAIM analysis (the QTAIM atoms in $CoSb_3$ are shown in Fig. 13.7, left) reveals only a small electron transfer from cobalt (iron) to antimony: for $CoSb_3$ the QTAIM charges are $Co^{+0.45}$ and $Sb^{-0.15}$ [31]. In combination the results provide a description of antimony-containing skutterudites as 3D frameworks with the composition T_4Sb_{12} per unit cell held together by covalent Sb-Sb and T-Sb bonds. Depending on the presence (and type) of the filler atoms, the framework may be neutral (non-filled skutterudites) or play the role of an anion (filled skutterudites).

The ionic interaction of the filler with the framework is indirectly confirmed by the difference electron density distribution around the M position in the iron-antimonides $M_xFe_4Sb_{12}$ of Na, Yb and La. In all independent crystallographic directions ([100], [110] and [111]) the charge density distribution is spherical (Fig. 13.8). In accordance with this all attempts to find an off-centre position of the M cation by crystal structure refinements at different temperatures up to 110 K failed [34, 35]. This was confirmed by the total energy calculations for the models with different off-centre deviations of sodium in $NaFe_4Sb_{12}$, where any deviation yields

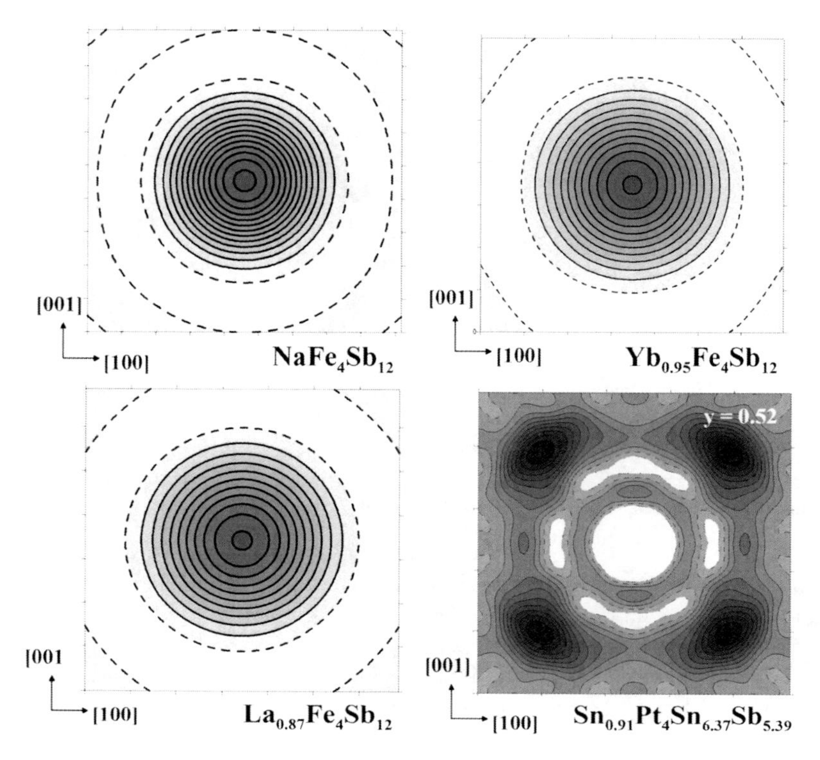

Fig. 13.8 Difference electron density distribution around the M position in the filled skutterudites with the fillers with different bonding abilities. The contour lines are drawn with the step of 5 e/Å^3 for $NaFe_4Sb_{12}$, 20 e/Å^3 for the rare-earth metal skutterudites and 1 e/Å^3 for the tin compound

an increase of the total energy of the lattice [32]. Plots of the isotropic displacement factor versus temperature from X-ray [35] and neutron diffraction data [36] do not show residual values at the extrapolation to zero temperature. This corroborates that the interaction between the fillers M and the framework for the mono-, two- and three-valent fillers is mainly ionic. However, the participation of the d electrons of Ba or La, or of the f electrons of other rare-earth metals in this interaction remains an open question. The detection of such interactions in real space may become possible applying the decomposition of the ELI into the partial contributions taking into account different ranges of the electronic DOS [36]. Using this procedure the spatial contributions of the f electrons of the rare-earth atoms to the interaction with their ligands in $La_7Os_4C_9$ (direct La-Os bonds) [37] or in EuT_2Ga_8 (T = Co, Rh, interaction of europium with the atoms of the second coordination shell) were visualized [38].

A completely different distribution of the difference electron density was found in the filled skutterudite $Sn_{0.91}Pt_4Sn_{6.37}Sb_{5.39}$ (Fig. 13.8, right bottom). Here, clear maxima of the charge density in the distance of 1 Å from the centre of the cavity

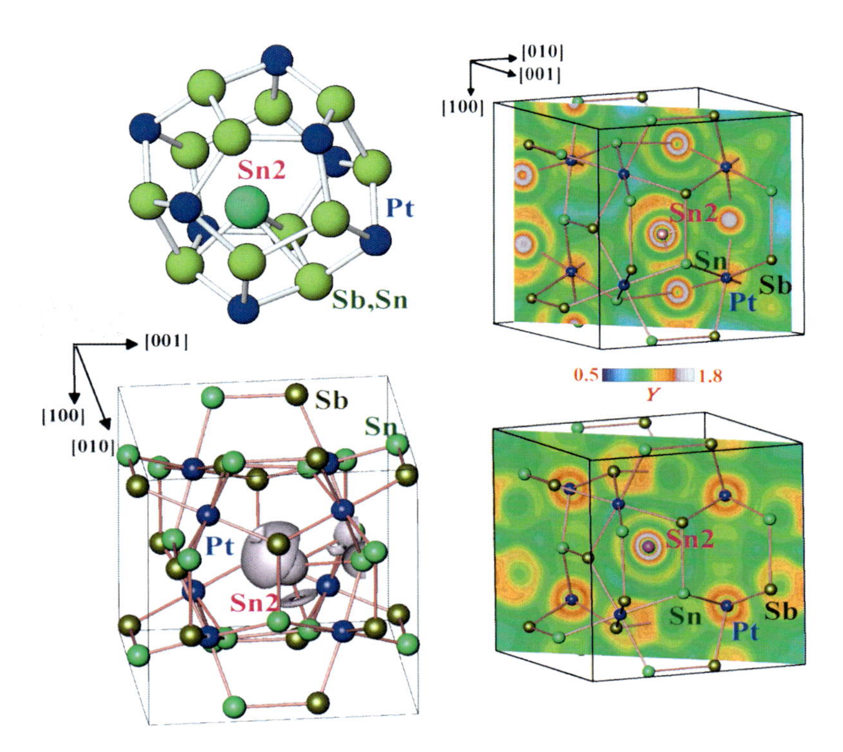

Fig. 13.9 Covalent interaction of the filler atom with the framework in $Sn_{0.91}Pt_4Sn_{6.37}Sb_{5.39}$: (*left top*) Tin atom located off-centre of the $(Sn,Sb)_{12}$ cage. (*left bottom*) ELI reveals direct interactions between the central atom with the tin and antimony atoms of the cage's wall as well as lone-pair like region on the opposite side of the filler atom. (*right*) Distribution of ELI around the filler atom in the hypothetical central position (*top*) and off-centre position (*bottom*)

were found. The complete analysis of the density distribution in this region was possible only after the identification of the filler atoms by NMR. The tin atom is located close to the wall of the cavity (Fig. 13.9, top left). Assuming an ordering of the central atoms in the off-centre positions, the distribution of the ELI around this position reveals direct bonds of the central tin atoms to the atoms of the wall as well as lone-pair like distributions on the opposite side of the central atom pointing into the cavity (Fig. 13.9, bottom left). This gives a clear picture of a tin atom, which is three-bonded to the wall and has a lone pair.

The ELI analysis is an appropriate tool to enhance the traditional analysis techniques of the charge density especially in cage compounds. Model calculations of ELI distributions for a tin atom in the centre and off-centre position (Fig. 13.9, right) reveal a high sensitivity of this functional to the bonding behaviour of atoms in the cages of a framework.

To conserve charge neutrality, tin atoms are also participating in the formation of the framework together with antimony. The distribution of the electron density in the vicinity of the framework positions was successfully described by two partially occupied Sn and Sb positions in full agreement with the different covalent radii of tin and antimony. The distribution of ELI confirms the covalent character of the atomic interactions within the framework.

In total, the understanding of the crystal structure and bonding in the filled skutterudite $Sn_xPt_4(Sn_{1-y}Sb_y)_{12}$ was only possible by a combined analysis of the charge density with NMR spectroscopy and calculations of the ELI. The four-valent filler tin is not acting as a cation in a skutterudite framework but is instead covalently bonded to the wall of the framework. Due to this bonding behaviour, $Sn_xPt_4(Sn_{1-y}Sb_y)_{12}$ does not exhibit notable thermoelectric activity.

13.3 Magnetic Materials

There has been an increasing interest in the study of the nature of the chemical bonding between metal atoms in dimers, cluster compounds and extended systems utilizing both experimental and theoretical methods [39]. Discussion about this topic, along with many interesting examples, can also be found in the earlier Chaps. 1, 10 and 12. We first discuss complexes containing two metal atoms in rather close proximity. Such systems comprise the smallest unit, and the studies then move to larger systems with longer range electronic communication between metal atoms leading finally to bulk magnetic systems.

The first complex to be discussed is a (non-magnetic) Co-complex, which exhibits a very unique bonding environment at its molecular center, Fig. 13.10 [40]. In simple terms, the complex is made from the chemical reaction of $Co_2(CO)_8$ and an alkynol leading to the addition of the triple bond to the $Co_2(CO)_8$ unit. This leads to a loss of a CO group from each Co and an apparent reduction of the triple bond to a double bond. The result is a strained center created by four atoms – the two Co-atoms and the two carbon-atoms from the saturated moiety – sitting at the corners of a slightly distorted tetrahedron. Several charge density studies have demonstrated the absence of a bcp between the two metal atoms when the two atoms are bridged by a common ligand. However, in this particular type of compound, the literature is full of conflicting observations [39]. There seems, however, to have been reached a consensus that there is no chemical bonding between two Co-atoms placed in geometries as this one, at least when topological analysis of the total electron density is used as a criterion for bonding interactions [41]. However, theoretical studies on the archetypal complex $Co_2(CO)_7$ has suggested that use of energy density criteria did in fact lead to indications of a stabilizing interaction between the two Co-atoms, which is not inferable from a topological analysis of the electron density [42].

Since the theoretical analysis [42] was carried out on an isolated molecule it was clearly of interest to get an experimental view point. The X-ray charge density study involved multipole modeling of single-crystal X-ray diffraction data collected at very low temperatures at the synchrotron beamline D3 at Hasylab on a complex in which the alkyne was 1-ethynyl-1-cyclohexanol. This crystal

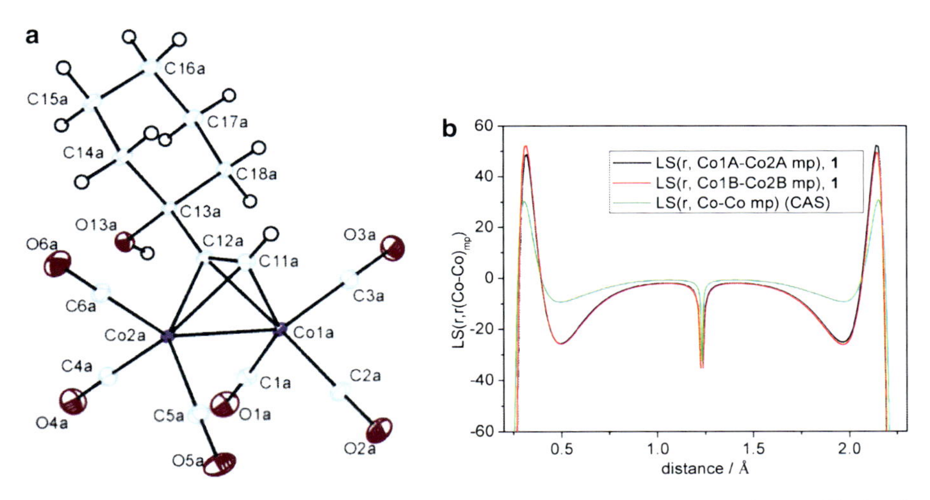

Fig. 13.10 (**a**) ORTEP drawing of the Co-dimer complex based on synchrotron data. (**b**) The local source, LS, evaluated at the bond midpoint along the Co-Co interatomic line

surprisingly contained two different molecules in the asymmetric unit which were identical except from the rotation of the cyclohexanol group by about 20° around a common axis. The experimental results were combined with high-level complete active space calculations on a gas-phase molecule of one of the two distinct molecules in the crystal structure. These revealed that the molecular complex more appropriately can be designated a singlet di-radical, in which the electronic structure has partial occupancy of Co-Co bonding and anti-bonding orbitals and no direct Co-Co bond. The theoretical result was entirely substantiated by the experimental charge density, which did not show any direct Co-Co bcp. On the other hand, the experimentally approximated energy density distribution both along the Co-Co line, and also along a line perpendicular to and crossing this line at its midpoint, displayed clear minima and close to negative value in the Co-Co region which may be interpreted as a signature of a stabilizing interaction. However, evaluation along the Co-Co line of the local source function [43] contributions to the electron density at the Co-Co midpoint showed a decrease in a region close to the midpoint, Fig. 13.10. It has been shown theoretically that the "drop" in the local source in such region comparatively decreases as the bonding between the atoms is included [44]. Similarly, the integrated source function from the Co-basins was close to zero and negative, showing that the Co atoms act as sinks for the density at the Co-Co midpoint. It has been established that the source function contribution to the density at a midpoint, or eventually at the bcp, increases to positive values from the neighboring metal atoms as this bonding interaction is increased in strength [44].

The study of the Co dimer complex revealed subtleties that required special care. The strained and rather uncommon bonding environment at the Co_2C_2-center was found to be difficult to describe properly based on the experimental data, and it was moreover obvious that the topological description in terms of the expected three

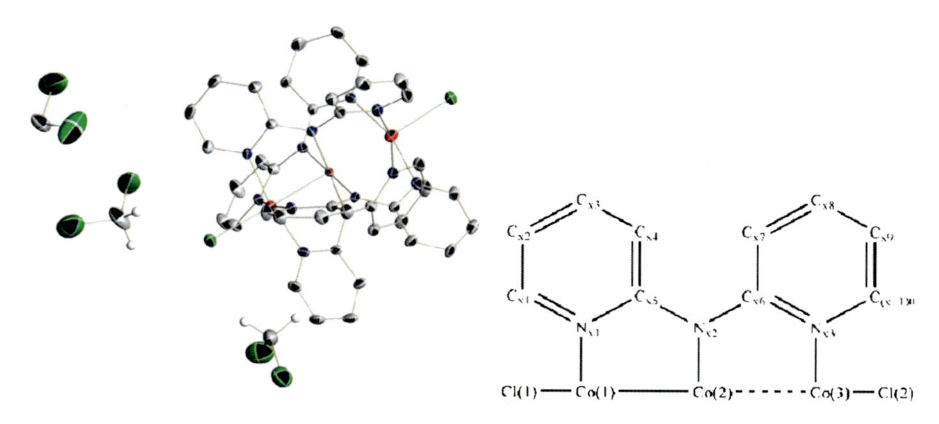

Fig. 13.11 (*left*) ORTEP drawing of the unsymmetrical molecule in the EMAC showing the three dichloromethane molecules. (*right*) Schematic showing the dipyridylamide ligands and the numbering scheme

bond critical points (bcp) and one ring critical point (rcp) within the constituting CoC_2-triangles was significantly model-dependent. This is due to the presence of a large electron-rich metal atom close to two carbon atoms, and while the theoretical density reproduce the two (non-straight) Co-C bond paths as well as the rcp at the centre of each of the CoC_2-planes, this was not clear from the experimental density. Modeling of dimer complexes clearly is very difficult, yet such complexes are only an entry into more complicated structures.

The next example is a significantly larger system with highly intriguing structural and magnetic properties. The system belongs to the family of extended metal atom chain compounds (EMAC's), which are comprised of linear chains of directly connected metal atoms bridged by different types of ligands and capped at the ends by a negative ion (often a halogen atom), Fig. 13.11 [45]. These systems exhibit large structural and magnetic changes with temperature and they have also been investigated as potential molecular scale electronic conductors (For a recent review of molecular electronics, see [46]). The first CD study of such systems concerned $Co_3(dpa)_4\ Cl_2 \cdot nCH_2Cl_2$ (dpa: dipyridylamide), where n is approximately 2.11, a value which is based on IAM refinements of the occupancy factors of three different solvent sites in the structure. There exists two different isomers of the complex $Co_3(dpa)_4Cl_2 \cdot nS$, either symmetrical (prefix *s*) or un-symmetrical (*u*) with respect to the Co-Co-Co bonding environment, and the isomer type is determined by the amount of solvent in the structure. The origin of the structural isomerism has been widely discussed, and it is not clear whether it is an intrinsic molecular effect or due to the surrounding environment. Over the past decades a wide range of studies on linear chain metal complexes have been carried out [47]. These studies have shown that the two types of molecular isomers of $Co_3(dpa)_4Cl_2 \cdot nS$ (i.e. *s* and *u*) are only observed in the solid state [48]. Thus, in solution only a single isomer is present (s–$Co_3(dpa)_4Cl_2 \cdot nS$), and this suggests that formation of isomerism is not due to

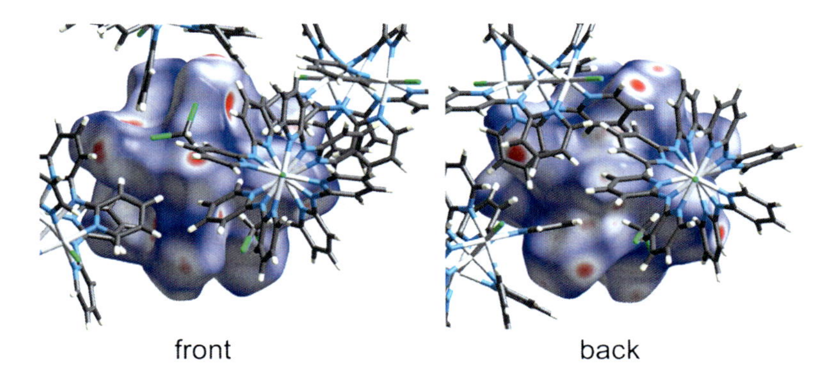

Fig. 13.12 Hirsfeld surfaces for the unsymmetrical Co3(dpa)2Cl2 complex showing the nearest neighbor environment around each Cl ligand. The view is along the Cl-Co-Co-Co-Cl axis of the complex within the surface, and only molecules with an atom within 5 Å of the Cl ligand are shown

an intrinsic molecular property, but determined by the crystal environment. Overall, the EMACs belong to a general class of systems where the potential energy surface (PES) exhibits a shallow minimum that is easily altered.

In order to investigate the effect of the crystal environment on the molecular entity, the intermolecular interactions in the crystal structures of the symmetric and unsymmetric complexes were scrutinized by the Hirshfeld surface analysis pioneered by Spackman and coworkers [49] (see Chap. 16). This analysis revealed large differences between the two solvates. In the symmetric system the two axial termini of the molecule have very similar intermolecular interactions, whereas in the unsymmetric system the two ends are very different, Fig. 13.12.

It is, however, difficult to judge whether the symmetry of the intermolecular interactions is the origin or an effect of the molecular isomerism. The metal-metal bonding in the EMACs structures has mainly been examined with qualitative molecular orbital models, but in these simplified models a specific set of orbitals is presumed to be responsible for both the bonding and the magnetic behavior. The use of simplified bonding schemes has an inherent danger of not capturing all the essential bonding features, and it was of interest to study these systems with density based methods. The charge density analysis of the unsymmetrical $Co_3(dpa)_2Cl_2$ complex showed that the qualitative picture of a $(Co_2)^{2+}$ dimer and an isolated Co^{2+} ion has some validity, but also that the chemical bonding in the molecule is complex. As an example clear bcps were located for both Co-Co interactions. Topological analysis gave atomic charges of $Co(1)^{+0.50}$, $Co(2)^{+0.77}$, and $Co(3)^{+1.36}$, and in general the metal-ligand interactions were more ionic for Co(3) than for Co(1) and Co(2). It was found that the Co(3) atom has a relatively low occupancy in the d_{xy} orbital and a relatively high occupancy in the $d_{x2\text{-}y2}$ orbital (z axis along the metal chain, x and y towards the nitrogen atoms) suggesting that these orbitals are significantly involved in the spin cross-over process responsible for the temperature dependent magnetism [47].

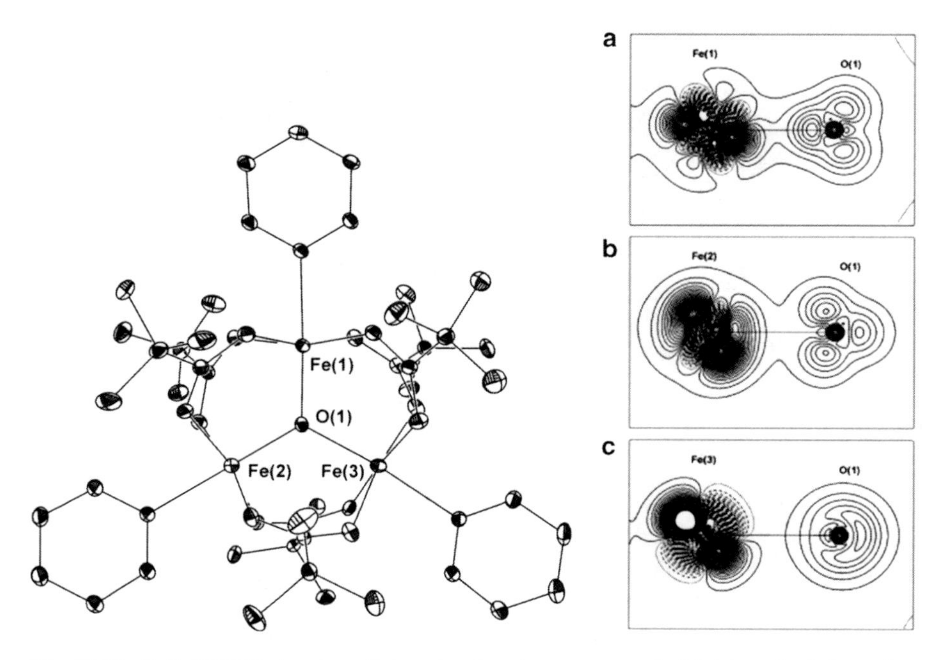

Fig. 13.13 (*left*) ORTEP drawing of the trinuclear iron compound; (*right*) static deformation density in the bonding planes perpendicular to the Fe_3O plane. Fe(2) is the divalent iron

Another class of structures exhibiting flat potential energy surfaces is the family of trinuclear oxo-centered carboxylate complexes [50]. These complexes are composed of a central μ^3-oxygen connected to three metal atoms within a plane. Each pair of metal atoms is bridged by two carboxylate groups above and below the central plane while the octahedral coordination sphere is completed by an apical ligand, Fig. 13.13. In the general case of a neutral compound with neutral apical ligands, the charges on the metal atoms add up to a total of +8 to neutralize the six carboxylate groups (−1) and the central O^{2-}. This gives a mixed valence compound with two Fe^{3+} atoms and one Fe^{2+} atom. If the molecule is fully symmetrical then the potential energy surface will have three minima of equal magnitude corresponding to localization of the extra electron on either of the iron centers. Depending on the barrier between the minima, electron transfer will be observed between the iron centers, and indeed the trinuclear systems have been studied as model compounds for electron transfer processes in enzymes [51]. However, the PES, which determines the location of this electron, can be perturbed by slight external or internal changes, such as temperature or re-crystallization in different solvents.

In the solvent-free $Fe_3O(piv)_6(py)_3$ (piv: pivalate, py: pyridine) system a multi-temperature structural study showed a clear correlation between the degree of disorder in the *t*-butyl groups of the pivalate-groups and the localization of the extra electron on the Fe-atoms [52]. The disorder in these groups, that are on the outer surface of the molecule, affects the PES to such an extent that the preferred location of the electron changes. The process is highly dynamic and on the time scale of a diffraction experiment a weighted average of occupancy of the extra electron on either Fe(1) or Fe(2) is observed (Fe(3) remains in the +3 state at all studied temperatures). As the temperature is lowered, the disorder in the *t*-Bu groups gradually vanishes and eventually below 20 K the system is completely localized, i.e. the extra electron is trapped at one particular iron site. Thus, below this temperature a well ordered state is present and the crystal is suitable for experimental charge density analysis.

The charge density in this system provided a clear surprise [53]. Since the central Fe_3O is completely planar it was expected that the central oxygen atom would exhibit an sp^2 type hybridization. However, it was found that O(1) exhibits an sp^3-like density deformation, Fig. 13.13b. CDs have also been determined in two other trinuclear carboxylate systems, one of which was also of mixed-valence (MV) type. Strikingly, the same sp^3-like density on the central O(1) was found also for this MV complex. On the other hand, in the oxidized complex the central oxygen was clearly sp^2 hybridized. The electronic features on the central oxygen were recently confirmed in a new study of an oxidized core [54] and the sp^3-like deformation therefore appears to be a fingerprint for the mixed valence systems. The study of three similar complexes provided additional insight that cannot be obtained from a study of a single complex. In particular, orbital population analysis revealed that the extra electron density on the Fe^{II} site is distributed primarily in the d_{yz} orbital (z axis towards O(1), y axis perpendicular to the Fe_3O-plane). Upon electron transfer electron density is moved to the d_{yz} orbital of the redox active Fe^{III} site. Presence of extra charge in the d_{yz} orbital correlates with a depletion of charge in the equatorial region and a decreased population of the d_{xy} orbital. This suggested that the equatorial ligands have a strong influence on the electron transfer process as was indeed observed to be the case for the disorder on the *t*-Bu groups.

Next we move to CD analysis of extended structures in the form of coordination polymers. As an extension of the dimer and trimer Co chains the extended polymer $Co_3(BDC)_4(C_4H_{12}N)_2$ $(DEF)_3$ (BDC: benzene dicarboxylate, DEA: diethylamine, DEF: diethylformamide) was studied [55]. This investigation was experimentally very challenging since only tiny crystals were obtainable and synchrotron radiation was imperative. When using conventional radiation it was not even possible to determine the structure of the crystal. The complex consists of three-atom chains of carboxylate bridged cobalt atoms interconnected by three unique BDC linkers, Fig. 13.14. The discrete Co-chain consists of a tetragonally distorted, octahedrally coordinated Co-atom at an inversion centre of the space group, and distorted tetrahedrally coordinated cobalt atoms at both ends of the chain. The central Co(2) bonds to six oxygen atoms of which two are also directly bonded to Co(1) while the remaining four oxygen atoms are part of bridging carboxylate groups from the BDC linkers. Thus, the terminal Co(1) are capped by an oxygen from a BDC linker as well as bonded to three oxygen atoms bonded to the central Co(2).

The magnetic properties of the compound are rather complex exhibiting several ordering schemes of different types [56]. Between temperatures of 85 K and 42 K

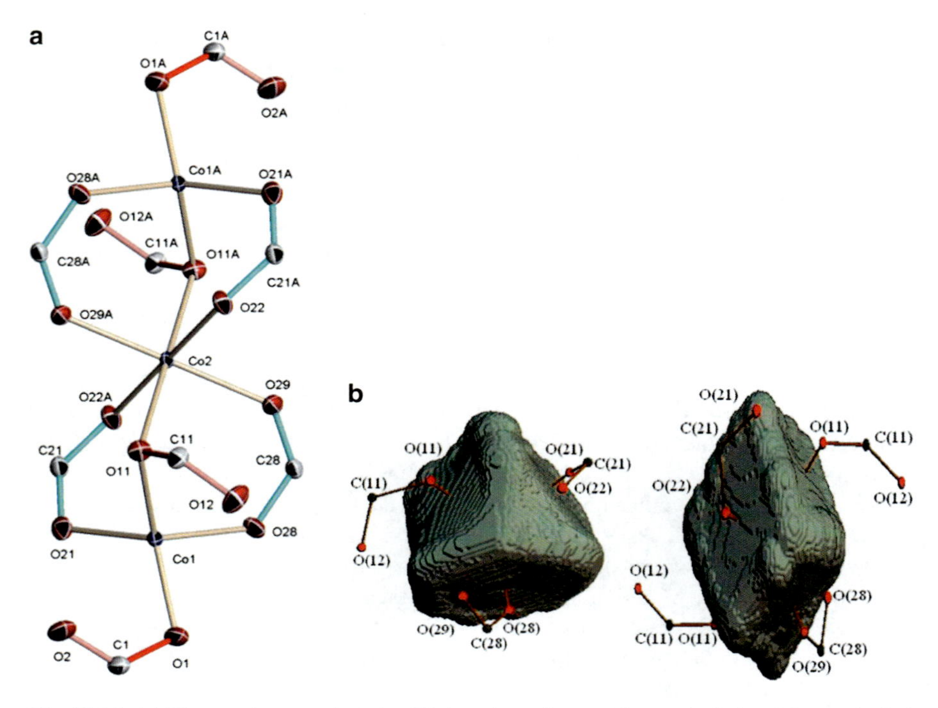

Fig. 13.14 (**a**) The metal atom sub chain. (**b**) Atomic surfaces of the tetrahedral and the octahedral Co atom

a ferromagnetic ordering is seen with an increase in χT. Cooling even further decreases the value of χT indicative of an antiferromagnetic ordering in this temperature regime and at very low temperature the slope of the χT curve changes again. Overall, it was concluded that there are a number of interaction pathways in the coordination polymer that cooperate to give the complex magnetic properties, and the CD study was dedicated to a description of the bond characteristics in the vicinity of the metal atoms. There was no sign of direct metal-metal bonding so the interaction must take place through a super-exchange pathway. The deformation density maps in the planes containing Co-O bonds indicated that all metal-ligand interactions, except the Co(1) – O(1) and Co(1) – O(11) bonds, are characterized by a negative metal region pointing towards a positive ligand region. The Co(1) – O(1) and Co(1) – O(11) interactions on the other hand seemed to have significant covalent contributions to the bonding with charge accumulation in the bond. The topological analysis of the density indicated that the two Co-O bonds to O(1) and O(11) are different in nature from the other metal ligand bonds as they present negative total energy densities at the bcp. The partly covalent nature of these two bonds suggests that they are the key electronic bridges in the magnetic interactions. Another important point of the study was that the crystal structure contains a formally negatively charged framework with cations and neutral molecules in the

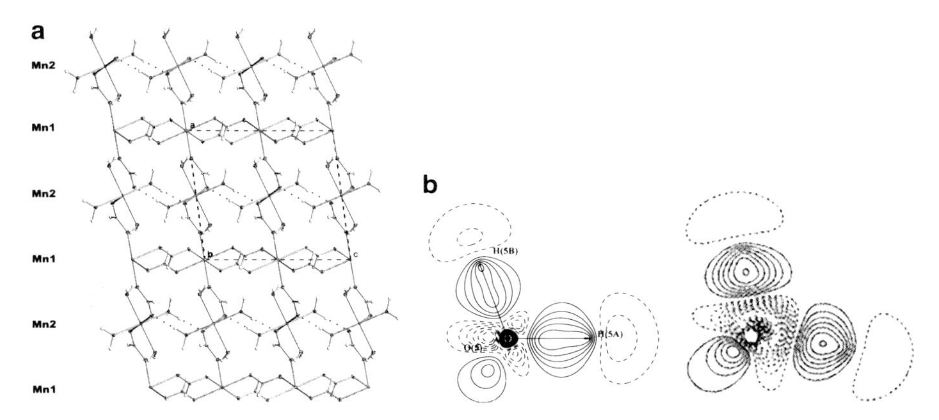

Fig. 13.15 (**a**) View of the framework structure in the infinite Mn coordination polymer. (**b**) Static deformation density. The left is from 10 K synchrotron data, the right from ab initio theory

voids. However, topological atomic charges sum to give the surprising result that the framework is actually close to neutral. Therefore, substantial charge transfer from the framework to the guest molecules must take place. The atomic charges suggest that qualitative conclusions based on formal charge counting, e.g. about guest inclusion properties, probably will be in error.

The CD studies of magnetic coordination polymers have also involved several Mn based structures. A benchmark study was carried out on the structurally simple, but magnetically complex system $[Mn(HCOO)_2(H_2O)_2]_\infty$. Data sets were measured on two different synchrotron sources as well as with conventional MoK_α radiation, and a detailed comparison of the data accuracy was carried out. Besides providing an example of the unique properties of synchrotron radiation the CD in this complex provided insights into the pathways for the magnetic orderings which occur at low temperature. The complex has one Mn-site where the Mn(1) ions are octahedrally coordinated to six formate oxygen atoms creating layers in the *bc*-plane, Fig. 13.15. The ions in these layers are interconnected with formate bridges. In the second site, Mn(2) ions are octahedrally coordinated with two bonds to formate oxygens and four bonds to water oxygens. These unlinked ions form separate layers which interpenetrate the layers formed by the Mn(1) ions. The two types of layers are connected with direct formate bridges between Mn(1) and Mn(2). Overall, the Mn(2) ions are isolated from each other, while the Mn(1) ions are connected.

The structure shows three magnetic transitions at 3.7, 1.7 and 0.6 K. Spin density measurements using polarized neutron diffraction enabled the derivation of magnetic moments on the two sites, which were estimated to be 0.38(2) μ_B on the Mn(1) ions and 1.73(2) μ_B on the Mn(2) ions [57]. It is somewhat surprising that the formal high spin Mn^{2+} ions show magnetic moments that are much smaller than the spin-only value of 5.92 μ_B. The isostructural Ni system also has been studied in detail [58]. In this system magnetic ordering is observed at 15.5 K and 3.75 K, and proton NMR was used to estimate magnetic moments of 2.38 μ_B for the Ni(1) ions and 0.38 μ_B for the Ni(2) ions [59]. The larger moment on the Ni(1) ions in the Ni system relative to the Mn(1) ions in the Mn system corroborates the higher ordering temperature of the M(1) lattice, but it is intriguing why the relative size of the moments of the M(1) and M(2) sublattices are interchanged. Even though the different 3d metals yield isostructural crystals, the subtle details of the metal ligand chemical bonding appear to change.

The CD study included magnetization measurements and here it was observed that the effective moment is 5.840(2) μ_B close to the free ion value. Orbital population analysis showed rather close to one electron in each d-orbital on both the Mn sites confirming that both ions to a first approximation are high spin Mn^{2+}. The CD and the magnetization measurements therefore clearly were at variance with the spin density study. The integrated atomic charges were slightly smaller than expected ($\sim +1.7$) possibly reflecting the covalent interactions responsible for the magnetic ordering. The covalent bonding contributions were apparent in the fact that the Mn^{2+} ions are not spherical. As mentioned above the study provided a benchmark for CD studies of coordination polymers. One important technical aspect was examination of deformation densities near the oxygen nuclei in water and carboxylate groups. From theory it is well established that the deformation features for such oxygen atoms should be negative, but in fact they are often observed to be positive probably due to slight errors in scale factors. In this study only the 10 K data set collected at a short wavelength synchrotron source exhibited features in quantitative agreement with ab initio theory (see Fig. 13.15b). It was found that the topological properties in the bcps are relatively more robust than subtle deformation density features close to the nuclei.

One of the original reasons for studying magnetic coordination polymers was their potential for obtaining spatial control of magnetic interactions. Since the reactants are preserved in the final product, coordination polymers to some extent can be compared to playing with LEGO. Thus, if one can systematically change the linker molecules, it may be possible to fine tune the magnetic superexchange interactions. However, this is easier said than done, since a change of linker molecules often will lead to quite substantial changes in the structure. However, in the case of two related Mn-complexes with either DMF (DiMethylFormamide) or DEF as coordinating solvent, $Mn_2(BDC)_2(DMF)_2$ and $Mn_3(BDC)_3(DEF)_2$, it was possible to obtain coordination polymers, which are structurally similar yet possess enough difference to fundamentally change the magnetic behavior [60]. Both these systems consist of chains of Mn atoms interconnected by bridging carboxylate groups along the a-axis. The carboxylate groups, or linkers, point in the two perpendicular directions to the Mn-chains thereby forming a three-dimensional network structure with significantly larger interchain than intrachain Mn-Mn separations, Fig. 13.16.

The two structures exhibit different temperature dependence of their magnetic properties as only the DEF structure shows any signs of ordering down to 2 K. This indicates that there are differences in the magnetic pathways in the two systems and an examination of the chemical bonding in the two systems was carried out using very low temperature synchrotron radiation data. Here it was clear that the

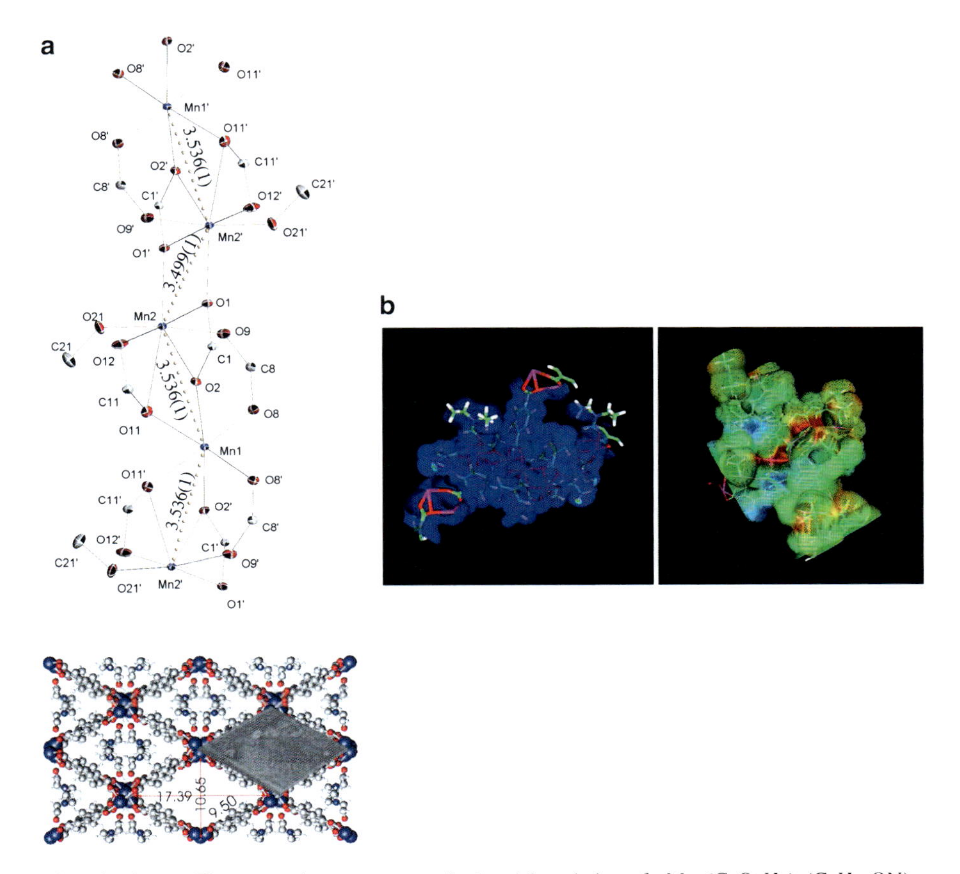

Fig. 13.16 (**a**) The crystal structure and the Mn chain of $Mn_3(C_8O_4H_4)_3(C_5H_{11}ON)_2$. (**b**) Experimentally determined electrostatic potential of $Mn_3(C_8O_4H_4)_3(C_5H_{11}ON)_2$ (*left*) and $Mn_2(C_8H_4O_4)_2(C_3H_7NO)_2$ (*right*)

DMF based complex, which shows no magnetic ordering, has predominantly ionic metal ligand interactions with completely spherical valence shells around the Mn atoms and atomic charges close to +2. In contrast the Mn atoms in the magnetic DEF based structure have highly anisotropic valence shells (from the analysis of the valence shell charge concentrations revealed by the Laplacian of the electron density [4]) and Mn atomic charges close to neutral. One of the very interesting differences between the two structures was found in the electrostatic potential in the nanoporous cavity. Here the changes in the atomic charges between the structures lead to significantly different electrostatic potentials, and thus if the nanoporous structures were used for gas storage, one would predict that they have very different inclusion properties.

The use of experimental electrostatic potentials for probing intermolecular interactions is indeed one of the most promising aspects of modern charge density

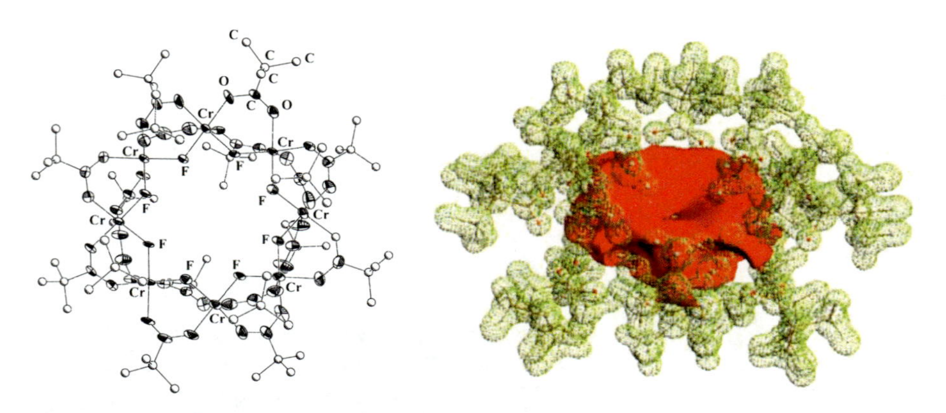

Fig. 13.17 (*left*) ORTEP drawing of the Cr-wheel complex without guest inclusion; (*right*) Isosurface of the experimentally determined electrostatic potential showing the values -0.54 eÅ^{-1} (*red*) and $+0.3$ eÅ^{-1}(*yellow*)

analysis. It has still not been used to a great extent, but one prominent example was the study of host-guest complexes based on the so called chromium wheel structures, Fig. 13.17 [61]. The Cr-ring host-structure has been successfully used to accommodate small organic structures and the inclusion complexes were found to be thermodynamically stable and structurally characterized [61]. The ring structure of the uncomplexed Cr-wheel was found to exhibit a slight twist and rotation of pivalate groups leading to a partial closure of one side of the molecule and an associated opening on the other side. The electron density of the 270 atom complex – which still ranks among the three largest molecular structures ever studied using the CD method – was determined from helium temperature synchrotron radiation data, and the corresponding electrostatic potential was used to rationalize the observed guest inclusion properties of the system, Fig. 13.17. The electrostatic potential in the molecule indicated a rather electronegative environment at the molecular center, which persisted all the way to the exterior along the molecular axis. This was fully in accordance with the observations that the guest molecules did not venture completely into the cavity but were fixed at the entrance with smaller molecules (DMF) able to penetrate further than larger (DMA, DiMethylAcetamide).

13.4 Gas Molecules Adsorbed in Metal Organic Framework Materials

Coordination polymers or metal organic frameworks (MOFs) have many other important properties than magnetism. They have also attracted attention due to their characteristic nano-sized porous structure [62, 63] which causes gas adsorption phenomena as well as their commercial application in gas separation [64], gas storage [65–67] and heterogeneous catalysis [68, 69]. To develop novel MOFs as

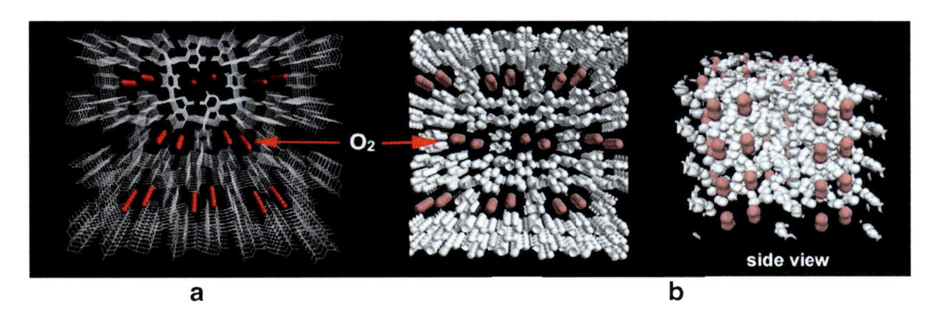

Fig. 13.18 (**a**) Crystal structure of CPL-1 with the adsorption of O_2. O_2 molecules are shown by *red balls*. Other atoms are shown by *connecting lines*. (**b**) MEM charge density of CPL-1 with the adsorption of O_2 as an equi-density contour surface. The equi-density level is 1.0 eÅ^{-3}. O_2 molecules are indicated in *red*

functional materials, precise structural information of the guest molecules and the host framework has been uncovered at the charge density level by the MEM [70]. These studies have successfully revealed the specific molecular arrangements in the nanopores as well as molecular interactions between the guest molecules and the host framework. Three representative studies of O_2, H_2 and C_2H_2 absorption will be described here since they are closely related to energy science.

Figure 13.18 shows the MEM charge density with O_2 molecules adsorbed in $[Cu_2(pzdc)_2(pyz)]_n$ (pzdc: pyrazine-2,3-dicarboxylate, pyz: pyrazine). This framework is also called CPL-1 (Coordination polymer 1 with a pillared-layer structure) and it has uniform nanochannels of dimension 4 Å × 6 Å [71, 72]. The MEM density gave a conclusive answer for the longstanding question of where the gas molecules are located inside the nanopore of the MOF and revealed novel aspects of the adsorbed gas molecules.

The MEM/Rietveld method provides an efficient self-consistent structure analysis method. At the start of the Rietveld analysis, it is necessary to have a primitive structural model. The MEM can provide useful information purely from observed structure factor data beyond the presumed primitive crystal structure model and it is this visualization ability of the MEM that is useful to construct a detailed structural model of novel nano-structured materials.

The flow chart of the method is shown in Fig. 13.19 with an example of the structure analysis procedure for the CPL-1 with the adsorption of O_2.

In the first step of analysis, the Rietveld refinement was carried out with a primitive structural model of only the CPL-1 framework determined by single-crystal X-ray diffraction [71], without assuming any guest molecules inside the nanopores. In this refinement, a rigid body model including restraints for intermolecular distances and angles was applied to maintain the intrinsic molecular forms.

The data used in the study was measured in a in-situ gas absorption powder diffraction experiment using the Large Debye-Scherrer Camera at BL02B2 in SPring-8 [73]. The result of the Rietveld fitting for the fully adsorbed specimen

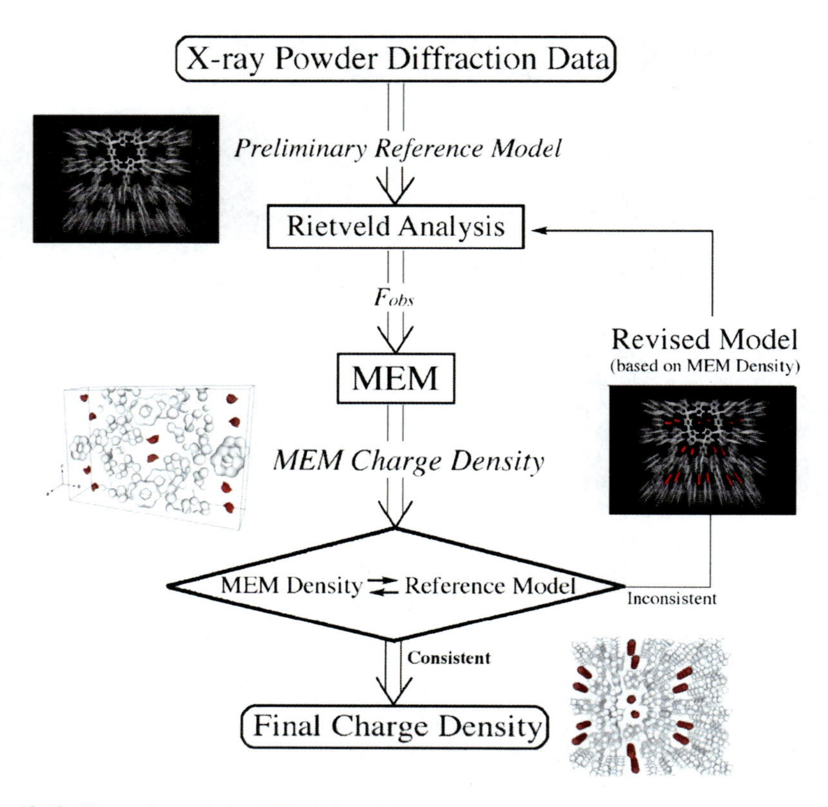

Fig. 13.19 Flow chart of the MEM/Rietveld method with structure modeling and imaging of charge density

is shown in Fig. 13.20a. The misfit between the observed and calculated powder patterns is largely due to the lack of guest gas molecules in the primitive structural model. The reliability factors based on the whole powder profile, R_{WP}, and the Bragg integrated intensities, R_I, were 16.4% and 44.9%, respectively. Despite the unsatisfactory fit to data, the integrated intensities of each reflection were evaluated from the observed diffraction patterns using the result of the Rietveld refinement.

The structure factors derived from these integrated intensities were used in the MEM analysis with the phases calculated from the structural model. The MEM calculation was carried out using a computer program, ENIGMA [74]. In the MEM calculation, the total number of electrons in the unit cell was given as a constraint including the number of electrons for adsorbed gas molecules. The number of adsorbed gas molecules was independently examined by the gas adsorption isotherm. The obtained three-dimensional MEM charge density distribution is the inserted figure at middle left in Fig. 13.19. The framework structure can be clearly seen. Despite the structural model without assuming guest molecules, two peaks of charge density were visualized inside the pores. This suggests that the diffraction intensities contain the information of the adsorbed O_2 molecule.

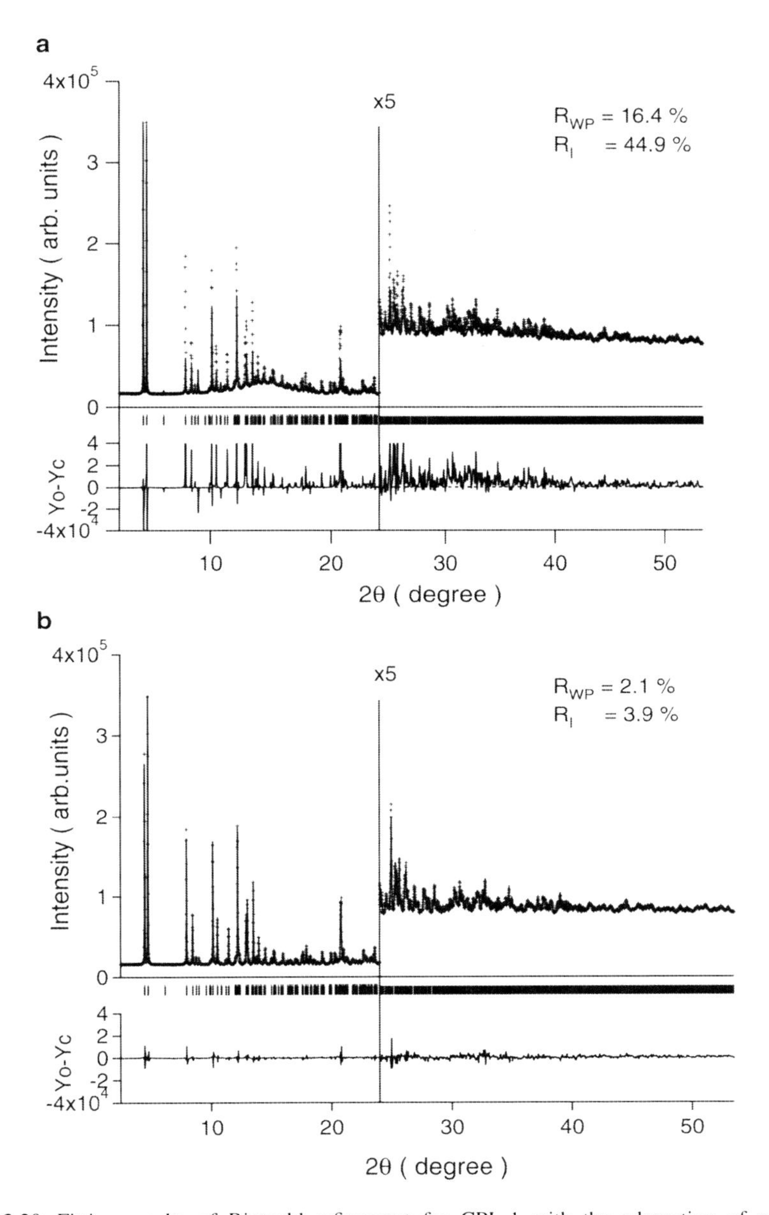

Fig. 13.20 Fitting results of Rietveld refinement for CPL-1 with the adsorption of oxygen. (**a**) Initial refinement using a structural model of CPL-1 framework without guest molecules. (**b**) Final refinement using the modified structural model of CPL-1 with oxygen

In the next step of the analysis, a modified structural model including O_2 molecules was applied in the Rietveld refinement. The results of the final Rietveld refinement are shown in Fig. 13.20b. The reliability factors R_{WP} and R_I were dramatically improved to 2.1% and 3.9%, respectively. A new set of structure factors were extracted from the observed data and used an input in MEM calculations. The features of the obtained MEM charge density were consistent with the structural model used in the Rietveld refinement and thus it was judged as a final result of the analysis.

The MEM charge density is shown in Fig. 13.18 together with the structural model. Here the dumbbell-shaped electron densities of O_2 molecules are clearly recognizable in the middle of the nanochannels forming a ladder structure. There are 15.8(1) electrons in total around the O_2 sites, which virtually agrees with an O_2 molecule. Therefore, the dumbbell-shaped electron densities are identified as O_2 molecules and it is found that one O_2 molecule is adsorbed per Cu atom without significant charge transfer between O_2 molecules and/or O_2 molecules and the pore wall. No electron density showing covalent chemical bonding is seen between the O_2 molecule and the pore walls. The interatomic distances between the O_2 molecule and the atoms on the pore walls are from 2.7 to 3.3 Å, which equals the van der Waals contact of these atoms. These observations support the idea that O_2 molecules are physisorbed in the nanochannels of CPL-1. The isotropic temperature factor B of the O_2 molecule was refined to be 4.1 $Å^2$, which shows that the O_2 molecules have large thermal vibrations in the nanochannels. However, the charge density distributions of adsorbed O_2 molecules were clearly separated showing that the molecules are trapped in their positions.

The state of the oxygen in the nanochannels is thought to be closer to a solid state than a liquid state. As shown here, the MEM charge density provides information on the molecular interaction, *i.e.* the ionicity of atoms and the charge transfer between the molecules.

The work on O_2 absorption in CPL-1 was the breakthrough, which lead to studies on absorption of other gasses in porous coordination polymers such as H_2 [75], N_2, Ar, CH_4 [76] and C_2H_2 [77]. Adsorption of H_2 molecules in the porous coordination polymers [66, 67, 78] is a promising solution to the problematic issue of hydrogen storage. To develop a rational synthesis strategy for porous coordination polymers for high performance hydrogen storage, the elucidation of the intermolecular interaction between H_2 molecules and the pore walls is essential.

The weak X-ray scattering amplitude of hydrogen makes it difficult to determine the hydrogen position by X-ray structure analysis. However, the MEM/Rietveld analysis combined with high brilliance synchrotron powder diffraction makes it possible to reveal the position of hydrogen atoms as well as the chemical bonding of hydrogen [75, 79]. Kubota et al. carried out the structure determination of adsorbed hydrogen in the nanochannel of CPL-1 and revealed that the H_2 molecules also produce a one-dimensional array in the nanochannel analogous to the O_2 molecule [75].

Distributions of the MEM charge density of CPL-1 with H_2 molecules are shown in Fig. 13.21. Small elongated shaped peaks in the electron distribution are observed

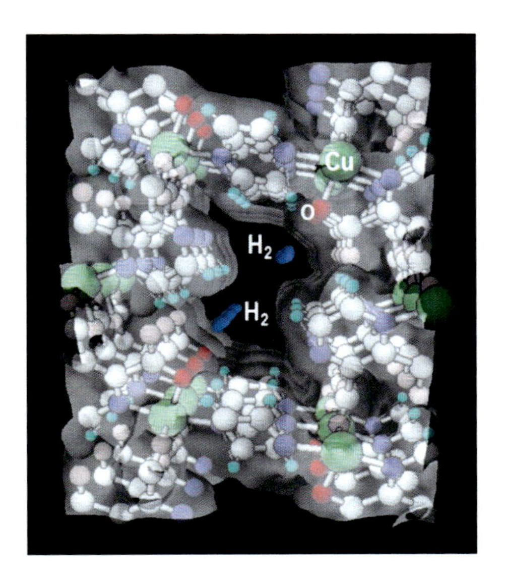

Fig. 13.21 Close-up view of MEM charge density distribution around an adsorbed H_2 molecule as an equi-density contour surface. The equi-density level is 0.11 $eÅ^{-3}$. The structural model of the CPL framework is superimposed on the charge density map

inside the nanochannels belonging to the H_2 molecules. The position of the H_2 molecule is close to that of the oxygen atom belonging to the carboxylate group of the framework. The H_2 molecule seems to be locked in the space formed by the oxygen atom and the hydrogen atom of pyrazine in the CPL framework.

The interatomic distances between the hydrogen atoms of the H_2 molecule and the nearest oxygen atom are 2.7(1) Å and 2.4(1) Å, respectively, which are comparable to the distance calculated from the van der Waals radii. The oxygen atom nearest to the H_2 molecule is forming a coordination bond with the Cu atom and it is negatively charged. Interestingly Rosi et al. showed a similar result in a different type of MOF material [80]. They observed that the site around Zn and O is one of the H_2 molecule binding sites by inelastic neutron scattering spectroscopy. This suggests that the metal and oxygen atoms also play an important role in hydrogen adsorption.

From a materials science perspective adsorbtion of acetylene (C_2H_2) is also attractive. Acetylene is one of the key starting material for many products in the petrochemical and electronic industries. In order to obtain highly pure C_2H_2 for the preparation of these materials, the separation of C_2H_2 from gas mixtures containing CO_2 impurities is an important subject. In addition to this, acetylene is well known as a highly reactive molecule and, therefore, it cannot be compressed above 0.2 MPa; otherwise it explodes without oxygen, even at room temperature. Thus MOF materials may open the door to safe storage of C_2H_2. The features of C_2H_2 adsorption were found to be quite different from that of other gases [77]. Figure 13.22a shows the MEM charge density of a C_2H_2 molecule in CPL-1.

The inter-layer dumbbell-shaped electron density of the C_2H_2 molecules shows that one C_2H_2 molecule was adsorbed per two Cu atoms, which is different from the

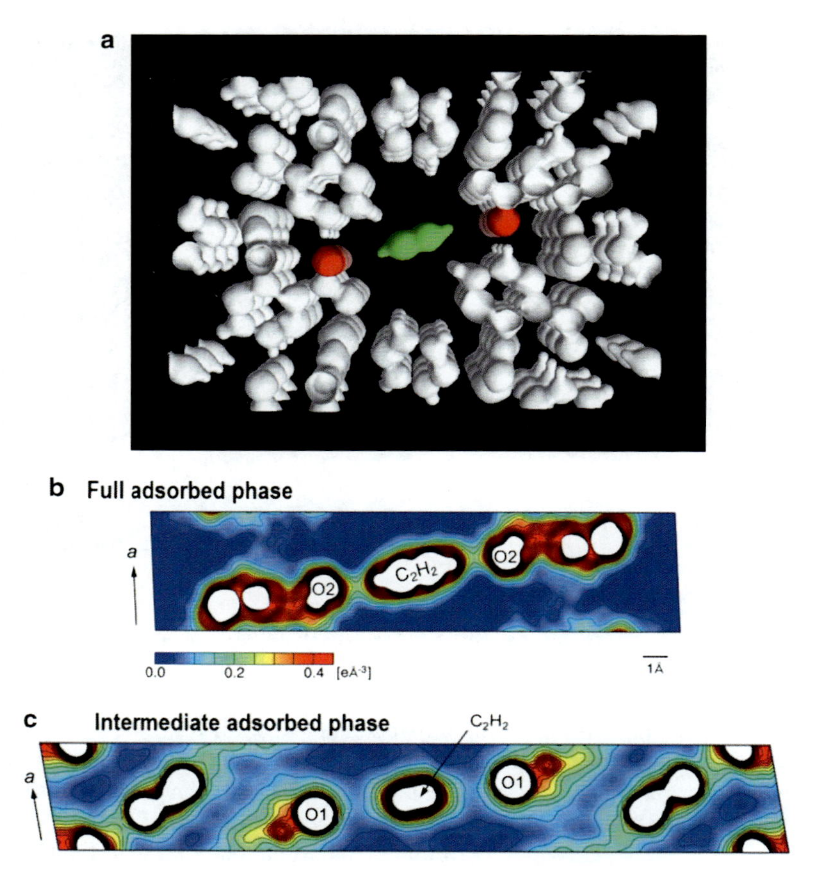

Fig. 13.22 MEM charge density distribution of CPL-1 with adsorbed acetylene. (**a**) Three-dimensional MEM map as an equal-density contour surface viewed from the nanochannel direction. The equi-density level is 1.0 eÅ^{-3}. The central C_2H_2 molecule and oxygen atoms are represented by *yellow-green* and *red*, respectively. MEM charge density sections containing the molecular axis of C2H2 and the a-axis of (**b**) fully adsorbed phase and (**c**) intermediate adsorbed phase. Contour lines are drawn from 0.00 to 1.00 eÅ-3 with intervals of 0.05 eÅ-3. Higher density regions are omitted in the figure

single element molecules, O_2, N_2 and H_2, where one molecule was absorbed per Cu atom. Interestingly, the C_2H_2 molecules align almost perpendicular the *a*-axis with an inclination of 75.6° and with an intermolecular distance of 4.78 Å, which means that C_2H_2 molecules are completely isolated from each other in the nanochannels of CPL-1. The section of electron density containing the molecular axis of C_2H_2 in Fig. 13.22b reveals that there is some charge density between the hydrogen atom of C_2H_2 and the oxygen atom of the carboxylate in the host framework. This indicates that O····H-C hydrogen bonding plays an important role in the stabilization of C_2H_2 in the nanochannel.

13.5 Concluding Remarks

In the present survey we have discussed charge density studies on a range of advanced materials with unique properties. Many of the studies are highly challenging and non-standard, and they include heavy atoms, large unit cells, high symmetry inorganic solids (with the problem of extinction), structural disorder, microcrystals and powders, and air sensitive compounds. Multi-temperature approaches, combined theory and experiment, and both synchrotron, neutron and conventional X-ray tube radiation have been employed. The charge density analysis focused both on application of thermally smeared densities, deformation densities, orbital populations, the quantum theory of atoms in molecules, electron localizability indices and electrostatic properties. The examples clearly document that charge density analysis is a strong tool for materials science, which can unlock secrets that are very hard to obtain with other methods. As both experimental and theoretical techniques continue to become more and more sophisticated, the importance of charge density information in materials science is bound to increase.

References

1. Coppens P (1997) X-ray Charge densities and chemical bonding. Oxford University Press, Oxford
2. (a) Coppens P, Koritzansky T (2001) Chemical applications of charge densities. Chem Rev 101:1583–1628; (b) Coppens P, Iversen BB, Larsen FK (2004) The use of synchrotron radiation in X-ray charge density analysis of coordination complexes Coord Chem Rev 249:179–195; (c) Coppens P (2005) Charge densities come of age. Angew Chem Int Ed 44:6810–6811
3. (a) Collins DM (1982) Electron-density images from imperfect data by iterative entropy maximization. Nature 298:49–51; (b) Sakata M, Sato M (1990) Accurate structure-analysis by the maximum-entropy method. Acta Crystallogr A 46:263–270; (c) Takata M, Nishibori E, Sakata M (2001) Charge density studies utilizing powder diffraction and MEM. Exploring of high Tc superconductors, C-60 superconductors and manganites. Z Kristallogr 216:71–86
4. Bader RFW (1990) Atoms in molecules: a quantum theory. Clarendon, Oxford
5. Spackman MA (1992) Molecular electric moments from X-ray-diffraction data. Chem Rev 92:1769–1797
6. Bader RFW, Gatti C (1998) A Green's function for the density. Chem Phys Lett 287:233–238
7. (a) Kohout M (2004) A measure of electron localizability. Int J Quantum Chem 97:651–658; (b) Kohout M (2007) Bonding indicators from electron pair density. Faraday Discuss 135: 43–54
8. Snyder GJ, Toberer ES (2008) Complex thermoelectric materials. Nature Mater 7:105–114
9. Slack GA (1995) New materials and performance limits for thermoelectric cooling. In: Rowe DM (ed) CRC handbook of thermoelectrics. CRC Press, Boca Raton
10. (a) Sales BC, Mandrus D, Williams RK (1996) Filled skutterudite antimonides: a new class of thermoelectric materials. Science 272:1325–1328; (b) Venkatasubramanian R, Siivola E, Colpitts T et al (2001) Thin-film thermoelectric devices with high room-temperature figures of merit. Nature 413:597–602; (c) Hsu KF, Loo S, Guo F et al (2004) Cubic $AgPb_mSbTe_{2+m}$: bulk thermoelectric materials with high figure of merit. Science 303:818–821

11. (a) Kovnir KA, Shevelkov AV (2004) Semiconducting clathrates: synthesis, structure and properties. Russ Chem Rev 73:923–938; (b) Mudryk Y, Rogl P, Paul C et al (2002) Thermoelectricity of clathrate I Si and Ge phases. J Phys Condens Mat 14:7991–8004
12. (a) Blake NP, Latturner S, Bryan JD et al (2001) Band structures and thermoelectric properties of the clathrates $Ba_8Ga_{16}Ge_{30}$, $Sr_8Ga_{16}Ge_{30}$, $Ba_8Ga_{16}Si_{30}$, and $Ba_8In_{16}Sn_{30}$. J Chem Phys 115:8060–8072; (b) Blake NP, Bryan JD, Latturner S et al (2001) Structure and stability of the clathrates $Ba_8Ga_{16}Ge_{30}$, $Sr_8Ga_{16}Ge_{30}$, $Ba_8Ga_{16}Si_{30}$, and $Ba_8In_{16}Sn_{30}$. J Chem Phys 114:10063–10074
13. Bryan JD, Bentien A, Blake NP et al (2002) Nonstoichiometry and chemical purity effects in thermoelectric $Ba_8Ga_{16}Ge_{30}$ clathrate. J Appl Phys 92:7281–7290
14. (a) Chakoumakos BC, Sales BC, Mandrus DG et al (2000) Structural disorder and thermal conductivity of the semiconducting clathrate $Sr_8Ga_{16}Ge_{30}$. J Alloys Compd 296:80–86; (b) Keppens V, Sales BC, Mandrus D et al (2000) When does a crystal conduct heat like a glass? Philos Mag Lett 80:807–812; (c) Iversen BB, Palmqvist AEC, Cox D et al (2000) Why are clathrates good candidates for thermoelectric materials? J Solid State Chem 149:455–458
15. Bentien A, Palmqvist AEC, Bryan JD et al (2000) Experimental charge densities of semiconducting cage structures containing alkaline earth guest atoms. Angew Chem Int Ed 40:3613–3616
16. (a) Nataraj D, Nagao N, Ferhat M et al Structure, high temperature transport, and thermal properties of $Ba_8Gs_xSi_{46-x}$ ($x = 10$ and 16) clathrates prepared by the arc melting method. J Appl Phys 93:2424–2428; (b) Paschen S, Carrillo-Cabrera W, Bentien A et al (2001) Structural, transport, magnetic, and thermal properties of $Eu_8Ga_{16}Ge_{30}$. Phys Rev B 64:214404; (c) Tournus F, Masenelli B, Mélinon P et al (2004) Guest displacement in silicon clathrates. Phys Rev B 69:035208
17. Sales BC, Chakoumakos BC, Jin R et al (2001) Structural, magnetic, thermal, and transport properties of $X_8Ga_{16}Ge_{30}$ (X = Eu, Sr, Ba) single crystals. Phys Rev B 63:245113
18. See e.g. (a) Zerec I, Keppens V, McGuire MA et al (2004) Four-well tunneling states and elastic response of clathrates. Phys Rev Lett 92:185502; (b) Suekuni K, Avila MA, Umeo K et al (2007) Cage-size control of guest vibration and thermal conductivity in $Sr_8Ga_{16}Si_{30-x}Ge_x$. Phys Rev B 75:195210; (c) Avila MA, Suekuni K, Umeo K et al (2006) Glasslike versus crystalline thermal conductivity in carrier-tuned $Ba_8Ga_{16}X_{30}$ clathrates (X = Ge,Sn). Phys Rev B 74:125109
19. Bentien A, Iversen BB, Bryan JD et al (2002) Maximum entropy method analysis of thermal motion and disorder in thermoelectric clathrate $Ba_8Ga_{16}Si_{30}$. J Appl Phys 91:5694–5699
20. (a) Christensen M, Iversen BB (2007) Host structure engineering in thermoelectric clathrates. Chem Mater 19:4896–4905; (b) Christensen M, Iversen BB (2008) Host-guest coupling in semiconducting $Ba_8Zn_8Ge_{38}$. J Phys B Condens Mater 20:104244; (c) Christensen M, Lock N, Overgaard J et al (2006) Crystal structures of thermoelectric n- and p-type $Ba_8Ga_{16}Ge_{30}$ studied by single crystal, multitemperature, neutron diffraction, conventional X-ray diffraction and resonant synchrotron X-ray diffraction. J Am Chem Soc 129:15657–15665
21. Bader RFW (2005) The quantum mechanical basis of conceptual chemistry. Monatshefte Chem 136:819–854
22. (a) Debye P, Scherrer P (1918) Phys Z 19:474–483; (b) Bragg WL, James RW, Bosanquet CH (1992) The distribution of electrons around the nucleus in the sodium and chlorine atoms. Philos Mag 44:433–449
23. Gatti C, Bertini L, Blake NP et al (2003) Guest-framework interaction in type I inorganic clathrates with promising thermoelectric properties: on the ionic versus neutral nature of the alkaline-earth metal guest a in $A_8Ga_{16}Ge_{30}$ (A = Sr, Ba). Chem Eur J 9:4556–4568
24. Chen G, Dresselhaus MS, Dresselhaus G et al (2003) Recent developments in thermoelectric materials. Int Mater Rev 48:45–66
25. Saber HH, El-Genk MS, Callita T (2007) Tests results of skudderudite based thermoelectric unicouples. Energy Convers Manag 48:555–567
26. Guloy AM, Ramlau R, Tang Z et al (2006) A guest-free germanium clathrate. Nature 443: 320–323

27. (a) Sharp JW, Jones EC, Williams RK et al (1995) Thermoelectric properties of $CoSb_3$ and related alloys. J Appl Phys 78:1013–1016; (b) Zhang L, Grytsiv A, Kerber M et al (2009) $M_mFe_4Sb_{12}$- and $CoSb_3$-based nano-skutterudites prepared by ball milling: kinetics of formation and transport properties. J Alloys Compd 481:106–115
28. Pauling L (1978) Covalent chemical bonding of transition metals in pyrite, cobaltite, skutterudite, millerite, and related minerals. Can Mineral 16:447–452
29. Lefebvre-Devos I, Lassalle M, Wallart X et al (2001) Bonding in skutterudites: combined experimental and theoretical characterization of $CoSb_3$. Phys Rev B 63:125110
30. Ohno A, Sasaki S, Nishibori E et al (2007) X-ray charge density study of chemical bonding in skutterudite $CoSb_3$. Phys Rev B 76:064119
31. Leithe-Jasper A, Borrmann H, Akselrud L et al (2009) Unpublished results
32. Leithe-Jasper A, Schnelle W, Rosner H et al (2004) Weak itinerant ferromagnetism and electronic and crystal structures of alkali-metal iron antimonides: $NaFe_4Sb_{12}$ and KFe_4Sb_{12}. Phys Rev B 70:214418
33. (a) Wagner FR, Bezugly V, Kohout M et al (2007) Charge decomposition analysis of the electron localizability indicator a bridge between the orbital and direct space representation of the chemical bond. Chem Eur J 13:5724–5741; (b) Kohout M, Wagner FR, Grin Y (2002) Electron localization function for transition-metal compounds. Theor Chem Acc 108:150–156
34. Schnelle W, Leithe-Jasper A, Schmidt M et al (2005) Itinerant iron magnetism in filled skutterudites $CaFe_4Sb_{12}$ and $YbFe_4Sb_{12}$: stable divalent state of ytterbium. Phys Rev B 72:020402
35. Schnelle W, Leithe-Jasper A, Rosner H et al (2008) Magnetic, thermal, and electronic properties of iron-antimony filled skutterudites MFe_4Sb_{12} ($M = Na, K, Ca, Sr, Ba, La, Yb$). Phys Rev B 77:094421
36. Koza MM, Capogna L, Leithe-Jasper A et al (2010) Vibrational dynamics of filled skutterudites $M_{1-x}Fe_4Sb_{12}$ ($M =$ Ca, Sr, Ba, and Yb). Phys Rev B 81:174302
37. Dashjav E, Prots Y, Kreiner G et al (2008) Chemical bonding analysis and properties of $La_7Os_4C_9$-A new structure type containing C^- and C^{-2}-units as Os-coordinating ligands. J Solid State Chem 181:3121–3129
38. Sichevych O, Kohout M, Schnelle W et al (2009) $EuTM_2Ga_8$ (TM = Co, Rh, Ir) - a contribution to the chemistry of the $CeFe_2Al_8$-type compounds. Inorg Chem 48:6261–6270
39. Leung P, Coppens P (1983) Generalized relations between D-orbital occupancies of transition-metal atoms and electron-density multipole population parameters from X-ray-diffraction data. Acta Crystallogr B 39:377–387; (b) Bianchi R, Gervasio G, Marabello D (2001) Experimental electron density in the triclinic phase of $Co_2(CO)_6(\mu\text{-}CO)(\mu\text{-}C_4O_2H_2)$ at 120 K. Acta Crystallogr B 57:638–645; (c) Bianchi R, Gervasio G, Marabello D (2001) An experimental evidence of a metal-metal bond in μ-carbonylhexacarbonyl[μ-(5-oxofuran-2(5 H)-ylidene-κC,κC)]-dicobalt(Co-Co)[$Co_2(CO)_6(\mu\text{-}CO)(\mu\text{-}C_4O_2H_2)$]. Helv Chim Acta 84:722–734; (d) Bianchi R, Gervasio G, Marabello D (1998) Experimental charge density study of the Mn-Mn bond in $Mn_2(CO)_{10}$ at 120 K. Chem Commun 1535–1536; (e) Bianchi R, Gervasio G, Marabello D (2000) Experimental electron density analysis of $Mn_2(CO)_{10}$: metal-metal and metal-ligand bond characterization. Inorg Chem 39:2360–2366; (f) Macchi P, Proserpio DM, Sironi A (1998) Experimental electron density in a transition metal dimer: metal-metal and metal-ligand bonds. J Am Chem Soc 120:13429–13435; (g) Macchi P, Garlaschelli L, Martinengo S et al (1999) Charge density in transition metal clusters: supported vs unsupported metal-metal interactions. J Am Chem Soc 121:10428–10429; (h) Farrugia LJ, Mallinson PR, Stewart B (2003) Experimental charge density in the transition metal complex $Mn_2(CO)_{10}$: a comparative study. Acta Crystallogr B 59:234–247
40. Overgaard J, Clausen HF, Platts JA et al (2008) Experimental and theoretical charge density study of chemical bonding in a Co dimer complex. J Am Chem Soc 130:3834–3843
41. Farrugia LJ (2005) Is there a Co-Co bond path in $Co_2(CO)_6(\mu\text{-}CO)(\mu\text{-}C_4H_2O_2)$? Chem Phys Lett 414:122–126
42. Finger M, Reinhold J (2003) Energy density distribution in bridged cobalt complexes. Inorg Chem 42:8128–8130

43. Gatti C, Bertini L (2004) The local form of the source function as a fingerprint of strong and weak intra- and intermolecular interactions. Acta Crystallogr A 60:438–449
44. Gatti C, Lasi D (2007) Source function description of metal-metal bonding in d-block organometallic compounds. Faraday Discuss 135:55–78
45. Poulsen RD, Overgaard J, Schulman A et al (2009) Effects of weak intermolecular interactions on the molecular isomerism of tricobalt metal chains. J Am Chem Soc 131:7580–7591
46. Carroll RL, Gorman CB (2002) The genesis of molecular electronics. Angew Chem Int Ed 41:4378–4400
47. (a) Cotton FA, Daniels LM, Jordan GT et al (1997) Symmetrical and unsymmetrical compounds having a linear Co_3^{6+} chain ligated by a spiral set of dipyridyl anions. J Am Chem Soc 119:10377–10381; (b) Clerac R, Cotton FA, Dunbar KR et al (2000) A new linear tricobalt compound with di(2-pyridyl)amide (dpa) ligands: two-step spin crossover of $[Co_3(dpa)_4Cl_2][BF_4]$. J Am Chem Soc 122:2272–2278; (c) Clerac R, Cotton FA, Daniel LM et al (2001) Tuning the metal-metal bonds in the linear tricobalt compound $Co_3(dpa)_4Cl_2$: bond-stretch and spin-state isomers. Inorg Chem 40:1256–1264; (d) Clerac R, Cotton FA, Daniels LM et al (2001) Linear tricobalt compounds with Di(2-pyridyl)amide (dpa) Ligands: temperature dependence of the structural and magnetic properties of symmetrical and unsymmetrical forms of $Co_3(dpa)_4Cl_2$ in the solid state. J Am Chem Soc 122:6226–6236
48. (a) Clerac R, Cotton FA, Jeffery SP et al (2001) Compounds with symmetrical tricobalt chains wrapped by dipyridylamide ligands and cyanide or isothiocyanate ions as terminal ligands. Inorg Chem 40:1265–1270; (b) Cotton FA, Murillo CA, Wang X (1999) Linear tricobalt compounds with di-(2-pyridyl)amide (dpa) ligands: studies of the paramagnetic compound $Co_3(dpa)_4Cl_2$ in solution. Inorg Chem 38:6294–6297
49. (a) Spackman MA, Byrom PG (1997) A novel definition of a molecule in a crystal. Chem Phys Lett 267:215–220; (b) Spackman MA, McKinnon JJ (2002) Fingerprinting intermolecular interactions in molecular crystals. CrystEngComm 4:378–392; (c) McKinnon JJ, Spackman MA, Mitchell MS (2004) Novel tools for visualizing and exploring intermolecular interactions in molecular crystals. Acta Crystallogr B 60:627–668; (d) McKinnon JJ, Jayatilaka D, Spackman MA (2007) Towards quantitative analysis of intermolecular interactions with Hirshfeld surfaces. Chem Commun 3814–3816; (e) Spackman MA, McKinnon JJ, Jayatilaka D (2008) Electrostatic potentials mapped on Hirshfeld surfaces provide direct insight into intermolecular interactions in crystals. CrystEngComm 10:377–388; (f) Spackman MA, Jayatilaka D (2009) Hirshfeld surface analysis. CrystEngComm 11:19–32
50. Dziobkowski CT, Wrobleski JT, Brown DB (1981) Magnetic and spectroscopic properties of $Fe^{II}Fe_2^{III}O(CH_3CO_2)_6L_3$, $L = H_2O$ or C_5H_5N - direct observation of the thermal barrier to electron-transfer in a mixed-valence complex. Inorg Chem 20:679–684
51. (a) Blow DM, Birktoft JJ, Hartley BS (1969) Role of a buried acid group in mechanism of action of chymotrypsin. Nature 221:337; (b) Overgaard J, Schiøtt B, Larsen FK et al (2001) The charge density distribution in a model compound of the catalytic triad in serine proteases. Chem Eur J 7:3756–3767
52. Wilson C, Iversen BB, Overgaard J et al (2000) Multi-temperature crystallographic studies of mixed-valence polynuclear complexes; Valence trapping process in the trinuclear oxo-bridged iron compound, $[Fe_3O(O_2CC(CH_3)_3)_6(C_5H_5N)_3]$. J Am Chem Soc 122:11370–11379
53. Overgaard J, Larsen FK, Schiøtt B et al (2003) Electron density distributions of redox active mixed valence carboxylate bridged trinuclear iron complexes. J Am Chem Soc 125:11088–11099
54. Overgaard J, Larsen FK, Timco GA et al (2009) Experimental charge density in an oxidized trinuclear iron complex using 15 K synchrotron and 100 K conventional single-crystal X-ray diffraction. Dalton Trans 664–671
55. Clausen HF, Overgaard J, Chen YS et al (2008) Synchrotron X-ray charge density study of coordination polymer $Co_3(C_8H_4O_4)_4(C_4H_{12}N)_2(C_5H_{11}NO)_3$ at 16 K. J Am Chem Soc 130:7988–7996

56. Poulsen RD, Bentien A, Christensen M et al (2006) Solvothermal synthesis, multi-temperature crystal structures and physical properties of isostructural coordination polymers, $2C_4H_{12}N^+[M3(C_8H_4O_4)_4]^{2-}\cdot 3C_5H_{11}NO$, M = Co, Zn. Acta Crystallogr B 62:245–254
57. Radhakrishna P, Gillon B, Chevrier G (1993) Superexchange in manganese formate dihydrate, studied by polarized-neutron diffraction. J Phys Condens Matter 5:6447–6460
58. (a) Takeda K, Kawasaki K (1971) Magnetism and phase transition in 2-dimensional lattices – $M(HCOO)_2\cdot 2H_2O$ (M = Mn,Fe,Ni,Co). J Phys Soc Jpn 31:1026–1036; (b) Burlet P, Burlet P, Rossat-Mignod J et al (1975) Magnetic-behavior of dihydrate formates $M(HCOO)_2\cdot 2H_2O$ of transition-metals M = Mn, Fe, Co, Ni. Phys Status Solidi 71:675–685 (1975); (c) Kageyama H, Khomskii DI, Levitin RZ et al (2003) Weak ferrimagnetism, compensation point, and magnetization reversal in $Ni(HCOO)_2\cdot 2H_2O$. Phys Rev B 67:224422; (d) Pierce RD, Friedberg SA (1971) Heat capacities of $Fe(HCOO)_2\cdot 2H_2O$ and $Ni(HCOO)_2\cdot 2H_2O$ between 1.4 and 20 K. Phys Rev B 3:934–942; (e) Hoy GR, Barros SDS, Barros FDS et al (1965) Inequivalent magnetic ions in dihydrated formates of Fe^{++} and Ni^{++}. J Appl Phys 36:936–937
59. Zenmyo K, Kubo H, Tokita M et al (2006) Proton NMR study of nickel formate di-hydrate, $Ni(HCOO)_2\cdot 2H_2O$. J Phys Soc Jpn 75:104704
60. (a) Poulsen RD, Bentien A, Chevalier M et al (2005) Synthesis, physical properties, multi-temperature crystal structure, and 20 K synchrotron X-ray charge density of a magnetic metal organic framework structure, $Mn_3(C_8O_4H_4)_3(C_5H_{11}ON)_2$. J Am Chem Soc 127:9156–9166; (b) Poulsen RD, Bentien A, Graber T et al (2004) Synchrotron charge-density studies in materials chemistry:16 K X-ray charge density of a new magnetic metal-organic framework material, $[Mn_2(C_8H_4O_4)_2(C_3H_7NO)_2]$. Acta Crystallogr A 60:382–389
61. Overgaard J, Iversen BB, Palii SP et al (2002) Host-guest chemistry of the chromium-wheel complex $[Cr_8F_8(tBuCO_2)_{16}]$: prediction of inclusion capabilities by using an electrostatic potential distribution determined by modeling synchrotron X-ray structure factors at 16 K. Chem Eur J 8:2775–2786
62. Yaghi OM, O'Keeffe M, Ockwig NW (2003) Reticular synthesis and the design of new materials. Nature 423:705–714
63. Kitagawa S, Kitaura R, Noro S (2004) Functional porous coordination polymers. Angew Chem Int Ed 43:2334–2375
64. Matsuda R, Kitaura R, Kitagawa S (2005) Highly controlled acetylene accommodation in a metal-organic microporous material. Nature 436:238–241
65. Noro S, Kitagawa S, Kondo M et al (2000) A new, methane adsorbent, porous coordination polymer $[\{CuSiF_6(4,4\text{'-bipyridine})_2\}_n]$. Angew Chem Int Ed 39:2082–2084
66. Férey G, Latroche M, Serre C et al (2003) Hydrogen adsorption in the nanoporous metal-benzenedicarboxylate $M(OH)(O_2C\text{-}C_6H_4\text{-}CO_2)$ ($M = Al^{3+}$, Cr^{3+}), MIL-53. Chem Commun 24:2976–2977
67. Rowsell JLC, Millward AR, Park KS et al (2004) Hydrogen sorption in functionalized metal-organic frameworks. J Am Chem Soc 126:5666–5667
68. Seo JS, Whang D, Lee H et al (2000) A homochiral metal-organic porous material for enantioselective separation and catalysis. Nature 404:982–986
69. Ohmori O, Fujita M (2004) Heterogeneous catalysis of a coordination network: cyanosilylation of imines catalyzed by a Cd(II)-(4,4'-bipyridine) square grid complex. Chem Commun 1586–1587
70. (a) Takata M, Umeda B, Nishibori E et al (1995) Confirmation by X-ray-diffraction of the endohedral nature of the metallofullerene $Y@C_{82}$. Nature 377:46–49; (b) Takata M, Nishibori E, Sakata M (2001) Charge density studies utilizing powder diffraction and MEM. Exploring of high Tc superconductors, C-60 superconductors and manganites. Z Kristallogr 216:71–86; (c) Takata M (2008) The MEM/Rietveld method with nano-applications - accurate charge-density studies of nano-structured materials by synchrotron-radiation powder diffraction. Acta Crystallogr A 64:232–245
71. Kondo M, Okubo T, Asami A et al (1999) Rational synthesis of stable channel-like cavities with methane gas adsorption properties: $[\{Cu_2(pzdc)_2(L)\}_n]$ (pzde = pyrazine-2,3-dicarboxylate; L = a pillar ligand). Angew Chem Int Ed 38:140–143

72. Kitaura R, Kitagawa S, Kubota Y et al (2002) Formation of a one-dimensional array of oxygen in a microporous metal-organic solid. Science 298:2358–2361
73. Takata M, Nishibori E, Kato K et al (2002) High resolution Debye-Scherrer camera installed at SPring-8. Adv X ray Anal 45:377–384
74. Tanaka H, Takata M, Nishibori E et al (2002) ENIGMA: maximum-entropy method program package for huge systems. J Appl Crystallogr 35:282–286
75. Kubota Y, Takata M, Matsuda R et al (2005) Direct observation of hydrogen molecules adsorbed onto a microporous coordination polymer. Angew Chem Int Ed 44:920–923
76. Kitaura R, Matsuda R, Kubota Y et al (2005) Formation and characterization of crystalline molecular arrays of gas molecules in a 1-dimensional ultramicropore of a porous copper coordination polymer. J Phys Chem B 109:23378–23385
77. Matsuda R, Kitaura R, Kitagawa S et al (2005) Highly controlled acetylene accommodation in a metal-organic microporous material. Nature 436:238–241
78. Dybtsev DN, Chun H, Kim K (2004) Rigid and flexible: a highly porous metal-organic framework with unusual guest-dependent dynamic behavior. Angew Chem Int Ed 43:5033–5036
79. Noritake T, Aoki M, Towata S et al (2002) Chemical bonding of hydrogen in MgH_2. Appl Phys Lett 81:2008–2010
80. Rosi NL, Eckert J, Eddaoudi M et al (2003) Hydrogen storage in microporous metal-organic frameworks. Science 300:1127–1129

Chapter 14
A Generic Force Field Based on Quantum Chemical Topology

Paul L.A. Popelier

14.1 Why Force Fields?

It is amazing that one simple master equation, the Schrödinger equation, describes with exquisite detail and accuracy all molecules and their behaviour. The power of this equation cannot be overestimated and the practical challenge of solving it neither, unfortunately. As a result, the energetics, structure, dynamics and properties of molecules and their assemblies remain locked in the Schrödinger equation unless a sufficiently powerful computer cluster unlocks them, by means of an efficient yet accurate algorithm. The most accurate *ab initio* methods, such as CCSD(T), scale unfavourably with system size, in spite of highly efficient computational schemes. The sustained progress in computer hardware (Moore's law) is not dramatic enough to enable such methods to be used for systems of real interest to a wide community of chemists, both in life science and materials science. If one is interested in properties requiring the sampling of configuration space then the situation is even direr. Born-Oppenheimer Molecular Dynamics (BOMD) [1] necessarily uses levels of theory much more approximate than the limiting case of CCSD(T) with a quadruple zeta basis set. An alternative technique, called Car-Parrinello MD [2], is entangled with Density Functional Theory (DFT) and a plane-wave basis set. Although the calculation effort per time step is lower than for BOMD more time steps are required. DFT scales very well with system size but at the expense of accuracy and in the absence of a clear path to improve the functional at hand. Moreover, hybrid functionals, which involve experimental input and possibly violate rigorous DFT conditions, continue to outperform "purer" functionals devoid from such input.

P.L.A. Popelier (✉)
Manchester Interdisciplinary Biocentre (MIB), 131 Princess Street, Manchester M1 7DN, UK

School of Chemistry, University of Manchester, Oxford Road, Manchester M13 9PL, UK
e-mail: pla@manchester.ac.uk

C. Gatti and P. Macchi (eds.), *Modern Charge-Density Analysis*,
DOI 10.1007/978-90-481-3836-4_14,

It is against this background of computational effort and uncertain future prospects that force fields prosper. Rooted in the 1930s they consist of an easy, direct, compact and intuitive relationship between energy and nuclear positions. If such a relationship can be captured by simple analytical expressions then the force field approach clearly represents a feasible route to carrying out nanosecond simulations of large systems. What is more, force fields are guaranteed a secure future because they are the only feasible methods for such problems, probably for decades to come. However, not all is well with their current status.

At the outset, force fields focused on mainly hydrocarbons and organic molecules. Later they were constructed to work for proteins and nucleic acids. All along, water force fields have enjoyed special attention and their own development given the utmost importance of water as a liquid. Liquid water, the Solvent of Life, has been modelled by more than fifty potentials since 1933 [3]. Nowadays force fields are available for many substances, from zeolites to pure metals, from carbohydrates to bioinorganic transition metals, as well as for a wide variety of solvents.

14.2 Why Look at Force Fields Again?

In spite of promising developments in *ab initio* molecular dynamics many simulations will only be possible by means of force fields in the foreseeable future. Even geometry optimisations of polypeptides and conformational searches of smaller (drug-like) molecules still take an inordinate amount of time when carried out by quantum mechanical calculations. This is why it is important to continue improving the potentials that underpin force fields.

The shape of presently used biomolecular force fields (e.g. AMBER [4], CHARMM [5]) was decided as far back as the 1980s. David Case, one of the developers of AMBER, wrote quite recently [6] that "*Without further research into the accuracy of force-field potentials, future macromolecular modeling may well be limited more by validity of the energy functions, particularly electrostatic terms, than by technical ability to perform the computations.*" This concern is echoed, even at textbook level, where it is stated (on p. 43 of Ref. [7]) that "*Obtaining a good description of the electrostatic interaction between molecules (or between different parts of the same molecule) is one of the big problems in force field work*". On the same page he lists four main deficiencies of modeling the electrostatic energy by fixed atomic charges.

The questionable state of affairs in the current performance of popular force fields such as AMBER is clearly illustrated by an important paper [8] on the conformational dynamics of "trialanine", a small peptide with two peptide bonds. This molecule has emerged as a paradigm in the study of conformational dynamics of a small peptide in aqueous solution, due to an exceptional amount of experimental and *ab initio* data. 2D-IR studies suggest that trialanine exists predominately in the so-called P_{II} conformation as confirmed by NMR, polarised Raman and FTIR experiments. New experimental data suggest that trialanine may also exist to a smaller amount in the helix-like α conformation and the extended β conformation. In spite of a wealth of data, interpreting spectroscopic experiments in the condensed phase is a notorious problem since conclusions are often ambiguous due to reliance on highly simplified models. This is why, one [8] resorted to current force fields to retrieve an independent and unbiased theoretical description. MD simulations were run with six different force fields: two parameterisations of AMBER, two of GROMOS, one of CHARMM and one of OPLS. The Ramachandran probability distribution plots from 20 ns MD simulations were *qualitatively* different between force fields. Indeed, even the minor modification between AMBER's "parm94" and "parm96" parameterisations significantly changed the population ratio of the conformational states. Furthermore, OPLS could not resolve P_{II} and β, and AMBER "parm94" significantly populated the α conformation. The conclusion was very disappointing, suggesting that commonly used force fields are not capable of correctly describing the (un)folding of a peptide.

The state of the art in aqueous ion simulations also reveals shortcomings. Although there are more than fifty water potentials [3], including polarisable, flexible and dissociable ones, the development of accurate aqueous ion potentials is limited. Many are *effective* potentials, incorporating the effects of polarisation implicitly as part of the parameterisation. These limitations have been noted when inhomogeneous systems are considered, such as interfaces and narrow channels [9, 10]. Those ion potentials are limited by the simple *ad hoc* models that have also been developed for water. The statement that *"correctly representing the electrostatics is still very much an art form, not a science . . . "* is true not only for water but also for ions [11].

In summary, the basic architecture of force fields has not changed much since the 1980s (or even before). Point charges still dominate the electrostatic interaction in spite of well-known inaccuracies at short range. If polarisation is added to the force field it is carried out following dated ideas. We propose to take full advantage of the increase in computing power by four orders of magnitude since the 1980s, in order to drastically alter the shape of a force field. This task is huge but the potential reward tempting. The driving force behind this long and challenging project is enhancing the realism of a force field. This should be achieved in a bottom-up fashion, from small to large molecules, and by systematic validation. In this chapter we lay out a strategy to accomplish a novel force field architecture.

14.3 Two Different Worlds?

The energy of a molecular system can be described either as a quantum mechanical energy partitioning (Eq. 14.1) or as a direct function of geometrical coordinates, as in current force fields (Eq. 14.2). Introducing ρ_1 as the first-order reduced density matrix (and its diagonal, the electron density ρ), and ρ_2 as the diagonal of the second-order reduced density matrix, one can write [12]:

$$E = \int \left[-\frac{1}{2} \nabla_{\mathbf{r}}^2 \rho_1(\mathbf{r}, \mathbf{r}') \right]_{\mathbf{r}'=\mathbf{r}} d\mathbf{r} + \int v(\mathbf{r}) \rho(\mathbf{r}) d\mathbf{r} + \sum_{\alpha} \sum_{\beta>\alpha} \frac{Z_\alpha Z_\beta}{R_{\alpha\beta}} + \iint \frac{1}{r_{12}} \rho_2(\mathbf{r}_1, \mathbf{r}_2) d\mathbf{r}_1 d\mathbf{r}_2 \tag{14.1}$$

where the terms are in order: the electronic kinetic energy, the nuclear-electron potential energy, the nuclear-nuclear potential energy and the electron-electron potential energy. In this equation, v($\mathbf{r}$) is the nuclear potential acting as an external potential on the electrons, Z the atomic number, $R_{\alpha\beta}$ an internuclear distance (where α and β refer to nuclei) and r_{12} an inter-electronic distance. This master equation is valid at Hartree-Fock level as well as at post-Hartree-Fock level. On the other hand, in an AMBER force field, the potential energy V is:

$$V = \sum_{bonds} k_r (r - r_0)^2 + \sum_{bonds} k_\theta (\theta - \theta_0)^2 + \sum_{torsion} \frac{1}{2} V_n [1 + \cos(n\varphi - \gamma)] + \sum_{i<j} \left(\frac{A_{ij}}{r_{ij}^{12}} - \frac{B_{ij}}{r_{ij}^{6}} + \frac{q_i q_j}{\varepsilon r_{ij}} \right) + \sum_{H-bonds} \left(\frac{C_{ij}}{r_{ij}^{12}} - \frac{D_{ij}}{r_{ij}^{10}} \right) \tag{14.2}$$

The first sum describes the energy associated with stretching covalent bonds, treated as a spring with a force constant k_r and reference bond length r_0. The second sum accounts for the deformation energy of valence angles θ, while the third sum represents the intrinsic deformation energy corresponding to variation in dihedral angles φ, where γ is phase angle. The fourth sum, which runs over atoms i and j, corresponds to the core repulsion, dispersion and electrostatic interactions, respectively. These interactions depend on various powers of the interatomic distance r_{ij}. For simplicity, Lennard-Jones parameters A and B are used here as well as the traditional point charges q, deplorably accompanied by the relative permittivity ε. The final sum is a dedicated hydrogen bond term, again introducing Lennard-Jones-like parameters C and D. Cross terms and anharmonic contributions are added in more elaborate potentials. The energy expression of CHARMM only differs by an extra term for improper dihedral angles. In spite of this tight similarity, the values of the parameters (e.g. k_r, A_{ij}, q_i) are *not* exchangeable between force fields.

It is clear that Eqs. 14.1 and 14.2 are very different; hence the subtitle "Two different worlds?" For example, there is no force field that contains a kinetic energy term, to the best of our knowledge. Yet, the physics of the quantum reality underpinning a force field introduces a kinetic term. As a second example, torsion barriers can of course be modelled directly by a dedicated term in Eq. 14.2 but these barriers should also naturally arise from Eq. 14.1. For this purpose Eq. 14.1 needs to be partitioned into intra- and interatomic contributions [13], as shown in Eq. 14.3 for a Hartree-Fock wavefunction,

$$E_{tot} = \sum_{A} E_{Kin}^{A} + \frac{1}{2}\sum_{A} E_{coul}^{AA} + \frac{1}{4}\sum_{A} E_{X}^{AA} + \frac{1}{2}\sum_{\substack{A,B \\ B\neq A}} E_{coul}^{AB} + \frac{1}{4}\sum_{\substack{A,B \\ B\neq A}} E_{X}^{AB} \quad (14.3)$$

where the first three terms are intra-atomic (running over atomic index A) describe the kinetic, Coulomb and exchange contributions. The latter two terms describe the Coulomb and exchange interaction terms, sums running over distinct pairs of atoms. One can then study how these five types of terms, that govern all interactions, cooperate to form a torsion potential [14]. Torsion barriers arise from a subtle interplay between the different types of physical terms. A torsion around a central CC bond is caused by a completely different interplay between terms than a torsion around a OO bond.

The main problem with force field construction is that it deteriorates to a fitting exercise, introducing a large number of parameters that satisfy (*ab initio*) observations made on a confined subset of molecules. The act of fitting itself is not problematic but that the parameter values can be devoid of physical reality is. For example, in Eq. 14.2, a number of point charges q_i are introduced as parameters. What guarantees that these parameters are indeed purely electrostatic? How can one be sure that they express exactly what they are supposed to express, given the Coulomb formula in which they appear? In fact, there is no such guarantee and this is the source of force field deficiencies. It is highly desirable to ensure that the electrostatic term indeed represents the electrostatics, nothing more and nothing less. Only then does the force field make contact with Eq. 14.1 and is it guaranteed that at least one term in the force field is correct and real. Work on force field improvement uncovers that the success of a force field depends on a delicate balance between the terms. If one term is improved then the overall performance may worsen. However, if terms are well-defined and physically justified then the false balance cannot upset.

Parameter redundancy in classical force fields also warrants care when decomposing energy differences into individual terms. The rotational barrier in ethane, for example, may be solely reproduced by a HCCH torsional term, solely by an H-H van der Waals repulsion or solely by H-H electrostatic repulsion. Different force fields will have slightly different balances of these terms, and while one force field may contribute a conformational difference primarily to steric interactions, another may have the major determining factor to be the torsional energy, and a third may "reveal" that it is all due to electrostatic interactions [15].

The way forward, we believe, is to reduce to number of parameters by only fitting what needs to be fitted. In particular, there is no need to fit the electrostatics because atomic electron densities can be obtained directly, from a suitable partitioning method. Atomic (point) charges, directly obtained from *ab initio* electron densities, can populate the Coulomb term in Eq. 14.2. The next step towards increased realism is to replace the point charges by multipole moments, which has been accomplished before in various ways (e.g. [16–19]). We have gone a step further and used machine learning to construct a direct relationship between a given atom's multipole moments and the coordinates of its neighbouring atoms. This novel methodology offers an alternative to the familiar polarisability tensors of intermolecular Rayleigh-Schrödinger perturbation theory.

The basic tenet of our approach is to move away from Eq. 14.2 and preserve the physical realism of Eq. 14.1 as much as possible. Our line of attack is *not to "tweak"* an existing force field. Instead, it is a concerted effort to put force field design on a firmer footing. Inevitably, this will require computer resources but these are available at low cost anyway.

14.4 The Cornerstones

Before we discuss the three cornerstones of our approach a few words on transferability are appropriate. Transferability is the foundation of any force field and can be regarded as the zeroth cornerstone. It is the basic assumption that information acquired from small molecules can be used to make predictions on large molecules. An example of a transferable property is the atomic dipole moment of an oxygen. This moment, calculated for the oxygen in the OH group of the isolated amino acid serine, must be very similar to that in the same oxygen in serine, now part of a protein. Without transferability the very idea of a force field would collapse because a force field constructed for a set of molecules could then only make "predictions" within that set. A good force field must be able to break out of the training set and make reliable predictions outside the original domain of applicability. Transferability lies at the heart of functional groups, whose existence makes chemistry a more systematic science. The fact that functional groups exist is a proof of a high degree of transferability. However, perfect transferability is unattainable [20]. If a force field is properly parameterised then it should automatically possess transferable parameters. This is paramount for a force field to be truly predictive. This is why it is also important to be able to detect transferability, discussed in detail below. Unfortunately, current force fields typically impose transferability rather than detect it.

Because we are concerned with an atomistic force field we need a definition of an atom, which is the first cornerstone of a force field. We adopted the solution provided by the "Quantum Theory of Atoms in Molecules" (QTAIM) [21], which rigorously defines atoms from the electron density using only the gradient of the electron density, $\nabla\rho$. In 2003 we introduced [22] the more appropriate name *Quantum Chemical Topology (QCT)* as an overarching term for all approaches (e.g. [23–26]) that, analogously to QTAIM, use a gradient vector field as the partitioning engine, but then operating on a function different from the electron density or its Laplacian. As explained in the Introduction of ref. [27], the central idea of QCT is using a gradient vector field as the partitioning engine for a property function of interest. The use of the gradient vector is the most distinct feature that bundles the topological

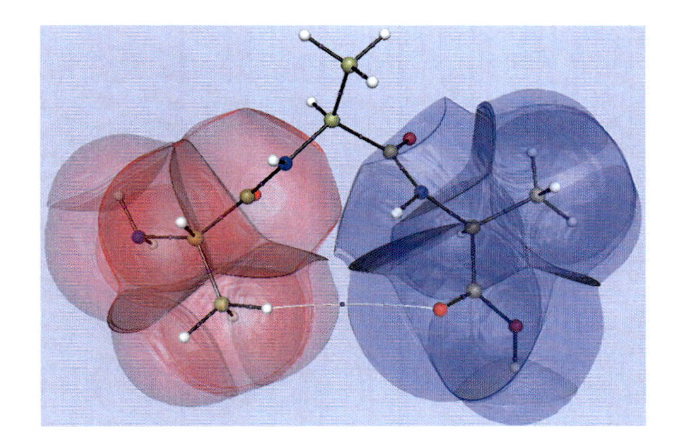

Fig. 14.1 Partitioning of trialanine into topological atoms. Each atom is malleable box, with specific shapes depending on the nuclear positions, nuclear charges and the number of electrons in the molecule

approaches mentioned above. Figure 14.1 shows how QCT partitions trialanine into (topological) atoms, when the partitioning engine is the gradient of the electron density. Such atoms have finite volumes due to their sharp boundaries. They do not overlap and leave no gaps between them.

Working with topological atoms is of course a choice since there is more than one way to define an atom in a molecule. There are five advantages associated with QCT partitioning. First, in the case where the partitioning engine is the gradient of the electron density, QCT is essentially based on the electron density, which is measurable and theoretically well-defined; it can be obtained regardless of the type of basis set used (Gaussian, plane-wave, Slater) and even in the absence of a basis set (e.g. Quantum Monte Carlo). Moreover, the electron density describes atoms regardless of whether they participate in *intra-* or *inter*molecular interactions. Secondly, QCT eliminates the problem of penetration (energy). Hence, there is no need for damping functions (which are cumbersome and formulated as late as 2000 for the electrostatic interaction [28]). Penetration energy arises from the overlap of interacting atoms and this overlap is absent in QCT. Thirdly, QCT atoms are rooted in quantum mechanics, which enhances their physical meaning. Fourthly, the multipolar expansion of topological atoms (see below) still converges at short range. Finally, QCT atoms can be visually represented due to their finite shape, which provides an intuitive understanding of interactions in complex situations, such as encountered in docking. An auxiliary conceptual advantage is the fact that their volume stems from the same volume integral as that yielding the multipole moments. A disadvantage relates to the computational expense and algorithm robustness of the atomic integration, which is still a topic of research.

Multipole moments constitute the second cornerstone of a realistic force field since at short range they provide a more accurate description of electrostatic interaction. Combining the QCT partitioning with spherical tensor multipole moments [29] led to a new way of evaluating the Coulomb interaction energy between atoms. Figure 14.2 shows two interacting oxygens in glycine, each endowed with a local axis system. Spherical tensors are perhaps less intuitive than Cartesian ones but

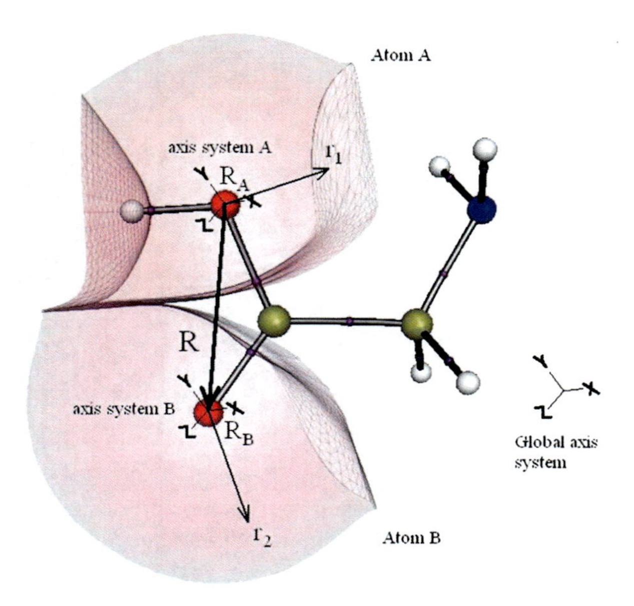

Fig. 14.2 Two interacting topological atoms *A* and *B*, each endowed with a local axis system. Each atom is described by its own set of multipole moments, expressed with respect to their axis systems

they are more compact since they are irreducible. For example, there are only five quadrupole moments instead of nine possible Cartesian ones (or six, after reduction due to symmetry). In general, for a multipole of rank ℓ there are $2\ell + 1$ multipole moments (i.e. components). This means that the whole electron density of a given atom can be compressed into only a handful of numbers (i.e. multipole components) sufficient to express the interaction with the electron density of another atom.

The use of multipole moments has a more general significance. They enable a separation of coordinates in $1/r_{12}$, which entangles both $\mathbf{r}_1$ and $\mathbf{r}_2$, each set of coordinates describing one of the interaction charge distributions. The separation is achieved by a well-known series expansion of $1/r_{12}$ into terms factorised (and hence separated) into $\mathbf{r}_1$ and $\mathbf{r}_2$. The advantage of this expansion is that multipole moments can be computed once and for all and then used in the calculation of the electrostatic interaction for any mutual orientation of the moments. The price paid for the series expansion is possible lack of convergence. If this occurs the expansion becomes useless. It should be emphasised that the multipole expansion can also be used in the context of the exchange term (last term in Eq. 14.2) because it also contains $1/r_{12}$. The corresponding multipole moments can be called *exchange moments*, which are given with explicit reference to molecular orbitals. We have demonstrated [30] that the corresponding expansion converges quickly. However, exchange moments based on canonical molecular orbitals are not transferable. Finding a way to transfer exchange moments, if it exists, is a major advance because it eliminates the use of empirical Lennard-Jones parameters. Treating repulsion in a manner fully integrated with the QCT treatment of Coulombic interactions (and polarisation) is conceptually superior to the current hybrid method.

In our early work we derived and implemented the first and second derivatives of a multipolar potential with respect to rigid body geometrical parameters [31]. Using Distributed Multipole Analysis (DMA) [18] we applied this approach to explore the potential energy surfaces of simple van der Waals complexes [32].

The third cornerstone is the use of machine learning to model polarisation and charge transfer. Originally we worked within the context of long range Rayleigh-Schrödinger perturbation theory and studied charge transfer along the hydrogen bond in the water dimer via QCT distributed polarisabilities [33]. Later we abandoned the perturbation approach in favour of capturing polarisation directly from supermolecular calculations. The main idea is to create a *direct mapping* between a given atomic multipole moment and the coordinates of its neighbouring atoms. In this approach there is no explicit polarisability (such as α for dipolar polarisability). Instead, the polarisability is implicitly accounted for in the mapping itself, which can be a set of trained weights if the machine learning method is a neural network. To make this procedure clearer, imagine generating thousands of dimer, trimer, tetramer, pentamer and hexamer water clusters. Monitoring the dipole moment of a single central water molecule in increasingly large water clusters shows that a plateau is reached after the hexamer [34]. Note that the dipole moment may be origin dependent if the water molecule is not neutral. Although it is nearly neutral in a cluster (see Sect. 14.5.3), this origin dependence does not affect the energy calculations as long as the origin is kept track of. The clusters are sampled from a Molecular Dynamics simulation that lacks polarisation. Then we train neural network to predict each atomic multipole moment of a central water molecule. The input of the neural net consists *solely* of the coordinates of the water molecules surrounding the central water. The output is a multipole moment of an atom in the central water. This method gives a *dynamic* multipolar representation of the water electron density without explicit polarisabilities. Secondly, there is no need to perform iterative calculations to self-consistency during the simulation. Thirdly, there is no need to include damping terms in order to avoid a polarisation catastrophe. Fourthly, dipolar (and higher rank) polarisation is treated on the same footing as charge transfer, unlike in a typical perturbational approach. This is because in our method charge transfer is just "monopolar polarisation", which is technically no different to dipolar (or higher ℓ) polarisation.

14.5 Building on the Cornerstones

The preliminary work at the outset of the force field project has been reviewed extensively in Ref. [35]. Here we reiterate salient points and expand the review to include more recent work, going beyond the previously published "gentle introduction" [36].

14.5.1 Convergence

As mentioned above, convergence needs to be carefully monitored since a diverging expansion is useless. In Sect. 7.4 of Stone's book [29] on intermolecular forces one comes across an erroneous statement. With reference to Voronoi polyhedra and topological atoms, it is claimed that *"neither are suitable for application to intermolecular forces because they have poor convergence properties."* Already in 1985 there was work that disproved this assertion [37]. Moreover, starting in 1996 [38], we have accumulated considerable evidence that topological atoms show favourable convergence, both in the electrostatic potential and the electrostatic interaction. The electrostatic potential was chosen as a starting point because it is a special case of electrostatic interaction, where one of the interacting partners is just a proton.

We computed [39] the exact atomic electrostatic potential and its value obtained via multipole expansion for molecules including molecular nitrogen, water, ammonia, imidazole, alanine and valine. The potentials were evaluated at the order of ten thousand grid points on the water-accessible surface. Excellent convergence behaviour for the C_α atoms in both alanine and valine is seen already for the octupole moment ($\ell = 3$), with a root mean square discrepancy of just above 0.1 kJ mol^{-1}. An advantage of the finite (i.e. sharply bounded) nature of topological atoms is that they have a *finite* convergence sphere. This means that the convergence condition associated with the multipolar expansion can be exactly obeyed, which is impossible for atoms that extend to infinity, as is the case in DMA.

Perhaps counter to intuition, this excellent convergence behaviour is compatible with the admittedly highly non-spherical shape of topological atoms. The cause [40] of the excellent convergence is the exponentially decaying electron density. The convergence of the electrostatic potential is dominated by the decay of the electron density inside a topological atom rather than by its shape.

The introduction of a *continuous* multipole expansion [41] expands the convergence region. The proposed method allows the electrostatic potential to be evaluated accurately at short range. The replacement of the classical multipole expansion by a moment representation, involving Bessel functions, enables the virtually exact computation of the atomic electrostatic potential at short range. With a reasonable number of pre-computed continuous moments one can predict the electrostatic potential right up to the boundary of the atom.

A second completely different method introduced [42] "inverse" multipole moments, enabling an expansion to converge everywhere. These moments are defined by negative powers of the magnitude of the position vector describing the electron density inside the atom. The method is fast since the multipole moments are computed beforehand. Once computed, they can generate a potential at any point of space. The computational cost is at best twice that of the classical multipole expansion. The drawback is the high rank necessary to get a convergent value of the potential at short range.

We also focused [43] on the Coulomb energy between atoms in *supermolecules.* Atom-atom contributions to the molecular intra- and intermolecular Coulomb energy were computed exactly, i.e. via a double integration (six-dimensional or 6D) over atomic basins, and by means of the spherical tensor multipole expansion, up to rank $L = \ell_A + \ell + 1 = 5$. The convergence of the multipole expansion was able to reproduce the exact interaction energy with an accuracy of 0.1–2.3 kJ mol^{-1} at $L = 5$ for atom pairs, each atom belonging to a different molecule in a given van der Waals complex, and for non-bonded atom-atom interactions in single molecules.

We then investigated [44] the convergence of both the energy and the geometry of a set of van der Waals complexes, with respect to the rank L of the multipole expansion. Although the QCT energies converge more slowly than the DMA ones, excellent agreement is obtained between the two methods at high rank ($L = 6$), both for geometry and energy. Subsequently we showed [45] how to distribute the multipole moments of a topological atom over extra non-nuclear positions. Such distribution is based on *shifting* a moment from its original expansion site to a new one. This distribution accelerates QCT multipolar convergence. The addition of extra sites is more beneficial to the convergence of the electrostatic interaction energy of *small* systems. However, in large systems excellent convergence was found for QCT *without* the introduction of extra sites, a surprising result.

The same distribution technology led to the gratifying result that the Coulomb interaction energy between atoms in a 1,3 relationship (e.g. the two oxygens in glycine, see Fig. 14.2) can be described by a *convergent* QCT multipole expansion [46]. The 1,4-interactions can also benefit from distribution to non-nuclear sites, if required in the first place. Unpublished work from our group shows that the electrostatic interaction between bonded atoms (i.e. 1,2-interactions) can be expanded by a convergent expansion, provided the original nucleus-centered moments are shifted by an appropriate amount. Finding this amount is cumbersome and has not been generalised to an arbitrary case. Note that 1,2-interactions between topological atoms has also been numerically computed [47] using the Kay, Todd, and Silverstone Bipolar Expansion [48]. This expansion does not have a convergence condition but this advantage comes at a price: its multipole moments are "r-dependent" functions, i.e. they cannot be calculated independently for each atomic basin, prior to the evaluation of the 1,2 Coulomb energy.

It is important to understand the convergence behaviour of the electrostatic interaction between topological atoms at a range longer than that occurring in the test molecules looked at in the work above. This is because the force field is initially set up for proteins, where distances of more than 20 Å are common. Evaluating high-rank multipole interactions is costly but fortunately this is only necessary locally. The majority of interatomic distances occurring in a protein are long-range and in that region point charges (and perhaps a dipole moment) suffice to represent the interaction exactly. For medium range we were able to set up a function [49] that relates atom type, interatomic distance, interaction rank L and energy error (of the multipole energy compared to the exact one). This paper answers question such as "*How close can two atoms get before their interaction energy diverges?*" and "*Which is the lowest (and therefore cheapest) rank necessary to achieve a certain error in energy?*"

14.5.2 Transferability

Over the years, the transferability of the properties of topological atoms has received much attention from the Bader group [20, 50] and others (e.g. [51–56]). There is also work [57] on transferability of pseudoatoms or multipole formalism [58] rooted in high-resolution X-ray crystallography. It is important to monitor transferability of the right physical property for it to be present in a way that actually matters to a force field. Ultimately, atomic interaction energy is the property that needs to be transferable because the energy of a system governs its structural and dynamical behaviour.

An early study [22, 59, 60], however, looked at atomic transferability from the point of view of intrinsic (non-interacting) atomic properties such as atomic energy, volume, population and the magnitude of the dipole, quadrupole, octopole and hexadecapole moment. For the first time, this work *computed* atom types, based on a statistical cluster analysis of 760 atoms taken from all natural amino acids and smaller derived molecules. This work [22] exposed weaknesses in the delineation of atom types in popular force fields such as AMBER.

Subsequently, we used the electrostatic potential as an indicator of how transferable an atom or functional group is between two molecules [61]. The potential acts as a "half-way house" between atomic multipole moments and interatomic interaction. The potential is a special case of interaction energy, in which the interaction partner is fixed as a proton. The potential has the advantage of incorporating the relative importance of the multipole moments. However, the necessity of a grid is a disadvantage since it introduces more decisions to be made and hence possible arbitrariness. The potential generated on a grid by the terminal aldehyde group in the biomolecule retinal was compared with the corresponding aldehyde group in smaller molecules derived from retinal. Similarly, we looked at the terminal amino group in the free amino acid lysine. This group turned out to be very little influenced by any part of the molecule further than two carbon atoms away. However, the aldehyde group in retinal was influenced by molecular fragments up to six carbon atoms away. We attributed this striking disparity to the fact that retinal contains a *conjugated* hydrocarbon chain but lysine not.

Finally, transferability was investigated [62] in terms of electrostatic interaction energy. Although a more informative assessment of transferability than by intrinsic atomic properties, this method invokes a "probe". This probe is an atom (or group of atoms) with which the atom of interest interacts. The atom of interest can now not be assessed for transferability by its intrinsic properties (i.e. multipole moments) but by its interaction with a probe. The necessity of this probe acts as a source of arbitrariness but is inevitable. Focusing on the water trimer and serine … $(H_2O)_5$ as pilot systems, it turned out that the atomic multipole moments in serine are more transferable than those of the water cluster, with the exception of the hydroxyl group. Also, in hydrated serine, the water cluster is more polarisable than serine.

14.5.3 *The Electrostatic Interaction Without Polarisation: Structure and Dynamics*

Geometries and intermolecular interaction energies of 27 hydrogen-bonded (DNA) base pairs predicted by QCT multipole moments (supplemented with a hard-sphere or Lennard-Jones potential) were compared [63] with those predicted by Merz-Kollman (MK) charges, Natural Population Analysis (NPA) and DMA. Globally, the AIM interaction energy curve follows the same pattern as that of MK, NPA and DMA. The MK model systematically underestimates the interaction energy and NPA shows undesirable fluctuations. A test of QCT on a DNA tetrad suggests that it is able to predict geometries of more complex nucleic acid oligomers than base pairs.

We also carried out a systematic comparison [64] of multipole moments versus point charges (AMBER, CHARMM, OPLS, MMFF and TAFF force fields) for pure water clusters, hydrated serine and tyrosine clusters. The reference geometries were generated at *ab initio* level. High rank multipole models ($L = 5, 6$) performed consistently better than point charge models in predicting the geometries. The $L = 5$ model is constantly most reliable in the vast majority of cases, although lower ranks are sometimes better for individual cases. Secondly, the quality of lower rank (multipole) models improves with increasing cluster size.

The same complexes were systematically analysed [65] in terms of QCT-partitioned supermolecular electron densities. Water molecules remain virtually neutral when present in the clusters studied. Observations were condensed in statements such as "the number of H-bonds donated by a water molecule determines its total molecular dipole" or "all oxygen atoms in water clusters are found to be more negatively charged than if they were in a free water molecule". The mechanism of change in dipole moment going from the gas phase to a cluster is also exposed. It is therefore possible to categorise types of water molecule with specific electronic characteristics based on H-bonds formed. This helps in the assignment of atom types.

Then we applied the exhaustive electrostatics offered by high-rank multipole moments in the context of Molecular Dynamics (MD) simulations of HF [66] at five temperatures between 203 and 273 K. This is made possible by the program DLMULTI [67], which is an offspring of the widely used program DLPOLY [68] developed at Daresbury Lab in the North-West of England. Long range interactions are represented by a high-rank *multipolar* Ewald summation, developed [67] and implemented by Leslie, using the spherical tensor formalism. Only two parameters in the Lennard-Jones part of the potential were adjusted to the experimental density and radial distribution functions. Agreement with experiment is excellent for the total energy and the density, and reasonable, with even overall performance, for the diffusion coefficient, the isobaric heat capacity and the thermal expansion coefficient.

MD simulations of liquid H_2O [69] at ambient conditions were carried out with the electrostatic interaction represented up to $L = 5$, and repulsion and dispersion via a Lennard-Jones potential. Simulations at a constant pressure of 1 atm, with 216 water molecules at 278, 288, 300, 308 and 318 K, showed that high-order multipolar interactions ($L = 5$) are essential to recover the typical features of a liquid-like structure. With only two adjustable parameters, σ (3.140 Å) and ε (0.753 kJ mol^{-1}) in the Lennard-Jones potential function, the density of the equilibrated system is 0.999 gcm^{-3}. Liquid simulations at four extra temperatures showed a maximum in the density at about 15°C, which is off by 11°C from the well-known experimental value of 4°C. The density of simulated water is within less than 1% of the experimental value, while the calculated energy of the liquid is within 3% of the experimental result. The experimental value of the self-diffusion coefficient, *D,* is underestimated by at least 32%. The value for C_p is overestimated by 28% and the thermal expansion coefficient α by 27%. This homogeneous error distribution is most encouraging given that simpler models such as TIP5P produced values for the same bulk properties (D, C_p and α) deviating from experiment by 14%, 62% and 145%, respectively.

A second simulation study [70] on liquid water explored the phase diagram more widely, now varying the pressure at 7 pressures between 1 and 10,000 atm and 17 temperatures between −35°C (238 K) and 90°C (363 K). The well-known maximum in the liquid's density at 4°C is reproduced at 6°C. Six bulk properties are calculated and found to deviate from experiment in a homogeneous manner, that is, without serious outliers, compared to several other potentials. Spatial distribution functions and the radial distribution functions are used to analyze the local water structure. At the lone pair side of a central water, neighboring waters form a continuous horseshoe-like distribution, with substantial narrowing in the central part. The latter feature is unique to the QCT multipolar potential since it is not found with SPC nor TIP5P [71]. Under high pressure, the local structure undergoes dramatic rearrangement and results in the collapse of second shell neighbors into the interstitial region of the first shell, which is in close agreement with experiment.

14.5.4 Polarisation

Originally we worked along the lines of the "Topological Partitioning of Electronic Properties (TPEP)" [72] approach, which combines QCT with long-range Rayleigh-Schrödinger perturbation theory. We investigated charge transfer response in a water dimer [33] involving distributed polarisability components. The *ab initio* computational approach of distributed response analysis was also used to quantify how electrons move across conjugated molecules in an electric field, in analogy to conduction [73]. The method promised to be valuable for characterizing the conductive behaviour of single molecules in electronic devices. Later we abandoned this approach for one involving machine learning. This was driven by the practical reason of the cumbersome retrieval of two-electron integrals necessary to compute the polarisability as well as by the conceptual reason of the breakdown of perturbation theory at short range.

The first demonstration of this novel technique involved HF dimer in vacuum [74]. High-rank multipole moments of a central molecule are predicted by a parsimonious neural network being fed with the relative position of its neighboring molecule. Excellent prediction errors are obtained, measured by direct comparison of multipole moments and by discrepancies in molecular interaction energy. Geometry optimization using neural net multipole moments yielded a potential interaction energy 5.2 kJ mol^{-1} lower than when using monomer moments on each molecule for exactly the same geometry optimized structure. The variation in time of a net molecular charge in each molecule of the HF dimer was also followed.

The next study [75] focused on intramolecular polarisation. Using glycine and N-methylacetamide as pilot systems, we show that neural networks can capture the change in electron density due to polarisation. After training, modestly sized neural networks successfully predict the atomic multipole moments from the nuclear positions of all atoms in the molecule. Accurate electrostatic energies between two atoms can be then obtained via a multipole expansion, inclusive of polarisation effects. As a result, polarisation is successfully modeled at short-range and without an explicit polarisability tensor. This approach puts charge transfer and multipolar polarisation on a common footing. Nonbonded atom-atom interactions in glycine cover an energy range of 948 kJ mol^{-1}, with an average energy difference between true and predicted energy of 0.2 kJ mol^{-1}, the largest difference being just under 1 kJ mol^{-1}. Very similar energy differences are found for NMA, which spans a range of 281 kJ mol^{-1}. The current proof-of-concept enables the construction of a new protein force field that incorporates electron density fragments that dynamically respond to their fluctuating environment. This work paves the way for the establishment of *flexible* multipole moments, to be used outside the restraint of the rigid body context, which continues to dominate multipolar potentials [76].

The neural networks were pushed to their limit looking [77] at water clusters up to the hexamer. Using thousands of dimer, trimer, tetramer, pentamer, and hexamer clusters sampled from a molecular dynamics simulation lacking polarisation, we trained neural networks to predict the atomic multipole moments of a central water molecule. The input of the neural nets consists *solely* of the coordinates of the water molecules surrounding the central water. This method gives a dynamic multipolar representation of the water electron density without explicit polarisabilities. Instead, the required knowledge is stored in the neural net. We showed that, due to the fitting method behind TIP3P, it is able to predict total Coulomb interaction energies that are more accurate than the polarisable model. However, this accuracy is due to a fortuitous cancelation of very inaccurate atom-atom interaction energies. This suggests that other empirical point charge models, fitted in a fashion similar to TIP3P, may again yield accurate total interaction energies for large clusters of water molecules but not for the correct reason.

The most recent study [78] deepened and perfected the treatment of polarisation for liquid water in terms of machine learning, in particular multi-objective optimisation. Here we focused on the balance between two competing objectives: accuracy, in terms of errors in Coulomb energy, and computing time. First, the predictive ability and speed of two additional machine learning methods, radial basis function neural networks (RBFNN) and Kriging [79, 80], are assessed with respect to water models based on neural networks (actually multilayer perceptrons). Compared to the latter, we find that RBFNNs are more accurate but take more computing time in predicting multipole moments. Kriging is even more accurate but needs most time. From the Pareto-front a user can choose which model is appropriate for his or her constraint on accuracy and computing time. A Kriging model for the water dimer yields multipole moments such that 90% of approximately a thousand test configurations yield an error of less than 1 kJ mol^{-1}. For the water pentamer (i.e. a central water with four waters around it) more than 50% correspond to an error of less than 4 kJ mol^{-1} ($\sim$1 kcal mol^{-1}). This study paved the way to MD simulations of liquid water with each water molecule being polarised by four nearest neighbours. Although this is technically still a classical simulation, the constant adjustment of the electronic structure of water makes it reminiscent of Car-Parrinello simulations, albeit at a fraction of the cost.

Finally, we comment on a concern that occasionally surfaces in the literature (e.g. [81, 82]), namely that the discrete partitioning of QCT may not allow for a straightforward construction of the total electron density, due to mismatches in the interatomic surfaces of transferred atoms. Indeed, we do not know if fragments leave gaps in a reconstructed electron density. This is because the machine learning methods do not focus on the shape of a topological atom. Instead, they focus on predicting the atom's multipole moments because they matter in the calculation of the electrostatic interaction between atoms. The atomic shape would be an indirect arbiter of the reliability of a model trained by machine learning. Even if gaps (or the opposite case of interatomic overlap) occurred, the error they introduce would have to be assessed through multipole moments or energy in order to be relevant for a force field. After all, a force field is about energy, not shape.

14.5.5 Alternative Methods: High-Resolution Crystallography

For space reasons this chapter could not report on non-QCT methods that share the goal of improving on electrostatics. However, given the prominence of experiment in this book it is perhaps appropriate to comment briefly on three methods proposed in X-ray crystallography. At the 2007 Gordon Research Conference on Electron Distribution and Chemical Bonding the independent work of three different group were presented: the approach of Volkov and Coppens [83] and co-workers, that of Jelsch and Lecomte [84] and co-workers, and the approach of Luger and Dittrich [85] and co-workers.

The first approach was applied for the first time [83] on protein-ligand interaction energies. The main idea behind this approach goes back to work by Koritsanszky et al. [57], who generated theoretical X-ray data from ab initio electron densities of isolated tripeptide molecules (in the gas phase). They fitted these densities to a model density obtained using the Hansen-Coppens formalism [58]. This is an established refinement method of high-resolution X-ray crystallography, already available in the late 1970s, and introduced as a modification to the rigid-pseudoatom scattering formalism proposed by Stewart [86]. This formalism describes a molecular (or crystalline) electron density as a superposition of aspherical atomic densities, by means of multipole moments. This formalism is responsible for partitioning the molecular electron density into atomic contributions, associated with so-called pseudoatoms. The partitioning results from a least-squares fit delivers fuzzy, interpenetrating atoms, which extend to infinity, in principle. The pseudoatoms, obtained for a sample consisting of a limited number of peptides, were shown to be highly transferable and fairly invariant under conformational changes. Subsequent work [87] developed this idea further, setting up a theoretical pseudoatom databank. The consistency achieved by this databank proved to be much better than that obtained with the AMBER99, CHARMM27, MM3, and MMFF94 force fields. This result corroborates the superiority of multipole moments over point charges. The multipolar expansion was truncated at the hexadecapolar level ($\ell_{\text{max}} = 4$) for the non-hydrogen atoms and at the quadrupolar level ($\ell_{\text{max}} = 2$) for hydrogen atoms. Finally, atom types were already present in this study but a "spawning procedure" for new families of chemically similar atoms was proposed later [83]. Volkov et al. [81] noticed large discrepancies (e.g. 69 kJ mol^{-1}) between energies obtained from the library of pseudoatoms and from theoretical structure factors calculated from Kohn–Sham wavefunctions. The paper stated that the "discrepancies were attributed to the limitation of the (Cartesian) multipole model when applied to short-range interactions", curiously without mentioning the convergence of the multipole expansion.

The second approach of Jelsch and Lecomte [84] and co-workers involves a library that also uses the Hansen-Coppens multipolar pseudo-atom model. The library is built from average multipole populations describing the electron density of chemical groups in all 20 natural amino acids. The main difference with the first approach is that the library values are now obtained from several small peptide (or amino acid crystal structures) refined against ultra-high-resolution X-ray diffraction data. The charge-density studies obtained from experimental diffraction data and quantum calculations generally agree well [88] although some dissimilarities appear for the polar atoms and valence populations of some atoms and when electrostatic properties are calculated. Compared to a spherical-atom model, the Experimental Library Multipolar Atom Model (ELMAM) delivers more accurate crystal structures, particularly in terms of thermal displacement parameters and bond distances involving hydrogen atoms.

The third approach of Luger and Dittrich [85] and co-workers, like the first approach, also describes the electron density via simulated data obtained from ab initio calculations of peptides in vacuo [89]. Their InvariomTool again invokes Hansen–Coppens pseudoatoms that are transferable from one molecule to another, called invarioms [90].

All these approaches have matured over many years and have revolutionised refinement in protein crystallography, delivering a "crispness" that scientists could only dream of a decade ago. The development of a QCT force field has still some way to go until it can be applied proteins, although work has started on the 46-residue protein crambin, a standard test case. The importance of giving reliable answers to simple but precise questions during this development cannot be overemphasized. For example, understanding the convergence behaviour of a multipole expansion, both from a theoretical and practical point of view, is vital to make solid progress. If neglected, one may encounter unpleasant surprises in the evaluation of interatomic energies, which can ultimately damage the reputation of a force field.

14.6 Conclusions

The viewpoint of our past work on a force field based on Quantum Chemical Topology (QCT) is laid out. We aim for a generic method to design a force field *ab ovo*, closer to quantum chemical reality than the popular classical force fields currently used. This is feasible given the abundance of current computing power. The approach is based on three cornerstones: a definition of an atom in a system, multipole moments and machine learning to capture how these moments change in response to a changing environment. When combined, one obtains malleable atoms that adjust their electronic structure to the coordinates of the neighbouring atoms. The force field delivers better accuracy at short-range and has been successfully used in molecular dynamics simulations. However, several future developments are desirable. One is the replacement of the empirical Lennard-Jones potential by one that is in line with the cornerstones. The fact that topological atoms do not overlap is an intriguing and unexploited fact in this context. A second development is the combination of the multipole moment treatment with classical expressions such as bonded terms. One reason that this is currently necessary is that Coulomb 1,2-interactions do not converge, although progress has been made with a method that shifts multipole moments. The other is the issue of transferability of so-called exchange moments. A further development is the automatic establishment of atom types throughout machine learning feature selection obfuscating the need for user intervention of user imposed transferability. In summary, several trends of a long-term project are coming together in an attempt to create a step change in the realism and hence reliability of force field design.

References

1. Payne MC, Teter MP, Allan DC, Arias TA, Joannopoulos JD (1992) Iterative minimization techniques for abinitio total-energy calculations - molecular-dynamics and conjugate gradients. Rev Mod Phys 64:1045–1097
2. Car R, Parrinello M (1985) Unified approach for molecular-dynamics and density-functional theory. Phys Rev Lett 55:2471–2474
3. Guillot B (2002) A reappraisal of what we have learnt during three decades of computer simulations on water. J Mol Liq 101:219–260

4. Cornell WD, Cieplak P, Bayly CI, Gould IR, Merz KM (1995) A 2nd generation force-field for the simulation of proteins, nucleic-acids, and organic-molecules. J Am Chem Soc 117:5179–5197
5. Brooks BR, Brooks CLI, MacKerell AJ, Ferguson DM, Spellmeyer DC, Fox T, Caldwell JW, Kollman PA, Mackerell AD, Nilsson L, Petrella RJ, Roux B, Won Y, Archontis G, Bartels C, Boresch S, Caflisch A, Caves L, Cui Q, Dinner AR, Feig M, Fischer S, Gao J, Hodoscek M, Im W, Kuczera K, Lazaridis T, Ma J, Ovchinnikov V, Paci E, Pastor RW, Post CB, Pu JZ, Schaefer M, Tidor B, Venable RM, Woodcock HL, Wu X, Yang W, York DM, Karplus M (2009) CHARMM: the biomolecular simulation program. J Comput Chem 30:1545–1614
6. Ponder JW, Case DA (2003) Force fields for protein simulations. Adv Protein Chem 66:27–85
7. Jensen F (2007) Introduction of computational chemistry, 2nd edn. Wiley, Chichester
8. Mu Y, Kosov DS, Stock G (2003) Conformational dynamics of trialanine in water. 2. Comparison of AMBER, CHARMM, GROMOS, and OPLS force fields to NMR and infrared experiments. J Phys Chem B 107:5064–5073
9. Roux B, Berneche S (2002) On the potential functions used in molecular dynamics simulations of ion channels. Biophys J 82:1681–1684
10. Chang TM, Dang LX (2006) Recent advances in molecular simulations of ion solvation at liquid interfaces. Chem Rev 106:1305–1322
11. Jungwirth P, Tobias DJ (2002) Ions at the air/water interface. J Phys Chem B 106:6361–6373
12. Parr RG, Yang W (1989) Density-functional theory of atoms and molecules. Oxford University Press, Oxford
13. Rafat M, Popelier PLA (2007) Long range behavior of high-rank topological multipole moments. J Comput Chem 28:292–301
14. Darley MG, Popelier PLA (2008) Role of short-range electrostatics in torsional potentials. J Phys Chem A 112:12954–12965
15. Jensen F (1999) Introduction to computational chemistry. Wiley, Chichester
16. Ren PY, Ponder JW (2003) Polarizable atomic multipole water model for molecular mechanics simulation. J Phys Chem B 107:5933–5947
17. Stone AJ (2008) Intermolecular potentials. Science 321:787–789
18. Stone AJ (1981) Distributed multipole analysis, or how to describe a molecular charge-distribution. Chem Phys Lett 83:233–239
19. Piquemal J-P, Gresh N, Giessner-Prettre C (2003) Improved formulas for the calculation of the electrostatic contribution to the intermolecular interaction energy from multipolar expansion of the electronic distribution. J Phys Chem A 107:10353–10359
20. Bader RFW, Becker P (1988) Transferability of atomic properties and the theorem of hohenberg and kohn. Chem Phys Lett 148:452–458
21. Bader RFW (1990) Atoms in molecules: a quantum theory, vol 22, International series of monographs on chemistry. Oxford Science, Oxford
22. Popelier PLA, Aicken FM (2003) Atomic properties of amino acids: computed atom types as a guide for future force-field design. Chem Phys Chem 4:824–829
23. Silvi B, Savin A (1994) Classification of chemical bonds based on topological analysis of electron localization functions. Nature 371:683–686
24. Tsirelson VG, Avilov AS, Lepeshov GG, Kulygin AK, Stahn J, Pietsch U, Spence JCH (2001) Quantitative analysis of the electrostatic potential in rock-salt crystals using accurate electron diffraction data. J Phys Chem B 105:5068–5074
25. Balanarayan P, Gadre SR (2003) Topography of molecular scalar fields. I. Algorithm and Poincare-Hopf relation. J Chem Phys 119:5037–5043
26. Blanco MA, Martín Pendás A, Francisco E (2005) Interacting quantum atoms: a correlated energy decomposition scheme based on the quantum theory of atoms in molecules. J Chem Theory Comput 1:1096–1109
27. Popelier PLA, Bremond EAG (2009) Geometrically faithful homeomorphisms between the electron density and the bare nuclear potential. Int J Quantum Chem 109:2542–2553
28. Freitag MA, Gordon MS, Jensen JH, Stevens WJ (2000) Evaluation of charge penetration between distributed multipolar expansions. J Chem Phys 112:7300–7306

29. Stone AJ (1996) The theory of intermolecular forces. Clarendon, Oxford
30. Rafat M, Popelier PLA (2007) Topological atom–atom partitioning of molecular exchange energy and its multipolar convergence. In: Matta CF, Boyd RJ (eds) The quantum theory of atoms in molecules: from solid state to DNA and drug design. Wiley-VCH, Weinheim, pp 121–140
31. Popelier PLA, Stone AJ (1994) Formulas for the 1st and 2nd derivatives of anisotropic potentials with respect to geometrical parameters. Mol Phys 82:411–425
32. Popelier PLA, Stone AJ, Wales DJ (1994) Topography of potential-energy surfaces for van-der-waals complexes. Faraday Discuss 97:243–264
33. in het Panhuis M, Popelier PLA, Munn RW, Angyan JG (2001) Distributed polarizability of the water dimer: field-induced charge transfer along the hydrogen bond. J Chem Phys 114:7951–7961
34. Handley CM, Popelier PLA (2008) The asymptotic behavior of the dipole and quadrupole moment of a single water molecule from gas phase to large clusters: a QCT analysis. Synth React Inorg, Met-Org Nano-Met Chem 38:91–99
35. Popelier PLA (2005) Quantum chemical topology: on bonds and potentials. In: Wales DJ (ed) Structure and bonding. Intermolecular forces and clusters. Springer, Heidelberg, pp 1–56
36. Popelier PLA, Rafat M, Devereux M, Liem SY, Leslie M (2005) Towards a force field via Quantum Chemical Topology. Lect Ser Comput Comput Sci 4:1251–1255
37. Cooper DL, Stutchbury NCJ (1985) Distributed multipole analysis from charge partitioning by zero-flux surfaces - the structure of HF complexes. Chem Phys Lett 120:167–172
38. Popelier PLA (1996) Integration of atoms in molecules: a critical examination. Mol Phys 87:1169–1187
39. Kosov DS, Popelier PLA (2000) Atomic partitioning of molecular electrostatic potentials. J Phys Chem A 104:7339–7345
40. Kosov DS, Popelier PLA (2000) Convergence of the multipole expansion for electrostatic potentials of finite topological atoms. J Chem Phys 113:3969–3974
41. Popelier PLA, Rafat M (2003) The electrostatic potential generated by topological atoms: a continuous multipole method leading to larger convergence regions. Chem Phys Lett 376: 148–153
42. Rafat M, Popelier PLA (2005) Atom-atom partitioning of intramolecular and intermolecular Coulomb energy. J Chem Phys 123:204103–204101,204107
43. Popelier PLA, Kosov DS (2001) Atom-atom partitioning of intramolecular and intermolecular Coulomb energy. J Chem Phys 114:6539–6547
44. Popelier PLA, Joubert L, Kosov DS (2001) Convergence of the electrostatic interaction based on topological atoms. J Phys Chem A 105:8254–8261
45. Joubert L, Popelier PLA (2002) Long range behavior of high-rank topological multipole moments. Mol Phys 100:3357–3365
46. Rafat M, Popelier PLA (2006) A convergent multipole expansion for 1,3 and 1,4 Coulomb interactions. J Chem Phys 124:144102
47. Martín Pendás AM, Blanco A, Fransisco E (2004) Two-electron integrations in the quantum theory of atoms in molecules. J Chem Phys 120:4581–4592
48. Kay KG, Todd HD, Silverstone HJ (1969) J Chem Phys 51:2363–2367
49. Rafat M, Popelier PLA (2007) Long range behaviour of high-rank topological multipole moments. J Comput Chem 28:832–838
50. Bader RFW, Bayles D, Heard GL (2000) Properties of atoms in molecules: transition probabilities. J Chem Phys 112:10095–10105
51. Mandado M, Grana AM, Mosquera RA (2002) Approximate transferability in alkanols. J Mol Struct-Theochem 584:221–234
52. Matta CF (2001) Theoretical reconstruction of the electron density of large molecules from fragments determined as proper open quantum systems: the properties of the oripavine PEO, enkephalins, and morphine. J Phys Chem A 105:11088–11101

53. Bohórquez HJ, Obregón M, Cárdenas C, Llanos E, Suárez C, Villaveces JL, Patarroyo ME (2003) Electronic energy and multipolar moments characterize amino acid side chains into chemically related groups. J Phys Chem A 107:10090–10097
54. Dittrich B, Scheins S, Paulmann C, Luger P (2003) Transferability of atomic volumes and charges in the peptide bond region in the solid state. J Phys Chem A 107:7471–7474
55. Alcoba DR, Ona O, Torre A, Lain L, Bochicchio RC (2008) Determination of energies and electronic densities of functional groups according to partitionings in the physical space. J Phys Chem A 112:10023–10028
56. Grabowsky S, Kalinowski R, Weber M, Foerster D, Carsten P, Luger P (2009) Transferability and reproducibility in electron-density studies - bond-topological and atomic properties of tripeptides of the type L-alanyl-X-L-alanine. Acta Crystallogr B 65:488–501
57. Koritsanszky T, Volkov A, Coppens P (2002) Aspherical-atom scattering factors from molecular wave functions. 1. Transferability and conformation dependence of atomic electron densities of peptides within the multipole formalism. Acta Crystallogr A 58:464–472
58. Hansen NK, Coppens P (1978) Electron population analysis of accurate diffraction data. 6. Testing aspherical atom refinements on small-molecule data sets. Acta Crystallogr A 34: 909–921
59. Popelier PLA, Aicken FM (2003) Atomic properties of selected biomolecules: quantum topological atom types of carbon occurring in natural amino acids and derived molecules. J Am Chem Soc 125:1284–1292
60. Popelier PLA, Aicken FM (2003) Atomic properties of selected biomolecules: quantum topological atom types of hydrogen, oxygen, nitrogen and sulfur occurring in natural amino acids and their derivatives. Chem Eur J 9:1207–1216
61. Popelier PLA, Devereux M, Rafat M (2004) The quantum topological electrostatic potential as a probe for functional group transferability. Acta Crystallogr A 60:427–433
62. Rafat M, Shaik M, Popelier PLA (2006) Transferability of quantum topological atoms in terms of electrostatic interaction energy. J Phys Chem A 110:13578–13583
63. Joubert L, Popelier PLA (2002) The prediction of energies and geometries of hydrogen bonded DNA base-pairs via a topological electrostatic potential. Phys Chem Chem Phys 4:4353–4359
64. Shaik MS, Devereux M, Popelier PLA (2008) The importance of multipole moments when describing water and hydrated amino acid cluster geometry. Mol Phys 106:1495–1510
65. Devereux M, Popelier PLA (2007) The effects of hydrogen-bonding environment on the polarization and electronic properties of water molecules. J Phys Chem A 111:1536–1544
66. Liem SY, Popelier PLA (2003) High-rank quantum topological electrostatic potential: molecular dynamics simulation of liquid hydrogen fluoride. J Chem Phys 119:4560–4566
67. Leslie M (2008) DL_MULTI - a molecular dynamics program to use distributed multipole electrostatic models to simulate the dynamics of organic crystals. Mol Phys 106:1567–1578
68. Smith W, Leslie M, Forester TR (2003) DLPOLY, a computer program. CCLRC, Daresbury Lab, Daresbury, Warrington
69. Liem SY, Popelier PLA, Leslie M (2004) Simulation of liquid water using a high-rank quantum topological electrostatic potential. Int J Quantum Chem 99:685–694
70. Liem SY, Popelier PLA (2008) Properties and 3D structure of liquid water: a perspective from a high-rank multipolar electrostatic potential. J Chem Theory Comput 4:353–365
71. Mahoney MW, Jorgensen WL (2000) A five-site model for liquid water and the reproduction of the density anomaly by rigid, nonpolarizable potential functions. J Chem Phys 112:8910–8922
72. Angyan JG, Loos M, Mayer I (1994) Covalent bond orders and atomic valence indexes in the topological theory of atoms in molecules. J Phys Chem 98:5244–5248
73. in het Panhuis M, Munn RW, Popelier PLA, Coleman JN, Foley B, Blau WJ (2002) Distributed response analysis of conductive behavior in single molecules. Proc Natl Acad Sci U S A 99:6514–6517
74. Houlding S, Liem SY, Popelier PLA (2007) A polarizable high-rank quantum topological electrostatic potential developed using neural networks: molecular dynamics simulations on the hydrogen fluoride dimer. Int J Quantum Chem 107:2817–2827

75. Darley MG, Handley CM, Popelier PLA (2008) Beyond point charges: dynamic polarization from neural net predicted multipole moments. J Chem Theory Comput 4:1435–1448
76. Price SL (2008) Computational prediction of organic crystal structures and polymorphism. Int Rev Phy Chem 27:541–568
77. Handley CM, Popelier PLA (2009) Dynamically polarizable water potential based on multipole moments trained by machine learning. J Chem Theory Comput 5:1474–1489
78. Handley CM, Hawe GI, Kell DB, Popelier PLA (2009) Optimal construction of a fast and accurate polarisable water potential based on multipole moments trained by machine learning. Phys Chem Chem Phys 11:6365–6376
79. Cressie N (1993) Statistics for spatial data. Wiley, New York
80. Rasmussen CE, Williams CKI (2006) Gaussian processes for machine learning. The MIT Press, Cambridge
81. Volkov A, Koritsanszky T, Coppens P (2004) Combination of the exact potential and multipole methods (EP/MM) for evaluation of intermolecular electrostatic interaction energies with pseudoatom representation of molecular electron densities. Chem Phys Lett 391:170–175
82. Walker PD, Mezey PG (1993) Molecular electron-density lego approach to molecule building. J Am Chem Soc 115:12423–12430
83. Dominiak PM, Volkov A, Li X, Messerschmidt M, Coppens P (2007) A theoretical databank of transferable aspherical atoms and its application to electrostatic interaction energy calculations of macromolecules. J Chem Theory Comput 3:232–247
84. Zarychta B, Pichon-Pesme V, Guillot B, Lecomte C, Jelsch C (2007) On the application of an experimental multipolar pseudo-atom library for accurate refinement of small-molecule and protein crystal structures. Acta Crystallogr A 63:108–125
85. Huebschle CB, Luger P, Dittrich B (2007) Automation of invariom and of experimental charge density modelling of organic molecules with the preprocessor program InvariomTool. J Appl Crystallogr 40:623–627
86. Stewart RF (1976) Electron population analysis with rigid pseudoatoms. Acta Crystallogr A 32:565–574
87. Volkov A, Li X, Koritsanszky T, Coppens P (2004) *Ab Initio* quality electrostatic atomic and molecular properties including intermolecular energies from a transferable theoretical pseudoatom databank. J Phys Chem 108:4283–4300
88. Volkov A, Gatti C, Abramov Y, Coppens P (2000) Evaluation of net atomic charges and atomic and molecular electrostatic moments through topological analysis of the experimental charge density. Acta Crystallogr A 56:252–258
89. Dittrich B, Huebschle CB, Luger P, Spackman MA (2006) Introduction and validation of an invariom database for amino-acid, peptide and protein molecules. Acta Crystallogr D 62:1325–1335
90. Dittrich B, Koritsanszky T, Luger P (2004) A simple approach to non spherical electron densities by using invarioms. Angew Chem Int Ed 43:2718–2721

Chapter 15
Frontier Applications of Experimental Charge Density and Electrostatics to Bio-macromolecules

Christian Jelsch, Sławomir Domagała, Benoît Guillot, Dorothee Liebschner, Bertrand Fournier, Virginie Pichon-Pesme, and Claude Lecomte

15.1 Introduction

Biocrystallography has evolved in recent years with the introduction of new methods and techniques at each stage of protein structure resolution. Data-collection techniques and data quality have been significantly improved thanks to third-generation synchrotron sources, automated data reduction and methodological improvements [1]. Moreover, protein model building and refinement can now in many cases be performed almost automatically, allowing the scientist to spend more time on structure analysis and interpretation. Some of these improvements arise from so-called high-throughput crystallography, which is dedicated to almost automated and rapid protein structure determination in the context of genomic projects. All these improvements, but in particular those in crystallogenesis and synchrotron-radiation sources, have another extremely important consequence: more and more protein or nucleic acid structures are being solved and refined at atomic or subatomic resolution.

Extension of charge density study methodologies from small molecules to biological macromolecules is a recent challenge. Macromolecules have specificities such as limited resolution of the X-ray diffraction data, their atoms display higher thermal motion and disorder, and the least-squares refinement method has a limited radius of convergence. It has become clear that the direct application of multipolar modelling to protein crystal structures, even at subatomic resolution, is conditioned

C. Jelsch (✉) • S. Domagała • B. Guillot • D. Liebschner • B. Fournier • V. Pichon-Pesme • C. Lecomte
Laboratoire de Cristallographie Résonance Magnétique et Modélisations (CRM2) CNRS, UMR 7036, Faculté des Sciences et Techniques, Institut Jean Barriol, Nancy University, BP 70239, 54506 Vandoeuvre-lès-Nancy, Cedex, France
e-mail: christian.jelsch@crm2.uhp-nancy.fr; SLAWDOM@chem.uw.edu.pl; benoit.guillot@lcm3b.uhp-nancy.fr; dorothee.liebschner@crm2.uhp-nancy.fr; bertrand.fournier@crm2.uhp-nancy.fr; Virginie.Pichon@crm2.uhp-nancy.fr; claude.lecomte@crm2.uhp-nancy.fr

C. Gatti and P. Macchi (eds.), *Modern Charge-Density Analysis*,
DOI 10.1007/978-90-481-3836-4_15, © Springer Science+Business Media B.V. 2012

to the prior evaluation of an electron density model that should be as close as possible to the final model [2].

On the other hand, electrostatic forces play an essential role in the process of protein folding and molecular recognition. While there has been significant progress and improved accuracy in the field of molecular mechanics applied to proteins in recent years, these methods still bear approximations, notably in the electrostatics modelling. The advantage of force-fields is their minimized computational effort required for calculations on macromolecules. However, one of the main issues in the development of next generation force fields is the reliable estimation of electrostatic interactions. The electrostatic component, in the traditional force fields, is calculated on the basis of an atom-centred point-charge model. This model is not advanced enough to describe aspherical features of atomic charge distributions. The charges modelling needs to be improved to be closer to the true molecular electron density.

The determination of charge densities and electrostatic properties directly from *ab initio* type approaches are not easily applicable because of exceedingly high computational cost when applied to macromolecules. Therefore, the use of average non-spherical charge density atomic parameters transferred from a database [3–6] is an efficient and fast approach to model accurately the electrostatic properties. The reconstructing of the electron density from transferable aspherical molecular fragments stored in a databank using the Hansen and Coppens [7] pseudo-atom model has been pursued by three research groups in the charge density community, including ours. Two databases (UBDB [6, 8] and Invariom [5]) were obtained from quantum calculations of molecules *in vacuo*, where theoretical structure factors are computed and used for a multipolar charge density refinement. The ELMAM library constructed in our laboratory is based on experimental charge densities.

Theoretical databases are more flexible: if a chemical group is missing, a quantum calculation can rapidly be performed. Crystallographic charge densities have the advantage of being experimental, but are subject to errors and incompleteness in the measurements and to atomic thermal motion. With diffraction data measured more routinely and accurately on synchrotron beam-lines at ultra low temperature using a helium cryostat, experimental charge densities will surely continuously improve with time. As ultra high resolution ($\sim$0.4 Å or better) can be achieved for the best crystals [9], the polarization of atoms (beyond transferability) is also observable experimentally. Therefore, both theoretical and experimental databases and studies of specific molecules are useful for an ever improved modelling of molecular charge densities and for the cross-validation of the two approaches.

15.2 Multipolar Refinement of Macromolecules

The Hansen and Coppens [7] pseudo-atom model cannot be directly applied to macromolecules because it requires the least squares estimation of at least 28 parameters per protein atom and therefore a huge number of structure factors modules. On the other hand, a protein is the repetition through space of similar peptide and side chains motives. This allows a systematic use of chemical equivalence constraints or restraints to decrease the number of variables over observations ratio. Furthermore, one can apply multipolar atomic scattering factors from a library built either from experimental or theoretical electron densities in peptides and small compounds. Such a library transferred model is a good starting point for further charge density refinement or for electrostatic potential calculation.

15.2.1 Electron Density Libraries and Transfer

The database transfer approach relies on the hypothesis that a same chemical group in different environments has, at first approximation, the same charge density. The experimental library of multipolar atom model (ELMAM), built in our group, contains average multipole populations describing the electron density of chemical groups in all 20 amino acids found in proteins [3, 4]. The library uses the Hansen and Coppens [7] multipolar pseudo-atom model to derive molecular electron density and electrostatic potential distributions. The library values are obtained from several small peptide or amino acid crystal structures refined against ultra-high-resolution X-ray diffraction data. The transferability principle postulates that the deformation charge density due to chemical bonding and electron lone pairs in a given chemical group is similar in different molecules and crystal contexts. The pseudo-atoms are highly transferable and fairly invariant with respect to dihedral rotations around single bonds in molecules [3–5, 10].

The library transfer is applied automatically in the MoPro software suite to peptide and protein structures measured at atomic resolution. The transferred multipolar parameters are kept fixed while the positional and thermal parameters are refined. This enables a proper deconvolution of thermal motion and valence electron density redistributions, even when the diffraction data do not extend to subatomic resolution. The use of the ELMAM databank has also a major impact on crystallographic structure modelling in the case of molecular and organic crystals at atomic resolution. Compared to a spherical atom model, the library transfer results in a more accurate crystal structure, notably in terms of thermal displacement parameters and bond distances involving H atoms [10]. Upon transfer, crystallographic statistics of fit are improved, particularly free R factors [11], and residual electron density maps are cleaner. It was also shown that the application of the database transfer approach, improves the precision and standard uncertainty of the Flack parameter and therefore the reliability of deducing the molecular chirality [12].

The very first attempt of application of multipolar modelling to a polypeptide compound, using transferred charge density parameters as initial guess, has been performed with success on the LBZ octapeptide [10]. Then, several applications on small proteins have been published: for instance on protein crambin [13] and on the scorpion *Androctonus australis Hector* protein toxin II [14].

Following this work, a first theoretical multipolar database (UBDB) has been proposed from quantum computations [6, 15, and related papers]. A second theoretical database (Invariom) for structural refinement of amino acid, oligopeptide and protein molecules was published by Luger and coworkers [5 and related papers]. These databases contain Hansen and Coppens [7] multipolar parameters obtained from charge density refinements against theoretical structure factors.

A charge density study on cyclosporine is an example of a refinement of the largest medium-sized macromolecular system in which the multipolar expansion was applied to every atom in the structure [16]. Two data sets were collected using synchrotron radiation at temperatures 5 and 90 K, and gave excellent statistics in terms of crystallographic descriptors. The 5 K temperature data set was selected to evaluate different multipolar modelling approaches. Three models have been thoroughly examined. In all three models, the same number of coordinate and thermal motion parameters was freely refined with the same constraints/restraints scheme. The multipolar parameters, including bond directed quadrupolar level for hydrogen atoms and up to hexadecapolar level for the remaining atoms, were freely refined in the first model. Some symmetry constraints and chemical equivalence constraints were introduced to reduce number of refined parameters. In the two further models, the charge density parameters were transferred from the UBDB [15] and Invariom [5] databases and kept fixed during the refinements. The mean average values and estimated standard deviations (esd) of the electron density parameters at the covalent bond critical points were calculated for all the models and compared with the published studies on the amino-acids [17]. A good agreement was obtained leading to discrepancies of $\rho(\mathbf{r}_{CP})$ within 3 esd's from the experimental and 9 esd's from the theoretical studies, respectively. Slightly higher variations were observed for the Laplacian function. Only minor differences for the electron density properties at the bond critical points for the three models occurred. The difference between the electron densities transferred from each database and the freely refined multipolar model was assessed on a grid for the same molecular geometry. The highest discrepancies were around 0.2 eÅ^{-3}, which was slightly less than the maximal residual densities after the multipolar refinements ($\sim$0.26 eÅ^{-3}). The main differences were observed in the region of the partially disordered water molecules and the methyl groups. Those evaluations proved the reliability and usability of both UBDB and Invariom databases in the charge density modelling of medium-sized systems.

Besides the experimental and two theoretical databases of multipolar atoms, a different quantum-chemical partitioning, the fuzzy density fragmentation method, has been introduced by Mezey et al. [18]. Another electron density modelling, which uses additional spherical scatterers on covalent bonds and electron lone pairs sites was also proposed for the crystallographic refinement of macromolecules [19, 20].

To the best of our knowledge, no protein charge density refinement based on theoretical database has been published yet. Comparison between the ELMAM experimental database and the theoretical UBDB database for the polypeptide backbone was discussed by Pichon-Pesme et al. [21] and by Volkov et al. [22].

The charge densities obtained from experimental diffraction data and quantum calculations agree generally well. The largest dissimilarities appear for the valence populations of the polar atoms especially, which might be a problem when electrostatic properties are calculated. The disagreement may partly be due to the fact that the experiment is performed on the crystalline state while the computations are performed *in vacuo*, leading to different results due to polarization effects [23].

15.2.2 The MoPro Software

With an increasing number of biological macromolecule structures solved at ultra-high resolution and with the advances of supramolecular chemistry, it has become necessary to design a charge density refinement program adapted to large systems. The software contains numerical tools of both small molecules and protein crystallography. The refinement program MoPro (Molecular Properties), dedicated to the charge density refinement at (sub)atomic resolution of structures ranging from small molecules to biological macromolecules was therefore developed [24, 25]. Besides the multipolar atom, alternative methods are also proposed, such as modelling bonding and lone pair electron density by virtual spherical atoms (bond charge model) [26].

Due to the large number of X-ray reflections in the case of macromolecules at ultra high resolution and to the complexity of the multipolar atom model, the algorithms were optimized to reduce the computation time. A large number of the normal matrix off-diagonal terms turn out to be very small in the case of large structures at very high resolution [27], the block diagonal approximation is thus particularly efficient. To refine large systems, the conjugate gradient algorithm was implemented. For proteins at atomic resolution, the program enables the transfer of multipolar parameters from the ELMAM charge density database. The program allows complex refinement strategies to be written and has numerous restraints, constraints and analysis tools for use in the structure and electron density analysis (http://www.crm2.uhp-nancy.fr/emqc).

15.2.3 Constraints on the Charge Density Refinement

Although it is possible, in principle, to determine the experimental electron density of a compound while refining all the multipole populations, this approach can be inappropriate, depending on the quality and resolution of the diffraction data, especially for macromolecules. Strong correlation between variables (κ and multipole population or thermal parameters) can occur, resulting in unstable refinement or unphysical parameter values, especially for non-centrosymmetric structures [28]. To decrease the large number of parameters needed to describe the electron density for larger systems and also to avoid the problems of over-fitting and high correlations between variables, different constraints can be imposed. Several types of constraints on the charge density can be applied in charge density programs [25, 29].

Symmetry constraints impose the multipoles which do not follow the local symmetry relation to have zero populations. For example, multipoles which are antisymmetric with respect to a symmetry mirror plane must be small. Several symmetries are available, the most common being mirrors, two-fold and three-fold axes, and the inversion centre. Symmetry restraints are also implemented in the program. The multipoles which do not follow the symmetry can be restrained to be equal to zero with an allowed standard deviation. The local symmetry of an atom is determined on the basis of its connectivity (the number and chemical type of connected atoms) and geometrical parameters describing bonds, valence angles and planarity of the considered atom and the first coordination shell.

Atoms with similar chemical surrounding can be considered, within the library transferability approximation, to have equivalent charge densities by constraining their parameters to be identical. Atoms equivalence constraints can apply to all the monopoles and multipoles, or only to higher multipole populations or to valence monopole populations only [25]. The contraction/expansion κ and κ' values of two or more atoms can also be constrained to be identical via the kappa equivalence constraints.

Analogously, it is possible to use (soft) restraints instead of (hard) constraints. Most of the constraints have their corresponding restraints available within the software MoPro. Due to the absence of core electrons, the positions and thermal factors of hydrogen atoms have a larger uncertainty. The atoms equivalence constraints are therefore especially well suited for hydrogen atoms, as their charge density parameters are more difficult to refine.

15.2.4 Optimal Local Axes and Generalized Experimental Databank

To define the real harmonic functions needed in the multipolar formalism, local axes systems associated to each atom of the molecular structure must be specified. The coordinate systems should be optimized to be able to assign the maximal symmetry to the multipoles and thus, to describe the electron density of an atom by using a minimal number of variables. In symmetry constrained refinements, the few remaining multipoles which are compatible with the local symmetry take generally significant values. Moreover, without symmetry constraints, the released multipoles remain small, which permits to analyze precisely the deviations from atomic pseudo-symmetry.

These local coordinates systems are centred on each atom of the structure and oriented with respect to two or three of their atomic neighbours. The following procedures are used to create the local coordinate system for an atom. The **u**, **v** and (if necessary) **w** normalized vectors represent the directions from the origin atom to its neighbours:

$$\mathbf{u} = (\mathbf{r}_1 - \mathbf{r}_0)/|\mathbf{r}_1 - \mathbf{r}_0|\,;\ \mathbf{v} = (\mathbf{r}_2 - \mathbf{r}_0)/|\mathbf{r}_2 - \mathbf{r}_0|\,;\ \mathbf{w} = (\mathbf{r}_3 - \mathbf{r}_0)/|\mathbf{r}_3 - \mathbf{r}_0| \quad (15.1)$$

where $\mathbf{r}_0$, $\mathbf{r}_1$, $\mathbf{r}_2$ and $\mathbf{r}_3$ are the vectors pointing the positions of the considered atom, and to the first, second and third neighbour atoms, respectively. The procedures use, at several steps, the vectorial product of two vectors to generate a third vector which is perpendicular. Thus an orthogonal basis set ($\mathbf{l}_1$, $\mathbf{l}_2$, $\mathbf{l}_3$) is acquired. In a final step, the $\mathbf{l}_i$ vectors are normalized.

In previous papers, including the one describing the ELMAM atomic transferable parameters library [4], the local axes used were always of type *XY*: *X* is directed towards a first atom, while the second atom defines the *XY* plane and *Z* is orthogonal to the plane. In the process of improvement of the experimental library and its extension from proteins/peptides to common chemical groups, new local axes systems are proposed. With careful choice of directions, the local coordinate system can take advantage of the spatial characteristics of certain multipoles, to model the bonding electron density derived from sp^2 and sp^3 hybrid orbitals, by the linear combination of only a few multipoles [25, 29].

New optimal local coordinate systems for the atoms in different environments were introduced using notably bisecting directions. In the case of the bisecting orientation *bXY*, the directions $\mathbf{l}_1$ and $\mathbf{l}_2$ are parallel to the sum and to the difference of the $\mathbf{v}$ and $\mathbf{u}$ normalized vectors, respectively. These two directions are already perpendicular and correspond to the inner and exterior bisecting directions of the ($\mathbf{u}$, $\mathbf{v}$) bivector. The third direction $\mathbf{l}_3$ is then readily defined by vectorial product $\mathbf{l}_3 = \mathbf{l}_1 \times \mathbf{l}_2$. Such a local axes system is for instance in accordance with the symmetry of a water molecule.

Automatic optimal local axes systems were implemented and tested within the MoPro crystallographic package. The application of symmetry constraints or restraints on the multipoles is more effective when the axes are in accordance with the local geometry of the atom [29] and can lead to improved crystallographic R_{free} residuals [11]. The new local axes systems, based on two or three atom neighbours, can also be usefully applied for the description of transition metal complexes.

A set of a seven different orientations for local axes, called: *XY*, *ZX*, *bXY*, *bZX*, *3ZX*, *3Zb* and *3bZ* is proposed (Fig. 15.1). The first four systems use two neighbour atoms while the *3ZX*, *3Zb* and *3bZ* systems need three atoms for a proper axes definition. These orientations are sufficient to describe the possible symmetries for systems containing up to six-coordinated atoms. The third atom is necessary for the proper definition of two perpendicular planes in the case of tetra- and penta-coordinated atoms and in the case of three-fold symmetry in trigonal prismatic configurations.

The type of local axes system is assigned according to the type and number of neighbours and the symmetry of the atom. The procedure of atoms selection used for axes definition depends on each local axes system and attempts to maximize the local symmetry. The symmetry is always considered within a certain tolerance of distances, angles and planarity deviations.

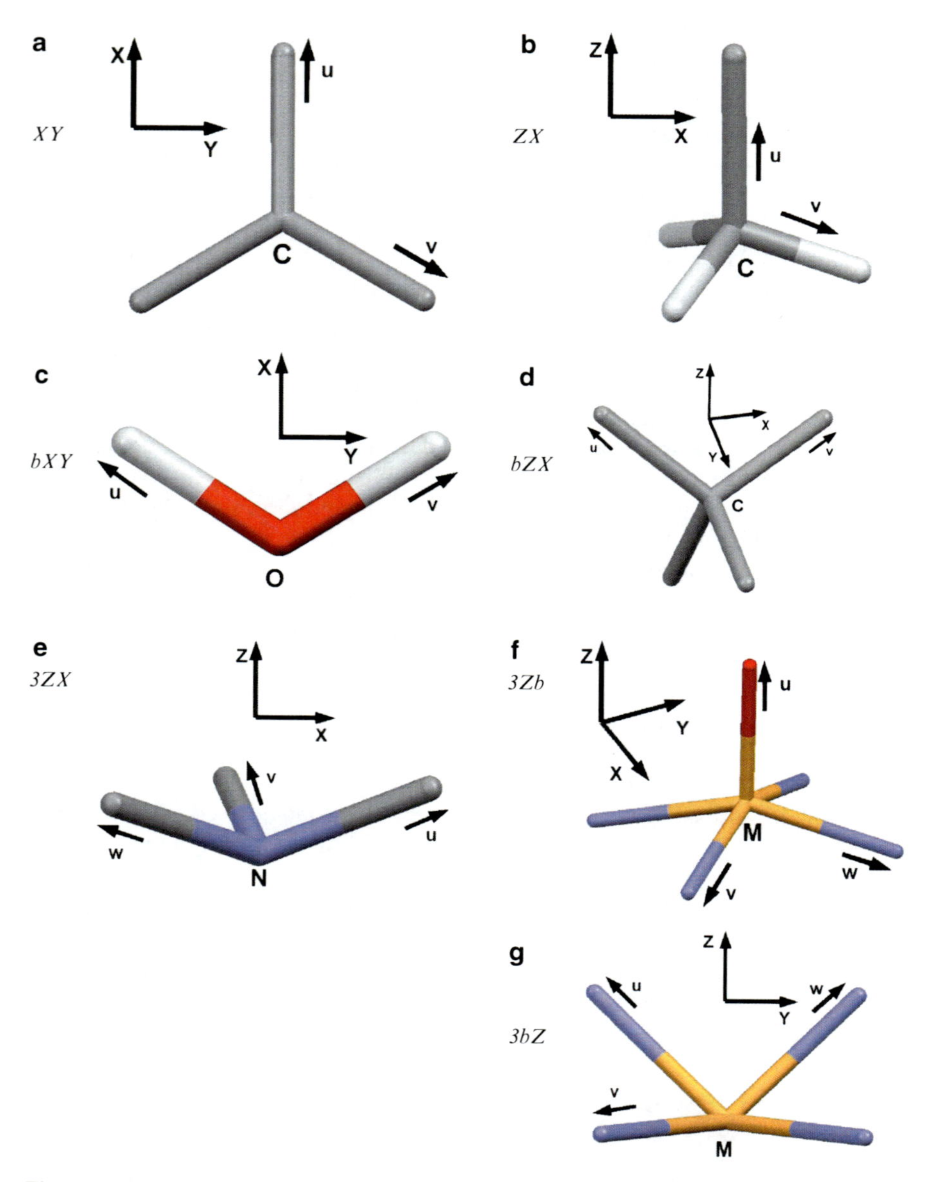

Fig. 15.1 Representation of the seven different local axes systems. The orientation of the *X*, *Y*, *Z* directions relative to the **u**, **v** and **w** vectors pointing towards the neighbouring atoms is shown

Another methodology, based on the approximate site symmetry and ligand field considerations, was proposed earlier [30] to choose an appropriate coordinate system for transition metal atoms. The local axes orientation at a transition-metal site is of importance in the interpretation of metal d-orbital populations and the analysis of chemical bonding. This could also be useful in the refinement of some metalloproteins.

Several libraries of multipolar atoms have been simultaneously built either from experimental data [3, 4] or from theoretical calculations [5, 6, and references herein]. An experimental multipolar atoms databank, extended from proteins to common chemical groups, is under construction and is based on the charge density refinement of 50 small compounds for which diffraction data are available from the literature or from the authors. Within the library transferability approximation, the polarization of atoms due to the immediate chemical environment is averaged over several atoms in the sample. Therefore, in the charge density refinement aimed at building the database, the equivalence and the symmetry constraints are applied and kept until the end of the refinement.

The consistent use of standard properly oriented local atomic axes is encouraged because:

1. axes for chemically similar atoms in different molecules and in different crystallographic environments can be similarly defined.
2. the number of multipolar parameters can be kept minimal when imposing local chemical symmetry constraints.
3. to build the generalized electron density library, the charge density parameters of several atoms from different compounds are averaged. All the atoms of a given chemical type need to have the same local axes definition.
4. atoms types in a new molecule need to be identified in a standardized protocol in order to transfer the multipole parameters from the library.
5. in a refinement, where the symmetry constraints are not applied, the use of optimal axes enables to highlight the deviations from the local pseudo-symmetry.

Optimal symmetry constraints can consequently be applied to atoms and a large number of multipole populations can be fixed to a zero value. The introduction of symmetry constraints in the multipolar refinement of small compounds allows the reduction of the number of multipolar parameters stored in the library and needed for the description of the atomic electron density.

15.3 Protein Structures at Ultra-High Resolution

Based on published results, the record resolutions for proteins are currently

- crambin at 0.54 Å resolution [13, 31]
- an antifreeze protein at 0.62 Å [32]
- hen egg white triclinic lysozyme at 0.65 Å [33]
- human aldose reductase (hAR) at 0.66 Å [34]
- the rubredoxin from Desulfovibrio gigas at 0.68 Å resolution [35]
- the rubredoxin W4L/R5S of Pyrococcus abyssi rubredoxin at 0.69 Å [36]
- a domain of syntenin at 0.73 Å [37]
- a hydrophobin at 0.75 Å [38]

Protein structures at resolutions between 1.0 and 0.5 Å allow detailed analysis of the protein structure on the atomic scale. Most of the H atoms then become visible in difference Fourier ($F_{obs}-F_{calc}$) maps. This is of extreme importance for understanding enzymatic catalysis and for protein-structure interpretation.

Diffraction data for a crambin crystal were measured to ultra-high resolution (0.54 Å) at low temperature $T = 100$ K by using short wavelength synchrotron radiation [31]. The charge density distribution of the protein crambin along the mainchain polypeptide was refined experimentally [13]. It was the first protein structure refined using a non-spherical, multipolar atom model and the transferability principle to accurately describe the molecular electron density distribution. The refined parameters agree within 25% with the transferable electron density library derived from accurate single crystal diffraction analyses of several amino acids and small peptides. The resulting electron density maps of redistributed valence electrons (deformation maps) compare quantitatively well with a high-level quantum mechanical calculation performed on a monopeptide. This study provided validation for experimentally derived parameters and demonstrated the feasibility of charge density analysis for the biological macromolecules for which crystals yield diffraction to subatomic resolution.

The crystal structure of a novel, lysine 49 PLA2 from *Agkistrodon acutus* venom was investigated at an ultra high resolution (0.8 Å) and reveals significant deformation density on many covalent bonds [39]. The trypsin of *Fusarium oxysporum* was revisited: crystallography at atomic resolution (0.8 Å) and quantum chemistry reveal details of catalysis [40]. A series of crystal structures of trypsin, containing either an autoproteolytic cleaved peptide fragment or a covalently bound inhibitor, were determined at atomic and ultra high resolution and subjected to *ab initio* quantum chemical calculations and multipole ELMAM databank transfer. Quantum chemical calculations reproduced the observed active site crystal structure with severe deviations from standard stereochemistry and indicated the protonation state of the catalytic residues. The charge distribution in the active site derived from the multipole databank transfer is in accordance with the *ab initio* calculations. The combined results confirmed the catalytic function of the active site residues and the two water molecules acting as the nucleophile and the proton donor. The crystal structures represent snapshots from the reaction pathway, close to a tetrahedral intermediate. The de-acylation of trypsin then occurs in true S_N2 fashion.

The crystal structure of the antifreeze protein RD1 was determined at an ultra-high resolution of 0.62 Å [32]. This 7-kDa globular protein from the Antarctic eel pout *Lycodichthys dearborni* belongs to type III of the four types of antifreeze proteins found in marine fishes living at subzero temperatures. The protein fold of RD1 comprises a compact globular domain with two internal tandem motifs arranged about a pseudo-dyad symmetry. Each motif of the "pretzel fold" includes four short β-strands and a 3_{10} helix (Figs. 15.1 and 15.2).

After extensive refinement that includes hydrogen atoms, significant residual electron densities associated with the electrons of peptides and many other bonds

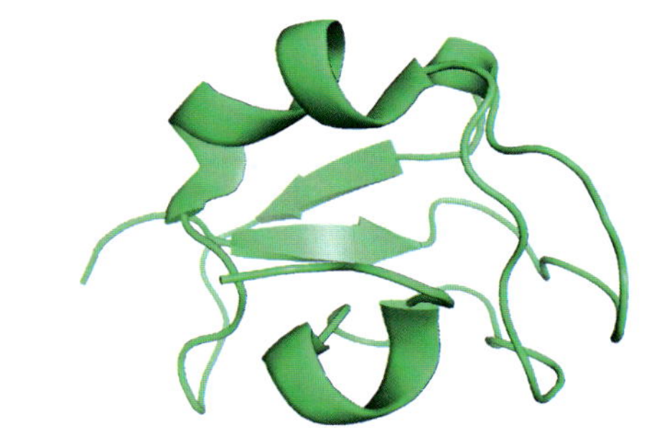

Fig. 15.2 Ribbon view of the antifreeze protein. The view was made with program Pymol [62]

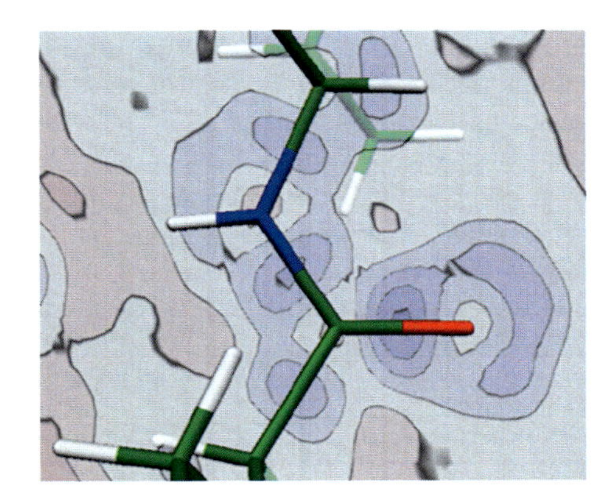

Fig. 15.3 Residual electron density map of the antifreeze protein averaged over the peptide bond planes of the protein main chain. Positive and negative contours are in pale blue and pale red respectively

could be visualized (Fig. 15.3). The averaged residual electron density is similar to that obtained for protein crambin [13].

The collection of diffraction data at ultra high resolution have also been reported for lysozyme, insulin and myoglobin which are macromolecules with molecular weights of 6–17 kDa [41]. The refinement of the protein structures using the Invariom multipolar atoms database is underway.

15.3.1 Charge Density Refinement of Human Aldose Reductase

The electron density and electrostatic potential in the human aldose reductase (hAR) holoenzyme complex with NADP + cofactor and IDD594 inhibitor were studied by density functional theory (DFT) and diffraction methods [42, 43]. Human aldose reductase, a member of the aldo-keto reductase superfamily, is a 36-kDa enzyme (316 residues) that catalyzes the NADPH-dependent reduction of diverse aldehydes to alcohols. As the first and rate-determining enzyme of the polyol pathway, hAR turns glucose into sorbitol. The accumulation of excess sorbitol resulting from diabetic hyperglycemia, plays a role in the development of diabetic complications. Thus, inhibition of hAR offers a potential treatment for these pathologies.

The 0.66 Å ultra-high resolution of the diffraction data and the low thermal-displacement parameters of the structure allow accurate atomic positions refinement and an experimental charge density analysis. Based on the X-ray structure, DFT calculations have been carried out on subsets of up to 711 atoms in the active site of the molecule. All calculations were performed at the experimental geometry with the DFT program SIESTA [44]. This software allows order-*N* calculations that scale linearly with the number of atoms in the system.

The charge density refinement of the protein was performed with the program MoPro by using the transferability principle and the database of charge density. Electrostatic potentials calculated from the ELMAM database, from DFT computations and from atomic point charges taken in the AMBER software dictionary are compared. The electrostatic complementary between the cofactor $NADP^+$ and the active site shows up clearly [42]. The anchoring of the inhibitor IDD594 is due mainly to hydrophobic forces and to only two polar interaction sites within the enzyme cavity. There is a remarkably short contact between the bromine atom of the inhibitor and the oxygen atom of THR113 side chain (Fig. 15.4). The electrostatic potentials calculated by X-ray and DFT techniques agree reasonably well. The potentials obtained directly from the database are in excellent agreement with the experimental ones. In addition, these results demonstrate the significant contribution of electron lone pairs and of atomic polarization effects to the host and guest recognition mechanism.

Crambin and human aldose reductase are the two only proteins for which multipolar refinements were reported in the literature [2, 13]. The crystallographic study by Guillot et al. [2] describes in details the methods used to perform a constrained charge density refinement of the most thermally stable parts of the hAR protein, including its active site, with a multipolar atom model. At first, a subset of amino acids lining the active site has been selected, mostly based on connectivity between residues and atomic thermal motion criteria. The bonding electron density is visible in Fourier residual maps for most of the bonds in this subset (Fig. 15.5). Then a high order IAM refinement has been performed. It has been shown in numerous charge density studies that the refinement of atomic coordinates and especially anisotropic displacement parameters against a high resolution shell allows to improve drastically the deformation density signal by achieving a first deconvolution between deformation density and thermal motion effects. In the hAR study, this refinement was performed in several high resolution ranges of decreasing sizes to ensure a proper convergence.

Then, the multipolar parameters from the ELMAM database were transferred to the protein atoms, followed by the multipolar refinement of the relevant atoms. The position of hydrogen atoms was constrained according to standard H-X bond distances obtained from neutron diffraction [45]. In order to reduce the number of refined parameters and thus avoiding any risk of over-fitting, a maximum number of symmetry and chemical equivalence constraints have been applied. In this particular study, the atom types nomenclature used in the ELMAM database was followed to

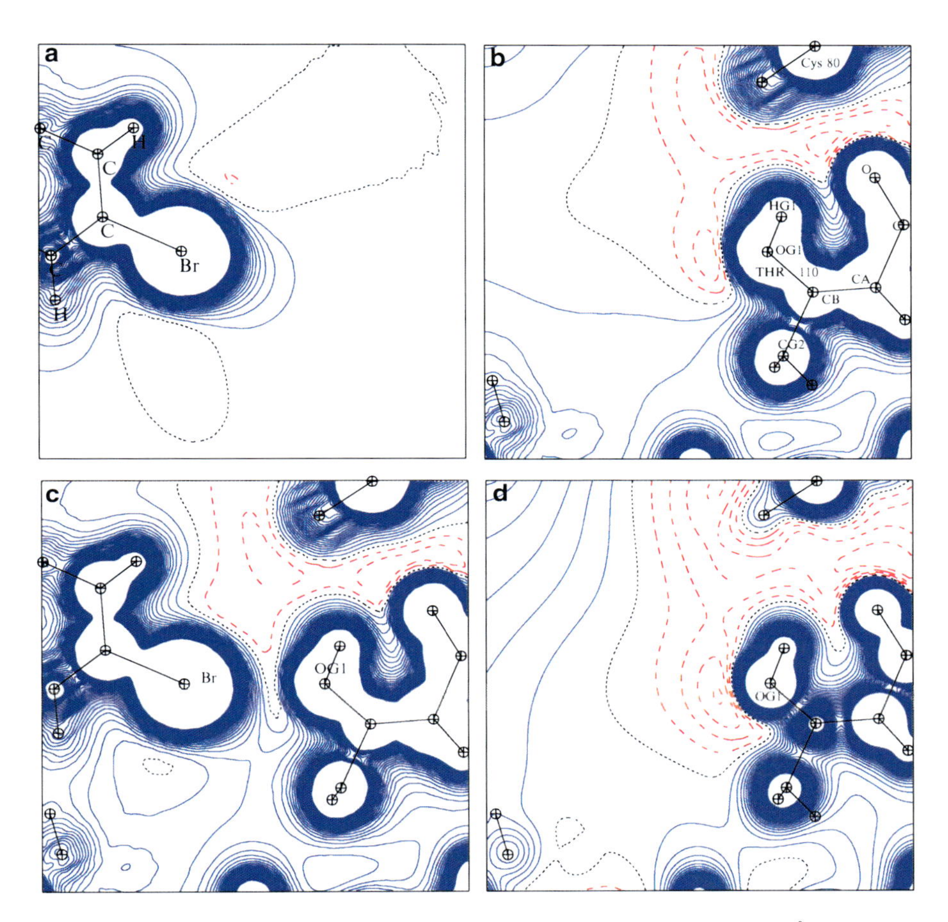

Fig. 15.4 View of the electrostatic potential around the short interaction ($d = 3.0$ Å) between a bromine atom of inhibitor IDD594 and the OG1 oxygen atom of Thr113 in hAR. (**a**) Around the bromine atom, DFT potential generated by the IDD594 inhibitor alone. (**b**) Around the hydroxyl group of OG1-Thr113, DFT potential generated by the protein active site, (**c**) DFT potential in the aldose reductase – inhibitor complex, (**d**) in the protein active site, potential computed after transfer from the experimental database ELMAM. The projection plane contains the bromine atom and the OG1-HG1 atoms of threonine 113. Blue and red contours ±0.05 e/Å are positive and negative potentials respectively

define the constraints scheme. Indeed, in the ELMAM database, only the multipoles following the chemical group symmetries have non-zero values. Thus, for the refinement of hAR charge density, the refined multipoles correspond to the non-zero multipoles defined in the library. In the same way, chemical equivalences constraints have been defined using the atom types in the ELMAM database. This way, 269 multipoles populations and 50 valence populations were refined in multipolar model of the protein. To ensure a good deconvolution between thermal

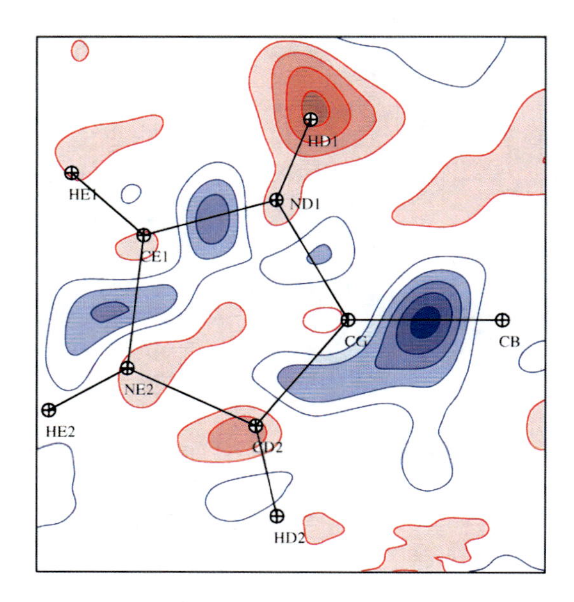

Fig. 15.5 Residual electron density map in the imidazole plane of the catalytic HIS110 of hAR. Positive and negative contours ± 0.05 e/Å^3 are in blue and red respectively. The histidine was here modelled as doubly protonated. The Fourier map displays however an electron deficit around atom HD1, indicating that atom ND1 bears no hydrogen atom

motion and deformation density effects, the structural and charge density parameters refinement have been performed alternately, up to total convergence. The final agreement factors are $R(F) = 8.74\%$ and $R_{\text{free}}(F) = 9.23\%$, which are respectively 1% and 0.7% better (in absolute value) than the initial (IAM) values.

The resulting static electron density maps are of good quality (Fig. 15.6), especially where the number of constrained chemically equivalent moieties is high. The refined static deformation density on peptide bonds shows little deviation when compared to the electron density transferred from ELMAM database.

After multipolar refinement the structural and thermal motion parameters of non-hydrogen atoms in the selected subset were refined, without application of any stereochemical restraints, in order to assess the effect of charge density modelling. As expected, the multipolar modelling led to a small, but almost systematic decrease of atomic equivalent thermal B factor. The difference of thermal mean square amplitudes along covalent bonds (rigid bond test) also tend, on average, to decrease.

The refinement of deformation density features also has an effect on the distribution of stereo-chemical parameters, like bond length or angles. For instance, the covalent bond lengths distribution shows a slight narrowing after the structure refinement using the ELMAM charge density modelling, when compared to a IAM unrestrained refinement.

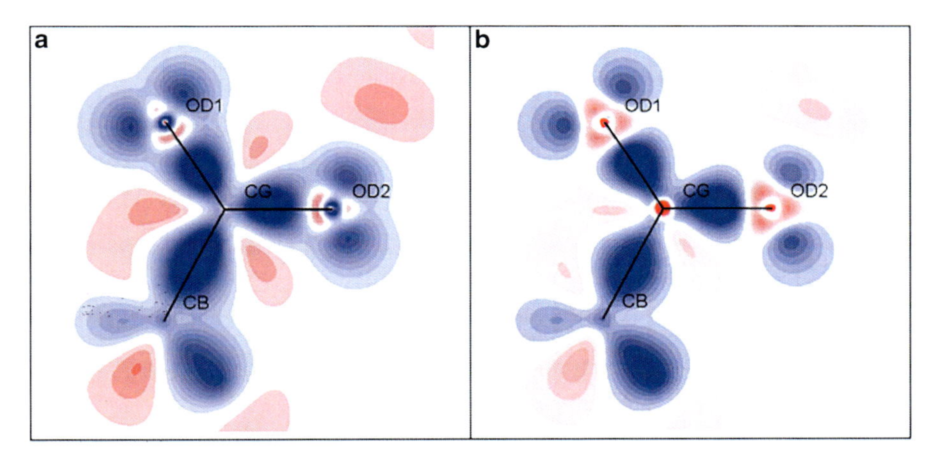

Fig. 15.6 Deformation electron density of the carboxylate group of residues ASP and GLU. (**a**) As defined in the ELMAM library. (**b**) After constrained refinement vs. the diffraction data of hAR. Positive contours are in *blue* and negative in *red*. Contour increment is ± 0.05 e/Å^3

15.3.2 Protein Electrostatics

Electrostatic properties of molecular systems are of major importance in analyzing the structure-function relationships of biological systems. Deriving these properties from the experimental multipolar charge distribution defined using the Hansen-Coppens formalism is routinely performed in small molecule cases. More methodological developments on electrostatic calculations can be found in another Chap. 11 of this book.

Despite significant improvements of experimental devices in X-ray diffraction, multipolar refinements for macromolecular systems are rarely possible because of the lack of sub-atomic resolution experimental data. To circumvent experimental limits, the transferability of multipolar parameters is a very reliable way to obtain an estimated non-spherical charge distribution model for macromolecular systems. The transferability principle was recently applied to several enzyme-inhibitor complexes using different atomic multipolar parameters databases. Non-spherical charge distribution models of enzyme-inhibitor complexes can be expected to retrieve more accurate description of electrostatic properties. Correct modelling of directionality of short range electrostatic interactions is crucial for drug-receptor recognition considering the major role of hydrogen bonding in biological systems.

Quantitative electrostatic properties can be calculated from charge distribution models. The total electrostatic interaction energy between the charge distributions of two chemical entities A and B, named respectively ρ_A and ρ_B, is given analytically by:

$$E_{elec} = \iint \rho_A \, \rho_B \; |r_A - r_B|^{-1} \, dr_A \, dr_B \tag{15.2}$$

This quantity is computed by 3D integration of the charge distribution ρ_A of the entity A, multiplied by the electrostatic potential φ_B of the entity B:

$$E_{elec} = \int \rho_A \; \varphi_B \; dr_A = \int \rho_B \, \varphi_A \, dr_B \tag{15.3}$$

The calculation of the integral is done in the VMoPro module, a properties calculation and visualization tool of the MoPro software [25] by summation on a non-regular grid. The integration method focuses on regions showing large variations of the function and uses a 5th order Taylor formula.

Although electrostatic interactions contribute only a part of the interaction energies between macromolecules, unlike dispersion forces, they are highly directional and therefore dominate the nature of molecular packing in crystals and in biological complexes. Electrostatics contribute therefore significantly to differences in inhibition strength among related enzyme inhibitors. In the study by Dominiak et al. [46], a wide range of complexes of influenza neuraminidases with inhibitor molecules (sialic acid derivatives and others) have been analyzed using charge densities from the UBDB transferable aspherical-atom data bank. The strongest interactions of the protein active site residues with inhibitors and the effect of some mutations on the electrostatic interactions were analyzed. The results are in agreement with the lower level of resistance of the mutated virus to glycerol-containing inhibitors compared to more hydrophobic derivatives.

In another study, the electrostatic energy interactions were analyzed for eight protein syntenin PDZ2 domain complexes with four amino-acid long peptides [6]. Two classes of peptides were investigated: type I with the sequence -(S/T)XΦ and type II with -ΦXΦ sequence, where X is any amino acid and Φ is a residue with hydrophobic side chains. The charge density parameters, were transferred from the UBDB databank [15] to the selected peptides and the 73-residue long fragments of the PDZ2 domain which were interacting with peptides in the groove. The solvent atoms and ions were omitted in the calculations. The electrostatic interaction energies were calculated from the charge density distributions with the exact potential/multipole moment (EPMM) method of Volkov et al. [47, 48] implemented in the XDPROP module of the XD package. The EPMM method combines numerical evaluation of the exact Coulomb integral for the short-range interactions between atoms within a certain limiting distance (here, taken as 4.5 Å) with the Buckingham type multipole approximation for the long-range interatomic interactions between atoms for which no charge density overlap occurs.

The results show that the major contribution to the electrostatic energy comes from the interaction with the P_0 and P_{-2} residues of the peptide (P_{-n} denotes the nth amino acid residue of the bound peptide counting from the C-terminal residue). The P_{-1} residue was shown to play a much smaller role. The charged P_{-3} residue contributes significantly to the total electrostatic interaction energy. Both class I and

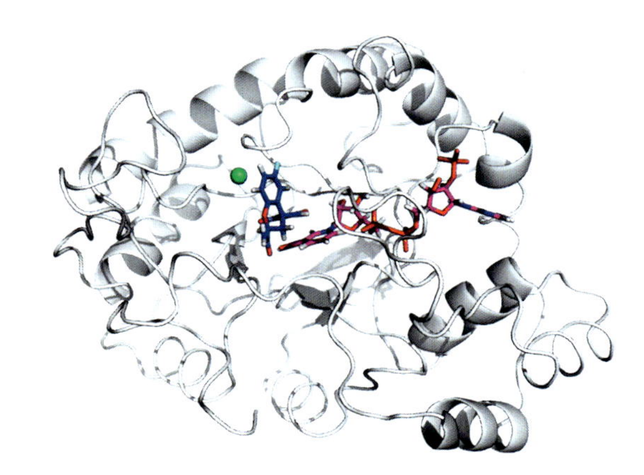

Fig. 15.7 Ribbon view of the human aldose reductase structure. The Fidarestat inhibitor and the NADP$^+$ cofactor are shown bound in the active site cleft. The view was made with Pymol [62]

II peptides were presented to be interacting with the same strength with the syntenin PDZ2 domain.

Another noticeable approach in protein/ligand interaction computations using the multipolar electron density modelling is the SIBFA polarizable molecular mechanics procedure, which is formulated and calibrated on the basis of quantum chemistry [49]. It embodies non classical effects such as electrostatic penetration, exchange-polarization, and charge transfer. This third-generation intermolecular potential is based on density fitting. The developed Gaussian electrostatic model relies on *ab initio*-derived fragment electron densities to compute the components of the total interaction energy. SIBFA offers the possibility of a continuous electrostatic model going from distributed multipoles to densities; it allows an inclusion of short-range quantum effects in the molecular mechanics energies. Recent applications are the docking of inhibitors to Zn-metalloproteins, namely metallo-β-lactamase, phosphomanno-isomerase, and the nucleocapsid of the HIV-1 retrovirus.

15.3.3 The Aldose Reductase – Fidarestat Complex

The charge distribution of human aldose reductase (hAR) complexes with inhibitors (Fig. 15.7) was modelled by transfer from the ELMAM database [50].

Compared to other inhibitors of hAR, fidarestat ((2*S*,4*S*)-6-fluoro-2′,5′-dioxospiro [chroman-4,4′-imidazolidine]-2-carboxamide) shows higher activity and selectivity. Furthermore, encouraging results are already highlighted: it normalizes erythrocytic sorbitol contents in neuropathic patients without significant side effects [51]. There is a stereospecificity in binding hAR: fidarestat ((2*S*,4*S*) stereoisomer) is more efficient than its (2*R*,4*S*) stereoisomer. The crystal structures of the complexes determined at 0.92 and 1.4 Å resolution by El-Kabbani et al. [52, 53] reveal that the binding modes of the two stereoisomers differ mainly in the orientation of the

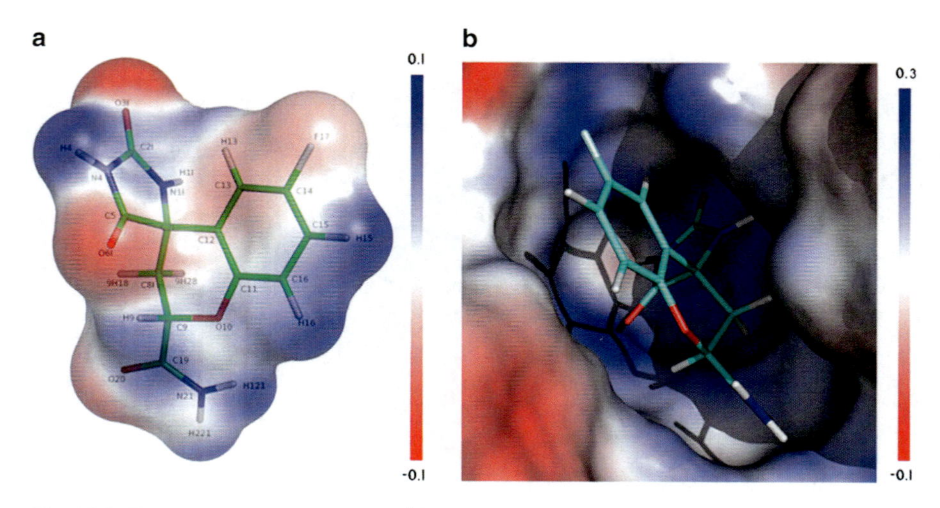

Fig. 15.8 Total electrostatic potential (e/Å) mapped on the solvent-excluded surface with a probe radius of 1.4 Å. The views were made with program Pymol [50]. (**a**) Around an isolated fidarestat molecule in its protein-bound conformation. (**b**) Around the active site generated by hAR and its cofactor

carbamoyl moiety in the active site, related to a configuration change of its C9 chiral carbon atom.

The electron density of fidarestat and of its (2*R*4*S*) stereoisomer were obtained from charge density analysis of Fidarestat crystals (Fig. 15.8a) [50]. The protein charge density was modeled by ELMAM database multipolar transfer. Accurate estimation of electrostatic potential in the active site was therefore derived from this detailed electron distribution. The three-dimensional total electrostatic potential was calculated with the VMoPro module. The potential, mapped on the solvent-excluded surface of the hAR structure, emphasizes the lock-key manner complementarity between the enzyme active site and fidarestat (Fig. 15.8b).

The total electrostatic interaction energies between the inhibitors (fidarestat and its (2R,4S) stereoisomer) and their neighboring residues in the hAR active site are summarized in Table 15.1. The summations of total electrostatic energies between the hAR active site and the inhibitors lead to a difference of -79 kJ.mol^{-1} in favor of fidarestat. The active site can be subdivided in three regions interacting each with one part of the fidarestat molecule (chroman ring, hydantoin moiety or carbamoyl moiety) (Fig. 15.9).

The electrostatic interaction between the ligands and the residues His110, Trp111, Tyr48, Trp20 and the NADP^{+} nicotinamide group is more attractive for fidarestat with a corresponding energy difference of -34 kJ·mol^{-1} compared to its stereoisomer. The total electrostatic interaction energies between the chroman ring and the residues Val47, Phe122 and Trp79 are quite similar with no significant difference.

The largest energy difference in the electrostatic interaction is observed between the carbamoyl moiety and its neighbors (Leu300, Ala299, Cys298 and Trp219) with

Table 15.1 Interactions between the inhibitors and the hAR active site residues. The corresponding intermolecular electrostatic interaction energies are given. The closest non-H atoms between the inhibitors and each residue are specified with the corresponding distances

Protein residue	Closest non-H atoms	Fidarestat Distance (Å)	$E_{elec,tot}$ (kJ/mol)	(2*R*,4*S*)-stereoisomer Distance (Å)	$E_{elec,tot}$ (kJ/mol)
His110[a]	NE2 . . . N4	2.75	−60	2.70	−52
Trp111[a]	NE . . . O6I	2.79	−25	2.79	−19
Tyr48[a]	OH . . . O3I	2.63	−47	2.50	−35
Trp20[b]	Ring . . . N1I	3.16	−46	3.17	−40
NADP+	NC4 . . . O3I	3.04	−20	2.93	−19
Subtotal			−199		−165
Phe122	CE1 . . . C16	3.83	1	3.70	0
Trp79	CZ3 . . . O6I	3.62	9	3.48	8
Val4	O . . . F17	3.21, 3.09	−5	3.11	−7
Subtotal			5		1
Leu300[a,c]	N . . . O20	2.95, 3.03	−34	2.80	−29
Cys298	CA . . . O20	3.44	6	3.00	38
Ala299[c]	N . . . O20	3.46, 3.55	1	3.72	5
Trp219[d]	CH2(CZ2) . . . N21	3.62	12	2.98	20
Subtotal			−15		34
Total			−209		−130

[a]These residues are involved in H-bonds with inhibitors. In this case, the closest heavy atoms distances correspond to Donor . . . Acceptor distances
[b]This interaction is a H-bond with an aromatic ring as π acceptor
[c]For some interactions with fidarestat, two closest distances are indicated because of structural disorder in the protein active site
[d]The closest carbon atom of Trp219 residue is different in the (2*R*,4*S*)-stereoisomer case and its label is indicated between parentheses

-49 kJ·mol^{-1} in favor of fidarestat. This major difference agrees with structural observation performed by El-Kabbani et al. [52, 53] with the predominant role being attributed to the orientation change of the carbamoyl moiety. Although the hydrogen bonds between the hydantoin moiety of the inhibitors and the active site residues are well preserved, slight conformational changes induce significant interaction energy difference which cannot be predicted by geometry considerations.

This example shows how the much highly detailed charge distribution description of enzyme-inhibitor complexes obtained by an ELMAM transfer provides useful insights to understanding electrostatic interactions in inhibitor-protein recognition and docking studies.

15.3.4 Application to the PfluDING Protein

The bacterial protein PfluDING is a member of the DING protein family, which has been identified in a wide range of organisms [54–56]. The name of these

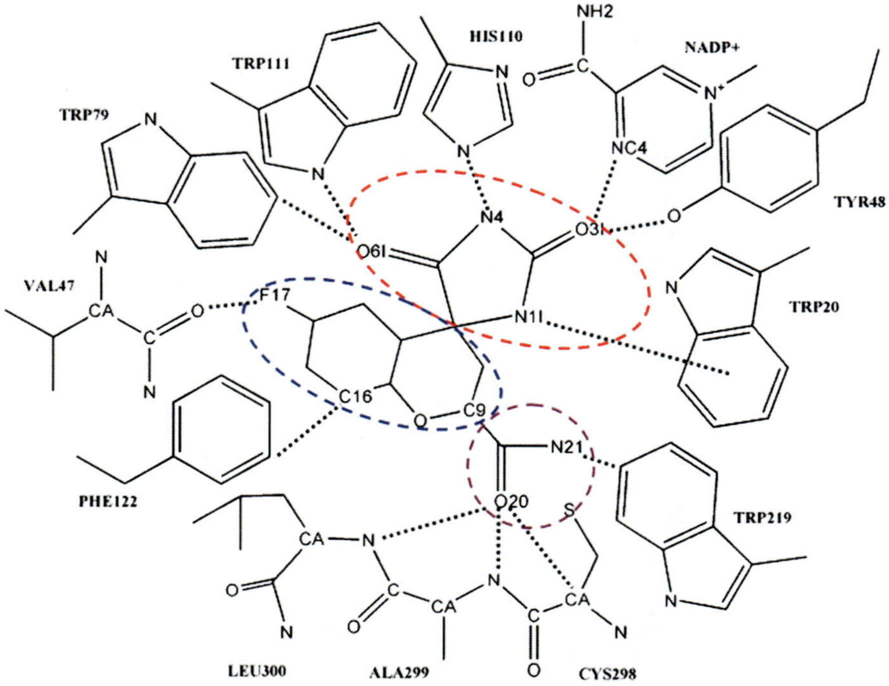

Fig. 15.9 Schematic diagram of the short contacts and hydrogen bonds network involving the fidarestat molecule in the hAR active site. Interactions between closest non-H atoms are indicated by *dash lines*. The different chemical moieties of fidarestat are specified by *coloured dash circles*: the hydantoin moiety in *red*, the chroman ring in *blue* and the carbamoyl moiety in *purple*

macromolecules comes from four conserved residues at the N-terminus. Although they occur in animals, plants and fungi, they are systematically absent from eukaryotic genome databases [57]. The PfluDING protein adopts a structure called "Venus flytrap", i.e. two globular domains hinge together and form a phosphate binding site. The phosphate ion is fixed to the protein via numerous hydrogen bonds [58, 59].

PfluDING has been crystallized at pH 4.5 and crystals at pH 8.5 were obtained by subsequent soaking. Diffraction experiments yielded resolutions of 0.98 and 0.88 Å for the crystal at pH 4.5 and 8.5, respectively. Final refinement steps of the bulk solvent [60], positional and thermal parameters in the binding cavity have been performed with MoPro.

The high resolution and the quality of the diffraction data permitted to locate hydrogen atoms in the phosphate binding site according to significant electron density peaks (Fig. 15.10). Thus, the phosphate binding mode could be determined. The phosphate oxygen atoms are involved in 11 hydrogen bonds to OH and NH groups of the protein and the only proton of the phosphate molecule forms a low

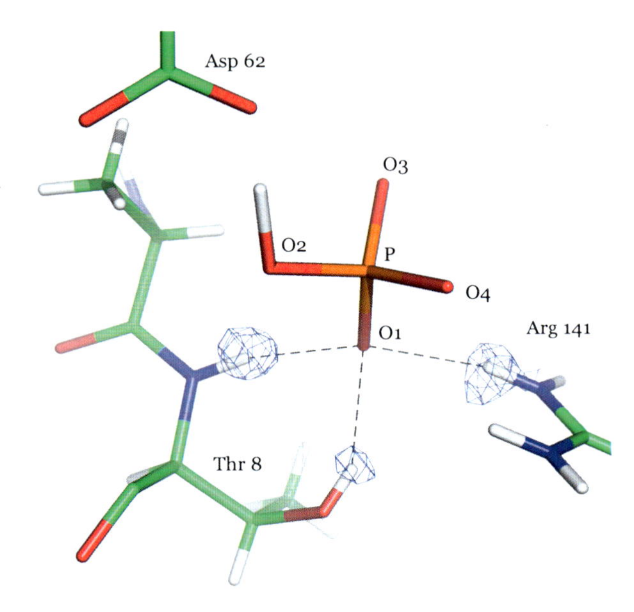

Fig. 15.10 Coordination of the phosphate oxygen atom O1 in the structure obtained at pH 4.5, and Fobs – Fcalc difference electron density map (*blue*, 2.7 σ contour level). The electron density of the omitted hydrogen atoms appears clearly. The view was made with Pymol [50]

barrier hydrogen bond with the carboxylate group of Asp 62 [61]. Surprisingly, the phosphate binding mode is identical at both pH values. Though, the protonation state of phosphate in solution varies according to its pH. At pH 4.5, phosphate is exceedingly monobasic ($H_2PO_4^-$), whereas it is dibasic (HPO_4^{2-}) at pH 8.5. As there was no change observed in the binding mode, PfluDING binds dibasic phosphate at the two pH values.

Since the atomic positions are precisely determined at sub-Angstrom resolution and hydrogen atoms could be placed in the active site, the precise electrostatic potential of the binding cave was calculated. Multipole parameters (up to octupoles for C, O, and N atoms and to dipoles for H atoms) and atomic charges were transferred from the ELMAM database to the final structural model. The covalent X-H bond lengths were elongated to standard distances obtained from neutron diffraction [45]. Afterwards, the electrostatic deformation potential was calculated using the VMoPro software (Fig. 15.11).

In the phosphate binding cleft, the computed deformation electrostatic potential turns out to be largely positive [61], contrarily to previous studies using punctual atomic charges modelling. Notably, around the negatively charged head of the aspartate, it is almost neutral. The positive electrostatic potential is due to the positive charges of the phosphate coordinating hydrogen atoms and to the arginine side chain in the binding site. This is in accordance with the ability of PluDING to attach negatively charged phosphate moieties.

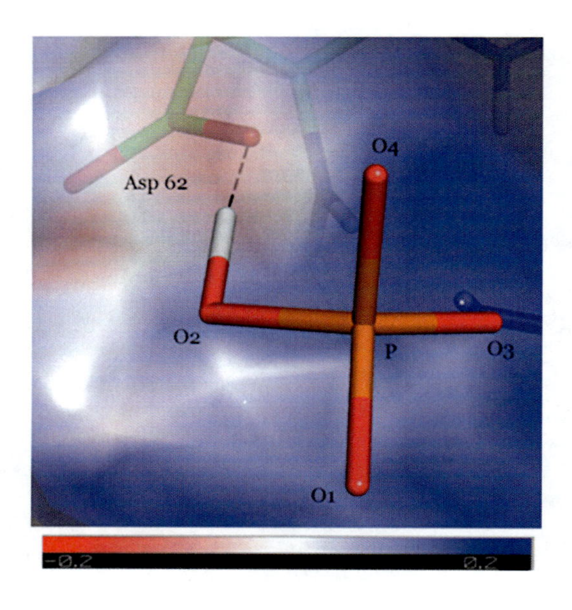

Fig. 15.11 Electrostatic potential (e/Å) in the phosphate binding cleft. The broken line shows the low barrier hydrogen bond interaction between the phosphate and the neighbouring aspartate group. Even around Asp 62, the potential is close to neutral. The view was made with Pymol [50]

15.4 Conclusion

The charge density of proteins at ultra high resolution can be refined when taking advantage of the repetition of chemical motifs along the polypeptide chain. The multipolar parameters can be constrained according to the numerous atoms chemical equivalence occurring in the polypeptide. The electron density in different secondary structure elements (alpha-helices, beta-strands) in proteins deserves also further analysis.

Substrate recognition and binding efficiency are directly related to the charge distributions of the ligand and of the enzyme active site. For the first time, the experimental charge density parameters of the ELMAM database have been used to interpret ligand binding in protein structures determined at atomic resolution: two hAR-ligand complexes and phosphate bound to a DING protein. The transfer of appropriate electron density parameters permits the accurate calculation of electrostatic properties otherwise inaccessible. This level of description can provide a novel view of electrostatic interactions to understand inhibitor-protein recognition and specificity in docking studies. Further work in this direction is encouraged.

Acknowledgements The ANR is gratefully acknowledged for grant NT05-3_41509 (programme blanc Physique). S.D. is grateful for a grant of Region Lorraine and Nancy University. B.F and D.L. are grateful for a PhD grant of the French Ministry of Research.

References

1. Petrova T, Podjarny A (2004) Very high resolution X ray structures of biological macromolecules (esp. proteins). Rep Prog Phys 67:1565–1605
2. Guillot B, Jelsch C, Podjarny A, Lecomte C (2008) Charge density analysis of a protein structure at subatomic resolution: the human aldose reductase case. Acta Crystallogr D64:567–588
3. Pichon-Pesme V, Lecomte C, Lachekar H (1995) On building a data bank of transferable experimental electron density parameters. Application to polypeptides. J Phys Chem 99:6242–6250
4. Zarychta B, Pichon-Pesme V, Guillot B, Lecomte C, Jelsch C (2007) On the application of an experimental multipolar pseudo-atom library for accurate refinement of small-molecule and protein crystal structures. Acta Crystallogr A63:108–125
5. Dittrich B, Hübschle CB, Luger P, Spackman MA (2006) Introduction and validation of an invariom database for amino-acid, peptide and protein molecules. Acta Crystallogr D62:1325–1335
6. Dominiak PM, Volkov A, Li X, Coppens P (2007) A theoretical databank of transferable aspherical atoms and its application to electrostatic interaction energy calculations of macromolecules. J Chem Theory Comput 3:232–247
7. Hansen NK, Coppens P (1978) Testing aspherical atom refinements on small-molecule data sets. Acta Crystallogr A34:909–921
8. Koritsánszky T, Volkov A, Coppens P (2002) Aspherical-atom scattering factors from molecular wave functions. 1. Transferability and conformation dependence of atomic electron densities of peptides within the multipole formalism. Acta Crystallogr A58:464–472
9. Birkedal H, Madsen D, Mathiesen RH, Knudsen K, Weber HP, Pattison P, Schwarzenbach D (2004) The charge density of urea from synchrotron diffraction data. Acta Crystallogr A60:371–381
10. Jelsch C, Pichon-Pesme V, Lecomte C, Aubry A (1998) Transferability of multipole charge density parameters: application to very high resolution oligopeptide and protein structures. Acta Crystallogr D54:1306–1318
11. Brünger AT (1992) Free R value: a novel statistical quantity for assessing the accuracy of crystal structures. Nature 355:472–475
12. Dittrich B, Strumpel M, Schäfer M, Spackman MA, Koritsánszky T (2006) Invarioms for improved absolute structure determination of light-atom crystal structures. Acta Crystallogr A62:217–223
13. Jelsch C, Teeter MM, Lamzin V, Pichon-Pesme V, Blessing RH, Lecomte C (2000) Accurate protein crystallography at ultra-high resolution: valence electron distribution in crambin. Proc Natl Acad Sci USA 97:3171–3176
14. Housset D, Pichon-Pesme V, Jelsch C, Benabicha F, Maierhofer A, David S, Fontecilla-Camps JC, Lecomte C (2000) Towards the charge density study of proteins: a room-temperature scorpion-toxin structure at 0.96 Å resolution as a first test case. Acta Crystallogr D56:151–160
15. Volkov A, Li X, Koritsanszky T, Coppens P (2004) Ab initio quality electrostatic atomic and molecular properties including intermolecular energies from a transferable theoretical pseudoatom databank. J Phys Chem A108:4283–4300
16. Dittrich B, Koritsánszky T, Luger P (2004) Dittrich charge density study on cyclosporine a. Angew Chem Int Ed 43:2718–2721
17. Matta CF, Bader RFW (2003) Proteins 52:360–399
18. Szekeres Zs, Exner TE, Mezey PG (2005) Fuzzy fragment selection strategies, basis set dependence, and HF – DFT comparisons in the applications of the ADMA method of macromolecular quantum chemistry. Int J Quantum Chem 104:847–860
19. Afonine P, Pichon-Pesme V, Muzet N, Jelsch C, Lecomte C, Urzhumtsev A (2003) Modelling of bond electron density by Gaussian scatters at subatomic resolution. CCP4 News Letter

20. Afonine PV, Grosse-Kunstleve RW, Adams PD, Lunin VY, Urzhumtsev A (2007) On macromolecular refinement at subatomic resolution with interatomic scatterers. Acta Crystallogr D63:1194–1197
21. Pichon-Pesme V, Jelsch C, Guillot B, Lecomte C (2004) A comparison between experimental and theoretical aspherical-atom scattering factors for charge density refinement of large molecules. Acta Crystallogr A60:204–208
22. Volkov A, Koritsánszky T, Li X, Coppens P (2004) Response to the paper a comparison between experimental and theoretical aspherical-atom scattering factors for charge-density refinement of large molecules, by Pichon-Pesme, Jelsch, Guillot & Lecomte. Acta Crystallogr A60:638–639
23. Fernandez-Serra MV, Junquera J, Jelsch C, Lecomte C, Artacho E (2000) Electron density in the peptide bonds of crambin. Solid State Commun 116:395–400
24. Guillot B, Viry L, Guillot R, Lecomte C, Jelsch C (2001) Refinement of proteins at subatomic resolution with MOPRO. J Appl Crystallogr 34:214–223
25. Jelsch C, Guillot B, Lagoutte A, Lecomte C (2005) Advances in proteins and small molecules. Charge density refinement methods using software MoPro. J Appl Crystallogr 38:38–54
26. Scheringer C (1980) The electron-density distribution in silicon. Acta Crystallogr A36: 205–210
27. Jelsch C (2001) Sparsity of the normal matrix in the refinement of proteins at atomic and subatomic resolution. Acta Crystallogr A57:558–570
28. El Haouzi A, Hansen NK, Le Hénaff C, Protas J (1996) The phase problem in the analysis of X-ray diffraction data in terms of electron density distributions. Acta Crystallogr A52:291–301
29. Domagała S, Jelsch C (2008) Optimal local axes and symmetry assignment for charge density refinement. J Appl Crystallogr 41:1140–1149
30. Coppens P, Sabino JR (2003) On the choice of *d*-orbital coordinate system in charge-density studies of low-symmetry transition-metal complexes. Acta Crystallogr A59:127–131
31. Lamzin VS, Morris RJ, Dauter Z, Wilson KS, Teeter MM (1999) Experimental observation of bonding electrons in proteins. J Biol Chem 274:20753–20755
32. Ko T, Robinson H, Gao Y, Cheng C, DeVries WA, Wang AHJ (2003) The refined crystal structure of an eel pout type III antifreeze protein RD1 at 0.62-Å resolution reveals structural microheterogeneity of protein and solvation. Biophys J 84:1228–1237
33. Wang J, Dauter M, Alkire R, Joachimiak A, Dauter Z (2007) Triclinic lysozyme at 0.65 Å resolution. Acta Crystallogr D63:1254–1268
34. Howard ER, Sanishvili R, Cachau RE, Mitschler A, Chevrier B, Barth P, Lamour V, Van Zandt M, Sibley E, Bon C, Moras D, Schneider TR, Joachimiak A, Podjarny A (2004) Ultrahigh resolution drug design I: details of interactions in human aldose reductase–inhibitor complex at 0.66 Å. Proteins 55:792–804
35. Chen CJ, Lin YH, Huang YC, Liu MY (2006) Crystal structure of rubredoxin from desulfovibrio gigas to ultra-high 0.68 Å resolution. Biochem Biophys Res Commun 349:79–90
36. Bönisch H, Schmidt CL, Bianco P, Ladenstein R (2005) Ultrahigh-resolution study on Pyrococcus abyssi rubredoxin. I. 0.69 Å X-ray structure of mutant W4L/R5S. Acta Crystallogr D61:990–1004
37. Kang BS, Devedjiev Y, Derewenda U, Derewenda ZS (2004) The PDZ2 domain of syntenin at ultra-high resolution: bridging the gap between macromolecular and small molecule crystallography. J Mol Biol 338:483–493
38. Hakanpää J, Linder M, Popov A, Schmidt A, Rouvinen J (2006) Hydrophobin HFBII in detail: ultrahigh-resolution structure at 0.75 Å. Acta Crystallogr D62:356–367
39. Liu Q, Huang Q, Teng M, Weeks CM, Jelsch C, Zhang R, Niu L (2003) The crystal structure of a novel, inactive, lysine 49 PLA2 from Agkistrodon acutus venom: an ultrahigh resolution, ab initio structure determination. J Biol Chem 278:41400–41408
40. Schmidt A, Jelsch C, Østergaard P, Rypniewski W, Lamzin VS (2003) Trypsin revisited: crystallography at (sub) atomic resolution and quantum chemistry revealing details of catalysis. J Biol Chem 278:43357–43362

41. Kalinowski R, Meents A, Luger P (2008) Application of the aspherical scattering formalism on the refinement of macromolecules. Acta Crystallogr A64:C570–C571
42. Muzet M, Guillot B, Jelsch C, Howard E, Lecomte C (2003) On the electrostatic complementarity in a human aldose reductase/NADP+/inhibitor complex as derived from first-principles calculations and ultra-high resolution crystallography. Proc Natl Acad Sci USA 100:8742–8747
43. Lecomte C, Guillot B, Muzet N, Pichon-Pesme V, Jelsch C (2004) Ultra-high-resolution X-ray structure of proteins. Cell Mol Life Sci 61:774–782
44. Sanchez-Portal D, Ordejon P, Artacho E, Soler JM (1997) Density-functional method for very large systems with LCAO basis sets. Int J Quantum Chem 65:453–461
45. Allen FH, Kennard O, Watson DG, Brammer L, Orpen AG, Taylor R (1987) Tables of bond lengths determined by X-Ray and neutron diffraction. Part 1. Bond lengths in organic compounds. J Chem Soc Perkin Trans II, S1–S19
46. Dominiak PM, Volkov A, Dominiak AP, Jarzembska KN, Coppens P (2009) Combining crystallographic information and an aspherical-atom data bank in the evaluation of the electrostatic interaction energy in an enzyme-substrate complex: influenza neuraminidase inhibition. Acta Crystallogr D65:485–499
47. Volkov A, Koritsánszky T, Coppens P (2004) Combination of the exact potential and multipole methods (EP/MM) for evaluation of intermolecular electrostatic interaction energies with pseudoatom representation of molecular electron densities. Chem Phys Lett 391:170–175
48. Volkov A, King HF, Coppens P (2006) Response to Spackman's comment on the paper 'on the calculation of the electrostatic potential, electric field and electric field gradient from the aspherical pseudoatom model'. J Chem Theory Comput 2:81–89
49. Gresh N, Cisneros GA, Darden TA, Piquemal JP (2007) Anisotropic, Polarizable Molecular Mechanics Studies of Inter- and Intramolecular Interactions and Ligand—Macromolecule Complexes. A Bottom-Up Strategy. J Chem Theory Comput 3:1960–1986
50. Fournier B, Bendeif EE, Guillot B, Podjarny A, Lecomte C, Jelsch C (2009) Charge density and electrostatic interactions of fidarestat, an inhibitor of human aldose reductase. J Am Chem Soc 131:10929–10941
51. Asano T, Saito Y, Kawakami M, Yamada N (2002) Fidarestat (SNK-860), a potent aldose reductase inhibitor, normalizes the elevated sorbitol accumulation in erythrocytes of diabetic patients. J Diabetes Complications 16:133–138
52. El-Kabbani O, Darmanin C, Schneider TR, Hazemann I, Ruiz F, Oka M, Joachimiak A, Schulze-Briese C, Tomizaki T, Mitschler A, Podjarny A (2004) Ultrahigh resolution drug design. II. Atomic resolution structures of human aldose reductase holoenzyme complexed with fidarestat and minalrestat: implications for the binding of cyclic imide inhibitors. Protein 55:805–813
53. El-Kabbani O, Darmanin C, Oka M, Schulze-Briese C, Tomizaki T, Hazemann I, Mitschler A, Podjarny A (2004) High-resolution structures of human aldose reductase holoenzyme in complex with stereoisomers of the potent inhibitor fidarestat: Stereospecific interaction between the enzyme and a cyclic imide type inhibitor. J Med Chem 47:4530–4537
54. Berna A, Bernier F, Scott K, Stuhlmüller B (2002) Ring up the curtain on ding proteins. FEBS Lett 524:6–10
55. Di Maro A, De Maio A, Castellano S, Parente A, Farina B, Faraone-Mennella MR (2009) The adp-ribosylating thermozyme from sulfolobus solfataricus is a ding protein. J Biol Chem 390:27–30
56. Pantazaki AA, Tsolkas GP, Kyriakidis DA (2008) A ding phosphatase in thermus thermophilus. Amino Acids 34:437–448
57. Berna A, Bernier F, Chabrière E, Perera T, Scott K (2008) Ding proteins; novel members of a prokaryotic phosphate-binding protein superfamily which extends into the eukaryotic kingdom. Int J Biochem Cell Biol 40:170–175
58. Morales R, Berna A, Carpentier P, Contreras-Martel C, Renault F, Nicodeme M, Chesne-Seck ML, Bernier F, Dupuy J, Schaeffer C, Diemer H, Van-Dorsselaer A, Fontecilla-Camps JC, Masson P, Rochu D, Chabrière E (2006) Serendipitous discovery and x-ray structure of a human phosphate binding apolipoprotein. Structure 14:601–609

59. Morales R, Berna A, Carpentierv P, Contreras-Martel C, Renault F, Nicodeme M, Chesne-Seck M-L, Bernier F, Dupuy J, Schaeffer C, Diemer H, Van-Dorsselaer A, Fontecilla-Camps JC, Masson P, Rochu D, Chabrière E (2007) Discovery and crystallographic structure of human apolipoprotein. Ann Pharm Fr 65:98–107
60. Fokine A, Urzhumtsev A (2002) Flat bulk-solvent model: obtaining optimal parameters. Acta Crystallogr D58:1387–1392
61. Liebschner D, Elias M, Moniot S, Fournier B, Scott K, Jelsch C, Guillot B, Lecomte C, Chabrière E (2009) Elucidation of the phosphate binding mode of ding proteins revealed by subangstrom x-ray crystallography. J Am Chem Soc 131:7879–7886
62. DeLano WL (2002) The PyMOL molecular graphics system. http://www.pymol.org

Chapter 16
Charge Densities and Crystal Engineering

Mark A. Spackman

16.1 Introduction

Modern research in crystal engineering spans a wide range of experimental and theoretical studies. A quick glance at the recent papers appearing in journals such as *Crystal Growth & Design* and *CrystEngComm* reveals topics as diverse as crystallization, crystal morphology, polymorphism, supramolecular architectures and inclusion compounds, fabrication of nanoparticles and thin films, crystal structure determination, crystal structure prediction, crystal structure analysis in terms of hydrogen bonds, halogen bonds, $\pi\cdots\pi$ interactions etc., and the measurement of bulk thermal, magnetic, optical and nonlinear optical (NLO) properties. All of these have a common focus on the engineering of crystalline materials with desirable physical and chemical properties, in the process exploiting the insight gained from an in depth understanding of the ways in which crystalline matter is constructed from component atoms, ions and molecules. This of course paraphrases Desiraju's widely cited definition of the field: "the understanding of intermolecular interactions in the context of crystal packing and in the utilization of such understanding in the design of new solids with desired physical and chemical properties" [1], and it is important to recognize that this definition involves two important goals: (i) understanding the nature of intermolecular interactions; and (ii) building on that knowledge through rational design of new crystalline solids. It will be evident from the wide range of topics mentioned above that it is impossible in this brief chapter to do justice to more than one or two. Therefore, we focus the discussion that follows on molecular crystals, and in particular the ways in which the present techniques and tools of charge density analysis – both theoretical and experimental – can assist in the realization of the first of Desiraju's goals.

M.A. Spackman (✉)
School of Biomedical, Biomolecular & Chemical Sciences, University of Western Australia, Crawley, WA 6009, Australia
e-mail: mark.spackman@uwa.edu.au

C. Gatti and P. Macchi (eds.), *Modern Charge-Density Analysis*,
DOI 10.1007/978-90-481-3836-4_16,

Before embarking on a summary of techniques and discussion of selected recent applications, it is worthwhile exploring an important philosophical point. In some of the earlier Chaps. (9–15) in the present book much of the emphasis has been on the use of topological analysis of the electron density – the quantum theory of atoms in molecules (QTAIM) [2] – for the detailed investigation of chemical bonding in molecules and crystals. Without doubt this approach has been highly successful in gaining enhanced understanding of chemical bonding in countless systems and environments, and this is amply demonstrated by the breadth of material that appeared in a recent monograph on QTAIM and applications [3]. However, despite considerable emphasis in that publication on the topological properties of hydrogen bonds, only a few pages of a single chapter [4] were devoted to the application of QTAIM to molecular crystals. In that chapter, Gatti commented that "QTAIM has proved to be a very powerful tool for isolating, detailing the weak intermolecular interactions responsible of molecular crystal formation and for quantitative characterization of the effect of these interactions on intramolecular bonding". Using the urea crystal by way of example, Gatti illustrated how QTAIM may be used to investigate the effects of crystal packing on the intramolecular bonds, especially in different parts of the molecule, to quantify the molecular dipole moment enhancement upon crystal formation, and to characterize the hydrogen bonds and the changes in atomic and molecular volumes upon crystallization. The properties employed in this type of analysis include those at bond critical points (bcps) as well as those obtained by integration over QTAIM atomic basins (see below).

But to what extent are tabulations of intermolecular bcp distances, electron densities and Laplacians informative for crystal engineering? Put another way, is it better to discuss intermolecular cohesion in terms of weak bonds between atoms in different molecules, or between entire molecular charge distributions? This question was addressed in considerable detail by Dunitz and Gavezzotti [5], and a number of the comments made in that review deserve highlighting for the present discussion. For example, Dunitz and Gavezzotti noted that "when the density at critical points is small, as it is between atoms in neighboring molecules in a crystal, the estimation of its exact value and that of its derivatives is fraught with uncertainties, which make reliable interpretation difficult". Furthermore, "it is not a question of whether we can find intermolecular bond paths, say between hydrogen atoms of neighboring phenyl groups in experimental density distributions. The question is whether and to what extent the information contained in such paths can be distinguished from experimental noise, from electron-density features inherent in the mere proximity of the atoms concerned, and from inadequacies in the multipole refinement model." These comments focus very much on the use of experimental charge densities derived from multipole refinements, but there is a deeper underlying issue clearly identified by Dunitz and Gavezzotti: "We see a danger that topological studies of the intermolecular charge density in crystals may put the emphasis in the wrong place; they will ascribe the intermolecular cohesion to interactions between pairs of atoms for which bond paths can be discerned, even if only weakly, rather than to interactions between the charge density distributions. The bond paths for such intermolecular contacts should be regarded as a secondary phenomenon (an epiphenomenon) accompanying the intermolecular attraction and caused by it."

It is with this dichotomy of viewpoints and approaches in mind that we embark on the remainder of this chapter, with a view to identifying those tools and techniques of charge density analysis that are currently finding useful application to questions in crystal engineering, those that show considerable promise for the future and, with the aid of a crystal ball, those that we envisage being of use but don't yet exist. To this end, Sect. 16.2 summarizes the key features of a selection of charge density-related tools in current use and, to illustrate their use by way of example, Sect. 16.3 describes their application to halogen bonding, a topic of considerable current interest and research in crystal engineering.

16.2 Charge Density Tools for Crystal Engineering

A decade ago, Seddon [6] outlined the tools available to the crystal engineer at that time: synthetic vectors (e.g., hydrogen bonding, inter-ring interactions, covalent bonds and Coulombic forces); computational chemistry; visualization tools for gaining insight into three-dimensional architecture; and the Cambridge Structural Database (CSD). An updated list of crystal engineering tools would include all of the above, most of which embody features of, or incorporate ideas from, charge density analysis, as well as the use of accurate X-ray diffraction experiments through experimental charge density analysis (which was a surprising omission from Seddon's list) and tools that have taken advantage of the power and availability of modern desktop and laptop computers: Hirshfeld surface analysis and related tools, and the calculation of accurate intermolecular interaction and crystal packing energies.

16.2.1 Theoretical Calculations

A number of other chapters in the present book deal with aspects of theoretical calculations of electron distributions in molecules and solids, and the reader is referred to those for detailed presentations of the most common methods and techniques. Of particular relevance to applications in crystal engineering are Chap. 2 (on charge densities from ab initio simulation of crystals), Chap. 9 (on charge density topological approaches), and Chap. 11 (on intermolecular interactions).

16.2.2 Experimental Charge Density Analyses

By far the most common strategy in experimental charge density analysis is multipole refinement, and this is discussed in some detail in Chap. 3 (on the deconvolution from thermal motion) and Chap. 5 (on different charge density models). Provided there is successful deconvolution of the electron distribution from nuclear motion (see Chap. 3), the outcome of these refinements is an analytical representation of the crystalline electron distribution in terms of nucleus-centred radial and angular functions. The QTAIM analysis of this electron distribution includes computation of properties at bond critical points (bcps), especially the electron density, ρ_b, its Laplacian, $\nabla^2\rho_b$, and the energy densities G_b, V_b and H_b, as well as those obtained by integration over QTAIM atomic basins, especially atomic volumes, $V(\Omega)$, charges, $q(\Omega)$, and dipoles, $\mu(\Omega)$. Unfortunately, the user-friendly nature of software such as XD [7] has encouraged the widespread publication of comprehensive tables of all of these properties – and more – as outcomes of many experimental charge density analyses. Although these are legitimate results of such analyses, by themselves they offer little insight into the issues of most concern in crystal engineering, and this is evident in the negligible uptake of these methods within the crystal engineering community.

However, that should not be interpreted to mean that the outcomes of charge density experiments are not having an impact on crystal engineering; the examples in Sect. 16.3 below provide compelling evidence that the opposite is in fact the case. Researchers in crystal engineering tend to think more in terms of the electrostatic nature of the constituent molecules in the crystal, especially molecular properties such as dipole moments and electrostatic potentials, and these can be readily obtained from experimental electron densities. Most importantly, experimental molecular charge distributions (including nuclei) can be used to compute electrostatic interaction energies, and these are discussed in the following Subsection.

16.2.3 Crystal Packing Energies

Crystal structure prediction, an important and extremely visible aspect of crystal engineering, requires addressing two key problems: (i) the accurate calculation of intermolecular, and hence lattice, energies; and (ii) the location of the relevant minima on the lattice energy hypersurface. The first of these is a problem common to many other facets of crystal engineering, especially the use of crystal packing energies to rationalize observed packing motifs and develop insight into the energetics associated with phenomena such as crystal polymorphism, and several authors have recently emphasized that the key to successful and meaningful analysis of crystal packing can only be obtained through an analysis of intermolecular energies of interaction [5, 8–10].

Researchers involved in the most recent evaluation of different methods of crystal structure prediction – the fourth blind test [11] – evaluated intermolecular energies in a variety of ways. The most impressive results in the blind test were obtained by Neumann et al. [12], who used a hybrid method involving DFT calculations to generate a tailor-made force field for each molecular structure, combined with an empirical dispersion correction expressed as a sum of atom-atom terms. The predicted crystal structures with the lowest energies agreed with observed experimental structures for all four compounds in the test. However, there is a catch: this approach is extremely computationally demanding and far from widely applicable at present. More common – and computationally less demanding – are approaches that use atomic charges or multipoles for the electrostatic term, typically combined with empirical exp-6 or off-the-shelf force fields for remaining terms in the energy expression, and these are much more attractive for routine application of an energy-based crystal packing analysis.

Instead of atomic charges or multipoles, Gavezzotti's PIXEL method (*cf.* 11.2.2.1) uses the entire molecular charge distribution, obtained by *ab initio* methods, to compute the electrostatic energy of interaction for a pair of molecules. As its name suggests, PIXEL is a numerical approach that breaks down the molecular electron distribution into 3D pixels (voxels) of charge, and the electrostatic energy is expressed as the sum of pairwise interactions between these elements on the two molecules. In the same manner, polarization energies, dispersion and repulsion energies are computed numerically; full details are provided in several key publications [9, 13, 14]. Importantly – and as emphasized by Gavezzotti many times – this approach quite deliberately avoids the concept of atom-atom potentials, commonly used in crystal structure prediction, and also specific atom-atom intermolecular interactions, commonly the object of topological analyses of experimental electron distributions. Equally important for crystal engineering purposes, this approach quickly yields results that compare favourably with *ab initio* results and, because the total energy is broken down into individual components with ready physical meaning, it is readily applied to the rationalization of organic crystal structures [15, 16]. Of course the quality of the results obtained by the PIXEL approach relies very much on the quality of the *ab initio* electron densities used in the calculations. Volkov, King and Coppens [17] undertook a detailed analysis of the dependence of *ab initio* electrostatic interaction energies as a function of basis sets and theoretical methods. Results for a set of glycine dimers were found to vary by several tens of kJ mol^{-1}, with stable results (i.e. close to benchmark values) obtained provided augmented Gaussian basis sets are used along with advanced correlation methods, including MP2.

Early efforts at using the results of experimental charge density analyses to calculate intermolecular energies of interaction typically employed atomic charges and multipole moments obtained directly from the nucleus-centred multipole expansion of the electron density. Examples include the work of Berkovitch-Yellin and Leiserowitz on crystal packing of amides [18] and carboxylic acids [19], work that in hindsight clearly had a crystal engineering focus, as well as our own early results [20]. More recently, electrostatic interaction energies from experimental charge density analyses have been made widely available in the XD package, where use is made of the EPMM numerical integration approach [21] (see Chap. 11) to compute E_{es}. Mention should also be made of the PAMOC package [22] which interfaces with the VALRAY multipole refinement package [23] and calculates electrostatic interaction energies using experimentally-derived atomic multipole moments. In both implementations, sets of atom-atom potentials are used to describe the repulsion and dispersion terms. Although the numerical approach in XD yields essentially an exact value for E_{es}, the use of atomic multipole moments in PAMOC has been augmented with the promolecule-promolecule electrostatic energy [24], yielding values for E_{es} within a few kJ mol^{-1} of the exact results of Volkov et al. [21].

16.2.4 Hirshfeld Surface Analysis

The Hirshfeld surface (HS) was originally devised for charge density analysis, as a means of identifying a molecule in a crystal [25] for the calculation of molecular dipole and quadrupole moments [26]. It encloses a molecule in a crystal, and is defined as the volume of space where the promolecule electron density exceeds that from all neighbouring molecules. It results in very close proximity of adjacent molecular volumes, but the volumes never overlap. The various tools associated with HS analysis have become increasingly popular in the discussion of intermolecular interactions in molecular crystals, with applications to a variety of topics in crystal engineering. This is largely due to the fact that the HS is defined by both the molecule and the proximity of its nearest neighbours, as well as the nature of the various atomic electron density functions (i.e. the relative extent of their radial distributions), and hence it effectively encodes information about close contacts between molecules in the bulk. And because the HS is smooth and differentiable, it is an ideal vehicle for visualizing intermolecular interactions in crystals, as well as the colour mapping of various properties on the surface.

Early applications in crystal engineering focused on the visualization of properties based on distance from the surface to nuclei interior and exterior to the surface (d_i and d_e, respectively) as well as functions of local surface curvature (shape index and curvedness) [27]. Perhaps more common in comparing crystal structures, especially polymorphs, is the 2D fingerprint plot, a histogram of the relative frequency of (d_i, d_e) pairs on the HS [28] and the breakdown of this information into pairs of atom types [29]. A convenient and up to date summary of the use of the various tools associated with HS analysis appeared recently [30], and the reader is referred to this for more complete descriptions, as well as citations to other key publications. These tools have been incorporated in the freely available CrystalExplorer package [31].

Of particular relevance to the work in the following section is the mapping of the molecular electrostatic potential (ESP) on Hirshfeld surfaces. As discussed in detail elsewhere [32], although a limited number of attempts had been made previously to relate molecular surface ESPs to crystal structure and packing motifs, the HS is an almost ideal vehicle for this purpose. This is because, by virtue of its definition, surfaces of adjacent molecules in crystals never overlap, unlike surfaces defined by the molecule alone, such as isosurfaces of the electron density. Recent examples [32] provide compelling evidence that the concept of "electrostatic complementarity" commonly invoked in discussing protein structure and binding is also commonplace in a wide variety of intermolecular contacts in crystals. Importantly, that work concluded that "even the relative magnitudes of the molecular ESP mapped on Hirshfeld surfaces correlate with electrostatic energies obtained by the PIXEL method for many different interactions". We pursue this point in some detail in the following section.

16.3 Applications to "Halogen Bonding" Interactions

The remainder of this chapter focuses on the ways in which the various approaches described above provide insight into "halogen bonding", the non-covalent interaction between halogen atoms as electrophiles and other halogen atoms, or electronegative atoms such as O and N. Halogen bonding has attracted considerable attention recently, especially as an alternative synthon to hydrogen bonding in crystal engineering (see, for example, the recent reviews by Metrangolo, Resnati and co-workers [33–35]). Our interest here is in evaluating the results and insight that can be obtained by applying the various tools and techniques of charge density analysis described in the preceding section to halogen bonded crystalline systems. This necessarily involves a comparison between the outcomes from various approaches, experimental and theoretical, where they are available. Although what follows represents an initial – and incomplete – attempt at such a comparison, it provides some incentive for more detailed future comparisons of the same kind.

Halogen bonding has been the subject of a small number of experimental charge density analyses. Three of these concern complexes of dihalotetrafluorobenzenes: 1,4-diiodotetrafluorobenzene, $C_6F_4I_2$ (F_4DIB) with (*E*)-1,2-bis(4-pyridyl)ethylene (bpe) [36], and with 4,4′-dipyridyl-*N*,*N*′-dioxide (bpNO) [37], and bpe with 1,4-dibromotetrafluorobenzene, $C_6F_4Br_2$ (F_4DBB) [38]. Most recently, the nature of the halogen trimer synthon has been investigated on the basis of charge density analysis of X-ray diffraction data on C_6Cl_6 [39]. We will examine the outcomes from three of these studies in some detail, where possible comparing them with the results from complementary theoretical charge density methods.

The complexes bpe·F_4DBB and bpe·F_4DIB provide an opportunity to explore the differences between Br···N and I···N interactions. These two experimental studies were based on X-ray diffraction data measured at 90 K, with electron density modelling performed with VALTOPO [40]. It is probably fair to say that the experimental electron distributions for these two complexes are representative of the quality of many present charge density analyses: the X-ray data extended to 1.22 $Å^{-1}$ and 1.15 $Å^{-1}$, and internal agreement factors, R_{int}, were 3.1% and 2.3% for

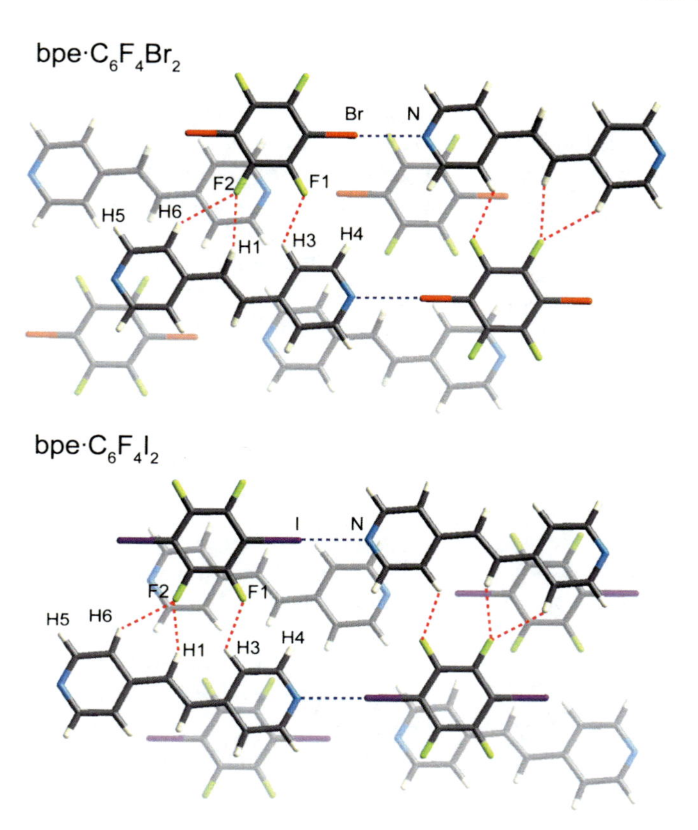

Fig. 16.1 Crystal packing diagrams for bpe·$C_6F_4Br_2$ and bpe·$C_6F_4I_2$. Selected atoms are labelled, and close intermolecular X···N and F···H contacts are indicated by *blue and red dashed lines*, respectively. The layers of molecules are essentially the same for the two structures, but they differ in the packing of one layer (*dark*) relative to that beneath it (*light*)

bpe·F_4DBB and bpe·F_4DIB, respectively. Nevertheless, it is worth bearing in mind that for heavy atoms such as I, and to a lesser extent Br, the Hartree-Fock core and valence monopole radial functions used in these studies may be sub-optimal, as they lack both electron correlation and relativistic effects. Furthermore, corrections for anomalous scattering can be considerable, for these elements amounting to ~0.3–0.5 e per atom for f' and 1.8–2.5 e for f''. Combined with the fact that multipole refinements on both complexes incorporated third and fourth order Gram-Charlier parameters on the halogen atoms, these considerations suggest that the electron distributions may not be of the quality typically obtained in studies on systems comprising only H, C, N and O atoms.

The crystal packing diagrams in Fig. 16.1 show that the two complexes form isostructural layers incorporating close X···N and F···H contacts, but they differ in the relationship between successive layers. Static electron densities derived from

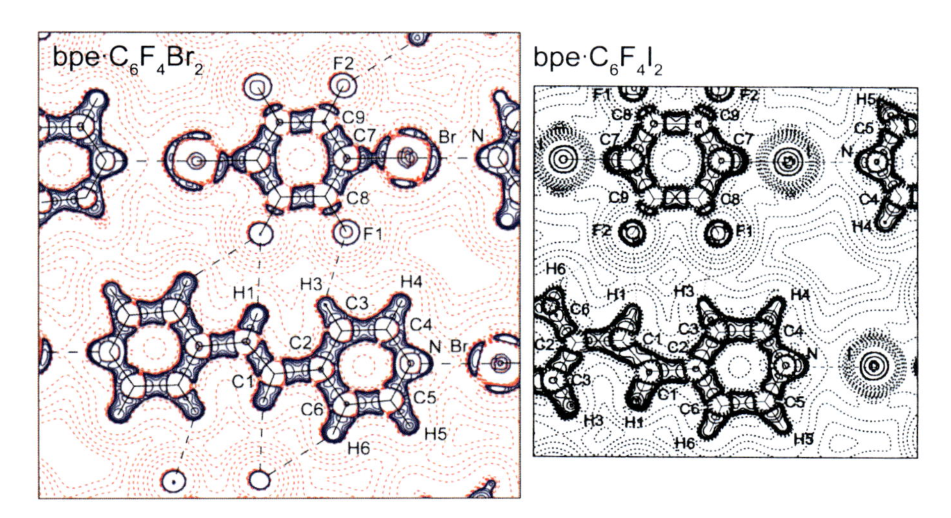

Fig. 16.2 Laplacian of the experimental electron distributions for bpe·$C_6F_4Br_2$ and bpe·$C_6F_4I_2$. Positive contours are *dashed/red* and negative contours *solid/blue* (Reproduced from Ref. [36] with kind permission of Wiley-VCH)

the multipole refinement were subjected to topological analysis, yielding numerous bcp properties for intra- and intermolecular interactions, as well as maps of the Laplacian of the electron density (ED) (Fig. 16.2).

What can these results tell us about the nature and strength of the intermolecular interactions in these crystals? In their review article Koritsanszky and Coppens [41] noted "If two reactants approach each other in a Lewis acid-base-type reaction, their relative orientation can be predicted by the Laplacian functions of their ED. Charge concentrations/depletions of one molecule can be considered to be complementary to depletions/concentrations of the other". This use of complementary concentrations and depletions in the Laplacian of the ED was pioneered by Bader and MacDougall [42], and a pattern of that kind appears to be evident in the region of the I···N interaction in bpe.F_4DIB (Fig. 16.2), but not in the Br···N interaction in bpe.F_4DBB This draws attention to the fact that the Laplacian in the valence region of the iodine atom in bpe.F_4DIB is negative everywhere, with no local charge concentrations, quite different from that in the valence region of the bromine atom in bpe.F_4DBB. However, it is worth noting in this context that this sort of qualitative analysis is not necessarily straightforward, because the correspondence between atomic shell structure of the Laplacian of the ED, commonly observed for second and third row atoms, is not as well-defined for atoms such as Br and I [43, 44]. Apart from suggesting the presence of bcps for the contacts F2···H6, F2···H1 and F1···H3 in both complexes, the Laplacian maps in Fig. 16.2 reveal little else about the intermolecular interactions in these two systems.

Both experimental charge density studies make use of the potential energy density at the bcp, V_b, to estimate energies associated with specific intermolecular

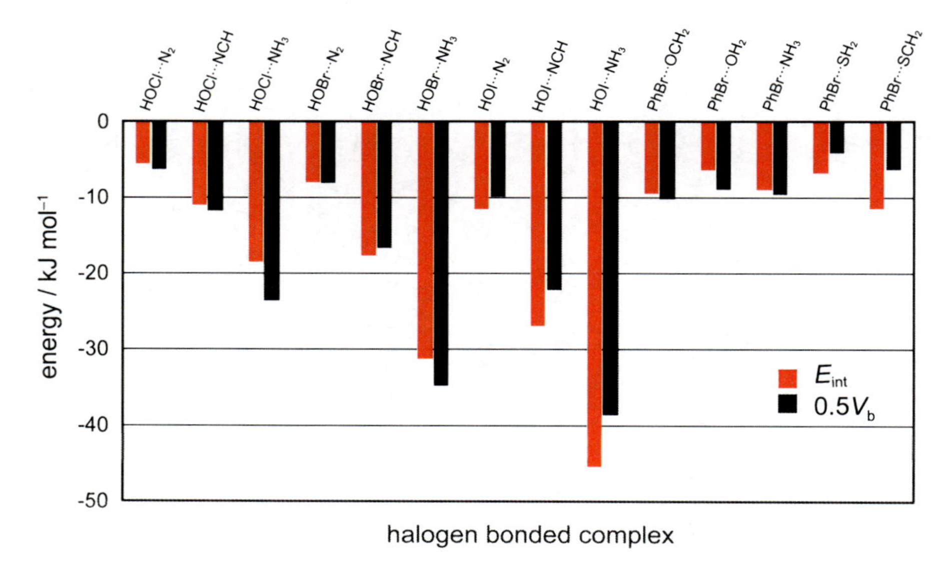

Fig. 16.3 Comparison between trends in *ab initio* interaction energy, E_{int}, and $0.5V_{\text{b}}$ for two series of halogen bonded complexes. Results for hypohalous acids are based on MP2 calculations with a high-level mixed basis set [49], and those for bromobenzene complexes are from MP2/aug-cc-pVDZ calculations [50]

interactions identified by bcps. This uses the relationship $E_{\text{int}} \approx 0.5V_{\text{b}}$, proposed by Espinosa, Molins and Lecomte [45] on the basis of topological analyses of experimental electron densities for a large number of X–H···O (X = O, N or C) hydrogen bonds (see 11.3.1). There is no reason to believe that this relationship will be universal and apply to a wide range of intermolecular interactions. In fact subsequent analysis of theoretical electron densities for a range of neutral hydrogen bonded dimers involving fluorine suggested $E_{\text{int}} \approx 0.42(2)V_{\text{b}}$, with a standard deviation of ~4 kJ mol^{-1} [46]. This relationship has been discussed in considerable detail by Gatti [47, 48], who provided several important caveats associated with its use that need to be borne in mind. To our knowledge, the relationship has not been tested for halogen bonded systems, and it is clear that any attempt to apply it to these systems to derive interaction energies from V_{b} would require calibration on a suitable set of halogen-bonded complexes. Nevertheless, there have been several recent ab initio QTAIM analyses of halogen bonding, for example those by Alkorta et al. [49] on complexes of hypohalous acids and nitrogen bases, and Lu et al. [50] on dimers involving bromobenzene, and bond critical point properties reported in those papers can be used to examine the applicability of the simple relationship $E_{\text{int}} \approx 0.5V_{\text{b}}$. Figure 16.3 compares these two quantities for 14 neutral complexes, and there is little doubt that both quantities exhibit the same basic trends. However, differences between them can be as large as 7 kJ mol^{-1} (for HOI···NH_3), and the rms deviation is 3.3 kJ mol^{-1}.

Table 16.1 Energies for molecular pairs (kJ mol^{-1}) in bpe.F_4DBB and bpe.F_4DIB

	PIXEL					QTAIM	
Molecular pair	E_{es}	E_{pol}	E_{dis}	E_{rep}	E_{tot}	$E_{int} \approx 0.5V_b$	Comment
bpe.F_4DBB							
aa/6.052	−10	−4	−33	19	−29	−11	2(Br···F2), F1···F1
aa/7.287	−8	−3	−29	20	−20	−7	2(Br···F1)
bb/6.052	−5	−1	−17	9	−15	−10	2(H3···H5), H6···H6
bb/7.287	−2	−1	−12	7	−8	−8	2(C2···C5)
ab/6.151	−5	−3	−25	14	−18	−5	C4··C9
ab/6.479	−6	−2	−14	10	−13	−13	H3···F1, H1···F2, H6···F2
ab/10.821	−35	−11	−15	51	−10	−28	Br···N
bpe.F_4DIB							
aa/6.267	−14	−3	−20	27	−10	−4	F2···F2; *no I···F2 bcp*
aa/9.448	−6	−2	−9	14	−3	–	*no I···F bcps*
bb/6.267	−7	−3	−19	10	−20	–	*no H···H bcps*
bb/9.448	−4	−2	−11	11	−5	−6	2(C4···C4)
ab/4.613	−20	−8	−40	54	−15	−11	2(C2··C8), C4···F2
ab/6.441	−10	−3	−16	14	−14	−16	H3···F1, H1···F2, H6···F2
ab/10.997	−77	−28	−24	109	−21	−37	I···N

The distance between centres of mass is indicated (in Å) and the label 'a' refers to molecules of F_4DBB or F_4DIB and 'b' refers to bpe molecules (see Figs. 16.4 and 16.5 for further details). PIXEL results are from Gavezzotti [51] and QTAIM results from Forni [38] and Bianchi et al. [36]

With this comparison in mind, we can estimate QTAIM interaction energies for pairs of adjacent molecules in bpe.F_4DBB and bpe.F_4DIB from experimental bcp properties, by summing over those bcps identified between the two molecules. Apart from the uncertainty inherent in using a factor of precisely 0.5 to relate E_{int} and V_b, this exercise implicitly assumes that for molecular pairs involving more than one intermolecular bcp the total interaction energy can be estimated by simply summing over the V_b values for individual interactions. The results are summarized in Table 16.1, accompanied by comments on the bcp interactions involved in the sum for each pair of molecules.

For comparison with the QTAIM estimates, Table 16.1 also includes the breakdown of energies reported by Gavezzotti in a recent application of the PIXEL method to halogen bonding in crystalline systems, including both bpe.F_4DBB and bpe.F_4DIB [51]. These energies are based on *ab initio* MP2/6-31 G** (MP2/DGDZVP for iodine) molecular electron densities, and we note that whereas the experimental results refer to crystal structures at 90 K (CSD refcodes IKUHUR and QIHCAL01), Gavezzotti's calculations were based on CSD refcodes IKUHUR03 (145 K) and QIHCAL (room temperature). Because of this some intermolecular separations differ significantly (e.g., the I···N distance is 2.780 Å at 90 K and 2.811 at RT) but, as emphasized by Gavezzotti, the distance dependence of the intermolecular energies for halogen bonded systems is relatively shallow. For consistency we label the pairs of molecules with the centre-of-mass distances used by Gavezzotti.

How then do the PIXEL total energies, E_{tot}, compare with QTAIM estimates of E_{int}? We see that for most molecular pairs E_{int} underestimates E_{tot}, especially for bpe.F_4DBB, but for both complexes the QTAIM estimate of the interaction energy for the halogen bonded pair exceeds the PIXEL estimate by ~17 kJ mol^{-1}. We suspect that there are two key reasons for these trends:

1. The QTAIM estimate depends on identifying all bcps between the molecules of interest, and failure to locate all bcps will lead to a systematic underestimate of E_{int}. Because the crystalline electron density between molecules is typically extremely flat, locating and characterizing all intermolecular bcps is non-trivial. As discussed by Gatti [48], the number and type of critical points in a crystal must satisfy the Morse equation, and failure to do so implies one or more critical points has been missed. Gatti also highlighted the case where two sets of cps for the urea crystal satisfy the Morse relation, but one did not include bcps and rcps associated with two unique N···N contacts. Neither of the charge density studies under consideration discusses the Morse equation, and Table 16.1 suggests that a number of bcps were either not reported or not identified in the study on bpe.F_4DIB. It is clear that identification of all cps in molecular crystals such as these will require the most careful cp search procedure possible.
2. Although Gavezzotti [51] suggested that the PIXEL value for the Br···N interaction may underestimate *ab initio* results by as much as 8 kJ mol^{-1}, the PIXEL energies for the X···N pairs in Table 16.1 are supported by independent computations. For example, Bianchi et al. reported a B3LYP interaction energy of 27 kJ mol^{-1} (uncorrected for BSSE) for the bpe.F_4DIB dimer, and Forni reported MP2/MIDI! and B3LYP/MIDI! interaction energies (corrected for BSSE) in the range 13–15 kJ mol^{-1} for the bpe.F_4DBB dimer, and 22 kJ mol^{-1} for the bpe.F_4DIB dimer. This suggests that the QTAIM results for the X···N interactions are systematically high, perhaps due to the inherent difficulties associated with multipole refinements involving heavy atoms such as these, coupled with the apparent need to model anharmonic thermal motion for these atoms.

Taken together, the comparison of QTAIM estimates with PIXEL energies might suggest that, with sufficient attention to the reduction of systematic errors that might compromise experimental charge densities, the QTAIM results based on V_b are capable of yielding meaningful experimental interaction energies. However, it is also clear that more careful studies and comparisons are required before this could become widely accepted and exploited. And a note of caution is also required: the energies estimated from values of V_b derived from multipole refined electron densities tend to differ little from those derived from an independent atom model (IAM), and they often agree within experimental error. For example, Bianchi et al. [36] also tabulated bcp properties from an IAM electron density for bpe.F_4DIB, and the resulting estimates of E_{int} based on this simple procrystal electron density differ by at most 2 kJ mol^{-1} from those listed in the QTAIM column in Table 16.1. This is less than the average error involved in using the approximation $E_{int} \approx 0.5V_b$, and it suggests more attention be paid to the question posed by us some time ago [52] and discussed in some detail by Gatti et al. [47], namely whether "experimental results differ from a simple reference model, that is the promolecular density, for weak interactions".

The final discussion in this section focuses on the mapping of the molecular ESP on Hirshfeld surfaces, the extent to which this conveys additional insight into the packing of molecules in these crystals, and their relationship with PIXEL results. For the homo- and heterodimers in these two complexes listed in Table 16.1, Figs. 16.4 and 16.5 show maps of the HF/MIDI! ESP mapped on Hirshfeld surfaces; in all cases the ESP is mapped over the range −0.025 au (red) through zero (white) to +0.025 au (blue). This mapping rapidly conveys the more obvious electrostatic nature of the two molecules. For bpe the hydrogen atoms along edges are electropositive while the remainder of the molecule is electronegative, substantially so in the vicinity of the pyridyl N atoms. In the case of F_4DBB and F_4DIB the fluorine atoms are electronegative, as is the region surrounding the C–Br and C–I bonds, while the regions above the phenyl ring and at the end of the C···X bonds are electropositive. The nature of the ESP around the C···X bonds is in complete agreement with numerous other reports, and is the direct result of the rearrangement of electron density upon bonding for these terminal halogen atoms (for example, see Price et al. [53], Saha et al. [54], Bujak et al. [55] and Bui et al. [39]). The greater electropositive (blue) area evident for F_4DIB, compared with F_4DBB, echoes other observations that the observed strength of halogen bonds Cl < Br < I is directly related to the increasingly electropositive region at the end of these terminal C–X bonds. This can actually be quantified, as the range of the ESP on the HS for F_4DBB is (−0.019, 0.105 au), compared with (−0.021, 0.158 au) for F_4DIB.

Detailed inspection of the pairs of molecular Hirshfeld surfaces mapped with the ESP in Figs. 16.4 and 16.5 shows that they can be readily correlated with the energies listed in Table 16.1. In a sense, these illustrations assist in rationalizing the energies for the various dimers, especially the breakdown into various components. To see this we elaborate on some of the more important observations, bearing in mind that only the electrostatic component of the PIXEL energies can be related directly to the ESP on the various surfaces:

1. For all molecular pairs in Figs. 16.4 and 16.5 there is evidence that the arrangement of the molecules relative to one another is such that complementary electrostatic regions (i.e. positive (blue) and negative (red)) are adjacent. This is quite obvious for the heterodimers (i.e. ab pairs), but careful inspection shows that it is also true for all aa and bb pairs, especially for the similar pairs aa/6.052 in Fig. 16.4 and aa/6.267 in Fig. 16.5. This electrostatic complementarity correlates nicely with the uniformly negative values of E_{es} in Table 16.1.
2. The X···N dimers (ab/10.821 and ab/10.997) involve close contact between complementary electrostatic regions on the two molecules, and in particular these contacts involve the most electronegative region on bpe molecules (the pyridyl N) and the most electropositive region on the tetrafluorohalobenzene molecules. This naturally leads to the largest electrostatic energy of all the molecular

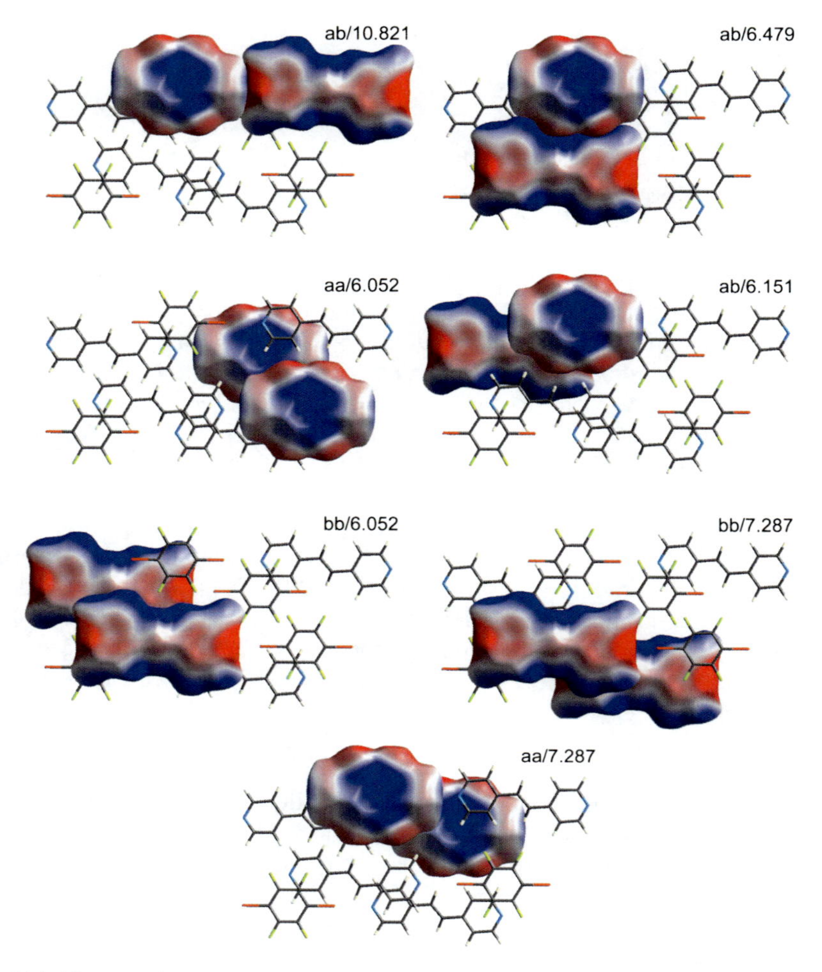

Fig. 16.4 Electrostatic potentials mapped on Hirshfeld surfaces of pairs of molecules for bpe·$C_6F_4Br_2$. The potential is mapped over the range −0.025 au (*red*) to +0.025 au (*blue*). The labels include the distance between centres of mass (in Å) and 'a' and 'b' refer to molecules F_4DBB and bpe, respectively

pairs, as evident in the PIXEL results. Moreover, the more positive electrostatic potential on the HS associated with the iodine atom (0.158 au compared with 0.105 au for Br) correlates with a much larger PIXEL electrostatic energy for this dimer.

3. The PIXEL polarization energy is typically only a few kJ mol^{-1}, except where two molecules are either very close together, as in the X···N interaction, or where there is substantial overlap of molecules with a strong electrostatic component, as in ab/4.613 in Fig. 16.5.

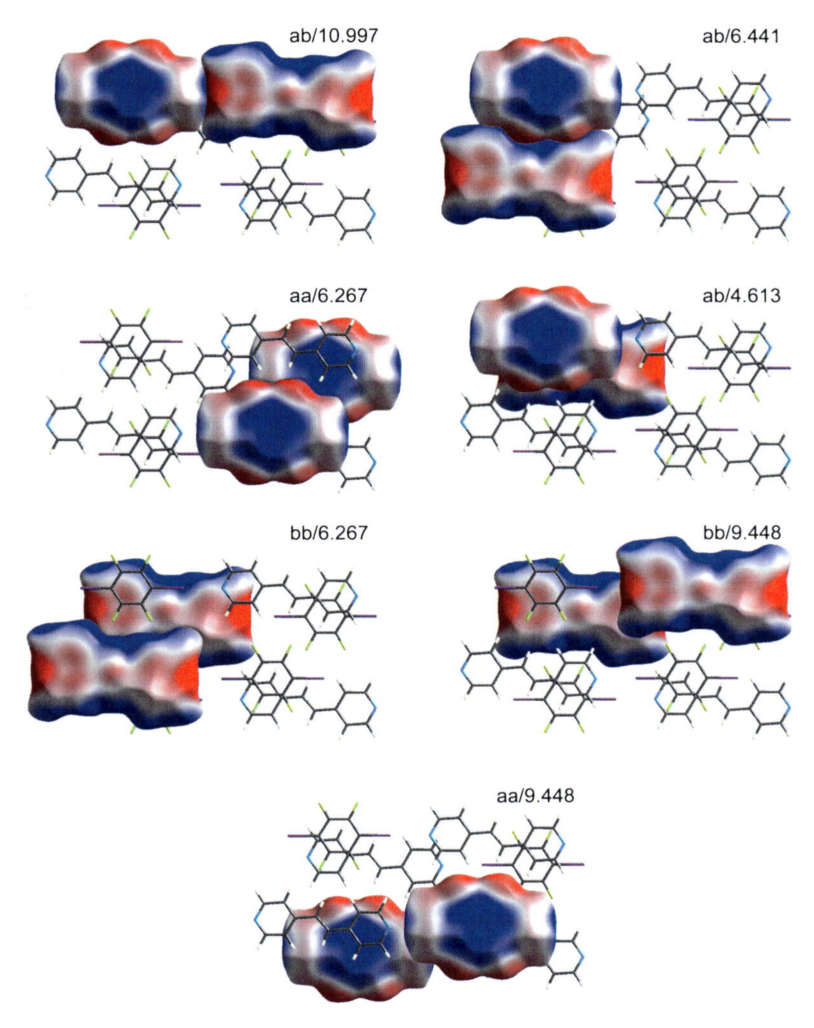

Fig. 16.5 Electrostatic potentials mapped on Hirshfeld surfaces of pairs of molecules for bpe·$C_6F_4I_2$. The potential is mapped over the range −0.025 au (*red*) to +0.025 au (*blue*). The labels include the distance between centres of mass (in Å) and 'a' and 'b' refer to molecules $C_6F_4I_2$ and bpe, respectively

4. The dispersion energy is always present, always attractive and always important, and quite clearly a function of the overlap between the two molecules. For example, the smallest value of E_{dis} for bpe.F_4DIB is −9 kJ mol^{-1} for the aa/9.448 pair, and this clearly involves the smallest overlap of molecules in Fig. 16.5. On the other hand, the largest E_{dis} for bpe.F_4DIB is −40 kJ mol^{-1} for the ab/4.613 pair, and this clearly involves the largest overlap of molecules in Fig. 16.5.

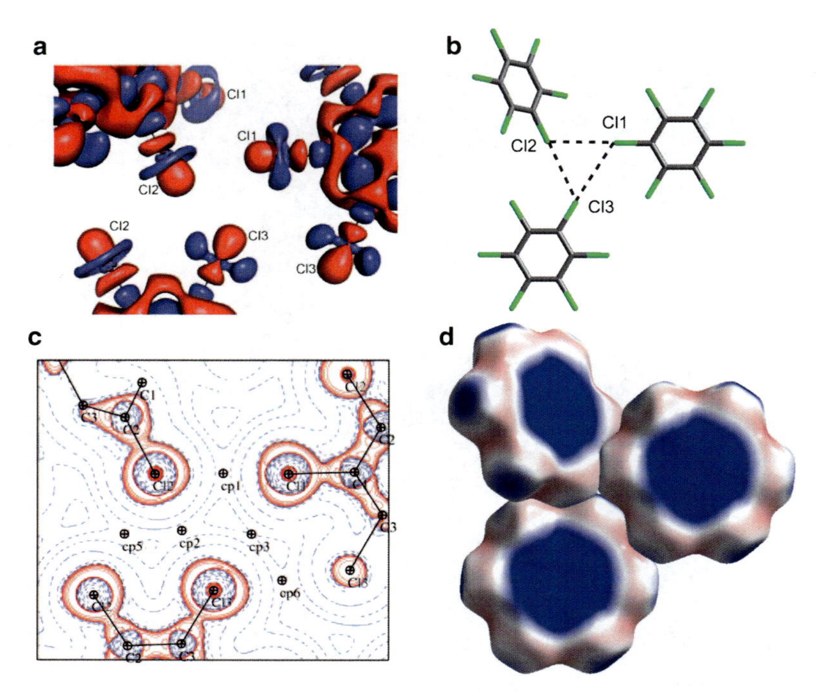

Fig. 16.6 Four different depictions of the halogen trimer synthon in C_6Cl_6: (**a**) experimental deformation electron density in the vicinity of the three C–Cl bonds involved (*red* represents a deficit of electron density relative to the promolecule, and *blue* an excess); (**b**) the three C_6Cl_6 molecules involved in the motif; (**c**) Laplacian and bcps associated with Cl···Cl contacts; (**d**) ESP mapped on Hirshfeld surfaces for the three molecules in (**b**). The ESP is mapped over the same range as in Figs. 16.4 and 16.5. Figs. 16.6a and c are reproduced from ref. [39] with kind permission of Wiley-VCH)

It would be extremely interesting to see if this overlap could be quantified by measuring the area of the touching Hirshfeld surfaces for particular contacts between pairs of molecules, perhaps as delineated by the curvedness function.

As a final example of charge density analyses applied to halogen bonding, in Fig. 16.6 we compare four different ways of depicting the halogen trimer motif in C_6Cl_6. Fig. 16.6a,c are reproduced from the experimental charge density analysis of Bui et al. [39], which made use of single crystal X-ray diffraction data measured at 100 K (data collected to 1.2 Å^{-1}, internal agreement factor $R_{\mathrm{int}} = 4.1\%$). The deformation electron density in Fig. 16.6a reveals the deformation characteristic of terminally bonded halogens, namely a deficit (red) in the C–Cl bond and beyond the Cl atom, and a surplus (blue) in a torus around the Cl atom. This rearrangement in turn gives rise to the characteristic features in the Laplacian map, Fig. 16.6c, which shows the "lock and key" arrangement of the local accumulations and deficits of electron density referred to previously (see Koritsanszky and Coppens [41]), but seen much more clearly in this case compared with either of bpe.F_4DBB or bpe.F_4DIB. Fig. 16.6b shows the way in which the three molecules involved in this motif are arranged relative to one another, and the molecular ESPs mapped on Hirshfeld surfaces are shown in Fig. 16.6d for the same arrangement of the three molecules. We have deliberately mapped the ESP in Fig. 16.6d over the same range as the maps in Figs. 16.4 and 16.5. In this way it becomes obvious that the range of ESP over the HS surface in C_6Cl_6 (-0.011 to $+0.042$ au) is considerably less than for either of F_4DBB or F_4DIB. The positive ESP on the HS at the end of the C–Cl bond is only 0.035 au, much less than noted above for the C–Br and C–I bonds (0.105 au and 0.158 au, respectively). This would be expected to result in a much smaller interaction energy for any of the pairs of molecules in Fig. 16.6. Published values of ρ_b (a maximum of 0.06 e Å^{-3}) and $\nabla^2\rho_b$ (maximum of 0.06 e Å^{-5}) in Bui et al. [39] suggest that E_{int} is at most $\sim$5 kJ mol^{-1}, in excellent agreement with typical PIXEL electrostatic energies for pairs of molecules obtained by Dunitz and Gavezzotti [56] in their analysis of 1,4-dichlorobenzene polymorphs.

16.4 Final Comments

The tools and approaches of charge density analysis will play an increasingly important role in addressing issues of concern in crystal engineering. This chapter has quite deliberately interpreted the term "charge density analysis" quite broadly, with the aim of demonstrating the wealth of complementary information that can be obtained from *ab initio* calculations (on molecules and the crystal) and X-ray diffraction measurements on single crystals, especially when the capabilities of modern graphics are exploited. Applications were discussed only for a small number of molecular crystals, with a view to seeing how present techniques are illuminating the topical issue of halogen bonding. We have not presented examples of intermolecular interaction energies calculated directly from the multipole refined electron density for these systems. That would have been ideal, and there are many examples in the literature where this has been reported, but not for the three crystalline systems that were the subject of the discussion. It needs to be recognized that these energies would necessarily refer to the electrostatic interaction between polarized molecules, and would actually be a mixture of E_{es} and E_{pol} in the PIXEL scheme (see the discussion in the Appendix of Suponitsky et al. [57] for further details of this subtle point). The QTAIM energies derived from V_b suggest it would be worthwhile to undertake a detailed study of the value of using QTAIM results such as this to infer interaction energies for a range of weak interactions including hydrogen bonds, especially if they were to be compared with PIXEL energies for the same pairs of molecules.

The HS maps in Figs. 16.4–16.6 provide further support for the assertion that electrostatic complementarity is an important driving force in the packing of molecules in crystals [32]. It is recognized that it is not possible for every pair of molecules to exhibit this pattern, but it is becoming increasingly clear that it is mandatory for those pairs with the largest interaction energies; further studies will be needed to confirm this. And it should be evident that using the ESP mapped on Hirshfeld surfaces greatly assists in the fuller appreciation of PIXEL energies, a point already emphasized by Wood et al. [58], who wrote in their study of the compression of the crystal structure of 3-aza-bicyclo(3.3.1)nonane-2,4-dione "This study has shown that the use of Hirshfeld surfaces and the PIXEL method together is a particularly effective combination for analyzing changes in crystal structures". We believe that the ability to compute both within the same software package would be extremely attractive.

References

1. Desiraju GR (ed) (1989) Crystal engineering: the design of organic solids. Elsevier, Amsterdam
2. Bader RFW (1990) Atoms in molecules: a quantum theory. Oxford University Press, Oxford
3. Matta CF, Boyd RJ (eds) (2007) The quantum theory of atoms in molecules. From solid state to DNA and drug design. Wiley-VCH, Weinheim
4. Gatti C (2007) Solid state applications of QTAIM and the source function: molecular crystals, surfaces, host-guest systems and molecular complexes. In: Matta CF, Boyd RJ (eds) The quantum theory of atoms in molecules. From solid state to DNA and drug design. Wiley-VCH, Weinheim, pp 165–206
5. Dunitz JD, Gavezzotti A (2005) Molecular recognition in organic crystals: directed intermolecular bonds or Nonlocalized bonding? Angew Chem Int Ed 44:1766–1787
6. Seddon KR (1999) Crystal engineering. A case study. In: Seddon KR, Zaworotko M (eds) Crystal engineering. The design and application of functional solids. Kluwer, Amsterdam, pp 1–28
7. Koritsánszky T, Mallinson P, Macchi P, Volkov A, Gatti C, Richter T, Farrugia L (2007) XD2006. A computer program package for multipole refinement, topological analysis of charge densities and evaluation of intermolecular energies from experimental or theoretical structure factors. http://xd.chem.buffalo.edu/
8. Dunitz JD, Gavezzotti A (2005) Toward a quantitative description of crystal packing in terms of molecular pairs: application to the hexamorphic crystal system, 5-methyl-2-[(2-nitrophenyl)amino]-3-thiophenecarbonitrile. Cryst Growth Des 5:2180–2189
9. Gavezzotti A (2005) Calculation of lattice energies of organic crystals: the PIXEL integration method in comparison with more traditional methods. Z Kristallogr 220:499–510
10. Novoa JJ, D'Oria E (2008) From bonds to packing: an energy-based crystal packing analysis for molecular crystals. In: Novoa JJ, Braga D, Addadi L (eds) Engineering of crystalline materials properties. State of the Art in modeling, design and applications. Springer, Dordrecht, pp 307–332
11. Day GM, Cooper TG, Cruz-Cabeza AJ, Hejczyk KE, Ammon HL, Boerrigter SXM, Tan JS, Della Valle RG, Venuti E, Jose J, Gadre SR, Desiraju GR, Thakur TS, van Eijck BP, Facelli JC, Bazterra VE, Ferraro MB, Hofmann DWM, Neumann M, Leusen F, Kendrick J, Price SL, Misquitta AJ, Karamertzanis PG, Welch GWA, Scheraga HA, Arnautova YA, Schmidt MU, van de Streek J, Wolf AK, Schweizer B (2009) Significant progress in predicting the crystal structures of small organic molecules – a report on the fourth blind test. Acta Crystallogr B 65:107–125
12. Neumann MA, Leusen FJJ, Kendrick J (2008) A major advance in crystal structure prediction. Angew Chem Int Ed 47:2427–2430
13. Gavezzotti A (2002) Calculation of intermolecular interaction energies by direct numerical integration over electron densities. 1. Electrostatic and polarization energies in molecular crystals. J Phys Chem B 106:4145–4154

14. Gavezzotti A (2003) Calculation of intermolecular interaction energies by direct numerical integration over electron densities. 2. An improved polarization model and the evaluation of dispersion and repulsion energies. J Phys Chem B 107:2344–2353
15. Gavezzotti A (2003) Towards a realistic model for the quantitative evaluation of intermolecular potentials and for the rationalization of organic crystal structures. Part I. Philosophy. CrystEngComm 5:429–438
16. Gavezzotti A (2003) Towards a realistic model for the quantitative evaluation of intermolecular potentials and for the rationalization of organic crystal structures. Part II. Crystal energy landscapes. CrystEngComm 5:439–446
17. Volkov A, King HF, Coppens P (2006) Dependence of the intermolecular electrostatic interaction energy on the level of theory and the basis set. J Chem Theory Comput 2:81–89
18. Berkovitch-Yellin Z, Leiserowitz L (1980) The role of coulomb forces in the crystal packing of amides. A study based on experimental electron densities. J Am Chem Soc 102:7677–7690
19. Berkovitch-Yellin Z, Leiserowitz L (1982) Atom-atom potential analysis of the packing characteristics of carboxylic acids. A study based on experimental electron density distributions. J Am Chem Soc 104:4052–4064
20. Spackman MA, Weber H-P, Craven BM (1988) Energies of molecular interactions from Bragg diffraction data. J Am Chem Soc 110:775–782
21. Volkov A, Koritsansky T, Coppens P (2004) Combination of the exact potential and multipole methods (EP/MM) for evaluation of intermolecular electrostatic interaction energies with pseudoatom representation of molecular densities. Chem Phys Lett 391:170–175
22. Barzaghi M (2006) PAMOC: properties of atoms in molecules and molecular crystals. http://www.istm.cnr.it/~barz/pamoc/
23. Stewart RF, Spackman MA, Flensburg C (2000) VALRAY – User's manual. 2.1 edn. Carnegie Mellon University & University of Copenhagen
24. Spackman MA (2006) The use of the promolecular charge density to approximate the penetration contribution to intermolecular electrostatic energies. Chem Phys Lett 418:158–162
25. Spackman MA, Byrom PG (1997) A novel definition of a molecule in a crystal. Chem Phys Lett 267:215–220
26. Whitten AE, Radford CJ, McKinnon JJ, Spackman MA (2006) Dipole and quadrupole moments of molecules in crystals: a novel approach based on integration over Hirshfeld surfaces. J Chem Phys 124:074106
27. McKinnon JJ, Spackman MA, Mitchell AS (2004) Novel tools for visualizing and exploring intermolecular interactions in molecular crystals. Acta Crystallogr B 60:627–668
28. Spackman MA, McKinnon JJ (2002) Fingerprinting intermolecular interactions in molecular crystals. CrystEngComm 4:378–392
29. McKinnon JJ, Jayatilaka D, Spackman MA (2007) Towards quantitative analysis of intermolecular interactions with Hirshfeld surfaces. Chem Commun 3814–3816
30. Spackman MA, Jayatilaka D (2009) Hirshfeld surface analysis. CrystEngComm 11:19–32
31. Wolff SK, Grimwood DJ, McKinnon JJ, Jayatilaka D, Spackman MA (2008) CrystalExplorer 2.1. http://hirshfeldsurface.net/CrystalExplorer
32. Spackman MA, McKinnon JJ, Jayatilaka D (2008) Electrostatic potentials mapped on Hirshfeld surfaces provide direct insight into intermolecular interactions in crystals. CrystEngComm 10:377–388
33. Metrangolo P, Meyer F, Pilati T, Resnati G, Terraneo G (2008) Halogen bonding in supramolecular chemistry. Angew Chem Int Ed 47:6114–6127
34. Metrangolo P, Neukirch H, Pilati T, Resnati G (2005) Halogen bonding based recognition processes: a world parallel to hydrogen bonding. Acc Chem Res 38:386–395
35. Metrangolo P, Resnati G, Pilati T, Biella S (2008) Halogen bonding in crystal engineering. Struct Bond 126:105–136
36. Bianchi R, Forni A, Pilati T (2003) The experimental electron density distribution in the complex of (E)-1,2-bis(4-pyridyl) ethylene with 1,4-diiodotetrafluorobenzene at 90 K. Chem Eur J 9:1631–1638

37. Bianchi R, Forni A, Pilati T (2004) Experimental electron density study of the supramolecular aggregation between 4,4′-dipyridyl-N, N′-dioxide and 1,4-diiodotetrafluorobenzene at 90 K. Acta Crystallogr B 60:559–568
38. Forni A (2009) Experimental and theoretical study of the Br···N halogen bond in complexes of 1, 4-dibromotetrafluorobenzene with dipyridyl derivatives. J Phys Chem A 113:3403–3412
39. Bui TTT, Dahaoui S, Lecomte C, Desiraju GR, Espinosa E (2009) The nature of halogen... halogen interactions: a model derived from experimental charge-density analysis. Angew Chem Int Ed 48:3838–3841
40. Bianchi R, Forni A (2005) VALTOPO: a program for the determination of atomic and molecular properties from experimental electron densities. J Appl Crystallogr 38:232–236
41. Koritsanszky TS, Coppens P (2001) Chemical applications of X-ray charge-density analysis. Chem Rev 101:1583–1627
42. Bader RFW, MacDougall PJ (1985) Toward a theory of chemical reactivity based on the charge density. J Am Chem Soc 107:6788–6795
43. Sagar RP, Ku ACT, Smith VH, Simas AM (1988) The Laplacian of the charge density and its relationship to the shell structure of atoms and ions. J Chem Phys 88:4367–4374
44. Shi Z, Boyd RJ (1988) The shell structure of atoms and the Laplacian of the charge density. J Chem Phys 88:4375–4377
45. Espinosa E, Molins D, Lecomte C (1998) Hydrogen bond strengths revealed by topological analyses of experimentally observed electron densities. Chem Phys Lett 285:170–173
46. Espinosa E, Alkorta I, Elguero J, Molins E (2002) From weak to strong interactions: a comprehensive analysis of the topological and energetic properties of the electron density distribution involving X-H center dot center dot center dot F-Y systems. J Chem Phys 117:5529–5542
47. Gatti C, May E, Destro R, Cargnoni F (2002) Fundamental properties and nature of CH··O interactions in crystals on the basis of experimental and theoretical charge densities. The case of 3,4-Bis(dimethylamino)-3-cyclobutene-1,2-dione (DMACB) crystal. J Phys Chem A 106:2707–2720
48. Gatti C (2005) Chemical bonding in crystals: new directions. Z Kristallogr 220:399–457
49. Alkorta I, Blanco F, Solimannejad M, Elguero J (2008) Competition of hydrogen bonds and halogen bonds in complexes of hypohalous acids with nitrogenated bases. J Phys Chem A 112:10856–10863
50. Lu YX, Zou JW, Wang YH, Jiang YJ, Yu QS (2007) Ab initio investigation of the complexes between bromobenzene and several electron donors: some insights into the magnitude and nature of halogen bonding interactions. J Phys Chem A 111:10781–10788
51. Gavezzotti A (2008) Non-conventional bonding between organic molecules. The 'halogen bond' in crystalline systems. Mol Phys 106:1473–1485
52. Spackman MA (1999) Hydrogen bond energetics from topological analysis of experimental electron densities: recognising the importance of the promolecule. Chem Phys Lett 301:425–429
53. Price SL, Stone AJ, Lucas J, Rowland RS, Thornley AE (1994) The nature of -Cl...Cl- intermolecular interactions. J Am Chem Soc 116:4910–4918
54. Saha BK, Nangia A, Jaskólski M (2005) Crystal engineering with hydrogen bonds and halogen bonds. CrystEngComm 7:355–358
55. Bujak M, Dziubek K, Katrusiak A (2007) Halogen... halogen interactions in pressure-frozen ortho- and meta-dichlorobenzene isomers. Acta Crystallogr B 63:124–131
56. Dunitz JD, Gavezzotti A (2002) Electrostatic energies in the 1,4-dichlorobenzene polymorph crystals: the role of charge density overlap effects in crystal packing analysis. Helv Chim Acta 85:3949–3964
57. Suponitsky KY, Tsirelson VG, Feil D (1999) Electron density based calculations of intermolecular energy: case of urea. Acta Crystallogr A 55:821–827
58. Wood PA, Haynes DA, Lennie AR, Motherwell WDS, Parsons S, Pidcock E, Warren JE (2008) The anisotropic compression of the crystal structure of 3-aza-bicyclo(3.3.1)nonane-2,4-dione to 7.1 GPa. Cryst Growth Des 8:549–558

Chapter 17
Electron Density Topology of Crystalline Solids at High Pressure

John S. Tse and Elena V. Boldyreva

17.1 Introduction

Electron density is a fundamental property of a system of interacting atoms. According to the Hohenberg-Kohn theorem the often used practical implementation of which is the Kohn-Sham DFT method, the ground state of a many-electron system is uniquely defined by a functional, which is the spatially dependent electron density. The exploitation of the topology of the electron density of a system helps to quantify the basic concept of chemical bonds. This theory, derived from principles of quantum mechanics, is known as the "Atoms-in-Molecules" theory (QTAIM) [1]. Over the last 20 years, QTAIM has been very successful for the elucidation of the structure, stability and reactivity of a wide variety of chemical systems [1–3]. A very powerful technique is the electron localization function (ELF) first introduced by Edgecombe and Becke [4] and further refined by others, notably by Silvi and collaborators [5]. The ELF is a convenient semi-quantitative index for the "measurement" of the covalency (thus electron localization at each point in space) between atom pairs [4, 6]. This principle is based on the pair probability function. It is reconciled that like-spin electrons occupy separate regions of space and the regions where an electron has a high probability of finding another electron of the same spin, are regions where the electrons are poorly localized, but where the probability is low the electrons are well localized. This method will be used extensively in the discussion. A comprehensive review on the theoretical

J.S. Tse (✉)
Department of Physics and Engineering Physics, University of Saskatchewan, Saskatoon, SK, Canada S7N 5E2
e-mail: John.Tse@usask.ca

E.V. Boldyreva
REC-008 Novosibirsk State University and Institute of Solid State Chemistry and Mechanochemistry SB RAS, ul. Kutateladze, 18, Novosibirsk 128, Russia
e-mail: eboldyreva@yahoo.com

C. Gatti and P. Macchi (eds.), *Modern Charge-Density Analysis*,
DOI 10.1007/978-90-481-3836-4_17, © Springer Science+Business Media B.V. 2012

background and the application of electron topological analysis to the elucidation of chemical bonding has been given in previous Chaps. (1 and 9) and will not be repeated here. This chapter is focused on the electron density topology of elemental, ionic and molecular systems under pressure. When atoms are compressed in the solid state, they are subjected to an additional confinement potential. The "chemical bonding" is very different from that at ambient conditions. In this review, no attempt was made to present all the systems which have been described in the literature. Instead, a number of selected examples are discussed in detail to illustrate the basic principles. An interested reader can find more examples in the original publications. Elementary, ionic, molecular solids have been studied at high pressures to a different extent, and therefore the number of examples, which could be selected to illustrate the effect of pressure on the electron density topology is not the same for different classes of compounds. The bonding changes along solid-state transitions are discussed also in Chap. 18.

The interaction between atoms in a crystalline solid is generally classified as covalent, ionic, metallic or van der Waals in nature. Typical examples of the respective classes are diamond, magnesium oxide, aluminum and solid benzene. In diamond, the carbon has a tetrahedral structure with sp^3 hybrid orbitals that interact with nearest neighbor atoms. In magnesium oxide, there is a charge separation between the magnesium cation and oxygen anion. The description of electron interaction in metals is still controversial. However, in general metals are regarded to possess itinerant electrons not localized on the atomic site. Finally, in molecular solids, such as benzene, the molecules are held together with weak van der Waals forces. A clear distinction on the classification of electron interactions, or bonding, however, is not as quite clear-cut when a solid is subjected to extreme external pressure. The success of the structural principles for solid materials is mainly built on our knowledge of the periodicity in the properties of atoms as revealed by the periodic table. The differences in the atomic size of the atoms, such as alkali and transition metal atoms, differentiate the chemical properties and, therefore, the structures of elemental solids and their compounds. Figure 17.1 shows the periodic trend of the atomic size of the elements. For example, the alkali elements with loosely bound electrons in the outmost *s* orbital have much larger atomic size as compare to the transition metal elements with much contracted (poorly screened) *d*-orbitals. When an atom is compressed, the electron density (electron per unit volume) increases and the differentiation between atoms becomes smaller (e.g. Fig. 17.1 at 100 GPa). Under extreme pressure (e.g. at 1,000 GPa or 1 TPa), the variation in electron density becomes less significant across the periodic table of the elements and the distinction between alkali and transition metals no longer exists. This observation is highly significant as it reveals that (i) the nature of "chemical" bonding will be very different under high pressure and consequently (ii) new crystal structures with novel properties will emerge. The anticipation of new chemistry and physics can be viewed in an energetic perspective. Figure 17.2 compares the (PV) work done on compressing a system to typical bond strengths. At low pressure <10 GPa, the compressive energy is comparable to a hydrogen bond. However, at pressure >100 GPa, the energy is large enough to break chemical bonds. Therefore,

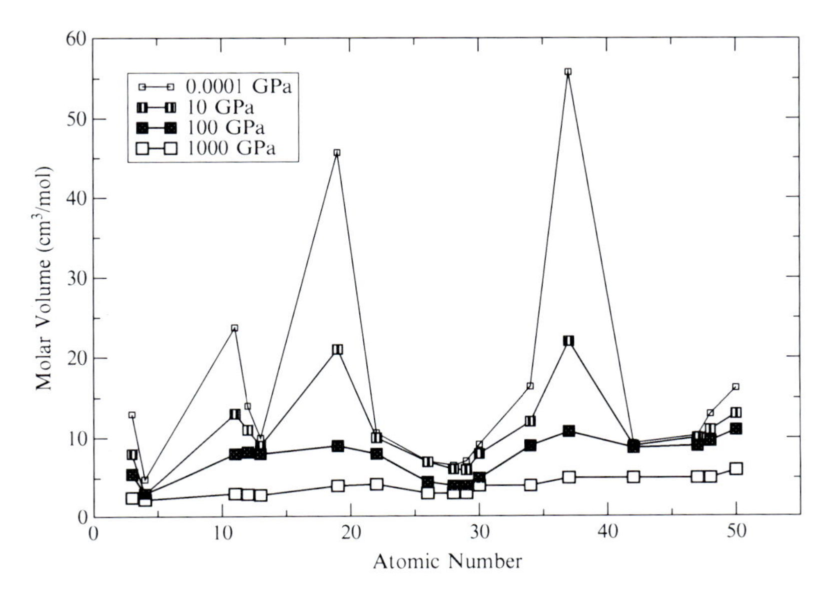

Fig. 17.1 Schematic depiction of the change of atomic size with applied pressure

Fig. 17.2 A comparison of estimated of PV work and typical bond energies

1 bar = 6.24×10^{-7} eV/Å^3
1 GPa = 6.24×10^{-3} eV/Å^3
100 GPa = 0.624 eV/Å^3
1000GPa = 6.24 eV/Å^3

E(H...O) = 0.1 – 0.5 eV/mol
E(O-H) = 5.1 3V/mol
E(N-H) = 4.7 eV/mol
E(C-H) = 4.3 eV/mol
E(Si-H) = 3.0 eV/mol
E(Sn-H) = 2.6 eV/mol

when a material is subject to extreme pressure, the mode of electron interaction is expected to change dramatically and the well established structural principles for chemical bonding will no longer be valid. New phenomena, sometimes exotic, are anticipated to occur [7]. For example, hydrogen, the simplest material is predicted to become a quantum and superconductive fluid at extreme pressure [8, 9]. In the past 10 years, advancements in synchrotron radiation diffraction with the diamond anvil cell technique have enabled the measurement of high quality diffraction patterns on powder samples, sometimes on single crystals, under high pressure and high temperature, to allow structural determination [10]. In hindsight, not unexpectedly, new and previously unobserved structures of simple elements were

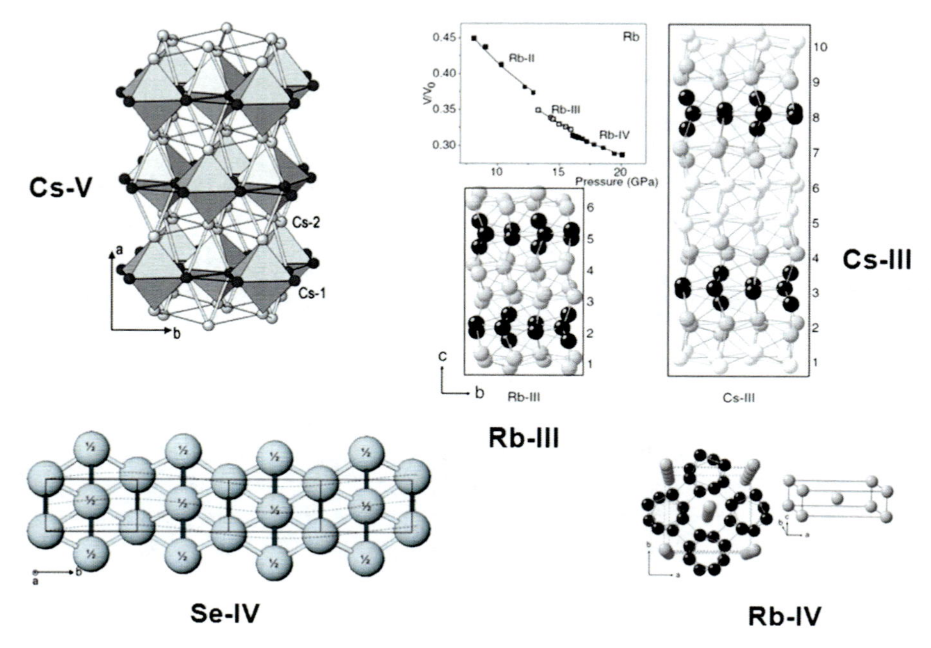

Fig. 17.3 A collection of *unusual* crystal structures for elemental solids illustrating the diversity of the structure types including open (Cs-V) [37], modulated (Cs-III [11b] and Rb-III [11b]), incommensurate (Se-IV) [14] and self-clathrated (Rb-IV) [12] structures (Figures are reproduced from references [11b, 12, 14, 37] with kind permission of The American Physical Society)

uncovered. Some of the unusual structures for elemental solids are collected in Fig. 17.3. These structures are complex including modulated (Cs-III [11], Rb-IV [12, 13]), incommensurate (Se-IV) [14], modulated incommensurate and even self-clathrate (Rb-IV) structural types [15]. Even group I alkali elements with a single valence electron, which are often regarded as "simple" metals, exhibit complicated pressure-induced transformation sequences and yield sophisticated structures. An interesting aspect is that instead of adopting more close-packed structures the observed structures consist of open framework and the bonding appears to be more directional. The well established structural principles may not be applicable to elucidate the nature of the bonding at high pressure [16]. It is highly desirable to examine the valence electron density distribution in these structures to gain insight into their structural stability.

17.2 Elemental Solids and Alloys

So how does the electron density distribution in a system change under pressure? As illustrated in Fig. 17.1, pressure has a dramatic effect on the atoms. At high pressure, as the electron density increases, so does the repulsion between valence electrons, either in the atom or between those bonding the neighbors, leading to orbital re-hybridization. As a result novel structural types with unusual "chemical bonds" are observed even in the elemental solids. Silicon (Si) presents a good example to illustrate the change in the charge density and the mode of chemical bonding [17]. At ambient pressure, Si-I has a diamond-type (*fcc*) structure. At 11 GPa, Si-I undergoes a strongly first order transition to a tetragonal structure (Si-II, *I41/amd*) [18] and then transforms to an orthorhombic structure (Si-XI, *Imma*) at 13 GPa [19]. Upon further compression, the structure changes successively to simple hexagonal, *sh*, (Si-V, *P6/mmm*) at 16 GPa [20], to the novel orthorhombic (Si-VI, *Cmca*) at 38 GPa [21], then to the hexagonal close-packed, *hcp*, (Si-VII, *P63/mmc*) at 42 GPa [20] and finally to a cubic close-packed structure (Si-X, *Fm3m*) at pressures higher than 79 GPa [22]. Figure 17.4 shows the evolution of the ELF [4] for the Si-I $\rightarrow$ Si-II $\rightarrow$ Si-XI phase transitions [17]. At zero pressure, the ELF has a very high value (maximum $\sim$0.92) and concentrates mostly on regions between the nearest Si atom pairs (Fig. 17.4). A high value of the ELF implies a small probability of finding a second electron with the same spin in its vicinity [4, 6, 23, 24]. Conversely, the probability of finding another electron with opposite spin from the reference electron is very high [6]. Therefore, the possibility of spin pairing leading to the formation of a chemical bond is also high. The topology of the ELF distribution indeed correctly reflects the presence of a tetrahedral (sp^3) covalent bond framework. The Si-I $\rightarrow$ Si-II transition at 11 GPa is accompanied by substantial "flattening" of the tetrahedral bonding framework along one of the cubic axes (Fig. 17.4). The crystal symmetry is distorted from cubic to tetragonal. The electron localization, however, is still very high, indicating that the strong four-coordinated directional covalent bonding network remains intact. An abrupt change in the ELF topology associated with the Si-II $\rightarrow$ Si-XI transition is observed (Fig. 17.4). The crystal structure changes from tetragonal to orthorhombic and continues the "flattening" of the Si 3-dimensional (3D) structure. The change in the nature of chemical bonding is clearly depicted by the ELF distribution. The "bonds" are now only found between Si atoms on the "flattened" planes – there is *no* electron localization between the planes! The "dimensionality" of the chemical bonding has reduced from a 3-D to a 2-D (sp^2) planar network. The Si-XI $\rightarrow$ Si-V transition pathway can be described as the displacement of Si (0,0,2z) (in the orthorhombic setting) in Si-XI along the z-axis from $z = 0.15$ to $z = ¼$ [18]. The change in ELF accompanying the transformation is remarkable (Fig. 17.4). The structural changes can be described as the consequence of the complete flattening of [102] planes in Si-XI. Accordingly, the ELF diminishes in these "inter-planar" regions, but there are no significant changes in the plane. This distortion weakens the interactions between the planes parallel to [102]. As a result, the Si chemical bond now becomes 1-D (*sp*) and runs along linear chains, parallel to the y-direction and lying on the [102] planes. An alternate view down an axis perpendicular to the [102] gives the *sh* structure of Si-V (Fig. 17.4). It is interesting to note that the usual way of presenting this structure is the stacking of parallel hexagonal planes which can be misleading in view of the nature of the chemical bonds.

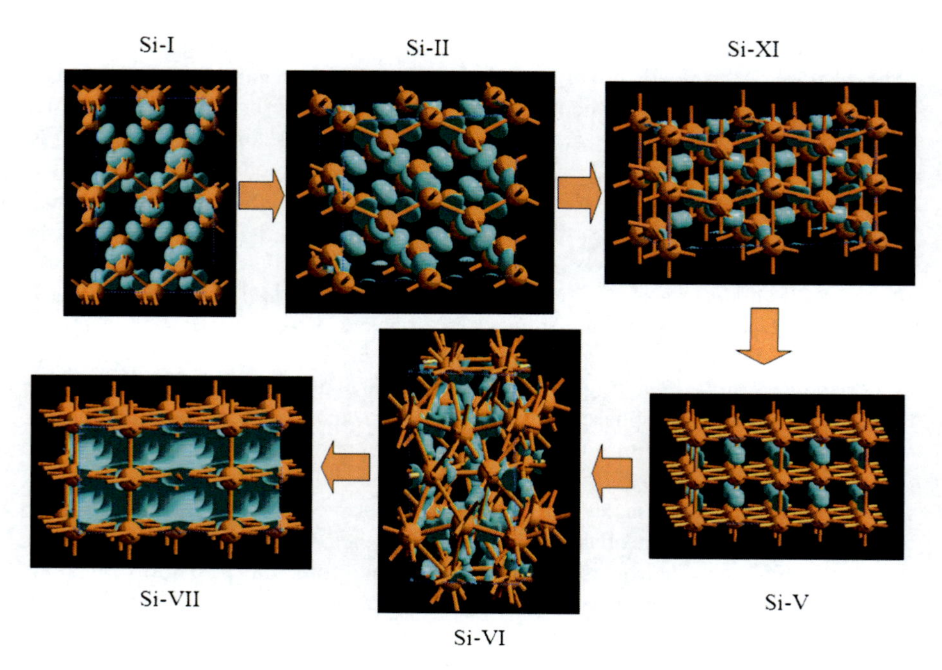

Fig. 17.4 Electron localization function (ELF) of different high pressure phases of silicon showing the chemical bonding change gradually from 3D (Si-I, II) → 2D (Si-XI) → 1D (Si-V) → 0D (Si-VI, VII). The ELF values for the iso-surface is 0.80 for Si-I, II, V, XI, and 0.7 for Si-VI and VII

The successive Si V → Si-VI → Si-VII transformations give perhaps the most interesting sequence. Inspection of the Si-V (*sh*) and Si-VII (*hcp*) structures would suggest a simplistic direct pathway by simple displacements of alternating hexagonal planes from ...AAAA... to ...ABAB... stacking. *Ab initio* phonon calculations predicted a soft phonon mode at 50 GPa corresponding to the shearing of the hexagonal plane. However, the direct transformation was not observed experimentally. Instead an intermediate phase with a complex but not close-packed structure (Si-VI *Cmca – oC16*) was identified [11]. Theoretical calculations confirmed that Si-VI is indeed a stable phase between Si-V and Si-VII [25]. The formation of Si-VI can be readily rationalized by comparing the changes in ELF for Si V → Si-VI (Fig. 17.4) and Si-V → Si-VII transformations. The direct transition from Si-V → Si-VII requires the disruption of the Si-Si bonds and therefore is energetically unfavorable. In contrast, although the Si-Si bonding is buckled and weakened (smaller ELF values), it is preserved in Si-VI (Fig. 17.4). The formation of Si-VI as an intermediate is possibly the lowest energy path for the distortion of the structure of Si-V under pressure. The very diffuse distribution and a low value (∼0.5 -free-electron like) of the ELF in Si-VII indicate that it is a simple metal and the "chemical bonding" may be regarded as 0-D. Further compression of Si-VII effectively forces the positive ions closer together and eventually leads to a cubic close-packed Si-X structure.

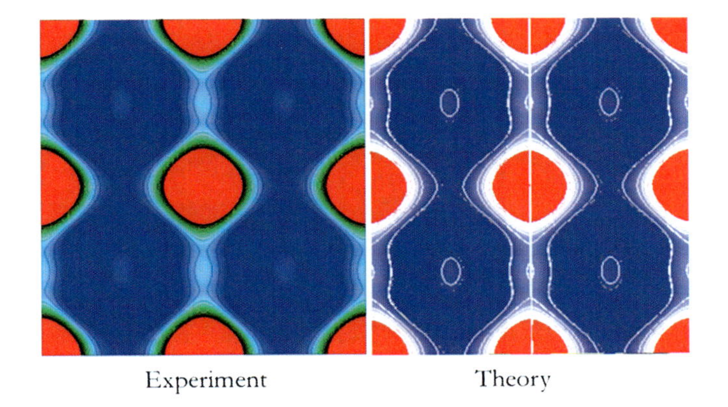

Fig. 17.5 Comparison of experimental and theoretical electron density maps of Si-V at 22 GPa. The light to dark colour shows increasing electron density

It is generally assumed that the coordination number increases with the denser structures. In the present case, the number of closest neighbors increases from 4 in Si-I to 6 in Si-II and Si-XI, 8 in Si-V and 12 in Si-VII. On the contrary, the ELF shows the number of chemical bonds reduced from 4 (tetrahedral) in Si-I to 2 (linear) in Si-V and 0 in Si-VII. Therefore, an increase in coordination number does not necessarily reflect stronger next-nearest neighbor interactions. The "bonding" topology changes with increasing pressure from 3-D to layered and eventually fully metalized. Accordingly, the number of valence electrons participated in the localized "chemical bonds" also decreases. Those "excluded" electrons then occupy spatially extended Si 3*d* orbitals and are responsible for the increasing metallic character. The electronic band structure of high-pressure Si can, therefore, be described as the simultaneous presence of itinerant electrons together with local pairs near the Fermi surface. As will be discussed below, these features have important consequences for the superconductivity. This description is consistent with calculated electronic band structures and with the Mulliken population analysis. The atomic orbital population in Si-XI is $s^{1.69}p^{2.6}d^{0.24}$, Si-V is $s^{1.64}p^{1.73}d^{0.61}$ Si-VI is $s^{1.60}p^{1.48}d^{0.82}$ (8*d*) and $s^{1.59}p^{1.52}d^{0.90}$ (8*f*) and Si-VII $s^{1.53}p^{1.36}d^{1.05}$. A gradual transfer of *p*-electrons into the *d*-orbitals in compressed Si is predicted. The predicted rehybrization in Si-V has been confirmed from electron density obtained from powder diffraction patterns using the method of maximum entropy (MEM) analysis (For an example of MEM analysis using powder diffraction pattern see Ref. [26]). The experimental and calculated charge density [27] of Si-V at 22 GPa are compared in Fig. 17.5. The charge density obtained from the powder diffraction pattern is in qualitative agreement with the theory. Most significantly, the predicted linear Si-Si linkages parallel to the crystallographic *c*-axes and the "chemical bond" between the Si atoms along the *longer* *c*-axes rather than in the *ab* plane with *shorter* Si-Si contacts are verified.

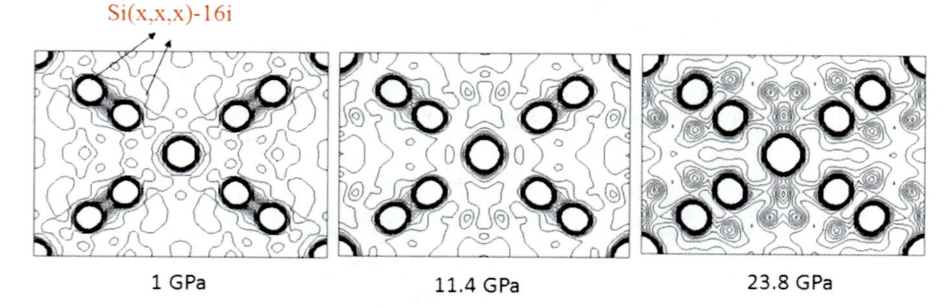

Fig. 17.6 Electron density maps derived from maximum entropy analysis of *in-situ* high pressure powder diffraction data of Ba_8Si_{46}. Only Si atoms in the [110] plane are shown [30]

The utilization of the spatially more extended *d*-orbital in Si containing compound at high pressure is a general phenomenon. The anomalous homothetic collapse in the volume in a Si clathrate doped with Ba (Ba_8Si_{46}) has been attributed to a similar effect [28, 29]. When compressed, the volume of Ba_8Si_{46} decreases until close to 16 GPa where a large volume reduction was observed. There is neither change in the space group of the new phase nor apparent discontinuity in the atomic positions. Using MEM analysis [26, 30], it is convincingly demonstrated that the collapse in the volume is the consequence of an abrupt re-hybridization of the Si atoms. Figure 17.6 shows the experimental electron density before and after the homothetic transition plotted in the [110] plane. It is clear from the figures that the electron density concentrated between two Si atom signifying a sp^3 Si-Si σ bond at 1 and 11.4 GPa is dispersed into the interstitial region and resembling the interactions of a pair of *d*-orbitals from each Si atom at 23.8 GPa. Charge density analysis shows unambiguously that the iso-structural phase transition observed at 16–17 GPa [9] is driven by a sudden change in the valence electron distribution as a result of *p-d* hybridization of the Si atoms.

An interesting example on the utilization of higher angular momentum (l) in the bonding at high pressure is highlighted by the formation of the K-Ag alloys under pressure [31]. The classical Miedema's rules [32, 33] state that alloy formation is favored between elements which have similar valence electron density and a large difference in electronegativity. Alkali metal K is, therefore, not expected to form alloys with noble and transition metals such as Ag. Nevertheless, two alloys with stoichiometries K_2Ag and K_3Ag were synthesized at high pressure. K_3Ag crystallizes in the BiF_3 structure type (space group $Fm3m$) with Ag forming a *fcc* sublattice and K occupying both octahedral and tetrahedral sites [34] (Fig. 17.7). A very unusual feature of this structure is the very short nearest neighbor K-K separation of 3.39 Å at 6.4 GPa. This distance is about 10% *shorter* than that observed in the *bcc* phase of elemental K at the same pressure. On the other hand, the first nearest neighbor distance of 5.54 Å between Ag atoms is exceptionally long. K_2Ag exists in the ω-phase structure type with the Ag atoms lying within the hexagonal planes [34]. The Ag-Ag distance in this plane is also 5.54 Å, identical to

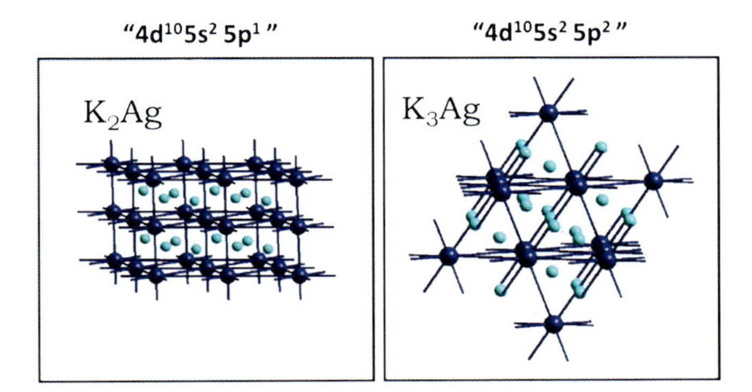

Fig. 17.7 Structural relationship between K_2Ag and K_3Ag viewed in a different perspective (Adapted from Ref. [31] with kind permission of The American Physical Society)

that found in the hexagonal planes of the *fcc* sublattice of K_3Ag. The planes of Ag atoms are stacked along the crystallographic c direction in a simple "...AAA..." sequence with an interlayer spacing of 3.605 Å at 6.1 GPa. In this structure, the K atoms are arranged in graphite-like sheets intercalated between the hexagonal Ag sheets. The K-K distance within these graphite-like sheets is 3.13 Å at 6.1 GPa, which is about 18% *shorter* than the closest interatomic distance in elemental K at the same pressure. Furthermore, the closest K-Ag distances are 3.39 Å and 3.62 Å for K_3Ag and K_2Ag, respectively, which are only slightly larger than the sum of the atomic radii of K and Ag of 3.30 Å at 6.4 GPa. On the basis of the structural information it might be conjectured that there are significant K-K and K-Ag bonding but the Ag-Ag interactions play only a minor role in the stabilization of the phases. This interpretation is quite illuminating in view of the fact that in most Zintl intermetallic phases, the interatomic distance between the electron-donating element is often shorter by 10% than in the elemental form [35]. If the K-Ag alloys are indeed Zintl type phases [36], this observation suggests that these structures are probably composed of K^+ ions and negatively charged Ag frameworks, *i.e.*, ionic closed shell-like K-K and Columbic K-Ag interactions but covalent Ag-Ag interactions. This description is in contradiction to the observed long Ag-Ag bonds.

To resolve this dichotomy, first-principles density functional (DFT) electronic calculations were performed [31] on K_2Ag and K_3Ag and corresponding hypothetical structures of the Ag sublattice with the K atoms removed, $\square_2Ag$ and $\square_3Ag$ ($\square$ denotes the site where a K^+ ion has been removed). The band structures of these compounds are compared in Fig. 17.8. In all cases, the Ag 4*d* orbitals form five completely filled flat bands and do not participate in the bonding. The next higher energy band in all the structures is predominantly Ag 4*s* in character and is completely occupied in the structure containing K. The highly dispersed and partially filled bands near the Fermi level of the K-Ag compounds are largely Ag 4*p* with only a minor K 4*s* component. There is no evidence supporting a K$s \rightarrow d$ hybridization in these compounds. The most striking feature common to the band

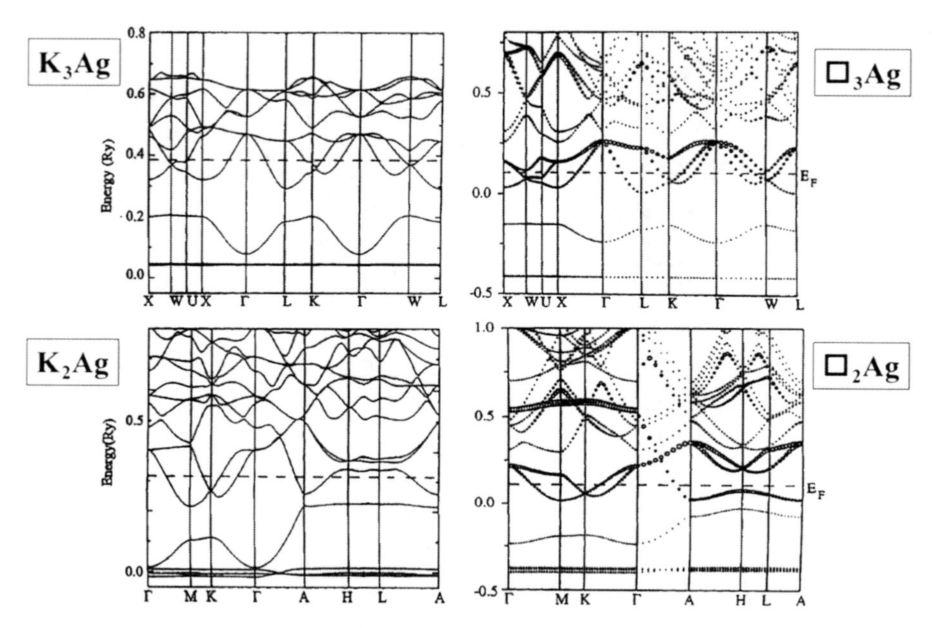

Fig. 17.8 Comparison of the electronic band structures of K_2Ag and K_3Ag and the corresponding hypothetical $\square_2Ag$ and $\square_3Ag$ with the K atoms removed from the respective structure [31]

dispersion diagrams for the corresponding structures of the K-Ag alloys with and without K is that the occupied bands are almost identical below the Fermi level both in terms of the number of bands and, *most importantly,* in the shape of the band dispersion. The similar pattern of band dispersions in the corresponding structures, K_2Ag and $\square_2Ag$, K_3Ag and $\square_3Ag$, indicates that the K atoms do not contribute appreciably to the occupied energy levels and have no significant effects on the bonding with and between the Ag atoms. This behavior of the electronic energy bands is consistent with Zintl's description [36] of alloy structures where in the K-Ag intermetallic phases there is almost complete electron transfer from K to Ag, completely filling the Ag 5*s* bands and partially occupying the Ag 5*p* bands. The K-Ag alloy structures can be described as a collection of K^+ situated within a negatively charged Ag framework with the effective electronic configuration for the Ag atom in K_2Ag as $4d^{10}5s^25p^1$ and $4d^{10}5s^25p^2$ in K_3Ag which is similar to the valence electron configuration for the Tl and Pb atoms, respectively.

A clear manifestation of orbital re-hybridization is observed in Cs-IV [37]. Cs-I has a *bcc* structure under ambient pressure [38]. It transforms to a *fcc* (Cs-II) structure at 2.3 GPa [39]. At 4.2 GPa, a second structural transformation occurred. The structure of this phase (Cs-III) was initially believed to be also *fcc* [40]. However, later experiment showed that it has a complex modulated structure [11]. The stability region of Cs-III [11] is very narrow and it transforms to a simple *bct* (body centered tetragonal) (Cs-IV) [41] structure at 4.3 GPa. Upon further compression to 10 GPa, Cs-IV transformed to an open orthorhombic *Cmca* structure

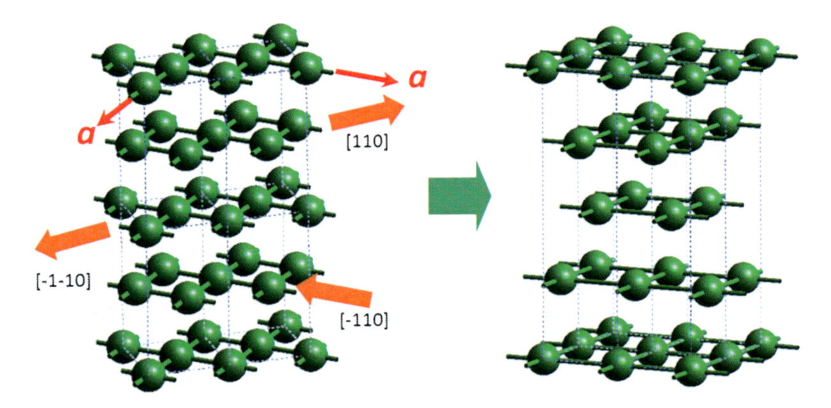

Fig. 17.9 The transformation from a *fcc* phase to the bct Cs-IV structure from the sliding of the Cs planes

[11, 37] and finally to a double hcp (dhcp) structure at 72 GPa [42]. The mechanisms for the Cs-II → Cs-III and the subsequent Cs III → Cs-IV transitions have been suggested resulting from 6*s* → 5*d* hybridization in Cs. The preference of occupying the 5*d* rather than 6*s* orbital in a confined potential in Cs has been predicted by Sternheimer [43]. It was shown from atomic calculations when a Cs atom is placed inside a potential well, such as at high pressure, the energetic order of the 6*s* and 5*d* orbital is reversed and the 6*s* and 5*d* may be hybridized. An earlier electronic calculation of the electron density of the Cs-IV structure revealed a novel sinusoidal pattern propagated between planes of Cs atoms [44]. The Cs-IV structure can be obtained from the precursor Cs-II *fcc* structure by translating consecutive 2D plans of Cs atoms in the [110] and [$\bar{1}$10] directions as shown in Fig. 17.9. To shed light on this sequence of transitions, band structure calculations were performed and the electron distribution in the high pressure phases was calculated [17]. Figure 17.10 shows the valence electron density for *fcc* Cs-II at two densities. It is obvious that the electrons are localized on the atom sites in Cs-II at low density but becoming more localized in the interstitial sites at high density. In an orbital description, the redistribution of valence electrons may be described as an electron transfer from the compact Cs 6s to the more diffuse Cs 5d. Enhanced overlaps between 5d orbitals of the diagonally positioned Cs atoms contribute to the stability of the crystal structure.

There are several important consequences to this description of orbital interactions which can be generalized for the examination of the structural motifs of high pressure structures. First, we observed that the "covalent bond" is not between nearest neighbors but rather between the next-nearest neighbor Cs atoms. Therefore, the concept of relating the number of short contacts to stability is not appropriate for the understanding of the structure of high-pressure structures. Second, the "bonding electrons" are localized on a 2D plane forming a square net following the $s \rightarrow d$ transformation [43]. In fact, the structure of Cs-IV can be described as ordered stacking of these planes with different orientations Fig. 17.9. The wave-like valence

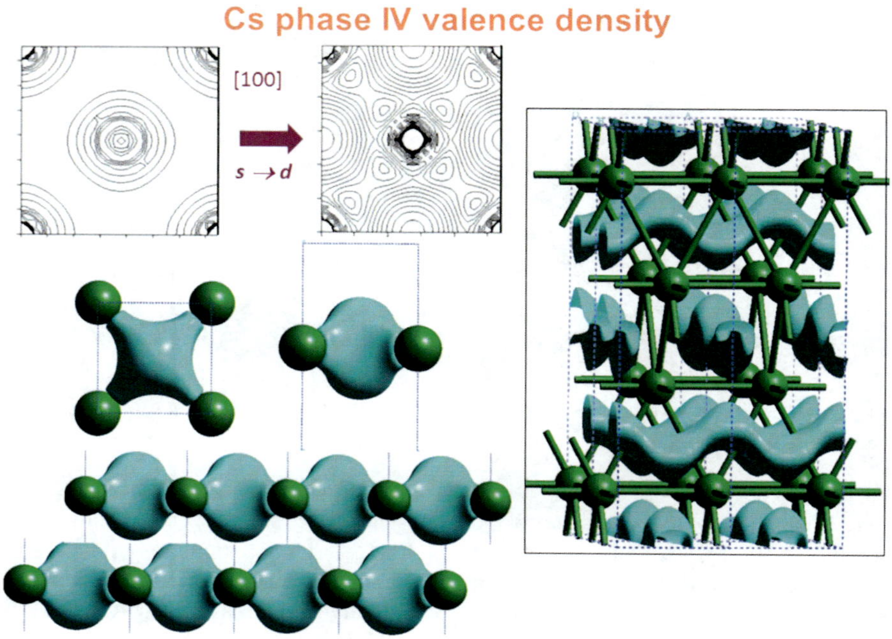

Fig. 17.10 Valence electron distribution in *fcc* Cs at high pressure showing the $s \rightarrow d$ transition and a schematic illustration of wave-like charge density predicted for Cs-IV

charge density in Cs-IV predicted in a band structure calculation [44] (Fig. 17.10) can be also easily explained by periodic displacements of these 2D square net (planes) in the $[\bar{1}10]$, $[\bar{1}\bar{1}0]$ and [110] directions perpendicular to the tetragonal c-axis from the precursor *fcc* Cs-II structure. The reason for the transformation from the *fcc* Cs-II to the complex and modulated Cs-III structure is not well established. It is believed that it is either driven by the $s \rightarrow d$ transition, or the system is very close to the electronic instability. ELF has the advantage of revealing fine details in the electron localization that may be obscured in the charge density. The crystal structure of Cs-III can be viewed alternatively as "... ABAB ..." stacking of two distinct *ac* planes at (y, ¼, 0.126) (1.68 Å) and (y, ¼, 0.375) (5 Å). ELF contours for the two planes are shown in Fig. 17.11. Both planes can be loosely described as originating from parallel square ladders in which the adjunct ladders are "sheared" against each other. An interesting pattern of electron localization emerges from the ELF topology of these planes. Although, electrons are found to be mostly localized on the atomic sites, however, at the center of the "squares" there is an unmistakable indication of weak electron localization. This unique feature is clearly absent in the region between the sheared square ladders. The "square-ladders" are then expected to be stabilized as a result of the small *d* overlaps. The 2D pattern of atom arrangement in the Cs-III is well known in Archimedean tiling [45]. This plane

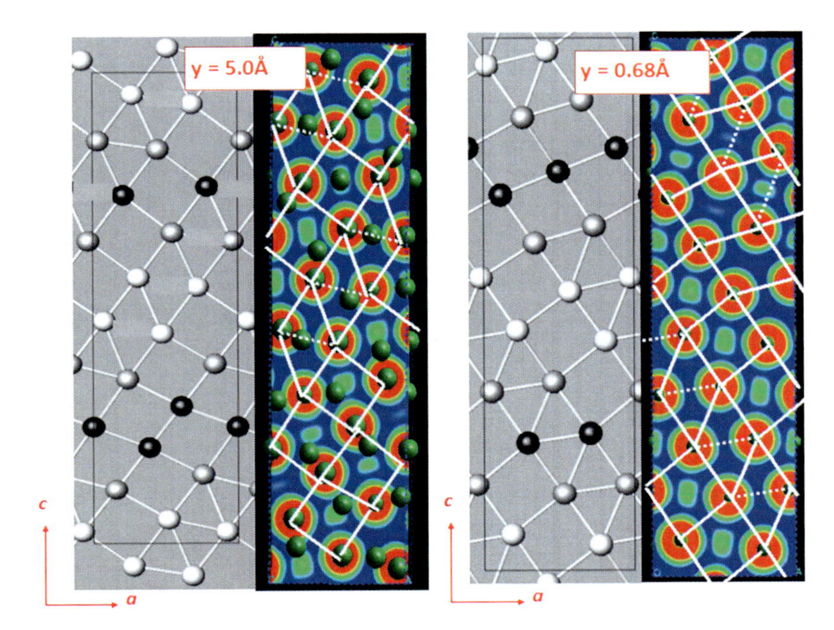

Fig. 17.11 Electron localization function plots of the Cs-III phase at two different Cs planes. The plot in *black grey* is the experimental structure. The *white dotted* line in the ELF contour plot showing the corresponding "triangular" arrangement of the Cs atoms corresponding to the experiment

is characterized by the Delaney symbol ($3^3 4^2$). This pattern has the next highest packing density for a 2D solid (Fig. 17.12) after the simple planar 2D square net.

Figure 17.13 shows the difference of the electron density of the atoms and the Cs-V structure. The charge distribution, to no surprise, is found to be concentrated between the two distinct planes of Cs atoms. Interestingly, the electron density shows that Cs atoms in the square 2D plane [40] have lost most of their valence electrons. Therefore, the Cs atoms in the Cs-V structure may be considered as self-disproportionate with valence electrons removed from the 2D square layer to the puckered layers.

Taken in all, Cs-III may be described as a transition metal with incomplete $6s \rightarrow 5d$ electron transfer. Due to the localization of electrons in the planes, the isotropic 3D metallic bonding in the precursor Cs-II became anisotropic and directional in Cs-III. The observed structure corresponds to the most efficient packing of atoms for a 2D solid. The description of the structure is consistent with the fact that Cs-III is the intermediate phase between Cs-II and Cs-IV. Following this interpretation of the structure, it can be easily shown that the atom arrangement in Cs-III can be derived directly from the distortion of a 2D square net of the *fcc* Cs-II structure. In hindsight, the intricate structures displayed in the high-pressure polymorphs of Cs (from Cs-III to Cs-V) are no more than the manifestation of

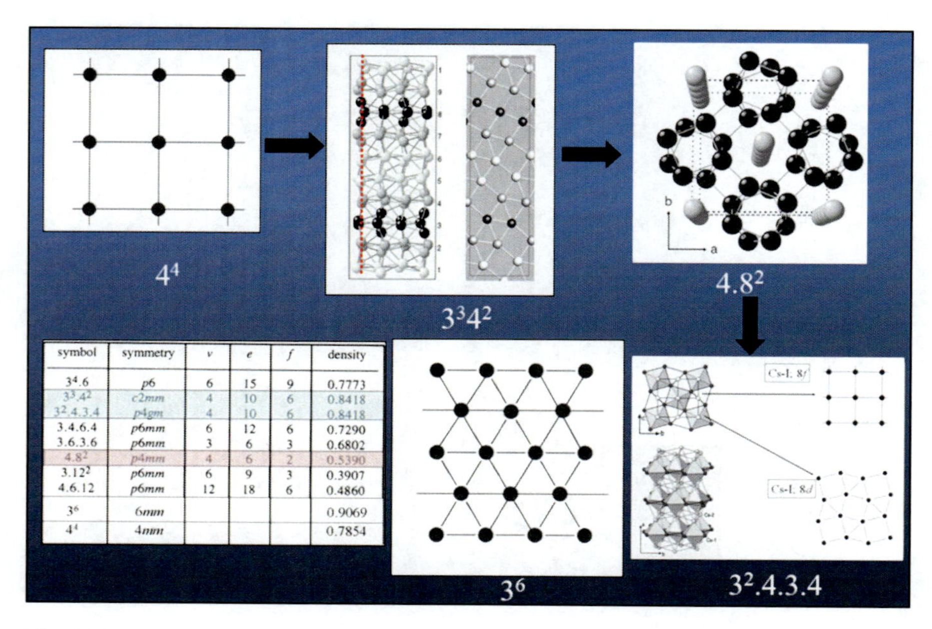

symbol	symmetry	v	e	f	density
$3^4.6$	*p*6	6	15	9	0.7773
$3^3.4^2$	*c*2*mm*	4	10	6	0.8418
$3^2.4.3.4$	*p*4*gm*	4	10	6	0.8418
3.4.6.4	*p*6*mm*	6	12	6	0.7290
3.6.3.6	*p*6*mm*	3	6	3	0.6802
4.8^2	*p*4*mm*	4	6	2	0.5390
3.12^2	*p*6*mm*	6	9	3	0.3907
4.6.12	*p*6*mm*	12	18	6	0.4860
3^6	6*mm*				0.9069
4^4	4*mm*				0.7854

Fig. 17.12 2D-Archimedean tiling of planar arrays of atoms with progressively higher packing density (see the inserted table) (Figures are reproduced from references [11b, 37] with kind permission of The American Physical Society)

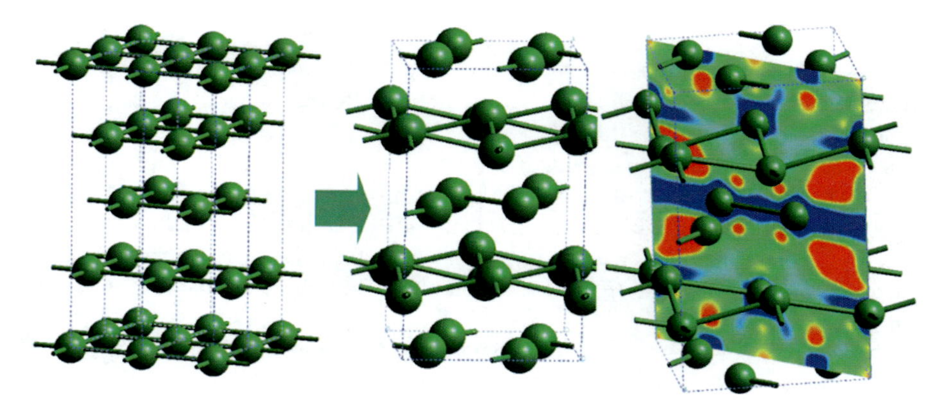

Fig. 17.13 Schematic representation of the transformation of Cs-IV to Cs-V structure and the electron density difference map of Cs-V relative to the Cs atoms

progressively higher density packing in 2-D as these structures follow the same pattern as the Archimedean tiling (Fig. 17.12) (or packing) of hard disks.

Among the alkali metals, Li has received special attention. With only a 2s valence electron, it was regarded as the simplest of all alkali metals. In reality, the structural changes and properties at high pressure are far more complex than expected. The ambient structure of Li is rather complex. Under ambient conditions, Li crystallizes in the body centered-cubic (*bcc*) structure [46]. Upon cooling, *bcc* Li goes through a martensitic transition to a rhombohedral 9*R* Li (*R*-3*c*) at around 75 K [47–49]. Upon compression, *bcc* Li transforms successively from *bcc* → face centered cubic (*fcc*) → *hR*1 (*R*-3*c*) → *cI*16 (*I*-43*d*) [50]. Surprisingly, the *fcc* and *cI*16 are found to be superconducting with a maximum critical temperature close to 17 K [51–53]. The observation of superconductivity is novel as it violates the traditional Matthias rule of superconductivity [54] that suggests main group elements with valence 1 should not be a superconductor. Moreover, it was found very recently that Li undergoes two successive structural transitions at 68 and 79 GPa [55]. Superconductivity was found to vanish at 68 GPa and instead, the two high-pressure phases exhibit semi-conducting properties indicating the opening of the band gap!

The near degeneracy of the 2*s* and 2*p* of Li suggests that hybridization of the two valence orbitals is likely. Theoretical calculations on the ELF of *bcc* bulk Li has revealed attractors localized (maximum ELF = 0.68) in the center of the octahedral cavities formed by the six nearest lithium atoms and coinciding with the maxima of the valence electron density [24, 56]. The electron localization becomes more significant in the *fcc* Li crystal. In addition to the presence of attractors in the octahedral sites, smaller attractor basins with ELF = 0.64 are located in the tetrahedral holes. This electron topology is indicative of multicenter bonding in Li crystals. A schematic representation of the frontier crystal orbitals of the *fcc* structure viewed down the (001) axis is shown in Fig. 17.14. Examination of the wave functions showed the presence of both *s*- and *p*- type characters. The *p*-π bonding interactions between Li atoms result in maxima of electron density above and below the Li-Li connecting lines located in the octahedral cavities (bound by the red square shown in Fig. 17.14).

The *fcc* Li transforms to a cubic *cI*16 structure via a transient hexagonal phase (*h*R1) at 38.9 GPa [51]. The transformation from a close-packed *fcc* structure to a more open *bcc* structure is surprising. The structural transition is driven by the softening of the transverse acoustic lattice vibration at the X symmetry point which corresponds to shearing between the hexagonal planes of Li in the *fcc* packing [57]. Thus, the structural transformation brings a 3D to a 2D atomic packing, as in the case of Si and Cs. The calculated density of states at the Fermi level shows that the contribution from Li 2*p* is higher than the 2*s* orbitals. A contour plot of the ELF of *cI*16 Li in the (110) plane is displayed in Fig. 17.15. The ELF reveals that electron localization at the interstitial sites becomes very prominent. Moreover, the "sheared ladders" packing of the Li, first observed in Cs-III (compare to Fig. 17.11), is evident. The lack of electron localization between adjacent rows of 1D Li square ladders leads to weaker interactions and is responsible for the "sheared" pattern.

Compared to other elements, the energy separation between the core 1s and valence 2s of 50 eV is relatively small [58]. Under pressure, the valence bandwidth is expected to increase. Therefore at very high packing density, the core 1s electrons potential might influence the electronic properties of Li. It was suggested that at pressures close to 100 GPa, dense Li may undergo a structural transition, possibly leading to a "paired-atom" phase with low symmetry and near-insulating

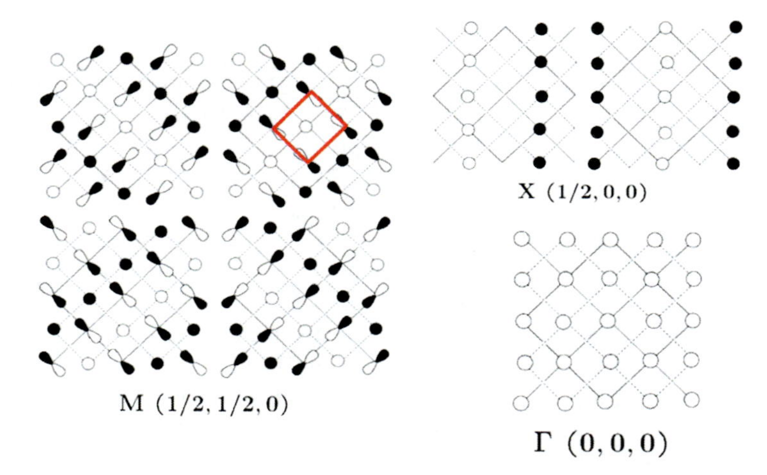

Fig. 17.14 Crystal orbital diagrams for bcc Li at Γ, X and M symmetry points. The red square highlights the Li-Li p-π bonding in the octahedral site (see text) (Reproduced from Ref. [56] with kind permission of John Wiley and Sons)

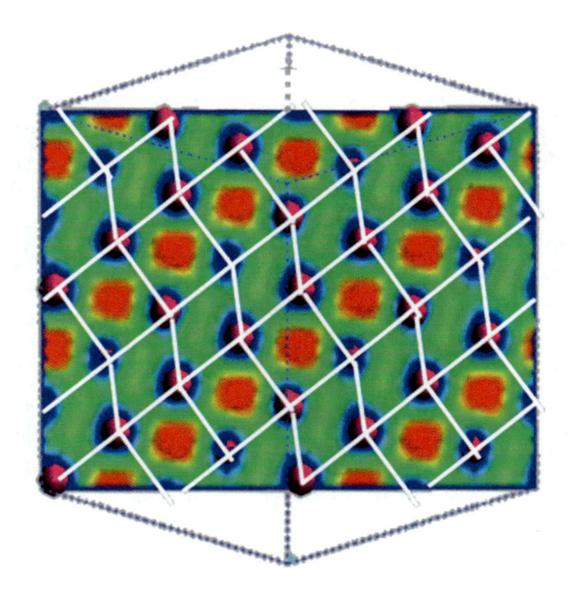

Fig. 17.15 Electron localization function of the (110) plane of the *cI*16 phase of Li

properties [59]. The predicted "paired-atom" structure features a very close contact between Li ion cores with electron localized in the interstitial region. Recent electrical resistance measurements [55] for Li up to 105 GPa revealed a significant increase in electrical resistivity and a change in its temperature dependence near 70–80 GPa. The experiment thus provided unambiguous evidence for a pressure-induced metal-to-semiconductor transition in Li. The structure of the corresponding

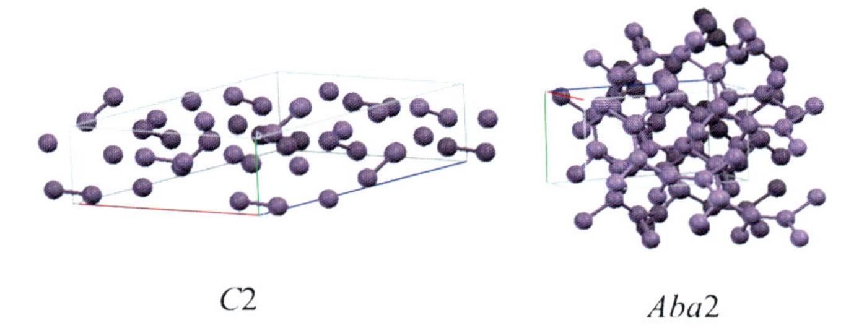

Fig. 17.16 Predicted insulating *C*2 and *Aba*2 high pressure phases of Li

Fig. 17.17 3D contour surface of the electron localization function (ELF value = 0.85) for the *C*2 and *Aba*2 phases of Li [60]

insulator phase(s) has (have) not been determined. The experimental observation was corroborated by theoretical calculations [60]. Using random search [61] and evolutionary algorithm [62, 63] methods with first-principles electronic structure calculations it was predicted that Li will transform from the metallic cubic *cI*16 phase to an insulating monoclinic *C*2 structure at 74 GPa [60] (Fig. 17.16). The *C*2 structure is the most stable phase up to 91 GPa, where it transforms to a second insulating orthorhombic *Aba*2 structure (Fig. 17.16) [60]. A common structural feature in *C*2 and *Aba*2 is the existence of alternating Li-Li bonds. Electronic band structures showed that an energy gap of *ca.* 0.3 eV is opened up in both structures [60]. The band gaps were found to increase with pressure, and get maximized at 85 GPa and 90 GPa for *C*2 and *Aba*2, respectively. The calculations show that the density of electronic states near the top of the valence band is dominated by Li 2*p* orbitals. The ELF iso-surfaces for the corresponding structure plotted at a value of 0.85 are shown in Fig. 17.17. A high ELF value indicates that electrons in both structures are strongly localized in the interstitial region. This is the combined result of dominant Li 2*p* contribution and a change in the mode of *p*-*p* interactions (*vide supra*).

To understand the origin of the anisotropic local structures of Li at high pressure, the electronic wave functions and their dependence upon Li–Li near-neighbor distance was examined with a hypothetical linear Li_4 chain model [64]. The results are summarized in Fig. 17.18. The total energy of the model system as a function of the Li–Li distance, $d_{Li\text{-}Li}$, is plotted in Fig. 17.18. At Li–Li distances >1.6 Å, the most stable structure is when the valence electrons occupy the *sp*-σ orbitals along the molecular axis. However, when Li–Li contact is shorter than 1.6 Å, the preferred ground state is with the valence electrons occupying the *p*π orbitals. Figure 17.18 also shows the contour plots of the total valence charge density at two Li-Li distances. At $d_{Li\text{-}Li} = 1.6$ Å, there is a discontinuity in the energy curve resulting from an abrupt change in the character of the chemical bonds. At longer distances, the bonding is predominantly of *sp*-σ type. The bonding changes to *p*π-type at shorter Li contacts. This change leads to density maxima no longer located directly between the atoms but rather situated above and below the "bond axis", that is, dispersed into the interstitial region. This qualitative description is in a complete agreement with the experimental observed changes in the nearest Li-Li contact distances (Fig. 17.18). The calculated Li-Li distance in the *cI*16 structure of 1.81 Å at 75 GPa shortened abruptly to 1.60 Å in the *C*2 structure at 78 GPa. The Li-Li contacts continue to become shorter on increasing pressure. The calculated nearest Li contact in the *Aba*2 structure at 95 GPa is 1.57 Å. It is also important to note that in the *bcc*, *fcc* and cI16 phase there is only one unique Li atom, and, hence, only one unique Li-Li distance in the crystal structure. The crystal symmetry is lowered in the high-pressure semi-conductor *C*2 and *Aba*2 phases with spiral Li-Li chains with alternating Li-Li distances. The presence of bond length alternation is a sign of Peierls distortion suggested in Ref. 57. It is not unreasonable to speculate that the opening of a band gap in the high-pressure phase is the consequence of the breaking of the degeneracy of Li electronic bands in the high-symmetry structure. This qualitative discussion accounts for both the prevalence of interstitial density maxima as well as the $s \rightarrow p$ rehybridization exhibited in elemental Li at high pressures. The experimentally observed and theoretically predicted semi-conducting phases of Li are in general agreement with the suggestion of an insulating symmetry breaking phase at high pressure. However, the predicted structures differ from the orthorhombic *Cmca* (oC8,α-Ga) suggested earlier [56, 59].

When the nearest neighbor Li-Li distance decreases further, the core 1*s* wave function is expected to overlap. At very high pressure, the core potential may affect the valence band significantly. A stable *fcc* structure ($Fd\bar{3}m$) of Li at 400 GPa has been predicted with the closest Li-Li contact of 1.30 Å [65]. Not unexpectedly, the ELF of this structure shows that valence electrons are almost completely localized in the octahedral cavities (Fig. 17.19). The re-emergence of a 3D closed-pack structure at very high density is not too surprising. At extreme pressure, pairing of electrons between the confined space of two Li atoms becomes highly unfavorable and the electrons prefer to be localized and paired in the open interstitial sites. This reduced the screening of the core charges between Li atoms. Eventually, the energetically favorable arrangement of the ions is a close-packed structure. The system is stabilized by the *PV* work and Coulomb interactions between localized

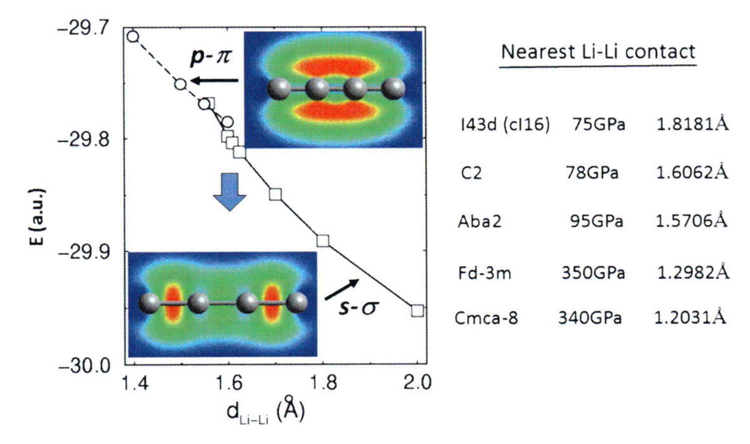

Fig. 17.18 Total energy as a function of Li ... Li distance for a hypothetical linear Li_4 molecule. Schematic representation of wavefunctions; note that the ground-state wavefunction for distances shorter than 1.5 Å consists entirely of Li–Li p-π interactions (Reproduced from Ref. [64] with kind permission of John Wiley and Sons)

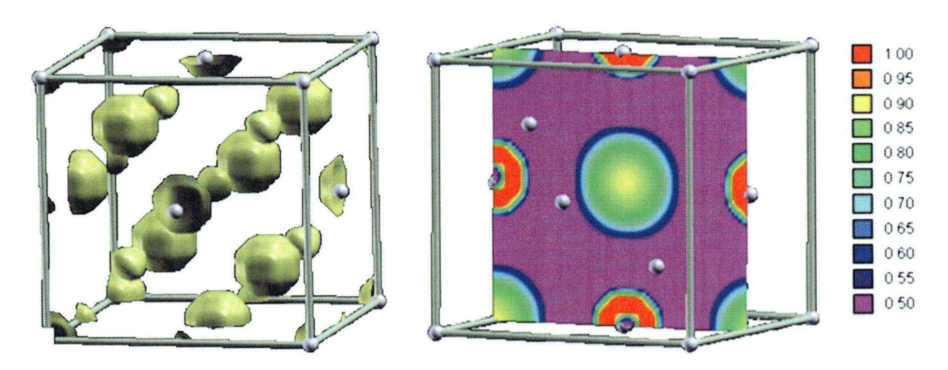

Fig. 17.19 The electron localization of the predicted high pressure *fcc* phase of Li at 380 GPa. Note the strong localization of electron pairs in the interstitial regions

electrons with the positively charged framework. This is a general phenomenon, and it explains why *all* known phases of elemental solids at very high pressure adopt close-packed structures, such as *hcp*, *dhcp* and *fcc*. For example, the high-pressure phase of Si at 90 GPa has a *fcc* structure [20] and that of Cs at 72 GPa is *dhcp* [42]. The core effect is evident in the $Fd\bar{3}m$ phase in which the dispersion of the 1*s* band is almost 10 eV (Fig. 17.20). For a comparison, the 1*s* band dispersion in the intermediate pressure "pairing" phases, for example, the *Aba*2 structure, is much smaller (<2 eV) (Fig. 17.20), Therefore, the core effect may not be as important at 100 GPa [66].

The localization of electrons in the interstitial sites has been reported in high pressure K at 25 GPa [67]. This phase has a Ni_2In type double hexagonal closed-

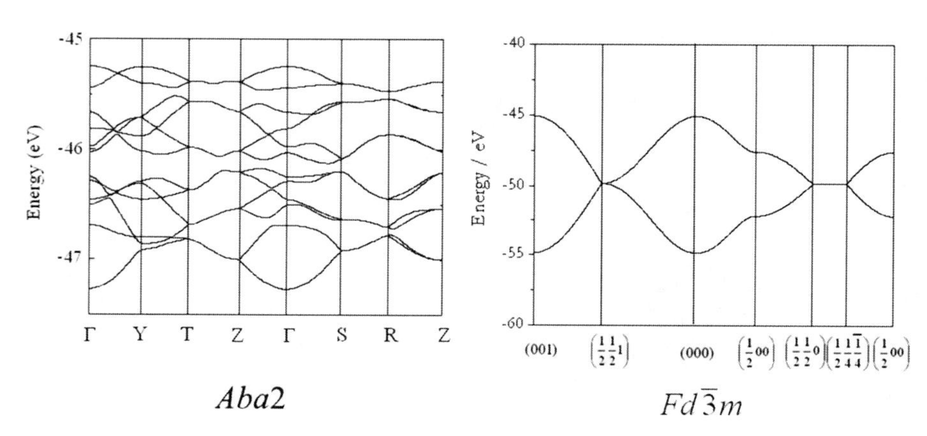

Fig. 17.20 Comparison of the electron band dispersions for Li at 100 GPa (*Aba*2) and at 380 GPa (*Fd*-3 *m*)

pack structure with the K atoms situated in the Ni sites. As already observed in high pressure phases of Si and Li (*vide infra*), ELF analysis of the calculated electron density reveals localization of electrons in the interstitial In sites. The very high ELF value (0.95) shows almost perfect pairing of the electrons. Therefore, the chemical bonding of the system can be considered as that of a pseudobinary ionic compound with the localized electrons acting as pseudoanions in the lattice formed from the K^+ cations. There is no information on the electrical properties of this novel phase.

On the other hand, theoretical calculations have predicted a high pressure hexagonal phase of Na (*P*6$_3$/*mmc*, h-P4) stable above 260 GPa, see Fig. 4 of Ref. [68]. A peculiar property of this phase is the charge localization resulting in complete separation of the valence electrons from the ionic cores and the system becomes an insulator with a calculated band gap >1.2 eV. The prediction of the transformation from a metal to an insulator was convincingly confirmed, as the originally dark sample became transparent, although at a much lower pressure of 200 GPa [68]. I this noteworthy that Na is predicted to have a *dhcp* (h-P4) structure at the highest pressure (>260 GPa to 1,000 GPa). The adoption of a close-packed hexagonal structure for the insulating state as a result of charge localization is entirely consistent with the discussion presented above.

Changes in the valence electron topology due to re-hybridization in elemental solids at high pressure are responsible for the observation of many new structures and expected properties, such as metal to insulator transition and superconductivity. Experimental studies have shown that regardless of the initial nature of the element, (*i.e.* metal or insulator) there will be a pressure range, in which an insulator → metal (or vice versa) transition may occur. This transition is caused by Peierls-like distortion from a high symmetry to lower symmetry structure. The occurrence of complex modulated, commensurate and incommensurate structure may be results of the charge density wave distortions. Moreover, the ubiquitous presence of

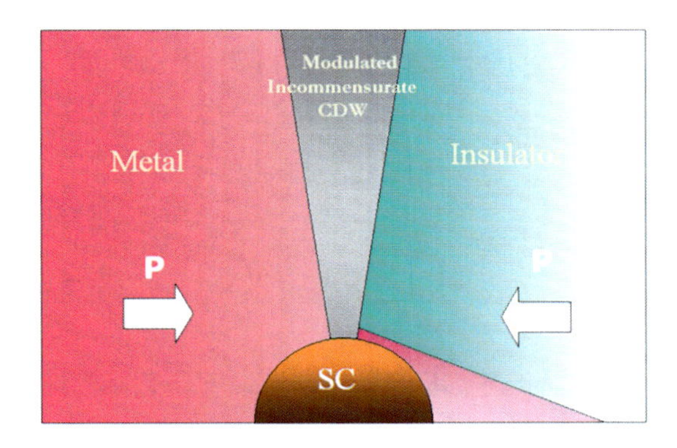

Fig. 17.21 A schematic illustration of the occurrence of electronic instability upon compression of a metal (*left*) and an insulator (*right*)

superconductivity may also be attributed to electronic instability in this pressure regime. A schematic representation of the pressure dependence of the electronic property is shown in Fig. 17.21.

17.3 Simple Ionic Solids

In an ionic solid the electric charges have already been transferred from the electropositive to the electronegative neutral atoms to generate cations and anions. There is only little perturbation to the charge density at low pressure. There are few experimental studies on the charge density topology of ionic solids at high pressure. However, QTAIM has been applied to elucidate the chemical bonding in a large number of solids under pressure [69]. An example is the theoretical analysis of the electron density topology (QTAIM) of several high pressure phases ($CaCl_2$, α-PbO_2 and pyrite) of silica (SiO_2) [70]. An interesting observation is that topological analysis of the highest pressure pyrite phase shows a (3,−1) critical point in the middle of each O-O "dumbell" in this structure at all pressures (Fig. 17.22). This is certainly not a bonding interaction as the calculated O-O interatomic distance is 2.385 Å at zero pressure and 2.043 Å at 260 GPa is much longer than the usual distances of 1.4–1.5 Å of peroxide bonds and more similar to nonbonding O-O distances in ionic oxides of 2.2 Å under ambient pressure. The corresponding electron density at the bond critical point is found to be very low *ca.* $0.22e$ Å^{-3} while $\nabla^2\rho$ is positive ($3.45e$ Å^{-5}), – these values suggest a closed-shell interaction (Fig. 17.22). It will be interesting to explore further the nature of the O...O interactions to determine whether it has a stabilizing or destabilizing effect. This will require the partition of the different interactions by an energy decomposition scheme as was discussed in detail in refs. [71–73].

An elegant demonstration of a study of charge density at high pressure is the experimental investigation of the chemical bonding in $Cs_2Au_2Br_6$ upon compression [74]. $Cs_2Au_2Br_6$ consists of linear $AuBr_2$ and planar $AuBr_4$ cluster units (Fig. 17.23). The charge density maps derived from MEM analysis of powder

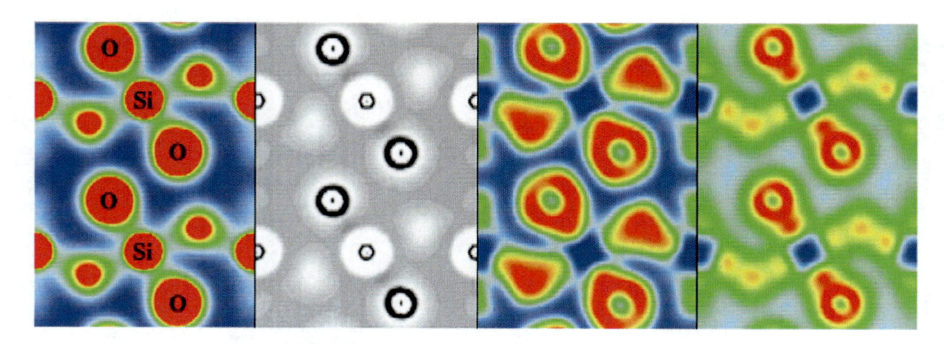

Fig. 17.22 LDA calculated (from *left* to *right*): total electron density (values from 0.11 to $1.5\,e\text{Å}^{-3}$ shown) of a hypothetical a high pressure silica phase, its Laplacian (-20 to $+20\,e\text{Å}^{-5}$), electron localization function (0–0.87), localized orbital locator (0.02–0.63) (Reproduced from Ref. [70] with kind permission of The American Physical Society)

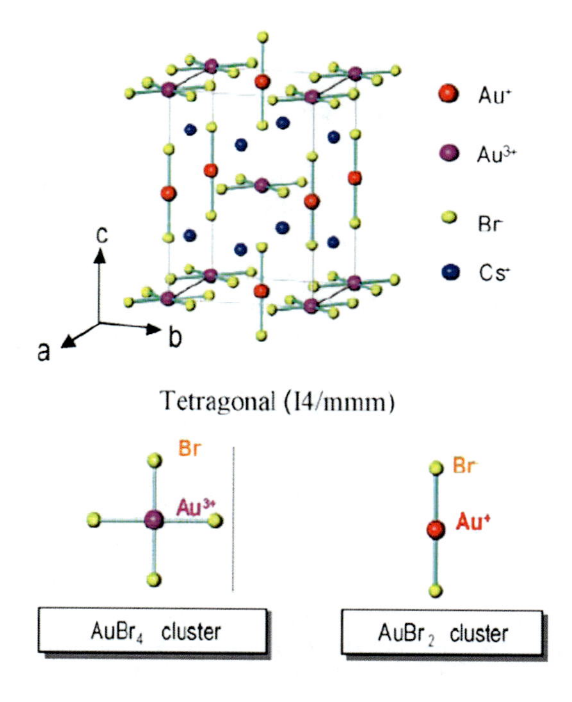

Fig. 17.23 The crystal structure of $Cs_2Au_2Br_6$. The $AuBr_4$ and $AuBr_2$ clusters are also shown (Reproduced from Ref. [74] with kind permission of Elsevier)

diffraction patterns at two pressures, 2.2 and 8.1 GPa of the $AuBr_2$ (110) plane and $AuBr_4$ (002) plane are shown in Fig. 17.24. At 2.2 GPa Au-Br covalent bonds are clearly seen in the charge distribution of the isolated $AuBr_2$ and $AuBr_4$ clusters. When this compound is compressed to 8.1 GPa, new covalent interactions are observed between the clusters as highlighted by the white arrows in Fig. 17.23.

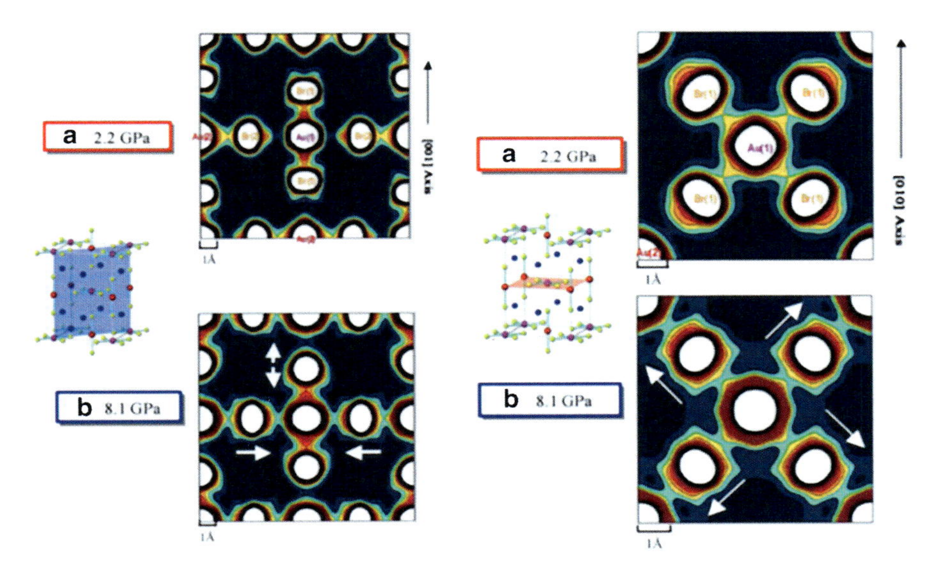

Fig. 17.24 Charge density of $Cs_2Au_2Br_6$ derived from maximum entropy analysis at 2.2 and 8.1 GPa. The contour lines drawn from 0.0 to 4.0 $e\text{Å}^{-3}$ with 0.2 $e\text{Å}^{-3}$ intervals. The white arrows highlight the inter- $AuBr_4$ interactions at high pressure (Reproduced from Ref. [74] with kind permission of Elsevier)

Therefore, pressure has modified the nature of the chemical bond and changes $Cs_2Au_2Br_6$ from a molecular complex of isolated $AuBr_2$ and $AuBr_4$ to a 2D network structure.

The minerals Ilmenites bearing transition elements have semi-conductive, ferro-electric and antiferromagnetic properties. In a recent study, the examination of the electron density distribution has helped to clarify the origin of electric conductivity and the anisotropy in $FeTiO_3$ at high pressure [75, 76]. For this purpose single crystal diffraction intensities obtained at 0.1 MPa and 8.2 GPa were used to derive the valence electron density maps from MEM analysis [26]. Density maps in the plane containing Fe, O, Ti atoms, as well as only Fe and Ti atoms, are shown in Fig. 17.25. The valence electron densities reveal significant covalent Fe-O and Ti-O bonding both at low and high pressure. The Fe-O-Ti linkages seemingly are not affected to any significant extent by pressure. In contrast, the electron density around the Fe and Ti atoms in the [011] plane parallel to the *c*-axis shows a remarkable change. At 0.1 MPa, the electron density located between the Fe atoms shows unambiguously Fe . . . Fe covalent interaction. This interaction, however, disappears at 8.2 GPa and, instead, is replaced by Fe-Ti bonds. Inspection of the electron densities suggests that the electric conduction in the direction parallel to the crystal *c* axis will increase with pressure. The enhancement in the conductivity can be interpreted assuming the probability of super-exchange of electron hopping between Fe and Ti to be higher due to the increased electron density between these two

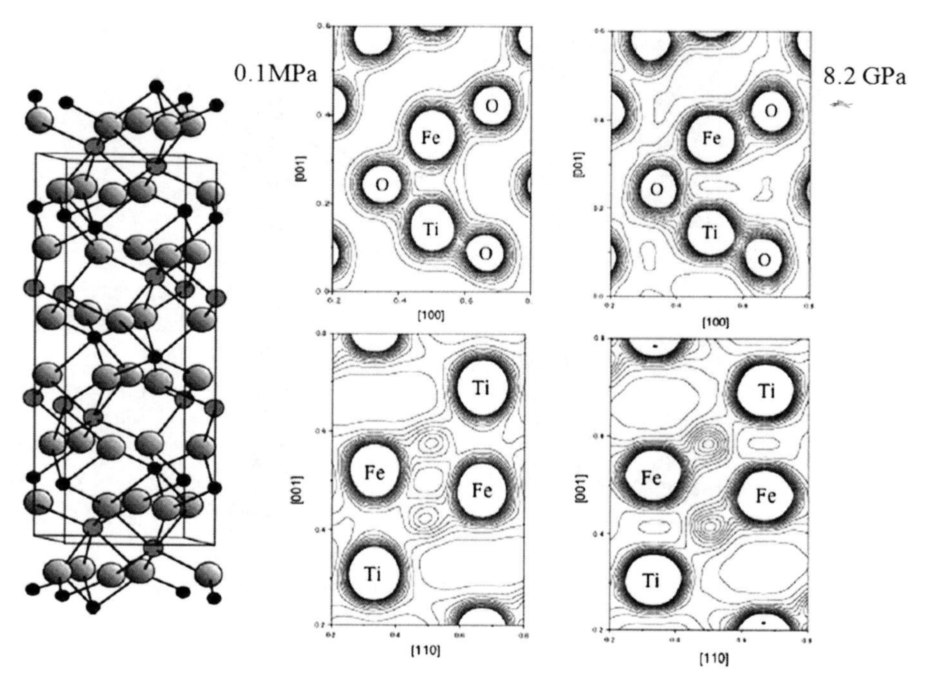

Fig. 17.25 The crystal structure of $FeTiO_3$ ilmenite (*left*) and the change of valence electron density (010) (*middle*) and (110) planes at 0.1 MPa and 8.2 GPa (Reproduced from Ref. [75] with kind permission of Springer)

atoms. The electron conduction, however, is dominant in the (001) plane due to the Fe-O-Ti linkage. Therefore, the study of the electron topology helps to elucidate the anisotropy of the electric conductivity of $FeTiO_3$ [76].

17.4 Molecular Solids

The pressure effect on the electron topology of molecular solids is not often as dramatic as the elemental solids. An early experimental study electron density at high pressure is the first use of the MEM technique to extract valence charge density to investigate the mechanism associates with the molecular → atomic (insulator → metal) phase transformation in molecular iodine at 19.8 GPa [77]. Results of the study are summarized in the electron density maps obtained at several pressures (Fig. 17.26). At low pressures (0.1 MPa and 7.4 GPa), molecular bonds between the nearest iodine atoms can be easily visualized. The interactions between neighbouring I_2 molecules characterized by the MEM determined charge densities ρ_{27} (0.16 $e/Å^3$) and $\rho_{21'}$ (0.07 $e/Å^3$) are quite weak. Upon compression

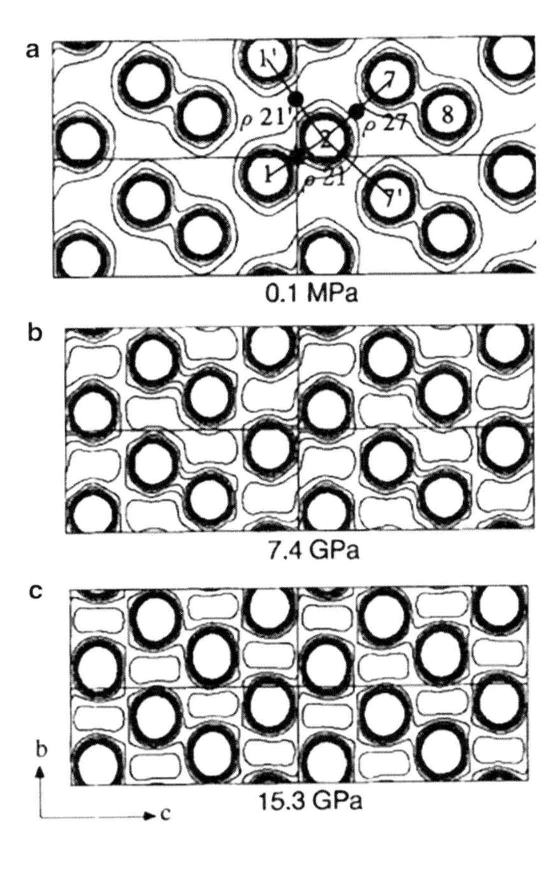

Fig. 17.26 Valence electron density maps of solid iodine derived from maximum entropy analysis of powder diffraction patterns as a function of pressure (Reproduced from Ref. [77] with kind permission of Taylor & Francis)

to 7.4 GPa, ρ_{27} and $\rho_{21'}$ increase to 0.36 e/Å^3 and 0.16 e/Å^3, respectively, and the I–I σ bond remains intact. When the pressure approaches the metal-insulator transition pressure (15.3 GPa), the optical gap is closed. The corresponding values of ρ_{27} and $\rho_{21'}$ are 0.38 e/Å^3 and 0.25 e/Å^3. Interestingly, the MEM electron density at the midpoint of two covalently bonded I atoms, ρ_{21}, remains fairly constant at 0.5 e/Å^3 over the entire pressure range. The results of the analysis show the mechanism for the insulator → metal transition is related to the increase of contact between neighbouring I_2 molecules a s result of the shorter $I_2 \ldots I_2$ distances upon compression. The sharing of electrons between the I_2 is the precursor step to the phase transition. The molecular (insulator) → atomic (metal) transition in I_2 has also been analyzed using the ELF [78]. As expected, a BCP was identified between the I atoms in the molecular phase and disappeared after the transition at 25 GPa. Inspection of the iso-surface with high ELF values provides now insight into the reason for the increase of the c/a ratio accompanying the phase transition. The repulsive interactions between the lone pairs of the I atoms above the conduction plane make the compression along the c-axis much stiffer than in the ab plane.

An interesting example on the change of electron density in a molecular solid is the magnetic → non-magnetic transition ($\delta \rightarrow \varepsilon$) in solid oxygen [79]. When solid oxygen is compressed at room temperature, a sequence of structural transformations was observed. At 5.4 GPa, rhombohedral (*R*-3 *m*) β-phase transforms to an orthorhombic (*Fmmm*) $\delta - O_2$ with a long-range anti-ferromagnetic order. At 9.6 GPa, δ-O_2 transforms into a monoclinic (*C2/m*) ε phase [80]. The ε phase is non-magnetic and is stable up to 97 GPa where it transforms into a metallic and superconducting ζ phase [81, 82]. The structures of the ε and δ were only solved by *in-situ* high pressure single crystal diffraction experiments very recently [81–83]. The ε phase reveals a remarkable monoclinic structure consisting of $(O_2)_4$ clusters. [81, 82] At 17.5 GPa the intermolecular distance within the cluster of 2.20 Å is significantly longer than the intramolecular distance of 1.20 Å [81, 82] which is almost identical to that of the free molecule. The very long intermolecular distance, however, raises important questions on the nature of the bonding between O_2 in the cluster. To understand the nature of chemical bonding in ε oxygen, this phase has been investigated through topological analysis of charge density using the QTAIM method [84].

With no magnetic moment in the ε-phase, the O_2 molecules must be spin-paired in the $(O_2)_4$ cluster. A zigzag chain configuration with a *Cmcm* space group [62, 85, 86] has been suggested to be the lowest energy structure from theoretical calculations. However, this structure was not observed in a recent experimental study. Rather, two $(O_2)_2$ units was found to "dimerized" forming $(O_2)_4$ clusters in a *C2/m* structure [81, 82]. To understand the nature of the chemical bonding, valence electron topology of the predicted *Cmcm* and observed *C2/m* have been examined from theoretical calculated electron densities. The (3,−1) bond critical points (BCP) obtained from QTAIM [1] analysis on the *C2/m* structure at 24 GPa are summarized in Table 17.1 and Fig. 17.28. As expected, the BCP's of the O_2 molecules have the largest value of local density $\rho(r_{cp})$ (∼0.55) and local energy $H(r_{cp})$ [87–89] (−1.10 a.u.) and are indicative of genuine O-O covalent bonds. Significantly, intracluster BCP's are found between O_2 molecules forming the $(O_2)_4$ cluster. The charge density at the critical point $\rho(r_{cp})$ ∼0.06 with a positive Laplacian and a negative $H(r_{cp})$ (−0.009 a.u.) are signs of covalent interactions between O_2 molecules albeit it is much weaker than that expected for a normal covalent bond [87, 88]. A positive Laplacian is perhaps surprising as it indicates "closed shell" interactions [1, 87–89] between the two O atoms of the adjacent O_2 molecules in the $(O_2)_4$ unit instead of the pairing of two "open-shell" spin-uncoupled orbitals forming a pair of two electron bonds as depicted in Fig. 17.28. If this were the case, the intermolecular distance would be close to that of a single σ-bond. The analysis, however, is consistent with the observed very long $O_2 \cdots O_2$ separation of about 2.34 Å at 10 GPa and 2.17 Å at 17 GPa. Results indicate that the O_2 molecules forming the rhomb-shaped clusters in the ε phase are stabilized via charge transfer, i.e. the donation of a *paired* π^* electron from one O_2 into the empty π^* orbital of the other. In this interpretation, the O_2 converted from a spin-uncoupled state of a free molecule into a spin-paired state (two electrons occupying one of the π^* orbital) to facilitate interactions with the neighbouring O_2 forming the cluster. It is noteworthy that the bond interatomic distance of 1.20(3) Å determined from

Table 17.1 Bond critical points for *C2/m* ε-O2 at 24 GPa (For the details of the structure see the caption of Ref. [24])

Bond	Type	$\rho(r_{CP})$	$\nabla^2\rho(r_{CP})$	λ_1	λ_2	λ_3	$H(r_{CP})$
O_3–O_3	Molecular	0.5501	−0.536	−1.555	−1.527	2.546	−1.109
O_1–O_2	Molecular	0.5468	−0.459	−1.518	−1.510	2.569	−1.087
$O_3 \ldots O_2$ (d1)	Intra-cluster	0.0622	0.231	−0.098	−0.091	0.421	−0.009
$O_3 \ldots O_1$ (d2)	Intra-cluster	0.0619	0.232	−0.098	−0.091	0.420	−0.009
$O_3 \ldots O_2$ (d3)	Inter-cluster	0.0276	0.099	−0.025	−0.024	0.148	0.003
$O_3 \ldots O_2$ (d4)	Inter-cluster	0.0194	0.089	−0.020	−0.019	0.128	0.003
$O_3 \ldots O_2$ (d5)	Inter-cluster	0.0163	0.080	−0.017	−0.017	0.122	0.004

Table 17.2 Bond critical points for a over-pressurized *C2/m* ε-O2 at 105 GPa (For the details of the structure see the caption of Fig. 17.27)

Bond	Type	$\rho(r_{CP})$	$\nabla^2\rho(r_{CP})$	λ_1	λ_2	λ_3	$H(r_{CP})$
O_3–O_3	Molecular	0.5882	−0.737	−1.702	−1.657	2.622	−1.253
O_1–O_2	Molecular	0.5866	−0.663	−1.659	−1.642	2.638	−1.238
$O_3 \ldots O_3$ (d5)	Inter-cluster	0.0652	0.291	−0.100	−0.093	0.484	−0.006
$O_3 \ldots O_2$ (d1)	Intra-cluster	0.0619	0.280	−0.094	−0.086	0.456	−0.005
$O_3 \ldots O_1$ (d2)	Intra-cluster	0.0619	0.282	−0.093	−0.084	0.459	−0.004
$O_3 \ldots O_2$ (d3)	Inter-cluster	0.0484	0.219	−0.066	−0.057	0.341	−0.001

the single crystal study at 17.6 GPa [80] is comparable to that of the free molecule of 1.207 Å but is slightly *longer* than 1.180 Å found in β-O_2 at 5.5 GPa [90] and 1.175 Å in δ-O_2 at 9.6 GPa and 297 K [91]. The longer O–O bond observed at even high pressure suggests that two valence electrons occupied a single π^* orbital, thus reducing the bond order and weaken the interatomic bond. The explanation offered here, of course, is also consistent with the non-magnetic character of ε-oxygen [92]. BCP's are also identified between oxygen atoms of different clusters. As shown in Table 17.1, the electron density at these inter-cluster BCPs (0.028) are much smaller. Moreover, both the Laplacian and local energy density $H(r_{cp})$ are positive indicating closed shell van der Waals interaction between the clusters [93]. Ring and cage CP's were also identified but they are of minor significance and will not be discussed here.

When ε-oxygen is compressed, the intra-cluster O_2 distance (d_1) and inter-cluster distance (d_3) decrease rapidly [81, 82]. Interestingly, the shape of the $(O_2)_4$ cluster distorts from a square prism to a diamond-shape prism with increasing pressure. The O···O···O bond angle decreases gradually from 88° at 20 GPa to 80° near the $\varepsilon \rightarrow \zeta$ transition at 97 GPa [81, 82]. The distortion leads to changes in the structure and the electron density topology. Table 17.2 and Fig. 17.27 report results of the QTAIM analysis on an over-pressurized *C2/m* structure at 105 GPa. As indicated by $\rho(r_{cp})$ and $\nabla^2\rho(r_{cp})$ the intra-cluster and inter-cluster interactions of 0.065 and 0.062, respectively, are now almost equivalent. The $(O_2)_4$ clusters in the *ab* plane start to connect to each other. At this pressure, the calculated electronic band structure shows that the *C2/m* is metallic but dynamically unstable. The linking of $(O_2)_4$ clusters may be the precursor for the $\varepsilon \rightarrow \zeta$ phase transition.

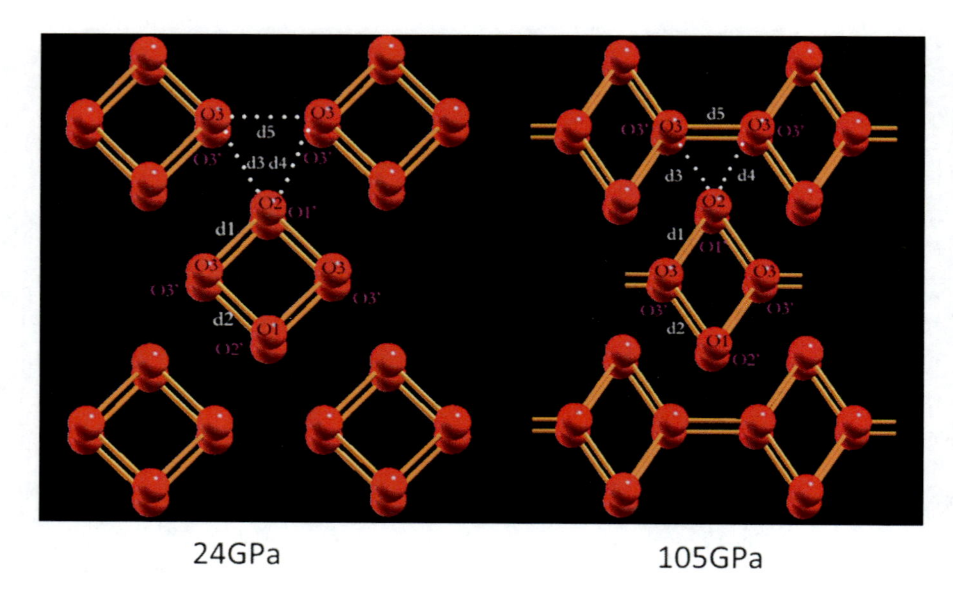

Fig. 17.27 Structure of the $(O_2)_4$ units of solid ε-O_2 at 24 and 105 GPa. The intra- (d_1) and intermolecular (d_3) O...O distances are, respectively, 2.06, 2.43 Å at 24 GPa and 1.95, 2.10 Å at 105 GPa

17.4.1 Extended Molecular Solids

The previous sections have illustrated to which extent the fine details of the electronic structure of inorganic solids (metals and minerals, as well as of simple inorganic molecular solids) under really extreme high-pressure conditions are now accessible through a combination of high-quality experimental work and theoretical quantum chemistry calculations. The studies of organic molecular crystals are still at their infancy. There seem to be no examples of direct observations of electron density in extended molecular solids yet, and the reason for this is to be sought in the difficulties of obtaining complete and accurate sets of diffraction data for the low-symmetry molecular crystals in a high-pressure cell. However, the techniques of collecting and processing data are progressing very quickly, and one may hope that the experimental data on the electron density in the extended molecular crystals become a reality in the near future. The usage of Ag-radiation can be useful for increasing the completeness of the data sets. Up to now, the analysis of chemical bonding is limited to the studies of the effect of pressure on molecular conformations and intermolecular contacts.

The pressures applied in the experiments are very modest as compared to the extremes described in the previous sections; they usually do not exceed 10 GPa, and many interesting effects are observed already in the 1 GPa range. In an agreement with the Fig. 17.1, this range of pressure is not sufficient for breaking or substantially modifying covalent bonds, but can affect hydrogen bonds and weaker non-covalent interactions. The discussion of the effect of pressure on the molecular crystals is often carried out under the assumption that the intramolecular geometry does not change. This assumption usually stems from the necessity to reduce the number of refined parameters, when experimental data are not sufficiently complete. However, the assumption turns out to be wrong to a larger or smaller extent, when better experimental data become available. The intramolecular bond lengths, valence and torsion angles in organic molecular crystals do vary with increasing pressure. These changes are interrelated with the anisotropic crystal structure compression, phase transitions and chemical reactions.

It has been well-known from spectroscopic experiments, that high pressures can induce considerable changes in the conformation of organic molecules in solutions. Some evidence on the pressure-induced conformational changes of organic molecules in crystals has been also reported based on spectroscopy and neutron diffraction (see [94] as an example). In the crystals of deuterated biphenyl the torsion angle between molecular phenyl rings is the modulated variable leading to two low-temperature modulated phases – they are strongly dependent on the pressure which by enhancing the crystal packing forces favors a coplanar conformation of the molecules [95]. Pressure has been shown to induce the ordering of molecular fragments, which at ambient pressure are disordered, as it takes place in the crystals of di(ethylenedithio)-tetrathiofulvene sesquiiodide [96].

A systematic X-ray diffraction study of the effect of pressure on the intra- and intermolecular bonds and interactions in organic crystals has been initiated by Katrusiak. In 1990 he has reported changes in the intramolecular geometry of 1,3-cyclohexanedione (CHD) in the course of a phase transition induced by hydrostatic pressure [97–99]. The high quality of the data made it possible to detect several significant changes in the molecular dimensions of the CHD molecule in the high-pressure phase as compared to the ambient-pressure one. It was evident, that the sequence of double and single bonds in the ambient-pressure conjugated system has reversed at high pressures, indicating that the H-atom involved in the intermolecular hydrogen bond has changed its donor and acceptor sites (Fig. 17.28). It is not straightforward to locate hydrogen atoms involved in hydrogen bonds at high pressure from the charge density topology, also due to the fact that the maximum of density associated to the H-atom may become hidden under the electron density tails of the H-acceptor and H-donor atoms, when hydrogen bond distances are largely reduced. However, there is often an indirect evidence of the position of hydrogen atoms provided by the lengths of the bonds between non-hydrogen atoms, torsion and valence angles. In the case of CHD, some systematic changes could be noted in the lengths of the bonds between the sp^2 atoms, which are particularly sensitive to the effects of pressure. For example, a bond elongation could be observed for the carbonyl group of CHD at high pressures, which was also consistent with this group being involved in the intermolecular hydrogen bond. The largest changes in the selected intramolecular bond length were 0.05–0.1 Å, the changes in the valence angles did not exceed 3–4 degrees. A further study of the related compound, 2-methyl-1,3-cyclopentanedione (MCPD) [100] (Fig. 17.29), has confirmed that not only significant changes in the orientation of molecules

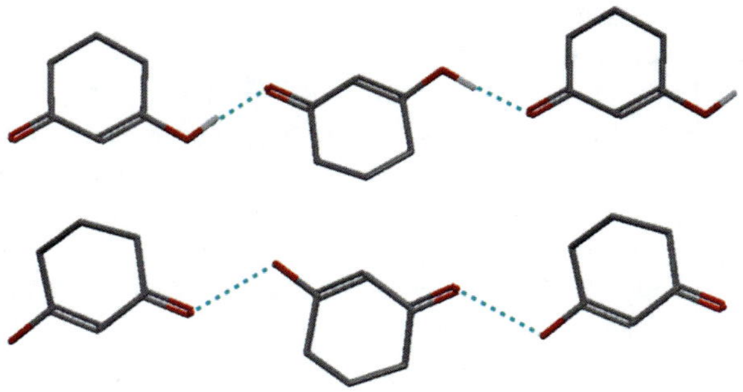

Fig. 17.28 Intra- and intermolecular changes in the 1,3-cyclohexanedione (*CHD*) induced by increasing hydrostatic pressure: the sequence of double and single bonds in the ambient-pressure conjugated system (*above*) has reversed at high pressures (*below*), indicating that the H-atom involved in the intermolecular hydrogen bond has changed its donor and acceptor sites [97–99]. OH...O hydrogen bonds are shown as dashed blue lines. The positions of H-atoms in the high-pressure phase are not shown, since they have been determined indirectly only from the distances between non-hydrogen atoms

Fig. 17.29 Intermolecular OH...O hydrogen bonds in the crystal structure of 2-methyl-1,3-cyclopentanedione (MCPD) [100] (shown as *dashed blue lines*)

and in the intermolecular hydrogen bonds networks can be reliably detected in a laboratory single-crystal X-ray diffraction experiment, but also the intramolecular distortions are measurable. Some systematic changes in bond lengths could be noted, in particular in the conjugated-bond moiety: bond O=C became longer by about 0.04 Å, while bond C–C shortened by 0.04 Å at 1.50 GPa and by 0.07 Å at 2.40 and 3.01 GPa. The elongation of the intramolecular C=O bond in the MCPD was consistent with the changes observed in the CHD crystals, and with the shortening of the intermolecular hydrogen bonds, in which the O atom of the carbonyl-group has been involved. The unsaturated bonds and the π–electron bond system were shown to be particularly sensitive to high pressures. A significant reduction in the atomic displacement parameters of all the non-hydrogen atoms versus pressure has been measured [100]. The first results obtained for a small series of organic compounds (CHD, MCPD, and dimedone, i.e. 5,5-dimethyl-1,3-cyclohexanedione [101]) have been summarized in [102]. In all the compounds the intramolecular changes were induced by cooperative hydrogen jumps from one site to another in the course of the phase transition, and correlated with the general distortion of the crystal structure. They could be interpreted in terms of the electrostatic interactions between the neighbouring molecules and the net atomic charges redistribution within a molecule.

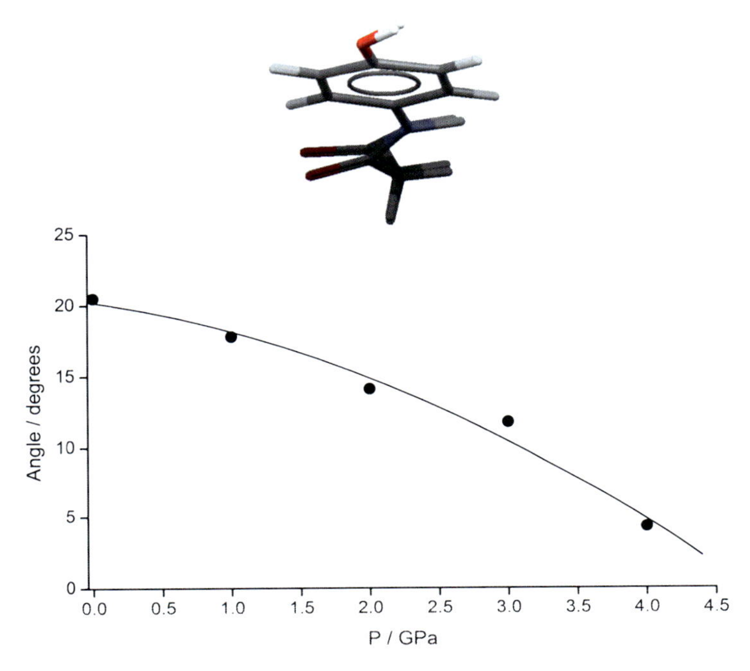

Fig. 17.30 Intramolecular changes in the crystals of paracetamol I with increasing hydrostatic pressure [103]: (**a**) an overlay plot of the molecules at ambient pressure and at 4/0 GPa; (**b**) the dihedral angle between the phenyl-ring and the acetamino-group *vs* pressure (Adapted from Ref. [161] with kind permission of IUCr Journals)

An evidence of the elongation of the carbonyl-group involved in the intermolecular hydrogen bond and of the considerable changes in the torsion angles in an organic molecule with increasing pressure, even if no phase transitions occurred, has been obtained for paracetamol, HO-C_6H_5-NH-C(=O)-CH_3 [103]: the length of the –C=O bond increased at about 0.023 Å at pressures about 4 GPa, and the dihedral angle between the phenyl-ring and the acetamino-group decreased at the same pressure from 22 to 4 degrees (Fig. 17.30). These distortions were related to a considerable shortening of intermolecular hydrogen bonds and to the anisotropic compression of the crystal structure with linear expansion along selected directions. The flattening of a paracetamol molecule is not just "better" for a more efficient close packing in the crystal, but is closely interrelated with the redistribution of the atomic net charges [104], which accompany any distortion of the hydrogen bond network in the crystal structure. An indirect indication at the pressure-induced proton transfer in the intermolecular hydrogen bonds in paracetamol is provided by the data on the changes in the Donohue angles [105], which are in the range 4–10 degrees on compression up to 4 GPa [103]. The shifts of the vibrational bands in the IR-spectra with increasing pressure can be a manifestation of the strengthening or loosening of the intermolecular hydrogen bonds, complementing the geometric data obtained from diffraction experiments. In complex crystal structures, a correlation of the frequency shifts (IR- or Raman- spectroscopy) and the changes in the interatomic distances (X-ray or neutron diffraction experiments) is not straightforward. Paracetamol can provide such an example: although both the N-O and the O-O distances in the NH...O and OH...O hydrogen bonds in paracetamol shorten with pressure, the vibrational frequency υ(NH) of the stretching vibrational shifts to the red with increasing pressure (as should be expected), whereas the vibrational frequency υ(OH) increases. A possible interpretation is that the –OH group not only donates a proton to the carbonyl –C=O group, but also accepts another proton from the –NH group [106].

An evidence of the measurable changes in the intramolecular geometry of a coordination compound, $[Co(NH_3)_5NO_2]Cl_2$, with increasing pressure has been obtained in [107]. The changes in the bond lengths $Co\text{-}NH_3$, $Co\text{-}NO_2$ and N-O were in the range of 0.015–0.025 Å, the O-N-O valence angle in the nitro ligand changed at about 1 degree and rotated with respect to the lines connecting the NH_3-ligands at about 3 degrees (Fig. 17.31). These changes were related to the continuous strengthening of the intermolecular NH...O hydrogen bonds with increasing pressure. The X-ray diffraction data agreed well with the results of measuring IR-spectra at high pressures and with the quantum chemistry calculations [108].

These examples show that already in the 1990s it was possible to obtain the X-ray diffraction data, which were complete and precise enough to measure reliably the subtle changes in the intramolecular distances and angles induced by increasing pressure. Nowadays, after a substantial progress in the hardware and software has been achieved (see [109] for a review), obtaining high-precision data became more straightforward, and much more examples of the experimental observations of the distortions in the intramolecular geometry could be mentioned. Research continues basically in those directions, which have been pioneered in the 1990s: studying the pressure-induced changes in the hydrogen bonds, rotation of ligands, conformational changes in organic molecules, transformation of non-covalent interactions into covalent bonds. The space limitations do not allow us to describe and even to list all the results published in the last decade. Therefore we shall mention only a few representative examples.

For carbon disulfide, the anisotropic structural distortion was followed up to 8 GPa, i.e. till the polymerization onset. The crystal structure was determined by direct methods (i.e. – without any assumptions based on modeling, just from the experimental diffraction data) from single-crystal X-ray diffraction at 295 K at two pressure points: 1.8 and 3.7 GPa (esd's in the lengths of C=S bond 0.001 Å). Molecular rearrangements have been rationalized by the close packing and equidistant S···S intermolecular interactions enforced by pressure (Fig. 17.32). Although only slight lengthening of the covalent double C=S bond has been observed till 3.7 GPa, the increase in the energy of the intermolecular S···S and C···S interactions revealed the possible reaction pathways of pressure-induced polymerization of CS_2 [110].

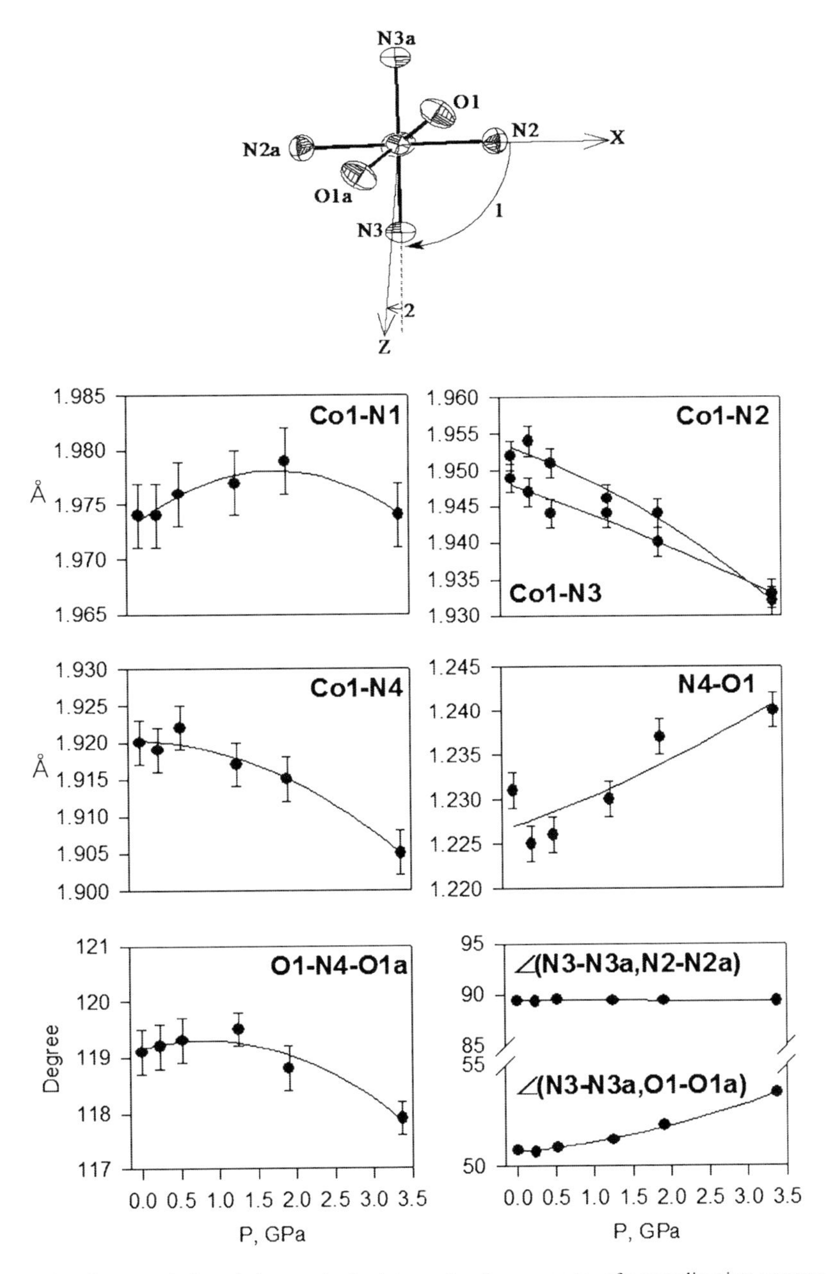

Fig. 17.31 Pressure-induced changes in the intramolecular geometry of a coordination compound, $[Co(NH_3)_5NO_2]Cl_2$ [107]: (*top*) a complex cation and numeration of the atoms, (*bottom*) the changes in the bond lengths $Co\text{-}NH_3$, $Co\text{-}NO_2$, N-O, and the O-N-O valence angle in the nitro ligand, as well as in the angles characterizing the rotation of the nitro-ligand in a complex cation [108] (Adapted from Ref. [161] with kind permission of IUCr Journals)

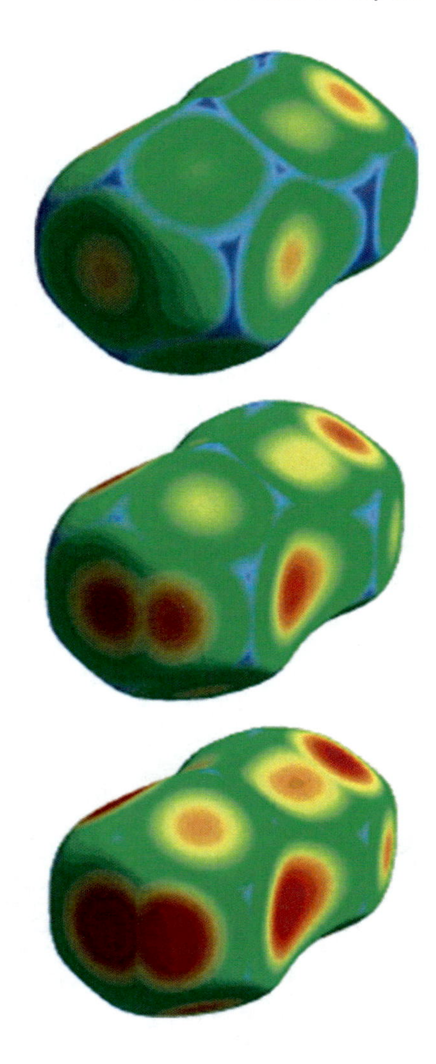

Fig. 17.32 Hirshfeld surfaces for the CS_2 molecule embedded in the crystal structure at 5.3 K/0.1 MPa (*top*), 295 K/1.8 GPa, and 295 K/3.7 GPa (*bottom*). The color scale on the surface represents the shortest distances from the surface element to the closest atoms outside or inside the surface (red corresponds to the shortest contacts, blue to the longest). The CS_2 molecule occupies the C_{2h}-symmetric site in the crystal; thus each contact is doubled (these along *C*2 or *Cs*) or quadrupled [110] (Reproduced from Ref. [110] with kind permission of The American Chemical Society)

Increasing efforts to understand the reaction mechanisms at the microscopic level, to set up and optimize synthetic approaches, are currently directed at carbon-based solids. A fundamental, but still unsolved, question concerns how the electronic excited states are involved in the high-pressure reactivity of molecular systems. The problem is being approached by the direct characterization of the electronic properties as a function of pressure by linear and nonlinear optical spectroscopy up to several GPa combined with theoretical calculations. Acetylene and ethylene (and their substituted derivatives) are probably the smallest molecules, for which the pressure-induced polymerization has been studied by a combination of theoretical and experimental techniques [111–115]. For benzene, the archetypal aromatic system, the measurement of two-photon excitation spectra in a molecular crystal under pressure, up to 12 GPa has been reported. Comparison between the pressure shift of the exciton line and the monomer fluorescence provided evidence for different compressibilities of the ground and first excited states. The structural excimers formed with increasing pressure, which involved molecules on equivalent crystal sites that were favorably arranged in a parallel configuration. These species represented the nucleation sites for the transformation of benzene into amorphous hydrogenated carbon [116]. The possibility of the dark-dimerization in the crystalline anthracene, phenanthrene, and chloroanthracene under hydrostatic pressure and under pressure combined with shear has been analyzed by theoretical DFT calculations and the experimental *in situ* studies of the photoluminescence [117–119].

Pressure-induced transformations between gauche-, gauche + and transoid conformations have been evidenced by X-ray single-crystal diffraction for 1,1,2-trichloroethane, and the energies of intermolecular interactions, conformational conversion, and the latent heat have been determined [120]. Angle-dispersive X-ray diffraction experiments on nitromethane single crystals were performed at room temperature as a function of pressure up to 19.0 GPa. The atomic positions were refined at 1.1, 3.2, 7.6, 11.0, and 15.0 GPa using the single-crystal data. The crystal structure was found to be orthorhombic, space group $P2_12_12_1$, with four molecules per unit cell, up to the highest pressure. In contrast, the molecular geometry undergoes an important change consisting of a gradual blocking of the methyl group libration about the C–N bond axis, starting just above the melting pressure and completed only between 7.6 and 11.0 GPa. Above this pressure, the orientation of the methyl group is quasi-eclipsed with respect to the NO bonds. This conformation allows the buildup of networks of strong intermolecular O$\cdots$H–C interactions mainly in the *bc* and *ac* planes, stabilizing the crystal structure. This structural evolution determines important modifications in the IR and Raman spectra, occurring around 10 GPa. The Raman and IR vibrational spectra as a function of pressure at different temperatures evidence the existence of a kinetic barrier for this internal rearrangement [121]. It is interesting to compare these results with the study of the effect of pressure on liquid nitromethane, which have been interpreted in terms of the distortions of the intramolecular geometry in the cyclic dimers [122].

As a recent example of an evidence of significant pressure-induced intramolecular distortions in a coordination compound one can mention the solid-state structural transformations occurring in species of the general formula $[Co_2(CO)_6(XPh_3)_2]$ (X = P, As) and studied by X-ray diffraction and *ab initio* theoretical modelling [123, 124]. These metal carbonyl dimers transform the conformation of carbonyls about the Co–Co bond from staggered to eclipsed when the volume is reduced. The phase transition is accompanied by *expansion* of metal-metal bonds, although the crystal packing as a whole becomes denser (Fig. 17.33). Long intermolecular interactions were shown to be converted into covalent bonds when the complex $[Cu_2(OH)(citrate)(guanidine)_2]^-$ is exposed to a pressure of 2.9 GPa; the coordination of the Cu centres changes from [4 + 1] to [4 + 2], but on increasing the pressure to 4.2 GPa some of the Cu centres become [4 + 1]-coordinated again [125].

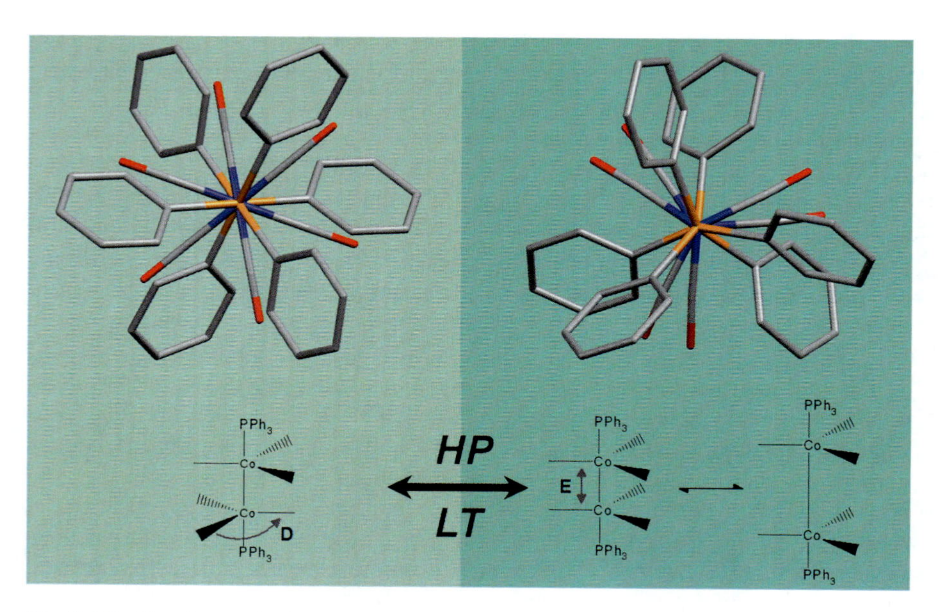

Fig. 17.33 The changes in the intramolecular geometry resulting from the solid-state structural transformations occurring in species of the general formula $[Co_2(CO)_6(XPh_3)_2]$ (X = P, As) and studied by X-ray diffraction and *ab initio* theoretical modelling [123, 124]. The conformation of carbonyls about the Co-Co bond is changed from staggered to eclipsed, and the metal-metal bonds *expand*, although the crystal packing as a whole becomes denser (Reproduced from Ref. [124] with kind permission of Springer)

Hexagonal ice is a prototypical H-bonded system. The electron and nuclei densities have been accurately measured by x-ray and neutron diffraction, respectively [126, 127]. In the earlier days, there was a controversy on the lengthening of the OH(D) bond (~1.0 Å) with respect to bound water molecule in other crystalline compounds. Unfortunately, the examination of the electron and nucleus densities failed to resolve this discrepancy. Instead, Floriano [128] et al., directly determined an interatomic O-D distance of 0.98 Å from the Fourier transform of high-angle (moment transfer) neutron powder diffraction data. This value is in complete agreement with an elaborate combined neutron single-crystal diffraction with vibrational spectroscopy analysis of *ca.* 0.98 Å. This observation highlights the difficulties on the location of hydrogen atom of a hydrogen bond from charge (nuclei) density maps evaluated from diffraction data. There has been no attempts to extract the electron density of ice polymorphs at high pressure. Nevertheless, electron topological analysis of calculated densities has been used to examine O–H and O–O bond critical points (BCP) and structural stability in ice II, VI, VIII, X and XI [129–132]. The results, in general, conform to the conventional description of hydrogen bonding. An interesting application of QTAIM to high pressure ices is the estimation of the metallic character of a bond. The bond metallicity index ξ is a local property, and is defined as the ratio of the charge density and its Laplacian,

$\xi\left(r_{bcp}\right) = \frac{\rho\left(r_{bcp}\right)}{\nabla^2\rho\left(r_{bcp}\right)}$ at the BCP [133]. It was shown that there was an enhancement in metallic character in the O–H bond of ice XI with the antifluorite structure at 5 TPa [134]. This observation is in apparent agreement with electronic band structure calculations [134] which show that ice XI becomes a metal at 1–1.5 TPa. However, caution may be exercised in examining the theoretical results, since the electron densities used in the QTAIM analysis were obtained from pseudopotential planewave calculations. The validity of ultrasoft pseudopotentials at these extreme pressures has not been verified thoroughly.

The only known example on the experimental determination of the valence electron density maps of an ice system under pressure is that of the xenon hydrate [135]. Clathrate hydrates are inclusion compounds in which guest molecules or atoms are trapped in cages formed by an ice-like host network of water molecules [136]. The combined Rietveld/maximum entropy method was used to derive the most probable charge density distribution at each pressure from the powder x-ray diffraction patterns from 0.16 to 1.98 GPa, within the stability region of the structure I hydrate. Structure-I hydrate has a cubic *Pm*3*m* symmetry [136] and characterized by small and large polyhedral cages It is shown that the charge density distribution of the encaged Xe atoms differs depending on the type of host cage at all pressures. Spherical electron density distributions were observed for the Xe atoms in the small cages while the atoms in the large cages showed longitudinal elongated electronic distributions. Along with the observed cage deformations, the change in electronic density distribution represents a clear indication that the guest-host interaction differs significantly for the occupation of the small and the large cages at high pressures. The results show that high quality valence charge density data can be extracted from high-pressure powder diffraction measurements, using even a rather limited range of diffraction angles. The theoretical justification of this is that the low-angle Bragg reflections are more sensitive to the valence electrons than those at high angles [135].

Even though the experimental determination of the charge density directly and/or from the difference maps remains a challenge, high quality structural data from single crystal diffraction can be obtained. In theses cases, not only the compression or expansion of hydrogen bonds can be measured quite precisely, but also provide reliable information on the shifts of a hydrogen atom between the donor and the acceptor atoms along a hydrogen bond. The latter can be obtained from indirect data (the changes in the corresponding intramolecular bonds and angles), or from spectroscopy. As a recent example one can mention a study of the oxalic acid dihydrate at pressures up to 5.3 GPa: considerable changes in the -C-O and -C=O bond lengths suggested that the intermolecular hydrogen-atom transfer between an oxalic acid and a water molecule in the crystal occurred with increasing pressure, resulting in the formation of ionic species, as has been predicted by periodic DFT calculations (Fig. 17.34) [137].

An interesting example has been provided by a study of L-alanine [138]. As other crystal structures of amino acids, the crystal structure of L-alanine is built by alanine zwitterions linked into head-to-tail chains (Fig. 17.35). Interestingly, in

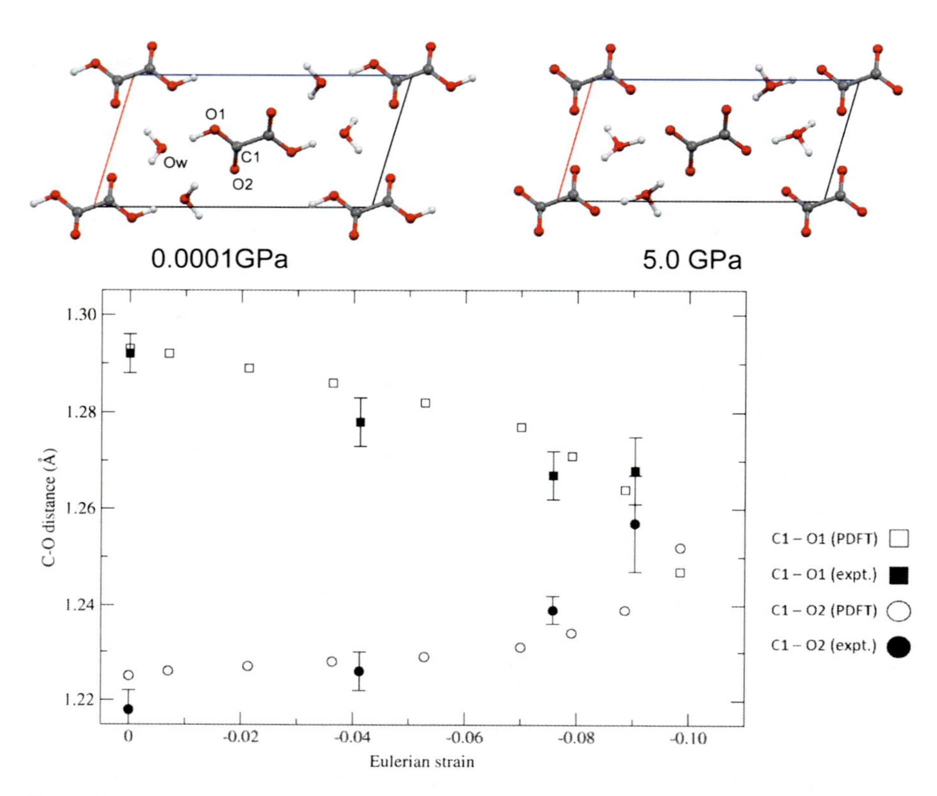

Fig. 17.34 Pressure-induced intermolecular proton transfer between neutral molecules to give ions in the oxalic acid dihydrate: (**a**) crystal packing (**b**) considerable changes in the -C–O and -C=O bond lengths [137] (Picture is courtesy of Dr. P. Macchi)

contrast to some other crystalline amino acids, at ambient conditions the two C–O bonds in the COO-group in L-alanine are not absolutely equal, what was confirmed also by several independent X-ray and neutron diffraction studies [139–145]. The longer of the two non-equal C–O bonds in the crystalline L-alanine is the one, which is involved in the formation of a N*–H*O* bond in the head-to-tail chain of zwitterions formed as a result of a proton transfer from a COOH head to a NH_2-tail giving zwitterions instead of molecules as a crystal structure is formed. The difference in the two C–O lengths may suggest that this proton transfer is not absolutely complete at any temperature at ambient pressure. The non-equality in the C–O bond lengths in the carboxylic group may serve as a measure of the completeness of the H*-atom transfer between the two alanine zwitterions within a chain. The difference in the lengths of the two intramolecular C–O bonds in the carboxylic group was shown to decrease with increasing pressure [138]. When the chain of zwitterions is compressed and the distance between the molecules shortens, the interaction between the carboxylic group and the amino group becomes stronger, and the proton shifts further to the amino group, so that polarization of the

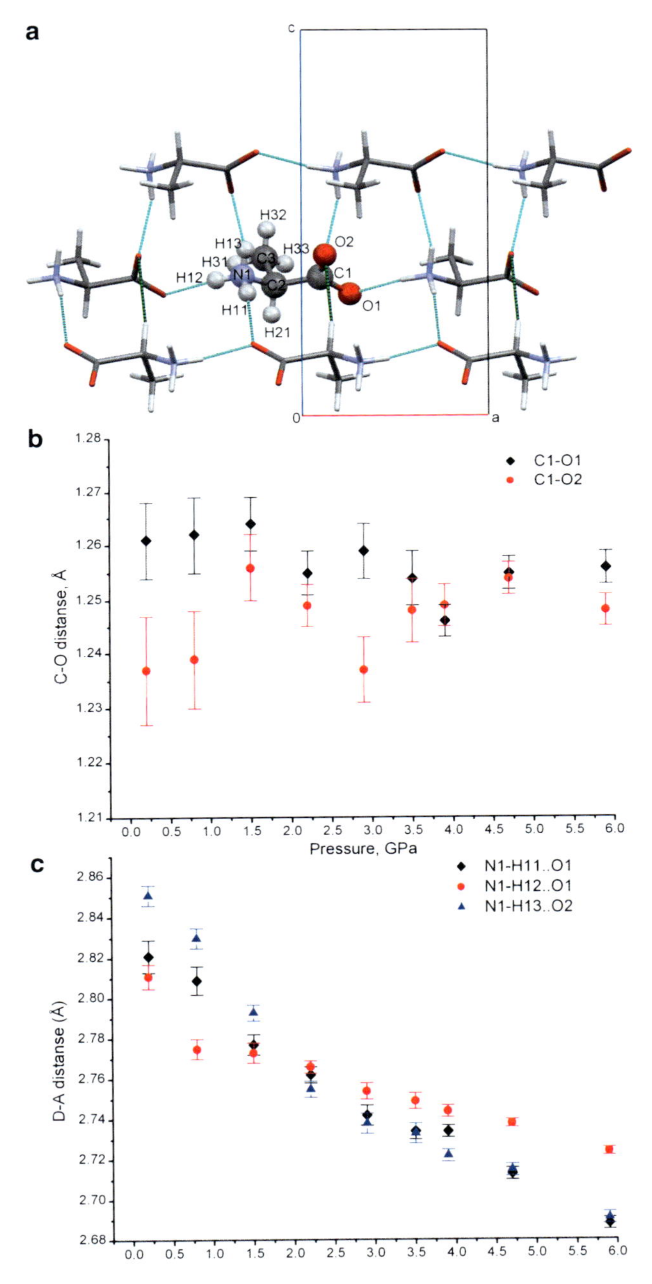

Fig. 17.35 Pressure-induced changes in the selected intra- and intermolecular bonds in the crystal of L-alanine: (**a**) crystal packing and numeration of atoms, hydrogen bonds are shown as blue dashed lines; (**b**) a decrease in the non-equivalence of the two -C–O bonds in a zwitterion; (**c**) different compression of the three types of the N-H . . . O hydrogen bonds, note that the H-bond linking zwitterions within a head-to-tail chain is compressed least of all [138] (Adapted from Ref. [161] with kind permission of IUCr Journals)

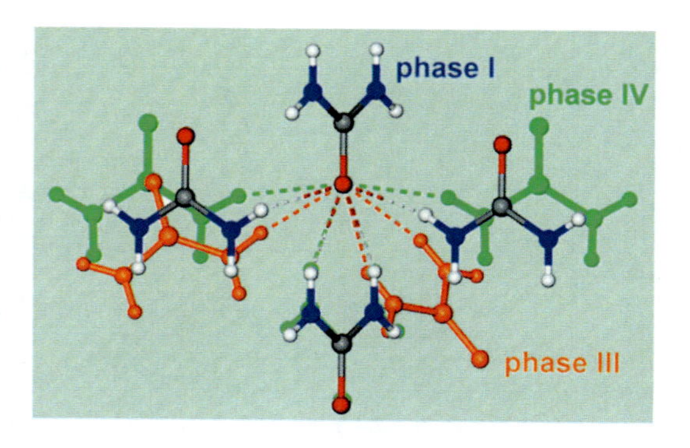

Fig. 17.36 Breaking and forming of hydrogen bonds with increasing pressure in the crystals of urea with increasing pressure: phase I transforms into phase III at 0.48 GPa, and then phase III transforms into phase IV at 2.80 GPa (Reproduced from Ref. [146] with kind permission of The American Chemical Society)

zwitterions increases. This effect can be to some extent compared with a similar, but a much stronger effect (a practically complete transfer of a proton between two neutral molecules to give ions, not just a small shift along a hydrogen bond linking two zwitterions) which has been observed at high pressure for the oxalic acid dihydrate [137].

One can also follow reliably breaking and forming of hydrogen bonds with increasing pressure. In the crystals of urea, hydrogen bonds NH . . . O are broken and restored, and their lengths changed by more than 1 Å, when exposed to high pressure (Fig. 17.36). At 0.48 GPa, on transformation from phase I (tetragonal space group $P42_1m$) to phase III (orthorhombic space group $P2_12_12_1$), the channel voids characteristic of phase I collapse, one of the NH . . . O bonds is broken, and the H-acceptor capacity of the oxygen atom is reduced from 4 to 3. Above 2.80 GPa, in phase IV (orthorhombic space group $P2_12_12$), the H-bonding pattern of phase I and fourfold H-acceptor oxygen are restored. The thermodynamic phase transitions in urea have been rationalized by a microstructural mechanism involving the interplay of pressure-induced molecular reorientations, with hydrogen bonds competing for access to lone-electron pairs of carbonyl oxygen, and by the increasing role of van der Waals interactions. None of phases I, III, and IV contain the hydrogen bond types most frequently encountered in urea co-crystals [146].

The changes in the hydrogen bonds have been shown to account for the phase transitions in the crystals of the β- and γ-polymorphs of glycine, and L-serine. A detailed analysis of the changes in the hydrogen bonds and other interatomic distances with increasing pressure helps to understand better, which structural distortions precede the phase transitions and eventually make them possible. The changes in the distances between N and O atoms in the NH . . . O hydrogen bonds in the β-polymorph throughout the phase transition into the β-polymorph suggest, that the phase transition occurs at the pressure point, when the NH . . . O hydrogen bond in a head to tail chain becomes bifurcated, that is when the two distances between the nitrogen atom of NH_3-tail and the two oxygen atoms of a neighboring COO-head become short enough, to allow the NH . . . O be classified as a hydrogen bond [147, 148]. The same holds also for the pressure-induced phase transition of the γ-form

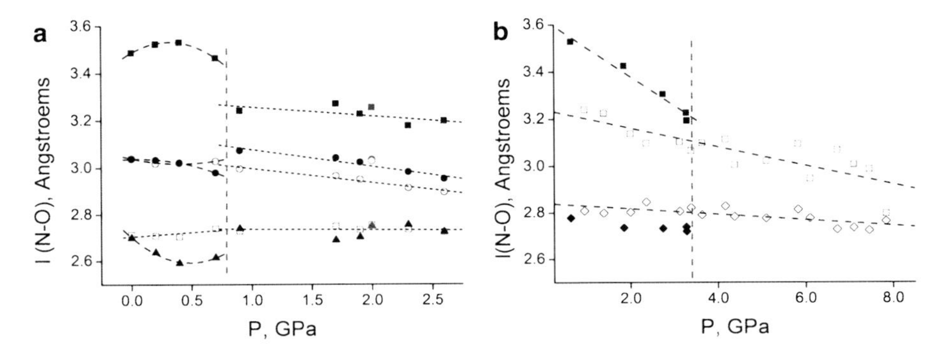

Fig. 17.37 The changes of the N-O distances in the NH ... O hydrogen bonds versus pressure in the β-polymorph (**a**) and in the γ-polymorph (**b**) of glycine. The pressures of the phase transitions are shown by dashed red vertical lines: (**a**) *black squares* – the long, *open squares* – the short NH ... O hydrogen bonds in the head-to-tail chain, *black* and *open circles* – the NH ... O bonds between the layers, triangles – the NH ... O bonds within a layer for the β- (before the phase transition) and for the β′- (after the phase transition) polymorphs; *red symbols* – the values for the α-polymorph; (**b**) *squares* – the long, and *rhombs* – the short NH ... O hydrogen bonds in the head-to-tail chain, *black symbols* – the γ-polymorph, *open symbols* – the δ-polymorph (Reproduced from Ref. [148] with kind permission of Taylor & Francis)

into the δ-polymorph (Fig. 17.37) [148–151]. Another interesting observation is that the distortion of some hydrogen bonds in the structure of β-glycine prior to the phase transition point is not monotonic, and the same hydrogen bonds change only slightly after the phase transition point [148]. The lengths of the NH ... O hydrogen bonds in the head to tail chains and of those linking the chains with each other in a layer become equal at the β to β′ phase transition point, and remain practically unchanged on further compression of the β′-polymorph, which is almost exclusively due to the compression of the hydrogen bonds between the layers [147, 148].

In contrast to the crystal structures of the glycine polymorphs, in L-serine the NH ... O hydrogen bonds linking the $^{+}NH_3$-tails and the COO^{-} - heads in a chain of zwitter-ions are bifurcated already at ambient pressure. The phase transitions induced by cooling or by increasing pressure in the crystals of L-serine are related not to the formation of the second NH ... O bond in the chain, but to the re-orientation of the $-CH_2OH$ side chains and a change in the type of the OH ... O hydrogen bonds linking either the two –OH groups, or an –OH and a –COO- group of the neighboring chains (Fig. 17.38) [152–155]. The phase transitions with increasing pressure are of a cascade-type: after a pronounced induction period the side chains rotate cooperatively, so that the whole head-to-tail chain expands jump-wise as a spring, and the types of hydrogen bonds formed by the side chain change – first from –OH ... OH to –OH ... O(CO), and then adding also an extra NH ... O bond. The single crystals are preserved intact during these transitions (Fig. 17.39) [156].

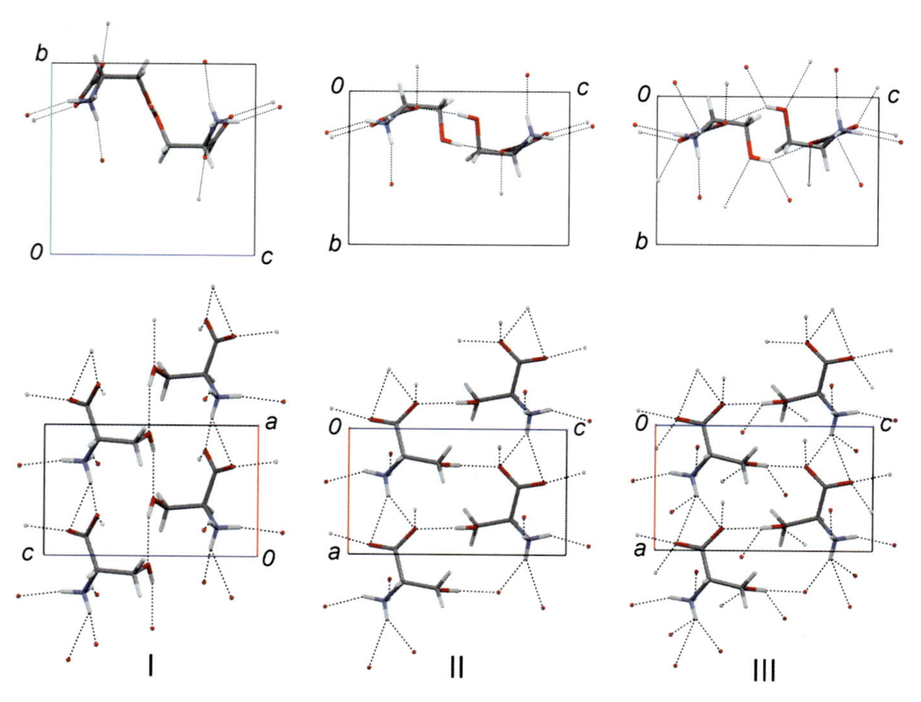

Fig. 17.38 The changes in the intermolecular hydrogen bonding (*dashed lines*) in the crystals of L-serine with increasing pressure [154]; phase II forms from phase I at 5.4 GPa, and phase III from phase II at ~8 GPa (Reproduced from Ref. [154] with kind permission of Elsevier)

17.5 Conclusions

This chapter aimed to illustrate at several selected examples that it is already possible to study the electron density topology of solids at high pressure, similar to how this is being almost routinely done at ambient conditions and at low temperatures. For inorganic systems and minerals such studies become quite common already nowadays. The high-pressure experimental studies of the electron density topology for molecular organic solids remain a challenge, but as the experimental techniques progress, this challenge will be surely met. There are already the first examples of theoretical papers treating the electron density topology in organic molecular crystals at high pressures [157]. The completeness of the experimental diffraction data is still low, to apply the same level of analysis, but other techniques – the calculation of Hirshfeld surfaces and fingerprints [110, 158, 159], the analysis of the intramolecular geometry and the intermolecular contacts, complementary Raman and IR-spectroscopy – are widely used, to follow the re-distribution of the electron density induced by pressure. The studies of the electron density topology can give us valuable information for the understanding of the nature of chemical bonding, solid-state reactivity and structure-properties relations. Extending the experimental

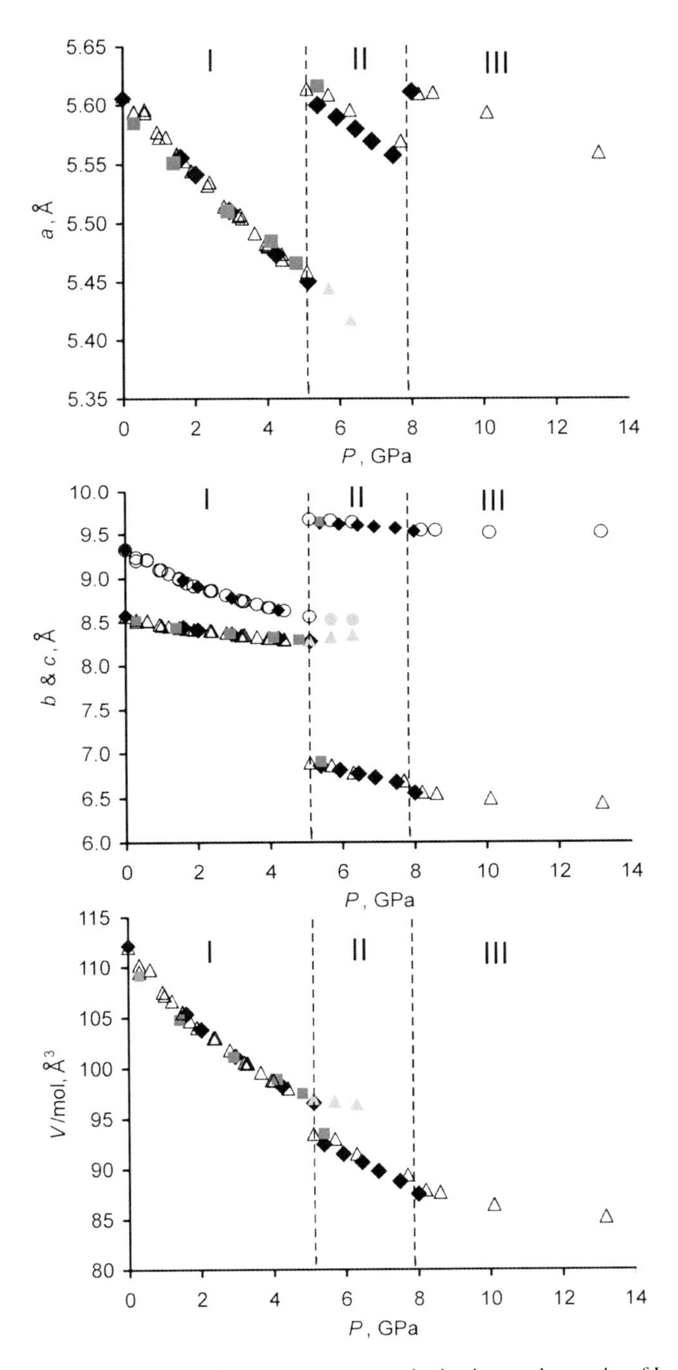

Fig. 17.39 Cell parameters and volume versus pressure in the three polymorphs of L-serine; *open symbols* – powder diffraction data, *black symbols* – single-crystal diffraction data, *gray symbols* – phase I partly preserved after the I–II transition in the powder sample [154]. Vertical dashed lines define the pressures at which phase transitions occur (Reproduced from Ref. [154] with kind permission of Elsevier)

conditions, at which the electron density topology studies are possible, we make it possible to explore new horizons in chemistry, since chemical bonding and intermolecular interactions at high-pressure may be very different from those, to which our ambient-pressure intuition has been used [16, 160].

References

1. Bader RFW (1990) Atoms in molecules: a quantum theory. Oxford University Press, Oxford
2. Bader RFW (1985) Atoms in molecules. Acc Chem Res 18:9–15
3. Bader RFW (1991) A quantum theory of molecular structure and its applications. Chem Rev 91:893–928
4. Becke A, Edgecombe K (1990) A simple measure of electron localization in atomic and molecular systems. J Chem Phys 92:5397–5403
5. Silvi B, Savin A (1994) Classification of chemical bonds based on topological analysis of electron localization functions. Nature 371:683–686
6. Burdett JK, McCirmick TA (1998) Electron localization in molecules and solids: the meaning of ELF. J Phys Chem A 102:6366–6372
7. Heine V (2000) As weird as they come. Nature 403:836–837
8. Bonev AA, Schwegler E, Ogitsu T, Galli G (2004) A quantum fluid of metallic hydrogen suggested by first-principles calculations. Nature 431:669–672
9. Babaev E, Sudbøø A, Ashcroft NW (2004) A superconductor to superfluid phase transition in liquid metallic hydrogen. Nature 431:666–668
10. Tse JS (2004) In: Sham TK (ed) Chemical application of synchrotron radiation, vol 2. Word Scientific, Singapore
11. (a) McMahon MI, Nelmes RJ, Rekhi S (2001) Complex crystal structure of cesium-III. Phys Rev Lett 87:255502; (b) Nelmes RJ, McMahon MI, Loveday JS, Rekhi S (2002) Structure of Rb-III: novel modulated stacking structures in alkali metals. Phys Rev Lett 88:155503
12. McMahon MI, Rekhi S, Nelmes RJ (2001) Pressure dependent incommensuration in Rb-IV. Phys Rev Lett 87:055501
13. Schwarz U, Grzechnik A, Syassen K, Loa I, Hanfland M (1999) Rubidium-IV: a high pressure phase with complex crystal structure. Phys Rev Lett 83:4085–4088
14. McMahon MI, Hejny C, Loveday JS, Lundegaard LF (2004) Confirmation of the incommensurate nature of Se-IV at pressures below 70 GPa. Phys Rev B 70:054101
15. McMahon MI, Nelmes RJ (2004) Chain "melting" in the composite Rb-IV structure. Phys Rev Lett 93:055501
16. Grochala W, Hoffmann R, Fengand J, Ashcroft NW (2007) The chemical imagination at work in very tight places. Angew Chem 46:3620–3642
17. Tse JS (2005) Crystallography of selected high pressure elemental solids. Z Kristallogr 220:521–530
18. McMahon MI, Nelmes RJ (1993) New high-pressure phase of Si. Phys Rev B 47:8337–8340
19. Olijnyk H, Sikka SK, Holzapfel WB (1984) Structural phase transitions in Si and Ge under pressures up to 50 GPa. Phys Lett A 103:137–140
20. Hu JZ, Spain IL (1984) Phases of silicon at high pressure. Solid State Commun 51:263–266
21. Hanfland M, Schwarz U, Syassen K, Takemura K (1999) Crystal structure of the high-pressure phase silicon VI. Phys Rev Lett 82:1197–1200
22. Duclos SJ, Vohra YK, Ruoff AL (1987) Hcp to fcc transition in silicon at 78 GPa and studies to 100 GPa. Phys Rev Lett 58:775–777

23. Savin A, Jepsen O, Flad J, Andersen OK, Preuss H, Von Schering HG (1992) Electron localization in solid-state structures of the elements: the diamond structure. Angew Chem Int Ed Engl 31:187–188
24. Silvi A, Gatti C (2000) Direct space representation of the metallic bond. J Phys Chem A 104:947–953
25. Tse JS, Klug DD, Patchkovskii S, Dewhurst JK (2006) Chemical bonding, electron-phonon coupling, and structural transformations in high-pressure phases of Si. J Phys Chem B 110:3721–3726
26. Takata M, Nishibori E, Sakata M (2001) Charge density studies utilizing powder diffraction and MEM. Exploring of high T_c superconductors, C_{60} superconductors and manganites. Z Kristallogr 216:71–86
27. Tse JS, Flacau R, Desgreniers S, Hanfland M, to be published
28. Kume T, Fukuoka H, Koda T, Sasaki S, Shimizu H, Yamanaka S (2003) High-pressure Raman study of Ba doped silicon clathrate. Phys Rev Lett 90:155503
29. San Miguel A, Merlen A, Toulemonde P, Kume T, Le Floch S, Aouizerat A, Pascarelli S, Aquilanti G, Mathon O, Le Bihan T, Itié JP, Yamanaka S (2005) Pressure-induced homothetic volume collapse in silicon clathrates. Europhys Lett 69:556–562
30. Tse JS, Flacau R, Desgreniers S, Iitaka T, Jiang JZ (2007) Electron density topology of high-pressure Ba_8Si_{46} from a combined Rietveld and maximum-entropy analysis. Phys Rev B 76:174109
31. Tse JS, Frapper G, Ker A, Rousseau R, Klug DD (1999) Phase stability and electronic structure of K-Ag intermetallics at high pressure. Phys Rev Lett 82:4472–4475
32. Miedema AP, de Châtel PF, de Boer FR (1980) Cohesion in alloys – fundamentals of a semi-empirical model. Phys B 100:1–28
33. Pettifor DG (1987) Electronic structure of simple metals and transition metals. Solid State Phys 40:43 Academic Press
34. Atou T, Hasegawa M, Parker LJ, Badding JV (1996) Unusual chemical behavior for potassium under pressure: potassium-silver compounds. J Am Chem Soc 118:12104–12108
35. Wells AF (1984) Structural inorganic chemistry. Clarendon, Oxford
36. Schäfer H (1985) On the problem of polar intermetallic compounds: the stimulation of E. Zintl's work for the modern chemistry of intermetallics. Annu Rev Mater Sci 15:1–42, and references therein
37. Schwarz U, Takemura K, Hanfland M, Syassen K (1998) Crystal structure of cesium-V. Phys Rev Lett 81:2711–2714
38. Weir CE, Piermarini GJ, Block S (1971) On the crystal structures of Cs II and Ga II. J Chem Phys 54:2768–2770
39. Hall HT, Merrill L, Barnet JD (1964) High pressure polymorphism in cesium. Science 146:1297–1299
40. Kennedy GC, Jayaman A, Newton RC (1962) Fusion curve and polymorphic transitions of cesium at high pressures. Phys Rev 126:1363–1366
41. Takemura K, Minomura S, Shimomura O (1982) X-Ray diffraction study of electronic transitions in cesium under high pressure. Phys Rev Lett 49:1772–1775
42. Takemura K, Shimomura O, Fujihisa H (1991) Cs(VI): a new high-pressure polymorph of cesium above 72 GPa. Phys Rev Lett 66:2014–2017
43. Sternheimer R (1950) On the compressibility of metallic cesium. Phys Rev 78:235–243
44. von Schnering HG, Nesper RK (1987) How nature adapts chemical structures to curved surfaces. Angew Chem Int Ed Engl 26:1059–1080
45. Darling D (2004) The universal book of mathematics: from Abracadabra to Zeno's paradoxes. Wiley, Hoboken
46. Barrett CS (1956) X-ray study of the alkali metals at low temperatures. Acta Crystallogr 9:671–677

47. Overhauser AW (1984) Crystal structure of lithium at 4.2 K. Phys Rev Lett 53:64–65
48. Smith HG (1987) Martensitic phase transformation of single-crystal lithium from bcc to a 9R-related structure. Phys Rev Lett 58:1228–1231
49. Schwarz W, Blaschko O (1990) Polytype structures of lithium at low temperatures. Phys Rev Lett 65:3144–3147
50. Hanfland M, Syassen K, Christensen NE, Novikov DL (2000) New high-pressure phases of lithium. Nature 408:174–178
51. Shimizu K, Ishikawa H, Takao D, Yagi T, Amaya K (2002) Superconductivity in compressed lithium at 20 K. Nature 419:597–599
52. Struzhkin VV, Eremets MI, Gan W, Mao H-K, Hemley RJ (2002) Superconductivity in dense lithium. Science 298:1213–1215
53. Deemyad S, Schilling JS (2003) Superconducting phase diagram of Li metal in nearly hydrostatic pressures up to 67 GPa. Phys Rev Lett 91:167001
54. Ashcroft NW (2002) Superconductivity: putting the squeeze on lithium. Nature 419:569–572
55. Matsuoka T, Shimizu K (2009) Direct observation of a pressure-induced metal-to-semiconductor transition in lithium. Nature 458:186–189
56. Rousseau R, Marx D (2000) Exploring the electronic structure of elemental lithium: from small molecules to nanoclusters, bulk metal, and surfaces. Chem Eur J 6:2982–2993
57. Tse JS, Ma Y, Tutuncu HM (2005) Superconductivity in simple elemental solids – a computational study of boron-doped diamond and high pressure phases of Li and Si. J Phys Condens Matter 17:S911–S920
58. Kirz JK, Attwood DT, Howells MR, Kenndy KD, Kim K-J, Kortright JB, Perera RC, Pianetta P, Riodan JC, Scofield JH, Stradling GL, Thompson AC, Underwood JH, Vaugh D, Williams GP, Winick H, Center for X-ray Optics (1986) X-ray data booklet. Lawrence Berkeley Laboratory, Berkeley
59. Neaton JB, Ashcroft NW (1999) Pairing in dense lithium. Nature 400:141–144
60. Yao Y, Tse JS, Klug DD (2009) Structures of insulating phases of dense lithium. Phys Rev Lett 102:115503
61. Pickard CJ, Needs RJ (2006) High-pressure phases of silane. Phys Rev Lett 97:045504
62. Oganov AR, Glass CW (2006) Crystal structure prediction using ab initio evolutionary techniques: principles and applications. J Chem Phys 124:244704
63. Yao Y, Tse JS, Tanaka K (2008) Metastable high-pressure single-bonded phases of nitrogen predicted via genetic algorithm. Phys Rev B 77:052103
64. Rousseau R, Uehara K, Klug DD, Tse JS (2005) Phase stability and broken-symmetry transition of elemental lithium up to 140 GPa. ChemPhysChem 6:1703–1706
65. Pickard CJ, Needs RJ (2009) Dense low-coordination phases of lithium. Phys Rev Lett 102:146401
66. Yao Y, Tse JS, Song Z, Klug DD (2009) Core effects on the energetics of solid Li at high pressure. Phys Rev B 79:092103
67. Marqués M, Ackland GJ, Lundegaard LF, Stinton G, Nelmes RJ, McMahon MI, Contreras-García J (2009) Potassium under pressure: a pseudobinary ionic compound. Phys Rev Lett 103:115501
68. Ma Y, Eremets M, Oganov AR, Xie Y, Trojan I, Medvedev S, Lyakhov AO, Valle M, Prakapenka V (2009) Transparent dense sodium. Nature 458:182–185
69. Martín Pendás A, Costales A, Blanco MA, Recio JM, Luaña V (2000) Local compressibilities in crystals. Phys Rev B 62:13970–13978; Calatayud M, Mori-Sánchez P, Beltrán A, Martín Pendás A, Francisco E, Andrés J, Recio JM (2001) Quantum-mechanical analysis of the equation of state of anatase TiO_2. Phys Rev B 64:1841113; Mori-Sánchez P, Marqués M, Beltrán A, Jiang JZ, Gerward L, Recio JM (2003) Origin of the low compressibility in hard nitride spinels. Phys Rev B 68:0641151–0641155; Marqués M, Flórez M, Recio JM, Santamaría D, Vegas A, García Baonza V (2006) Structure, metastability, and electron density of Al lattices in light of the model of anions in metallic matrices. J Phys Chem B 110:18609–18618

70. Oganov AR, Gillan MJ, Price GD (2005) Structural stability of silica at high pressures and temperatures. Phys Rev B 71:064104
71. Pendás AM, Francisco E, Blanco MA, Gatti C (2007) Bond paths as privileged exchange channels. Chem Eur J 13:9362–9371
72. Bader RFW (1998) A bond path: a universal indicator of bonded interactions. J Phys Chem A 102:7314–7323
73. Bader RFW (2009) Bond paths are not chemical bonds. J Phys Chem A 113:10391–10396
74. Sakata M, Itsubo T, Nishibori E, Moritomo Y, Kojima N, Ohishi Y, Takata M (2004) Charge density study under high pressure. J Phys Chem Solids 65:1973–1976
75. Yamanaka T, Komatsu Y, Nomori H (2007) Electron density distribution of FeTiO3 ilmenite under high pressure analyzed by MEM using single crystal diffraction intensities. Phys Chem Miner 34:307–318
76. Viswanath RP, Seshadri AY (1994) The ferroelectric characteristics in Fe-Ti-O system. Solid State Commun 92:831–833
77. Fujihisa H, Fujii Y, Takemura K, Shimomura O, Nelmes RJ, McMahon MI (1996) Pressure dependence of the electron density in solid iodine by the maximum-entropy method. High Press Res 14:335–340
78. Silvi B (2004) Phase transition in iodine: a chemical picture. J Phys Chem Solids 65:2025–2029
79. Schiferl D, Cromer DT, Schwalbe LA, Mills RL (1983) Structure of 'orange' $^{18}O_2$ at 9.6 GPa and 297 K. Acta Crystallogr B 39:153–157
80. Nicol M, Hirsch KR, Holzapfel HB (1976) Oxygen phase equilibria near 298 K. Chem Phys Lett 68:49–52
81. Lundegaard LF, Weck G, McMahon MI, Desgreniers S, Loubeyre P (2006) Observation of an O_8 molecular lattice in the ε phase of solid oxygen. Nature 443:201–204
82. Fujihisa H, Akahama Y, Kawamura H, Ohishi Y, Shimomura O, Yamawaki H, Sakashita M, Honda K (2006) O_8 Cluster structure of the epsilon phase of solid oxygen. Phys Rev Lett 97:085503
83. Weck G, Desgreniers S, Loubeyre P, Mezouar M (2009) Single-crystal structural characterization of the metallic phase of oxygen. Phys Rev Lett 102(25):255503
84. Tse JS, Yao Y, Klug DD, Desgreniers S (2008) Bonding in the ε-phase of high pressure oxygen. J Phys Conf Ser 121(Part 1):012006
85. Neaton JB, Ashcroft NW (2002) Low–energy linear structures in dense oxygen: implications for the ε-phase. Phys Rev Lett 88:20550
86. Burkhard M, Hemley RJ (2006) Crystallography: solid oxygen takes shape. Nature 443:150
87. Gatti C (2005) Chemical bonding in crystals: new directions. Z Kristallogr 220:399
88. Cremer D, Kraka E (1984) Chemical bonds without bonding electron density. Angew Chem Int Ed Engl 23:627–628
89. Cremer D, Kraka E (1984) A description of the chemical bond in terms of local properties of electron density and energy, in conceptual approaches in quantum chemistry – models and applications. Croatica Chem Acta 57:1259–1281
90. Schiferl D, Cromer DT, Mills RL (1981) Structure of O_2 at 5.5 GPa and 299 K. Acta Crystallogr B37:1329–1332
91. Freiman YA, Jodl HJ (2004) Solid oxygen. Phys Rep 401:1–228
92. Goncharenko IN (2005) Evidence for a magnetic collapse in the epsilon phase of solid oxygen. Phys Rev Lett 94:205701
93. Bone RGA, Bader RFW (1996) Identifying and analyzing intermolecular bonding interactions in van der Waals molecules. J Phys Chem 100:10892–10911
94. Sim PG, Klug DD, Ikawa S, Whalley E (1984) Effect of pressure on molecular conformations. 4. The flattening of trithiane as measured by its infrared spectrum. J Am Chem Soc 106:502–508

95. Cailleau H, Girard A, Messager JC, Delugeard Y, Vettier C (1984) Influence of pressure on structural phase transitions in p-polyphenyls. Ferroelectrics 54:257–260
96. Molchanov VN, Shibaeva RP, Kachinski VN, Yagubski EB, Simonov VI, Vainstein BK (1986) Molecular and crystal structure of an organic superconductor β-(BEDT-TTF)$_2$I$_3$. Dokl Akad Nauk SSSR 286:637–640 [Sov Phys Dokl, 31, 6 (1986)]
97. Katrusiak A (1990) High-pressure X-ray diffraction studies of organic crystals. High Press Res 4:496–498
98. Katrusiak A (1990) High-pressure X-ray diffraction study on the structure and phase transition of 1,3-cyclohexanedione crystals. Acta Crystallogr B 46:246–256
99. Katrusiak A (1992) Stereochemistry and transformation of -OH-O = hydrogen bonds. I. Polymorphism and phase transition of 1,3-cyclohexanedione crystals. J Mol Struct 269:329–354
100. Katrusiak A (1991) High-pressure X-ray diffraction study of 2-methyl-1,3-cyclopentanedione crystals. High Press Res 6:155–167
101. Katrusiak A (1991) High-pressure X-ray diffraction study of dimedone. High Press Res 6:265–275
102. Katrusiak A (1991) High-pressure X-ray diffraction studies on organic crystals. Cryst Res Technol 26:523–531
103. Boldyreva EV, Shakhtshneider TP, Vasilchenko MA, Ahsbahs H, Uchtmann H (2000) Anisotropic crystal structure distortion of the monoclinic polymorph of acetaminophen at high hydrostatic pressures. Acta Crystallogr B 56(2):299–309
104. Binev IG, Vassileva-Boyadjieva P, Binev YI (1998) Experimental and ab initio MO studies on the IR spectra and structure of 4-hydroxyacetanilide (paracetamol), its oxyanion and dianion. J Mol Struct 447:235–246
105. Katrusiak A (2003) Macroscopic and structural effects of hydrogen-bond transformations: some recent directions. Crystallogr Rev 9:87–89
106. Boldyreva EV, Shakhtshneider TP, Ahsbahs H, Uchtmann H, Burgina EB, Baltakhinov VP (2002) The role of hydrogen bonds in the pressure-induced structural distortion of 4-hydroxyacetanilide crystals. Polish J Chem 76:1333–1346
107. Boldyreva EV, Naumov DYu, Ahsbahs H (1998) Distortion of crystal structures of some Co^{III} ammine complexes. III. Distortion of crystal structure of [Co(NH$_3$)$_5$NO$_2$]Cl$_2$ at hydrostatic pressures up to 3.5 GPa. Acta Crystallogr B 54:798–808
108. Boldyreva EV, Burgina EB, Baltakhinov VP, Burleva LP, Ahsbahs H, Uchtmann H, Dulepov VE (1992) Effect of high pressure on the infra-red spectra of solid nitro- and nitrito- cobalt (III) ammine complexes. Ber Bunsengesell Phys Chem 96:931–937
109. Katrusiak A (2008) High-pressure crystallography. Acta Crystallogr A 64:135–148
110. Dziubek KF, Katrusiak A (2004) Compression of intermolecular interactions in CS$_2$ crystal. J Phys Chem B 108:19089–19092
111. Bernasconi M, Chiarotti GL, Focher P, Parrinello M, Tosatti E (1997) Solid-state polymerization of acetylene under pressure: Ab initio simulation. Phys Rev Lett 78:2008–2011
112. Santoro M, Ciabini L, Bini R, Schettino V (2003) High-pressure polymerization of phenylacetylene and of the benzene and acetylene moieties. J Raman Spectrosc 34:557–566
113. Schettino V, Bini R, Ceppatelli M, Ciabini L, Citroni M (2005) Chemical reactions at very high pressure. Adv Chem Phys 131:105–242
114. Mugnai M, Pagliai M, Cardini G, Schettino V (2008) Mechanism of the ethylene polymerization at very high pressure. J Chem Theory Comput 4:646–651
115. Mediavilla C, Tortajada J, Baonza VG (2009) Modeling high pressure reactivity in unsaturated systems: application to dimethylacetylene. J Comput Chem 30:415–422
116. Citroni M, Bini R, Foggi P, Schettino V (2008) The role of excited electronic states in the high-pressure amorphization of benzene. Proc Natl Acad Sci USA 105:7658–7663
117. Politov AA, Chupakhin AP (2009) Phenanthrene crystals investigation under high pressure and shear conditions. Proc NSU Ser Phys 4:55–58

118. Politov AA, Chupakhin AP, Tapilin VM, Bulgakov NN, Druganov AG (2010) To mechanochemical dimerization of anthracene. Crystalline phenanthrene under high pressure and shear conditions. J Struct Chem 51:1064–1069
119. Tapilin VM, Bulgakov NN, Chupkhin AP, Politov AA, Druganov AG (2010) On the mechanism of mechanochemical dimerization of anthracene. Different possible reaction pathways. J Struct Chem 51:635–641
120. Bujak M, Podsiadło M, Katrusiak A (2008) Energetics of conformational conversion between 1,1,2-trichloroethane polymorphs. Chem Commun 37:4439–4441
121. Citroni M, Datchi F, Bini R, Vaira MD, Pruzan P, Canny B, Schettino V (2008) Crystal structure of nitromethane up to the reaction threshold pressure. J Phys Chem B 112:1095–1103
122. Boldyreva EV, Burgina EB, Baltakhinov VP, Stoyanov E, Zhanpeisov N, Zhidomirov GM (1993) Effect of high pressure on vibrational spectrum of nitromethane molecules: changes in the kinematics of the vibrations as a result of the decrease in the distance between molecules. J Molec Struct 296:53–59
123. (a) Casati N, Macchi P, Sironi A (2005) Staggered to eclipsed conformational rearrangement of $[Co_2(CO)_6(PPh_3)_2]$ in the solid state: an x-ray diffraction study at high pressure and low temperature. Angew Chem Int Ed 44:7736–7739; (b) Casati N, Macchi P, Sironi A (2009) Molecular crystals under high pressure: theoretical and experimental investigations of the solid-solid phase transitions in $[Co_2(CO)_6(XPh_3)_2]$ (X = P, As). Chem Eur J 15:4446–4457
124. Macchi P (2010) Ab initio quantum chemistry and semi-empirical description of solid state phases under high pressure: chemical applications. In: Boldyreva EV, Dera P (eds) High-pressure crystallography. Springer, Dordrecht
125. Moggach SA, Galloway KW, Lennie AR, Parois P, Rowantree N, Brechin EK, Warren JE, Murrie M, Parsons S (2009) Polymerisation of a Cu(II) dimer into 1D chains using high pressure. CrystEngComm 11:2601–2604
126. Goto A, Hondoh T, Mae A (1990) The electron density distribution in ice I_h determined by single-crystal x-ray diffractometry. J Chem Phys 93:1412–1416
127. Kuhs WF, Lehmann MS (1983) The structure of the ice Ih by neutron diffraction. J Phys Chem 87:4312–4313
128. Floriano MA, Klug DD, Whalley E, Svensson EC, Sears VF, Hellman ED (1987) Direct determination of the intramolecular O–D distance in ice Ih and Ic by neutron diffraction. Nature 329:821–823
129. Jenkins S, Kirk SR, Ayers PW (2007) Topological transitions between ice phases. In: Kuhs W (ed) Physics and chemistry of ice. Royal Society of Chemistry, Cambridge, pp 249–256
130. Jenkins S, Kirk SR, Ayers PW (2007) The importance of O–O bonding interactions in various phases of ice. In: Kuhs W (ed) Physics and chemistry of ice. Royal Society of Chemistry, Cambridge, pp 256–265
131. Jenkins S, Kirk SR, Ayers PW (2007) The chemical character of very high pressure ice phases. In: Kuhs W (ed) Physics and chemistry of ice. Royal Society of Chemistry, Cambridge, pp 265–272
132. Marques M, Ackland GJ, Loveday JS (2009) Nature and stability of ice X. High Press Res 29:208–211
133. Jenkins S (2002) Direct space representation of metallicity and structural stability in SiO solids. J Phys Condens Matter 14:10251
134. Hama J, Shiomi Y, Suito K (1990) Equation of state and metallization of ice under very high pressure. J Phys Condens Matter 2:8107
135. Flacau R, Tse JS, Desgreniers S (2008) Electron density topology of cubic structure I Xe clathrate hydrate at high pressure. J Chem Phys 129:244507
136. Ripmeester JA, Ratcliffe CI, Klug DD, Tse JS (1994) Molecular perspectives on structure and dynamics in clathrate hydrates. Annals N Y Acad Sci 715:161–176
137. Casati N, Macchi P, Sironi A (2009) Hydrogen migration in oxalic acid di-hydrate at high pressure? Chem Commun 2679–2681

138. Tumanov NA, Boldyreva EV, Kurnosov AV, Quesada Cabrera R (2010) Pressure-induced phase transitions in L-alanine, revisited. Acta Crystallogr B 66:458–471
139. Barthes M, Bordallo HN, Denoyer F, Lorenzo J-E, Zaccaro J, Robert A, Zontone F (2004) Micro-transitions or breathers in L-alanine? Eur Phys J B 37:375–382
140. Destro R, Marsh RE, Bianchi R (1988) A low-temperature (23 K) study of L-alanine. J Phys Chem 92:966–973
141. Destro R, Bianchi R, Gatti C, Merati F (1991) Total electronic charge density of L-alanine from X-ray diffraction at 23 K. Chem Phys Lett 186:47–52
142. Destro R, Soave R, Barzaghi M (2008) Physicochemical properties of zwitterionic L- and DL-alanine crystals from their experimental and theoretical charge densities. J Phys Chem B 112:5163–5174
143. Dunitz JD, Ryan RR (1966) Refinement of the L-alanine crystal structure. Acta Crystallogr 21:617–618
144. Lehmann MS, Koetzle TF, Hamilton WC (1972) Precision neutron diffraction structure determination of protein and nucleic acid components. I. The crystal and molecular structure of the amino acid L-alanine. J Am Chem Soc 94:2657–2660
145. Wilson CC, Myles D, Ghosh M, Johnson LN, Wang W (2005) Neutron diffraction investigations of L- and D-alanine at different temperatures: the search for structural evidence for parity violation. New J Chem 29:1318–1322
146. Olejniczak A, Ostrowska K, Katrusiak A (2009) H-bond breaking in high-pressure urea. J Phys Chem C 113:15761–15767
147. Tumanov NA, Boldyreva EV, Ahsbahs H (2008) Structure solution and refinement from powder or from single-crystal diffraction data? Pro and contras: an example of the high-pressure β-polymorph of glycine. Powder Diffr 23:307–316
148. Boldyreva EV (2009) Combined X-ray diffraction and Raman spectroscopy studies of phase transitions in crystalline amino acids at low temperatures and high pressures. Selected examples. Phase Transit 82:303–321
149. Boldyreva EV, Ivashevskaya SN, Sowa H, Ahsbahs H, Weber H-P (2004) Effect of high pressure on crystalline glycine: a new high-pressure polymorph. Doklady Phys Chem 396:111–114
150. Boldyreva EV, Ivashevskaya SN, Sowa H, Ahsbahs H, Weber H-P (2004) Effect of hydrostatic pressure on the gamma-polymorph of glycine: a phase transition. Mater Struct 11:37–39
151. Boldyreva EV, Ivashevskaya SN, Sowa H, Ahsbahs H, Weber H-P (2005) Effect of hydrostatic pressure on the γ-polymorph of glycine 1. A polymorphic transition into a new δ-form. Z Kristallogr 220:50–57
152. Moggach SA, Allan DR, Morrison CA, Parsons S, Sawyer L (2005) Effect of pressure on the crystal structure of L-serine-I and the crystal structure of L-serine-II at 5.4 GPa. Acta Crystallogr B 61:58–68
153. Drebushchak TN, Sowa H, Seryotkin YuV, Boldyreva EV, Ahsbahs H (2006) L-serine III at 8.0 GPa. Acta Crystallogr E 62:4052–4054
154. Boldyreva EV, Sowa H, Seryotkin YuV, Drebushchak TN, Ahsbahs H, Chernyshev VV, Dmitriev VP (2006) Pressure-induced phase transitions in crystalline l-serine studied by single-crystal and high-resolution powder X-ray diffraction. Chem Phys Lett 429:474–478
155. Moggach SA, Marshall WG, Parsons S (2006) High-pressure neutron diffraction study of L-serine-I and L-serine-II, and the structure of L-serine-III at 8.1 GPa. Acta Crystallogr B 62:815–825
156. Kolesnik EN, Goryainov SV, Boldyreva EV (2005) Different behavior of the crystals of L- and DL-serine at high pressure. Transitions in L-serine and the stability of the phase of DL-serine. Doklady Phys Chem 404:169–172
157. Zhurova EA, Tsirelson VG, Zhurov VV, Stash AI, Pinkerton AA (2006) Chemical bonding in pentaerythritol at very low temperature or at high pressure: an experimental and theoretical study. Acta Crystallogr B 62:513–520

158. Wolff SK, Grimwood DJ, McKinnon JJ, Jayatilaka D, Spackman MA (2007) CrystalExplorer, version 2.1. Tech. Rep. University of Western Australia. http://hirshfeldsurfacenet.blogspot.com
159. Wood PA, McKinnon JJ, Parsons S, Pidcock E, Spackman MA (2008) Analysis of the compression of molecular crystal structures using hirshfeld surfaces. CrystEngComm 10:368–376
160. Boldyreva EV, Dera P (eds) (2010) High-pressure crystallography. Springer, Dordrecht
161. http://journals.iucr.org/

Chapter 18
Bonding Changes Along Solid-Solid Phase Transitions Using the Electron Localization Function Approach

Julia Contreras-García, Miriam Marqués, Bernard Silvi, and José M. Recio

18.1 Introduction

The characterization of chemical changes induced by thermodynamic variables in crystals is of capital interest in a variety of scientific areas, from fundamentals in solid state chemistry to applications in materials science. Among these variables, pressure constitutes a key parameter for precise tuning of interatomic distances, thus controlling the electronic structure and virtually all the interatomic interactions that determine materials properties. Indeed, pressure tuning usually enables a more rapidly and clean optimization of properties than either chemical agents or thermal effects, which introduce greater disorder and undesirable experimental by-side products. A deep knowledge of the complex interplay between structural, energetic, and bonding aspects involved in the system is required, should we contribute to the global understanding of the behavior of solids under high pressure conditions. These correlations may evolve softly within the same crystalline structure or sharply between different polymorphs if a solid-solid phase transition occurs.

In general, detailed information on the geometry and phase stability dependence on pressure (and temperature) of a given crystal structure can be obtained from

J. Contreras-García (✉) • J.M. Recio
MALTA-Consolider Team and Departamento de Química Física, Universidad de Oviedo, E-33006 Oviedo, Spain
e-mail: contrera@lct.jussieu.fr; mateo@fluor.quimica.uniovi.es

M. Marqués
SUPA, School of Physics and Centre for Science at Extreme Conditions, The University of Edinburgh, Mayfield Road, Edinburgh EH9 3JZ, UK
e-mail: mmarques@staffmail.ed.ac.uk

B. Silvi
Laboratoire de Chimie Théorique (UMR-CNRS 7616), Université Pierre et Marie Curie, 3 rue Galilée, 94200 Ivry sur Seine, France
e-mail: silvi@lct.jussieu.fr

C. Gatti and P. Macchi (eds.), *Modern Charge-Density Analysis*,
DOI 10.1007/978-90-481-3836-4_18,

X-ray and/or neutron diffraction experiments, and also predicted from computer assisted simulations. Inter-phase phenomena are more elusive to experiments and should usually resort to martensitic-like or molecular dynamics models describing the mechanism of the transformation, whose validity ultimately relies on the comparison with the available data of the observed phase transition properties.

In both intra- and inter-phase phenomena in solids, chemical bonding appears as a central topic towards which many efforts are currently invested in order to reconcile the traditional picture derived from the Lewis theory with the outcome of first principles quantum-mechanical methodologies [1]. Quantitative and rigorous formalisms based on the topological analysis of scalar fields as the electron density or the electron localization function (ELF) have been successfully used in the molecular realm, but their application in the solid state has been less frequent (see Ref. [2] for a review). ELF represents a suitable tool to study chemical changes induced by pressure in crystalline solids due to its ability to decompose the space in meaningful chemical fragments. This function provides foundation for a rich characterization of the chemistry of compounds that merits to be exploited in detail in the context of Lewis theory in the solid state.

The main applications of ELF to solid state studies have been centered in the description of intra-phase phenomena and the characterization of bonding. It was first applied by Savin et al. [3] in a seminal paper that established the relationship of ELF with the Pauli repulsion principle and adapted the function to DFT methodologies (see below). This formulation was then used to analyze pair localization in diamond and other group-IV related compounds. Following applications of ELF to solid-state were mainly related to the characterization of bonding in intermetallic and Zintl phases [4, 5], whereby ionic, covalent and delocalized interactions may occur simultaneously [5]. In 2000, Silvi and Gatti [6] envisaged the study of numerous metals in order to establish the common features that characterize metallic behavior. They ended up with the concept of localization window, which, by means of the difference between the ELF at the maxima and the first order saddle point or bond interconnection point, quantifies the delocalization across the crystal.

Not much work has been devoted to track chemical bonding changes across solid-solid phase transformations using topological analysis. We should point out that two main types of modifications occur: related to crystal packing (physical), and related to bonding changes (chemical). Within the former, the capacity of pressure to promote higher atomic coordinations in crystalline solids is one of its most outstanding features. This fact leads to densification processes of fundamental interest in areas ranging from planetary sciences to materials engineering. The electronic changes associated with a more effective atomic packing is an issue that needs to be addressed if a complete characterization of the densification process is desired. In addition, the phase transition may be accompanied by a change in the general bonding pattern of the solid. In this case, understanding the process of bond formation and rupture becomes of crucial interest. The ELF ability to unveil the nature of these two types of changes in solid-solid phase transitions is at the heart of the subject of this chapter. In all the cases, special attention will be paid to show how

ELF is able to provide: (i) insight into the differences in the chemical nature of the parent and daughter phases and (ii) a detailed track of the bonding reconstruction process along the transformation mechanism.

18.2 ELF Description of Bonding in Solids

In the upcoming paragraphs, we will review the main concepts associated with the topological analysis of the ELF gradient field. First, we will focus on the description that ELF provides of bonding, emphasizing the most relevant differences between molecular and solid topologies. Then, we will go deeper into the analysis of bond formation, which holds the key for understanding bond reconstruction across transition paths. The main conclusion to be drawn from these sections is that the spatial distribution of critical points of ELF is crucial for understanding the changes in the bonding pattern, whereas the ELF values at these points become essential in order to gain deeper insight into electronic fluctuation and bond orders.

18.2.1 Static Characterization of Bonding

The Electron Localization Function (ELF) was originally designed by Becke and Edgecombe to identify "localized electronic groups in atomic and molecular systems" [7]. It relies, through its kernel, to the laplacian of the conditional same spin pair probability scaled by the homogeneous electron gas kinetic energy density:

$$\chi_\sigma(\mathbf{r}) = \frac{D_\sigma(\mathbf{r})}{D_\sigma^0(\mathbf{r})}, \quad ELF = \eta(\mathbf{r}) = \frac{1}{1+\chi(\mathbf{r})^2}, \tag{18.1}$$

in which

$$D_\sigma(\mathbf{r}) = t_\sigma(\mathbf{r}) - \frac{1}{4}\frac{|\nabla\rho_\sigma(\mathbf{r})|^2}{\rho_\sigma(\mathbf{r})} \tag{18.2}$$

is the difference of the actual definite positive kinetic energy density $t_\sigma(\mathbf{r})$ and the von Weizsäcker kinetic energy density functional [8], whereas

$$D_\sigma^0(\mathbf{r}) = \frac{3}{5}(6\pi^2)^{2/3}\rho_\sigma^{5/3}(\mathbf{r}) \tag{18.3}$$

is the kinetic energy density of the homogeneous electron gas. This formulation enabled its calculation from Kohn-Sham orbitals [3, 5, 9]. Orbital-based interpretations of ELF have been proposed by Burdett [10] and more recently by Nalewajski et al. [11] who considered the non additive interorbital Fisher information. Another route pioneered by Dobson [12] explicitly considers the pair functions. It has been independently developed by Kohout et al. [13] and by one of us [14], allowing the extension of ELF to correlated wave functions [15].

From a simple statistical viewpoint, the concept of electron density localization at a given position **r** relies on the standard deviation of the electron density integrated over a sampling volume $V(\mathbf{r})$ encompassing the reference point and containing a given quantity of matter, in other words a given charge q. The smaller the standard deviation is, the higher is the localization. Instead of the standard deviation, it is advantageous to use its square, the variance σ^2, which can be expressed as the expectation value of the variance operator [16]:

$$\begin{aligned} \langle \hat{\sigma}^2(\bar{N}[V(\mathbf{r})]) \rangle &= \bar{\Pi}(V(\mathbf{r}), V(\mathbf{r})) - \bar{N}(V(\mathbf{r}))\,(\bar{N}(V(\mathbf{r})) - 1) \\ &= \bar{\Pi}(V(\mathbf{r}), V(\mathbf{r})) - q^2 + q \end{aligned} \tag{18.4}$$

in which $\bar{N}(V(\mathbf{r})) = q$ and $\bar{\Pi}(V(\mathbf{r}), V(\mathbf{r}))$ are respectively the one particle and two particle densities integrated over the sample $V(\mathbf{r})$. In the expression of the variance given above, only $\bar{\Pi}(V(\mathbf{r}), V(\mathbf{r}))$ is function of the position and therefore $-q^2 + q$ can be regarded as a constant and ignored. The integrated pair density is the sum of an opposite spin contribution, $2\bar{\Pi}^{\alpha\beta}(V(\mathbf{r}), V(\mathbf{r}))$ almost proportional to q^2 and of a same spin contribution $\bar{\Pi}^{\alpha\alpha}(V(\mathbf{r}), V(\mathbf{r})) + \bar{\Pi}^{\beta\beta}(V(\mathbf{r}), V(\mathbf{r}))$. The expectation value of the variance of the integrated opposite spin pair density,

$$\begin{aligned} \langle \hat{\sigma}^2(\bar{\Pi}^{\alpha\beta}(V(\mathbf{r}), V(\mathbf{r}))) \rangle &= \bar{\Pi}^{\alpha\alpha}(V(\mathbf{r}), V(\mathbf{r}))\bar{\Pi}^{\beta\beta}(V(\mathbf{r}), V(\mathbf{r})) \\ &\quad + \bar{N}^{\alpha}(V(\mathbf{r}))\bar{\Pi}^{\beta\beta}(V(\mathbf{r}), V(\mathbf{r})) \\ &\quad + \bar{N}^{\beta}(V(\mathbf{r}))\bar{\Pi}^{\alpha\alpha}(V(\mathbf{r}), V(\mathbf{r})) + \bar{N}^{\alpha}(V(\mathbf{r}))\bar{N}^{\beta}(V(\mathbf{r})) \\ &\quad - (\bar{\Pi}^{\alpha\beta}(V(\mathbf{r})))^2 \end{aligned} \tag{18.5}$$

also depends upon the integrated same spin pair densities. The integrated same spin pair density has numerically been shown proportional to a function of the position, say $c_\pi(\mathbf{r})$, times $q^{5/3}$. In the limit $q > 0$, the ratio

$$\frac{\bar{\Pi}(V(\mathbf{r}), V(\mathbf{r}))}{q^{5/3}}$$

tends to the spin pair composition, $c_\pi(\mathbf{r})$, a local function independent of the size of the sample [14]. The ability of this function to localize "electronic groups" can be illustrated by a very simple example in which two α and two β spin electrons are confined in a box of volume Ω as represented in Fig. 18.1. For the sake of simplicity we assume the electron density probability to be uniform, i. e. $\rho(\mathbf{r}) = 4/\Omega$ without spin polarization ($\rho^\alpha(\mathbf{r}) = \rho^\beta(\mathbf{r}) = 2/\Omega$), such as the opposite spin pair functions,

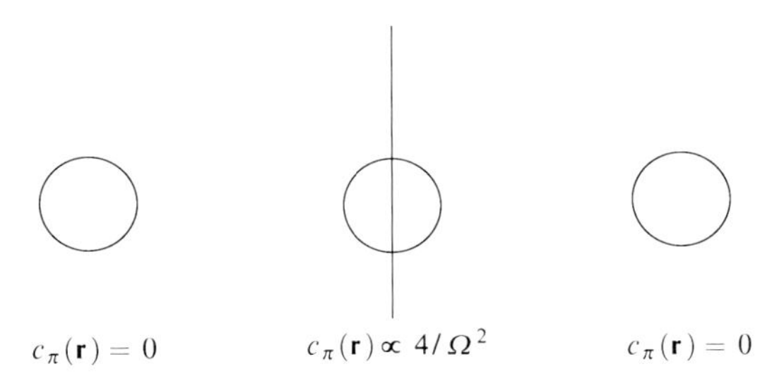

Fig. 18.1 Each half space contains two of opposite spin electrons and the density is assumed to be constant. The population $\bar{N}(\mathbf{r})$ of the sampling volume being denoted by q, the antiparallel pair population of the sample is also constant and equal to $\frac{q^2}{2}$. The spin pair composition is identically zero when the sample is entirely in the same half space, it reaches its maximum when the sample crosses the boundary

$\Pi^{\alpha\beta}(\mathbf{r}_1, \mathbf{r}_2) = \Pi^{\beta\alpha}(\mathbf{r}_1, \mathbf{r}_2) = 4/\Omega^2$ are constant. This model enables to consider two localization cases. On the one hand, the opposite spin pairs are delocalized over the box and the same spin pair functions are constant: $\Pi^{\alpha\alpha}(\mathbf{r}_1, \mathbf{r}_2) = \Pi^{\beta\beta}(\mathbf{r}_1, \mathbf{r}_2) = 2/\Omega^2$ and therefore $c_\pi(\mathbf{r})$ is also constant. On the other hand, each opposite spin pair occupies one half of the box such as:

$$\Pi^{\alpha\alpha}(\mathbf{r}_1, \mathbf{r}_2) = \Pi^{\beta\beta}(\mathbf{r}_1, \mathbf{r}_2) = \begin{cases} 0 & \mathbf{r}_1, \mathbf{r}_2 \in \text{same half box} \\ 4/\Omega^2 & \mathbf{r}_1, \mathbf{r}_2 \in \text{different half boxes} \end{cases} \tag{18.6}$$

It follows that

$$c_\pi(\mathbf{r}) \begin{cases} = 0 & \mathbf{r} \notin \text{the boundary} \\ \propto 4/\Omega^2 & \mathbf{r} \in \text{the boundary} \end{cases} \tag{18.7}$$

which enables to locate the boundary between the two opposite spin pair regions.

For Hartree-Fock wavefunction, it can be easily demonstrated [14] that:

$$c_\pi(\mathbf{r}) \approx \chi_\sigma(\mathbf{r}) \tag{18.8}$$

The ELF itself is further obtained through the transformation of $\chi_\sigma(\mathbf{r})$ into a lorentzian function

$$\eta(\mathbf{r}) = \frac{1}{1 + \chi_\sigma^2(\mathbf{r})} \tag{18.9}$$

so that it tends to 1 in those regions where the localization is high and to small values at the boundaries between such regions.

A partition of the 3D space is achieved by applying the dynamical system theory [17] to the ELF gradient field [1, 18]. This yields basins of attractors which can be thought of as corresponding to atomic cores, bonds, and lone pairs and therefore recovering the Lewis picture of bonding and the electronic domains of the valence shell electron pair repulsion (VSEPR) approach. Moreover, it has been recently shown that the electrostatic repulsions between the ELF basins provide a justification of the VSEPR rules [19].

The core basins surround nuclei with atomic number $Z > 2$ and are labeled C(A) where A is the atomic symbol of the element. The union of the valence basins encompassing a given core C(A) constitutes the valence shell of atom A. A valence basin may be shared by several valence shells, this is a generalization of Lewis's fourth postulate "Two atomic shells are mutually interpenetrable" [20]. The valence basins are characterized by the number of atomic valence shells to which they participate, or in other words by the number of core basins with which they share a boundary. This number is called the synaptic order. Thus, there are monosynaptic, disynaptic, trisynaptic basins, and so on. Monosynaptic basins, labeled V(A), correspond to the lone pairs of the Lewis model, and polysynaptic basins to the shared pairs of the Lewis model. In particular, disynaptic basins, labeled V(A, X) correspond to two-centre bonds, trisynaptic basins, labeled V(A, X, Y) to three-centre bonds, and so on [21].

In the context of the ELF analysis, the concept of domain is very important because it enables definition of chemical units within a complex system. The topological concept of domain has been introduced in chemistry by P. Mezey in order to recognize functional groups within organic molecules [22]. In the context of the ELF analysis, a f-localization is a subset of the 3D space satisfying the following requirements:

1. for any point P of the domain $\eta(\mathbf{r}_P) \geq f$
2. between any two points of the domain there exists a path entirely contained in the domain.

The f-localization domains are volumes bonded by an external $\eta(\mathbf{r}) = f$ isosurface which can be either filled or hollowed if they surround other f-localization domains. A localization domain surrounds at least one attractor; in this case it is called irreducible. If it contains more than one attractor, it is reducible. An irreducible domain is a subset of a basin whereas a reducible one is the union of subsets of different basins. Except for atoms and linear molecules, the irreducible domains are always filled volumes whereas the reducible ones can be either filled volumes or hollowed volumes. Upon the increase of the value of $\eta(\mathbf{r})$ defining the bounding isosurface, a reducible domain splits into several domains each containing less attractors than the parent domain. The reduction of localization occurs at turning points which are first order saddle points located on the separatrix of two basins involved in the parent domain. These saddle points are called *basin interconnection points* often abbreviated as *bips* [23]. Ordering these turning points (localization nodes) by increasing $\eta(\mathbf{r})$ enables tree-diagrams to be built, reflecting the hierarchy of the basins and providing the connectivity of the different fragments of the system [24, 25].

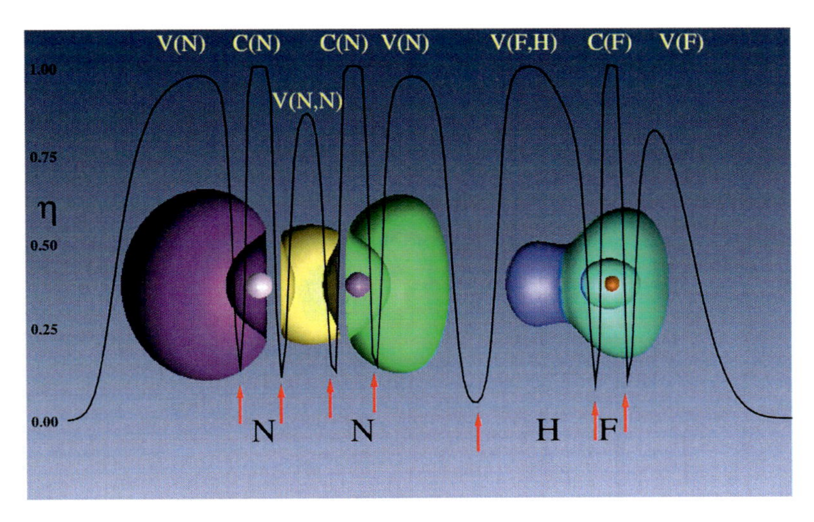

Fig. 18.2 ELF localization domains of $FH\cdots N_2$ and ELF profile along the C_∞ axis. The isosurface value is $\eta(r) = 0.75$. The *red arrows* points to the basin interconnection points

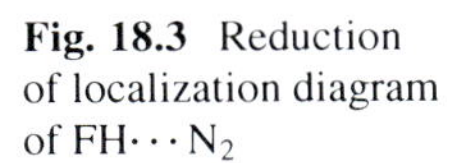

Fig. 18.3 Reduction of localization diagram of $FH\cdots N_2$

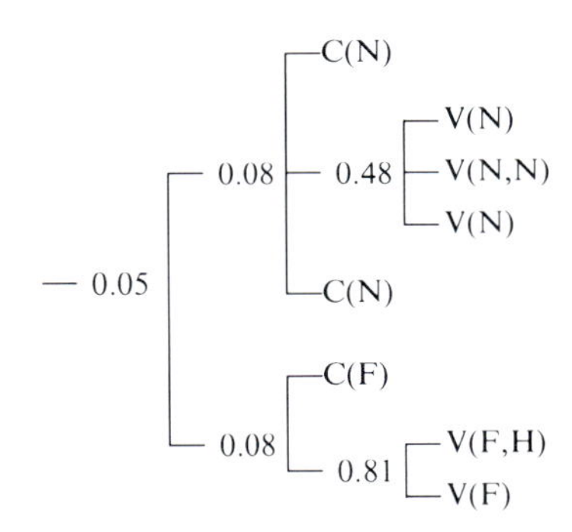

Figure 18.2 illustrates the different concepts introduced above on the example of the weakly bonded $FH\cdots N_2$ complex. The valence shell of the external nitrogen atom is the union of the V(N) monosynaptic basin represented in magenta and of the V(N,N) disynaptic basin in yellow. This latter also contributes to the valence shell of the proton acceptor nitrogen. The V(F, H) basin light blue is shared by the valence shell of both fluorine and hydrogen and is therefore disynaptic whereas V(F), in turquoise blue belongs only to the fluorine valence shell. The reduction of localization diagram, Fig. 18.3, clearly shows that in the $FH\cdots N_2$ complex the two moieties remain distinct chemical units. A first bifurcation occurs for $\eta(\mathbf{r}) = 0.05$ which splits the parent domain into two filled children corresponding to FH and N_2. These latter become hollowed at c.a. $\eta(\mathbf{r}) = 0.08$ which corresponds to the core valence separation. Each valence domain is further separated at higher ELF values. There are several factors that draw a clear line between the topological analyses of

ELF in the molecular and the crystalline realms. To begin with, an infinite number of critical points is expected due to the periodicity of the solid. Secondly, the presence of all types of critical points of rank 3 is ensured:

$$n_{(3,-3)} \geq 1, \qquad n_{(3,-1)} \geq 3, \qquad n_{(3,+1)} \geq 3, \qquad n_{(3,+3)} \geq 1 \tag{18.10}$$

being n the number of maxima (3,−3), *bips* or first order saddle points (3,−1), ring or second order saddle points (3,+1) points, and minima or cage points (3,+3). Furthermore, the Morse relationship followed by these sets is equal to zero:

$$n_{(3,-3)} - n_{(3,-1)} + n_{(3,+1)} - n_{(3,+3)} = 0. \tag{18.11}$$

As far as critical points of lower dimensionality are concerned, they are not observed in the solid state since the collapse of degenerated critical points is imposed by symmetry [1]. This gives rise to the main algorithmic difference between solids and molecules [19]. A great concentration of critical points cluster in small volumes that hampers exhausting the solutions by mere formation of dimers, triads, etc. Moreover, all solid basins have a well defined finite volume due to periodicity, which is not the case in molecules. On the other hand, very small and flat/steep profiles combinations are found in the solid state that once again complicates their algorithmic implementation. A methodology that overcomes these drawbacks has been used throughout most of the following examples [26, 27]. It is based on a core-valence separation algorithm [26] whose applicability has been tested across a wide variety of crystalline systems (see below).

Two of the quantities that are most commonly calculated in solids are the volume (Ω_i) and the population ($\bar{N}$) of the basin:

$$\Omega_i = \int_{\Omega_i} d\mathbf{r}, \quad \bar{N}(\Omega_i) = \int_{\Omega_i} \rho(\mathbf{r}) d\mathbf{r}. \tag{18.12}$$

These integrations have been able to prove the clear relationship between ELF and electronic pairing, by showing that the population associated to basins follow the expected values and tendencies from the Aufbau principle and the VSEPR theory [28].

Given the fact that topological regions are non-overlapping and space filling in crystalline solids, the so obtained basin contributions present the interesting property of recovering the value of the system volume when added up. Furthermore, since the topological analysis of the ELF gradient field [1, 18] yields basins that can be associated to Lewis entities, the integration of properties over ELF basins assigns chemically meaningful properties to bonds, lone pairs, etc. As shown in Fig. 18.4, ELF is able to clearly represent and differentiate between the different types of solids. Nitrogen, as a molecular solid, shows well differentiated chemical ELF objects that are identified as N_2 molecules. At both ends of the N_2 molecular axis, we find the lone pairs, V(N), which surround the nitrogen nuclei, N. The cylindrical basin in between two nitrogen atoms represents the N-N bond, V(N,N).

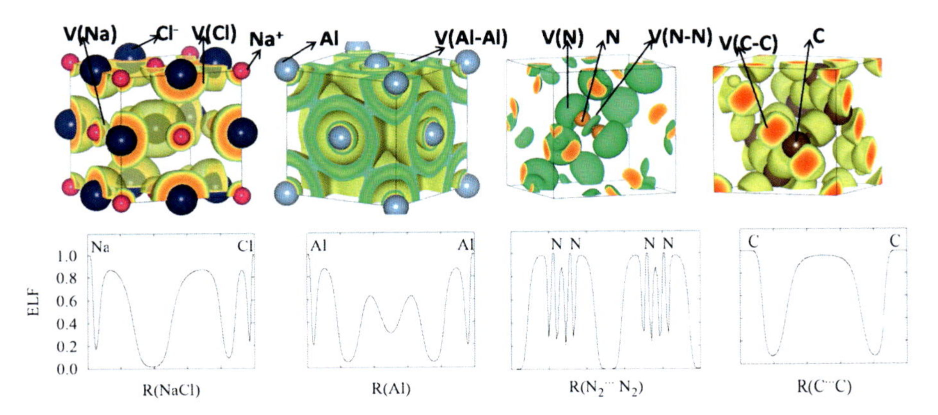

Fig. 18.4 3D ELF isosurfaces (*top*) and 1D ELF profiles along the bonding direction (*bottom*) of representative solids. From *left* to *right*, NaCl rocksalt, Al (fcc), molecular N_2 and C-diamond. Labels stand for elements and names of basins (see text)

A covalent solid like diamond displays a continuous 3D-network of tetrahedrically coordinated carbons, C, bound together by means of single-bond basins, V(C,C). Ionic compounds like NaCl are characterized by the absence of bond basins, only showing closed-shell basins whose shape approach that of a sphere. As far as metals are concerned, their ELF profile distinguishes core regions behaving as cations and valence basins without high ELF values, around 0.5 in the case of Al.

18.2.2 The Dynamic Process of Bond Formation

The concepts of catastrophe theory, that enable to localize the regions of topological instability, are usually complemented among molecular examples by the quantitative analysis of the degree of bond formation (i.e. delocalization) by means of variance and covariance populations [1]. However, their calculation is usually limited in solid state simulations due to the difficult access to high order density matrixes. Hence, we have followed a different approach here, based on our studies of the process of bond formation and the relationship between ELF and localized molecular orbitals. Following the numerous orbital interpretations present in the literature [10–14], the value of ELF at the *bip* of two approaching isolated fragments can be directly related to electron delocalization via the overlap (S) of the relevant orbitals involved in the interaction. As the molecules come closer, new overlaps arise between fragments and the Pauli principle at work demands antisymmetrization of the corresponding orbitals [29]. Following the Heitler-London approach for the orthogonalization of two interacting orbitals, let us say Φ_i and Φ_j, the density and the excess of kinetic energy resulting from the interference are given, respectively [107] by

$$\rho = \frac{1}{1+S^2}\left(\phi_i^2 + \phi_j^2 + 2S\phi_i\phi_j\right) \tag{18.13}$$

$$\Delta t = t_p = \frac{-S^2}{2(1+S^2)}\left(\nabla\phi_i^2 + \nabla\phi_j^2\right) + \frac{S}{1+S^2}\left(\nabla\phi_i\nabla\phi_j\right). \tag{18.14}$$

The results of the interference are easily analyzed at the bond middle point, $R/2$, although it is apparent along the whole bonding region. For a homonuclear diatomic molecule, A_2, the change in the electron density and kinetic energy, $\Delta\rho$ and Δt, with respect to the isolated atoms, ρ_A and t_A, at this point are given by:

$$\Delta\rho\left(\frac{R}{2}\right) = \frac{S(1-S)}{1+S^2}\rho_A\left(\frac{R}{2}\right) > 0 \tag{18.15}$$

$$t_p\left(\frac{R}{2}\right) = \frac{-2S(1+S)}{1+S^2}t_A\left(\frac{R}{2}\right) < 0, \tag{18.16}$$

where we have taken into account the contragradience of the orbitals at $R/2$, $t_B(R/2) = -t_A(R/2)$ [30]. These relationships highlight the main characteristics of the exchange contribution in the bonding region: density increases and electrons become slower. As overlap (or pressure) increases, these modifications along the interaction line result in a decrease of the ratio $t_p/\rho^{5/3}$ (and hence a rise of the ELF value) at those places across the intermolecular region where it was earlier negligible, such as first order saddle points or *bips*.

The relationship between this analysis of delocalization and bond order clearly comes to light if we take into account that the exchange contribution dictated by the Pauli principle is the main energetic term involved in the formation of covalent bonds [31–33]. Indeed, if we analyze the relationship between our ELF analysis of a homonuclear binary molecule and its Wiberg index, B [34]:

$$B = 4\left(\frac{S(1+S)}{1+S^2}\right)^2, \tag{18.17}$$

it can be seen that the term $S(1+S)/(1+S^2)$ is exactly the same as that for the Pauli kinetic energy density term in Eq. 18.16 at the (3,−1) critical point. Hence, the ELF value at this point, that we shall call EDI from now onwards, (after "ELF delocalization index"), is not only able to characterize the process of bond formation in the system, but also provides an idea of the delocalization within it by means of one electron properties. Indeed, the relationship between the EDI and fluctuation has been implicitly used in the literature in the bifurcation diagrams [23–25] we saw before, as well as in the delocalization index by Silvi and Gatti (see Eq. 18.18 below) [6].

18.3 ELF Description of Phase Transitions

Along this section we will apply the concepts developed above in order to understand the changes in electronic structure induced by pressure in a wide variety of solid-solid phase transitions. We will follow thereto a classification criterion

Table 18.1 Classification of phase transitions according to bonding reconstruction and to changes in bonding nature (inspired in Buerger classification)

Type	Example
I. With changes in coordination	
Covalent	α-cristobalite $\leftrightarrow$ stishovite
	α-$LiAlH_4$ $\leftrightarrow$ β-$LiAlH_4$
Ionic	BeO-*B*3 $\leftrightarrow$ BeO-*B*1
II. With changes in chemical bonding	
Molecular $\rightarrow$ Metallic	$I_2 \rightarrow 2I$ (with atomization)
	ε-O_2 $\rightarrow$ ζ-O_2 (without atomization)
Molecular $\rightarrow$ Covalent	CO_2-II $\leftrightarrow$ CO_2-VI (Pressure)
	α-$TmGa_2$ $\leftrightarrow$ β-$TmGa_2$ (Temperature)
Metallic $\rightarrow$ pseudo-ionic	K-fcc $\rightarrow$ K-hp4
	Na-cI16 $\rightarrow$ Na-oP8 $\rightarrow$ Na-hp4
III. Disorder involved	
Electric (electronic)	Ferroelectric $PbTiO_3$ $\leftrightarrow$ paraelectric $PbTiO_3$
	Antiferroelectric $PbZrO_3$ $\leftrightarrow$ paraelectric $PbZrO_3$
Incommensuration (atomic)	P-sc $\rightarrow$ P-IV
	Sn-I $\rightarrow$ Sn-II

established by Buerger [35], which sets the main mechanistic sources of difference in the nature of the chemical changes taking place. Hence, we will make a first differentiation considering those transitions where the bonding nature is maintained (type I) and those where a drastic change takes place (type II). Type II transitions are thus characterized by the appearance of new electronic characteristics in the solid. Metallization would constitute a typical example. In type I instead, the main source of phase reorganization under pressure is the achievement of a more efficient atomic packing by means of coordination changes. Important differences will arise depending on the type of bonding pattern. Finally, we will try to shed some light into processes of disorder, either electronic or atomic, that we name as type III transitions. Type III transformations could obviously fit into some of the above categories, but we believe that their category eases the conceptual understanding of the relationship between their localization pattern and their properties.

Some well-known examples of each of these transitions are organized in Table 18.1 and will be reviewed later. This classification has a two-fold aim: (i) it will help us to illustrate to which extent ELF is able to identify changes in very different cases, and (ii) it will hopefully ease the reading of the text, highlighting the differences and similarities between the transitions analyzed so far. We will pay especial attention to the understanding of the electronic changes that accompany the changes in coordination, as well as to track their consequences on the system properties and their identification by means of the ELF topology. We will see that all the relevant chemical changes are suffered by the valence electrons and reflected by the outer core ones, whereas inner cores remain virtually untouched.

18.3.1 Type I: Changes in Coordination

As a general principle in the pressure loading of solids, it is to be noted that a shortening of nearest neighbor distances occurs. This shortening is accompanied by a raise in the chemical potential (μ) of the solid. Since the chemical potential always increases with pressure (at constant temperature), volume is reduced to keep μ as low as possible. The process continues up to a stage where the increase in connectivity becomes energetically competitive with the shortening of distances, and eventually a phase transition to a denser structure occurs. It is interesting to know if the high pressure coordination appears from the beginning or emerges at a later stage. The main goals here are to gain insight in the process of creation of these new atomic coordinations, and to investigate how the bond reconstruction correlates with the energetic profile and the atomic displacements involved in the transformation.

Analysis of atomic coordinations in a covalent solid is straightforward. It suffices to count the number of bonds surrounding a given core. However, coordination counting can sometimes become a hard task, especially when we are dealing with low symmetry ionic patterns. In these cases, establishing the limit of what belongs and what does not to the active sphere of coordination becomes a matter of threshold flavor. It then becomes interesting to analyse spatial location of Outer Core Maxima (OCM) which reflect in an indirect and subtle way the coordination of the solid and even enable to understand phase relationships [36]. OCM are ELF attractors of the most external core shell of an atom or ion. In the case of soft ions (usually anions) the OCM are disposed along the nearest neighbors directions and reflect the polarization of the ion. If we analyze the outer core of a hard ion instead, the OCM are disposed so as to occupy the interstitial voids in between the bonding basins. This feature recovers the ligand opposed core charge concentrations (LOCCC) within the laplacian field, whose maxima were found to point away from the ligands instead of towards them [37]. It will be shown that their spatial arrangement is not opposed, but they are disposed in order to minimize the repulsion with the valence basins. Thus, they form dual polyhedrons: attractors adopt the direction of the center of the faces of the surrounding polyhedron, e.g. when the coordination is octahedral, they adopt a dual cubic shape. This means that the valence shell electron pair repulsion theory also holds for all electrons in the system following a more general principle that would be called EPR (*electron pair repulsion*). All in all the ELF maxima highlight the distribution of bonds and voids, providing an alternative look at coordination. This type of analysis will help to face, at the end of the chapter, the problem of discarding between different transition mechanisms.

18.3.1.1 Covalent

The analysis of covalent packing efficiency upon pressurization encloses a great potentiality in materials design. This is the case of $LiAlH_4$ [38], where analysis of the change in Al coordination from 4 to 6 in the pressure induced $\alpha \rightarrow \beta$ transition enabled to predict the hydrogen storage potentialities of the quenched β phase due to its efficient six-folded packing. Another example are hard substances of high energy densities, most commonly obtained from the quenching of high pressure phases.

A very illustrative example of a covalent compacting transition is the emergence of hexacoordinated Si after compression of silica polymorphs where a tetrahedral environment is initially found. Thermodynamics and kinetics aspects of these transformations have been the subject of a number of studies [39–43], although it is still needed to shed some light into the evolution of the chemical bonding network from low to high Si coordination.

Since the space group of α-cristobalite ($P4_12_12$) is a subgroup of that of stishovite, it is very convenient to use a transition pathway of this symmetry under the martensitic approach [39, 40, 42, 44]. The transformation from Si four-fold to Si six-fold coordinated can be characterized by two sets of movements, the atomic displacement and the unit cell strain, which is mainly a reduction of the c/a ratio. The opening of O-Si-O angles creates low-energy passage ways for the Si atoms to move from four-fold to six-fold bonding, and under further compression (or thermal activation), Si atoms move to their final location while the positions of the O atoms slightly adjust to achieve the octahedral coordination.

The changes in coordination along the transition path are easily followed in terms of bonds and OCM [36]. The analysis of the critical points in α-cristobalite reveals that valence electrons are invested in the formation of four single Si-O bonds (0.9 electrons) around each Si. The eccentric position of the bond maxima highlights the tension of the structure. Each oxygen is two-fold coordinated to Si and posseses another voluminous lone pair basin that holds 6.8 electrons. As has already been found for a number of SiO_2 [45] and Al_2SiO_5 polymorphs [46], this triangular arrangement of valence basins around oxygen responds to VSEPR principles. Although the tetrahedral coordination of Si atoms is easily inferred from ELF bonds, it can also be tracked down to the Si outer core according to the EPR theory above (see α-cristobalite in Fig. 18.5). The L shell is divided into four OCM that are located forming a tetrahedron, dual to the oxygens' positions.

A clearly different pattern arises for stishovite. Lone pairs are formed by two basins that occupy axial positions with respect to the equatorial Si–O bonds. Two different sets of Si-O bonds arise (two axial and four equatorial Si–O bonds corresponding to the octahedral-like coordination), holding 1.1 and 1.6 electrons respectively. Silicon cores are now surrounded by eight OCM (see stishovite in Fig. 18.5). This OCM distribution clearly highlights the inefficiency of the ligand opposed principle. Instead of obtaining an octahedral distribution, outer core electrons form a dual polyhedron (i.e. a cube). Preliminary calculations reveal that the change in coordination from α-cristobalite to stishovite takes place in one step [36], with the transition state within the four-folded step (Fig. 18.5). The analysis of OCM along the transition not only enables to locate this change in coordination but also provides insight into the shape of the transition path. Comparison of the four-fold structure that preludes the transformation with the cristobalite parent phase reveals a high deformation that is in agreement with it being the transition state. The

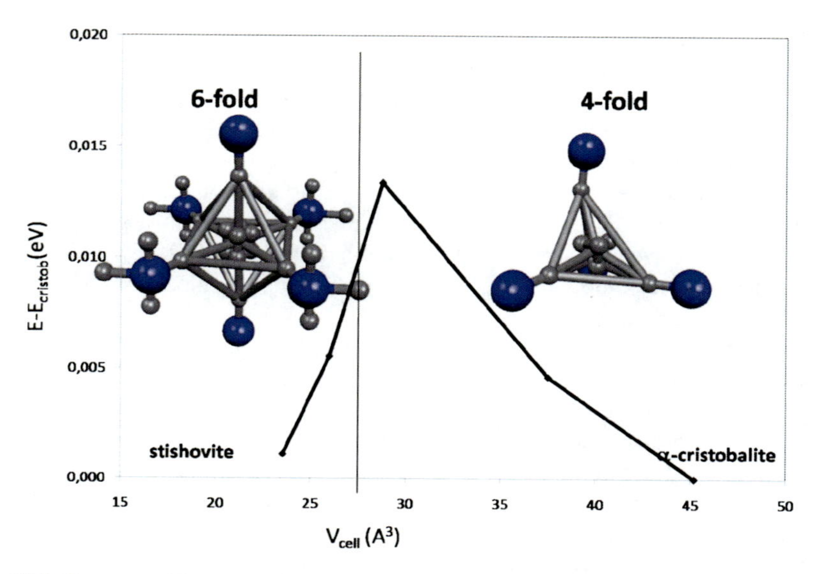

Fig. 18.5 Energy profile across the α-cristobalite $\rightarrow$ stishovite transition. The relationship between OCMs and coordination is represented in the parent phases. Blue spheres represent O cores, whereas Si cores are at the center of the polyhedra. Grey spheres highlight the position of ELF maxima. The vertical line separates the two regions with different ELF patterns

subsequent six-fold intermediate phase instead involves a small reconstruction with respect to stishovite that explains the fast stabilization within the six-fold region.

18.3.1.2 Ionic

To illustrate the analysis of atomic coordination in ionic crystals, we consider the $B3 \rightarrow B1$ phase transition in BeO. The first feature of the ELF analysis is that the electron pairs are distributed in both $B1$ and $B3$ structures forming closed atomic electronic shell around the nuclei: one shell for beryllium (the core K(Be)-shell) and two for oxygen (the core K(O)-shell and the valence L(O)-shell). As expected, there are no electrons associated with bonding basins. Hence, ionic coordination can be tracked by the OCM of either the anion or the cation. Following the changes in OCM arrangements of the anion throws also light into the polarization reorganization along the transition.

We will follow the transformation of oxygen from four-folded tetrahedral ($B3$) to six-folded octahedral ($B1$) at around 100 GPa in BeO [47–49]. Using an orthorhombic *Imm*2 unit cell [50] to monitor the transition mechanism, a small distortion occurs in the cell parameters and the Be cation is displaced from the (0,0,0.25) crystallographic position in the four-fold $B3$ phase to (0,0,0.00) in the six-fold $B1$ one [48]. Hence, z_{Be} becomes the most appropriate transformation coordinate under

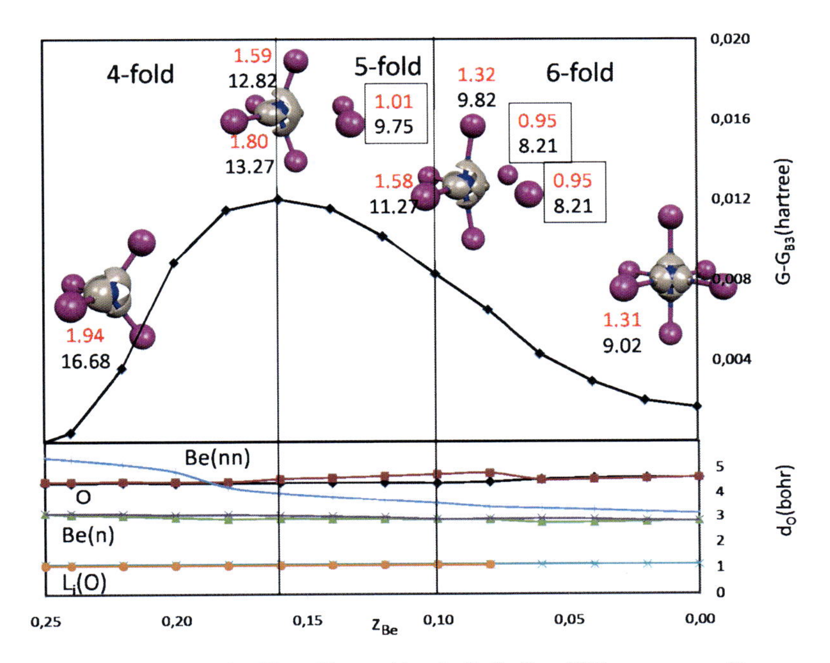

Fig. 18.6 Changes across the $B3 \rightarrow B1$ transition in BeO. *Top*: Gibbs energy profile *versus* z_{Be} at $p < p_t$. *Bottom*: distances from oxygen to oxygen valence attractors (L_i(O)), nearest (Be(n)) and next nearest (Be(nn)) Be atoms, and O nearest atoms. Characteristic chemical patterns for each topological region are shown along the Gibbs profile with oxygen in blue and Be in purple. Isosurfaces in brown are represented at the *bip* value. One new oxygen basin emerges at $z_{Be} = 0.18$ associated to the transition state and at $z_{Be} = 0.10$, when the final coordination is reached. Charge (*red*) and volume (*black*) of the new basins are given in the insets

the generalized Landau theory [51], and a normalized transformation coordinate in the [0, 1] range can be defined as $\xi = [z_{Be} - z_{Be}(B3)]/[z_{Be}(B1) - z_{Be}(B3)]$.

The change in coordination is reflected by an increase in oxygen OCM from 4 to 6 (2 to 3 sets of symmetrically non equivalent basins respectively) [48]. This topological reorganization is easily visualized in Fig. 18.6 as due to two new Be atoms approaching from the next nearest Be coordination shell to the nearest Be coordination shell. The topological change proceeds across three domains of structural stability:

1. $z_{Be} > 0.18$: four-fold coordination
2. $0.18 > z_{Be} > 0.10$: five-fold coordination
3. $z_{Be} < 0.10$: six-fold coordination

First, the new pair of approaching Be atoms induces the appearance of a new oxygen OCM at around $z_{Be} = 0.18$ (see Fig. 18.6), *i.e.* close to the transition state. This new OCM is located along the plane that bisects the angle formed by the O atom and the two approaching Be atoms. As the Be atoms approach the oxygen

($z_{Be} = 0.10$), this initial OCM is divided into two ones corresponding to the new O-Be interactions. The new oxygen L-shell basins, very small in electron population and volume at the beginning, evolve across the transition so that they resemble the other four. The $B1$ conformation is achieved at $z_{Be} = 0.0$ ($\xi = 1$), when all three sets of OCM become equivalent and perfectly show the new coordination in the crystal (Fig. 18.6) Due to this increase in the number of basins, the process is classified as polymorphic according to bond evolution theory [52, 53]. However, this should not mislead the reader towards the thinking that it is associated with a chemical change. The OCM, as reflected by the small value of the difference between the maximum ELF value ($\eta(max)$) and the ELF value at the *bip* ($\eta(bip)$), $\Delta\eta = \eta(max) - \eta(bip)$ [48], form just one meaningful chemical unit, the L shell [21, 54]. Hence, their number is mainly relevant to crystal coordination but not to the chemical nature of the system. In fact, the value of $\Delta\eta$ is the best way to account for the small changes in electronic reorganization, since it provides insight into the deformation of each OCM. Since the polarity of a bond is usually pictured as an oriented deformation of the electron density concentration towards neighbouring atoms, the greater the $\Delta\eta$ value, the more polarized the anion is. This phenomenon is easily visualized when the $B1$ and $B3$ phases in Fig. 18.6 are compared. It can be seen how the six OCM basins in the $B1$ phase almost reconstruct a spherical surface surrounding the anion and leaving a very small space in between, whereas deformation from sphericity in the $B3$ phase is clearly visible and noticeably larger.

Let's now relate all these changes to the changes in the atomic environments and the energetic profile. We observe that O-Be and O-O curves show continuous and soft trends except, perhaps, for the pronounced decrease of the distance of the two approaching Be atoms just before the transition state is reached. This behavior is also related to a catastrophic change in the localization pattern as well as to changes in the energetic profile. It can be seen in Fig. 18.6 that the transition state ($z_{Be} = 0.16$) is not far away from the corresponding topological change. Hence, it is possible to relate the energetic toll of the transition to the microscopic changes unveiled by ELF. The most expensive reconstructive step of the transition comes from the approach of the second Be sphere of coordination to form part of the first one.

18.3.2 *Type II: Changes in Bond Nature*

The analysis of solid-solid phase transitions involving dramatic changes in the localization pattern has become a major topic over the last years thanks to the availability of new high pressure techniques [55, 56]. Extremely multifarious polymorphic sequences have been observed in very simple solids whose underlying electronic reorganization is still not well understood. Furthermore, these transitions enclose an incredibly rich potential in the understanding of chemical bonding due to the great range of conformations it provides. Indeed, it is believed that these experiments hold the key for future developments in both physics and chemistry, so that understanding the microscopic factors that drive these changes and comprehend the thermodynamical and dynamic evolution is crucial. A compacting principle is found here to be of general applicability to the microscopic factors determining the progression of these polymerizations. The minimization of the chemical potential is clearly facilitated by the "weakest" Lewis entity (most voluminous), which changes its nature to provide a more efficient packing of the structure. Lone pairs and multiple bonds (more voluminous than the underlying sigma ones) are good examples of "weak" Lewis entities that enable the formation of new bonds following this compacting principle. This fact is also in agreement with those molecular principles that predict a decreasing in the bond order of non-saturated molecules upon compression. Along the next sections we will review how ELF analysis developed in Sect. 18.2.2 can provide insight into this compacting principle in some of the most representative examples of this high pressure induced polymorphism.

18.3.2.1 From Molecular to Covalent

We will approach here the changes in the cohesion of a solid from purely long-range electrostatic and van der Waals interactions between fragments (molecular phases) to covalency. Following the compacting principle along this polymerization, some intermolecular interactions are expected to strengthen as the intramolecular bond weakens. One of the major concerns will be to identify the active-pressure chemical entity and to analyze whether this process is synchronous or not. In order to analyze these points, we will resort to the EDI across the intermolecular region (see above).

As a first illustration of this scenario, we will track the polymerization of CO_2-II [57] into CO_2-VI under pressure (Fig. 18.7). At large cell volumes (Region A), the molecular phase is the stable polymorph, i.e. the multiple C=O bond of different fragments is favored over an extended single bond network. Hence, molecular units that respect the *in vacuo* geometries, separated by great distances between them, are observed. Indeed, the inspection of the intramolecular bond in Fig. 18.7 reveals a ring shape for the bond attractors, a finger print of double bonds in solid state. The intermolecular forces that retain the CO_2 molecules together in this range (named A in Fig. 18.7) are long range and virtually no overlap between the orbital fragments occurs: the main source of stabilization is the quadrupole-quadrupole interactions that arise from a favorable relative orientation of approaching quadrupoles (i.e. the 'T' and the slipped parallel planar CO_2 arrangements). The analysis of the charge flow upon compression shows that electrons go from the lone pairs to the bond, supporting the molecular interpretation of the solid [29]. Since EDI remains at a negligible value between molecules [29], the compression of the solid leads to a flow of valence electrons toward inner molecular valence basins. As the volume is further compressed, the approach of surrounding molecules induces chemical changes in the molecular structure (region B of Fig. 18.7). The EDI value in the intermolecular region between intermolecular valencies rises (see inset of Fig. 18.7), highlighting the drastic contribution of electronic exchange to polymerization [58, 59]. As the distance between fragments becomes smaller, a charge transfer contribution appears between the virtual orbitals (π^*) of the neighboring CO_2 molecule. The net effect

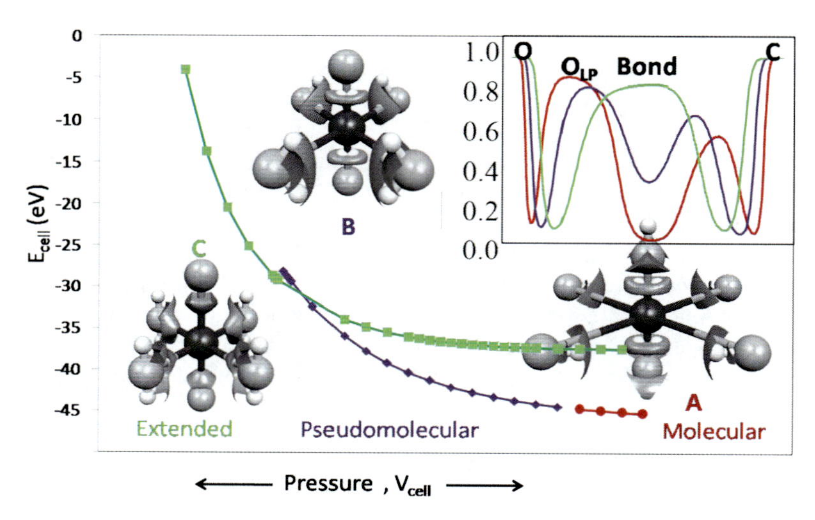

Fig. 18.7 Total energy *versus* volume curve for CO_2-II (*molecular* and *pseudomolecular*) and CO_2-VI (*extended*). Characteristic chemical patterns for each interacting region are shown along the curve. Capital letters represent different interaction regimes: region A (*red*) stands for strictly molecular, B (*blue*) for secondary interactions range, and C (*green*) for extended covalent network (see text). Black, grey, and white balls stand for carbon core, oxygen core, and oxygen valence attractors, respectively. Inset displays evolution of the ELF profile upon compression along the O–C direction of bond formation. A normalized (*blind*) distance has been used in the inset to highlight topological changes. Maxima in the inset represent oxygen nucleus (*O*), oxygen lone pair (O_{LP}), and carbon nucleus (*C*). Colors of the curves follow the same pattern as above

of this transfer is the weakening of the intramolecular C–O bond, as observed in the loss of its annular shape (characteristic of the strictly molecular phases) and in the decrease of its electron population (see CO_2 pictures in Fig. 18.7). The flowing intramolecular charge from the π bonds is redirected towards the intermolecular region for the formation of new σ bonds that allow a favorable compacting of the structure. The rupture of intramolecular bonds and the creation of new intermolecular bonds is carried out in a synchronous manner, when the EDI value between CO_2 fragments is high enough to allow charge flow from the double C–O bond into the intermolecular lone pair region.

In fact, as pressure is induced, the density reorganization of the lone pairs gives rise to new lone pair maxima oriented towards the approaching molecule that represent the "secondary interactions" (see basins in Fig. 18.7). The emerging basin can be considered as a prelude to a future bond, giving rise to a $2+4$ carbon coordination. This fact would explain the incipient stabilization of a six-fold coordinated carbon (stishovite-like) at high pressure, as proposed by Iota et al. [60], and would be in agreement with another report by Iota et al. [57], who state that phase II is a precursor of six-fold phase VI, instead of being of four-fold coordinated phase V, as was previously believed.

As pressure keeps on rising, the new attractors progressively approach the intermolecular direction. The phase transition towards covalent six-fold CO_2 is characterized by the collapse of the molecular C–O bond maxima onto the internuclear line (Region C of Fig. 18.7). This final stage corresponds to the complete formation of a covalent network, where all multiple bonds have been substituted by an extended single bond framework. Simultaneously, oxygen lone pairs are split into distinct chemical units, some of which ultimately become new C-O single bonds.

Similar packing routes, where lone pairs are reinvested in bond formation, have been found by other authors in the polymerization of intermetallic compounds. For example, Leoni et al. [61] analyzed the parent structures of the $\alpha \rightarrow \beta$ transition of $BaAl_2Ge_2$. The high temperature β modification presents a 2D(Al, Ge)-arrangement that transforms upon cooling into the 3D(Al, Ge)-network of the α phase by reinvestment of the Ge lone pairs in the formation of new interlayer bonds.

18.3.2.2 From Molecular to Metallic

As illustrated by the pioneer work of Bridgman [62], metallization at high pressure is a very interesting field demanding new experimental and theoretical insights. Regarding the analysis of chemical bonding, Savin [63] studied the atomization of molecular iodine and employed the analysis of the ELF to unveil some of the structural and electronic characteristics of the transition. In order to account for the metallic nature of the final phase it is desirable to recall the index introduced by Silvi and Gatti [6]:

$$\Delta\eta^{val} = \eta^{val}_{(3,-3)} - \eta^{val}_{(3,-1)} \tag{18.18}$$

Taking into account that the homogeneous electron gas has an ELF value of 0.5, valence electrons of metals should deviate very slightly from this quantity and small $\Delta\eta$ values are expected for metallic compounds. Let us emphasize here that the delocalization concept is able to provide a reliable measure of metallization that identifies metallic phases regardless of the structural changes across the transition. As an example, we will see a completely different route of metallization without atomization. Raman studies of the pressure induced metallization of oxygen ($\varepsilon \rightarrow \zeta$) unveil the persistence of the O_2 vibron across the transition. The ε phase is formed by $(O_2)_4$ clusters in a monoclinic *C2/m* cell [64, 65]. At pressures over 95 GPa, the ε structure experiments an isostructural phase transition towards a metallic phase [66], ζ-O_2. At least, two main questions are interesting about the ε phase of oxygen, which are clearly related to the chemical bonds responsible for the molecular clustering: (i) the nature of the forces that stabilize the ε phase and give rise to a cluster structure not found in its family, and (ii) the changes associated to the $\varepsilon \rightarrow \zeta$ transition that preserve the O–O bond while they provide the phase with metallicity.

The analysis of ELF in the ε phase reveals that O_2 units are far from being equivalent to their *in vacuo* counterparts. As expected from the long O–O bond ($d_{O\text{-}O} = 2.07$ Å, far from the molecular value of 1.20 Å), O atoms are held together

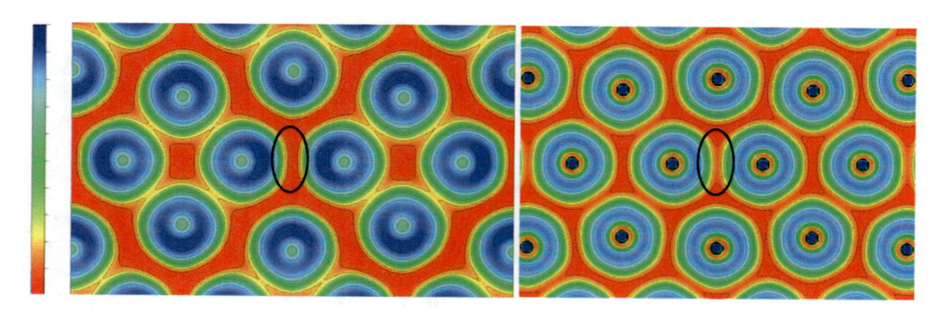

Fig. 18.8 2D ELF isosurfaces for the ε phase (*left*) and ζ phase (*right*) of crystalline oxygen. Circles highlight the change from absence (ε) to new intercluster connections in the ζ phase. The ELF values range from 0 (*red*) to 1 (*blue*)

by only one attractor, which highlights the single bond character of this bond. This is in agreement with the population of 5.03 electrons found in the lone pair, which points at a resonance with the ionic excited state and an electron transfer from the double bond to the lone pair. Contrary to earlier ideas [67], no covalent bonds stabilize the $(O_2)_4$ clusters, but a very large delocalization takes place between them instead with $\eta(bip) = 0.2$ (see Fig. 18.8 (left)). This result is also in concordance with the analysis of the laplacian at the electron density critical point [68]. The positive sign of the laplacian but negative electronic energy density points to an incipient intracluster O–O bond [69].

At around 95 GPa, an isostructural phase transition towards the metallic ζ phase [66] takes place associated to the metallization of the phase. Due to the isostructural nature of the transition, the transformation from the ε to the ζ phase occurs spontaneously upon relaxation. A discontinuity in the a and b cell parameters appears that explains the first order nature of the transition in spite of the nearly soft trend of the $p-V$ curve. The structural analysis informs that the new structure induces a change in the electronic distribution: an increasing electronic delocalization across the layers appears that is responsible for the metallicity of this compound. Increasing values at the bip points between neighboring clusters give rise to new basin connections (see circles in Fig. 18.8 (right)) yielding small $\Delta\eta^{val}$ values (see Eq. 18.18).

A second interesting feature arises from the analysis of the dependence of the ELF topology with pressure in this ζ-phase. Valence electrons are distributed over the solid, the lone pair contribution coalesces in the internuclear line, giving a perfect sp re-hybridization, whereas extra basins of very small charge appear in the intercluster region. New bonds are formed between the O_2 layers, intra and intercluster. This reorganization gives rise to a very flat ELF profile and to delocalization along the lone pair layers. This topological transition towards a very delocalized and unstable phase could be associated with the superconducting phase found by Shimizu et al. [70] at 100 GPa and 0.6 K.

18.3.2.3 From Metallic to Pseudo-ionic

It is nowadays becoming strikingly clear that the Wigner-Seitz model of metals is not applicable when metals are compressed. A deviation from the "ideal" metallic behavior is observed at intermediate pressures, manifested by the apparition of open and incommensurate structures. This challenging state of matter has for example been explained in terms of Peierls distortions or $s \rightarrow p$, $s \rightarrow d$ electronic transitions. However, these explanations lack of predictive character since they are not based in the intrinsic electronic structure of these phases. Hence, the understanding of their real space electronic distribution and behavior upon compression becomes crucial. Potassium adopts the bcc crystal structure at ambient pressure and upon compression up to around 11 GPa it transforms to the fcc structure. Upon further compression several host-guest structures have been identified. What is more interesting is that, occasionally, a hexagonal commensurate phase appears in this pressure range. Analysis with ELF of this new phase reveals that the delocalization channels present in metallic K have completely disappeared, and have been substituted by new pseudo-anions that occupy the cell voids resembling an ionic-like electronic structure [71]. The high localization of these valence electrons in K double hexagonal closed packed (dhcp) is highlighted by the high ELF value of the maxima (0.95), that reveals their true Lewis pair nature. Indeed, integration of the charge density inside the basin corresponding to the ELF valence attractor leads to 1.76 electrons, very close to the electron pair value [71]. These valence attractors, in opposition to localized pairs found at low pressure (mainly bonds and lone pairs), adopt very singular shapes that occupy the voids of the metallic structure. The metal atom completely loses its valence electrons, which are relocated as coreless pseudoanions. Several direct and indirect proofs of the "real" nature of these pseudoanions can be found. In spite of sharing the same space group and atomic positions of the dhcp, the new observed hexagonal phase shows a very different atomic arrangement as a consequence of a much lower c/a ratio (close to 1.35) than the ideal dhcp structure ($c/a = 3.266$). However, if we assume that the structure incorporates "real" anions, its crystal structure turns out to be the same as in many of the ionic compounds of the same family. For example, cation arrays with $c/a \simeq 1.3$ are found under the Ni_2In-type structures of several dialkali metal monochalcogenides and in their corresponding oxides (Na_2S ($c/a = 1.31$), Rb_2Te ($c/a = 1.29$), Na_2SO_4 ($c/a = 1.34$), K_2SO_4 ($c/a = 1.37$)). Intuitively, chalcogen ions occupy in these structures the same positions as the "pseudoanions" of our phase. Indeed, this is in agreement with the observation of Hyde and Andersson [72] that the space of the unit cell occupied by a lone pair is of the same order as an oxide or fluoride anion. Furthermore, the above analysis allows to understand that the closed packing of the crystal is preserved if pseudoanions are taken into account. These observations fall into the anions in metallic matrices model (AMM), where the crystal structure is interpreted in terms of a metallic matrix acting as a host lattice for the non-metallic atoms [73, 74].

Further experimental proof of the pseudo-ionic nature of metals at intermediate pressures has been found in Na (see Fig. 18.9 for 2-D ELF plots). It transforms

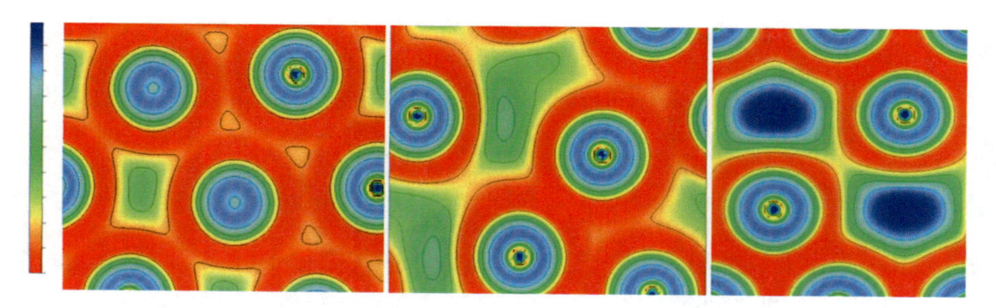

Fig. 18.9 Electron localization function (*ELF*) for the (from *left* to *right*) (110) plane of cI16, (040) plane of oP8, and (110) plane of hP4 structures of Na. The ELF values range from 0 (*red*) to 1 (*blue*)

from a high-reflecting free-electron metal at ambient conditions to a oP8 phase with a significantly reduced reflectivity at 118 GPa. Finally, it turns into a wide gap insulator ($E_g = 2.1$ eV) with unusual strong Raman activity having the hP4 structure [75]. Analysis of the Raman active mode of E_{2g} symmetry reveals that it involves the stretching along the metal-pseudoanion direction, giving an experimental and measurable proof of the new state of matter we are facing [76].

18.3.3 *Type III: Disorder Involved*

18.3.3.1 Paraelectric ↔ Ferroelectric

(Anti)ferroelectric arrangements in metals are usually perturbed by temperature, leading to paraelectric phases when temperature rises. The analysis of the consequence of these electronic transitions in real space has been carried out by Seshadri et al. [77, 78] by use of the electron localization function. The changes in electronic localization that accompany the transition are related to crucial changes in the lone pairs structure and arrangement. They are disposed in a pseudo closed shell manner in the paraelectric structure whereas they form localized lobes in the (anti)ferroelectric one, permitted by the low symmetry of the low-temperature phase. Furthermore, the relative spatial disposition of lone pairs enables to understand the ferroelectric or antiferroelectric nature of the low temperature phase. In $PbZrO_3$ [78] and $PbTiO_3$ [77] lobes pairs are arranged in an opposed manner, so that the polarization of the structure cancels out, but the arrangement of the lobes is able to suggest its antiferroelectric nature. In $PbTiO_3$ instead, the lone pair disposition is even along the crystal, suggesting a greater tendency to ferroelectric distortion.

These changes induced in the lone pair shapes have been related to an admixture with *p* orbitals [78] in agreement with the rules by Hyde and Anderson that establish that both the counter-anion and the degree of covalency are important for the manner in which the lone pairs dispose themselves.

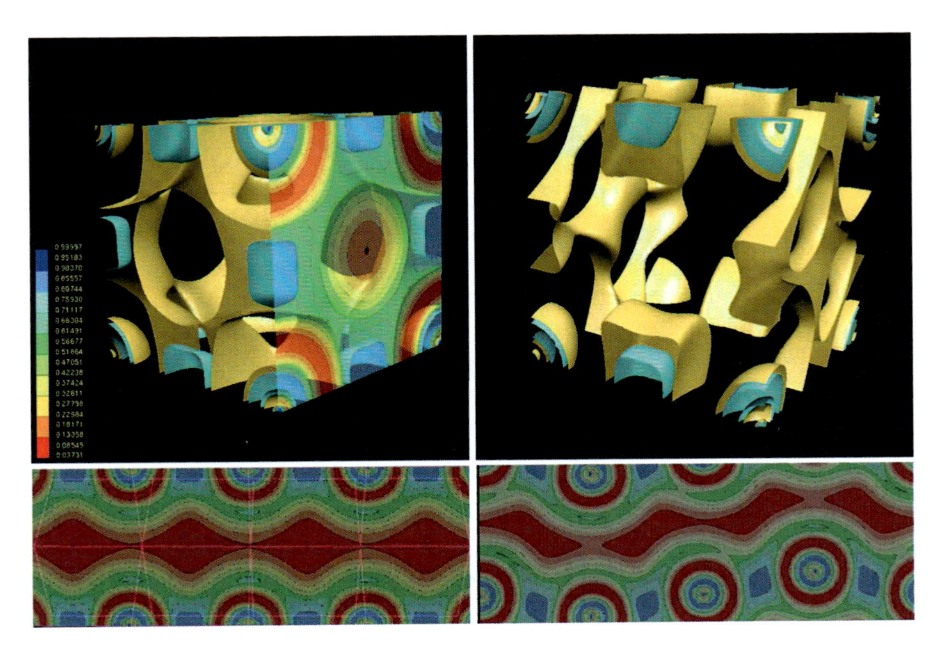

Fig. 18.10 ELF 3D and 2D isosurfaces of simple cubic, simple hexagonal and *Cmcm* structures of P at high pressure. The represented isosurfaces correspond to the values 0.6 (*yellow*) and 0.8 (*blue*), and the ELF scale varies between 0 (*red*) and 1 (*dark blue*). Isosurfaces and ELF across the (001) plane of the simple cubic phase (*top-left*), isosurfaces of the simple hexagonal phase (*top-right*), ELF modulation along the *c* axis of the fcc structure at the fcc → *Cmcm* transition pressure under orthogonal symmetry (*bottom-left*), and ELF modulation along the *c* axis of the *Cmcm* phase. (*bottom-right*)

18.3.3.2 Incommesuration

As stated above, the availability of new high pressure techniques has led to the discovery of a number of complicated phases when metals are subjected to pressure. Among them, incommesurate modulated phases, where an incommensurate modulation leads to the destruction of the 3D crystalline periodicity, have recently called much attention [79, 80]. We will center our analysis in two main questions concerning these phases: (i) which is the nature of the chemical bonding in these phases and (ii) what is the reason underlying their stability. One of the most interesting cases of incommensurate phases has been that of phase P-IV of phosphorous [81], which due to its large modulation amplitude (with atomic displacements $\simeq$15% of the unit cell length!, even larger than that observed in Te-III [82]) constitutes a perfect example in order to analyze the chemical bonding [83]. P-IV phase lies between two phases with extremely simple crystal structures, simple cubic and simple hexagonal. The simple cubic phase shows connected ELF isosurfaces across the whole crystal, consistent with a metallic nature (see yellow surface in Fig. 18.10, top-left). The electronic structure of the simple hexagonal phase instead, shows a stratification in metallic *ab* layers connected by localized bonds along the *c* axis (Fig. 18.10, top-right). The average structure of the incommensurate phase is intermediate between those of the two much simpler structures [84]. "Covalent" bonds appear in the approximant intermediate phase with stronger localization along the bonding directions and partial rupture along the *c* axis [83]. A closer look at this progressive electron localization along the *c* axis provides insight into the stability of the modulated phase. The appearance of the disorder is found in a coupled phonon-lattice parameter distortion that is related to the localization and concomitant opening of a pseudogap. The maxima along the bonding P-P direction in the fcc phase would coalesce at the commensurate → incommesurate transition pressure. The reasons of the instability of such a bonding pattern are highlighted in Fig. 18.10 (bottom), where it becomes clear that the modulation of the *Cmcm* phase boosts the bonding in the *ab* plane every two *ab* atomic planes.

Further proof of the localization trend that stabilizes these intermediate phases is found in the modulated incommensurate phase II of antimony [85]. Comparison between the electronic structure of Sb-II with that of a rhombohedral structure greater in energy and previously proposed by Ormeci and Rosner, showed that the incommensurate structure was further stabilized by the presence of guest-guest bonds that were lacking in the periodic structure. It is the point of view of the authors that incommensurate phases can be understood as an intermediate stage between two stable bonding patterns. At the transient pressures, none of the ending bonding patterns is completely achieved, so that a deformation is favored in order to promote greater bond order in one direction in detriment to the others.

18.4 From ELF Analysis to Discarding Mechanisms

Selecting the appropriate mechanism of a solid-solid phase transition is one of the toughest tasks due to the intrinsic complexity of resolving the most stable pathway at a given condition of pressure and temperature. The high number of involved variables inhabilitates a practical complete optimization of the active space. Hence, new global optimization routes have to be found. Recent advances in metadynamics [86, 87] and evolutionary algorithms [88] seem to indicate that we are nowadays closer to being able to predict phases and mechanisms from a theoretical point of view. However, in all these cases the computation is still very demanding. Therefore, simple approaches that go deeper into the microscopic nature of the problem and enable a first approach to discarding different routes, claim to be looked for. A guiding general rule analogous to the VSEPR of molecules that enables to understand the reasons underlying the structure and the stability of phases is still lacking in the solid state. The search could be established in the direction of electron localization and chemical bonding in the crystalline realm. However, mechanistic studies are usually based on the calculated energetic profiles, a property that is difficult to generalize and simulate by simple models. A number of properties of the cell, the transformation itself and the interatomic interactions constitute very powerful tools in order to understand why a certain mechanism is favored over the others. Since the analysis of OCMs is able to show the voids that atomic reorganization leaves across the transition path and highlights the relationship between the ending phases, we will envisage the discarding of transition mechanisms from the understanding of phase reconstruction in a paradigmatic case: the zircon → scheelite transition of $ZrSiO_4$ [89, 90].

According to relative atomic displacements and cell distortions, transitions are usually classified as either reconstructive or displacive. Displacive mechanisms should involve little phase reconstruction, so that they usually have a negligible activation barrier and generally occur easily and fast. The ending structures are very similar, show same atomic coordinations and have a group-subgroup relationship. In this way, the transition occurs by mere distortions without relevant atomic diffusions. In $ZrSiO_4$, the space group of the high pressure scheelite-type lattice is a subgroup of that of the low pressure zircon (also called reidite) structure, and the four-fold and eight-fold oxygen coordinations of, respectively, Si and Zr, are present in both structures. These two distinctive features would point towards the displacive nature of the transition [91]. However, a number of experimental results points toward a reconstructive conversion, where deep changes in the first coordination sphere take place. The transition displays a volume collapse around 10% and pressure dependent thermal activation barriers for the direct transition around 1,000 K at 10 GPa [92–94] and 300 K at 20–23 GPa [95, 96], and 1,273 K at 0 GPa for the reverse scheelite → reidite transformation [97]. The question emerging from these observations is simple: why the direct, displacive-like, mechanism is not able to reproduce the experimental results? Is the direct path hindered over reconstructive ones? Why? We will try to answer these questions below. OCM will show to be crucial to unveil crystal reorganization [89] and to provide once again insight into ionic interactions.

18.4.1 Displacive or Reconstructive?

Let us try to unveil if there is any bond reconstruction along the so-called displacive mechanism, with the aim at understanding why this path is not able to reproduce the experimental results of the transition. Following the EPR theory, the OCM of Zr are disposed pointing to the "spare" space in both zircon and scheelite (Fig. 18.11). The eight-folded sphere of coordination of Zr is disposed forming two cages (distorted squared-based pyramids in the vertical axes of the figure), so that the OCM of Zr occupy the voids in between these cages. Hence, two OCM (up and down) are found in each polymorph. The main difference arise with respect to the oxygen coordinations: in the zircon case, oxygens share the same SiO_4 tetrahedron, whereas in the scheelite, they belong to different silicon subunits. During the transition, ZrO_8 and SiO_4 polyhedra go from an alternated disposition, forming chains through edge sharing, to a distribution where ZrO_8 and SiO_4 are linked by corners. This environment change requires a reconstructive flip of silicon tetrahedra

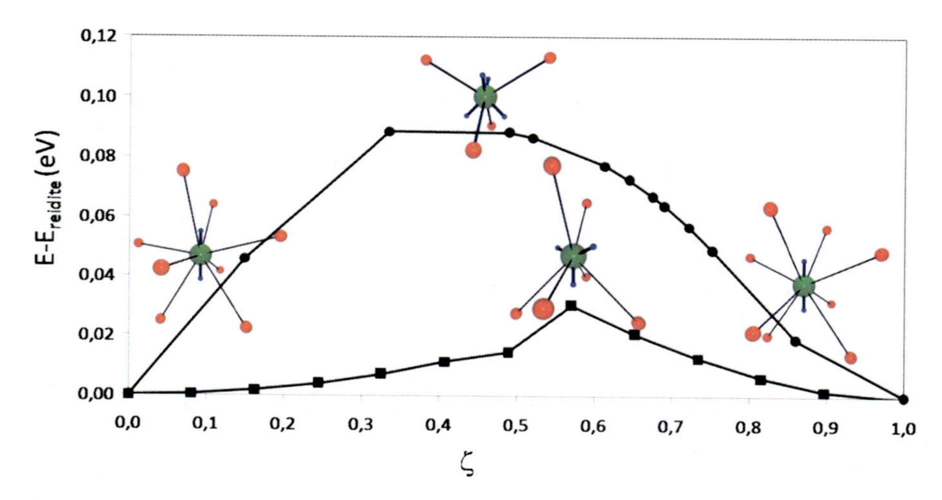

Fig. 18.11 Changes across the $ZrSiO_4$ reidite → scheelite transition. Gibbs energy profile *versus* the transformation coordinate for the pseudo-monoclinic (*bottom*) and tetragonal (*top*) pathways. Chemical patterns for the ending structures and the transition states of each path are shown, highlighting the reconstruction along each of them thanks to the OCM. Zr in *green*, O in red and OCM in *blue*

that determines the location of the transition state. Indeed, it is the tilting of the SiO_4 tetrahedra to pass from the sharing with the ZrO_8 bidisphenoid of a O-O edge in zircon to a O corner in scheelite that hinders the phase transformation at ambient conditions. The analysis of the transition shows that OCM go from two in the ending phases to four in the transition state of the so-called displacive mechanism (top curve of Fig. 18.11). These OCM occupy a pseudo-tetrahedral disposition revealing that half of the oxygens have left their original position, complementary to that of the remaining oxygens. Hence, it involves the simultaneous rupture of four Zr-O bonds and highlights the reconstructive nature of the mechanism. In other words, we can assert that the same metal coordinations of group-subgroup related phases is not sufficient to classify a transition as displacive. Given the fact that the direct mechanism is reconstructive, the following question arises: Is it nonetheless the energetically favored mechanism? Is it possible to find another pathway that implies less reconstruction?

18.4.2 Discarding Mechanisms

In addition to the direct tetragonal mechanism, Kusaba et al. [98] suggested a lower symmetry pathway. According to these authors, the transition would occur along a quasi-monoclinic symmetry mechanism. Each of these two mechanisms would in principle give rise to very different descriptions of the bonding pattern associated to the phase transformation. Hence, the analysis of the electronic structure given by ELF can provide deeper insight into the changes and bond reconstruction involved in each of the pathways.

In opposition to the tetragonal pathway, the eight-fold coordination is maintained most of the time along the monoclinic pathway. The transition state is characterized by a sudden exchange in two of the oxygens around Zr. Indeed, two Zr OCM highlight the position of the outgoing ligands (Fig. 18.11) whereas two oxygens remain at their position. One of the O_4 cages along with its Zr OCM remain untouched (Fig. 18.11), to be compared with the enormous phase reconstruction of the tetragonal path that should give rise to a very high energetic toll.

All in all, the monoclinic pathway favors the correct electron localization for a smaller reorganization along the pathway, whereas the tetragonal one, supports a new Zr coordination. Moreover, the number of broken bonds along the latter clearly stands for the suitability of the monoclinic description, where the coordination remains close to that of the stable parent phases at all points. This topological analysis enables to rule out the zircon $\rightarrow$ scheelite pathway as (i) displacive (ii) tetragonal. Indeed, these results are in agreement with energetic and strain criteria [89]. A very high activation barrier (236 kJ/mol) an greater lattice strains accompany the tetragonal path, whereas a lower activation energy (80 kJ/mol) is required for the monoclinic alternative.

18.5 Conclusions

Across this chapter we have examined the ability of ELF to track the microscopic changes underlying phase transitions. Three main groups of such transformations have been followed: type I, which only imply a change in the atomic coordination, type II, which are accompanied by alterations in the nature of chemical bonding, and are hence associated to drastic changes in the solid properties, and type III, which involve atomic or electronic disorder. From a chemical point of view, type I transitions are characterized by a change in the number of bonds. Since we have mainly focused on pressure-induced phase transitions, these changes in coordination are often associated to increases in the coordination number for a more efficient packing of the structure. We have seen that an important principle arises here, that of the *Electronic Pair Repulsion*: the Valence Shell Electron Repulsion Theory was found to be applicable to all the electrons of the system (therefore, EPR theory). This principle is reflected in the crystal in the spatial distribution of the outer core maxima (OCM), whose distribution is able to unveil in a very subtle way the coordination of the system. If the OCM of a polarizable atom are analyzed, they reflect coordination by forming a polyhedron with its vertexes directed towards the neighboring atoms. If we look at hard atoms instead, their OCM form a dual polyhedron with respect to the neighbors. The analysis of OCM distribution in the crystal provides clear insight into the effective coordination in the solid and into crystal reconstruction processes.

We have analyzed in depth two bonding types: covalent and ionic. In both cases the ELF analysis is able to unveil not only the changes in coordination by means of the OCM, but it is also able to track chemical changes. In covalent cases, as that of the α-cristobalite $\rightarrow$ stishovite phase transition, the new electron pairs are readily localized by ELF maxima. In ionic patterns, as was our example of the BeO $B3 \rightarrow B1$ phase transition, the slight changes across the transition path are mainly associated to polarity changes, which have been shown to be related to the relative position (and ELF value) of the ELF maxima and *bips*. These concepts have been applied at the end of the chapter in order to discard mechanisms in the zircon $\rightarrow$ reidite phase transition of $ZrSiO_4$. The group-subgroup relationship between the parent and daughter phases, along with the same cation coordination displayed in both phases, suggests a displacive mechanism according to common analysis. The analysis of bond reconstruction by the means of ELF sheds light into real bond reorganization taking place and identifies that both, the tetragonal and the monoclinic transition mechanisms, are reconstructive.

As far as phase transitions with a fundamental bonding change are concerned (type II), we have analyzed some of the most troublesome chemical bonding changes in current high pressure physics. Most generally, we believe that the polemic analysis of the nature of the phases involved has been centered on the geometry of the compounds, whereas little efforts have been paid to unveil their underlying electronic structures. The broadening of the ELF scope we have introduced has allowed us to prove that the great changes in electronic properties observed across these reactions can be tracked back to the microscopic structure of the ending phases. As an example of the ability of ELF to relate electronic structure and macroscopic properties, we have followed the transition from molecular to metallic $(O_2)_4$ clusters. The new dynamical approach to bond formation under ELF where no high order density matrices are required has enabled us to follow the processes where new covalent bonds appear. A compacting principle has been found to be operative in all cases, stating that the most voluminous basins (lone pairs and multiple bonds) are reinvested in the formation of new bonds that enable a better packing under pressure. This has been the case of CO_2 polymerization, where our study confirms that the transformation follows a synchronic weakening of the intramolecular (C=O) double bond and the birth of a new intermolecular C–O bond controlled by the oxygen lone pairs. ELF has been also used to get deeper insight into high pressure phases of metals, where new localized patterns have been recently found. The analysis of pairing reveals that these phases and their properties can be understood from an ionic scheme where valence electrons detach from the atoms and form interstitial pseudo-anions.

Finally, some observations have been made on disordered structures. Due to their current relevance in high pressure physics, incommesurate structures have been illustrated using P-IV as a hot example, and the consequences of modulation tracked back to the localization pattern. Some ideas on the reasons underlying incommesurate stability have been put forward. All in all, we have shown that atomic displacements, transition energy and chemical bonding are different faces of the same coin, and can all be used and combined in the complex practice of proposing and analyzing phase transition mechanisms.

Acknowledgments The authors want to thank L. Contreras-García for proof reading the manuscript. Financial support from Spanish MEC and FEDER programs (MAT2006-13548-C02-02) and MALTA-Consolider Ingenio 2010 Program (Project CSD2007-00045) is gratefully acknowledged. JCG thanks the Fulbright program for a Ruth Lee Kennedy travel grant.

Appendix: Computational Methods

First-principles total-energy calculations were carried out within the density-functional theory (DFT) formalism with a plane-wave pseudopotential approach, as implemented in the Vienna *ab initio* simulation package [99]. We have used both, LDA and GGA levels of calculation using standard exchange-correlation functionals as Perdew-Wang ones [100], and the projector augmented wave (PAW) all-electron description of the electron-ion-core interaction [101]. Brillouin zone integrals were approximated using the method of Monkhorst and Pack [102], and the energies converged to 1 meV with respect to k-point density and 0.2 meV with respect to planewave cutoff. Upon compression, we calculated the total energy (E) at a number of selected values of the volume (V) for each lattice, relaxing the atomic coordinates and lattice parameters subjected to the constraints of symmetry and volume conservation. All structural relaxations were performed via a conjugate-gradient minimization of the total energy using the Methfessel-Paxton method. For the final calculation of the optimized crystal structures the tetrahedron method with Blöchl correction was used. The variation with hydrostatic pressure of the lattice parameters and atomic coordinates has been obtained by means of numerical and standard equations of state fittings to the sets of computed (E, V) points [103]. This procedure also provides $G(p)$ curves in the static approximation (zero temperature and neglecting zero point vibrational contributions).

Analysis of ELF topologies along the transition pathways has been possible thanks to an automated and efficient code developed by the authors [26, 27]. The algorithm is able to completely characterize the topology induced by ELF in solids, including identification and characterization of all critical points and basin integration. It is based on the fact that this topology presents two well-differentiated regions. On the one hand, the valence, which can be determined following previous crystalline topological methods [104]. On the other hand, the core, whose sphericity holds the key for designing new automated algorithms. In order to ensure a reliable and quantitative analysis of the ELF topology, all-electron wavefunctions are required [105]. To this end, the `VASP` optimized structures were recalculated with the `CRYSTAL98` code [106] using the same exchange and correlation functionals as in the pseudopotential calculations.

References

1. Silvi B, Savin A (1994) Classification of chemical bonds based on topological analysis of electron localization functions. Nature 371:683–686
2. Gatti C (2005) Chemical bonding in crystals: new directions. Z Kristallogr 220:399–457
3. Savin A, Jepsen O, Flad J, Andersen L, von Schnering HG, Preuss H (1992) Electron localization in solid-state structures of the elements: the diamond structure. Angew Chem Int Ed Engl 31:187–188
4. Fässler TF (2003) The role of non-bonding electron pairs in intermetallic compounds. Chem Soc Rev 32:80–86
5. Savin A, Nesper R, Wengert S, Fässler T (1997) ELF: the electron localization function. Angew Chem Int Ed Engl 36:1809–1832
6. Silvi B, Gatti C (2000) Direct space representation of the metallic bond. J Phys Chem A 104:947–953
7. Becke AD, Edgecombe KE (1990) A simple measure of electron localization in atomic and molecular systems. J Chem Phys 92:5397–5403
8. von Weizsäcker CF (1935) Zur Theorie der Kernmassen. Z Phys 96:431–458
9. Savin A, Becke AD, Flad J, Nesper R, Preuss H, von Schnering HG (1991) A new look at electron localization. Angew Chem Int Ed Engl 30:409–412
10. Burdett JK, McCormick TA (1998) Electron localization in molecules and solids: the meaning of ELF. J Phys Chem A 102:6366–6372
11. Nalewajski RF (2005) Electron localization function as information measure. J Phys Chem A 44:10038–10043
12. Dobson JF (1991) Interpretation of the Fermi hole curvature. J Chem Phys 94:4328–4332
13. Kohout M, Pernal K, Wagner FR, Grin Y (2004) Electron localizability indicator for correlated wavefunctions. I. Parallel-spin pairs. Theor Chem Acc 112:453–459
14. Silvi B (2003) The spin-pair compositions as local indicators of the nature of the bonding. J Phys Chem A 107:3081–3085
15. Matito E, Silvi B, Duran M, Solà MJ (2006) Chem Phys 125:024301
16. Silvi B (2004) How topological partitions of the electron distributions reveal delocalization. Phys Chem Chem Phys 6:256–260
17. Abraham RH, Marsden JE (1994) Foundations of mechanics. Addison Wesley, Reading
18. Haussermann U, Wengert S, Hoffmann P, Savin A, Jepsen O, Nesper R (1994) Localization of electrons in intermetallic phases containing aluminum. Angew Chem Int Ed Engl 33:2069–2073
19. Martín Pendás A, Francisco E, Blanco MA (2008) Electron-electron interactions between ELF basins. Chem Phys Lett 454:396–403
20. Lewis GN (1916) The atom and the molecule. J Am Chem Soc 38:762–785
21. Silvi B (2002) The synaptic order: a key concept to understand multicenter bonding. J Molec Struct 614:3–10
22. Mezey PG (1994) Quantum-chemical shape - new density domain relations for the topology of molecular bodies, functional-groups, and chemical bonding. Can J Chem 72:928–935
23. Kohout M, Wagner F, Grin Y (2002) Electron localization function for transition-metal compounds. Theor Chem Acc 108:150–156
24. Calatayud M, Andrés J, Silvi B (2001) The hierarchy of localization basins: a tool for the understanding of chemical bonding exemplified by the analysis of the VOx and VOx + $(x = 1 - 4)$ systems. Theoret Chem Acc 105:299–308
25. Savin A, Silvi B, Colonna F (1996) Topological analysis of the electron localization function applied to delocalized bonds. Can J Chem 74:1088–1096
26. Contreras-García J, Martín Pendás A, Silvi B, Recio JM (2009) Computation of local and global properties of the electron localization function topology in crystals. J Chem Theor Comp 5:164–173

27. Contreras-García J, Martín Pendás A, Silvi B, Recio JM (2008) Useful applications of the electron localization function in high-pressure crystal chemistry. J Phys Chem Solids 69:2204–2207
28. Kohout M, Savin A (1996) Atomic shell structure and electron numbers. Int J Quantum Chem 60:875–882
29. Contreras-García J, Martín Pendás A, Silvi B, Recio JM (2009) Bases for understanding polymerization under pressure: the practical case of CO_2. J Phys Chem B 113:1068–1073
30. Goddard WA, Wilson CW Jr (1972) The role of kinetic energy in chemical binding. Theor Chim Acta (berl) 26:211–230
31. Bickelhaupt FM, Baerends EJ (2000) Kohn-sham density functional theory: predicting and understanding chemistry. Rev Comput Chem 15:1–86
32. Cooper DL, Ponec R (2008) A one-electron approximation to domain-averaged Fermi hole analysis. Phys Chem Chem Phys 10:1319–1329
33. Ruedenberg K (1962) The physical nature of the chemical bond. Rev Modern Phys 34: 326–376
34. Wiberg KB (1968) Application of the pople-santry-segal CNDO method to the cyclopropyl-carbinyl and cyclobutyl cation and to bicyclobutane. Tetrahedron 24:1083–1096
35. Buerger J (1951) Phase transformation in solids. Wiley, New York
36. Gracia L, Contreras-García J, Beltrán A, Recio JM (2009) Bonding changes across the α-cristobalite $\rightarrow$ stishovite transition path in silica. High Press Res 29:93–96
37. Bytheway I, Gillespie RJ, Tang TH, Bader RFW (1995) Core distortions and geometries of the difluorides and dihydrides of Ca, Sr, and Ba. Inorg Chem 34:2407–2414
38. Vajeeston P, Ravindran P, Vidya R, Fjellvag H, Kjekshus A (2003) Huge-pressure-induced volume collapse in $LiAlH_4$ and its implications to hydrogen storage. Phys Rev B 68:212101
39. Huang LP, Durandurdu M, Kieffer J (2006) Transformation pathways of silica under high pressure. Nat Mater 5:977–981
40. Klug DD, Rousseeau R, Uehara K, Bernasconi M, Page YL, Tse JS (2001) Ab initio molecular dynamics study of the pressure-induced phase transformations in cristobalite. Phys Rev B 63:104106
41. Liu J, Topor L, Zhang J, Navrotsky A, Liebermann RC (1996) Calorimetric study of the coesite-stishovite transformation and calculation of the phase boundary. Phys Chem Miner 23:11–16
42. Silvi B, Jolly LH, Darco PJ (1992) Pseudopotential periodic Hartree-Fock study of the cristobalite to stishovite phase transition. Mol Struct 92:1–9
43. Tsuchida Y, Yagi T (1990) New pressure-induced transformations of silica at room temperature. Nature 347:267–269
44. O'Keeffe M, Hyde BG (1976) Cristobalites and topologically-related structures. Acta Crystallogr B 32:2923–2936
45. Gibbs GV, Cox DF, Boisen MB, Downs RT, Ross NL (2003) The electron localization function: a tool for locating favourable proton docking sites in the silica polymorphs. Phys Chem Miner 30:305–316
46. Burt JB, Gibbs GV, Cox DF, Ross NL (2006) ELF isosurface maps for the Al_2SiO_5 polymorphs. Phys Chem Miner 33:138–144
47. Cai YX, Wu S, Xu R, Yu J (2006) Pressure-induced phase transition and its atomistic mechanism in BeO: a theoretical calculation. Phys Rev B 73:184104
48. Contreras-García J, Martín Pendás A, Recio JM (2008) How the electron localization function quantifies and pictures chemical changes in a solid: the B3 $\rightarrow$ B1 pressure induced phase transition in BeO. J Phys Chem B 112:9787–9794
49. Park CJ, Lee SJ, Ko YJ, Chang KJ (1999) Theoretical study of the structural phase transformation of BeO under pressure. Phys Rev B 59:13501–13504
50. Miao MS, Lambrecht WRL (2005) Universal Transition state for high-pressure zinc blende to rocksalt phase transitions. Phys Rev Lett 94:225501
51. Dmitriev VP, Rochal SB, Gufan YM, Toledano P (1988) Definition of a transcendental order parameter for reconstructive phase transitions. Phys Rev Lett 60:1958–1961

52. Krokidis X, Noury S, Silvi B (1997) Characterization of elementary chemical processes by catastrophe theory. J Phys Chem A 101:7277–7282
53. Polo V, Andres J, Castillo R, Berski S, Silvi B (2004) Understanding the molecular mechanism of the 1,3-dipolar cycloaddition between fulminic acid and acetylene in terms of the electron localization function and catastrophe theory. Chem Eur J 10:5165–5172
54. Mori P, Recio JM, Silvi B, Sousa C, Martín Pendás A, Luaña V, Illas F (2002) Rigorous characterization of oxygen vacancies in ionic oxides. Phys Rev B 66:075103
55. Eremets MI, Hemley RJ, Mao H, Greforyanz E (2001) Semiconducting non-molecular nitrogen up to 240 GPa and its low-pressure stability. Nature 411:170–174
56. Hemley RJ (2000) Effects of high pressure on molecules. Annu Rev Phys Chem 51:763–800
57. Yoo C, Kohlmann H, Cynn H, Nicol MF, Iota V, Bihan TL (2002) Crystal structure of pseudo-six-fold carbon dioxide phase II at high pressures and temperatures. Phys Rev B 65:104103
58. Contreras-García J, Recio JM (2009) From molecular to polymeric CO_2: bonding transformations under pressure. High Press Res 29:113–117
59. Morokuma K (1971) Molecular orbital studies of hydrogen bonds. III. C=O...H–O hydrogen bond in $H_2CO \ldots H_2O$ and $H_2CO \ldots 2H_2O$. J Chem Phys 55:1236–1244
60. Iota V, Yoo C, Klepeis J, Jenei Z, Evans W, Cynn H (2007) Six-fold coordinated carbon dioxide VI. Nat Mater 6:34–38
61. Leoni S, Carrillo-Cabrera W, Schnelle W, Grin Y (2003) $BaAl_2Ge_2$: synthesis, crystal structure, magnetic and electronic properties, chemical bonding, and atomistic model of the $\alpha \rightarrow \beta$ phase transition. Solid State Sci 5:139–148
62. Bridgman PW (1935) Theoretically interesting aspects of high pressure phenomena. Rev Mod Phys 7:1–33
63. Savin A (2004) Phase transition in iodine: a chemical picture. J Phys Chem Solids 65:2025–2029
64. Falconi S, Ackland GJ (2006) Ab initio simulations in liquid caesium at high pressure and temperature. J Phys Rev B 73:184204
65. Militzer B, Hemley R (2006) Crystallography: solid oxygen takes shape. Nature 443:150–151
66. Weck G, Loubeyre P, LeToullec R (2002) Observation of structural transformations in metal oxygen. Phys Rev Lett 88:035504
67. Ma Y, Oganov AR, Glass CW (2007) Structure of the metallic ε-phase of oxygen and isosymmetric nature of the ε-ζ phase transition: Ab initio simulations. Phys Rev B 76:064101
68. Tse JS, Yao Y, Klug DD, Desgreniers S (2008) Bonding in the e-phase of high pressure oxygen. J Phys Conf Ser 121:012006
69. Espinosa E, Alkorta I, Elguero J, Molins E (2002) From weak to strong interactions: a comprehensive analysis of the topological and energetic properties of the electron density distribution involving XH ... FY systems. J Chem Phys 117:5529–5542
70. Shimizu K, Suhara K, Ikumo M, Eremets MI, Amaya K (1998) Superconductivity in oxygen. Nature 393:767–769
71. Marqués M, Ackland GJ, Lundegaard LF, Contreras-García J, McMahon MI (2009) Potassium under pressure: a pseudobinary ionic compound. Phys Rev Lett 103:115501
72. Hyde BG, Anderson S (1989) Inorganic crystal structures. Wiley, New York
73. Marqués M, Flórez M, Recio JM, Santamaría-Pérez D, Vegas A, García Baonza V (2006) Structure, Metastability, and electron density of Al lattices in light of the model of anions in metallic matrices. J Phys Chem B 110:18609–18618
74. Vegas A, Santamaría-Pérez D, Marqués M, Flórez M, García Baonza V, Recio JM (2006) Anions in metallic matrices model: application to the aluminium crystal chemistry. Acta Crystallogr B 62:220–227
75. Ma Y, Eremets M, Oganov AR, Xie Y, Trojan I, Medvedev S, Lyakhov AO, Valle M, Prakapenka V (2009) Transparent dense sodium. Nature 458:182–185
76. Marqués M, Santoro M, Guillaume C, Gorelli F, Contreras-García J, Goncharov AF, Gregoryanz E (2011) Optical and electronic properties of dense sodium. Phys Rev B 83: 184106-1–184106-7

77. Seshadri R, Baldinozzi G, Felsera C, Tremel W (1999) Visualizing electronic structure changes across an antiferroelectric phase transition: Pb_2MgWO_6. J Mater Chem 9:2463–2466
78. Seshadri R (2001) Visualizing lone pairs in compounds containing heavier congeners of the carbon and nitrogen group elements. Proc Indian Acad Sci (Chem Sci) 113:487–496
79. McMahon MI, Nelmes RJ (2004) Incommensurate crystal structures in the elements at high pressure. Z Kristallogr 219:742–748
80. McMahon MI, Nelmes RJ (2006) High-pressure structures and phase transformations in elemental metals. Chem Soc Rev 35:943–963
81. Akahama Y, Kobayashi M, Kawamura H (1999) Simple-cubic → simple-hexagonal transition in phosphorus under pressure. Phys Rev B 59:8520–8525
82. McMahon MI, Hejny C (2003) Pressure-Induced magnetization in FeO: evidence from elasticity and Mössbauer spectroscopy. Phys Rev Lett 93:215502
83. Marqués M, Ackland GJ, Lundegaard LF, Falconi S, Hejny C, McMahon MI, Contreras-García J, Hanfland M (2008) Origin of incommensurate modulations in the high-pressure phosphorus IV phase. Phys Rev B 78:054120
84. Fujihisa H, Akahama Y, Kawamura H, Ohishi Y, Gotoh Y, Yamawaki H, Sakashita M, Takeya S, Honda K (2007) Incommensurate structure of phosphorus phase IV. Phys Rev Lett 98:175501
85. Ormeci A, Rosner H (2004) Electronic structure and bonding in antimony and its high pressure phases. Z Kristallogr 219:370–375
86. Glass CW, Oganov AR, Hansen N (2006) USPEX: evolutionary crystal structure prediction. Comput Phys Commun 175:713–720
87. Martonak R, Laio A, Bernasconi M, Ceriani C, Raiteri P, Parrinello M (2005) Z Kristallogr 220:489–498
88. Winkler B, Pickard CJ, Milman V, Thimm G (2001) Systematic prediction of crystal structures. Chem Phys Lett 337:36–42
89. Flórez M, Marqués M, Contreras-García J, Recio JM (2009) Quantum-mechanical calculations of zircon to scheelite transition pathways in $ZrSiO_4$. Phys Rev B 79:104101
90. Marqués M, Contreras-García J, Flórez M, Recio JM (2008) On the mechanism of the zirconreidite pressure induced transformation. J Phys Chem Solids 69:2277–2280
91. Lang M, Zhang F, Lian J, Trautmann C, Neumann R, Ewing RC (2008) Irradiation-induced stabilization of zircon ($ZrSiO_4$) at high pressure. Earth Planet Sci Lett 269:291–295
92. Liu LG (1979) High-pressure phase transformations in baddeleyite and zircon, with geophysical implications. Earth Planet Sci Lett 44:390–396
93. Ono S, Tange Y, Katayama I, Kikegawa T (2004) Equations of state of $ZrSiO_4$ phases in the upper mantle. Am Mineral 89:185–188
94. Reid AF, Ringwood AE (1969) Newly observed high pressure transformations in Mn_3O_4, $CaAl_2O_4$, and $ZrSiO_4$. Earth Planet Sci Lett 6:205–208
95. Knittle E, Williams Q (1993) High-pressure Raman spectroscopy of $ZrSiO_4$: observation of the zircon to scheelite transition at 300 K. Am Mineral 78:245–252
96. van Westrenen W, Frank MR, Hanchar JM, Fei Y, Finch RJ, Zha CS (2004) In situ determination of the compressibility of synthetic pure zircon ($ZrSiO_4$) and the onset of the zircon-reidite phase transition. Am Mineral 89:197–203
97. Kusaba K, Syono Y, Kikuchi M, Fukuoka K (1985) Shock behavior of zircon: phase transition to scheelite structure and decomposition. Earth Planet Sci Lett 72:433–439
98. Kusaba K, Yagi T, Kikuchi M, Syono Y (1986) Structural considerations on the mechanism of the shock-induced zircon-scheelite transition in $ZrSiO_4$. J Phys Chem Solids 47:675–679
99. Kresse G, Furthmuller J (1996) Efficient iterative schemes for ab initio total-energy calculations using a plane-wave basis set. Phys Rev B 54:11169–11186
100. Perdew JP, Wang Y (1992) Accurate and simple analytic representation of the electron-gas correlation energy. Phys Rev B 45:13244–13249
101. Kresse G, Joubert D (1999) From ultrasoft pseudopotentials to the projector augmented-wave method. Phys Rev B 59:1758–1775

102. Monkhorst HJ, Pack JD (1976) Special points for Brillouin-zone integrations. Phys Rev B 13:5188–5192
103. Blanco MA, Francisco E, Luaña V (2004) GIBBS: isothermal-isobaric thermodynamics of solids from energy curves using a quasi-harmonic Debye model. Comput Phys Commun 158:57–72
104. Otero-de-la-Roza A, Martín Pendás A, Blanco MA, Luaña V (2009) Critic: a new program for the topological analysis of solid-state electron densities. Comput Phys Commun 180:157–166
105. Kohout M, Savin A (1997) Influence of core-valence separation of electron localization function. J Comput Chem 18:1431–1439
106. Saunders VR, Dovesi R, Roetti R, Causá M, Harrison NM, Orlando R, Zicovich-Wilson CM (1998) CRYSTAL98 user's manual. University of Torino, Torino
107. Contreras-García J, Recio JM (2010) Electron delocalization and bond formation under the ELF framework. Theor Chim Acta 128:411–418

Chapter 19
Multi-temperature Electron Density Studies*

Riccardo Destro, Leonardo Lo Presti, Raffaella Soave, and Andrés E. Goeta

19.1 Introduction

To obtain accurate experimental electron densities $\rho(\mathbf{r})$ it is mandatory to perform a reliable deconvolution of the static electron density distribution (EDD) from thermal motion. To this end, single-crystal X-ray diffraction (XRD) experiments, especially on molecular crystals, have to be performed at low temperature. Under these conditions enough high-order diffraction data – which contain information on the core electrons – become measurable, allowing an appropriate treatment of the thermal motion. Chapter 3 deals with this problem; it describes atomic displacements in crystals in the harmonic approximation and beyond, and comments on recently developed methods that combine experimental and theoretical information to achieve deconvolution. In the present chapter, the effects of temperature on the experimental electron densities are examined from a more general perspective.

A firm tenet in the charge density community, apropos of temperature, is "the lower the better" [1]. This recommendation derives from at least three reasons: (i) more reliable determination of the static $\rho(\mathbf{r})$; (ii) lower values of the corre-

* To their great sorrow, Riccardo Destro, Leonardo Lo Presti and Raffaella Soave announce the death of Dr. Andrés E. Goeta on July 29th, 2011.

R. Destro (✉) • L. Lo Presti
Department of Physical Chemistry and Electrochemistry, Università degli Studi di Milano, Via Golgi 19, 20133 Milan, Italy
e-mail: riccardo.destro@unimi.it; leonardo.lopresti@unimi.it

R. Soave
Istituto di Scienze e Tecnologie Molecolari (ISTM), Italian National Reseearch Council (CNR), Via Golgi 19, 20133 Milan, Italy
e-mail: raffaella.soave@istm.cnr.it

A.E. Goeta
Chemistry Department, Durham University, South Road, Durham, DH1 3LE UK
e-mail: a.e.goeta@durham.ac.uk

C. Gatti and P. Macchi (eds.), *Modern Charge-Density Analysis*,
DOI 10.1007/978-90-481-3836-4_19, © Springer Science+Business Media B.V. 2012

lation coefficients between positional/thermal parameters and electron population coefficients of generalized scattering factors; (iii) higher precision, i.e. smaller estimated standard deviations (esd's), of the refined parameters, and hence of all the electrostatic properties derived from $\rho(\mathbf{r})$. However, what is the quantitative gain in accuracy and precision when collecting data below, say, the liquid nitrogen temperature? Furthermore, is it possible to establish an upper limit on the temperature, above which experiments aimed at a reliable determination of the electron density are not worth the effort? Possible answers to these questions are critically examined in the next Section, through examples taken from the pertinent literature.

A change in the temperature can also give rise to a significant change in the interactions between the molecules. The lowering of T causes the reduction of the overall thermal motion in the crystal, decreasing the amplitudes of both the molecular and the lattice vibration modes. Moreover, the intermolecular (interatomic) potential due to the crystal field changes because of the lattice shrinking[1] [2] and, at the same time, less and less states become thermally accessible in the crystal. The reduction of the vibration amplitudes makes the molecules approach each other, strengthening all the interactions (both attractive and repulsive) between them. The change of the potential and the reduction of the thermally accessible regions of the potential energy surface may drive the molecules into energy minima that correspond to a conformation or packing different from that observed at room temperature. In the last case, a structural phase transition occurs, considerably altering the intermolecular interactions. If the experimental EDD is available above and below the transition temperature T_{tr}, it is possible to gain insight into the observed structural changes by relating them to the changes of the scalar field $\rho(\mathbf{r})$ itself. To this end, the Quantum Theory of Atoms in Molecules (QTAIM) [3, 4] illustrated in Chaps. 1, 9 and 12 provides useful topological tools to describe the nature of the chemical interactions and the corresponding changes which occur as a function of T. Moreover, a change in the temperature can vary the population of electronic states. This is particularly well-known in the field of spin crossover (SCO) materials, which are molecular compounds with the ability to switch between a paramagnetic high spin state (HS) and a diamagnetic low-spin ground state (LS) [5].

Special attention is required when comparing the experimental EDD above and below T_{tr}, because in most cases the $\rho(\mathbf{r})$ obtained above T_{tr} shows intrinsically lower precision than the $\rho(\mathbf{r})$ obtained at $T < T_{tr}$, for the reasons mentioned above. In addition further difficulties may arise, e.g. both phases [6] or at least the high-T one [7] may be disordered, there may be hysteresis effects [8], twinning [6, 9], or coexistence of phases [10]. All these problems hamper or even prevent the determination of accurate experimental electron densities, even at very low T. However, if enough single-crystal XRD data are available to at least solve the structure with satisfactory accuracy, such a structure may serve as a starting point

[1] In negative thermal expansion materials (such as ZrW_2O_8 or HfW_2O_8) [2] the effect of the T lowering is just the opposite, i.e. it implies an expansion, rather than a shrinking, of the lattice. See also Chap. 17.

for theoretical periodic quantum calculations in both the high-T and the low-T phases. QTAIM can then be applied to the theoretical $\rho(\mathbf{r})$ above and below T_{tr} and the transition can be followed topologically. Such solid-state simulations may be performed with the CRYSTAL or the ESPRESSO codes described in Chap. 2 and with other publicly available software [11, 12]. On the other hand, when both experimental and theoretical results are available, it is important to compare them on the same grounds. For instance, it could be useful to project the theoretical EDD onto a set of Fourier-calculated structure factor amplitudes, which can then be examined by a conventional multipole model and easily compared with the experiment-derived ones.

Sections 19.3 and 19.4 deal with the T-dependent phenomena described above. In Sect. 19.3, the interplay between $\rho(\mathbf{r})$ and other physical properties as a function of T is considered, and some cases are presented for which the EDD is available (experimentally or theoretically) both above and below the transition temperature. Sect. 19.4 will then be devoted to the changes in $\rho(\mathbf{r})$ associated with spin transitions, i.e. changes linked to an intra-ionic move of electrons between non-bonding and anti-bonding orbitals and structurally reflected in the metal to ligand distances.

19.2 The Role of Temperature in Accurate Experimental Charge-Density Studies

Multi-temperature studies of crystalline structures can be carried out for a variety of purposes, which include the accurate analysis of molecular thermal motion in crystals [13, 14], the explicit search for structural phase transitions [15, 16], the investigation of the effect of T on the *intra-* [17, 18] and *inter-* [19] molecular interactions and on hydrogen bonding [17, 18, 20], and the search for the best conditions to accurately evaluate the electron density.

In early investigations, when the experimental analysis of EDD was still a challenge, one of the main targets was the determination of the so-called difference Fourier synthesis, defined as:

$$\Delta\rho = 1/\mathrm{V}\sum_{\mathrm{hkl}}\Delta\mathrm{F}\,(\mathrm{hkl})\exp\left[-\mathrm{i}\phi_{\mathrm{hkl}}\right] = 1/\mathrm{V}\sum_{\mathrm{hkl}}\left(|\mathrm{F_o}\,(\mathrm{hkl})| - |\mathrm{F_c}\,(\mathrm{hkl})|\right)\exp\left[-\mathrm{i}\phi_{\mathrm{hkl}}\right] \tag{19.1}$$

where V is the unit cell volume, $|\mathrm{F_o(hkl)}|$ and $|\mathrm{F_c(hkl)}|$ are the observed and calculated structure factor amplitudes, respectively, and ϕ_{hkl} is the phase of $\mathrm{F_c}$. As the structure factors were computed using spherical-atom form factors, they could not account for the local deformation of the electron charge cloud due to chemical bonding. Provided that all the atoms were correctly included in the structure

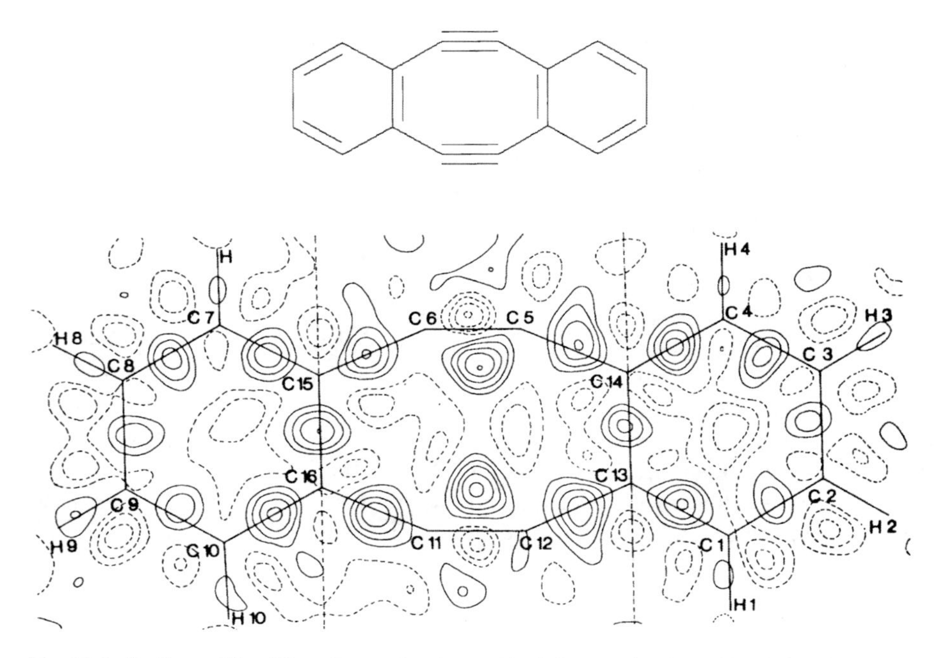

Fig. 19.1 Sections of the ΔF syntheses showing residual electron density at $T = 113$ K. The three sections of the composite, separated by chain-dotted lines, are defined by the planes through the benzene rings and the plane through the triple bonds. Contour levels at intervals of 0.05 e·Å^{-3}. *Solid lines* positive; *dashed lines* negative; zero contours omitted

(i.e. there were no missing, no misassigned nor misplaced atoms), the $\Delta\rho$ maps revealed just those features of $\rho(\mathbf{r})$ that are related to the bonding. These features were indeed present in the observed data, but were not modelled by the structure factors calculated with spherical-atom form factors. Before the era of the multipole formalisms [21–23], the interpretation of such maps was the main method employed to extract experimental information on the chemical bonds from XRD data. Because the F_0 magnitudes depend on the thermal motion in the crystal through the Debye-Waller factors, the lower the data collection temperature, the more informative the corresponding $\Delta\rho$ maps. When *sym*-dibenzo-1,5-cyclooctadiene-3,7-diyne, for example, was studied at $T = 290$, 218 and 113 K [24], it was found that the most interesting difference-Fourier features concern the central eight-membered ring, and particularly the region close to the two triple bonds, where both negative and positive residues are clearly visible (see Fig. 19.1). While the negative holes were ascribed to the inadequacy of the spherical atom scattering factor formalism, the positive peaks were rationalized by invoking an asymmetrical distribution of the electron charge density of the formal triple bonds, in agreement with the results of theoretical calculations. These features were less and less apparent when the same map was obtained from data sets collected at 218 K and 290 K (see Fig. 5 in Ref. [24]), indicating that the lowering of T was mandatory to achieve both qualitative and quantitative information on the chemical bonds.

19.2.1 The Problems of Deconvolution and Correlations Between Parameters

An alternative $\Delta\rho$, the so-called static model deformation density, is nowadays preferred to that obtained by difference Fourier synthesis. This static $\Delta\rho$ is defined [25] as

$$\Delta\rho = \rho_{\text{model}} - \rho_{\text{free atom}} \tag{19.2}$$

where ρ_{model} is the sum of model density functions, which are usually atom-centred multipole functions as in the formalism developed by Stewart [21] or by Hansen and Coppens [23]; $\rho_{\text{free atom}}$ is the reference spherical density of the free atoms held at the same positions as the multipole atoms. Like the earlier difference Fourier map, this $\Delta\rho$ also maps the electronic rearrangement occurring upon formation of chemical bonds, but now the atoms are "frozen" at well defined positions, i.e. the thermal motion is supposed to have been completely deconvoluted from the deformation density (see also Chap. 3).

Deformation densities are not the only tool – in the experimental charge density field – to investigate chemical bonding properties: the development of the QTAIM [3, 4] has triggered a great deal of effort to obtain reliable *total* $\rho(\mathbf{r})$'s with the help of the above mentioned multipole models. Accurate $\rho(\mathbf{r})$'s are actually the main target of modern low-T XRD experiments, and the well-known severe requirements for the evaluation of high-quality total $\rho(\mathbf{r})$ from single-crystal XRD data are continuously investigated [26]. Optimal sample quality [27], high redundancy, large number and high precision of measurements, especially at high angle, are all needed for a proper treatment of the thermal motion, to reduce the least-squares correlations between parameters and to avoid slow convergence and poor-quality (in terms of precision) of the final multipole coefficients. Unfortunately, it is not always possible to accomplish these goals with the available experimental data, particularly because the number of laboratories equipped with in-home cryogenic systems capable to reach very low temperatures ($T < 50$ K) is quite limited.

Two questions are of particular relevance in this context: (i) are the multipole parameters transferable? If a precise and accurate set of multipole coefficients is obtained from a low-T XRD experiment, with what confidence can such parameters be employed to model the EDD of the same substance at another (higher) temperature? (ii) How low must the temperature of the data collection be to achieve a satisfactory deconvolution of the thermal motion from the static deformation density? The next subsections present some attempts to answer these questions.

19.2.2 Transferability of Low-T Multipole Parameters on High-T Structures

In the absence of T-dependent chemical or physical transformations, the *static*[2] molecular EDD is not influenced by the temperature. As a consequence, properties like the atomic charges, the molecular electrostatic moments, and the nature of the *intramolecular* chemical bonds (to mention but a few), are expected not to vary by changing T. Hence, if accurate multipole parameters are known from low-T experiments, it is reasonable to assume that they could be employed at higher temperatures on the same substance, for instance to get a better description of its thermal motion. Furthermore, the topological properties of $\rho(\mathbf{r})$ should also be transferable between similar molecules, according to the well-known QTAIM [3] assumption of atoms as building blocks of chemical systems. This implies that the multipole parameters should be, at least to some extent, transferable also between chemically similar substances, even when they are studied at different temperatures.

The pioneering work by Pratt-Brock, Dunitz and Hirshfeld in 1991 explored this possibility [28]. They transferred the atomic charge-deformation parameters obtained by applying the Hirshfeld formalism [22] on extensive XRD data (up to $\sin\vartheta/\lambda = 1.26\,\text{Å}^{-1}$) of perylene at 83 K to naphtalene and anthracene. The main scope of that work was to improve the temperature-dependent ADP's that were originally derived by conventional least-squares analyses from XRD data of moderate resolution (up to $\sin\vartheta/\lambda = 0.65\,\text{Å}^{-1}$) at temperatures in the range 92–239 K for naphtalene and 94–295 K for anthracene. Although the authors admitted that it was not possible to answer with complete confidence whether the "use of transferred deformation densities has actually led to *improved* values of the atomic ADP's", they were among the first to develop a method for transferring experimentally derived deformation densities between chemically related molecules.

Several other recent works have nowadays assessed the transferability of multipole parameters. Mata et al., for instance, obtained the EDD of L-histidinium dihydrogen orthophosphate orthophosphoric acid from X-ray and neutron diffraction data at $T = 120$ K; then, they collected another set of diffraction data at $T = 294$ K on the same substance and successfully interpreted them using the low-T EDD [29]. Furthermore, transferability between chemically related substances is the underlying idea of the data bank of experimental aspherical atom parameters, applicable to charge density studies of macromolecules of biological interest (such as polypeptides) [30–32]. This topic is discussed in some detail in Chap. 15.

[2]The "static" EDD corresponding to an energy minimum of the potential energy surface at $T = 0$ K is just a *model*, i.e. a purely theoretical concept. Nevertheless, it supplies sensible and reliable information on the nature of the chemical interactions in molecules, being characteristic of a *specific* chemical system.

19.2.3 Setting an Upper Limit on the Temperature?

Once an exhaustive, accurate and adequately phased single-crystal XRD data set has been obtained, the main problem, even if the experiment was performed at low temperature, is to assess whether the final experimental EDD is free of errors due to an insufficient treatment of thermal motion.

Several experimental works (mostly on organic or organometallic substances) have been devoted to this topic; here we comment only on a few of them. Sometimes the multi-temperature data collections have been carried out with a single experimental setup, one of the main focuses being just the deconvolution problem [33, 34]; sometimes the performances of different experimental setups (including synchrotron X-ray sources [35–38]) for an accurate EDD determination have been examined. In all these works the XRD data sets have been collected at T not higher than ~200 K.

Two recent studies by Overgaard et al. [37, 38] compare the structural and electronic properties of polynuclear organometallic complexes as obtained from synchrotron data at T < 20 K and from conventional X-ray sources at 100 K. For one complex, $Co_2(CO)_6(HC{\equiv}CC_6H_{10}OH)$ [37], the two data sets were compared on the basis of the EDD properties at all bond critical points (bcp's). The authors report that "this approach indicates quite different densities from the two data sets", mostly due to significant differences in the carbonyl bonds, for which a small shift (~0.03 Å towards C) in the position of the bcp was found for the synchrotron data set. A detailed comparison with the theoretical electron density showed that the synchrotron data were to be preferred, even though the conventional data provided results closer to theory for certain integrated properties (e.g. the atomic charges on Co). Similarly, preference to 15 K synchrotron data was given in the case of another complex $[(Fe_3O(HCOO)_6(HCOO)_3]^{2-}{\cdot}H_2O{\cdot}2(\alpha\text{-}CH_3NC_5H_5)^+$, [38]), for which relevant discrepancies between the EDD results from the two data sets were also found, particularly in $\rho(\mathbf{r})$ and its Laplacian at the bcp's. All this did not prevent the authors from concluding that "two independent high-resolution studies, including one using an intense synchrotron source in combination with very low temperatures, lead to comparable results" [38]. The conclusion that "the errors in the synchrotron data are smaller than those for the conventional data" [37] is probably the consequence of both the lower temperature of the data collection and the very high intensity available with the synchrotron radiation.

Messerschmidt et al. [36] compared the $\rho(\mathbf{r})$-derived electrostatic properties from four different experimental setups at three temperatures between 15 and 100 K. Two data sets were collected with conventional Mo $K\alpha$ radiation under N_2 gas stream at $T = 100$ K on a SMART three-circle diffractometer with APEX CCD area detector and on an EXCALIBUR2 four-circle diffractometer with a SAPPHIRE CCD area detector, respectively. The third data set was measured with synchrotron radiation at $T = 15$ K (Huber four-circle diffractometer, closed-cycle Helium cryostat), while the fourth set was obtained at $T = 25$ K with a Huber four-circle diffractometer equipped with standard sealed-tube Mo $K\alpha$ radiation, a

Table 19.1 Comparison of electronic parameters (d_n) at different temperatures[a]

	100 K	135*a* K	170*a* K	205*a* K
100 K		2.9	4.7	9.5
135*a* K	3.3		2.8	7.7
170*a* K	5.4	3.1		7.9
205*a* K	11.0	9.1	8.7	

[a]The variance-covariance matrix entry in the top row is used. (Reproduced from Ref. [34] with kind permission of The American Chemical Society).

closed-cycle Helium cryostat and an APEX detector. Good agreement was found, on average, among all four settings in terms of both the topological and the electrostatic properties. However, some unexpected dissimilarities in the electrostatic potential from the EXCALIBUR2 data set were reported. These disparities were attributed to differences in the integration software, but their origin could not be clearly explained. In any case, it was concluded that "in general, the lowest possible temperature should be used for charge-density determinations of organic compounds to obtain the optimum I/σ ratios".

A recent paper by Oddershede and Larsen [34] examines systematically the extent of the deconvolution of the thermal motion from the static density as temperature is increased. The authors undertook a study of the EDD of naphtalene at $T = 100$, 135, 170 and 205 K. Because of an unexpected reaction between the original sample and its coating epoxy glue, only the 100 K data obtained from this sample were used in the subsequent analysis. The higher temperature data collections were all performed on a second sample, coated with baby oil, and were labeled as 135*a* K, 170*a* K and 205*a* K, respectively. The refined atomic multipole coefficients p from the various data sets were compared by calculating their distance d_n in parameter space as obtained at two different temperatures, T_1 and T_2:

$$d_n = \left[\frac{1}{n}(p_{T_1} - p_{T_2})^t V^{-1} (p_{T_1} - p_{T_2})\right]^{1/2} \tag{19.3}$$

where n is the number of independent refined parameters and V is the variance-covariance matrix at either T_1 or T_2. If the differences between p_{T1} and p_{T2} were due to random errors only, d_n is expected to be close to 1. In Table 19.1 (Table 3 in the original paper) the values of d_n for all the possible parameter comparisons among the four reliable data collections are reported.

Note that the distance is lower when the variance-covariance matrix of the higher temperature is used, because the variances increase with increasing temperature. The authors recognize that "multipole populations exhibit some variation with temperature", but state that only the 205 K parameters disagree systematically from the rest. They conclude that "the deconvolution of the thermal motion from the electron density has been achieved for the data measured below 200 K". This conclusion appears somewhat arbitrary, because it is not clear which value of d_n is to be adopted as a benchmark for assessing an acceptable similarity of the electronic parameters. In addition, this statement lacks evidence coming from measurements at very low T: if the trend shown in the Table would be maintained down to 70, 40 or 20 K, the threshold for satisfactory deconvolution might have to be lowered to perhaps 100 K. Anyhow, the authors verified that the main features of the static experimental $\rho(\mathbf{r})$ were virtually identical to those obtained by theoretical calculations.

As regards the inorganic substances, it is well known that for a given temperature the ADP's of heavier atoms are significantly smaller than those of light elements such as carbon, nitrogen, oxygen and, of course, hydrogen. So, when only heavy elements are present, the deconvolution problem should be easily faced and, at least in principle,[3] accurate EDD studies do not need to be performed at low T. It appears that for inorganic compounds systematic investigations of the extent of deconvolution as a function of T have not been performed, but several room temperature studies of the static $\rho(\mathbf{r})$ have been reported [39–43]. In these cases short-wavelength or synchrotron radiation was used to collect a significant number of accurate high-order reflections[4] [44, 45]. However, while the agreement between the topological properties coming from synchrotron data sets and those obtained by theoretical calculations is generally good, the "experimental bcp properties, generated with conventional low energy X-ray diffraction data for several rock forming minerals, were found overall to be in poorer agreement with the theoretical properties" [46].

The evident conclusion is that no single upper limit of T for accurate experimental EDD determinations can be set, even for a specific, restricted class of compounds. Rather, there is general consensus on the significant advantages resulting from performing single-crystal XRD experiments at T as low as possible, especially when synchrotron radiation is used. Clearly, the quality of experimental $\rho(\mathbf{r})$ depends not only on T, but on several other features of the experimental setup as well, such as, for example, the beam homogeneity or the performances of the detector, together with the data collection and reduction strategies. Therefore, although very low temperatures ($T < 50$ K) are to be preferred, EDD's obtained at higher temperatures are not necessarily of lower quality. It is certainly true that the precision of the EDD-derived properties is generally higher in experiments performed at lower T.

[3]In fact, this is not always possible because of the occurrence of other phenomena which are to be handled with great care, such as anisotropic X-ray absorption and extinction, atomic anharmonic motion, and so on. Moreover, because of the large number of electrons in inorganic materials, the valence density contributes significantly to the diffracted intensities even at higher $\sin\vartheta/\lambda$ (not so for organic substances). Accordingly, measures carried out at higher 2ϑ are needed to perform a satisfactory deconvolution.

[4]Some pioneering works devoted to the determination of the electron density distribution were performed in the past at room temperature on home diffractometers, and actually dealt with very few (<20) structure factor amplitudes [44] or needed the replacement of lowest-angle, highly-extinction-affected reflections by data from X-ray diffraction powder measurements [45].

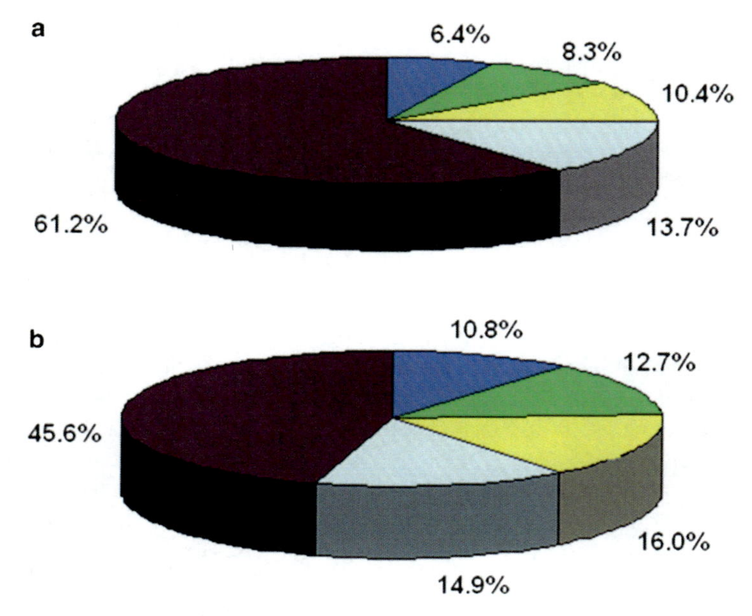

Fig. 19.2 Statistics of the measured intensities at (**a**) 23 K, (**b**) 115 K: ■ I < 1σ(I); ■ 1σ(I) < I < 3σ(I); ■ 3σ(I) < I < 6σ(I); ■ 6σ(I) < I < 10σ(I); ■ I > 10σ(I)

19.2.4 Going Beyond the Convolution Problem: Comparison Between the Charge Density Distributions at T = 23 and 115 K from a Crystal of L-Alanine

All the works discussed so far are lacking a quantitative estimate of the gain in accuracy and precision for the electronic parameters when XRD experiments are performed at very low T. We obtained such an estimate in a test case, the aminoacid L-alanine, for which we collected high-resolution data of comparable quality, from the same crystal, at $T = 115$ K (hereinafter IT) and $T = 23$ K (LT). Full details of the LT crystal structure and charge density analysis, based on a slightly different data set, have already been reported [47–49].

As the temperature is lowered from 115 to 23 K, the number of very significant diffracted intensities ($I/\sigma(I) > 10$) increases from ~50 to ~61% (Fig. 19.2) and the number of very weak measures ($I < \sigma(I)$) decrease from ~11 to ~6%. Similarly, the number of unobserved data ($I < 0$) at 23 K is 1/3 of the corresponding quantity at 115 K.

A comparison of the electron population parameters (C_{plm}'s) obtained at the end of the multipole refinement (based on the Stewart formalism [21]) shows that, on average, the esd's of the C_{plm}'s for the LT are ~0.7 times those at 115 K. The C_{plm}'s at the two temperatures are very similar, with an average difference of about 1 pooled esd (hereinafter σ_p); the largest deviation in terms of σ_p between the two

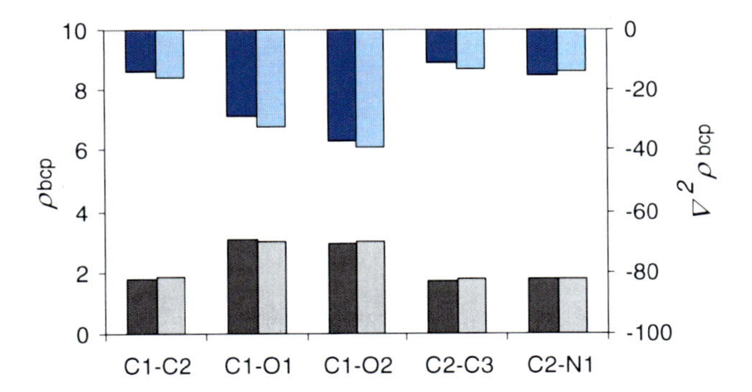

Fig. 19.3 Electron density and Laplacian at the bcp's. The *lower bars* show $\rho(\mathbf{r})$ values at the bcp's (*dark gray* for LT set and *light gray* for IT set), while the *upper bars* display $\nabla^2\rho(\mathbf{r})$ values at the same bcp's (*dark blue* for LT and *light blue* for IT)

sets is 3.7, and is observed for the monopole population of a hydrogen atom of the $-CH_3$ group.

As for the QTAIM properties of the experimental EDD, the LT esd's are again smaller than the IT esd's by a factor of ~0.7. The topological properties at the two temperatures are very similar (see Fig. 19.3). The largest deviation between EDD values at the bcp (ρ_{bcp}) of the C–C, C–O and C–N bonds is 1.6 σ_p for the C1–C2 bond, and the average difference is 1.0 σ_p. Almost identical deviations are observed for the EDD Laplacian at the bcp ($\nabla^2\rho_{bcp}$). For the two C–O bonds, whose lengths differ by ~0.02 Å [47, 49], it is found that the LT ρ_{bcp} value of the C1–O2 bond is smaller by 2.2 σ_p than the ρ_{bcp} value of the shorter C1–O1 bond, while the corresponding difference at IT amounts to 0.5 σ_p only. This suggests that, thanks to the correspondingly lower esd's, the LT data allow a finer discrimination between the chemical features of the two carboxylic oxygen atoms.

Figure 19.4 shows the experimentally derived molecular electrostatic potential $\Phi(\mathbf{r})$ from both data sets, mapped with the program MolIso [50] onto the isodensity surfaces of 0.00675 e Å^{-3}. The overall features of $\Phi(\mathbf{r})$ are largely comparable, and both maps exhibit a large charge separation, as expected. The minima of $\Phi(\mathbf{r})$ are equal within 1 esd, amounting to −464(49) kJ mol^{-1} and −486(72) kJ mol^{-1} for the LT and IT data set, respectively. The O2 atom is closest to the minimum in both cases (at distances of 1.148 and 1.159 Å, respectively). Although these findings suggest that the study of the molecular electrostatic potential does not require temperatures as low as 23 K, the esd's for this extensive property are significantly smaller at 23 K, making the experimental estimate of $\Phi(\mathbf{r})$ more reliable at 23 than at 115 K.

The IT and LT integrated QTAIM properties are also markedly similar. The largest and average differences between the QTAIM volumes of the four atom groupings in the alanine zwitterion (COO^-, NH_3^+, CH and CH_3) amount to 1 and 0.4 Å^3, respectively. The similarity is even more pronounced if the QTAIM group charges are taken into account: LT and IT estimates never differ by more than $1\sigma_{23K}$.

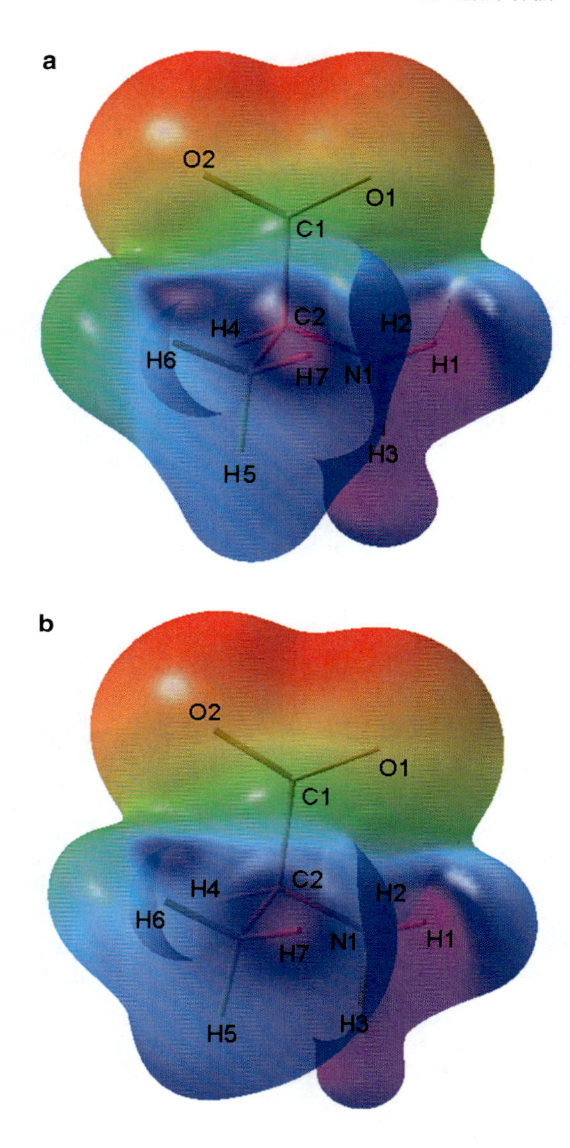

Fig. 19.4 Molecular electrostatic potential mapped on the isodensity surface of 0.00675 eÅ^{-3} for the alanine molecule in the crystal. (**a**) LT and (**b**) IT. The colour scheme ranges from *red* (negative) *via green* (neutral) to *blue* (positive) with values in the range $-410 \rightarrow +456$ kJmol^{-1}

The dipole moments of L-alanine, based on QTAIM partitioning of the LT and IT experimental EDD's, are listed in Table 19.2; all components are expressed in the inertial reference system of coordinates. The following features can be noted: (i) the values from the LT data set are slightly but systematically more precise than those obtained from the IT set; (ii) the difference between the magnitudes of $\boldsymbol{\mu}$ for the in-crystal estimates is larger than for the extracted molecule, but the difference of 0.7 D is scarcely significant in terms of standard deviation; (iii) the polarization of the electron density appears slightly more pronounced from the IT data; (iv) the

Table 19.2 Experimental molecular dipole moment (μ, Debye) from QTAIM partitioning of the EDD in zwitterionic L-alanine. First row: in-crystal values; second row: values for molecules extracted from the crystal, based on a cutoff of $\rho = 0.001$ atomic units (au)

	LT	IT
$\lvert\mu\rvert$	12.4(4)	13.1(5)
	12.1(4)	12.5(5)
μ_x	−12.2(4)	−12.9(5)
	−12.0(4)	−12.3(5)
μ_y	2.0(4)	1.8(4)
	1.9(3)	2.1(5)
μ_z	0.9(2)	1.5(2)
	0.9(2)	1.1(2)

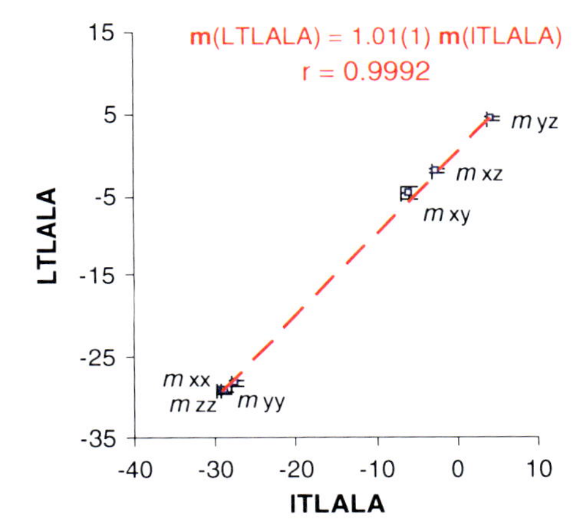

Fig. 19.5 Linear correlation between experimental QTAIM values of the components of the in-crystal second moment (in debye Å) of the LT and IT data sets. The s.u.'s are reported as vertical/horizontal bars

direction of the in-crystal dipole moment is in both cases roughly parallel to the principal axis of the molecule and to the **c** axis of the crystal, the angle between the latter and the dipole vector amounting to 1.0 and 0.1° for the LT and IT data set, respectively.

Individual components of the molecular second moments are graphically correlated in Fig. 19.5. The linear correlation between the experimental in-crystal QTAIM values from the IT and LT sets shows a slope of 1.01(1), indicating quantitative agreement between the values of the tensor components at the two temperatures.

Experimental estimates of the electrostatic interaction energy, E_{es}, for the three hydrogen-bonded molecular pairs in the aminoacid crystal at the two temperatures are summarized in the fourth column of Table 19.3 [see Fig. 4 of Ref. [49] for a graphical display of the three pairs A, B and C]. The values of E_{es} were obtained from the multipolar X-ray $\rho(\mathbf{r})$'s through the highly accurate EP/MM method [52, 53].

Within experimental error, the two data sets predict the same values of the electrostatic interaction energies, but again we find that the estimates from the IT data set are less precise than those obtained from the 23 K data.

This test case shows that the difference between the results obtained at the two temperatures is very modest, but the comparison proves that working at

Table 19.3 Electrostatic contribution (E_{es}, kJ mol^{-1}) to the total interaction energies for the three hydrogen-bonded molecular pairs of L-alanine

Pair[a]	Data set	d_{CM} (Å)[b]	E_{es}[c]
A	LT	5.7939	−149(10)
	IT	5.7913	−153(17)
B	LT	4.1541	−170(9)
	IT	4.1600	−184(12)
C	LT	5.1401	−25(4)
	IT	5.1360	−24(6)

[a]Symmetry operations relating the second molecule of a pair to the parent one (at x, y, z) are: A (x, y, −1 + z); B (1/2 + x, 1/2 − y, 1 − z); C (3/2 − x, −y, −1/2 + z)
[b]Distance between centres of mass of the molecules in a pair
[c]Obtained from the experimental $\rho(\mathbf{r})$ adopting the EP/MM method as implemented in the PAMoC code [51]. The esd's of the MM term are in parentheses

really low temperature ($T < 30$ K) increases the precision of the EDD and its derived properties. This is particularly important when studying intermolecular or interatomic low-density regions (e.g. metal-metal interactions in polynuclear coordination complexes), where it is crucial to assess the physical significance of subtle, very flat EDD features.

19.3 Interplay Between Electron Density and *T*-Driven Changes in Physicochemical Properties

In recent decades, exceptionally fast and accurate XRD data collections have become feasible at synchrotron facilities and home diffractometers equipped with very sensitive area detectors even for microcrystalline samples. Meanwhile, the increasing power of modern computers and the development of robust and fast algorithms for periodic calculations have made available various software packages [11, 12, 54] for accurate simulation of various physical properties in the solid state, such as magnetic coupling constants, optical properties, electronic structure, and so on. These developments facilitated correlation of the T-driven changes in such properties (either measured or theoretically estimated) with EDD, and elucidation of the interplay between lattice, electronic and magnetic degrees of freedom within the material. Nowadays, this field of study is of increasing importance and hopefully it will disclose in the next future new perspectives in crystal engineering and in the search for advanced materials. In the literature, two different approaches are followed to merge the information on EDD with that on other properties: in one approach, $\rho(\mathbf{r})$ is determined only at very low T by single-crystal XRD to provide insight into the physicochemical nature of *intramolecular* covalent or coordinative interactions [55, 56]. In a second approach, the changes in the *intermolecular* $\rho(\mathbf{r})$ are examined as a function of T and, if necessary, they are correlated with the corresponding changes of other electronic or structural parameters [57, 58]. In the following discussion, two examples based on the latter approach will be presented.

The mutual interplay between crystal structure, $\rho(\mathbf{r})$, electron configuration and thermal motion was studied for the Kondo material CeB_6 [57]. The experimental EDD was obtained at four temperatures between 100 and 298 K and investigated using a non-conventional atomic orbital analysis [59, 60]. In particular, it was found that the lowering of T forces a charge transfer of $4f$ electrons from Ce to the B–B bonds connecting the B_6 octahedra, resulting in a significant enhancement of the anharmonic motion of the metal ion. Enhanced anharmonicity raises the entropy and provides the driving force for the electron transfer. The authors claim that "changes in crystal structure, electron density, electron configuration and anharmonic vibration are found to be closely correlated with one another". Eventually, they conclude that "multidimensional X-ray diffraction analysis in which temperature, time and energy are added as experimental variables to the three dimensions of space will be a very fascinating field in science".

A recent study follows the strengthening of intermolecular interactions in pentaerythritol and corresponding changes in the intermolecular electron density, both due to a lowering of temperature or to an increase of pressure [58]. This substance forms hydrogen-bonded layers, which come closer at low T or high p. At the same time, the intermolecular interaction energies become more negative, while topological and energetic indicators at the bcp's of $\rho(\mathbf{r})$ signal the enhancement of the interlayer attractive interactions. On the other hand, the interactions within each molecular layer are found to become weaker on the same grounds. As a take-home message it can be learned from this work that it is possible to *quantitatively* relate changes in charge density within the unit cell with changes in the *observed* crystal packing as a function of T (or p).

As pointed out in the introduction of this chapter, a lowering of T may sometimes induce significant conformational and packing rearrangements of the asymmetric unit, i.e. structural phase transitions may occur. If an accurate $\rho(\mathbf{r})$ is available (experimentally or theoretically) both above and below the transition temperature T_{tr}, structural phase transitions can be characterized quantitatively in terms of the observed changes in intermolecular (or atom-atom) interactions, energetics, and electrostatic moments, thus providing insight into the mechanism[5] of the phase transition. The next Sections deal with this topic, showing a few examples of single-crystal-to-single-crystal phase transitions induced by a change in temperature and studied from the EDD point of view.

[5]In the literature, the mechanism of solid-solid phase transitions is usually associated to thermodynamics or structural considerations rather than kinetics, which would imply the knowledge of nucleation and growth dynamics of the new phase in the bulk. Accordingly, hereinafter the term "mechanism" is intended in the pure applicative meaning of "way by which atoms displace across the transition".

19.3.1 Electron Density Analysis Across Structural Phase Transitions: The Case of Organic Charge-Transfer Complexes

So far only a few studies have been devoted to the characterization of structural phase transitions through EDD analysis, probably because the requirements for an accurate determination of the experimental $\rho(\mathbf{r})$ are very demanding. Organic charge-transfer complexes are the most studied substances from this point of view. These materials are composed of flat molecules, which invariably pile up in parallel stacks. Since the 1970s they attracted great interest as low-dimensional electric conductors. Their electronic properties are heavily T-dependent, and the detailed knowledge of the changes in intermolecular interactions at the atomic level (hence of the EDD) in a wide T range is mandatory to explain their transport properties exhaustively. The literature concerning this topic is very rich, but we refer here to those studies which are explicitly aimed at the experimental and/or theoretical $\rho(\mathbf{r})$ analysis above and below T_{tr}.

A first example is the charge-transfer complex anthracene-tectracyanobenzene (A–TCNB), which undergoes a single-crystal-to-single-crystal second-order, order-disorder phase transition at $T_{tr} = 206$ K. This system was studied in great detail during the 1980s by means of both experimental and theoretical techniques; the disordered phase above T_{tr} was interpreted as "dynamically disordered" in an early EDD study performed in 1980 [61]. The paper reports the experimental EDD for both the molecules in the complex, obtained at seven different temperatures above and below T_{tr}. The high-T and low-T phases were assigned to the space groups Cm and $P2_1/a$, respectively. In Fig. 19.6, an excerpt of Fig. 1 in Ref. [61] is reported, showing the EDD plots of anthracene obtained as direct Fourier summations over the observed structure factors amplitudes F_o's. The large, smeared contours in $\rho(\mathbf{r})$ around the C4–C4M and C7–C7M bonds of anthracene above T_{tr}, mirrored by equally large and elongated thermal ellipsoids on the atoms C4, C4M, C7 and C7M were originally interpreted as the effect of librational motion around the normal to the molecular plane, i.e. as a sign of dynamic disorder. These findings were heavily criticized in a later work [62], which proposed the symmetry C2/m as a valuable alternative for the high-T phase. A static model for the disordered anthracene was also proposed, in which anomalous, large thermal ellipsoids arise because of the superposition of two molecules, inclined at $\pm 6°$ with respect to the mirror plane in Cm. Subsequent works [63–65] maintained the C2/m symmetry for the high-T phase, but at the same time confirmed that the orientational disorder of anthracene is better described as dynamic, in agreement with the original proposal [61]. As pointed out by Larsen [15], "sometimes it is difficult to disentangle in a crystallographic study the results of molecular disorder", and "data collected at different temperatures may be of help in obtaining structural information by discerning between statistical and dynamic disorder".

The tetrathiofulvalene–chloranil complex (TTF–CA) is a key example of a T-driven neutral-ionic phase transition of the first-order [66]. It takes place at

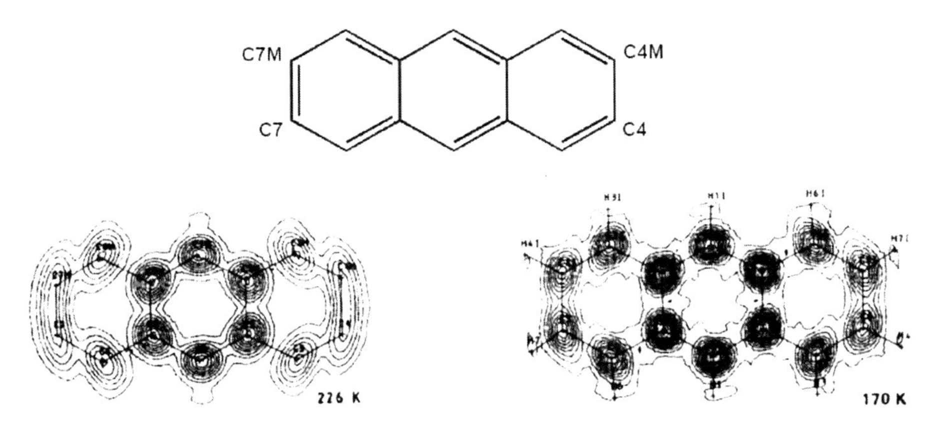

Fig. 19.6 Excerpt of Fig. 1 of Ref. [61]. Electron density, F_0, plots for the A molecules of A: TCNB at the indicated temperatures. The picture at left contains examples of the high-T phase; that at right displays examples from the low-T phase. Plots are contoured at 1e Å^{-3} levels (Reproduced from Ref. [61] with kind permission of The American Institute of Physics)

$T_{tr} = 81$ K, involving a reduction of the crystal symmetry from $P2_1/n$ (high-T phase) to Pn (low-T phase) and abrupt changes in the amount of charge transfer, optical and transport properties. The EDD of this complex [67], especially the corresponding QTAIM descriptors, was studied theoretically as a function of T using plane-wave calculations based on experimental neutron-diffraction geometries obtained at $T = 40$, 90 and 300 K. A direct relationship was found between the relative variations of the potential energy density at the intermolecular bcp's and the intermolecular distances. The lowering of T causes an increase in the strength of all the intermolecular interactions; shorter contacts are strengthened more, on average, than the longer ones. Moreover, the amount of the charge transfer was studied as a function of T both topologically and with other methods; results were found to be in close agreement with each other and indicated a charge transfer of about 0.45 e at 300 K and 0.65 e at 40 K, occurring from TTF (donor) to CA (acceptor). The combination of an EDD study with electronic structure calculations gives a valuable insight into the transition mechanism. The authors claim that their "results on TTF-CA show that HOMO-LUMO overlap and hydrogen bonds dominate the intermolecular interactions in both high- and low-temperature structures and drive the molecular deformation and reorientation occurring" at the neutral-ionic transition.

More recently, high-resolution single-crystal XRD experiments on the same system, both above ($T = 105$ K) and below ($T = 15$ K) T_{tr}, were performed to accurately characterize the intra- and interstack intermolecular interactions and the amount of charge transfer directly from the experiment [68]. In particular, the charge transfer q_{CT} was related to the observed changes in the intermolecular interactions and in the covalent bond lengths of the TTF and CA molecules. The amount of q_{CT}, estimated through the integration of the observed $\rho(\mathbf{r})$ inside each atomic basin, was found to be as large as 0.21 (2)e at 105 K and 0.77 (2)e at 15 K. On going from 105 to 15 K, the TTF C=C bond lengths increase, whereas the lengths of the C–S bonds decrease (although the latter shortening is barely significant in terms of the experimental esd's). In CA, on the other hand, the single C–C bond distances decrease, while the C=C, C=O and C–Cl distances increase. These findings were rationalized in terms of the charge transfer from the TTF HOMO into the CA LUMO, the latter being π antibonding for both C=O and C–Cl bonds. It should be noted, however, that these conclusions assume that the thermal motion has been properly deconvoluted at both temperatures.

As regards the intermolecular interactions, the topological analysis of the experimental EDD both above and below T_{tr} showed that "the C–H···O intrastack contacts are the strongest interactions and are reinforced in the low-T phase, whereas the strength of the Cl···Cl interactions do not change across the transition". The authors claim that in the low-T phase "the creation of pairs of donor-acceptor dimers clearly shows up, with an increase of the intradimer charge density". The statement that the dimer "is clearly identified" by plotting the $\rho(\mathbf{r}) = 0.065$ e Å^{-3} isocontour in the unit cell, appears to be somewhat arbitrary. The authors conclude that "accurate X-ray measurement and electron density modelling is a necessary tool to directly investigate intermolecular interactions and tiny charge transfers" and acknowledge that "agreement between theory and experiment is not fully satisfactory and requires more accurate calculations and modelling". Indeed, plane wave theoretical DFT calculations are not able to reproduce the observed charge transfer across the transition and disagree in the position of some intermolecular cp's with respect to the experimental results.

Another theoretical study on a similar complex, 2,6-dimethyltetrathiofulvalene-p-chloranil (DMTTF–CA), was recently performed [69], and led to substantially similar conclusions.

19.3.2 Tracking the Temperature Dependence of the EDD Topology Near a Phase Transition: The Case of KMnF$_3$

Several interesting physical properties of perovskite-type inorganic materials, such as piezoelectricity, dielectric susceptibility, and magneto-transport properties, are related to distortions from the ideal, cubic structure with P m$\bar{3}$m symmetry. The perovskite $KMnF_3$, for example, undergoes a weak first-order phase transition, in which the room temperature cubic phase is tetragonally distorted (I4/mcm) below $T_{tr} = 186$ K. The transition involves the alternate rotation of the MnF_6 octahedra about the C_4 cubic axes. Ivanov et al. [70] have performed single-crystal XRD experiments at 190, 240 and 298 K with the aim of following the effect of T on the experimental EDD topology immediately preceding the phase transition. In this study, important experimental requirements have been fulfilled: the sample was grown to a sphere, the reciprocal lattice was explored up to very high resolution to account for anharmonic thermal displacements, and absorption and secondary extinction corrections were properly considered. Therefore, the obtained results appear very reliable in terms of accuracy, although the reported λ_1 and λ_2 curvatures (see Table 3 in Ref. [70]), which are very close to zero and scarcely significant in terms of esd's, may cast some doubts on the critical points classification. The authors demonstrated that the chemical bonding network started to adjust 50–60°C above T_{tr}. At 298 K only cation–anion (Mn–F and K–F) bond paths are evident, with a cage critical point lying on the C_3 axis along the Mn–K vector (Fig. 19.7). As T is lowered, the cage critical point disappears and the ring critical point (rcp) on the (110) plane moves towards the threefold axis; at the same time, a new Mn–K bcp appears in the neighbourhood of the rcp. At $T = 240$ K these two points are only 0.04 Å apart, i.e. the system is close to a topological catastrophe according to Bader's QTAIM [3] (Fig. 19.7). When the temperature is lowered further (Fig. 19.7), the bcp and the rcp veer away from each other (at $T = 190$ K they are 0.35 Å apart) and the topological indicators of the Mn–K interaction clearly indicate its strengthening. At the same time, the K–F interaction becomes weaker. The authors interpreted their findings as a precursor effect, i.e. as a reconstruction of the bonding network of the system as it approaches the phase transition temperature. It is worth noting that, although in this study the EDD was not obtained both above and below $T_{tr,}$ its detailed topological analysis as a function of T provided insight at atomic resolution into the bonds rearrangement near the phase transition itself.

19.3.3 The Elusive Phase Transition of the Fungal Metabolite Austdiol Below 70 K

Sometimes, the occurrence of structural phase transitions can be revealed indirectly from low-T EDD studies, as illustrated by the following example.

Austdiol ($C_{12}H_{12}O_5$, $P2_12_12$, $Z = 4$ [71]; see Fig. 19.8 for the molecular formula and the atom numbering scheme) is the main toxic component of a mixture of substances produced in mouldy maize meal by the fungus *Aspergillus Ustus*. A single-crystal XRD data collection aimed at the accurate determination of the EDD was performed at very low T (20 K) several years ago. Intensities, profile shapes (at azimuthal angle $\Psi = 0$), redundancy and resolution of both data sets seemed of satisfactory quality for an electron density study. Although at first glance there was no evidence of a structural transformation (e.g. cell distortions, changes in the systematic extinction rules, appearance of satellite peaks, and so on), the ADP's of the heavy atoms looked very anomalous at the end of the spherical atom refinement (Fig. 19.8). All the ellipsoids were elongated in the same direction, almost parallel to the **b** axis (which is the longest one). These unexpected features did not depend on the $\sin\vartheta/\lambda$ cutoff adopted in the refinement, i.e. both the high-order and the low-order reflections led to the same, clearly unphysical, ADP shapes. At room temperature, the thermal ellipsoids assumed a normal shape, with U_{ij} values identical to those determined at the same T before the low-T data collections.

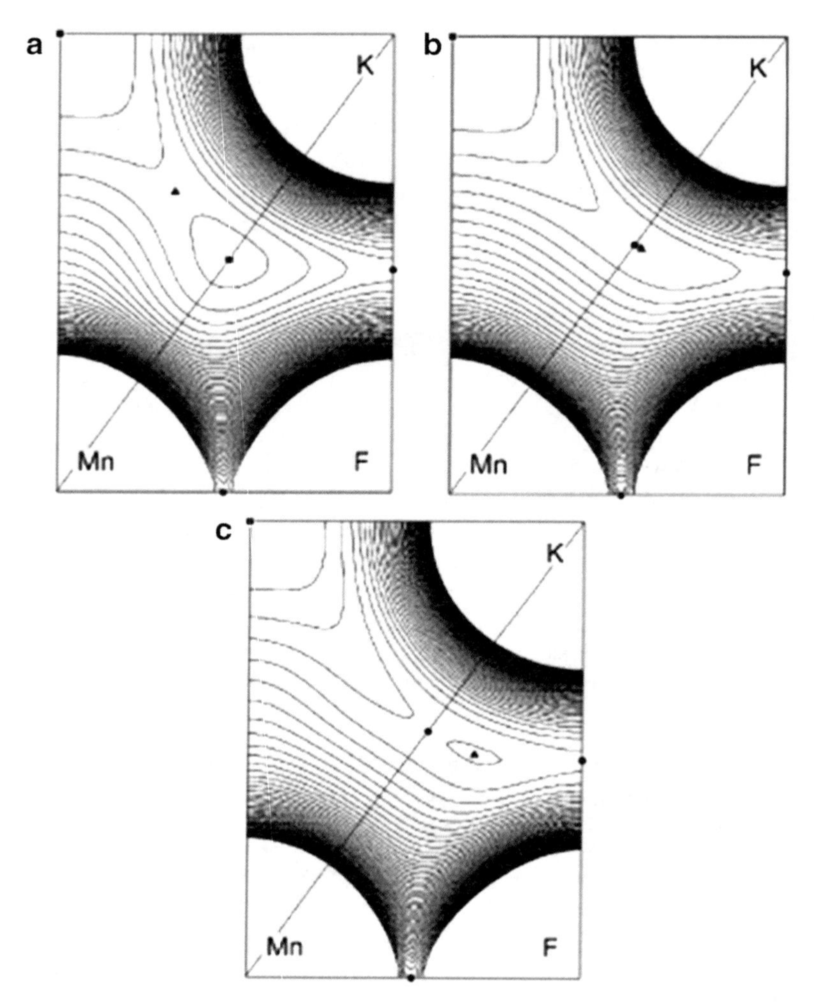

Fig. 19.7 Static model of the total electron density in the (110) plane of $KMnF_3$ at (**a**) 298, (**b**) 240 and (**c**) 190 K. Contour interval: 0.01 e·Å^{-3}. The critical points of the electron density are denoted as follows: filled circles: (3,–1), bond; filled triangles: (3,+1), ring; filled square: (3,+3), cage (Reproduced from Ref. [70] with kind permission of the International Union of Crystallography)

An accurate multi-temperature study was then carried out on several samples of austdiol, crystallized from various solvents, to understand the origin of the observed anomaly in the molecular thermal motion. Eventually, it was concluded that it was due to some kind of fully reversible structural transformation affecting the observed diffraction intensities below $T = 70$ K. Indeed, several azimuthal scans, performed at $T = 62$–64 K on intense, low-order reflections, showed a clear splitting of the profiles when the azimuthal angle Ψ was different from zero (Fig. 19.9); the same

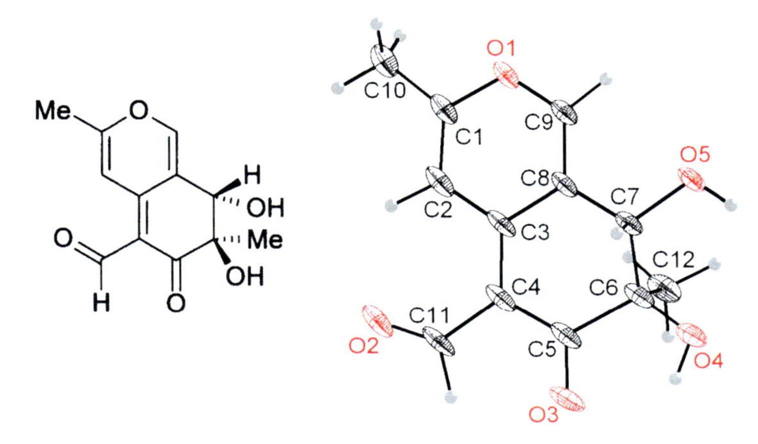

Fig. 19.8 Austdiol ($C_{12}H_{12}O_5$): molecular formula and connectivity, with the atom numbering scheme, as obtained from the single-crystal X-ray diffraction experiment at $T = 20$ K (see text). The thermal ellipsoids are drawn at the 75% probability level. The ball-and-stick picture was realized with the Diamond v3.2a graphical software, © Klaus Brandenburg (1997–2009) Crystal Impact GbR, Bonn, Germany

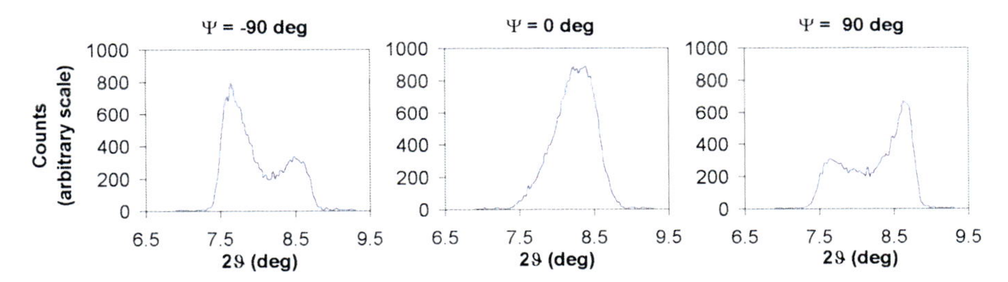

Fig. 19.9 Azimuthal scans for the −1 0 1 reflections of austdiol at $T = 62$ K

profiles appeared perfectly normal in shape when the azimuthal scan was performed at room temperature. Having determined $T \sim 64$ K as the possible temperature for the beginning of the phase transformation, the accurate EDD study was based on data measured at slightly higher temperature, i.e. at $T = 70$ K [72].

This test case shows that it is necessary to verify the accuracy of an EDD analysis not only in terms of local EDD descriptors, especially when anomalous ADP's, or surprising changes in electrostatic moments, are found. In fact, this kind of unusual findings is likely to be the consequence of an inadequate multipole model and/or of wrong data treatment [73] or, as in this case, they may signal an unexpected and otherwise elusive structural transformation.

19.4 Charge Density Studies of Temperature-Dependent Phenomena: Spin-Crossover and Associated Changes in $\rho(\mathbf{r})$

19.4.1 Introduction

As stated at the beginning of this chapter, changes in temperature may not only have a direct effect on atomic positions, and hence on molecular and crystal structure, but may also alter the features of the electron density distribution $\rho(\mathbf{r})$ by inducing changes in the population of certain atomic orbitals. Spin crossover (SCO) in transition metal complexes is a typical example of this.

When a transition metal ion is subject to a ligand field the degeneracy of the five atomic d-orbitals is broken and different spin states become possible. The final spin state of the metal will depend on the energies and populations of the non-degenerate d-orbitals. In the frequent case of an octahedral ligand field, the *lobes* of the $d_z{}^2$ and $d_x{}^2{}_{-y}{}^2$ orbitals are oriented along metal-ligand axes, leading to an increase in their energy due to repulsion from the electrons in the ligand orbitals. By contrast, it is the *nodes* of the d_{xy}, d_{xz} and d_{yz} orbitals that lie along the metal-ligand axes, resulting in a decrease in energy of these orbitals. The difference in energy (Δ) between the destabilised orbitals (the e_g set) and the stabilised ones (the t_{2g} set) depends on the nature of the bonded ligands. Spin-pairing two electrons in the same orbital requires a pairing energy (P). The relative magnitude of Δ and P governs the electron distribution within the orbitals (Fig. 19.10). If $\Delta < P$ then Hund's Rule is obeyed, i.e. all five d-orbitals will be singly occupied prior to spin-pairing of electrons in the t_{2g} set; maximum spin multiplicity is achieved. This case is referred to as high-spin (HS). Conversely, when $\Delta > P$, the orbital occupation is governed by the Aufbau Principle, allowing for full occupation of the t_{2g} set prior to population of the e_g set; spin multiplicity is at a minimum. This case is then referred to as low-spin (LS). The spin state of a metal ion can then be controlled through a modification of the ligands around the metal, i.e. through a change in the ligand field strength. With appropriately chosen ligands, the spin crossover phenomenon can be observed: an intra-ionic electron transfer between the e_g and t_{2g} orbitals producing a change in the spin state. This is usually induced by a perturbation to the sample environment, such as a change in pressure or as a result of light irradiation. However, the most common stimulus to induce this effect is a change in temperature, the low spin state being inherently more stable at low temperatures and the high spin state being the more stable at higher temperatures. Although SCO is markedly more common in d^6 iron(II) complexes, it is also possible in any transition metal that has an electronic configuration between d^4 and d^7.

HS and LS configurations are structurally and electronically very different and can easily be distinguished from the metal-ligand bond distances: population of the anti-bonding e_g orbitals and depopulation of the non-bonding t_{2g} orbitals will lengthen them, as has been observed numerous times [74]. Accompanying

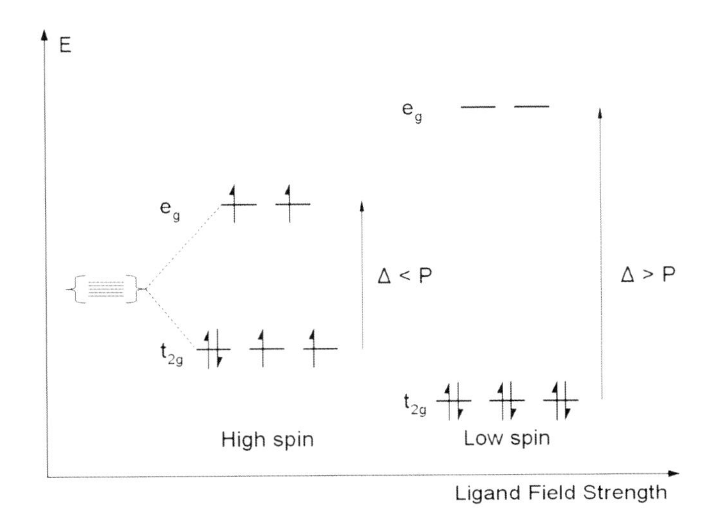

Fig. 19.10 Schematic representation of the electronic configuration of a d^6 metal ion in an octahedral ligand field, in the HS and LS states

differences in the electron density distribution have not been studied as extensively, but have been analysed on the basis of high resolution diffraction data for some materials in either the LS or the HS state, in particular for porphyrins and phthalocyanines [75–80]. As in the case of a structural phase transition, obtaining the experimental $\rho(\mathbf{r})$ above and below a spin transition would improve our understanding of the relationship between the changes in the electronic and molecular structure, by relating them to the changes of the scalar field $\rho(\mathbf{r})$ itself.

However, performing such an experiment would introduce additional effects inherent to the temperature difference. This difficulty can be circumvected by making use of the thermal hysteresis frequently present in these systems. This would allow to obtain $\rho(\mathbf{r})$ for the HS as well as the LS states of a complex at the same temperature. The challenge here is to find a suitable sample, one that does not introduce other intrinsic difficulties, such as the presence of another heavy metal. The material must also form high quality crystals and present a transition with a wide hysteresis loop centered at a temperature low enough to allow this kind of study. To our knowledge, no such high resolution diffraction study has been published to date. Milen & Maslen published in 1988 electron density studies at 295 K and 120 K of an iron(III) spin crossover complex showing a gradual transition centred at ~200 K [81]. Much more recently Pillet et al. [82] reported electron density studies of the light induced metastable high spin state of $Fe(phen)_2(NCS)_2$. However, the work reported by Legrand et al. [83] on the electron density studies at 15 K of the LS and quenched HS states of $Fe(btr)_2(NCS)_2 \cdot H_2O$, comes closest to the scenario mentioned above.

19.4.2 Charge Density Analysis of $Fe(btr)_2(NCS)_2{\cdot}H_2O$

The compound $Fe(btr)_2(NCS)_2{\cdot}H_2O$ (btr: 4,4'-bis-1,2,4-triazole) undergoes an abrupt, complete spin transition with a hysteresis width of 21 K ($T_{1/2\downarrow} = 123.5$ K, $T_{1/2\uparrow} = 144.5$ K) as revealed by magnetic susceptibility and Mössbauer spectroscopy measurements [84]. Initial structural information on this compound was reported at room temperature [84], however, Pillet et al. carried out a careful structural study of the thermal HS-LS transition, showing that it occurs through a domain nucleation and growth process [85]. The crystal structure shows layers parallel to the *bc* plane, formed by each Fe(II) centre coordinating to the terminal N-atoms of 4 btr ligands in the basal plane. The octahedral coordination of each Fe(II) is completed by two NCS groups in *trans* position, predominantly aligned along the *a*-axis.

An electron density distribution study on this compound, both in the LS and HS state, was reported by Legrand et al. [83]. The results reported are not from data collected at a temperature within the hysteresis loop, which is too high to allow proper deconvolution of the thermal smearing effects from the deformation density. However, the authors collected high quality data of the LS and quenched HS states at 15 K. Thermal quenching was achieved by flash freezing the crystal in a cold stream of He gas. This procedure allowed comparison of the electron density in different electronic states at the same temperature. Differences in the atomic displacement parameters are then likely to only represent the modification of vibrational modes and amplitudes due to the structural changes induced by the spin transition.

The quenched HS state is a metastable state but relaxation of metastable HS states is, in general, very slow at temperatures below 30 K, in particular for highly cooperative systems. In this case, the rate of relaxation was slow enough to allow collection of high resolution diffraction data from the quenched HS state sample and hence, obtain a full characterisation of this state by the analysis of the electron density distribution.

The static deformation densities in the mean plane of the triazole ring in both the HS and LS states show characteristics similar to those observed in a study of a btr crystal [86]. In fact, identical deformation density features are observed for both spin states with the exception that, in the HS state, there is a displacement of density towards the metal-coordinated nitrogen of the triazole's N–N bond.

As expected, the electron density around the metal centre shows very different features depending on the spin state (Fig. 19.11). In the LS state, negative deformation density is found in all six Fe-N coordination bonds; this is interpreted as a depopulation of the $d_{x^2-y^2}$ and d_{z^2} atomic orbitals with respect to the isolated Fe atom. Similarly, along the diagonal directions, corresponding to the d_{xy}, d_{xz} and d_{yz} atomic orbitals, positive deformation density can be observed. These features are in agreement with the aforementioned splitting of the Fe 3d orbitals into the e_g and t_{2g} energy levels, when the metal sits in an octahedral environment. Figure 19.11 shows that the nitrogen lone pairs of the btr and NCS groups are directed towards the Fe(II) ion, with a maximum of deformation density along the Fe-N axes. The

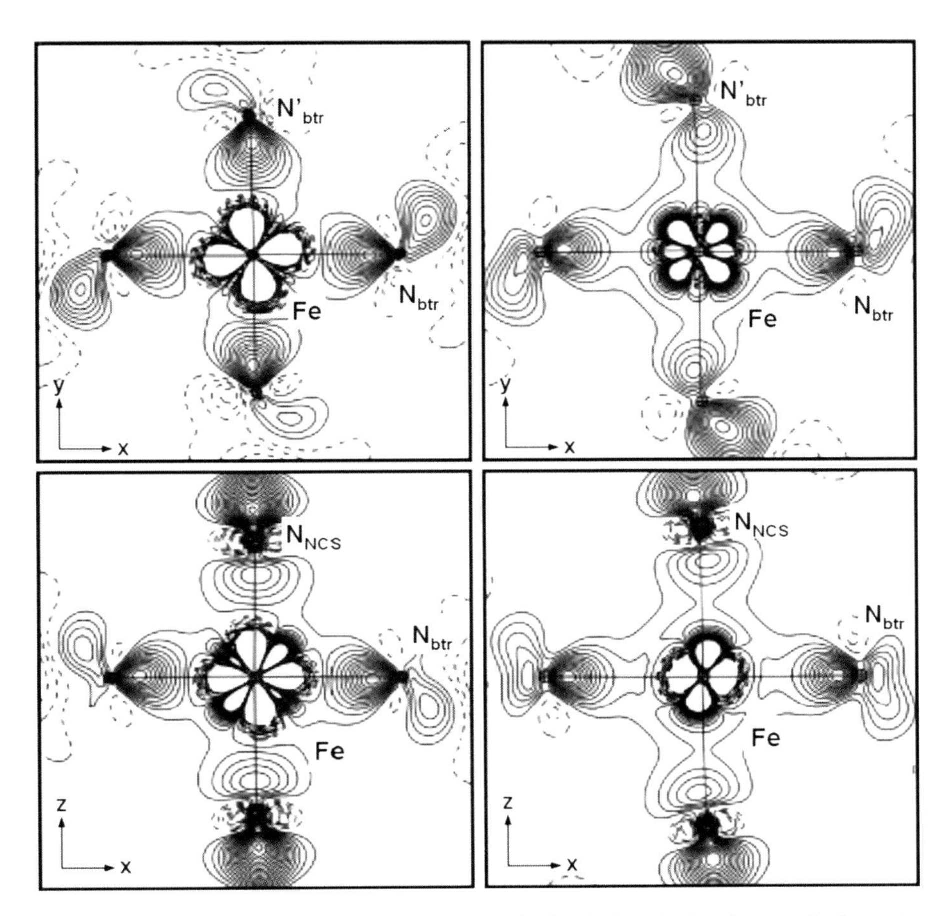

Fig. 19.11 Static deformation density in the triazole-Fe basal plane (*top*) and perpendicular to the basal plane (*bottom*) for the LS (*left*) and HS (*right*) states. Contour levels at 0.05 e/Å^3; positive contours shown as solid lines and negative as dashed lines (Adapted from [83] with kind permission of The American Chemical Society)

deformation density corresponding to the NCS nitrogen lone pair is smaller and more diffuse in directions perpendicular to the Fe-N axis compared to those of the btr nitrogen atoms. In general, the nitrogen deformation density extends to the Fe site, indicating a significant covalent character of the Fe-N bonds in the LS state.

In the HS state, some of the negative deformation density in the direction of the basal btr nitrogen atoms is retained. Positive deformation density is accumulated in the axial direction though, in contrast to the LS state (Fig. 19.11), indicating a large electron redistribution among the Fe 3d orbitals. As for the LS state, the deformation density corresponding to the lone pairs of the coordinating N-atoms extends towards the Fe site, indicating some covalent character of the considerably longer Fe-N bonds. The electron density features of the btr and NCS ligands are very similar in both spin states, in agreement with the SCO effect being a charge transfer within the metal atom, hence no significant changes in the ligands are expected.

Multipolar electron density studies on transition metal complexes (or any other fitting of the charge distribution by analytical density functions) allow direct calculation of the 3d orbital populations [87]. This could not be more suitable than for the case of Fe(II) spin crossover complexes, where not only the differences in those populations is pronounced but also where they are so closely related to the magnetic properties. The relationships between populations of spherical harmonic functions on metal atoms and *d*–orbital occupancies have been derived within the limits of crystal field theory, which neglects covalent interactions between the metal atom and ligands [88]. Hence, orbital population calculations are most reliable when the metal and ligand orbitals do not strongly overlap. As discussed above this assumption is not strictly valid for the present compound as there may be significant covalent character present in the Fe-N bonds. However, the approximation has been proven to be sufficient when a moderate overlap is present [89] and it can clearly detect the differences between high and low spin states as well as small values of orbital occupancies ascribed to σ-donation and π-back-donation [75–80]. Hence, it is expected to also yield enlightening results for the present case.

A comparison of the orbital populations calculated from the multipolar model of the electron density, for the HS and LS states, with the Crystal Field theory ideal orbital populations for Fe(II) in an octahedral ligand field, are shown in Table 19.4. The populations calculated for the LS state agree well with those of a d^6-Fe(II) metal centre having all electrons in the t_{2g} orbitals and none in the e_g orbitals. Nonetheless, the crystal field destabilized $d_{x^2-y^2}$ and d_{z^2} orbitals exhibit some population, attributed mainly to σ-donation from the NCS and btr ligands due to a high metal-ligand σ overlap. Table 19.4 shows that the amount of electron donation from the ligands to the metal is the same, 0.1 *e* by each NCS group and 0.1 *e* by each triazole ring. In the (metastable) HS state, all 3d orbitals are more uniformly populated, consistent with theoretical predictions. Part of the differences in populations between experiment and theory in both the LS and HS states have been ascribed by the authors of this work to π back-bonding. However, another reason for the discrepancy can be the deviations from perfect octahedral symmetry: different ligands, M–N bond distances and N–M–N angles; the latter deviating from right angles. Hence, it can be expected that the theoretical orbital occupancies will not necessarily be those ideally predicted by crystal field theory. Even with differences much smaller than the ones observed in this case, the asphericity of the electron density around the metal resulted in noticeable changes in orbital populations for the case of $Cr(CO)_6$ [90]. Nevertheless, taking into account the errors associated with the diffraction data and the possible differences in expected theoretical values, the experimental results reflect the differences anticipated for LS and HS states as suggested by the theoretical predictions of crystal field theory.

Topological analysis of the charge distribution [3] has yielded values of the electron density at the Fe–N bond critical points (bcp's) that mirror the differences in bond distances, i.e. they are systematically higher in the LS state where shorter and stronger Fe–N bonds are observed. However, a better characterization of metal-

Table 19.4 Atomic orbital populations from experimental X-ray data on $Fe(btr)_2(NCS)_2{\cdot}H_2O$ [83]

	d_{xy}	d_{xz}	d_{yz}	$d_{x^2-y^2}$	d_{z^2}	Total 3d population
LS	1.50	2.22	1.94	0.40	0.20	6.26
HS	1.59	0.94	1.17	0.95	1.49	6.14
LS (theo)	2	2	2	0	0	6
HS (theo)	1.33	1.33	1.33	1	1	6

ligand bonding may be given in terms of the kinetic $G(\mathbf{r})$ and potential $V(\mathbf{r})$ energy densities at the bcp's, where a larger degree of covalent character is associated with an excess of potential energy over the kinetic energy density. Both quantities can be obtained through the Abramov approximation [91]. In the present case, the potential energy density of the Fe-N bonds dominates the total energy density, while the normalised kinetic energy density $G(\mathbf{r})/\rho(\mathbf{r}) > 1$, indicating an intermediate type of bond [92], with some covalent character.

The Laplacian of the electron density indicates the location of local charge concentration ($\nabla^2\rho < 0$). In this case, the Laplacian distribution in the FeN_4 planes clearly shows minima close to the N atoms and towards the Fe ion, corresponding to the electron concentration of the N lone pairs. However, the Laplacian distribution around the Fe ion shows different features depending on the spin state. The authors of this work have found that in the LS state, isovalues of the Laplacian adopt a cubic shape with faces perpendicular to the Fe-N bonds. The actual values of the Laplacian show charge deficit along the bonds, i.e. on the faces of the cube, and concentration of charge as far as possible from these, i.e. on the vertices of the cube. Such a distribution is expected for octahedrally coordinated d^4–d^7 transition metals subject to a strong ligand field, such as an Fe(II) ion in the LS state, due to the electron density and lack of density associated to the t_{2g} and e_g orbitals respectively [93, 94]. In contrast, charge concentrations in the HS state are located along the z axis and along the diagonal directions of the basal plane, fully consistent with the 3d orbital populations described before.

The example summarized above shows that the experimental electron density distribution in a spin crossover complex is consistent with the predictions of ligand field theory, commonly used to explain the observed physical properties of these complexes, and with information obtained by other techniques.

19.4.3 Electron Density Rearrangements in Spin Crossover Systems

The features of the electron density above and below the SCO transition for the most commonly studied mononuclear pseudo-octahedral iron(II) SCO complexes have been described in Sect. 19.4.2. However, there are other examples of SCO transitions that involve more considerable rearrangements of the electron density. Although no formal electron density distribution studies have been published to date on any of these systems, it is worth describing structural results from a few of these cases.

19.4.3.1 Spin Crossover Associated to a Reversible Change of Coordination Number

Given the intraionic nature of the spin crossover, the most important changes on the electron density due to such a transition are observed around the metal, although always keeping its coordination number. However, in 2007 Guionneau et al. reported thermal and light induced spin crossover transitions on $FeL(CN)_2{\cdot}H_2O$ (L:2,13-dimethyl-6,9-dioxa-3,12,18-triazabicyclo[12.3.1]octadeca-1(18),2,12,14,16-pentaene) [95]. The magnetic transitions on this complex are associated to a large modification of the electron density and bonding features around the metal, as the metal changes from 7-coordinate in the high spin state ($FeN_3O_2C_2$) to 6-coordinate (FeN_3OC_2) in the low spin state. The material undergoes a thermal spin transition between 250 K and 150 K, with different degrees of completeness, depending on the cooling rate. Above 250 K the structure is such that the iron(II) is in a pentagonal bipyramidal environment with the metal lying on a two-fold axis. The symmetry restricts the Fe-O distances to be the same. Below the transition, there is a loss of symmetry associated to a Fe–O bond breaking, since the Fe–O bond lengths of 2.334(1) Å at room temperature become 2.243(1) Å and 3.202(1) Å. It is obvious then the large degree of change in electron density distribution around the metal, involving the change of coordination from 7 to 6. In pseudo-octahedral iron(II) complexes, it has been shown that the volume of the coordination polyhedron decreases from 13.0(5) $Å^3$ in HS to 10.0(5) $Å^3$ in LS [96]. The drastic and large change in molecular structure in the present case, is also demonstrated by the volume of the coordination polyhedron around the metal decreasing from 17.0(1) $Å^3$ to 10.0(1) $Å^3$, more than double the standard case.

Another remarkable change of this complex molecular structure after the transition from HS to LS is the inversion of the $O{-}CH_2{-}CH_2{-}O$ ethylene ring conformation. This is shown for example by the distance ethylene carbon atoms to the mean plane defined by the two oxygen atoms and the metal. In the HS state, for symmetry reasons, the distances to the plane are identical (±0.353 Å) and the two carbon atoms are located on opposite sides of the mean plane. In contrast, in the LS state these carbon atoms are located on the same side of the mean plane and at different distances to it (0.872 Å and 0.303 Å).

The inversion of the ethylene conformation and the transformation of the coordination sphere induce significant variations of the intermolecular interactions. In particular, changes of the hydrogen bonds involving the water of solvation, the ethylene carbon atoms and more significantly, of those H-bonds involving the oxygen atoms belonging to the coordination sphere. In the HS state, these two oxygen atoms do not participate in any intermolecular interaction however, in the LS state the oxygen atom that leaves the coordination sphere forms a hydrogen bond with a methyl group of a neighbouring molecule (O . . . H = 2.33 Å, OCH = 154°). The large changes in the intermolecular contacts are noteworthy since they play an essential role in the propagation of the spin crossover across the solid.

19.4.3.2 Spin Crossover in Fe(II) Phosphine Complexes

One of the most interesting features of the SCO effect is the feasibility of addressing information reversibly in the solid state by light irradiation, according to the well-known Light-Induced Excited-Spin-State Trapping (LIESST) effect and reverse-LIESST effect. However, trapping and keeping the system in the metastable HS state generally requires very low temperatures. The HS → LS relaxation process, largely a quantum mechanical tunnelling process, is known to be influenced among other factors, by the differences in metal-ligand bond lengths and differences in energies between the two states. This process has been widely studied, mostly on SCO complexes involving Fe(II) centres with nitrogen donor ligands, as in the previous examples. However, an alternative to nitrogen donor ligands is the use of phosphine ligands.

$[Fe(dppen)_2X_2]\cdot 2S$ type compounds [dppen = cis-1,2-bis(diphenylphosphino) ethylene; X=Cl, Br; S=$CHCl_3$, CH_2Cl_2, $(CH_3)_2CO$] have been shown to undergo both thermal-, pressure- and light induced spin transitions [97–100]. Single crystal structure determinations have been carried out above and below the SCO transition temperatures for $[Fe(dppen)_2Cl_2]\cdot 2(CH_3)_2CO$ [97] and $[Fe(dppen)_2Br_2]\cdot 2CHCl_3$ [100]. Both cases have revealed a substantial shortening of the Fe–P bond, an average of 0.28(1) Å, upon conversion from HS to LS, larger than the more common ~0.20 Å contraction in Fe–N bonds [74]. The decrease of the Fe–Halide bond length is an order of magnitude smaller at an average of 0.033(1) Å. The variation of the bond angles in the FeP_4Hal_2 fragment is also relatively small, as indicated by the ~15% change in the value of the octahedron distortion parameter Σ [96] upon the transition, in comparison to values of 30–55% for FeN_6 coordination. Thus, the main structural change upon a spin crossover transition in these complexes is the Fe–P bond lengths. From measurements of the HS → LS relaxation curves at different temperatures of the metastable HS state after LIESST, Wu et al. [100] have estimated the temperature independent tunneling rate and correlated them with the transition temperatures $T_{1/2}$ for four FeP_4Hal_2 complexes. Comparison of this correlation with that of Hauser [101] for FeN_6 spin crossover complexes, shows that for a given $T_{1/2}$ value, the FeP_4Hal_2 complexes exhibit a much slower rate of low-temperature tunneling than the FeN_6 complexes. They ascribe this result to the larger bond length changes observed between the HS and LS states.

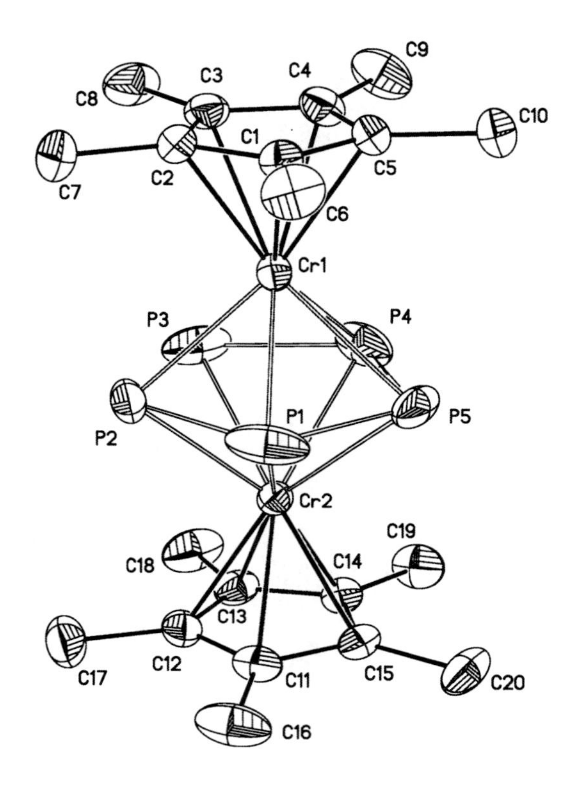

Fig. 19.12 ORTEP plot of the cation of TD1 at 170 K. Ellipsoids at 50% probability. H atoms have been omitted for clarity. (Reproduced from Ref. [7] with kind permission of The Royal Society of Chemistry)

19.4.3.3 Abrupt Rearrangement of the Electron Density in a Dinuclear Cr(II) Spin Crossover System

As mentioned above most structural studies reported on spin crossover concern octahedrally coordinated Fe(II) complexes. However, at least three cases have been reported that include a d^4 metal ion as the magnetic centre [7, 102–104] and only one of them, a triple decker dinuclear Cr(II) complex, shows a metal environment that is not octahedral [7, 102].

$[(Cp^*)Cr(\mu^2{:}\eta^5 - P_5)Cr(Cp^*)]^+$ $(SbF_6)^-$ (TD1), where Cp^* is a pentamethyl-cyclopentadienyl ion, undergoes a sharp SCO at approximately 23 K. The molar susceptibility and the structural results between 290 and 170 K are consistent with 2 non-interacting d^4 Cr^{2+} ions. At these temperatures the asymmeric unit of TD1 (in space group *Fddd*) contains a cation, shown in Fig. 19.12, and an anion. At 290 K the five-membered phosphorus ring (^5P ring) shows an almost continuous ring of electron density. This is attributed to librational disorder, as the electron density separates into discrete atomic entities as the temperature is lowered (Fig. 19.13). The Cr–Cr separation gradually decreases from 3.1928(5) Å at 290 K to 3.1588(4) Å at 170 K. The P–P distances and Cr-P-Cr angles are all similar within this temperature range.

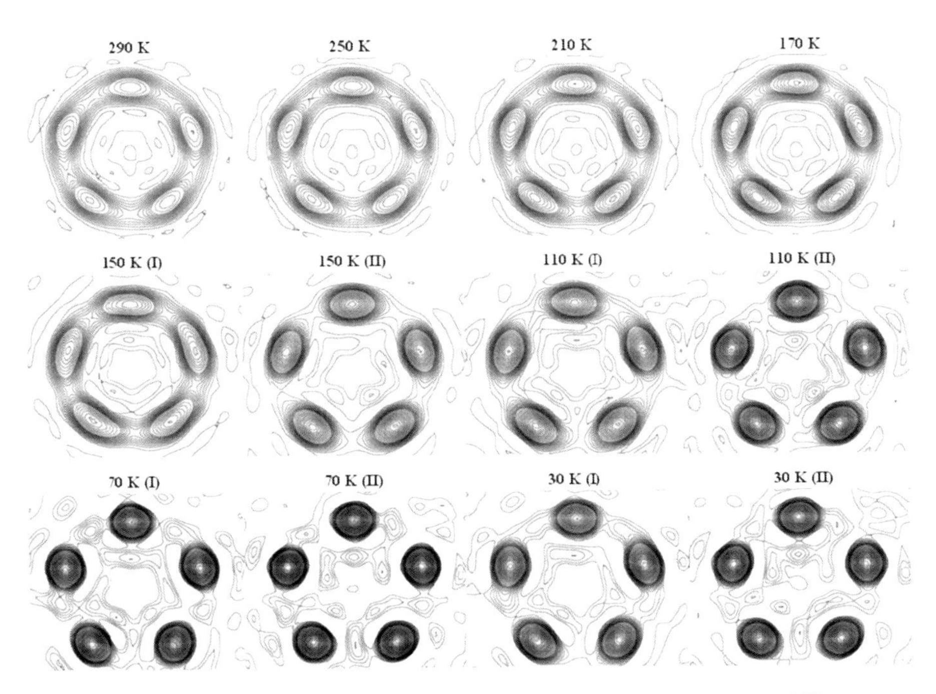

Fig. 19.13 Electron density maps (Fobs) of the five-membered phosphorus ring at different temperatures. Below 160 K there are two crystallographically independent cations in the asymmetric unit, (I) and (II) (Reproduced from Ref. [7] with kind permission of The Royal Society of Chemistry)

A structural phase transition with a change of symmetry is observed at 160 K, from space group *Fddd* to *I2/a*, in parallel with a change shown in the magnetic behaviour between 150 and 25 K. This results in a structure containing 2 independent cations per asymmetric unit. At 150 K the Cr-Cr separations are 3.1201(8) Å for cation I and 3.1657(7) Å for cation II, clearly showing that the geometries of the 2 cations, although similar, are inequivalent. The libration of the ^5P rings is markedly different (Fig. 19.13), cation I still showing a large libration of the ^5P ring but a reduced Cr-Cr distance while cation II, where the libration of the ^5P ring has been dramatically decreased, shows less change in the Cr-Cr separation.

From 150 to 25 K, the Cr-Cr separation decreases for both cations, the change for cation I (~ 0.23 Å) being 4 times that of cation II. The dramatic change between 150 and 25 K in cation I suggests a strong AF interaction, while cation II could still be considered as containing 2 non-interacting Cr^{2+} ions. At 25 K the Cr-Cr separation in cation I is 2.886(2) Å, 0.222 Å smaller than the value of 3.108(2) Å for cation II. All P–P distances and Cr-P-Cr angles are still similar within each cation but clearly different between the 2 symmetry independent cations. The average values at 25 K are 2.225(3) Å and 74.7(2)° for cation I and, 2.200(7) Å and 79.4(3)° for cation II respectively.

The unit cell parameters from 290 K down to 25 K had shown an approximately linear decrease in the volume with temperature. However, they show a sharp change between 20 K and 18 K. Furthermore, a sudden change in the observed intensity of a few monitored reflections can be seen at 24 K, clearly indicating a sudden rearrangement of the electron density. These results are all in accordance with the sharp magnetic transition at 23 K [102], given that small differences in measured transition temperatures can be attributed to differences in temperature calibrations for the different instruments.

No alteration of the symmetry is observed below 25 K however, at 12 K the Cr-Cr separations of the 2 independent cations again become quite similar, at 2.782(2) and 2.798(2) Å. If an AF interaction is assumed between the Cr ions with the shorter Cr–Cr distance at 25 K, it is evident that at 12 K both pairs of Cr ions present an AF interaction. This would make the material diamagnetic, as is observed experimentally [102].

The rearrangement of the electron density below the transition at 23 K is considerable. Below this transition the two ^5P rings become again very similar. However, the P-P distances within each ^5P ring are no longer equivalent. Each ^5P ring distorts and the P-P bonds can be grouped in two short, two medium and a long one. The ^5P rings could then be pictured as consisting of two entities: one group of 3 phosphorus atoms separated by an average distance of 2.165(5) Å and a group of 2 phosphorus atoms separated by an average distance of 2.386(6) Å. The average remaining P-P distance is intermediate at 2.265(10) Å. Concurrently with the ^5P ring distortion the Cr ions move towards the longest P–P bond. Hence, the elongation of a P–P bond in each ring below the magnetic transition could be seen as due to the pronounced η^2-coordination to the Cr ions and/or as a means for the Cr ions to maximise their antiferromagnetic interaction by way of a superexchange interaction through the P atoms. The shift of electron density from a P–P bond to the Cr–P contacts would improve the magnetic coupling between the metal ions. Additionally, at 12 K the Cr ions are found in positions very similar to those formed at 25 K by the cation already showing the suggested AF interaction, with Cr-P-Cr angles in both cations of 74.2(3)°.

This explanation for the shortening of the Cr–Cr distance differs from that suggested by Hughes et al. [102], who explained the magnetic interaction in terms of metal-metal bonding. Their theory would imply that metal $d_z{}^2$ orbitals from the Cr atoms are overlapping through the 'hole' in the ^5P ring, forming a M–M sigma bond. The ideal positions for these metal orbitals to overlap are above and below the centroid of the ^5P ring. Hence, this hypothesis does not explain the movement of the Cr atoms away from these ideal positions towards 2 of the phosphorus atoms, after the transition at 23 K.

The observed results may also be interpreted as a two-step spin crossover, a process that has been observed before in dinuclear compounds [105–107]. Given that two non-interacting Cr^{2+} ions with $S = 1$ will explain the value of 4.1 μ_B for the effective magnetic moment between RT and 160 K, the spin state of a Cr^{2+} ion with $S = 1$ can be called high spin (HS). Two (or four) non-interacting Cr^{2+} ions with $S = 0$, i.e. in a low spin state (LS), will explain then a diamagnetic material below 23 K. But at 25 K, the material can be thought of as having the two Cr^{2+} ions of one cation as non-interacting with S = 1 (HS, longer Cr-Cr separation) and the two Cr^{2+} ions of the second independent cation as non-interacting with S = 0 (LS, shorter Cr-Cr separation). At all times, within each molecule, the behaviour of each Cr_2P_5 unit is symmetrical, e.g. the distance from each Cr to the centroid of the ^5P rings is the same within experimental error. Hence, instead of having dinuclear units in a HS-LS state, TD1 presents between 160 and 25 K a dinuclear unit in which the Cr^{2+} ions are both in a high spin state, i.e. HS-HS, and a dinuclear unit in which the Cr^{2+} ions are both in a low spin state, i.e. LS-LS. However, explaining the behaviour of TD1 using this theory, would not account for the change in geometry of the ^5P rings below 23 K.

This test case highlights the relevance of performing careful analyses of the electron density distribution. Information obtained simply by following changes in atomic positions through structural transitions cannot provide a complete understanding of magnetic behaviour. A full analysis of the experimental electron density distribution could aid the understanding of the underlying magnetic transition pathways. However, as previously stated, the experiments require the use of very low temperatures and their complexity can be increased by inherent problems such as twinning and the presence of heavy elements, which intrinsically limits this approach. In these cases, combining modern theoretical methods with measured structural data may still provide a route to understanding the correlations between $\rho(\mathbf{r})$ and physical observables.

References

1. Koritsanszky T, Coppens P (2001) Chemical applications of X-ray charge-density analysis. Chem Rev 101:1583–1627
2. Mary TA, Evans JSO, Vogt T, Sleight AW (1996) Negative thermal expansion from 0.3 to 1050 Kelvin in ZrW_2O_8. Science 272:90–92
3. Bader RFW (1990) Atoms in molecules: a quantum theory, vol 22, International series of monographs on chemistry. Oxford Science, Oxford
4. Matta CF, Boyd RJ (eds) (2007) The quantum theory of atoms in molecules: from solid state to DNA and drug design. Wiley-VCH, Weinheim
5. Gütlich P, Garcia Y, Goodwin HA (2000) Spin crossover phenomena in Fe(II) complexes. Chem Soc Rev 29:419–427
6. Hostettler M, Birkedal H, Gardon M, Chapuis G, Schwarzenbach D, Bonin M (1999) Phase-transition-induced twinning in the 1:1 adduct of hexamethylenetetramine and azelaic acid. Acta Crystallogr B 55:448–458
7. Goeta AE, Howard JAK, Hughes AK, O'Hare D, Copley RCB (2007) Structural-magnetic correlations of the first dinuclear spin crossover d^4 system. J Mater Chem 17:485–492
8. Samulon EC, Islam Z, Sebastian SE, Brooks PB, McCourt MK Jr, Ilavsky J, Fisher IR (2006) Low-temperature structural phase transition and incommensurate lattice modulation in the spin-gap compound $BaCuSi_2O_6$. Phys Rev B 73:100407
9. Yang G, Shang P, Jones IP, Abell JS, Gough CE (1993) Monoclinic phase transition and twinning in $Bi_2Sr_2CaCu_2O_y$ single crystals. Phys Rev B 48:16873–16876

10. Lee HJ, Kim KH, Kim MW, Noh TW, Kim BG, Koo TY, Cheong S-W, Wang YJ, Wei X (2002) Optical evidence of multiphase coexistence in single crystalline (La, Pr, Ca)MnO_3. Phys Rev B 65:115118
11. Blaha P, Schwartz K, Madsen GKH, Kvasnicka D, Luitz J (2001) WIEN2K, an augmented plane wave + local orbitals program for calculating crystal properties. Karlheinz Schwarz Techn. Universitat, Wien
12. Kresse G, Marsman M, Furthmüller J (2009) VASP the GUIDE. Universität Wien, Wien
13. Capelli SC, Albinati A, Mason SA, Willis BTM (2006) Molecular motion in crystalline naphtalene: analysis of multi-temperature X-ray and neutron diffraction data. J Phys Chem A 110:11695–11703
14. Bolotina NB, Hardie MJ, Speer RL Jr, Pinkerton AA (2004) Energetic materials: variable-temperature crystal structures of γ- and ε-HNIW polymorphs. J Appl Crystallogr 37:808–814
15. Larsen FK (1995) Diffraction studies of crystals at low temperatures – crystallography below 77 K. Acta Crystallogr B 51:468–482
16. Lo Presti L, Invernizzi D, Soave R, Destro R (2005) Looking for structural phase transitions in the colossal magnetoresistive thiospinel $FeCr_2S_4$ by a multi-temperature single-crystal X-ray diffraction study. Chem Phys Lett 416:28–32
17. Seliger J, Žagar V, Gotoh K, Ishida H, Konnai A, Amino D, Asaji T (2009) Hydrogen bonding in 1,2-diazine-chloranilic acid (2:1) studied by a ^{14}N nuclear quadrupole coupling tensor and multi-temperature X-ray diffraction. Phys Chem Chem Phys 11:2281–2286
18. Piccoli PMB, Koetzle TF, Schultz AJ, Zhurova EA, Stare J, Pinkerton AA, Eckert J, Hadzi D (2008) Variable temperature neutron diffraction and X-ray charge density studies of tetraacetylethane. J Phys Chem A 112:6667–6677
19. Thomas LH, Cole JM, Wilson CC (2008) Orientational disorder in 4-chloro-nitrobenzene. Acta Crystallogr C 64:o296–o302
20. Parkin A, Adam M, Cooper RI, Middlemiss DS, Wilson CC (2007) Structure and hydrogen bonding in 2,4-dihydroxy-benzoic acid at 90, 100, 110 and 150 K; a theoretical and single-crystal X-ray diffraction study. Acta Crystallogr B 63:303–308
21. Stewart RF (1976) Electron population analysis with rigid pseudoatoms. Acta Crystallogr A 32:565–574
22. Hirshfeld FL (1977) A deformation density refinement program. Isr J Chem 16:226–229
23. Hansen NK, Coppens P (1978) Testing aspherical atom refinements on small-molecule data sets. Acta Crystallogr A 34:909–921
24. Destro R, Pilati T, Simonetta M (1977) The structure and electron density of *sym*-dibenzo-1,5-cyclooctadiene-3,7-diyne by X-ray analysis at three different temperatures. Acta Crystallogr B 33:447–456
25. Coppens P, Becker PJ (1995) International tables for crystallography, vol C. Kluwer Academic Publisher, Dordrecht, p 628
26. Zhurov VV, Zhurova EA, Pinkerton AA (2008) Optimization and evaluation of data quality for charge density studies. J Appl Crystallogr 41:340–349
27. Destro R, Loconte L, Lo Presti L, Roversi P, Soave R (2004) On the role of data quality in experimental charge-density studies. Acta Crystallogr A 60:365–370
28. Pratt Brock C, Dunitz JD, Hirshfeld FL (1991) Transferability of deformation densities among related molecules: atomic multipole parameters from perylene for improved estimation of molecular vibrations in naphthalene and anthracene. Acta Crystallogr B 47:789–797
29. Mata I, Espinosa E, Molins E, Veintemillas S, Maniukiewicz W, Lecomte C, Cousson A, Paulus W (2006) Contributions to the application of the transferability principle and multipolar modelling of H atoms: electron-density study of L-histidinium dihydrogen orthophosphate orthophosphoric acid. I. Acta Crystallogr A 62:365–378
30. Pichon-Pesme V, Lecomte C, Lachekar H (1995) On building a data bank of transferable experimental electron density parameters: application to polypeptides. J Phys Chem 99:6242–6250

31. Jelsch C, Pichon-Pesme V, Lecomte C, Aubry A (1998) Transferability of multipole charge-density parameters: application to very-high resolution oligopeptide and protein structures. Acta Crystallogr D 54:1306–1308
32. Koritsanszky T, Volkov A, Coppens P (2002) Aspherical-atom scattering factors from molecular wave functions. 1. Transferability and conformation dependence of atomic electron densities of peptides within the multipole formalism. Acta Crystallogr A 58:464–472
33. Li N, Maluandes S, Blessing RH, Dupuis M, Moss GR, DeTitta GT (1994) High-resolution X-ray diffraction and *ab initio* quantum chemical studies of glycouril, a biotin analog. J Am Chem Soc 116:6494–6507
34. Oddershede J, Larsen S (2004) Charge density study of naphtalene based on X-ray diffraction data at four different temperatures and theoretical calculations. J Phys Chem A 108:1057–1063
35. Macchi P, Proserpio DM, Sironi A, Soave R, Destro R (1998) A test of the suitability of CCD area detectors for accurate electron-density studies. J Appl Crystallogr 31:583–588
36. Messerschmidt M, Scheins S, Luger P (2005) Charge density of (−)-stricnine from 100 to 15 K, a comparison of four data sets. Acta Crystallogr B 61:115–121
37. Overgaard J, Clausen HF, Platts JA, Iversen BB (2008) Experimental and theoretical charge density study of chemical bonding in a Co dimer complex. J Am Chem Soc 130:3834–3843
38. Overgaard J, Larsen FK, Timco GA, Iversen BB (2009) Experimental charge density in an oxidized trinuclear iron complex using 15 K synchrotron and 100 K conventional single-crystal X-ray diffraction. Dalton Trans 664–671
39. Lippmann T, Schneider JR (2000) Topological analyses of cuprite, Cu_2O, using high-energy synchrotron-radiation data. Acta Crystallogr A 56:575–584
40. Kirfel A, Lippmann T, Blaha P, Schwarz K, Cox DF, Rosso KM, Gibbs GV (2005) Electron density distribution and bond critical point properties for forsterite, Mg_2SiO_4, determined with synchrotron single crystal X-ray diffraction data. Phys Chem Miner 32:301–313
41. Kirfel A, Krane H-G, Blaha P, Schwarz K, Lippmann T (2001) Electron density distribution in stishovite, SiO_2: a new high-energy synchrotron-radiation study. Acta Crystallogr A 57:663–677
42. O'Toole NJ, Streltsov VA (2001) Synchrotron X-ray analysis of the electron density in CoF_2 and ZnF_2. Acta Crystallogr B 57:128–135
43. Hansen NK, Schneider JR, Yelon WB, Pearson WH (1987) The electron density of beryllium derived from 0.12 Å γ-ray diffraction data. Acta Crystallogr A 43:763–769
44. Scheringer C (1980) The electron-density distribution in silicon. Acta Crystallogr A 36:205–210
45. Spackman MA, Hill RJ, Gibbs GV (1987) Exploration of structure and bonding in stishovite with fourier and pseudoatom refinement methods using single crystal and powder X-ray diffraction data. Phys Chem Miner 14:139–150
46. Gibbs GV, Cox DF, Rosso KM, Kirfel A, Lippmann T, Blaha P, Schwarz K (2005) Experimental and theoretical bond critical point properties from model electron density distributions for earth materials. Phys Chem Mater 32:114–125
47. Destro R, Marsh RE, Bianchi R (1988) A low-temperature (23 K) study of L-alanine. J Phys Chem 92:966–973
48. Destro R, Bianchi R, Gatti C, Merati F (1991) Total electronic charge density of L-alanine from X-ray diffraction at 23 K. Chem Phys Lett 186:47–52
49. Destro R, Soave R, Barzaghi M (2008) Physicochemical properties of zwitterionic L- and DL-alanine crystals from their experimental and theoretical charge densities. J Phys Chem B 112:5163–5174
50. Hübschle CB, Luger P (2006) MolIso – a program for colour-mapped iso-surfaces. J Appl Crystallogr 39:901–904
51. Barzaghi M (2002) PAMoC Online user's manual. CNR-ISTM, Institute of Molecular Science and Technologies, Milano. http://www.istm.cnr.it/pamoc/

52. Volkov A, Coppens P (2004) Calculation of electrostatic interaction energies in molecular dimers from atomic multipole moments obtained by different methods of electron density partitioning. J Comput Chem 25:921–934
53. Volkov A, Koritsanszky T, Coppens P (2004) Combination of the exact potential and multipole methods (EP/MM) for evaluation of intermolecular electrostatic interaction energies with pseudoatom representation of molecular electron densities. Chem Phys Lett 391:170–175
54. Dovesi R, Saunders VR, Roetti C, Orlando R, Zicovich-Wilson CM, Pascale F, Civalleri B, Doll K, Harrison NM, Bush IJ, D'Arco Ph, Llunell M (2006) CRYSTAL06 user's manual. Università di Torino, Torino
55. Poulsen RD, Bentien A, Christensen M, Iversen BB (2006) Solvothermal syntesis, multi-temperature crystal structures and physical properties of isostructural coordination polymers, $2C_4H_{12}N^+$-$[M_3(C_8H_4O_4)_4]^{2-}\cdot 3C_5H_{11}NO$, $M = $ Co, Zn. Acta Crystallogr B 62:245–254
56. Poulsen RD, Bentien A, Chevalier M, Iversen BB (2005) Synthesis, physical properties, multitemperature crystal structure, and 20 K synchrotron X-ray charge density of a magnetic metal organic framework structure, $Mn_3(C_8O_4H_4)_3(C_5H_{11}ON)_2$. J Am Chem Soc 127:9156–9166
57. Tanaka K, Ōnuki Y (2002) Observation of 4*f* electron transfer from Ce to B_6 in the Kondo crystal CeB_6 and its mechanism by multi-temperature X-ray diffraction. Acta Crystallogr B 58:423–436
58. Zhurova EA, Tsirelson VG, Zhurov VV, Stash AI, Pinkerton AA (2006) Chemical bonding in pentaerythritol at very low temperature or at high pressure: an experimental and theoretical study. Acta Crystallogr B 62:513–520
59. Tanaka K (1988) X-ray analysis of wavefunctions by the least-squares method incorporating orthonormality. I. General formalism. Acta Crystallogr A 44:1002–1008
60. Tanaka K (1993) X-ray analysis of wavefunctions by the least-squares method incorporating orthonormality. II. Ground state of the Cu^{2+} ion of bis(1,5-diazacyclooctane)copper(II) nitrate in a low-symmetry crystal field. Acta Crystallogr B 49:1001–1010
61. Stezowski JJ (1980) Phase transition effects: a crystallographic characterization of the temperature dependency of the crystal structure of the 1:1 charge transfer complex between anthracene and tetracyanobenzene in the temperature range 297 to 119 K. J Chem Phys 73:538–547
62. Boeyens JCA, Levendis DC (1984) Static disorder in crystals of anthracene-tetracyanobenzene charge transfer complex. J Chem Phys 80:2681–2688
63. Luty T, Munn RW (1984) Theory of phase transitions in charge-transfer complexes: anthracene-tetracyanobenzene. J Chem Phys 80:3321–3327
64. Lefebvre J, Odou G, Muller M, Mierzejewski A, Luty T (1989) Characterization of an orientational disorder in two charge-transfer complexes: anthracene-tetracyanobenzene (A–TCNB) and naphthalene-tetracyanobenzene (N–TCNB). Acta Crystallogr B 45:323–336
65. Lefebvre J, Ecolivet C, Bourges P, Mierzejewski A, Luty T (1991) The structural phase transition in anthracene-TCNB. Phase Transit 32:223–234
66. Dressel M (2007) Ordering phenomena in quasi-one-dimensional organic conductors. Naturwissenschaften 94:527–541
67. Oison V, Katan C, Rabiller P, Souhassou M, Koenig C (2003) Neutral-ionic phase transition: a thorough *ab initio* study of TTF–CA. Phys Rev B 67:035120
68. García P, Dahaoui S, Katan C, Souhassou M, Lecomte C (2007) On the accurate estimation of intermolecular interactions and charge transfer: the case of TTF-CA. Faraday Discuss 135:217–235
69. Oison V, Rabiller P, Katan C (2004) Theoretical investigation of the ground-state properties of DMTF–CA: a step toward the understanding of charge transfer complexes undergoing the neutral-to-ionic phase transition. J Phys Chem A 108:11049–11055
70. Ivanov Y, Nimura T, Tanaka K (2004) Electron density and electrostatic potential of $KMnF_3$: a phase-transition study. Acta Crystallogr B 60:359–368
71. Lo Presti L, Soave R, Destro R (2003) The fungal metabolite austdiol. Acta Crystallogr C 59:o199–o201

72. Lo Presti L, Soave R, Destro R (2006) On the interplay between CH···O and OH···O interactions in determining crystal packing and molecular conformation: an experimental and theoretical charge density study of the fungal secondary metabolite austdiol ($C_{12}H_{12}O_5$). J Phys Chem B 110:6405–6414
73. Whitten AE, Turner P, Klooster WT, Piltz RO, Spackman MA (2006) Reassessment of large dipole moment enhancements in crystals: a detailed experimental and theoretical charge density analysis of 2-methyl-4-nitroaniline. J Phys Chem A 110:8763–8776
74. Gütlich P, Goodwin HA (eds) (2004) Spin crossover in transition metal compounds. Topics in current chemistry, vols 233–235. Springer, New York and references therein
75. Lecomte C, Chadwick DL, Coppens P, Stevens ED (1983) Electronic structure of metalloporphyrins. 2. Experimental electron density distribution of (meso-tetraphenylporphinato) iron(III) methoxide. Inorg Chem 22:2982–2992
76. Coppens P, Li L (1984) Electron density studies of porphyrins and phthalocyanines. III. The electronic ground state of iron(II) phthalocyanine. J Chem Phys 81:1983–1993
77. Tanaka K, Elkaim E, Liang L, Zhu NJ, Coppens P, Landrum J (1986) Electron density studies of porphyrins and phthalocyanines. IV. Electron density distribution in crystals of (meso-tetraphenylporphinato) iron(II). J Chem Phys 84:6969–6978
78. Lecomte C, Blessing RH, Coppens P, Tabard A (1986) Electron density studies of porphyrins and phthalocyanines. 5. Electronic ground state of iron(II) tetraphenylporphyrin bis(tetrahydrofuran). J Am Chem Soc 108:6942–6950
79. Elkaim E, Tanaka K, Coppens P, Scheidt RW (1987) Low temperature study of bis(2-methylimidazole)(octaethylporphinato) iron(III) perchlorate. Acta Crystallogr B 43:457–461
80. Li N, Coppens P, Landrum J (1988) Electron density studies of porphyrins and phthalocyanines. 7. Electronic ground state of bis(pyridine)(meso-tetraphenylporphinato) iron(II). Inorg Chem 27:482–488
81. Milne AM, Maslen EN (1988) Electron density in the spin crossover complex trans-[N, N'-ethylenebis(salicylideneaminato)]bis(imidazole)iron(III) perchlorate. Acta Crystallogr B 44:254–259
82. Pillet S, Legrand V, Weber HP, Souhassou M, Létard JF, Guionneau P, Lecomte C (2008) Out-of-equilibrium charge density distribution of spin crossover complexes from steady-state photocrystallographic measurements: experimental methodology and results. Z Kristallogr 223:235–249
83. Legrand V, Pillet S, Souhassou M, Lugan N, Lecomte C (2006) Extension of the experimental electron density analysis to metastable states: a case example of the spin crossover complex $Fe(btr)_2(NCS)_2 \cdot H_2O$. J Am Chem Soc 128:13921–13931
84. Vreugdenhil W, Van Diemen JH, De Graaf RAG, Haasnoot JG, Reedijk J, Van Der Kraan AM, Kahn O, Zarembowitch J (1990) High-spin $\leftrightarrow$ low-spin transition in $[Fe(NCS)_2(4,4'\text{-bis-}1,2,4\text{-triazole})_2](H_2O)$. X-ray crystal structure and magnetic, mössbauer and epr properties. Polyhedron 9:2971–2979
85. Pillet S, Hubsch J, Lecomte C (2004) Single crystal diffraction analysis of the thermal spin conversion in $[Fe(btr)_2(NCS)_2]\cdot H_2O$: evidence for spin-like domain formation. Eur J Phys B 38:541–552
86. Legrand V (2005) Cristallographie et photo-cristallographie haute résolution de matériaux à transition de spin: propriétés structurales, électroniques et mécanismes de conversion. PhD thesis, University Henri Poincaré, Nancy I
87. Holladay A, Leung P, Coppens P (1983) Generalized relations between d-orbital occupancies of transition metal atoms and electron density multipole population parameters from X-ray diffraction data. Acta Crystallogr A 39:377–387
88. Stevens ED, Coppens P (1979) Refinement of metal d-orbital occupancies from X-ray diffraction data. Acta Crystallogr A 35:536–539
89. Coppens P (1997) X-Ray charge densities and chemical bonding, International union of crystallography texts on crystallography series. Oxford University Press, Oxford

90. Rees B, Mitschler A (1976) Electronic structure of chromium hexacarbonyl at liquid nitrogen temperature. 2. Experimental study (X-ray and neutron diffraction) of σ and π bonding. J Am Chem Soc 98:7918–7924
91. Abramov YA (1997) On the possibility of kinetic energy density evaluation from the experimental electron density distribution. Acta Crystallogr A 53:264–272
92. Bader RFW, Essen H (1984) The characterization of atomic interactions. J Chem Phys 80:1943–1960
93. Bianchi R, Gatti C, Adovasio V, Nardelli M (1996) Theoretical and experimental (113 K) electron density study of lithium bis(tetramethylammonium) hexanitrocobaltate(III). Acta Crystallogr B 52:471–478
94. Macchi P, Sironi A (2003) Chemical bonding in transition metal carbonyl clusters: complementary analysis of theoretical and experimental electron densities. Coord Chem Rev 238–239:383–412
95. Guionneau P, Le Gac F, Kaiba A, Sánchez Costa J, Chasseau D, Létard JF (2007) A reversible metal-ligand bond break associated to a spin-crossover. Chem Commun 3723–3725
96. Guionneau P, Marchivie M, Bravic G, Létard JF, Chasseau D (2004) Structural aspects of spin crossover. Example of the $[Fe^{II}L_n(NCS)_2]$ complexes. In: Gütlich P, Goodwin HA (eds) Topics in current chemistry, vol 234. Springer, Berlin, pp 97–128
97. Cecconi F, Di Vaira M, Midollini S, Orlandini A, Sacconi L (1981) Singlet ↔ quintet spin transitions of iron(II) complexes with a P_4Cl_2 donor set. X-Ray structures of the compound $FeCl_2(Ph_2PCH = CHPPh_2)_2$ and of its acetone solvate at 130 K and 295 K. Inorg Chem 20:3423–3430
98. Di Vaira M, Midollini S, Sacconi L (1981) Low spin and high spin six-coordinate iron(II) complexes with a P_4Cl_2 donor set. X-Ray structures of $FeCl_2[(Ph_2PCH_2CH_2)_2PPh]_2{\cdot}2(CH_3)_2CO$ and of $FeCl_2(Me_2PCH_2CH_2PMe_2)_2$. Inorg Chem 20:3430–3435
99. McCusker JK, Zvagulis M, Drickamer HG, Hendrickson DN (1989) Pressure-induced spin-state phase transitions in $Fe(dppen)_2Cl_2$ and $Fe(dppen)_2Br_2$. Inorg Chem 28:1380–1384
100. Wu C-C, Jung J, Gantzel PK, Gütlich P, Hendrickson DN (1997) LIESST effect studies of iron(II) spin-crossover complexes with phosphine ligands: relaxation kinetics and effects of solvent molecules. Inorg Chem 36:5339–5347
101. Hauser A (1991) Intersystem crossing in Fe(II) coordination compounds. Coord Chem Rev 111:275–290
102. Hughes AK, Murphy VJ, O'Hare D (1994) Synthesis, X-ray structure and spin crossover in the triple-decker complex $[(\eta^5\text{-}C_5Me_5)Cr(\mu^2{:}\eta^5\text{-}P_5)Cr(\eta^5\text{-}C_5Me_5)]^+[A]^-$ ($A = PF_6$, SbF_6). J Chem Soc Chem Commun 163–164
103. Sim PG, Sinn E (1981) First manganese(III) spin crossover and first d^4 crossover. Comment on cytochrome oxidase. J Am Chem Soc 103:241–243
104. Halepoto DM, Holt DGL, Larkworthy LF, Jeffery Leigh G, Povey DC, Smith GW (1989) Spin crossover in chromiun(II) complexes and the crystal and molecular structure of the high spin form of bis[1,2-bis(diethylphosphino)ethane] di-iodochromiun(II). J Chem Soc Chem Commun 1322–1323
105. Real JA, Gaspar AB, Muñoz MC, Gütlich P, Ksenofontov V, Spiering H (2004) Bipyrimidine-bridged dinuclear iron(II) spin crossover compounds. In: Gütlich P, Goodwin HA (eds) Topics in Current Chemistry, vol 233. Springer, Berlin, pp 167–193 and references therein
106. Amoore JLM, Kepert CJ, Cashion JD, Moubaraki B, Neville SM, Murray KS (2006) Structural and magnetic resolution of a two-step full spin-crossover transition in a dinuclear iron(II) pyridil-bridged compound. Chem Eur J 12:8220–8227
107. Kaiba A, Shepherd HJ, Fedaoui D, Rosa P, Goeta AE, Rebbani N, Létard JF, Guionneau P (2010) Crystallographic elucidation of purely structural, thermal and light-induced spin transitions in an iron(II) binuclear complex. Dalton Trans 39:2910–2918

Chapter 20
Transient Charge Density Maps from Femtosecond X-Ray Diffraction

Thomas Elsaesser and Michael Woerner

20.1 Introduction

X-ray diffraction represents a key method for determining the atomic structure of condensed matter under equilibrium conditions. X-rays interact with electrons most of which are bound to atoms. As a result, the x-ray diffraction pattern, the superposition of waves scattered from the spatial electron distribution, provides direct access to the spatial arrangement of atoms and allows for measuring interatomic distances with a precision of a small fraction of a chemical bond length [1, 2]. The intensity of the different diffraction peaks is determined by the structure factor which is proportional to the Fourier transform of the 3-dimensional electron density. Thus, sectional views of the electronic charge distribution, i.e., charge density maps (CDMs), can be derived from the diffraction patterns [3].

In the solid phase, equilibrium structures of an impressive range of inorganic and organic crystalline materials have been determined by x-ray diffraction. Both Bragg diffraction of monochromatic x-rays and Laue diffraction of x-rays in a broad spectral range have been applied to record a multitude of diffraction peaks and derive both the time-averaged atomic structure and CDMs with a spatial precision of up to 1 pm. The diffraction patterns predominantly reflect the positions of heavier atoms with a substantial number of electrons, whereas light components, in particular hydrogen atoms, are more difficult to locate [4].

Nonequilibrium processes underlying physical, chemical or biological function are frequently connected with structure changes on a multitude of time scales. In crystal lattices, atomic motions and structural rearrangements are determined by the multidimensional interatomic potential energy surfaces and their changes which – vice versa – are related to the electronic structure. In recent years, there have been substantial efforts to develop time-resolved x-ray diffraction methods that can map

T. Elsaesser (✉) • M. Woerner
Max-Born-Institut für Nichtlineare Optik und Kurzzeitspektroskopie, D-12489 Berlin, Germany
e-mail: elsasser@mbi-berlin.de; woerner@mbi-berlin.de

C. Gatti and P. Macchi (eds.), *Modern Charge-Density Analysis*,
DOI 10.1007/978-90-481-3836-4_20, © Springer Science+Business Media B.V. 2012

such structural dynamics directly. A key approach is the pump-probe technique in which excitation by an ultrashort optical pulse initiates the structural change and a short hard x-ray pulse is diffracted from the excited sample. The measurement of diffraction patterns for different delay times of the x-ray pulse relative to the excitation pulse provides a sequence of 'snapshots' from which the momentary atomic positions and/or charge distributions can be derived.

Diffraction experiments with picosecond x-ray probe pulses from synchrotrons have given insight into transient structures being formed or existing on time scales longer than 70 ps [5–8]. However, the time scales relevant for the most elementary processes are in the ultrafast domain below 1 ps [9], and, thus, a picosecond time resolution is insufficient for observing basic atomic motions in a time resolved way. In recent years, there has been substantial progress in x-ray diffraction with a femtosecond time resolution [10–13] and first applications to molecular crystals have been reported [14]. Femtosecond x-ray diffraction provides much more direct structural information than femtosecond optical spectroscopies which give insight into transient changes of the dielectric function.

So far, most x-ray diffraction studies with a femtosecond time resolution have concentrated on Bragg scattering from bulk or nanolayered single crystals. Changes of the angular position and/or intensity have been measured for single or a small number of Bragg peaks. Reversible atomic motions, e.g., phonon oscillations [15–18], and irreversible changes of crystal structure such as photoinduced melting [13, 19] have been observed and transient geometries, e.g., changes in the separation of lattice planes have been derived. Information on changes of electronic structure and/or charge distributions has remained very limited, mainly because of the small number of diffraction peaks analyzed.

Very recently, we have performed the first diffraction studies of crystalline powders with a femtosecond time resolution, using hard x-ray pulses from a laser-driven plasma source for probing photoinduced structure changes [20]. This extension of the Debye-Scherrer method into the ultrafast time domain provides a large number of diffraction rings *simultaneously* and, thus, generates much more extended sets of transient diffraction data. The experimental results allow for determining transient CDMs, in addition to changes of atomic positions and lattice geometries. In this article, we review progress in this new area of femtosecond x-ray diffraction. We focus on prototype experiments with a laser-driven femtosecond x-ray source in which structural dynamics of the ionic hydrogen bonded material ammonium sulfate [$(NH_4)_2SO_4$] were addressed. We discuss transient charge density maps derived from the diffraction data and reveal a so far unknown structure transformation which is connected with a concerted transfer of electrons and protons in the crystal lattice.

The article is organized as follows. In Sect. 20.2, we briefly discuss the experimental methods for x-ray generation in laser-driven plasma sources and the femtosecond Debye-Scherrer method. The equilibrium properties of ammonium sulfate and the time-resolved diffraction data are presented in Sect. 20.3, followed by a discussion of transient CDMs and structural changes (Sect. 20.4). Conclusions and a brief outlook are given in Sect. 20.5.

20.2 Experimental Techniques of Femtosecond X-Ray Diffraction

In ultrafast x-ray diffraction, excitation by a femtosecond pump pulse initiates structural dynamics which are monitored by hard x-ray probe pulses diffracted from the excited sample. Measurements for different pump-probe delays give a sequence of diffraction patterns from which the momentary structure of the sample can be inferred [10, 11]. This method requires optical and hard x-ray pulses synchronized with an accuracy of better than 100 fs. This condition is fulfilled in laser-driven x-ray diffraction experiments where pump and probe pulses are both derived from a single amplified laser system.

Our experimental setup (Fig. 20.1a, b) implements a laser-driven plasma source for hard x-ray pulses with a high kilohertz repetition rate. This source offers a highly stable x-ray output consisting of 100 fs pulses with a total flux of up to 5×10^{10} x-ray photons/s from a very small emitting area. The femtosecond Ti:sapphire laser system provides sub-50 fs pulses at 800 nm of up to 5 mJ energy per pulse with a repetition rate of 1 kHz. A minor part of the laser output is reflected by a beamsplitter and generates the pump pulses. Both the laser fundamental at 800 nm and frequency converted pulses in a wide spectral range from the mid-infrared to the ultraviolet can be applied. The interaction length l in the sample is limited by the absorption coefficient at the pump wavelength and – in powder samples – by elastic scattering of the pump light. In the latter case, typical values are $l \leq 50$ μm for a pumping wavelength of 400 nm.

20.2.1 Laser-Driven Plasma Source for Hard X-Ray Pulses

The irradiation of metal targets by femtosecond laser pulses with a peak intensity higher than 10^{16} W/cm^2 results in the generation of a plasma that emits radiation in the hard x-ray range. The key mechanism is the acceleration of free electrons into the target by the very high electric field of the pulses and the generation of characteristic radiation and Bremsstrahlung by interactions with target atoms. More precisely, electrons from the plasma are accelerated up to kinetic energies of several tens of keV away from and then back onto the target in a single cycle of the laser field [21]. Such energetic electrons ionize Cu atoms in the target by generating K-shell holes. Recombination of electrons into the unoccupied K-shell is connected with the emission of characteristic Cu K_α fluorescence into the full solid angle 4π. The time structure of the emitted x-ray burst is ultrashort for an ultrashort driving laser field and an interaction length in the target of 10–20 μm.

In our setup (Fig. 20.1a), the major part (95%) of the laser output is focused onto a moving 20 μm thick Cu tape target to generate probe pulses at the characteristic Cu K_α photon energy of 8.05 keV (wavelength 0.154 nm) [22, 23]. The target is placed in a vacuum chamber (pressure 10^{-3} mbar), together with a moving plastic tape that

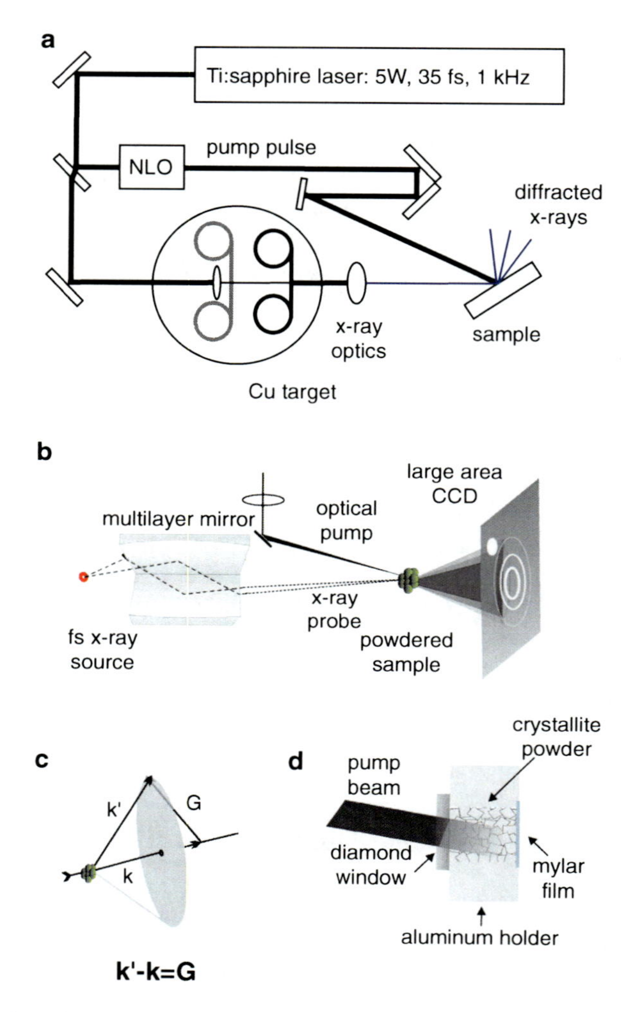

Fig. 20.1 (**a**) Setup for femtosecond optical pump/hard x-ray probe experiments. To generate x-rays in a plasma source working at a 1 kHz repetition rate, the main part of the output of an amplified Ti:sapphire laser is focused onto a 20 μm thick copper tape target placed in a vacuum chamber. The x-rays emitted in forward direction are focused onto the sample by multilayer x-ray optics. The diffraction signals are detected with a CCD detector. The sample is optically excited by a second part of the laser output at 800 nm or by frequency-converted (NLO) pump pulses. The diffracted x-ray signals are measured as a function of pump-probe delay. (**b**) Scheme of a powder diffraction experiment with polycrystalline samples. A multitude of diffracted Debye-Scherrer rings is detected with a large-area CCD detector. (**c**) Schematic illustration of the k-vector geometry for powder diffraction. (**d**) Scheme of the sample holder for femtosecond powder diffraction studies

protects the entrance window of the chamber against deposition of Cu debris. The x-rays emitted in forward direction leave the chamber through an exit slit sealed by a second plastic tape. The x-ray source has a very small diameter of 10 ± 2 μm as

Table 20.1 Femtosecond x-ray plasma source

Laser pulse energy	5 mJ
Repetition rate	1 kHz
Laser pulse duration	35 fs
Laser intensity	$\leq 10^{18}$ W/cm^2
Cu K_α photon energy	8.05 keV
X-ray photons emitted in 4π	5×10^{10} photons/s
Source diameter	10 ± 2 μm
X-ray pulse duration	100 fs
X-ray photons on sample	$\approx 10^6$ photons/s
X-ray spot diameter on sample	30–200 μm

the kinetic energy of the electrons is insufficient to move away from the interaction region. The very small source diameter facilitates imaging of the emitted x-rays by multilayer x-ray optics used in a transmission geometry.

Details of the most advanced version of this kilohertz plasma source have been presented in Ref. [22]. In Table 20.1, some of the key parameters are summarized. The driving laser, the x-ray source, the setup for generating pump pulses and the diffraction experiment fit onto a conventional experimental table (area ≈ 5 m^2), similar to all-optical femtosecond experiments.

20.2.2 Femtosecond X-Ray Powder Diffraction

In a powder sample of randomly oriented crystallites, a particular set of lattice planes with the Miller indices (hkl) and the reciprocal lattice vector **G**(hkl) show all orientations in the full solid angle 4π. In x-ray diffraction (Fig. 20.1c), a subset of crystallite orientations is selected via the Bragg condition for the wavevectors **k** of the incoming x-ray beam, **k**′ of the diffracted x-rays, and the reciprocal lattice vector **G**(hkl): $\mathbf{G}(\mathrm{hkl}) = \mathbf{k}' - \mathbf{k}$ with $|\mathbf{k}| = |\mathbf{k}'|$ for elastic scattering. This condition defines a cone of reciprocal lattice vectors **G**(hkl) contributing to the diffracted signal and, thus, a diffraction cone defined by **k** and **k**′ with an opening angle 2θ. Different sets of lattice planes give rise to different diffraction cones as illustrated in Fig. 20.1b. On a planar x-ray detector, one detects diffraction rings with a diameter determined by 2θ.

The number of photons scattered per unit time into a particular ring is given by $N_{scatt} = P_{scatt}/(hc/\lambda)$ with

$$P_{\mathrm{scatt}} = I_0 \frac{r_e^2 V \lambda^3 M(hkl) F_{hkl}^2}{4 v_a^2} \left(\frac{1 + \cos^2(2\theta_{hkl})}{2 \sin \theta_{hkl}} \right), \qquad (20.1)$$

where I_0 is the incident x-ray intensity and P_{scatt} is the scattered x-ray power, c is the velocity of light, λ is the x-ray wavelength, r_e is the classical electron radius, $V = \rho_{\mathrm{powder}}/\rho_{\mathrm{material}} V_{\mathrm{crystal}}$ is the effective illuminated volume with $\rho_{\mathrm{powder}}/\rho_{\mathrm{material}}$

≈0.8 for usual powders, $M(hkl)$ is the multiplicity of the reflection hkl, v_a is the unit cell volume and F_{hkl} is the structure factor for x-ray scattering [1]. The structure factor is proportional to the Fourier transform of the 3-dimensional density of electronic charge with respect to the reciprocal lattice vector **G**(hkl). The last term of Eq. 20.1 (in parentheses) is the Lorentz polarization factor. For a wide range of materials, the diffraction efficiency into a single Debye-Scherrer diffraction ring has values from 10^{-3} to 10^{-5}.

In our setup (Fig. 20.1b), a multilayer x-ray mirror in transmission geometry focuses the x-ray pulses onto the powder sample, resulting in a x-ray flux of 10^6 photons/s within a 200 μm (FWHM) spot on the sample [24]. Details of the sample geometry are shown in Fig. 20.1d. The powder sample of 1 mm diameter and 250 μm thickness is contained in a metallic sample holder. The powder is sealed by a thin polycrystalline diamond front window and a sub-10 μm thick mylar foil on the rear side. The high thermal conductivity of the diamond window supports the transport of thermal excess energy originating from the optical excitation of the powder to the metallic sample holder which serves as a heat sink. The diamond window is optically transparent in a wide range from the far-infrared up to the ultraviolet. The x-rays diffracted from the sample are detected with a large-area CCD detector and up to approximately 20 Debye-Scherrer rings are recorded simultaneously as a function of the time delay between pump and probe pulses.

Changes of the atomic arrangement and/or CDMs are derived from the measured changes of the diffracted intensity $\Delta I/I_0 = (I - I_0)/I_0$ and of the diffraction angle $\Delta\Theta/\Theta_0$, I_0 and Θ_0 being the diffracted intensity and the diffraction angle for the unexcited sample. A fundamental limit of the dynamic range over which such quantities can be measured is set by the photon counting statistics in the x-ray detection process. For an incoming x-ray flux of 10^6 photons/s on the sample and a diffraction efficiency of 10^{-3} and 10^{-5}, one detects between 10 and 1,000 diffracted x-ray photons per second. For typical data acquisition times of 20 min, the total number of counts is between $N_{tot} = 1.2 \times 10^4$ and 1.2×10^6. The statistical error $\Delta N/N_{tot} = 1/\sqrt{N_{tot}}$ has values between approximately 10^{-2} and 10^{-3}. Such numbers set a limit for the smallest $\Delta I/I_0$ that can be measured in a time resolved experiment. The time-resolved data presented in Sect. 20.3 were measured with an integration time on the CCD detector of 300–600 s for each individual value of the time delay and up to ten delay scans were averaged.

20.3 Ultrafast Structural Dynamics of Elementary Chemical Processes

Photoinduced electron and proton transfer are among the most elementary chemical processes that occur in many (macro)molecular systems and govern their function. Examples range from photosynthesis to artificial molecular arrays for the separation of charges and/or storage of hydrogen. The reaction pathways and dynamics of condensed phase electron and proton transfer are determined by the interplay of molecular potential energy surfaces with the response of the environment to the change in charge distribution and/or molecular geometry [25, 26]. Ultrafast optical spectroscopies have provided substantial insight into the dynamics and mechanisms of such processes. In most cases, however, transient charge distributions and molecular geometries generated on a femtosecond time scale have not been characterized.

Both Bragg diffraction from molecular single crystals and Debye-Scherrer diffraction from polycrystalline powder samples are of interest for mapping transient molecular structure directly and for deriving time-dependent CDMs. The first femtosecond Bragg diffraction study of molecular single crystals has addressed the structural response of a crystalline environment consisting of polar molecules to a dipole change of photoexcited chromophores [14]. The photoinduced intramolecular charge transfer in a small fraction of excited 4-(diisopropylamino)benzonitrile (DIABN) chromophores induces a rotational reorientation of the surrounding unexcited molecules which lowers the dipole energy. The time evolution of this dipole solvation process is governed by the formation of the charge transfer state, i.e., the environment follows the dipole change without any measurable delay. Due to the collective nature of the response, this process gives rise to strong changes in the x-ray diffraction pattern.

In the following, we concentrate on the first femtosecond diffraction study of powders consisting of molecular crystallites. Application of the time-resolved Debye-Scherrer method described in Sect. 20.2.2 provides detailed insight into ultrafast changes of the electronic charge distribution and the concomitant changes of crystal structure.

20.3.1 The Model System Ammonium Sulfate [$(NH_4)_2SO_4$]

In the crystalline phase, ammonium sulfate (AS) forms an ionic structure held together by both Coulomb forces between anions and cations, i.e., the Madelung energy, and intermolecular hydrogen bonds between the ammonium and sulfate groups. The material belongs to a prototype class of molecular ferroelectrics with physical properties that are understood only in part. At T = 300 K, AS crystallizes in an orthorhombic structure belonging to the space group Pnam with four $(NH_4)_2SO_4$ entities per unit cell (Fig. 20.2a–c) [27–29]. The dimensions of the unit cell are a = 0.778 nm, b = 1.064 nm, and c = 0.599 nm and the material is paraelectric. Under equilibrium conditions, AS undergoes a structural phase transition at T = 223 K into a ferroelectric phase of lower symmetry and the space group $Pna2_1$. A second phase transition occurs at a higher temperature of $T = 423$ K retaining the space group Pnam. Here, a contraction of the unit cell [29] and pronounced changes of the dielectric function [30–34] have been found. Stationary x-ray and neutron diffraction experiments show that both phase transitions are connected with a rearrangement of the atoms within the unit cell. The microscopic

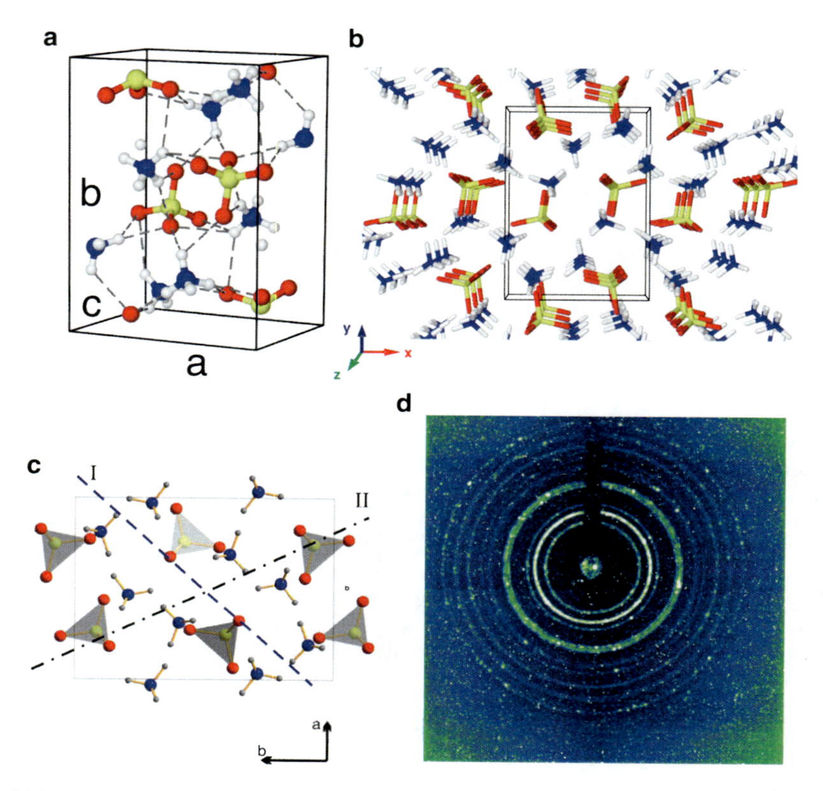

Fig. 20.2 (**a**) Unit cell of ammonium sulfate (*left*) with sulfate SO_4^{2-} and ammonium NH_4^+ groups (*yellow*: sulfur atoms, *red*: oxygen, *blue*: nitrogen, *gray*: hydrogen). The *dashed lines* indicate hydrogen bonds. (**b**) View of the crystal structure into the z direction. (**c**) Specific lattice planes defined by the z direction and the line linking the oxygen atoms 1 and 2 of opposite sulfate groups (I, *dashed line*) and by the z axis and the line linking the hydrogen atoms of opposite ammonium groups (II, *dash-dotted line*). In Sect. 20.4, transient charge density maps for such planes are presented. (**d**) Steady-state Debye-Scherrer diffraction pattern of ammonium sulfate powder recorded with the femtosecond x-ray source

driving forces of and the lattice motions involved in the phase transitions have, however, not been identified so far.

For the powder diffraction experiments, commercially available high-purity AS powder was ground in a heated mortar to avoid contamination with water. The resulting grain size was 5–10 μm. The processed powder was sealed in the sample holder shown in Fig. 20.1. In a first series of measurements, steady-state diffraction patterns of the AS powder were recorded with the output of the femtosecond x-ray source and the large-area CCD detector of the setup shown in Fig. 20.1. A ring pattern measured with a fixed orientation of the powder sample and a CCD integration time of 420 s is displayed in Fig. 20.2d. One clearly resolves up to 20 Debye-Scherrer rings originating from different sets of lattice planes in

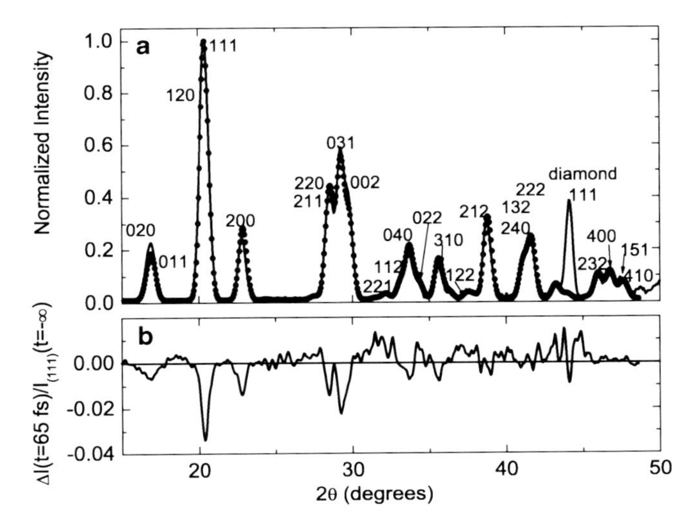

Fig. 20.3 (**a**) Measured (*solid line*) and calculated (*dots*) intensity profile of the Debye Scherrer ring pattern of AS as a function of 2θ (θ: diffraction angle). The intensities are normalized to that of the (111) peak. (**b**) Intensity change of the rings observed 65 fs after excitation of the sample

the crystallites. The total number of crystallites contributing to this pattern is of the order of 10,000. In this finite ensemble, larger crystallites make a stronger contribution to the diffraction signal than smaller ones. As the orientations of the larger crystallites are not perfectly random, one observes a nonuniform intensity of particular diffraction rings, i.e., areas of enhanced intensity. Such inhomogeneities are avoided when the powder sample is rotated in the incident x-ray beam.

In Fig. 20.3a, the diffracted intensities integrated over the individual rings are plotted as a function of the diffraction angle 2θ (solid line). Details of data processing and the integration procedure have been discussed in Ref. [20]. An intensity of 1 at the maximum of a ring corresponds to a number of counts of the order of 1.5×10^5 for a 420 s integration time per ring. This steady-state diffraction pattern was analyzed by calculating the structure factors for the different rings using the atomic positions derived from neutron diffraction studies of AS [27] and the atomic form factors. The result (dots in Fig. 20.3a) is in excellent agreement with the measured AS pattern. The additional experimental peak at $2\theta = 44$ degrees represents the (111) reflection of the polycrystalline diamond window.

20.3.2 *Femtosecond Powder Diffraction from Ammonium Sulfate*

The femtosecond experiments address *nonequilibrium* structure changes that are induced by photoexcitation of AS [35]. A powder sample at T = 300 K is excited via 3-photon absorption of 50 fs pulses at 400 nm. The penetration depth of the pump pulses into the sample is determined by the elastic scattering length of approximately 40 μm, rather than the optical penetration depth of 200 μm. The deposition of pump energy results in a temperature increase of the AS powder which was analyzed with a calibrated thermocamera. For the pump energy of 75 μJ per pulse, one finds a maximum temperature increase of 40 K. The structural dynamics initiated by the pump pulse were mapped by diffracting ultrashort Cu K_α pulses from the excited sample. The angular position and the intensity of the 20 Debye-Scherrer diffraction rings shown in Figs. 20.2d and 20.3a were measured simultaneously as a function of the pump-probe delay. On the 40 μm interaction length, the group velocity mismatch, i.e., the difference of propagation times between the optical pump and the hard x-ray probe pulses, is negligible compared to the duration of the x-ray pulse of 100 fs, the latter mainly determining the time resolution of the experiment.

In Fig. 20.3b, the change of intensity of the AS diffraction rings measured at a pump-probe delay of 65 fs is plotted as a function of diffraction angle. Pronounced intensity changes are observed on a number of rings, the changes $\Delta I(t = 65\,fs)$ reaching values of up to several percent of $I_{(111)}(t = -\infty)$, the steady-state intensity diffracted into the (111) ring. The time evolution of the intensity changes was studied in an extended series of measurements. A subset of results is presented in Fig. 20.4 where the time dependent change of intensity is plotted for three selected rings for delay times between −0.5 and 0.5 ps and on a longer picosecond time scale. The transients display strong oscillations with a period of 650 fs, corresponding to the frequency of the A_g phonon of AS of 50 cm^{-1}. The change of the (111) intensity (bottom panel) displays a very fast rise within the first 100 fs, demonstrating the very high time resolution of the experiment. The dashed line represents a fit through the data points, giving the time-integrated cross-correlation function of the optical pump and x-ray probe pulses. The time derivative of this curve (solid line, upper part) is proportional to the cross-correlation function of a 130 fs width. This result points to a duration of the x-ray probe pulses of $\leq$ 100 fs. All changes of diffracted intensity decay on a slower time scale beyond 20 ps, i.e., the underlying changes of AS structure are fully reversible. It is important to note that the angular positions of all diffraction rings remain unchanged over the time range of Fig. 20.4. For the change $\Delta\theta_{002}$ of the position of the (002) reflection relative to the angular position $\theta_{111}^{\mathrm{diamond}}$ of the (111) diamond reflection one estimates an upper limit $\Delta\theta_{002}/\theta_{111}^{\mathrm{diamond}} < 0.005$ for delay times $t \leq 15$ ps. Thus, changes in the average positions of the heavier atoms S, N, and O in the unit cell are minor and the size (volume) and shape of the unit cell remain unchanged.

20.4 Transient Charge Density Maps and Structure Changes

The large set of powder diffraction data allows for deriving transient 3-dimensional CDMs of AS and for analyzing the changes of the atomic arrangement that result from photoexcitation. The static AS crystal structure was determined with high

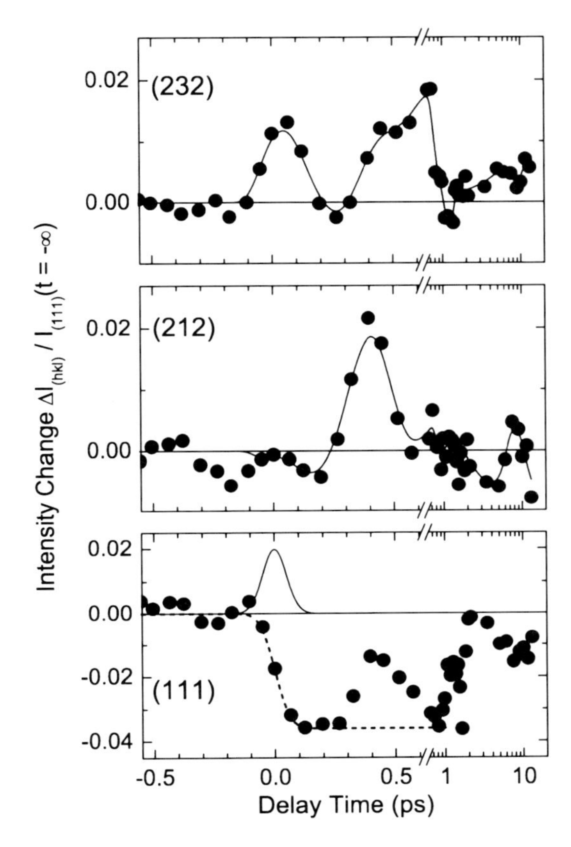

Fig. 20.4 Changes in intensity (*circles*) of particular Debye-Scherrer rings as a function of time delay between the optical pump and x-ray probe pulses. The thin solid lines are to guide the eye. Bottom panel: The *dashed line* represents a fit to the data giving the time-integrated cross-correlation function of optical pump and x-ray probe pulses. The time derivative of this fit is the cross-correlation function plotted in the upper part of the panel

accuracy by neutron diffraction [27]. The lattice positions of atoms derived there and the respective atomic scattering factors serve as a basic input of the analysis presented in the following.

The diffracted intensity or number of photons diffracted per unit time into a particular ring is proportional to the square of the structure factor, i.e., $N_{\text{scatt}} \propto C|F_{hkl}|^2$ with C containing the Lorentz polarization factor and the multiplicity (cf. Eq. 20.1). To derive the complex structure factor $F_{hkl} = |F_{hkl}|\exp(i\phi_{hkl})$ and the spatially resolved charge density from the diffraction data, one needs to solve the 'phase problem', i.e., determine ϕ_{hkl}. In general, recursive methods of data analysis are applied to address this issue [1].

In our time-resolved diffraction experiments, we exploit the interference of x-rays diffracted from the small fraction of excited (modified) unit cells and from the majority of unexcited ones within a single crystallite, i.e., the x-rays diffracted from the unexcited cells serve as a local oscillator enhancing the sensitivity of the experiment. The unexcited AS powder at T = 300 K displays inversion symmetry (space group Pnam). In other words, each crystallite is a racemate of AS molecules, containing for each molecule also its mirror image. Since femtosecond photoexcitation does not select one type of enantiomeres, there exists for each unit cell which contains a photoexcited AS molecule of *arbitrary shape*, a corresponding inverted unit cell containing the mirror image of the photoexcited species. Thus, when averaging over the crystallite inversion symmetry is preserved even after an arbitrary photoinduced structure change. As a result, F_{hkl} averaged over a single crystallite is and remains real-valued with $\phi_{hkl} = 0$ or π and keeps the phase of the initial crystal structure.

The change of diffracted intensity is given by:

$$\frac{\Delta I(\theta_{hkl}, t)}{I_{(111)}(t = -\infty)} = C_{hkl} \frac{\left|\eta \left(F_{hkl}^{ex} + F_{hkl}^{ex*}\right)/2 + (1-\eta) F_{hkl}^{gr}\right|^2 - \left|F_{hkl}^{gr}\right|^2}{I_{(111)}(t = -\infty)} \quad (20.2)$$

where η is the fraction of excited unit cells in the sample and F_{hkl}^{ex}, F_{hkl}^{ex*} and F_{hkl}^{gr} are the structure factors of the excited unit cells, their mirror image, and the unexcited unit cells, respectively. For weak excitation as in our experiments, i.e., $\eta \ll 1$, this expression can be simplified by keeping only terms linear in η:

$$\eta \Delta F = \eta \mathrm{Re}\left(F_{hkl}^{ex} - F_{hkl}^{gr}\right) = \frac{\Delta I(\theta_{hkl}, t)}{2 C_{hkl} \left|F_{hkl}^{gr}\right|} \quad (20.3)$$

This equation directly connects the observed changes in the intensity to the change of the structure factor. All quantities in the denominator on the r.h.s. of the equation are known, i.e., a measurement of $\Delta I(\theta_{hkl}, t)$ gives the transient structure factor $\mathrm{Re}\left(F_{hkl}^{ex}\right)$ if the value of η is known.

From Eq. 20.3 the change $\Delta\rho$ of the 3-dimensional charge density is calculated:

$$\eta \Delta\rho(x, y, z, t) = \frac{\eta}{abc} \sum_{hkl} \Delta F_{hkl}(t) \cos\left[2\pi\left(\frac{hx}{a} + \frac{ky}{b} + \frac{lz}{c}\right)\right] \quad (20.4)$$

Averaging over the crystallite gives the real part of the structure factor change $\mathrm{Re}\left(\Delta F_{hkl}^{ex}\right)$. The calculated $\Delta\rho = (\Delta\rho(x, y, z, t) + \Delta\rho(-x, -y, -z, t))/2$ is the average of the charge density change and the spatially inverted charge density change being proportional to the CDMs of the modified unit cells $(\rho_{ex}(x, y, z, t) + \rho_{ex}(-x, -y, -z, t))/2$ with $\rho_{ex}(x, y, z, t)$, $\rho_{ex}(-x, -y, -z, t) > 0$. The charge density map $(\rho_{ex}(x, y, z, t) + \rho_{ex}(-x, -y, -z, t))/2$ represents an average over one crystallite and, thus, does not tell us how many molecules per excited unit cell and which particular molecule in an excited unit cell are structurally modified. Since $\rho_{ex}(x, y, z, t) > 0$ and $\rho_{gr}(x, y, z, t) > 0$, however, a negative $\Delta\rho(x, y, z, t)$ can exclusively occur at spatial positions where initially atoms (electronic charge) are present and positive peaks in the difference map $\Delta\rho(x, y, z, t)$ prove definitely the occurrence of charge at the respective spatial positions in one of the excited unit cells.

To calculate the absolute value of $\Delta\rho$, the fraction of excited unit cells η needs to be determined in the experiment, taking into account boundary conditions set by the physical process under study. For instance, a transfer of electrons over a distance larger than the atomic radius results in a decrease of charge density at the original site and an increase at the new position. The charge integral at the new site has a value identical to the elementary charge e or a multiple of it:

$$\int_{\mathrm{V_{CT}}} \eta\Delta\rho(x, y, z, t)\, dx\, dy\, dz = N \cdot e, \tag{20.5}$$

where N is the number of electrons involved in the transfer process, and V_{CT} is the volume of the charge-accepting unit. This boundary condition allows for a calibration of η. In our experiments, η has a value of 0.06.

Equilibrium and transient CDMs of AS are presented in Fig. 20.5 for two particular planes in the crystal structure. The first plane (upper row in Fig. 20.5 and dashed-dotted line II in Fig. 20.2c) contains the z axis and the line connecting the hydrogen atoms of two adjacent ammonium groups. The second plane (lower row in Fig. 20.5 and dashed line I in Fig. 20.2c) contains the z axis and the line connecting the oxygen atoms 1 and 2 of opposite sulfate groups. The equilibrium CDMs were calculated using the atomic positions determined by neutron diffraction [27] and the known form factors of the different atoms.

The CDMs in Fig. 20.5 show a channel-like geometry of enhanced electron density along the z axis. Such feature occurs at the spatial position (x,y,z) = (0,0,z) and (0,5a,0,5b,z) in the unit cell with local maxima of electron density at (0,0,0), (0,0,0.5c), (0.5a,0.5b,0) and (0.5a,0.5b,0.5c). In the AS equilibrium structure (Fig. 20.2), there are no lattice atoms at those positions. The channel region displays a strong lateral confinement with a diameter in the xy plane of approximately $2a_B$, a_B being the Bohr radius of the hydrogen atom. This finding points to a stabilization of the channel geometry by (positively charged) protons which move from the neighboring NH_4^+ groups into positions between the oxygen atoms 1 and 2 of adjacent SO_4^{2-} groups (inset of Fig. 20.6a). The formation of this hydrogen bonded new geometry requires a concerted charge and proton transfer. This process occurs within the first 100 fs after photoexcitation, as is evident from the data in Fig. 20.3b and the very fast rise of the time-dependent changes of diffracted x-ray intensity in Fig. 20.4.

The upper row of Fig. 20.5 shows CDMs in the plane II containing the channel and the original positions of the protons on the NH_4^+ groups. At 260 fs, one observes the strong increase of charge density in the channel region, in agreement with the corresponding map for plane I. At 450 fs, however, this feature has disappeared and one finds a strong increase of charge density close to the initial positions of the protons. At 640 fs, there is a slight spatial displacement of this charge density increase with respect to the initial proton positions (open circles). This slight displacement of the NH_4^+ groups results in a negative $\Delta\rho$ at the initial proton position. The time-dependent CDMs point to an oscillation of charge density in the channel which is connected with a concerted relocation of charge and protons

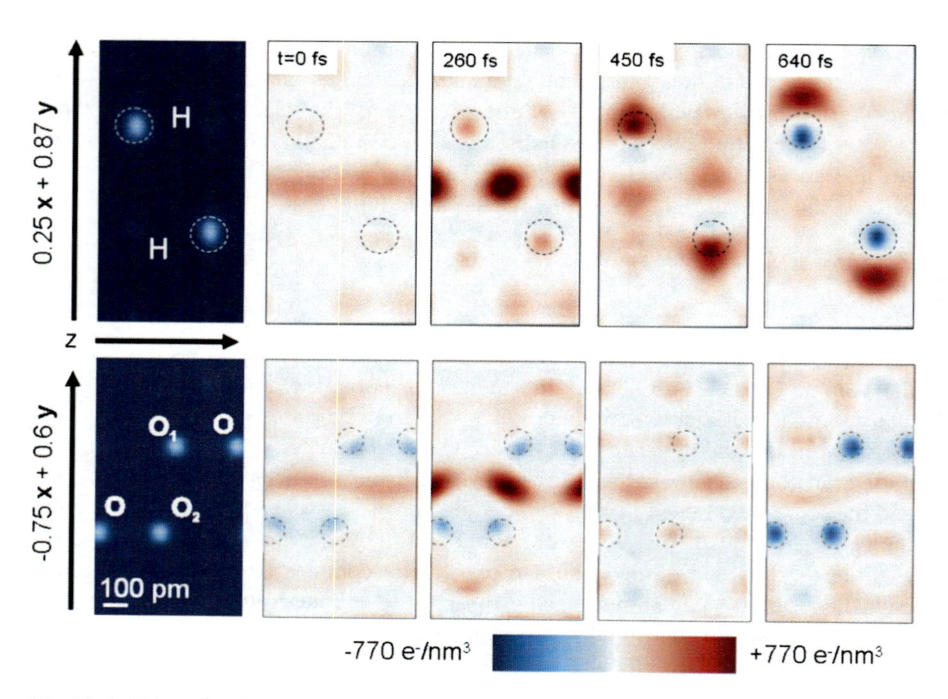

Fig. 20.5 Charge density maps derived from the diffraction data for different time delays and the planes indicated in Fig. 20.2c. *Upper row*: Steady state charge density (*left panel*) and transient changes $\Delta\rho(x, y, z, t)$ of charge density for different pump-probe delays t in plane II that includes the z axis and the line connecting the hydrogen atoms of opposite NH_4^+ groups, marked with a *dashed circle* (*dash-dotted line* in Fig. 20.2c). *Lower row*: Same for plane I containing the z axis and the line linking the oxygen atoms 1 and 2 of opposite sulfate groups (*dashed line* in Fig. 20.2c)

between the original and the new proton positions in the channel. This behavior is borne out more clearly when integrating the charge density over the channel region. In Fig. 20.6a, the integrated charge change Δq is plotted as a function of delay time, clearly displaying periodic oscillations with the frequency of 50 cm^{-1} of the A_g phonon. This lattice mode modulates the O_1–O_2 distance (inset of Fig. 20.6a) periodically and induces a periodic charge and proton transfer between the channel and the original site of the proton on the NH_4^+ group. Eventually the oscillations decay by vibrational dephasing of the lattice mode.

The overall charge in the plane defined by the channel and the line connecting the proton sites on two neighboring NH_4^+ groups is constant during the oscillation (Fig. 20.6b). Concomitantly, the charge on the two SO_4^{2-} groups close to the channel remains depleted (Fig. 20.6c), demonstrating again that the sulfur atoms represent the main photoinduced supply of charge for the channel formation.

A comment should be made on the spatial resolution of the CDMs presented here. The fundamental resolution limit set by the momentum transfer is given by $\Delta k = \lambda/4\pi \sin\theta$ which has a value of the order of 30 pm for Cu K_α radiation and a diffraction angle $\theta = 25$ degrees. The finite angular resolution of our powder

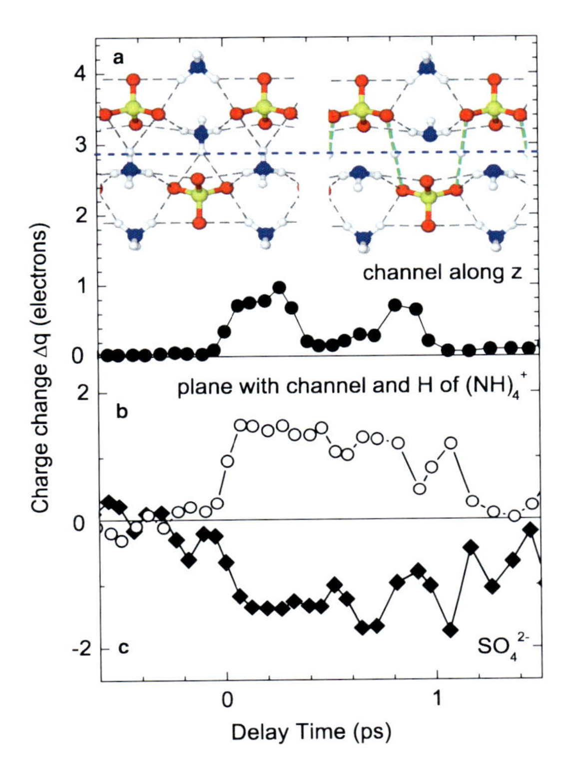

Fig. 20.6 (**a**) Total charge change in the channel region (in units of the elementary charge) as a function of pump-probe delay. Inset: Crystal structure along the z direction. The dashed line marks the channel region. (**b**) Change of electronic charge integrated in plane II. (**c**) Same for the two SO_4^{2-} groups close to the electron channel. The transients in (**b**) and (**c**) display the same time evolution with opposite signs of the charge changes

diffraction setup represents another limitation which has been discussed in detail in Ref. [20]. The relative angular resolution is approximately $\Delta\theta/\theta = 0.02$ which – for the AS structure – translates into an uncertainty of the spacing of lattice planes between 10 and 20 pm. To test the spatial resolution experimentally, we measured static diffraction patterns of the ferro- and paraelectric phase of AS single crystals with our setup. Upon the phase transition, the position of a subgroup of oxygen atoms of the SO_4^{2-} groups changes by 30 pm [27]. This change is manifested in our diffraction patterns via changes of diffracted intensity on a number of peaks, demontrating the high spatial resolution of the experiment. We conclude that the spatial resolution of the CDMs is of the order of 30–40 pm. The relative uncertainty of the CDMs and the charge changes Δq is estimated from the scattering of their values for different delay times (Figs. 20.5 and 20.6) and is of the order $\delta(\Delta q)/\Delta q \approx 0.1$.

20.5 Conclusions

We have demonstrated ultrafast x-ray powder diffraction using an optical pump/x-ray probe scheme with a time resolution of 100 fs. The Cu K_α probe pulses were generated in a laser-driven plasma source working at a 1 kHz repetition rate.

In our prototype experiments, photoinduced structural dynamics were studied in crystallites of ammonium sulfate, an ionic hydrogen bonded molecular material. Transient changes of diffracted intensity were measured simultaneously on up to 20 diffraction rings. The transient diffraction patterns display pronounced intensity changes with unchanged angular positions of the individual diffraction rings. From such data, transient charge density maps were derived in a quantitative way. Pronounced changes of the spatially resolved charge density demonstrate the occurrence of a photoinduced concerted electron and proton transfer into an initially unoccupied part of the elementary cell. This new channel-like geometry displays a lateral confinement determined by the proton's size and is stabilized by a change of the hydrogen bond pattern. For a few picoseconds, both the electronic charge and the protons undergo a periodic relocation between the new and the initial proton positions driven by the 50 cm^{-1} phonon of ammonium sulfate.

Our results demonstrate the potential of femtosecond powder diffraction for studying changes of geometric and electronic structure in a wide range of polycrystalline inorganic and organic materials. For the comparably light sulfur, oxygen, and nitrogen atoms such as in ammonium sulfate, changes of diffracted intensity are measured with high accuracy for signal integration times of the order of 30 min and even proton positions can be inferred from the data. For enhancing the spatial resolution in the derived charge density maps, diffraction of femtosecond x-ray pulses at shorter wavelengths will be essential. In laser-driven plasma sources, other target materials, e.g., Mo, should allow for generating such pulses. Another option are accelerator based femtosecond sources providing a sufficient hard x-ray flux such as free electron lasers. We envisage a broad range of applications of this novel technique, in particular for studying materials with correlated electron and spin systems and biomolecular systems.

Acknowledgements We would like to acknowledge the important contributions of our coworkers Z. Ansari and M. Zamponi to the results reported here. This work has been supported in part by the Deutsche Forschungsgemeinschaft through the Priority Program 1134.

References

1. Warren BE (1969) X-ray diffraction. Dover, New York
2. Als-Nielsen J, McMorrow D (2001) Elements of modern x-ray physics. Wiley, Sussex
3. Coppens P (1997) X-ray charge densities and chemical bonding. Oxford University Press, New York
4. Parkin A, Seaton CC, Blagden N, Wilson CC (2007) Designing hydrogen bonds with temperature-dependent proton disorder: the effect of crystal environment. Cryst Growth Des 7:531–534
5. Neutze R, Wouts R, Techert S, Davidsson J, Kocsis M, Kirrander A, Schotte F, Wulff M (2001) Visualizing photochemical dynamics in solution through picosecond x-ray scattering. Phys Rev Lett 87:195508-1–195508-4
6. Collet E, Lemee-Cailleau MH, Buron Le Comte M, Cailleau H, Wulff M, Luty T, Koshihara SY, Meyer M, Toupet L, Rabiller P, Techert S (2003) Laser-induced ferroelectric structural order in an organic charge-transfer crystal. Science 300:612–615

7. Coppens P, Vorontsov II, Graber T, Gembicky M, Kovalevsky AY (2005) The structure of short-lived excited states of molecular complexes by time-resolved x-ray diffraction. Acta Crystallogr A 61:162–172
8. Lorenc M, Hebert J, Moisan N, Trzop E, Servol M, Buron-Le Cointe M, Cailleau H, Boillot ML, Pontecorvo E, Wulff M, Koshihara S, Collet E (2009) Successive dynamical steps of photoinduced switching of a molecular Fe(III) spin-crossover material by time-resolved x-ray diffraction. Phys Rev Lett 103:028301-1–028301-4
9. For a recent overview: Corkum P, de Silvestri S, Nelson K, Riedle E, Schoenlein R (eds) (2009) Ultrafast phenomena XVI. Springer, Berlin
10. Rousse A, Rischel C, Gauthier JC (2001) Colloquium: femtosecond x-ray crystallography. Rev Mod Phys 73:17–31
11. von Korff Schmising C, Bargheer M, Woerner M, Elsaesser T (2008) Real-time studies of reversible lattice dynamics by femtosecond x-ray diffraction. Z Kristallogr 223:283–291
12. Rischel C, Rousse A, Uschmann I, Albouy PA, Geindre JP, Audebert P, Gauthier JC, Förster E, Martin JL, Antonetti A (1997) Femtosecond time-resolved x-ray diffraction from laser-heated organic films. Nature 390:490–492
13. Siders CW, Cavalleri A, Sokolowski-Tinten K, Toth C, Guo T, Kammler M, von Hoegen MH, Wilson KR, von der Linde D, Barty CPJ (1999) Detection of nonthermal melting by ultrafast x-ray diffraction. Science 286:1340–1342
14. Braun M, von Korff Schmising C, Kiel M, Zhavoronkov N, Dreyer J, Bargheer M, Elsaesser T, Root C, Schrader TE, Gilch P, Zinth W, Woerner M (2007) Ultrafast changes of molecular crystal structure induced by dipole solvation. Phys Rev Lett 98:248301-1–248301-4
15. Sokolowski-Tinten K, Blome C, Blums J, Cavalleri A, Dietrich C, Tarasevitch A, Uschmann I, Förster E, Kammler M, Horn-von-Hoegen M, von der Linde D (2003) Femtosecond x-ray measurement of coherent lattice vibrations near the Lindemann stability limit. Nature 422: 287–289
16. Bargheer M, Zhavoronkov N, Gritsai Y, Woo JC, Kim DS, Woerner M, Elsaesser T (2004) Coherent atomic motions in a nanostructure studied by femtosecond x-ray diffraction. Science 306:1771–1773
17. von Korff Schmising C, Bargheer M, Kiel M, Zhavoronkov N, Woerner M, Elsaesser T, Vrejoiu I, Hesse D, Alexe M (2007) Coupled ultrafast lattice and polarization dynamics in ferroelectric nanolayers. Phys Rev Lett 98:257601-1–257601-4
18. Johnson SL, Beaud P, Vorobeva E, Milne CJ, Murray ED, Fahy S, Ingold G (2009) Directly observing squeezed phonon states with femtosecond x-ray diffraction. Phys Rev Lett 102:175503-1–175503-4
19. Lindenberg AM et al (2005) Atomic-scale visualization of inertial dynamics. Science 308: 392–395
20. Zamponi F, Ansari Z, Woerner M, Elsaesser T (2010) Femtosecond powder diffraction with a laser-driven hard x-ray source. Opt Express 18:947–961
21. Brunel F (1987) Not-so-resonant, resonant absorption. Phys Rev Lett 59:52–55
22. Zamponi F, Ansari Z, von Korff Schmising C, Rothhardt P, Zhavoronkov N, Woerner M, Elsaesser T, Bargheer M, Trobitzsch-Ryll T, Haschke M (2009) Femtosecond hard X-ray plasma sources with a kilohertz repetition rate. Appl Phys A 96:51–58
23. Zhavoronkov N, Gritsai Y, Bargheer M, Woerner M, Elsaesser T, Zamponi F, Uschmann I, Förster E (2005) Microfocus Cu K_{α} source for femtosecond x-ray science. Opt Lett 30:1737–1739
24. Bargheer M, Zhavoronkov N, Bruch N, Legall H, Stiel H, Woerner M, Elsaesser T (2005) Comparison of focusing optics for femtosecond x-ray diffraction. Appl Phys B 80:715–719
25. Jortner J, Bixon M (eds) (1999) Electron transfer – from isolated molecules to biomolecules. Adv Chem Phys 106
26. Elsaesser T, Bakker HJ (eds) (2002) Ultrafast hydrogen bonding dynamics and proton transfer processes in the condensed phase. Kluwer, Dordrecht
27. Schlemper EO, Hamilton WC (1966) Neutron-diffraction study of the structures of ferroelectric and paraelectric ammonium sulfate. J Chem Phys 44:4498–4509

28. Ahmed S, Shamah AM, Kamel R, Badr Y (1987) Structural changes of $(NH_4)_2SO_4$ crystals. Phys Status Solidi (a) 99:131–140
29. Ahmed S, Shamah AM, Ibrahim A, Hanna F (1989) X-ray studies of the high temperature phase transition of ammonium sulphate crystals. Phys Status Sol (a) 115:K149–K153
30. Syamaprasad U, Vallabhan CPG (1981) Observation of a high-temperature phase transition in $(NH_4)_2SO_4$ from dielectric studies. J Phys C Solid State Phys 14:L865–L868
31. Desyatnichenko AV, Shamshin AP, Matyushkin EV (2004) Dielectric dispersion in crystal $(NH_4)_2SO_4$. Ferroelectrics 307:213–219
32. Shen RH, Chen YC, Shern CS, Fukami T (2009) Conductivity and dielectric relaxation phenomena in $(NH_4)_2SO_4$ single crystal. Solid State Ion 180:356–361
33. Kim J-L, Lee KS (1996) The origin of the nonexistence of prototypic paraelastic phase in $(NH_4)_2SO_4$. J Phys Soc Jap 65:2664–2669
34. Sobiestianskas R, Banys J, Brilingas A, Grigas J, Pawlowski A, Hilczer B (2007) Dielectric properties of $(NH_4)_3$ $H(SO_4)_2$ crystals in room- and high-temperature phases. Ferroelectrics 348:75–81
35. Woerner M, Zamponi F, Ansari Z, Dreyer J, Freyer B, Prémont-Schwarz M, Elsaesser T (2010) Concerted electron and proton transfer in ionic crystals mapped by femtosecond X-ray powder diffraction. J Chem Phys 133:064509-1–064509-8

Chapter 21
Charge Density and Chemical Reactions: A Unified View from Conceptual DFT

Paul A. Johnson, Libero J. Bartolotti, Paul W. Ayers, Tim Fievez, and Paul Geerlings

21.1 Overview

While it is well known that the electron density provides key information about structural properties of molecules and materials [1, 2], it is less appreciated that the electron density, along with its response to perturbations, contains key information about chemical reactivity. Using the electron density and its various response functions to understand and predict chemical reactivity is the goal of conceptual (or chemical) density functional theory (DFT) [3, 4]. The mathematical structure of conceptual DFT is similar to classical thermodynamics with state functions, total differentials, variational principles, and Legendre transforms [5–7]. This "thermodynamic" approach provides an alternative perspective on the role of the electron density in quantum mechanics that recalls the classic pre-DFT work on ideal and real electron gasses [8–11].

The key idea in conceptual DFT is that the response of a system to perturbations determines its reactivity [12, 13]. If a system reacts favourably to a perturbation (i.e., the energy goes down, or at least does not increase very much) then this indicates that the system will react favourably with a certain class of reagents. Differentials of the energy may thus be interpreted as reactivity indicators.

P.A. Johnson • P.W. Ayers
Department of Chemistry, McMaster University, Hamilton, ON, Canada L8S4M1
e-mail: johnsopa@mcmaster.ca; ayers@mcmaster.ca

L.J. Bartolotti
Department of Chemistry, East Carolina University, Greenville, NC 27858, USA
e-mail: bartolottil@mail.edu

T. Fievez • P. Geerlings (✉)
Eenheid Algemene Chemie (ALGC), Vrije Universiteit Brussel (VUB), Pleinlaan 2, 1050 Brussel, Belgium
e-mail: tfievez@rub.vub.ac.be; pgeerlin@vub.ac.be

C. Gatti and P. Macchi (eds.), *Modern Charge-Density Analysis*,
DOI 10.1007/978-90-481-3836-4_21, © Springer Science+Business Media B.V. 2012

In this chapter we will systematically present the elements of conceptual DFT. In Sect. 21.2 we will show how responses of the electronic energy to changes in the number of electrons and the external electrostatic potential can be interpreted as chemical reactivity indicators. The first-order response of the energy to changes in external potential is none other than the electron density, so in this context the electron density's role in determining the molecular electrostatic potential and internuclear forces naturally arises. Together with the electrostatic potential, responses of the electron density predict regioselectivity. In open electronic systems like a molecule in a solvent or a defect in a crystal, the number of electrons in the system fluctuates about its mean value, so it is impossible to precisely define the number of electrons in the system. Performing a Legendre transform allows us to rewrite all the key equations from Sect. 21.2 in terms of a chemical potential. The resulting reactivity indicators are presented in Sect. 21.3. The Hohenberg-Kohn Theorem says that we can determine all properties of any system, including its reactivity, from the electron density. A different Legendre transform, this one with respect to the external potential, allows us to reformulate the key equations in terms of the electron density and its responses. This density-based approach is presented in Sect. 21.4. Almost every reactivity indicator in conceptual DFT fits into one of these three pictures: the closed-system picture (using the number of electrons), the open-system picture (using the chemical potential), and the electron-preceding picture (using the electron density). Some other types of reactivity indicators are discussed in Sect. 21.5. Some extensions to the theory and open problems are presented in Sect. 21.6. A few representative applications are presented in Sect. 21.7. Common criticisms of conceptual DFT are addressed in Sect. 21.8. Concluding remarks are in Sect. 21.9.

21.2 Reactivity Indicators in the Closed System Picture

As mentioned earlier, reactivity indicators are associated with the response of the system to perturbations. When a reagent approaches the system of interest, the number of electrons in the system changes due to electron transfer and the external potential that the electrons in the system feel also changes. Specifically the electrons in the system now feel not only an attraction to the nuclei in the system, but also an attraction to the nuclei of the attacking reagent and a repulsion to the electrons of the attacking reagent. The exact form of the effective external potential (system + environment) has been discussed at some length but the details are not relevant to the present discussion [14–20].

The change in the system's energy when the number of electrons and the external potential change is given by [5]:

$$dE\left[v\left(\mathbf{r}\right);N\right]=\left(\frac{\partial E}{\partial N}\right)_{v(\mathbf{r})}dN+\int\left(\frac{\delta E}{\delta v\left(\mathbf{r}\right)}\right)_{N}\delta v\left(\mathbf{r}\right)d\mathbf{r} \tag{21.1}$$

This is the total differential of the energy. The coefficient of dN, in analogy to classical thermodynamics, is called the electronic chemical potential [5],

$$\mu = \left(\frac{\partial E}{\partial N}\right)_{v(\mathbf{r})} \tag{21.2}$$

Because the chemical potential measures the energetic benefit of adding electrons to the system, it is clearly related to the electronegativity. In fact, in conceptual DFT we define the electronegativity as $-\mu$. Often we use the working approximation based on the finite difference approximation [5, 21]

$$\mu \approx -\frac{I + A}{2} \tag{21.3}$$

where I is the ionization potential and A is the electron affinity. In this approximation the chemical potential is simply -1 times Mulliken's electronegativity [22]. In addition to its utility as an electronegativity measure, the chemical potential has been used to measure the intrinsic strength of Lewis acids and bases. (The "intrinsic strength" is the acid/base strength in the absence of solvent and other system-dependent effects [23].)

The coefficient of $\delta v(\mathbf{r})$ in Eq. 21.1 is none other than the electron density.

$$\rho(\mathbf{r}) = \left(\frac{\delta E}{\delta v(\mathbf{r})}\right)_N \tag{21.4}$$

The notation in this equation is called a functional derivative. Recall that a function, $f(\mathbf{r})$, is a rule for mapping a scalar or vector quantity $\mathbf{r}$ to a real number, $f(\mathbf{r})$. Similarly, a functional is a rule for mapping a *function* $f(\mathbf{r})$ to a real number, $F[f]$. The derivative/gradient of a function is a rule for mapping infinitesimal changes in the argument of the function, $d\mathbf{r}$, to infinitesimal changes in the value of the function,

$$df = \nabla f(\mathbf{r}) \cdot d\mathbf{r} = \sum_{i=1}^{d} \frac{\partial f(\mathbf{r})}{\partial r_i} dr_i \tag{21.5}$$

Similarly, the derivative of a functional, $F[f]$, is a rule for mapping infinitesimal changes in a function, $\delta f(\mathbf{r})$, to infinitesimal changes in the value of the functional, dF, and is denoted[1]

$$dF = \int \frac{\delta F[f]}{\delta f(\mathbf{r})} \delta f(\mathbf{r}) \, d\mathbf{r} \tag{21.6}$$

[1] The relationship between Eqs. 21.5 and 21.6 will be more clear if one imagines the integral as a Riemann sum or, alternatively, conceives of the change in the function, $\delta f(r)$, as a vector in a Hilbert (or, more generally, a Banach) space.

For example, every choice for the number of electrons, N, and the external potential, $v(\mathbf{r})$, produces a unique value for the energy, $E[v(\mathbf{r});N]$. Thus the energy is a function of the number of electrons, and a functional of the external potential. The functional derivative can be evaluated by determining the change in the value of a functional when the underlying function is changed at $\mathbf{r}$:

$$\left(\frac{\delta E\left[v_0;N_0\right]}{\delta v\left(\mathbf{r}\right)}\right)_N = \left.\frac{\partial E\left[v_0\left(\mathbf{r}'\right)+\varepsilon\delta\left(\mathbf{r}'-\mathbf{r}\right);N_0\right]}{\partial\varepsilon}\right|_{\varepsilon=0}$$
$$= \lim_{\varepsilon\to 0}\frac{E\left[v_0\left(\mathbf{r}'\right)+\varepsilon\delta\left(\mathbf{r}'-\mathbf{r}\right);N_0\right]-E\left[v_0\left(\mathbf{r}'\right);N_0\right]}{\varepsilon} \tag{21.7}$$

This notation means "the functional derivative of the energy with respect to the external potential evaluated for N_0 electrons and the external potential $v_0\left(\mathbf{r}\right)$". When it's clear that a function depends on $\mathbf{r}$, we will often not show this dependence explicitly.

The chemical potential and the electron density are themselves functions of the number of electrons and functionals of the external potential. We can express these dependencies through total differentials

$$d\mu = \left(\frac{\partial\mu}{\partial N}\right)_{v(\mathbf{r})} dN + \int\left(\frac{\delta\mu}{\delta v\left(\mathbf{r}\right)}\right)_N \delta v\left(\mathbf{r}\right) d\mathbf{r} \tag{21.8}$$

$$\delta\rho\left(\mathbf{r}\right) = \left(\frac{\partial\rho\left(\mathbf{r}\right)}{\partial N}\right)_{v(\mathbf{r})} dN + \int\left(\frac{\delta\rho\left(\mathbf{r}\right)}{\delta v\left(\mathbf{r}'\right)}\right)_N \delta v\left(\mathbf{r}'\right) d\mathbf{r}' \tag{21.9}$$

The derivatives in these equations define second-order reactivity indicators. The $\mathbf{r}$-independent second-order reactivity indicator was identified by Parr and Pearson as a measure of the chemical hardness [24],

$$\eta = \left(\frac{\partial\mu}{\partial N}\right)_{v(\mathbf{r})} = \left(\frac{\partial^2 E}{\partial N^2}\right)_{v(\mathbf{r})} \approx I - A \tag{21.10}$$

The approximation I–A is based on the same quadratic energy model that gave rise to the Mulliken chemical potential (Eq. 21.3). This definition of the chemical hardness has been used to elucidate the hard/soft acid/base principle [23, 25–30] and to derive the maximum hardness principle [31–36]. The maximum hardness principle states that the stability of molecular systems increases with their chemical hardness. Ergo, transition states tend to be associated with low hardness and the most stable products of a reaction tend have the highest hardness. While this principle is not universally valid [35, 37, 38], it represents one of the great successes of conceptual DFT because the principle could not be formulated or understood by other means.

The $\mathbf{r}$-dependent derivatives in Eqs. 21.8 and 21.9 are equal; they define the Fukui function [39, 40]

$$f(\mathbf{r}) = \left(\frac{\partial \rho(\mathbf{r})}{\partial N}\right)_{v(\mathbf{r})} = \left(\frac{\delta \mu}{\delta v(\mathbf{r})}\right)_N = \frac{\partial \delta E}{\partial N \delta v(\mathbf{r})}$$
$$\approx \frac{\rho_{N+1}(\mathbf{r}) - \rho_{N-1}(\mathbf{r})}{2} \tag{21.11}$$

The Fukui function is one of the key regioselectivity indicators in conceptual DFT [41, 42]. Notice that the Fukui function captures the response of the electron density to changes in electron number. In the molecular orbital (MO) picture, electrons are removed from the highest occupied molecular orbital (HOMO) and added to the lowest unoccupied molecular orbital (LUMO). This suggests that the Fukui function captures the essence of frontier molecular orbital (FMO) theory but with corrections for orbital relaxation [40, 43–45] and electron correlation. It also supports our initial assertion that responses of the electron density to perturbations provide key information about chemical reactivity. Once again, the approximate formula for the Fukui function is consistent with the Mulliken approximation for the chemical potential.

Finally the two-point (nonlocal) reactivity indicator in Eq. 21.9 is identified as the linear response kernel [7]

$$\chi(\mathbf{r},\mathbf{r}') = \left(\frac{\delta \rho(\mathbf{r})}{\delta v(\mathbf{r}')}\right)_N = \left(\frac{\delta^2 E}{\delta v(\mathbf{r})\, \delta v(\mathbf{r}')}\right)_N \tag{21.12}$$

This term models the polarization of the electron density by the change in external potential. In fact the electric dipole polarizability is simply

$$\alpha = \iint \frac{xx' + yy' + zz'}{3} \chi(\mathbf{r},\mathbf{r}')\, d\mathbf{r} d\mathbf{r}' \tag{21.13}$$

There is unfortunately no Mulliken type approximation for this quantity, which explains why it is rarely used. However, there is clearly chemically relevant information encapsulated in this term. Fortunately, with some clever tricks to be described in the next section, we can approximately capture these effects.

Third and higher order reactivity indicators are rarely considered [12, 46–49]. The most notable exception is the dual-descriptor [46, 50–52]:

$$f^{(2)}(\mathbf{r}) = \left(\frac{\partial^2 \rho(\mathbf{r})}{\partial N^2}\right)_{v(\mathbf{r})} = \left(\frac{\partial f(\mathbf{r})}{\partial N}\right)_{v(\mathbf{r})}$$
$$= \left(\frac{\delta \eta}{\delta v(\mathbf{r})}\right)_N = \frac{\partial^2 \delta E}{\partial N^2 \delta v(\mathbf{r})}$$
$$\approx \rho_{N+1}(\mathbf{r}) - 2\rho_N(\mathbf{r}) + \rho_{N-1}(\mathbf{r}) \tag{21.14}$$

The dual-descriptor also describes FMO effects and is especially important when a molecule acts as both an electron donor and an electron acceptor (e.g., pericyclic reactions) [53–56]. The approximation in the last line of Eq. 21.14 is once again a Mulliken type approximation.

Note that, except for the global (**r**-independent) quantities (μ, η, etc.), all reactivity indicators of any order can be reduced to the electron density or one of its derivatives with respect to N and $v(\mathbf{r})$. This electron density and its response provide the essential information needed for studies of chemical reactivity.

21.3 Reactivity Indicators in the Open System Picture

If one could measure the number of electrons in a liquid or solid solution, then one would observe that the number of electrons fluctuates around its mean value due to the interchange of electrons with the "solvent".[2] While we cannot speak of the number of electrons with any precision, the chemical potential in Eq. 21.2 is well-defined. This suggests the following Legendre transform [5, 6]

$$\begin{aligned} \Omega\,[v;\mu] &= E\,[v;N] - N\left(\frac{\partial E}{\partial N}\right)_{v(\mathbf{r})} \\ \Omega\,[v;\mu] &= E\,[v;N\,(\mu)] - N\,(\mu)\,\mu \end{aligned} \tag{21.15}$$

The state function in the open system picture, the grand potential, satisfies a variational principle analogous to the DFT variational principle

$$\Omega\,[v;\mu] = \min_{\{\rho(\mathbf{r})\geq 0\}} \Omega_V\,[\rho] \tag{21.16}$$

where

$$\Omega_v\,[\rho] = E_v\,[\rho] - \mu \int \rho\,(\mathbf{r})\,d\mathbf{r} \tag{21.17}$$

and $E_v\,[\rho]$ is the universal energy functional from DFT [57].

Instead of adjusting the number of electrons, in the open system picture we adjust the chemical potential. In practical terms, the chemical potential can be tuned by adjusting the acidity of the environment. For example, increasing the acidity of a molecule's surroundings decreases the chemical potential of the system as a whole

[2]This is obvious from the fact that the number of electrons in, for example, a solvated ion is usual not an integer. However, any measurement of the number of electrons will always produce an integer answer.

(molecule + surroundings) and the average number of electrons in the molecule decreases because it donates electrons to the increasingly acidic solvent.

Just as before, chemical reactivity indicators arise from the total differential of the state function,

$$d\Omega = \left(\frac{\partial \Omega}{\partial \mu}\right)_{v(\mathbf{r})} d\mu + \int \left(\frac{\delta \Omega}{\delta v(\mathbf{r})}\right)_{\mu} \delta v(\mathbf{r})\, d\mathbf{r} \tag{21.18}$$

where

$$\left(\frac{\partial \Omega}{\partial \mu}\right)_{v(\mathbf{r})} = -N \tag{21.19}$$

$$\left(\frac{\delta \Omega}{\delta v(\mathbf{r})}\right)_{\mu} = \rho(\mathbf{r}) \tag{21.20}$$

Once again the electron density emerges as the key carrier of chemical information. The response of the electron density to changes in μ and $v(\mathbf{r})$ likewise emerge as local reactivity descriptors.

Differentiating Eqs. 21.19 and 21.20 gives

$$dN = \left(\frac{\partial N}{\partial \mu}\right)_{v(\mathbf{r})} d\mu + \int \left(\frac{\delta N}{\delta v(\mathbf{r})}\right)_{\mu} \delta v(\mathbf{r})\, d\mathbf{r} \tag{21.21}$$

$$d\rho(\mathbf{r}) = \left(\frac{\partial \rho(\mathbf{r})}{\partial \mu}\right)_{v(\mathbf{r})} d\mu + \int \left(\frac{\delta \rho(\mathbf{r})}{\delta v(\mathbf{r}')}\right)_{\mu} \delta v(\mathbf{r}')\, d\mathbf{r}' \tag{21.22}$$

The $\mathbf{r}$-independent (global) derivative is identified as the chemical softness [58],

$$S = \left(\frac{\partial N}{\partial \mu}\right)_{v(\mathbf{r})} = \left[\left(\frac{\partial \mu}{\partial N}\right)_{v(\mathbf{r})}\right]^{-1} = \frac{1}{\eta} = \left(\frac{\partial^2 \Omega}{\partial \mu^2}\right)_{v(\mathbf{r})} \tag{21.23}$$

It is merely one over the chemical hardness. The $\mathbf{r}$-dependent (local) derivative is called the local softness [58],

$$\begin{aligned} s(\mathbf{r}) &= -\left(\frac{\delta N}{\delta v(\mathbf{r})}\right)_{\mu} = \left(\frac{\partial \rho(\mathbf{r})}{\partial \mu}\right)_{v(\mathbf{r})} \\ &= \frac{\partial \delta \Omega}{\partial \mu \delta v(\mathbf{r})} = \left(\frac{\partial \rho(\mathbf{r})}{\partial N}\right)_{v(\mathbf{r})} \left(\frac{\partial N}{\partial \mu}\right)_{v(\mathbf{r})} = Sf(\mathbf{r}) \end{aligned} \tag{21.24}$$

It measures the change in electron density due to a change in chemical potential. It is, unsurprisingly, proportional to the Fukui function. The two-point (non-local) derivative is called the softness kernel [7],

$$s\left(\mathbf{r},\mathbf{r}'\right) = -\left(\frac{\delta \rho\left(\mathbf{r}\right)}{\delta v\left(\mathbf{r}'\right)}\right)_{\mu} = \left(\frac{\delta^2 \Omega}{\delta v\left(\mathbf{r}\right) \delta v\left(\mathbf{r}'\right)}\right)_{\mu} \tag{21.25}$$

It reflects the change in density due to a change in external field and is related to the dielectric response. The softness kernel is related to the linear response kernel by the Berkowitz-Parr relation [7]

$$\chi\left(\mathbf{r},\mathbf{r}'\right) = Sf\left(\mathbf{r}\right) f\left(\mathbf{r}'\right) - s\left(\mathbf{r},\mathbf{r}'\right) \tag{21.26}$$

Sometimes the linear response kernel is approximated from Eq. 21.26 by omitting, or crudely approximating, the softness kernel [59–61].

As in the closed-system picture, third- and higher-order derivatives are rarely considered. The most notable exception is the second-order local softness [49, 54–56], which is the analogue of the dual descriptor (cf. Eq. 21.14) in the open-system picture,

$$s^{(2)}\left(\mathbf{r}\right) = \left(\frac{\partial^2 \rho\left(\mathbf{r}\right)}{\partial \mu^2}\right)_{v(\mathbf{r})} = \frac{f^{(2)}\left(\mathbf{r}\right)}{\eta^2} + \left(\frac{\partial \eta}{\partial N}\right)_{v(\mathbf{r})} \frac{f\left(\mathbf{r}\right)}{\eta^3} \tag{21.27}$$

In both the closed system and open system pictures, the density and its derivatives encapsulate the key information about where and how a molecule interacts with an approaching reagent. The **r**-independent (global) reactivity indicators, on the other hand, contain information about the propensity of a molecule to react but give no information about where or how the molecule will interact with a reagent. Taken together, all of the important chemical effects are captured: electron transfer, electrostatic interactions, frontier orbital interactions, mutual polarization, etc.

21.4 Reactivity Indicators in the Electron-Preceding Picture

In the closed and open system pictures the external potential is a variable and variations in the external potential induce changes in the electron density. This is called the electron-following picture, because the electrons adapt to changes in the external potential [62, 63]. The Hohenberg-Kohn theorem indicates that just as changes in external potential give rise to well-determined changes in electron density, changes in the electron density can be used to infer the underlying change in the external potential [57]. This latter perspective is called the electron-preceding picture, because we envision the electron density changing and then the external potential changing in response to the change in electron density. To derive reactivity indicators, in the electron-preceding picture, we again perform a Legendre transform of the energy, this time with respect to the external potential

$$F[\rho] = E[v;N] - \int \left(\frac{\delta E}{\delta v(\mathbf{r})}\right)_N v(\mathbf{r})\, d\mathbf{r}$$
$$= E[v[\rho];N[\rho]] - \int \rho(\mathbf{r})\, v[\rho;\mathbf{r}]\, d\mathbf{r} \tag{21.28}$$

where

$$N[\rho] = \int \rho(\mathbf{r})\, d\mathbf{r} \tag{21.29}$$

$F[\rho]$ is called the Hohenberg-Kohn functional [57]. Formally, performing the Legendre transformation on $E[v; N]$ with respect to $v(\mathbf{r})$ defines a functional $F[\rho; N]$, but because N itself is a functional of $\rho(\mathbf{r})$, we prefer the more succinct notation, $F[\rho]$. (This notation also avoids the thorny mathematical issues associated with ill-defined "constrained functional derivatives" [64–67].)

Notice that Eq. 21.28 only applies when the density is v-representable. If one considers a non-v-representable density (that is, a density that is not the ground state for any external potential), then $v[\rho]$ is not defined and $F[\rho]$ cannot be computed from Eq. 21.28. In this case, one needs to use the variational principle for $F[\rho]$, namely [68]

$$F[\rho] = \sup_{v(\mathbf{r})} E[v;N] - \int \rho(\mathbf{r})\, v(\mathbf{r})\, d\mathbf{r} \tag{21.30}$$

The supremum in Eq. 21.30 can be replaced by a maximum for v-representable densities [68, 69].

Once again, total differentials of the state function (here $F[\rho]$) are associated with reactivity indicators. The first order differential is

$$dF = \int \frac{\delta F}{\delta \rho(\mathbf{r})} \delta \rho(\mathbf{r})\, d\mathbf{r} \tag{21.31}$$

$$\frac{\delta F}{\delta \rho(\mathbf{r})} = \mu - v(\mathbf{r}) \tag{21.32}$$

In the electron-preceding picture, the external potential replaces the electron density as the key first-order reactivity indicator. The second order reactivity indicator arises from the differential

$$d[\mu - v(\mathbf{r})] = \int \frac{\delta(\mu - v(\mathbf{r}))}{\delta \rho(\mathbf{r}')} \delta \rho(\mathbf{r}')\, d\mathbf{r}' \tag{21.33}$$

and defines the hardness kernel [7],

$$\eta(\mathbf{r},\mathbf{r}') = \frac{\delta(\mu - v(\mathbf{r}))}{\delta \rho(\mathbf{r},\mathbf{r}')} = \frac{\delta^2 F}{\delta \rho(\mathbf{r})\, \delta \rho(\mathbf{r}')}. \tag{21.34}$$

The hardness kernel is the inverse of the softness kernel if the softness kernel is viewed as an integral operator [7]:

$$\delta\left(\mathbf{r}-\mathbf{r}''\right)=\int \eta\left(\mathbf{r},\mathbf{r}'\right)s\left(\mathbf{r}',\mathbf{r}''\right)d\mathbf{r}' \tag{21.35}$$

Integrating both sides of Eq. 21.35 with respect to $\mathbf{r}''$ gives the famous Chattaraj-Cedillo-Parr identities [70]

$$\begin{aligned} 1 &= \int \eta\left(\mathbf{r},\mathbf{r}'\right)s\left(\mathbf{r}'\right)d\mathbf{r}' \\ \eta &= \int \eta\left(\mathbf{r},\mathbf{r}'\right)f\left(\mathbf{r}'\right)d\mathbf{r}' \end{aligned} \tag{21.36}$$

All of the information needed to describe chemical reactivity theory can be computed from the electron density and the related responses. This is clearly true for local reactivity information, because the key regioselectivity indicators are the electron density and its derivatives with respect to N and/or μ. One of the most useful features of the electron-preceding picture, however, is that it provides the key equations by which the global reactivity indicators can be deduced from $\mathbf{r}$-dependent quantities. For example, the chemical potential can be extracted from the asymptotic decay of $\delta F/\delta\rho(\mathbf{r})$ (cf. Eq. 21.32). The chemical hardness can be obtained from the Fukui function and the hardness kernel, Eq. 21.36. The global softness can then be obtained from $S = 1/\eta$ or, alternatively, by using the sequential relations that couple the global softness, local softness, and the softness kernel:

$$S=\int s(\mathbf{r})\,d\mathbf{r}=\iint s\left(\mathbf{r},\mathbf{r}'\right)d\mathbf{r}d\mathbf{r}'. \tag{21.37}$$

If desired, the Mulliken-style formulae for μ and η (Eqs. 21.3 and 21.10, respectively) can be determined by extracting the ionization potential and the electron affinity from the asymptotic decay of the electron density [71–75] and the Fukui function [41], respectively. There are similar equations for the higher-order global reactivity indicators in terms of the density and its responses [12, 48, 49]. The electron density and its response to perturbations, then, readily provide *all* the information that is needed to provide a complete description of chemical reactivity.

Notice that the reactivity indicators in all three pictures are closely related to each other [7, 12, 48, 76]. This indicates that the results of one's analysis will not depend on which picture one chooses. (This is directly analogous to the freedom of choosing an ensemble in statistical mechanics.) Therefore, one can use whichever picture is most natural or most computationally facile for the system of interest. In practice, most previous work in conceptual DFT was performed using either the closed system or open system pictures. The electron-preceding picture, while mathematically elegant, is computationally challenging. The first functional derivative of $F[\rho]$ requires a method for determining the external potential directly from the electron density. Programs that perform this task are computationally demanding, but widely available [77–82]. The mathematical formulae for evaluating the second functional derivative (i.e., the hardness kernel) from the output of a Kohn-Sham DFT calculation have been worked out, but seemingly never implemented [83, 84].

21.5 Derived Chemical Reactivity Indicators

All of the fundamental chemical reactivity indicators arise from one of the derivatives in one of the pictures. In some cases, however, it is useful to derive special combinations of these indicators that are particularly relevant for an important chemical problem. The most famous of these "derived indicators" is the electrophilicity measure of Parr et al. [85]

$$\omega = \frac{\mu^2}{2\eta} \tag{21.38}$$

The electrophilicity measures the change in energy when a system with chemical potential μ and hardness η is brought into contact with a perfect electron donor (that is, an electron reservoir with zero chemical potential and zero hardness). As such, this electrophilicity indicator measures the inherent electrophilicity of a substance [86, 87].

There are many more derived indicators, some of which have detailed mathematical derivations like the electrophilicity and some of which were proposed for wholly utilitarian reasons. Among "mathematical" derived indicators we may mention the nucleofugality [88], the electrofugality [88], the Ξ indicator [89, 90], and the point charge response [15, 91]. Among "heuristic" derived indicators are the relative electrophilicity and nucleophilicity [92] and the leaving-group indicators from the Campodonico and Contreras group [93–95]. (There are many, many more derived indicators, but most of them are conceived as easy-to-compute approximations to one of the standard reactivity indicators.) Some indicators lie between these two extremes, for example the family of philicity indicators including, most notably, the local electrophilicity [87]

$$\omega(\mathbf{r}) = \omega f(\mathbf{r}), \tag{21.39}$$

the multiphilic descriptor [96, 97], the local hardness [98–102], and the family of indicators associated with the comprehensive decomposition analysis of stabilization energy (CDASE) [103]. Another "intermediate" reactivity indicator is the modified electrophilicity/nucleophilicity measure called the electron-donating and electron-accepting power [104].

It should be emphasized that nothing is lost by restricting oneself only to the fundamental derivative-based reactivity indicators. It is merely that sometimes a "derived indicator" provides a more convenient package for the key reactivity phenomena. Derived reactivity indicators tend to be especially important for quantitative structure activity relationships (QSAR) and quantitative structure property relationships (QSPR) because, in favourable cases, the effect one is trying to predict may be captured by a single "derived indicator", allowing one to avoid the difficulties attendant to multivariate regression.

21.6 Extensions

21.6.1 Including Inter-nuclear Repulsion

In conceptual DFT, the focus is usually on the electronic energy and its Legendre transforms. However, in chemistry, we are often more interested in the Born-Oppenheimer potential energy surface (PES) because of its direct correspondence to thermodynamic and kinetic properties. It is a small modification of the preceding treatment to include the nuclear-nuclear repulsion [15]. For the closed-system, and open system pictures, we just add the nuclear-nuclear repulsion to the state functions

$$U\left[v;N\right] = E\left[v;N\right] + V_{nn}\left[v\right] \tag{21.40}$$

$$W\left[v;\mu\right] = \Omega\left[v;\mu\right] + V_{nn}\left[v\right] \tag{21.41}$$

where V_{nn} is the nuclear-nuclear repulsion energy

$$\begin{aligned} V_{nn} &= \frac{1}{2}\sum_{\substack{\alpha=1\\ \alpha\neq\beta}}^{Natoms}\sum_{\beta=1}^{Natoms}\frac{Z_\alpha Z_\beta}{\left|\mathbf{R}_\alpha-\mathbf{R}_\beta\right|} \\ &= \frac{1}{32\pi^2}\iint\limits_{\mathbf{r}\neq\mathbf{r}'}\frac{\nabla^2 v\left(\mathbf{r}\right)\nabla^2 v\left(\mathbf{r}'\right)}{\left|\mathbf{r}-\mathbf{r}'\right|}d\mathbf{r}d\mathbf{r}' \end{aligned} \tag{21.42}$$

The derivation in the electron-preceding picture is somewhat more complicated, but the resulting state function has the intuitively expected form,

$$\mathrm{F}\left[\rho\right] = E\left[v;N\right] - \int\left(\rho\left(\mathbf{r}\right)+\rho_{nuc}\left(\mathbf{r}\right)\right)v\left(\mathbf{r}\right)d\mathbf{r}, \tag{21.43}$$

where

$$\rho_{nuc}\left(\mathbf{r}\right) = -\sum_{\alpha=1}^{Natoms} Z_\alpha\delta\left(\mathbf{r}-\mathbf{R}_\alpha\right) = \tfrac{-1}{4\pi}\nabla^2 v\left(\mathbf{r}\right) \tag{21.44}$$

is the charge density of the nuclei. The minus sign in Eq. 21.44 is required because our charge density convention chose the electron density as a positive distribution.

One advantage of including the internuclear repulsion is that the electrostatic potential [105, 106] emerges as one of the fundamental reactivity indicators. Taking the total differential of the closed-system PES we get [89]

$$dU = \mu dN + \int (\rho(\mathbf{r}) + \rho_{nuc}(\mathbf{r}))\delta v(\mathbf{r})\, d\mathbf{r} \tag{21.45}$$

When a reagent is very far away, the electrostatic interaction with the molecule dominates. Therefore, for very distant reagents, it is reasonable to approximate $\delta v(\mathbf{r})$ as a point charge (or a collection of point charges). Choosing $\delta v(\mathbf{r})$ to be a single positive point charge,

$$\delta v(\mathbf{r}) = \frac{-1}{|\mathbf{r} - \mathbf{R}|} \tag{21.46}$$

gives a special class of reactivity indicators. Inserting Eq. 21.46 into Eq. 21.45 demonstrates that the first of this class of reactivity indicators is simply the molecular electrostatic potential

$$\Phi(\mathbf{r}) = -\int \frac{\rho(\mathbf{r})}{|\mathbf{r} - \mathbf{R}|} d\mathbf{r} + \sum_{\alpha=1}^{Natoms} \frac{Z_\alpha}{|\mathbf{R} - \mathbf{R}_\alpha|} \tag{21.47}$$

Inserting Eq. 21.46 into expressions for the second-order differentials gives rise to additional reactivity indicators such as the Fukui potential [107]

$$v_f(\mathbf{r}) = \int \frac{f(\mathbf{r})}{|\mathbf{r} - \mathbf{R}|} d\mathbf{r} \tag{21.48}$$

and the polarization potentials [108]

$$\begin{aligned} V_\chi(\mathbf{R},\mathbf{R}') &= \int \frac{1}{|\mathbf{r} - \mathbf{R}|} (\chi(\mathbf{r},\mathbf{r}')) \frac{1}{|\mathbf{r}' - \mathbf{R}'|} d\mathbf{r}\, d\mathbf{r}' \\ \tilde{V}_\chi(\mathbf{R},\mathbf{R}') &= V_\chi(\mathbf{R},\mathbf{R}') + \frac{1}{|\mathbf{R} - \mathbf{R}'|} \end{aligned} \tag{21.49}$$

Similar expressions are obtained in the open-system picture [108].

21.6.2 Condensed Reactivity Indicators

It is often useful to obtain a "coarse-grained" description of the local and non-local reactivity indicators. For example, when the molecule of interest and the reagent are far away from each other, the interaction between them is insensitive to detailed point-by-point variations of the electron density, the Fukui function, etc.. In such cases, it is sufficient to condense the reactivity indicators to the atomic level [109], and speak of "atomic Fukui functions" and "atomic populations." This "coarse-grained" description provides a compact representation of the most important reactivity features of the molecule, and is often used to simplify the analysis of conceptual DFT [43, 110–118]. The condensed reactivity indicators also, by construction, provide a faithful representation of molecular interactions when the molecule and the reagent are far apart. In particular, the condensed reactivity indicators capture the essential electrostatic effects, for example,

$$\Phi(\mathbf{R}) \approx \sum_{\alpha=1}^{Natoms} \frac{q_\alpha}{|\mathbf{R} - \mathbf{R}_\alpha|} \tag{21.50}$$

$$v_f(\mathbf{R}) \approx \sum_{\alpha=1}^{Natoms} \frac{f_\alpha}{|\mathbf{R} - \mathbf{R}_\alpha|} \tag{21.51}$$

In fact, these expressions may be used as the definition of the atomic charges and the atom-condensed Fukui functions because the best choice for these quantities is the one that minimizes the error in Eqs 21.50 and 21.51.

More commonly, condensed reactivity indicators are computed in one of two inequivalent ways, based on atomic partitions of unity. An atomic partition of unity is a sum of weight operators, $\hat{w}_\alpha(\mathbf{r})$, that satisfy

$$\hat{1} = \sum_{\alpha=1}^{N_{\text{atoms}}} \hat{w}_\alpha(\mathbf{r}) \tag{21.52}$$

$$0 \leq \hat{w}_\alpha(\mathbf{r}) \leq 1 \tag{21.53}$$

Usually the weight operators are simple functions, as in Bader's atoms-in-molecules partitioning [119–121] and in Hirshfeld-style partitioning [122–125], for example. The weights are nonmultiplicative operators (as in Mulliken partitioning) [126–129]. For simplicity, only the case where the atomic weight operator is a function is considered here; in this case the atomic weight function, $w_\alpha(\mathbf{r})$, represents the probability that an electron at $\mathbf{r}$ is associated with atom α. The electron density of an atom in a molecule can then be defined as

$$\rho_\alpha(\mathbf{r}) = w_\alpha(\mathbf{r})\,\rho(\mathbf{r}) \tag{21.54}$$

and the resulting atomic population is

$$p_\alpha = \int \rho_\alpha(\mathbf{r})\,d\mathbf{r}. \tag{21.55}$$

Inserting the atomic partition of unity into Eq. 21.11 gives either

$$f(\mathbf{r}) = \left(\frac{\partial \left(\sum_{\alpha=1}^{N_{\text{atoms}}} w_\alpha(\mathbf{r}) \right) \rho(\mathbf{r})}{\partial N} \right)_{v(\mathbf{r})} \tag{21.56}$$

or

$$f(\mathbf{r}) = \sum_{\alpha=1}^{N_{\text{atoms}}} w_\alpha(\mathbf{r}) \left(\frac{\partial \rho(\mathbf{r})}{\partial N} \right)_{v(\mathbf{r})} \tag{21.57}$$

depending on where the partition of unity is inserted. Both cases lead to natural definitions for the atomic Fukui functions, which are then integrated to obtain condensed values,

$$f_\alpha = \left(\frac{\partial p_\alpha}{\partial N} \right)_{v(\mathbf{r})} = \int \left(\frac{\partial \rho_\alpha(\mathbf{r})}{\partial N} \right)_{v(\mathbf{r})} d\mathbf{r} \tag{21.58}$$

$$f_\alpha = \int w_\alpha(\mathbf{r}) f(\mathbf{r}) d\mathbf{r} = \int w_\alpha(\mathbf{r}) \left(\frac{\partial \rho(\mathbf{r})}{\partial N} \right)_{v(\mathbf{r})} d\mathbf{r} \tag{21.59}$$

corresponding to Eqs. 21.56 and 21.57, respectively. Although this ambiguity was already noted in the foundational paper of Yang and Mortier [109], it was mostly ignored until recently [116]. Almost all of the previous work has used Eq. 21.58, because atomic populations are standard output from quantum chemistry programs; this is called the response-of-molecular-fragment approach [116]. A few papers have adopted the approach based on Eq. 21.59, which is called the fragment-of-molecular-response approach [112, 114, 116, 121]. The fragment-of-molecular-response approach is arguably better, as it is more closely linked to the probabilistic interpretation of $w_\alpha(\mathbf{r})$ and it requires only one numerical integration to evaluate the reactivity indicator; the response of molecular fragment approach typically requires multiple numerical integrations.

There are many different atomic partitions of unity, each with its advantages and drawbacks. Different population analysis methods give different condensed reactivity indicators, so their ambiguity is unavoidable.

21.6.3 Spin Reactivity Indicators

Most accurate DFT calculations use the spin density instead of the total density as the fundamental quantity [130, 131]. Unsurprisingly, conceptual DFT can also be generalized into a spin-resolved form [132–136]. In spin-resolved conceptual DFT, the number of electrons is split into two variables, N_α and N_β. The N_α up-spin electrons are bound by the external potential $v_\alpha(\mathbf{r})$ and the N_β down-spin electrons are bound by the external potential $v_\beta(\mathbf{r})$. While $v_\alpha(\mathbf{r}) = v_\beta(\mathbf{r})$ in most cases, this is not true in external magnetic fields. The spin-resolved state functions in the closed-system, open-system, and electron-preceding pictures are as follows

$$\begin{aligned} E\left[v_\alpha,v_\beta;N_\alpha,N_\beta\right] &\Leftrightarrow E\left[v;\mathrm{N}\right] \\ \Omega\left[v_\alpha,v_\beta;\mu_\alpha,\mu_\beta\right] &\Leftrightarrow \Omega\left[v;\mu\right] \\ F\left[\rho_\alpha,\rho_\beta\right] &\Leftrightarrow F\left[\rho\right] \end{aligned} \tag{21.60}$$

Here $\rho_\alpha(\mathbf{r})$ and $\rho_\beta(\mathbf{r})$ are the spin densities.

Often it is preferable to express the spin-dependence of DFT in terms of the total density and the magnetization density, which are normalized as follows

$$\begin{aligned} N &= \int \rho(\mathbf{r})\, d\mathbf{r} = \int \left(\rho_\alpha(\mathbf{r}) + \rho_\beta(\mathbf{r})\right) d\mathbf{r} \\ N_S &= \int m(\mathbf{r})\, d\mathbf{r} = \int \left(\rho_\alpha(\mathbf{r}) - \rho_\beta(\mathbf{r})\right) d\mathbf{r} \end{aligned} \tag{21.61}$$

The state functions are easily re-expressed in terms of these new variables through the linear transformations [137]

$$\begin{aligned} \begin{bmatrix} N \\ N_S \end{bmatrix} &= \begin{bmatrix} 1 & 1 \\ 1 & -1 \end{bmatrix} \begin{bmatrix} N_\alpha \\ N_\beta \end{bmatrix} \\ \begin{bmatrix} \rho(\mathbf{r}) \\ m(\mathbf{r}) \end{bmatrix} &= \begin{bmatrix} 1 & 1 \\ 1 & -1 \end{bmatrix} \begin{bmatrix} \rho_\alpha(\mathbf{r}) \\ \rho_\beta(\mathbf{r}) \end{bmatrix} \\ \begin{bmatrix} v(\mathbf{r}) \\ v_m(\mathbf{r}) \end{bmatrix} &= \begin{bmatrix} \frac{1}{2} & \frac{1}{2} \\ \frac{1}{2} & \frac{-1}{2} \end{bmatrix} \begin{bmatrix} v_\alpha(\mathbf{r}) \\ v_\beta(\mathbf{r}) \end{bmatrix} \end{aligned} \tag{21.62}$$

With these definitions, the entire edifice of conceptual DFT can be extended to the spin-resolved case [113, 132–136, 138–144]. In fact, using the matrix/vector notation in Eq. 21.62, all of the fundamental equations are directly transferable, it is only that the reactivity indicators become "decorated" with vector notation [138].

Spin-resolved DFT is especially important for reactions between molecules with unpaired electrons, i.e., radicals. In these cases the spin-reactivity indicators describe the most favourable spin-transfer processes [140, 145, 146]. Spin-reactivity indicators are also useful for describing excited electronic states. By changing N_S while keeping the total number of electrons constant, we can access the lowest excited states for each value of the M_S quantum number. This fact can be used to provide hints about excited state reactivity.

21.6.4 Excited State Reactivity Indicators

A thorough and rigorous treatment of reactivity indicators and excited states is very difficult [54, 147–153]. Partly, this is because elementary DFT is a theory for electronic ground states [57, 154]. Generalizations of DFT to excited states have been made [153, 155–163] and the extensions of these approaches to conceptual DFT have been considered by a few researchers [152, 153, 164]. The overall structure of the theory is very similar to the ground state formulation, but now there are additional indices indicating which excited state's reactivity is being scrutinized. For reactivity indicators that depend on changes in the number of electrons, there is an additional index indicating which excited state of the $N-1$ or $N+1$ electron system is obtained after electron-transfer [153].

21.6.5 Derivative Discontinuities

To this point, we have assumed that the system of interest has a positive temperature. In this case, the energy is a smooth function of the number of electrons, and differentiation with respect to the number of electrons is always well-defined [165, 166]. Most quantum chemistry calculations are performed for molecules at absolute zero temperature. In this case, a plot of the energy versus the number of electrons is a sequence of straight lines [165, 167]. For non-integer numbers of electrons, the derivative of the energy with respect to the number of electrons is just -1 times the ionization potential of the system with the next largest integer number of electrons. I.e.,

$$\begin{aligned} \mu &= \left(\frac{\partial E}{\partial N}\right)_{v(\mathbf{r})} = E\left[v; \lceil N \rceil\right] - E\left[v; \lfloor N \rfloor\right] \\ &= -I\left[v; \lceil N \rceil\right] \\ 0 &= \left(\frac{\partial^n E}{\partial N^n}\right)_{v(\mathbf{r})} \end{aligned} \tag{21.63}$$

Note that all of the higher-order derivatives are zero. When the number of electrons is an integer, the chemical potential is discontinuous and therefore ill-defined, with reasonable values in the range

$$-I \leq \mu \leq -A \tag{21.64}$$

At an integer, N_0, the hardness is infinite, and has the expression [168]

$$\eta = (I - A)\,\delta\,(N - N_0) \tag{21.65}$$

Notice that the Parr-Pearson approximation of the hardness enters naturally into this rigorous formula.

The derivative discontinuity also has implications for the Fukui function and other higher-order reactivity indicators. The Fukui function at an integer splits into two functions depending on whether the derivative is taking from above or from below [39, 41]

$$\begin{aligned} f^{+}(\mathbf{r}) &= \rho\left[v; N_0+1, \mathbf{r}\right] - \rho\left[v; N_0, \mathbf{r}\right] \\ f^{-}(\mathbf{r}) &= \rho\left[v; N_0, \mathbf{r}\right] - \rho\left[v; N_0-1, \mathbf{r}\right] \end{aligned} \tag{21.66}$$

The Fukui function from above indicates the most favourable places to add electrons to the molecule and therefore indicates sites of high nucleophilicity. The Fukui function from below indicates the most favourable places to remove electrons from the molecule and therefore indicates sites of high electrophilicity. In practice, it is often convenient to assume a small positive temperature so that all of the derivatives exist. (This is especially useful for the open-system and electron-preceding pictures.[3]) With this assumption, all of the preceding expressions in this chapter are correct. At zero temperature, the equations usually still hold, but the reactivity indicators must be "decorated" with superscript +'s and −'s to indicate the direction of differentiation.

To introduce a temperature into a Kohn-Sham DFT calculation, one may simply use the standard Mermin formulation and choose the Kohn-Sham orbital occupation numbers according to the Fermi-Dirac distribution function [169]. This is not perfectly rigorous because the exchange-correlation functional also depends on the temperature, but it is usually good enough.

21.6.6 Nuclear Reactivity Indicators

In the preceding analysis, changes in the number of electrons and changes in the chemical potential were always performed at constant external potential. That is, the preceding analysis is based on vertical electron attachment and electron removal processes. Changing the number of electrons, however, usually also affects the molecular geometry. (I.e., vertical and adiabatic electron attachment/removal is inequivalent.) In classical frontier molecular orbital theory, removing an electron from a diatomic molecule in which the HOMO is a binding orbital will cause the bond length to increase. Similarly, adding an electron to a diatomic molecule in which the LUMO is a bonding orbital will cause the bond length to decrease.

[3] Unfortunately, it is difficult to formulate the open-system picture at zero temperature because the Legendre transform from the number of electrons to the chemical potential is no longer well-defined. The electron-preceding picture may be formulated, but the hardness kernel has derivative discontinuities in both variables and is therefore difficult to work with.

It is desirable to find a DFT-based approach to this same problem. DFT-based indicators that reflect how the forces on atomic nuclei, $\mathbf{F}_\alpha$, change with respect to changes in electron number or chemical potential are called nuclear reactivity indicators [44, 118, 170]. These forces can, in turn, be used to determine the molecular Hessian (force constant matrix) and the changes in bond length using the equivalence between the force-based "compliance" approach and the conventional approach based on bond lengths [171–175].

The most commonly used nuclear reactivity indicators are the nuclear Fukui function [44],

$$\begin{aligned}\boldsymbol{\phi}_\alpha &= \left(\frac{\partial \mathbf{F}_\alpha}{\partial N}\right)_{v(\mathbf{r})} = \int f(\mathbf{r}) \frac{\partial}{\partial \mathbf{R}_\alpha}\left(\frac{Z_\alpha}{|\mathbf{R}_\alpha - \mathbf{r}|}\right) d\mathbf{r} \\ &= Z_\alpha \int f(\mathbf{r}) \frac{(\mathbf{R}_\alpha - \mathbf{r})}{|\mathbf{R}_\alpha - \mathbf{r}|^3} d\mathbf{r}\end{aligned} \tag{21.67}$$

and the nuclear local softness [44],

$$\begin{aligned}\boldsymbol{\sigma}_\alpha &= \left(\frac{\partial \mathbf{F}_\alpha}{\partial \mu}\right)_{v(\mathbf{r})} = \int s(\mathbf{r}) \frac{\partial}{\partial \mathbf{R}_\alpha}\left(\frac{Z_\alpha}{|\mathbf{R}_\alpha - \mathbf{r}|}\right) d\mathbf{r} \\ &= Z_\alpha \int s(\mathbf{r}) \frac{(\mathbf{R}_\alpha - \mathbf{r})}{|\mathbf{R}_\alpha - \mathbf{r}|^3} d\mathbf{r}\end{aligned} \tag{21.68}$$

One-sided variants of these indices are obtained by replacing $f(\mathbf{r})$ and $s(\mathbf{r})$ by their one-sided variants, $f^\pm(\mathbf{r})$ and $s^\pm(\mathbf{r})$, respectively [44]. The nuclear Fukui function from below, $\boldsymbol{\phi}_\alpha^-$, represents the change in the force on atom α, $\mathbf{F}_\alpha$, when the number of electrons in the system is decreased. This allows us to determine which of the bonds to atom α strengthen (shorten) and which weaken (lengthen) upon electron transfer from the molecule. In the molecular-orbital picture, similar information could be obtained by studying the nodal structure (ergo the bonding/antibonding character) of the highest occupied molecular orbital, but $\boldsymbol{\phi}_\alpha^-$ also includes information about orbital relaxation and electron correlation that is missed in the simple molecular-orbital description. Similarly, $\boldsymbol{\phi}_\alpha^+$, can be used to determine which bonds strengthen/weaken upon electron transfer to the system under scrutiny.

As an alternative to the one-sided nuclear reactivity indices, one can use the nuclear dual descriptor or the nuclear local hypersoftness,

$$\boldsymbol{\phi}_\alpha^{(2)} = \left(\frac{\partial^2 \mathbf{F}_\alpha}{\partial N^2}\right)_{v(\mathbf{r})} = Z_\alpha \int f^{(2)}(\mathbf{r}) \frac{(\mathbf{R}_\alpha - \mathbf{r})}{|\mathbf{R}_\alpha - \mathbf{r}|^3} d\mathbf{r} \tag{21.69}$$

$$\boldsymbol{\sigma}_\alpha^{(2)} = \left(\frac{\partial^2 \mathbf{F}_\alpha}{\partial \mu^2}\right)_{v(\mathbf{r})} = Z_\alpha \int s^{(2)}(\mathbf{r}) \frac{(\mathbf{R}_\alpha - \mathbf{r})}{|\mathbf{R}_\alpha - \mathbf{r}|^3} d\mathbf{r}, \tag{21.70}$$

which contain similar information. Notice that all of the nuclear reactivity indicators can be evaluated in any coordinate system for the nuclei. It is often more convenient to use internal (bond length, bond angle, dihedral angle) coordinates for evaluating the nuclear reactivity indicators, instead of the atomic coordinates we have, for simplicity, considered here.

Essentially the entire mathematical edifice of DFT reactivity theory can be extended into the "nuclear reactivity" framework [176–181]. There have also been numerous applications of the nuclear reactivity indicators, which are often appealing because they are easy to compute from the output of standard quantum chemistry packages. Partly because of this ease of computation, nuclear reactivity indicators have been widely applied by non-chemists to understand chemical phenomenon as diverse as predicting/understanding crystal polymorphs of drug molecules [182–186] and interpreting the fragmentation pathways of high explosives [187].

21.6.7 Related Topics: Local Thermodynamics, Electron Pair Visualization, Information Theory, etc

The field commonly known as conceptual DFT is vast, and there is no way to capture its full breadth here. For example, we have not mentioned relativistic effects on the conceptual DFT indicators or even on the electron density itself [188, 189].

There are other electron-density-based indicators of electronic structure and molecular reactivity that we have omitted. Most of these indicators are relevant to conceptual DFT, but have for various reasons remained on the periphery of the discipline. For example, we have not mentioned the local temperature and the local entropy [101, 190–192]. These quantities extend the thermodynamic analogy considered here, but the "local thermodynamics" approach to density-functional theory is far less developed. Important outstanding issues include the ambiguity in the definition of the local temperature and the fact that the standard definitions for local temperature and local entropy do not comprise a Legendre-transform pair, in seeming violation of the principles of thermodynamics. The interested reader is referred to a recent review for more details [193].

The local temperature (sometimes called the nighness indicator) [193, 194] and the local entropy [195] both reveal regions of charge concentration in atoms and molecules, facilitating the development of a Lewis-like electron-pair structure. There are other density-based tools that reveal similar shell structure, including the electron localization function (ELF) [196–198], the Laplacian of the electron density [199–201], and the electronic stress tensor [202–209]. The latter two indicators are key ingredients of Bader's quantum theory of atoms in molecules (QTAIM) [119]. QTAIM has developed in parallel with conceptual DFT, but there has been very little cross-over between the disciplines. Typically QTAIM is more concerned with molecular electronic structure and conceptual DFT is more concerned with molecular chemical reactivity.

There has also been an immense amount of work on information-theoretic (i.e., maximum-entropy) methods for interpreting chemical reactivity and molecular electronic structure, chiefly by Roman Nalewajski [123, 124, 172, 210–223]. Like most of the other topics mentioned in this section, those developments have yet to penetrate into the mainstream of research activity in conceptual DFT, but they are certainly worthy of further study.

21.6.8 Alternative Formalisms: Conceptual DFT from Variational Principles and Perturbation Theory

The results in this chapter can be developed from many different perspectives. We have adopted the thermodynamic analogy; this was the most important approach, historically, and is arguably the easiest approach, both pedagogically and mathematically. There are other approaches which make tighter links to the wavefunction-based formulation of quantum mechanics, like the approach of reactivity indicators through variational principles [15, 34, 70, 112] or through perturbation theory [12, 13, 83]. In the perturbation-theoretic approach, the fundamental state function for the system is expanded in a functional Taylor series. For example, the energy is expanded as

$$\begin{aligned}
\Delta E =& E\left[v+\Delta v; N+\Delta N\right] - E\left[v;N\right] \\
=& \sum_{k=1}^{\infty} \frac{1}{k!}\left(\frac{\partial^k E}{\partial N^k}\right)(\Delta N)^k \\
&+ \sum_{k=1}^{\infty} \frac{1}{k!} \int \left(\frac{\partial^k}{\partial^k N}\left(\frac{\delta E}{\delta v(\mathbf{r}_1)}\right)_N\right)_{v(\mathbf{r})} \Delta v(\mathbf{r}_1)\, d\mathbf{r}_1 \\
&+ \sum_{k=1}^{\infty} \frac{1}{2!k!} \iint \left(\frac{\partial^k}{\partial^k N}\left(\frac{\delta^2 E}{\delta v(\mathbf{r}_1)\,\delta v(\mathbf{r}_2)}\right)_N\right)_{v(\mathbf{r})} \Delta v(\mathbf{r}_1)\,\Delta v(\mathbf{r}_2)\, d\mathbf{r}_1 d\mathbf{r}_2 \\
&+ \cdots \\
&+ \sum_{k=1}^{\infty} \frac{1}{k!} \iint \cdots \int \left(\frac{\delta^k E}{\delta v(\mathbf{r}_1)\cdots\delta v(\mathbf{r}_k)}\right)_N \Delta v(\mathbf{r}_1)\cdots\Delta v(\mathbf{r}_k)\, d\mathbf{r}_1 \ldots d\mathbf{r}_k
\end{aligned} \tag{21.71}$$

All of the information about the molecule being reacted with is encapsulated in the coefficients of ΔN and $\Delta v(\mathbf{r})$; these define the reactivity indicators for the system. The characteristics of the attacking reagent (i.e., the relative magnitude of ΔN and $\Delta v(\mathbf{r})$) control which terms in Eq. 21.71 are most important. As a result, different reactivity indicators will be more relevant in different situations. For example, the electrostatic potential is more relevant for hard reagents (where $\|\Delta v(\mathbf{r})\|$ is large compared to $|\Delta N|$), but the Fukui function is more relevant for soft reagents. Although the third and higher-order terms in the series may not be quantitatively small, an abundance of experience suggests that they rarely affect the qualitative trends [13]. The reader is referred to Ref. [13] for a mathematical perspective on the Taylor series, including the (very mild) conditions that are required for its convergence.

21.7 Applications

The applications of this density-based approach to chemical reactivity are so numerous and so diverse as to defy any attempt to summarize them. Suffice it to say that this DFT-based chemical reactivity theory has been applied across the full range of chemistry, including all of the "traditional" chemical disciplines (physical, inorganic, organic, and biological chemistry) and also most, if not all, of chemistry's sister fields (environmental science, materials science, pharmaceutical science, condensed-matter physics, atomic and molecular physics, etc.). Here we will choose only a few very specific applications, which we have chosen as a way to demonstrate the key points in preceding exposition. A useful list of references to the "early" applications of DFT reactivity theory (up to roughly the year 2002) can be found in the 2003 review of Geerlings, De Proft, and Langenaeker [4]. The field has matured significantly since then, with notable progress on redox systems [224–227], sterically hindered systems [145, 228, 229], molecules with multiple reactive sites [90, 230], and surface chemistry [231–234]. In addition, a family of external potential-based reactivity indicators has begun to emerge [15, 91, 108, 235, 236].

To demonstrate the breadth of application, we have selected a few representative examples. The first type of examples include the primary regioselectivity indicators in DFT reactivity theory: the Fukui function and the electrostatic potential. These indicators provide complementary information, and in Sect. 21.7.1 we hope to provide some insight into the nuanced way that experts use these powerful indicators. When a reagent serves as both an electrophile and a nucleophile, the dual descriptor is a more appropriate indicator. This is treated in Sect. 21.7.2, using an example of pericyclic reactions. Finally, in Sect. 21.7.3, we treat the nuclear Fukui function. Although the nuclear Fukui function is a reactivity indicator derived from the closed-system picture, it is an "honorary member" of the electron-preceding (or "density") picture because it reveals how changes in electronic structure (due to adding or removing electrons) lead to changes in nuclear positions (due to forces on the nuclei).

21.7.1 Regioselectivity from the Electrostatic Potential and the Fukui Function

21.7.1.1 Nucleophilic Attack on SCN^-. Hardness and Softness in Regioselectivity

The thiocyanate anion, SCN^-, is a simple example that serves to demonstrate many different features of a DFT-based chemical reactivity theory [13, 15, 230, 237]. The thiocyanate anion is an ambident nucleophile, which can bind through either the sulfur atom (forming thiocyanate compounds) or the nitrogen atom (forming isothiocyanate complexes). The sulfur atom tends to be the preferred site when the attacking electrophile is soft; the nitrogen atom tends to be the preferred site when the attacking electrophile is hard.[4] When thiocyanate reacts with a soft electrophile, charge transfer is significant, and so electrophilic attack tends to occur at the site that is most willing to donate electrons—that is, the site where the Fukui function is the largest. The isosurface of the Fukui function is plotted in Fig. 21.1a; the Fukui function isosurface bulges outwards on the sulfur atom, indicating that soft reagents will attack here. The inner isosurface encapsulates regions where the Fukui function is negative; this is an orbital relaxation effect, and indicates that the σ-bond in SCN^- becomes stronger following electrophilic attack [112, 238].

Small, highly charged ions tend to be hard, and such ions resist electron transfer and favour ionic bonding. In such cases the reactivity is dominated by electrostatic effects, and so the electrophile attacks at the most negatively charged site, that is, the place where the electrostatic potential is most negative. In SCN^-, this is the nitrogen atom (Fig. 21.1b). Figure 21.1, b thus clearly indicate a difference in regioselectivity depending on the soft/hard character of the attacking electrophile (e.g. H^+ (hard) or Ag^+ (soft)).

Note that the plot of the Fukui function is proportional to the local softness, so the closed-system and open-system pictures yield the same conclusions for regioselectivity.

[4]Of course, matters are more complicated. A soft electrophile whose highest occupied orbital(s) do not have the appropriate size/symmetry to interact strongly with the sulfur atom may prefer to bind through the nitrogen atom. Similarly, a hard electrophile for which the charge on the active site is small (a rare, but not impossible occurrence) may prefer to bind through the sulfur atom. Finally, in some cases a molecular rearrangement subsequent to the initial attack may change the binding pattern. Complications like these arise in any qualitative theory of chemical reactivity, but the DFT-based approach is perhaps unique in that they can be addressed within the context of the theory, although capturing such effects usually require difficult to compute higher-order reactivity indicators.

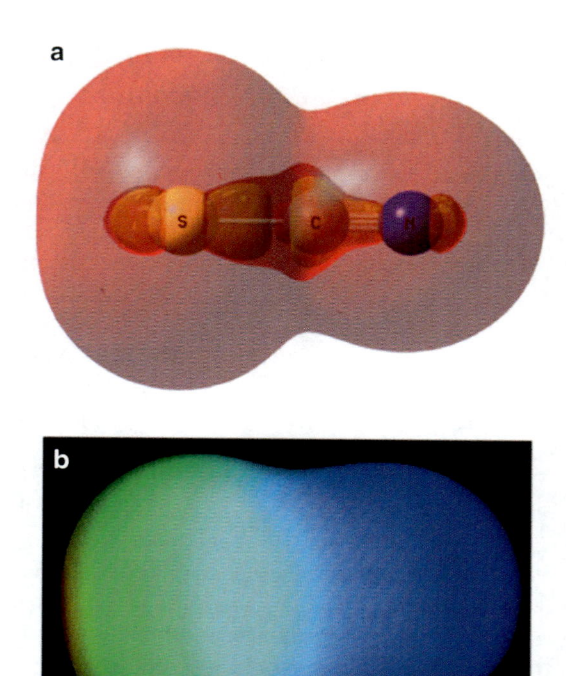

Fig. 21.1 (**a**) Isosurface map of the Fukui function from below, $f^{-}(\mathbf{r})$, for SCN^{-}. Notice that the Fukui function bulges outwards near the sulfur atom, indicating that soft reagents will prefer to bind here. (**b**) The electrostatic potential is plotted on the molecular van der Waals surface with *blue* values indicating small (negative) electrostatic potentials and *red* values indicating large (positive) electrostatic potentials

21.7.1.2 Electrophilic Attack on Propylene. The Power of Condensed Reactivity Indicators. Regioselectivity for Two Centres of Comparable Hardness/Softness: The Fukui Function and MEP for Propylene

The Fukui function is usually the correct reactivity indicator for predicting the regioselectivity of soft molecules, while the electrostatic potential is usually the correct reactivity indicator for predicting the regioselectivity of hard molecules [23, 26, 27, 89, 239–241]. Exceptions exist [190, 242]. As an example we consider the electrophilic attack of propylene, $H_2C{=}CH_2CH_3$, which occurs at the electron-rich double bond. The Fukui function from below, $f^{-}(\mathbf{r})$, is plotted in Fig. 21.2a, and it indicates that propylene most readily donates electrons from the vicinity of the double bond [42]. The same result could be obtained by examining where the highest occupied molecular orbital is large; $|\phi_{\mathrm{HOMO}}(\mathbf{r})|^2$ is plotted in Fig. 21.2b. As noted before, the Fukui functions resemble the frontier orbital densities in most cases, because orbital relaxation and electron correlation rarely changes the qualitative structure of the Fukui function. This result follows from the Yang-Parr-Pucci relations [40, 44],

Table 21.1 Condensed Fukui function from below, f_α^-, for propylene [42]

Atom	C1	C2	C3	H1	H2	H3	H4	H5	H6
f_α^-	0.280	0.216	0.072	0.088	0.073	0.039	0.073	0.093	0.076

$$f^-(\mathbf{r}) = \left(\frac{\partial\rho(\mathbf{r})}{\partial N}\right)^-_{v(\mathbf{r})} = |\phi_{\mathrm{HOMO}}|^2 + \sum_{i=1}^{\mathrm{HOMO}}\left(\frac{\partial|\phi_i(\mathbf{r})|^2}{\partial N}\right)^-_{v(\mathbf{r})} \tag{21.72}$$

$$f^+(\mathbf{r}) = \left(\frac{\partial\rho(\mathbf{r})}{\partial N}\right)^+_{v(\mathbf{r})} = |\phi_{\mathrm{LUMO}}|^2 + \sum_{i=1}^{\mathrm{HOMO}}\left(\frac{\partial|\phi_i(\mathbf{r})|^2}{\partial N}\right)^+_{v(\mathbf{r})} \tag{21.73}$$

and the fact that, if orbital relaxation upon electron donation (Eq. 21.72) or electron acceptance (Eq. 21.73) is small [40, 44, 45], then the terms in the summations are likewise small.

Looking at Fig. 21.2a, b, one cannot say definitively which of the two double bonded carbon atoms will be attacked by the electrophile. (If anything, one would guess that carbon 2 was attacked.) To obtain a semi-quantitative measure of the relative reactivity of the two carbon atoms, we can use the condensed Fukui function, as shown in Table 21.1. The condensed Fukui function is larger on carbon one by .06*e*, clearly indicating that carbon 1 is most susceptible to attack, in accord with the Markovnikov rule.

We can also inspect the electrostatic potential, as shown in Fig. 21.2c. It is difficult to make any statement about preferred reactivity from this figure, though it can be argued (based on slight differences in atomic charges, ca. 0.01*e*, or small differences in electrostatic potential) that the electrostatic potential shades ever-so-slightly towards the terminal carbon [243]. However, because electrophilic attack on a double bond is almost always associated with extensive charge donation to the nucleophile, this process is best described by the Fukui function. The near electrostatic equivalence of the two carbons helps explain why, when propylene encounters a hard nucleophile (like a metal in a high oxidation state), it can actually coordinate simultaneously through both centres, forming an η^2 complex.

21.7.1.3 TeF_5^- When to Ignore the Electrostatic Potential

Just because the electrostatic potential and the Fukui function predict reactivity at two different sites doesn't mean that both sites will actually be reactive. Consider the tellurium pentafluoride anion, TeF_5^-, a commonly cited exemplar of VSEPR theory that is used as a fluorine donor in organometallic catalysis [244, 245]. The Tellurium atom is formally in the +4 oxidation state, and a plot of the electrostatic potential on the molecular van der Waals surface (Fig. 21.3a) reveals that, indeed, the Tellurium atom is positively charged, while the Fluorine atoms are negatively charged. But there seems to be no experimental evidence of electrophiles attacking the Fluorine

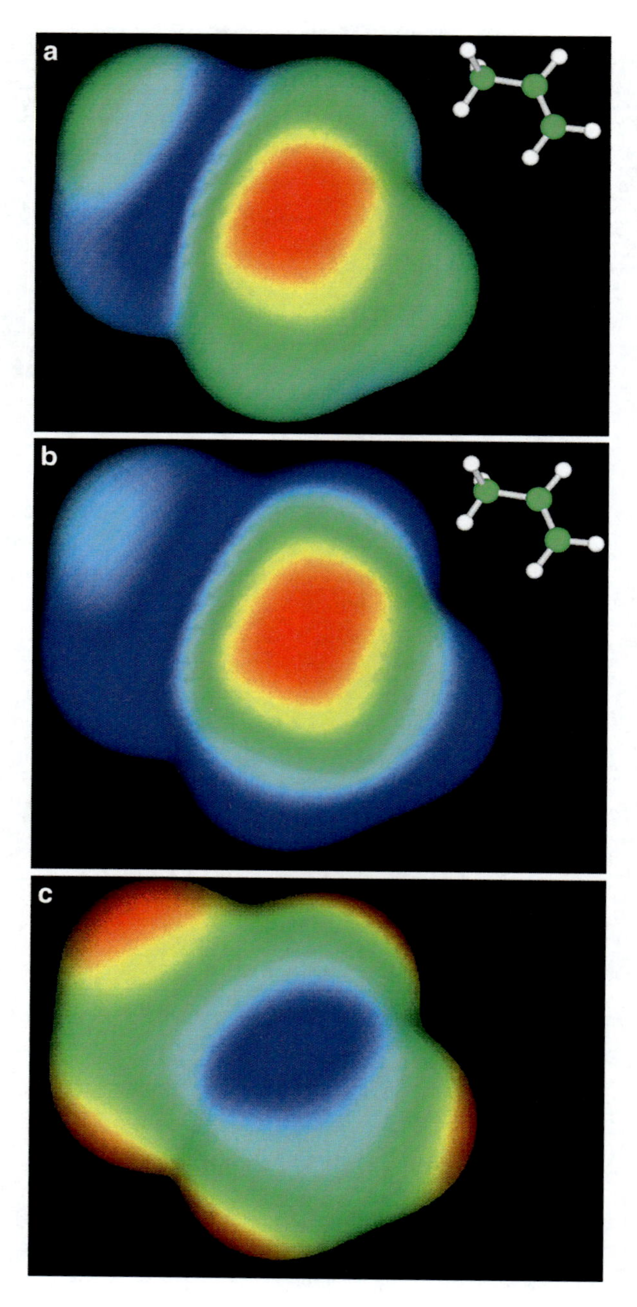

Fig. 21.2 The (**a**) Fukui function from below, $f^{-}(\boldsymbol{r})$, (**b**) the highest-occupied orbital density, $|\phi_{HOMO}(\boldsymbol{r})|^2$, and (**c**) the molecular electrostatic potential are plotted on the molecular van der Waals surface of propylene. Electrophilic attack is predicted to occur at the double bond. The largest (most positive) values of the functions are *red*; the smallest (most negative) values of the functions are *blue*

atoms in this ligand, or in the related teflate $OTeF_5^-$ ligand.[5] The Fukui function from below (Fig. 21.3b) indicates that Tellurium, despite its positive charge, is the best electron donor in the system. Electrophiles invariably attack TeF_5^- at the Tellurium atom. Because electron-rich compounds like this one are willing electron donors, their reactivity is usually controlled by the Fukui function, which governs electron donation. In this case, the fact that tellurium is at the bottom of the periodic table also means that the ligand is quite soft, enhancing charge transfer even further.

TeF_5^- is just one—albeit one of the most extreme—examples of a case where electrophilic attack occurs on a highly positively charged centre. Many other electron-rich compounds also react on a positively charged atom that possesses a lone pair (inert pair effect). The positive charge on the phosphorus atom in PF_3 makes this ligand a weak σ-donor, but its fascinating coordination chemistry nonetheless features binding through the positively charged phosphorous atom [246–249]. By contrast, because the NF_3 molecule is quite hard, and does not readily donate electrons, so it has a much more elaborate chemistry (including attack at the Fluorine atoms by cations, although attack at the nitrogen atom can still happen) [250, 251].

21.7.2 *Pericyclic Reactions and the Woodward-Hoffmann Rules*

For molecules that can both donate and accept electrons, the dual descriptor—which represents both the electron-donating and electron-accepting capability of the molecule at the same time, often provides a clearer description of molecular reactivity than the Fukui function alone. This is especially true when a molecule is simultaneously donating and accepting electrons, as in pericyclic reactions. In such cases, one needs to align the nucleophilic/basic regions of one molecule ($f^{(2)}(\mathbf{r}) > 0$) with the electrophilic/acidic regions of its partner ($f^{(2)}(\mathbf{r}) < 0$), and vice versa. This approach has proved effective for the full class of pericyclic reactions [52–55, 252, 253], and also for some other compounds that serve as both donors and acceptors of electrons [56].

A particular clear example is the 4 + 2 cycloaddition reaction of butadiene with ethene, which is rendered in Fig. 21.4 [54]. Notice that the electrophilic regions of ethene nicely align with the nucleophilic regions of butadiene; this is characteristic of an allowed reaction. Had we instead drawn the interactions between ethene,

---

[5]The Fluorines do not react because the Fluorine atoms hold their charge very tightly and, in TeF_5^-, they are nowhere nearly as negative as their formal -1 charge might lead one to anticipate. The Hirshfeld charges on the Fluorine atoms range from -26 to -35 and the Mulliken charges range from -46 to -53. The least charged Fluorine atom is the "exposed" atom at the top of the deformed square pyramid, which one might otherwise expect to be the most reactive.

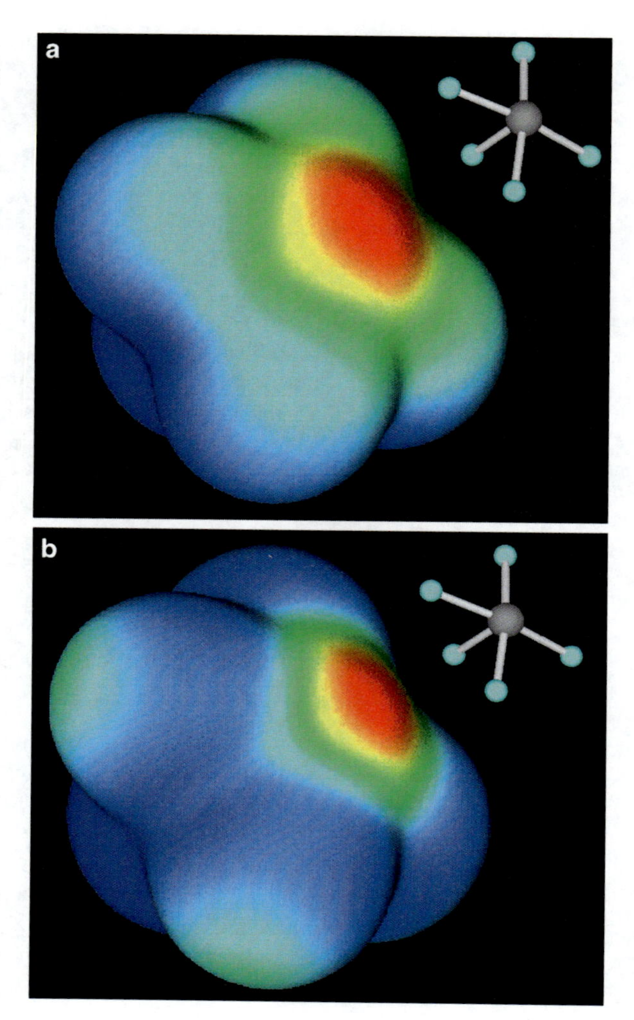

Fig. 21.3 The (**a**) molecular electrostatic potential, and (**b**) Fukui function from below are plotted on the molecular van der Waals surface of TeF_5^-. The largest (most positive) values of the functions are red; the smallest values of the functions are blue. Electrophilic attack occurs at the Te atom

or butadiene, with itself, it would be clear that suprafacial-suprafacial attack was unfavourable (not allowed) in those cases. DFT reactivity theory is often criticized, compared to molecular orbital theory, because it usual requires performing computations: it is not perceived as a "purely conceptual" tool. Figure 21.4, however, is simply the result of a back-of-the-envelope calculation. Indeed, in cases like this one, where the frontier orbitals can be reasonably approximated by Hückel-type calculations, adequate approximations to the Fukui function and the dual descriptor can also be obtained [54].

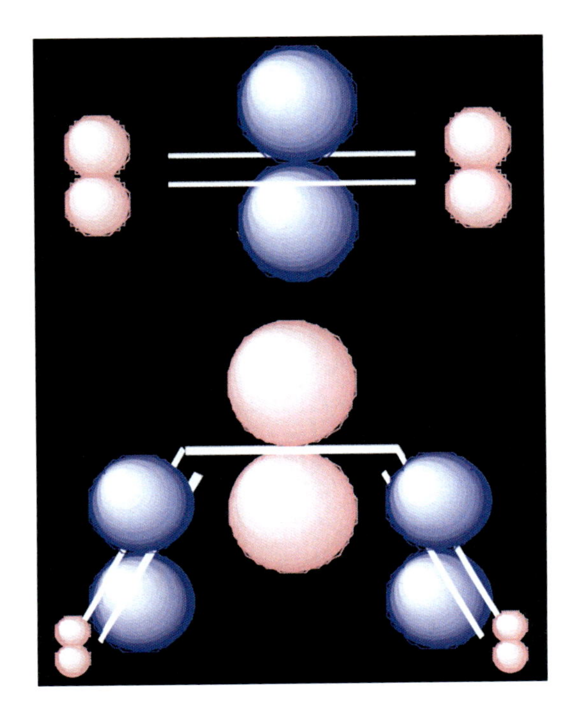

Fig. 21.4 The dual descriptors for ethene and butadiene, stacked in the reactive suprafacial-suprafacial arrangement. A "litmus" colour scheme is used, indicating how acidic regions (*pink*, $f^{(2)}(\mathbf{r}) > 0$) and basic regions (*blue*, $f^{(2)}(\mathbf{r}) < 0$) of the molecules match in a complimentary way, making the reaction favourable (Figure adapted from Ref. [54] with kind permission of Wiley VCH)

21.7.3 The Effects of Electron Transfer on the Bonds in H_2O: The Nuclear Fukui Function

As mentioned in Sect. 21.6.6, the nuclear Fukui functions measure the change in force on an atomic nucleus due to electron transfer from ($\boldsymbol{\phi}_\alpha^-$) or to ($\boldsymbol{\phi}_\alpha^+$) the molecule. The Fukui functions and the associated nuclear Fukui functions for the water molecule are plotted in Fig. 21.5a–d, respectively. Notice that even though the highest occupied molecular orbital in H_2O is usually labeled as a nonbonding orbital, removing an electron from water results in a force that tends to a stretching of the O-H bonds and an opening of the HOH angle, in line with the linearization of the molecule upon ionization. Similarly, adding an electron to H_2O induces forces that tend to stretch the bonds and decrease the angle because the lowest unoccupied molecular orbital has antibonding character [143, 181]. Like the reactivity indicators in the electron-preceding picture, the nuclear Fukui functions capture how changes in electron density induce changes in the nuclear positions (encapsulated in the external potential).

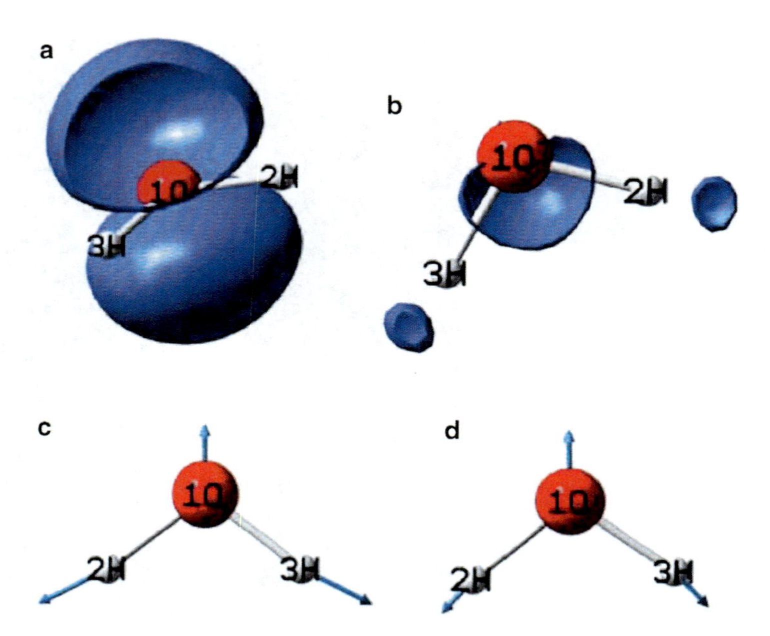

Fig. 21.5 Isosurfaces of (**a**) the Fukui function from below, $f^{-}(\boldsymbol{r})$, and (**b**) the Fukui function from above, $f^{+}(\boldsymbol{r})$, are plotted for the water molecule. The resulting nuclear Fukui functions, capturing the forces on the atomic nuclei associated with (**c**) removing electrons, $\boldsymbol{\phi}_{\alpha}^{-}$, or (**d**) adding electrons, $\boldsymbol{\phi}_{\alpha}^{+}$, are also rendered (Figure adapted from Ref. [143] with kind permission of The American Institute of Physics)

21.8 Misunderstandings, Criticisms, and Perspectives on Conceptual DFT

21.8.1 What Is the Purpose and Philosophy of Conceptual DFT?

In this section, we wish to address some of the most common criticisms and misunderstandings about conceptual DFT, and also to express our personal perspective about the discipline. In this first subsection, however, we wish to provide a broad overview of the goals, objectives, and philosophy of conceptual DFT.

Conceptual DFT is based on three fundamental precepts:

Observability: Our understanding of chemical observations should be based on quantum mechanical observables.

Universality: The tools we use to understand the results of a quantum mechanical calculation should not depend on the type of calculation that is performed.

Rigour: The tools we use to understand chemistry should fit into a well-defined mathematical framework. While practical calculations may not (and probably will not) exploit this framework, the existence of a systematic mathematical foundation for an interpretative technique is reassuring.

We do not claim that conceptual DFT is the only theory based on the preceding axioms, though it is probably the most popular such theory. Nor do we claim that every theory must be based on these axioms: many scientists prefer theories based on other considerations. In particular, many experimental and computational chemists favour reactivity indicators that are easy to obtain and use. While most of the fundamental reactivity indicators of conceptual DFT are not difficult to obtain and use, there are many chemical phenomena that are easier to explain with molecular orbital theory and/or the theory of resonance (valence bond theory). On the other hand, there are phenomena (strongly correlated systems where there are tens of thousands of significant Slater determinants, heavy-atom systems, etc.) that are much easier to explain with conceptual DFT.

Before briefly commenting on each of the foundational principles of conceptual DFT, we would like to recall a quote of Mulliken from 45 years ago [254],

> ... with old-fashioned chemical concepts, which at first seemed to have their counterparts in MQM [Molecular Quantum Mechanics], the more accurate the calculations became the more the concepts tended to vanish into thin air. So we have to ask, should we try to keep these concepts—do they still have a place—or should they be relegated to chemical history. Among such concepts are electronegativity ..., hybridization, population analysis, charges on atoms, even the idea of orbitals ....

This problem has been reiterated many times in the intervening years (see, for example, Ref. [255]), and the problem has arguably been exacerbated by the increasing power of computers: as we improve the treatment of electron correlation in a molecular-orbital theory calculation, the number of Slater determinants explodes, so that it is difficult to find any simple interpretation for the wavefunction. Similarly, as we increase the accuracy of a valence-bond calculation, the number of resonance structures (and especially "absurd" ionic structures) grows rapidly, and the structures that are commonly used in discussions of "electron pushing" and "hybridization" are often found to be quantitatively unimportant. Finally, accurately including the effects of relativity requires using four-component (or at least two-component) wavefunctions; such wavefunctions are difficult to interpret.

In our opinion, the only sure way to avoid concepts that "vanish into thin air" as the level of computation increases or the type of calculation changes is to base our understanding on observables. This is a very common viewpoint among those who use the electron density and related reactivity tools to understand chemistry, and its most eloquent and vociferous advocate is certainly Richard Bader, whose quantum theory of atoms in molecules also relies upon this perspective [256–258]. This is a very uncommon viewpoint among those who use molecular orbital theory, as molecular orbitals are unobservable, owing to the (very useful, but totally arbitrary) phase of molecular orbitals, which can even be chosen to take purely imaginary values. Conceptual DFT is based only on quantities that can, in principle, be observed: chiefly responses of the energy and electron density to changes in the number of electrons and external electrostatic potential.

Explaining chemical phenomena in terms of quantum mechanical observables forces rigorous thinking, grounded in reality, and we believe that it is useful for this reason. This pattern of reasoning is unfamiliar to most chemists and it is not always easy, even for these authors. Indeed, we sometimes begin by understanding a phenomenon with orbitals; then we use our orbital understanding to find observable quantities that capture key features. E.g., the inspiration for our approach to the Woodward-Hoffmann rules [54, 252] came from molecular orbital theory.

As Mulliken noted, because traditional molecular-orbital theory and valence-bond theory concepts are based on interpreting the wavefunction, they vanish into thin air when correlation is treated at a high level. The complexity of the wavefunction explodes exponentially as computed properties approach the exact answer. However, the nature of the computed properties (like the electron density and the other reactivity indicators of conceptual DFT) does not change; only the values adjust. The most accurate wavefunctions—like those from explicitly correlated methods, quantum Monte Carlo, or from including relativistic effects by the Dirac equation—seem entirely independent of any orbital model. It is difficult to imagine making a molecular-orbital theory study starting from a quantum Monte Carlo wavefunction. By contrast, the key reactivity indicators of conceptual DFT are readily accessed from any type of wavefunction, because the nature of the observables is independent of the complexity of the underlying wavefunction. While there has been some consideration of accurately correlated wavefunctions [112, 238, 259, 260] and multi-component relativistic treatments [261], most conceptual DFT calculations are based on independent particle models like Hartree-Fock or Kohn-Sham DFT. It is only in the rare cases where electron correlation and/or relativistic effects beyond scalar relativity are *qualitatively important* that going beyond simple and readily interpretable Slater determinant wavefunctions is critical. This helps explain, in our mind, the unquestioned utility of molecular orbital theory.

Perhaps the greatest advantage of conceptual DFT is that it provides a mathematically rigorous, formally exact, approach to understanding chemical changes. Such an understanding can be based, for example, on the perturbation expansion in Eq. 21.71 [13]. Each term in Eq. 21.71 has a well-defined interpretation, and there are no mathematical issues (though there are practical issues) associated with the expansion. Using Eq. 21.71, one can understand exactly which contributions to the physical interaction energy control chemical reactivity. There is a stark contrast with molecular-orbital theory and valence-bond theory, where concepts begin to "vanish" as higher-order corrections are made. It also allows us to treat electrostatics, polarization, and electron-transfer on equal footing, something that is rarely done (though it is possible) in orbital-based approaches.

To us, the conceptual DFT philosophy is useful and appealing; we prefer the firmly grounded reasoning that working with observables requires and we like knowing that we can refine our calculations without compromising the utility of our interpretative tools. Exploiting conceptual DFT requires a trade-off, however, because it is often easier to work with molecular orbitals or resonance structures.

21.8.2 Why Is It Called Conceptual DFT?

The authors are often asked why this approach to chemical reactivity is called "conceptual DFT." Obviously all of the key reactivity indicators in conceptual DFT can be evaluated (and often are) from wavefunction calculations. One of the key ideas in conceptual DFT is differentiation with respect to the number of electrons. This requires considering fractional numbers of electrons, and it is not, for example, conventional to write a wavefunction for a 10.032 electron system. Another key idea in conceptual DFT is the Legendre transform, and one does not usually think about Legendre transforms in wavefunction theory. In chemistry, the development of Legendre transforms and techniques for dealing with fractional electron number were driven by investigations of the mathematical structure of DFT, motivated by a desire to develop more accurate density functionals. A fringe benefit of these studies was the development of conceptual DFT.

Most of the key *concepts* in conceptual DFT originate in DFT itself, and not the extension of these principles to chemical reactivity. For this reason, many authors prefer other names for this field, like "chemical DFT," "density-functional chemical reactivity theory," or (most explicitly) "chemical reactivity from DFT." We do believe that it is useful to consider this subject as an area of inquiry within density-functional theory because of the key role played by the electron density and its response functions. Even the reactivity indicators that are, superficially, not electron-density-based are closely related to the electron density. For example, the chemical potential arises as the Lagrange multiplier used to enforce the normalization of the electron density in the variational principle.

21.8.3 What Is the Goal of Conceptual DFT?

Density-functional theorists are often disappointed by the "sloppy reasoning" that is common in conceptual DFT. It is true that the level of mathematical and physical rigour that characterizes conceptual DFT is usually much lower than the rest of density-functional theory. Sometimes this is inexcusable, but often it is a natural consequence of the fact conceptual DFT strives to find heuristic, qualitative, models for chemical phenomena. The mathematical foundation of conceptual DFT is firm and deep, but these mathematical foundations are rarely explicitly exploited because it is often impractical to do so. (The authors nonetheless find it reassuring that such a foundation exists.)

In addition, compared to molecular-orbital and valence-bond theorists, researchers in conceptual DFT have traditionally been more engaged in endeavours like quantitative structure-activity relationships (QSAR), quantitative structure-property relationships (QSPR), and molecular similarity. From the standpoint of most theoretical chemists, such studies abuse statistics and give scant chemical insight. From the standpoint of a scientist immersed in those fields, approaches based on conceptual DFT reactivity indicators are based on much firmer physical foundations than the usual models, which are frequently over-parameterized and rarely give any chemical insight at all.

The breadth of conceptual DFT runs from mathematical derivations that are more intricate and formal than most aspects of theoretical chemistry, to interpretative studies within the purview of the molecular-orbital and valence-bond approaches, to statistical correlations between reactivity indicators and molecular properties and reactivity. Many papers contain two or more of these aspects, and the link between the aspects is sometimes tenuous. Sometimes a formal derivation is used to motivate a practical approximation, which elucidates a phenomenon, which suggests an even less rigorous QSPR study. While we have focused mostly on the first two aspects in this review, readers must bear in mind that much of the development of conceptual DFT has been motivated by QSPR and QSAR.

21.8.4 Is Molecular-Orbital Theory Linked to Conceptual DFT?

As has been repeatedly stressed in the conceptual DFT literature, most of the predictions of molecular orbital theory emerge very easily from conceptual DFT. The frontier molecular orbital densities that emerge are equal to the Fukui functions in the limit where electron correlation is weak, orbital relaxation is negligible, and the speed of light is infinite [40]. Similarly, atom and group charges, commonly used in MO-based investigations, appear in conceptual DFT as condensed reactivity indicators. Ergo, the most commonly used reactivity indicators in molecular orbital theory appear as zeroth-order approximations to reactivity indicators in conceptual DFT. In fact, the effectiveness of molecular-orbital theory may be explained by noting that, when electron correlation, orbital relaxation, and non-scalar relativistic effects are neglected, the dominant contributions to the interaction energy expression in Eq. 21.71 involve the frontier MO densities and the electrostatic potential of the electron density.

Because conceptual DFT is based on observables, all information about the phase of the wavefunction is lost. In MO theory, if the phase of two orbitals can be chosen so that they constructively interfere, then this indicates a favourable interaction. (But one must choose the phases of the orbitals carefully: if one chooses the phases incorrectly, one can obtain an unfavourable destructive interference instead.) It is only very recently that conceptual DFT has been able to address phenomena where the MO-based explanation makes explicit use of the phases of the frontier orbitals [54, 252, 253], and this is still a relatively unexplored topic. Because Eq. 21.71 includes all of chemistry, such phenomena are certainly within the purview of conceptual DFT. However, it seems that these types of chemical interactions are, and perhaps always will be, easier to explain with MO theory. Conceptual DFT is probably most useful for phenomena that feature net electron transfer between reagents; electron rearrangements without electron transfer are not impossible to explain, but they are more difficult.

21.8.5 *Can Conceptual DFT Be Used for "Back of the Envelope" Calculations?*

Usually conceptual DFT is used as a post-computational tool for expressing results in chemical language. It can be used in a back-of-the-envelope fashion, however.

Back-of-the-envelope electronegative equalization can be used to predict atomic populations. This is much easier in the simplest case where the hardness kernel is diagonal, or at least tridiagonal, but the general problem of solving a linear equation for a $N_{\text{atoms}} \times N_{\text{atoms}}$ matrix is easier than the matrix eigenvalue problem of the same size that one needs to solve in a Hückel-type calculation.

If one does perform a simple Hückel or extended-Hückel calculation, then the Fukui functions can be estimated directly from the frontier MOs, the hardness and electronegativity can be estimated directly from the frontier molecular orbital energies, and the leading term in the expression of the polarizability kernel can be estimated from the frontier orbitals and their energies. Other reactivity indicators of conceptual DFT can be similarly approximated. So any "back of the envelope" molecular orbital theory calculation contains the information needed for a simple conceptual DFT analysis. The authors often do such calculations, but rarely publish them (but see Ref. [54]). Instead, we prefer to do full calculations, with orbital relaxation and electron correlation included (at least approximately). Even though these effects are very rarely qualitatively important, by allowing these effects to contribute, conceptual DFT reactivity indicators provide a check on the reliability of simple MO-based models.

The charge sensitivity analysis of Roman Nalewajski is another conceptual DFT tool that is amenable to hand-calculation [118, 180, 262]. Nalewajski's communication theory of the chemical bond is too [213, 215].

21.8.6 *Is Conceptual DFT a Predictive Theory?*

Most conceptual DFT calculations are interpretative, not predictive. Indeed, prediction is extremely challenging for any theory. Within a family of molecules, insights gleaned from some members of the family can be used to predict properties of other molecules in the family. This sort of study is not uncommon in conceptual approaches to chemistry, including MO-based, VB-based, and DFT-based studies. Such results are relatively unchallenging, requiring merely a well-chosen reactivity indicator and a carefully chosen family of molecules.

Predicting the values of properties from reactivity indicators is much more challenging, but the applications of conceptual DFT in QSAR and QSPR suggests that the reactivity and properties of molecules can be predicted with reasonable accuracy.

The most demanding test of a theory is whether it can explain, and ideally predict, entirely new chemical phenomenon. The first effect that was derived from conceptual DFT is the maximum hardness principle [31–33]. The maximum hardness principle has proved broadly useful for understanding the relative stability of different molecules and different conformations of the same molecule. Another success of conceptual DFT is the elucidation of redox induced electron transfer (RIET), where oxidation (reduction) of a molecule is coupled to reduction (oxidation) of a fragment within the molecule [263]. This phenomenon is very easy to explain in conceptual DFT: it happens whenever the Fukui function is negative on a portion of the molecule. Mathematical analysis using conceptual DFT allows one to predict what characteristics a molecule must have to exhibit RIET [112, 263, 264]. By contrast, an explanation based on MOs is relatively complicated because oxidizing a molecule should decrease all of the atomic populations in proportion to the highest occupied orbital density, $|\phi_{\mathrm{HOMO}}|^2$, on the atoms. RIET is an orbital relaxation effect [263], and such effects seem easier to explain using conceptual DFT. Importantly, the conceptual DFT treatment of RIET preceded the experimental pursuit of this phenomenon [265, 266].

21.8.7 Why Not Consider Both Reagents?

In the MO-based approach to describing the reactivity between two molecules, one usually examines both reagents at the same time and searches for a favourable matching between the energies, shapes, and phases of their frontier orbitals. By contrast, conceptual DFT usually does not consider the attacking reagent explicitly. This works because the detailed character of the attacking reagent rarely matters: a rough description often suffices. For example, most nitrogen-containing bases are subject to electrophilic attack on the nitrogen, regardless of the identity of the electrophile. Most boron-containing acids are subject to nucleophilic attack on the boron, regardless of the identity of the nucleophile. With almost any Brønsted-Lowry acid, ammonia is more reactive than pyridine, which is more reactive than pyrrole. Soft electrophiles (nucleophiles) almost always react where the appropriate Fukui function is large; hard electrophiles (nucleophiles) almost always react in places where the electrostatic potential is highly positive (negative).

As exemplified in Sect. 21.7, in conceptual DFT we use the properties of reactive molecule and the attacking reagent to determine which terms in the interaction energy expression (Eq. 21.71) are likely to dominate; this leads to an appropriate choice for a reactivity indicator. When it is not clear how to choose the best chemical reactivity indicator, one uses several reactivity indicators to tease out the subtleties of the chemical reactivity profile. Alternatively, in troublesome cases, it is becoming increasingly common to use the same types of two-reactant approaches that are common in molecular-orbital theory. So far, most of the two-reagent work in conceptual DFT uses the dual descriptor [51, 54, 55, 96].

In conceptual DFT, the two-reagent picture was pioneered by Berkowitz, who explained how the overlap between Fukui functions is analogous to the overlap between the HOMO and the LUMO in MO theory and, moreover, revealed how the magnitude of this interaction is modulated by the chemical hardness [107]. This model was extended by Anderson et al., who started with a formal expression for the interaction energy between reagents and showed how the balance between electrostatic interactions and electron-transfer effects is modulated by the hardness, the charge, and condensed Fukui function of the attacking reagent's active site. They then used their model to explain the reactivity preferences of a family of molecules that is not easily be described using frontier MO theory [89, 90, 267].

At second order, the expression for the interaction energy between two reagents is

$$\Delta U_{AB}[\Delta N] = \left(T_s^{\text{non}-\text{add}}\left[\rho_A^0, \rho_B^0\right] + E_{xc}^{\text{non}-\text{add}}\left[\rho_A^0, \rho_B^0\right]\right) + \left[\begin{array}{l} \iint \frac{\left(\rho_{A,\text{nuc}}(\mathbf{r}) + \rho_A^0(\mathbf{r})\right)\left(\rho_{B,\text{nuc}}(\mathbf{r}') + \rho_B^0(\mathbf{r}')\right)}{|\mathbf{r}-\mathbf{r}'|} d\mathbf{r} d\mathbf{r}' \\ +2\left(E_A^{(\text{pol})}\left[\Phi_B^0\right] + E_B^{(\text{pol})}\left[\Phi_A^0\right]\right) \\ -\Delta N_A \int \left(f_A(\mathbf{r})\Phi_B^0(\mathbf{r}) - f_B(\mathbf{r})\Phi_A^0(\mathbf{r})\right) d\mathbf{r} \\ -(\Delta N_A)^2 \iint \frac{f_A(\mathbf{r}) f_B(\mathbf{r}')}{|\mathbf{r}-\mathbf{r}'|} d\mathbf{r} d\mathbf{r}' \end{array}\right] + \left\{\begin{array}{l} \Delta N_A\left(\mu_A^0 - \mu_B^0\right) + \frac{(\Delta N_A)^2}{2}\left(\eta_A^0 + \eta_B^0\right) \\ -E_A^{(\text{pol})}\left[\Phi_B^0\right] - E_B^{(\text{pol})}\left[\Phi_A^0\right] \end{array}\right\} \tag{21.74}$$

This expression is highly reminiscent of the expressions that occur in density-functional embedding theory [14, 18–20]. The first group of terms captures primarily steric effects, and the last group of terms captures the effects of charge transfer and the energy cost of polarizing (ergo deforming) the fragment densities away from their ground electronic state. The middle terms are electrostatic, including the electrostatic interaction energy between the isolated fragments and the changes in electrostatic interaction energy due to polarization of the fragments' electron density and electron-transfer between the fragments [268].

Though it is still uncommon, sometimes three or even four reagents are treated simultaneously in conceptual DFT. Such considerations, for example, are very helpful for elucidating the HSAB principle [23, 28, 29].

21.8.8 When Should One Use Conceptual DFT? When Is Caution Warranted?

To summarize, conceptual DFT is especially useful for describing chemical reactions in which orbital relaxation, electron correlation, or relativistic effects are believed to be important. More generally, conceptual DFT is very useful when the reaction may be broadly characterized as being of acid-base type.

While conceptual DFT, just like molecular-orbital and valence-bond theory, can be used to provide a complete picture of chemical reactivity, there are cases where conventional chemical tools are perhaps preferable. For example, when the molecular-orbital elucidation of a chemical reaction mechanism relies upon the phase of the frontier molecular orbitals, the conceptual DFT approach often requires third-order reactivity indicators, and is thus relatively complicated. Conceptual DFT for photochemical reactions is still a developing field [269, 270], but the molecular-orbital description of such reactions is fully mature. When performing "back of the envelope calculations," conceptual DFT is highly effective for atomic populations (using electronegativity equalization). However, for back-of-the envelope calculations, most of the DFT reactivity indicators contain no more information than their molecular orbital theory counterparts, and so it is more direct to couch one's arguments in molecular orbital theory.

In both molecular orbital theory and valence bond theory, the same mathematical structure can be used to elucidate both molecular electronic structure and chemical reactivity. Conceptual DFT, on the other hand, is traditionally a theory only of chemical reactivity. There are density-based approaches to molecular electronic structure (e.g., Bader's quantum theory of atoms in molecules [119] and Nalewajski's communication theory of the chemical bond [213, 215]), but the conceptual tools used by these approaches do not seem to fit in the same mathematical framework as the chemical reactivity indicators. At present, conceptual DFT is a theory of chemical reactivity, while molecular-orbital theory and valence-bond theory can be considered true theories of molecules.

21.9 Conclusion

The electron density and its response to perturbations encapsulate all of the information that is needed to understand and predict chemical reactivity; the mathematical formalism of density-functional theory can be used to extract this information. Just as the change in thermodynamic state functions can be evaluated by differentiating with respect to the variables that define the state of the system, so also the molecular changes associated with chemical reactivity can be evaluated by differentiating with respect to the variables that define the molecular structure. In thermodynamics one may choose to define the system in different ways, based on the experimental setup; the same is true in the DFT-based approach to chemical reactivity. To a quantum chemist, the most useful of these representations is probably the closed-system picture (Sect. 21.2), because quantum chemists specify their system based on the external potential (molecular geometry) and the number of electrons. To a condensed-matter theorist, the most useful representation is probably the open-system picture (Sect. 21.3), because condensed-matter theorists typically specify their system based on the external potential and the chemical potential (Fermi level). To an X-ray spectroscopist, who can measure the electron density and determines the nuclear positions from the peaks in the electron density, the electron-preceding picture may be the most useful (Sect. 21.4). All three pictures are equivalent; the choice of representation is coloured by convenience and personal taste. Indeed, one often uses "derived" reactivity indicators that do not belong to a specific "picture" of chemical reactivity theory (Sect. 21.5). A few "complications"—fine points and extensions of the basic theory—were covered in Sect. 21.6. One of these—the nuclear reactivity indicators—provide information about the changes in molecular geometry upon electron transfer to or from the system. Other extensions allow one to study the excited-state reactivity and spin-transfer processes. Condensed reactivity indicators provide a convenient coarse-grained description of chemical reactivity that contains the chemists' concept of partial charges, but extends it to quantities that measure the susceptibility of atoms in a molecule to electron transfer, polarization, and other perturbations.

Although the mathematical formalism in the DFT-based chemical reactivity theory is sometimes very complicated, applying the theory is usually easy. There are simple, and effective, "working formulas" for most of the key reactivity indicators. By using these working formulae, we can represent molecular reactivity in a simple pictorial way. A few representative examples were presented in Sect. 21.7.

The strengths of the DFT-based approach to chemical reactivity are its theoretical rigour, its generality, and its reliance on experimentally observable quantities, the electron density itself being a prominent example. Unlike other conceptual approaches to chemical reactivity, the complete family of DFT reactivity indicators encapsulates all of the information about the chemical reactivity of a system. Also unlike these approaches, the DFT reactivity indicators can be computed from any type of quantum chemistry method, even computational methods like quantum Monte Carlo or massive configuration interaction, for which orbital-based interpretations are very challenging. The DFT reactivity indicators are also experimentally accessible, at least in principle. These factors, combined with the immense practical utility of the DFT-based approach, continue to drive the progress of the field.

Acknowledgments Paul Johnson thanks the governments of Ontario and Canada for OGS-M and CGS-M graduate fellowships. Paul Ayers acknowledges research support for NSERC, Sharcnet, and the Canada Research Chairs. Paul Geerlings acknowledges the Fund for Scientific Reasearch-Flanders (FWO) and the Free University of Brussels (VUB) for continuous support to his group. He thanks all present and past group members for many years of intellectually stimulating work in the area of Conceptual DFT. Tim Fievez thanks the FWO-Flanders for a Predoctoral Fellowhip as Aspirant. Lee Bartolotti acknowledges financial support from RENCI@ECU.

References

1. Coppens P (2005) Charge densities come of age. Angew Chem Int Ed 44:6810–6811
2. Koritsanszky TS, Coppens P (2001) Chemical applications of X-ray charge-density analysis. Chem Rev 101:1583–1627

3. Parr RG, Yang W (1989) Density-functional theory of atoms and molecules. Oxford University Press, New York
4. Geerlings P, De Proft F, Langenaeker W (2003) Conceptual density functional theory. Chem Rev 103:1793–1873
5. Parr RG, Donnelly RA, Levy M, Palke WE (1978) Electronegativity: the density functional viewpoint. J Chem Phys 68:3801–3807
6. Nalewajski RF, Parr RG (1982) Legendre transforms and Maxwell relations in density functional theory. J Chem Phys 77:399–407
7. Berkowitz M, Parr RG (1988) Molecular hardness and softness, local hardness and softness, hardness and softness kernels, and relations among these quantities. J Chem Phys 88: 2554–2557
8. Fermi E, Amaldi E (1934) Le orbite oos degli elementi. Accad Ital Rome 6:117–149
9. Fermi E (1928) A statistical method for the determination of some atomic properties and the application of this method to the theory of the periodic system of elements. Z Phys 48:73–79
10. Dirac PAM (1930) Note on exchange phenomena in the Thomas atom. Proc Camb Philos Soc 26:376–385
11. Thomas LH (1927) The calculation of atomic fields. Proc Camb Philos Soc 23:542–548
12. Senet P (1996) Nonlinear electronic responses, Fukui functions and hardnesses as functionals of the ground-state electronic density. J Chem Phys 105:6471–6489
13. Ayers PW, Anderson JSM, Bartolotti LJ (2005) Perturbative perspectives on the chemical reaction prediction problem. Int J Quantum Chem 101:520–534
14. Cortona P (1991) Self-consistently determined properties of solids without band-structure calculations. Phys Rev B 44:8454–8458
15. Ayers PW, Parr RG (2001) Variational principles for describing chemical reactions. Reactivity indices based on the external potential. J Am Chem Soc 123:2007–2017
16. Ayers PW (2000) Atoms in molecules, an axiomatic approach. I. Maximum transferability. J Chem Phys 113:10886–10898
17. Wesolowski TA (2004) Quantum chemistry 'without orbitals' – an old idea and recent developments. Chimia 58:311–315
18. Wesolowski TA, Warshel A (1994) Ab-initio free-energy perturbation calculations of solvation free-energy using the frozen density-functional approach. J Phys Chem 98:5183–5187
19. Wesolowski TA, Warshel A (1993) Frozen density-functional approach for Ab-initio calculations of solvated molecules. J Phys Chem 97:8050–8053
20. Vaidehi N, Wesolowski TA, Warshel A (1992) Quantum-mechanical calculations of solvation free-energies – a combined ab initio pseudopotential free-energy perturbation approach. J Chem Phys 97:4264–4271
21. Parr RG, Bartolotti LJ (1982) On the geometric mean principle for electronegativity equalization. J Am Chem Soc 104:3801–3803
22. Mulliken RS (1934) A new electroaffinity scale: together with data on states and an ionization potential and electron affinities. J Chem Phys 2:782–793
23. Ayers PW, Parr RG, Pearson RG (2006) Elucidating the hard/soft acid/base principle: a perspective based on half-reactions. J Chem Phys 124:194107
24. Parr RG, Pearson RG (1983) Absolute hardness: companion parameter to absolute electronegativity. J Am Chem Soc 105:7512–7516
25. Chattaraj PK, Lee H, Parr RG (1991) HSAB principle. J Am Chem Soc 113:1855–1856
26. Gazquez JL, Mendez F (1994) The hard and soft acids and bases principle: an atoms in molecules viewpoint. J Phys Chem 98:4591–4593
27. Mendez F, Gazquez JL (1994) Chemical-reactivity of enolate ions – the local hard and soft acids and bases principle viewpoint. J Am Chem Soc 116:9298–9301
28. Ayers PW (2005) An elementary derivation of the hard/soft-acid/base principle. J Chem Phys 122:141102
29. Ayers PW (2007) The physical basis of the hard/soft acid/base principle. Faraday Discuss 135:161–190

30. Nalewajski RF (1984) Electrostatic effects in interactions between hard (soft) acids and bases. J Am Chem Soc 106:944–945
31. Pearson RG (1987) Recent advances in the concept of hard and soft acids and bases. J Chem Educ 64:561–567
32. Pearson RG (1999) Maximum chemical and physical hardness. J Chem Educ 76:267–275
33. Parr RG, Chattaraj PK (1991) Principle of maximum hardness. J Am Chem Soc 113: 1854–1855
34. Ayers PW, Parr RG (2000) Variational principles for describing chemical reactions: the Fukui function and chemical hardness revisited. J Am Chem Soc 122:2010–2018
35. Pearson RG, Palke WE (1992) Support for a principle of maximum hardness. J Phys Chem 96:3283–3285
36. Zhou Z, Parr RG (1989) New measures of aromaticity: absolute hardness and relative hardness. J Am Chem Soc 111:7371–7379
37. Torrent-Sucarrat M, Luis JM, Duran M, Sola M (2004) The hardness profile as a tool to detect spurious stationary points in the potential energy surface. J Chem Phys 120:10914–10924
38. Torrent-Sucarrat M, Luis JM, Duran M, Sola M (2002) Are the maximum hardness and minimum polarizability principles always obeyed in nontotally symmetric vibrations? J Chem Phys 117:10561–10570
39. Parr RG, Yang WT (1984) Density functional approach to the frontier-electron theory of chemical reactivity. J Am Chem Soc 106:4049–4050
40. Yang WT, Parr RG, Pucci R (1984) Electron density, Kohn-Sham frontier orbitals, and Fukui functions. J Chem Phys 81:2862–2863
41. Ayers PW, Levy M (2000) Perspective on "density functional approach to the frontier-electron theory of chemical reactivity" by Parr RG, Yang W (1984). Theor Chem Acc 103:353–360
42. Ayers PW, Yang WT, Bartolotti LJ (2009) Fukui function. In: Chattaraj PK (ed) Chemical reactivity theory: a density functional view. CRC Press, Boca Raton, pp 255–267
43. Langenaeker W, Demel K, Geerlings P (1991) Quantum-chemical study of the Fukui function as a reactivity index. 2. Electrophilic substitution on mono-substituted benzenes. J Mol Struct THEOCHEM 80:329–342
44. Cohen MH, Ganduglia-Pirovano MV (1994) Electronic and nuclear chemical reactivity. J Chem Phys 101:8988–8997
45. Bartolotti LJ, Ayers PW (2005) An example where orbital relaxation is an important contribution to the Fukui function. J Phys Chem A 109:1146–1151
46. Fuentealba P, Parr RG (1991) Higher-order derivatives in density-functional theory, especially the hardness derivative. J Chem Phys 94:5559–5564
47. Geerlings P, Proft FD (2008) Conceptual DFT: the chemical relevance of higher response functions. Phys Chem Chem Phys 10:3028–3042
48. Cardenas C, Echegaray E, Chakraborty D, Anderson JSM, Ayers PW (2009) Relationships between third-order reactivity indicators in chemical density-functional theory. J Chem Phys 130:244105
49. Ayers PW, Parr RG (2008) Beyond electronegativity and local hardness: higher-order equalization criteria for determination of a ground-state electron density. J Chem Phys 129:054111
50. Morell C, Grand A, Toro-Labbé A (2006) Theoretical support for using the delta f(r) descriptor. Chem Phys Lett 425:342–346
51. Morell C, Grand A, Toro-Labbé A (2005) New dual descriptor for chemical reactivity. J Phys Chem A 109:205–212
52. Sablon N, de Proft F, Geerlings P (2009) Reformulating the Woodward-Hoffmann rules in a conceptual density functional theory context: the case of sigmatropic reactions. Croat Chem Acta 82:157–164
53. De Proft F, Chattaraj PK, Ayers PW, Torrent-Sucarrat M, Elango M, Subramanian V, Giri S, Geerlings P (2008) Initial hardness response and hardness profiles in the study of Woodward-Hoffmann rules for electrocyclizations. J Chem Theory Comput 4:595–602

54. Ayers PW, Morell C, De Proft F, Geerlings P (2007) Understanding the Woodward-Hoffmann rules using changes in the electron density. Chem Eur J 13:8240–8247
55. Morell C, Ayers PW, Grand A, Gutierrez-Oliva S, Toro-Labbé A (2008) Rationalization of Diels-Alder reactions through the use of the dual reactivity descriptor delta f(r). Phys Chem Chem Phys 10:7239–7246
56. Cardenas C, Rabi N, Ayers PW, Morell C, Jaramillo P, Fuentealba P (2009) Chemical reactivity descriptors for ambiphilic reagents: dual descriptor, local hypersoftness, and electrostatic potential. J Phys Chem A113:8660–8667
57. Hohenberg P, Kohn W (1964) Inhomogeneous electron gas. Phys Rev B 136:864–871
58. Yang WT, Parr RG (1985) Hardness, softness, and the Fukui function in the electron theory of metals and catalysis. Proc Natl Acad Sci 82:6723–6726
59. De Proft F, Geerlings P, Liu S, Parr RG (1998) Variational calculation of the global hardness and the Fukui function via an approximation of the hardness kernel. Pol J Chem 72: 1737–1746
60. Fuentealba P (1998) Reactivity indices and response functions in density functional theory. J Mol Struct THEOCHEM 433:113–118
61. Simon-Manso Y, Fuentealba P (1998) On the density functional relationship between static dipole polarizability and global softness. J Phys Chem A 102:2029–2032
62. Nakatsuji H (1974) Common nature of the electron cloud of a system undergoing change in nuclear configuration. J Am Chem Soc 96:24–30
63. Nakatsuji H (1974) Electron-cloud following and preceding and the shapes of molecules. J Am Chem Soc 96:30–37
64. Parr RG, Bartolotti LJ (1983) Some remarks on the density functional theory of few-electron systems. J Phys Chem 87:2810–2815
65. Gal T (2001) Differentiation of density functionals that conserves the normalization of the density. Phys Rev A 63:049903
66. Gal T (2007) The mathematics of functional differentiation under conservation constraint. J Math Chem 42:661–676
67. Gal T (2002) Functional differentiation under conservation constraints. J Phys A 35: 5899–5905
68. Lieb EH (1983) Density functionals for Coulomb systems. Int J Quantum Chem 24:243–277
69. Levy M, Perdew JP (1985) The constrained search formulation of density functional theory. NATO ASI Ser B 123:11–30
70. Chattaraj PK, Cedillo A, Parr RG (1995) Variational method for determining the Fukui function and chemical hardness of an electronic system. J Chem Phys 103:7645–7646
71. Levy M, Parr RG (1976) Long-range behavior of natural orbitals and electron density. J Chem Phys 64:2707–2708
72. Morrell MM, Parr RG, Levy M (1975) Calculation of ionization potentials from density matrixes and natural functions, and the long-range behavior of natural orbitals and electron density. J Chem Phys 62:549–554
73. Katriel J, Davidson ER (1980) Asymptotic behavior of atomic and molecular wave functions. Proc Natl Acad Sci 77:4403–4406
74. Ahlrichs R, Hoffmann-Ostenhof M, Hoffmann-Ostenhof T, Morgan JD III (1981) Bounds on the decay of electron densities with screening. Phys Rev A 23:2106–2117
75. Hoffmann-Ostenhof M, Hoffmann-Ostenhof T (1977) "Schrodinger inequalities" and asymptotic behavior of the electron density of atoms and molecules. Phys Rev A 16:1782–1785
76. Baekelandt BG, Cedillo A, Parr RG (1995) Reactivity indexes and fluctuation formulas in density- functional theory – isomorphic ensembles and a new measure of local hardness. J Chem Phys 103:8548–8556
77. Freed KF, Levy M (1982) Direct 1st principles algorithm for the universal electron- density functional. J Chem Phys 77:396–398
78. Zhao Q, Parr RG (1993) Constrained-search method to determine electronic wave functions from electronic densities. J Chem Phys 98:543–548

79. Zhao Q, Morrison RC, Parr RG (1994) From electron-densities to Kohn-Sham kinetic energies, orbital energies, exchange-correlation potentials, and exchange- correlation energies. Phys Rev A 50:2138–2142
80. van Leeuwen R, Baerends EJ (1994) Exchange-correlation potential with correct asymptotic-behavior. Phys Rev A 49:2421–2431
81. Wu Q, Yang WT (2003) A direct optimization method for calculating density functionals and exchange-correlation potentials from electron densities. J Chem Phys 118:2498–2509
82. Colonna F, Savin A (1999) Correlation energies for some two- and four-electron systems along the adiabatic connection in density functional theory. J Chem Phys 110:2828–2835
83. Senet P (1997) Kohn-Sham orbital formulation of the chemical electronic responses, including the hardness. J Chem Phys 107:2516–2524
84. Ayers PW (2001) Strategies for computing chemical reactivity indices. Theor Chem Acc 106:271–279
85. Parr RG, Von Szentpaly L, Liu SB (1999) Electrophilicity index. J Am Chem Soc 121: 1922–1924
86. Chattaraj PK, Sarkar U, Roy DR (2006) Electrophilicity index. Chem Rev 106:2065–2091
87. Chattaraj PK, Maiti B, Sarkar U (2003) Philicity: a unified treatment of chemical reactivity and selectivity. J Phys Chem A 107:4973–4975
88. Ayers PW, Anderson JSM, Rodriguez JI, Jawed Z (2005) Indices for predicting the quality of leaving groups. Phys Chem Chem Phys 7:1918–1925
89. Anderson JSM, Melin J, Ayers PW (2007) Conceptual density-functional theory for general chemical reactions, including those that are neither charge nor frontier-orbital controlled. I. Theory and derivation of a general-purpose reactivity indicator. J Chem Theory Comput 3:358–374
90. Anderson JSM, Melin J, Ayers PW (2007) Conceptual density-functional theory for general chemical reactions, including those that are neither charge- nor frontier-orbital-controlled. 2. Application to molecules where frontier molecular orbital theory fails. J Chem Theory Comput 3:375–389
91. Cedillo A, Contreras R, Galvan M, Aizman A, Andres J, Safont VS (2007) Nucleophilicity index from perturbed electrostatic potentials. J Phys Chem A 111:2442–2447
92. Roy RK, Krishnamurti S, Geerlings P, Pal S (1998) Local softness and hardness based reactivity descriptors for predicting intra- and intermolecular reactivity sequences: carbonyl compounds. J Phys Chem A 102:3746–3755
93. Campodonico PR, Aizman A, Contreras R (2006) Group electrophilicity as a model of nucleofugality in nucleophilic substitution reactions. Chem Phys Lett 422:340–344
94. Campodonico PR, Andres J, Aizman A, Contreras R (2007) Nucleofugality index in alpha-elimination reactions. Chem Phys Lett 439:177–182
95. Campodonico PR, Perez C, Aliaga M, Gazitua M, Contreras R (2007) Electrofugality index for benhydryl derivatives. Chem Phys Lett 447:375–378
96. Padmanabhan J, Parthasarathi R, Elango M, Subramanian V, Krishnamoorthy BS, Gutierrez-Oliva S, Toro-Labbé A, Roy DR, Chattaraj PK (2007) Multiphilic descriptor for chemical reactivity and selectivity. J Phys Chem A 111:9130–9138
97. Padmanabhan J, Parthasarathi R, Subramanian V, Chattaraj PK (2006) Chemical reactivity indices for the complete series of chlorinated benzenes: solvent effect. J Phys Chem A 110:2739–2745
98. Chattaraj PK, Roy DR, Geerlings P, Torrent-Sucarrat M (2007) Local hardness: a critical account. Theor Chem Acc 118:923–930
99. Torrent-Sucarrat M, De Proft F, Geerlings P, Ayers PW (2008) Do the local softness and hardness indicate the softest and hardest regions of a molecule? Chem Eur J 14:8652–8660
100. Ayers PW, Parr RG (2008) Local hardness equalization: exploiting the ambiguity. J Chem Phys 128:184108
101. Ghosh SK, Berkowitz M, Parr RG (1984) Transcription of ground-state density-functional theory into a local thermodynamics. Proc Natl Acad Sci 81:8028–8031

102. Berkowitz M, Ghosh SK, Parr RG (1985) On the concept of local hardness in chemistry. J Am Chem Soc 107:6811–6814
103. Bagaria P, Saha S, Murru S, Kavala V, Patel BK, Roy RK (2009) A comprehensive decomposition analysis of stabilization energy (CDASE) and its application in locating the rate-determining step of multi-step reactions. Phys Chem Chem Phys 11:8306–8315
104. Gazquez JL, Cedillo A, Vela A (2007) Electrodonating and electroaccepting powers. J Phys Chem A 111:1966–1970
105. Politzer P, Truhlar D (1981) Chemical applications of atomic and molecular electrostatic potentials. Plenum, New York
106. Gadre SR, Kulkarni SA, Shrivastava IH (1992) Molecular electrostatic potentials – a topographical study. J Chem Phys 96:5253–5260
107. Berkowitz M (1987) Density functional-approach to frontier controlled reactions. J Am Chem Soc 109:4823–4825
108. Ayers PW, Liu SB, Li TL (2009) Chargephilicity and chargephobicity: two new reactivity indicators for external potential changes from density functional reactivity theory. Chem Phys Lett 480:318–321
109. Yang WT, Mortier WJ (1986) The use of global and local molecular parameters for the analysis of the gas-phase basicity of amines. J Am Chem Soc 108:5708–5711
110. Fuentealba P, Perez P, Contreras R (2000) On the condensed Fukui function. J Chem Phys 113:2544–2551
111. De Proft F, Van Alsenoy C, Peeters A, Langenaeker W, Geerlings P (2002) Atomic charges, dipole moments, and Fukui functions using the Hirshfeld partitioning of the electron density. J Comput Chem 23:1198–1209
112. Ayers PW, Morrison RC, Roy RK (2002) Variational principles for describing chemical reactions: condensed reactivity indices. J Chem Phys 116:8731–8744
113. Chamorro E, Perez P (2005) Condensed-to-atoms electronic Fukui functions within the framework of spin-polarized density-functional theory. J Chem Phys 123:114107
114. Tiznado W, Chamorro E, Contreras R, Fuentealba P (2005) Comparison among four different ways to condense the Fukui function. J Phys Chem A 109:3220–3224
115. Sablon N, De Proft F, Ayers PW, Geerlings P (2007) Computing Fukui functions without differentiating with respect to electron number. II. Calculation of condensed molecular Fukui functions. J Chem Phys 126:224108
116. Bultinck P, Fias S, Alsenoy CV, Ayers PW, Carbó-Dorca R (2007) Critical thoughts on computing atom condensed Fukui functions. J Chem Phys 127:034102
117. Olah J, Van Alsenoy C, Sannigrahi AB (2002) Condensed Fukui functions derived from stockholder charges: assessment of their performance as local reactivity descriptors. J Phys Chem A 106:3885–3890
118. Nalewajski RF (1995) Chemical reactivity concepts in charge sensitivity analysis. Int J Quantum Chem 56:453–476
119. Bader RFW (1990) Atoms in molecules: a quantum theory. Clarendon, Oxford
120. Popelier PLA (2000) Atoms in molecules: an introduction. Pearson, Harlow
121. Bulat FA, Chamorro E, Fuentealba P, Toro-Labbé A (2004) Condensation of frontier molecular orbital Fukui functions. J Phys Chem A 108:342–349
122. Hirshfeld FL (1977) Bonded-atom fragments for describing molecular charge densities. Theor Chim Act 44:129–138
123. Nalewajski RF, Parr RG (2000) Information theory, atoms in molecules, and molecular similarity. Proc Natl Acad Sci 97:8879–8882
124. Parr RG, Ayers PW, Nalewajski RF (2005) What is an atom in a molecule? J Phys Chem A 109:3957–3959
125. Bultinck P, Van Alsenoy C, Ayers PW, Carbó-Dorca R (2007) Critical analysis and extension of the Hirshfeld atoms in molecules. J Chem Phys 126:144111
126. Mulliken RS (1955) Electronic population analysis on LCAO-MO molecular wave functions 1. J Chem Phys 23:1833

127. Mulliken RS (1955) Electronic population analysis on LCAO-MO molecular wave functions 2. J Chem Phys 23:1841
128. Mulliken RS (1955) Electronic population analysis on LCAO-MO molecular wave functions 4. J Chem Phys 23:2343
129. Mulliken RS (1955) Electronic population analysis on LCAO-MO molecular wave functions 3. J Chem Phys 23:2338
130. Von Barth U, Hedin L (1972) A local exchange-correlation potential for the spin polarized case I. J Phys C 5:1629–1642
131. Rajagopal AK, Callaway J (1973) Inhomogeneous electron gas. Phys Rev B 7:1912–1919
132. Ghanty TK, Ghosh SK (1994) Spin-polarized generalization of the concepts of electronegativity and hardness and the description of chemical-binding. J Am Chem Soc 116:3943–3948
133. Galvan M, Vela A, Gazquez JL (1988) Chemical-reactivity in spin-polarized density functional theory. J Phys Chem 92:6470–6474
134. Vargas R, Galvan M, Vela A (1998) Singlet-triplet gaps and spin potentials. J Phys Chem A 102:3134–3140
135. Vargas R, Galvan M (1996) On the stability of half-filled shells. J Phys Chem 100: 14651–14654
136. Galvan M, Vargas R (1992) Spin potential in Kohn Sham theory. J Phys Chem 96:1625–1630
137. Garza J, Vargas R, Cedillo A, Galvan M, Chattaraj PK (2006) Comparison between the frozen core and finite differences approximations for the generalized spin-dependent global and local reactivity descriptors in small molecules. Theor Chem Acc 115:257–265
138. Perez P, Chamorro E, Ayers PW (2008) Universal mathematical identities in density functional theory: results from three different spin-resolved representations. J Chem Phys 128:204108
139. Chamorro E, Perez P, Duque M, De Proft F, Geerlings P (2008) Dual descriptors within the framework of spin-polarized density functional theory. J Chem Phys 129:064117
140. Pinter B, De Proft F, Van Speybroeck V, Hemelsoet K, Waroquier M, Chamorro E, Veszpremi T, Geerlings P (2007) Spin-polarized conceptual density functional theory study of the regioselectivity in ring closures of radicals. J Org Chem 72:348–356
141. Chamorro E, Perez P, De Proft F, Geerlings P (2006) Philicity indices within the spin-polarized density-functional theory framework. J Chem Phys 124:044105
142. Chamorro E, De Proft F, Geerlings P (2005) Hardness and softness reactivity kernels within the spin-polarized density-functional theory. J Chem Phys 123:154104
143. Chamorro E, De Proft F, Geerlings P (2005) Generalized nuclear Fukui functions in the framework of spin-polarized density-functional theory. J Chem Phys 123:084104
144. De Proft F, Chamorro E, Perez P, Duque M, De Vleesschouwer F, Geerlings P (2009) Spin-polarized reactivity indices from density functional theory: theory and applications. RSC periodical specialist report. Chem Model 6:63
145. Ugur I, De Vleeschouwer F, Tuzun N, Aviyente V, Geerlings P, Liu SB, Ayers PW, De Proft F (2009) Cyclopolymerization reactions of diallyl monomers: exploring electronic and steric effects using DFT reactivity indices. J Phys Chem A 113:8704–8711
146. Melin J, Aparicio F, Galvan M, Fuentealba P, Contreras R (2003) Chemical reactivity in the N, N-S, ν(r) space. J Phys Chem A 107:3831–3835
147. Fuentealba P, Simon-Manso Y, Chattaraj PK (2000) Molecular electronic excitations and the minimum polarizability principle. J Phys Chem A 104:3185–3187
148. Chattaraj PK, Poddar A (1999) Molecular reactivity in the ground and excited electronic states through density-dependent local and global reactivity parameters. J Phys Chem A 103: 8691–8699
149. Chattaraj PK, Poddar A (1999) Chemical reactivity and excited-state density functional theory. J Phys Chem A 103:1274–1275
150. Chattaraj PK, Poddar A (1998) A density functional treatment of chemical reactivity and the associated electronic structure principles in the excited electronic states. J Phys Chem A 102:9944–9948

151. Ayers PW, Parr RG (2000) A theoretical perspective on the bond length rule of Grochala, Albrecht, and Hoffmann. J Phys Chem A 104:2211–2220
152. Morell C, Labet V, Grand A, Ayers PW, De Proft F, Geerlings P, Chermette H (2009) Characterization of the chemical behavior of the low excited states through a local chemical potential. J Chem Theory Comput 5:2274–2283
153. Ayers PW (2001) Variational principles for describing chemical reactions. PhD dissertation, Department of Chemistry, University of North Carolina at Chapel Hill
154. Lieb EH (1985) Density functionals for coulomb systems. NATO ASI Ser B 123:31–80
155. Singh R, Deb BM (1999) Developments in excited-state density functional theory. Phys Lett 311:48–94
156. Theophilou AK, Gidopoulos NI (1995) Density-functional theory for excited-states. Int J Quantum Chem 56:333–336
157. Theophilou AK (1979) Energy density functional formalism for excited-states. J Phys C 12:5419–5430
158. Ayers PW, Levy M (2009) Time-independent (static) density-functional theories for pure excited states: extensions and unification. Phys Rev A 80:012508
159. Nagy A, Levy M (2001) Variational density-functional theory for degenerate excited states. Phys Rev A 63:052502
160. Levy M, Nagy A (1999) Variational density-functional theory for an individual excited state. Phys Rev Lett 83:4361–4364
161. Levy M, Nagy A (1999) Excited-state Koopmans theorem for ensembles. Phys Rev A 59:1687–1689
162. Levy M (1999) On time-independent density-functional theories for excited states. In: Proceedings of the 1st international workshop electron correlation and material properties. Rhodes, Greece, pp 299–308
163. Gorling A (1996) Density-functional theory for excited states. Phys Rev A 54:3912–3915
164. Nagy A (2005) Hardness and excitation energy. J Chem Sci 117:437–440
165. Perdew JP, Parr RG, Levy M, Balduz JL Jr (1982) Density-functional theory for fractional particle number: derivative discontinuities of the energy. Phys Rev Lett 49:1691–1694
166. Zhang YK, Yang WT (2000) Perspective on "density-functional theory for fractional particle number: derivative discontinuities of the energy". Theor Chem Acc 103:346–348
167. Yang WT, Zhang YK, Ayers PW (2000) Degenerate ground states and fractional number of electrons in density and reduced density matrix functional theory. Phys Rev Lett 84: 5172–5175
168. Ayers PW (2008) The continuity of the energy and other molecular properties with respect to the number of electrons. J Math Chem 43:285–303
169. Mermin ND (1965) Thermal Properities of the inhomogeneous electron Gas. Phys Rev 137:A1441–A1443
170. Nalewajski RF (1992) On geometric concepts in sensitivity analysis of molecular charge distribution. Int J Quantum Chem 42:243–265
171. Brandhorst K, Grunenberg J (2007) Characterizing chemical bond strengths using generalized compliance constants. Chem Phys Chem 8:1151–1156
172. Nalewajski RF (2003) Electronic structure and chemical reactivity: density functional and information-theoretic perspectives. Adv Quantum Chem 43:119–184
173. Jones LH, Ryan RR (1979) Interaction coordinates and compliance constants. J Chem Phys 52:2003–2004
174. Swanson BI, Satija SK (1977) Molecular vibrations and reaction pathways – minimum energy coordinates and compliance constants for some tetrahedral and octahedral complexes. J Am Chem Soc 99:987–991
175. Decius JC (1963) Compliance matrix and molecular vibrations. J Chem Phys 38:241–248
176. Cardenas C, Chamorro E, Galvan M, Fuentealba P (2007) Nuclear Fukui functions from nonintegral electron number calculations. Int J Quantum Chem 107:807–815
177. Cardenas C, Lamsabhi AM, Fuentealba P (2006) Nuclear reactivity indices in the context of spin polarized density functional theory. Chem Phys 322:303–310

178. Nalewajski RF (2000) Coupling relations between molecular electronic and geometrical degrees of freedom in density functional theory and charge sensitivity analysis. Comput Chem 24:243–257
179. De Proft F, Liu SB, Geerlings P (1998) Calculation of the nuclear Fukui function and New relations for nuclear softness and hardness kernels. J Chem Phys 108:7549–7554
180. Nalewajski RF, Korchowiec J, Michalak A (1996) Reactivity criteria in charge sensitivity analysis. In: Nalewajski RF (ed) Density functional theory IV: theory of chemical reactivity. Springer, Berlin, pp 25–141
181. Balawender R, De Proft F, Geerlings P (2001) Nuclear Fukui function and Berlin's binding function: prediction of the Jahn-Teller distortion. J Chem Phys 114:4441–4449
182. Li TL, Ayers PW, Liu SB, Swadley MJ, Aubrey-Medendorp C (2009) Crystallization force-a density functional theory concept for revealing intermolecular interactions and molecular packing in organic crystals. Chem Eur J 15:361–371
183. Li TL (2007) Understanding the polymorphism of aspirin with electronic calculations. J Pharm Sci 96:755–760
184. Li TL (2006) Understanding the large librational motion of the methyl group in aspirin and acetaminophen crystals: insights from density functional theory. Cryst Growth Des 6: 2000–2003
185. Li TL, Feng SX (2005) Study of crystal packing on the solid-state reactivity of indomethacin with density functional theory. Pharm Res 22:1964–1969
186. Feng SX, Li TL (2005) Understanding solid-state reactions of organic crystals with density functional theory-based concepts. J Phys Chem A 109:7258–7263
187. Swadley MJ, Li TL (2007) Reaction mechanism of 1,3,5-trinitro-s-triazine (RDX) deciphered by density functional theory. J Chem Theory Comput 3:505–513
188. Eickerling G, Reiher M (2008) The shell structure of atoms. J Chem Theory Comput 4:286–296
189. Kiewisch K, Eickerling G, Reiher M, Neugebauer J (2008) Topological analysis of electron densities from Kohn-Sham and subsystem density functional theory. J Chem Phys 128:044114
190. Ghosh SK, Berkowitz M (1985) A classical fluid-like approach to the density-functional formalism of many-electron systems. J Chem Phys 83:2976–2983
191. Nagy A, Parr RG (1994) Density-functional theory as thermodynamics. Proc Indian Acad Sci Chem Sci 106:217–227
192. Nagy A, Parr RG (2000) Remarks on density functional theory as a thermodynamics. J Mol Struct THEOCHEM 501:101–106
193. Ayers PW, Parr RG, Nagy A (2002) Local kinetic energy and local temperature in the density-functional theory of electronic structure. Int J Quantum Chem 90:309–326
194. Ayers PW (2005) Electron localization functions and local measures of the covariance. J Chem Sci 117:441–454
195. Chattaraj PK, Chamorro E, Fuentealba P (1999) Chemical bonding and reactivity: a local thermodynamic viewpoint. Chem Phys Lett 314:114–121
196. Becke AD, Edgecombe KE (1990) A simple measure of electron localization in atomic and molecular-systems. J Chem Phys 92:5397–5403
197. Silvi B, Savin A (1994) Classification of chemical-bonds based on topological analysis of electron localization functions. Nature 371:683–686
198. Savin A, Becke AD, Flad J, Nesper R, Preuss H, Vonschnering HG (1991) A New look at electron localization. Angew Chem 30:409–412
199. Bader RFW, Gillespie RJ, Macdougall PJ (1988) A physical basis for the VSEPR model of molecular geometry. J Am Chem Soc 110:7329–7336
200. Gillespie RJ, Bytheway I, Dewitte RS, Bader RFW (1994) Trigonal bipyramidal and related molecules of the main-group elements – investigation of apparent exceptions to the VSEPR model through the analysis of the Laplacian of the electron-density. Inorg Chem 33: 2115–2121

201. Malcolm NOJ, Popelier PLA (2003) The full topology of the Laplacian of the electron density: scrutinising a physical basis for the VSEPR model. Faraday Discuss 124:353–363
202. Bader RFW (1980) Quantum topology of molecular charge-distributions. 3. The mechanics of an atom in a molecule. J Chem Phys 73:2871–2883
203. Bartolotti LJ, Parr RG (1980) The concept of pressure in density functional theory. J Chem Phys 72:1593–1596
204. Szarek P, Sueda Y, Tachibana A (2008) Electronic stress tensor description of chemical bonds using nonclassical bond order concept. J Chem Phys 129:094102
205. Szarek P, Tachibana A (2007) The field theoretical study of chemical interaction in terms of the rigged QED: new reactivity indices. J Mol Model 13:651–663
206. Tachibana A (2005) A new visualization scheme of chemical energy density and bonds in molecules. J Mol Model 11:301–311
207. Tachibana A (2004) Spindle structure of the stress tensor of chemical bond. Int J Quantum Chem 100:981–993
208. Ichikawa K, Tachibana A (2009) Stress tensor of the hydrogen molecular ion. Phys Rev A 80:062507
209. Ayers PW, Jenkins S (2009) An electron-preceding perspective on the deformation of materials. J Chem Phys 130:154104
210. Nalewajski RF, Parr RG (2001) Information theory thermodynamics of molecules and their Hirshfeld fragments. J Phys Chem A 105:7391–7400
211. Ayers PW (2006) Information theory, the shape function, and the Hirshfeld atom. Theor Chem Acc 115:370–378
212. Nalewajski RF (2008) Use of fisher information in quantum chemistry. Int J Quantum Chem 108:2230–2252
213. Nalewajski RF (2006) Probing the interplay between electronic and geometric degrees-of-freedom in molecules and reactive systems. Adv Quantum Chem 51:235–305
214. Nalewajski RF (2005) Partial communication channels of molecular fragments and their entropy/information indices. Mol Phys 103:451–470
215. Nalewajski RF (2004) Communication theory approach to the chemical bond. Struct Chem 15:391–403
216. Nalewajski RF (2003) Information theoretic approach to fluctuations and electron flows between molecular fragments. J Phys Chem A 107:3792–3802
217. Nalewajski RF (2003) Information principles in the theory of electronic structure. Chem Phys Lett 372:28–34
218. Nalewajski RF, Switka E (2002) Information theoretic approach to molecular and reactive systems. Phys Chem Chem Phys 4:4952–4958
219. Nalewajski RF, Switka E, Michalak A (2002) Information distance analysis of molecular electron densities. Int J Quantum Chem 87:198–213
220. Nalewajski RF (2000) Entropic measures of bond multiplicity from the information theory. J Phys Chem A 104:11940–11951
221. Borgoo A, Godefroid M, Indelicato P, De Proft F, Geerlings P (2007) Quantum similarity study of atomic density functions: insights from information theory and the role of relativistic effects. J Chem Phys 126:044102
222. Sen KD, De Proft F, Borgoo A, Geerlings P (2005) N-derivative of Shannon entropy of shape function for atoms. Chem Phys Lett 410:70–76
223. Borgoo A, Godefroid M, Sen KD, De Proft F, Geerlings P (2004) Quantum similarity of atoms: a numerical Hartree-Fock and information theory approach. Chem Phys Lett 399: 363–367
224. Moens J, Jaque P, De Proft F, Geerlings P (2008) The study of redox reactions on the basis of conceptual DFT principles: EEM and vertical quantities. J Phys Chem A 112:6023–6031
225. Moens J, Geerlings P, Roos G (2007) A conceptual DFT approach for the evaluation and interpretation of redox potentials. Chem Eur J 13:8174–8184
226. Moens J, Roos G, Jaque P, Proft F, Geerlings P (2007) Can electrophilicity act as a measure of the redox potential of first-row transition metal ions? Chem Eur J 13:9331–9343

227. Moens J, Jaque P, De Proft F, Geerlings P (2009) A New view on the spectrochemical and nephelauxetic series on the basis of spin-polarized conceptual DFT. Chem Phys Chem 10:847–854
228. Liu SB (2007) Steric effect: a quantitative description from density functional theory. J Chem Phys 126:244103
229. Torrent-Sucarrat M, Liu SB, De Proft F (2009) Steric effect: partitioning in atomic and functional group contributions. J Phys Chem A 113:3698–3702
230. Anderson JSM, Ayers PW (2007) Predicting the reactivity of ambidentate nucleophiles and electrophiles using a single, general-purpose, reactivity indicator. Phys Chem Chem Phys 9:2371–2378
231. Li TL, Liu SB, Feng SX, Aubrey CE (2005) Face-integrated Fukui function: understanding wettability anisotropy of molecular crystals from density functional theory. J Am Chem Soc 127:1364–1365
232. Cardenas C, De Proft F, Chamorro E, Fuentealba P, Geerlings P (2008) Theoretical study of the surface reactivity of alkaline earth oxides: local density of states evaluation of the local softness. J Chem Phys 128:034708
233. Calatayud M, Tielens F, De Proft F (2008) Reactivity of gas-phase, crystal and supported V2O5 systems studied using density functional theory based reactivity indices. Chem Phys Lett 456:59–63
234. Sablon N, De Proft F, Geerlings P (2009) Molecular orbital-averaged Fukui function for the reactivity description of alkaline earth metal oxide clusters. J Chem Theory Comput 5: 1245–1253
235. Liu SB, Li TL, Ayers PW (2009) Potentialphilicity and potentialphobicity: reactivity indicators for external potential changes from density functional reactivity theory. J Chem Phys 131:114106
236. Liu SB, Schauer CK, Pedersen LG (2009) Molecular acidity: a quantitative conceptual density functional theory description. J Chem Phys 131:164107
237. De Proft F, Martin JML, Geerlings P (1996) Calculation of molecular electrostatic potentials and Fukui functions using density functional methods. Chem Phys Lett 256:400–408
238. Melin J, Ayers PW, Ortiz JV (2007) Removing electrons can increase the electron density: a computational study of negative Fukui functions. J Phys Chem A 111:10017–10019
239. Klopman G (1968) Chemical reactivity and the concept of charge and frontier-controlled reactions. J Am Chem Soc 90:223–234
240. Gazquez JL (1997) The hard and soft acids and bases principle. J Phys Chem A 101: 4657–4659
241. Melin J, Aparicio F, Subramanian V, Galvan M, Chattaraj PK (2004) Is the Fukui function a right descriptor of hard-hard interactions? J Phys Chem A 108:2487–2491
242. Loos R, Kobayashi S, Mayr H (2003) Ambident reactivity of the thiocyanate anion revisited: Can the product ratio be explained by the hard soft acid base principle? J Am Chem Soc 125:14126–14132
243. Suresh CH, Koga N, Gadre SR (2001) Revisiting Markovnikov addition to alkenes via molecular electrostatic potential. J Org Chem 66:6883–6890
244. Petragnani N, Stefani HA (2005) Advances in organic tellurium chemistry. Tetrahedron 61:1613–1679
245. Huheey JE, Keiter EA, Keiter RL (1993) Inorganic chemistry: principles of structure and reactivity. HarperCollins, New York
246. Drews T, Rusch D, Seidel S, Willemsen S, Seppelt K (2008) Systematic reactions of Pt(PF3)(4). Chem Eur J 14:4280–4286
247. Kruck T (1967) Trifluorophosphine complexes of transition metals. Angew Chem Int Ed 6:53–67
248. Kruck T, Baur H (1965) Synthesis of tetrakis(trifluorophosphine)-platinum(0) and tetrakis(trifluorophosphine)-palladium(0). Angew Chem Int Ed 4:521

249. Basolo F, Johnston RD, Pearson RG (1971) Kinetics and mechanism of substitution reactions of the tetrakis(trifluorophosphine) complexes of nickel(0) and platinum(0). Inorg Chem 10:247–251
250. Pei KM, Liang J, Li HY (2004) Gas-phase chemistry of nitrogen trifluoride NF_3: structure and stability of its M^+-NF_3 (M = H, Li, Na, K) complexes. J Mol Struct 690:159–163
251. Borocci S, Bronzolino N, Giordani M, Grandinetti F (2006) Ligation of Be^+ and Mg^+ to NF_3: structure, stability, and thermochernistry of the Be^+-(NF_3) and Mg^+-(NF_3) complexes. Int J Mass Spectrom 255:11–19
252. De Proft F, Ayers PW, Fias S, Geerlings P (2006) Woodward-Hoffmann rules in conceptual density functional theory: initial hardness response and transition state hardness. J Chem Phys 125:214101
253. Jaque P, Correa JV, Toro-Labbé A, De Proft F, Geerlings P (2010) Regaining the Woodward–Hoffmann rules for chelotropic reactions via conceptual DFT. Can J Chem 88:858–865
254. Mulliken RS (1965) Molecular scientists and molecular science: some reminiscences. J Chem Phys 43:S2–S11
255. Head-Gordon M (1996) Quantum chemistry and molecular processes. J Phys Chem 100:13213–13225
256. Bader RFW (1994) Why define atoms in real-space. Int J Quantum Chem 49:299–308
257. Bader RFW, Nguyen-Dang TT (1981) Quantum-theory of atoms in molecules – Dalton revisited. Adv Quantum Chem 14:63–124
258. Matta CF, Bader RFW (2006) An experimentalist's reply to "what is an atom in a molecule?". J Phys Chem A 110:6365–6371
259. Melin J, Ayers PW, Ortiz JV (2005) The electron-propagator approach to conceptual density-functional theory. J Chem Sci 117:387–400
260. Ayers PW, Melin J (2007) Computing the Fukui function from ab initio quantum chemistry: approaches based on the extended Koopmans' theorem. Theor Chem Acc 117:371–381
261. Sablon N, Mastalriz R, De Proft F, Geerlings P, Reiher M (2010) Relativistic Effects on the Fukui Function. Theoret Chem Acc 127:195–202
262. Nalewajski RF, Korchowiec J (1997) Charge sensitivity approach to electronic structure and chemical reactivity. World Scientific, Singapore
263. Ayers PW (2006) Can one oxidize an atom by reducing the molecule that contains It? Phys Chem Chem Phys 8:3387–3390
264. Senet P, Yang M (2005) Relation between the Fukui function and the Coulomb hole. J Chem Sci 117:411–418
265. Min KS, DiPasquale AG, Rheingold AL, White HS, Miller JS (2009) Observation of redox-induced electron transfer and spin crossover for dinuclear cobalt and iron complexes with the 2,5-Di-tert-butyl-3,6-dihydroxy-1,4-benzoquinonate bridging ligand. J Am Chem Soc 131:6229–6236
266. Miller JS, Min KS (2009) Oxidation leading to reduction: redox-induced electron transfer (RIET). Angew Chem Int Ed 48:262–272
267. Dewar MJS (1989) A critique of frontier orbital theory. J Mol Struct THEOCHEM 59: 301–323
268. Ayers PW, unpublished
269. Jaque P, Toro-Labbé A, Geerlings P, De Proft F (2009) Theoretical study of the regioselectivity of 2 + 2 photocycloaddition reactions of Acroleins with olefins. J Phys Chem A 113:332–344
270. Pinter B, De Proft F, Veszpremi T, Geerlings P (2008) Photochemical nucleophilic aromatic substitution: a conceptual DFT study. J Org Chem 73:1243–1252

Index

C. Gatti and P. Macchi (eds.), *Modern Charge-Density Analysis*,
DOI 10.1007/978-90-481-3836-4, © Springer Science+Business Media B.V. 2012

D

I

K

L

M

Q